Proceedings of the 15th

HYPERVELOCITY IMPACT SYMPOSIUM (HVIS2019)

THE AMERICAN SOCIETY OF MECHANICAL ENGINEERS
Two Park Avenue * New York, N.Y. 10016

ISBN: 978-0-7918-8355-6

Proceedings of the 15th Hypervelocity Impact Symposium (HVIS2019)

This proceedings contains papers and extended abstracts from the 2019 Hypervelocity Impact Symposium (HVIS 2019) held in Destin, Florida on April 16–20, 2019. This was the fifteenth symposium since the re-initiation of symposia related to hypervelocity impact in 1986. The proceedings for the previous fourteen symposia have been published as separate volumes of the International Journal of Impact Engineering and Procedia Engineering. Papers in this proceedings address advancements in the basic understanding of hypervelocity impact physics, related phenomenology, and engineering applications. Each paper in this proceedings has undergone peer review by experts in their respective fields of research.

The opening keynote address, following the tradition that was started in 1989, was given by the recipient of the Hypervelocity Impact Society's Distinguished Scientist Award, Dr. Dennis Grady (from Applied Research Associates). Keynote lectures on other days of the symposium were delivered by Dr. Werner Arnold (from MBDA-TDW) and Mr. Peter Sable (from Marquette University). The banquet speaker was Dr. Richard P. ("Dick") Hallion (Trustee of Florida Polytechnic University) who gave a presentation entitled "Inventing Hypersonic Flight".

A total of 96 papers were presented at the symposium. Among these were 32 papers selected for a special issue of the International Journal of Impact Engineering which can be found at https://www.sciencedirect.com/journal/international-journal-of-impact-engineering/special-issue/10MTGP5W4VJ. Extended abstracts of these papers appear in this proceedings. The first paper in this volume is a summary of the lecture given by Dr. David Crawford (from Crawford Technical Services), Distinguished Scientist Awardee from the 2017 HVIS Symposium held April 24-28 in Canterbury, Kent, UK. A total of 91 papers and abstracts appears in this proceedings.

I would like to thank the following individuals and groups who volunteered their time and expertise to make this symposium a reality: the Hypervelocity Impact Society board of directors, society committee chairs and committee members, the technical program chair, and the members of the HVIS 2019 technical program committee. Without their hard work, this symposium would not have been possible. Plans are to continue the Hypervelocity Impact Symposia, and we hope that the papers from future symposia can also become part of the archival literature.

David Littlefield
HVIS 2019 Symposium Chair
Editor, Proceedings of the 15th Hypervelocity Impact Symposium (HVIS2019)

HVIS 2019 Organizing Committee

Symposium Chair:
  David Littlefield, *The University of Alabama at Birmingham*

Technical Program Chair:
  Geremy Kleiser, *Air Force Research Laboratory*

Exhibits Chair:
  William Reinhart, *Sandia National Laboratories*

Conference Coordination:
  Coordinator: Michelle Wiginton, *Missouri S&T*
  Secretary: Tammy Mace, *Missouri S&T*

Board of Directors:
  President: William Schonberg, *Missouri S&T*
  Secretary-Treasurer: James Wilbeck, *Dynetics*
  Past President: Todd Bjerke, *ASA/LT*
  David Littlefield, *The University of Alabama at Birmingham*
  Mark Burchell, *University of Kent*
  Kathryn Harriss, *University of Kent*
  William Reinhart, *Sandia National Laboratories*
  Shannon Ryan, *Defense Science and Technology Organization*

Committee Chairs:
  Membership Committee: Brooke Myers Corbett, *Institute for Defense Analyses*
  Nominations Committee: Andrew Tonge, *Army Research Laboratory*
  Awards Committee: Josh Miller, *The University of Texas at El Paso*
  Education Outreach Committee: John Borg, *Marquette University*
  Communications Committee: James Walker, *Southwest Research Institute*
  Site Selection Committee: Gene Hertel, *Sandia National Laboratory*

Technical Program Committee:
  Andrew Tonge, *Army Research Laboratory*
  Angela Diggs, *Air Force Research Laboratory*
  Brett Sorenson, *Army Research Laboratory*
  Casey Uhlig, *Army Research Laboratory*
  David Crawford, *Crawford Technical Services*
  David Lambert, *Air Force Research Laboratory*
  David Price, *Atomic Weapons Establishment*
  Eric Christiansen, *National Aeronautics and Space Administration*
  Gene Hertel, *Sandia National Laboratories*
  James Cazamias, *Army Research Laboratory*
  James Walker, *Southwest Research Institute*
  John Niederhaus, *Sandia National Laboratories*

Josh Miller, *The University of Texas at El Paso*
Joel Williamsen, *Institute for Defense Analyses*
Kent Danielson, *Engineer Research and Development Center*
Michael Hopson, *Naval Surface Warfare Center*
Michael Burkett, *Los Alamos National Laboratory*
Philip Mulligan, *Missouri S&T*
Rohan Banton, *Army Research Laboratory*
Scott Alexander, *Sandia National Laboratories*
Sidney Chocron, *Southwest Research Institute*
Shannon Ryan, *Defense Science and Technology Organization*
Sikhanda Satapathy, *Army Research Laboratory*
Tracy Vogler, *Sandia National Laboratories*
Todd Bjerke, *ASA/LT*
Wayne Chen, *Purdue University*

HVIS 2019 Exhibitors

CertaSIM, LLC
Specialised Imaging
Cordin Scientific Imaging
AMETEK/VISION
HADLAND/SHIMADZU/IX
Amotronics
NAC
Photron
PCO Americas
Coherent Solutions Limited

CONTENTS

PROCEEDINGS OF THE 2019 HYPERVELOCITY IMPACT SYMPOSIUM

Proceedings of the 2019 Hypervelocity Impact Symposium
HVIS2019
April 14-19, 2019, Destin, FL, USA
HVIS2019-DSA2017

Coils, Codes and Comets: My Attempts to Partially Understand Some Particular Aspects of Hypervelocity Impacts

David A. Crawford*

Crawford Technical Services, LLC, Winona, Minnesota, USA

Abstract

At the time of this writing in 2019, I have been studying hypervelocity impacts for about 34 years. I attended my first Hypervelocity Impact Symposium in 1992. In this paper, I attempt to show what I've been up to all this time but also show some of the things I've learned along the way. I've singled out three topics that stand out in my mind as milestones in my career: the impact of Comet Shoemaker-Levy 9 on Jupiter in 1994, the development of adaptive mesh refinement for the CTH hydrocode in 1998-2001 and my ongoing studies of hypervelocity impact-generated magnetic fields from 1985 to the present day.

Keywords: Shoemaker-Levy 9, Jupiter, adaptive mesh refinement, AMR, CTH, magnetism, paleomagnetism, Moon

1. Introduction

The impact of Comet Shoemaker-Levy 9 on Jupiter in July, 1994 was the largest, most energetic impact event on a planet ever witnessed. Because it broke up during a close encounter with Jupiter in 1992, it was bright enough to be discovered more than a year prior to impact, allowing the scientific community an unprecedented opportunity to assess the effects such an event would have. Many excellent observations were made from Earth-based telescopes, the Hubble Space Telescope (HST) and the Galileo spacecraft en route to Jupiter. In 1993 and 1994, my Sandia colleagues and I performed simulations prior to the impact that compared well qualitatively with many of the observations. After the impact event, we used the observations in conjunction with CTH computational simulations to determine the sizes of the fifteen fragments that made discernible impact features on the planet and from that made an estimate of the total size and energy of the impact event.

In 1998-2001, I implemented an adaptive mesh refinement capability into CTH. The adaptivity is block-based with refinement and un-refinement occurring in an isotropic 2:1 manner. The capability, like CTH, is designed to run on serial, multiprocessor and massively parallel platforms. By 2001, we could run large, parallel 3D AMR calculations that realized about an order of magnitude decrease in CPU requirements over non-AMR counterparts. The fastest available computers at the time required 3D calculations to be of a heroic nature even with AMR. The advance of computing hardware in the past decade, however, has enabled routine 3D calculations with AMR to become a staple of very large CTH calculations. The surprising thing, however, is the much greater performance advantage we're seeing on certain types of AMR problems – in some cases, by two or three orders of magnitude. Credit for this, I think is partly due to the unique approach we use in CTH for the definition of refinement indicators. Rather than a more automated approach, we require analysts to construct indicators in their input decks using a simple construct of operators, filters, database variables and thresholds.

* Corresponding author. Tel.: 1-507-459-0659
E-mail address: dave_crawford@q.com.

The origin and evolution of the Moon's magnetic field has been a major question in lunar science ever since Luna 1 made the first magnetic measurements in the vicinity of the Moon in 1959. Orbital measurements show that the magnetic field at the surface of the Moon has local scale lengths on the order of 1-100 km but correlation with specific geologic features remains elusive. The source of the anomalies from an internal core dynamo or from exogenic processes such as impact, remains a key question. In my earliest work, I showed that impact-generated magnetic fields showed promise in explaining the magnetism of lunar samples particularly and perhaps the "jumbled" magnetic state of the lunar surface today. In my most recent work, I show that craters at nearly all scales may leave behind remnant magnetic fields observable at the surface or in samples. For at least five large basins on the Moon, simulated anomalies are similar enough to orbital magnetic field data for us to conclude there is an alternative explanation for many lunar anomalies that doesn't require the presence of a lunar dynamo. At least some anomalies may be associated with magnetization acquired by rocks in the presence of the transient magnetic field produced by the impact itself.

2. Numerical simulations of the impact of Comet Shoemaker-Levy 9 on Jupiter

In early July, 1992, periodic comet Shoemaker-Levy 9 broke up during a close encounter with Jupiter (Fig. 1). For a brief two-year period, about 20 large fragments and associated debris followed one last orbit about Jupiter before striking the planet at an estimated velocity of 60 km/s. The largest fragments entered the Jovian atmosphere during the week of July 16-22, 1994. Although the impact sites were located just beyond the limb of Jupiter and were not directly visible from Earth, the Galileo spacecraft was positioned for direct viewing of the impact sites. While impact phenomena were not spatially resolved by the spacecraft, its timing, spectral and luminosity data are invaluable for comparison with analytical and numerical models. Fireballs and plumes generated by the impacts were visible in line-of-sight from Earth within a minute and the impact locations themselves rotated into view within 7-20 minutes [1]. The wealth of data provided by this fortuitous event gave us an opportunity to assess models of meteoroid entry into planetary atmospheres and to estimate the size of the Shoemaker-Levy 9 parent body based on observations of the radiated light flux observed by the Galileo spacecraft and by Earth-based telescopes.

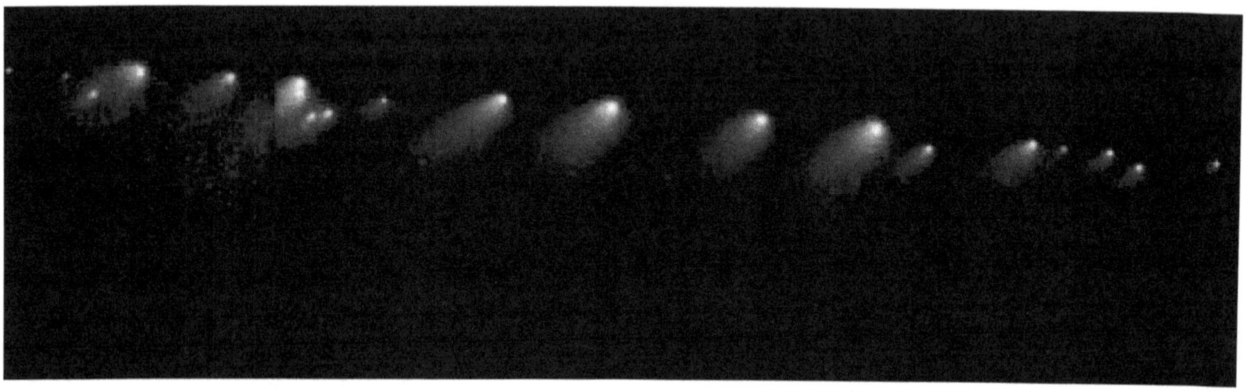

Fig. 1. Comet Shoemaker-Levy 9 soon after its discovery in 1993 (Hubble Space Telescope Science Institute: see [1]). The fragments were labeled A through W. Some disappeared from view (e.g. J, M and P1) and some split (e.g. Q1, Q2).

Simulations performed prior to the impact of Comet Shoemaker-Levy 9 using the CTH shock-physics code [2] compare well qualitatively with many of the observations [3-5]. Nevertheless, good quantitative models of the event were required in order to extract useful information from observations of the events. Fortunately, models were strongly constrained by the excellent data collected (Fig. 2), especially: (1) the direct light-flux observations made by the Galileo spacecraft [6]; (2) the pattern of dark ejecta the impacts left in the Jovian stratosphere which varied considerably in albedo and lateral extent [1, 7]; and (3) the Earth-based telescope observations of the infrared impact light flashes [8]. By using a semi-analytical meteoroid entry model [9] as initial conditions for plume evolution models we could investigate the Shoemaker-Levy 9 events in the context of matching visible and infrared light fluxes (Fig. 3). We performed simulations for ten test cases with comet fragments ranging in size from 250 to 1250 m in diameter. In all cases, we assumed the comets were of density 0.95 g/cm^3 or 0.25 g/cm^3 with an equation-of-state (including melting, vaporization, dissociation and ionization) appropriate for water ice. A two-dimensional representation, symmetric to the 45° entry angle, was used to simulate the first few minutes of fireball evolution in CTH. Output light curves were computed using opacity appropriate for the Jovian atmosphere [10] and compared against Galileo spacecraft observations (Fig. 3).

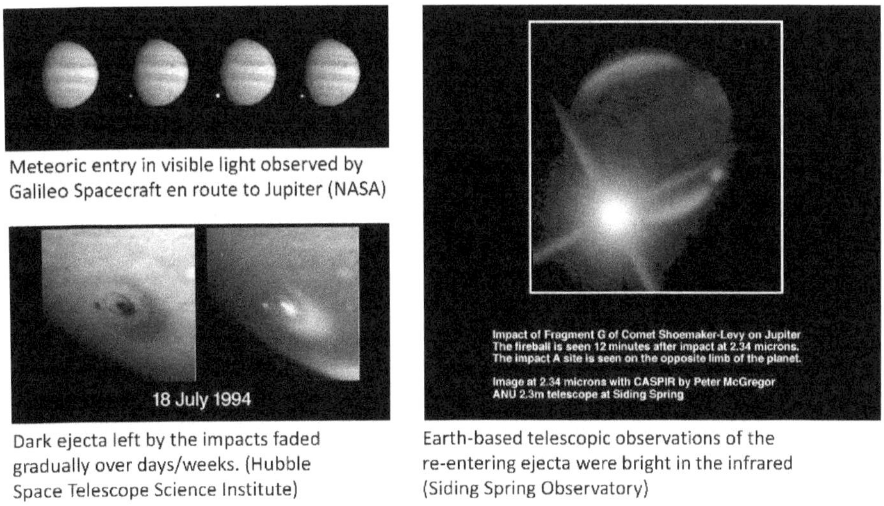

Fig. 2. Representative data collected during the impacts of Comet Shoemaker-Levy 9 on Jupiter in 1994 [1,6-8,11-14].

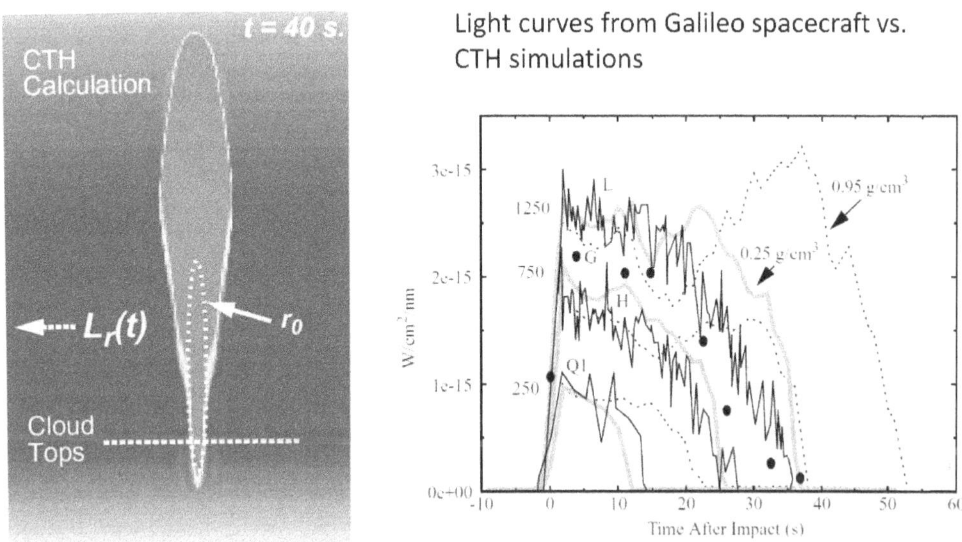

Fig. 3. A modified version of CTH was run with input from a semi-analytical entry model we developed [9] and using the Jovian atmospheric opacity model of [10]. Output light curves (L_r) were compared to Galileo spacecraft observations of the L, H and Q1 impacts (black lines) and the G impact (black dots) [6] to determine impactor size and density. CTH light curves from 250, 750 and 1250 m impactors are shown. The best matching light curves resulted from low density (0.25 g/cm^3) impactors (gray curves).

Table 1. Shoemaker-Levy 9 fragment size estimates (from [9]).

Fragment*	HST Class†	2.3 μm Peak (Jy)‡	12 μm Peak (Jy)‡	Diameter (m)♦
A	2a	20	1,200	450
B	3			(50)
C	2a	25		380
D	3	4		200
E	2a	100	1,550	610
G	1	460		1,000•
H	2a	100	2,400	660•
K	1	460		1,000•
L	1	1,000	12,700	1,270•
N	3	0.04		45•
Q2	3			(50)
Q1	2b	90	1,700	600•
R	2b	70	990	530
S	2c	120		640
W	2c	60		490•
Parent	-	-	-	1,400¶

*Fragments F, P2, T, U and V produced no discernible impact features. Fragments J, M and P1 faded from view before impact [7]. The letters I and O were not used.

†Based on first view of the impact site with the Hubble Space Telescope [7]. Class 1 = large dark feature (>10,000 km radius), Class 2 = medium dark feature (4000-8000 km radius), Class 3 = small dark feature (<3000 km radius).

‡Observed light flux of the main infrared event seen from Earth at 2.3 and 12 μm [8, 11-13].

♦Diameter is estimated from a least-squares analysis incorporating Galileo light-flux observations (marked with • symbol and calibrated against CTH light-output calculations) and assuming r^3 dependence of the peak flux (for a given wavelength) observed from Earth-based telescopes [8]. Values in parentheses are estimates assuming diameter equivalence within HST class 3. Uncertainty (1 σ) of individual fragment diameter is 15% for fragments A, E, H, L, Q1 and R and 30% for fragments B, C, D, G, K, N, Q2, S and W. Best fitting fragment density is about 0.25 g/cm³.

¶Assuming a density of 0.5 g/cm³ before breakup in 1992. The total volume of the fragments would make a sphere 1760 m in diameter (at 0.25 g/cm³). Uncertainty of these estimates is dominated by uncertainties of the largest fragments (L, K and G), about 25%.

Table 1 shows our estimated sizes for the Shoemaker-Levy 9 fragments that produced discernible features on the planet. The fragment diameter is estimated from a least-squares analysis incorporating Galileo light-flux observations (calibrated against CTH light-output calculations) and assuming r^3 dependence (mass scaling) of the peak flux for a given wavelength observed from Earth-based telescopes [8, 11-13]. For fragments that produced the smallest impact features on the planet, we assumed a constant diameter within the site classification derived from HST imagery [7]. For most cases where sufficient overlap between the Galileo and Earth-based observations exists, the r^3 scaling law holds fairly well. Normal variations due to density heterogeneity among the fragments is expected. There are a couple of notable exceptions, however. Fragments Q1 and W were relatively small by Galileo light flux estimates but of moderate size based on Earth-based observations. This could be explained if they were denser than the other fragments (e.g. fully dense silicate vs. fluffy, muddy water ice).

The fragments of Shoemaker-Levy 9 that hit Jupiter with discernible effect have a total volume that would make a 1.8±0.5 km diameter sphere at an average density of 0.25 g/cm³. This is consistent with the total volume of dusty debris seen on the planet after the impacts, corresponding approximately to a 1-km diameter sphere [14], assuming a typical debris particle density of (1-2 g/cm³). With a pre-breakup density of 0.5 g/cm³ [15-16], the parent body had a diameter of 1.4±0.4 km. With these parameters, the Shoemaker-Levy 9 family pummeled Jupiter with the energy equivalent of 300 Gigatons of TNT.

3. The development of adaptive mesh refinement (AMR) in CTH

Adaptive mesh refinement (AMR) has been used for improving computational resolution when resources are limited and has been used for hyperbolic problems on an experimental basis for years. For a mature Eulerian multi-material shock-physics code family like CTH and its predecessors, adaptivity was a natural next step in code development. By 1998, CTH had been running on the largest parallel machines in the world. At around that time, I performed a scaling study by adding AMR to a simple 2D ideal gas hydrocode and determined we should be able to realize an order of magnitude improvement in performance on many large parallel 3D calculations if we adopted AMR for CTH. I proposed this to Gene Hertel, who was the CTH project lead at the time. With his go ahead and support from Paul Yarrington who was my manager at the time, we started what turned into a three-year effort to put AMR into CTH.

In order to achieve adaptivity yet retain the man-years of effort that have been expended on the development of physics routines, the AMR in CTH is block-based, where each block consists of a small patch of cells (typically 8x8x8 or 10x10x10). In order to facilitate its implementation on massively parallel platforms, each block communicates with its neighbors using a generalization of the message-passing paradigm developed for multiprocessor CTH. A maximum 2:1 resolution difference across block boundaries is strictly enforced. Block refinement and un-refinement is isotropic. In multiprocessor AMR-CTH calculations, blocks are distributed across processors and CPU loading is optimized.

By 2001, we could run large, parallel 3D AMR calculations that realized about an order of magnitude decrease in CPU requirements over non-AMR counterparts (see Fig. 4a). The fastest available computers at the time required 3D calculations to be of a heroic nature even with AMR. The advance of computing hardware in the past decade, however, has allowed AMR to become a staple routine of very large CTH calculations. Extrapolating the curve in Figure 4a and assuming a doubling of CPU memory and performance every 18 months would suggest that AMR calculations would improve their performance advantage over non-AMR calculations by about a factor of two every ten years. Today, therefore, we expect to see a 30-40 times performance advantage for AMR vs. non-AMR. The surprising thing, however, is the much greater performance advantage we're seeing on other types of AMR problems – in some cases, by two or three orders of magnitude. Credit for this, I think is partly due to the unique approach we use in CTH for the definition of refinement indicators. Rather than an automated approach, we require analysts to construct indicators in their input decks using a simple construct of operators, filters, database variables and thresholds. While this is more work for them, it allows maximum flexibility and puts more control of computational resources into their hands. The analyst is, after all, the most likely person to know where to focus refinement to achieve their goals.

Figure 4b shows an example of an AMR calculation which exhibits a 200-300 times performance advantage over its non-AMR counterpart. It is a simulation of the hypothesized formation of the Moon by the impact of a Mars-sized body on the proto-Earth. Because of the importance of the large volume of vapor produced during the impact, the most effective adaptive meshing strategy is to maintain equal mass zoning as much as possible. In this way, CTH with AMR can realize some of the efficiencies of Lagrangian mesh free methods such as SPH (the typical method used on this type of problem) while capturing features more accurately represented by Eulerian methods (e.g. shocks in gases).

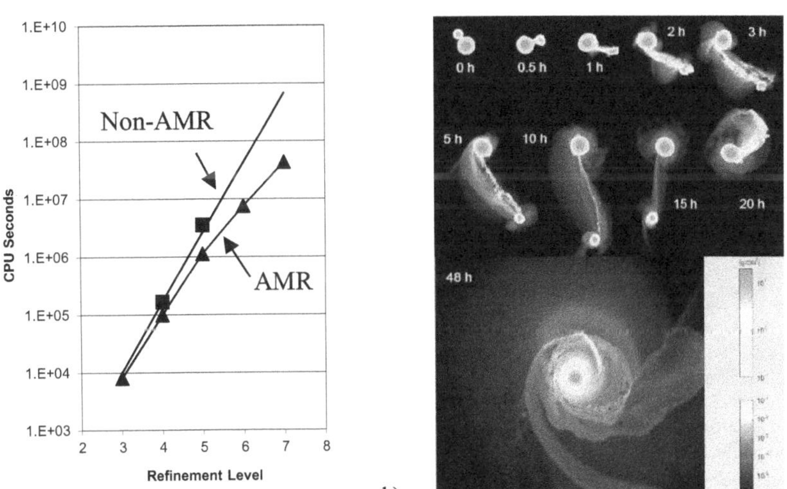

a) b)

Fig. 4. a) CPU timing study comparing AMR vs. Non-AMR methods on a heroic problem in 2001. At highest resolution, the AMR calculation outperformed the Non-AMR calculation by a factor of 10. b) Recent high-resolution AMR-CTH simulation of a Mars-sized object striking the proto-Earth in a candidate Moon forming impact (from [19]). Time after impact (in hours) is shown for each panel. This AMR calculation required approximately 40 million zones, including ghost cells and was run on 256 processors for six days. The non-AMR equivalent would need at least 20 billion zones – nearly 500 times the memory and 200-300 times the CPU requirements of the AMR calculation.

4. Studying magnetic and electric fields (and separated electric charge) caused by hypervelocity impacts

The origin and evolution of the Moon's magnetic field has been a major question in lunar science ever since Luna 1 made the first magnetic measurements in the vicinity of the Moon in 1959. Orbital measurements show that the magnetic field at the surface of the Moon has local scale lengths on the order of 1-100 km. While this could suggest a correlation with impact craters, most lunar magnetic anomalies don't appear to correlate with known geologic structures, including impacts. Nevertheless, the question of lunar magnetism has been a strong motivator for this work from the time when I started studying impact-generated magnetic fields to the present day [20-25]. And now, with my most recent work, I can explain why magnetic anomalies don't generally correlate with impact structures even if produced by impact events. Because lunar materials have relatively low thermoremanence susceptibility, impacts produce magnetic fields observable in orbital magnetic field data only for large and relatively rare impact craters - large lunar basins greater than ~200 km diameter. If occurring in thick units with higher thermoremanence susceptibility, craters as small as 50-100 km may produce orbital anomalies but this would also be a relatively rare occurrence [25].

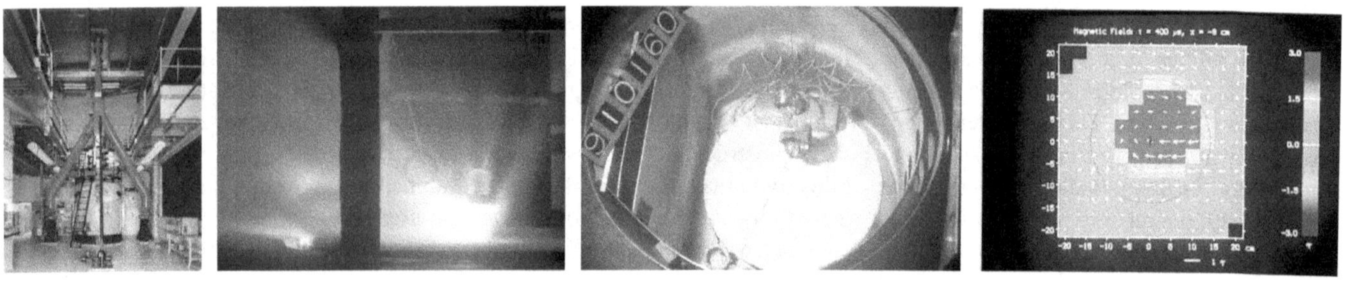

Fig. 5. From left-to-right: the NASA Ames Vertical Gun Range; a time exposure of a hypervelocity impact into powdered dolomite with magnetic search coils position above and below the target surface; a mu-metal shielded target was used in later experiments to reduce the ambient magnetic field to lunar-like conditions (~450 nT) with carefully wrapped and shielded magnetic search coils designed to reduce electrostatic noise; a map of the magnetic field observed 400 micro-seconds after the vertical impact of a 1/4 inch Al sphere at 5 km/s.

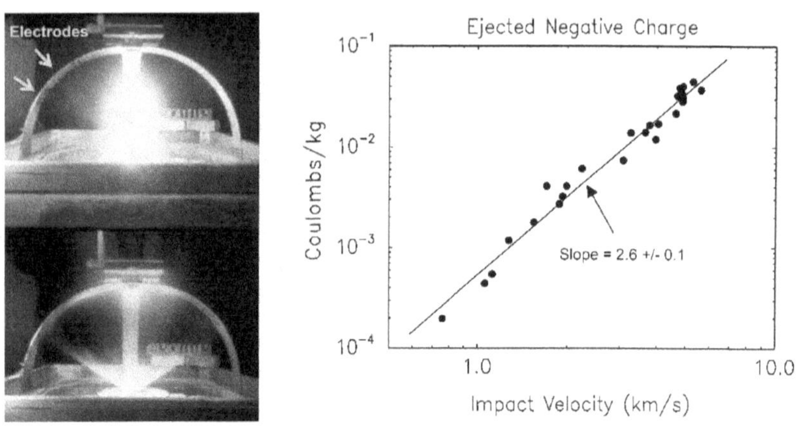

Fig. 6. Apparatus used to measure electrostatic charge separation during a hypervelocity impact event (left) at the NASA AVGR. The moment-of-impact is shown in the top frame, ejecta curtain development shown in the bottom frame. The total charge carried out of the crater by the ejecta scales linearly with projectile mass and has a $v^{2.6}$ dependence on velocity (right) [23].

In 1985, during my junior year at Brown University, Pete Schultz introduced me to the question of impact-generated magnetic fields. Starting then, and during my graduate studies with Pete thereafter, I pursued the study of impact-generated magnetic fields by primarily using the experimental facilities of the NASA Ames Vertical Gun Range (AVGR) (Fig. 5). These studies culminated with the unambiguous mapping of impact-generated magnetic fields during hypervelocity impact which we published in the HVIS proceedings in 1992 [22]. After this work was completed, I decided to attempt an understanding of the strong electrostatic noise signals that had always bedeviled our magnetic measurements. It was really these measurements of the electrostatics (how static electric charges were created and moved about the target during a hypervelocity impact) that informs our understanding of the magnetic field production mechanism. In the publication of the 1998 HVIS Proceedings [23], we showed how the quantity of electric charged produced during hypervelocity impact scaled with the mass and velocity of the impactor (Fig. 6) and how impact-generated magnetic fields should increase linearly with the radius of the projectile. In the 20 years since that publication, I tinkered with theoretical models based on plasma sheath theory to better understand the charging mechanism. This culminated in the publication in the 2015 HVIS Proceedings of a model implemented into CTH and validated against the earlier experimental data [24]. This model was general enough to allow us to extrapolate to large natural crates on the Moon. Combined with modifications to CTH to compute 3D magnetic fields and capture remanence, we now have the ability to predict magnetic anomalies from natural impact events on the Moon (Figs. 7 and 8) and elsewhere throughout the solar system.

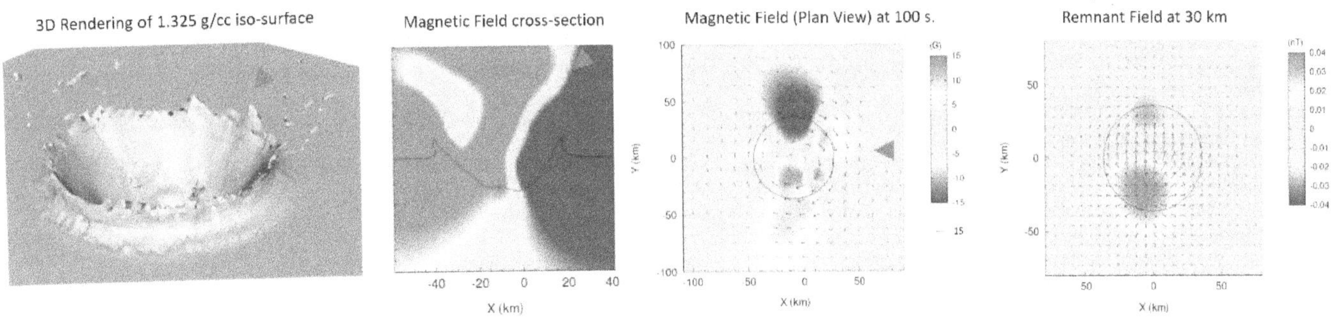

Fig. 7. CTH simulation of a 72-km central-peak crater on the Moon 100 seconds after impact. From left-to-right: 3D rendering of the 1.325 g/cc density iso-surface, Y-component of the magnetic field in cross section, all components of the magnetic field (horizontal components depicted with arrows, vertical component with color) looking from above. Magnetic field strength shown in Gauss (1G = 10^{-4} Tesla). Pre-impact projectile trajectory indicated by the blue triangles and final crater rim indicated by circles. At far right, the remnant field that would be observed today at 30 km altitude. (modified from [25]).

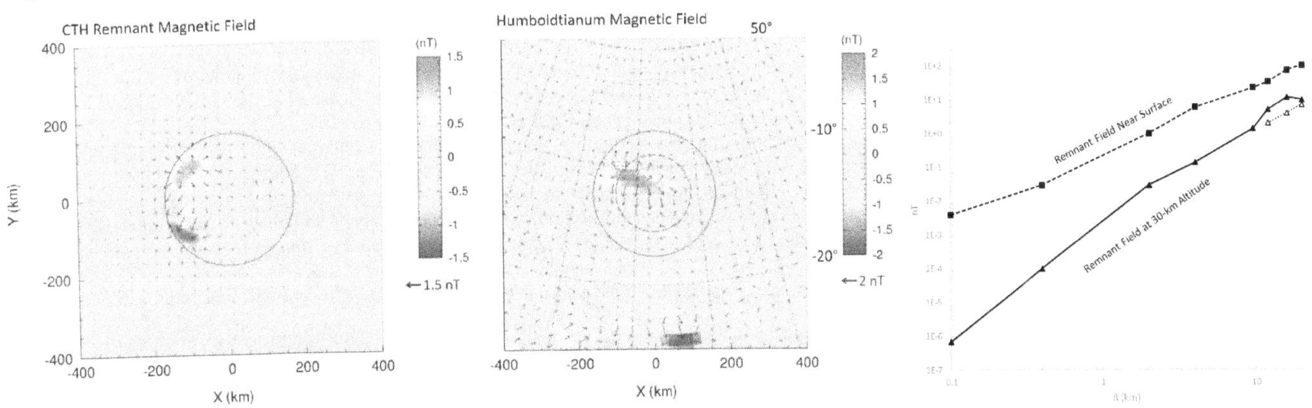

Fig. 8. From left-to-right: CTH simulation of a 340-km crater, the magnetic anomaly of the 322-km lunar basin Humboldtianum, and a plot of remnant magnetic field intensity near the surface and at 30-km altitude for a range of projectile radii. Only the largest projectiles (10 km radii and larger) which produce basins larger than 200 km in diameter will produce strong enough anomalies to be observable from orbit with present technology. (modified from [25])

Acknowledgements

I am grateful to the following individuals with whom I have co-authored papers and technical reports during the course of my hypervelocity impact career: Olivier Barnouin, Amy Barr, Ray Bell, Mark Boslough, Aaron Brundage, Robert Butler, Robin Canup, Mark Cintala, Millie Elrick, Eric Harstad, Dave Hensinger, Gene Hertel, Marlin Kipp, Charles Mader, Sharon Petney, Sue Phelps, Betty Pierazzo, Stephanie Quintana Bouchey, Allen Robinson, Kevin Ruggirello, Bob Schmitt, Pete Schultz, Shane Schumacher, Jim Simmons, Angela Stickle, Paul Taylor, Tim Trucano and Kelly Wrobel.

I want to single out several of the individuals mentioned above. Foremost, I want to thank my friend and colleague and PhD advisor, Pete Schultz. He continues to inspire me to pursue impact-generated magnetic fields and other questions of planetary impact science. Secondly, I want to recognize some of my many colleagues at Sandia. I worked with Mark Boslough extensively on Shoemaker-Levy 9. He was a principal influence during my earliest days at Sandia. I am grateful to several core members of the CTH development team: Gene Hertel, Bob Schmitt, Eric Harstad and Paul Taylor. Gene in particular has been a personal mentor and a tireless advocate of CTH when he was the team lead in the 1990s right up to the present day.

References

[1] Hammel, H. B. et al., 1995, HST Imaging of Atmospheric Phenomena Created by the Impact of Comet Shoemaker-Levy 9, *Science, 267*, 1288-1296.

[2] McGlaun, J. M, Thompson S. L. and Elrick, M. G., 1990, CTH - A Three-Dimensional Shock-Wave Physics Code, *International Journal of Impact Engineering, Vol. 10*, p. 351.

[3] Boslough, M. B., Crawford, D. A., Trucano, T. G. and Robinson, A. C., 1994, Watching for Fireballs on Jupiter, *EOS Transactions, American Geophysical Union, Vol. 75, Issue 27, pp. 305-310.*

[4] Crawford, D. A., Boslough, M. B., Trucano, T. G. and Robinson, A. C., 1994, The impact of comet Shoemaker-Levy 9 on Jupiter, *Shock Waves, 4*, pp. 47-50.

[5] Crawford, D. A., Boslough, M. B., Trucano, T. G. and Robinson, A. C., 1995, The Impact of Comet Shoemaker-Levy 9 on Jupiter, *International Journal of Impact Engineering*, pp. 253-262.

[6] Chapman, C. R., 1996, Galileo observations of the impacts, in *The Collision of Comet Shoemaker-Levy 9 and Jupiter*, K. S. Noll, H. A. Weaver and P. D. Feldman, eds., Cambridge Univ. Press, 121-132.

[7] Hammel, H. B., 1996, HST imaging of Jupiter shortly after each impact: Plumes and fresh sites, in *The Collision of Comet Shoemaker-Levy 9 and Jupiter*, K. S. Noll, H. A. Weaver and P. D. Feldman, eds., Cambridge Univ. Press, 111-120.

[8] Nicholson, P. D., 1996, Earth-based observations of impact phenomena, in *The Collision of Comet Shoemaker-Levy 9 and Jupiter*, K. S. Noll, H. A. Weaver and P. D. Feldman, eds., Cambridge Univ. Press, 81-109.

[9] Crawford, D. A., 1997, Comet Shoemaker-Levy 9 Fragment Size Estimates: How big was the Parent Body?, in *Near-Earth Objects, The United Nations International Conference,* John L. Remo, ed., Annals of the New York Academy of Sciences, Vol 822, pp. 155-173.

[10] Nemtchinov, I. V. et al., 1996, Spectral Opacities and Ablation Rates for Recognized Meteoroid Materials, Interim Report 001 on task 1.0 under the Sandia National Laboratories Contract AS-6889, Institute for Dynamics of Geospheres, Russian Academy of Sciences, Moscow, 136 pp.

[11] Lagage, P. O. et al., 1995, SL-9 fragments A, E, H, L, Q1 collision on to Jupiter: Mid-infrared light curves, *Geophys. Res. Lett., 22*, 1773.

[12] Takeuchi, S. et al., 1995, Near-IR imaging observations of the cometary impact into Jupiter: Time variation of radiation from impacts of fragments C, D and K, *Geophys. Res. Lett., 22*, 1581.

[13] McGregor, P. et al. 1996, CASPIR observations of the collision of Comet Shoemaker-Levy 9 with Jupiter, *Icarus, 121*, 361-388.

[14] West, R. A., 1996, Particulate matter in Jupiter's atmosphere from the impacts of Comet P/Shoemaker-Levy 9, in *The Collision of Comet Shoemaker-Levy 9 and Jupiter*, K. S. Noll, H. A. Weaver and P. D. Feldman, eds., Cambridge Univ. Press, 269-292.

[15] Scotti, J. V. and Melosh, H. J., 1993, Tidal Breakup and Dispersion of P/Shoemaker-Levy 9: Estimate of Progenitor Size, *Nature, 365*, 7333.

[16] Asphaug, E. and Benz, W., 1994, Density of Comet Shoemaker-Levy 9 deduced by modelling breakup of the parent 'rubble pile', *Nature, 370*, 120-124.

[17] Crawford, D. A., Taylor, P. A., Bell, R. L. and Hertel, E. S., 2002, Adaptive Mesh Refinement in the CTH Shock Physics Hydrocode, *New Models and Hydrocodes for Shock Wave Processes in Condensed Matter,* Edinburgh, U.K., May 19-24, 6 pages.

[18] Canup, R. M., Barr, A. C. and Crawford, D. A., 2013, Lunar-forming impacts: High-resolution SPH and AMR-CTH simulations, *Icarus, 222*, pp. 200-219.

[19] Crawford, D.A., 2011, *42nd Lunar and Planetary Sci. Conf.*, Abs. #2112 (http://www.lpi.usra.edu/meetings/lpsc2011/pdf/2112.pdf).

[20] Crawford, D. A. and Schultz, P. H., 1988, Laboratory observations of impact-generated magnetic fields, *Nature, 336*, pp. 50-52.

[21] Crawford, D. A. and Schultz, P. H., 1991, Laboratory Investigations of Impact-Generated Plasma, *Journal of Geophysical Research, 96*, pp. 18,807-18,817.

[22] Crawford, D. A. and Schultz, P. H., 1993, The Production and Evolution of Impact-Generated Magnetic Fields, *International Journal of Impact Engineering, 14*, pp. 205-216.

[23] Crawford, D. A. and Schultz, P. H., 1999, Electromagnetic Properties of Impact-Generated Plasma, Vapor and Debris, *International Journal of Impact Engineering, Vol.* 23, pp. *169-180.*

[24] Crawford, D. A., 2015, Computational modeling of electrostatic charge and fields produced by hypervelocity impact, *Proceedings of the 13th Hypervelocity Impact Symposium, Procedia Engineering, Vol. 103*, pp 89-96.

[25] Crawford, D. A., 2019, Simulations of Magnetic Fields Produced by Asteroid Impact: Possible Implications for Planetary Paleomagnetism, *Proceedings of the 15th Hypervelocity Impact Symposium*, this volume.

Proceedings of the 2019 Hypervelocity Impact Symposium
HVIS2019
April 14-19, 2019, Destin, FL, USA
HVIS2019-017

Extended Abstract+

Mesoscale modeling and debris generation in hypervelocity impacts

Stephanie N. Q. Bouchey* and Jeromy T. Hollenshead

Sandia National Laboratories, PO Box 5800 Mail Stop 1185, Albuquerque, NM 87185-1185
SAND2019-5602 A

1. Introduction

Material fragmentation after a hypervelocity impact is of interest to predictive electro-optical and infrared (EO/IR) modeling. Successful comparisons with data require that submicron fragments are generated in such impacts; however, experimental data has so far been unable to produce fragments of this scale [e.g., 1-3]. This effort investigated the generation of predicted debris from hypervelocity impact of a sphere on a flat, semi-infinite plate. It is hypothesized that explicit modeling of grains, especially in the presence of void and varying grain properties, may lead to differences in predicted strain rates (locally higher) associated with the grain boundaries. Such an effect may lead to smaller predicted fragments sizes than when using the traditional bulk modeling approach and may provide improved understanding of fragmentation modeling in hypervelocity impacts. Comparisons of predicted strain rates at failure (a proxy for fragment size) and material temperature were made between simulations run using a bulk modeling approach and a mesoscale grain modeling approach.

2. Input development

Computational modeling was performed at Sandia National Laboratories using the shock physics analysis package, CTH [4-5]. The current investigation was based on a two-dimensional (2D) description of the problem and was comprised of two suites of simulations. The first suite tested the standard bulk (isotropic, homogeneous) structure and was the nominal case with which to compare the results of the second study, which explicitly modeled mesoscale grains. The nominal problem design was an aluminum sphere impacting a semi-infinite, flat aluminum plate target at a velocity of 4 km/s. Each material was modeled with the SESAME equation of state, a Johnson-Cook strength model, a Johnson-Cook fracture model, and a Grady-Kipp fragmentation model.

Although the bulk model was developed first, many of the system characteristics (e.g. resolution, problem geometry) were driven by the expected requirements for the mesoscale model. High resolution through the grains necessitated exceptionally small simulation domains in order to incorporate the sphere-on-plate geometry. Additionally, the target was designed to be much larger than the projectile to mitigate edge and boundary effects. The projectile was nominally 7 mm in diameter (500 grains across), and the total target area was 2.1 cm wide by 4.1 cm deep. To fill the entire target area with grains would be prohibitively expensive. Thus, a smaller "grains region" of the target was modeled to be 3.5 mm thick and 7 mm wide (14 mm wide when mirrored about the axis). In the bulk model, this small target area was modeled as a single material that matched the material properties of the larger surrounding target matrix area. In both cases, the target matrix area was modeled with a single, bulk material.

+ Full paper can found at https://www.sciencedirect.com/journal/international-journal-of-impact-engineering/special-issue/10MTGP5W4VJ

* Corresponding author. Tel.: +1-505-284-8128.
E-mail address: snquint@sandia.gov.

Traditional model descriptions within CTH are typically macro-scale and do not include any explicit grain structure representation. Consequently, a procedure was necessary to model and initialize grains within CTH. A MATLAB Monte Carlo simulation of 2D grain growth using a Q-state Potts model generated a unit cell of grains. The current configuration allows for eight different grain types in both the target and projectile (16 total materials). Each grain type was randomly distributed within the unit cell, and the entire cell was then scaled such that the average grain size was ~14 μm in diameter. The 14 μm size was estimated from Nakai and Itoh [6], based on the short axis of a characteristic aluminum grain. Finally, this scaled unit cell was tiled repeatedly to generate the impact geometry (projectile and target grains areas), and the result was written out pixel by pixel in a form (diatom) that could be read by CTH.

3. Bulk model

The bulk model suite simulated a projectile striking a target within a larger target matrix. This model suite enabled the following studies: resolution, impact velocity, projectile size, and projectile shape. A custom post-processing script produced histograms and cumulative distribution functions (CDFs) for strain rate at failure and material temperature. For brevity, only histograms will be shown here.

First, an appropriate resolution needed to be defined. For the resolution study, a single projectile size was evaluated across a range of resolutions from 0.875 to 14.0 μm at the nominal impact velocity of 4 km/s. Convergence occurred for material temperature but not the strain rate at failure. This result is not surprising as the Grady-Kipp model used here is dependent on resolution. Rise time of the initial peak in strain rate histograms was chosen as an indication of convergence. Based on these results, the 3.5 μm was chosen as an acceptable mesh resolution.

Next, the projectile size study assessed the effects of projectile curvature and overall size relative to the domain. Histograms and CDFs were evaluated at equivalent times for each model. The histogram plots for strain rate at failure and temperature are shown in Fig. 1 for the projectile. Material temperature was similar between all projectile sizes and appears to be related to geometry scaling. Conversely, strain rate at failure varied significantly between projectile sizes, with smaller projectiles manifesting higher strain rates. In the projectile, the histogram of strain rate at failure exhibited a bimodal shape. The distance between the peaks in each strain rate histogram appears to be at least partially driven by transit time across the object (i.e., by a characteristic length scale, which is projectile diameter in this case). Strain rate at failure was also recorded in the semi-infinite target. There, the strong bimodality seen in the projectile was absent, which suggests that the bimodality may be due primarily to projectile shape. In both the projectile and target, however, onset of the strain rate histogram was related to projectile size.

Three additional simulations tested projectile shape by impacting a cube (face-on), rather than a sphere, into the target at 4 km/s. Results for the projectile are shown in Fig. 2. The lack of bimodality in Fig. 2a is evidence that bimodality is a product of projectile curvature (as was suggested in the previous projectile size study). A more step-like strain rate at failure histogram was apparent, especially in the larger cube impacts. The size of the cube, however, did not influence temperature; this result was also seen in the sphere impacts.

A final set of bulk model simulations explored the effect of increasing impact velocity from the nominal 4 km/s to 6 km/s. This impact velocity increase resulted in a shift to higher strain rates and temperatures, but overall shape and bimodality (in the projectile) were not significantly affected. Impact velocity clearly controls temperature in the projectile, although temperature histograms were nearly unchanged by projectile size for impacts at the same speed.

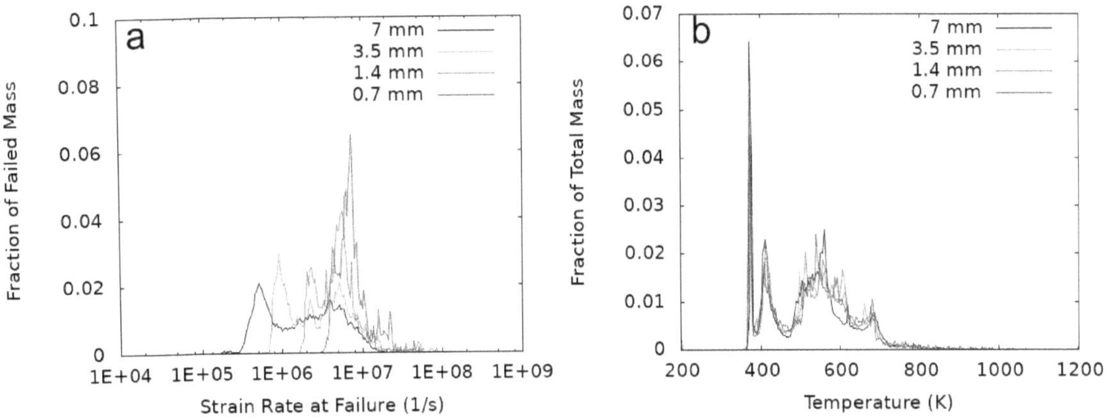

Fig. 1. Projectile size study results. Histograms for (a) strain rate at failure and (b) material temperature for a sphere impacting at 4 km/s.

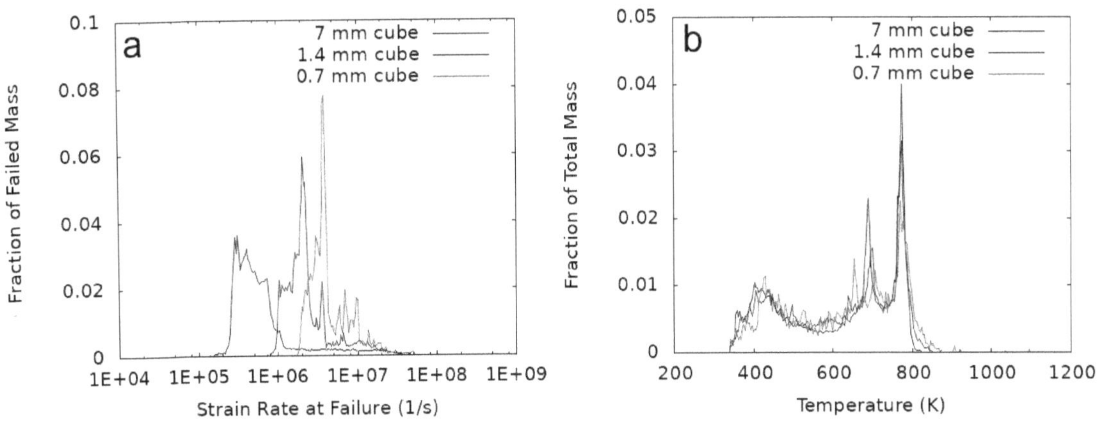

Fig. 2. Projectile shape study results. Histograms for (a) strain rate at failure and (b) material temperature for a face-on cube impacting at 4 km/s.

4. Mesoscale model

In the mesoscale model, the projectile and "grains region" of the target were filled with individual grains, which were described in CTH as separate materials. Eight grain types were randomly distributed through the projectile by the tile method described in Section 2. Similarly, eight (separate) grain types were randomly distributed and tiled through the target grains region. The surrounding target matrix was defined as a single, bulk material. This modeling approach enabled the following studies: fracture strength distribution, yield strength distribution, and porosity. Because each grain type is a separate material, a separate histogram and CDF was produced for each grain type. The histograms/CDF results of each grain type were then averaged for ease of comparison with the isotropic simulation. An initial "consistency case" where all grain types were the modeled as the same material produced the closest approximation to the bulk simulations and served as the standard of comparison for the rest of the simulations (Fig. 3, black line).

The first deviation from the consistency case modified fracture strength. Aluminum alloys vary in fracture strength, and so to simulate this variation in the mesoscale model, a normal distribution of plausible fracture strengths, centered on the fracture strength of Aluminum 1100 was applied to the grains in both the projectile and the target. Results for this study can be found for the projectile in Fig. 3 (green line). An increase in the lowest strain rates compared to the grains consistency result was observed, but the amount of material at higher strain raters remained similar between the two tests. Additionally, temperature remained largely unchanged between the fracture study results and the consistency result.

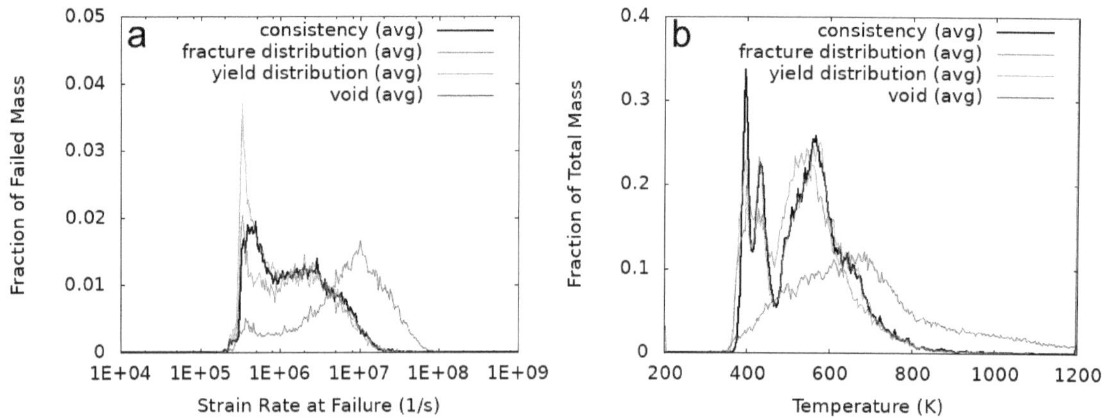

Fig. 3. Summary of mesoscale study results. Histograms for (a) strain rate at failure and (b) material temperature for a sphere impacting at 4 km/s. The insertion of material interfaces (through void) drastically increases strain rate at failure and material temperature.

Another important material property is yield strength. In a process similar to that described above, a distribution of reasonable yield strength values for aluminum was determined by varying parameters in the Johnson-Cook strength equation [7-8]. This distribution was then incorporated into the grain-type materials of both the target and the projectile. Results indicate that changing the strength of the grains reduced the amount of lower strain rates of failure that were produced, but it did not have a substantial effect on higher strain rates at failure or material temperature (Fig. 3, blue line).

Finally, the last suite of simulations explored the role of porosity (void space) in the grain structure. To test the upper bounds of porosity effects, all grains of a single grain-type were removed from the calculation. The resulting void space caused interfaces between grains that allowed several sets of shock and rarefaction waves to pass through the grains. Such behavior dampened the overall shockwave but resulted in locally higher strain rates at failure and material temperatures. This local increase in both parameters resulted in substantial deviation from the grains consistency study (Fig. 3, purple line). The strain rate at failure histogram nearly lost its bimodal shape in the projectile and shifted to higher strain rates in both the projectile and the target. This shift indicates that the projectile size was no longer the dominant property controlling the strain rate at failure. Instead, individual grains became more important.

5. Summary

This work compared the effects of modeling grain structure in hypervelocity impact simulations. Comparisons of strain rate at failure (fragment size) and material temperature were made between a suite of simulations performed with the standard bulk modeling structure and one in which individual grains were modeled. Smaller fragments or higher temperatures are needed to match EO/IR signatures from observed impacts. Results from the various studies described herein indicate that strain rate at failure is influenced primarily by projectile size, impact velocity, and material porosity. Material temperature is predominantly influenced by impact velocity and porosity; not by projectile size. Changes to the material properties within grains tended to affect lower strain rates only, but material interfaces (here, manifested as material porosity) drastically increased strain rate at failure and material temperatures. Higher strain rates are likely to produce smaller debris fragments, which, along with hot debris may help provide evidence supporting the generation of sub-micron fragments currently required by many EO/IR predictive models to successfully compare with observed hypervelocity impacts. Future work will focus on extending the study to three dimensions, assessing more realistic grain aspect ratios, and simulating other types of interfaces such as inclusions and dislocations.

Acknowledgements

The authors wish to thank Torch Technologies (specifically Jim Burke, Steven Henke, and Karen Hirsch) for supporting this work. This paper describes objective technical results and analysis. Any subjective views or opinions that might be expressed in the paper do not necessarily represent the views of the U.S. Department of Energy or the United States Government. Sandia National Laboratories is a multimission laboratory managed and operated by National Technology & Engineering Solutions of Sandia, LLC, a wholly owned subsidiary of Honeywell International Inc., for the U.S. Department of Energy's National Nuclear Security Administration under contract DE-NA0003525.

References

[1] Glenn, L. A. and Chudnovsky, A., 1986. Strain-energy effects on dynamic fragmentation, Journal of Applied Physics, vol. 59, no. 4, p. 1379.

[2] Grady, D. E. and Kipp, M. E., 1997. Fragmentation properties of metals, International Journal of Impact Engineering, vol. 20, no. 1-5, pp. 293-308.

[3] Grady, D. E., 2008. Fragment size distributions from the dynamic fragmentation of brittle solids, International Journal of Impact Engineering, vol. 35, no. 12, pp. 1557-1562.

[4] McGlaun, J.M., Thompson, S.L., Elrick, M.G., 1990. CTH: A Three-Dimensional Shock Wave Physics Code, International Journal of Impact Engineering, 10(1-4), pp. 351-360.

[5] Hertel, E.S. Bell, R.L., Elrick, M.G., Farnsworth, A.V., Kerley, G.I., McGlaun, J.M., Petney, S.V., Silling, S.A., Taylor, P.A., Yarrington, L., 1993. "CTH: A Software Family for Multi-Dimensional Shock Physics Analysis," Proceedings of the 19th International Symposium on Shock Waves, pp. 377-382.

[6] Nakai, M., Itoh, G., 2014. The Effect of Microstructure on Mechanical Properties of Forged 6061 Aluminum Alloy, Materials Transactions 55, pp. 114-119.

[7] Johnson, G.R., Cook, W.H., 1983. "A Constitutive Model and Data for Metals Subjected to Large Strains, High Strain Rates and High Temperatures," Seventh International Symposium on Ballistics, The Hague, Netherlands.

[8] Silling, S.A., 1996. "CTH Reference Manual: Viscoplastic Models," SAND91-0292, Sandia National Laboratories, Albuquerque, NM.

Proceedings of the 2019 Hypervelocity Impact Symposium
HVIS2019
April 14-19, 2019, Destin, FL, USA
HVIS2019-033

Hypervelocity penetration of granular silicon carbide from mesoscale simulations

Brian J. Demaske* and Tracy J. Vogler

Sandia National Laboratories, Livermore, CA 94540, USA

Abstract

Penetration of gold rods into SiC powder targets at velocities of 1 to 3 km/s are investigated using mesoscale simulations. The range of impact velocities is chosen to coincide with previous penetration experiments and represents a new regime over which to test the applicability of mesoscale simulations of granular materials. Both 2D and 3D geometries of the combined penetrator and powder system are considered. Analysis of the penetration depth histories at various impact velocities shows the penetrator undergoes an initial transient period of rapid deceleration within the first several microseconds before converging to a steady state characterized by jumps in the penetration velocity on the order of a few hundred meters per second. Steady-state penetration velocities obtained from 2D and 3D simulations agree well with one another, but lie below those computed using hydrodynamic theory, which indicates a non-zero strength for the simulated powders over this range of impact velocities. For comparable initial powder densities, 3D simulations predict steady-state penetration velocities in good agreement with those measured in penetration experiments on pre-compacted SiC powder specimens.

Keywords: hypervelocity penetration; mesoscale simulations; silicon carbide; granular materials

Nomenclature

v_p	impact velocity (km/s)
L	penetrator length (mm)
P_{d}	penetration depth (mm)
$\dot{P}_{\mathrm{d}}$	instantaneous penetration velocity (km/s)
u	steady-state penetration velocity (km/s)

1. Introduction

Mesoscale simulations of powders, in which individual grains are explicitly resolved, have shown considerable success in modeling the shock compaction behavior of granular metals [1-3] and ceramics [4-8] as well as low-velocity penetration of sand [9]. However, mesoscale simulations have yet to be applied to the regime of hypervelocity penetration. Anderson et al. [10] performed a series of experiments in which pre-compacted SiC powder specimens were launched at velocities of 1–3 km/s at stationary gold rods. The experiments were designed to determine the effect of varying levels of target pre-damage on the penetration response of SiC. In this sense, the powder was chosen to represent the limiting state of a pre-damaged SiC target as it is the raw material from which all SiC targets were made [10]. Flash X-rays taken at different time delays were used to monitor the progress of the gold rod through the powder specimens. It was found that the penetration depth exhibited steady-state (or

* Corresponding author. Tel.: +1-925-294-1138.
E-mail address: bjdemas@sandia.gov.

constant-velocity) motion for impact velocities > 1.5 km/s. Measured steady-state penetration velocities were found to increase linearly with impact velocity, which is a well-known property of many types of projectile and target materials for this range of impact velocities [11]. Moreover, the penetration velocities were found to lie below those predicted by hydrodynamic theory thereby indicating a non-zero strength of the powder under penetration.

In this work, a mesoscale simulation approach is adopted to model the penetration response of a gold projectile into granular SiC. Continuum simulations utilizing a linear compaction model and pressure-dependent strength model for the powder have already been performed by Holmquist et al. [12]. In that work, parameters for the compaction and strength models for the SiC powder were obtained by fitting the simulations to results of the penetration experiments discussed above. By contrast, in this work, there is no fitting to experiment and material models for the target and projectile are assumed to be simple with parameters derived from basic material properties. Both 2D and 3D geometries of the problem are considered. Mesoscale simulations using the 2D geometry are computationally much cheaper to perform, so it is desirable to compare results to those obtained for the full 3D geometry. Penetration depth histories are calculated for different initial densites and configurations of the simulated powder beds over the full range of impact velocities from experiment. Steady-state penetration velocities obtained from 2D and 3D simulations are compared to one another as well as to results of experiments [10].

2. Simulation methods

The dynamics of a gold rod penetrating a bed of SiC powder is simulated using the finite-volume Eulerian mechanics code CTH [13]. In the experiments in Ref. [10], the SiC powder specimens were contained in an Al sleeve with an inner diameter of 18 mm and length of 35 mm. The powders had an estimated grain size of ~1-3 μm and, prior to the penetration experiments, were pre-compacted to ~73% of the theoretical maximum density (TMD) of SiC. The penetrator was a gold rod with a diameter of 1 mm and length of 70 mm. Our aim is to reproduce the initial conditions of experiments as closely as possible. However, it is not possible to reproduce the small grain size and large size of the powder compacts simultaneously. Assuming a spherical grain size, one powder compact would contain roughly a trillion grains, which would lead to an intractable number of computational cells in the simulations even for a very rough mesh resolution of one cell per grain. Thus, the grain size in simulation is increased by one-hundred-fold relative to experiment, i.e., the simulated grain size range is 100-300 μm.

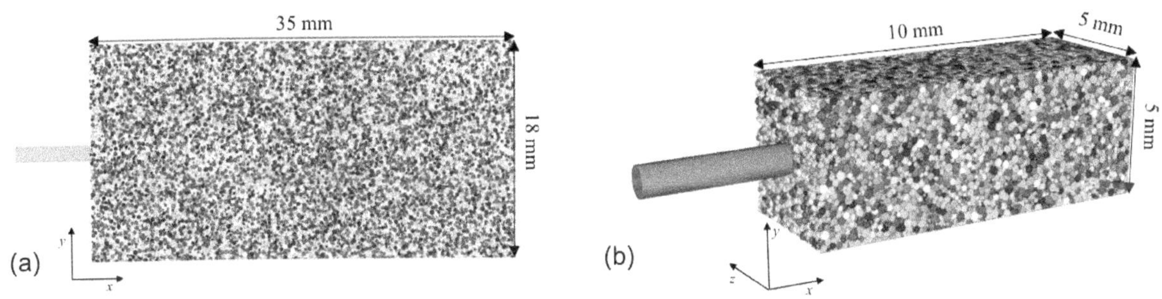

Fig. 1. Illustrations of the (a) 2D and (b) 3D geometries used in the mesoscale simulations. Only the portion of the penetrator initially part of the mesh is shown.

As stated in the Introduction, we consider both 2D and 3D geometries of the combined projectile and powder system. Illustrations of the 2D and 3D simulation geometries are shown in Fig. 1. Grains making up the SiC powder bed are modeled as spheres in the 3D geometry, whereas they are modeled as circles in the 2D geometry with infinite extent along the z-axis. Due to the mesoscale treatment of the powder, it does not make physical sense to use an axisymmetric geometry for the 2D simulations as this would lead to toroidally-shaped grains. Powder bed dimensions for the 2D geometry are 35×18 mm^2, the same as the powder specimens in experiment, whereas powder dimensions for the 3D geometry are $10 \times 5 \times 5$ mm^3. Due to computational limitations, it was not possible to go to larger dimensions for the 3D geometry. In both geometries, the impact direction is taken as the $+x$-axis. The penetrator is treated as a slab in the 2D simulations with a thickness of 1 mm along the y-axis, whereas it is treated as a cylinder in the 3D simulations with a diameter of 1 mm. Its length along the x-axis is 70 mm. The penetrator is positioned to the left of the powder bed along the x-axis and centered along the y-axis and z-axis, where necessary.

Initial grain arrangements are generated using the granular package of the molecular dynamics (MD) code LAMMPS [14]. Each grain is treated as a finite-size particle that interacts elastically with neighboring particles via a Hertzian force field. Damping and tangential forces are ignored. Particles are inserted at random locations inside the simulation domain and assigned a diameter from a uniform distribution of 100-300 μm. By varying the number of particles, the overall volume fraction of the powder bed can be controlled. The insertion step does not consider the finite size of the particles, so initial particle arrangements are highly-overlapped. Once inserted, each particle is given a small velocity with a random orientation. To reduce the large forces resulting from the high-degree of particle overlap, a step-size-limited MD run is performed until the overlap is reduced below a threshold value. Grain beds with high volume fractions can be produced by setting this threshold to be a significant fraction of the minimum particle diameter. Afterwards, velocities are reinitialized and particles trajectories are evolved for the time it takes for an average particle to traverse several mean free paths. The initialization procedure outlined here is similar to previous work [4, 5] except here initial grain locations are random instead of ordered and particles are free to overlap.

The final particle locations from the MD runs are used as input for the initial powder bed in the mesoscale simulations performed using CTH. All grains are treated as having the properties of fully-dense polycrystalline SiC. The penetrator and grains are described by a Mie-Grüneisen equation of state (EOS) for gold and SiC, respectively. Following previous mesoscale simulation work for shock compaction of granular WC [4, 5, 7], the strength behavior of the SiC grains is described by a constant von Mises yield surface with a yield strength taken from results of planar impact experiments [15]. This is the simplest approach to modeling strength and results from these simulations will help determine whether more sophisticated models are necessary in future work. At the melt temperature of 3100 K [16], the strength of SiC goes to zero. The strength of the gold penetrator is described by a rate-independent Steinberg-Guinan-Lund (SGL) viscoplastic model, which includes effects of strain-hardening, a pressure- and temperature-dependent shear strength and thermal softening with melting described by a Lindemann law [17, 18]. Fracture is modeled using a simple cap on the hydrostatic tensile pressure. Fracture strengths for gold and SiC are taken from spall strength data from planar impact experiments. The parameters for the EOS and strength models used in the CTH simulations are provided in Table 1. Values for the yield strength and melting temperature for gold listed in Table 1 are the corresponding values at zero pressure, room temperature and zero plastic strain.

Table 1. Parameters for Mie-Grüneisen equation of state and strength models for silicon carbide and gold used in CTH calculations.

Material		Silicon carbide	Gold
Solid density	ρ_s (g/cm^3)	3.227	19.240
Bulk sound speed	C_0 (km/s)	8.29	3.056
Hugoniot shock velocity slope	S	1.21	1.572
Grüneisen coefficient	Γ	0.95	2.97
Specific heat	C_v (J/g·K)	0.862	0.127
Yield strength	Y (GPa)	8.0 [15]	0.02*
Poisson ratio	ν	0.168 [19]	0.42
Fracture strength	σ_f (GPa)	0.65 [20]	3.3 [21]
Melting temperature	T_m (K)	3100 [16]	1970*

*initial values at zero pressure, room temperature and zero plastic strain

At the start of a CTH simulation, the penetrator is given an initial uniform velocity v_p along the $+x$-axis. Periodically during the simulation, location and volume fraction data for cells containing penetrator material are output. The penetration depth $P_d(t)$ at time t is defined as the position along the x-axis where the cumulative mass of penetrator material summed from right to left exceeds the corresponding mass for a 0.1 mm long section of the undeformed projectile. Cell sizes of 16.667 μm and 25 μm are used in the 2D and 3D simulations, respectively. Decreasing the cell size below these values was found to not have a significant effect on the trajectory of the penetrator. Plots of material interfaces, see, for example, Fig. 2, are used to visualize the penetrator and powder bed at different instants in time during the simulation.

3. Simulation results

Snapshots from a 2D mesoscale simulation of a gold slab penetrating a SiC powder bed are shown in Fig. 2. The penetrator has an initial velocity $v_p = 2.088$ km/s and the powder bed has an initial volume fraction of 54.8%, which is considerably less than ~73% for the powder compacts used in experiment [10]. The snapshot in Fig. 2(a) is taken at an early stage of the penetration process, while the snapshot in Fig. 2(b) is taken at a much later stage when the penetrator has almost reached the far end of the powder bed. Once the penetrator contacts the powder bed, a compaction wave forms around the head of the penetrator and propagates radially outward ahead of the penetrator into the powder bed. The curved front of the compaction wave can be seen clearly in both snapshots. Grains nearest to the penetrator exhibit the most deformation with the amount of deformation decreasing as one moves farther away from the penetrator. As the penetrator propagates into the powder bed, it starts to flow perpendicular to the impact direction and forms a mushroom-like head, which is clearly visible in Fig. 2(a). Penetrator material is deposited along the powder bed interface as it makes its way through the powder bed. Cracks formed from failure of intergranular contacts can be seen in Fig. 2(b).

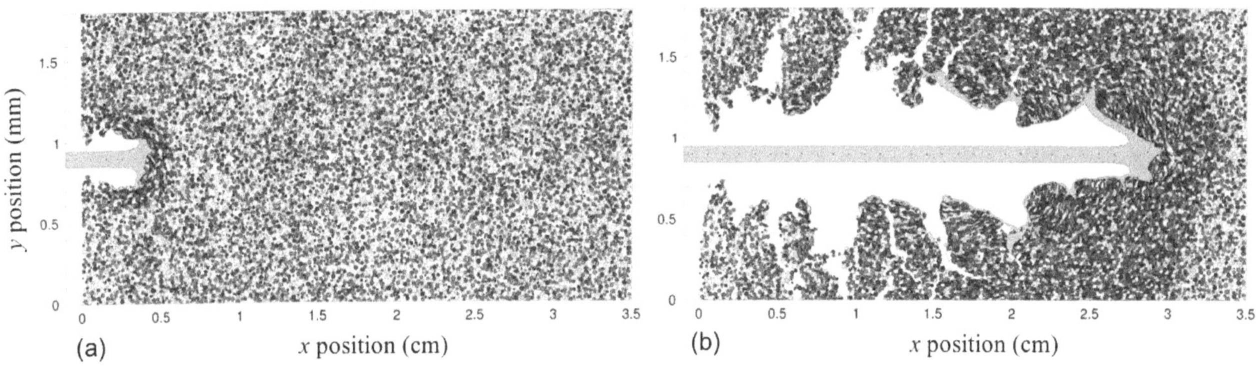

Fig. 2. Snapshots at (a) 2.7 μs and (b) 19.9 μs from a 2D mesoscale simulation for an impact velocity of 2 km/s and initial powder bed volume fraction of 54.8%. The penetrator is colored gray, while the grains are multi-colored. Only the portion of the penetrator inside the powder bed is shown.

The time-history of the penetration depth $P_d(t)$ and penetrator length $L(t)$ are plotted in Fig. 3(a). Both plots exhibit nearly linear behavior from the start of penetration to the end of the simulation, at which point the penetrator has reached a depth of ~33 mm. The initial length of the penetrator is 70 mm and, over this time interval, its length has decreased to ~56 mm. The penetration depth data in Fig. 3(a) is numerically differentiated with respect to time to compute the instantaneous penetration velocity $\dot{P}_d$**Error! Bookmark not defined.**(t), which is plotted in Fig. 3(b). At time zero, the penetration velocity equals the impact velocity v_p, but decreases rapidly as the penetrator contacts the first several grains in the powder bed. After several microseconds, $\dot{P}_d(t)$ converges to a steady-state characterized by fluctuations on the order of a few hundred meters per second about a constant level of ~1.5 km/s. Periods of increasing penetration velocity occur when small finger-like projections propagate ahead of the main interface with the powder bed, while periods of decreasing velocity occur when these projections change direction or slow their growth.

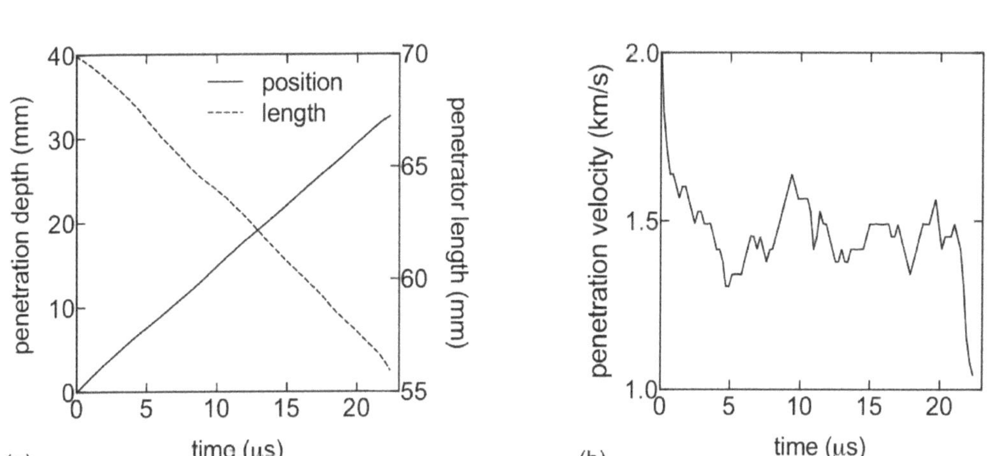

Fig. 3. Time-history of (a) penetration depth and penetrator length and (b) penetration velocity from a 2D mesoscale simulation at an impact velocity of 2.088 km/s. The powder bed has an initial volume fraction of 54.8%.

Simulations using the 2D geometry are performed for v_p = 1-3 km/s to determine the penetration response over the full range of impact velocities used in experiments [10]. For each value of v_p, ten runs are performed for different initial powder bed configurations with roughly the same initial volume fraction of 55% to determine what effect, if any, the initial grain arrangement has on the trajectory of penetrator. Initial volume fractions for the different powder bed configurations vary from 54.7% to 55.4% due to grain diameters being assigned randomly during the powder bed initialization process. The trajectories averaged over all ten grain arrangements are plotted in Fig. 4(a) along with the corresponding envelopes that bound the extrema trajectories. At early times, trajectories obtained for different powder bed configurations are all very similar as evident by the narrow envelope widths in the plots. However, as the penetrator propagates deeper into the powder bed, the envelope widths increase indicating more variation in the trajectories. Corresponding time-history plots of the instantaneous penetration velocities are shown in Fig. 4(b). As in Fig. 4(a), the grain-averaged responses are given by the solid curves, while the corresponding envelopes bound extrema trajectories. Although some overlap in the envelopes occurs for runs at different impact velocities, the grain-averaged penetration velocities increase monotonically with increasing v_p.

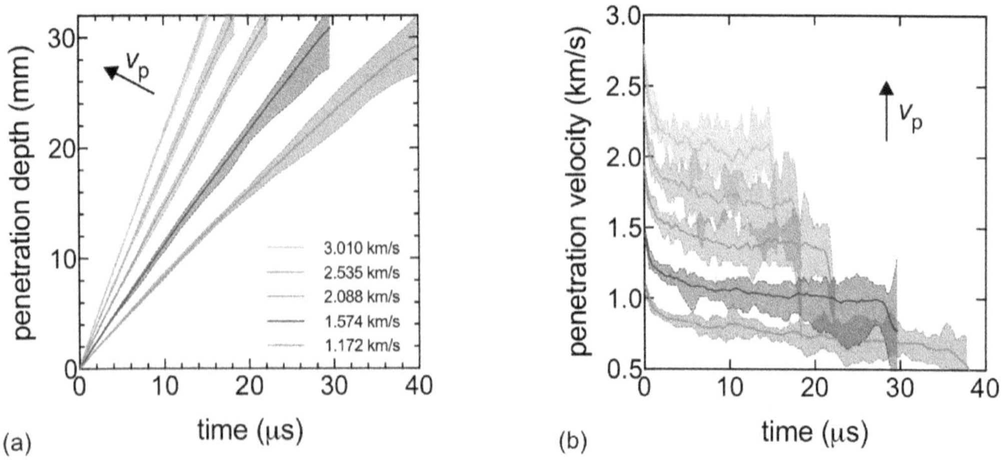

Fig. 4. Time-histories of (a) penetration depths and (b) penetration velocities from 2D mesoscale simulations for impact velocities v_p = 1-3 km/s. Envelopes bound the extrema trajectories for 10 runs with different initial powder configurations with the mean responses given by the solid curves.

Simulations using the full 3D geometry are carried out for v_p = 1-3 km/s as was done for the 2D geometry above. However, due to the high computational cost of the 3D simulations, only five initial powder bed configurations are used instead of the ten used in 2D. As in the 2D simulations, the random nature in which grain diameters are assigned during the powder initialization process leads to some variation in the initial volume fractions of the powder bed configurations used in the 3D simulations. $P_d(t)$ and $\dot{P}_d(t)$ data for each of the 3D simulations are calculated and averaged over the different powder bed configurations in an identical manner to the 2D simulations. Overall, the 3D simulations exhibit much of the same general behavior that was found for the 2D simulations, including an initial transient period of rapid deceleration when the penetrator makes contact with the powder bed followed later by a period of steady-state penetration behavior. However, for the 3D simulations there is less variation in the penetrator trajectories among the different powder bed configurations.

Grain-averaged responses for the instantaneous penetration velocity $\dot{P}_d(t)$ are plotted in Fig. 5 for both 2D and 3D geometries at v_p = 2.088 km/s. The approach to steady-state behavior takes considerably longer in 2D than in 3D. However, once a steady state is reached the penetraion velocities in both cases are similar. A dependence on the simulation geometry was also observed in the time required for formation of a quasi-steady compaction wave in piston-driven mesoscale simulations of granular WC [7]. The authors observed that it took roughly 16 grain diameters for the 2D simulations to achieve a quasi-steady stress state behind the compaction wave front, whereas it took only 1.8 grain diameters in 3D. In the penetration simulations in this work, a compaction wave is generated in the powder due to the motion of the penetrator, see Fig. 2. Therefore, it is not surprising that the transient penetration velocity is dependent of the geometry of the system, as similar processes must be governing the approach to steady-state behavior in both cases.

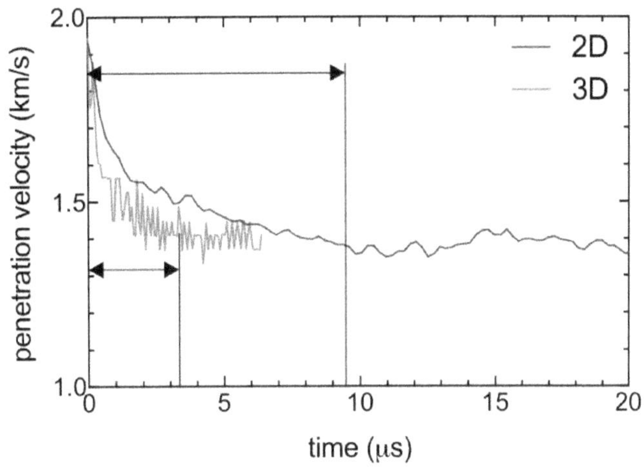

Fig. 5. Grain-averaged pentration velocities for 2D and 3D simulations for an impact velocity of 2.088 km/s. Arrow widths denote the time required for the penetration velocity to approach steady-state behavior.

The steady-state penetration velocity u for a given value of v_p is calculated by fitting a line to the grain-averaged $P_d(t)$ data once the penetrator has achieved steady-state behavior as discussed in the preceeding paragraph. The u-v_p data for both 2D and 3D simulations are plotted in Fig. 6 along with the response from hydrodynamic theory for a target of the same density. Hydrodynamic theory assumes the penetrator and target are incompressible and have zero strength. No data point for $v_p = 1.172$ km/s is plotted for the 2D simulations as steady-state penetration was not achieved in this case. The calculated u-v_p data for the 2D and 3D simulations lie very close to one another indicating that there is almost no difference in the penetration behavior in the steady-state regime. The u-v_p data from the simulations exhibit linear behavior in agreement with the penetration experiments [10]. The equations for the linear fits are given by $u = 0.6971v_p - 0.0710$ km/s and $u = 0.7358v_p - 0.1262$ km/s for the 2D and 3D simulations, respectively. Results from the 2D and 3D simulations lie below the hydrodynamic response, which indicates the simulated powders exhibit non-zero strength under penetration.

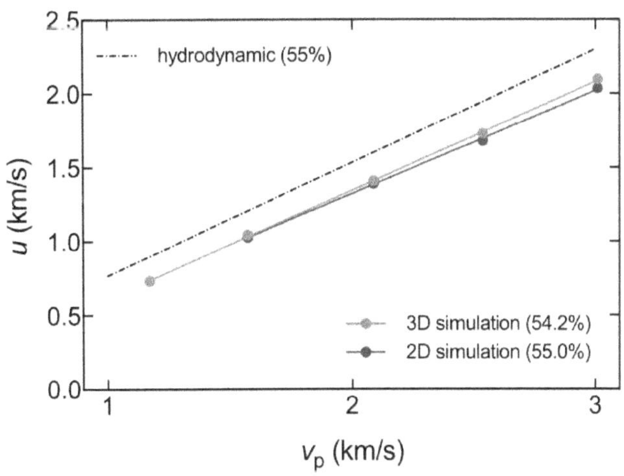

Fig. 6. Steady-state penetration velocity u versus impact velocity v_p from (a) 2D and (b) 3D mesoscale simulations for an initial powder bed volume fraction of ~55%. The penetration response from hydrodynamic theory is plotted for comparison.

To determine the effect of the initial powder density on the penetration response, penetration simulations using the 3D geometry are carried out for v_p = 1-3 km/s using powder beds with initial densities up to 73% of the TMD of SiC, which corresponds to roughly the density of the pre-compacted powder specimens used in the penetration experiments of Ref. [10]. In contrast to data obtained from simulations for initial powder densities of ~55%, $P_d(t)$ data obtained from simulations at higher initial powder bed densties start to show significant nonlinearity at later times, especially when v_p is low. In these cases, u is calculated by fitting a line to the $P_d(t)$ data over the time interval where it displays approximately linear behavior, as was done in experiment [10]. Independent of the initial density of the powder bed, it is found that u increases linearly with increasing v_p as was found for the 55%-dense powder beds. The linear fits to the simulation data are plotted in Fig. 6(a). From the figure, it is clear that u decreases as the initial density of the powder bed increases, which is expected as more densely-packed powder beds should be more resistant to penetration.

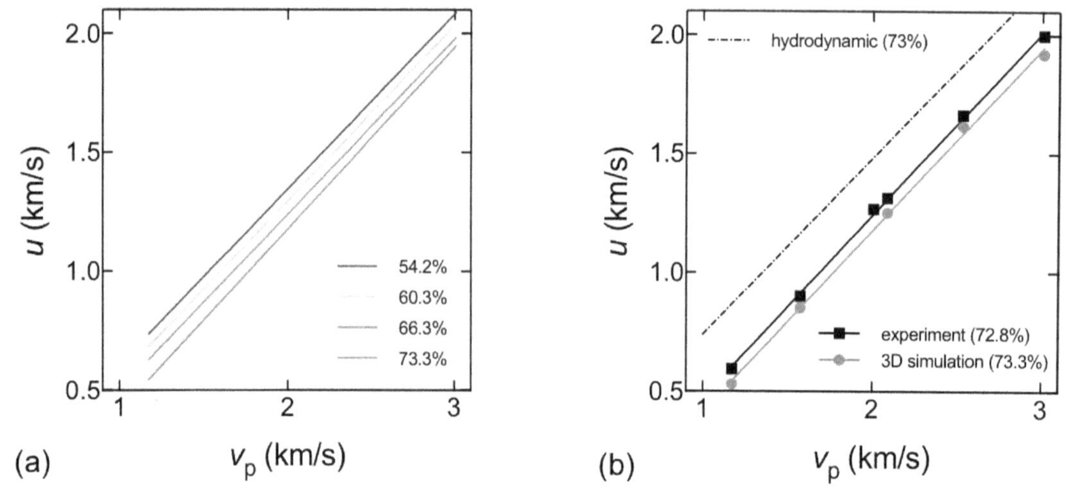

Fig. 7. Steady-state penetration velocity u versus impact velocity v_p from 3D mesoscale simulations for (a) different initial volume fractions of powder bed and (b) at a volume fraction of ~73% to allow comparison with results from penetration experiments [10] and hydrodynamic theory.

Measured u-v_p data from penetration experiments are plotted in Fig. 7(b) along with results from 3D simulations. The volume fractions for the powder specimens used in experiment are 72.8%, which is slightly lower than the average volume fraction of 73.3% for the powder beds used in the simulations. Overall, steady-state penetration velocities predicted by simulations are in good agreement with those measured in experiment over the entire range of impact velocities investigated. The linear fit to the 3D simulation data is given by $u = 0.7639v_p - 0.3506$ km/s, while the fit to the experimental data is given by $u = 0.7606v_p - 0.2760$. The slopes of the two fits agree extremely well, while simulations predict a y-intercept that is slightly more negative than experiment. In experiment, the effects of target strength are said to be largely contained in the y-intercept with target strength increasing for more negative y-intercepts [10]. Therefore, as the slopes obtained in experiment and simulation are nearly the same, we conclude that the simulated powder beds exhibit a slightly higher strength than the powder specimens used in experiment. This may be due in part to the 0.5% higher initial density for the powder beds used in the simulations.

4. Conclusions

Mesoscale simulations were performed of a gold rod impacting a SiC powder bed at velocities of 1-3 km/s. Such simulations represent the first application of the mesoscale modeling approach to the hypervelocity regime of penetration. Both 2D and 3D variants of the problem were investigated. It was found that for comparable initial powder densities simulations performed using the two geometries gave similar steady-state penetration velocities. However, the approach to steady state behavior differed between the two geometries with 2D simulations exhibiting a significantly longer transient period compared to the 3D simulations. Steady-state penetration velocities obtained from 3D simulations are in good agreement with results of penetration experiments on pre-compacted SiC powder specimens [10], which demonstrates the applicability of the mesoscale modeling approach to this regime of hypervelocity penetration problems. By tuning the strength of the SiC grains and/or by improving the description of intergranular contact, it may be possible to further improve the agreement with experiment.

Acknowledgements

Sandia National Laboratories is a multimission laboratory managed and operated by National Technology & Engineering Solutions of Sandia, LLC, a wholly owned subsidiary of Honeywell International Inc., for the U.S. Department of Energy's National Nuclear Security Administration under contract DE-NA0003525.

References

[1] Williamson, R. L., 1990, "Parametric studies of dynamic powder consolidation using a particle-level numerical model," Journal of Applied Physics, 68(3), pp. 1287-1296.

[2] Benson, D. J., 1994, "An analysis by direct numerical simulation of the effects of particle morphology on the shock compaction of copper powder," Modelling and Simulation in Materials Science and Engineering, 2(3A), p. 535.

[3] Benson, D. J., Do, I., and Meyers, M. A., 2002, "Computational Modeling of the Shock Compression of Powders," AIP Conference Proceedings, 620(1), pp. 1087-1092.

[4] Borg, J. P., and Vogler, T. J., 2008, "Mesoscale calculations of the dynamic behavior of a granular ceramic," International Journal of Solids and Structures, 45(6), pp. 1676-1696.

[5] Borg, J. P., and Vogler, T. J., 2009, "Aspects of simulating the dynamic compaction of a granular ceramic," Modelling and Simulation in Materials Science and Engineering, 17(4), p. 045003.

[6] Borg, J. P., Vogler, T. J., and Fraser, A., 2009, "A review of mesoscale simulations of granular materials," AIP Conference Proceedings, 1195(1), pp. 1331-1336.

[7] Borg, J. P., and Vogler, T. J., 2013, "Rapid compaction of granular material: characterizing two- and three-dimensional mesoscale simulations," Shock Waves, 23(2), pp. 153-176.

[8] Derrick, J. G., LaJeunesse, J. W., Davison, T. M., Borg, J. P., and Collins, G. S., 2018, "Mesoscale simulations of shock compaction of a granular ceramic: effects of mesostructure and mixed-cell strength treatment," Modelling and Simulation in Materials Science and Engineering, 26(3), p. 035009.

[9] Borg, J. P., and Vogler, T. J., 2008, "Mesoscale simulations of a dart penetrating sand," International Journal of Impact Engineering, 35(12), pp. 1435-1440.

[10] Anderson, C. E., Behner, T., Orphal, D. L., Nicholls, A. E., and Templeton, D. W., 2008, "Time-resolved penetration into pre-damaged hot-pressed silicon carbide," International Journal of Impact Engineering, 35(8), pp. 661-673.

[11] Orphal, D. L., and Anderson, C. E., 2006, "The dependence of penetration velocity on impact velocity," International Journal of Impact Engineering, 33(1), pp. 546-554.

[12] Holmquist, T. J., and Johnson, G. R., 2008, "The failed strength of ceramics subjected to high-velocity impact," Journal of Applied Physics, 104(1), p. 013533.

[13] McGlaun, J. M., Thompson, S. L., and Elrick, M. G., 1990, "CTH: A three-dimensional shock wave physics code," International Journal of Impact Engineering, 10(1), pp. 351-360.

[14] Plimpton, S., 1995, "Fast Parallel Algorithms for Short-Range Molecular Dynamics," Journal of Computational Physics, 117(1), pp. 1-19.

[15] Vogler, T. J., Reinhart, W. D., Chhabildas, L. C., and Dandekar, D. P., 2006, "Hugoniot and strength behavior of silicon carbide," Journal of Applied Physics, 99(2), p. 023512.

[16] Rumble, J. R., 2018, CRC Handbook of Chemistry and Physics, 99th Edition, CRC Press.

[17] Steinberg, D. J., Cochran, S. G., and Guinan, M. W., 1980, "A constitutive model for metals applicable at high-strain rate," Journal of Applied Physics, 51(3), pp. 1498-1504.

[18] Steinberg, D. J., and Lund, C. M., 1989, "A constitutive model for strain rates from 10−4 to 106 s−1," Journal of Applied Physics, 65(4), pp. 1528-1533.

[19] Carnahan, R. D., 1968, "Elastic Properties of Silicon Carbide," Journal of the American Ceramic Society, 51(4), pp. 223-224.

[20] Dandekar, D. P., 2005, "Spall Strength of Silicon Carbide Under Normal and Simultaneous Compression-Shear Shock Wave Loading," International Journal of Applied Ceramic Technology, 1(3), pp. 261-268.

[21] Robbins, D. L., Gehr, R. J., Harper, R. W., Rupp, T. D., Sheffield, S. A., and Stahl, D. B., 2000, "Laser-driven MiniFlyer induced gold spall," AIP Conference Proceedings, 505(1), pp. 1199-1202.

Proceedings of the 2019 Hypervelocity Impact Symposium
HVIS2019
April 14-19, 2019, Destin, FL, USA
HVIS2019-052

Extended Abstract+

Statistical Distributions of the Elastic Moduli of Particle Aggregates at the Mesoscale

Gerald Pekmezi [a]*, David Littlefield [a], Bruno Chareyre [b]

[a] Department of Mechanical Engineering, The University of Alabama at Birmingham, Birmingham, Alabama, USA
[b] Univ. Grenoble Alpes, CNRS, Grenoble INP, 3SR, Grenoble, France

1. Introduction

The current work is part of an ongoing effort to couple the Discrete Element Method (DEM), used as a mesoscale method, with Statistical Mechanics in order to construct a framework for the modeling, homogenization, and uncertainty quantification of particulate geomaterials. The aspect of the framework tackled in this work, is the quantification of stress and deformation of heterogenous DE models. Such quantification is necessary to allow DEM to "speak continuum". DEM is a discrete particle method, wherein the concept of "stress" does not fit naturally, but where each particle pair is considered to interact through discrete forces. Additionally, DEM is an explicit method, and equivalent continuum deformation must be defined in the context of particle kinematics.

2. Computational Approach

2.1. Surface Reconstruction of Particle Aggregate Boundaries with Alpha Shapes

The definition of a boundary for free-floating particles in space, also known as a "point cloud", requires some sort of surface reconstruction algorithm. Serendipitously, surface reconstruction from point clouds has been an area of intensive research and progress in recent years. This has been spurred by rapid developments in 3D scanning and photogrammetry. There are several approaches used to extract an exterior surface from a point cloud.

Discrete Element Models do pose an additional challenge for surface reconstruction algorithms, the discrete particles are actually spheres with radii, surface area, and volume. This complication can be resolved by using "weighted" alpha shapes, that account for the radii of the DEM particles. The alpha-shape associated with a set of points is a generalization of the concept of the convex hull [1].

For each real number α, define the concept of a generalized disk of radius $1/\alpha$ as follows:

- If $\alpha = 0$, it is a closed half-plane (convex hull)
- If $\alpha > 0$, it is closed disk of radius $1/\alpha$
- If $\alpha < 0$, it is the closure of the complement of a disk of radius $-1/\alpha$

Weighted alpha shapes on a regular triangulation of the DEM particles can be used to construct a Laguerre-Voronoi Diagram (a.k.a. Power Diagram) of the boundary [2]. The LV Diagram essentially is a map on the DE model boundary of the boundary areas belonging to each particle determined by the weighted alpha shape algorithm to be a boundary particle

---

+ Full paper can found at https://www.sciencedirect.com/journal/international-journal-of-impact-engineering/special-issue/10MTGP5W4VJ

* Corresponding author. Tel.: +1-205-996-2261; fax: +1-205-975-7217.
E-mail address: gpekmezi@uab.edu.

2.2. *Surface Reconstruction for Stress and Deformation Quantification*

A highlight of this micromechanical boundary implementation is that the boundary particles may become interior particles and vice-versa. While this makes micromechanical boundaries more versatile, it does increase computation complexity for original-to-deformed mapping schemes such as the deformation gradient. This complexity is herein resolved by introducing into the computation a fixed rectangular grid which is "fitted" to the particle aggregate boundary at either configuration. The interpolated regular surface grid makes it possible to compute stress and deformation by providing matching reference points for the current as well as the reference configuration.

The fitting of this regular grid is done by 1) translating the coordinates of the LV Diagram into spherical coordinates, 2) translating the coordinates of a surface grid from a tessellation of a unit cube into spherical coordinates, 3) linear interpolation of the radii of the unit cube from the LV Diagram, 4) conversion of the interpolated spherical coordinates back to Cartesian coordinates.

2.3. *Deformation Gradient from Particle Kinematics*

The macroscopic deformation gradient is defined by the volume average of the macro-scale deformation:

$$\overline{\mathbf{F}} = \frac{1}{V}\int_V \mathbf{F}dv \tag{1}$$

where the deformation gradient is defined as:

$$\mathbf{F} = \nabla \mathbf{x} \tag{2}$$

Application of the divergence theorem, gives:

$$\overline{\mathbf{F}} = \frac{1}{V}\int_{\partial V} \mathbf{x} \otimes \mathbf{N}ds \tag{3}$$

where $\mathbf{N}$ is the unit normal vector at the outer boundary for the reference configuration. Defining the area vector as

$$\mathbf{A_q} = \int_{\partial V} \mathbf{N}ds \tag{4}$$

then for a discrete setting Equation (4) is equivalent to:

$$\overline{\mathbf{F}} = \frac{1}{V}\sum_{q=1}^{Q} \mathbf{x}_q \otimes \mathbf{A_q} \tag{5}$$

which may be described as the volume average of the tensor products of the area vectors in the reference configuration and the particle position vectors in the current configuration.

2.4. *Love-Weber Average of the Cauchy Stress Tensor*

The volume average of the Cauchy stress tensor may be defined as:

$$\overline{\boldsymbol{\sigma}}_{ij} = \frac{1}{V}\int_V \boldsymbol{\sigma}_{\mathbf{ik}} \frac{\partial \mathbf{x_k}}{\partial \mathbf{x_j}} dv \tag{6}$$

Using the chain rule and applying the divergence theorem, Equation (6) becomes

$$\overline{\boldsymbol{\sigma}}_{ij} = \frac{1}{V}\int_{\partial V} \boldsymbol{\sigma}_{ik}\mathbf{x}_j\mathbf{N}_k dS - \frac{1}{V}\int_V \frac{\partial \boldsymbol{\sigma}_{ik}}{\partial \mathbf{x}_k}\mathbf{x}_j dv \tag{7}$$

Incorporating the balance of momentum equation into the second term of Equation (7), gives:

$$\overline{\boldsymbol{\sigma}}_{ij} = \frac{1}{V}\int_{\partial V} \boldsymbol{\sigma}_{ik}\mathbf{x}_j\mathbf{N}_k dS - \frac{1}{V}\int_V (\rho\dot{\mathbf{v}}_i - \boldsymbol{\gamma}_i)\mathbf{x}_j dv \tag{8}$$

If $\mathbf{f}_i^{ext}$ is the exterior force for all points on the boundary, then:

$$\overline{\boldsymbol{\sigma}}_{ij} = \frac{1}{V}\int_{\partial V} \mathbf{f}_i^{ext}\mathbf{x}_j dS - \frac{1}{V}\int_V (\rho\dot{\mathbf{v}}_i - \boldsymbol{\gamma}_i)\mathbf{x}_j dv \tag{9}$$

From Equation (9), it is obvious that the average Cauchy stress tensor is composed of two parts. The first integral on the right side of Equation (9) represents the static component of the stress, involving the forces applied at the boundary. The second integral is an inertial term, representing the acceleration of each material point. For static and quasi-static analyses, the second term may be ignored.

For the case of a particle aggregate with P interior particles and Q boundary particles, Equation (9) may be written as:

$$\overline{\boldsymbol{\sigma}}_{ij} = \frac{1}{V}\sum_{q=1}^{Q} \mathbf{f}_i^{ext,q}\mathbf{x}_j^q - \frac{1}{V}\sum_{q=1}^{P+Q}\int_{V_q} (\rho\dot{\mathbf{v}}_i - \boldsymbol{\gamma}_i)\mathbf{x}_j dV \tag{10}$$

For a particle aggregate in equilibrium when subjected to external forces the second term vanishes, giving the classical Love-Weber formula of the average Cauchy stress tensor [3, 4]:

$$\overline{\boldsymbol{\sigma}}_{ij}^{LW} = \frac{1}{V}\sum_{q=1}^{Q} \mathbf{f}_i^{ext,q}\mathbf{x}_j^q \tag{11}$$

2.5. Simulating Eglin Sand

In this work, a well-graded fine sand known in various settings as either Eglin sand, or Quikrete sand 1961, is used to illustrate the approach and utility of the methodology developed. Eglin sand is chosen for one important reason, it is a sand that has been tested micro-experimentally and had its micro-mechanical characteristics described statistically [5-10]. This, along with triaxial test data [11], makes it possible to evaluate the quality of both the mesoscale as well as the macroscale model. This is perhaps not essential to the current effort, which is mostly geometric and analytical in nature, however it is crucial to the evaluation of the overall framework the current effort is part of.

The Particle Size Distribution (PSD) of Eglin sand is used to build a DEM packing of $1\text{x}10^8$ particles, hereafter referred to as the "source" model. The source model is then sampled thousands of times to build mesoscale DEMs of a set size. All simulations in the current work were carried out on DEMs of 1000 particles each, sampled from the source particle model.The ensemble used here consists of 22000 DEMs. Each realization is confined to three different mean stress targets of 5MPa, 20MPa, and 50MPa.

Each of these models is then loaded along three different paths using both the uniform displacement as well as the uniform traction boundary conditions. The loading paths used are:

1) Isotropic Compression – the forces/velocities on the boundary particles are increased until a volumetric strain of about 0.5% is achieved. The slope of the mean stress vs. volumetric strain curve is equal to the bulk modulus.
2) Uniaxial Stress – the forces/velocities on the boundaries are adjusted to achieve an axial strain of about 0.8%, with a constant lateral stress. The slope of the principal stress difference vs. axial strain is equal to the Young's modulus.
3) Deviatoric Stress – the forces/velocities on the boundaries are adjusted to achieve a principal stress difference increase while maintaining a constant mean stress with a target deviatoric strain of about 0.6%. The slope of the deviatoric stress vs. deviatoric strain curve is equal to twice the shear modulus.

The stresses and strains are computed from the Love-Weber averages of the Cauchy stress and the averaged deformation gradient as indicated in the preceding sections. Each of the three moduli: the bulk modulus, the Young's modulus, and the shear modulus, were evaluated for each realization in the ensemble at confinements of 5MPa, 20 MPa, 50 MPa, 100MPa, 200MPa, and 400 MPa.

References

[1] H. Edelsbrunner, D. Kirkpatrick, R. Seidel, On the shape of a set of points in the plane, IEEE Transactions on information theory 29(4) (1983) 551-559.

[2] F. Aurenhammer, Power diagrams: properties, algorithms and applications, SIAM Journal on Computing 16(1) (1987) 78-96.

[3] A.E.H. Love, A treatise on the mathematical theory of elasticity, at the University Press1920.

[4] J. Weber, Recherches concernant les contraintes intergranulaires dans les milieux pulvérulents, Bulletin de Liaison des Ponts-et-chaussées 20 (1966) 1-20.

[5] H. Luo, W.L. Cooper, H. Lu, Effects of particle size and moisture on the compressive behavior of dense Eglin sand under confinement at high strain rates, International Journal of Impact Engineering 65 (2014) 40-55.

[6] V. Nardelli, M. Coop, J. Andrade, F. Paccagnella, An experimental investigation of the micromechanics of Eglin sand, Powder Technology 312 (2017) 166-174.

[7] D.M. Cole, Laboratory observations of frictional sliding of individual contacts in geologic materials, Granular Matter 17(1) (2015) 95-110.

[8] C. Sandeep, K. Senetakis, Grain-scale mechanics of quartz sand under normal and tangential loading, Tribology International 117 (2018) 261-271.

[9] F. Wang, B. Fu, H. Luo, S. Staggs, R.A. Mirshams, W.L. Cooper, S.Y. Park, M.J. Kim, C. Hartley, H. Lu, Characterization of the Grain-Level Mechanical Behavior of Eglin Sand by Nanoindentation, Experimental Mechanics 54(5) (2014) 871-884.

[10] K. Senetakis, M.R. Coop, M.C. Todisco, The inter-particle coefficient of friction at the contacts of Leighton Buzzard sand quartz minerals, Soils and Foundations 53(5) (2013) 746-755.

[11] B. Martin, O. Cazacu, Experimental and theoretical investigation of the high - pressure, undrained response of a cohesionless sand, International Journal for Numerical and Analytical Methods in Geomechanics 37(14) (2013) 2321-2347.

Proceedings of the 2019 Hypervelocity Impact Symposium
HVIS2019
April 14-19, 2019, Destin, FL, USA
HVIS2019-060

Extended Abstract+

Validation of an Improved Contact Method for Multi-Material Eulerian Hydrocodes in Three-Dimensions

Kenneth C. Walls[a] and David L. Littlefield[a]*

[a]*University of Alabama at Birmingham, 1075 13th Street South, Birmingham, Alabama 35294, USA*

1. Introduction

Realistic and accurate modeling of contact for problems involving large deformations and severe distortions presents a host of computational challenges. Due to their natural description of surfaces, Lagrangian finite element methods are traditionally used for problems involving sliding contact. However, problems such as those involving ballistic penetrations, blast-structure interactions, and vehicular crash dynamics, can result in elements developing large aspect ratios, twisting, or even inverting. For this reason, Eulerian, and by extension Arbitrary Lagrangian-Eulerian (ALE), methods have become popular. However, additional complexities arise when these methods permit multiple materials to occupy a single finite element.

Multi-material Eulerian formulations in computational structural mechanics are traditionally approached using mixed-element thermodynamic and constitutive models. These traditional approaches treat discontinuous pressure and stress fields that exist in elements with material interfaces by using a single approximated pressure and stress field. However, this approximation often has little basis in the physics taking place at the contact boundary and can easily lead to unphysical behavior. This work presents a significant departure from traditional Eulerian contact models by solving the conservation equations separately for each material within each computational element and then imposing inequality constraints associated with contact to the solutions for each material with the appropriate tractions included. The advantages of this method have been demonstrated with several computational examples which are fully detailed in [6] and only partially described here. These examples are limited to contact between two materials. Future work will examine contact between three or more materials within an element. This work concludes by drawing a comparison between the method put forth in this work and traditional treatment of multi-material contact in Eulerian methods.

2. Research Approach

The theory developed in this work represents a significant departure from traditional treatment of contact in an Eulerian framework. It essentially represents a generalization of a technique used in Lagrangian formulations and implements it in an Eulerian framework in order to avoid the mesh distortion drawbacks of a Lagrangian method while also removing issues related to mixture theory associated with an Eulerian approach. In this work, mixed cell thermodynamic and constitutive models traditionally used in an Eulerian framework are not used. Rather, the governing equations are solved for each material individually and then specific contact constraints are imposed. After these constraints are enforced the traction, which is traditionally handled implicitly in Eulerian methods, is determined explicitly. This results in a set of coupled equations that can be approximated using an uncoupled system. The following sections briefly describe the finite element formulations and contact constraints used in this work. Readers can reference the article by Walls and Littlefield in [6] and a dissertation by Walls [1] for more complete descriptions of the conservation equations, interface tracking formulation, finite element formulation, contact constraints, and other background information about the work presented here. The formulations developed in this work

+ Full paper can found at https://www.sciencedirect.com/journal/international-journal-of-impact-engineering/special-issue/10MTGP5W4VJ

* Corresponding author. Tel.: +1-205-934-8343.
E-mail address: kwalls71@uab.edu.

have been implemented in the multi-material finite element code ALEAS (Arbitrary Lagrangian-Eulerian Adaptive Solver) in order to examine the advantages of this approach.

In this work, the conservation equations are solved using a finite element formulation that makes use of two operator splits. The first operator split is traditionally performed in Eulerian formulations and is solved without accounting for the traction, **t**. The second operator split is unique to this work and enforces contact by determining the traction and applying the Signorini contact constraints. This second operator split is valid everywhere, but since the traction is zero except in elements where contact occurs it is only performed in elements where the contact constraints are not identically satisfied. In this work the values for density, ρ, and specific internal energy, e, are considered to be piecewise constants, while the velocity function, **v**, is a piecewise linear function. A visual representation of the finite element approach used in this work is shown in Figure 1.

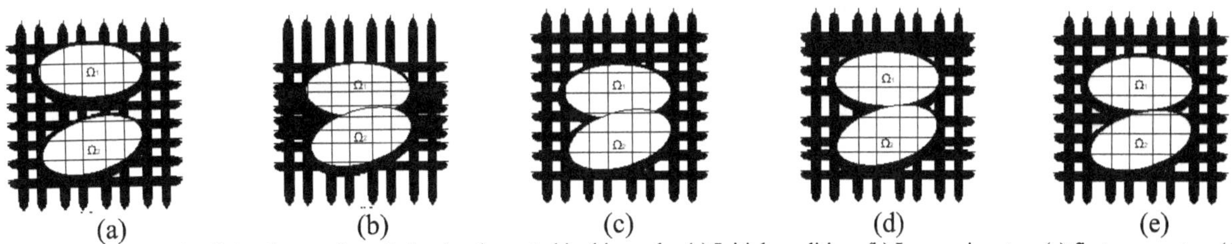

Figure 1: Representation of the finite element formulation implemented in this work. (a) Initial condition, (b) Lagrangian step, (c) first remap step, (d) contact enforcement step, and (e) second remap step. Steps (d) and (e) are unique to this work.

3. Discussion

There are several advantages to using the form of the conservation equations described in this work. Since this formulation treats each material separately, and there are no quantities associated with the whole cell, such as temperature, pressure, or strain rate; there is no need for the use of traditional cell mixture theory involving mixed cell thermodynamic, strength, and fracture algorithms, all of which typically contribute to significant amounts of error in simulations where contact is important. Additionally, the form of the conservation equations used in these formulations guarantees element-level conservation of mass, momentum, and energy. This allows for the amount of material being transported between adjacent elements to be determined easily. One other feature that distinguishes the approach used in this work is that each material has its own momentum balance and, consequently, separate densities, velocities, energies, and stresses are updated for each material.

The main complication of this method is a result of the ability of each material to have its own velocity field. Because of this property it is possible for advection volumes to become complicated in shape or even disjoint. This difficulty is accounted for by using interface tracking, however not in the traditional sense. In a typical multi-material Eulerian contact formulation interface tracking is used to determine the contents of an advection volume whose shape is known. However, in this formulation the contents of the advection volume are known, and thus the shape of the advection volume can be determined.

The advantages of using the formulation presented in this work over traditional Eulerian methods for classes of problems where large deformations and high strains are present, but contact is important are evident and can be seen in the results of a set of validation problems described below.

4. Results

In order to demonstrate the effectiveness of the method presented in this work, a set of validation problems have been developed. These test cases have been developed to demonstrate the superiority of the method described in this work to a traditional Eulerian formulation as well as determine the applicability of the newly developed method in cases that are handled well by traditional Eulerian methods with mixture theory. The simulations presented here were carried out in the multi-material hydrocode ALEAS. Simulations using the method developed for this work were carried out using the multi-material momentum (MMM) option, while comparison runs using mixture theory were run using the single-material momentum (SMM) option. Two validation cases are described in this extended abstract. Interested readers should refer to [1] and [6] for a more complete validation of the method presented here.

Validation 1: Long Rod Penetration of Oblique Plates

As a means of validating the advantages of the method implemented in ALEAS MMM this work used the three test conditions presented by Caglıostro, et al. [4] which utilized data from Fugelso and Taylor [3] to validate the MESA code. A more complete description of the problem setup can be found in [6]. The initial velocities and orientations of the rods and plates were designed in such a way that the rod impact velocity relative to the plate was 1.29 km/s at an obliquity of 65° for yaws of -9.3°, 0°, and +10.3°. The penetration of oblique plates by projectiles of various yaw serves as a means of demonstrating the shortcomings of traditional Eulerian formulations with mixture theory. Due to the increased interaction between materials as the rod passes through the plate an artificial slowing of the rod compared to experimental data is often seen due to material bonding when mixture theory is employed. This result is not desirable, especially for applications such as multi-layered armor or other events where the ability to accurately model a secondary or further subsequent impact is important.

(a) (b)

Figure 4: Results of -9.3° yaw case for (a) ALEAS MMM and (b) ALEAS SMM at 100μs

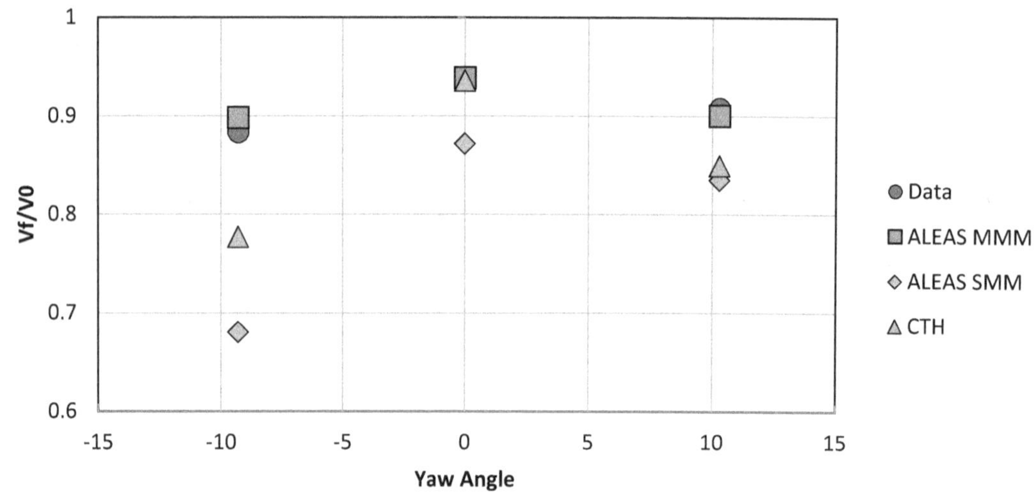

Figure 5: Comparison of normalized exit velocity versus yaw angle

Results of the -9.3° yaw case is shown in Figure 4 and the relative exit velocity versus yaw angle is summarized in Figure 5 for all cases. ALEAS MMM showed significant improvements over ALEAS SMM and CTH which use mixture theory. Both of the mixture theory runs show a significant artificial slowing of the penetrator due to material bonding. In the CTH runs this effect is mitigated by the use of an ad hoc fracture model, which has not yet been implemented in ALEAS. Figure 4 (b) illustrates the shortcomings of the mixture theory formulation. As the penetrator passes through the plate large chunks of plate material bond to the rod. This is non-physical and is not seen in the ALEAS MMM formulation shown in Figure 4 (a).

Validation 2: Long Rod Penetration of Semi-infinite Targets

Another class of impact and penetration events considered for the validation of ALEAS was the penetration of tungsten alloy long rod projectiles into semi-infinite steel targets. This class of problems is frequently used as a validation of hydrocodes. The setup for the simulations consists of a tungsten-heavy-alloy cylindrical penetrator with diameter of 0.5 cm and length of 5 cm impacting a semi-infinite hardened steel target at initial velocities of 500, 1000, 1500, 2000, and 3000 m/s.

The role of contact in this class of problems is less important due to hydrodynamic erosion, so traditional Eulerian formulations that use mixture theory often perform very well. As such, it was desirable to ensure that the new method developed in this work was capable of obtaining reasonable results. As can be seen in Figure 6, the depth of penetration for the 3000 m/s impact case showed good agreement between both methods, however, the ALEAS SMM case showed greater amounts of erosion of the impactor along the crater wall, while material continued to slide down the wall and collected at the base of the crater at late time in the ALEAS MMM case. Due to the hydrodynamic erosion that takes place in the rod at this velocity a trail of material should be present along the length of the crater and is seen experimentally. This serves to demonstrate a limitation of the method developed in this work. Since contact is not an important factor in long rod penetration of semi-infinite targets the method developed for this work was not expected to provide perfect agreement, however, when comparing penetration efficiency, the primary measure for assessing these simulations, the agreement between the new contact method and mixture theory proved to be adequate, however the added computation cost makes the method developed in this work less appealing for this type of problems where sliding contact is not a significant factor.

(a) (b)

Figure 6: Results of the penetration of L/D=10 tungsten alloy rod into steel target with V_0=3000m/s at 50 μs for (a) ALEAS MMM and (b) ALEAS SMM

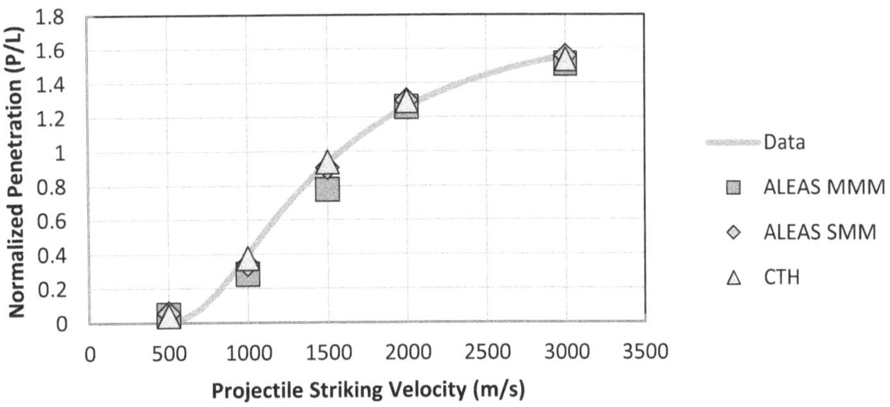

Figure 7: Comparison of normalized depth of penetration (P/L) for ALEAS MMM, ALEAS SMM, CTH, and fit to experimental data [5] for L/D=10 tungsten alloy rods into steel targets

5. Conclusions

This study has validated a new method for the treatment of multi-material contact-impact problems in Eulerian hydrocodes used for computational structural mechanics. Contact constraints have been added to a multi-material Eulerian formulation using numerical approximations to impose the contact inequalities using a finite element model. The conservation equations have also been solved separately for each material in the problem space. As a result, a separate velocity field is assigned to each contacting body, leading to a tremendous improvement over traditional Eulerian contact methods that make use of mixture theory for problems where large deformations are present, but contact is an important factor. The advantages of using the method presented in this work have been shown in a series of validation problems, introduced here and fully described in [6].

References

[1] Walls, K.C. (2017). An Improved Contact Method for Multi-material Eulerian Hydrocodes. Ph.D. Thesis. The University of Alabama at Birmingham.

[2] Banerjee, B. (2005). An Evaluation of Plastic Flow Stress Models for the Simulation of High-temperature and High-strain-rate Deformation of Metals. *arXiv.org:cong-mat/0512466.*

[3] Fugelso, E., Taylor, J. (1978). Evaluation of Obliquity and Yaw for U 0.75 wt% Ti Penetrators. *Los Alamos National Laboratory Report LA-7402-MS.*

[4] Cagliostro, D., Mandell, D., Schwalbe, L., Adams, T., & Chapyak, E. (1989) MESA 3-D Calculations of Armor Penetration by Projectiles with Combined Obliquity and Yaw. *Proceedings of the Hypervelocity Impact Symposium.* San Antonio, TX, December 12-14, 1989.

[5] Park, B., Leavy, R., & Niederhaus, J. (2013). Penetration of Rod Projectiles in Semi-Infinite Targets: A Validation Test for Eulerian X-FEM in ALEGRA. *Sandia National Laboratories Report SAND2013-1863.* March 12, 2013.
[6] Walls, K.C. and Littlefield, D.L. (2020) Validation of an Improved Contact Method for Multi-Material Eulerian Hydrocodes in Three-Dimensions. International Journal of Impact Engineering, Volume 137, March 2020, 103444.

Proceedings of the 2019 Hypervelocity Impact Symposium
HVIS2019
April 14-19, 2019, Destin, FL, USA
HVIS2019-078

AN ACCURATE SPH SCHEME FOR HYPERVELOCITY IMPACT MODELING

Anthony Collé[a,b]* , Jérôme Limido[a] , Thomas Unfer[a] and Jean-Paul Vila[b]

[a]*Impetus AFEA, 6 rue du Cers, 31330 Grenade sur Garonne, France*
[b]*Institut de Mathématiques de Toulouse, Institut National des Sciences Appliquées de Toulouse, 135 Avenue de Rangueil, 31400 Toulouse, France*

Abstract

We focus in this paper on the use of a meshless numerical method called Smooth Particle Hydrodynamics (SPH), to solve fragmentation issues as Hyper Velocity Impact (HVI). Contrary to classical grid-based methods, SPH does not need any opening criteria which makes it naturally well suited to handle material failure. Nevertheless, SPH schemes suffer from well-known instabilities questioning their accuracy and activating nonphysical processes as numerical fragmentation. Many stabilizing tools are available in the literature based for instance on dissipative terms, artificial repulsive forces, stress points or Particle Shifting Techniques (PST). However, they either raise conservation and consistency issues, or drastically increase the computation times. It limits then their effectiveness as well as their industrial application. To achieve robust and consistent stabilization, we propose an alternative scheme called γ -SPH-ALE. Firstly implemented to solve Monophasic Barotropic flows, it is secondly extended to the solid dynamics. Particularly, based on the ALE framework, its governing equations include advective terms allowing an arbitrary description of motion. Thus, in addition of accounting for a stabilizing low-Mach scheme, a PST is implemented through the arbitrary transport velocity field, the asset of ALE formulations. Through a nonlinear stability analysis, CFL-like conditions are formulated ensuring the scheme conservativity, robustness, stability and consistency. Besides, stability intervals are defined for the scheme parameters determining entirely the stability field. Its implementation on several test cases reveals particularly that the proposed scheme faithfully reproduces the strain localization in adiabatic shear bands, a precursor to failure. By preventing spurious oscillations in elastic waves and correcting the so-called tensile instability, it increases both stability and accuracy with respect to classical approaches.

Keywords: Smoothed Particle Hydrodynamics; Arbitrary Lagrangian Eulerian; Non-Linear Stability; Tensile Instability; Strain Localization;

Nomenclature

h, Δx, $\boldsymbol{x}$	smoothing length and particle spacing and position (m)
$\boldsymbol{v}, \boldsymbol{v}_0, \boldsymbol{w}$	material velocity, arbitrary transport velocity field and relative ALE velocity (m/s)
e	specific internal energy ($\frac{\text{J}}{\text{kg}}$)
W, $\boldsymbol{\nabla} W$, $\boldsymbol{A}$	Kernel function, kernel function gradient and renormalized kernel function gradient (m^{-4})
Greek symbols	
ρ, ω	density ($\frac{\text{kg}}{\text{m}^3}$) and volume ($\text{m}^3$)
$\boldsymbol{\Phi}$	vector of the conserved variables in $\mathbb{R}^4$
α, γ	artificial viscosity and low-Mach scheme parameters
Subscripts	
i, j	defines particles among the particle domain P
Superscripts	
k, l	defines the coordinate among x,y,z axis

* Corresponding author. Tel.: +33 5 61 74 86 82.
E-mail address: anthony@impetus.fr.

1. Introduction

In Aeronautics, Defense and Space, the dynamic fragmentation is induced by extreme loadings involved in events like Hypervelocity Impacts (HVI). Regarding the experimental limitations to replicate the real impact configurations, numerical tools are then crucial to study structures vulnerability. However, combining large deformations as well as numerous interface productions, this mechanism is particularly challenging from a numerical point of view. Lagrangian finite element methods (FE) are classically used to handle the numerical modeling of these events. They propose to reproduce the material failure process thanks to mesh opening techniques as element erosion, node splitting or XFEM. However, these approaches induce very long computation times. Indeed, it is due to the large number of elements needed to limit the mass lost through erosion, or to the complexity of the opening algorithms. Moreover, they experience severe energy conservation issues, whereas it is crucial in this framework. Among the Lagrangian approaches the meshless methods appear as a relevant alternative to prevent such shortcomings. We focus in this paper on the meshless method called Smooth Particle Hydrodynamics (SPH).

SPH is a Lagrangian particle method used to solve partial differential equations (PDE). Originally developed by Lucy [1] for astrophysical problems and by Gingold & Monaghan [2-4] for hydrodynamic applications (incompressible free surface flows), the method was then adapted by Libersky & Petschek [5] and Benz [6,7] to the solid mechanics by including material models. The computational domain is discretized in a set of interpolation points interpreted as particles interacting between themselves through kernel functions and carrying material properties. Based on a set of moving interpolation points then, it disregards any connectivity between its elements. Thus, it does not need any opening criteria which makes it naturally well suited to handle material failure. In addition of its conceptual simplicity, robustness, ability to handle large arbitrary deformations, to not produce diffusive interfaces and for its monolithic nature, SPH is particularly attractive. Still, SPH schemes suffer from well-known instabilities as lack of interpolation completeness, nonphysical oscillations, tensile instability or particle clumping. Questioning their accuracy, they also activate nonphysical processes as numerical fragmentation. Since its emergence, SPH benefits from many researches aiming to improve its accuracy. Thus, various stabilizing tools are available in the literature based for instance on high order kernels [8-10], additional diffusive terms [2,11,12], artificial repulsive forces [13,14] or Particle Shifting Techniques (PST) [15-18]. However, they also either raise conservation and consistency issues, or drastically increase the computation times. Thus, even if today most of the commercial transient dynamic software propose SPH like methods, its instabilities as well as its long computation times still limit its industrial application.

To achieve robust and consistent stabilization, we focus on a different SPH framework introduced by Vila [19] based on Arbitrary Lagrangian Eulerian (ALE) considerations. The idea is to combine and benefit from both Eulerian and Lagrangian movement descriptions. Indeed, the Eulerian description is generally preferred to handle large deformations, while a Lagrangian one for interface tracking. In practice this combination relies on an arbitrary transport velocity field $\boldsymbol{v}_0$, which by a smart choice, allows to increase the stability and robustness of the scheme without degrading its consistency. Generally combined to Riemann Solvers (RSALE) [17], such formulation improves the scheme accuracy, but at the cost of an increased solving complexity (solving algorithm) and low-Mach limitations [20,21]. Regarding this late aspect, in the Finite Volume (FV) framework some authors [22,23] proposed a stabilizing Low-Mach scheme. It relies on a corrective velocity term added to the continuity and momentum equations. A non-linear stability analysis provided them stability conditions (CFL-like) insuring robust and accurate computations whichever flow regime. Clayer et al. [24] adapted this Low-Mach scheme to the SPH framework and achieved similar convincing conclusions.

We propose in this work an alternative scheme called γ-SPH-ALE relying on the combination of the SPH-ALE formalism and the FV low-Mach scheme. Its governing equations and the corresponding stability conditions are exposed in section 2. Firstly, detailed for a hydrodynamic version, they are extended then to a solid formulation. Finally, in section 3, the proposed γ-SPH-ALE scheme is evaluated through several validation cases (both hydrodynamic and solid).

2. γ-SPH-ALE: A new meshless scheme

In this section the proposed γ-SPH-ALE scheme is detailed. The general SPH framework is firstly exposed and secondly apply to the ALE context. The governing equations of the proposed scheme and the corresponding stability conditions are thirdly exposed in the hydrodynamic version. They are then extended to a solid formulation.

Note that in the following Ω denotes the whole material domain, P defines the particle domain among which i describes one particle. We set V_i its neighboring particle set, and we take $j \in V_i$. Finally, $n \in \mathbb{N}$ defines the time step number and $d \in \mathbb{N}^*$ the space dimension. In this section we refer to [19,25] for more details on SPH and SPH-ALE.

2.1. SPH basics

SPH relies on approximation features allowing to discretize conservation laws on a particle set $(x_i, \omega_i)_{i\in P}$ (respectively the particle position and volume). These approximations are achieved thanks to kernel functions $W(\|\boldsymbol{x}_i - \boldsymbol{x}_j\|, h) = W_{ij}$ depending on the smoothing length h (half radius of its compact support) and the distance between two particles. The kernel gradient can also be evaluated as $\boldsymbol{\nabla} W_{ij} = grad_x W_{ij}$ allowing to define then the smoothed particle approximation of a function f and of its derivative

$$\Pi^h(f)_i = \sum_{j\in P} \omega_j f_j W_{ij} \text{ and } \boldsymbol{\nabla}\Pi^h(f)_i = \sum_{j\in P} \omega_j f_j \boldsymbol{\nabla} W_{ij} \tag{1}$$

As introduced above, classical SPH suffers from a lack of interpolation completeness which can be compensated by using renormalized kernels. Vila [9] proposed to replace $\boldsymbol{\nabla} W_{ij}$ by $\boldsymbol{A}_{ij} = B_{ij}\boldsymbol{\nabla} W_{ij}$ where $\boldsymbol{B}_{ij} = (\boldsymbol{B}_i + \boldsymbol{B}_j)/2$ and B is the renormalization matrix (introduced by Johnson et al. [8] and Vila [9]) defined as

$$B_i = E_i^{-1} \text{ and } \forall (k,l) \in [\![1,..,d]\!]^2, E_i^{kl} = \sum_{j\in P} \omega_j \left(\boldsymbol{x}_j^l - \boldsymbol{x}_i^l\right) \boldsymbol{\nabla} W_{ij}^k \tag{2}$$

We set in the following $\boldsymbol{n}_{ij} = \boldsymbol{A}_{ij}/\|\boldsymbol{A}_{ij}\|$.

To ensure the conservative property of the scheme, alternative derivative operators are used in practice and combined to symmetric kernels such that $W_{ij} = W_{ji}$ and $\boldsymbol{\nabla} W_{ij} = -\boldsymbol{\nabla} W_{ji}$

$$\boldsymbol{D}_h f_i = \sum_{j\in P} \omega_j \left(f_j - f_i\right) \boldsymbol{A}_{ij} \text{ and } \boldsymbol{D}_h^* f_i = \sum_{j\in P} \omega_j \left(f_j + f_i\right) \boldsymbol{A}_{ij} \tag{3}$$

Note that, as proved in [19], operator $\mathbf{D}_h$ (3) combined to renormalization strongly approximates the gradient operator under the condition $\Delta x/h = O(1)$ (where Δx is the particle spacing). It is a crucial point to ensure the scheme consistency. In standard SPH this convergence property depends on the ratio $\Delta x/h$ and is only enforced for particular values [22]. Classically $h = 1.3\Delta x$ in 3D.

2.2. ALE Framework

The idea is now to apply these approximation features to a conservation law in ALE formalism. We consider then the following law

$$L_{v0}(\boldsymbol{\Phi}) + div[F_E(\boldsymbol{\Phi}) - \boldsymbol{\Phi}\otimes \boldsymbol{v}_0] = S(\boldsymbol{\Phi}) \text{ with } L_v: \boldsymbol{\Phi} \to \frac{\partial \boldsymbol{\Phi}}{\partial t} + \sum_{k=1,..,d} \frac{\partial (v^k \boldsymbol{\Phi})}{\partial x^k} \tag{4}$$

Where $\boldsymbol{v}_0 \in \mathbb{R}^3$ is a regular transport field, $\boldsymbol{\Phi} \in \mathbb{R}^4$ the vector of the conserved variables, $\boldsymbol{F}_E = (F_E^k)_k^4$ the Eulerian flux vectors, $\boldsymbol{S} = (S^k)_k^4$ the source term and L_{v0} the transport operator associated to $\boldsymbol{v}_0$. Considering the continuum mechanics equations, we have

$$\boldsymbol{\Phi} = \begin{pmatrix} \rho \\ \rho \boldsymbol{v} \\ \rho e \end{pmatrix}, \boldsymbol{F}_E = \begin{pmatrix} \rho \boldsymbol{v} \\ -\boldsymbol{\sigma} + \rho \boldsymbol{v}\otimes \boldsymbol{v} \\ -\boldsymbol{\sigma}.\boldsymbol{v} + \rho e \boldsymbol{v} \end{pmatrix} \text{ and } \boldsymbol{S} = \begin{pmatrix} 0 \\ \rho \boldsymbol{f} \\ \rho \boldsymbol{f}.\boldsymbol{v} \end{pmatrix} \tag{5}$$

Where $\rho, \boldsymbol{v}, e$ and $\boldsymbol{\sigma}$ are respectively the density, material velocity, specific total energy and stress tensor. Thus, to approximate (4) considering the moving particle set $(x_i, \omega_i)_{i\in P}$, we apply the SPH discretization process detailed in [19] based on the weak formulation of (4) providing the following equation

$$\frac{d(\omega_i \boldsymbol{\Phi}_i)}{dt} + \omega_i \sum_{k=1,..,d} \boldsymbol{\nabla}_h^{k,*} [F_E^k(\boldsymbol{\Phi}) - v_0^k . \boldsymbol{\Phi}] = 0 \tag{6}$$

Where $-\boldsymbol{\nabla}_h^*$ is the dual of an operator $\boldsymbol{\nabla}_\mathrm{h}$ approximating the derivative operator $\boldsymbol{\nabla}$.

2.3. Application to the Barotropic Euler Equations

The idea is to choose $\boldsymbol{\nabla}_\mathrm{h}$ in order to stay conservative and consistent. Several choices are admissible (as $\mathbf{D}_h$ (3) like in [19]), and we propose to work with the following scheme discretized in space

$$\frac{dx_i}{dt} = \boldsymbol{v}_{0i} \tag{7}$$

$$\frac{d\omega_i}{dt} = \omega_i \sum_{j \in P} \omega_j \langle (\boldsymbol{v}_{0j} - \boldsymbol{v}_{0i}), \boldsymbol{A}_{ij} \rangle \tag{8}$$

$$\frac{d(\omega_i \rho_i)}{dt} = -\omega_i \sum_{j \in P} \omega_j \left(\rho_j + \rho_i\right) \langle \boldsymbol{w}_{ij}, \boldsymbol{A}_{ij} \rangle \tag{9}$$

$$\frac{d(\omega_i \rho_i \boldsymbol{v}_i)}{dt} = -\omega_i \sum_{j \in P} \omega_j \left[\left(\rho_j \boldsymbol{v}_j + \rho_i \boldsymbol{v}_i\right) \langle \boldsymbol{w}_{ij}, \boldsymbol{A}_{ij} \rangle + p_{ij} \boldsymbol{A}_{ij} - \boldsymbol{\pi}_{ij} \right] \tag{10}$$

With

$$\boldsymbol{w}_{ij} = \frac{1}{2}\left(\boldsymbol{w}_i + \boldsymbol{w}_j\right) - \boldsymbol{\Gamma}_{ij} \tag{11}$$

$\boldsymbol{w}$ corresponds to the relative ALE velocity defined by $\boldsymbol{w} = \boldsymbol{v} - \boldsymbol{v}_0$. We also define $p_{ij} = p_i + p_j$, $\boldsymbol{\Gamma}_{ij}$ corresponds to the stabilizing term coming from the FV low-Mach scheme and $\boldsymbol{\pi}_{ij}$ corresponds to the artificial viscosity. $\langle .,. \rangle$ defines the usual scalar product on $\mathbb{R}^d$.

A vectorial version of Monaghan's stabilizing artificial viscosity [2] is proposed. It depends on a parameter α to adjust in order to minimize the amount of artificial diffusion induced in the scheme. c_0 is the material sound speed and $\rho_{ij} = \left(\rho_i + \rho_j\right)/2$

$$\boldsymbol{\pi}_{ij} = \alpha_{ij}\left(\boldsymbol{v}_j - \boldsymbol{v}_i\right) \text{ with } \alpha_{ij} = \alpha c_0 \rho_{ij} \left\| \boldsymbol{A}_{ij} \right\| \tag{12}$$

The coupling between the SPH version of the FV Low-Mach scheme and the SPH-ALE formalism is achieved by the term $\boldsymbol{\Gamma}_{ij}$ through the mean relative ALE velocity $\boldsymbol{w}_{ij}$

$$\boldsymbol{\Gamma}_{ij} = \gamma_{ij}\left(p_j - p_i\right)\boldsymbol{n}_{ij} \text{ with } \gamma_{ij} = \frac{\gamma}{2 c_0 \rho_{ij}} \tag{13}$$

Note that, according to (7-10) the choice of γ is totally independent from the choice of the movement description (meaning from the choice of $\boldsymbol{v}_0$). Also disregarding $\boldsymbol{\Gamma}_{ij}$ and $\boldsymbol{\pi}_{ij}$, the classical Lagrangian SPH formulation ($\boldsymbol{w} = \boldsymbol{0}$) with constant masses $m_i = \omega_i \rho_i$ is recovered. With this ALE formulation particles are moved following the velocity field $\boldsymbol{v}_0$ (7). Such field is said to be arbitrary in the sense that, as suggested by Vila [19], a smart choice of $\boldsymbol{v}_0$ could increase both stability and robustness of the calculations without its consistency. Various methods are available in the literature as Monaghan's XSPH method [3] or PST [15-18]. They consist in reorganizing the particle set during calculations (quasi-lagrangian mode) to prevent instabilities such as tensile instability [13] or the formation of Lagrangian particle structures [16]. The idea is to set $\boldsymbol{v}_{0i} = \boldsymbol{v}_i + \boldsymbol{\delta v}_i$ where $\boldsymbol{\delta v}_i$ stores the regularizing technique. The integration is performed thanks to a second order Runge-Kutta Heun (RK2H) time integration scheme defined as

$$\begin{cases} \bar{\Theta}^{n+1} = \Theta^n + \Delta t \dfrac{d\Theta^n}{dt} \\ \Theta^{n+1} = \dfrac{1}{2}\left[\Theta^n + \bar{\Theta}^n + \Delta t \dfrac{d\bar{\Theta}^{n+1}}{dt}\right] \end{cases} \text{ Where } \Theta = (\boldsymbol{x}, \omega, \omega\rho, \omega\rho\boldsymbol{v})^t \tag{14}$$

2.4. Nonlinear Stability Analysis

In order to ensure that the proposed scheme is conservative, robust, stable and consistent, stability conditions on the scheme parameters are exhibited thanks to a nonlinear stability analysis. The proof is briefly presented but not detailed in this paper, we refer to the work of Grenier et al. [22], Lavalle et al. [23] and Couderc et al. [26] for a similar process in a FV framework.

According to the equation set (7-10), and considering that $\boldsymbol{A}_{ij} = -\boldsymbol{A}_{ji}$, $\boldsymbol{w}_{ij} = \boldsymbol{w}_{ji}$ (11,13) and $\boldsymbol{\pi}_{ij} = -\boldsymbol{\pi}_{ji}$ (12) we can show that the proposed scheme is indeed conservative. Also, according to (8) the volume evolution is controlled by the imposed field $\boldsymbol{v}_0$ which provides volume bounds $(\omega_{min}, \omega_{max}) \in \mathbb{R}^{+*}$. Now, thanks to (9), and considering $(\omega_{min}, \omega_{max})$, we can find a CFL $N_\rho \in \mathbb{R}^{+*}$ which, insures the existence of density bounds $(\rho_{min}, \rho_{max}) \in \mathbb{R}^{+*}$ and achieves the expected robustness. Regarding the scheme stability, an energy balance is performed. The idea is to provide a CFL condition ensuring a control on the system total energy $\mathcal{E}$. We define $\mathcal{E} = \frac{1}{2}\rho\|\boldsymbol{v}\|^2 + \varphi$ where $\varphi = \rho \int_{\rho_0}^{\rho} \frac{p(r)}{r^2} dr$ is the total free energy and corresponds to the barotropic pressure. Through this energy balance, the production terms are exhibited and evaluated. It allows to highlight a CFL $N_E \in \mathbb{R}^{+*}$ insuring that

$$\begin{aligned} &For\ \frac{\Delta t c_0}{h} \leq N_E,\ \exists \mathcal{E}_T \in \mathbb{R}^{+*}, \exists(\alpha_{min}, \alpha_{max}, \gamma_{min}, \gamma_{max}) \in \mathbb{R}^{+,4}\ such\ that \\ &\forall(\alpha,\gamma) \in [\alpha_{min}, \alpha_{max}]x[\gamma_{min}, \gamma_{max}], \forall t \in [0,T], \mathcal{E}(t) \leq \mathcal{E}_T \end{aligned} \quad (15)$$

It worth noticing that these stability conditions are defined regardless the expression of $\boldsymbol{v}_0$ revealing its arbitrary property previously introduced. However, they still depend on a geometrical constant meaning that the stability is ensured while the particle distribution stays regular enough. Then it is crucial to be allowed to implement a conservative and consistent regularizing technique through $\boldsymbol{v}_0$ such as PST. Considering now the volume, density and energy bounds (15), we can finally achieve the weak consistency of the proposed scheme thanks to a Lax-Wendroff like theorem. We refer to [19] where Vila details the employed technique.

2.5. Extension to Solids dynamics

The proposed γ-SPH-ALE scheme is then directly extended to account for material strength. To do so the momentum equation (10) is enhanced by the deviatoric stress tensor contribution. To account for material strength models, we use then discrete formulations of strains, to describe the isotropic linear elasticity (through Hooke's law), and rotation, to insure the stresses objectivity (through Jaumann's derivative). Working in an ALE framework, a discretization of the advection component of path-dependent variables, such as stresses, is also considered. Thanks to the Von Mises yield stress criterion and a flow stress model, we are then able to describe the material plastic flow. Aiming to handle the fragmentation process, we also consider a description of the material damage growth, a direct consequence of its deformation and a precursor to its failure. Finally, we implement an initial damage distribution as well to account for material heterogeneity determining its failure behavior (presence of structural defects such as inclusions or cavities). See [18,27] for similar detailed processes.

3. Validation

In order to evaluate the capability of γ-SPH-ALE, we propose to investigate three test cases: isentropic shock tube, cylinder fragmentation and hypervelocity impact.

3.1. Isentropic Shock Tube

The isentropic shock tube is a classical academic test case with a known exact solution. Many configurations and data are available in the literature allowing then to efficiently evaluate γ-SPH-ALE. We focus here on a simple barotropic shock tube using a linear equation of state (EOS) (21). We choose the case of a pressure discontinuity generated by the following initial density field (22,23) (where L and R define the left and right states).

$$p = c_0^2(\rho - \rho_0)\ and\ \begin{cases} \rho_L = 1100\ kg/m^3\ and\ \|\boldsymbol{v}_L\| = 0\ m/s \\ \rho_R = 1000\ kg/m^3\ and\ \|\boldsymbol{v}_R\| = 0\ m/s \end{cases} \quad (16)$$

The fluid is taken as having a referential density ρ_0=1000kg/m3 and a sound speed c_0=1466m/s. The tube is taken with a length Lx=0.1m and a width Ly=0.01m. The discontinuity is initially defined at x=0.05m. All computations are performed with cubic B-splines kernel functions and a fixed smoothing length. The exact solution used as a reference is given by an exact 1D Riemann solver with 2000 cells. SPH computations are performed in 2D with periodic boundary conditions to mimic a 1D case. We present the results given by standard SPH (with Monaghan's artificial viscosity such as α=β=1.5), by an ALE SPH HLLC Riemann solver with a MUSCL scheme (RSALE), and by the γ-SPH-ALE scheme (used with α=0.0). The initial spacing is set to Δx=2.10-4m, giving 50.000 particles. Note that for standard SPH the values of the artificial viscosity parameters correspond to a threshold below which the scheme is unstable (significant overshoots).

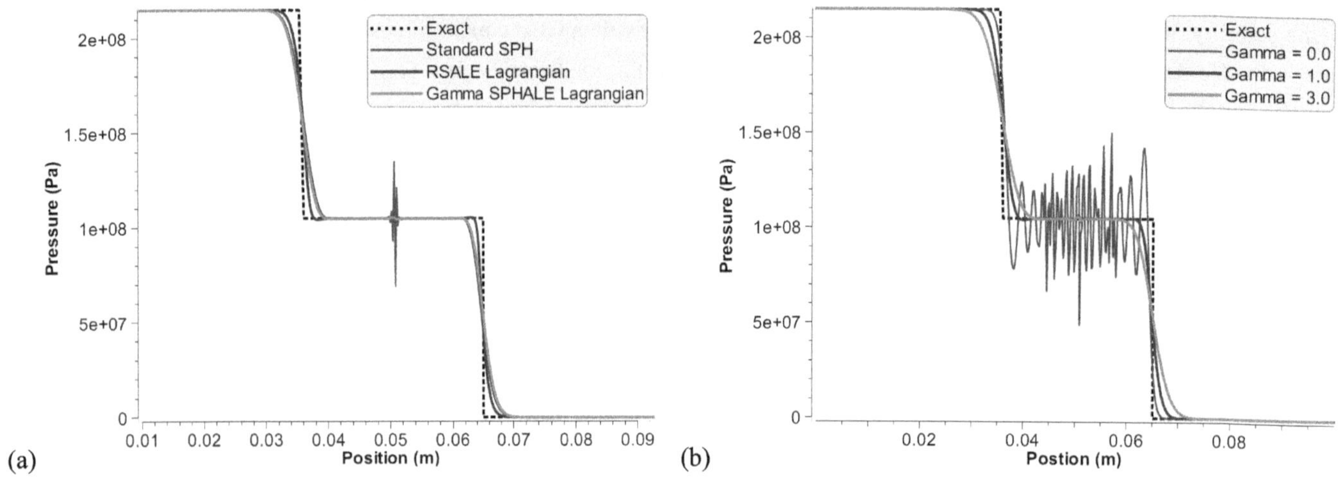

Fig. 1. **Shock Tube** Pressure fields comparisons at $t = 10^{-5}s$ (a) for several schemes (γ=1.0 for Gamma SPHALE), (b) for γ-SPH-ALE with several values of γ.

As depicted on Fig.1a the discontinuity generates a left rarefaction wave and a right shock wave. The solution is then characterized by three pressure levels (2.14928 Pa, 1.04938 Pa, 0 Pa). Fig. 1a shows the comparison between the exact solution, standard SPH, RSALE and γ-SPH-ALE with γ=1.0. Standard SPH, RSALE and γ-SPH-ALE are all able to reproduce the pressure jumps. While capturing correctly the pressure jumps, standard SPH produces significant oscillations at the discontinuity initial position both RSALE and γ-SPH-ALE manage to prevent. Note that these oscillations cannot be prevented even with more artificial viscosity. Also, RSALE manages to capture the jumps more efficiently than standard SPH and γ-SPH-ALE, generating narrower pressure variations. However, RSALE also produces overshoots the proposed scheme completely damps.

Regarding the choice of γ, Fig. 1a results are obtained with the value corresponding to the threshold below which overshoots appear. Fig. 1b shows the results for different γ. We can see that increasing its value stabilizes the scheme by damping the oscillations in a very efficient way, but it also diffuses the shocks (larger jumps). γ-SPH-ALE is then able to faithfully reproduce the shocks and the use of γ allows to easily control its stability and accuracy without the Riemann solver complexity. Note that both RSALE and γ-SPH-ALE (for a given value of γ) produce results identical to the one presented on Fig. 1b (Lagrangian description of motion i.e. $\boldsymbol{v}_0 = 0$) in Eulerian description of motion ($\boldsymbol{v}_0 = 0$) or with PST (ALE description of motion i.e. $\boldsymbol{v}_0 = arbitrary$).

3.2. Hypervelocity Impact

Hypervelocity Impacts are events occurring at velocities around 1 to 10 km/s. They are particularly challenging for space vehicles design when generated by orbiting debris. Regarding the experimental limitations (to reach the impact velocity), providing reliable predictive modelling is then crucial to evaluate the vehicle vulnerability. We focus here on the case of an Aluminium sphere (diameter 9.53 mm) impacting Aluminium plates with different thickness at 6.7 km/s. The experiment is described in [26]. Two cases are reproduced: 1- thickness = 0.8 mm, 2- thickness = 4.039 mm. The γ-SPH-ALE scheme is implemented in full 3D with an Elastic perfectly Plastic material model as well as a Mie-Gruneisen equation of state. The spalling process is managed by a pressure threshold above which the deviatoric stress is relaxed (only pressure remains). Such simple criterion is used as the idea is to evaluate the γ-SPH-ALE scheme ability to reproduce the cloud of debris. The constitutive parameters are the following: $\rho_0 = 2703$ kg/m3, Young modulus 76 GPa, Poisson ratio 0.3, Yield stress 260 MPa, Linear Hugoniot Slope coefficient 1.338, Gruneisen gamma 2.0. The numerical results are achieved with $1{,}5.\,10^6$ SPH particles.

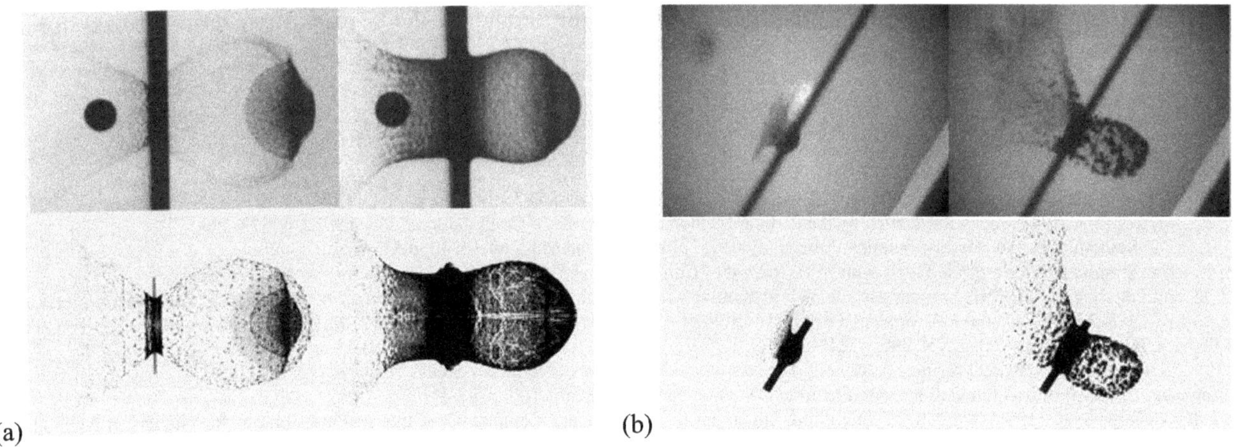

Fig. 2. **Hypervelocity Impact** Comparison between experiments (top) and γ-SPH-ALE (bottom) (a) normal impact Thickness = 0.8mm (left), Thickness = 4.039 mm (right), (b) oblique impact.

Fig. 2a shows the experimental and numerical results for both configurations described above. The experimental results are achieved thanks to X-ray flashes. We can see that in both configurations the numerical model is able to faithfully reproduce the cloud of debris characteristic features: an ejecta veil on the plate front side, an expending debris cloud "bubble" on the plate rear side and a disc-like structure inside the expending bubble (appearing clearly on case thickness 0.8 mm).

A similar configuration is now studied and corresponds to an oblique HVI. The sphere diameter is 3 mm with a velocity 4050 m/s and the plate thickness is 2 mm. Both elements still are in Aluminium (same constitutive parameters). However, the plate is now 32° tilted with respect to the sphere impact direction. Fig. 2b-top shows the experimental results at different times achieved by Thiot Ingenerie (introduced in [29]), and Fig. 2b-bottom shows the γ-SPH-ALE numerical results full 3D with $6{,}5.\,10^6$SPH particles. We can see that the proposed numerical model is again able to faithfully reproduce the cloud of debris shape.

4. Conclusion

We have proposed a new mesh-less scheme called γ-SPH-ALE based on the combination of both SPH-ALE formulation [14] and FV low-Mach scheme [19,20]. PDE are written in ALE formalism (4) allowing for a simple treatment of the arbitrary velocity field $\boldsymbol{v}_0$, ruling the particle motion (7). A stabilizing term extracted from the FV low-Mach scheme is also introduced by means of a velocity corrective term (11,13) (depending on a parameter γ). Proportional to the pressure gradient it improves the state variables evaluation. An artificial viscosity (12) (depending on a parameter α) is combined with this γ velocity to ensure the global stability of the scheme. The arbitrary property of $\boldsymbol{v}_0$ also allows to implement a PST increasing both accuracy and stability while maintaining the conservative and consistency property of the scheme.

We evaluated the performances of the γ-SPH-ALE scheme (for both hydrodynamic and solid contexts) on two test cases: isentropic shock tube and hypervelocity impact. For the shock tube case the γ-SPH-ALE scheme managed to faithfully reproduce the pressure field by efficiently damping the spurious oscillations (in both Lagrangian and Eulerian descriptions). Regarding the HVI cases, γ-SPH-ALE managed to faithfully reproduce the cloud of debris for both normal and oblique configurations.

Future works will be the enhancement of material models and fracture modelling to handle more complex materials and configurations.

Acknowledgements

We want to firstly thank the Institut de Mathématiques de Toulouse, the Institut National des Sciences Appliquées de Toulouse as well as the Université Paul Sabatier for providing the necessary literature resources to conduct the researches through online review access. Besides, we also thank particularly Impetus AFEA for providing the numerical material needed for the investigations (workstations with Impetus Solver®).

References

[1] Lucy, L. B., 1977. A numerical approach to the testing of the fission hypothesis, Astronomical Journal 82, p. 1013-1024.
[2] Monaghan, J., Ginglod, R. A.,1983. Shock Simulation by the Particle Method SPH. Journal of Computational Physics 52, p. 374-389.
[3] Monaghan, J., 1992. Smoothed Particle Hydrodynamics, Annual review of astronomy and astrophysics 30, p.543-574.
[4] Monaghan, J., 1994. Simulating Free Surface Flows with SPH. Journal of Computational Physics 110, p. 399-406.
[5] Libersky, L.D. and Petschek, A.G., 1991. Smooth particle hydrodynamics with strength of materials, in *"Advances in the Free-Lagrange Method Including Contributions on Adaptive Gridding and the Smooth Particle Hydrodynamics Method"* Trease, H.E. and Fritts, M.F. and Crowley, W.P. Editor. Lecture Notes in Physics, Berlin Springer Verlag, vol. 395, p. 248-257.
[6] Benz, W., 1990. Smooth Particle Hydrodynamics: A Review. The Numerical Modelling of Nonlinear Stellar Pulsations, p. 269-288.
[7] Benz, W., Asphaug, E., 1994. Impact Simulations with Fracture. I. Method and Tests. Icarus 107, p. 98-116.
[8] Johnson, G. R., Beissel, S. R., 1996. Normalized Smoothing Functions for SPH Impact Computations. International Journal for Numerical Methods in Engineering 39, p. 2725-2741.
[9] Vila, J. P., 1998. "Methodes particulaires regularisees. Developpements recents et nouvelles applications". ESAIM Proceedings 3 p. 131-146.
[10] Frontiere, N., Raskin, C. D., Owen, J. M., 2017. CRKSPH: A Conservative Reproducing Kernel Smoothed Particle Hydrodynamics Scheme. Journal of Computational Physics 332, p. 160-209.
[11] Marrone, S., Antuono, M., Colagrossi, A., Colicchio, G., Touze, D. L., Graziani, G., 2011. Computer Methods in Applied Mechanics and Engineering 200, p. 1526-1542.
[12] Antuono, M., Colagrossi, A., Marone, S., 2012. Numerical diffusive terms in weakly-compressible SPH schemes. Computer Physics Communications 183, p. 2570-2580.
[13] Monaghan, J., 2000. SPH Without a Tensile Instability. Journal of Computational Physics 159, p. 290-311.
[14] Swegle, J., Hicks, D., Attaway, S., 1995. Smoothed Particle Hydrodynamics Stability Analysis. Journal of Computational Physics 116, p. 123-134.
[15] R. Xu, R., Stansby, P., Laurence, D., 2009. Accuracy and stability in incompressible SPH (ISPH) based on the projection method and a new approach. Journal of Computational Physics 228, p. 6703-6725.
[16] Lind, S., Xu, R., Stansby, P., Rogers, B., 2012. Journal of Computational Physics 231, p. 1499-1523.
[17] Oger, G., Marrone, S., le Touze, D., de Leffe, M., 2016. SPH Accuracy Improvement through the Combination of a Quasi-Lagrangian Shifting Transport Velocity and Consistent ALE Formalism. Journal of Computational Physics 313, p. 76-98.
[18] Zhang, C., Hu, X. Y., Adams, N. A.,2017. A generalized transport-velocity formulation for Smoothed Particle Hydrodynamics. Journal of Computational Physics 337, p. 216 – 232.
[19] Vila, J. P., 1999. On Particle Weighted Methods and Smooth Particle Hydrodynamics. Mathematical Models and Methods in Applied Sciences 09, p. 161-209.
[20] Guillard, H., Viozat, C., 1999. On the behavior of upwind schemes in the low Mach number limit. Computers and Fluids 28, p. 63-86.
[21] Degond, P., Tang, M., 2011. All Speed Scheme for the Low Mach Number Limit of the Isentropic Euler Equations. Communications in Computational Physics 10, p. 1-31.
[22] Grenier, N., Vila, J. P., Villedieu, P., 2013. An Accurate low-Mach Scheme for a Compressible Two-fluid Model Applied to Free-surface Flows. Journal of Computational Physics 252, p. 1-19.
[23] Lavalle, G., Vila, J. P., Blanchard, G., Laurent, C., Charru, F., 2015. A Numerical Reduced Model for Thin Liquid Films Sheared by a Gas Flow. Journal of Computational Physics 301, p. 119-140.
[24] Clayer, M., Lacome, J. L., Limido, J., Vila, J. P., 2014. "Pressure Evaluation Improvement for Euler Isentropic SPH Scheme", Proceedings of the 9th international SPHERIC workshop, Paris, France.
[25] Lanson, N. and Vila, J.P., 2008. Renormalized Meshfree Schemes I: Consistency, Stability, and Hybrid Methods for Conservation Laws. SIAM Journal on Numerical Analysis 46, p. 1912-1934.
[26] Couderc, F., Duran, A., Vila, J. P., 2017.An explicit asymptotic preserving low Froude scheme for the multilayer shallow water model with density stratification. Journal of Computational Physics 343, Supplement C, p. 235-270.
[27] Huerta, A., Casadei, F., 1994. New ALE applications in nonlinear fast transient solid dynamics. Engineering Computations 11, p. 317-345.
[28] Piekutowski, A. J., 1993. Characteristics of debris clouds produced by hypervelocity impact of aluminum spheres with thin aluminum plates. International Journal of Impact Engineering 14, p. 573-586.
[29] Plassard, F., Mespoulet, J., Hereil, P., 2011. "Hypervelocity impact of aluminium sphere against aluminium plate: experiment and LS-DYNA correlation", 8th European LS-DYNA Users Conference, Strasbourg, France.

Proceedings of the 2019 Hypervelocity Impact Symposium
HVIS2019
April 14-19, 2019, Destin, FL, USA
HVIS2019-090

Analytically derived space time-based boundary condition (STBC) to account for stress wave propagation in a heterogeneous micromechanical model at hypervelocity impact

Zhiye Li[a], Somnath Ghosh[b]*

[a] *Department of Civil and Environmental Engineering, Stanford University, Stanford, CA 94305, United States*
[b] *Department of Civil Engineering, Johns Hopkins University, Baltimore, MD 21218, United States*

Abstract

Recent years have seen a surge in research on material and structural response of composites using the homogenization based hierarchical modeling method. The microstructural representative volume element (RVE) is a small micro-region for which the volume average of variables is the same as those for the entire body. Representations of the microstructure are used for micromechanical simulations in determination of effective material properties by homogenization. Conventionally, periodic boundary conditions (PBC) are applied on the RVE boundary. However, when the heterogeneous microstructure is under very high strain rate loading conditions ($10^5\ s^{-1} - 10^7 s^{-1}$), periodic boundary conditions (PBC) do not accurately represent the effect of stress wave propagation. Improper boundary conditions can lead to significant error in homogenized material properties. In order to increase the accuracy of the homogenization model, this study introduces a new space-time dependent boundary condition (STBC) for a 3D microscopic RVE subjected to high strain rate deformation in explicit FEM simulation by using the characteristics method of traveling waves. The advantages of the STBC are discussed in comparison with time-dependent averaged results of examples using PBC. The proposed STBC offers significant advantages over conventional PBC in the RVE-based analysis of heterogeneous materials.

Keywords: stress wave propagation, inhomogeneity, micro-mechanics, high strain rate, hypervelocity impact, space-time dependent boundary condition

Nomenclature

c	elastic wave speed
T	time
L	length
X_i	node coordinates, i =1, 2, 3.
V	volume
C_{ijkl}	components of the fourth-order stiffness tensor

1. Introduction

The study of composites and their structural applications (e.g. commercial vehicle and protective armor) under high strain rate loading has many advantages. S-glass fiber reinforced composites containing magnesium aluminosilicate or borosilicate fibers are particularly noteworthy and durable because of their high stiffness and strength to weight ratio, impact resistance, and durability under extreme temperature or corrosive environments. However, the understanding of dynamic mechanisms and assistance for material structural design are far from integrated. This lack of integration is due to the deficiency of microscale experimental

* Corresponding author. fax: +1 410 516 7473.
E-mail address: sghosh20@jhu.edu (S. Ghosh).

thermo-mechanical data on controlled high-speed loading, insufficient attempts to elucidate the effect of stress waves propagating among heterogeneities in the RVE-microsecond range, and lack of applications to hierarchical multiscale modeling. Robust modeling, accounting for microstructural details, is an indispensable ingredient of the material-by-design model in order to reduce the composite plate thickness and penetration depth and to increase the penetration resistance force and energy absorption ratio of energy absorption and weight.

It has long been recognized that the phenomenon of substantial deformation under extreme loading conditions varies according to strain, strain rate, temperature, and loading history. One significant mechanism resulting in this phenomenon is the propagation of the stress wave which universally affects theoretical and computational models with various constitutive laws, damage mechanisms, and failure modes. The stress wave is the most fundamental element for determining the difference between the processes within bodies subjected to high rate loads and the processes in static or quasi-static situations.

2. Microstructural representative volume element (RVE)

The representative volume element or RVE is defined as a micro-scale sub-domain, on which a volume average of variables yield the macro-scale model variables [1-6]. The choice of the RVE depends on the material property of interest and can vary from one class of properties to another. Statistically equivalent RVEs have been developed from detailed characterization studies of non-uniformly distributed micro-structures in [7-8]. In this study, the RVE of a uniformly distributed unidirectional composite, shown in Fig 1, is simulated to demonstrate the shockwave motion in a hypervelocity problem. In Fig 1 (c) the RVE is represented by a cuboidal matrix containing a cylindrical fiber [9].

2.1. Periodic boundary conditions(PBCs) limitations at hypervelocity

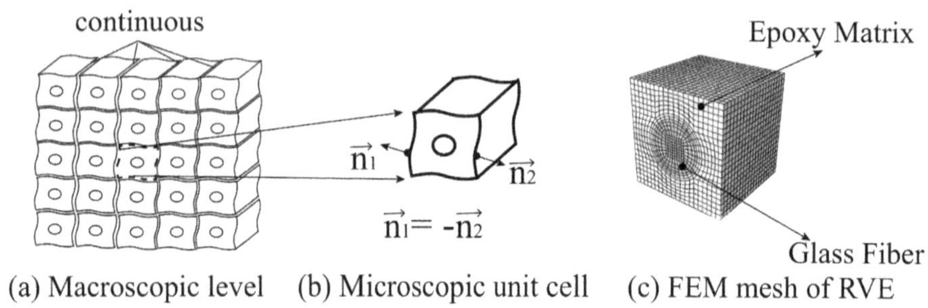

Fig.1. Representation of a continuous unidirectional composite domain and the RVE simulation model: (a) the microstructure with periodically repetitive unidirectional cylindrical fibers, (b) RVE represented by a microstructural unit cell, and (c) finite element model of the RVE. The materials of the RVE are epoxy matrix and glass fiber. [5]

Assuming periodicity of the RVE, the FEM model in Fig 1 (c) is subjected to incremental periodic boundary conditions. Following procedures in [11, 12], the macroscopically applied strain increment on the RVE ΔE_{ij} is obtained by decomposing the displacement increment ΔU_i on the RVE boundary into an RVE-averaged term and a periodically perturbed term, expressed as:

$$\Delta U_i = \Delta E_{ij} X_j + \Delta \breve{U}_i \tag{1}$$

The periodic displacement component $\Delta \breve{U}_i$ is equal for corresponding nodes on opposite faces of the RVE, e.g. nodes n1 and n2 in Fig 1 (b). Accordingly, the total displacements at the corresponding node-pair (n1, n2) are related in terms of the macroscopic strain increment as:

$$(\Delta U_i)_{n1} - (\Delta U_i)_{n2} = \Delta E_{ij}((X_j)_{n1} - (X_j)_{n2}) \tag{2}$$

Where $(X_j)_{n1}$ and $(X_j)_{n2}$ are coordinates of the node-pair on the RVE boundary. Details of implementation in a FEM simulation for dynamic simulations are given in [5, 9-10].

However, periodicity in the response variables necessitated by the application of Eq. (2) depends on the material state and the applied loading conditions. This is not always true in geometrically repeated microstructures. [10] gives a mathematical proof that, when an elastic wave enters through one side of the unit cell, periodicity requires that an identical wave simultaneously exits the opposite side. Thus this kinematic constraint is generally valid only for quasi-static problems. In other words, it will add an unrealistic incident wave into the RVE for a hypervelocity problem.

2.2. Bloch boundary conditions and its limitations at hypervelocity

The Bloch wave theory for time-harmonic waves in periodic lattice structures is an alternative method for solving PBCs at hypervelocity. The Bloch condition has been used in the study of photonic bandgaps [13] and electromagnetic waves [14]. However, there are some limitations in the assumptions of the Bloch theorem so that, in this study, the Bloch-periodic wave assumptions are not resorted to establish a micromechanics FEM model for the high strain rate problem. The limitations are:

- In this study, the microscopic FEM model is under continuum mechanics assumptions. Unlike how photons are analogized to electrons [15], the continuous material cannot be analogized to electrons directly.
- Similar to periodic boundary conditions, Bloch boundary conditions induce an incident wave on both of the responding points on the surface of FEM models. They omit the un-periodic history of the incident wave propagating into the research domain, interacting with inhomogeneities and finally reaching a quasi-static Periodic Bloch wave state. The un-periodic process is the main phenomenon in the high rate FEM model of this paper.
- Bloch conditions limit the mathematical format of the loading condition function. Only time harmonic wave loading can be applied with this boundary condition. This is very different from the high strain rate tension/compression or shear problem in the current study.
- The boundary conditions needs to be calculated for every frequency. One possible way is to decompose the input into a superposition series of waves with different frequencies. When applying the boundary conditions to the FEM model, the boundary conditions are the superposition of the Bloch condition of all frequencies obtained from the previous step. This requires huge computational work to tune the constraint equations in the FEM model for every particular deformation input.

2.3. Space-time dependent boundary conditions (STBC)

Consider a composite microstructure formed by spatial repetition of a two phase composite RVE, as shown in figure Fig 3(a). If material 1 and material 2 are both epoxy in Table 1, the analytical solution is plotted in Fig 2(a). If material 1 is epoxy, and material 2 is glass in Table 1, the analytical solution is plotted in Fig 2(b). The stress wave propagation process in a 1-D infinite domain shown in Fig 2(a) and Fig 2(b) is correctly performed with novel space-time boundary conditions (STBCs) in Fig 3(b) for extreme high strain rate impact or blast.

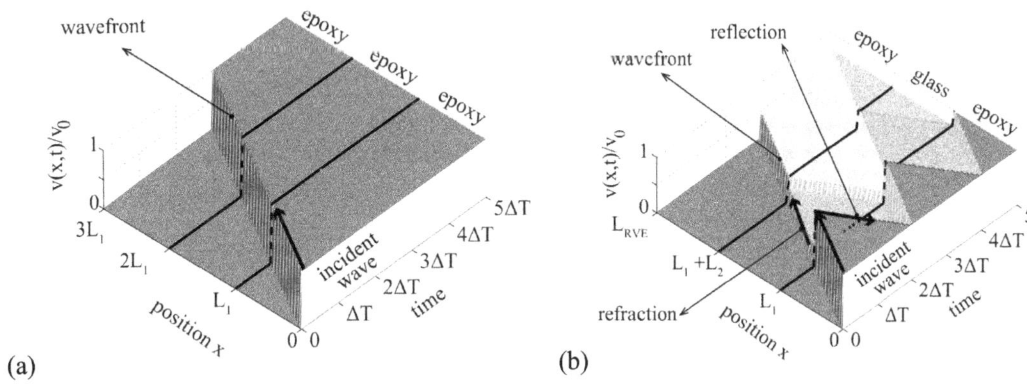

Fig. 2. Analytical solution of velocity ratio-time-position function along local position ($0 \le x \le L_{RVE}$). In this figure, $T_1=L_1/c_1$, $T_2=L_2/c_2$, and $T_1 = T_2 = \Delta T$. (a) homogenous RVE, (b) heterogeneous RVE.

Table 1. Elastic material properties of the constituent phases of the periodic medium. [5]

Material/Property	Young's Modulus (GPa)	Poisson Ratio	Density(g/cc)
EPON 828 epoxy	3.0	0.4	1.00
Borosilicate glass	7.4	0.26	2.00

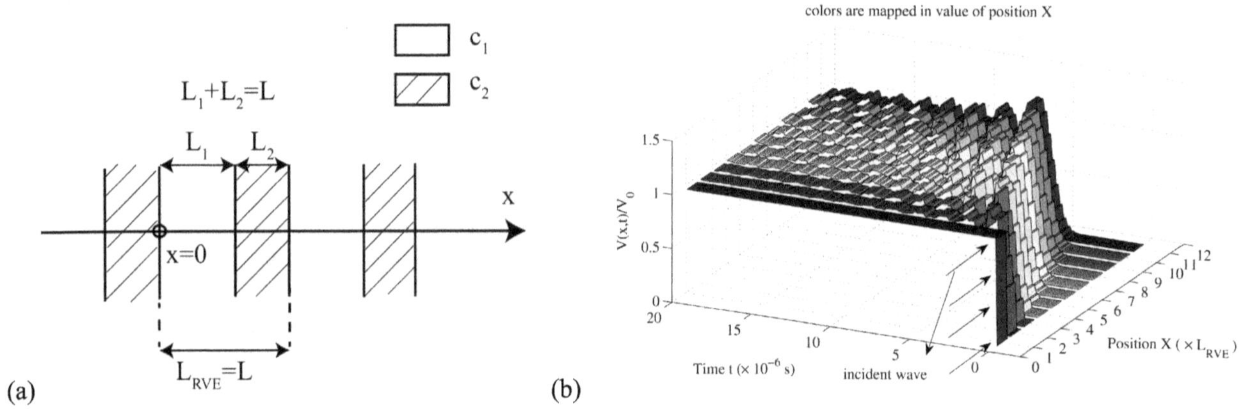

Fig. 3. (a) One-dimensional periodic structure of a two-material system in an infinite medium showing a unit cell. (b) Analytical solution of velocity ratio-time-position function at RVE boundary (N $\times L_{RVE}$), N = 1, 2, 3....

Space-time boundary conditions (STBCs) are implemented in the commercial code ABAQUS/Explicit® (2014) using multi-point constraint equations. For nodes on the surfaces shown in Fig 1 (b), the total displacements at the corresponding node-pairs (n1, n2) are constrained in terms of the macroscopic strain increment as:

$$(\Delta U_i)_{n2} - (\Delta U_i)_{n1} = \Delta E_{ij}\left[\Delta R[t - t^c\left(\left(X_j\right)_{n2}\right)] - \Delta R[t - t^c\left(\left(X_j\right)_{n1}\right)]\right]\Delta X_j \quad (3)$$

$$\Delta X_j = \left(X_j\right)_{n2} - \left(X_j\right)_{n1} \quad (4)$$

2.1. Comparison of the result of PBC and STBC

This section discusses the results of models with PBC and STBC. The single unit cell RVE with PBCs is subject to simple compression (x-axis longitudinal wave) with overall strain rate of $10^6 s^{-1}$. Other models compared with this model are subjected to the same amplitude of incident wave. A 1 × 5 unit cell RVE containing 59,200 elements is setup as a reference example of how the stress wave is propagating in a large domain. It is also constrained by PBCs. In this model, unit cell 1 and unit cell 2 marked in Fig 4(a) are studied and compared with the single fiber RVE with STBC for unit cell 1 and unit cell 2, respectively [10].

Since the essential output data which needs to be input for the homogenization model [4,16-17] is volume-averaged stress-strain, the objective of the paper is to minimize the volume-averaged stress-strain curve difference between the results of specific unit cells in the 1×5 RVE PBC model (e.g. unit cell 1 or unit cell 2 in Fig 4(a)) and the result of the 1×1 RVE model (11,840 elements) by applying STBCs instead of conventional PBCs. The volume-averaged von-Mises stress and effective strain are respectively defined as:

$$\Sigma_{vM} = \sqrt{\frac{3}{2}\boldsymbol{\Sigma}^{dev}:\boldsymbol{\Sigma}^{dev}} \quad (5)$$

$$e_{eff} = \sqrt{\frac{2}{3}\mathbf{e}^{dev}:\mathbf{e}^{dev}} \tag{6}$$

where $\mathbf{\Sigma}^{dev}$ and $\mathbf{e}^{dev}$ are the deviatoric components of the volume-averaged stresses and strains respectively, expressed as:

$$\mathbf{\Sigma} = \frac{1}{V^{RVE}}\int_{V^{RVE}} \boldsymbol{\sigma}\, dV \tag{7}$$

$$\mathbf{e} = \frac{1}{V^{RVE}}\int_{V^{RVE}} \boldsymbol{\varepsilon}\, dV \tag{8}$$

Here, $\mathbf{\Sigma}$ and $\mathbf{e}$ are the microscopic stresses and strains in the RVE volume.

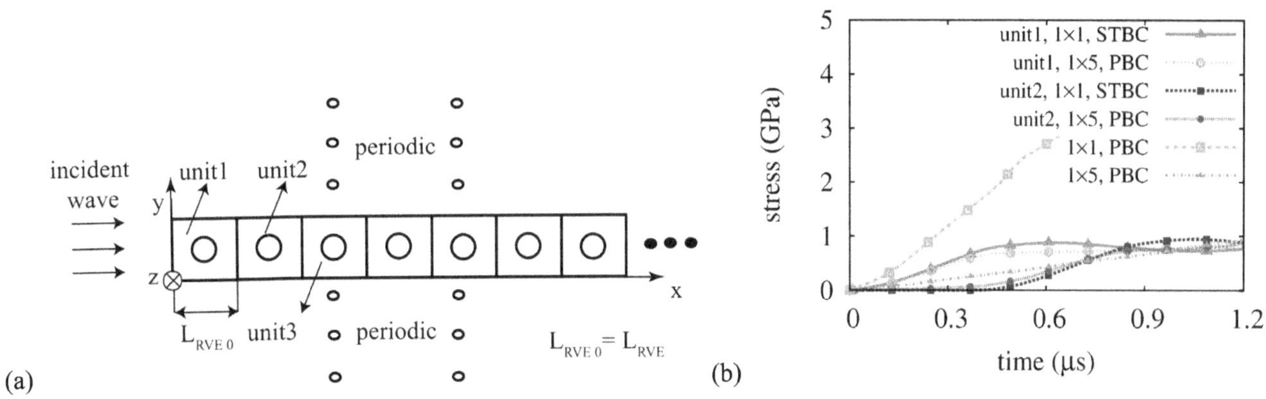

Fig. 4. (a) Schematic diagram of the analysis model investigated by comparing results using RVE 0, RVE 1 and RVE 2. (b) plot of volume-averaged von-Mises stress Σ_{vM} as a function of time. The overall strain-rate is $\bar{d}_{11}$=$10^7 s^{-1}$, in compression.

Fig 4(b) is the volume-averaged von-Mises stress versus time plot. Two unit cells from the 1 × 5 unit cell RVE are studied in Fig 4(a), with unit cell 1 having volume V_1 and unit cell 2 with volume V_2. At t = 0, V_1 = V_2 due to structural periodicity. The legend of these two curves are "unit 1, 1 × 5, PBC" and "unit 2, 1 × 5, PBC", respectively. In addition to the two studied unit cells, the volume-averaged curve of the whole 1 × 5 RVE (V=5 ×V_1=5 ×V_2) is also plotted with the legend as "1 × 5, PBC" to show the overall material performance of the large RVE. The three plots prove that the homogenized stress strain behavior of the unit cell in different positions and the homogenized result of the whole model are different functions of time. The unit 1 and 2 curves are compared with the 1×1 RVE with STBC for unit cell 1 (the legend is "unit 1, 1 × 5, STBC") and unit cell 2 (the legend is "unit 2, 1 × 5, STBC") for the same time frame. For comparison, the 1×1 RVE in Fig. 1 (c) with PBC is calculated under the same loading condition as with the "1 × 5, PBC" example. The corresponding stress-time curve is labeled as "1 × 1, PBC". Fig 4(b) shows that both the unit cell 1 and unit cell 2 averaged stress-time results from the 1 × 5 PBC model are very different from the 1 × 1-PBC result. Meanwhile, the 1 × 1 STBC model for unit 1 and unit 2 are very close to the 1 × 5-PBC model for unit 1 and unit 2, respectively.

The primary reason STBC can improve the accuracy of high strain rate deformation simulations is that STBC can simulate the evolution of effective volume fractions. The effective volume fraction is the stressed fiber volume divided by the stressed RVE volume. The effective volume fraction, unlike the constant microstructure dependent volume fraction, changes during stress wave propagation. In high strain rate deformation, the time span is extremely short. With shrinking time span, the effect of evolving effective volume fraction is increasingly considerable [9-10].

Fig. 5 depicts the corresponding homogenized relative stiffness component C_{1111}/E_{matrix} as a function of averaged strain ϵ_{11}, as well as the evolution of the effective stressed fiber volume fraction. The unidirectional cylindrical single fiber RVE is solved under tension for the different loading strain rates with PBCs and STBCs. The static fiber volume fraction is equal to the geometric fiber volume fraction of the undeformed RVE, which is a constant equal to 0.2. The dynamic volume fraction is the stressed fiber volume over the total stressed volume.

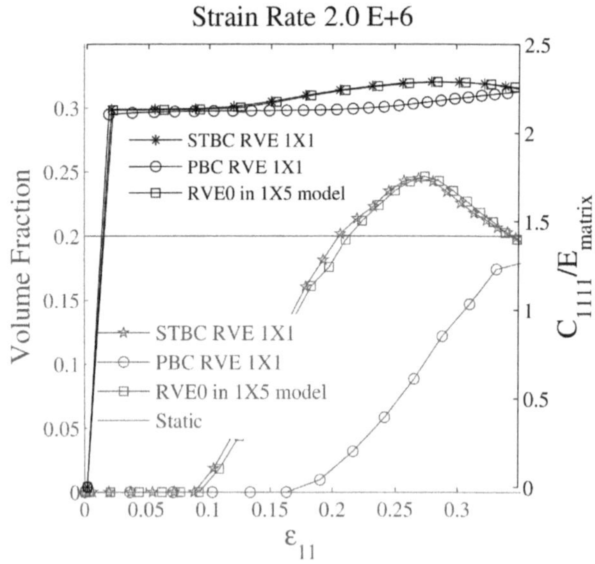

Fig. 5. Homogenized relative stiffness component C_{1111}/E_{matrix} and effective volume fraction as a function of ϱ11 at strain rate of $10^6 s^{-1}$ [9].

In this study, the threshold von Mises stress of a stressed element is chosen as 260 MPa (lower than 10% of the historic maximum local von Mises stress value). From Fig 5, the results of the single fiber RVE with STBCs show good agreement with the results of unit 1 in the large model in Fig 4(a), while curves of the single fiber RVE with PBCs indicate a different evolution from the previous two. All volume fractions are zero in the early stage followed by rapid growth. The effective volume fractions of the STBCs model and of the large model grow at a faster rate. Unlike the single fiber RVE with PBCs curves, which only increase and saturate below the static volume fraction horizontal line, these two curves increase higher than the static volume fraction and then decrease and saturate to a value which is very close but slightly lower than the static value. This is because there is a small decrease of the deformed static geometric fiber volume fraction from the undeformed value [9-10]. Before the single fiber RVE with PBCs curves branch away from the other two, there is no obvious difference in all homogenized relative stiffness components. However, during the rapid change of the effective volume fractions, the homogenized relative stiffness components of the single fiber RVE with PBCs disagree with and, concurrently, grow slower than the other two results. The coincidence indicates that in a considerable beginning stage of a high strain rate deformation problem the evolution of dynamic effective volume fraction leads to the change of homogenized or averaged stress-strain behavior of composite RVEs [9-10]. Fig 4(a) shows evidence that STBCs provide a convincing evolution process of stress wave propagation and dynamic material behavior.

3. Conclusion

This study focuses on understanding and adopting the elastic wave propagation effect in the microscopic RVE model of periodically repetitive arrayed heterogeneities. The first stage of this project is observation and acquisition of the characterization of stress wave propagation in a 1-D analytical solution. Fig. 2(a) summarized a classical solution of the elastic wave traveling in a single material phase. Fig 2(b) summarized the reflection/refraction equation at a perfectly bonded interface of two elastic materials. Combining the background of the characteristics of the method of traveling waves and the reflection/refraction equation, the analytical solutions of elastic wave under nonhomogeneous boundary conditions and under periodic boundary conditions are derived. A Matlab® code to generate the analytical solution with respect to the different material parameters and volume fractions is developed. Comparison studies of localized and globalized results show that the STBC has exhibited improved accuracy and time-saving advantage by integrating the effect of shockwave motion and effective volume fraction evolution. The improvement of STBC on the FEM model is more significant in high strain rate simple tension/ compression and in shear deformation.

Acknowledgements

This work is sponsored by the Center for Materials in Extreme Dynamic Environments (CMEDE) of the Army Research Laboratory and was accomplished under Cooperative Agreement Number W911NF-12-2-0022. The sponsorship is gratefully acknowledged. Computational support of this work has been provided by the Homewood High Performance Compute Cluster (HHPC) at Johns Hopkins University and the Maryland Advanced Research Computing Center (MARCC).

References

[1] Drugan, W.J. and Willis, J.R., 1996. A micromechanics-based nonlocal constitutive equation and estimates of representative volume element size for elastic composites. Journal of the Mechanics and Physics of Solids, 44(4), pp.497-524.

[2] Kouznetsova, V., Geers, M.G., Brekelmans, W.M., 2002. Multi-scale constitutive modelling of heterogeneous materials with a gradient enhanced computational homogenization scheme. Int. J. Numer. Methods Eng. 54 (8), 1235–1260.

[3] Kanit, T., Forest, S. , Galliet, I. , Mounoury, V. , Jeulin, D. , 2003. Determination of the size of the representative volume element for random composites: statistical and numerical approach. Int. J. Solids Struct. 40 (13), 3647–3679.

[4] Jain, J.R. , Ghosh, S. , 2008. Homogenization based 3d continuum damage mechanics model for composites undergoing microstructural debonding. J. Appl. Mech. 75 (3) . 031011–031011

[5] Li, Z., Ghosh, S., Getinet, N. and O'Brien, D.J., 2016. Micromechanical modeling and characterization of damage evolution in glass fiber epoxy matrix composites. Mechanics of materials, 99, pp.37-52.

[6] Zhang, X., Li, Z., Ghosh, S. and O'Brien, D.J., 2016. Parametric Homogenization Based Continuum Damage Mechanics Model for Composites. In Proceedings of the American Society for Composites: Thirty-First Technical Conference.

[7] Swaminathan, S. , Ghosh, S. , 2006. Statistically equivalent representative volume el- ements for unidirectional composite micro-structures: Part ii-with interfacial debonding. J. Compos. Mater. 40 (7), 605–621 .

[8] Swaminathan, S. , Ghosh, S. , Pagano, N. , 2006. Statistically equivalent representative volume elements for unidirectional composite.

[9] Zhiye Li, "Micromechanical Studies of Glass Fiber Reinforced Epoxy Matrix Composites Undergoing Deformation and Damage at High Strain Rates." 2018.

[10] Li, Z., Ghosh, S., 2019. Developing Space-Time Dependent Boundary Conditions for Composite RVEs at High Strain-Rate. International Journal of Solids and Structures, https://doi.org/10.1016/j.ijsolstr.2019.02.019

[11] Pellegrino, C. , Galvanetto, U. , Schrefler, B. , 1999. Numerical homogenization of pe- riodic composite materials with non-linear material components. Int. J. Numer. Methods Eng. 46 (10), 1609–1637 .

[12] Segurado, J. , Llorca, J. , 2002. A numerical approximation to the elastic properties of sphere-reinforced composites. J. Mech. Phys. Solids 50 (10), 2107–2121.

[13] Ho, K., Chan, C., Soukoulis, C., 1990. Existence of a photonic gap in periodic dielectric structures. Phys. Rev. Lett. 65 (25), 3152–3155.

[14] Zhang, Z., Satpathy, S., 1990. Electromagnetic wave propagation in periodic structures: Bloch wave solution of maxwell's equations. Phys. Rev. Lett.65 (21), 2650–2653.

[15] Kushwaha, M.S., Halevi, P., Dobrzynski, L. and Djafari-Rouhani, B., 1993. Acoustic band structure of periodic elastic composites. Physical review letters, 71(13), p.2022.

[16] Li, Z., Zhang, X., Ghosh, S. and O'Brien, D.J., 2019. Adiabatic heating and damage formation of composite associated with high-velocity impact . Proceedings of the 2019 Hypervelocity Impact Symposium. #109

[17] Xiaofan Zhang, Zhiye Li, Somnath Ghosh, and Daniel J. OBrien. Parametrically Homogenized Continuum Damage Mechanics(PHCDM) Model for Composites from Micromechanical Analysis." submited to TMS 2019 Conference Proceedings: 2019 TMS Annual Meeting & Exhibition, 2019.

Proceedings of the 2019 Hypervelocity Impact Symposium
HVIS2019
April 14-19, 2019, Destin, FL, USA
HVIS2019-091

Extended Abstract+
Mesoscale Modeling of Al/Ni Composites

Grant Smith[a,*], Scott Bardenhagen,[a] John Nairn[b]

[a]*Wasatch Molecular Incorporated, 825 N. 300 W. Ste W003, Salt Lake City, UT 84013, USA*
[b]*Oregon State University, 112 Richardson Hall, Corvallis OR 97331, USA*

1. Introduction

Granular composite formulations involving metal/metal [1,2] (intermetallic) reactions have been investigated for a number of applications. Interfaces between metallic particles have been attributed particular importance due to their ability to form localized hotspots, where the local temperature of the granular composite may significantly exceed the global temperature, leading to a localized reaction initiation that may initiate a global reaction.[3] Understanding of the response of metal/metal granular composites to mechanical loading can be greatly facilitated by *mesoscale* modeling in which the underlying composite structure of the components is clearly resolved. For the Al/Ni composite discussed in this study, this involves explicitly resolving Al and Ni particles and intermetallic (Al/Al, Ni/Ni and Al/Ni) interfaces. In this work we utilize an idealized but realistic model for an Al/Ni composite using EOS, constitutive and interfacial models informed by molecular dynamics (MD) simulations. The Al/Ni microstructures are subjected to uniaxial compressive loading up to 30 GPa, shock loading up to 70 GPa (2000 m/s) and impact loading at 2000 m/s, and the mechanical and thermal response of the composite are analyzed.

2. Main text

2.1. Methodology and Systems Simulated

Continuum mesoscale simulations of the Al/Ni microstructures were carried out using the Lagrangian based Material Point Method (MPM) [4] that is capable of simulating elastic and plastic responses, contact (e.g., friction between phases) and large deformations. The OSParticulas [5] package was employed in all simulations and a newly developed contact methodology [6,7] was utilized.

In modeling a granular composite at the mesoscale, such as Al/Ni considered here, representative volume elements, or RVEs, are usually considered. An RVE must be large enough to contain a representative distribution of the phases present in the composite (i.e., Al and Ni particles). Furthermore, properties of each material as well as interfacial properties that describe contact between these phases are needed before mesoscale simulations can be performed. A modest-sized 2D system consisting of randomly packed ellipses representing Al and Ni particles was chosen for the sake of computational expediency, allowing for investigation of a large set of mesoscale parameters and loading conditions. The randomly packed Al/Ni system was compressed to yield microstructures of 90.3%, 97.3% and 100% of TMD (maximum theoretical density).

Uniaxial (plane strain) compression was applied to the 90.3% of TMD, 97.3% and 100% of TMD Al/Ni composites. Shock loading (plane strain) up to particle velocities of 2000 m/s was applied to the 97.3% and 100% (2000 m/s only) TMD composites. Uniaxial loading was applied via a piston moving from right to left at 100 m/s) resulting in a strain rate of 1.4 x

+ Full paper can found at https://www.sciencedirect.com/journal/international-journal-of-impact-engineering/special-issue/10MTGP5W4VJ

* Corresponding author. Tel.: +1-801-746-9258.
E-mail address: gsmith@wasatchmolecular.com.

10^6/s. Shock loading was applied by setting the left wall velocity instantaneously to various particle velocities representative of shock loading.

2.2. PVT Behavior of Al/Ni

Fig. 1a shows the pressure as a function of strain for uniaxial compression of Al/Ni at three porosities. The composite exhibits behavior typical of granular materials. Specifically, an elastic region at small strains is followed by a region with little dependence of pressure on volume followed by a region of stiffening behavior. The intermediate (low slope) region is associated with the "squeezing out" of void space. After void space is removed the stiffness of the material resembles that of the metals of which it is comprised. Fig. 1b shows a comparison of pressure vs. strain for shock and uniaxial compression for the 97.3% Al/Ni composite. For a given strain the pressure is higher on the Hugoniot (shock) than on the isentrope (uniaxial compression), particularly at higher strains, as expected.

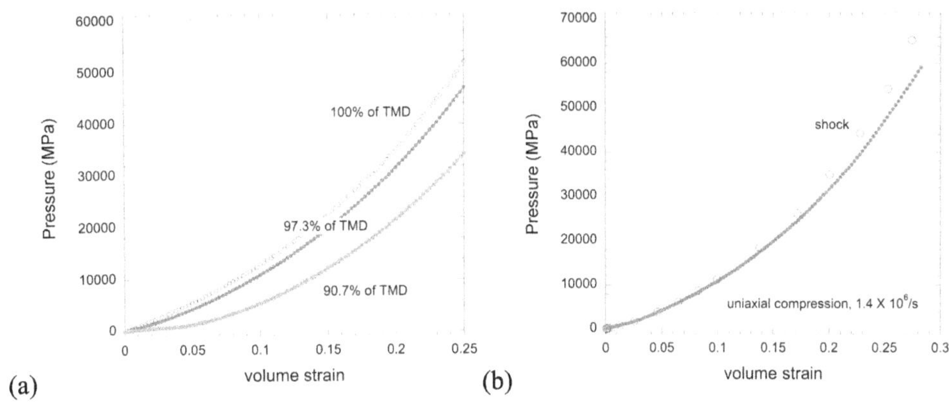

Fig. 1. (a) Pressure vs. strain from uniaxial compression for three porosities. (b) Pressure vs. strain from uniaxial compression and shock loading of the 97.3% of TMD composite.

2.3. Influence of Particle-Particle Friction on Uniaxial Compression

In order to gain insight into the influence of particle-particle frictional contact on Al/Ni response to uniaxial compressive loading, we investigated the behavior of the Al/Ni microstructures when perfect interfaces are employed. The perfect interface (PI) model has a single velocity field and does not allow sliding motion of the particles with respect to each other. However, heating due to mechanical effects and plastic flow still occur. Fig. 2a shows a comparison of pressure vs. strain from uniaxial strain compression of the 97.3% of TMD Al/Ni microstructure with and without frictional sliding. As is found for Al/Ni with frictional interfaces (multimaterial model), the material with perfect interfaces exhibits behavior typical of granular material, specifically a "void collapse" regime with a small pressure vs. strain slope following an initial low-strain elastic regime, albeit with a somewhat stiffer response. This result indicates that plastic deformation allows for the filling of voids upon compression, but not as efficiently as when combined with frictional particle-particle contact/sliding. After void collapse has occurred there is very little difference in the mechanical response of the frictional sliding and perfect interface models.

Fig. 2b shows a comparison of mean temperature vs. stress for the frictional interface and PI models for the 97.3% of TMD Al/Ni microstructure subjected to uniaxial compression. For a given stress, the PI model shows a smaller temperature rise than observed for the frictional model due to the absence of frictional heating. Fig. 2c shows the temperature profile at 15 GPa for the PI model for the 90.7% of TMD Al/Ni composite. The temperature distribution is much more homogeneous than was found for the frictional model at the same compression shown in Fig. 2d. In particular, the hotspots associated with void collapse and particle-particle boundaries are no longer present.

The treatment of particle-particle contact influences the mechanical and more strongly the thermal response of the low porosity Al/Ni granular composites to uniaxial compressive loading. It is worth noting, however, that under no circumstances for the uniaxial compressive loading scenarios studied do we observe hotspot formation sufficient (or even nearly sufficient) to lead to melting of the aluminum phase.

2.4. Influence of Particle-Particle Friction and Voids on Temperature Distribution under Shock Loading

Fig. 3 shows the temperature distribution in Al/Ni composites subjected to shock at a particle velocity of 2000 m/s for two microstructures (97.3% of TMD and 100% of TMD) and two models ("frictional heating" and "no heat". The "no heat" model allows frictional sliding but without heat generation. For all cases the mean temperature is around 1400 K. A heterogeneous distribution of temperatures is observed for all cases, which is much broader than observed under uniaxial compression. The coolest regions (blue) are nickel and the hotter regions (green-yellow) are aluminum. For the 97.3% of TMD microstructure with frictional heating (Fig. 3a) hotspots as hot as 3300 K can be observed, centred primarily on collapsed voids. Only a few hotspots approaching 3000 K remain when frictional heating is neglected (Fig. 3b). When voids are removed (Fig. 3c) there are only few regions at particle-particle boundaries with temperatures approaching 3000 K. Finally, in the absence of voids and frictional heating no hotspots approaching 3000 K are observed (Fig. 3d). These results indicate that frictional effects associated with the collapse of voids at particle-particle boundaries under shock loading can lead to local temperatures much higher than the mean temperature. Note that even the highest local temperature observed with a particle velocity of 2000 m/s, around 3300 K, remains below the melting point of aluminum at 65-70 GPa. However, it appears that sufficiently high void content could result in larger voids as well as the interaction of locally high temperature regions (coalescence of hotspots) that could lead to melting of aluminum and/or initiation of Al/Ni reactions.

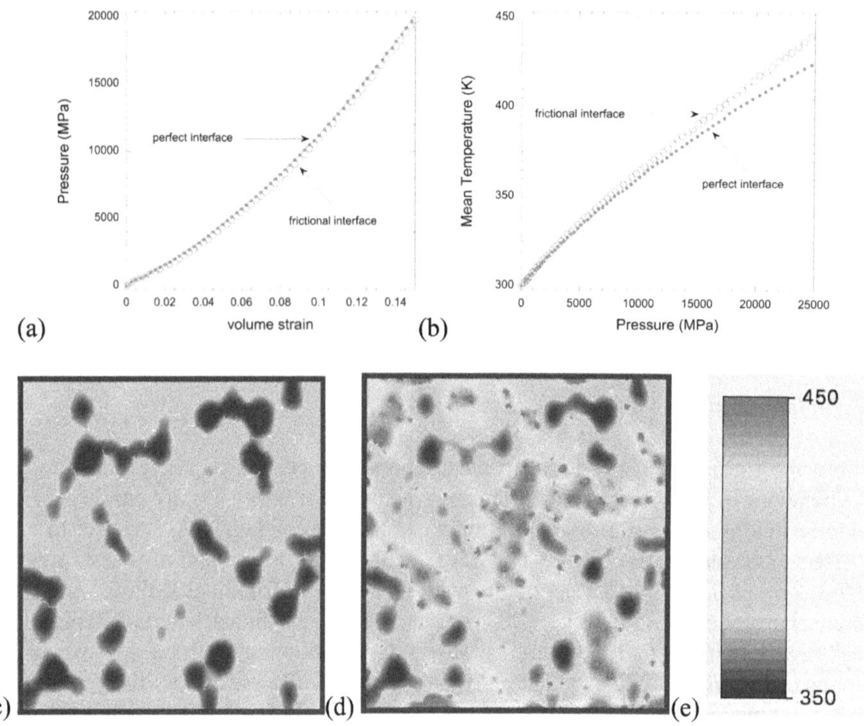

Fig. 2. (a) Pressure vs. strain from uniaxial compression of the 97.3% of TMD Al/Ni composite for the frictional interface and perfect interface (PI) models. (b) Mean temperature vs. pressure from uniaxial compression of the same microstructure using the same models. (c) Temperature profile for the 90.7% of TMD composite at 15 GPa using the perfect interface model. (d)The temperature profile for the 90.7% of TMD composite at 15 GPa with friction. (e) The temperature scale (in K).

2.5. Conclusions

For the low porosity Al/Ni granular composites investigated in this study, the distributions of temperatures and stresses during uniaxial strain compression were observed to be highly heterogeneous, reflecting the heterogeneous microstructure of the material. Plastic deformation, mechanical work, and particle/particle friction led to modest increase of temperature and hotspots unixial compressive loading (isentropic) up to 30 GPa. For shock loading, higher pressures, much higher temperatures and greater heterogeneity was observed for a given volume strain. the importance of frictional effects and voids on mean temperature and particularly hotspot formation for the low porosity composites studied was clearly evident. However, none of the uniaxial compression and shock loading scenarios investigated led to sufficient local heat to cause Al melting at the resulting pressure state, thought to be a key step in the initiation of Al-Ni reactions. However, subsequent events to shock loading, such as fragmentation and unconstrained impact, could initiation Al/Ni reactions.

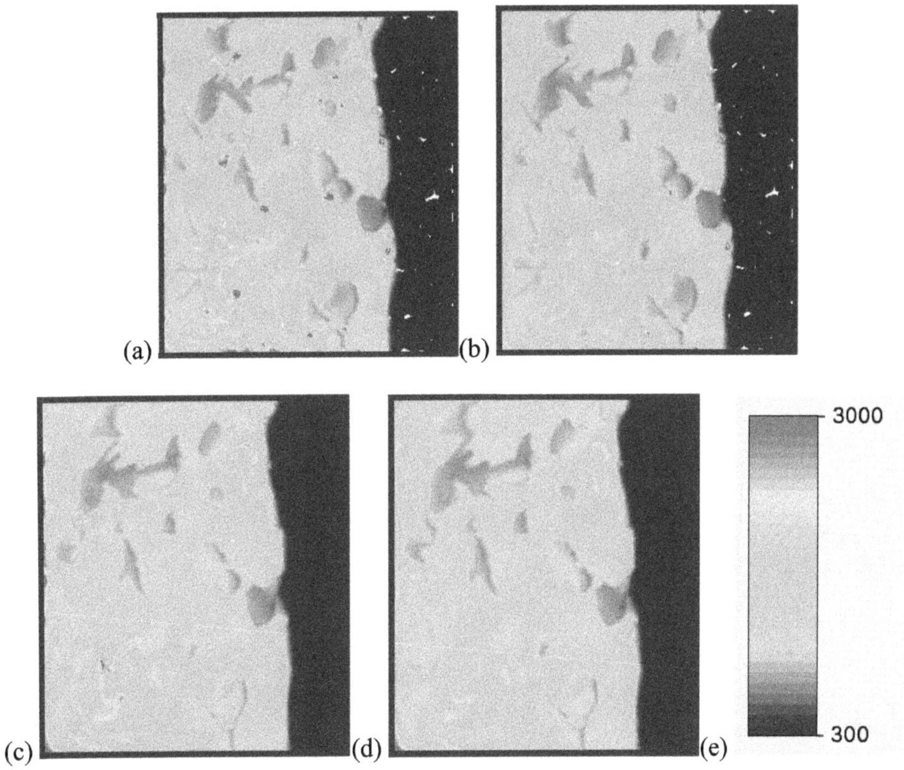

Fig. 3. Temperature distribution for Al/Ni composite subjected to shock at a particle velocity of 2000 m/s. (a) 97.3% of TMD with frictional heating. (b) 97.3% of TMD witout frictional heating. (c) 100% of TMD with frictional heating. (d) 100% of TMD without frictional heating. (e) The temperature scale (in K).

Acknowledgements

Authors would like to acknowledge the support by the Air Force Research Laboratory through the STTR program (contract numbers FA8651-16-P-0245 and FA8651-17-C-0076). This work has been approved for public release (case reference number 96TW-2019-0222). Opinions, interpretations, conclusions, and recommendations are those of the authors and are not necessarily endorsed by the United States Government.

References

[1] Du, S.W., Thadhani, N.N., 2009. "Impact Initiation of Pressed Al-Based Intermetallic-Forming Powder Mixture Compacts", in *Proc., 16th APS Topical Conference on Shock Compression of Condensed Matter. American Physical Society* (M. Elert, M. D. Furnish, W. W. Anderson, W. G. Proud, and W. T. Butler, eds.), vol. 54, (Nashville, TN), pp. 470-473, Woodbury, N.Y.: AIP Press.

[2] Fischer, S.H., Grubelich, M.C., 1996. "A Survey of Combustible Metals, Thermites, and Intermetallics for Pyrotechnic Applications", 32nd Joint Propulsion Conference and Exhibit. Lake Buena Vista, FL, U.S.A.

[3] Jette, F.X., Higgins, A.J., Goroshin, S., Frost, D.L. Charron-Tousignant, Y., Radulescu, M.I., Lee, J.J., 2011. "In-situ measurements of the onset of bulk exothermicity in shock initiation of reactive powder mixtures", J. Appl. Phys. 109, 084905.

[4] Sulsky, D., Zhou, S.J., Schreyer, H.L., 1995. "Application of a particle-in-cell method to solid mechanics", Computer Physics Communications 87, 236-252.

[5] osupdocs.forestry.oregonstate.edu/index.php/OSParticulas

[6] Nairn, J.A., Bardenhagen, S.G., Smith, G.D., 2018. "Generalized contact and improved frictional heating in the material point method," Computational Particle Mechanics 5, 285-296.

[7] Nairn, J.A., Hammerquist, C.C., Smith, G.D., "New MPM Contact Methods for Improved Accuracy, Large-Deformation Problems, and Proper Null-Space Filtering," Computer Methods in Applied Mechanics and Engineering, awaiting clearance (2019).

Proceedings of the 2019 Hypervelocity Impact Symposium
HVIS2019
April 14-19, 2019, Destin, FL, USA
HVIS2019-109

Extended Abstract+

Adiabatic heating and damage formation of a composite associated with high-velocity impact

Zhiye Li[a], Xiaofan Zhang[b], Daniel J.O'Brien[c], Somnath Ghosh[b]*

[a] *Department of Civil and Environmental Engineering, Stanford University, Stanford, CA 94305, United States*
[b] *Department of Civil Engineering, Johns Hopkins University, Baltimore, MD 21218, United States*
[c] *Composite and Hybrid Materials Branch, U.S. Army Research Laboratory*

1. Introduction

This work aims to develop a physically-based multiscale model incorporating material heterogeneities in order to study multi-physics damage and failure of S-glass fiber reinforced epoxy composites under high-velocity impact.

Mechanical response of S-glass fiber reinforced composites is of interest in several automotive and defensive applications. Most of the studies available in the literature have considered static or quasi-static loading. Loading strain-rates on the order of 10^4 s^(-1)–10^5 s^(-1) are of primary interest in impact loading scenariosbut are often neglected in conventional studies. The mechanical response of composites at high strain rates are drastically different from those in the quasi-static regime. The failure and damage modes that occur in the high-strain-rate regimes are not captured by quasi-static analysis and experiments. Hence, in the present study, we consider the effects of hypervelocity impact on the mechanical response of fiber reinforced composites.

Micromechanical analysis of unidirectional fiber composites is performed in conjunction with the phenomenon of stress wave propagation in representative volume elements (RVEs) subjected to high strain rate deformation. The deformation and failure response of the RVE model undergoing fiber and matrix damage as well as cohesive interfacial cracking is simulated. The rate-dependent nonlocal continuum damage model (CDM) is validated for the DER 353 Epoxy donut impact experiment. From the split Hopkins bar (SHPB) high-speed impact experiment, thermal softening became more prominent as the strain rate increased. Adiabatic heating is observed in the donut-shaped specimens and implemented in the thermo-mechanical damage model. The dynamic rate-dependent cohesive models are validated by cruciform experiments and micro-droplet experiments. With model parameters calibrated from experiments, characterization of failure properties and microstructure criteria is studied in realistic microstructure models by using a Voronoi cell characteristic method, e.g., crack nucleation, growth and spacing decrease.

1.1. Computational epoxy model calibration with high strain rate SHPB

Donut-shaped specimen experiments using the SHPB have been conducted in [1] for characterizing the damage and thermal properties of the DER 353 Epoxy material. The shape of the donut specimen and the experimental set-up are shown in Fig 1. In this figure, the rigid bottom plate is fixed while the rigid top plate is pushing down. A time-dependent displacement function is recorded from the experiment and applied to the FE simulation. Stress-strain response of the specimen was then calculated using the reaction force and the cross-sectional area assuming no change in the density of the material. The true strain was calculated from the imposed uniform plate displacement and instantaneous specimen thickness. A quarter symmetric FE model of an epoxy

+ Full paper can found at https://www.sciencedirect.com/journal/international-journal-of-impact-engineering/special-issue/10MTGP5W4VJ

* Corresponding author. fax. +1 410 516 7473.
E-mail address: sghosh20@jhu.edu (S. Ghosh).

donut is developed to obtain a volumetrically-averaged stress-strain curve to compare with experimental data. The overall strain rate of the experiment is 12,000 per second. Coefficients of friction between the sample and two plates are the same, $\mu = 0.01$. The geometric data of the specimen is given in [1].

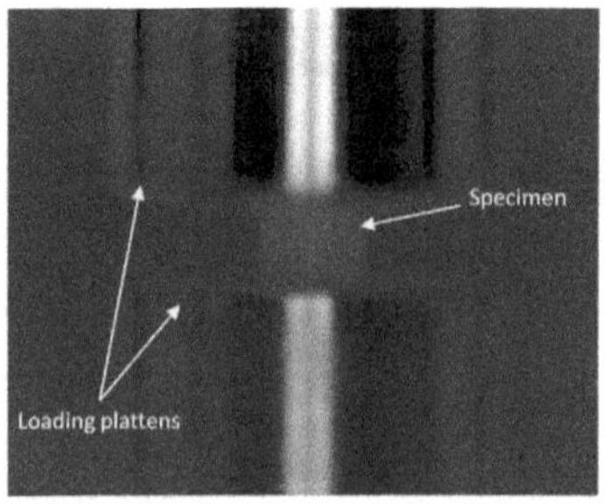

Fig. 1. The donut-shaped DER353 epoxy specimen prior to a compression test. [1]

In Fig. 2, comparison of the FE simulation results with the experimental data shows that CDM with adiabatic heating can capture the stress softening behavior of the DER 353 Epoxy donut-shaped specimens under high-speed impact. Although the FEM curves do not fit with the experimental curves exactly, the trend can be captured within a tolerable error range.

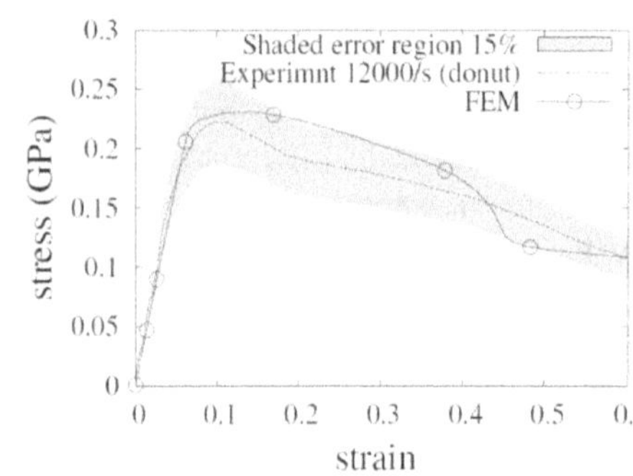

Fig. 2. Comparison of the FE simulation results with experimental data for the donut specimen: macroscopic stress-strain plots. [6]

2. Damage and failure in single fiber RVE and Random fibers RVE

A typical 3D finite element model of the RVE for a uniformly distributed unidirectional fiber-reinforced composite is shown in Figure 3(a). A randomly distributed unidirectional fiber RVE composite is shown in Figure 3(b). The RVE is enforced to deform in a periodic manner by applying periodic boundary conditions at nodes on opposite faces of its boundary.

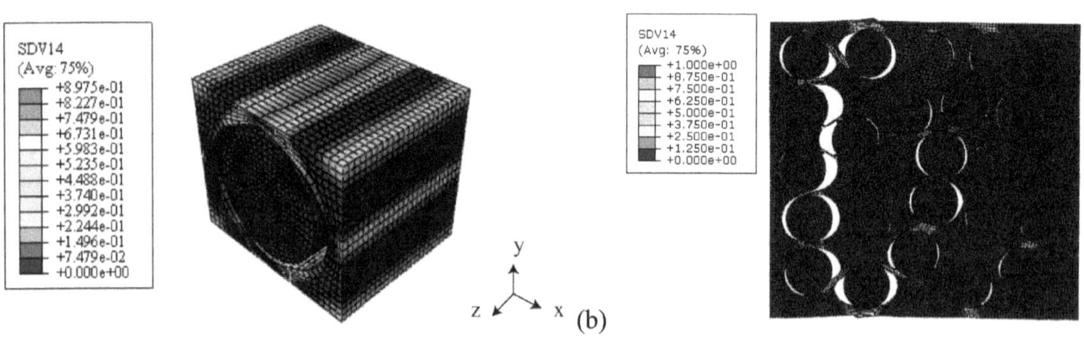

(a) (b)

Fig. 3. (a) Contour of damage variable D in Eq. (8) for a 3D RVE for multiple unidirectional fibers under tension at $e_{eff} = 0.2$. Overall strain rate is $10^5\ s^{-1}$. (b) Contour of damage variable D in Eq. (8) for a 3D RVE for multiple unidirectional fibers under tension. Overall strain rate is $10^4\ s^{-1}$.

The constitutive law of the fiber is the nonlocal continuum damage model (CDM) from [4], and the model for the matrix of the RVE is the SHPB experimentally-calibrated model from Section 2 of the article. Fiber-matrix interfacial debonding in this study is simulated using a bilinear function based cohesive zone model (bilinear CZM) [4]. Fig. 3 (b) shows debonding at material interfaces near the RVE boundary, contributing to the primary failure mechanisms of the overall RVE. This is different from the observations of the damage variable contours of the single fiber RVE in Fig. 3 (a). This difference is caused by the periodic boundary condition (PBC) applied on the surface nodes of the RVE and the stress wave propagation induced by the PBC.

3. Conclusion

As extended work of [4], this study develops an experimentally calibrated and validated 3D finite element model for simulating strain-rate dependent deformation and damage behavior in representative volume elements of S-glass fiber reinforced Der353 epoxy matrix composites. This study considers adiabatic heating and thermal expansion under high strain rate deformation. Calibration of the high-speed impact mechanical-thermal constitutive model is calibrated using a dynamic compression experiment, using an adapted split Hopkins bar (SHPB). The experimental detail and the resultant curves are published in [1]. Finally, this study examines the effect of microstructural morphology on the evolution of damage and its path. It is useful to optimize the microstructure in the material-by-design process. These models can also be a foundation for work in hierarchical multi-scale modeling [3, 5] to develop the bilateral multiscale analysis.

Acknowledgments

This work is sponsored by the Center for Materials in Extreme Dynamic Environments (CMEDE) of the Army Research Laboratory and was accomplished under Cooperative Agreement Number W911NF-12-2-0022. The sponsorship is gratefully acknowledged. Computational support of this work has been provided by the Homewood High-Performance Compute Cluster (HHPC) at Johns Hopkins University and the Maryland Advanced Research Computing Center (MARCC).

References

[1] S. Tamrakar, 2018. Characterization of s-glass epoxy composite interface under various rates of loading, Ph.D. dissertation, University of Delaware.
[2] Li, Z., Ghosh, S, 2019. Analytically derived space time-based boundary condition (STBC) to account for stress wave propagation in heterogeneous micromechanical model at hypervelocity impact. Proceedings of the 2019 Hypervelocity Impact Symposium. #090
[3] Xiaofan Zhang, Zhiye Li, Somnath Ghosh, and Daniel J. OBrien. Parametrically Homogenized Continuum Damage Mechanics(PHCDM) Model for Composites from Micromechanical Analysis." TMS 2019 Conference Proceedings: 2019 TMS Annual Meeting & Exhibition, 2019.
[4] Li, Z., Ghosh, S., Getinet, N. and O'Brien, D.J., 2016. Micromechanical modeling and characterization of damage evolution in glass fiber epoxy matrix composites. Mechanics of materials, 99, pp.37-52.
[5] Zhang, X., Li, Z., Ghosh, S. and O'Brien, D.J., 2016. Parametric Homogenization Based Continuum Damage Mechanics Model for Composites. In Proceedings of the American Society for Composites: Thirty-First Technical Conference.
[6] Li, Z., 2018. Micromechanical Studies of Glass Fiber Reinforced Epoxy Matrix Composites Undergoing Deformation and Damage at High Strain Rates. Ph.D. dissertation, Johns Hopkins University.
[7] Li, Z., Ghosh, S., 2019. Developing Space-Time Dependent Boundary Conditions for Composite RVEs at High Strain-Rate. International Journal of Solids and Structures, https://doi.org/10.1016/j.ijsolstr.2019.02.019
[8] Lemaitre, J. and Chaboche, J.L., 1994. Mechanics of solid materials. Cambridge university press.
[9] Z. Bazant and G. Pijaudier-Cabot. Nonlocal continuum damage, localization instability and convergence. J. Appl. Mech., 55(2):287–293, 1988.
[10] Siviour CR, Jordan JL (2016) High strain rate mechanics of polymers: a review. J Dyn Behav Mater 2:15–32. https://doi.org/10.1007/s40870-016-0052-8
[11] Chen WW, Song B (2011) Split Hopkinson (Kolsky) bar: design, testing and applications. Springer, New York
[12] Gama B, Lopatnikov SL, Gillespie JW (2004) Hopkinson bar experimental technique: a critical review. Appl Mech Rev 57:223. https://doi.org/10.1115/1.1704626

Proceedings of the 2019 Hypervelocity Impact Symposium
HVIS2019
April 14-19, 2019, Destin, FL, USA
HVIS2019-111

Extended Abstract+

Higher-order finite elements for lumped-mass explicit modeling of high-speed impacts

Robert S. Browning[a], Kent T. Danielson[b]*, Ph.D., Mark D. Adley[b], Ph.D., P.E.

[a]Geotechnical & Structures Laboratory, U.S. Army Engineer Research and Development Center
3909 Halls Ferry Road, CEERD-GSM, Vicksburg, MS 39180-6199 USA

[b]Geotechnical & Structures Laboratory, U.S. Army Engineer Research and Development Center
3909 Halls Ferry Road, CEERD-GMR, Vicksburg, MS 39180-6199 USA

1. Introduction and background

Classical finite element analysis (FEA) continues to be a primary computational method of choice for most solid mechanics applications and the explicit method is significantly used in the defense industry for high-speed impact analysis. The explicit lumped-mass approach, without a stiffness matrix, is well suited for rapidly changing/high rate short duration applications, but can produce distinct nuances and severely affect element performances differently than in typical static/implicit methods. In contrast to automatic tetrahedral meshing approaches applied to the entire volume, hexahedral-dominant modeling methods (e.g., [1-3]) attempt to mesh with all hexahedral elements and then automation with other element types is applied only to regions where the Hex mesher has trouble. The methods typically use wedge and/or pyramid elements to transition from hexahedral elements to fill volumes with other types. This results in models that typically have much fewer elements than with "all-Tet" approaches

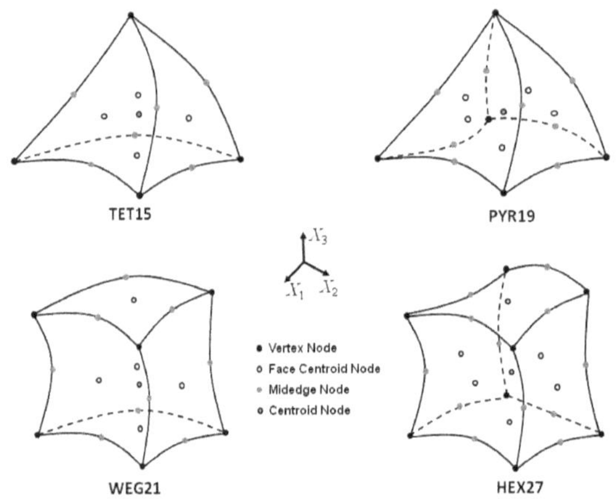

+ Full paper can found at https://www.sciencedirect.com/journal/international-journal-of-impact-engineering/special-issue/10MTGP5W4VJ

* Corresponding author. Tel.: +1-601-634-2039; fax: 1-601-634-2642.
E-mail address: Kent.T.Danielson@usace.army.mil.

and generally contain the more desirable hex elements in large regions of the mesh. In this paper, the four 2nd order element types depicted in Fig. 1 are used, which authors [4-6] have found to mass lump well for explicit methods and have well-defined contact

Fig. 1. 15-node tetrahedron, 19-node pyramid, 21-node wedge, and 27-node hexahedron type finite element topologies used in this paper; serendipity element versions are obtained by omitting the nodes on all faces and at the centroid

tractions that reliably perform well both with nearly incompressible materials and in flexural applications. Higher-order elements avoid many of the problems of 1st order ones, but at an increased computational cost per element. Second-order elements seem to be a "sweet spot" in terms of accuracy and computational cost for impact analyses. These 2nd order elements eliminate the need for hourglass control and provide higher resolution with fewer elements, since a single element can inherently capture curvature and flexural modes. Second order elements can also simplify meshing, since they are less prone to volumetric locking associated with near incompressibility, such as what occurs in metal plasticity, than certain types of 1st order elements. Methods to avoid severe locking at the element level permit unstructured meshing, e.g., typical all-tet methods or hexahedral-dominant methods. These technologies were developed within the in-house meshing (ProMesher), parallel analysis (*ParaAble*), and visualization (PenView) codes as well as placed into popular production software for meshing (Cubit), parallel analysis (EPIC), and visualization (ParaView). The IMPETUS AFEA [7] and LS-DYNA [8] explicit finite element codes have also implemented similar elements. As part of this paper, the authors will also present some initial promising findings of similar performances for the natural extension of these works to a 19-node 2nd order pyramid element, which can additionally facilitate meshing.

2. Finite element formulation

2.1. Discrete equations of motion

The basic form of the Lagrangian nonlinear equations is derived as that in standard texts, e.g., [9-12]. The motion is described by the current position of any point, at time, t, related to the undeformed position, $\mathbf{X}$, using the displacements, $\mathbf{u}$, from the undeformed configuration. The virtual work for large strain/deformation dynamic equilibrium at any time is defined with the body forces per unit mass, $\mathbf{b}$, e.g., gravity accelerations, the applied surface tractions, $\hat{\mathbf{t}}$, and the inertia forces all moving through virtual displacements, $\delta\mathbf{u}$, from the deformed configuration. It is spatially discretized into a mesh of standard displacement-based isoparametric Galerkin elements; their general formulations are found in the literature for many different element types, e.g., [9-12], and for those herein, in references [4-6]. In addition, an augmented version of the 13/14-node pyramid elements [13] to include all face and the body centroid nodes is used. Defining h as the shape function and q as a generalized degree of freedom, the general elemental equation for node, I, at time, t, is

$$\left[\int_{V_o} \rho_o \ddot{u} h_I dV_o\right] \delta q_I = \left[\int_{V_o} \rho_o b_q h_I dV_o + \int_{A_d} \hat{t}_q \ddot{u} dA_d + \int_{V_o} \sigma : \partial h_I / \partial x \, dV_d\right] \delta q_I \tag{1}$$

where δ is the variational operator designating a virtual quantity, the subscript d refers to the deformed configuration, each dot refers to differentiation with respect to time, ρ_d is the current mass density, and A_d and V_d are the deformed configuration surface area and volume, respectively. σ are Cauchy (true) stresses. The displacement boundary conditions are satisfied by explicitly controlling the motion of each appropriate nodal point by user model input at runtime. The acceleration, body force, and applied traction components with the q subscript are the values in the direction of the generalized degree of freedom. The virtual quantities are virtual nodal displacements that are now factored out, so that the elemental equation for each degree of freedom is defined by the expressions within the square brackets of Eq. (1). Large deformations are accommodated by the incremental transient explicit time integration procedures.

2.2. Numerical methods

Either "Full" (exact for ideal shape) quadrature rules are used for compressible materials or "Selective Reduced" (SR appended to the names in Fig. 1) integration for nearly incompressible materials to address volumetric locking [13, 14], whereby deviatoric and pressure terms use "Full" and a lower-order "Reduced" quadrature, respectively. Following [4-6], the higher-order elements considered herein use a row summation scheme for mass lumping, whereby all of the components in the row are summed and placed on the diagonal. A central difference scheme is used for temporal integration, e.g., [8, 15-19] without mass scaling, damping, or other special procedures. A simple 1-D wave propagation approach approximates a critical time increment, using a shortest distance between pairs of adjacent nodes in each element and fastest material sound wave speed of the corresponding element to compute the time required to transverse that distance. Artificial viscosity (pressure) is added to the pressure in explicit methods [8, 15-19] to prevent element collapse in the presence of high-velocity gradients and to quiet truncation frequency oscillations or "ringing".

3. Examples

3.1. Cylindrical bar impact

Here, a cylindrical bar impacting a rigid frictionless wall at 227 m/s is modeled with a quarter symmetric mesh and definition similar to that in [3, 9]. The bar is 3.24×10^{-2} m long with a 3.2×10^{-3} m radius. The copper isotropic hardening plasticity model uses a modulus of elasticity of 1.17×10^{11} Pa, Poisson's ratio of 0.33, yield stress of 4.00×10^{8} Pa, tangent plastic modulus of 1.00×10^{8} Pa, and mass density of 8930 kg/m^3. Figure 2 shows undeformed and deformed plots of the hexahedral-dominant model using TET15SR, PYR19SR, and WEG21SR elements placed along the bar centerline at the impact zone and HEX27SR elements placed elsewhere. The model predictions of 7.3612451 mm and 21.465609 mm for the final radius and length, respectively, compare favorably with the results in [3, 9].

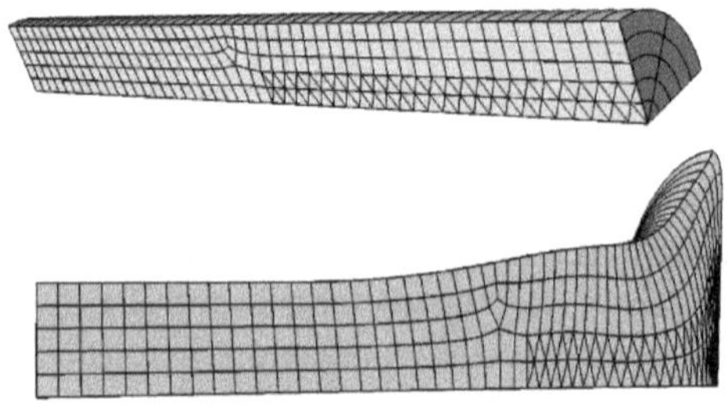

Fig. 2. Deformed unstructured mesh plots for bar impact problem using Quarter-symmetry 2nd Order Hex-Dominant mesh of HEX27SR, WEG21SR, PYR19SR, and TET15SR type elements.

3.2. Penetration into concrete using cavity expansion functions

This example models the penetration of a solid ogive-nose projectile into a concrete half-space using cavity expansion penetration resistance functions, as in [20] with penetration test data [21]. The projectile has a 3.0 CRH ogive-nose, shank length of 169.5 mm, nose length of 33.7 mm, and shank diameter of 20.3 mm. It is made of 4340 R_c 45 steel and modeled with a power law hardening elastic-plastic model with modulus of elasticity of 206.8 GPa, Poisson Ratio of 0.32, yield stress of 1.207 GPa, hardening constant of 382 MPa, hardening exponent of 0.266, plastic modulus of 141 MPa, and mass density of 7810 kg/m^3. The concrete has an unconfined strength of 58.4 MPa, mass density of 2320 kg/m^3, and a cavity expansion S-factor of 9.037.

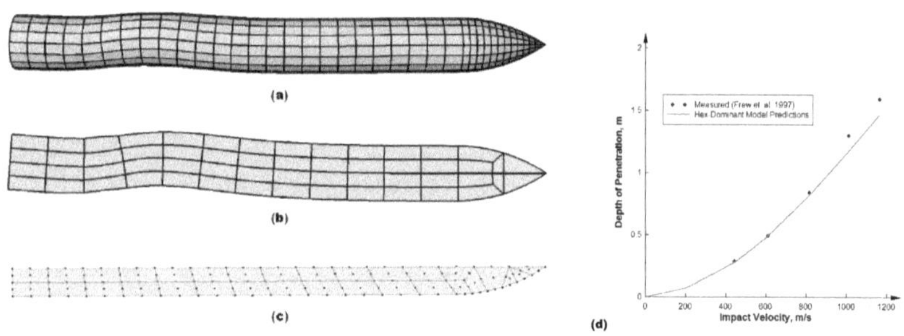

Fig. 3. Model plots for (a) Full-symmetry deformed 1st order brick mesh [33]; (b) Half-symmetry deformed 2nd Order Hex-Dominant mesh; (c) Undeformed Half-symmetry 2nd Order 2-D mesh Automatically created by a Paving Algorithm; and (d) test and predicted results for normal impact.

Figure 3 shows (a) the 1st order model of [21]; (b) a 2nd order Hex-Dominant model requiring only a little user intervention; and (c) a 2nd order 2-D mesh automatically generated by a paving approach on the half cross-section. The three-dimensional model in Fig. 3(b) was created by simply revolving a user-created 2nd order 2-D mesh about the projectile axis to create a model of HEX27SR, WEG21SR, and TET15SR elements. The 2-D mesh in Fig. 3(c) was created automatically from the CAD geometry, which is especially beneficial for large tradespace or shape optimization applications that need to create numerous meshes. The automation, however, places elements with a single node on the axis of revolution. These are easily and automatically identified and accommodated in a 3-D mesher by splitting the quadrilaterals into two triangles, and any triangle with a single node on the axis can now be automatically generated into a series of PYR19 elements by the rotation of the 2-D mesh. As seen in Fig. 3, the Hex-Dominant model predictions are in good agreement with the normal impact predictions and test data of the examples in [20-21], as well as with the 1st order hexahedron model for a non-normal impact velocity of 500 m/s with an Angle of Attack (AOA) of 5.0 degrees. For non-normal impact, the projectile models both predict very similar final positions at about a 0.27 m depth and a double permanent bending in the aft end. The 1st order model used default hourglass control parameters and the 2nd order model, of course, represents hourglass modes naturally without resorting to methods of artificial control.

4. Concluding remarks

The results show that the 2nd order elements have distinct advantages for certain applications, especially for flexure, curved geometries, and unstructured meshes with nearly incompressible materials, and are thus useful for 2nd order Hex-Dominant approaches that can greatly simplify and automate meshing. The ability to use unstructured combinations of the different element types in explicit dynamics codes provides an important modeling alternative to all-Hex or all-Tet meshing. The Hex-Dominant meshes are typically more computationally intensive than corresponding all-Hex models, but typically much less so than corresponding all-Tet models. The ability to automate meshing of reasonable sizes can significantly reduce analysts effort and is quite beneficial for applications like shape optimization.

Acknowledgements

Permission to publish was granted by Director, Geotechnical and Structures Laboratory.

References

[1] Meshkat S., Talmor D., 2000. Generating a mixed mesh of hexahedra, pentahedra and tetrahedra from an underlying tetrahedral mesh. Int J Numer Meth Eng 49:17–30

[2] Owen S., Saigal S., 2000. H-Morph: an indirect approach to advancing front hex meshing. Int J Numer Meth Eng 49:289–312

[3] Abaqus, 2018. CAE, Theory, Benchmarks, and Examples Manuals. Dassault Systèmes Simulia Corp., Providence, RI, USA

[4] Danielson K., Adley M., Williams T., 2016. Second-order finite elements for Hex-Dominant explicit methods in nonlinear solid dynamics. Finite Elements Anal Des 19:63–77.

[5] Danielson K., O'Daniel J., 2011. Reliable second-order hexahedral elements for explicit methods in nonlinear solid dynamics. Int J Numer Meth Eng 85(9):1073–1102. doi: 10.1002/nme.3003

[6] Danielson, K., 2014. Fifteen Node Tetrahedral Elements for Explicit Methods in Nonlinear Solid Dynamics. Comput Meth Appl Mech Eng 272:160-180, ISSN 0045-7825, http://dx.doi.org/10.1016/j.cma.2014.01.012

[7] IMPETUS AFEA, 2012. SOLVER, User Documentation, version 3.0 beta, IMPETUS AFEA, Flekkefjord, Norway

[8] LS-DYNA, 2012. KEYWORD USER'S MANUAL VOLUME I, Version 971 R6.1.0. Livermore Software Technology Corporation, Livermore, CA

[9] Zienkiewicz O., Taylor R., Zhu J., 2005. The finite element method: its basis and fundamentals; The finite element method for solid and structural mechanics, sixth edition. Elsevier Butterworth-Heinemann, Burlington, MA

[10] Bathe K-J., 1996. Finite Element Procedures. Prentice-Hall, Englewood Cliffs, New Jersey

[11] Belytschko, T., Liu, W., Moran, B., 2000. Nonlinear Finite Elements for Continua and Structures. John Wiley & Sons, New York

[12] Hughes T. 1987. The finite element method linear static and dynamic finite element analysis. Prentice-Hall, Englewood Cliffs, NJ

[13] Felippa, C., 2004. "A compendium of FEM integration formulas for symbolic work", Engineering Computations, Vol. 21 Issue: 8, pp.867-890, https://doi.org/10.1108/02644400410554362

[14] Nagtegaal J., Parks D., Rice J., 1977. On numerically accurate finite element solutions in the fully plastic range. Comput Meth Appl Mech Eng 4:153-177

[15] Presto User's Guide Version 4.16 ., 2012. Report SAND2010-3112, Sandia National Laboratories, Albuquerque, NM

[16] Zywicz E., Lin J., 2012. DYNA3D: A nonlinear, explicit, three-dimensional finite element code for solid and structural mechanics: version 12.1. Lawrence Livermore National Laboratory Report, LLNL-SM-599533. Livermore, CA

[17] Benson D., 1992. Computational methods in Lagrangian and Eulerian hydrocodes. Comput Meth Appl Mech Eng 99:235-394

[18] Taylor L., Flanagan D., 1989. PRONTO 3D: A three-dimensional transient solid dynamics program. Report SAND 87-1912, Sandia National Laboratories, Albuquerque, NM

[19] Johnson G., Beissel S., Gerlach C., Holmquist T., Stryk R., 2010. User Instructions for the 2010 Version of the EPIC Code, Southwest Research Institute, Final Report, Contract W56HZV-06-C-0194 with U.S. Army RDECOM-TARDEC.
[20] Warren T., Tabbara M., 1997. Spherical cavity-expansion forcing function in PRONTO 3D for applications to penetration problems. SAND97-1174, Sandia National Laboratories, Albuquerque, NM.
[21] Frew, D., Hanchak, S., Green, M., Forrestal, M., 1998. Penetration of concrete targets with ogive-nose steel rods. Int J Impact Engng 21(6):489-497

Proceedings of the 2019 Hypervelocity Impact Symposium
HVIS2019
April 14-19, 2019, Destin, FL, USA
HVIS2019-003

Analysis of the distribution of BAD generated during the normal penetration of a variable cross-section EFP on RHA

Boyang XING [a], Yunhui HOU [b], Zhenyan GUO [c],Dongjiang ZHANG [b], Liang CHEN [d], Yongliang Yang [a], Jianhua Luo [a], Rongzhong LIU [a], Rui GUO [a,]*

[a]*School of Mechanical Engineering, Nanjing University of Science and Technology, Nanjing 210094, Jiangsu, China*
[b]*Xi'an Modern Control Technology Research Institute, Xi'an 710065, Shanxi, China*
[c]*School of Electronic and Optical Engineering, Nanjing University of Science and Technology, Nanjing, 210094, China*
[d]*The Seventh Design Department, Aerospace Science and Technology Corporation, the Seventh Research Institute, Chengdu 610100, Sichuan, China*

Abstract

The purpose of this study is to analyse how the thickness of Rolled Homogeneous Armor (RHA) and impact velocity of an Explosively Formed Projectile (EFP) influence the middle mass behind-armor debris (BAD) when a variable cross-section EFP penetrates RHA normally. Numerical simulation is adopted, the thickness of RHA varies from 10mm to 70mm, and the impact velocity of the EFP varies from 1650m/s to 1860m/s. The results indicate that: (1) when the impact velocity of the EFP is 1650m/s and the thickness of RHA varies from 10mm to 70mm, p_{1g} of the RHA and EFP decreases with increasing H_0. The thin target could be used to produce a large proportion of the middle mass BAD from RHA (including BAD from the EFP and BAD from the RHA and EFP). (2) When the impact velocity of the EFP varies from 1650m/s to 1860m/s and the thickness of the RHA is 40mm, p_{1g} of the RHA is less than 50%, p_{1g} of the EFP is more than 70%, and p_{1g} of the RHA and EFP is more than 50%.

Keywords: Explosively Formed Projectile; variable cross-section characteristic; behind-armor debris; mass and number distribution; numerical simulation

Nomenclature

p_{1g}	$p_{1g} = N_{1g} / N_{total}$.
N_{1g}	The number of BAD whose weight is between 0.1 g and 1 g.
N_{total}	The number of BAD whose weight is more than 0.1 g.
H_0	The thickness of RHA.
v_0	The impact velocity of EFP.

1. Introduction

The function of an EFP is not only to penetrate the armor but also to generate BAD to destroy objects behind the armor. So it is significant to predict how the mass and number distribution of BAD varies with thickness of the RHA (H_0) and impact velocity of the EFP (v_0) [10-11].

References [1-3] studied the aftereffect of EFP penetration into a target of finite thickness by the theoretical method, but they all neglected the variable cross-section characteristic of the EFP. Reference [4] studied the behind-armor effects caused by an EFP impact by numerical simulation, while references [5-6] studied that by experiment, but none analyzed the distribution of BAD from the target and EFP. Theoretical studies on the distribution of BAD are all based on a penetrator whose main part is a slender

* Corresponding author. Tel.: +86-02584317972
E-mail address: guorui@njust.edu.cn

and equal cross section cylindrical rod [8-9] and it could not be applied directly to the situation examined in this study. There are also few papers that analyze how H_0 and v_0 effect the distribution of BAD generated during the normal penetration of a variable cross-section EFP on RHA.

This study examines how H_0 and v_0 influence the middle mass BAD, with masses varying from 0.1g to 1g, when a variable cross-section EFP penetrates RHA. H_0 is varied from 10mm to 70mm and v_0 from 1650m/s to 1860m/s.

2. Model verification and establishment

2.1. Model verification

Dalzell [4] pointed out that Smooth Particle Hydrodynamics (SPH) offered significant advantages over traditional Eulerian and Lagrangian methods when he used AUTODYN-3D™ to simulate behind-armor debris formed by the perforation of an EFP through a steel target.

The shape and size of the EFP in the experiment of reference [5] are shown in the X-ray photo of the lower part of Fig. 1. This EFP normally penetrates a 20mm-thick 45# steel target at an impact velocity of 2120m/s, with the length and width of the target both being 200mm. The shape and size of the EFP in the numerical simulation model are shown in the upper part of Fig. 1. The numerical simulation model is shown in Fig. 2, where the red part is a fixed constraint applied to the boundary of the target. The particle size of the EFP and target are both 0.5mm. The material of the EFP and target are copper and 45# steel, respectively. Equation of state of the EFP and target are linear and shock, respectively. Constitutive equations of the EFP and target are both Johnson Cook. Accurate parameters can be found in references [5-6], since the two references used the same EFP warhead.

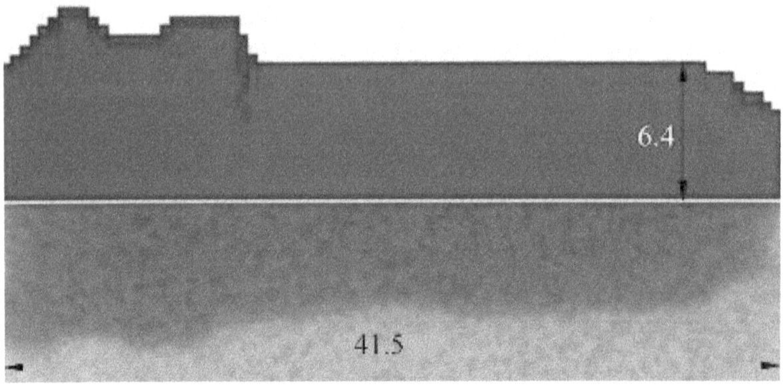

Fig. 1. The shape of EFP (unit: mm)

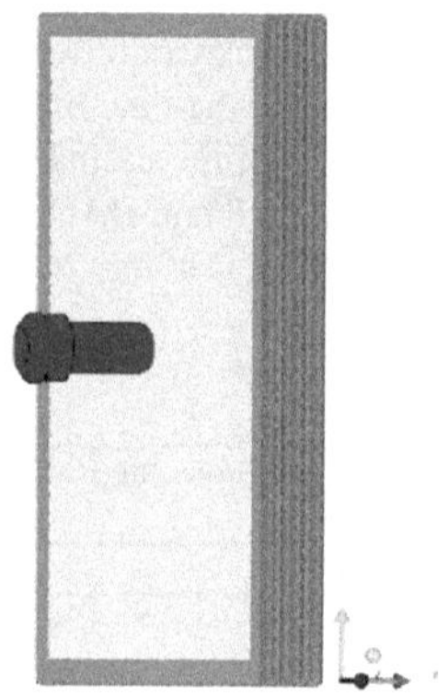

Fig. 2. Numerical simulation model

Shapes of the BAD clouds at a time of 80μs and 95μs in the numerical simulation are shown in the upper part of Fig. 3, and X-ray photos taken at the corresponding time given in reference [5] are shown in the lower part of Fig. 3. It can be seen that the BAD clouds obtained by numerical simulation are in good agreement with those obtained in reference [5]. So the numerical simulation method is credible.

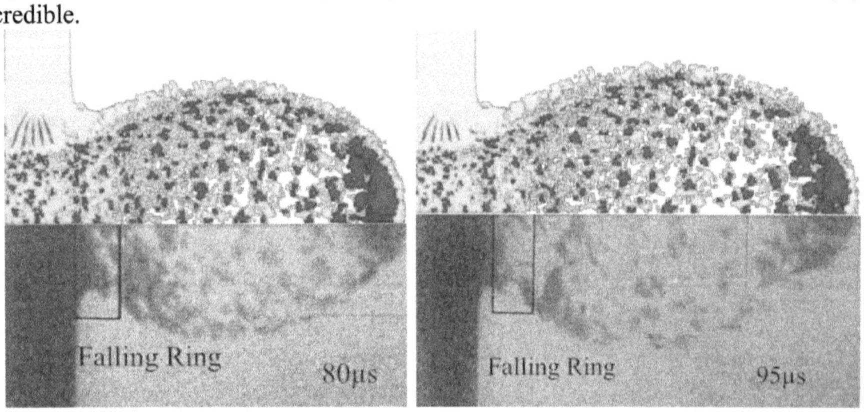

Fig. 3. BAD clouds at different times

The radial and head expansion velocities of the BAD are 531m/s and 1690m/s respectively, at a time of 95μs in the numerical simulation model. At the same time, radial and head expansion velocities of the BAD are 525m/s and 1678m/s, respectively, in reference [5]. Deviations between results in the numerical simulation model and reference [5] are both less than 2%, so the numerical simulation method is credible.

2.2. Model establishment

Laser high-speed photography of a typical EFP warhead, used in this study, is shown in the upper part of Fig. 4, while the equivalent model constructed in the numerical simulation is shown in the lower part of Fig. 4. The maximum radius is 28.5mm and the maximum length is 98mm.

The numerical simulation model of BAD generated by RHA subjected to normal penetration of the typical EFP is shown in Figure 5. The blue part is the typical EFP, the green part is the target, and the red part is a fixed constraint applied to the boundary of the target. In order to reduce influence of the boundary effect, the length and width of the target are both 1000 mm, which are more than 17 times the maximum diameter of the typical EFP.

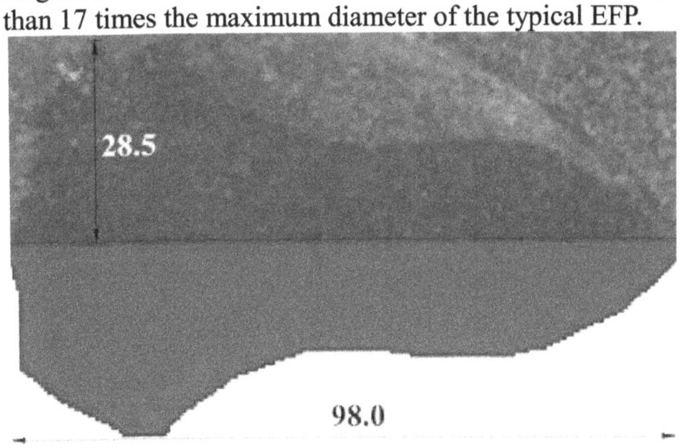

Fig. 4. Experiment result and simulation model (unit: mm)

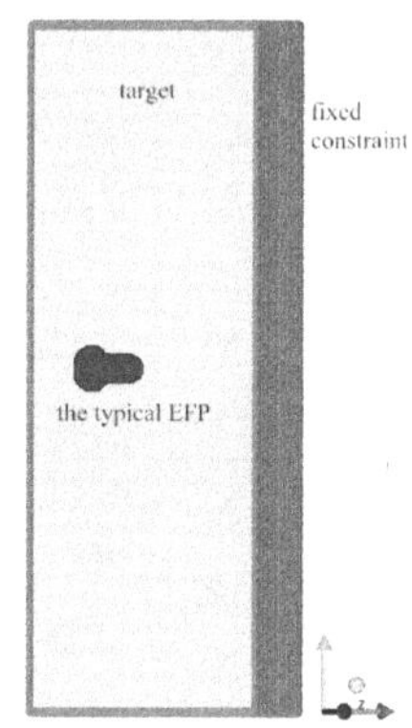

Fig. 5. Simulation model of the typical EFP

The SPH method in AUTODYN-3D™ was adopted in this study. Equations of state of the RHA and EFP are Linear and Gruneisen, respectively. Constitutive equations of the RHA and EFP are both Johnson-Cook. Accurate parameters can be found in references [7-8]. We set the particle size to 0.4 mm, 0.5 mm and 0.6 mm. We found that the influence of particle size on the number of BAD fragments whose mass is more than 0.1 g is less than 10%, the influence of particle size on the mass of BAD is less than 10%, and the influence of particle size on the size of the BAD cloud is less than 10%. Therefore, we set the particle size for the RHA and EFP to 0.5mm in order to reduce the calculation time. A fixed constraint was applied to the boundary of the target. An example numerical simulation model for the typical EFP penetrating 70mm-thick RHA at 1650m/s is introduced in detail in references [10-11]. The numerical simulation model in this study and in references [10-11] are all the same. The mass of BAD are calculated by the theoretical model in reference [10], and the size of the hole in target are calculated by the theoretical model in reference [11]. Results obtained by the theoretical model have been compared with results obtained by numerical simulation in references [10-11], and they both demonstrated that the numerical simulation model is credible.

3. The mass and number distribution

3.1. Effect of RHA thickness

RHA thickness is one of the main factors that influences the mass and number distribution of BAD, so we examined the typical EFP impacting RHA normally with target thickness varying from 10mm to 70mm and a EFP velocity v_0 of 1650m/s. The BAD from the RHA is shown in Fig 6 (a), the BAD from the EFP is shown in Fig 6 (b), and the BAD from both the RHA and EFP is shown in Fig 6 (c). The Y-coordinate is the proportion of debris whose mass is less than its X-coordinate. Since the debris whose mass is less than 0.1g does not generate much damage, only the debris whose mass is more than 0.1g is taken into consideration. The initial point of every curve is the proportion of debris whose mass is between 0.1g and 0.2g.

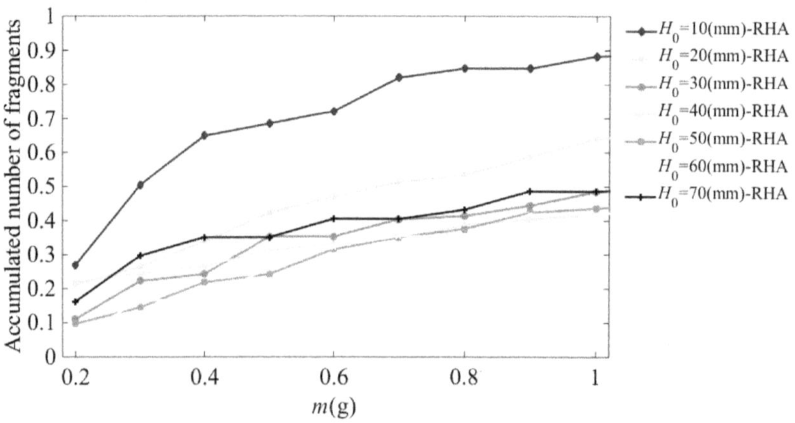

(a)

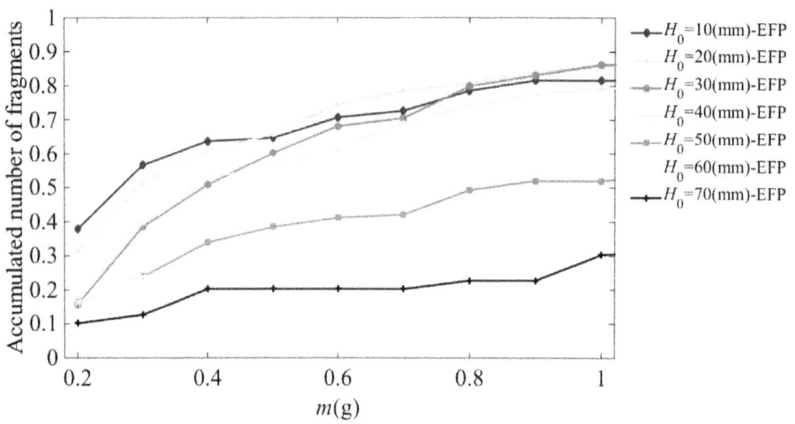

(b)

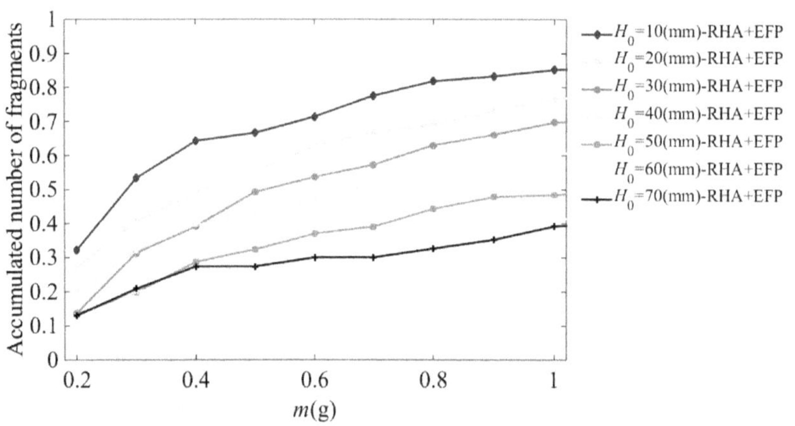

(c)

Fig. 6. (a) BAD generated by the RHA and (b) BAD generated by the EFP and (c) BAD generated by both the RHA and EFP

It can be seen from Fig 6 (a) that p_{1g} of the RHA is less than 50% when H_0 varies from 30 mm to 70 mm, but p_{1g} of the RHA is nearly 90% when H_0 is 10 mm. Since v_0 is the same, the strength of the shock wave is the same when the EFP impacts the target.

The shock wave reflects between two faces of the target and the strength decreases with increasing H_0. So p_{1g} of the RHA is large when H_0 is 10 mm and p_{1g} of the RHA is small when H_0 is more than 30 mm. As such, the thin target could be used to produce a larger proportion of middle mass BAD from the RHA.

It can be seen from Fig 6 (b) that p_{1g} of the EFP is 50% or less when H_0 varies from 50 mm to 70 mm, but p_{1g} of the EFP is 80% or more when H_0 is between 10 mm and 40 mm. The residual length of the EFP decreases with increasing H_0. The shock wave reflects between the head and end of the EFP, so the strength of it decreases with increasing H_0. So p_{1g} of the EFP is large when H_0 is less than 40 mm and p_{1g} of the EFP is small when H_0 is more than 50 mm. As such, the thin target could be used to produce a larger proportion of middle mass BAD from the EFP.

It can be seen from Fig 6 (c) that p_{1g} of the RHA and EFP is only 50% or less when H_0 varies from 50 mm to 70 mm, while p_{1g} of the RHA and EFP is more than 80% when H_0 is 10 mm. So, the thin target could be used to produce a larger proportion of middle mass BAD from the RHA and EFP.

In general, when v_0 is 1650m/s and H_0 varies from 10mm to 70mm, p_{1g} of the RHA and EFP decreases with increasing H_0.

3.2. Effect of EFP impact velocity

The EFP impact velocity is one of the main factors that influences the mass and number distribution of BAD, so we examined the typical EFP impacting RHA normally with target thickness of 40mm and v_0 varying from 1650m/s to 1860m/s. BAD from the RHA is shown in Fig 7 (a), BAD from the EFP is shown in Fig 7 (b), and BAD from both the RHA and EFP is shown in Fig 7 (c).

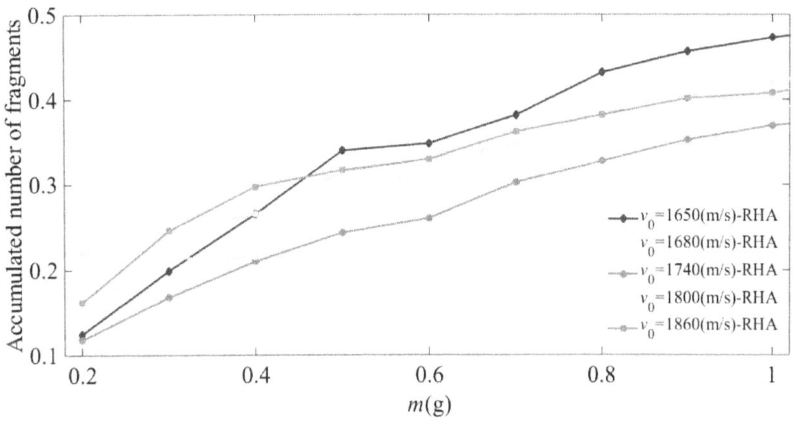

(a)

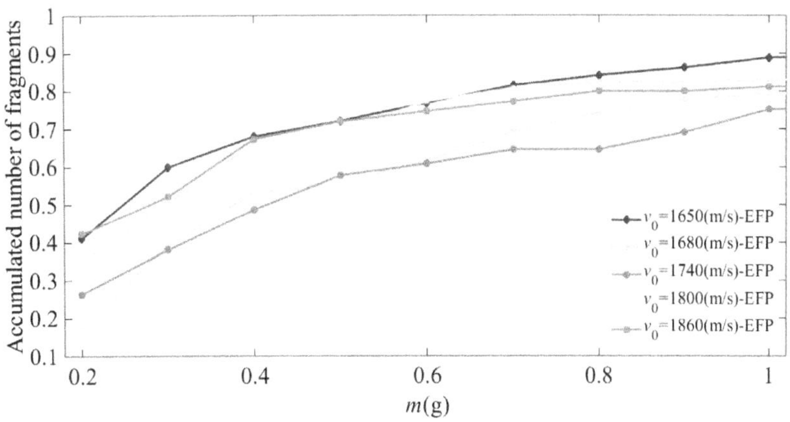

(b)

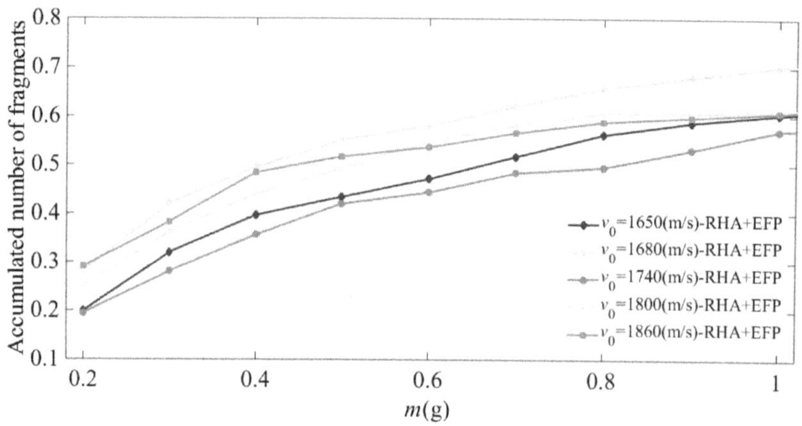

(c)

Fig. 7. (a) BAD generated by the RHA and (b) BAD generated by the EFP and (c) BAD generated by both the RHA and EFP

It can be seen from Fig 7 (a) that p_{1g} of the RHA is less than 50% when v_0 varies from 1650m/s to 1860 m/s. With increasing v_0, p_{1g} of the RHA first decreases, then increases. For H_0 of 40mm, p_{1g} of the RHA is always less than 50% when v_0 varies from 1650m/s to 1860 m/s.

It can be seen from Fig 7 (b) that p_{1g} of the EFP is more than 70% when v_0 varies from 1650m/s to 1860 m/s. For H_0 of 40mm, p_{1g} of the EFP is always more than 70% when v_0 varies from 1650m/s to 1860 m/s.

It can be seen from Fig 7 (c) that p_{1g} of the RHA and EFP is more than 50% when v_0 varies from 1650m/s to 1860 m/s. For H_0 of 40mm, p_{1g} of the RHA and EFP is always more than 50% when v_0 varies from 1650m/s to 1860 m/s.

In general, v_0 has a little influence on p_{1g} of the RHA (as well as p_{1g} of the EFP, and p_{1g} of the RHA and EFP) since the deviation between maximum and minimum of v_0 is 210 m/s, only 11% of the maximum of v_0. As v_0 varies from 1650m/s to 1860m/s and for H_0 of 40mm, p_{1g} of the RHA is less than 50%, p_{1g} of the EFP is more than 70%, and p_{1g} of the RHA and EFP is more than 50%.

4. Conclusions

The distribution of BAD generated during the normal penetration of a variable cross-section EFP on RHA under different H_0 (from 10mm to 70mm) and v_0 (from 1650m/s to 1860m/s) is simulated by AUTODYN-3D™, and the following conclusions are drawn:

(1) For v_0 of 1650m/s and H_0 varying from 10mm to 70mm, p_{1g} of the RHA and EFP decreases with increasing H_0. The thin target could be used to produce a larger proportion of middle mass BAD from the RHA (including BAD from the EFP, and BAD from the RHA and EFP).

(2) For v_0 varying from 1650m/s to 1860m/s and H_0 of 40mm, p_{1g} of the RHA is less than 50%, p_{1g} of the EFP is more than 70%, and p_{1g} of the RHA and EFP is more than 50%.

Acknowledgements

This work was financially supported by the National Natural Science Foundation of China (Grant No. 11372136; 21875109) and Postgraduate Research & Practice Innovation Program of Jiangsu Province (2019).

References

[1] Kim H S, Arnold W, Hartmann T, Rottenkolber E., Klavzar A., 2011. "A model for behind armor debris from EFP impact," Proceedings of the 26th International Symposium on Ballistics. Miami, Florida. USA, pp. 1410-1419.

[2] Rui LI, Zhengxiang HUANG, Xudong ZU, Qiangqiang XIAO, Xin JIA,2018. Theoretical analysis on the spallation of EFP vertical penetration target , Explosion and Shock Waves 5,p.1039. (in Chinese)
[3] Xianfeng ZHANG, Huiwu CHEN, Youshou ZHAO, 2006. Investigation of process and after effect of EFP penetration into target of finite thickness. Explosion and Shock Waves 4, p.323. (in Chinese)
[4] Dalzell M W, Hazell P J, Meulman J H, 2002. "Modelling behind-armour debris formed by the perforation of an EFP through a steel target," Proceedings of the 20th International Symposium on Ballistics. Orlando, Florida. USA,pp. 786-793.
[5] Xin WANG, Jianwei JIANG, Shuyou WANG, Mei LI, 2018. Experimental Research on Debris after Explosively-formed Projectile Penetrating into Steel Target. Acta Armamentarii 39(7), p.1284. (in Chinese)
[6] Yangyang WANG, Jianwei JIANG, Jiayu MENG, Jianbin MEN, 2016. "Effect of add-on explosive reactive armor on EFP penetration," Proceedings of the 29th International Symposium on Ballistics. Edinburgh, Scotland, UK, pp. 2395-2406.
[7] Wu J, Liu J, Du Y. Experimental and numerical study on the flight and penetration properties of explosively-formed projectile, 2007. International Journal of Impact Engineering 34(7), p.1147.
[8] Yossifon G, Yarin A L, 2002. Behind-the-armor debris analysis. International Journal of Impact Engineering, 27(8), p.807.
[9] Yarin A L, Roisman I V, Weber K, Hohler V, 2000. Model for ballistic fragmentation and behind-armor debris. International Journal of Impact Engineering, 24(2), p.171-201.
[10] Boyang Xing, Rongzhong Liu, Dongjiang Zhang, Liang Chen, Yunhui Hou, Rui Guo, 2018. Modelling on the mass of behind-armor debris generated by armor steel plate subjected to normal penetration of variable cross-section EFP. Explosion and Shock Waves: http://kns.cnki.net/kcms/detail/51.1148.O3.20180719.1120.020.html. (in Chinese)
[11] Boyang Xing, Dongjiang Zhang, Guo Zhenyan, Yunhui Hou, Rui Guo, Rongzhong Liu, Liang Chen, Hao Zhou, Yongliang Yang, Jianhua Luo, Analysis of the mass of behind-armor debris generated by RHA subjected to normal penetration of variable cross-section EFP, Defence Technology (2018), doi: https://doi.org/10.1016/j.dt.2018.10.006.

Proceedings of the 2019 Hypervelocity Impact Symposium
HVIS2019
April 14-19, 2019, Destin, FL, USA
HVIS2019-004

Calculation of jet characteristics from hydrocode analysis

Justin C. Sweitzer[a]*, Nicholas Peterson[b], and Scott Hill[a]

[a]*Practical Energetics Research, Inc. Huntsville, AL, 35802*
[b]*US Army CCDC-AvMC, Redstone Arsenal, AL 35878*

Abstract

The penetration performance of a shaped charge jet is affected strongly by factors such as straightness, stretch rate, and breakup time. Straightness is related to manufacturing tolerances, assembly techniques, and system integration features. Stretch rate and breakup time are controllable features of charge design. A higher stretch rate is desirable for short standoff performance. The stretch rate is easily altered by a change of explosive or modification of the angle with which the detonation wave sweeps the liner surface, however, an increased stretch rate generally results in a decreased breakup time. Many of the recent gains in shaped charge performance have been made possible by increasing the effective breakup time of the jet.

Several models exist for calculating breakup time. They include analytic models, such as Chou & Carleone's dimensionless strain rate model, and empirical or semi-empirical models such as Walsh's theory and those proposed by Pearson, et al. These models can be applied to raw hydrocode calculation data and used to determine a Jet Characterization (JC) file. The JC file can then be used to perform further calculations, such as Penetration Versus Stand Off (PVSO) curves.

This paper details adaptation of the Chou & Carleone model for predicting breakup time using hydrocode data. The hydrocode is used to determine the physical parameters of the jet which are then extrapolated back to a virtual origin for breakup time calculation. This results in a model that is design independent, relying on hydrocode determination of jet variables. The model implementation will be discussed, and comparisons of predicted jet characteristics will be made to test data for several charge geometries.

Keywords: Shaped charges, Breakup, Fragmentation

1. Introduction

Many empirical relationships for the breakup time of a shaped charge jet have been derived, including those put forward by Hirsch[1–4], Pfeffer[5], Haugstad[6], and Haugstad & Dullum[7]. Hirsch[1] related the breakup time to a simple ratio of jet diameter to plastic wave velocity. This model provides reasonable estimates of breakup time within a very narrow range of charge geometries, but lacks a dependence on strain rate.[8]

Pfeffer[5] performed numeric simulation to empirically arrive at a relationship with inverse dependence on strain rate and a direct relationship with the ratio of jet radius to shock velocity in the jet material. This model neglects the jet strength of Hirsch's model (in the form of plastic wave velocity), and thus also finds limited application. Haugstad[6] and Haugstad & Dullum[7]

* Corresponding author. Tel.: +1-256-867-1222.
E-mail address: justin.sweitzer@per-hq.com.

formulate relationships within a continuum mechanics framework that depend on a hydrodynamic description of the jet strength, with an initial yield stress supplemented by an artificial viscosity term.

Walsh[9] developed a dimensionless 'instability parameter', which includes yield strength, density, initial stretch rate, and jet radius. Breakup time is correlated to this parameter through a critical value, which is material dependent. Walsh's results have been further developed in recent history. In Baker, et al[10], correlations are developed building off of a previous study[11] focused on molybdenum liners, to describe average particle lengths, number of particles, and breakup times. Their metric of choice is the cubic root of the derivative of accumulated mass with respect to jet velocity, and they show this parameter to be closely related to the Walsh parameter.

Chou & Carleone[12] developed an analytic relationship[13,14] between a dimensionless breakup time and dimensionless initial strain rate. The formula is expressed

$$\bar{t}_b = 3.75 - 0.125\bar{\eta}_0 + \frac{1}{\bar{\eta}_0} \quad (1)$$

The dimensionless breakup time is the ratio of breakup time t_b multiplied by plastic wave velocity, $C_p = \sqrt{Y/\rho}$, where Y and ρ are yield strength and density, and initial jet radius r_0, or $\bar{t}_b = t_b C_p / r_0$. The dimensionless strain rate is $\bar{\eta}_0 = \dot{\varepsilon}_0 r_0 / C_p$, where $\dot{\varepsilon}_0$ is the initial strain rate. This model has also been shown to closely approximate breakup time across a wide range of charge geometries and liner materials[8]. They provide correlations including an empirical correction factor[15] in later works, which is reviewed by Walters & Summers[8,16].

Hennequin[17] proposed a model for a conical copper shaped charge introducing a 'fragment shape index' which is determined experimentally. The model represents elements of Hirsch's and Chou & Carleone's developments with correction terms based on the average shape of jet particles. The calculated results represent a marginal improvement over Chou & Carleone, while requiring empirical characterization of the shape index. The simplicity and accuracy of the Chou and Carleone model, and the wide availability of the required mechanical data for liner materials lead to the selection of this model for integration with hydrocode calculation in this work.

2. Hydrocode Integration

Shaped charge design is largely approached today through iterative simulation/test cycles. Hydrocode analysis is used to identify trends in desired performance traits and design candidates are prototyped and tested to provide validation and improve model fidelity. One of the challenges of this design approach is the calculation of jet breakup times. Hydrocode simulations are capable of closely predicting jet characteristics such as velocity and mass gradients, but diverge from experiment when estimating material failure. Finite element simulations introduce artificial sources of jet instability through mesh resolution, and failure models tend to produce inaccurate estimates of breakup at high strain rates.[18]

A combined approach leveraging the hydrocode predictions of jet characteristics with the Chou & Carleone breakup model can easily be implemented, allowing breakup time to be treated as a design variable. In the current work, the ALE3D hydrocode[19] is used to simulate the jet formation, and the jet characteristics are extracted through the Visualization Toolkit (VisIt)[20,21]. An external Python code is then used to calculate the breakup times. The output of the Python code is in the format used by Dyna East's WASP program, which can be used to predict penetration performance.

The Chou & Carleone model predicts breakup time as a function of local strain rate, and does not provide particle-by-particle estimates of length. In order to divide the jet into discrete particles, a stochastic natural fragmentation model[22–24] is used to determine the average particle length. This model was derived for the case of a one-dimensional rod stretching uniformly in tension. The mean length is

$$\bar{l} = \phi\sqrt{\frac{\pi}{2}}l_1 + (1-\phi)1.48l_2, \qquad \phi = \mathrm{erf}\left(\sqrt[4]{\frac{Y\dot{\varepsilon}r_0^3}{\rho c^3}}\right) \quad (2)$$

In the above equation, c is the material sound speed, and l_1, l_2 are length scales for brittle and ductile fracture models. The parameter ϕ is an estimate of dynamic ductility, and is the probability of a brittle-like fracture completing before plastic deformation can relieve stress in a rapidly propagating crack. The length scales are

$$l_1 = \sqrt[4]{\frac{\sigma^2}{\rho}\frac{Yc}{r_0\dot{\varepsilon}^3}}, \qquad l_2 = \sqrt[3]{\frac{3\sigma}{2}\frac{Y}{\rho\dot{\varepsilon}^2}}, \qquad \sigma = 4.5\sqrt{\frac{2Y}{\rho\dot{\varepsilon}^2\gamma^3}}$$

The γ in the above set of equations is the Mott physics-based[25] distribution parameter, and can be estimated from the Johnson-Cook constitutive model[26] as

$$\gamma \sim \frac{160Bn\varepsilon^{n-1}}{A + B\varepsilon^n}$$

More complete details of the model derivation can be found in the references[22–24].

3. Hydrocode Integration

The first step of this approach is a 2D, axisymmetric hydrocode simulation. Where available, the Johnson-Cook constitutive model is preferred for the liner material, as the Mott γ is calculated from the parameters. The Johnson-Cook parameters are held constant, and not applied as statistical distribution functions. The validation studies reported in latter sections of this paper were performed with a mesh density of 3 zones/mm. No mesh resolution sensitivity analysis has been performed at this point.

The hydrocode model is allowed to run until the jet has completely formed, and radial components of velocity have approached zero. Data is extracted from the simulation at this point, and post-processing begins. In order to ensure that the calculations are less dependent on the time from which the jet characterization data is extracted, the jet is moved to a common reference frame. Initially, the virtual origin was used for this purpose, but for waveshaped and non-conical charge geometries this approximation does not hold well. Instead, the reference frame was chosen as the point where the jet tip is two charge diameters from the initial liner face. The choice of reference frame is somewhat arbitrary, in that it was chosen based on a very limited-scope sensitivity study using a single warhead geometry. The jet is moved to this location by allowing the velocity gradient to determine the new coordinates and conserving mass by increasing jet radius, treating each differential segment of jet as a cylinder. This choice of reference frame is somewhat in conflict with the Chou-Carleone assumptions, which are based around the initial state immediately after collapse of the jet element.

Again, the virtual origin would be the ideal location to satisfy these assumptions, but the wide variance of the error in the virtual origin approximation for non-conical liners prohibits its use here. An additional challenge with this approach is that the jet is stretching in free flight, converting some of the kinetic energy into internal energy. Translating the jet in space and conserving mass and velocity, as has been done here, therefore does not conserve energy. Calculation accuracy and timestep independence would be improved by a more thorough description of jet kinematics, such as presented in Babkin, et al[27,28].

With the jet in the common frame of reference, Eq. 2 is used to determine the average particle length, and the number of particles is calculated by dividing the total length by particle length. The jet is segmented into the total number of particles using the velocity gradient as the binning factor. This produces shorter particles toward the front of the jet, where the strain rate is higher, and longer particles toward the back where the strain rate is lower. The breakup time is then calculated using Eq. 1 for each individual particle. Because the particle size calculation is performed assuming fracture occurs around the reference frame, the results reflect a material with an average breakup length of two charge diameters. Performing the breakup time calculation provides an estimate of actual strain-to-failure. The final step in the process, then, is to stretch the particle out to the calculated breakup time.

4. Results

The first comparison made here is to data collected for the Viper warhead[29]. Viper is a 65 mm diameter, 44° conical copper shaped charge. Figure 1 below shows the calculated accumulated mass versus velocity plots alongside the experimental data. The calculations show a slower tip speed than the experimental results, but generally follow along with the experimental data. The experimental data shows 79 particles down to an axial velocity of 2 km/s. By comparison, the calculations result in 75 particles at roughly the same velocity. Scaled cumulative jet lengths are compared in Figure 2a, which indicates that the calculated particle lengths are very close to the experimental particle lengths. Divergence occurs at the tip and tail in the same locations as the cumulative mass profiles diverge.

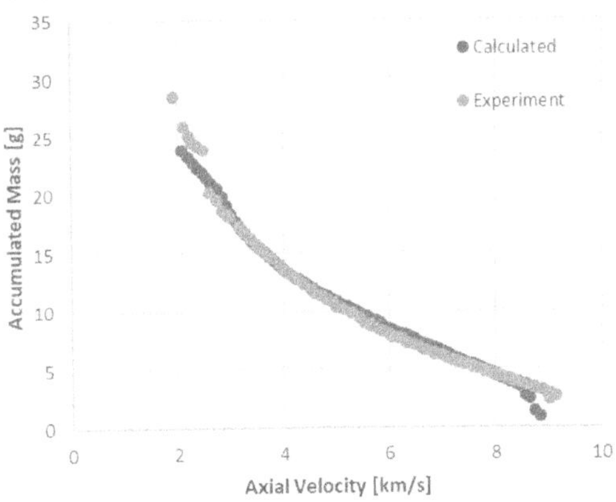

Figure 1. Accumulated mass versus velocity for Viper charge

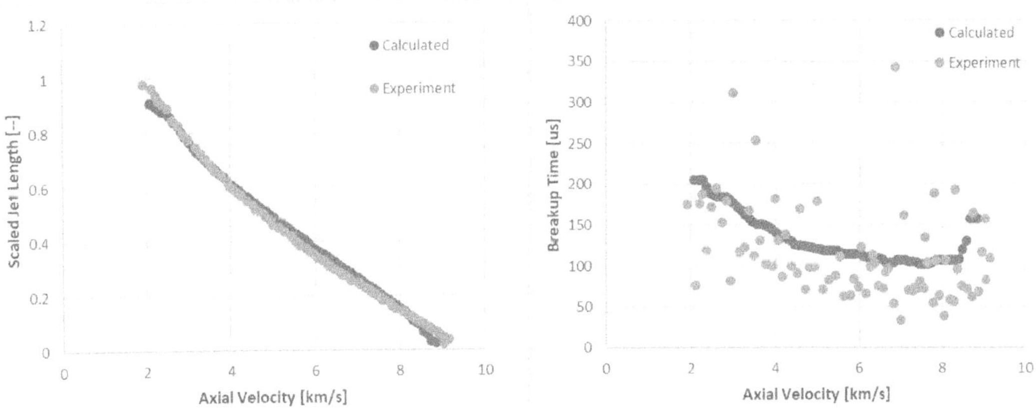

Figure 2. Scaled accumulated length versus velocity (a) and breakup time (b) for Viper charge

Figure 2b shows the breakup time comparison. The average calculated breakup time is 132 us, compared to the experimental result of 125 us. The experimental breakup time is estimated by the distance from the tail of the ith particle to the tip of the i+1th particle divided by the velocity difference. Across the range of axial velocities, the calculated breakup time is near the top of the range of observed values.

A second comparison is made for a 60 mm, waveshaped molybdenum cone[11]. This charge is designed to produce a fast, low mass jet. The Steinberg-Guinan strength model was used for molybdenum, as opposed to the Johnson-Cook constitutive equation for copper. The accumulated mass profile in Figure 3 shows more divergence between the hydrocode and experimental results. This may be a result of constitutive model inaccuracies for the Steinberg-Guinan model employed for molybdenum in ALE3D. It could also arise from tolerance stackups in the manufacturing process. The calculated values of jet length in Figure 4a are accordingly shorter than the experimental values, but the calculated breakup times in Figure 4b are in line with the observed values.

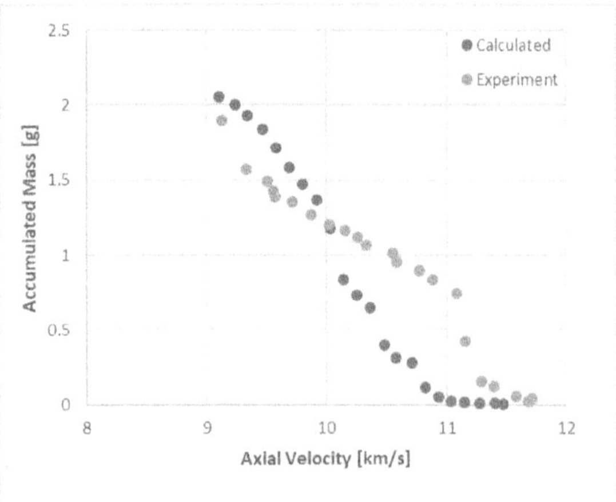

Figure 3. Accumulated mass versus velocity for 60 mm Waveshaped Mo Cone

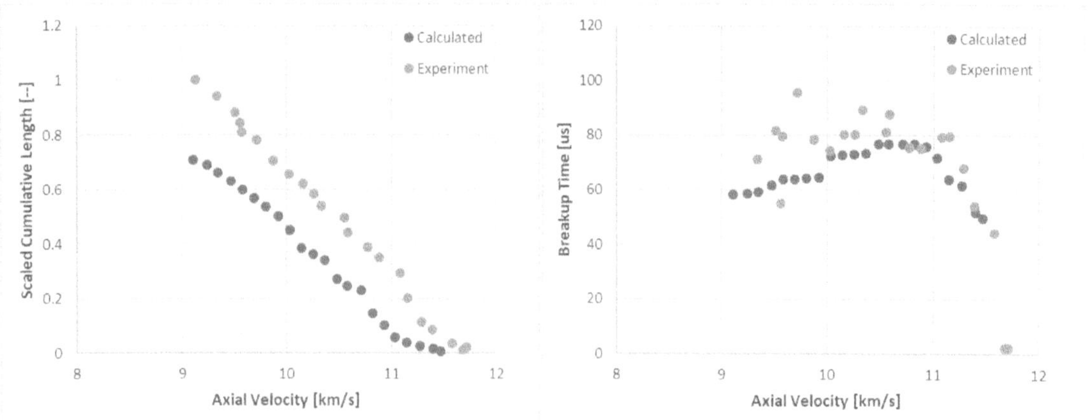

Figure 4. Scaled acc. length (a) and breakup time (b) vs velocity for 60 mm Waveshaped Mo Cone

A final comparison is made for NHN1-2, which is a 165 mm, waveshaped copper liner with experimental data from Pham, et al[30]. The accumulated mass profile in Fig 5 shows good agreement between the hydrocode and experimental data, with some minor overprediction of tip speed. There is also small divergence at the tail, where the experimental data shows that the last two particles are comparatively large. The scaled accumulated length profile depicted in figure 6a also indicates good agreement with the experimental data. The calculated breakup times in Fig. 6b show reasonable agreement with the experimental data, which appears to have been linearized in the source report[30]. The calculated average breakup time is 334 us in comparison to the experimental 352 us.

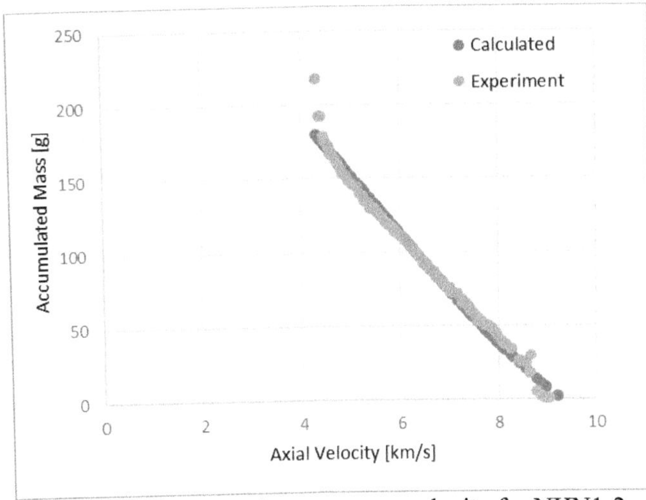

Figure 5. Accumulated mass versus velocity for NHN1-2

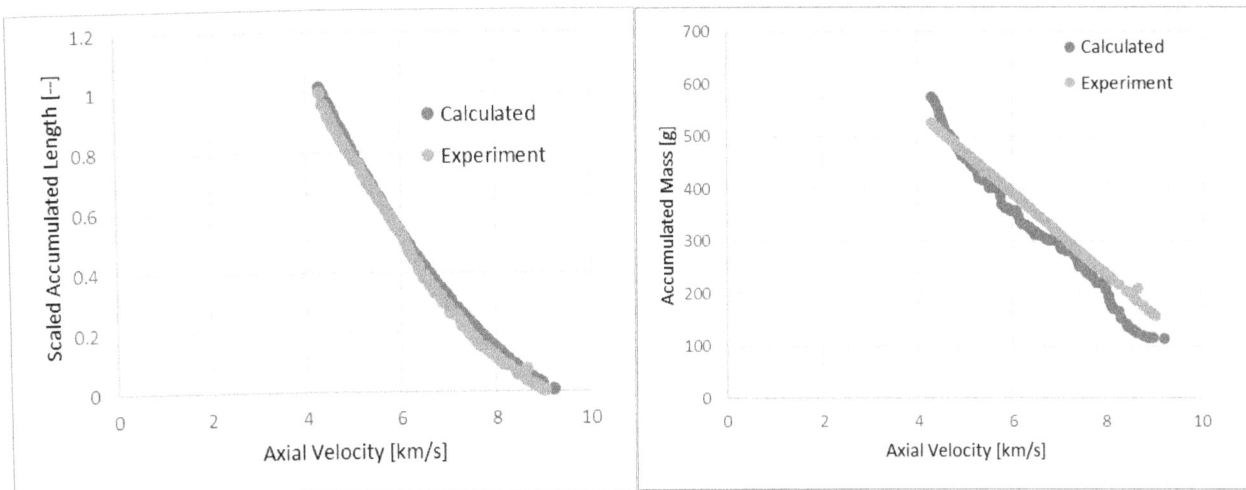

Figure 6. Scaled acc. length (a) and breakup time (b) vs velocity for NHN1-2.

A summary of average characteristics observed versus calculated appears in Table 1.

Table 1. Summarized comparisons.

	Tip Speed		Scaled Length		Average Breakup		Number of Particles	
	km/s		--		us		--	
	Observed	Calculated	Observed	Calculated	Observed	Calculated	Observed	Calculated
Viper	9.02	8.84	1	0.91	125	132.4	79	75
Mo Cone	11.7	11.5	1	0.71	60.1	66.7	22	23
NHN1-2	9.03	9.2	1	1.07	352	334	93	97

5. Conclusions

The simulation approach tends to produce effective estimates of breakup time for a straight wall, conical liner and waveshaped molybdenum and copper liners. More extensive validation of the model predictions will be undertaken, including calculations for additional liner materials and charge geometries (where experimental data is available). Additionally, more complete descriptions of jet kinematics[27,28] will be implemented in the post-processing code for translating the jet between formation, reference, and final breakup locations. This is expected to further reduce the dependence on the timestep and improve the accuracy of the calculations.

Currently the Chou-Carleone model is implemented in the post-processing code as the breakup time relationship because of the simplicity and availability of input parameters. Additional models will be implemented alongside this option and evaluated for consistency and accuracy of results. The ability to perform shaped charge simulation with extensive jet characterization as the output is becoming more necessary in modern iterative design processes.

In addition, more formal sensitivity analyses should be performed regarding the choice of reference frame and mesh density in the hydrocode calculation.

6. References

1. Hirsch, E. A Formula for the Shaped Charge Jet Breakup-Time. *Propellants, Explos. Pyrotech.* **4**, 89–94 (1979).
2. Hirsch, E. A Model Explaining the Rule for Calculating the Break-up Time of Homogeneous Ductile Metals. *Propellants Explos.* **6**, 11–14 (1981).
3. Hirsch, E. The Effect of the Liner Metallurgical State on the shaped charge jet break-up time. *Propellants, Explos. Pyrotech.* **15**, 166–176 (1990).
4. Hirsch, E. Scaling of the shaped charge jet break-up time. *Propellants, Explos. Pyrotech.* **31**, 230–233 (2006).
5. Pfeffer, G. Détermination par simulations numériques de l'état et des lois de fragmentation des jets de charges creuses. in *Proc. 5th Int. Symp. on Ballistics.* (1980).
6. Haugstad, B. On the Break-Up of Shaped Charge Jets. *Propellants Explos. Pyrotech.* **8**, 119–120 (1983).
7. Haugstad, B. S. & Dullum, O. S. Optimization of shaped charges. *J. Appl. Phys.* **55**, 100–106 (1984).
8. Walters, W. P. & Summers, R. L. A Review of Jet Breakup Time Models. *Propellants Explos. Pyrotech.* **18**, 241–246 (1993).
9. Walsh, J. M. Plastic instability and particulation in stretching metal jets. *J. Appl. Phys.* **56**, 1997–2006 (1984).
10. Baker, E. L., Pham, J. & Vuong, T. *An Empirical Shaped Charge Jet Breakup Model.* (2014).
11. Baker, E. L., Daniels, A., Pham, J., Vuong, T. & Defisher, S. JET BREAK-UP CHARACTERIZATION OF MOLYBDENUM SHAPED CHARGE LINERS. in *2003 Society for Experimental Mechanics Conference* (2003).
12. Chou, P. C. & Carleone, J. The stability of shaped-charge jets. *J. Appl. Phys.* **48**, 4187–4195 (1977).
13. Chou, P. C. & Flis, W. J. Recent Developments in Shaped Charge Technology. *Propellants, Explos. Pyrotech.* **11**, 99–114 (1986).
14. Chou, P. C., Tanzio, C. A., Carleone, J. & Cicarelli, R. D. *SHAPED CHARGE JET BREAKUP STUDIES USING RADIOGRAPH MEASUREMENT AND SURFACE INSTABILITY CALCULATIONS.* (1977). doi:10.1109/CSMR.2012.24
15. Chou, P. C., Grudza, M., Liu, Y. F. & Ritzman, Z. Shaped-Charge Jet Break-Up Formula with Metal Anisotropy. in *Proceedings of the 13th International Symposium on Ballistics (Vol 2)* (1992).
16. Walters, W. P. & Summers, R. L. The Particulation of a Shaped Charge Jet for Face-Centered-Cubic Liner Materials. *Arl-Tr-114* 60 (1993).
17. Hennequin, E. Modelling of the shaped charge jet break-up. *Propellants, Explos. Pyrotech.* **21**, 181–185 (1996).
18. Petit, J., Jeanclaude, V. & Fressengeas, C. Breakup of Copper shaped-charge jets: Experiment, numerical simulations, and analytical modeling. *J. Appl. Phys.* **98**, 123521 (2005).
19. Nichols (ed), A. *User's Manual for ALE3D: An Arbitrary Lagrange/Eulerian 3D Code System.* (2007).
20. Childs, H. VisIt: An End-User Tool for Visualizing and Analyzing Very Large Data. (2013).
21. *VisIt Python Interface Manual.* (2013).
22. Sweitzer, J. C. & Banish, R. M. A closed form, energy-based theory of dynamic fragmentation. *J. Appl. Phys.* **123**, 075901 (2018).
23. Sweitzer, J. C., Hill, S. D. & Banish, R. M. Energy based distribution for naturally fragmenting warheads. in *California: Joint Classified Bombs/Warheads & Ballis-tics Symposium Monterey* 1–4 (2016).

24. Sweitzer, J. C. Material Sections and Natural Fragmentation Size Distributions in Heterogenous Shells: A Dissertation. (University of Alabama in Huntsville, 2016).
25. Mott, N. F. Fragmentation of Shell Cases. *Proc. R. Soc. A Math. Phys. Eng. Sci.* **189**, 300–308 (1947).
26. Johnson, G. R. & Cook, W. H. A constitutive model and data for metals subjected to large strains, high strain rates and high temperatures. *7th International Symposium on Ballistics* 541–547 (1983). doi:10.1038/nrm3209
27. Babkin, A. V., Ladov, S. V., Marinin, V. M. & Fedorov, S. V. Characteristics of inertially stretching shaped-charge jets in free flight. *J. Appl. Mech. Tech. Phys.* **38**, 171–176 (1997).
28. Babkin, A. V, Ladov, S. V, Marinin, V. M. & Fedorov, S. V. Regularities of the stretching and plastic failure of metal shaped- charge jets. *J. Appl. Mech. Tech. Phys.* **40**, 571–580 (1999).
29. Fermen-Coker, M. *A 2-D Axisymmetric Model of a Stretching Jet in ALEGRA*. (2004).
30. Pham, J. D., Baker, E. L. & Defisher, S. *SHAPED CHARGE JET FLASH RADIOGRAPH DIGITIZATION*. (2005).

Proceedings of the 2019 Hypervelocity Impact Symposium
HVIS2019
April 14 - 19, 2019, Destin, FL, USA

HVIS2019-050

Defeating Modern Armor and Protection Systems

Markus Graswald *, Raphael Gutser, Jakob Breiner, Florian Grabner, Timo Lehmann, and Andrea Oelerich **

TDW GmbH, Hagenauer Forst 27, 86529 Schrobenhausen, Germany

*** WTD 91, Schießplatz 1, 49716 Meppen, Germany*

Abstract

An open source research and vulnerability study of main battle tanks and their protections systems revealed that current anti-tank weapons may not be suited to defeat modern threats. One example is the novel T-14 tank being developed and tested in the Russian army with its combined hard-kill and soft-kill active protection system AFGANIT / SHTORA, its new reactive armor MALACHIT as well as improved multi-component passive armor. Additionally, modern active protection systems currently developed in, e.g., Israel, the United States, and Germany feature also multi-sensor and multi-effector systems with drastically improved detection and intercept ranges, short system reaction times as well as protection against multiple threats attacking simultaneously and / or from similar directions. While known effectors and concepts may overcome fielded active protections systems, they are probably not suited in defeating such modern and even future systems. Countermeasures relying on high engagement velocities through improved kinetic energy projectiles or hypervelocity penetrators may provide a potential solution. Another promising concept generates directed, far-distance electromagnetic effects defeating sensors and communications systems of modern main battle tanks. After such a mission kill, a following salvo attack through an anti-tank or modern multi-role weapon will eventually lead to a catastrophic kill. Feasibility studies of these mobile electromagnetic effectors have already shown their high potential.

Keywords: Main battle tank; active protection system; reactive / passive armor; kinetic energy / hypervelocity penetrator; high power electromagnetic.

1 Introduction

The enduring competition between weapon effectiveness and platform protection has already entered a new act. The threat on main battle tanks imposed by various effectors such as anti-tank missiles (anti-tank (guided) missile, ATM / ATGM) or shoulder-launched rockets (rocket propelled grenade, RPG) with tandem-shaped charge warheads (e.g., TOW 2A, PARS3 LR), kinetic energy (KE) projectiles (e.g., M829 or DM 63), and top-attack munition with explosively formed projectile (EFP) warheads (e.g., TOW 2B or SMArt) has been drastically increased over the last decades [1]. This lead to recent developments with T-14 tanks (built on the standardized ARMATA track vehicle platform) in Russia using novel and / or improved active protection systems (APS), explosive reactive armor (ERA), and passive protection. Table 1 compares Israeli main battle tank MERKAVA 4 with Russian tanks T-90 and T-14 showing significant improvements in terms of both fire power and protection systems of the latter. It needs to be noted that data refers to 2013 sources and may be dependent also upon environmental conditions, i.e., weather, day vs. night time. The greatest challenge, however, is offered through the new generation of active protection system being developed.

This paper presents major results of a comprehensive survey on trends in developments on new main battle tanks and their protection systems. Focussing on active protection systems, a classification scheme is provided and countermeasures assessed. Two promising concepts, a combined kinetic energy effector and a mobile high power electromagnetic effector, are introduced in detail along with simulation and experimental results of subsystems.

2 Technological trends in the field of main battle tank protection

The novel T-14 main battle tank currently tested in the Russian army was first fully shown at a parade in Moscow on May 9th, 2015. After this initial rate, an ambitious series production of 2000+ units shall be completed in the early 2020s. Although not all details on its capabilities are known, it uses a comprehensive protection concept with indirect and direct measures: [1, 3]

- Reducing or changing signatures through stealth technologies in infrared and radar spectrum such as special coatings, improved heat isolation, absorbing materials, cooling of exhaust gases, as well as an active, electromagnetic mine protection system

* Corresponding author: Email markus.graswald@mbda-systems.de, Tel. +49 (8252) 99-7264.

Table 1. Overview comparing Israeli and Russian main battle tanks. [2]

Technical data	MERKAVA 4	T-90 (A,S,M)	T-14
Main weapon, caliber in mm	120	125	125 ... 152
Firing speed per min	4	8	10 ... 12
Detection range in m	4500	>5000	>5000
Maximum engagement range in m	4000 ... 6000	4000 ... 5000	7000 ... 8000
Mine protection	Passive	Active and passive	Active
Active protection system	TROPHY	SHTORA	AFGANIT
Reactive protection	Yes	Yes	Yes
Armor equivalent in mm	700 ... 750	800 ... 830	>900
Engine power in PS	1500	1130	1200 ... 2000
Maximum mass in t	65	46.5 ... 48	50 ... 55
Maximum velocity in km/h	60 ... 65	60 ... 65	80 ... 90
Range in km	500	500	>500

- Avoiding hits by soft kill active protection systems like SHTORA-1
- Avoiding both hits and penetration by hard kill active protection systems like ARENA, DROZD, or novel AFGANIT
- Reducing the effectiveness through explosive reactive armor like KONTAKT-5, RELIKT, or novel MALACHIT
- Reducing the system impact through passive, multi-component ballistic protection based on novel high-strength steels such as 44S-SW combined with non-metallic materials like ceramics, aramid fabrics, and plastics, e.g., PU (polyurethane) or PE (polyethylene), providing an RHA (rolled homogeneous armor) equivalent of more than 900 mm in the front section

Active protective systems of main battle tanks and armored vehicles are typically classified into soft or hard kill systems. Their functionality is visualized in Figure 1. Based on a threat detected by sensors, a fire control solution will be determined dependent upon threat analysis, its predicted trajectory, and the counter measure. A soft kill system distracts the incoming threat either trough infrared (IR) emitters, jammer, or decoy/flares, or interrupts the line-of-sight by smoke shells, while a hard kill system is designed to destroy the threat in a distinct and safe distance to the tank through deployed or distributed blast/fragmentation projectiles, shaped charges or (multi) EFP, or directed high explosive (HE) charges. [1]

In a comprehensive literature and online search, a total of 26 active protection systems have been found and categorized [3, 5]. Besides various Russian systems named earlier, SASLON/ZASLON as active and NOZH and DUPLET as reactive, Ukrainian made protection systems are supposed to be highly effective. Israeli Trophy APS installed on MERKAVA MK4 tanks are known as combat proven in several armed conflicts. In NATO countries like the United States and Germany, a number of both soft kill and hard kill active protection systems are developed and mounted onto main battle tanks or armored vehicles such as PUMA and STRYKER. Turkey and South Korea develop their own systems called AKKOR and KAPC for future integration on their main battle tanks ALTAY and K2, respectively.

Table 2 shows a classification scheme of APS with their protection against threat types, their sensor and effector principle, and system aspects along with typical examples and essential data of Russian, Israeli, and American systems. It differentiates between classical state-of-the-art systems like DROZD or SHTORA and modern or future systems like AFGANIT or TROPHY since engagement concepts and weapon systems overcoming them may differ significantly. It reveals that modern and potential future developments

- use a combination of sensors with different physical principles, long detection ranges, virtually no dead zones, and capabilities to track multiple threats simultaneously
- rely on (a combination of) hard kill effectors with increased engagement ranges, and
- provide drastically reduced system reaction times from threat detection to countermeasure interaction

This allows the defeat of multiple – also high velocity – threats attacking simultaneously or successively from the same or different directions. Further development trends concentrate on both sensors and countermeasures and indicate to use active protection systems through cooperative engagements for other or even unprotected vehicles as well. Besides new and modern tanks, classical tanks like T-72 or T-90 may also be upgraded with modern APS.

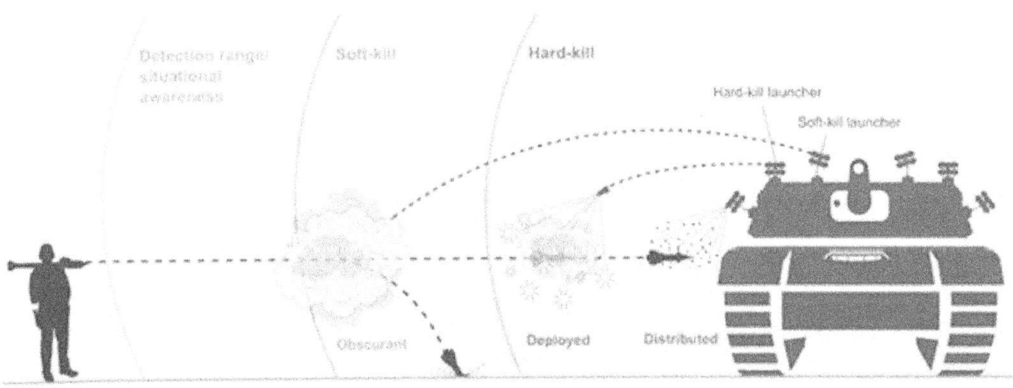

Figure 1. Principle of active protection systems visualized for hard and soft kill effectors. [4]

Table 2. Classification of active protection systems*. [3]

Classification	Parameter	Classical APS	Modern APS
Protection against	Single threats	ATM, ATGM, RPG	Most ATM, ATGM, RPG; in part KE
	Multiple threats	No	Yes, simultanously and/or from different attack directions (AFGANIT, TROPHY, IAAPS, IRON FIST)
	Threat velocities	70 . . . 700 m/s, up to 1200 m/s (DROZD-2)	1700 m/s (AFGANIT 1), 3500 m/s (AFGANIT 2)
Threat detection	Sensor type	Mostly radar	AESA, laser (RUS); combinations of AESA, EO, IR, EW (TROPHY, IRON FIST, FSAP, IAAPS), also integrated into turret (AFGANIT, ARENA-E)
	Zone	Limited in Az, El (especially from top)	360 deg Az (typically), also 180 deg El (AFGANIT)
	Detection range	5 . . . 75 m (typically)	Up to 10 km (AFGANIT, IAAPS)
Tracking and fire control	Sensor type	Mostly radar	Mostly AESA, especially for hard kill
Effector principle	Hard kill	Mostly fragmenting or blast projectiles (ADS)	Partly (multi) EFP (TROPHY, FSAP, AFGANIT 2) or MG / cannon (AFGANIT 1), also through blast (FSAP, IRON FIST)
	Soft kill	IR (SHTORA-1, IAAPS), decoy/ flares/smoke (POMALS, SHTORA-1)	
	Engagement range	1 . . . 10 m (typically)	Up to 35 m
System aspects	Reaction time	>1 s	<1 ms (AFGANIT), may also depend upon threat velocity
			Combinations of APS on one platform (AFGANIT / SHTORA-1)

Various approaches have been identified to overcome active protection systems as visualized through an onion model adapted for single or salvo engagements of effectors in Figure 2. These opportunities can also be combined for being more effective and sustainable. The analysis revealed that existing anti-tank weapons and engagements principles exploiting sensor-dead zones and using saturation attacks may still be suited against classical tanks, while they might be outdated for current and future developments of active protection systems. Potential solutions meeting these challenges are, therefore, provided through a combined kinetic energy effector relying on super or hyper velocities and a mobile high power electromagnetic effector aimed at defeating APS sensors. They are introduced in the following subsections. In addition, anti-tank or multi-role missiles with tandem or new multi-effect warheads may still be effective against explosive reactive and passive armor. They can also be used against a wider target set consisting of personnel, unarmored vehicles, and light infrastructures.

3 Combined penetrator effector

Besides high explosive anti-tank missiles, kinetic energy projectiles like DM 63 or M829 provide a high threat potential against classical main battle tanks. They are typically fired from 120 mm caliber guns and reach high velocities $\gg$ 1000 m/s

*Acronyms not introduced earlier: active electronically scanned array (AESA), electro-optical (EO), infrared (IR), electronic warfare (EW), azimuth (Az), elevation (El), machine gun (MG).

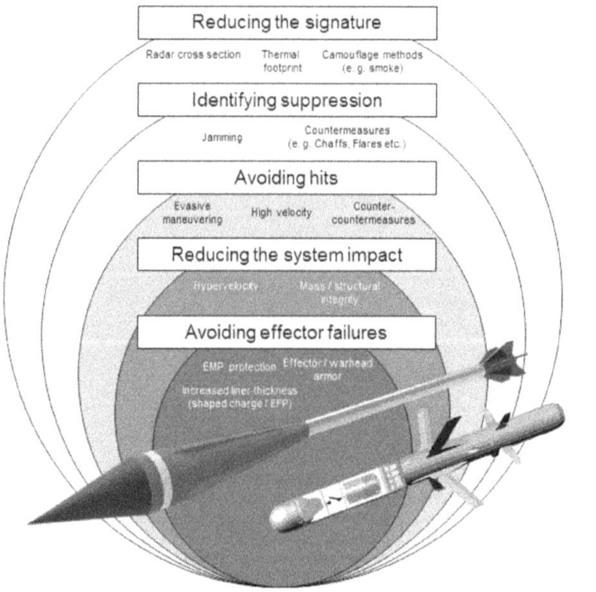

Figure 2. Adapted onion model showing various opportunities for effectors to overcome active protection systems that can be achieved through single or salvo engagements.

to penetrate both reactive and passive armor. Modern tanks with active protection systems, improved reactive and passive armor may, however, drastically decrease their performance.

Supplementing a kinetic energy penetrator with a precursor shaped charge provides an incredible excess velocity so that the EFP may be fired well outside the APS engagement range hitting unrecognized the explosive reactive armor and initiating it. The concept is visualized in Figure 3. An EFP charge may also serve as a deception device for APS, so that the KE penetrator survives and eventually engages the main target effectively. The precursor charge consists of a high strength metal such as tantal, a high-performance octogene-based high explosive charge like a KS33 or P31 with a reactive detonation wave shaper and a shock-hardened inline (electronic safe-and-arm device, ESAD) fuze system embedded into light weight titanium structure. A sub-caliber, high L/D tungsten penetrator is located behind and may be further improved through reactive materials. This cost-efficient and robust lethal package can be integrated into various systems like a hyper-velocity missile or an armor piercing fin-stabilized discarding sabot (APFSDS) munition.

The EFP precursor charge was designed meeting desired velocities of >800 m/s, L/D ratios of 5, and fin-stabilization measures with the hydrocode SPEED [6]. Its effectiveness against various targets was modeled using a 3D Eulerian mesh as well. Results of a representative explosive reactive armor module consisting of Comp B (RDX/TNT) applying the History Variable Reactive Burn (HVRB) initiation model are shown in Figure 4. They indicate that the EFP can penetrate inert flyer plates and initiate the explosive layer at least with a deflagration or low-order detonation reaction. Assuming a sufficient timely stand-off, the ERA target should be cleared for the succeeding penetrator killing the main target with its remaining passive armor.

4 High power electromagnetic effector

A mobile high power electromagnetic (HPEM) effector integrated into an anti-tank missile as displayed in Figure 5 seems interesting for defeating active protection systems sensors and communication devices of main battle tanks. Early studies generating electromagnetic effects through explosive devices date back several decades, e.g., [7, 8, 9] and recent feasibility studies of have already shown their high potential [10, 11]. The concept consists of an explosively driven flux compression generator (FCG), an electrical opening switch and the subsequent shaping of the electrical pulse into an high frequency (HF) radio signal (e.g., by applying a magnetron) that is radiated through an antenna system. This enables a compact design for integration into mobile applications such as missiles, shoulder-launched munitions, or artillery shells. The following section describes the design of an FCG and several proof-of-principle firing tests as well as a HF generator and directed antenna investigating the vulnerability of electronic components.

4.1 Proof-of-principle tests on prototype FCGs

The working principle of a flux compression generator can be explained by using the transient Maxwell equations. Its main purpose is generating a short-time high level electric power peak by converting chemical energy into kinetic energy – released by detonating a high explosive charge – and eventually into electrical power. This changes the magnetic field and induces an Eddy current similar to an electrical motor. The high-explosive driven magnetic flux compression generator shown in

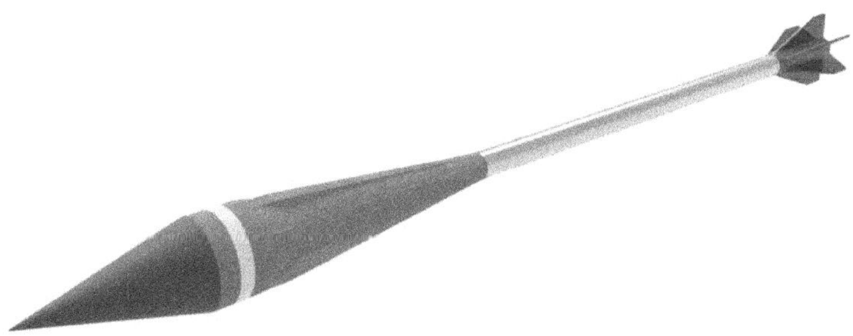

Figure 3. Rendered artist impression of a high velocity effector combining an EFP precursor charge with a kinetic energy penetrator.

(a) Before penetration. (b) After penetration and HE initation.

Figure 4. 3D SPEED hydrocode simulation of an EFP attacking a representative ERA target.

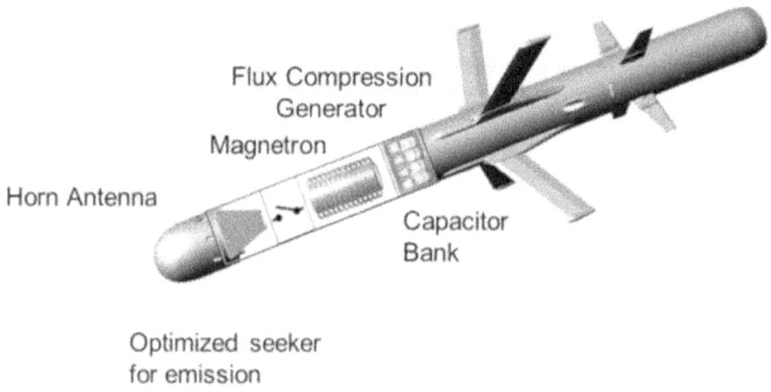

Figure 5. Directed HPEM missile designed against active protection systems of main battle tanks.

Figure 6 consists of a capacitor bank, a stator coil (solenoid), an armature filled with a HE charge, and a load switch. This system operates through these three major steps:

- Seeding current: An initial magnetic flux density is required to start the process which can then be compressed by the system. This flux density has to be provided conventionally by an appropriate high power source such as a Marx Generator or a capacitor bank that generates the seeding current.
- Flux compression: After charging the solenoid with the seeding current and establishing a high magnetic flux density within the system, the conventional power generator is removed from the system and the detonation process starts. The expansion of the armature creates an electric contact with the coil/solenoid compressing the magnetic flux in a very short time. This results in an instantaneous increase of the magnetic energy density within the system allowing the generation of currents of several 100 kA.
- Power conversion: The generated power needs to be transmitted to a load such as a HF generator with an antenna structure. For the flux compression process, the load switch needs to be closed in order to enable a high current for flux compression. This short circuit switch needs to be opened in a short time transmitting the power to the load. An explosive electric opening switch is a possible option to realize fast switching.

Prototypes of high explosive driven generators, a test setup and diagnostics including a Rogowski probe for current measurements were developed for proof-of-principle trials at Schrobenhausen site. Circuits for the flux compression generator and detonator initiation were galvanically separated. Test setups with FCG prototypes are displayed in Figure 7 (with capacitor bank not shown). Figure 8 shows a successfully measured output current from the FCG (red line) prototype and shunt (blue line) during detonation. The detonation of the FCG was timed to the positive peak of the pulse generator resulting in an output current of several hundreds of kA in the microsecond regime. This complies to a current amplification with a factor of approx. 30 that was basically predicted through the system simulation.

FCGs can be further optimized through increasing the seeding current, the forming of the magnetic field and the subsequent timely trigger for initiating the high explosive charge. An appropriate design of the armature and solenoid, the type of high explosive charge, and a precise manufacturing and assembly process are also important. Another development step is the interface design between the current power output and HF generator to fulfil the radiation characteristic requirements. This current to HF signal transformation depends also on the targets of interest and operational endgame settings.

4.2 Vulnerability tests of representative components

Since microcontrollers and FPGAs serve as subsystem controllers in a variety of applications such as sensoring, navigation, and other data processing, a malfunction of these devices lead usually to a critical system state and may cause overall system failures as well. Their damage levels are typically classified as follows:

- Level 1: disturbance of core functionalities of electronic components while the interference signal is active

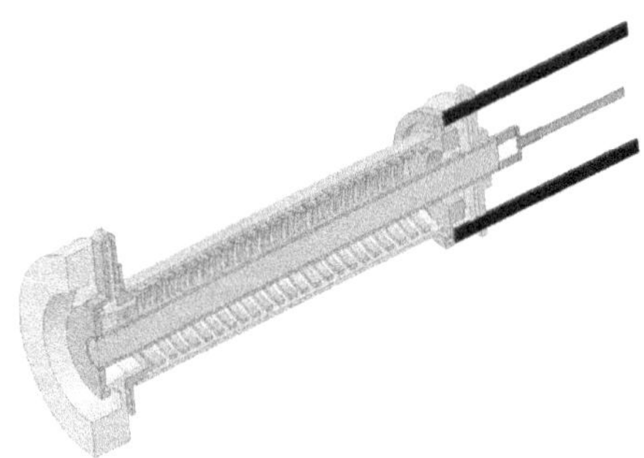

Figure 6. 3D view of a flux compression generator including detonator and cables to the capacitor bank.

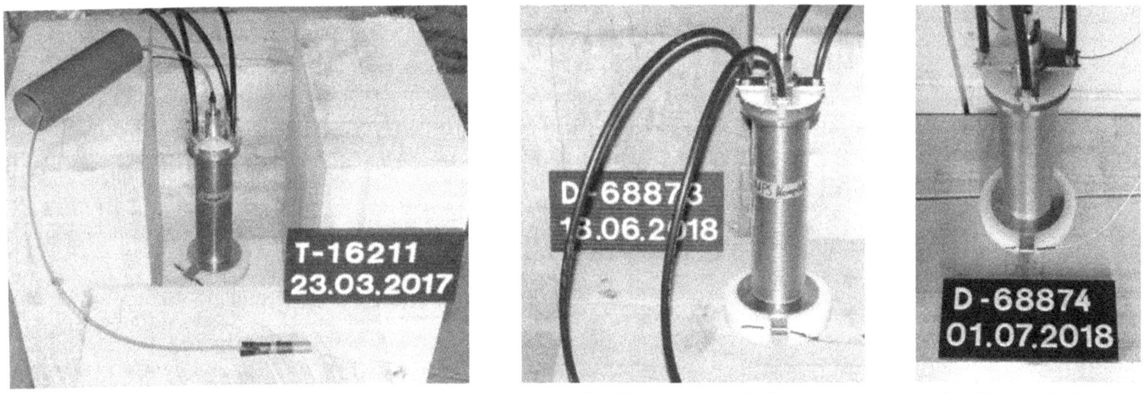

(a) Test setup T16211. (b) Test setup T68873. (c) Setup T68874.

Figure 7. Testing the flux compression generator prototype.

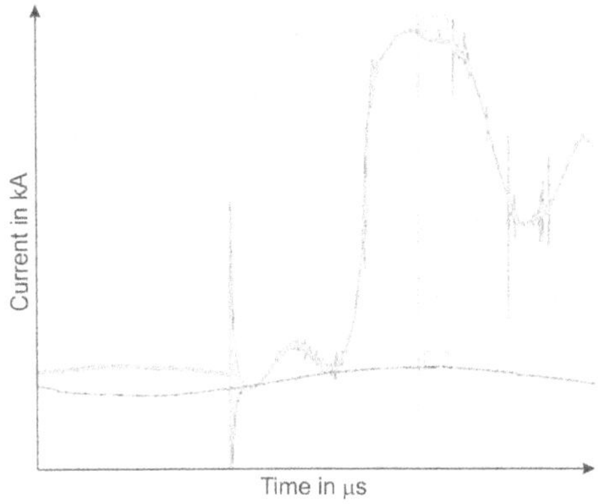

Figure 8. Rogowski probe measurement of current generated by the flux compression generator prototype (red line) and capacitor bank shunt current (blue line).

- Level 2: temporary disturbance of core components until system reboot
- Level 3: permanent damages of components – their exchange is required for proper operations

Besides these electronic components, transmission lines such micro strip lines and coaxial lines are considered as well since they are used in integrated circuits and for connecting all electronic devices transmitting signals in the high frequency domain. These lines are modeled in COMSOL Multiphysics [12] and eventually exposed to electrical fields. Figure 9a shows a coaxial line modeled through three concentric circles, a dielectricum between conductor lines, and ports at conductor ends: electrical field lines pointing from the inner to the outer line, magnetic field lines passing in inconcentric circles parallel to both lines, and a pointing vector pointing into the plane. The model implementation is initially simulated using a test signal and without outer influences for verifying typical line characteristics such as damping depending upon signal frequency or impedance with analytical calculations and experimental data. Figure 9b shows a good correlation of simulation results with analytical and literature data except in the lower frequency domain.

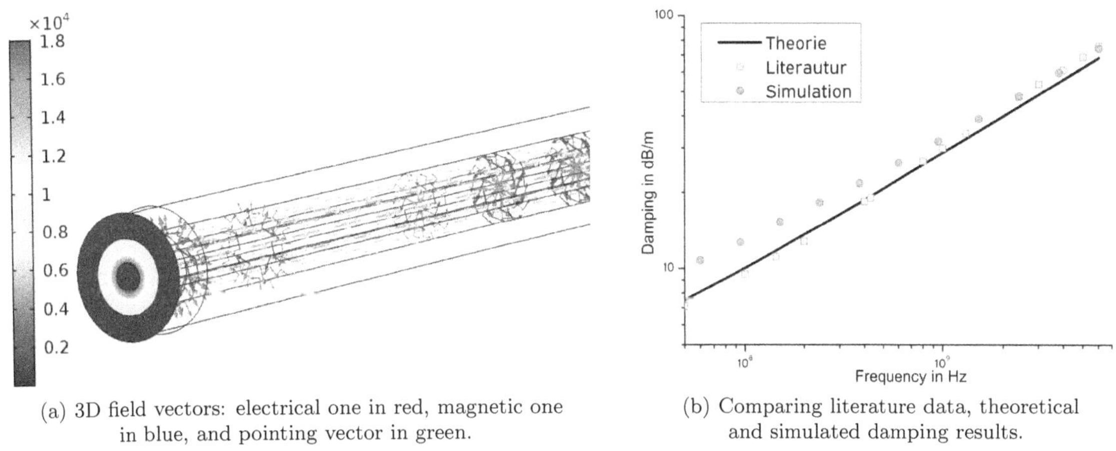

(a) 3D field vectors: electrical one in red, magnetic one in blue, and pointing vector in green.

(b) Comparing literature data, theoretical and simulated damping results.

Figure 9. Simulation of a coaxial cable with COMSOL Multiphysics.

Shielding electronics against electromagnetic fields was also investigated through simulations. Figure 10a displays a copper Faraday cage that provides a perfect shielding. Openings such as slits and grits are mostly required for practical reasons. They are also used in radar sensor systems and communication devices and allow a transmission of electromagnetic signals as exemplified in Figures 10b and 10c.

A representative microcontroller (LPC2103FBD48) is also modeled in a COMSOL Multiphysics simulation [12] and its response on externally generated electromagnetic fields evaluated. The induced current into the component is simulated and the resulting damage evaluated through the given CMTI (Common mode Transient Immunity) value. Figure 11 shows the microcontroller model and the simulated induced current at a given field strength. If the component dependent CMTI value measured through means of surge protection measurement (SPM) is exceeded, the system will usually shutdown. Permanent component damages can be assumed by exceeding this value by a factor of 5.

In an experimental setup for subcomponent tests, a continuous and a pulsed magnetron were used for generating HF signals that were transmitted through a horn antenna. Figure 12 shows a simulated far field and gain of a typical horn antenna with CST Microwave Studio [13]. The electrical field strength is measured robustly with Ddot sensors. The effectiveness of the emitted signal is eventually evaluated against commonly used FPGAs such as an Altera MAX10 and Xilinx Spartan6 as well as microcontrollers like an Intel PIC18.

Development boards with these two embedded FPGAs and a microchip were used for vulnerability tests at the anechoic chamber located at Schrobenhausen site. The test setup is shown in detail in Figure 13. Each board uses a different set of input and output devices, e.g., LEDs, switches, VGA port, GPIO pins. Hundreds of tests of these board were performed. During each test campaign, the variation on supply voltages besides a logic signal in the domain of kilohertz have been measured. Parameters varied include the distance and direction of the antenna to the component relating to the field strength as well as pulse and idle times of the interference signal.

The experimental investigation shows a variety of results. The transmitted wave form of the signal, e.g., either a continuous RF signal or a pulsed rectangular wave signal, remains the same in the measured signal of the boards. While a dependency of the results on pulse power periods and their power-to-idle ratio could not be observed, the electrical field strength and angle of attack are significant parameters. Changing this angle of attack of the antenna to the component, it was proven that individual angles exist where sudden increases or decreases of the signal voltage measured occur. Parameter settings for best and worst signal deflections were eventually identified. Damage levels observed vary widely between all three components investigated. While a level 1 damage was mostly observed, a level 2 was only noticed infrequently and a level 3 not at all. This might be changed through applying higher field strengths or through shorter wave lengths providing more efficient disturbances at a given field strength. These experimental results will be used for verifying a simulation of the electronic components modeled in LT Spice.

5 Conclusions

The ongoing development of a new generation of main battle tanks with its improved protection systems provides a challenge for existing anti-tank effectors. The study identified vulnerabilities of modern and future active protection systems. Several potential counter measures were assessed and two promising concepts down selected, i.e., a combined penetrator effector relying on excess velocities and a high power electromagnetic effector for defeating sensor and communication devices. The theoretical and experimental investigations performed enable detailed concept designs for different system applications. In future, it is planned to build prototypes of complete effector systems for performing experimental tests against relevant military targets based on operational scenarios and associated endgame settings.

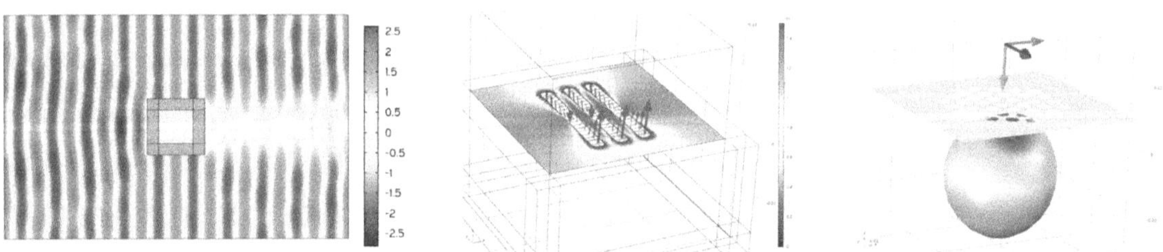

(a) Faraday cage. (b) Slits. (c) Grit.

Figure 10. Assessment of geometric shielding measures against electromagnetic radiation.

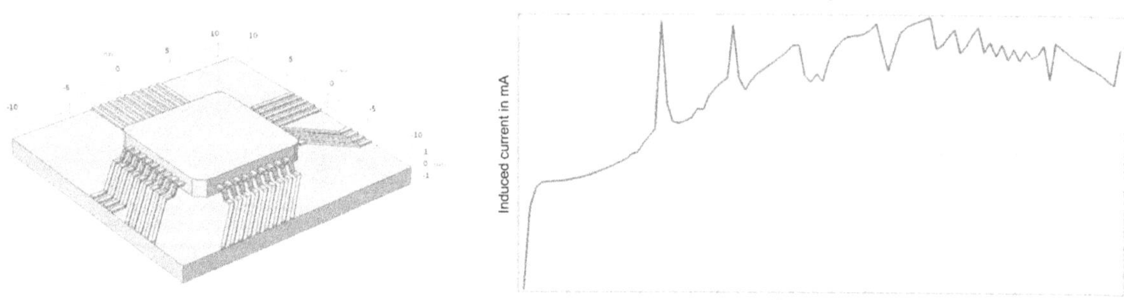

(a) COMSOL model. (b) Current vs. frequency at a given field strength.

Figure 11. Vulnerability of a microcontroller evaluated through COMSOL Multipysics simulations.

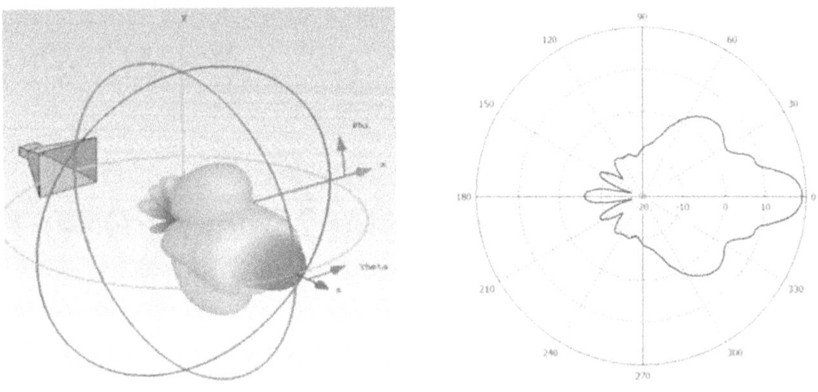

(a) 3D far field. (b) Antenna gain.

Figure 12. Horn antenna simulation providing an directional electromagnetic effect.

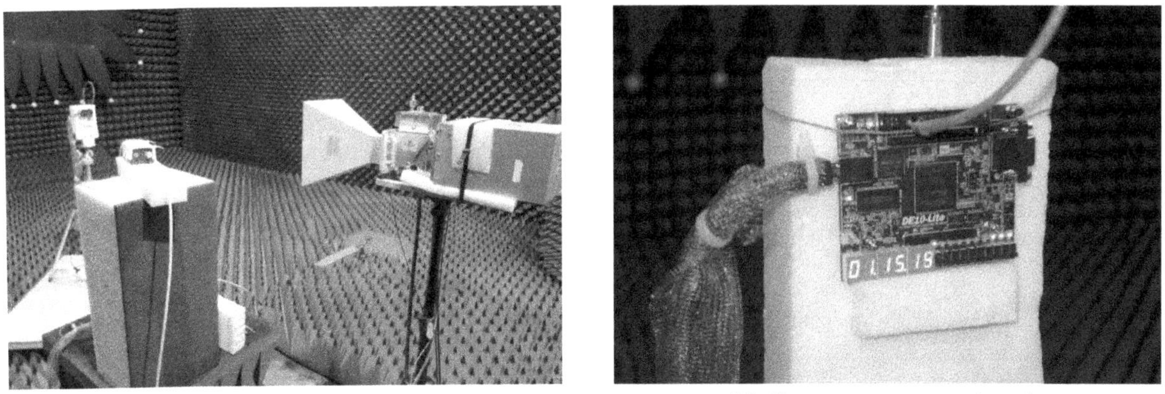

(a) Site overview. (b) Close-up view on test board.

Figure 13. Test setup with test board, video camera, horn antenna, and magnetron (from left to right) in an anechoic chamber.

Acknowledgements

The authors gratefully acknowledge the financial support by Meppen proving ground WTD 91, GF-440, and BAAINBw K1.5.

References

1. R. Holthaus. Plattformschutz – Hard- und Softkill-Systeme. *Strategie & Technik*, Nov 2011.
2. http://youinf.ru/armata-vs-mercava/. Accessed Nov 2015.
3. M. Graswald. *Recherche und Klassifikation aktiver Schutzsysteme inklusive des Gesamtschutzes moderner Kampfpanzer*. Number TDW-TN-EN-15-0017. TDW GmbH, Schrobenhausen, Feb 2016.
4. J. E. Staff. Developments in APS technologies. https://ihsmarkit.com/research-analysis/developments-in-active-protection-system-aps-technologies.html, Oct 2017. Accessed Nov 2018.
5. M. Graswald and R. Gutser. *Recherche zum Gesamtschutz moderner Kampfpanzer und Wirksystemtechnologien gegen Aktivschutzsysteme/ UAV-Schwärme*. Number TDW-TN-EN-17-0012. TDW GmbH, Schrobenhausen, Dec 2017.
6. N. N. *SPEED v 3.1 Users Manual*. Numerics GmbH, Petershausen, 2018.
7. C. M. Fowler and L. L. Altgilbers. *Magnetic Flux Compression Generators: a Tutorial and Survey*. Los Alamos National Laboratory, 1975.
8. E. Kleinschmidt. *Studie zur Erzeugung elektro-magnetischer Pulse (EMP) mit Sprengladungen (Teil 1)*. Number TDW-TN-PE4-96-0043. TDW GmbH, Schrobenhausen, 1996.
9. E. Kleinschmidt. *Studie zur Erzeugung elektro-magnetischer Pulse (EMP) mit Sprengladungen (Teil 2)*. Number TDW-TN-PE4-96-0043. TDW GmbH, Schrobenhausen, 1996.
10. V. K. Chernyshev. Superpower explosive magnetic energy sources. In *Proceedings of the International Workshop High Energy Density Hydrodynamics*, Novosibirsk, Russia, 2004.
11. P. Appelgren. Modeling of a Small Helical Magnetic Flux-Compression Generator. *IEEE Transactions on Plasma Science*, 36(5), 2008.
12. N. N. *COMSOL Multiphysics version 5.3*. COMSOL AB, Stockholm, Sweden, 2017.
13. N. N. *CST EMC Studio User Manual*. Dassault Systems, Vélizy-Villacoublay, France, 2017.

Proceedings of the 2019 Hypervelocity Impact Symposium
HVIS2019
April 14-19, 2019, Destin, FL, USA
HVIS2019-069

Evaluation of Critical Ricochet Angles for 25mm APDS-T Projectile on Metallic Targets - Modeling and Verification

Alon Weiss[a]*, Asaf Borenstein[a], Vitaly Paris[a], Moshe Ravid[b], Nimrod Shapira[b]

[a]*Plasan Sasa Ltd., M.P. Merom Hagalil, 13870, Israel*
[b]*Rimat Advanced Tech. Ltd., Hod Hasharon, 45264, Israel*

Abstract

This paper is a study on ricochet characteristics of metal plates hit by 25 mm APDS-T projectile. A series of ballistic tests has been carried out on three different armor steel types chosen to present a range of hardness. Each armor steel target was impacted at varying angles of incidence until ricochet was observed. This investigation is not only expected to help in understanding the phenomenon of projectile ricochet but also to provide some useful data that can be used for calibration and validation of the finite element (FE) models. In general, the ricochet angle was found to decrease with an increase in target hardness due to different failure mechanism.

Keywords: critical ricochet angle; 25mm APDS-T projectile; armor steel target; numerical simulation

---

1. Introduction

Ricochet is a phenomenon of projectile deflection that may be observed when striking highly oblique targets. Critical ricochet angle is defined here as a transition angle from regime of ricochet to regime of penetration for sufficiently thick plates or perforation for thinner plates. The critical ricochet angle may depend on material properties of both target and projectile (density, strength, failure, etc.), geometry, target thickness, projectile L/D ratio, shape, and the impact velocity [1].

An extensive review of the ricochet of compact projectiles impacting oblique targets is provided in Ref. [2]. Major interest has been shown in a description of limiting states of ballistic behavior, such as critical velocities and angles of obliquity for embedment or ricochet, and in various fragmentation forms of the strikers. Together with the concept of the ballistic limit, these parameters are essentially structural/geometrical rather than continuum mechanical properties. Much of the research in ricochet centers is on compact projectiles or on rigid projectiles. A much smaller fraction of the work is focused on medium- and long-rod ricochet, where a different phenomenology often manifests itself. Tate first described ricochet using a simplified two-dimensional hydrodynamic model [3]. Rosenberg et al. [4] concerned the origins of the asymmetric force that acts upon the eroding tip of the impinging rod. Several more authors examined the ricochet problem combined efforts to produce new results [5]–[12].

Significant amount of experimental data of 25 mm APDS-T projectile, including tests of oblique impact (see Fig 1) was conducted by Plasan in the recent years. Most of these results were obtained with combinations of materials such as aluminum/steel and ceramics/steel. Oblique impact data for armor steels are limited to HH steel and to obliquity angles in the perforation regime. Even though the oblique steel plates are not as weight effective as aluminum/steel and ceramics/steel combinations, they may be a part of effective passive armor combinations as well as element of active armor protective solutions.

* Corresponding author. Tel.: +972-73-2619482; fax: +972-4-6809001.
E-mail address: alonw@plasan.com.

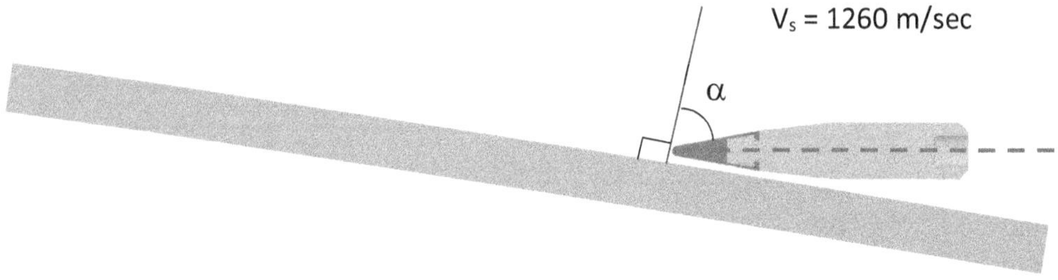

Fig.1. Illustration of plate obliquity (Nato angle α) between the armor steel target and the 25 mm APDS-T projectile. α was varied from one shot to another.

Currently we investigate the effect of thickness and strength of steel plates on critical ricochet/perforation transition angle. To rely on the 3-D FE models used for such applications, it is important to validate the models for similar conditions and to make some tuning if needed. The need to validate the materials models was a primary motivation for currently reported tests results. In addition to model validation, relative performance of various types of steel and failure behavior of both armor plates and projectiles in regime of interaction was verified. In order to validate the models for armor steels targets against the 25 mm APDS-T projectile in the regime of perforation/ricochet transition, a series of ballistic tests were performed and reported.

2. Experimental details

The experiments were conducted using tungsten-alloy projectile 25x137 mm APDS-T at a striking velocity in the range of 1260 m/s. The total mass of the projectile is 149 gr with a WA core mass of 102 gr. Its penetration capabilities in semi-infinite RHA target is approximately 45 mm at 1260 m/s. Projectile velocities are measured at impact by a flight velocity measurement system. The gun was interfaced to a photographic system capable of producing high speed photographs of the impact event at oblique impact. Two high-speed cameras (Phantom® V710, 50,000 fps) continuously recorded the impact and penetration events. One camera was situated horizontally perpendicular to the trajectory of the projectile within the target and the other was placed vertically perpendicular to the target. A schematic presentation of the test set-up is shown in Fig 2.

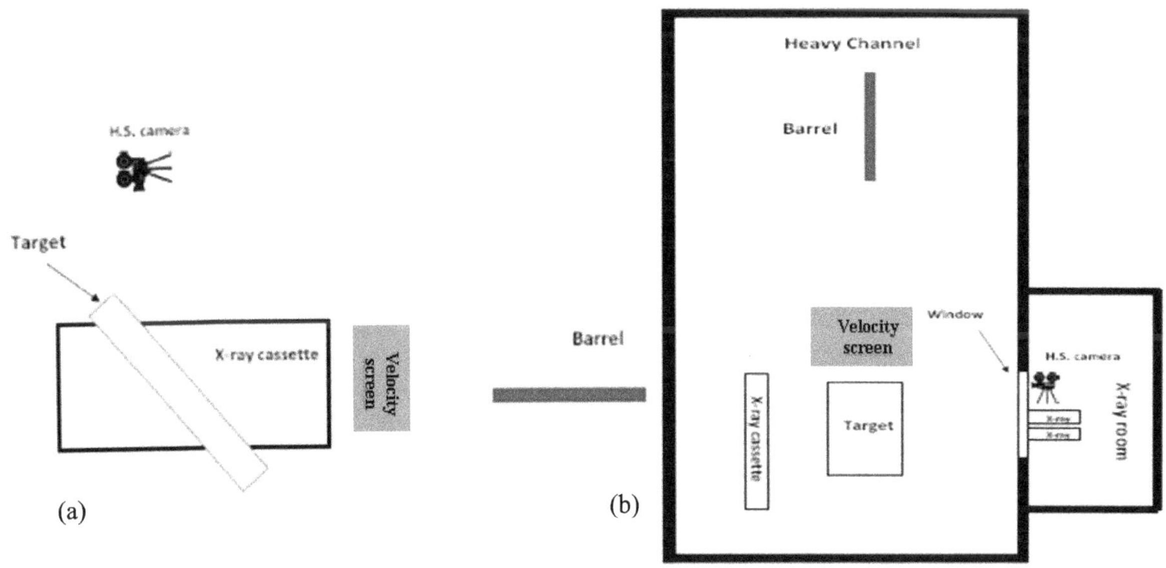

Fig. 2. Schematic illustration of the experimental set-up; (a) side view, (b) top view.

Table 1. Summary of test results

Material	Nominal hardness (*BHN*)	Obliquity angle (*degrees*)	Result
UHH	600	81	ricochet
UHH	600	79	ricochet
UHH	600	78	ricochet
UHH	600	77	penetration
UHH	600	76.7	penetration
HH	500	79.8	penetration
HH	500	81.3	ricochet
HH	500	80.3	penetration
HH	500	81	ricochet
RHA	370	82	ricochet
RHA	370	80.8	penetration
RHA	370	82	ricochet
RHA	370	80.5	penetration

3. Results and discussions

Oblique impact tests with 25 mm APDS-T and 10 mm thick armor steel plates of 3 types (RHA, HH and UHH) were carried out to evaluate their ricochet capabilities and to provide a data for calibration and validation of the FE models. Plate obliquity (NATO angle α) was measured before each shot. Pitch and yaw of the projectile were measured for each shot to verify stability of projectile flight and exact impact. α was varied from one shot to another in such manner that at least two ricochets and two perforations were obtained for each type of armor steel. Test results are detailed in Table 1 and Fig 3.

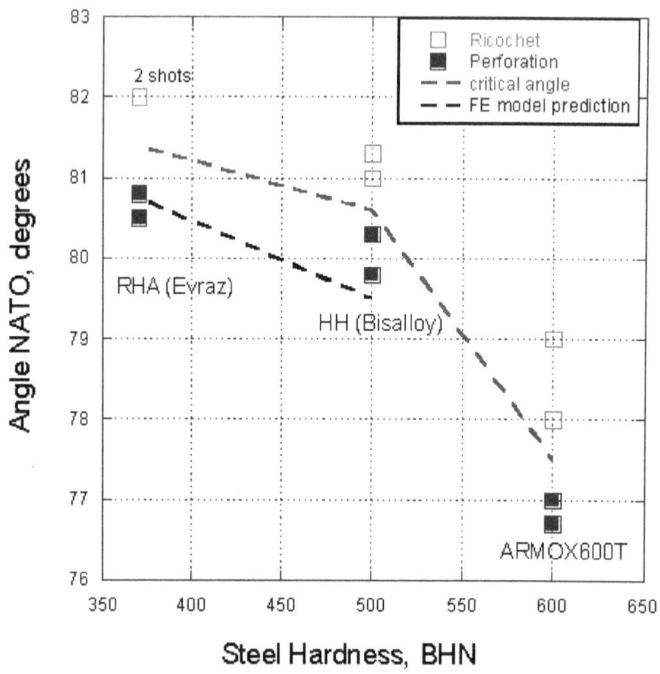

Fig. 3. Critical ricochet angles of 25 mm APDS-T for 3 types of armor steel plates.

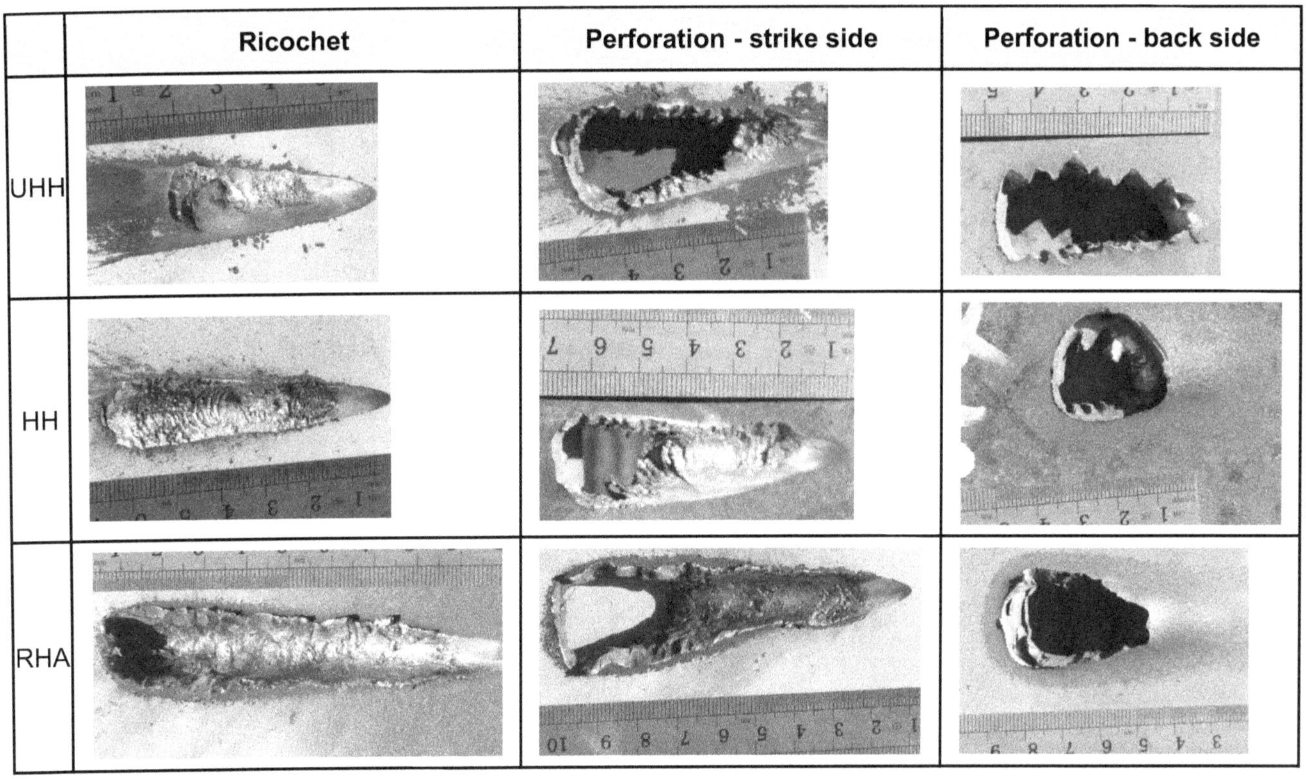

Fig. 4. The penetration crater obtained at both ricochet and perforations angles for 3 types of armor steel plates: UHH, HH and RHA.

. In both cases (ricochet and perforation) the projectile was shattered and deflected upwards. The hardness values measured are typical values for these armor steel grades. There may be some differences in the hardness between steel target plates tested and those used for materials models in the prediction analyses. The obtained results allowed estimating the ricochet/perforation transition angle (α50 in analogy to V50). The transition angle for UHH is approximately 78 degrees versus 80.6 and 81.3 degrees for HH and RHA, respectively.

Apparently, the transition angle is decreasing with increasing of the steel target hardness; it is in agreement with theory and FE simulations predictions. Increasing of nominal hardness from 500 to 600 BHN had about twice stronger effect on the transition angle than increasing from 370 to 500 BHN.

Analysis of the high-speed videos and visual examination of the targets (see Fig 4) revealed also different qualitative behavior of the 3 types of armor steel:

- The penetration crater obtained at ricochet angles just above the transition angle is largest for RHA, moderate for HH and much smaller for UHH.
- The holes in the penetrated RHA and HH plates are small relatively to UHH and most likely were created by shear failure after significant local shear deformation.
- The holes in the UHH plates have prolonged jagged shape.

The 25 mm APDS-T projectiles were shattered in brittle manner after the impact with significant fragments dispersal and without visible difference between the 3 types of armor steel.

Figure 5 display the actual test results on different thicknesses of HH plates. Fig 6 derived the line of sight weight to obtain the critical ricochet angle. Figure 6 show the effect of plate thickness on areal weight at critical ricochet angle for 25 mm APDS-T projectiles. Apparently there is an optimal thickness of HH steel. The optimum is about 8 mm for 25 mm APDS-T projectiles striking at 1260 m/s. Possibly the optimum thickness for ricochet is affected by projectile strength and velocity.

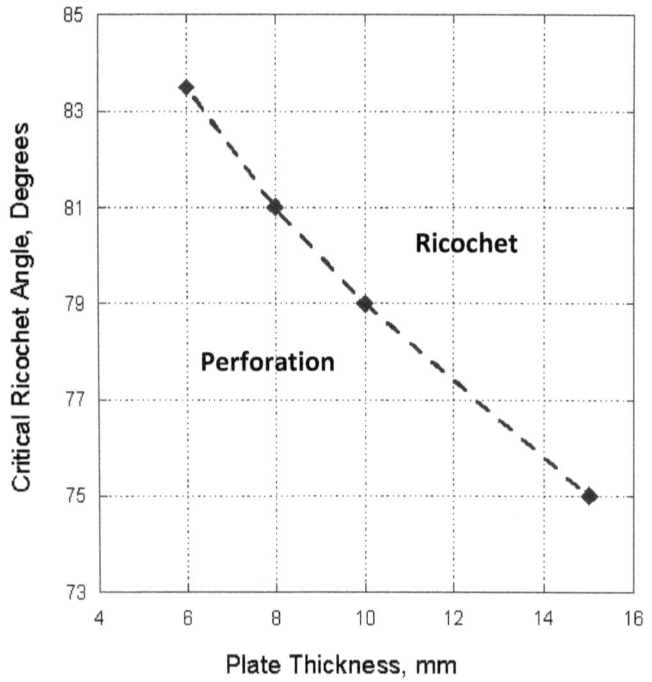

Fig. 5. Critical ricochet angles of 25 mm APDS-T as a function of HH steel plate thickness.

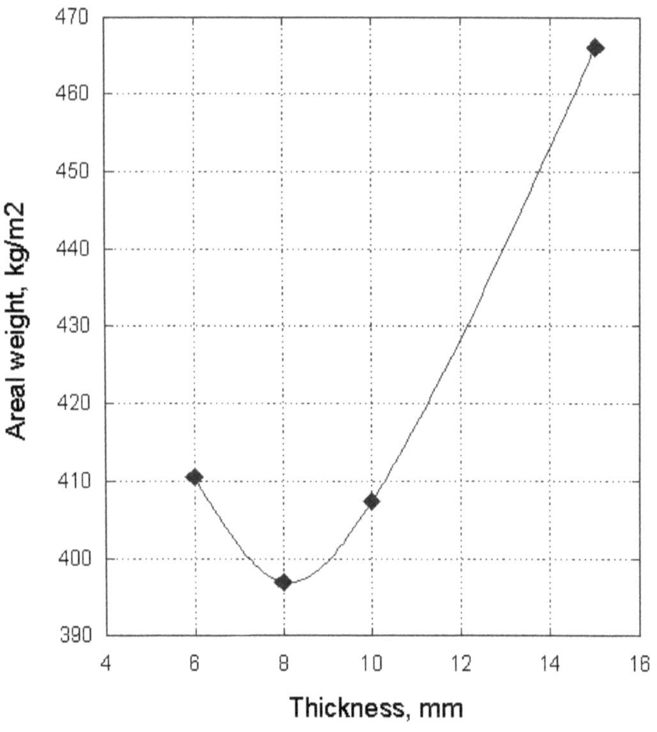

Fig. 6. Areal weight at critical ricochet angle as a function of HH plate thickness.

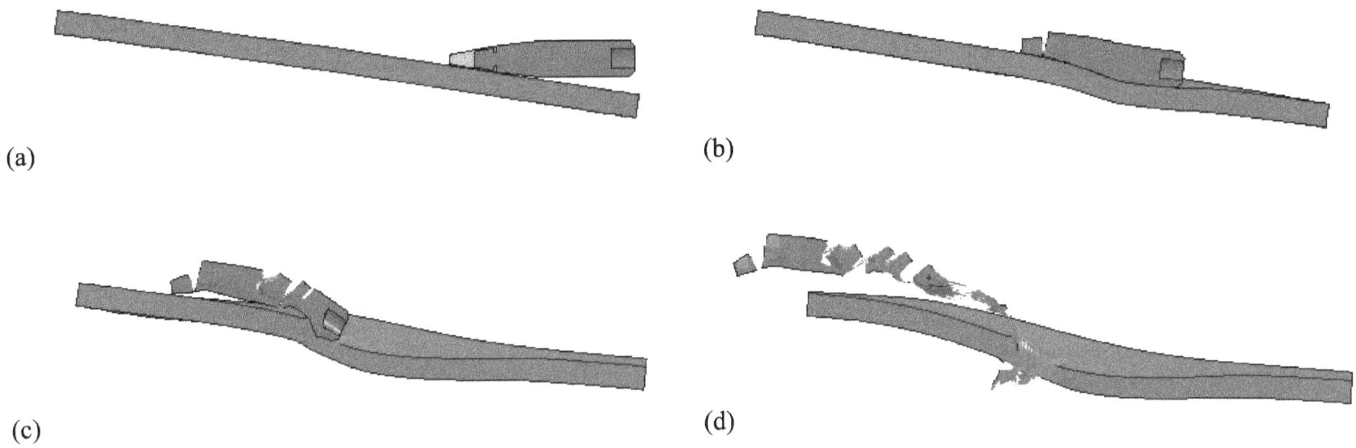

Fig. 7. Numerical results at optimum thickness of 8 mm HH steel plate, $\alpha = 81°$: (a) Time = 0, (b) Time = 40 μs, (c) Time = 88 μs and (d) Time = 136 μs.

4. Numerical simulations

We will present a brief outline of the simulation results for the effect of plate thickness on areal weight at critical ricochet angle. Here, due to limited available space, we shall outline just an illustrative part of the complete numerical research. Further analysis of the numerical simulation will be given in our next paper. Numerical simulations were carried out using LS-DYNA commercial finite element (FE) software. All simulations were conducted assuming half symmetry problems. The modeled plate was circular with 100 mm diameter. The target and projectile were modeled by brick elements; the average element size used for the projectile and the target was 0.5 mm

Numerical results at the optimum thickness of 8 mm HH steel plate against 25x137 mm APDS-T projectile with a striking velocity of 1260 m/s are shown in Fig. 7. In this case, the effect of plate thickness is explained by balance of two competing mechanisms. For thicker plates interaction is mainly by erosion of both projectile and steel plate and deflection of small fragments. Normal load component decreases with increasing angle leads to increase of the fraction of deflected projectile components thereby improving the efficiency. For thinner plates the effect of back face deformation and failure becomes dominant. This leads to bulging in the plate and fracturing of the projectile with subsequent perforation by tail segment of the projectile. Still, the effect of thickness on the performance is much less pronounced for the APDS-T projectiles than for the long rod penetrators. Moreover, there is still more work to be done in calibrating the numerical models in order to better predicts for the actual ballistic test results.

5. Concluding remarks

Oblique impact tests with 25 mm APDS-T and 10 mm thick armor steel plates of 3 types (RHA, HH and UHH) were carried out to evaluate their ricochet capabilities and to provide a data for calibration and validation of the FE models.

The obliquity angle was varied from shot to shot in such manner that at least two ricochets and two perforations were obtained for each steel type.

The obtained results allowed estimating the ricochet/perforation transition angles.

The tests demonstrate an inverse trend between transition angles to target hardness. As verified with theory and FE simulation predictions - when target hardness is increased, the transition angle decreases. Increasing of nominal hardness from 500 to 600 BHN had about twice-stronger effect on the transition angle than increasing from 370 to 500 BHN.

In addition, we observed an optimal thickness at 8 mm HH steel plate. Possibly the optimum thickness for ricochet is affected by projectile strength and velocity.

It is recommended, in future work, to validate experimentally the most interesting and promising cases (e.g. optimum thickness for ricochet of HH plate), additional materials (e.g. Aluminum) may be considered as well.

References

[1] Zukas, Jonas A., 1990. High Velocity Impact Dynamics, John Wiley & Sons. Ltd, New York, U.S.A.

[2] Goldsmith W. 1999. Non-Ideal Projectile Impact on Target, International Journal of Impact Engineering 22, p. 95.

[3] Tate, A. 1979. A Simple Estimate of the Minimum Target Obliquity Required for the Ricochet of a High Speed Long Rod Projectile, Journal of Physics D Applied Physics 12, p. 1825.

[4] Rosenberg, Z., Yeshurun, Y. and Mayseless, M. 1989. "On the ricochet of long-rod projectiles," Proceedings of the 11th International Symposium on Ballistics. Brussels, Belgium. pp. 501-506.

[5] Lee, W., Lee, H. J. and Shin, H. 2002. Ricochet of a Tungsten Heavy Alloy Long-Rod Projectile from Deformable Steel Plates, Journal of Physics D Applied Physics 35, p. 2676.

[6] Rosenberg, Z., Ashuach, Y. and Dekel, E. 2007. More on the Ricochet of Eroding Long Rods - Validating the Analytical Model with 3D Simulations International Journal of Impact Engineering 34, p. 942.

[7] Jo, J. H. and Lee, Y. S. 2012. A Study of Ricochet Phenomenon for Inclined Impact of Projectile, International Journal of Modern Physics. Conference Series 6, p. 443.

[8] Senf, H., Rothenhauser, H., Scharpf, F., Poth, A. and Pfrang, W. 1981. "Experimental and numerical investigation of the ricocheting of projectiles from metallic targets," Proceedings of the 6th International Symposium on Ballistics, Orlando, FL, U.S.A.

[9] Yaziv, D., Mayseless, M. and Reifen, Y. 2001. "The penetration process of long rods into thin metallic targets at high obliquity," Proceedings of the 19th International Symposium on Ballistics. Interlaken, Switzerland. pp. 1249-1255.

[10] Abdul Majid, M. S., Birch, R. S. and Jais, M. I. 2006. "The effects of varying target hardness on ricochet angle," Proceedings of AIP Conference, Volume 909. pp. 85-89.

[11] Koene, L. 2016. "Modelling Bullet Ricochet from Wooden Targets," Proceedings of the 29th International Symposium on Ballistics. Edinburgh, Scotland, UK. pp. 1952-1963.

[12] Seidl, M., Wolf, T. and Nuesing, R. 2017. "Numerical Investigations on Ricochet of a Spin-stabilized Projectile on Differently Shaped Target Surfaces," Proceedings of the 30th International Symposium on Ballistics. Long Beach, CA. pp. 2299-2311.

Proceedings of the 2019 Hypervelocity Impact Symposium
HVIS2019
April 14-19, 2019, Destin, FL, USA
HVIS2019-093

Effect of Liquid Parameters on Protective Performance of a Liquid Composite Target Subjected to Jet Impact

Tan Yaping, Jia Xin*, Huang Zhengxiang, Cai Youer, Zu Xudong

School of Mechanical Engineering, Nanjing University of Science & Technology,Nanjing 210094, P.R. China

Abstract

In order to study the influence of liquid parameters on the protective performance of liquid composite targets (LCT), based on the theory of interaction between the jet and the LCT, three dimensionless numbers - C, G, and V - are obtained by dimensional analysis in this paper. These 3 dimensionless parameters represent the compressibility, inertia, and viscosity of the liquid, respectively. The empirical formula, $P/H = 0.346C^{1.251}G^{-0.712}V^{0.036}$, was obtained by fitting experimental data of the static depth of penetration (DOP) experiment which can predict the residual depth of penetration (RDOP) of the jet penetrating the LCT. It turns out that the 2 dimensionless parameters - C and G - which characterize the compressibility and inertia of the liquid, plays a decisive role in the protection performance of the LCT, while the influence of liquid viscosity is small. In addition, according to the research results of this paper, the protective performance of the LCT can be improved by selecting a liquid with high sound velocity, high viscosity, and low density.

Keywords: Liquid composite target(LCT); Density of liquid; Sound velocity of liquid; Viscosity of liquid; Dimensionless numbers

1. Introduction

In past decades, scholars have done a lot of experimental, theoretical and simulation studies on the process of jets penetrating liquid. Held et al. [1][2][3] studied the penetration process of jets formed from copper, aluminum, and tantalum in water using experimental methods. On the basis of the Szendrei equation, a new equation of the crater growing process caused by a shaped charge jet penetrating into the water was proposed, which is called the Szendrei-Held equation. The excellent anti-jet penetration ability of liquid-filled confined structures (LFCS) was verified by White et al. by experiments [4]. In the X-Ray experiment and numerical simulation study of Anderson et al. [5] it was found that the tail segment of the jet would break because of the interference of the liquid in the process of the jet penetrating the LFCS. The simulation results of Wang, Haifu, et al. [6] showed that the penetration performance of a jetting penetrator in water is significantly affected by the stand-off (S/O). With the same mass ratio of charge to liner, Zhang X et al. [7] studied the penetration properties of a jet formed by collapse of the liner with different materials in water through experiments and simulations. The results showed that the jet formed by a W-Cu alloy has stronger penetration capability. A theoretical model of interaction between a diesel-filled confined structure and a jet was presented by Zhen-Yu Gao et al. [8] for the first time, and which was slightly modified by Zhang Xian et al. [9] by considering the propagation direction of a shock wave in the liquid. Zu XD et al. [10] studied the effect of liquid molecular weight on the protective performance of the LCT through experiments. Residual penetration tests on a multicellular LCT with different kinds of liquids were carried out by Tan Y P et al. [11]. It turned out that both liquid density and viscosity affect the protective performance of the LCT, while liquid viscosity has no significant effect on the protective performance of the LCT.

* Corresponding author. Tel.: +86 13913858955; fax: +86 025-84315649.
E-mail address:jiaxinnjust@hotmail.com

The above research shows the excellent anti-jet penetration ability of the LCT, and reveal the interference mechanism and influencing factors of the liquid on the jet penetration ability in the LCT. However, the existing theoretical models are complex and difficult to calculate, and the analysis of liquid parameters is limited to a qualitative description without quantitative analysis, which limits their application.

In the present paper, 16 control parameters of a jet penetrating the LCT are extracted based on the virtual origin method and interaction mechanism between the jet and LCT. A dimensionless formula on penetration depth is proposed using 3 dimensionless parameters by dimensional analysis. The 3 dimensionless parameters are C, G, and V, which characterize the compressibility, inertia, and viscosity of the liquid. Not only can the existing theory be verified by this empirical formula, but also the RDOP of a jet penetrating the LCT can be predicted easily, quickly and accurately. In addition, this paper provides reliable suggestions for the selection of liquid parameters in the LCT. Protective performance of the LCT can be improved by selecting a liquid with high sound velocity, high viscosity, and low density.

2. Theory of Jet penetrating the confined fluid

Based on the theoretical model proposed by Zhen-Yu Gao et al. [8] and Zhang Xian et al. [9], the basic physical quantities and their dimensions and units involved in the process of jet penetration into a LCT are shown in Table 1.

Table 1 Physical quantities of jet penetration into a LCT

Macro object	Physical name	Symbol	Dimension [ ]	Unit { }
Jet	Tip velocity of jet	v_j	LT^{-1}	m/s
	Tail velocity of jet	v_s	LT^{-1}	m/s
	Tip diameter of jet	d_j	L	m
	Tail diameter of jet	d_s	L	m
	Density of jet	ρ_j	ML^{-3}	kg/m^3
	Sound velocity of jet	c_j	LT^{-1}	m/s
Confined structure	Height of cavity	h_t	L	m
	Sound velocity of shell	c_t	LT^{-1}	m/s
	Density of shell	ρ_t	ML^{-3}	kg/m^3
	The internal diameter of cavity	d_t	L	m
Liquid	Density of liquid (DL)	ρ_l	ML^{-3}	kg/m^3
	Sound velocity of liquid (SVL)	c_l	LT^{-1}	m/s
	Viscosity of liquid (VL)	μ_l	$ML^{-1}T^{-1}$	Pa • s
Others	Position coordinates of virtual origin	z_0	L	m
	Time coordinates of virtual origin	t_0	T	s
	S/O	H	L	m

The target physical quantity RDOP is determined by equation (1)

$$P = f\left(v_j, v_s, d_j, d_s, \rho_j, c_j, h_t, c_t, \rho_t, d_t, \rho_l, c_l, \mu_l, z_0, t_0, H\right) \tag{1}$$

Taking ρ_l 、 c_l and H as basic quantities, a dimensionless analysis based on Eq. (1) leads to

$$\frac{P}{H} = f\left(\frac{v_j}{c_l}, \frac{v_s}{c_l}, \frac{d_j}{H}, \frac{d_s}{H}, \frac{\rho_j}{\rho_l}, \frac{c_j}{c_l}, \frac{h_t}{H}, \frac{c_t}{c_l}, \frac{\rho_t}{\rho_l}, \frac{d_t}{H}, 1, 1, \frac{\mu_l}{\rho_l c_l H}, \frac{z_0}{H}, \frac{t_0 c_l}{H}, 1\right) \tag{2}$$

Eq. (2) can be transformed to

$$\frac{P}{H} = f\left(\frac{v_j}{c_l}, \frac{v_s}{v_j}, \frac{d_j}{H}, \frac{d_s}{H}, \frac{\rho_j}{\rho_l}, \frac{c_j}{v_j}, \frac{h_t}{H}, \frac{c_t}{v_j}, \frac{\rho_t}{\rho_j}, \frac{d_t}{H}, \frac{\mu_l}{\rho_j v_j H}, \frac{z_0}{H}, \frac{t_0 v_j}{H}\right) \tag{3}$$

When only the liquid type of the LCT is variable in the experiment, then $\pi_1=\frac{v_s}{v_j}$, $\pi_2=\frac{d_j}{H}$, $\pi_3=\frac{d_s}{H}$, $\pi_4=\frac{c_j}{v_j}$, $\pi_5=\frac{h_t}{H}$, $\pi_6=\frac{c_t}{v_j}$, $\pi_7=\frac{\rho_t}{\rho_j}$, $\pi_8=\frac{d_t}{H}$, $\pi_9=\frac{z_0}{H}$ and $\pi_{10}=\frac{t_0 v_j}{H}$ all are constant, and Eq. (3) can be further simplified to

$$\frac{P}{H}=f(\mathrm{C},\mathrm{G},\mathrm{V}) \tag{4}$$

where $\mathrm{C}=\frac{v_j}{c_l}$, $\mathrm{G}=\frac{\rho_j}{\rho_l}$, $\mathrm{V}=\frac{\rho_j v_j H}{\mu_l}$, characterize the compressibility, inertia, and viscosity of the liquid, respectively.

For the same shaped charge jet, the virtual origin coordinates and jet parameters all are equal, and the length of the jet when it strikes the target is only related to the S/O. Therefore, the left side of Eq. (4) essentially represents the normalized RDOP of the jet. The 3 dimensionless parameters representing the compressibility, inertia, and viscosity of the liquid are independent of each other so the normalized RDOP function can be written as an empirical power law formula as follows:

$$\frac{P}{H}=k\cdot\mathrm{C}^{\alpha}\mathrm{G}^{\beta}\mathrm{V}^{\gamma} \tag{5}$$

3. Experiment and Results Analysis

In this paper, the exponents of Eq.5 are obtained by fitting the experimental results of measuring the sound velocity, density, and viscosity of liquids and RDOP.

3.1. Liquid and rheological experiments

Five Newtonian fluids - water, diesel, PEG200, PEG400, and PPG6000 - were stdied in the experiment. The test method and conditions are the same as those of Newtonian fluids in reference [11].

3.2. LCT and Residual Penetration Experiments

The LCT specimen which was used the experiment is shown in Fig. 1, and the site layout is shown in Fig. 2 which is the same as those in reference [11].

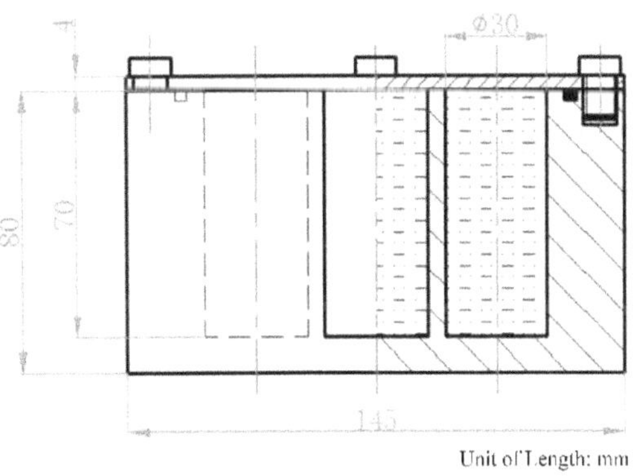

Fig.1 Structure sketch of LCT

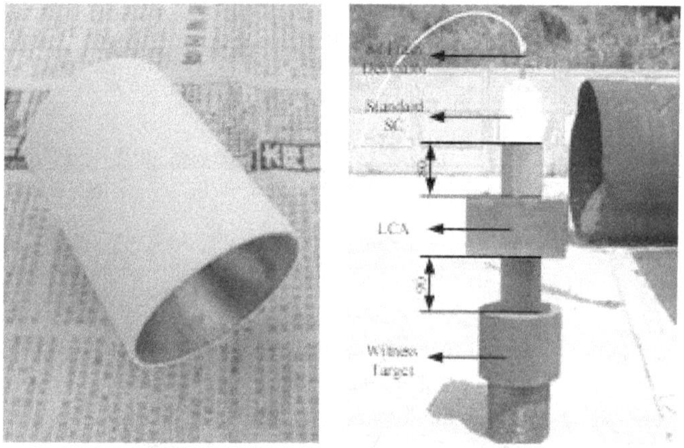

(a) (b)

Fig.2 Testing stand: (a) – Φ56 mm standard SC; (b) – DOP experiment setup

The same Φ56mm shaped charge as in reference [8] is used in the study. And the virtual origin coordinates of the jet formed by the Φ56mm shaped charge are zo=28.85mm, to=8.8μs. The tip velocity, tip diameter, tail velocity and tail diameter of the jet are 6.51mm/μs, 2mm, 1.189mm/μs and 14mm, respectively. The S/O, H is 80mm, and the distance between the LCT and the witness target is 90mm.

3.3. Results and Analysis

3.3.1 Fitting of Empirical Formula Index

Sound velocity, density, and viscosity of the five liquids and the DOP experiment results are shown in Table 2.

Tab. 2 Sound velocity, density, and viscosity of liquid and results of RDOP experiments

Liquid / Parameters	PEG200	PPG6000	Diesel	Water	PPG400
ρ_l (g/cm^3)	0.98	0.98	0.85	1.03	1.19
c_l (km/s)	1.597	1.278	1.377	1.547	1.519
μ_l (Pa·s)	4.77e-2	1.13	5e-3	8.937e-4	13
P (mm)	64	75.5	75	79.5	68

According to the data of the first four groups, Eq. (6) can be fit as follows:

$$\frac{P}{H} = 0.346\mathrm{C}^{1.251}\mathrm{G}^{-0.712}\mathrm{V}^{0.036} \tag{6}$$

The predicted RDOP calculated by substituting the data of the fifth group, Diesel, into Eq. (6) is 71.06mm. It is shown that the empirical formula can accurately predict the RDOP of the jet which penetrated the LCT.

3.3.2 Discussion on 3 Influencing Factors

From reference [8], it is stated that, when the jet starts penetrating the liquid of the LCT, the fluid will move towards the wall of the cavity first, then when the high-speed shock wave reflects from the wall, the fluid will move towards the axis of the jet. This reverse flow of liquid is called 'reflux'. The reflux liquid interacts with the jet, and this transverse cutting effect of the liquid on the jet leads to a decrease in penetration ability of the jet.

Taking water as an example, the RDOP can be calculated by Formula (7) without considering the reflux of water – i.e. no radial reflux occurs in the liquid- proposed by Xiao, Qiang-Qiang [12]:

$$\frac{P}{P_0}=\left[1-\left(\delta_r^*\right)^N\right]\left(1-\xi e^{-\eta\frac{1}{g^*}}\right) \tag{7}$$

where P_0 is the DOP of the continuous jet, δ^*_r is a dimensionless radial drift distance of jet particles, g^* is a dimensionless spacing between fractured jet particles, and N, ξ, η are empirical constants which are 1.5, 0.2 and 2.

With the setup illustrated in Fig 2, considering the liquid as solid-like material, without considering the reflux effect, and the target material is 4mm steel +70mm water (no reflux) +10mm steel, the RDOP is P=185mm, obtained by substituting experimental parameters into Eq. (7). However, the experimental result is 79mm. It fully illustrates the influence of liquid reflux on the subsequent penetration ability of the jet.

Exponents of the 3 dimensionless parameters in Eq. (6) show that the sound velocity and density of liquid play a significant role in the residual penetration capacity of the jet, while the viscosity has a relatively small influence. It is noteworthy that the RDOP of the jet decreases with increasing sound velocity and viscosity of the liquid, and increases with increasing liquid density. In order to further analyze the influence of liquid parameters on the subsequent penetration ability of the jet, the compressibility, inertia, and viscosity of the liquid were analyzed on the basis of the empirical formula, Eq. (6).

Fig. 3 shows the RDOP calculated by Eq. (6) when the density and viscosity of the liquid are 1000kg/m^3 and 8×10^{-4} Pa·s, respectively, while the sound velocity of the liquid ranges from 800m/s to 1800m/s. It turns out that the RDOP decreases with increasing sound velocity of the liquid; that is, the more compressible the liquid, the greater the RDOP of the jet. Meanwhile, it can be seen from Fig. 3(a) that the change rate of RDOP decreases with increasing SVL.

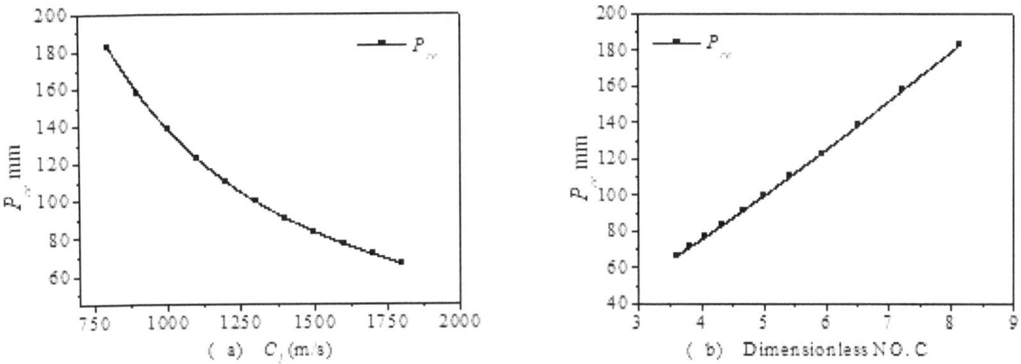

Fig.3 Effect of SVL on RDOP

Fig. 4 shows the range and interval length of the interference velocity calculated by reference [8] when the density and viscosity of the liquid are 1000kg/m^3 and 8×10^{-4} Pa·s, respectively, and the sound velocity of the liquid ranges from 800m/s to 1800m/s. It turns out that with increase of the SVL, the velocity range of the disturbed jet moves to the tip of the jet, and the length of the interval increases gradually. According to the theory of jet penetration, the penetration ability of the jet is proportional to the length of jet, and the penetration ability of jet in the high-speed section is stronger. Therefore, with the increase of length of the interval of the interference velocity and movement of the interference velocity to the high-speed section, the subsequent penetration ability of jet will be reduced and the RDOP will be eventually decreased. In addition, as shown in Fig. 4(b), with the increase of the SVL, the change of the length of the interference velocity interval becomes less, which will also lead to a gradual change in the RDOP. The rule obtained by the empirical formula is in good agreement with that obtained by reference [8].

Fig. 5 shows the RDOP calculated by Eq. (6) when the SVL and VL are 1547m/s and 8×10-4 Pa·s, respectively, while the DL ranges from 700kg/m^3 to 2000kg/m^3. It turns out that the RDOP increases with increasing DL, that is, the stronger the inertia of liquid, the greater the RDOP of the jet. This is contrary to the density rule in classical penetration theory. When the jet penetrates the LCT, the fundamental reason for the decrease of the jet penetration ability is not the erosion consumption caused by the density of the target, but the fracture of the jet caused by the lateral interference of the liquid reflux in the closed structure. With the increase of the DL, the radial velocity of the liquid under impact decreases which results in a delay of the backflow time, so the interference position of the liquid moves to the tail of the jet. It may even occur that the liquid returns to the axis of penetration after the jet completely penetrates the LCT. Meanwhile, it can be seen that the change rate of the RDOP is not obvious with the increase of liquid density from Fig. 5(a).

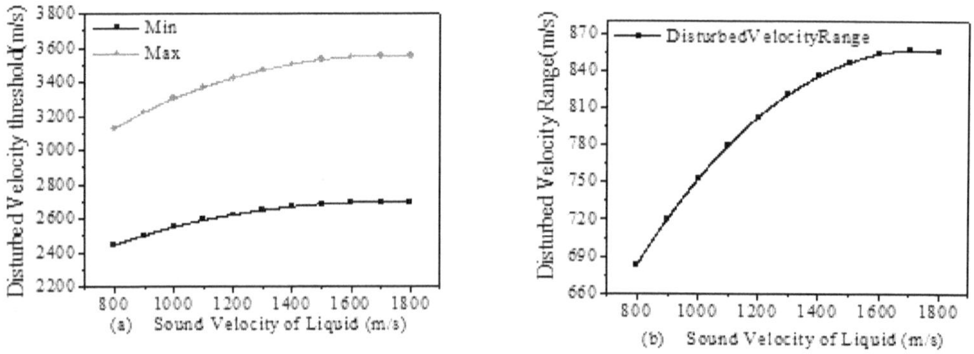

Fig. 4 Effect of SVL on Interference Velocity

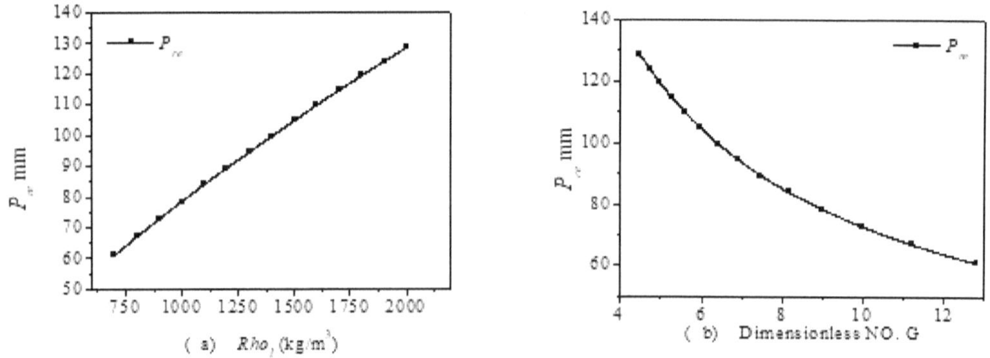

Fig. 5 Effect of DL on RDOP

Fig. 6 shows the range and interval length of the interference velocity calculated by reference [8] when the SVL and VL are 1547m/s and 8×10-4 Pa·s, respectively, while the DL ranges from 700kg/m3 to 2000kg/m3. It turns out that with the increase of liquid density, the velocity range of the disturbed jet moves to the tail of the jet, and the length of the interval decreases gradually. Therefore, with the decrease of length of the interference velocity interval and the movement of the interference velocity interval to the low-velocity section, the subsequent penetration capacity of the jet is enhanced, and the RDOP will eventually increase, which is completely consistent with the conclusion of the dimensional analysis. In addition, as shown in Fig. 6(b), the length of the interference velocity interval increases with increasing DL. However, it should be noted that the disturbed jet is closer to the low-speed tail section for these circumstances, so the change rate of the RDOP does not necessarily increase.

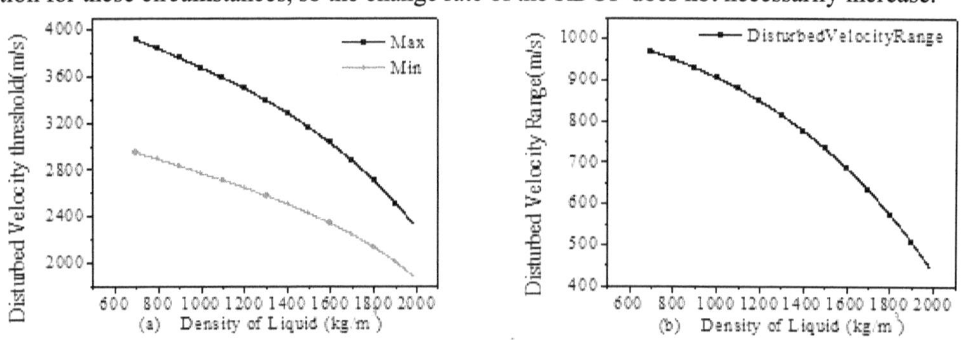

Fig. 6 The influence of DL on the Interference Velocity

Fig. 7 shows the RDOP calculated by Eq. (6) when the SVL and DL are 1547m/s and 1000kg/m3, respectively, while the VL ranges from 0 Pa·s to 15 Pa·s. It turns out that the RDOP decreases with increasing VL, that is, the more viscous the liquid is, the smaller the RDOP of the jet. This is due to the existence of no-slip conditions so that a thin layer of liquid will adhere to the outer surface of the jet after the liquid reflux, which will continue to interfere with the jet, eventually resulting in its fracture, thus leading to the results shown in Fig. 7.

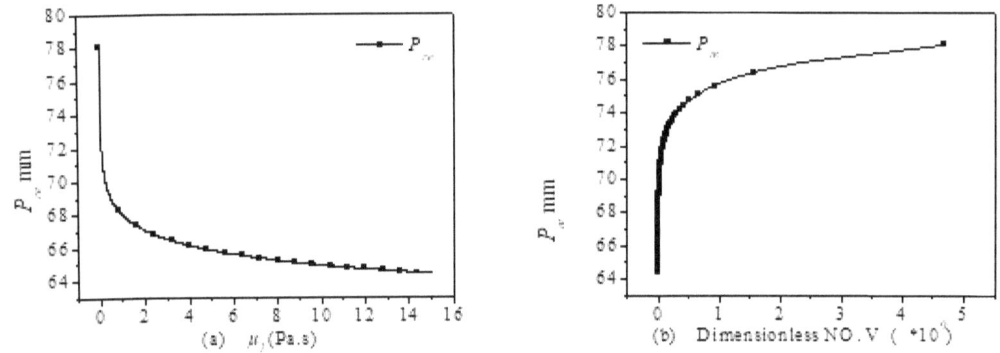

Fig. 7 Effect of VL on RDOP

According to analysis of the 3 factors, the protective performance of liquid composite armor can be improved by using a liquid with high sound velocity, low density, and high viscosity.

4. Conclusions

(1) The interaction mechanism between a jet and LCT, the penetration process of the same shaped charge, with the same S/O, LCT structure and other layouts for different types of liquids, are studied by dimensional analysis. Three dimensionless parameters - C, G and V - are obtained to characterize the compressibility, inertia, and viscosity of the liquid, respectively. An empirical formula, $P/H = 0.346C^{1.251}G^{-0.712}V^{0.036}$, for predicting the residual penetration capability of a jet is obtained through experimental fitting and verification. The formula shows that the key factor determining the residual penetration ability of the jet is the sound velocity and density of the liquid, while the effect of viscosity is small. The higher the sound velocity and viscosity of the liquid, the better the protection performance of the LCT, while the higher the liquid density, the worse the protection performance of the LCT.

(2) Based on the three parameters of sound velocity, density and viscosity of water, this paper quantitatively analyzes the influence of these three factors through the empirical formula calculation and verifies the relevant conclusions in references [8] and [11]. At the same time, it is further revealed that the radial reflux of the liquid is the fundamental reason for the decrease in the penetration ability of the jet when it penetrates the LCT.

(3) The research results of this paper point out the direction for designers of liquid composite armor when choosing a liquid. In order to obtain a liquid composite armor with good protection performance, a liquid with high sound speed, low density, and high viscosity should be chosen.

Acknowledgments

This paper is supported by the National Natural Science Foundation of China (Grant No. 11602110) and the Postgraduate Research & Practice Innovation Program of Jiangsu Province (Grant No. KYCX18_0456).

References

[1] Held M, Backofen J E. Penetration of shaped charge into water[C]. Proceedings of the 12th international symposium on ballistics, San Antonio, TX, 30 Oct.-1 Nov 1990, 1.

[2] Held M. Verification of the equation for radial crater growth by shaped charge jet penetration [J]. International journal of impact engineering, 1995, 17(1): 387 398.

[3] Held M, Huang N S, Jiang D, et al. Determination of the crater radius as a function of time of a shaped charge jet that penetrates water [J]. Propellants, explosives, pyrotechnics, 1996, 21(2): 64-69.

[4] White J J, Wahll M J. Shaped charge jet interactions with liquids[C]. Proceedings of the 6th international symposium on ballistics, ADPA Orlando, Florida, 1981: 554-561.
[5] Andersson G, Karlsson S, Watterstam A. Shaped charge jet interaction with confined water[C]. Proceedings of the 12th international symposium on ballistics, 1998: 183-190.
[6] WANG Hai-fu, JIANG Zeng-rong, YU Wei-min, LIU Zong-Wei. Behavior of Jetting Penetrator Charge Operating Underwater [J]. Journal of Beijing University of Technology, 2006, 26(3): 189-192.
[7] Zhang X, Wu C, Huang F. Penetration of shaped charge jets with tungsten–copper and copper liners at the same explosive-to-liner mass ratio into water [J]. Shock Waves, 2010, 20(3): 263-267.
[8] Zhen-Yu Gao, Zheng-Xiang Huang, et al. Theoretical Study of a Diesel-Filled Airtight Structure Unit Subjected to Shaped Charge Jet Impact. Propellants, Explosives, Pyrotechnics. 2016.
[9] Zhang Xian, Zu Xudong, et al. Analysis of Liquid-filled unite cell structure subjected to shaped charge jet impact [J]. Explosion and Shock Waves, 2017, 37(6):1101-1106.
[10] Zu X D, Huang Z X, Xiao Q Q, et al. Study on Cell Structure Liquid Composite Armour Subjected to Shaped Charge Jet Impact[C]. Proceedings of the 28th International Symposium on Ballistics, Atlanta, Georgia, USA, September 22-26, 2014: 1325-1332
[11] Tan Y P, Huang Z X, et al. Experiment study on anti-penetration performance of liquid composite armor to shaped-charged jet [C]. Proceedings of the 30th International Symposium on Ballistics, Long Beach, CA, United States, USA, September 2-2, 2014: 1968-1976
[12] Xiao, Qiang-Qiang, Huang, Zheng-Xiang, Zu, Xu-Dong, et al. Influence of Drift Velocity and Distance Between Jet Particles on the Penetration Depth of Shaped Charges[J]. Propellants Explosives Pyrotechnics, 2016, 41(1):76-83.

Proceedings of the 2019 Hypervelocity Impact Symposium
HVIS2019
April 14-19, 2019, Destin, FL, USA
HVIS2019-094

Effects of Additional Body on Jet Velocity of Hyper-cumulation

XU Mengwen[a], JIA Xin[a]*, HUANG Zhengxiang[a]

[a]*School of Mechanical Engineering, Nanjing University of Science and Technology, Nanjing 210094, Jiangsu, China*

Abstract

In this paper, with different cone angles of the liner, the effects of the additional body's material, thickness, and diameter on the velocity are studied by AUTODYN-2D. The results show as follows: 1. The additional body can increase the top velocity of the jet but has little effect on the tail velocity. 2. When the cone angle is larger, the velocity increases more, about 36%, and the velocity gradient is greater at the head of the jet. 3. There is little influence of material yield strength, but there is a large correlation with density. With the increase of material density, the velocity and velocity gradient increase. 4. The larger the density of the additional body, the smaller the optimum thickness. When the thickness is not less than the optimum thickness, an increase in the diameter of the additional body can improve the velocity and stability of the jet. Consequently, this kind of hyper-cumulation structure with a flat plate additional body is more suitable to improve the classical liner with a large cone angle. By modifying the design of the shaped charge in this way, the velocity of the jet can be improved effectively, enhancing the penetration power compared with the classical one.

Keywords: hyper-cumulation; additional body; velocity; numerical simulation.

Nomenclature

$2r$	the truncated diameter of the liner (mm)
V	velocity (m/s)
Greek symbols	
δ	main line thickness (mm)
2α	cone angle of line (°)
Subscripts	
j-tip	head of the jet
j-tail	end of the jet

1. Introduction

The shaped-charge jet is widely used in military and civil fields due to its excellent penetration ability. Nowadays, a variety of new armors are used as additional armors, for which made the classical jet cannot meet the penetration requirements. In order to enhance the penetration ability of the jet, a structure called the hyper-cumulation shaped charge was proposed by a Russian scholar named Minin et al. [1-2] Compared with the traditional jet, they defined that the hyper-cumulation jet's diameter is larger than that of the clubber. S. V. Fedorov et al. [3-5] developed combined shaped-charge liners in the form of a truncated ellipsoid of variable thickness, allowing the formation of the compact elements' velocity to have a large improvement. Another simple but widely studied hyper-cumulation structure is obtained by adding an additional circular body with a slightly larger diameter on the cutoff of the truncated liner, as shown in Fig.1. Chinese scholar Wang Cheng et al. [6] studied the hyper-cumulation structure with an additional body made of a titanium alloy. They concluded that hyper-cumulation could obviously improve the overall

* Corresponding author. Tel.: 13913858955.
E-mail address: jiaxin@mail.just.edu.cn.

velocity of the jet and quality utilization of the line. Li Qingxin [7] summarized that a high-speed jet in good shape can be obtained using copper, tantalum, and tungsten as additional bodies.

There is no doubt that the combination of the truncated liner and the additional body can affect the formation of the jet. Therefore, it is necessary to study which parameters of the additional body plays the most important role. In this paper, a shaped charge with the combination of a flat plate additional body and a truncated liner is studied. By varying the cone angle of the liner, the material, the diameter, and the thickness of the additional body, numerical simulations of jet formation are carried out by AUTODYN-2D. The factors and results are discussed about the formation of the jet.

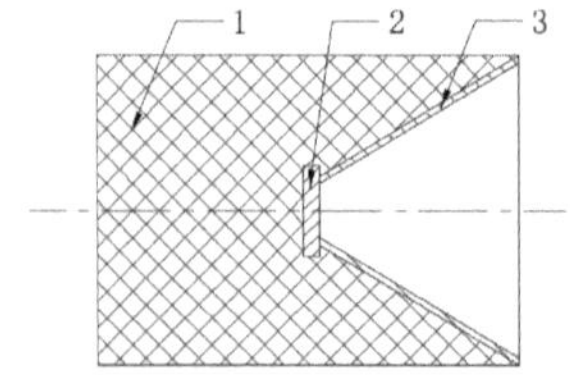

1-explosive; 2-additional body; 3-liner

Fig. 1. The structure of the hyper-cumulation.

2. Numerical simulation model

2.1. Computational model

The hyper-cumulation model presented in this paper is axisymmetric, so the simulation software AUTODYN-2D was used. The explosive detonation and the collapse of the liner are large deformation movements. The Euler method takes the flow field as the description object to study the motion of points in space, which has great advantages in analyzing the large deformation process. Therefore, the high-precision solver Euler-2D Multi-material is used for numerical simulation.

The schematic of hyper-cumulation is shown in Fig.2. The liner's cone angle and the additional body's material, thickness, and diameter were varied, while other structure data remained the same, as shown in Fig.2, with units of "mm".

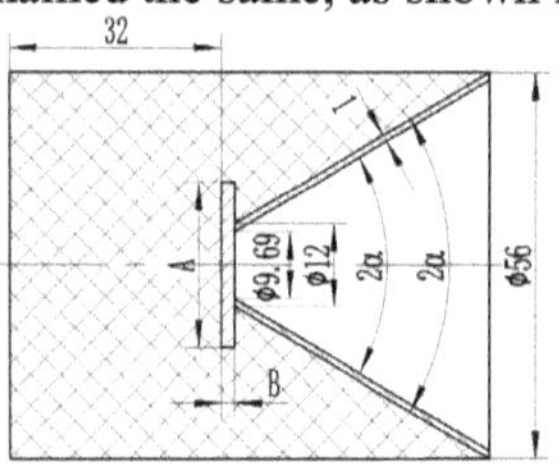

Fig. 2. The structure of the hyper-cumulation with data.

In the hyper-cumulation, the liner material was ALUMINUM, and the additional body material was ALUMINUM, TI6%AL4%V, COPPER and TUNGSTEN ALLOY, respectively. The single cone liner angle is usually between 50 to 70 degrees, so 50°, 60°, and 70° were selected in this paper. If the additional body is thinner than the liner, the collapse velocity of the additional body will be faster than the liner, interfering with jet formation. So the additional body we designed was 2mm or 4mm thick, which was two or four times the thickness of the liner. The diameter of the additional body was 2r or 4r. '2r' was the truncated diameter of the liner, about 12mm. According to the above design, we obtained 48 kinds of structures. The model of the numerical simulation is shown in Fig. 3.

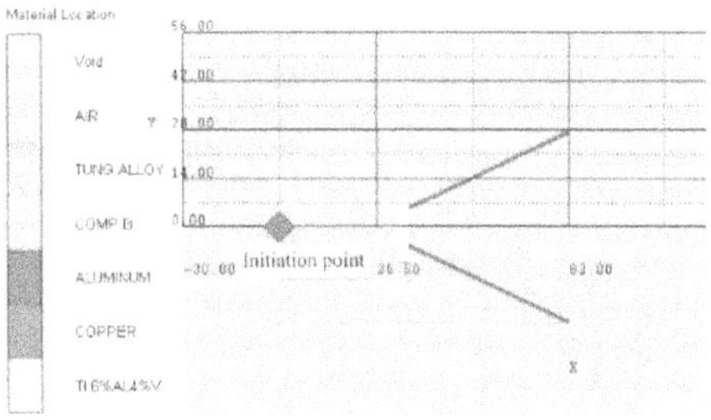

Fig. 3. The model of numerical simulation.

2.2. Material model and parameters

In the numerical simulation, the materials of explosive, air, liner and additional body were involved. All materials were selected from the standard material library of AUTODYN. The explosive was COMP-B, and the equation of state was JWL, which can accurately describe the pressure, volume and energy characteristics of detonation products in the process of explosion driving. For ALUMINUM, the equation of state was Shock, and the strength model was von-Mises. For TI6%AL4%V, the equation of state was Shock, and the strength model was Steinberg Guinan. For COPPER, the equation of state was Shock, and the Piecewise JC strength model was used. For TUNGSTEN ALLOY, the equation of state was Shock, and the strength model was Johnson Cook. Parameters of each material are shown in Tables 1 and 2.

Table 1. Material model and equation of state of COMP-B

Material	$\rho(g/cm^3)$	$A(GPa)$	$B(GPa)$	R_1	R_2	W	$V_{C-J}(km/s)$	$E_{C-J}(10^7\ KJ/m^3)$
COMP-B	1.717	524	7.678	4.2	1.1	0.34	7.98	0.85

Table 2. Material model and parameters for the equation of state of different materials

Material	$\rho(g/cm^3)$	γ	$C_1(10^3\ m/s)$	S_1	$S_2(m/s)$	$E(10^7\ kPa)$	$Y(10^5\ kPa)$
ALUMINUM	2.71	2.1	5.38	1.337	0.00	2.67	2.9
TI6%AL4%V	4.419	1.23	5.13	1.028	0.00	4.19	13.3
COPPER	8.9	2	3.958	1.497	0.00	4.64	1.20
TUNG.ALLOY	17.0	1.54	4.029	1.237	0.00	16.0	15.06

It can be seen in Table 2 that the density of the 4 kinds of materials is approximately a multiple of the liner density. TI6%AL4%V and COPPER have almost the same Young's modulus *E*; TI6%AL4%V and TUNGSTEN ALLOY have a much higher yield strength *Y*.

3. Numerical analysis

3.1. Overall analysis of results

In the simulation, the initiation point is (0,0), shown in Fig.3. The calculation stops when the jet length is stretched to twice the standoff distance. If the material of the additional body is ALUMINUM, the structure is similar to the traditional one. Therefore, the jet formed by the structure of an ALUMINUM additional body can be regarded as a comparative benchmark.

The numerical simulation results of the jet head and tail velocities of 48 structures are shown in Fig. 4. Compared to the benchmark (ALUMINUM, a), the additional body has a great influence on the head velocity of the jet but has little effect on the tail velocity. The Figs.4, a, b, c, d represent different structures of the additional body, shown in Table 3.

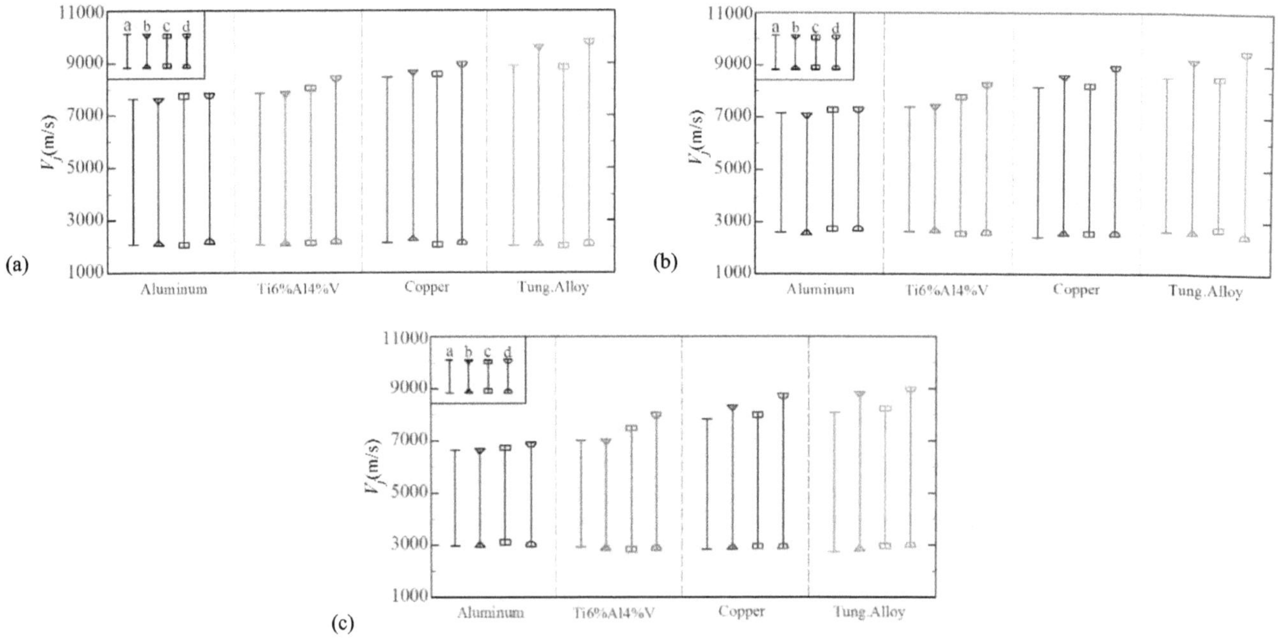

Fig. 4. Jet head and tail velocities when the cone angle is (a) 50 degrees, (b) 60 degrees and (c) 70 degrees.

Table 3. Structures of additional body

Structures	Diameter(mm)	Thickness(mm)
a	12	2
b	24	2
c	12	4
d	24	4

3.2. Effect of the additional body on jet formation of the liner with different cone angles

When the cone angle of the liner is set as 50°, 60°, and 70°, Fig. 4 shows that with the same cone angle, the benchmark(with the material of ALUMINUM and the structure of a), which is equivalent to the traditional one, has the lowest tip velocity. When the material is TUNGSTEN ALLOY and the structure is d, the tip velocity of the jet is the highest. The data is shown in Table 4 according to Fig. 4. With a good design of the additional body, $V_{j\text{-}tip}$ can be increased by about 30%. The last column in Table 4 reveals that the larger the cone angle of the liner, the more $V_{j\text{-}tip}$ can be increased with a good additional body design.

Table 4. Effect of additional body on the increase of $V_{j\text{-}tip}$ with different cone angles

2α (°)	$V_{j\text{-}tip}$		(Max-Min)/Min
	Minimum(m/s)	Maximum(m/s)	
50	7642	9893	0.2945
60	7165	9562	0.3345
70	6628	9056	0.3663

3.3. Effect of material on tip velocity and the shape of the jet

In Fig. 4, comparing these three graphs with different structures and materials for the additional body, the change of V_{j-tip} is the same. Therefore, we chose to analyze the effect of additional body structures and materials at 60° . The V_{j-tip} of the jet is shown in Table 5.

Table 5. The V_{j-tip} (m/s) of the jet with different designs for the additional body

Structures \ Materials	ALUMINUM	TI6%AL4%V	COPPER	TUNG.ALLOY
a	7165	7395	8184	8519
b	7185	7481	8627	9209
c	7316	7875	8291	8584
d	7370	8326	8981	9562

As we know, Young's modulus is a physical quantity describing the deformation resistance of solid materials. But during the process of the liner collapse, we assume that the material is a fluid. So, we can consider that Young's modulus doesn't affect velocity. The yield stress in a fluid is the shear stress which enables the fluid to begin flowing. This reveals when detonation waves can deform the additional body.

The yield stresses of the four kinds of materials are sorted from small to large as COPPER, ALUMINUM, TI6%AL4%V and TUNGSTEN ALLOY. With the increase of yield stress, there is no obvious change of jet velocity under the same structure. However, with the increase of density, the jet velocity increases. Above all, it is speculated that the density is the principle factor affecting jet velocity.

When the size of the additional body is $\phi 24mm \times 4mm$ (structure d), V_{j-tip} is the highest. Fig.5 shows the velocity distribution of the jet at the axis under the structure d. The jet velocity gradient with ALUMINIUM tends to decrease, while with the other three materials, it tends to decrease first and then increase.

From Fig. 5(b), comparing the other three materials' results with the ALUMINUM additional body, when the position coordinates are more than 160mm (where the head location of the jet is), the velocity gradient of the jet increases, and the greater the material density, the greater the velocity gradient. It can be concluded that the additional body only affects the velocity near the head of the jet, and the affected location and length are basically the same.

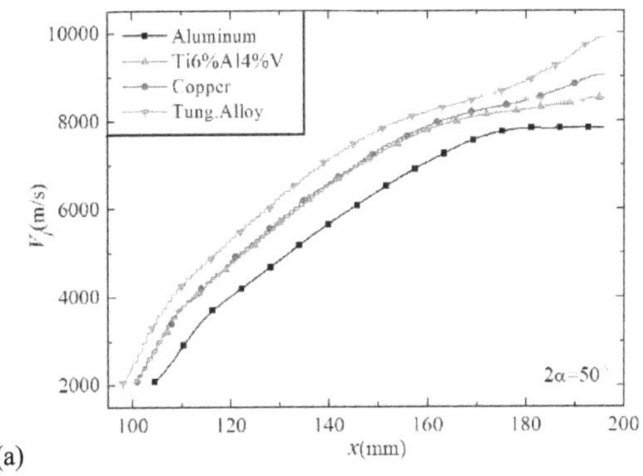

(a)

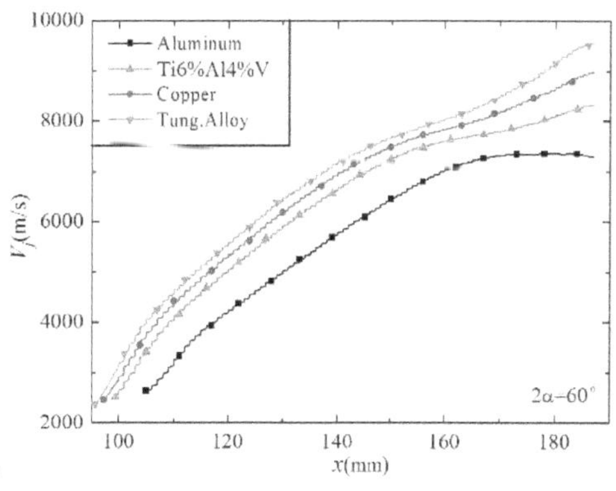

(b)

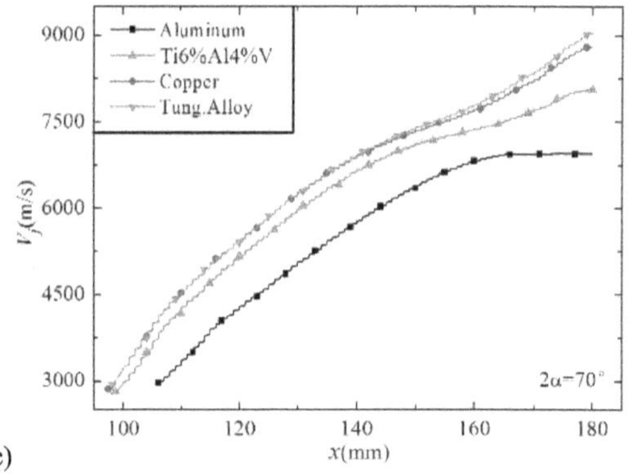

(c)

Fig. 5. The velocity distribution of the jet along the axis.

In addition, with different kinds of materials, the shapes of the jets are different. The results are shown in Table 6, with $2\alpha = 60°$ and the structure d. Case 1 represents the deformation of the liner when the calculation time is 10us. Case 2 represents the calculation stopped when the jet length is stretched to twice the standoff distance. When the material is ALUMINUM, the material accumulates at the head of the jet forming a thick head. With increasing material density, the jet becomes slender. Accordingly, we know that the additional body can change the collapse process near the top of the liner and affect the formation of the jet head.

Table 6. The shapes of the jet with different materials for the additional body in two cases

Cases \ Materials	ALUMINUM	TI6%AL4%V	COPPER	TUNG.ALLOY
1	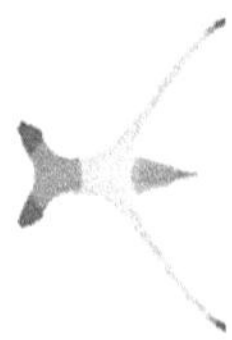			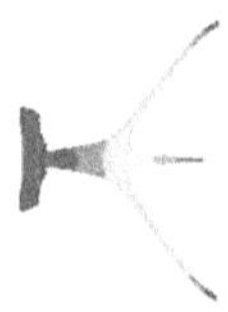
2				

3.4. Effect of the structure size of the additional body

In this paper, we changed the diameter and thickness of the disc-shaped additional body. Increasing the diameter and thickness can improve its resistance to deformation and transfer more detonation energy to the additional body. From Table 5, comparing the results of structures a and c when the materials are ALUMINUM and TI6%AL4%V, V_{j-tip} increases with increasing thickness but changes little when the material is COPPER or TUNG.ALLOY. However, when the diameter increases to 24mm in structures b and d, V_{j-tip} is improved with all 4 materials.

Therefore, a preliminary observation is that increasing the anti-deformation ability of the additional body can improve the V_{j-tip}. Ideally, the additional body is a rigid body. Under the condition of smaller thickness, a small diameter additional body made of denser materials has higher deformation resistance. Once the thickness reaches a certain value, which is related to the density and diameter, further increasing the thickness will not improve deformation resistance. When the diameter of the additional body increases, meaning more detonation wave energy is transferred to the additional body, V_{j-tip} increases. That is, the additional body transfers more energy to the jet. It is also found that the jet head is not easy to fracture when the diameter of the additional body is increased under the same thickness. But the above conclusions need more simulation and experimental data to verify, which will be our next research direction.

4. Conclusions

A series of models of hyper-cumulation were calculated by AUTODYN-2D. The influence of the cone angle of the liner, the material, and the structure of the additional body on the jet were investigated by the control variable method. The conclusions are as follows:

1. The additional body has a great influence on the head velocity of the jet, but has little effect on other characteristics of the jet.
2. The larger the cone angle of the liner, the greater the jet tip velocity can be improved with proper additional body design, by as much as 36%.
3. The increase of jet head velocity with an additional body is independent of material yield strength but has the greatest correlation with density. With the increase of material density, the velocity and the velocity gradient of the jet head also increase. The affected location and length of the jet are basically the same with different materials.
4. The larger the density of the additional body, the smaller the optimum thickness. When the thickness is not less than the optimum thickness, increasing the diameter of the additional body can improve the velocity and stability of the jet head.

Acknowledgements

The authors gratefully acknowledge the support from the National Natural Science Funds for Distinguished Young Scholar of China under the Grant 11602110 and Jiangsu Province Graduate Research and Practice Innovation Program (No.KY CX18_0470).

References

[1] Minin V F , Minin O V , Minin I V . Physics hypercumulation and comdined shaped charges[C]// International Conference on Actual Problems of Electronics Instrument Engineering. IEEE, 2013.
[2] Minin V F , Minin O V, Minin I V. Some Possibilities of Hyper-cumulative Regime of Jet Formations[J]. Applied Mechanics and Materials, 2015, 782:42-48.
[3] Fedorov S V , Bayanova Y M , Ladov S V . Numerical analysis of the effect of the geometric parameters of a combined shaped-charge liner on the mass and velocity of explosively formed compact elements[J]. Combustion Explosion & Shock Waves, 2015, 51(1):130-142.
[4] Fedorov, S. V . Numerical simulation of the formation of shaped-charge jets from hemispherical liners of degressive thickness[J]. Combustion, Explosion, and Shock Waves, 2016, 52(5):600-612.
[5] Fedorov S V , Ladov S V , Nikol'skaya, Ya. M, et al. Formation of a high-velocity particle flow from shaped charges with a liner consisting of a hemisphere and a degressive-thickness cylinder[J]. Combustion, Explosion, and Shock Waves, 2017, 53(4):479-482.
[6] Jun-Lei S , Ying-Bin L , Xiao-Yan H U , et al. Numerical Simulation of Effect of Material of Additional Liner on the Performances of Hypercumulation[J]. Chinese Journal of Explosives & Propellants, 2017. Doi: 10.14077/j.issn.1007-7812.2017.01.014
[7] Qing-Xin LI , Zhi-Jun W , Li C , et al. Simulation Research of a Super Shaped Charge Structure[J]. Journal of Ordnance Equipment Engineering, 2016.

Proceedings of the 2019 Hypervelocity Impact Symposium
HVIS2019
April 14-19, 2019, Destin, FL, USA
HVIS2019-100

Study on jet formation and penetration of double-layer sub-caliber shaped charge

Yu-ting Wang[a], Qiang-qiang Xiao[a] *,Zheng-xiang Huang[a], Xu-dong Zu[a], Bin Ma[a]

[a]*School of Mechanical Engineering, Nanjing University of Science & Technology, 210094, Jiangsu, China*

Abstract

The ordinary shaped charge has a low kill-radius because of its structure. In order to adapt to the different battlefields, this paper proposed the double-layer sub-caliber shaped charge, which is modified from an ordinary shaped charge with insensitive explosive appended to its outer layer. In this paper, a study of the Ø56 mm shaped charge was conducted as a benchmark case. The simulations were carried out using AUTODYN for jet formation and penetration of different axial thicknesses of additional insensitive explosive and the additional height of the charge, with the aim to obtain the optimal size of double-layer sub-caliber shaped charge. The results show that the optimum axial thickness of additional insensitive explosive is $\Delta R/R_0 = 0.8$, and the optimum size of the additional height of charge is $\Delta L/L_0 = 0.4$.

Keywords: double-layer sub-caliber shaped charge; jet; formation; penetration

Nomenclature	
ΔR	Axial thickness of additional insensitive explosive
R_0	Radius of the benchmark shaped charge
ΔL	Additional height of charge
L_0	Height of the benchmark shaped charge

1. Introduction

The ordinary shaped charge has a low kill-radius because of its structure. In order to adapt to the different battlefields, that is, to increase the kill-radius of shaped charge while also satisfying the penetration requirements of the armor-piercing index, double-layer sub-caliber shaped charge can be used. A double-layer sub-caliber shaped charge is a shaped charge with insensitive explosive appended to the outer layer of the shaped charge. Compared with the ordinary shaped charge, the double-layer sub-caliber shaped charge can increase the charge quantity and reduce the sensitivity of ammunition while the armor-piercing index is also satisfied.

Researchers have carried out numerous studies on detonation wave propagation in double-layer charges [1-4]. Zhang X F [5] obtained characteristic parameters of the sandwich shaped charge using AUTODYN, conducted a numerical study on the jet forming of a typical shaped charge with a single structure charge and a sandwich charge, and calculated the parameters of the sandwiched shaped charge jets of different explosive. The results show that, compared with the single structure charge, the jet head speed of the sandwiched shaped charge is increased by 20%. Hussain T [6,7] compared the ballistic performance of Double-layer shaped charges (DLSC) with the ordinary shaped charge (OSC) by using the hydrodynamic software LS-DYNA, which found that OSC may have some advantages over DLSC, especially in smaller lengths. They also found that the shaped charges in which an explosive with higher Chapman-Jouguet pressure is utilized in the inner core, yield more penetration. Wang

* Corresponding author. Tel.: 13851796765.
E-mail address: xiaowawj@njust.edu.cn.

Z[8] performed a theoretical analysis on detonation wave propagation in a double-layer shaped charge (DLSC), and compared between jet formations in DLSC and OSC using AUTODYN. The results show that the improved detonation velocity ratio and radial charge percentage of outer-to-inner layer charges are conducive to the formation of a convergent detonation wave, which contributes to enhancement of jet top velocity in DLSC. The thickness and mass percentages of liner flowing into the jet in DLSC closely follows an exponential distribution along the radial direction, but the percentages in DLSC and the mass of the effective jet, which have a significant influence on the penetration depth, are lower than those in the OSC with the outer layer charge.

Researchers have previously focused on the difference of jet formation and penetration between the ordinary shaped charges and double-layer shaped charges whose calibres remain the same. The study of the double-layer sub-caliber shaped charge has not been conducted. In this paper, we research the jet formation and the penetration of double-layer sub-caliber shaped charges with different axial thickness of additional insensitive explosive and additional height of charges using the finite element code AUTODYN, and obtained the optimal size of a double-layer sub-caliber shaped charge.

2. Numerical Simulation Model

The structure of the double-layer sub-caliber shaped charge is shown in Fig2.1. The multi-material Euler solver AUTODYN was adopted to conduct the simulation of jet formation. The initiation point is located at the center of the end surface of the charge. The initiation mode is indirect to ensure that the detonation of the inner explosive triggers the initiation of the external explosive. The optimum thickness of the additional insensitive explosive was obtained through numerical simulations, with $\Delta R/R_0$ = 0, 0.2, 0.4, 0.6, 0.8 and 1.0 respectively. To gain the optimum size of additional height of charge, we then simulated the working conditions of $\Delta L/L_0$ with values of 0, 0.2, 0.4, 0.6, 0.8 and 1.0 respectively under the circumstances of the optimum axial thickness of additional insensitive explosives.

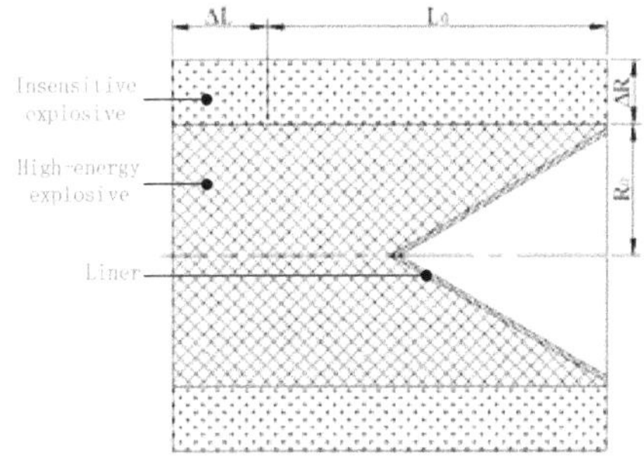

Fig 2.1 Double-layer sub-caliber shaped charge

3. Material Model and Parameters

The high-energy explosive used in this study is 8701, and the insensitive explosive is Composition B. The material model used is high explosive detonation, and the Jones-Wilkins-Lee (JWL) equation of state is used.

$$p = A\left(1-\frac{w}{R_1 V}\right)e^{-R_1 V} + B\left(1-\frac{w}{R_2 V}\right)e^{-R_2 V} + \frac{w}{V}E$$

where p is pressure, V is the relative volume, E is the internal energy, A and B are material parameters. R_1, R_2 and w are constants. Specific material parameters are shown in Table 3.1.

Table 3.1 The calculated parameters of 8701 and Composition B

Material	ρ(g/cm³)	D(m/s)	e_0(KJ/m3)	P_{CJ}(GP)	A(GP)	B(GP)	ω	R_1	R_2
8701	1.717	8350	8.5E+6	29.66	618.4	6.9	0.38	4.3	0.87

Composition B[9]	1.717	7980	8.5E+6	29.5	524.23	7.678	0.34	4.2	1.10

The material of the liner is copper, and the target material is carbon steel C45. The material model is Johnson-Cook and the equation of state is SHOCK.

$$p\begin{cases} p_H+\Gamma\rho\left(e-e_H\right),u\geq 0 \\ \rho_0 c_0 u,u>0 \end{cases}$$

With

$$p_H=\frac{\rho_0 c_0^2 u\left(1+u\right)}{\left[1-\left(\lambda-1\right)u\right]^2},\quad e_H=\frac{1}{2}\frac{p_H u}{\rho_0\left(1+u\right)}$$

where Γ is the Grüneisen coefficient, ρ_0 is the initial density, $\mu=\rho/\rho_0-1$. λ and c_0 are constants, determined by the following shock wave experimental relationship: $D=\lambda v+c_0$, D is the shock wave velocity, v is the velocity of the particle after the wave.

Specific material parameters of Cu and 45 steel are shown in Table 3.2

Table 3.2 The calculated parameters of Copper and 45#

Material	ρ_0 (g/cm^3)	Green coefficient	C1 (m/s)	S_1	T_s (K)	Shear modulus (GPa)	Yield Strength (MPa)	T_m (K)
Cu	8.930	2.02	3940	1.489	383	47.7	120	1790
45 steel	7.83	2.17	4569	1.33	408	76.99997	352	1760

4. The influence of the axial thickness of additional insensitive explosive

4.1. Shaped charge jet forming

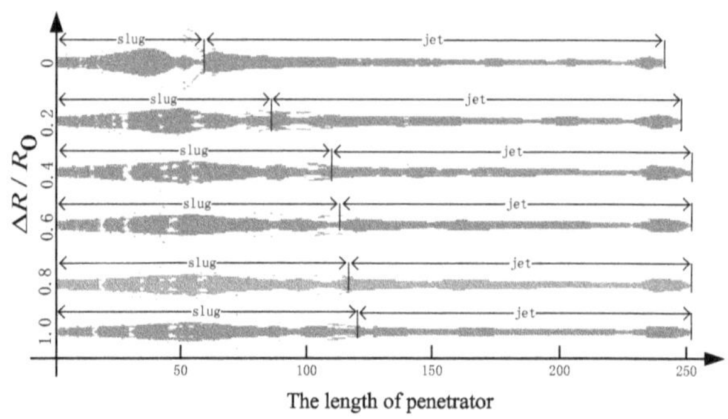

Fig 4.1 The morphology of penetrator at 50us while the axial thickness of additional insensitive explosive varies

Table 4.1 The related data of penetrator while the axial thickness of additional insensitive explosive varies

$\Delta R/R_0$	The jet top velocity (m/s)	The jet tail velocity(m/s)	The jet length (mm)	The maximum energy of Cu(MJ)
0	6074	1273	183.5	0.125
0.2	6183	1972	160	0.167
0.4	6282	2717	140	0.190
0.6	6275	2998	132	0.197
0.8	6283	3022	134	0.198
1.0	6278	3023	130	0.199

Fig. 4.1 shows the penetrator morphology at 50us when the axial thickness of the additional insensitive explosive varies. The penetrator is composed of jet and slug, where the last collapse unit of the liner was regarded as the boundary of jet and slug, as shown at Fig4.1. It can be seen that with the increase of the axial thickness of the additional insensitive explosive the length of the jet decreases gradually, the length of slug increases, and the maximum diameter of the slug decreases. We can see that the slug of the double-layer sub-caliber shaped charge has tensile capacity, but the back segment of slug shows serious breakage.

Figure 4.2 was drawn by using the related data of the jet while the axial thickness of additional insensitive explosive was varied. As shown in Fig 4.2, the jet tip velocity increases with the axial thickness of additional insensitive explosive until $\Delta R/R_0 = 0.4$. The jet tail velocity and the maximum energy of copper increase with the axial thickness of additional insensitive explosive until $\Delta R/R_0 = 0.6$. Initially the length of jet increases, when the axial thickness of additional insensitive explosive increase between $\Delta R/R_0 = 0$ and $\Delta R/R_0 = 0.6$, then it is almost a constant value.

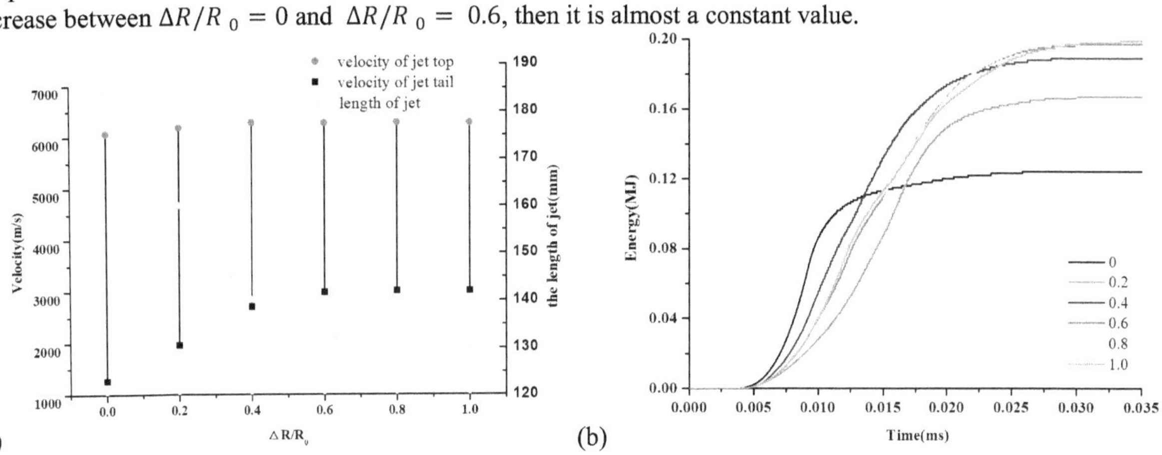

(a) (b)

Figure 4.2 (a) The parameters of jet (b) The energy of the copper

4.2. Penetration

In order to analyze the penetration ability of the penetrator when the axial thickness of additional insensitive explosive varies, numerical simulations were conducted at stand-off = 80 mm, 160 mm and 280 mm. It is well known that the slug of an ordinary shaped charge jet has no benefit to penetration because of its low velocity and bulky shape. But the slug of a double-layer sub-caliber shaped charge jet can increase the penetration depth because of the increase of the tail velocity of the jet and the velocity gradient of the front part of the slug, which shows that slug has certain stretching ability and penetration ability. The penetration ability of the jet decreases due to the decrease in the length of the jet. The penetration data is shown in Table 4.2.

Table 4.2 Data of Penetration

$\Delta R/R_0$	Penetration depth by penetrator when stand-off=80mm	Penetration depth by penetrator when stand-off=160mm	Penetration depth by jet when stand-off=160mm	Penetration depth by penetrator when stand-off=280mm	Penetration depth by jet when stand-off=280mm
0	232.1mm	251.1mm	242.1mm	225.7mm	218.1mm
0.2	222.3mm	282.9mm	262.1mm	289.4mm	237.1mm
0.4	216.9mm	264.1mm	223.1mm	316.7mm	239mm
0.6	192.9mm	271.7mm	230mm	324.9mm	239.1mm
0.8	200.1mm	270.9mm	218.1mm	342.8mm	255.2mm
1.0	204.8mm	309.1mm	235.1mm	299.7mm	228.1mm

The total penetration depth of the penetrator while axial thickness of the additional insensitive explosive varies is shown in Figure 4.3(a). It can be seen from the figure that the penetration depth of the double-layer sub-caliber shaped charge is smaller than the penetration depth of the ordinary shaped charge at stand-off = 80mm, and reaches minimum value at $\Delta R/R_0 = 0.6$.

The decrease in penetration depth can be attributed to the reduction in the length of jet and the slug not being fully stretched, therefore the slug cannot reach the bottom of the penetration hole so the slug has no contribution to the penetration depth.

When stand-off = 160mm, the penetration depth of double-layer sub-caliber shaped charge is larger than the penetration depth of ordinary shaped charge, and it reaches the extreme value at $\Delta R/R_0 = 0.2$, and maximum value at $\Delta R/R_0 = 1.0$. The penetration depth of the double-layer sub-caliber shaped charge remains basically constant between $\Delta R/R_0 = 0.4$ and $\Delta R/R_0 = 0.8$, and it is higher than the penetration depth of ordinary shaped charge.

While stand-off = 280mm, the penetration depth of double-layer sub-caliber shaped charge increases when $\Delta R/R_0$ increases, and reaches maximum value at $\Delta R/R_0 = 0.8$. The penetration depth increases most sharply at $\Delta R/R_0 = 0.2$.

To find out the difference between the penetration process of the double-layer sub-caliber shaped charge and the ordinary shaped charge, the penetration depth of the penetrator and the penetration depth of the jet were both considered, as shown in Fig. 4.3(b). It can be seen from the figure that the trend of the penetration depth of the jet is consistent with that of the penetration depth of the penetrator. The difference between the penetration depth of penetrator and the penetration depth of jet is the penetration ability of slug. The penetration ability of slug increases with $\Delta R/R_0$, until $\Delta R/R_0 = 0.6$. The variation tendency of the jet tail velocity is the same as the penetration ability of slug.

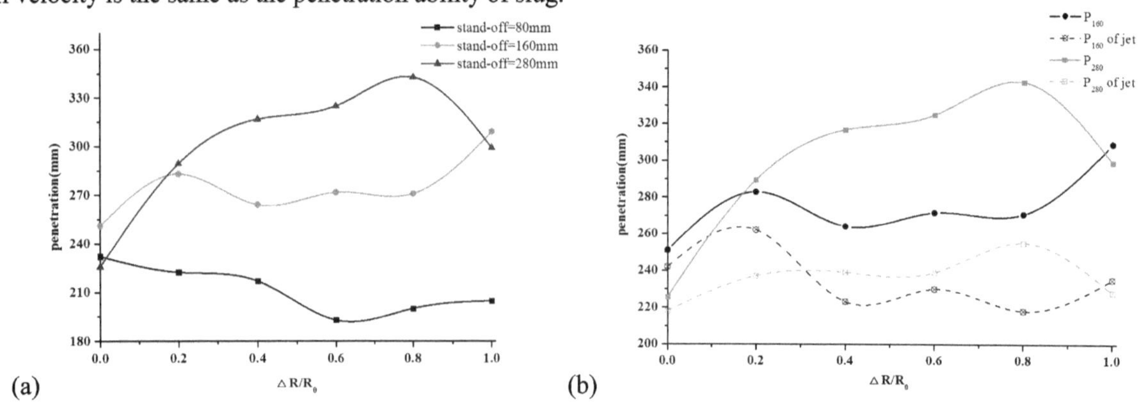

Figure 4.3 Data of penetration (a) penetration depth of penetrator (b) Penetration depth of penetrator and jet at stand-off=160,280mm.

4.3. Summary

Table 4.3 penetration error to $\Delta R/R_0 = 0$

$\Delta R/R_0$	Penetration error of penetrator when stand-off=80mm	Penetration error of penetrator when stand-off=160mm	Penetration d error of jet when stand-off=160mm	Penetration error of penetrator when stand-off=280mm	Penetration error of jet when stand-off=280mm
0	/	/	/	/	/
0.2	-4.2%	12.7%	8.3%	28.2%	8.7%
0.4	-6.5%	5.2%	-7.8%	40.3%	9.6%
0.6	-16.9%	8.2%	-5.0%	44.0%	9.6%
0.8	13.8%	7.9%	-9.9%	51.9%	17.0%
1.0	11.8%	23.1%	2.9%	32.8%	4.6%

The penetration error for the penetration of $\Delta R/R_0 = 0$ at different stand-off is shown at Table 4.3. It can be seen from Table 4.3 that, at stand-off = 80mm, the axial additional insensitive explosive is not conducive to increasing the penetration depth. When stand-off = 160mm, the penetration depth by the penetrator of the double-layer sub-caliber shaped charge reaches the maximum value at $\Delta R/R_0 = 1.0$, resulting in an increase of 23.1% compared to the penetration depth by the penetrator of the ordinary shaped charge. While stand-off = 280mm, the penetration depth by the penetrator of the double-layer sub-caliber shaped charge reaches the maximum value at $\Delta R/R_0 = 0.8$, giving a 51.9% increase compared to the penetration depth of the ordinary shaped charge. The improvement in penetration for stand-off = 280mm is far greater than for stand-off = 160mm, and in this paper, we consider the penetration depth of the penetrator as the most important property of shaped charge jet. Therefore, the optimum axial thickness of the additional insensitive explosive is $\Delta R/R_0 = 0.8$.

5. The influence of the additional height of charge

5.1. Shaped charge jet forming

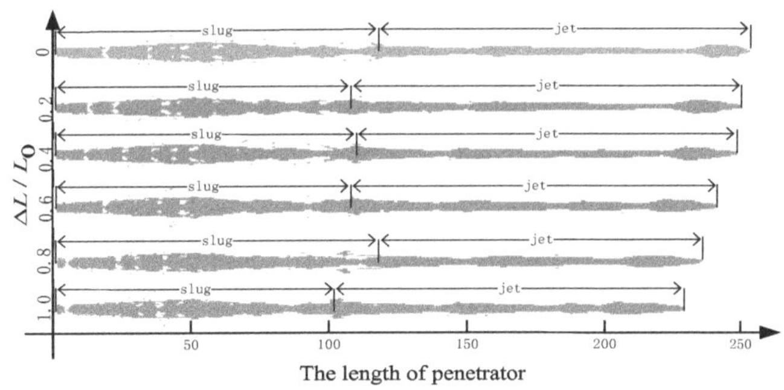

Fig 5.1 The morphology of penetrator at 50us when the additional height of charge varies

In this section, to determine the optimum size of the additional height of the charge, we simulated the working conditions of $\Delta L/L_0$ with values of 0, 0.2, 0.4, 0.6, 0.8 and 1.0 respectively at the optimum additional thickness, that is $\Delta R/R_0 = 0.8$. Fig. 5.1 shows the morphology of penetrator at 50us when the additional height of charge varies. The last collapse unit of the liner was regard as the boundary of jet and slug, as shown at Fig5.1. It can be seen from the figure that, while the height of the double-layer explosive charge increases, the length of the jet varies, the length of the penetrator decreases, the front segment of slug has tensile capability, the velocity of the front of slug varies and the phenomenon of necking occurs. Also, the back segment of the slug breaks seriously.

Table 5.1 The related data of the penetrator at 50us while the additional height of charge varies

$\Delta L/L_0$	The jet top velocity(m/s)	The jet tail velocity (m/s)	The jet length(mm)	The maximum energy of Cu(MJ)
0	6283	3022	134	0.198
0.2	6570	2956	143	0.225
0.4	6776	2984	139	0.249
0.6	6921	3194	132	0.266
0.8	7050	3148	118	0.276
1.0	7117	3232	124	0.292

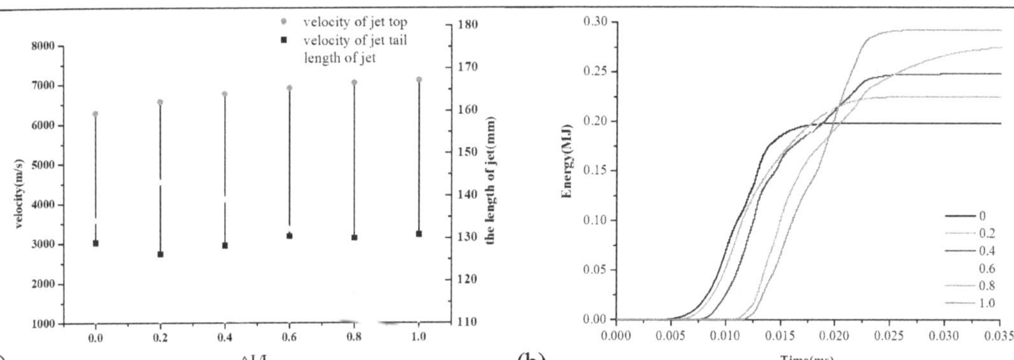

(a) (b)

Figure 5.2 (a) The parameters of jet (b) The energy of the copper

Figure 5.2 was drawn by using the related data of the jet while the additional height of the charge varies. It can be clearly seen from Fig 5.2 that the jet tip velocity and the maximum energy of Cu increase when the additional height of charge increases. The jet tail velocity increases while the additional height of charge increases, and the jet tail velocity reaches the maximum value at $\Delta L/L_0 = 0.6$, then the tail velocity of the jet decreases slightly. The length of the jet varies with the additional height of charge, and it reaches the maximum value at $\Delta L/L_0 = 0.4$. In addition, the length of the jet reaches the minimum value at $\Delta L/L_0 = 0.8$.

5.2. Penetration

In order to analyze the penetration ability of the penetrator while the additional height of the charge varies with the axial thickness of additional insensitive explosive held constant at $\Delta R/R_0 = 0.8$, numerical simulations were conducted at stand-off = 80 mm, 160 mm and 280 mm. The penetration data is shown in Table 5.2. Since the penetrator has not been fully stretched, only the penetration depth of penetrator was collected when stand-off=80mm.

Table 5.2 Data of Penetration

$\Delta L/L_0$	Penetration depth by penetrator when stand-off=80mm	Penetration depth by penetrator when stand-off=160mm	Penetration depth by jet when stand-off=160mm	Penetration depth by penetrator when stand-off=280mm	Penetration depth by jet when stand-off=280mm
0	200.1mm	270.9mm	218.1mm	342.8mm	255.2mm
0.2	204mm	294.5mm	219.2mm	359.6mm	268mm
0.4	208.8mm	310.4mm	216mm	355.4mm	282.7mm
0.6	200mm	305.5mm	188.3mm	365.7mm	277.8mm
0.8	198.7mm	270mm	170.7mm	375.4mm	217.8mm
1.0	214mm	276mm	194mm	394.7mm	270mm

The penetration depth of the penetrator for varying additional heights of the charge is shown in Figure 5.3(a). It can be seen from the figure that the change in total penetration depth is not obvious when the additional height of the double-layer sub-caliber explosive charge varies at stand-off = 80mm. The penetration depth reaches the extreme value at $\Delta L/L_0 = 0.4$ and the maximum value at $\Delta L/L_0 = 1.0$.

When stand-off = 160mm, the penetration depth of the double-layer sub-caliber shaped charge increases with the height of the additional charge between $\Delta L/L_0 = 0$ and $\Delta L/L_0 = 0.4$, and it decreases with the height of the additional charge between $\Delta L/L_0 = 0.6$ and $\Delta L/L_0 = 0.8$, then it increases at $\Delta L/L_0 = 1.0$.

When stand-off = 280mm, the penetration depth increases when $\Delta L/L_0$ increases.

The penetration depth of the penetrator and the penetration depth of the jet are shown in Fig. 5.3(b). It can be seen from the figure that the changing trend of the penetration depth of the jet is consistent with that of the total penetration depth of the penetrator, except when $\Delta L/L_0 = 0.8$ at stand-off = 160mm. When stand-off = 160mm, the penetration ability of slug increase with $\Delta L/L_0$ until $\Delta L/L_0 = 0.6$, then it begins to decrease. When stand-off = 280mm, the penetration ability of slug decreases with $\Delta L/L_0$ until $\Delta L/L_0 = 0.4$, then it begins to increase, and slug has greatest penetration ability at $\Delta L/L_0 = 0.8$.

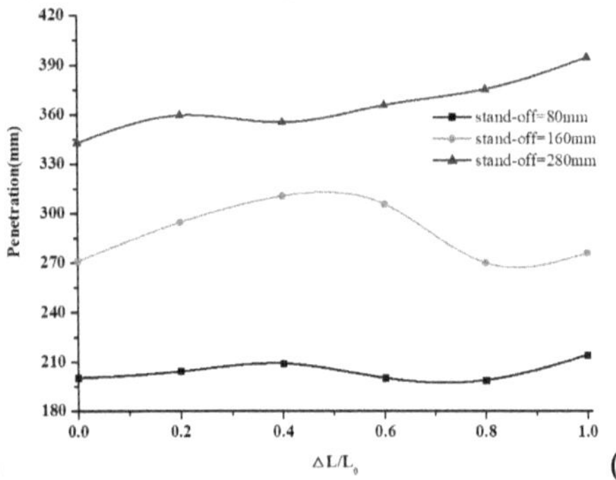

(a)

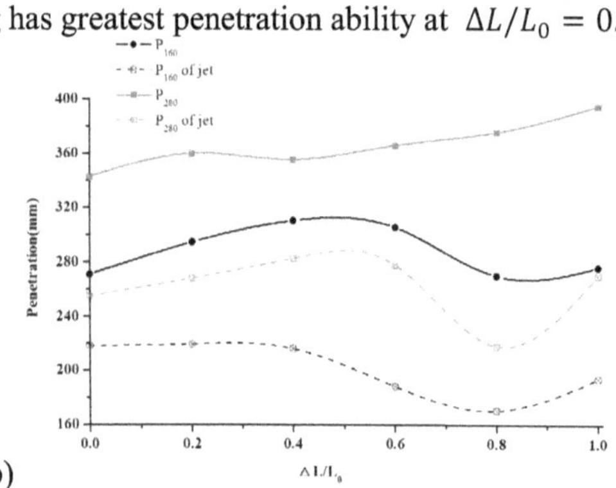

(b)

Figure 5.3 Data of penetration (a) penetration depth of penetrator (b) Penetration depth of penetrator and jet at stand-off=160,280mm

5.3. Summary

Table 5.3 penetration error to $\Delta L/L_0 = 0$

$\Delta L/L_0$	Penetration error of penetrator when stand-off=80mm	Penetration error of penetrator when stand-off=160mm	Penetration d error of jet when stand-off=160mm	Penetration error of penetrator when stand-off=280mm	Penetration error of jet when stand-off=280mm
0	/	/	/	/	/
0.2	1.9%	8.7%	0.5%	4.9%	5.0%
0.4	4.3%	14.6%	-1.0%	3.7%	10.8%
0.6	0%	12.8%	-13.7%	6.7%	8.9%
0.8	-0.7%	-0.3%	-21.7%	9.5%	-14.7%
1.0	6.9%	1.9%	-11.1%	15.1%	5.8%

The penetration error for the penetration with $\Delta L/L_0 = 0$ at different stand-off distances is collected in Table 5.3. Based on a comprehensive analysis of Table 5.3, at stand-off = 80mm the additional height of double-layer sub-caliber explosive charge can hardly change the penetration depth. When stand-off = 160mm, the penetration depth of the penetrator for the double-layer sub-caliber shaped charge reaches the maximum value at $\Delta L/L_0 = 0.4$, and it increases by 14.6%. When stand-off = 280mm, the penetration depth of the penetrator increases when $\Delta L/L_0$ increases, and it increase by 15.1% at $\Delta L/L_0 = 1.0$. The improvement in penetration depth of the penetrator when stand-off = 160mm and 280mm is almost the same. Considering the utilization of the explosive, we think the optimum additional height of the charge is $\Delta L/L_0 = 0.4$.

6. Experiments

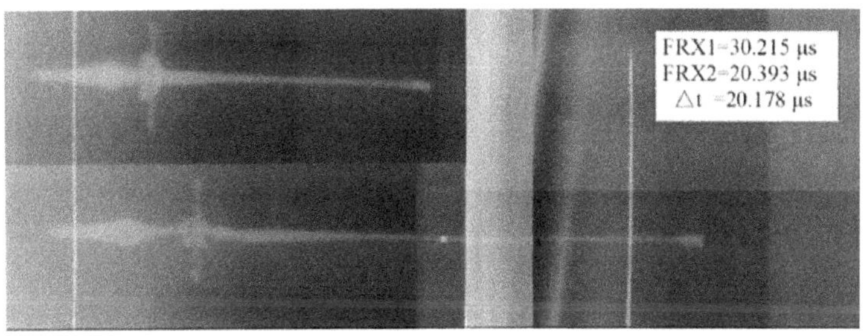

Fig 6.1 Double flash X-ray exposures of the shaped charge jet at 30 and 50 us after initiation

To verify the validity of the simulations described in this paper, we compared experimental results with simulation results for the benchmarked shaped charge. As Fig. 6.1 illustrates, the jet tip and tail velocity were measured by the X-ray system. The jet tip velocity was 6510 m/s, and jet tail velocity was 1189 m/s. The length of the jet was approximately 167.5 mm at 30 us and 291.5 mm at 50 us after initiation. [10] Depth of penetration (DOP) experiments at stand-off = 80 mm, 160 mm and 280 mm were executed to analyze the jet penetration ability of Ø56mm. [11]

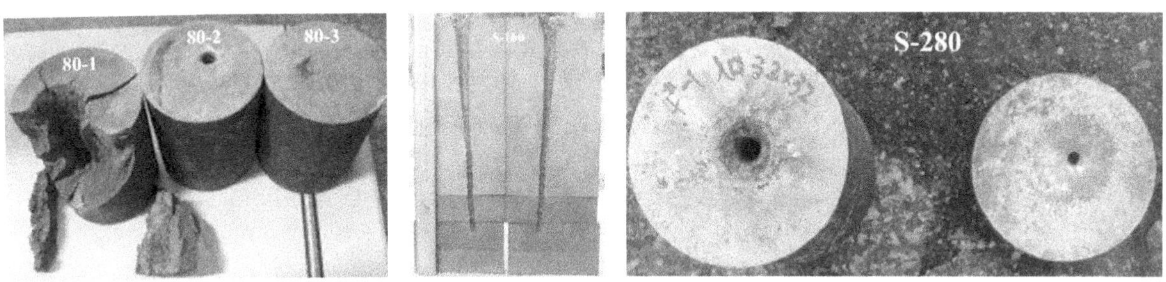

(a) (b) (c)

Fig 6.2 Target of DOP experiment at different stand-off (a) stand-off=80mm, (b) stand-off=160mm, (c) stand-off=280mm

Table 6.1 Data of experiment and simulation of Ø56 mm shaped charge jet

	Data of jet			Data of penetration		
	The jet tip velocity (m/s)	The jet tail velocity(m/s)	The jet length at 50us(mm)	Stand-off =80mm	Stand-off =160mm	Stand-off =280mm
experiment	6510	1189	219.2	201	246	217
simulation	6109.5	1298	183.5	232.1	251.1	225.7
error	6.15%	9.16%	16.28%	15.47%	2.07%	4.01%

A comparison of the data from the experiments and simulations for the Ø56 mm shaped charge jet is shown in Table 6.1. In the table, both the jet tip and the jet tail velocity are the averages between 30us and 50us because of the measurement method employed in the experiment. From Table 6.1, it can be seen that the simulations described in this paper can produce accurate jet velocities and penetration depths at stand-off=160 and 280mm. For stand-off=80mm a deviation of the jet length and penetration depth can be seen between the simulation and the experiment, but the difference is still within an acceptable range. The measured penetrations of Ø56 mm shaped charge at stand-off=80mm, 160 mm, 280mm, as well as the X-ray radiographs of the jets formed by Ø56 mm shaped charge, demonstrate that the research of this article is credible.

7. Conclusion

In this paper, we research the influence of the axial thickness of additional insensitive explosive and the additional height of charge on the jet formation and the penetration depth of a double-layer sub-caliber shaped charge using the finite element code AUTODYN. According to the simulation results, it can be shown that, for the benchmarked Ø56 mm shaped charge, the optimum axial thickness of additional insensitive explosive for the double-layer sub-caliber shaped charge is $\Delta R/R_0 = 0.8$, and the optimum size of the additional height of the double-layer sub-caliber shaped charge is $\Delta L/L_0 = 0.4$.

Acknowledgements

This study is supported by the National Natural Science Foundation of China (Grant Nos. 11702144).

References

[1] Held, M. 1995. Behavior of Dual Composition Explosive, EUROPYRO, Tours, France, June: 185–191.

[2]Itoh S, Liu Z, Nagano S, et al. Visualization of an overdriven detonation phenomenon in a high explosive[J]. Journal of Flow Visualization and Image Processing, 1999, 6(4).

[3]Stuivinga M, Verbeek H J, Carton E P. The double explosive layer cylindrical compaction method[J]. Journal of Materials Processing Technology, 1999, 85(1-3): 115-120.

[4]Kato H, Murata K, Itoh S, et al. Application of overdriven detonation in high density explosive to shaped charge[C]//23rd International Symposium on Ballistics, Tarragona, Spain. 2007: 223-230.

[5] Zhang X F, Ding J, Zhao X. Numerical simulation of double-layer shaped charge[J]. Explosives. Shock Waves, 2009, 30: 63-67.

[6] Hussain T, Yan L, Feng--lei H. Numerical simulation of double-layer shaped charges and comparison with ordinary shaped charges[C]//Modeling, Simulation, and Applied Optimization (ICMSAO), 2015 6th International Conference on. IEEE, 2015: 1-5.

[7] HUSSAIN T, YAN L I U, HUANG F, et al. An Analysis of Double-layer Shaped Charges[C]//29th International Symposium on Ballistics. 2016.

[8]Wang Z, Jiang J W, Wang S Y, et al. Jet Formation and Penetration Study of Double-Layer Shaped Charge[J]. Journal of Energetic Materials, 2018, 36(2): 152-168.

[9] Murphy M J, Lee E L, Weston A M, et al. Modeling shock initiation in composition B[R]. Lawrence Livermore National Lab., CA (United States), 1993.

[10] Jia X, Huang Z, Zu X, et al. Experimental study on the performance of woven fabric rubber composite armor subjected to shaped charge jet impact[J]. International Journal of Impact Engineering, 2013, 57: 134-144.

[11] Bin Ma, Research on the Coupling Characteristics between the Strong Magnetic Field and the Shaped Charge[D]. Nanjing University of Science & Technology.2018.

Proceedings of the 2019 Hypervelocity Impact Symposium
HVIS2019
April 14-19, 2019, Destin, FL, USA
HVIS2019-110

3D Printed Conical Shaped Charge Performance

Phillip Mulligan[a*], Catherine Johnson[b], Jason Ho[c], Cody Lough[c], and Edward Kinzel[d]
[a]Assistant Research Professor, Missouri University of Science and Technology, 1006 Kingshighway, Rolla MO 65409, *pmulligan@mst.edu
[b]Assistant Professor, Missouri University of Science and Technology
[c]Research Assistant, Missouri University of Science and Technology,
[d]Associate Professor, University of Notre Dame

Abstract

A Conical Shaped Charge (CSC) is a versatile device utilized in construction, mining, petroleum and defense industries. The geometry and material structure of the metal liner play an integral role in the CSC performance. The performance of CSC liners has been relatively well-characterized for liners manufactured via hydroforming, hydraulic pressing, or turning on a CNC lathe. With advancements in Additive Manufacturing (AM) CSC liners can be 3D printed with metal powders. AM can provide significant design freedom in terms of realizing better properties through introduced hierarchic structuring or anisotropy. However, it is unclear as to how metal liners produced with Selective Laser Melting (SLM), will influence the conical shaped charge's performance. This paper explores the performance, relative to the penetration of steel plates, of CSCs using 3D printed metal liners benchmarked against machined liners. The metal liners were printed with SLM parameters that were optimized to maximize the print density. The metal liner dimensions (thickness, height, and outer diameter) were designed using the recommended ratios of the liner's inner diameter presented by Virgil (1988). The 3D printed metal liners are compared to a CNC machined liner, with the same dimensions. The comparison enables the evaluation of how 3D printing a liner influences penetration performance. The results indicate conical shaped charges could utilize 3D printed liners. These results open a wide range of performance design opportunities that cannot be achieved via conventional manufacturing and justify the current increased cost associated with additive manufacturing metal components. Future work will continue to explore how print density, printed material, and advanced geometries modify the conical shaped charge performance.

Keywords: Conical Shaped Charges, Additive Manufacturing, 3D printing, Select Laser Melting

Nomenclature

A	Position of Detonation wave on liner surface
C	Collision point
D	Internal diameter of conical shaped charge liner
M	Mass
P	Penetration depth
V	Velocity
Subscripts	
C	Casing
E	Explosive
L	Liner
p	Initial cone
j	Jet

1. Introduction

The penetration performance of Conical Shaped Charges (CSC) is a function of the liner material properties and geometry [1]. Exploration of the CSC performance has been limited by manufacturing considerations, specifically the practicality of machining or forming the CSC liners. Additive Manufacturing (AM) has advanced to the point where metals can be deposited with properties exceeding cast metals. AM techniques provide the possibilities for complex geometric configurations (e.g., lightweight truss structures). In addition, the engineering properties of the part can be varied spatially (e.g., reduction of local yield stress by not fully fusing powder). A first step toward realizing the potential of AM for shaped charge liners is to benchmark CSC liners created with AM against machined liners. This paper presents a simple comparison of the penetration performance of CSC liners produced with Selective Laser Melting (SLM) Stainless Steel 304L and CNC liners with identical geometries machined from cold worked stainless steel 304L rod. The selection of 304L is not optimal for CSC liners. However, the use of 304L was based on material availability for the SLM system at the time of this research.

2. Conical Shaped Charge

A hollow charge, i.e., explosives with a hollow cavity at the end opposite the initiation, focuses the energy of the detonation products into a high-intensity localized force. This phenomenon is commonly referred to as the Munroe effect and typically generate a crater approximately one-half the diameter of the cavity. When the hollow cavity is lined with a material such as glass, metal, ceramics, pressed powders, etc. the liner material forms a jet similar to that demonstrated in Fig 1a shortly after detonation. The resulting jet can have a velocity in excess of 8 km/sec [2].

The jet velocity is a function of the liner mass and internal angle of the cone, as well as the shape of the shockwave interacting with the liner and quantity of explosives [3]. Substantial research has been conducted into optimizing the performance of CSCs. For example, Sandia's Shaped Charge Analysis Program (SCAP) is a numerical code that can analytically model and optimize the design of a CSC. Figure 1b, shows optimized results from this program, parameterized in terms of the internal diameter of the liner, *D*. The same study showed that the CSC has the best performance when the internal angle is between 42 and 45 degrees.

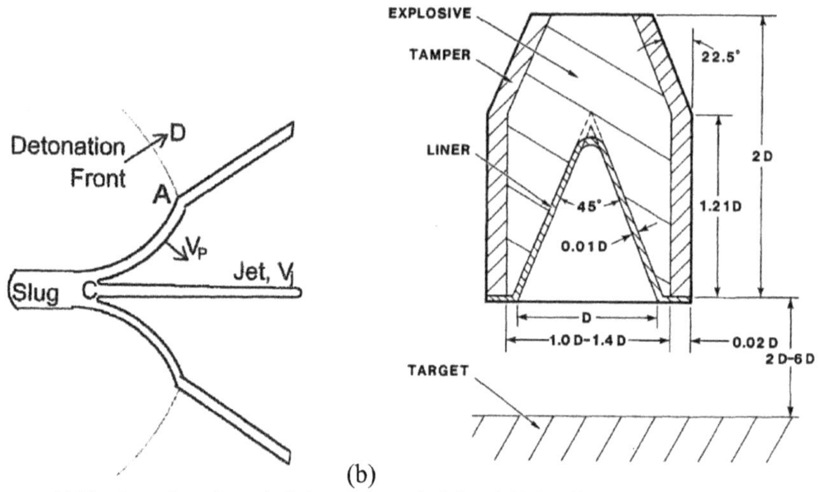

Fig. 1. Conical Shaped Charge geometry (a) The formation of a conical shaped charge jet [4] and (b) Sandia optimized conical shaped charge configuration relative to the internal diameter, *D*, of the cone [1].

The distance the CSC is from the target is also critical. The jet is fully accelerated once it leaves the collapse point. The standoff distance is required to permit time and space for the jet to stretch. Too little space and the penetrator is short whereas too much space and the jet particulates and literal drift decreases penetration. The ideal distance from the conical shaped charge to the target is 2-6 *D*. The penetration performance is also analyzed as a function of charge diameters, see Fig 2a. The penetration shown in Fig 2a is the penetration of cold rolled copper and steel plates (traditionally manufactured liners) that were machined to the desired geometry. Figure 2a does not factor in the conical shaped charge penetration into multiple steel plates, which is a common technique for analyzing penetration because the penetration can be easily assessed without needing to cut the sample in half. However, it is known that the penetration of a conical shaped charge into a layered target will not have equal penetration to the same device penetrating a single target of a larger thickness. This is due to the boundary between the plate layers, that would not be present in a single target, enabling the energy of the jet and slug to escape the cavity [3]. A well designed and constructed layered target will decrease penetration slightly (~3-5%) due to the introduction of additional free surfaces. Additionally, in some steels, thin plates can be harder, which increases penetration resistance, thus decreasing penetration resistance.

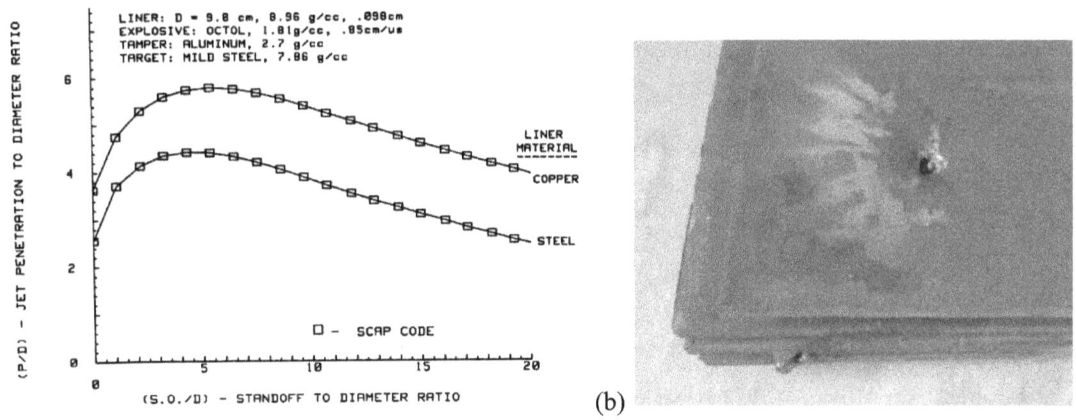

(a) (b)

Fig. 2. (a) Conical shaped charge penetration as a function of charge diameter standoff [1] and (b) the penetration of a pressed powder copper conical shaped charge [5].

Pressed powder conical shaped charge liners have been shown to have similar penetration to liners formed from sheet metal when fired into a solid target, even exceeding traditional liners in some cases [5]. When fired into a target made up of layered plates the deposited liner material can be seen in-between the plate layers, see Fig 2b. The amount of powder deposited between each plate can be associated with the spacing between each target plate. Traditional liners have been observed to deposit material between the layers of multiple plate targets, but the metal deposition wasn't observed until the last layers corresponding to the deepest penetration. Despite the jet from a pressed powder liner losing mass between each plate layer, the overall penetration was observed to be similar to the traditional liner [5].

3. Conical Shaped Charge Manufacturing and Methodology

The SLM process is a powder-bed additive manufacturing technique in which 3D metal parts are produced layer by layer. During SLM, a laser scans over a metal powder bed, locally heating the powder so that it is fused to adjacent material, see Fig 3a. The laser traces paths defined by a computer following part geometry with pre-set process parameters (laser power, scan-speed, hatch spacing, and layer thickness). The process parameters can be used to control the material properties. The resulting melt pools can be seen in Fig 3b. The general appearance is analogues to placoid scales. It was noted that the bond between the melt pools might fail during the jet formation process and the liners may behave similarly to pressed powder liners (this effect could be deliberately introduced by depositing material with less laser energy). The samples in Fig 3b were mounted in bakelite, polished to 0.05 micron using diamond suspension and then electrolytically etched using 60/40 Nitric acid. The material that will be utilized for this test series is 304L stainless steel. While it is understood that copper would be ideal for the test, the pyrophoric nature or reactive metals such as copper makes depositing self-passivating stainless steel powder a safer choice for the university environment.

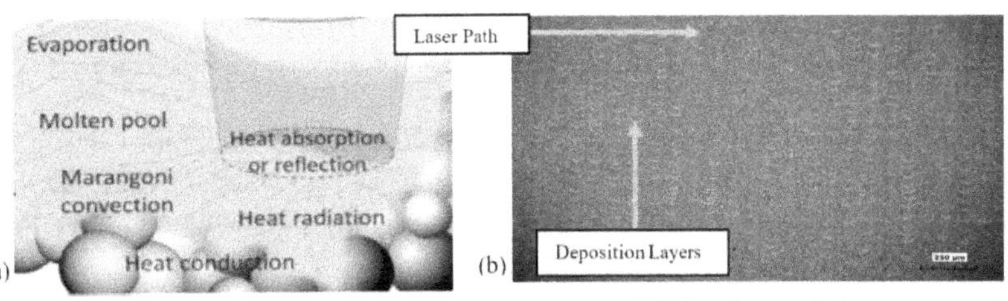

(a) (b)

Fig. 3. Heat transfer and molten pool dynamics during powder-based additive manufacturing [6] (a) and an optical micrograph of the cross-sections of laser tracks in 304L stainless steel part produced by SLM (without layer-to-layer rotation) illustrating the molten pool (b).

For the research presented in this paper, a Renishaw AM250 SLM system is used. The laser power was set to 200 watts with an exposure time of 75 microseconds. These properties are selected based on prior experience to maximize the density (minimize porosity) in 304L. Previous tests have shown that these parameters produce 304L with a yield strength of 428±7 MPa and an ultimate strength of 598±5 MPa. The strength of the 3D printed material exceeds both the average yield strength and ultimate strength of AISI 304L stainless steel, 210 MPa, and 564 MPa respectively [7] because of the smaller grain structure associated with rapid cooling in SLM. The traditional liners were machined using CNC manufacturing. Due to manufacturing limitations, the CNC liners have a thicker base than the 3D printed liners. No post-processing was performed on either liner to obtain a precise final shape or surface finish.

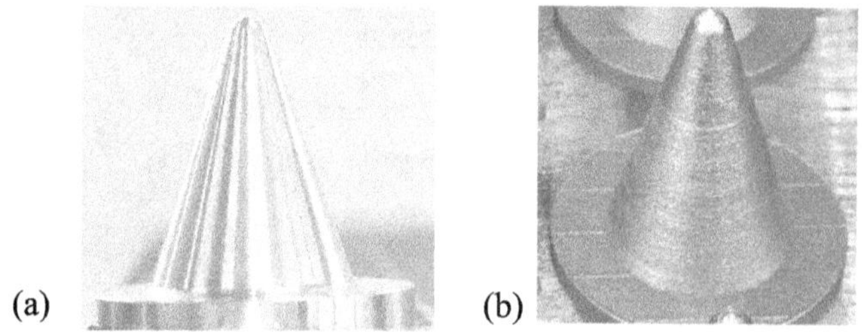

Fig. 3. Conical shaped charges utilized in this project CNC liner stainless steel (a) and 3d printed 304l stainless steel [8] (b).

To evaluate the performance of the 3D printed conical shaped charge liner the geometry of the liner was based on the diameter ratios identified with Sandia's SCAP presented in Fig 1b. The diameter of the liner was back calculated to ensure the liner would fit within commercially available piping and require a charge weight that was below the site limit. The final liner diameter (D) was set to 3.05 cm. The resulting liner geometry is shown in Fig 5a. The internal angle was set to 42 degrees. Each charge was packed with 193 grams of C-4 and a 10-gram stinger (PETN) coupled to a detonator was used to initiate the CSC. The overall test configuration can be seen in Fig 5b. Three replications were conducted for each test liner material. It is understood that the optimized configuration would utilize a single target plate and casing material with greater tensile strength than PVC. However, the objective of the research presented in this paper was not to identify the maximum penetration of an optimized design, but rather to explore the viability of 3D printed liners and provide a comparison to a traditionally manufactured liner. This configuration was chosen because of the ease of penetration analysis and fragmentation concerns to the test facility. It should be noted the spacing between target plates for the research presented in this paper was approximately 0.08 cm.

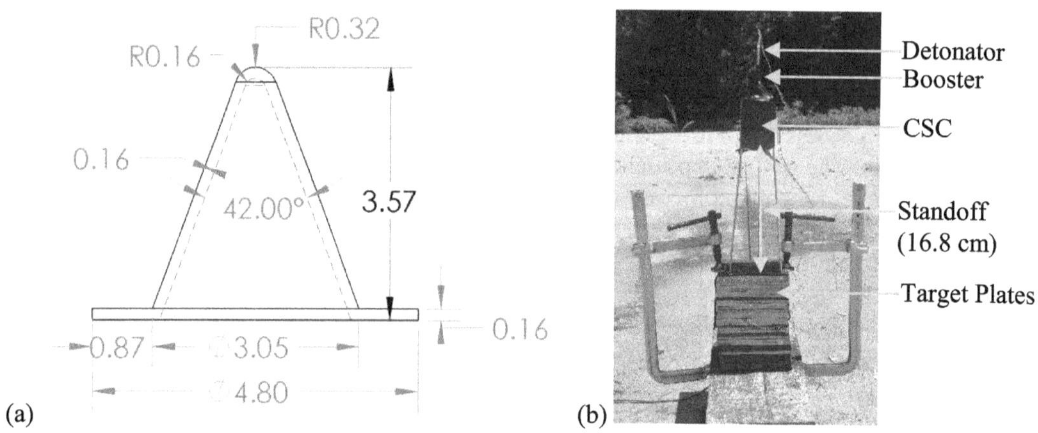

Fig. 5. (a) The conical shaped charge liner dimensions (cm) [9] and (b) the corresponding casing, standoff, explosive geometries.

The test setup was simulated using Sandia's Hydro-code CTH to explore the jet formation, slug geometry, and predicted penetration. The simulation was run in a 2D domain and a flat 0.025 cm mesh. The Equations of State (EOS) and Material Models are shown in Table 1. Discards were used for material with densities lower than 0.001g/cm and temperatures above 1 electron-volt (11,604 degrees Kelvin).

Table 1: CTH Part EOS and Material Models

Part	EOS		Material Model	
Liner	Mie-Gruneisen	304_SS	Steinberg	304_SS
Explosive	History Variable Reactive Burn	COMPC4		
Casing	Mie-Gruneisen	PMMA	Johnson Cook	LEXAN
Targets	Mie-Gruneisen	V-250_STEEL	Johnson Cook	STEEL

The simulation indicates that the liner geometry will produce a jet similar to that predicted in Fig 1a. The initial jet velocity is 9 km/sec, and the configuration forms a slug, see Fig 6a. The slug separates from the jet and produces the shape shown in Fig 6b traveling at 0.6 km/sec. The CSC was predicted to penetrate eight plates with 0.08 cm spacing, and the slug became lodged in the first few plates, Fig 6c.

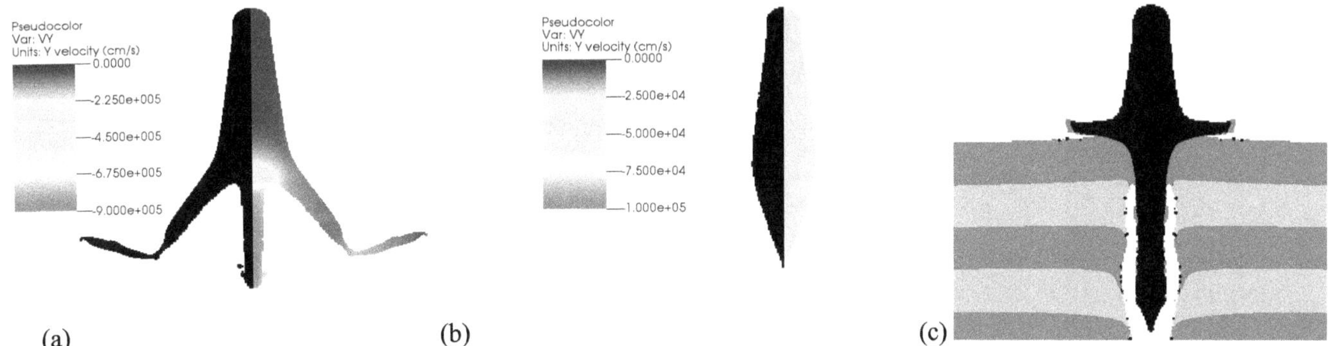

Fig. 6. CTH results illustrating the jet formation (a), slug geometry (b), and predicted slug penetration (c).

4. Experimental Results and Discussion

The objective of this test was to examine if the 3D printed conical shaped charge liners could perform the same a tradition liner under the same test conditions. Each test was examined for the depth of penetration and the penetration hole diameter for each plate. The penetration data was recorded for the last plate fully penetrated. The plate immediately following the last fully penetrated plate typically had some indentation or penetration, an approximation was made of the penetration depth into the final plate. Figure 7 presents the penetration depth as a ratio of total penetration and liner diameter (P/D) versus the liner mass to explosive mass ratio (M_L/M_E). There were slight differences in the final liner to explosive mass ratios. However, when evaluating the penetration data, these slight differences did not seem to have much influence on the conical shaped charge liners penetration. The average penetration of the two liners appear to be similar, with the CNC liner slightly outperforming the 3D printed liner, see Fig 7. The average penetrations of the two liners were within 0.27 cm (i.e., less than one plate thickness).

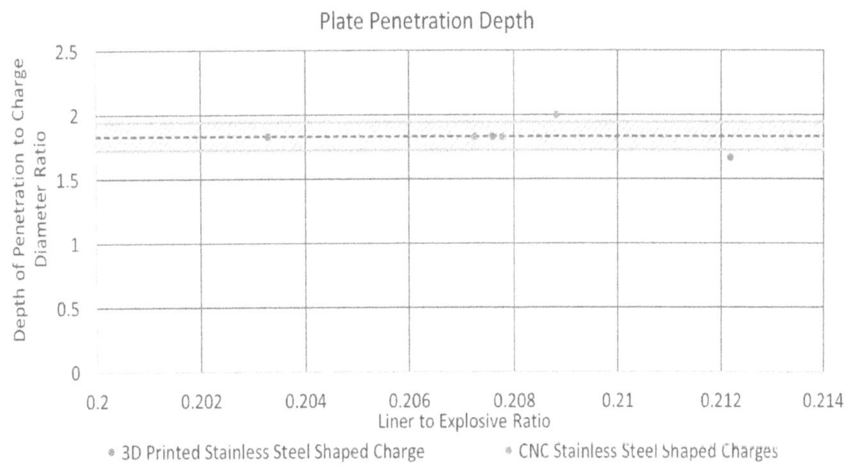

Fig. 7. Analysis of the 3D printed and CNC stainless steel shaped charge penetration into 0.635 cm plates of mild steel.

The overall penetration is not representative of the penetrations shown in Fig 2a. This can be attributed to a non-optimized charge design. For example, slight variances in the PVC tubing diameter, detonator placement, and packing density have are known to influence penetration performance [3]. Additionally, the mass of the PVC casing (M_C) provides little confinement for the gas expansion, resulting in more energy loss out the sides of the charge rather than an increased acceleration to the CSC jet [3]. Furthermore, the stainless steel liners may have formed a highly particulate jet that was not represented in the SCAP code and the standoff distances are based on "rules-of-thumb" rather than from an optimization study. While it is worth noting the importance of an optimized charge design and test setup for achieving maximum penetration of a liner, the objective of this research was to perform a top-level analysis of the 3D printed CSC liners.

In addition to the total depth of penetration the penetration hole diameter was monitored to see how the penetration varied from plate to plate. The hole diameter of each plate penetrated for the two liners show an overall trend of decreasing hole diameter as the jet penetrates further into the target plates. However, the 3D printed liner produced a slightly larger hole, than the CNC liner, on the first initial plates (plates 1-4), but produces a smaller diameter hole for plates 5-8, see Fig 8.

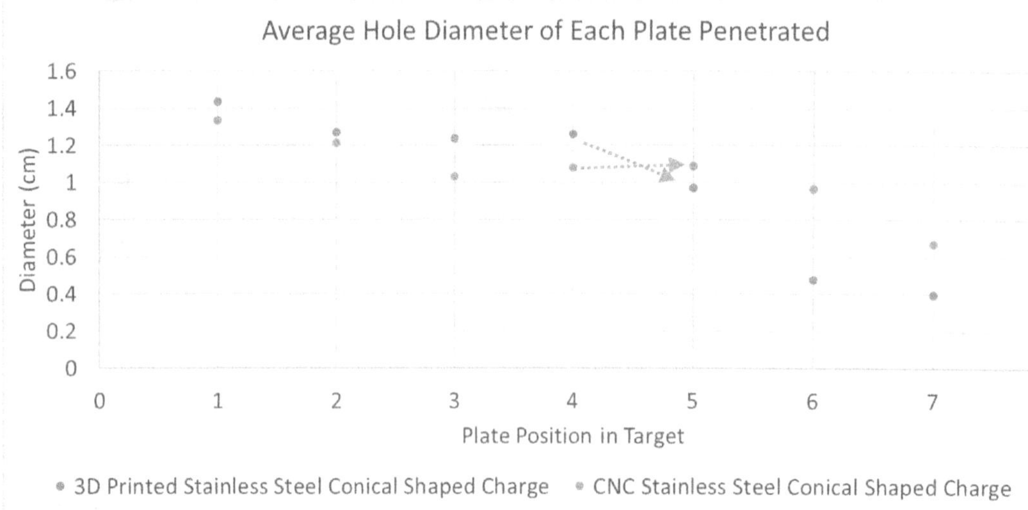

Fig. 8. Average penetration hole diameter for each plate in order of penetration

The larger diameter for the 3D printed conical shaped charges can be correlated to the presence of the slug that was recovered in the first four target plates, see Fig 9. The CNC liners did not produce a slug that was observed in the target plates. The absence of a slug indicates the slug was able to penetrate through to plates 5-8 and produced the larger hole diameters than the 3D printed liners. The penetration of the slug from the 3D printed liner is similar to the predicted penetration and geometry from the CTH simulations.

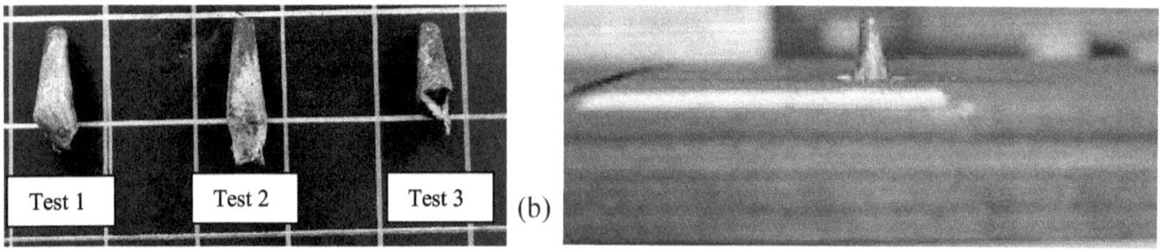

(a) (b)

Fig. 9. (a) Slugs found after 3D printed conical shaped charge test and (b) a slug caught after 2 plates of penetration.

In addition to the depth of penetration and the diameter of the penetration holes for each plate, it was noted that the 3D printed liners appear to have deposited significantly more material between each target plate layer. The material deposited between each layer was formed into thin sheets, see Fig 10. During the examination of the plate penetration, it was observed that these thin sheets of stainless steel could be removed as one piece. The CNC liners did deposit some material, but the deposition was not observed until after Plate 3, whereas the 3D printed liner deposited material immediately after penetrating the first plate and the quantity of material deposited from the 3D printed liners was significantly higher than the CNC liners. Figures of the front of each plate are shown in Appendix A: Representative Conical Shaped Charge Penetration by Plate.

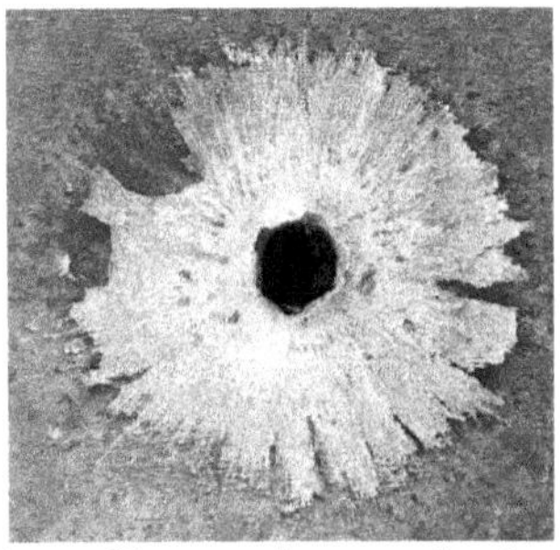

Figure 10. Representative deposition of stainless steel material observed between the target plate layers of the 3d printed conical shaped charge test.

5. Summary and Conclusions

The performance of the 3D printed liners are promising and indicate that 3D printed conical shaped charge liners can have similar penetration performance as a traditional liner. The penetration of both liners is much lower than the predicted penetration, shown in Fig 2a. This can be attributed to a non-optimized test setup. The increased strength of the 3D printed liner appears to have resulted in a slug that was recoverable in the first several plates, that was not present with the CNC liners. The presence of a slug and the slugs penetration depth of the 3D liners influenced the penetration hole diameters and is likely the reason the CNC liners had a slightly higher average penetration depth. The 3D printed liners had a similar performance to the performance predicted with CTH. The 3D printed liner material exhibited material deposition between layers similar to pressed powder liners.

Further testing is needed to quantify if the melt pools delaminated resulting in a highly particulate jet and the material deposition between the plates, as well as explore if an optimized CSC design and test setup illuminate additional performance differences between the two liner manufacturing processes. Future testing will explore flash x-ray imaging of the jet formation to determine how the jet and slug form in comparison to one another. While further testing is needed, the need for the CNC liners to have a thicker base than the 3D printed liners is an illustration of how 3D printing conical shaped charge liners could expand the innovation of conical shaped charges.

6. Bibliography

[1] M. G. Vigil, "Optimized Conical Shaped Charge Design Using the SCAP Code," Albuquerque, 1988.

[2] W. P. Walters, "The Shaped Charge Concept, Part I. Introduction," Ballistics Research Lab, ABERDEEN PROVING GROUND, MARYLAND, 1990.

[3] W. P. Walters and J. A. Zukas, Fundamentals of Shaped Charges, Baltimore: CMCPress, 1998.

[4] P. W. Cooper, Explosives Engineering, Wiley-VCH, 1996, p. 387.

[5] Abdo GM and Elshenawy T, "Density Effect of the Compacted Copper-Tungsten Shaped Charge Powder Liners on its Penetration Performance," *Powder Metallurgy & Mining* , vol. 6, no. 2, 2017.

[6] T. DebRoy, H.L. Wei, J.S. Zuback, T. Mukherjee, J.W. Elmer, J.O. Milewski, A.M. Beese, A. Wilson-Heid, A. De, W. Zhang, "Additive manufacturing of metallic components – Process, structure and properties," *Progress in Materials Science,* p. 112–224, 2018.

[7] A. A. S. Metals, "AISI Type 304 Stainless Steel," ASM Aerospace Specification Metals, 2018. [Online]. Available: http://asm.matweb.com/search/SpecificMaterial.asp?bassnum=mq304a. [Accessed 2019].

[8] Ho J., Lough C., Mulligan P., Kinzel E., Johnson C., "Additive Manufacturing of Liners for Shaped Charges," in *Proceedings of the 29th Annual International Solid Freeform Fabrication Symposium – An Additive Manufacturing Conference*, 2018.

[9] Ho J., Lough C., Mulligan P., Kinzel E., Johnson C., "Additive Manufacturing of Liners for Shaped Charges: Practicality and Performance," in *Proceedings of the International Society of Explosive Engineers 45th Annual Conference on Explosives and Blasting Technique*, Nashville TN, 2019.

[10] T. Elshenawy, G. Abdo, A. Elbeih, "High Penetration Performance of Powder Metallurgy Copper-Tungsten Shaped Charge Liners," *Central European Journal of Energetic Materials,* vol. 15, no. 4, pp. 610-628, 2018.

A. Appendix: Representative Conical Shaped Charge Penetration by Plate

	CNC Conical Shaped Charge	3D Printed Conical Shaped Charge
Plate 1	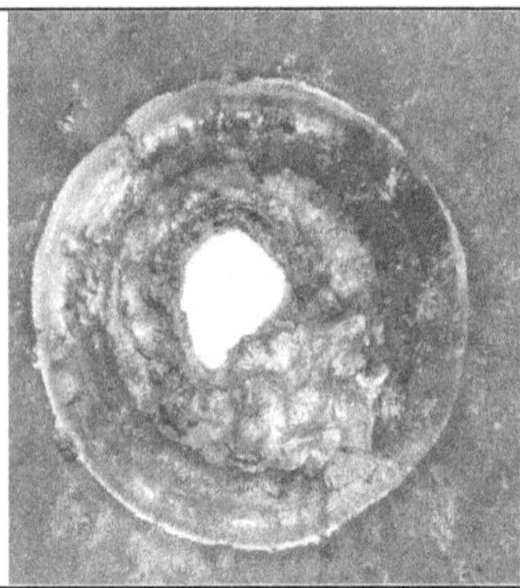	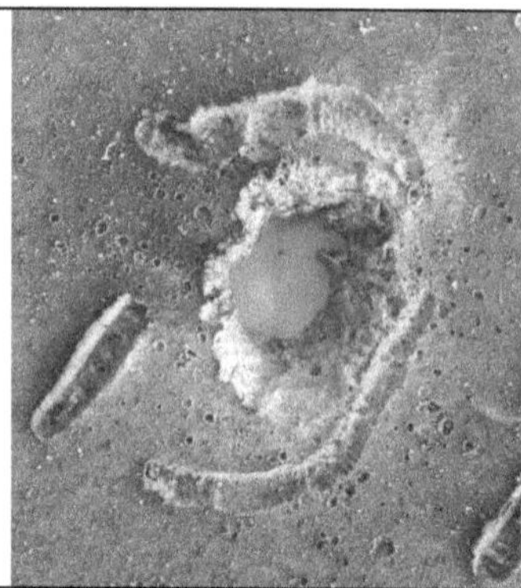
Plate 2	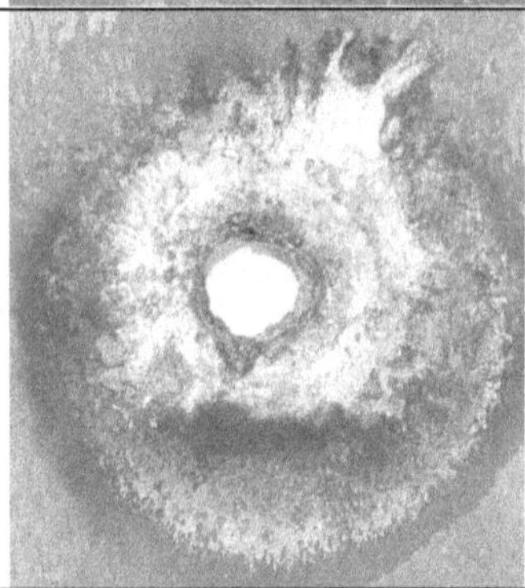	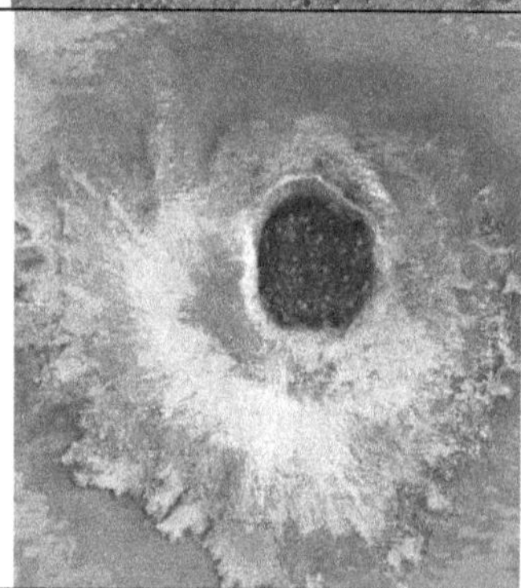
Plate 3	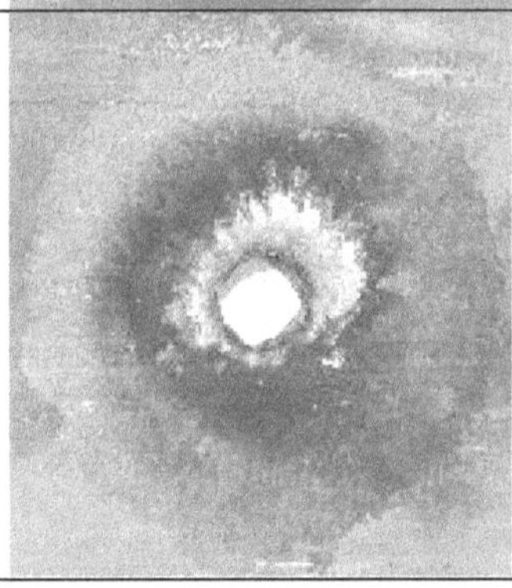	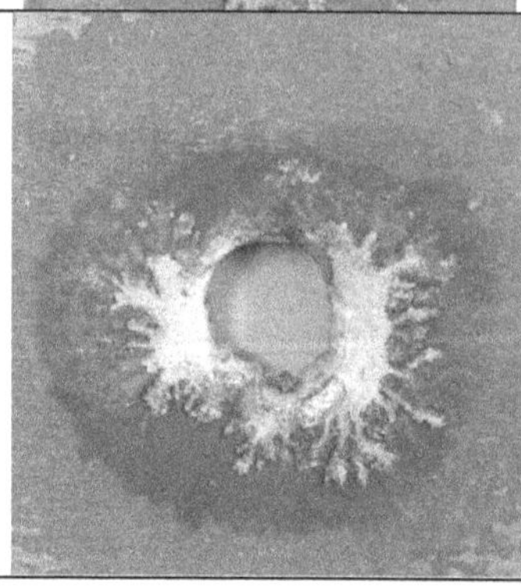

Plate 4	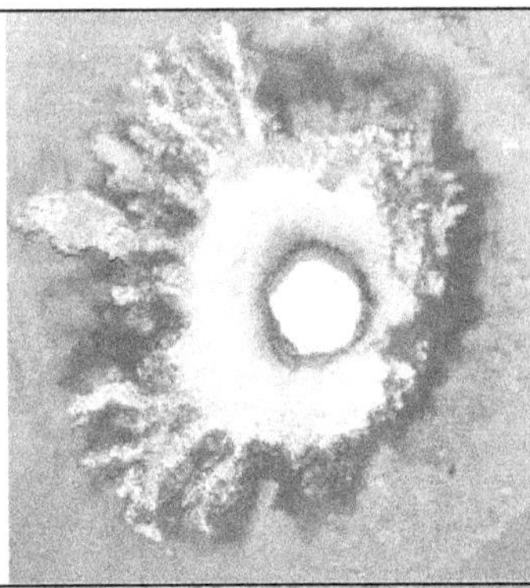	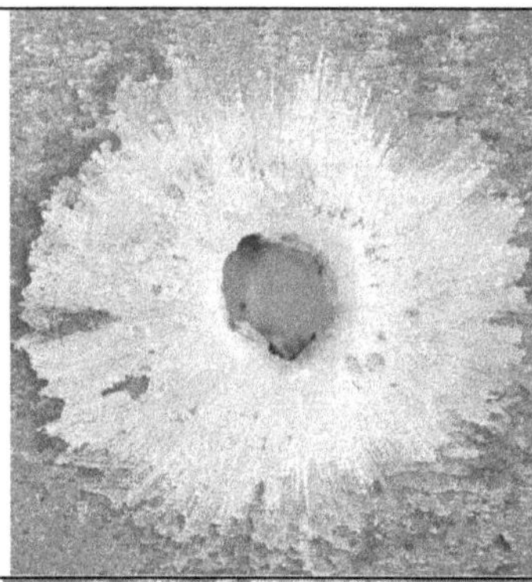
Plate 5	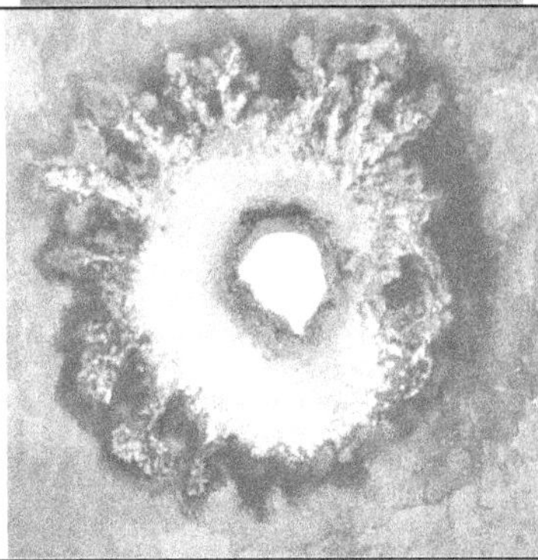	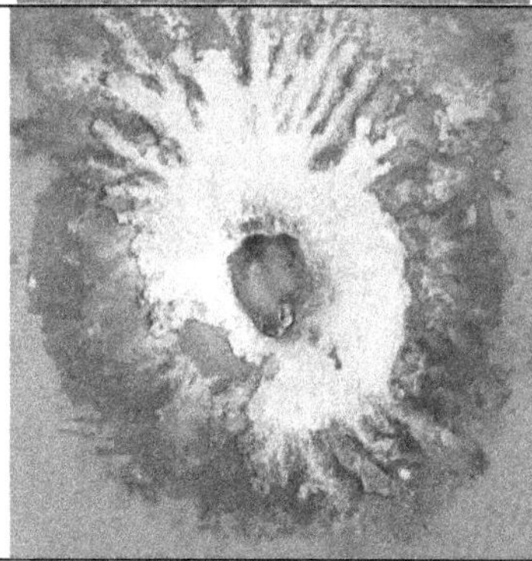
Plate 6		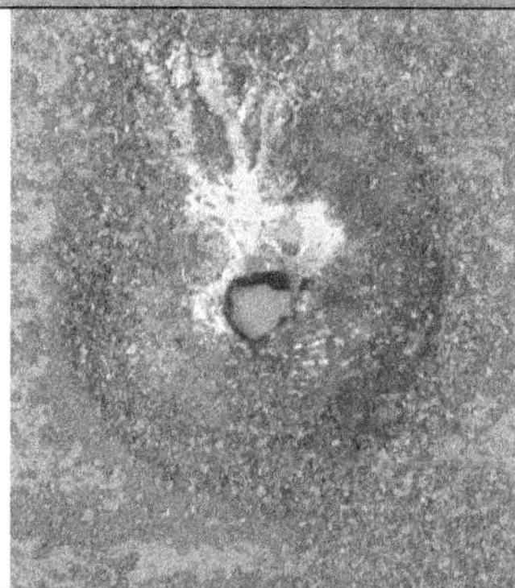
Plate 7		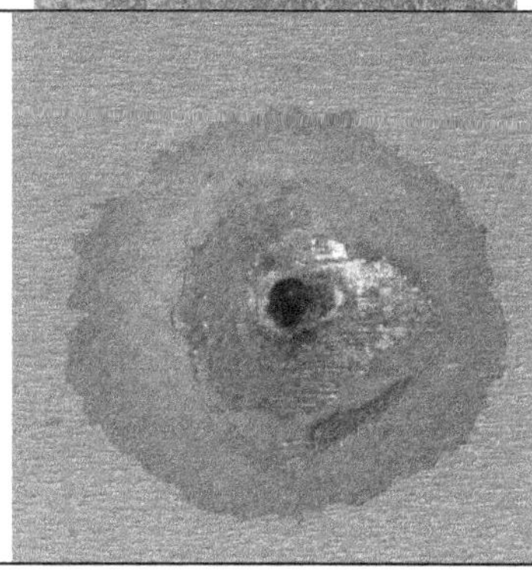

Plate 8	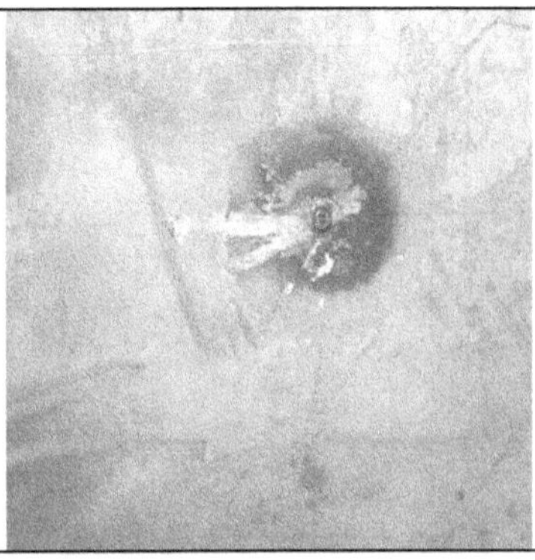	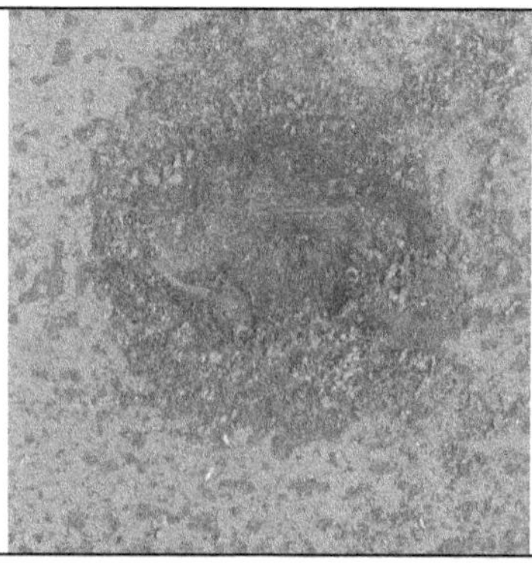

Proceedings of the 2019 Hypervelocity Impact Symposium
HVIS2019
April 14-19, 2019, Destin, FL, USA
HVIS2019-028

Extended Abstract+

Momentum transfer in hypervelocity cratering of meteorites and meteorite analogs: Implications for asteroid deflection

George J. Flynn[a*], Daniel D. Durda[b], Mason J. Molesky[c], Brian A. May[c], Spenser N. Congram[c], Colleen L. Loftus[c], Jacob R. Reagan[c], Melissa M. Strait[c], Robert J. Macke[d]

[a]*SUNY-Plattsburgh, 101 Broad St., Plattsburgh, NY 12901 USA*
[b]*Southwest Research Institute, 1050 Walnut Street, Suite 300, Boulder, CO 80302 USA,*
[c]*Alma College, Alma, MI 48801 USA*
[d]*Vatican Observatory, V-00120 Vatican City*

Abstract

Asteroid porosity ranges from 0 to >50%, with most >20%, and some asteroids exhibit a water feature in their reflection spectra. Porosity and hydration are expected to influence the momentum transferred in hypervelocity collisions. We conducted a series of measurements of the post-impact momentum, characterized by a factor β, the ratio of the total linear momentum acquired by the target to the momentum of the impactor. We measured β for anhydrous meteorites, samples of their asteroidal parent bodies, spanning a wide range of porosities: 7 samples of the CV3 carbonaceous chondrite Northwest Africa (NWA) 4502 (2.1% porosity), 7 samples of the ordinary chondrite NWA 869 (6.4% porosity), and 4 samples of the ordinary chondrite Saratov (15.6% porosity), as well as 2 samples of terrestrial pumice (80% porosity). We also measured hydrous meteorite analog targets, including 2 samples of terrestrial serpentine (17.9% porosity) and 4 samples of terrestrial montmorillonite (51.5% porosity), the two clay minerals that dominate the composition of the hydrous CI carbonaceous chondrite meteorites, as well as 4 samples of hydrous meteorite analog material prepared by powdering and hydrating an anhydrous carbonaceous chondrite. We found that for both anhydrous and hydrous samples β decreased with increasing porosity, consistent with hydrocode modeling. The β for each target type was >2 demonstrating that crater ejecta makes a significant contribution to recoil in hypervelocity collisions.. The β values we measured for the anhydrous samples are larger, with β = 3.55 for NWA 4502, 2.69 for NWA 869, 2.10 for Saratov, and 2.15 for pumice, than results from hydrocode modeling for 10 km/s impacts into relatively strong, porous rock targets. The momentum enhancement by ejecta (β - 1) for the moderate porosity (17.9%) hydrous serpentine targets (β = 4.70), the highly porous (51.55% porosity) hydrous montmorillonite targets (β = 2.79), and the intermediate porosity (~26%) CI-analogs (β = 2.99) are much larger than β value for anhydrous targets of similar porosity, indicating jetting of water vapor could significantly affect deflection of hydrous asteroids and comets in natural or human-induced collisions.

Keywords: hypervelocity cratering; meteorites; asteroids; kinetic impact deflection, hydrous targets.

---

1. Introduction

The response of asteroids or comets to hypervelocity impact influences the morphology and size of craters, the orbital change resulting from collisions, fragmentation in catastrophic collisions, and the size-frequency distribution and the speed distribution of dust produced by the impact. The physical properties of asteroids and comets are important in understanding the change to their orbits from natural or human-induced collisions. The extent of this alteration is important for asteroid or comet orbital evolution modeling and for attempts to deflect an asteroid or comet on a collision course with Earth using a kinetic impactor or a

+ Full paper can found at https://www.sciencedirect.com/journal/international-journal-of-impact-engineering/special-issue/10MTGP5W4VJ

* Corresponding author. Tel.: +1-518-564-3163.
E-mail address: flynngj@plattsburgh.edu.

surface/subsurface nuclear explosion. The momentum transferred to the target in a hypervelocity impact is characterized by a factor β, the ratio of the total linear momentum acquired by the target to the linear momentum of the impactor, given by:

$$\beta = (m_p v_p + p_e)/m_p v_p = M_t V_t / m_p v_p \qquad \text{(Equation 1)}$$

where m_p is the mass of the projectile, v_p is the speed of the projectile, p_e is the magnitude of the momentum of the crater ejecta opposite the recoil direction, M_t is the mass of the target, and V_t is the change in speed of the target due to the impact. Two major factors contribute to β: the momentum of the impactor and the recoil from crater ejecta produced in the impact.

Porosities inferred for nearly two dozen asteroids range from nearly zero to >50%, with most >20% and with a mean of ~30% [1]. Asteroids exhibit a diversity of mineralogies. Reflection spectra show the inner half of the main-belt is dominated by S-type asteroids, believed to be similar to ordinary chondrite meteorites, a link firmly established when samples from S-type asteroid Itokawa were shown to have mineralogy and mineral chemistry identical to those of thermally metamorphosed LL ordinary chondrites [2]. The outer half of the main belt is dominated by darker asteroids, mainly classified as C-type, which are believed to be similar in composition to carbonaceous chondrite meteorites. Reflectance spectra of some show a 0.7-μm feature, indicating the presence of hydrous minerals [3]. These dark hydrous asteroids are similar to the hydrous CI or CM carbonaceous chondrite meteorites, while the dark, anhydrous asteroids are likely similar to the anhydrous carbonaceous chondrites, mainly CV or CO meteorites. The recovery of meteorites spanning the range of mineralogies seen among the asteroids demonstrates that members of each type occur in the Near Earth Object (NEO) population. Mineralogy, porosity and water content each effect the response to hypervelocity impact. The average velocity of crater ejecta decreases with decreasing material strength and with increasing porosity [4], and the recoil of liquid water targets is more than 10 times that of dry rocks [5].

At the 2017 Hypervelocity Impact Symposium, we reported hypervelocity recoil results from samples of three anhydrous meteorites spanning a range of porosities, including five cratering impacts of the low porosity CV3 carbonaceous chondrite Northwest Africa (NWA) 4502, seven cratering impacts of the intermediate porosity NWA 869 L3-6 ordinary chondrite, and one cratering impact of the high porosity L4 ordinary chondrite Saratov [6]. For each target type the β values were tightly clustered except for one cratering impact into NWA 4502, which gave a β value more than three times the other NWA 4502 impacts. We suggested that impactor struck a vein of hydrous weathering product. We have extended that work to include more samples of NWA 4502 and Saratov and hydrous targets of montmorillonite, serpentine, and laboratory hydrated CI-analog material.

1.1. Samples and Experimental Procedures

We have measured the momentum transfer in hypervelocity cratering for 18 anhydrous meteorite targets spanning a range of porosities from 2.1% to 15.6%, and 2 samples of terrestrial pumice (80% porosity) to extend anhydrous target measurements to higher porosity than is sampled by meteorites. We also measured momentum transfer for hydrous meteorite analogs, including 2 samples of terrestrial serpentine (18% porosity) and 4 samples of terrestrial montmorillonite (55% porosity), the two clay minerals that dominate in the hydrous CI meteorites [7]. These CI carbonaceous chondrites are rare and generally unavailable in large enough pieces for cratering measurements, but they are represented in the NEO population. For example Bennu, the target of NASA's OSIRIS-REx sample return, is a primitive, hydrous, carbon-rich asteroid with bulk density of 1.26 g/cm^3 and porosity of 40% ± 10% [8]. This suggests similarity to the CI meteorite Orgueil, which has a bulk density of 1.5 g/cm^3 [9]. To investigate the cratering recoil of CI-like targets we prepared hydrous meteorite analog targets. Fragments of NWA 4502 were crushed to a fine powder and laboratory hydrated in slightly alkaline water at high temperature in a pressure bomb for months, using a procedure intended to mimic, on a shorter time scale, the hydration process that occurs on asteroid parent bodies [10].

The measurement procedures for the grain and bulk densities, unconfined compressive strengths and recoil speeds are described in Flynn et al. [6]. Each target was suspended on a light nylon string about 2 m in length in the target chamber of the two-stage light gas gun at the NASA Ames Vertical Gun Range (AVGR). The AVGR Chamber was pumped to ~0.5 Torr to minimize atmospheric interference with the projectiles. The gun launched a nylon sabot containing a 1/16th or 1/8th inch diameter Al-sphere. After the sabot was stripped from the projectile the speed was measured using a laser-interrupt system, and the projectile was imaged to verify its structural integrity. Projectile speeds ranged from ~4 km/s to ~5.9 km/s, comparable to the mean collision speed in the main-belt [11]. To minimize the effect of gas emitted by the gun, a mylar disk was located downstream from the sabot stripper, and a paper disk was placed over the exit port of the gun chamber. We confirmed that there was no detectable target recoil when the gun was fired with a normal powder load but no projectile. High-speed video cameras were located in the viewing ports of the AVGR chamber, and we determined β by measuring the recoil speed (V_t) of the target relative to the 1x1inch grid pattern and combining this with the target mass, impactor mass and speed using Equation 1. In four cases the targets split into roughly equal fragments, with only a small mass of additional material. In these cases we determined the speed of each fragment and β was determined from the sum of the linear momentum transferred to each fragment.

2. Results

The typical mass of the crater ejecta is hundreds of times the mass of the impactor, thus even a relatively low mean ejecta speed produces a significant recoil momentum.

2.1 Anhydrous Targets

NWA 4502 is a minimally weathered (W1) and minimally shocked (S2), oxidized CV3 carbonaceous chondrite. Porosities of two of NWA 4502 whole stones are 3.0% and 1.2% (mean porosity 2.1% ± 0.9%), which is much lower than typical, since Macke [12] reported a mean porosity of 17.7% for oxidized CV3 falls. This low porosity is consistent with the observations that the matrix of NWA 4502 is unusual, being "rather compact" [13], since most meteorite porosity is within the matrix as cracks or as voids between the grains. The unconfined compressive strengths of four samples of NWA 4502 ranged from 18.9 to 58.8 MPa, with a mean of 32.9 ± 17.8 MPa, comparable to the unconfined compressive strength of the CV3 meteorite Allende [14]. Five of the seven cratering impacts into NWA 4502 targets produced β values in the narrow range from 3.02 to 3.93, with a mean β of 3.55 ± 0.38, consistent with the relatively narrow range of β values for other types of targets. They produced ejecta masses, the difference between the initial target mass and the masses of the largest fragments, ranging from 3.0 g to 18.0 g, with a mean ejecta mass of 10.3 g. In each case the ejecta mass was more than 500 times the mass of the impactor. Two NWA 4502 shots produced remarkably different values of β, 8.95 and 11.72, with consistent values from both camera angles. This suggests these impactors struck a different material than the other five NWA 4502 cratering impacts. Although NWA 4502 was collected in a dry desert, terrestrial weathering can still occur relatively rapidly [15], so we suspected impact into hydrous weathering veins, which are prominent on the surfaces of our NWA 4502 stones, gave rise to these higher β values [6]

NWA 869 has a weathering grade W1, indicating that there are only small oxide rims around metal and troilite and small oxide veins, and a shock stage S3, indicating it was only weakly shocked (15 to 20 GPa). We measured the porosities of six fragments of NWA 869. The mean grain density of these six samples of NWA 869, 3.58 ± 0.08 g/cm^3, the same as the mean value of 3.58 ± 0.01 g/cm^3 reported by Macke [12] for L-chondrite falls, consistent with minimal effects from terrestrial weathering, which lowers the grain density. The mean bulk density of these six samples, 3.36 ± 0.04 g/cm^3, consistent with the mean value of 3.30 ± 0.01 g/cm^3 reported by Macke [12] for L-chondrite falls. The porosity values ranged from 2.7% to 10.2%, with a mean of 6.4% ± 2.8%, slightly lower than typical for L-chondrite falls (8.0% ± 0.3% [12]). The unconfined compressive strengths of the six samples of NWA 869 ranged from 52 to 114 MPa, with a mean value of 87.4 ± 25.6 MPa, comparable to the range reported for L-type ordinary chondrites [16]. All seven of the cratering impacts into NWA 869 targets produced β values in the narrow range from 1.94 to 3.74, with a mean β of 2.69 ± 0.42. They produced ejecta masses ranging from <0.5 g to 5.0 g, with a mean ejecta mass of 2.3 g, much lower than the ejecta masses produced in the NWA 4502 cratering impacts.

Saratov is an L4 ordinary chondrite, showing minimal weathering and shock (S2). We measured porosities of two Saratov samples of 15.2% and 16.1%, giving a mean porosity of 15.6% ± 0.6%, much in the form of interstitial voids rather than cracks. Saratov has unusually high porosity among ordinary chondrites, almost double the mean of 8.0% + 0.3% for L-chondrite falls [12]. All four of the Saratov cratering impacts produced β values in the range from 1.49 to 2.84, a mean β of 2.10 ± 0.60.

To extend our anhydrous targets to higher porosity we impacted samples of highly vesicular, rough-textured terrestrial pumice. We cut ten rectangular blocks, averaging about 1 g with volumes ~2 cm^3, from an unshot pumice sample. These had densities ranging from 0.45 to 0.62 gm/cm^3, with a mean density of 0.538 ± 0.059 g/cm^3. Taking literature values of 2.4 to 2.8 g/cm^3 as representative of non-porous samples, we estimate the porosity of these pumice targets to be in the range of 77 to 82%. The two pumice impacts produced β values of 2.3 and 2.0, giving a mean β of 2.15 ± 0.21.

2.2 Hydrous Targets

To investigate the effect of hydrous minerals on cratering recoil we used targets of serpentine and montmorillonite, the two clay minerals that dominate the structure of CI meteorites [7]. We measured the porosities of two serpentine samples as 17.0 and 18.8%, giving a mean porosity of 17.9% ± 1.3%, and two samples of montmorillonite as 51.60% and 51.35%, giving a mean porosity of 51.5% ± 0.2%. A 61.96 g serpentine sample was immersed in water for two days before the shot in an attempt to add liquid water to the target. Its mass increased to 64.89 g, adding almost 3 g of water. However, we could not determine how much water was lost during the more than half-hour interval while the AVGR chamber was pumped to ~0.5 torr before the shot. These two serpentine impacts gave β values of 4.40 and 4.93, with a mean β of 4.7 ± 0.4. The values are quite similar for the two samples, indicating that the immersion in water did not change the β value, possibly because of water evaporation during pump down of the AVGR chamber. Four montmorillonite impacts gave β values in the range from 2.10 to 3.34, with a mean β of 2.8 ± 0.5. Both of these β values are significantly higher than we measured for anhydrous targets of similar porosities.

Labroratory Hydrated NWA 4502 was prepared by crushing the CV3 NWA 4502 using a procedure shown by Jones and Brearly [10] to produce CI-like material. We confirmed the production of clays in our processing by infrared spectroscopy, but we cannot ensure complete alteration of anhydrous silicates to clays. The mean density of the resulting samples was ~1.8 g/cm^3. If the targets were fully hydrous then, using the mean grain density of the CI meteorites (2.42 g/cm^3 [12]), the porosity would be ~26%. Any remaining crystalline silicates would increase the mean grain density, thus giving even higher porosity. We performed four cratering impacts of these CI-like targets. These very weak targets could not be hung directly from the ceiling of the chamber, since they were so weak they would come apart under their own weight if the support line were attached to the top. So, we suspended each of these laboratory hydrated CI stimulant targets in a hair net sling of negligible mass compared to the target. The sling supported the bottom of the target, but it did not interfere with the incoming projectile or with the recoil speed of the target. This provided the necessary support without interfering with the projectile or recoil. The CI-analog targets were impacted at ~5 km/s. These four targets gave β values of 3.58, 3.83, 2.36, and 2.17, giving a mean β of 2.99 ± 0.84.

3. Discussion and Conclusions

Figure 1 shows the mean β for each of our stone target types as well as the value for low porosity river rock. Two trends are clearly evident. First, β decreases with increasing porosity, consistent with results of hydrocode modeling [17]. Second, the recoil enhancement due to crater ejecta for each of the hydrous targets studied is significantly greater than the value for an anhydrous target of similar density. These results suggest hydration substantially enhances momentum transfer, likely by jetting of water vaporized in the hypervelocity cratering event. The enhancement of β for hydrous targets has important implications for orbital evolution from natural collisions and momentum requirements for kineic or nuclear deflection of hydrous asteroids and icy comets. The β values we measured on anhydrous porous targets are significantly larger than predicted by hydrocode modeling of moderately strong porous rocks, since SPH modeling for 10 km/s cratering impacts into rocks (100 MPa strength) gave β <1.2 over this porosity range [17]. Our values for porous hydrous targets are even higher, but effects of water vaporization have not been modeled. For each stone target, over the full range of porosity investigated, the ratio of the momentum transferred by the crater ejecta to the momentum of the projectile (β – 1) is >1. Thus, momentum enhancement by ejecta must be considered in design of kinetic impact deflection missions and modeling of the alteration of orbits by collisions.

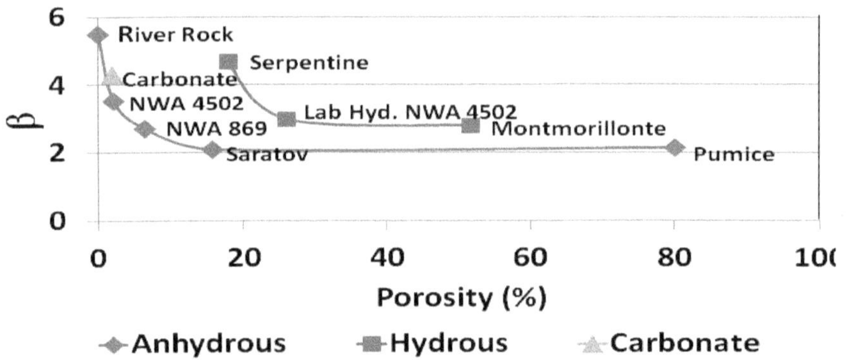

Fig. 1. Mean porosity vs. β for three types of anhydrous meteorite targets, NWA 4502, NWA 869 and Saratov, as well as terrestrial pumice and the river rock from Housen and Holsapple [18], as well as three hydrous CI-analog materials, serpentine, laboratory hydrated NWA 4502, and montmorillonite. Lines show the trend in β for anhydrous and hydrous targets.

References

[1] Britt, D.T., D. Yeomans, K. Housen, G. Consolmagno, 2002. Asteroid density, porosity, and structure, in Asteroids III, U. of Arizona Press, 485-500.
[2] Nakamura, T. and 21 co-authors , 2011, Itokawa Dust Particles, Science, 333, 1113-1116.
[3] Campins, H. and 5 co-authors, 2018. Compositional Diversity Among Primitive Asteroids, in Primitive Meteorites and Asteroids, Elsevier, 345-369.
[4] Michikami, T. and 3 co-authors, 2007. Ejecta velocity distribution for impact cratering experiments, Planetary and Space Science, 55, 70-88.
[5] Holsapple, K.A., K.R. Housen, 2012. Momentum transfer in asteroid impacts. I. Theory and scaling, Icarus, 221, 875-887..
[6] Flynn, G. J. and 7 co-authors, 2017, Hypervelocity impact cratering of chondritic meteorites, Procedia Engineering, 204:146-153.
[7] Bass, M.N., 1971. Montmorillonite and serpentine in Orgueil meteorite, Geochim Cosmochim. Acta, 35, 139-147.
[8] Chesley, S. R. and 15 co-authors, 2014. Orbit and bulk density of the OSIRIS-REx target Asteroid (101955) Bennu, Icarus, 235, 5-22.
[9] Britt, D.T. and G.J. Consolmagno, 2003. Stony meteorite porosities and densities, Meteorit. Planet. Sci., 38, 1161-1180.
[10] Jones, C.L., A. Brearley, 2006. Experimental aqueous alteration of the Allende meteorite, Geochim.. Cosmochim. Acta, 70, 1040-1058.

[11] Bottke, W.F., M. C. Nolan, R.Greenberg, and R.A. Kolvoord, 1994. Velocity distributions among colliding asteroids, Icarus, 107, 255–268.
[12] Macke R.J., 2010. Survey of meteorite physical properties: Density, porosity and magnetic susceptibility. Ph.D. dissertation. University of Central Florida.
[13] Sapah, M.S., A.N. Krot, Y. Amelin, 2013. Mineralogy and Petrography of Refractory Inclusions in NWA 4502, Lunar Planet. Sci. Conf. XLIV, Abs. #1036.
[14] Cotto-Figueroa, D. and 8 co-authors, 2016. Scale-dependent measurements of meteorite strength, Icarus, 277, 73-77.
[15] Al-Kathiri, A., B.A. Hofmann , A.J.T. Jull , E. Gnos, 2015. The Weathering of meteorites from Oman:, Meteoritics & Planet. Sci., 40, 1215-1239.
[16] Flynn, G.J., G.J. Consolmagno, P. Brown, R.J. Macke, 2018. Physical properties of the stone meteorites, Chemie der Erde, 78, 269-298.
[17] Bruck Syal, M., J.M. Owen, P.L. Miller, 2016. Deflection by kinetic impact: Sensitivity to asteroid properties, Icarus, 269, 50-61.
[18] Housen, K. R. and K. A. Holsapple , 2012. Deflecting Asteroids by Impacts: What is Beta?, Lunar and Planet. Sci. Conf.XLIII, Abst. #2539).

Proceedings of the 2019 Hypervelocity Impact Symposium
HVIS2019
April 14-19, 2019, Destin, FL, USA
HVIS2019-032

Extended Abstract[+]

Simulations of Magnetic Fields Produced by Asteroid Impact: Possible Implications for Planetary Paleomagnetism

David A. Crawford*

Crawford Technical Services, LLC
Winona, Minnesota, USA

1. Introduction

The origin and evolution of the Moon's magnetic field has been a major question in lunar science ever since Luna 1 made the first magnetic measurements in the vicinity of the Moon in 1959. Orbital measurements show that the magnetic field at the surface of the Moon has local scale lengths on the order of 1-100 km. While this could suggest a correlation with impact craters, most lunar magnetic anomalies don't appear to correlate with known geologic structures, including impacts [1]. However, the magnetic field produced by impact events are spatially and temporally complex. Add in the complexity of remanence acquisition (localized regions of heating/cooling and/or shock that can produce remanence in the presence of a magnetic field) and we have the potential for a complex pattern to emerge. Wieczorek et al. [1] showed just how such complexity may play out. In their simulations, some lunar magnetic anomalies may be caused by regions of concentrated magnetic materials associated with fragments of the South Pole-Aitken impactor, especially if the impactor was differentiated with an iron core. More recently, Oliveira et al. [2] showed that magnetic anomalies associated with five large lunar basins may be caused by impact melt sheets that cooled in the presence of an early lunar dynamo. In this paper we will look at an alternative explanation for many lunar anomalies that doesn't require the presence of a lunar dynamo. At least some lunar anomalies may be associated with a deeper, thicker yet more varied region of magnetization acquired by rocks that became hot and cooled rapidly enough during crater formation to have acquired the transient magnetic field produced by the impact itself.

In prior work, we developed a model of electric charge separation and associated electric and magnetic fields produced during hypervelocity impacts ([3] and references therein). The model, referred to as the CTH electrostatic charge model, was implemented into the CTH shock physics hydrocode [4] and validated against experimental vertical impacts (0.3-0.5 cm aluminum projectiles, 1-6 km/s). In this work, we extend the length scale of our simulations to include a range of possible asteroid impacts on planetary surfaces to help provide a more complete picture of the magnetic field environment during natural crater formation. We have added the ability to magnetize materials as they cool through the Curie point. This was added to CTH as an extension to the Brittle Damage with Localized thermal softening (BDL) model [5]. With these new capabilities, we can predict the magnetic field throughout time and throughout the entirety of the spatial domain and explore the complex patterns of magnetization that may appear, especially associated with basin-scale impacts on the Moon.

+ Full paper can found at https://www.sciencedirect.com/journal/international-journal-of-impact-engineering/special-issue/10MTGP5W4VJ

* Corresponding author. Tel.: +1-507-459-0659.
E-mail address: dave_crawford@q.com.

2. Calculation of magnetic fields from lunar impacts

We performed eight CTH simulations of lunar impact events for projectile radii of 0.1-20 km, producing lunar craters with diameters of 4.4-660 km. We used spherical projectiles with an impact velocity of 20 km/s, 45° from horizontal. The target was a half space initially in hydrostatic equilibrium (gravitational acceleration appropriate for the Moon was used). It had a density at the surface of 2650 kg/m^3. For the projectile, we used the ANEOS description for dunite with molecules [6] and for the target, we used the ANEOS description for SiO_2 with molecules [7]. We used adaptive mesh refinement, with an initial resolution equal to 10 zones across the projectile radius, and ran the calculations on the 128-core cluster at Crawford Technical Services. It took approximately two days for each calculation to run to completion.

The Brittle Damage with Localized thermal softening (BDL) model with parameters appropriate for basalt was used to describe constitutive behavior of the target [5]. The BDL model starts with the damage model of Collins et al. [8], where the yield strength of fully damaged material follows a Coulomb friction law and thermal softening of bulk material occurs approaching the melting point. We extended the model by estimating statistical crack spacing based on strain rate, a determination of shear heating within cracks and heat loss via conduction away from cracks. With this extension, thermal softening can occur as the temperature within a crack approaches the melt temperature. By temporarily weakening materials, the temperature rise near cracks plays a role in the development of the complex crater form but can also lead to magnetization of materials as they cool below the Curie point.

We used the CTH electrostatic charge model, in conjunction with the computational tree-solving algorithm we originally wrote for self-gravity [9], to calculate the magnetic fields produced during the simulations. While vaporization of both the projectile (dunite) and target (SiO_2) materials occurs, the most significant ion produced is Mg+ from the dunite projectile.

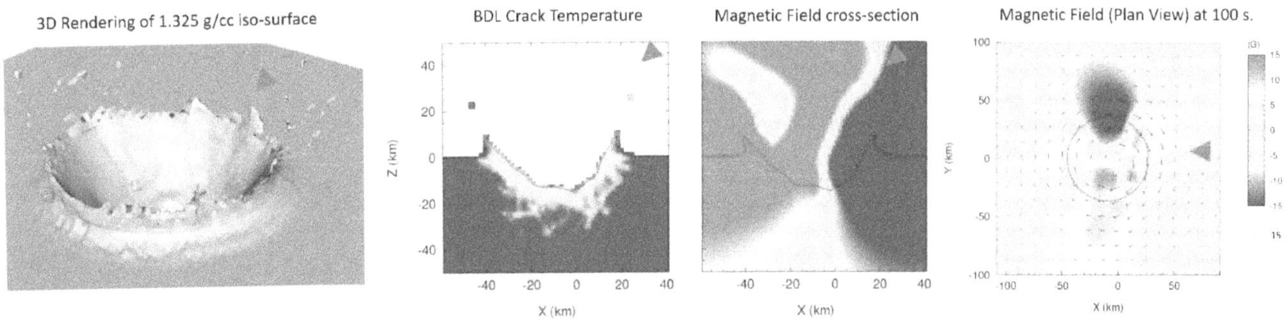

Fig. 1. CTH simulation of a 72-km central-peak crater on the Moon 100 seconds after impact. From left-to-right: 3D rendering of the 1.325 g/cc density iso-surface, crack temperature from the BDL model, Y-component of the magnetic field in cross section, and all components of the magnetic field (horizontal components depicted with arrows, vertical component with color) looking from above. Magnetic field strength shown in Gauss (1G = 10^{-4} Tesla). Pre-impact projectile trajectory indicated by the blue triangles and final crater rim indicated by circles.

Figure 1 shows a simulation of a 2-km radius projectile striking the Moon at 20 km/s, 45° from horizontal. Strong magnetic fields were produced throughout the target region. At 20 seconds into the simulation, field strengths were approximately 10^{-2} Tesla with the largest fields confined within and below the crater or in the free space above the target. At 100 seconds (shown in the figure), field strength declined by about an order of magnitude (10^{-3} T) but extended significantly beyond the crater rim and was of similar magnitude both above and below the crater. By the end of crater formation (~644 seconds), the magnetic field had declined a further two orders of magnitude (10^{-5} T) and was strongest in the region just below the crater.

Crawford and Schultz [10] predicted magnetic fields generated by hypervelocity impact scale linearly with impactor radius, a significant implication for natural impact crater events where the length scale can be 10^4 -10^6 times larger than the laboratory scale. According to this estimate, the impact of a 1-km asteroid could generate a 0.03 Tesla magnetic field lasting for about 100 seconds (~1000 times Earth's surface field). As shown in Figure 2a, CTH analyses of lunar impact events using the electrostatic charge model also show the transient magnetic field has a linear dependence on projectile size for impactor radii smaller than 1-2 km (crater diameters smaller than 40-60 km). For larger projectiles, the maximum field continues to increase with projectile size, but at a slower rate. However, the magnetic fields predicted by CTH in this study are about three orders of magnitude lower than prior work [3, 10] would imply. The biggest contributor to the discrepancy comes from the different dominant ion in the simulations. The Mg+ ion (E_i = 7.65 eV) has a higher ionization energy than the Ca+ ion (E_i = 6.11 eV) which was the dominant ion in the prior work. According to the Saha equation this translates to a 400-fold reduction in ionization at 3,000 K [3].

3. Magnetization of target materials

By explicitly computing the thermal properties in the vicinities of cracks, at least in a statistical sense, the BDL model already has most of the capabilities needed to estimate magnetization. The dominant magnetic carrier of lunar materials is thought to be iron [11] with a Curie point of 1043 K. If material in the vicinity of a crack heats above the Curie point and then subsequently cools, it will retain a magnetization with a magnitude proportional to the magnetic field at that time. For this study, we added a capability to the BDL model to record the magnetic field (magnitude and orientation) at the time when materials in the crack vicinity dropped below the Curie point. We also record the local ratio of the thermal zone thickness to the crack spacing – a direct measure of the volume fraction of magnetized material. With this recorded information and assuming reasonable values of material thermoremanence susceptibility (χ_{trm}) and saturation remnant magnetization, we can post-process to create intensity maps of magnetic anomalies that may be present on the Moon today.

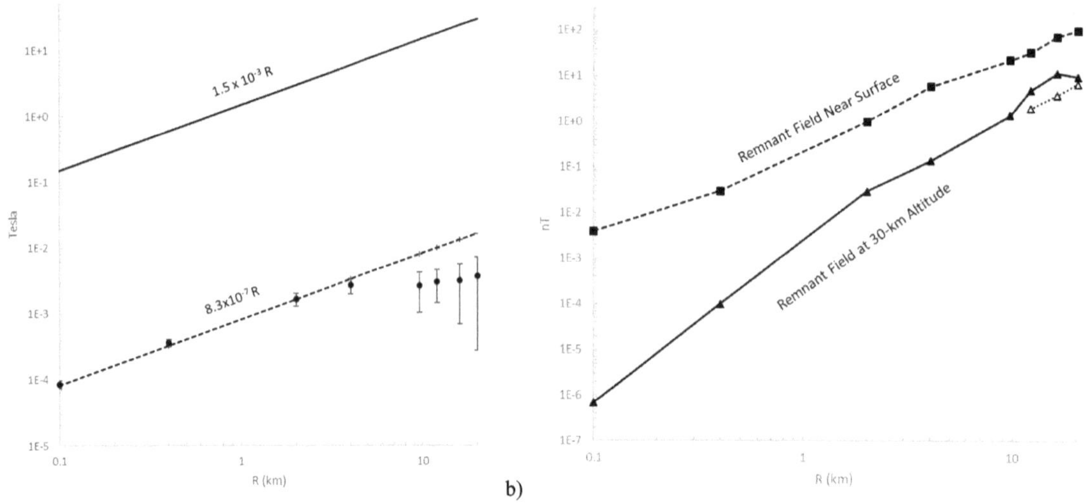

Fig. 2. a) Magnetic field strength in Tesla vs. projectile radius computed by CTH using the electrostatic charge model. The fields were measured in CTH at four points on the target surface 50*R* from the crater center (up-range, down-range, left and right). The vertical bars show the range of values measured. Small projectiles follow a linear trend (dashed line) as predicted by prior work [10] (solid line), although the field strength is about 3 orders of magnitude lower (see text). b) Remnant magnetic field predicted by the simulations if measured near the surface (squares) and if measured at 30-km orbital altitude (triangles). For these plots we use a value for χ_{trm}=3x10^{-5}, which is reasonable for pristine feldspathic highland rocks [1].

4. Discussion

For the range of crater sizes studied, the remnant magnetic field near the lunar surface is anywhere from one to four orders of magnitude larger than if measured at 30-km altitude (see Fig 2b). Not surprisingly, the surface field is dominated by nearby materials, typically in the crater wall and materials uplifted to form the crater rim. These materials are also more likely to be more magnetically susceptible (e.g. mare basalts, mafic impact-melt breccias or materials containing meteoric iron). However, surface remnant fields are also more likely to be disturbed, even erased, by subsequent impact events over the course of lunar history.

Assuming χ_{trm} of 3x10^{-5}, the predicted remnant fields at 30-km altitude produced signals comparable to lunar orbital observations (typically a few nT) only for large basin-sized events – crater diameters greater than ~200 km. This helps explain why lunar magnetic anomalies don't generally correlate with impact craters. Craters of this size are relatively rare. Smaller craters may have magnetic anomalies if they formed in regions of higher magnetic susceptibility, such as mare basalts or impact breccias or regions containing relatively high proportions of meteoric iron, but such regions would have to be 10's of km thick to produce an observable field at orbital altitude. Again, we expect such situations to be relatively rare.

In Figure 3, we show a comparison between 30-km altitude remnant field predictions from CTH and the magnetic anomalies associated with two lunar basins: Nectaris and Serenitatis – two of the five basins studied by Oliveira et al. [2]. The magnetic field data are from Purucker and Nicholas [12] graphed using Lambert azimuthal equal-area projection with grid lines overlain for every 10° of latitude and longitude. In each case the crater rim is indicated by the solid circle with additional features such as outer rings and inner depressions as identified in Neumann *et al.* [13] shown by dashed circles. The Serenitatis plot also shows the location of the Serenitatis North basin (210 km diameter).

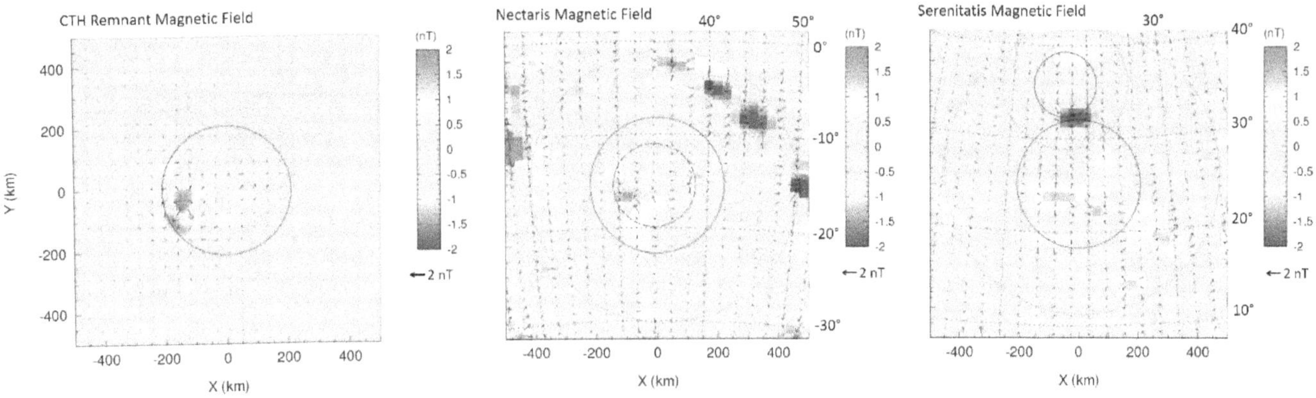

Fig. 3. CTH simulation of a 420-km crater compared with two similarly sized lunar basins: Nectaris (440 km) and Serenitatis (416 km).

As a first cut, the remnant field comparisons in Figure 3 are remarkably good. The strength of the simulated fields and their spatial extents are comparable to the observations. The comparison potentially demonstrates that transient impact-generated magnetic fields may be a viable additional mechanism for the magnetization of lunar rocks. In future work we hope to use simulations with different impact angles, speeds, impactor compositions and geometries to see if we can glean more information about the specific impactor properties that formed each of these basins.

5. Conclusions

We show that impacts occurring on the Moon in pristine feldspathic highland rocks with relatively low thermoremanence susceptibility produce magnetic anomalies observable in orbital magnetic field data only for relatively rare events - large lunar basins greater than ~200 km diameter. If occurring in thick units with higher thermoremanence susceptibility, craters as small as 50-100 km may produce orbital anomalies. Craters at nearly all scales may leave behind remnant magnetic fields observable at the surface or in samples. For at least some large basins on the Moon, these simulated anomalies are similar enough to orbital magnetic field data for us to conclude there is an alternative explanation for these lunar anomalies that doesn't require the presence of a lunar dynamo. At least some lunar anomalies may be associated with a deep and thick region of magnetization acquired by rocks in the presence of the transient magnetic field produced by the impact itself.

References:

[1] Wieczorek, M. A.,Weiss, B. P. and Stewart, S. T., 2012, An Impactor Origin for Lunar Magnetic Anomalies, *Nature, Vol. 335,* pp. 1212-1215.
[2] Oliveira, J. S., Wieczorek, M. A., and Kletetschka, G., 2017, Iron abundances in lunar impact basin melt sheets from orbital magnetic field data, *Journal of Geophysical Research: Planets*, 122, pp. 2429–2444.
[3] Crawford, D. A., 2015, Computational modeling of electrostatic charge and fields produced by hypervelocity impact, *Proceedings of the 13th Hypervelocity Impact Symposium, Procedia Engineering, Vol. 103*, pp 89-96.
[4] McGlaun, J. M, Thompson S. L. and Elrick, M. G., 1990, CTH - A Three-Dimensional Shock-Wave Physics Code, *International Journal of Impact Engineering, Vol. 10*, p. 351.
[5] Crawford, D. A. and Schultz, P. H., 2013, A Model of Localized Shear Heating with Implications for the Morphology and Paleomagnetism of Complex Craters, *Large Meteorite Impacts and Planetary Evolution V*, Sudbury, Canada.
[6] Canup, R. M., Barr, A. C. and Crawford, D. A., Lunar-forming impacts: High-resolution SPH and AMR-CTH simulations, *Icarus, 222*, pp. 200-219 (2013).
[7] Melosh, H. J., 2007. A hydrocode equation of state for SiO2, *Meteoritics and Planetary Science*, 42, pp. 2079–2098.
[8] Collins, G.S., Melosh, H. J. and Ivanov, B. A., 2004, *Meteoritics and Planetary Science*, 39:2, pp. 217-231.
[9] Crawford, D. A., 2010, Parallel N-Body Gravitational Interaction in CTH for Planetary Defense and Large Impact Simulations, *Proceedings of the 11th Hypervelocity Impact Symposium.*
[10] Crawford, D. A. and Schultz, P. H., 1999, Electromagnetic Properties of Impact-Generated Plasma, Vapor and Debris, *International Journal of Impact Engineering, Vol.* 23, pp. *169-180.*
[11] Fuller, M. and Cisowski, S. M., 1987, Lunar Paleomagnetism, *Geomagnetism*, vol. 2, pp. 307–455.
[12] Purucker, M. E. and Nicholas, J. B., 2010, A New Map of the Internal Magnetic Field of the Moon and its Implications, *Lunar and Planetary Science Conference,* Abstract #2291.

[13] Neumann, G. A., Zuber, M. T., Wieczorek, M. A., Head, J. W., Baker, D. M. H., Solomon, S. C., Smith, D. E., Lemoine, F. G., Mazarico, E., Sabaka, T. J., Goossens, S. J., H. Melosh, H. J., Phillips, R. J., Asmar, S. W., Konopliv, A. S., Williams, J. G., Sori, M. M., Soderblom, J. M., Miljković, K., Andrews-Hanna, J. C., Nimmo, F., Kiefer, W. S., 2015, Lunar impact basins revealed by Gravity Recovery and Interior Laboratory measurements, *Science Advances*, Vol 1, No. 9, e1500852.

Proceedings of the 2019 Hypervelocity Impact Symposium
HVIS2019
April 14-19, 2019, Destin, FL, USA
HVIS2019-038

Extended Abstract+

Impact Modeling for the Double Asteroid Redirection Test Mission

Emma S.G. Rainey[a], Angela M. Stickle[a], Andrew F. Cheng[a], Andrew S. Rivkin[a], Nancy L. Chabot[a], Olivier S. Barnouin[a], Carolyn M. Ernst[a], and the AIDA/DART Impact Simulation Working Group

[a]Johns Hopkins University Applied Physics Laboratory, 11100 Johns Hopkins Rd, Laurel, MD 20723, USA
**Corresponding author. Emma.Rainey@jhuapl.edu*

1. Introduction

The Asteroid Impact Deflection Assessment (AIDA) collaboration is a joint ESA-NASA planetary defense collaboration that will include the first full-scale test of an asteroid deflection by kinetic impactor [1]. The AIDA collaboration comprises two independent spacecraft, the NASA-sponsored Double Asteroid Redirection Test (DART) and the ESA-led Hera. In September 2022 the DART spacecraft will impact the secondary member of the binary asteroid system 65803 Didymos (Didymos-B) at a speed of ~6.7 km/s and mass ~500 kg. The resulting period change in the orbit of Didymos-B will be measured using Earth-based observations. Hera will arrive post-impact and perform detailed measurements to characterize Didymos-B.

Post-impact, the measured period change of Didymos-B will be used to compute the momentum transferred to the asteroid by the DART spacecraft. It is expected that the momentum change induced by DART could be larger than the incoming spacecraft momentum due to ejecta escaping from Didymos-B. This momentum enhancement effect has been studied using both models and experiments [e.g., 2-4] and is commonly parameterized by a momentum enhancement factor β, which is defined as the ratio of the momentum change of the target post-impact to the momentum of the impactor. Studies of kinetic impactors have shown that β varies significantly with impact velocity, impact geometry, target internal structure and material properties [e.g., 5,6]. The properties of Didymos-B are poorly known and the local conditions of the specific impact site on the body influence the impact event; these factors result in high uncertainty in the expected momentum transfer of the DART impact. Early modeling efforts have found that β could reasonably be as low as ~1 or greater than 5 in end-member cases [7,8]. Initial studies have also included a benchmarking campaign [9] to characterize the variability in results from different numerical impact models used by the community, which showed that when using similar material properties the impact models produced similar results for crater size and momentum transfer. The DART impact will provide rare and valuable planetary-scale validation data for these numerical impact models, increasing confidence in the use of numerical models for design of kinetic impactor missions.

2. Material Property Sensitivity Study

In this paper we present results from a sensitivity study of the effect of bulk material properties on the DART impact outcome using homogeneous asteroid models in 2D axisymmetry. This purposes of this study were (1) to determine a range of potential outcomes for the DART impact, (2) to determine which model inputs are significant predictors of the DART impact outcome, and (3) to reduce the parameter space required for future, more complex simulations. The use of 2D models allows for a large number of fast-running simulations at high resolution, but does not allow for variable impact geometries or the complex shape of the DART spacecraft. 3D simulations will be used in the future to explore these effects.

+ Full paper can found at https://www.sciencedirect.com/journal/international-journal-of-impact-engineering/special-issue/10MTGP5W4VJ

* Corresponding author. Tel.: +1-240-228-4860.
E-mail address: Emma.Rainey@jhuapl.edu.

2.1. Impact Simulation Setup

All simulations were run using CTH, a shock physics Eulerian hydrocode developed by Sandia National Laboratories [10]. The DART spacecraft was modeled as a homogeneous aluminum sphere or cylinder (see Section 2.2) with mass 586 kg and closing speed 5.975 km/s. These values were based on the DART baseline trajectory design as of 7 January 2019, which has since been updated. The updated trajectory results in a slightly higher closing speed and lower arrival mass, but these changes are minor enough that they are not expected to affect the conclusions of this study. The asteroid was modeled as a homogeneous basalt sphere with radius 81.5 m. The CTH mesh had a constant grid cell size of 1.86 cm, greater than 20 cells per projectile radius and sufficient to fully resolve relevant physical effects [9].

In all simulations the aluminum spacecraft was modeled using the SESAME equation of state [11] and the Johnson-Cook strength model [12]. For the asteroid, the equation of state was varied (see Section 2.2). In all cases the strength model used for the target was the Brittle Damage with Localized Thermal Softening (BDL) model [13,14]. The BDL model includes a pressure-dependent yield and brittle damage model similar to that of Collins *et al.* [15], in which the plastic strain at failure increases across the brittle and ductile regimes and the yield strength of fully damaged material follows a Coulomb friction law. The model also incorporates a pressure-dependent melt curve [16].

2.2. Run Matrix

A list of the impactor and target properties varied and the range of values for each property are presented in Table 1. The "cylinder with wide panels" impactor included a thin, wide disk to approximate the effect of wide solar panels in 2D axisymmetry. 10% of the total impactor mass was contained in the disk, approximately equal to the relative mass of the DART solar panels. The impactor was modeled either as fully dense aluminum (2.7 g/cm^3) or aluminum with 90% microporosity (0.27 g/cm^3), which yields a density similar to the volume-averaged density of the DART spacecraft. Three different equations of state were used for the asteroid material: the SESAME alpha quartz model with density scaling to represent basalt [17], an ANEOS basalt model [18], and a Mie-Gruneisen basalt model [19]. The target porosity was varied between 0-40%, and was modeled as microporosity using the p-alpha model [20] both with and without crush curve parameters [21]. Nine of the BDL strength model inputs were also varied, with the parameter ranges chosen to encompass values for weak sediment as well as strong rock.

Table 1. List of parameters varied in sensitivity study

	Parameter	Range of Values
Impactor Properties	Impactor shape	Sphere, cylinder, cylinder with wide panels
	Average density	0.27 - 2.7 g/cc
Target Properties	Equation of state	SESAME, Aneos, Mie-Gruneisen
	Yield strength at low pressure	0.1-100 MPa
	Tensile strength	0.1-100 MPa
	Yield strength at infinite pressure	300 MPa – 3.5 GPa
	Microporosity	0-40%, with and without crush curve
	Coefficient of internal friction	1-2 (intact material), 0.1-0.8 (damaged material)
	Failure strain	0.01-0.05 (brittle regime), 0.05-0.1 (ductile regime), 0.1-0.3 (plastic regime)
	Poisson ratio	0.2-0.4
	Melting temperature at zero pressure	1200-1800 K
	Average crack spacing	0.5-5 m
	Damage at failure	0.7-0.95

The run matrix was designed using JMP Design of Experiments Statistical Software [22]. In a design of experiments approach, experimental parameters are varied simultaneously and systematically, which enables the influence of each parameter as well as multi-factor parameter interactions to be determined with a relatively small number of experiments. For this study, we used the JMP Custom Design tool to select 28 runs with different combinations of 17 CTH input parameters varied. In this study we chose not to restrict any part of the parameter space, which means that some input combinations described improbable material behavior for an asteroid (e.g., highly porous material that is also strong).

All simulations were run for at least 0.5 second simulation time or until crater growth had stopped as determined by inspection. For each simulation the momentum transfer efficiency factor β was calculated by summing the total vertical momentum of all ejecta produced during the cratering process. Due to the significant expansion of the ejecta field over of the course of the simulations, it was not possible for the CTH domain size to be large enough to contain all the ejecta produced. The model domain was kept compact for efficiency, and fixed tracers were placed in each cell above the impact plane in order to track the momentum flux of the ejecta as a function of time. The crater width and depth were also measured at the end of each simulation, with the crater width measured parallel to the x-axis at the impact plane, and the crater depth measured parallel to the y-axis as the distance between the impact plane and the crater bottom.

3. Results

The calculated β and crater width/depth ratios for all CTH runs were analyzed using JMP to determine which inputs had highest statistical significance for predicting the CTH simulation results. The results showed that β is influenced by a relatively small number of target strength parameters, while crater morphology is influenced by the same target strength parameters as well as impactor shape. The parameters with highest statistical significance for predicting β were target porosity, yield strength at low pressure, tensile strength, average crack spacing, and coefficient of internal friction. A fit model for β constructed from these variables alone has $R^2 = 0.88$ when compared with actual values of β. The inputs with highest significance for crater width and width/depth ratio are porosity, yield strength at low pressure, tensile strength, average crack spacing, and impactor shape. Fit models constructed using these variables alone to predict crater dimensions have $R^2 > 0.90$.

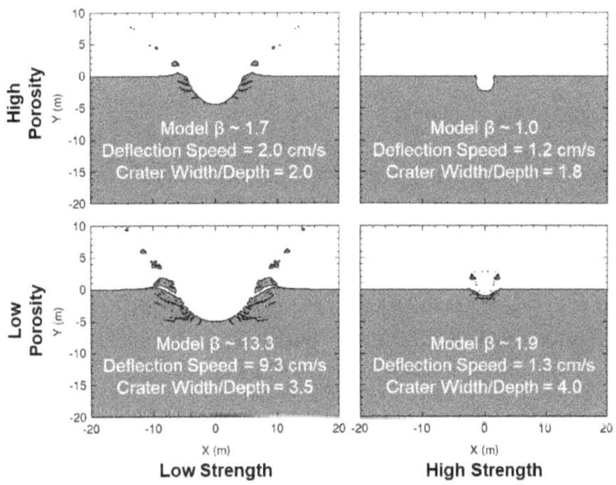

Fig. 1. Impact craters and calculated values of β, target deflection speed, and final crater width/depth ratio from end-member CTH simulations with high and low porosity and strength values. High target porosity results in a crater that has a lower width-to-depth ratio compared with low porosity targets. Note that the end-member cases shown in the upper right and lower left of the figure have unrealistic parameter combinations for asteroid materials (high porosity and strong, low porosity and weak).

End-member results for cases with high and low values of porosity and strength are shown in Figure 1. At the extremes, β was found to be as low as 1.0 and as high as 9.1. Note that these extremes resulted from simulations with combinations of material properties unrealistic for real asteroids (high porosity with high strength, low porosity with low strength). The realistic range of expected β values from the DART impact is likely much narrower. As shown in Figure 1, a moderate value of β between 1.5-2 can occur for either a porous or non-porous target. However, these cases differ in their deflection speeds (due to mass differences) and crater size and morphology. This suggests that characterization of the DART crater post-impact, as planned by Hera, may be useful for constraining the properties of Didymos-B along with the measured deflection.

4. Conclusions

The results of this study show that for realistic combinations of asteroid material properties, the DART impact would be expected to produce a β between 1.5 and 2. In addition, the outcome of the DART impact is expected to be primarily driven by a relatively small number of material properties, in particular strength and porosity. Note that the number of significant inputs found here is smaller than in the sensitivity study carried out by Bruck Syal *et al.* [5], who considered a much broader range of kinetic impactor masses and velocities beyond the specific case of the DART impact.

It should also be noted that to correctly capture the fracture processes leading to crater excavation and ejecta production, simulations should be run in 3D, and any impact outcome predictions based on 2D simulations should be treated as highly uncertain. However, the relative importance of the large number of potential material model inputs is likely the same in 2D and 3D models. Results from this study can therefore be used to greatly reduce the parameter space required to span the range of potential DART impact outcomes for future 3D modeling efforts, in which each simulation will be much more computationally expensive. In addition to 3-D simulations for homogeneous targets, follow-on studies currently underway by the DART Impact Simulation Working Group will compare results from impact models into homogeneous targets and rubble pile targets, and will incorporate realistic models for the DART spacecraft geometry into the impact simulations.

Acknowledgements

This study was funded by NASA for the Double Asteroid Redirection Test (DART) Phase C-D, contract NNN06AA01C to Johns Hopkins University Applied Physics Laboratory

References

[1] Cheng, A. F., Rivkin, A.S., Michel, P., Atchison, J., Barnouin, O., Benner, L., Chabot, N. L., Ernst, C., Fahnestock, E. G., Kueppers, M., Pravec, P., Rainey, E., Richardson, D. C., Stickle, A. M., Thomas, C., 2018. AIDA DART asteroid deflection test: Planetary defense and science objectives, Planetary and Space Science 157, p. 104.

[2] Walker, J. D., Chocron, S., 2011. Momentum enhancement in hypervelocity impact, International Journal of Impact Engineering 38, p. A1.

[3] Holsapple, K. A., Housen, K. R., 2012. Momentum transfer in asteid impacts I. Theory and scaling, Icarus 221, p. 875.

[4] Michel, P., Jutzi, M. 2014, Hypervelocity impacts on asteroids and momentum transfer I. Numerical simulations using porous targets, Icarus 229, p. 247.

[5] Bruck Syal, M., Owen, J. M., Miller, P. L., 2016. Deflection by kinetic impact: Sensitivity to asteroid properties, Icarus 269, p. 50.

[6] Owen, J. M., 2017. Modeling kinetic impactors on a rubble pile asteroid, IAA Planetary Defense Conference, Tokyo, Japan.

[7] Stickle, A. M., Atchison, J. A., Barnouin, O. S., Cheng, A. F., Crawford, D. A., Ernst, C. M., Fletcher, Z., Rivkin, A. S., 2015. Modeling momentum transfer from kinetic impacts: Implications for redirecting asteroids, Procedia Engineering 103, p. 577.

[8] Stickle, A. M., Rainey, E. S. G., Bruck Syal, M., Owen, J. M., Miller, P., Barnouin, O. S., Ernst, C. M., and the AIDA Impact Simulation Working Group, 2017. Modeling impact outcomes for the Double Asteroid Redirection Test (DART) mission. Procedia Engineering 204, p. 116.

[9] Stickle, A. M., Bruck Syal, M., Cheng, A., Collins, G., Gisler, G., Guldemeister, N., Heberling, T., Luther, R., Michel, P., Owen, J. M., Rainey, E. S. G., Rivkin, A. S., Rosh, T., Wunnemann, K., 2020. Benchmarking impact hydrocodes in the strength regime: Implications for modelling deflection by kinetic impactor, Icarus 338, https://doi.org/10.1016/j.icarus.2019.113446.

[10] McGlaun, J. M., Thompson, S. L., Elrick, M. G., 1990. CTH: A three-dimensional shock wave physics code, International Journal of Impact Engineering 10, p. 351.

[11] Kerley, G. I., 1987. Theoretical equation of state for aluminum, International Journal of Impact Engineering 5, p. 441.

[12] Johnson, G. R., Cook, W. H., 1983. "A Constitutive Model and Data for Metals Subjected to Large Strains, High Strain Rates, and High Temperatures," Proceedings of the 7th International Symposium on Ballistics, The Hague, p. 541.

[13] Crawford, D., Schultz, P., Quintana, S., 2013. A model of localized shear heating with implications for the morphology and paleomagnetism of complex craters, Large Meteorite Impacts and Planetary Evolution V, Ontario, Canada.

[14] Schultz, P. H., Crawford, D. A., 2016. Origin and implications of non-radial Imbrium Sculpture on the Moon, Nature 535, p. 391.

[15] Collins, G. S., Melosh, H. J., Ivanov, B. A., 2004. Modeling damage and deformation in impact simulations, Meteoritics and Planetary Science 39, p. 217.

[16] Senft, L. E., Stewart, S. T., 2009. Dynamic fault weakening and the formation of large impact craters, Earth and Planetary Science Letters 287, p. 471.

[17] Kerley, G. I., 1999. "Equations of State for Composite Materials," Kerley Publishing Services report KPS99-4, December 1999.

[18] Pierazzo, E., Artemieva, N. A., Ivanov, B. A., 2005. Starting conditions for hydrothermal systems underneath Martian craters: hydrocode modelling, in: Kenkmann, T., Horz, F., Deutsch, A. (Eds.), Large Meteorite Impacts III. In: Geological Society of America Special Paper, vol. 384, p. 443.

[19] Ahrens, T., 1995. Rock physics and phase relations: a handbook of physical constants, American Geophysical Union.

[20] Herrmann, W., 1969. Constitutive equation for the dynamic compression of ductile porous materials, Journal of Applied Physics 40, p. 2490.

[21] Jutzi, M., Benz., W., Michel, P., 2008. Numerical simulations of impacts involving porous bodies I: Implementing sub-resolution porosity in a 3-D SPH hydrocode, Icarus 198, p. 242.

[22] http://www.jmp.com

Proceedings of the 2019 Hypervelocity Impact Symposium
HVIS2019
April 14-19, 2019, Destin, FL, USA
HVIS2019-049

Extended Abstract+

Size scaling of crater size, ejecta mass, and momentum enhancement due to hypervelocity impacts into 2024-T4 and 2024-T351 aluminum

James D. Walker*, Sidney Chocron, and Donald J. Grosch

Engineering Dynamics Department, Mechanical Engineering Division, Southwest Research Institute, 6220 Culebra Rd., San Antonio, TX 78238, U.S.A.

1. Introduction

Momentum enhancement occurs when impactors strike objects at hypervelocities due to the formation of crater ejecta whose departure from the impact body impart more momentum to the impacted body. In previous work the momentum enhancement caused when metals, rock, and pumice were impacted have been examined [1-7]. Momentum enhancement is quantified by β, which is the ratio of the resulting target momentum by the impactor momentum. By quantifying momentum enhancement it is possible to make informed decisions about the use of hypervelocity impactors to deflect celestial bodies such as asteroids or comet nuclei.

An interesting feature of momentum enhancement is that it has been observed that the size scaling of momentum enhancement is not linear for impacts into consolidated materials. This interesting feature of the momentum enhancement event has been explored, as it is an intriguing result. The new two-stage light gas gun facility at SwRI was used to extend the data for momentum enhancement of aluminum targets to higher speeds (above 5 km/s impact speed) and larger scales (3.0-cm diameter impactors) to compare with smaller impactor results of Denardo and Nysmith [8]. This data allows a further examination of the size-scale behavior to new regimes.

In the following, the Denardo and Nysmith data for 1100-O is included in the plots for comparison, though our focus is comparing the 2024-T4 results to our new 2024-T351 large-scale impact results.

2. Experimental Impacts into Aluminum

Experiments were performed with SwRI's large two-stage light gas gun, which has a 11.4 cm pump tube inner bore and a 3.8 cm launch tube inner bore (Fig. 1 left). Al 6061-T6 spheres of 3.0-cm diameter were launched into 10.16-cm-thick Al 2024-T351 blocks (Fig. 1 right). The target blocks were nominally 30.5 cm square. Each aluminum block was weighed to an accuracy of 0.1 gram prior to being impacted. A mild steel plate, 2.54-cm thick and nominally 30.5-cm square was placed behind the aluminum block. The aluminum block/steel plate target is chained directly to a heavy steel ballistic pendulum (Fig. 2 left). The pendulum is located in a large target tank and has a 25-cm diameter thru hole directly behind the target (Fig. 2 right). A bore-mounted laser is used to position the pendulum/target assembly. A mirror on the face of the target is used to confirm zero-degrees obliquity (adjustments are made until the laser shines back onto itself). Shot numbers for the shots reported here are sequential from LGG 53 to LGG 61.

A high-speed video camera was positioned to view the flight path prior to impact. The camera is used to measure the velocity of the projectile. It also records the post-impact ejecta exiting the target and allows measurement of the velocity of

+ Full paper can found at https://www.sciencedirect.com/journal/international-journal of impact-engineering/special-issue/10MTGP5W4VJ

* Corresponding author. Tel.: +1-210-522-2051; fax: +1-210-522-6290.
E-mail address: james.walker@swri.org.

some of these particles. Impact speed values were also measured from the flash x-rays and a set of pre-impact lasers.

Fig. 1. Two-stage light gas gun at SwRI (left) and 3.0-cm-diameter aluminum sphere with sabot opened up and petals laid side by side (right).

Fig. 2. Post shot image of 10.16-cm-thick Al 2024-T351 plate attached to steel pendulum (left, LGG 53) and pre-test image showing pendulum swing inside target tank (right, LGG 52).

A rotational potentiometer was placed at the pivot point of the pendulum to record the pendulum swing. The potentiometer is calibrated prior to each test using a calibrated digital inclinometer held on the surface of the pendulum. An initial nominally 90-degree value (the value when the pendulum is at rest) is used as one calibration point and then a chain hoist is used to pull the pendulum up to around 70-degrees (Fig. 2 right) as a second calibration point. On the way back down to 90-degrees, a point at around 80-degrees is checked to confirm the calibration.

The maximum swing, as measured by the rotational potentiometer, was used to determine the momentum enhancement β. Specifically, the equation used was

$$\beta = \frac{m}{m_p}\sqrt{2gl(1-\cos(\theta-\theta_{bd}))}, \tag{1}$$

where m is the mass of the pendulum and target, m_p is the mass of the projectile, $g = 980$ cm/s is the acceleration of gravity, $l = 83.82$ cm is the distance from the rotation axis of the pendulum to the strike location on the pendulum, θ is the maximum angle of the swing and θ_{bd} is the blow-down test swing angle (LGG 56). For LGG 53 through LGG 59, $m = 809.2$ kg, and for LGG 60 and LGG 61, $m = 834.8$ kg, reflecting the fact that two 10.16-cm-thick aluminum targets were stacked back-to-back due to the higher impact speeds. To determine the amount of the swing due to the gun-launch gases, a blow-down test (LGG 56) was performed where no sabot/projectile was launched but the gun was fired with the same powder load, hydrogen pump-tube pressure, and atmospheric pressure in the flight path and target chambers as were used in LGG 53 and LGG 54. The resulting swing was $\theta_{bd} = 3.65°$. The same value of θ_{bd} was always used in determining β as additional blow-down experiments to reflect other tests powder loads and pump tube pressures, used to achieve different projectile speeds, were not performed.

The mass of the aluminum target block was measured before and after the test to provide the amount of material lost to the ejecta. This data was then normalized by the projectile mass and is shown in the last column of the table.

Table 1. Testing summary

Test No.	Impact speed (km/s)	Ball Mass (g)	Target Pre-Test Weight (g)	Target Post-Test Weight (g)	Crater Dia (cm)	Crater Depth (cm).	Swing (°)	β	Target Mass loss divided by projectile mass
LGG-53	5.04	38.44	26,762	26,329	14.3 15.2	5.11	14.67	2.30	11.26
LGG-54	5.06	38.44	26,831	26,254	15.1 14.1	4.78	14.94	2.35	15
LGG-56		0	26,807	26,807	0	0	3.65		
LGG-57	4.15	38.43	26,807	26,442	12.7 13.3	4.45	12.33	2.20	9.5
LGG-58	4.18	38.45	26,753	26,430	12.7 11.4	4.29	11.39	1.95	8.4
LGG-59	5.47	38.43	26,854	26,226	14.6 15.2	5.16	20.95	3.32	16.34
LGG-60	5.46	38.43	26,813	26,204	14.3 15.2	5.21	15.55	2.36	15.85
LGG-61	5.77	38.45	26,895	26,166	16.3 15.2	5.44	16.97	2.50	18.96

3.0 Discussion of Results

A major objective of this work was to see if the surprising size scaling behavior observed in small size experiments performed at NASA Ames in the 1960s [8] continues to larger sizes which can be reached with the SwRI large two-stage light gas gun. The largest projectiles that the Ames facilities launched were 1.27-cm diameter spheres. We are able to more than double the diameter and increase the mass by 13.2 times beyond the largest previous test data on this topic.

3.1 Mass of the Ejecta

Examination of the target mass loss data implies that we are undergoing a transition in the fragment formation process. The ejecta mass normalized by the projectile mass is shown in Fig. 3 (left). The plot is in terms of the impact speed squared times the square root of the projectile diameter. The 2024 ejecta mass data falls onto straight lines, passing through the origin. The upper line is for the smaller projectiles from the Denardo/Nysmith data. The second line is for our larger projectile data. (Also shown is a gray line through the 1100-O target data, which we are not discussing). The fact that our new data with a 3.0-cm-diameter sphere does not fall on the line of the other data implies that somewhere in the vicinity of a 1.27-cm-diameter impactor, the functional behavior of the ejecta mass has changed.

The smaller diameter impactors behave with a term that, when multiplied by a density, have units of fracture toughness. If the same size scaling occurred for the 3.0-cm-diameter projectile as occurred for the smaller projectiles, then the ejecta mass would have been *twice* what it was. Thus, a significant change in behavior has occurred, with much less mass being liberated than we would have expected based on the small scale behavior.

3.2 Momentum Enhancement

The momentum enhancement for the larger projectiles scales on impactor size to the 0.45 power, and is shown in Fig. 3 (right). It shows the same nonlinear scaling behavior for our larger projectiles as is seen in the original Denardo/Nysmith data. The fact that there is no change in behavior in the momentum enhancement nonlinear size scaling is quite intriguing since we demonstrated in our tests a quantitative change in the ejecta mass scaling, discussed in the previous paragraph.

There was one anomalous point of high momentum enhancement (LGG 59). In this test the chain broke and the aluminum block ended up on the floor of the test chamber. This test demonstrates that two bodies in contact but weakly connected can lead to large momentum enhancements, perhaps up to double as is seen with elastic impacts of different sizes. Because of the fixture failure, the value is not representative of the aluminum material impacts (though dividing the result by two gives a lower bound on the momentum enhancement), but it still provides insight on possible effects of impacting bodies comprised on consolidated pieces of material in weak contact.

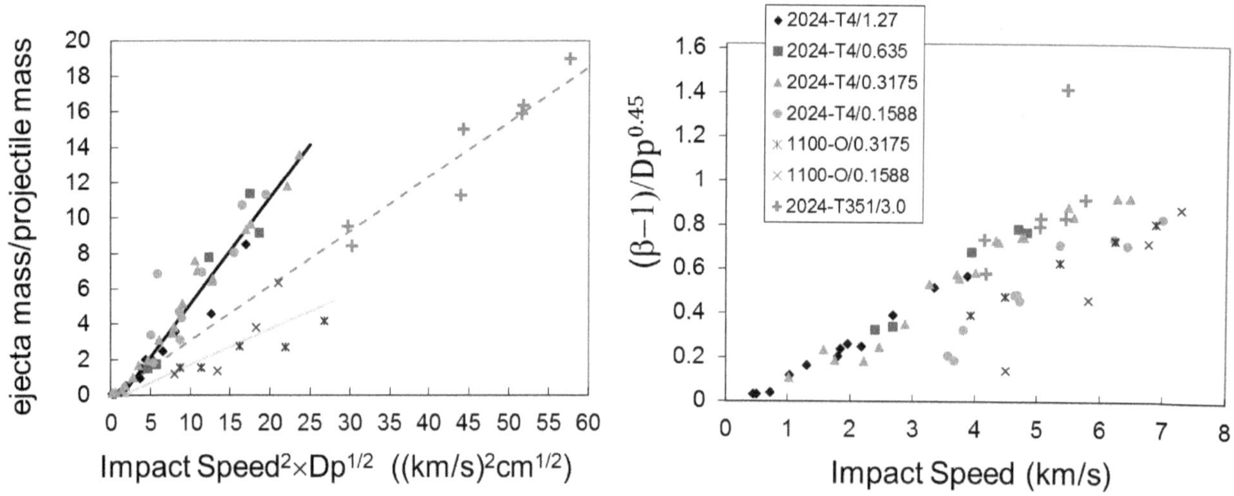

Fig. 3. Ejecta mass normalized by projectile mass as a function of impact speed squared time square root of projectile diameter (left). Momentum enhancement $\beta - 1$divided by projectile diameter to the 0.45 power as a function of impact speed (right). The legend shows the alloy / projectile diameter in cm. New 3.0-cm-diameter projectile data from this work indicated by the blue plus (+) signs. Smaller diameter projectile data from Denardo and Nysmith [8].

4. Summary and conclusions

Impact tests were performed from 4 to 5.77 km/s of 3.0-cm-diameter aluminum sphere impacting 2024-T351 aluminum targets. As presented at the Hypervelocity Impact Symposium and described in the full paper submitted to the conference issue of the International Journal of Impact Engineering, it was seen that there are no size-scaling effects for the crater depth. The slight size scaling effect in the crater diameter directly correlated with the nonlinear size-scaling effect in the ejecta mass. It was also shown that there is a qualitative change in the ejecta mass somewhere in the vicinity of 1.27-cm-diamter impactors, in that less than this the ejecta mass scales with a square root of the impactor diameter but for larger diameter spheres there is no longer a size dependence as the ejecta mass scales with impact velocity squared. This is likely due to some sort of failure saturation effect, where the size scale is large enough that material damage involving intrinsic length or time is able to completely occur. Finally, we showed that the 0.45 power dependence on the projectile diameter for the momentum enhancement $\beta - 1$ continues up through the 3.0-cm-diameter, even though there is a qualitative and quantitative change in the ejecta mass behavior. This work was funded by Southwest Research Institute's internal research and development program.

References

[1] Holsapple, K. A., 2004. About Deflecting Asteroids and Comets, in "*Mitigation of Hazardous Comets and Asteroids*" M. J. S. Belton, T. H. Morgan, N. H. Samarasinha, D. K. Yeomans, Editors. Cambridge Univ. Press, Cambridge, pp. 113-140.

[2] Walker, J. W., Chocron, S., 2011. Momentum Enhancement in Hypervelocity Impact, Int. J. Impact Engng 38, p. A1-A7, doi:10.1016/j.ijimpeng.2010.10.026.

[3] Walker, J. W., Chocron, S., 2012. "Role of Target Strength in Momentum Enhancement," Shock Compression of Condensed Matter-2011, M. L. Elert, W. T. Buttler, J. P. Borg, J. L. Jordan, T. J. Vogler, Editors. AIP Press, Melville, NY, pp. 1019-1022.

[4] Walker, J. D., Chocron, S., Durda, D. D., Grosch, D. J., Movshovitz, N., Richardson, D. C., Asphaug, E., 2013. Momentum Enhancement from Aluminum Striking Granite and the Scale Size Effect, Int. J. Impact Engng 56, pp. 12-18.

[5] Walker, J. D., Chocron, S., Durda, D. D., Grosch, D. J., Movshovitz, N., Richardson, D. C., Asphaug, E., 2013. Scale Size Effect in Momentum Enhancement, Procedia Engineering 58, pp. 240-250.

[6] Walker, J. D., Chocron, S., 2015. Damage Modeling, Scaling, and Momentum Enhancement for Asteroid and Comet Nucleus Deflection, Procedia Engineering 103, pp. 636-641.

[7] Chocron, S., Walker, J. D., Grosch, D. J., Carpenter A. J., Durda D. D., Housen K. R., 2017. Hypervelocity Impact on Pumice: Scale Effects on Experiments and Simulations. Procedia Engineering 204, pp. 154-161.

[8] Denardo, B. P., Nysmith, C. R., 1964. "Momentum Transfer and Cratering Phenomena Associated with the Impact of Aluminum Spheres into Thick Aluminum Targets at Velocities to 24000 ft/s," Proceeding of the AGARD-NATO Specialists, Vol. 1: The Fluid Dynamic Aspects of Space Flight, Gordon and Breach Science Publishers, New York, 1964, pp. 389-402.

Proceedings of the 2019 Hypervelocity Impact Symposium
HVIS2019
April 14-19, 2019, Destin, FL, USA
HVIS2019-059

Hypervelocity Impact on Concrete and Sandstone: Momentum Enhancement from Tests and Hydrocode Simulations

Sidney Chocron [a*], James D. Walker[a], Donald J. Grosch[a], Stephen R. Beissel[a],

Daniel D. Durda[b], Kevin R. Housen[c]

[a]*Engineering Dynamics Department, Southwest Research Institute, 6220 Culebra Rd., San Antonio, TX 78238, U.S.A.*
[b]*Southwest Research Institute, 1050 Walnut St., Suite 300, Boulder, Colorado 80302 USA*
[c]*The Boeing Co. (Ret.), MS ST-50, Seattle, Washington 98124 USA*

Abstract

Concrete, sandstone, and, in a previous round of experiments, pumice, were tested under hypervelocity impact at SwRI. Aluminum spheres with diameters of 1 and 1.75 in were shot at a velocity of approximately 2 km/s using a 50-mm conventional powder gun. The targets were mounted on a swing so that the momentum enhancement could be measured. The size effect, i.e. comparing momentum enhancement generated by the small and large projectiles, was of particular interest in this project. The targets were also scaled, although for sandstone we were limited by the natural geometry of the rocks. The results from the experiments show a clear size effect for the concrete while sandstone did not show any size effect, possibly because of experimental artifacts. The sandstone behavior was investigated with computations using the EPIC hydrocode. The porosity and compressive strength of the sandstone used in the impact tests were measured and reported. The rock is very similar to one reported and extensively tested by Lawrence Livermore Laboratory in 1974. Two material models (Holmquist-Johnson Concrete and Johnson-Holmquist-Beissel) were fit to the data from LLL. The momentum enhancement predicted by the code is reported for different parameter studies.

Keywords: Momentum enhancement, hypervelocity impact, ejecta, concrete, sandstone, pumice

1. Introduction

Southwest Research Institute has been involved for many years in both testing and computations of momentum enhancement for impacts at hypervelocity [1]–[8]. After impact and perforation by a high-velocity projectile, the object impacted may release ejecta in the opposite direction of the impact. Simple momentum conservation dictates that the object impacted will acquire the momentum of the impactor plus that of the ejecta, i.e. the momentum of the object impacted is "enhanced" by the ejecta momentum. This enhancement is of interest for the planetary defense community as it increases the efficiency of the payload in a mission aimed at deflecting asteroids using kinetic impactors.

For the last conference, we presented hypervelocity impact on pumice rock, a low-density and high porosity material. As expected, momentum enhancement on this material was relatively small and did not show a clear size effect when impactor and target were scaled.

In this paper, hypervelocity impact tests of aluminum spheres on concrete and sandstone are reported as well as the momentum enhancement obtained during the tests. First, the materials will be described as well as the test set-up. Then the results from tests and computations will be summarized and followed by analysis.

* Corresponding author: schocron@swri.edu

2. Materials

The projectiles procured for all the tests were made of Al6061-T6 spheres (www.precisionballs.com) with diameters of 2.54 cm (1 in, 24 gram) and 4.45 cm (1.75 in, 120.4 gram). Impact velocities ranged from 1.97 to 2.06 km/s. The concrete material, from a high-strength concrete mix (brand name Quickrete®), was acquired at a local store of a large hardware chain. The concrete was molded with cylindrical geometry using Sonotube® preform. The targets were 39.4-cm (15.5 in) diameter and 50.8-cm (20-in) long for the small projectile and 61-cm (24-in) diameter and 58.4-cm (23-in) long for the large projectile. The sandstone was also acquired locally at Keller Material Ltd. (9388 Corporate Dr., San Antonio, TX 78514). The geometry of sandstone targets was not under SwRI control as it is a natural material. The typical size for the small targets was 45×45×45 cm^3 (18×18×18 in^3) and, for the large targets, 100×80×40 cm^3 (40×30×15 in^3). Note that for the large sandstone target shots the shot line was the short direction, i.e. the thickness of the target was 40 cm (15 in).

The sandstone density was measured to be 2.12 g/cm^3 using two of the cylindrical specimens cut from the rock with waterjet. The porosity was also determined by measuring the volume of two specimens soaked in water. After extracting them from the water, the specimens were dried in an oven, pulverized, and the volume of the particles measured again. The porosity calculated was 20.8%. Additionally, six compression tests were performed in a universal testing machine providing a strength for the sandstone between 25 and 40 MPa. Note that this rock is anisotropic as it is layered. Compression tests with the stress applied in the direction perpendicular to the layers (or grain) resulted in a strength of ~37 MPa, while compression tests along the plane of the grain provided a strength of ~27 MPa. The rock was found to be similar enough to a sandstone described by Duba et al. [9]. Hence, the pressure vs. density as well as the failure envelope used for the computations were assumed the ones published by Duba.

Figure 1a) shows the pressure vs. volumetric strain derived from the pressure vs. density curve published by Duba and labeled as "LLL". The same plot contains fits to Holmquist-Johnson concrete (HJC) model [10] and to Johnson-Holmquist-Beissel (JHB) model [11] that will be discussed later in the paper. Figure 1b) reproduces the failure envelope as a function of the hydrostatic pressure from Duba. The black continuous line was obtained when virgin sandstone material was tested, while the black dashed line was obtained for material pre-pressurized to 700 MPa, i.e. it had less porosity than the virgin material. Again, curve fits to HJC and JHB are presented. Note that the models cannot fit the complexity of the virgin material failure envelope. It plateaus around 200 MPa and at 500 MPa increases very fast to meet again with the pre-pressurized rock data at 800 MPa. The curve fits for the failure envelope favored the pre-pressurized data over the virgin material as it was believed that high pressure behavior was dominant in this hypervelocity impacts.

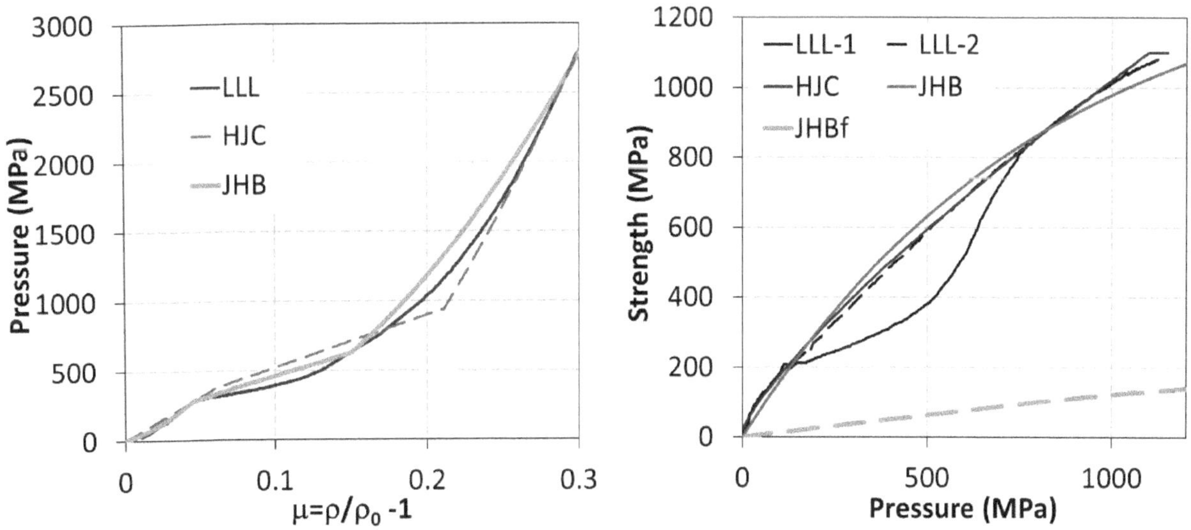

Figure 1: Sandstone properties [9] a) Pressure vs. volumetric strain, b) Failure envelope. Curve fits for numerical models are also presented

3. Methods

Experiments were performed with SwRI's 50-mm powder-gun. The concrete target was placed on a swing supported by four steel wires, as shown in Figure 2a. Note the chains surrounding the frontal part of the concrete target to provide additional confinement and avoid lateral spall of large pieces. Typical lengths for the wires were 105-120 cm (41-47 in). Two Phantom v7.11 high-speed video cameras were used to see the impact face and track the motion of the target. Witness marks on the swing and the target helped performing software-automated tracking. An example of the horizontal (x) and vertical (y) displacement of one of the fiducials placed on the target is shown in Figure 2b for a concrete test (number 14). The blue continuous line is the x-position vs. time for the fiducial for a full swing; the red continuous line provides the y-position vs. time. The dashed lines are the theoretical displacements as calculated from the simple swing theory, and using as initial velocity v_0 the one measured for the initial 100 ms after the impact. For the x-position, the equation used was:

$$x(t) = v_0 \sqrt{\frac{L}{g}} \sin\left(\sqrt{\frac{g}{L}} t\right)$$

where v_0 is the initial velocity immediately after impact, g the gravity constant, L the length of the pendulum, and t the time. Note that the trajectory recorded follows nicely the theoretical prediction for this test. In general theory and experiments match well for concrete targets. With sandstone targets the maximum swing is sometimes overpredicted by the theory. The "final" momentum, i.e. the momentum used to compute momentum enhancement, was calculated using v_0. and the initial mass of the package (targets+chains+swing).

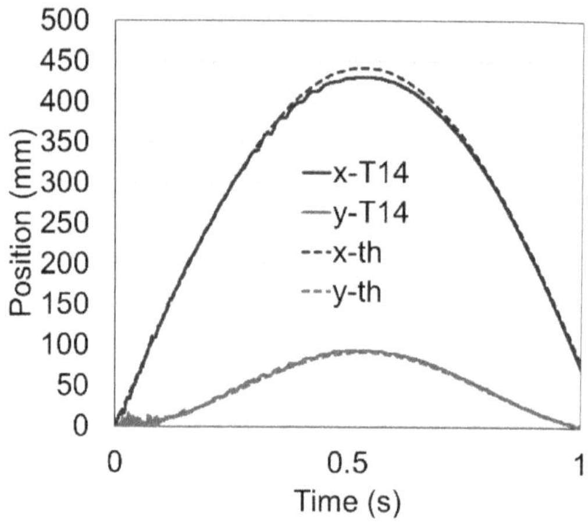

Figure 2. a) Picture of test set-up. b) Position vs. time history recorded during the test (continuous lines) and predicted by theory (dashed)

The sandstone target set-up was very similar with the main difference being that chicken wire was used around the target with the intent of avoiding ejecta from lateral and back surfaces. This was properly implemented for the small targets. Unfortunately, during the analysis of the results it was realized that for the large targets the chicken wire covered partially the front of the target. I.e., for the large targets, momentum enhancement may have been underestimated because of some ejecta from the front being captured by the chicken wire. An additional caveat with the large sandstone targets was that the shot line thickness was too small to prevent spall from the back of the target. Plates of laminated wood were added to contain the spall on the back of the large sandstone targets.

4. Results from Tests

The results from the concrete tests are summarized in Table 1. The small spheres were shot against the small targets and the large spheres were shot against the large targets. Target geometry is described in the Materials section. Concrete targets averaged a momentum enhancement of β=2.19 for the small projectiles and 2.71 for the large projectiles, hence showing a significant size-effect. Note that test 12 has an asterisk because this shot was off-axis by 10-cm (4 in) due to an alignment problem. If test 12 is discarded the enhancement increases from 2.19 to 2.22, i.e. not a significant increase. This off-axis test gives us an idea of the effect of the edges on momentum enhancement. Centered shots happened at a distance of almost 20 cm (8 in) from the edge while the off-axis test happened at 10 cm from one edge but 40 cm from the other. Overall the momentum enhancement effect was probably within the scatter of the data.

Figure 3 shows all the concrete targets after the tests. Note that tests 10 and 11 did not have the confinement provided by the chains on the front of the target. Without the confinement a large piece sheared-off on the bottom-right of the target upon impact for test 10 (small projectile). For the large projectile, test 11, the event was much more violent and basically destroyed the whole front of the target. Interestingly the momentum enhancement measured for the unconfined test (11) was larger than the other two tests with the same projectile.

Crater morphology was relatively repeatable for all the tests that were laterally confined. Radial cracks extended all the way to the edges of the target and impact crates were somewhat consistent from impact to impact.

Table 1. Summary of momentum enhancement results for impact on sandstone and concrete

Material	Test No.	Ball Dia. (in)	Impact Velocity (km/s)	Swing Initial Velocity (m/s)	Momentum (kg m/s)	β
Concrete	10	1	2.05	0.5	105.3	2.14
Concrete	12	1	2.06	0.46	101.7	2.06*
Concrete	13	1	2.06	0.54	118.4	2.40
Concrete	16	1	2.06	0.486	106.8	2.16
					Average	2.19
Concrete	11	1.75	1.95	1.5	672.1	2.86
Concrete	14	1.75	1.98	1.31	604.3	2.53
Concrete	15	1.75	1.99	1.42	655.6	2.74
					Average	2.71

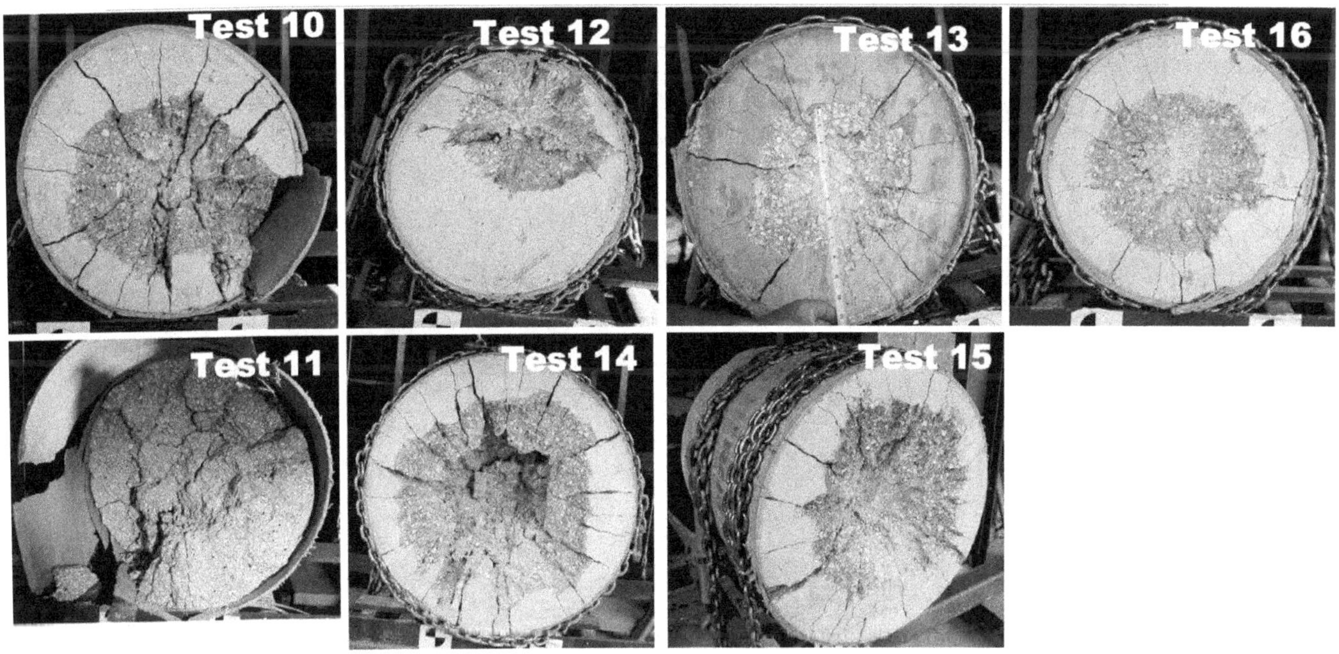

Figure 3. Concrete targets after shots at 2 km/s with an Al6061 sphere: 2.54 cm or 1 inch (top row) and 4.45 cm or 1.75 (bottom row)

The results for the sandstone targets are summarized in Table 2. The average momentum enhancement for the small projectiles was 2.10 while for the large projectile it was found to be 1.98. Interestingly, test 18 has less cracks and damage, and also less momentum enhancement. This may be related to the direction of the grain in the rock as the strength in the direction perpendicular to the planes of the rock is larger than the other two directions.

For the large shots, the asterisk on test 20 indicates that it was not used to find the average. The large targets were actually thinner in the shot line direction than the smaller targets because thicker targets were not available for the tests. The first test on the large target produced large spall on the back of the target. This result was considered invalid because ejecta on the same direction of the shot reduces the momentum enhancement.

The most puzzling result from these series of tests is that the momentum enhancement for the large impactors was smaller than for the small impactors. Two possible reasons for the reduction in the momentum enhancement are speculated. The first one is that the targets are too thin for this large projectile and do not provide enough self-confinement in the rock. Less confinement is also less pressure during the impact which results in less strength for the rock and less the elastic rebound. All these effects should reduce the momentum enhancement. The second reason is that momentum of the ejecta may have been artificially reduced because of the chicken wire wrapping in part the front of the target. Movies do show some ejecta particles being trapped by the wire.

Table 2. Summary of momentum enhancement results for impact on sandstone

Material	Test No.	Ball Dia. (cm)	Impact Velocity (km/s)	Swing Initial Velocity (m/s)	Momentum (kg m/s)	β
Sandstone	17	2.54	2.03	0.4	97.88	2.01
Sandstone	18	2.54	2.05	0.25	84.56	1.72
Sandstone	19	2.54	2.06	0.4	122.40	2.48
					Average	2.10
Sandstone	20	4.44	1.97	0.66	411.71	1.74*
Sandstone	21	4.44	1.99	0.695	467.93	1.95
Sandstone	22	4.44	1.99	0.76	483.06	2.02
					Average	1.98

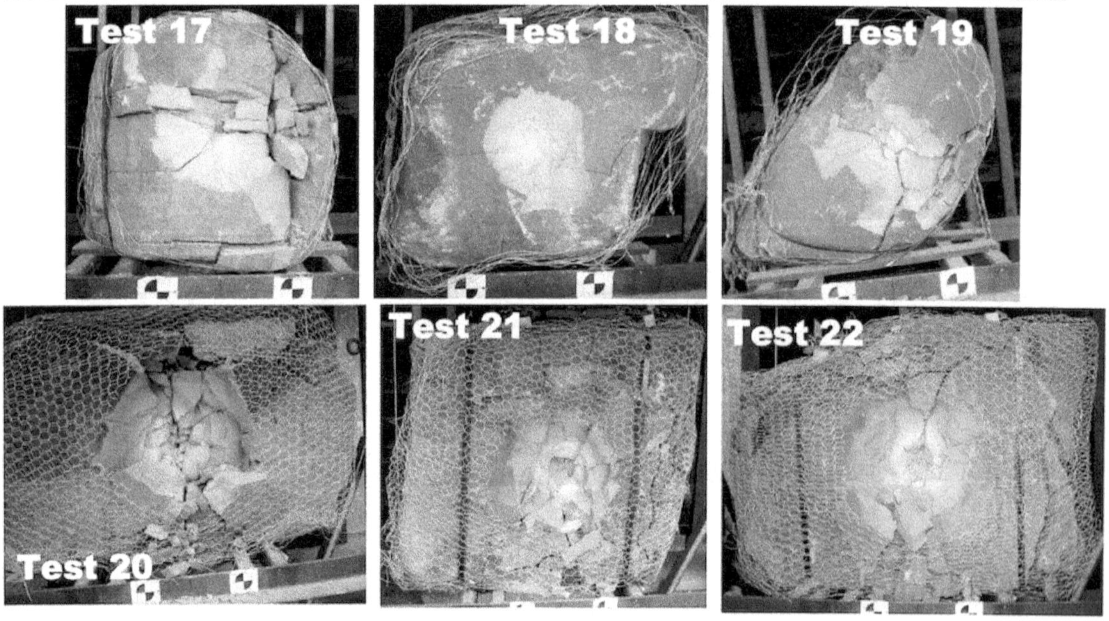

Figure 4. Sandstone targets after shots at 2 km/s with an Al6061 sphere: 2.54 cm or 1 inch (top row) and 4.45 cm or 1.75 (bottom row)

5. Computer Simulations

Computer simulations were performed with EPIC, a Lagrangian hydrocode with the capability of converting elements to particles when the mesh gets severely distorted, see Johnson et al. [12]. The initial geometry and mesh are shown in Figure 5a. The brick elements in the target (1.84-cm side) are subdivided in 24 tetrahedral elements each. The typical size of the side of the tet in the projectile was 0.8 cm. The thickness of the target in the shot line direction was 61 cm, and the side of the target 46 cm. The Al6061-T6 ball was 1-inch diameter impacting at 2 km/s for these runs.

When trying to reproduce numerically the momentum enhancement there are two parts that are critical: the material model of the target and the particle conversion algorithm. For the material model, the important physics to capture are probably: crush-up of porosity, pressure dependence of strength, strength dependence on damage, non-linearity of equation of state, and, possibly, bulking of the material upon failure and strain-rate effects. The Holmquist-Johnson concrete (HJC) model [10] can capture all of the above physics except the bulking of the material. Note that the strength in HJC model only decreases slightly when being fully damaged. The Johnson-Holmquist-Beissel (JHB) model [11], does potentially capture all the physics, and, in particular, the crush-up and the bulking after failure.

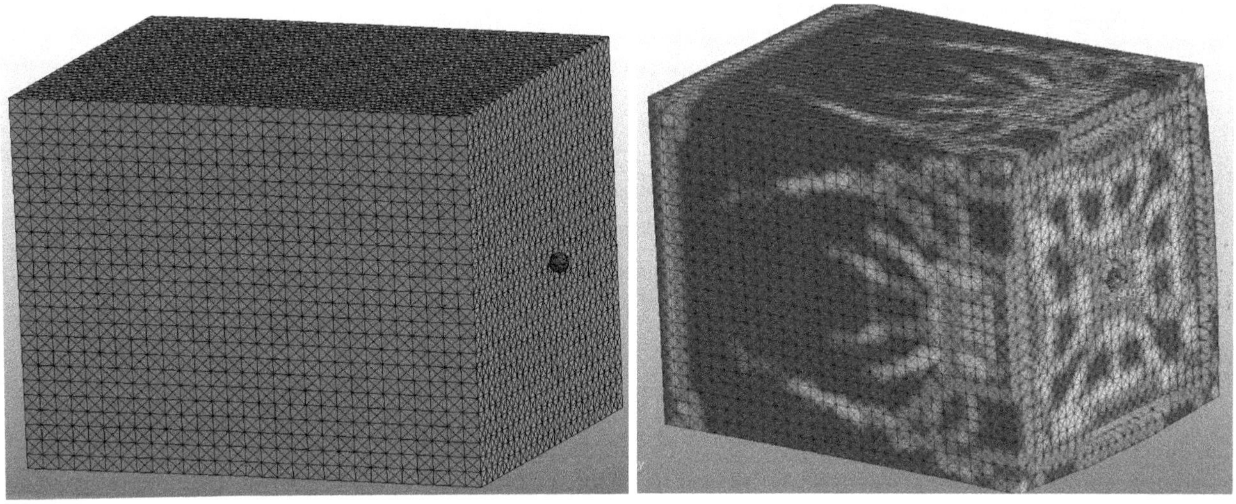

Figure 5. a) Initial geometry for the EPIC computations. b) Effective plastic strain for last step (200 ms), note the particles flying away of target.

Several computations were performed using the HJC and the JHB material models shown in Figure 1. The curve fits for the pressure vs. volumetric strain matched reasonably well the curve from the tests published by Lawrence Livermore although the full crush-up point for both models was adjusted to allow for a better fit on the fully crushed material. For the failure envelope, as mentioned previously, priority was on matching the high-pressure part of the curve. Hence, the "anomalous" crush-up plateau was ignored when doing the curve fit. Note also the failed material strength for the JHB model, much lower than the intact. For the HJC model, strength of the fully damaged material is very similar to the intact strength. The low strength of the fully damaged material (labeled as JHBf) on the JHB model is not based on data, but just a choice to see its effect on the momentum enhancement.

Table 3 summarizes the computations performed using the HJC model. The baseline run (sandstone04) used properties as close as possible to the ones found in the LLL report. The damage constants D1=0.04 and D2=1 were, nevertheless, from the HJC paper as these constants were unknown. The hydrostatic tensile strength used for the rock was 0.4 MPa and was calculated from the uniaxial strength in the LLL report. EPIC uses by default a plastic strain of 0.4 to transform elements into particles.

With the above material properties, the prediction of the code for momentum enhancement was 1.046, very far from the momentum enhancement of 2 seen in the experiments. The crater depth was well predicted by the code but the diameter was underpredicted (6 cm vs. 20 cm). Note that the crack pattern shown in Figure 5b does look realistic with radial cracks reaching the edges of the target.

A parametric exercise was performed to understand the effect of the different HJC variables on the momentum enhancement. The key variables identified were tensile strength and damage parameters. Decreasing the tensile strength of the sandstone by a factor of 10 did not increase significantly the momentum of the rock. Similarly decreasing the damage parameter, i.e. forcing full damage of the rock earlier, did not increase the momentum enhancement.

As these parameters did not affect the momentum enhancement, the attention was turned to the particle conversion algorithm. Initially, the plastic strain for conversion from elements to particles was decreased from 0.4 to 0.05 and then 0.005, but the effect was opposite to the one desired. EPIC also has the option of an element conversion formulation called the "combined particle-element algorithm" that does not have the zero energy modes of the default formulation. Using the new conversion formulation, the momentum enhancement only increased slightly.

The JHB model for the sandstone provided a significant improvement of the momentum enhancement (1.276) with respect to HJC, and closer but still very far from the one observed in the experiments. Additionally, the crater diameter and depth were also properly predicted by the JHB model. Computations are still ongoing to investigate further if this model has the potential to provide the momentum enhancement expected.

Table 3. Parametric computations performed with the HJC model in EPIC

Computation ID	β Computation	Parametric Study Performed
sandstone04	1.046	Baseline HJC
sandstone05	1.047	Decreased tensile strength by factor of 10
sandstone06	1.035	Decreased d1 by a factor of 10
sandstone07	1.053	Decreased d2 by a factor of 10
sandstone08	1.005	Reduced plastic strain to failure for particle conversion to 0.05
sandstone09	1.002	Reduced plastic strain to failure for particle conversion to 0.005
sandstone10	1.060	Used combined particle-element algorithm
sandstone15	1.276	JHB

4.0 Conclusions

Sandstone rocks and concrete were impacted by an Al6061-T6 sphere at two different scales. The concrete specimens, where it is easy to control the geometry, did show a clear size effect with momentum enhancement increasing at larger scales. The sandstone did not show any scale effect, but the large targets may not have had the right geometry or set-up to capture the effect. Numerical simulations with EPIC showed that the two materials models chosen (Holmquist-Johnson concrete and Johnson-Holmquist-Beissel) underpredicted the momentum enhancement. A parametric study changing some of the key material properties and the element conversion formulation did not improve significantly the results.

References

[1] J. D. Walker *et al.*, "Momentum enhancement from aluminum striking granite and the scale size effect," *Int. J. Impact Eng.*, vol. 56, pp. 12–18, Jun. 2013.

[2] J. D. Walker and S. Chocron, "Role of target strength in momentum enhancement," in *AIP Conference Proceedings*, 2012, vol. 1426.

[3] J. D. Walker *et al.*, "Scale size effect in momentum enhancement," *Procedia Eng.*, vol. 58, pp. 240–250, 2013.

[4] J. D. Walker and S. Chocron, "Damage modeling, scaling and momentum enhancement for asteroid and comet nucleus deflection," in *Procedia Engineering*, 2015, vol. 103.

[5] J. D. Walker and S. Chocron, "Momentum enhancement in hypervelocity impact," in *International Journal of Impact Engineering*, 2011, vol. 38, no. 6.

[6] J. D. Walker and S. Chocron, "Momentum Enhancement from Kinetic Impactors and Conventional Explosives," in *International Seminar on Nuclear War and Planetary Emergencies: 43rd Session, Erice, Italy, 19-24 August 2010.*

[7] D. D. Durda, S. Marchi, D. J. Grosch, S. Chocron, J. D. Walker, and K. R. Housen, "Impact Experiments with Iron-Nickel Targets: Momentum Enhancement and Crater Morphology," in *Lunar and Planetary Science XLVIII, 2017.*

[8] S. Chocron, J. D. Walker, D. J. Grosch, A. J. Carpenter, D. D. Durda, and K. R. Housen, "Hypervelocity Impact on Pumice: Scale Effects on Experiments and Simulations," *Procedia Eng.*, vol. 204, pp. 154–161, 2017.

[9] A. G. Duba, A. E. Abey, R. P. Bonner, H. C. Heard, and R. N. Schock, "High-Pressure Mechanical Properties of Kayenta Sandstone, Lawrence Livermore Laboratory Report UCRL-51526," 1974.

[10] T. J. Holmquist, G. R. Johnson, and W. H. Cook, "A computational constitutive model for concrete subjected to large strains, high strain rates, and

high pressures," in *Proceedings of the 14th International Symposium on Ballistics, Quebec City, Canada, September 1993*, 1993.

[11] G. R. Johnson, T. J. Holmquist, and S. R. Beissel, "Response of aluminum nitride (including a phase change) to large strains, high strain rates, and high pressures," *J. Appl. Phys.*, vol. 94, no. 3, pp. 1639–1646, 2003.

[12] G. R. Johnson, S. R. Beissel, and R. a. Stryk, "A generalized particle algorithm for high velocity impact computations," *Comput. Mech.*, vol. 25, no. 2–3, pp. 245–256, 2000.

Proceedings of the 2019 Hypervelocity Impact Symposium
HVIS2019
April 14-19, 2019, Destin, FL, USA
HVIS2019-022

BULKING AS A MECHANISM IN THE FAILURE OF ADVANCED CERAMICS

Brendan M. L. Koch[1], Calvin Lo[1], Tomoko Sano[2], James David Hogan[1*]

[1] *Department of Mechanical Engineering, the University of Alberta, Edmonton, AB T6G 2R3, Canada*
[2] *Weapons and Materials Research Directorate, Army Research Laboratory, Aberdeen Proving Ground, MD 21005*

Abstract

Failure in brittle materials is characterized by crack growth and fracture, processes which involve an increase in the volume of a sample to accommodate these cracks. This process is called bulking and it is known to be an important factor in the failure of materials such as ceramics, stone, and concrete. While volumetric strains are obtainable under quasi-static conditions, under dynamic conditions technical challenges have stood in the way of obtaining multi-dimensional strain data that would allow for assessment of bulking under the sort loading conditions that would simulate a high velocity impact. Advances in digital-image-correlation and ultra-high-speed-photography have however opened up the capacity to obtain this higher dimensional data. This data in turn has prompted an assessment of prior theory to produce a framework through which stress-strain behavior can be expressed in terms of changes to multiple elastic constants simultaneously. This presentation offers initial results in quasi-static and dynamic experiments and discusses the implications for brittle material behavior and crack evolution phenomenon under a variety of conditions.

Keywords: ceramic, digital-image-correlation, bulking, microcracks, failure

Nomenclature

D	Elastic property damage
E	Young's modulus
Greek symbols	
ν	Poisson's ratio
Subscripts	
0	Initial value
1	Stiffness damage
2	Poisson's ratio damage
3	Residual strain
Eff	Effective value

1. Introduction

During a kinetic impact event, there is a complex interplay of factors that determine whether or not a projectile is stopped or not by an interposing material. As velocities increase all materials will ultimately fail, but from a terminal ballistics and armor design perspective how a material fails becomes key. Because of the mechanics of brittle failure giving them particular ability to resist kinetic penetrator impacts[1]–[7], advanced ceramics such as alumina, boron carbide, and silicon carbide have received considerable interest in the literature. Rather than failing in a localized shear the way ductile metals do, brittle ceramics fracture and fragment in a conoidal volume that better distributes the forces and energy. In further contrast to metals, ceramics

* Corresponding author. Tel.: +1-780-492-1733.
E-mail address: jdhogan@ualberta.ca.

front of the projectile. This process further dissipates energy and con-tributes to eroding and blunting the projectile, reducing its penetrative capacity, and potentially, leading to projectile defeat[10], [11]. Ceramics also demonstrate features such as strain-rate dependent hardening up to their Hugoniot elastic limit[12], and this means that these particular properties increase in importance up to the hypervelocity impact regime.

Current models of the failure process of ceramics involve the growth of microcracks from pre-existing defects, such as pores or inclusions[13]–[18]. These are typically referred to as "wing-crack" models, due to the way that secondary cracks in the direction of applied stress branch off of the main crack body like wings. Under an applied compressive stress, the defects transform macroscopic compression into local tension, which causes the cracks to grow[13], [14]. Eventually these growing cracks begin to interact and coalesce with each other, resulting in fracture and fragmentation[18]–[20].

A major usage of the microcrack model is in modelling the failure of ceramics. Among these models, a typical way of representing the macroscopic effects of the failure is by the inclusion of a damage term, a factor that decreases the apparent stiffness of a material in order to capture the effects of microcrack growth[18], [21]. The formulation used is based around damage changing the effective Young's modulus, of the form:

$$E_{eff} = E_0(1 - D) \tag{1}$$

Here the effective Young's modulus, E_{eff}, is equal to the nominal Young's modulus, E_0, reduced by a damage term, D. A key component to the microcrack models is that the cracks grow under loading, which requires the total volume of a given sample to increase in order to accommodate this growth. This process, called dilatancy or bulking, requires measurement of 2D or 3D strains. This is because cracks cannot expand perpendicular to a compressive force as the compression would close those cracks, and thus the volume expansion has to occur in the directions orthogonal to the largest applied force. Dilatancy effects that alter the Poisson's ratio of a material have been observed in brittle materials such as coal or concrete[22]–[26], so there is physical evidence for this sort of behavior.

Experimental research into these sorts of effects on advanced ceramics under high loading rates and in impact scenarios have thus far not occurred due to a lack of adequate tools to measure 2D and 3D strains that would reveal these effects. With new advances in technology allowing for the probing of these multidimensional strains via digital-image-correlation (DIC)[27], we are better able to assess the influence of bulking upon cracking and failure mechanisms[28]. In particular we will examine how brittle materials such as alumina or cermets behave in both quasi-static and dynamic loading conditions through the use of high speed cameras and DIC. With both a variety of materials and loading conditions, we discover behaviors previously not discussed in literature, particularly within modelling work where it is assumed that changes to apparent elastic constants from microcracking are irreversible and proceed only in one direction. With this experimental data, we extend the microcrack models to account for damage mechanisms beyond a purely stiffness altering phenomenon to a multidimensional phenomenon that influences multiple elastic constants, and use the evolution of these constants to extrapolate cracking mechanisms occurring within an opaque material. This approach allows for the identification of various physical phenomena such as crack closure or cracking induced anisotropy.

2. Experimental Set Up

2.1. Experimental Tools – Qausi-Static vs. Dynamic

Initial experiments performed to study bulking phenomenon in brittle material used an MTS 810 quasi-static testing machine to compress cuboids machined to dimensions of 3.5 mm X 2.7 mm X 2.5 mm to the point of failure. During this process, high resolution images were taken with Promon cameras to capture the surface deformation for digital-image-correlation purposes. By matching one dimensional stress-strain data from the MTS machine to the two-dimensional strain data from DIC, the axial and lateral stress strain curves for samples under uniaxial loading can be determined. This allows for the analysis of changes in Young's modulus and Poisson's ratio with increasing strain. A sample image still from a quasi-static experiment is seen in Figure 1, which is representative of both quasi-static and dynamic testing configurations.

In addition to quasi-static testing, additional characterization was done using a Kolsky bar testing apparatus to obtain stiffness and strength values and a Shimadzu HPV-X2 high-speed camera to obtain images for DIC at 1 to 2 million frames per second, depending on material used and the duration of input strain pulses, which were typically on the order of tens of microseconds, with significant events typically being confined to periods of time of less than ten microseconds. The Kolsky bar used incident and transmitted bars 12.7 mm in diameter made of maraging steel (Service Steel America C 350) with a Young's modulus of 200 GPa, Poisson ratio of 0.29, yield strength of 2.68 GPa, and a density of 8100 kg/m^3. A projectile made from the same steel was launched using compressed gas to strike an incident bar, which then transmitted the strain pulse into a sample of alumina, with another bar receiving the remaining stress pulse from the event. For the experiments used, a 1 mm thick disc of 99% tin was placed between the striker and incident bars in order to produce desired pulse ramp rate and duration. This set up allows for strain rates of 10^2 to 10^3 s^{-1} to be produced and stress equilibrium achieved.

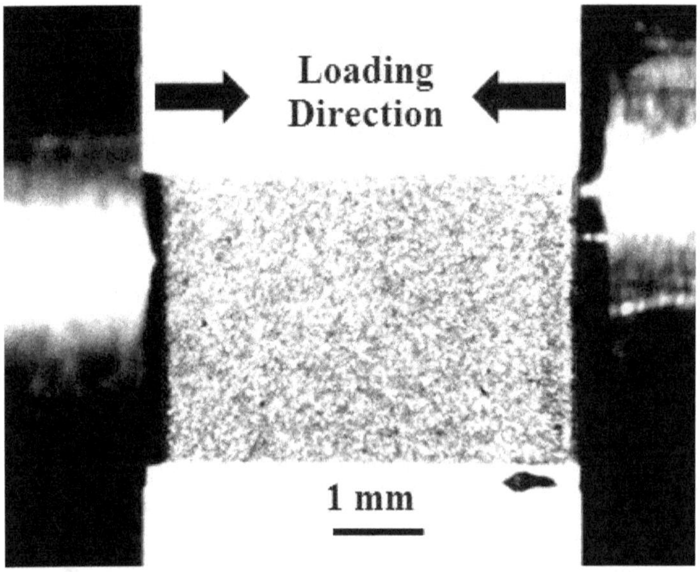

Figure 1 – Alumina sample showing the speckle pattern applied for DIC purposes. Image is representative of both quasi-static and dynamic testing configurations.

2.2. Materials Used

The primary materials studied were aluminum oxide (alumina), with CoorsTek AD85 and AD995 being the types of alumina studied. These product names correspond to an 85% and 99.5% purity of alumina, respectively. These materials were chosen due to alumina being a common ceramic used in body armor applications due to its high strength to weight ratio and good economic suitability for wide usage. Being able to compare two materials that are chemically very similar but that have a variety of microstructural differences also allows for the assessment in differences in behavior arising from said differences.

An additional material used was a titanium-aluminum-alumina cermet composite material made by Luminant under the brand name of TitanMade®. This is included to examine further variations in how the overall category of brittle material can exhibit variations in an overall cracking and fragmentation failure.

2.3. Loading vs. Unloading

One of the key questions of crack growth is that of how damage accumulates in loading and unloading. If damage continues to accumulate during the unloading phase then this has different implications for failure models than if damage remains constant or even decreases after peak stress is reached. For quasi-static conditions this can be achieved by loading a sample to a stress less than is required to cause failure, but high enough that local damage in the form of surface cracking and chipping can be observed, at which point the load is released and differences in stiffness and Poisson's ratio are looked for. Under dynamic conditions unloading is more difficult to achieve as the sample typically fails and fragments and thus cannot be unloaded in the same way as in quasi-static experiments. When the sample fails its ability to resist deformation from stress decreases and thus the stress sustained will decrease in an unloading event. Alternatively, with careful set up of the Kolsky bar it is possible to put in a stress pulse that does not cause the sample to fail on the first pass, putting it through at least one full load-unload cycle. Both of these examples of loading and unloading under dynamic conditions have been observed.

2.4. *Held Loading*

Finally, the implications of damage suggest the possibility of an extended subcritical load being able to cause failure. Microcrack damage formulations typically assume that damage in the form of a reduction of stiffness begins to accumulate before peak stress is reached. A lower stiffness produces more strain for the same applied stress, which if damage accumulates according to strain would increase the amount of damage, until failure occurred. Oppositely, if damage accumulates according to stress then no additional damage would occur. This task is simple for quasi-static loading as it only involves compressing to a subcritical failure value and holding that stress, where for dynamic testing this requires an extremely long pulse that maintains a constant stress state for an extended period. This type of pulse is possible with a sufficiently long striker and careful pulse shaping and is explored in this paper.

3. Experimental Results

Experiments began with simple load to failure quasi-static testing to validate DIC testing methods and confirm the accuracy of manufacturer supplied elastic properties. For the alumina those experiments confirmed linear elastic behavior followed by sudden catastrophic failure, while the cermet tested demonstrated a continuous loss of stiffness and increase in Poisson's ratio before undergoing brittle failure, as seen in Figure 2.. AD995 has a similar behaviour to AD85, but because of greater failure strength being on the same plot as the cermet would obscure the non-linear behaviour of the cermet. These continuous changes in elastic constants suggest that for the cermet the damage is accumulating linearly with strain.

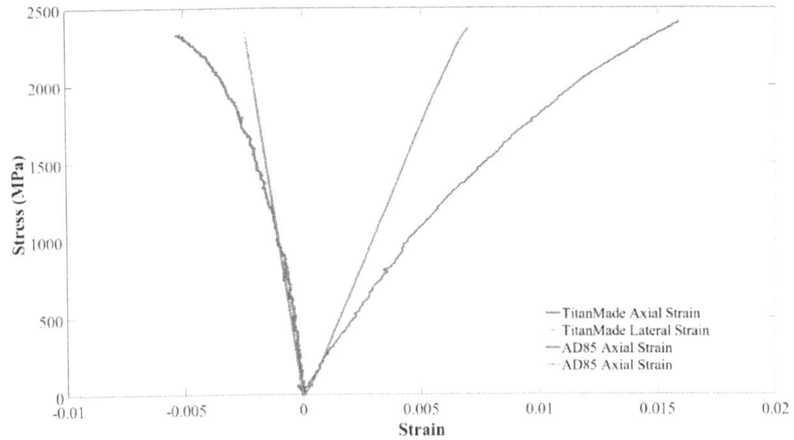

Figure 2 – Quasi-static stress-strain data for TitanMade® cermet and AD85 alumina, showing axial and lateral strains. For TitanMade the axial and lateral strains are non-linear, while for AD85 the strains are highly linear under these conditions.

Beyond simple quasi-static tests to failure, a variety of dynamic tests and more complex quasi-static tests such as load-unload and held loading were also performed, producing more complex stress-strain curves. These results frequently ran counter to the expectations generated by modelling papers, and Figure 3 is typical of the complexities seen, showing the stress-strain curves for an AD-995 sample driven to failure in the Kolsky bar. Fragmentation of the material ends DIC analysis before the pulse recorded by the strain gauges on the Kolsky bar return to a zero stress state, hence why the curves cut off.

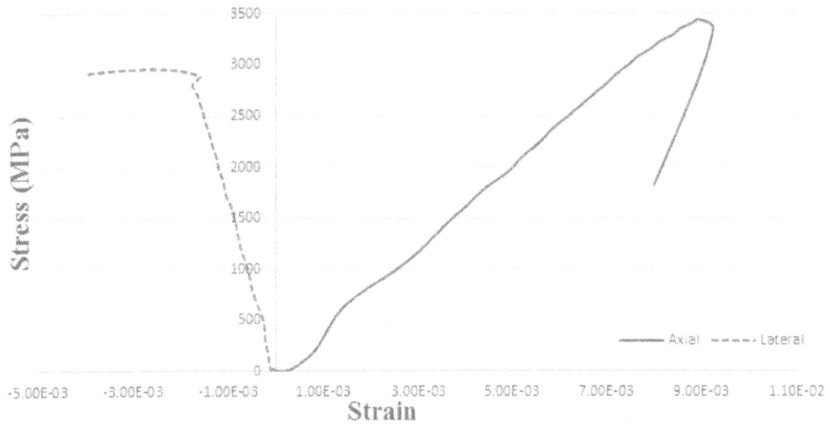

Figure 3 – A stress-strain plot showing the behaviour of axial and lateral strains for a representative sample of AD995 compressed in the Kolsky bar apparatus.

The largest divergence from expected results is on the portion of the curve where the axial stress is decreasing from its peak value. Rather than strain continuing to increase as stress decreases as is typically expected, the axial strain decreased as the stress decreased. The slope of the unloading portion of the curve is steeper than the loading portion, suggesting the possibility of an increased stiffness, which should be impossible by current understanding of brittle failure. For the lateral strains, the Poisson's ratio is slowly increasing until it abruptly rapidly increases right before failure makes further tracking impossible via DIC. The final slope of the lateral vs. axial strain curve is actually negative, and while negative Poisson's ratio values are permitted by elasticity, it is tempting to simply declare these elements noise or evidence of a failed experiment. Rather than discard the data in Figure 3 and other experiments like it, a more fruitful use is to go back to the physical processes assumed to be behind microcrack theory and damage accumulation, and to see if actual physical processes can be described that produce these sorts of stress-strain curves.

4. Interpretation of Results

4.1. Expansion of Damage Terms

The results seen in Figure 3 are complex as they involve non-linear behaviors on both the stress-strain curve and the axial vs. lateral strain curves, showing non-constant Young's modulus and Poisson's ratio that evolve with the applied stress states, and do not follow the expectations of models such as those of Paliwal and Ramesh[18] or Hu *et al.*[21]. Fortunately prior theory has a number of insights to guide analysis. To begin with, Nemat-Nasser and Obata[19] proposed a microcrack theory of bulking in brittle materials based off of the assumption that individual microcracks follow the dynamics of macroscopic wing cracks seen in other materials[29]. A wing crack of the type observed in macroscopic brittle materials can be seen in Figure 4A, with a highly simplified form that reduces all elements to purely orthogonal components seen in Figure 4B.

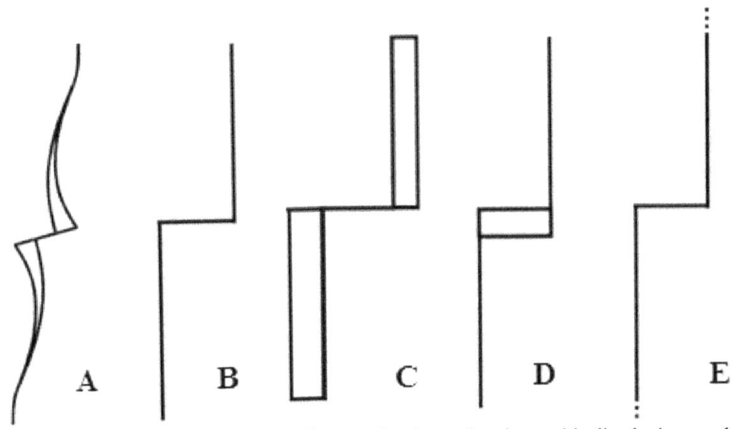

Figure 4 – Simplified wing-crack failure mechanisms, showing an idealized wing crack (A), a simplified crack (B), crack sliding (C), crack face opening (D), and crack extension (E).

The key insight of the Nemat-Nasser and Obata model[19] is that they attribute three additional factors to the stress-strain relationship aside from the standard elasticity components. The first of these factors is crack sliding where the faces of an initial crack slide against each other, widening the cracks branching off as wings, depicted in a simplified form in Figure 4C. The second factor is crack opening, where the initial crack faces separate away from each other, seen in Figure 4D. The third factor is crack extension, where the wing cracks elongate in the direction of applied stress, as shown in Figure 4E. Of these factors, crack sliding and opening are formulated as not being directly influenced by applied stress but by the growth of the cracks altering crack size distributions, while crack elongation requires an applied stress. While all three of these factors interact with each other, in a most idealized state only crack sliding and crack face opening influence the volume the crack occupies. By combining these insights with the sort of damage accumulation formulation seen in Equation 1, it is possible to produce an expanded damage formulation that captures the complex behaviour seen in experiment and provides a bridge between experiment and modelling.

The first feature we would expect in an expanded damage formulation is a replication of the behaviour seen in Equation 1, with both linear elasticity and a change in stiffness expressed. Next we know that we have a change in internal volume that would be expressed by a change in Poisson's ratio. Finally, we have the assumption of permanent strains both laterally and axially that should appear even when stress is zero, caused by crack face sliding and opening. This leads to the requirement of three factors to account for axial, lateral, and permanent changes. Working from basic elasticity, these can be derived for a uniaxial stress state, giving two equations, one for each direction.

$$\varepsilon_{11} = \frac{\sigma_{11}}{E} + \frac{\sigma_{11} D_1}{E(1+D_1)} + D_3 \tag{2}$$

$$\varepsilon_{22} = \frac{-v\sigma_{11}}{E} - \frac{v(D_1 - D_2)\sigma_{11}}{E(1+D_1)} - v(1 + D_2)D_3 \tag{3}$$

Equations 2 and 3 show formulation for axial (ε_{11}) and lateral strains (ε_{22}) respectively, showing the three new damage terms. The first, D_1, is the same as the D term in Equation 1 and corresponds to changes in Young's modulus (E). Where this term differs is that for nomenclature purposes it is allowed to be essentially any value, with a complete loss of stiffness corresponding to -1 rather than 1 in Equation 1. The first new term is D_2, which corresponds to changes in Poisson's ratio (v) due to damage. If damage results in a loss of internal volume the D_2 term will manifest as a negative value, while an increase in internal volume will manifest as a positive value. Finally there is D_3, which corresponds to permanent strains caused by damage. This last term is included for completeness, as advanced ceramics typically undergo catastrophic failure and disintegrate once cracking is initiated, but for other types of brittle materials such as concrete cracks can grow without catastrophic failure[26], [30]. The D_3 term also provides a method for describing unloading curves where extrapolation would not bring it back to zero strain under zero applied stress.

The reason for allowing all three damage values to vary in both positive and negative ways is that their interactions can produce effects that would previously not have been mathematically permitted under prior damage models where D can only vary between 0 and 1[18]–[21]. To demonstrate the full breadth of possibility, imagine for example a material that begins with cracks present that is compressed. Under this situation some of the material is being compressed into void space rather pressing against other material, and it thus manifests as a lower apparent stiffness. As cracks are closed along the direction of the stress, the stiffness would increase, presumably to a maximum value equal to the value of a pristine sample. While this crack closure was occurring, the Poisson's ratio would be less than for a pristine value as the material could move into void spaces rather than expanding around the free surfaces.

Now let the sample behave as if it were pristine for some time before the internal cracks begin to grow parallel to the loading direction, reducing the apparent stiffness and increasing the apparent Poisson's ratio. Before failure occurs the experiment goes into an unload phase and the cracks that had been compressed shut begin to rebound. The slope of the stress-strain curve is now greater than during the load sequence as the measured strain has a greater decrease with decreasing stress than the pristine sample would suggest, even while the sample is still in a compressive state. This produces an apparent increase in stiffness, and thus D_1 is assigned a positive value in order to keep nomenclature straight. This rebound effect could also produce a positive D_3 value as the material reaches zero stress with a greater total length than it started with due to crack volume expansion. In this hypothetical, D_2 may also exhibit unintuitive behaviors such as producing a negative total Poisson's ratio, which would indicate that the damage has a significant anisotropic character and has resulted in auxetic behavior.

This hypothetical experiment shows how all three damage values can be combined together to determine how cracks, which may be microscopic and internal to an opaque material, are evolving by their influence on macroscopic elastic constants. The mathematical construction also allows for cracks to close as well as open. A complex stress-strain curve can thus be broken into multiple phases where the combination of damage terms derived from observed macroscopic behavior allows for the inference of microscopic crack evolutions.

4.2. Quasi-Static Cermet Analysis

The quasi-static cermet experiment seen in Figure 2 serves to illustrate the behaviour one would expect from dilatancy theory in a material that accumulates damage easily. The axial and lateral strains follow quadratic curves, suggesting that in addition to a linear elasticity term the D_1 and D_2 terms both are linear functions of strain, producing a continuous decrease in stiffness and a continuous increase in Poisson's ratio. Performing a damage analysis produces Figure 5, which confirms the linear nature of damage accumulation, with stiffness decreasing by 40% and Poisson's ratio increase by 100%. This corresponds to the classical microcrack model assumptions with crack growth deteriorating stiffness and increasing the volume of the sample via the internal volume of the cracks needing to be accommodated. This simple but illustrative case allows for contrast with the more complex dynamic behavior of AD-995.

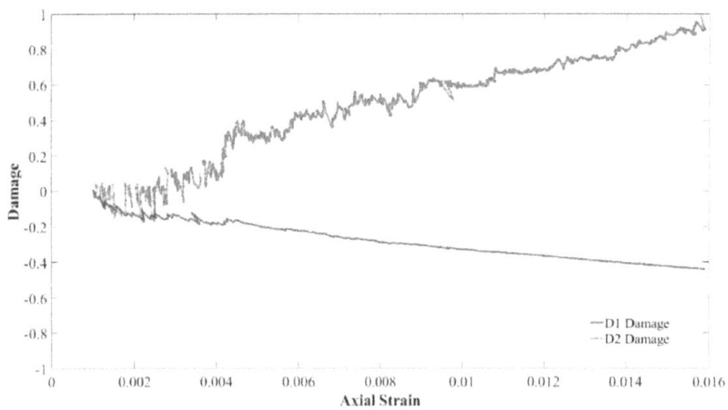

Figure 5 – D_1 (stiffness) and D_2 (Poisson's ratio) damage for TitanMade cermet, showing that both are linear functions of axial strain.

4.3. Dynamic AD-995 Analysis

In contrast with the cermet, the experiment shown in Figure 3 has stress-strain is primarily linear for the majority of the incident pulse, but once peak stress is reached the curve begins to bend downward as stiffness is lost. Comparing the axial versus lateral strains at these times, the lateral stress strongly diverges from linear behavior in a manner that suggests that cracks are opening up internally. Thus during the unloading phase, we can regard the steeper slope not as evidence of a higher actual stiffness, but as evidence that the crack faces are separating even while still under macroscopic compression. Extrapolating the unload line to zero stress produces a virtual D_3 value of 0.006 strain. The increase in apparent stiffness and Poisson's ratio and a permanent compressive strain suggests that the sample, had it not fragmented, would have been permanently deformed. For advanced ceramics this is a non-physical behavior due to unstable crack growth, but it is illustrative as in another material such as concrete it is possible for the axial length to shorten and the sides buckle outward by both the compression and growth of internal cracks. When taken together with the changes to Poisson's ratio and appearance of a permanent strain, the apparent increase in Young's modulus becomes a sensible behavior based on engineering strain. The data further shows that because the sample fragmented, this suggests that further damage accumulation was occurring during the unloading sequence, and that ultimate failure can occur at less than peak strain.

5. Conclusions

In this paper we have explored the use of digital-image-correlation and ultra-high-speed-photography in studying bulking in brittle materials by being able to capture 2D strains for events occurring in microseconds. A variety of ceramics and cermets were studied under both quasi-static and dynamic conditions, with variations in response arising from both differences in material and differences in strain rates. Under dynamic conditions the 2D strains could become complex and failed to conform to predicted responses. By taking into consideration the expected microscopic phenomena, damage to elastic constants could be expanded from Young's modulus to also include changes to Poisson's ratio and permanent strains. By considering these damage terms simultaneously, failure mechanisms such as crack closure or opening can be discerned, providing information on the dynamic evolution of crack growth. These new insights helped to make sense of previously complex experimental data by providing a way to evaluate damage behaviors simultaneously, providing explanation for behaviors that could not be explained in prior models. This insight into crack behavior can lead to improvements in input parameters and validation methods for simulations and modelling, and can also aid in guiding design for protective systems against high velocity impacts.

Acknowledgements

This research was sponsored by the Army Research Laboratory and was accomplished under Cooperative Agreement Number W91NF-16-2-0079. The views and conclusions contained in this document are those of the authors and should not be interpreted as representing the official policies, either expressed or implied, of the U.S. Army Research Laboratory or the U.S. Government. The U.S. Government is authorized to reproduce and distribute reprints for Government purposes notwithstanding any copyright notation herein.

References

[1] D. A. Shockey, A. H. Marchand, S. R. Skaggs, G. E. Cort, M. W. Burkett, and R. Parker, "Failure Phenomenology of Confined Ceramic Targets and Impacting Rods," *Int. J. Impact Eng.*, vol. 9, pp. 263–275, 1990.

[2] M. L. Wilkins, "Mechanics of penetration and perforation," *Int. J. Eng. Sci.*, vol. 16, no. 11, pp. 793–807, 1978.

[3] M. Wilkins, C. Honodel, and D. Sawle, "APPROACH TO THE STUDY OF LIGHT ARMOR.," California Univ., Livermore. Lawrence Radiation Lab., 1967.

[4] W. W. Chen, A. M. Rajendran, B. Song, and X. Nie, "Dynamic Fracture of Ceramics in Armor Applications," *J. Am. Ceram. Soc.*, vol. 90, no. 4, pp. 1005–1018, Apr. 2007.

[5] J. D. Clayton, "Penetration resistance of armor ceramics: Dimensional analysis and property correlations," *Int. J. Impact Eng.*, vol. 85, pp. 124–131, Nov. 2015.

[6] T. J. Moynihan, S.-C. Chou, and A. L. Mihalcin, "Application of the Depth-of-Penetration Test Methodology to Characterize Ceramics for Personnel Protection:," Defense Technical Information Center, Fort Belvoir, VA, Apr. 2000.

[7] F. Cui, G. Wu, T. Ma, and W. Li, "Effect of Ceramic Properties and Depth-of-penetration Test Parameters on the Ballistic Performance of Armour Ceramics," *Def. Sci. J.*, vol. 67, no. 3, p. 260, Apr. 2017.

[8] D. R. Curran, L. Seaman, T. Cooper, and D. A. Shockey, "Micromechanical model for comminution and granular flow of brittle material under high strain rate application to penetration of ceramic targets," *Int. J. Impact Eng.*, vol. 13, no. 1, pp. 53–83, 1993.

[9] J. D. Hogan, L. Farbaniec, N. Daphalapurkar, and K. T. Ramesh, "On Compressive Brittle Fragmentation," *J. Am. Ceram. Soc.*, vol. 99, no. 6, pp. 2159–2169, Jun. 2016.

[10] A. Krell and E. Strassburger, "Order of influences on the ballistic resistance of armor ceramics and single crystals," *Mater. Sci. Eng. A*, vol. 597, pp. 422–430, Mar. 2014.

[11] A. Krell, E. Strassburger, T. Hutzler, and J. Klimke, "Single and Polycrystalline Transparent Ceramic Armor with Different Crystal Structure," *J. Am. Ceram. Soc.*, vol. 96, no. 9, pp. 2718–2721, Sep. 2013.

[12] D. E. Grady, "Shock-wave compression of brittle solids," *Mech. Mater.*, vol. 29, no. 3, pp. 181–203, 1998.

[13] H. Horii and S. Nemat-Nasser, "Compression-Induced Microcrack Growth in Brittle Solids," *J. Geophys. Res.*, vol. 90, pp. 3105–3125, Mar. 1985.

[14] H. Deng and S. Nemat-Nasser, "Microcrack arrays in isotropic solids," *Mech. Mater.*, vol. 13, pp. 15–36, 1992.

[15] E. N. Landis and S. P. Shah, "The influence of microcracking on the mechanical behavior of cement based materials," *Adv. Cem. Based Mater.*, vol. 2, no. 3, pp. 105–118, 1995.

[16] F. Longy and J. Cagnoux, "Plasticity and Microcracking in Shock-Loaded Alumina," *J. Am. Ceram. Soc.*, vol. 72, no. 6, pp. 971–979, 1989.

[17] L. Farbaniec, J. D. Hogan, and K. T. Ramesh, "Micromechanisms associated with the dynamic compressive failure of hot-pressed boron carbide," *Scr. Mater.*, vol. 106, pp. 52–56, Sep. 2015.

[18] B. Paliwal and K. T. Ramesh, "An interacting micro-crack damage model for failure of brittle materials under compression," *J. Mech. Phys. Solids*, vol. 56, no. 3, pp. 896–923, Mar. 2008.

[19] S. Nemat-Nasser and M. Obata, "A microcrack model of dilatancy in brittle materials," *J Appl Mech*, vol. 55, no. 1, pp. 24–35, 1988.

[20] H. Deng and S. Nemat-Nasser, "Microcrack Interaction and Shear Fault Failure," *Int. J. Damage Mech.*, vol. 3, no. 1, pp. 3–37, Jan. 1994.

[21] G. Hu, J. Liu, L. Graham-Brady, and K. T. Ramesh, "A 3D mechanistic model for brittle materials containing evolving flaw distributions under dynamic multiaxial loading," *J. Mech. Phys. Solids*, vol. 78, pp. 269–297, May 2015.

[22] V. L. Shkuratnik, P. V. Nikolenko, and A. E. Koshelev, "Stress dependence of elastic P-wave velocity and amplitude in coal specimens under varied loading conditions," *J. Min. Sci.*, vol. 52, no. 5, pp. 873–877, Sep. 2016.

[23] V. R. Shea and D. R. Hanson, "Elastic Wave Velocity and Attenuation as Used to Define Phases of Loading and Failure in Coal," *Int. J. Rock Mech. Min. Sci.*, vol. 25, no. 6, pp. 431–437, 1988.

[24] K. K. Phani and D. Sanyal, "Critical reevaluation of the prediction of effective Poisson's ratio for porous materials," *J. Mater. Sci.*, vol. 40, no. 21, pp. 5685–5690, Nov. 2005.

[25] M. Rempe, T. Mitchell, J. Renner, S. Nippress, Y. Ben-Zion, and T. Rockwell, "Damage and seismic velocity structure of pulverized rocks near the San Andreas Fault: DAMAGE STRUCTURE OF PULVERIZED ROCKS," *J. Geophys. Res. Solid Earth*, vol. 118, no. 6, pp. 2813–2831, Jun. 2013.

[26] C. Poinard, E. Piotrowska, Y. Malecot, L. Daudeville, and E. N. Landis, "Compression triaxial behavior of concrete: the role of the mesostructure by analysis of X-ray tomographic images," *Eur. J. Environ. Civ. Eng.*, vol. 16, no. sup1, pp. s115–s136, Jun. 2012.

[27] F. Hild and S. Roux, "Digital Image Correlation: from Displacement Measurement to Identification of Elastic Properties - a Review," *Strain*, vol. 42, no. 2, pp. 69–80, May 2006.

[28] J. Han and B. Pan, "A novel method for measuring discontinuous deformation in digital image correlation based on partition and dividing strategy," *Eng. Fract. Mech.*, vol. 204, pp. 185–197, Dec. 2018.

[29] W. F. Brace and E. G. Bombolakis, "A note on brittle crack growth in compression," *J. Geophys. Res.*, vol. 68, no. 12, pp. 3709–3713, Jun. 1963.

[30] Z. P. Bazant and M. H. Hubler, "Theory of cyclic creep of concrete based on Paris law for fatigue growth of subcritical microcracks," *J. Mech. Phys. Solids*, vol. 63, pp. 187–200, Feb. 2014.

Proceedings of the 2019 Hypervelocity Impact Symposium
HVIS2019
April 14-19, 2019, Destin, FL, USA
HVIS2019-025

Extended Abstract+

Deformation and Acceleration of Zn and Cu Liners under Explosive Shock Loading

Puwadet Sutipanya[a], Fumikazu Saito[a], Mitsuyori Nakashita[b], Takashi Ogino[b], Yutaka Takizawa[b], Yohsuke Okada[b], Masahiko Natsubori[b] *

[a] National Defense Academy, 1-10-20, Hashirimizu, Yokosuka 239-8686, Japan
[b]*CHUGOKU KAYAKU CO., LTD., 2530 Iwasaki, Yoshii, Takasaki, Gunma, 370-2131, Japan*

1. Introduction

Shaped charge jets (SCJs) are of interest in the field of ballistics because of their excellent penetration ability. The various characteristics of an SCJ, such as density, temperature and length of the jet, are important when estimating the expected penetration depth. Therefore, these characteristics have been widely researched [1-5]. Zernow studied the density deficit associated with an in-flight jet using densitometry [1]. This prior work determined that the 15-20% density deficit of a Cu jet (as reported by Jamet [2]) includes a temperature contribution of 3.5%, an experimental error of 5% and a so-called "vacancy avalanche" contribution of 6.5%. Uhlig and Hummer measured the conductivity and temperature of in-flight Cu SCJs by employing a novel magnetic diffusion technique and five color radiometry. The resulting conductivity values indicated that the tip of the SCJ had a bulk temperature on the order of 1200 K, which is in the vicinity of the melting point of Cu (1350 K) [3]. Hertel *et al.* developed an analytical model for the SCJ formation process. CTH calculations utilizing their analytical model accurately predicted the thermodynamic characteristics of the SCJ [4, 5]. Huang quantified the density of an SCJ and reported the density distribution based on evaluating the x-ray absorption coefficient of a metal jet [6], and found that the density of a powder metal jet decreases toward the jet tip as the jet elongates. These prior reports indicate that observations of various jet features under high temperature conditions (close to the melting point) can be a useful means of assessing the density deficit of an SCJ.

The purpose of the present study was to characterize a metal jet generated from a conical liner and to evaluate the relationship between the melting point of the metal and the density of the jet under high-temperature conditions. Both zinc ($T_{melt} \cong 693$ K [7]) and copper ($T_{melt} \cong 1350$ K [8]) were used as the metals in this work, and images of the resulting metal jets were acquired using a flash x-ray system. The physical quantities of the jets, such as mass and apparent density, were determined based on *in situ* mass analysis in conjunction with x-ray image processing (*i*-MAX).

2. *i*-MAX Method

Recently, our group reported a novel method of *in situ* mass analysis based on flash x-ray image processing, which can be used to estimate the mass of lightweight ceramic targets, including those made of mullite, silicon nitride and alumina [9]. To extend this technique to fragments obtained from other materials using the same principle, we attempted to demonstrate that the estimated mass (M) of an SCJ could be determined from an x-ray image using the equation

---

+ Full paper can found at https://www.sciencedirect.com/journal/international-journal-of-impact-engineering/special-issue/10MTGP5W4VJ

* Corresponding author. Tel.: +81-046-841-3810; fax: +81-046-844-5912.
E-mail address: ed18010@nda.ac.jp

$$M = \rho \sum t_i \cdot S(t_i) \tag{1}$$

where ρ is the density of a standard sheet (termed a reference sheet) of the material used as the liner, t_i is the thickness of the ith sequence obtained by analysis of the decomposition area of the SCJ image, and $S(t_i)$ is the area of the ith sequence thickness. This last parameter is estimated using the equation

$$S(t_i) = \frac{B(t_i)}{A(t_i)} S_{stand} \tag{2}$$

where S_{stand} is the constant area of the standard sheet, $A(t_i)$ is the of x-ray intensity at the irradiated area of the ith sequence thickness (that is, the peak height of a normal distribution curve), and $B(t_i)$ is an optimization factor for the peak height of the ith sequence thickness. Therefore, the mass of the SCJ can be estimated from the following equation.

$$M = \rho \sum \left(t_i \cdot \frac{B(t_i)}{A(t_i)} S_{stand} \right) \tag{3}$$

3. Experimental setup

The experiments were performed using a flash x-ray system with an output voltage of 100-120 VAC to generate soft x-rays. In-flight images of the SCJs were acquired using imaging plates (IPs, Fujifilm) that were highly sensitive to x-rays. X-ray images of the zinc and copper SCJs were obtained immediately after the jets were generated (at a standoff distance of 4 CD (Charge Diameter) and after break up of the jets (at a standoff distance of 25 CD). The SCJ masses were estimated by employing standard zinc and copper reference sheets having various thickness along the jet trajectory direction, as shown in Figure 1. The compact shaped charges each had a conical zinc or copper liner with a diameter of 30 mm and thickness of 2 mm, together with a number 6 electronic detonator and a Pentolite explosive charge. All experiments were performed in ambient air for the sake of simplicity.

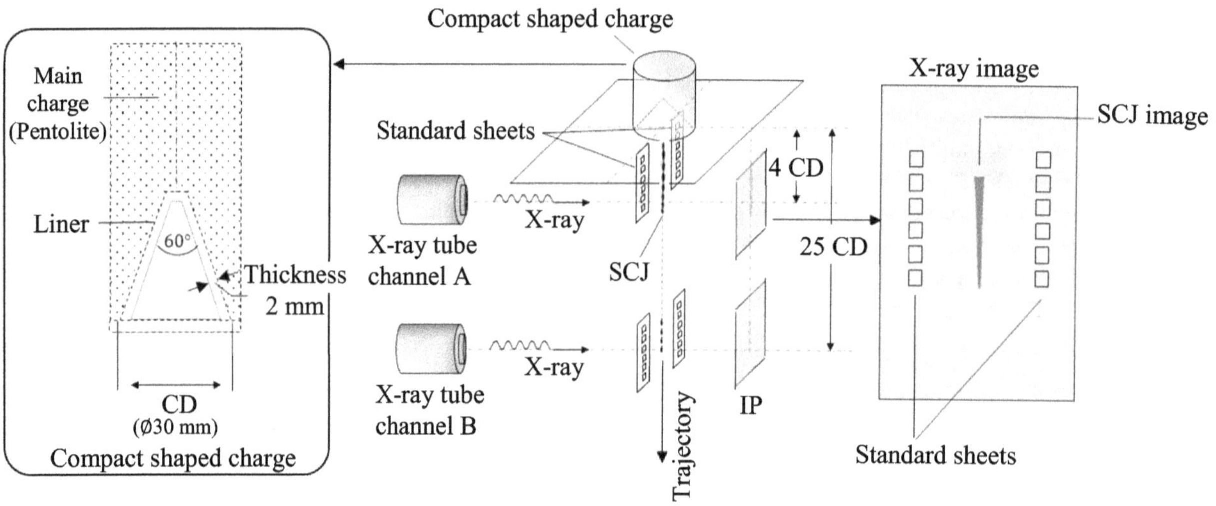

Fig. 1. Schematics showing the experimental apparatus and setup.

4. X-ray image analysis

4.1. Stretch and break up of the SCJs

Figures 2(a) and 2(b) respectively show x-ray radiographs of zinc and copper SCJs acquired at standoff distances of 4 CD and 6 CD. The shadow images generated by the zinc and copper standard sheets having various thickness placed in the trajectory direction indicate the difference in blackness due to variations in the extent of x-ray transmission. The images were calibrated based on the intensity of horizontal and vertical lines throughout the background area, as indicated to the side of the figures. The zinc SCJ in Figure 2(a) shows stretching of the jet along the trajectory direction, while Figure 2(b) pictures the copper SCJ, which is broken up and separated into small fragments. Figures 2(c) and 2(d) are enlarged images of the head portions of the zinc and copper SCJs, respectively. We defined the leading jet of the zinc SCJ as the region 4 mm from the tip, similar to the length of the copper leading jet. The rear part of the zinc leading jet in Figure 2(c) exhibits a comet-like tail of small particles which is quite unique compared to the morphology of the copper leading jet in Figure 2(d). The leading jet of the zinc SCJ remains attached to the following jet, while the copper leading jet is broken up and show an appearance like a droplet. These results demonstrate that, although the SCJs were generated at a close position, both SCJs were obviously distinct.

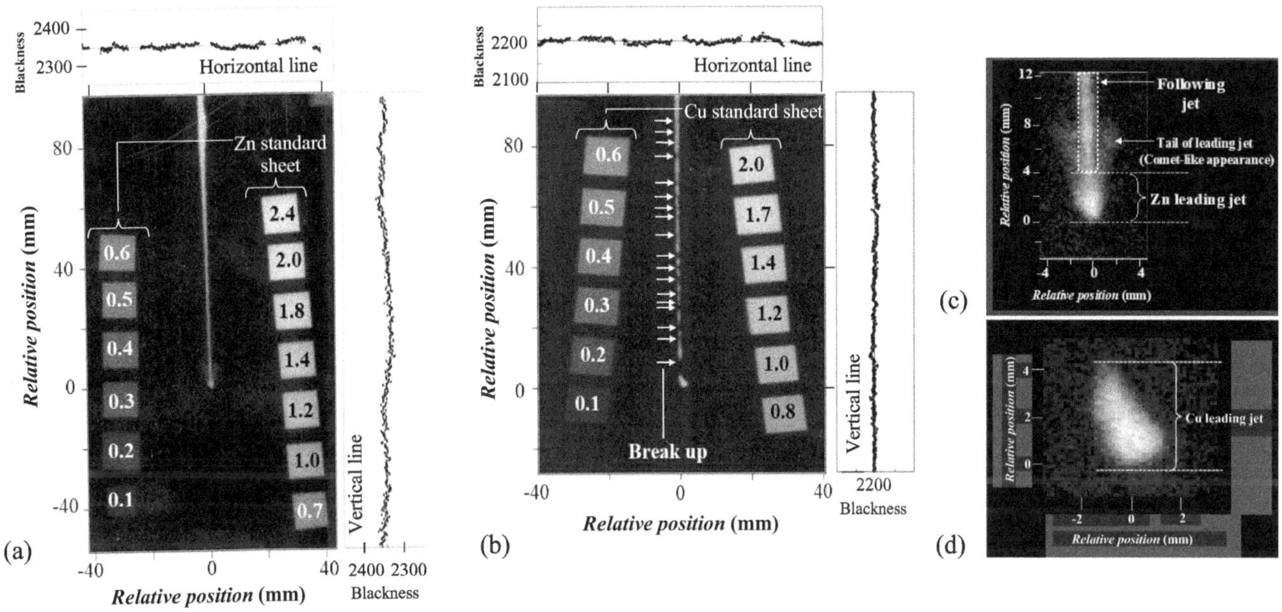

Fig. 2. X-ray radiographs of (a) the zinc SCJ (showing stretching) and (b) the copper SCJ (showing break up) acquired in the 4 CD and 6 CD standoff regions, respectively. The figures display the shadow images of zinc and copper standard sheets having various thickness placed in the trajectory direction. Beside each figure the corrected blackness intensity results are shown in the horizontal and vertical directions. Enlarged x-ray images showing the different features of the jet tips of (c) the zinc SCJ (which, due to stretching, is divided into a leading jet, leading jet tail (with a comet-like appearance) and following jet) and (d) the copper SCJ (which is broken up and shows an appearance like a droplet).

4.2. Estimation of the physical quantities of the jets

An analytical method termed *in situ* mass analysis from x-ray image processing (*i*-MAX) was utilized to estimate the mass and apparent density of each SCJ in this study. The masses determined for the zinc and copper leading jets were (34 ± 2) mg and (68 ± 3) mg, respectively. The overall mass distributions of both the zinc and copper SCJs showed the same tendency to decrease toward the tip of the jet, with the exception of the leading jet. The density distribution of the zinc SCJ gradually decreased toward the leading jet as it stretched. In contrast, the density distribution of the copper SCJ varied depending on the appearance of a necking region, and the minimum apparent density was observed 30 mm from the leading jet.

4.3. The evidence for crater spreading

Figures 3(a) and 3(b) show the craters on witness plates created by the impact of the zinc and copper SCJs, respectively. These plates were placed at a distance of 27 CD from the initial position of the shaped charge. The craters caused by the zinc SCJ impact are relatively shallow and widely dispersed relative to the position of the deepest crater, as shown in Figure 3(a). In contrast, the craters in Figure 3(b) appear in a line, showing that the copper SCJ deviated from its original ballistic direction. Spraying of the zinc SCJ onto the witness plate is also clearly evident, suggesting that the zinc was in a liquid state and/or was composed of a cloud of fine particles (that is, a dust). The evidence from these witness plates is in good agreement with the x-ray images provided in Figure 2(a).

(a) 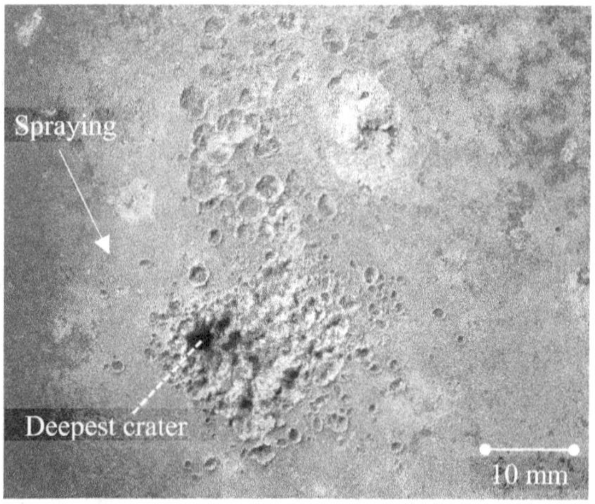

(b) 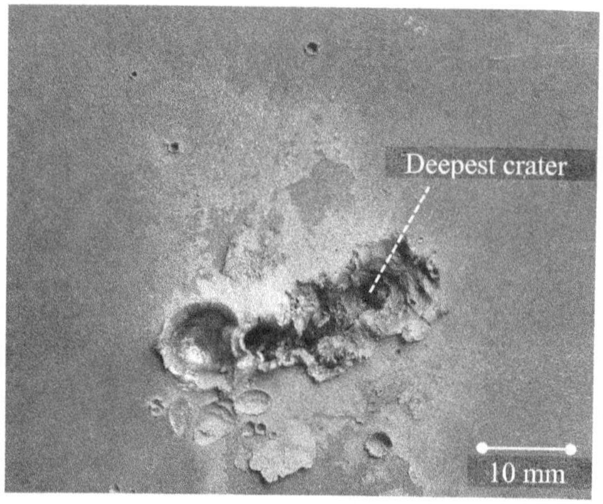

Fig. 3. Photographs of witness plates at a standoff distance of 27 CD after impact by (a) the zinc SCJ and (b) the copper SCJ. Figure (a) shows the wide dispersal of many craters relative to the position of the deepest crater. Overall, the craters are relatively shallow and there is evidence of spray from the zinc SCJ around the crater boundaries. The craters in (b) are relatively deep and appear as a small cluster.

Acknowledgements

The authors would like to acknowledge funding for this project through Japan funding agency. The authors would also like to acknowledge the assistance of personnel at the Chugoku Kayaku Company, Ltd. during the experimental setup.

References

[1] Zernow, L., 1997. The Density Deficit in Stretching Shaped Charge Jets, International Journal of Impact Engineering 20, p. 849.
[2] Jamet, F., 1976. Measurements of Densities in Shaped Charge Jets and Detonation Waves, American Society for Non-Destructing Testing.
[3] Uhlig, W., Hummer, C., 2013. "In-Flight Conductivity and Temperature Measurements of Hypervelocity Projectiles," Proceedings of the 12th Hypervelocity Impact Symposium, pp. 48-57.
[4] Sable, P., Helminiak, N., Gullerud, A., Harstad, E., Hollenshead, J., Hertel, E., 2017. "Characterizing In-Flight Temperature of Shaped Charge Penetrators in CTH," Proceeding of the 14th Hypervelocity Impact Symposium, pp. 375-382.
[5] Helminiak, N. S., Sable, P., Harstad, E., Gullerud, A., Hollenshead, J., Hertel, E., 2017. "Characterizing In-Flight Temperature of Explosively Formed Projectiles in CTH," Proceeding of the 14th Hypervelocity Impact Symposium, pp. 178-185.
[6] Huang, H., 2013. "Characterization of Powder Metal Jet and Penetration Calculation," Proceedings of the 12th Hypervelocity Impact Symposium, pp. 471-478.
[7] Wang, L., Xian, A., 2003. Density Measurement of Liquid Indium and Zinc by the γ-ray Attenuation Method, High Temperatures – High Pressures 35, p. 659.
[8] Brillo, J., Egry, I., 2003. Density Determination of Liquid Copper, Nickel, and their Alloys, International Journal of Thermophysics 24, p. 1155.
[9] Saito, F., Kawai, N., Tamura, H., 2012. "Observation of Mass and Velocity of Projectile Fragments Produced by Hypervelocity Impact with Lightweight Ceramic Targets," American Institute of Physics Conference Proceedings, Shock compression of Condensed Matter, pp. 891-894.

Proceedings of the 2019 Hypervelocity Impact Symposium
HVIS2019
April 14-19, 2019, Destin, FL, USA
HVIS2019-041

Modelling the Fracture of High-Hardness Armour Steel in Taylor Rod-on-Anvil Experiments

Brodie McDonald[a,b], Shannon Ryan[b], Stephen J. Cimpoeru[b], Nathan Edwards[c], Adrian Orifici[a]*

[a]*School of Engineering, RMIT University, PO Box 71, Bundoora, Victoria 3083, Australia.*
[b]*Defence Science and Technology Group, 506 Lorimer Street, Fishermans Bend, Victoria 3207, Australia*
[c]*Department of Mechanical and Product Design Engineering, Swinburne University of Technology, John Street, Hawthorn, Victoria 3122, Australia*

Abstract

A series of Taylor rod-on-anvil experiments have been performed to validate the predicted impact velocity fracture threshold and fracture mode of a high hardness armour steel (HHA) obtained through explicit finite element simulations. Experimentally, the rods exhibited principal shear failure, a condition that can be closely linked to adiabatic shear band (ASB) formation in high strength steel. Using a stress triaxiality and Lode angle dependent failure strain criterion (Basaran 3D fracture locus), calibrated from quasi-static mechanical characterisation tests, the simulations were unable to predict the onset of fracture observed in experiments. As such, a strength-fading criterion is proposed using a phenomenological description to capture the loss of load-carrying capacity resulting from ASB formation. The ASB criterion is based on an exponential fit to experimentally-observed instability strains measured at different average stress triaxialities in a series of tests on inclined cylindrical and modified flat-hat specimens. With the prediction of ASB formation the material strength is reduced to model the thermal softening experienced in the shear band, and fracture of the material (in the form of element erosion) remains controlled by the Basaran fracture model. Incorporating the ASB-based criterion, the numerical models were found to accurately predict both the impact velocity fracture threshold, as well as the general appearance of the observed principal shear fracture. The proposed criterion enables the effects of ASB formation to be captured in an impact simulation with little increase in computational cost.

Keywords: Taylor rod-on-anvil; Finite element simulations; Adiabatic shear bands; Armour steel

1. Introduction

The Taylor rod-on-anvil experiment [1] involves a planar impact between a cylindrical test specimen and a large, high strength anvil causing plastic deformation and possible fracture of the test sample under varying impact velocities. Coupled with finite element simulations, Taylor rod-on-anvil experiments have been widely utilized to assess the predictive capacity of constitutive models in reproducing various aspects of dynamic material behaviour. Numerous plasticity models have been evaluated against Taylor impact experiments for the prediction of deformed rod profiles [2,3], free surface velocity-time response [4] and full-field surface strain evolution [5]. In contrast, evaluation of fracture modelling against Taylor anvil experiments is less common and the complexity of unique failure modes and 3D fracture propagation increases computational requirements.

Tensile splitting and petalling fracture of Aluminium AA6061-T6 was first modelled by Anderson et al. [6] using the Johnson-Cook (JC) plasticity and fracture models [2,7] calibrated against cylindrical notched tensile experiments. While the fracture mode was qualitatively reproduced, the impact velocity to induce fracture was significantly over-estimated. Chocron et al. [8] re-modelled Anderson et al.'s experiments and, utilizing a novel Lode angle dependent extension to the JC plasticity and fracture model, successfully reproduced the experimental fracture initiation velocity and petalling behaviour. Rakvåg et al. [9]

* Corresponding author.
E-mail address: brodie.mcdonald@dst.defence.gov.au

modelled the deformation and fracture of tool steel with three different yield strength values: 395, 1203, and 1660 MPa. Tensile splitting and spiral shear failure of the two lower strength materials was well captured by a stochastic Cockcroft-Latham fracture model [10] calibrated against un-notched tensile samples tested quasi-statically. For the highest strength steel however, which experienced principal shear failure caused by adiabatic shear band (ASB) formation and significant fragmentation at higher impact velocities, qualitative reproduction of the failure modes was less satisfactory. The Cockcroft-Latham fracture model, which is dependent on the tensile principal stress ($\langle\sigma_1\rangle$), is unable to accurately predict fracture under the negative stress triaxialities experienced with ASB formation.

This paper examines the response of HHA steel in Taylor rod-on-anvil experiments and presents a computational modelling approach to capturing the deformation and fracture behaviour in finite element simulations. Sections 2 and 3 present a new mechanical testing and constitutive modelling approach to characterise adiabatic shear band formation exhibited by HHA steel under compressive and shear loading as a function of stress triaxiality. Section 4 summarises the setup and results of a series of Taylor rod-on-anvil experiments. Numerical modelling of the tests is presented in Section 5 which highlights the accuracy of the new constitutive model in predicting the impact velocity fracture threshold and failure modes of HHA steel during Taylor rod-on-anvil experiments.

2. Mechanical Characterization

McDonald et al. [11] characterised the plasticity and ductile fracture behaviour of a Class 1 high hardness armour steel (HHA), as per MIL-DTL-46100E [12], under tensile loading across a range of loading conditions including elevated temperatures and strain rates. Additional mechanical testing has been performed in compression and shear to include zero and negative stress triaxiality conditions. Characterisation within this region follows the method described by Herzig et al. [13]. Inclined cylindrical samples [13] (Figure 1a), were utilised to characterise fracture behaviour during uniaxial compression and combined compression-shear loading. Inclination angles of 90°, 85° and 80° were used to vary the stress state within the sample. Material behaviour in shear was characterised using flat-hat specimens [13] (Figure 1b) with several modifications to those originally proposed by Clos et al. [14]. The modified flat-hat shear specimen was designed with an optimised gauge section geometry to minimise changes in stress state during deformation and encourage failure initiation away from material boundaries. The gauge section morphology and computational design approach are inspired by Roth and Mohr [15] but inverted vertically to produce a proportional shear stress state under the compressive loading of the sample.

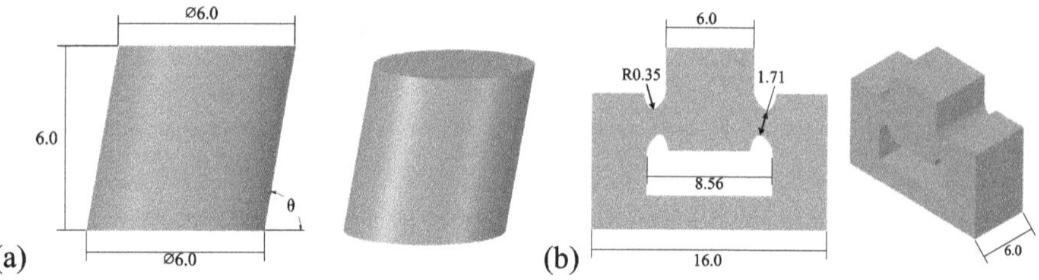

Figure 1. (a) Inclined compression specimens, θ= 90°, 85°, 80° (b) Modified flat-hat shear specimen.

Each inclined cylindrical and modified flat-hat specimen type was tested at 0.001 s^{-1} in a universal test machine. Up to the load capacity of the test frame (100 kN) none of the inclined compression samples failed, see Figure 2a, demonstrating significant quasi-static ductility in negative stress triaxialities. The shear samples were fractured with DIC data showing strain localised in the gauge section throughout the test. Dynamic tests were performed using split Hopkinson pressure bars (SHPB) at average strain rates between 2300 s^{-1} and 2500 s^{-1} (Figure 2b). All specimens were fractured (excluding one 90° sample) at axial strains significantly lower than that sustained quasi-statically. The inclined compression samples failed with a slanted fracture mode characteristic of adiabatic shear band formation [16]. The axial strain to failure was heavily influenced by the angle of inclination, suggesting a stress state dependence in the onset of failure.

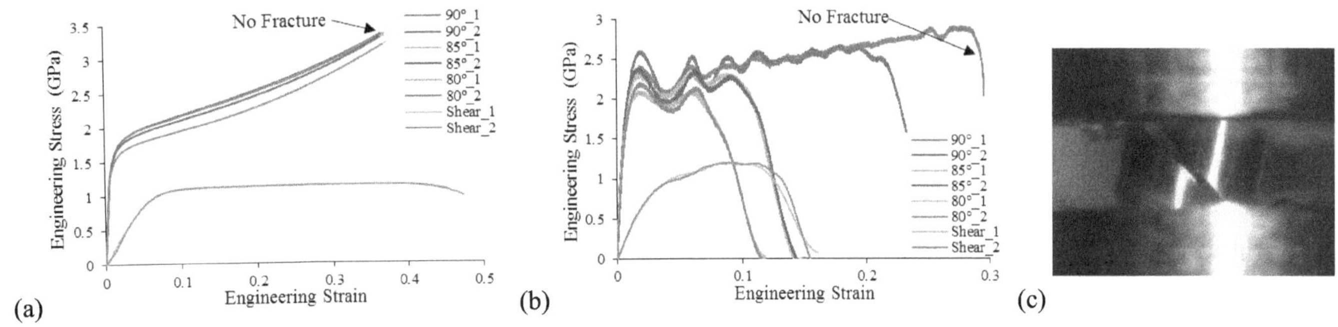

Figure 2. Engineering stress-strain response of inclined compression and shear specimens. (a) Quasi-static (b) Dynamic (c) Slanted failure mode of 80° sample.

3. Constitutive Material Modelling

Plasticity behaviour of the HHA steel was modelled using a linear equation of state and a modified Johnson-Cook (MJC) strength model with combined Voce-Ludwik strain hardening and two-part strain rate hardening. The J3-dependent generalised yield surface (GYS) proposed by Sengoz [17] was utilised to capture the pronounced tension/compression strength differential exhibited by the HHA steel. (see [18] for more details of the plasticity model and its calibration). Ductile fracture was modelled by the accumulation of plastic strain (ε_{pl}) at the current stress state and loading conditions described by Equation 1:

$$D = \int \frac{\dot{\varepsilon}_{pl}}{\varepsilon_f} dt \tag{1}$$

Where the damage variable D, is equal to 1 at fracture resulting in element erosion and ε_f is the fracture strain of the material at any proportional stress state and loading condition as given by Equation 2. ε_f is a function of stress triaxiality (η), the normalised third deviatoric stress invariant (ξ), plastic strain rate ($\dot{\varepsilon}_{pl}$) and temperature (T).

$$\varepsilon_f = f(\eta, \xi)\ g(\dot{\varepsilon}_{pl})\ h(T) \tag{2}$$

Basaran [19] proposed a function $f(\eta, \xi)$ that describes the effect of stress state on material ductility given by Equation 3:

$$\begin{aligned} f(\eta, \xi) = &\left[\frac{1}{2}(D_1^- + D_1^+) - D_1^0 + \frac{1}{2}\{D_2^- \exp(-D_3^-\eta) + D_2^+ \exp(-D_3^+\eta)\} - D_2^0 \exp(-D_3^0\eta)\right]\xi^2 \\ &+\left[\frac{1}{2}(D_1^+ - D_1^-) + \frac{1}{2}\{D_2^+ \exp(-D_3^+\eta) - D_2^- \exp(-D_3^-\eta)\}\right]\xi + D_1^0 + D_2^0 \exp(-D_3^0\,\eta) \end{aligned} \tag{3}$$

Where D_i^+, D_i^-, D_i^0 are nine empirical constants used to fit the fracture locus to experimental data. In the current modelling approach, a symmetric locus shape is enforced by making $D_i^+ = D_i^-$. Eq. (3) describes an exponential decay function for the influence of stress triaxiality as seen in the Johnson-Cook model [7], while the influence of the normalised third deviatoric invariant is defined by an asymmetric parabola with a minimum at an ordinate of 0, representing plane strain. The dependencies of the failure strain on strain rate and temperature are taken directly from the MJC model [20]. The Basaran 3D fracture locus was calibrated across a range of stress states from uniaxial tension to high stress triaxiality plane strain and was shown in [21] to accurately reproduce the fracture behaviour of HHA steel under near-field air-blast loading when failure was a result of biaxial tension (i.e. $\eta \approx 0.66$), albeit with significant Lode-angle effects.

Dynamic fracture under low stress triaxialities can be closely linked to the formation of adiabatic shear bands, which are a thermo-mechanical instability producing narrow bands of high temperature, highly strained material with limited load-bearing capability. At a microstructural level, ASBs in themselves do not constitute an actual fracture mode, as crack initiation and propagation throughout the band is caused by void nucleations elongating in shear and linking together along the shear plane [22]. To model the plasticity effects of adiabatic shear band formation in finite element simulations (strain localization, temperature increases, large deformation) requires extremely fine discretization to resolve the narrow band and limit element distortion. This level of computational expense quickly become impractical for the simulation of structural problems [22].

A simplified approach to capturing the macro-scale load carrying effects of ASB is proposed via an instability criterion which reduces the strength of an element with the prediction of ASB formation. Plastic strain within the shear band continues to develop and the eventual erosion of the element is modelled by the Basaran ductile fracture model. The ASB instability criterion is based on stress state, using a three parameter (C_1, C_2, C_3) empirical function from Hancock and Mackenzie [23]:

$$\varepsilon_f = C_1 + C_2 \exp(C_3 \eta) \tag{6}$$

Numerical simulations of each dynamic characterisation test were run without a fracture model to record the stress triaxiality (η) and plastic strain (ε_{pl}) history in the sample. The onset of failure is identified by the divergence of the experimental and numerical stress-strain signals and used to define the end of plastic strain and stress triaxiality development (see Figure 4). The critical element at the onset of adiabatic shear band formation is identified with the highest plastic strain and most adverse stress triaxiality history and used to define a plastic strain-weighted average stress triaxiality ($\tilde{\eta}$) for each sample type by Equation 7.

$$\tilde{\eta} = \int_0^{\varepsilon_f} \eta(\varepsilon_{pl})\, d\varepsilon_{pl} \tag{7}$$

For the current specimens, the critical element is in the centre of the sample for the 90° compression sample, at opposing corners of the 85° and 80° compression samples and in the lower portion of the gauge section in the shear sample. The stress state and plastic strain evolution to failure for the critical element of each specimen type is given in Figure 3a and the resulting ASB instability criterion in Figure 3b.

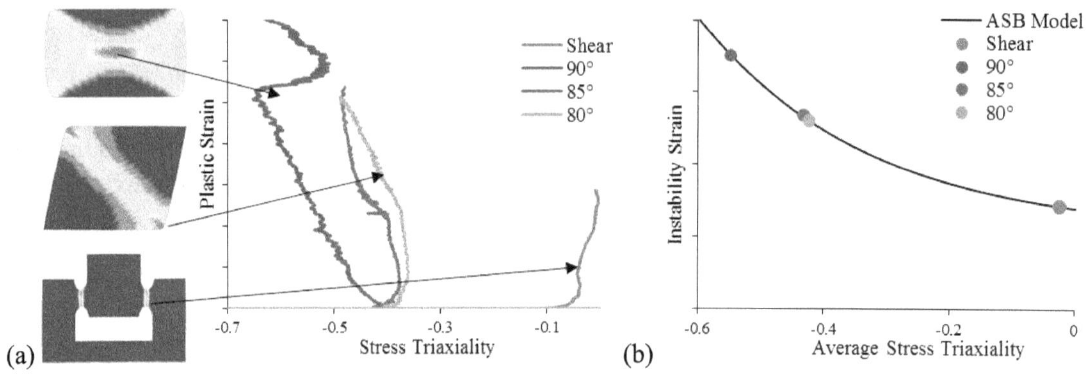

Figure 3. Stress state up to fracture in inclined compression and shear specimens. (a) Stress state evolution in critical elements. (b) ASB facture model fit to the average stress triaxiality measured in each specimen type.

To verify the ASB model calibration, the dynamic mechanical tests are re-simulated with the Basaran fracture model and the combined Basaran and ASB instability criterion activated. A comparison of the experimental and numerically predicted response of the 85° inclined compression sample is given in Figure 4 alongside the specimen appearance at the onset of fracture and following complete fracture along the shear band (both utilising the ASB criterion). While the onset of instability at an engineering strain of around 0.1 reflects the experimental response (identified by divergence between the two numerical stress signals), the rate of stress drop-off is under-estimated by the present numerical model, suggesting the transition from ASB formation to fracture initiation occurs more rapidly in the real material. Nevertheless, the current ASB criterion provides a novel approach to capturing the shear localisation, allowing the Basaran ductile fracture model to predict failure under dynamic compressive loading with significantly greater accuracy.

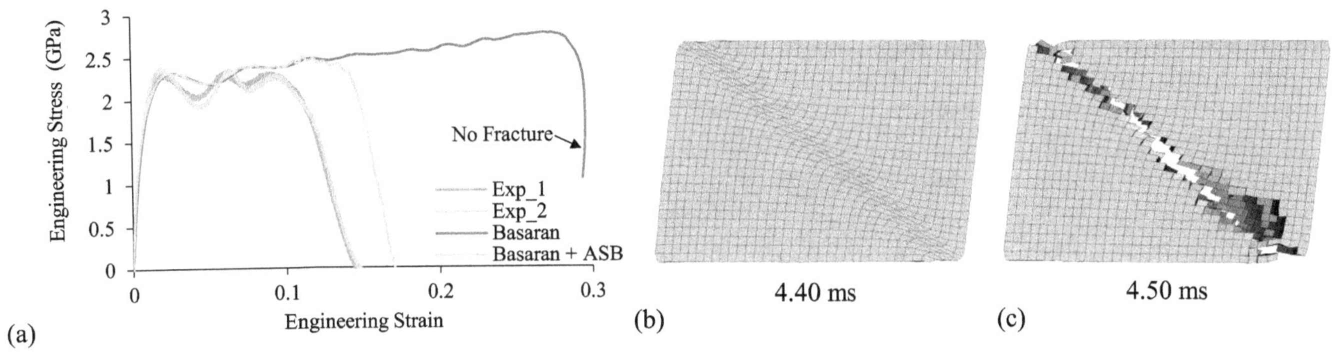

Figure 4. Numerical (FEM) response of 85° inclined compression sample including ASB fracture model. (a) Engineering stress-strain response. (b) Sample appearance at the onset of fracture and (c) Following complete through-thickness fracture.

The constitutive model parameters for the HHA steel are given in Figure 5a. Additionally, the 3D fracture locus defined by the Basaran model (Eq. 3) is plotted in Figure 5b together with the ASB formation criterion (Eq. 7), highlighting the significant difference in material response predicted between the two models. The ASB criterion is restricted to the calibration envelope of stress triaxialities between -0.6 and 0.0, and outside of this range no instability will be predicted.

Parameter		Value (unitless)
Basaran Fracture Model	D_1^-, D_1^+	0.06
	D_2^-, D_2^+	3.94
	D_3^-, D_3^+	2.46
	D_1^0	0.13
	D_2^0	0.55
	D_3^0	2.51
	D_4	-0.0035
	D_5	1.40

(a)

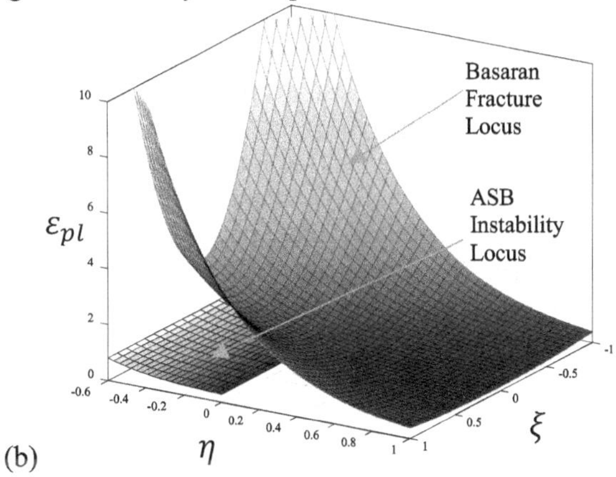

(b)

Figure 5. Instability and fracture criteria for HHA steel. (a) Constitutive model parameters. (b) Instability and fracture locus in terms of stress triaxiality (η) and normalised third deviatoric stress invariant (ξ).

4. Taylor Rod-on-anvil Experiments

A series of Taylor rod-on-anvil impact experiments were conducted in order to generate a range of large deformation and subsequent fracture results with increasing impact velocity (four intact and four fractured conditions were achieved). The experiments were performed at South West Research Institute using the apparatus, instrumentation and data processing described in detail by Bigger et al. [5]. The HHA steel was machined into cylinders with 6.35 mm diameter and 31.75 mm length and prepared with a stochastic speckle pattern to allow high speed 3D digital image correlation (DIC) throughout the experiment.

5. Taylor Rod-on-Anvil Simulations

The rod-on-anvil impact experiments were numerically simulated using the explicit finite element code LS-DYNA® to assess the accuracy of the constitutive material model for HHA steel described in Section 3. The focus of this work is on the prediction of fracture, but for an extensive evaluation of the plasticity modelling, see [18]. The experiment was modelled in 3D using hexahedral elements with an edge length ranging from 0.1 mm at the center of the sample to 0.2 mm at the outside surface. The maraging steel anvil was modelled as an elastic material with a mesh size of 0.25 mm and a yield-limited friction coefficient of 0.1 as suggested by Rakvåg et al. [9]. The results of the experiments and numerical simulations are plotted in Figure 6 in terms of the maximum diameter and rod length across all intact impact velocities, highlighting the accuracy of the HHA plasticity model. Experimentally, the HHA steel was fractured at an impact velocity of 207 m/s and remained intact in another test at 208 m/s, suggesting that the impact velocity threshold for fracture initiation is close to this range.

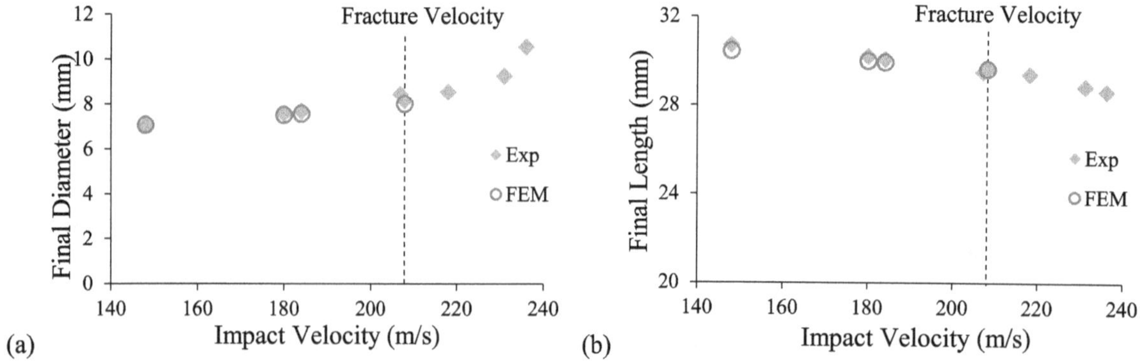

Figure 6. Experimental (Exp) and numerical (FEM) results of Taylor rod samples. (a) Maximum rod diameter (b) Rod length.

With the combined ASB instability criterion and Basaran fracture model, activated element erosion is predicted for impact velocities above 207 m/s, reflecting the experimental fracture initiation threshold. However, for a perfectly planar impact, together with a perfectly homogenous description of the rod material, the numerical simulations do not accurately reproduce the experimentally observed principal shear failure mode. Heterogenization can be introduced to the model via the inclusion of random defects in the meshed part (e.g. [24]), through the incorporation of probabilistic material properties (e.g. [9]), or via the introduction of a small degree of pitch/yaw in the impacting rod (e.g. [25]). For this study a 1° pitch is applied to the Taylor rod sample. While the effect of this rotation on sample deformation is negligible (>2% change in deformation for any node) and the velocity for fracture initiation is unaffected, it allows fracture to initiate locally and propagate through the sample more realistically. A comparison of the fracture modes predicted at an impact velocity of 207 m/s is given in Figure 7 for a perfectly planar impact and an impact with a 1° pitch, using the Basaran and ASB failure criteria.

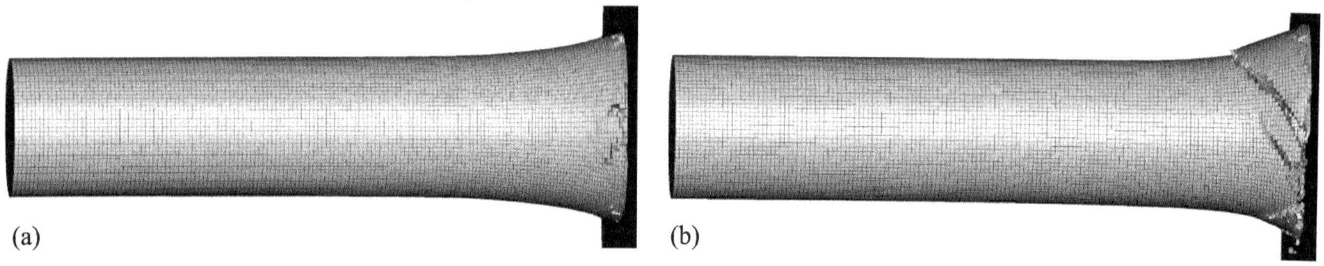

Figure 7. Taylor rod appearance following impact at 207 m/s demonstrating the effect of minor heterogeneity in the impact event. (a) 0° pitch. (b) 1° pitch.

6. Evaluation

The final shape of the Taylor rod sample tested experimentally at 207 m/s was recreated from orthogonal profiles and is compared to the predicted appearance (1° pitch included) using the ASB instability criterion and Basaran fracture model in Figure 8a and b. With the angular perturbation, the numerical model reproduces the principal shear failure mode seen experimentally, ejecting a single large fragment with an approximately 45° fracture surface. While the size of the fragment is larger and a small amount of impact face erosion is predicted in the numerical model, the agreement is considered very good. By comparison, using the Basaran ductile fracture model only (Figure 8c), a maximum damage value (D) of 0.3 is predicted at the experimental fracture initiation velocity. Indeed, for this failure model, an impact velocity of 500 m/s is required to produce the first element erosion and the rod appearance bears no resemblance to the experiments.

Figure 8. Comparison of the experimental and numerical HHA Taylor rod appearance following fracture. Impact velocity given in brackets. (a) Experimental (207 m/s) (b) Basaran + ASB instability criteria (207 m/s) (c) Basaran fracture model (500 m/s).

To further examine the ASB formation and fracture process, the stress state history is given in Figure 9 for an element at the root of the fracture path responsible for the large fragment separation. The strength fading initiated by the ASB criterion can be clearly seen at a plastic strain of 0.3. Prior to ASB formation in shear, the stress triaxiality and normalised third deviatoric stress invariant (NTDSI) in the element are around -0.3 and -1 respectively, which indicates essentially uniaxial compression. Following the onset of ASB localisation, seen graphically in Figure 9b, the stress state in the element shifts significantly as the loading around the softened material is redistributed. From uniaxial compression, the stress state transitions to plane strain tension where the predicted ductility is significantly lower and within a region of the Basaran fracture locus which was extensively characterised and validated in [11]. Final fracture of the element occurs at a plastic strain of 0.4 with over 85% of the damage accumulation occurring following ASB formation.

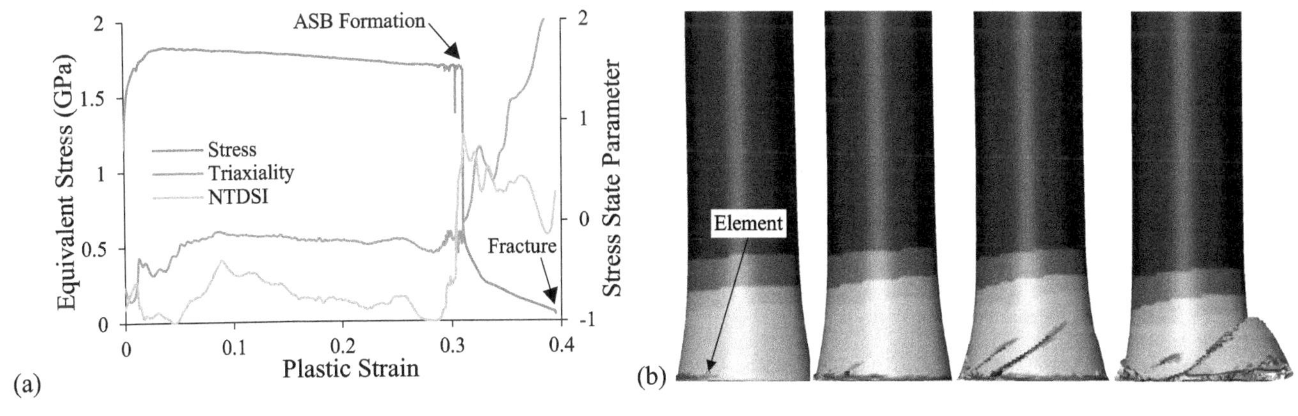

Figure 9. Fracture of Taylor rod at 207 m/s. (a) Stress state history at crack root. (b) ASB formation and propagation (plastic strain contour, $\varepsilon_{pl,max}$ =1).

7. Discussion

While the results of this evaluation of the ASB criterion and associated numerical modelling approach appear promising, several aspects of the model require development to allow further applicability in impact problems. Additional mechanical characterisation is required to assess the strain rate dependence of ASB formation across the range of specimen types and calibrate a strain rate extension to the criterion with a cut-off rate, below which ASB formation is inhibited as seen for other materials [26]. The current ASB criterion was characterised for triaxialities between -0.6 and 0.0 but did not consider the effects of lode angle on ASB formation. The development and testing of additional specimen types such as plane strain compression and combined tensile-shear flat-hat specimens will extend the characterisation envelope and provide a physical basis for the range of stress states under which ASB will develop. Further validation of the ASB criterion is required for impact problems involving the development of ASB and fracture such as penetration by blunt projectile [27] or explosively formed projectiles [28].

8. Conclusions

This experimental and numerical investigation evaluated the fracture behaviour for a HHA steel during Taylor rod-on-anvil impact. A series of mechanical characterisation experiments were performed to evaluate the failure of HHA steel under zero and negative stress triaxiality loading with the use of inclined cylinders and a new flat-hat shear specimen design. The material showed high ductility under quasi-static loading which was significantly affected at dynamic loading rates by the formation of adiabatic shear bands in all specimen types. An inverse numerical modelling approach was utilized to characterise the average stress triaxiality and plastic strain at the onset of ASB localization in each specimen type and calibrate a strength fading criterion (referred to as the ASB criterion) to capture the loss of load-bearing capacity with ASB formation. Material fracture was modelled using the stress triaxiality and Lode dependent Basaran ductile fracture model calibrated in [11].

A series of numerical simulations were conducted in LS-DYNA® to assess the performance of the proposed ASB criterion and fracture model in the prediction of fracture behaviour in Taylor rod-on-anvil experiments. In the simulations 1° of pitch was applied to the rod in order to induce a degree of heterogeneity, which is necessary to accurately reproduce the experimentally observed behaviour. At the onset of fracture, the experimental rods demonstrated principal shear failure on approximately 45° planes from the face of the cylinder. At the low and negative stress triaxialities experienced in Taylor rod samples, the Basaran 3D fracture model was unable to reproduce these failure modes. However, inclusion of the additional ASB instability criterion resulted in very good prediction, both qualitatively in the appearance of the fracture mode, and quantitatively in the prediction of the threshold impact velocity required to induce fracture of the rod. The proposed ASB criterion provides an effective means of representing the macro effect of ASB formation in high strength steels when the use of highly refined mesh configurations or multi-scale modelling approaches are not feasible.

Acknowledgements

The authors are grateful to the Impact Dynamics Laboratory at the University of New South Wales for conducting the mechanical characterisation experiments. The authors would also like to acknowledge the Computational Mechanics Section at the South West Research Institute for conducting the instrumented Taylor rod-on anvil experiments.

References

[1] Taylor G. The use of flat-ended projectiles for determining dynamic yield stress. I. Theoretical considerations. Proc R Soc A Math Phys Eng Sci 1948;194:289–99.
[2] Johnson GR, Cook WH. A constitutive model and data for metals subjected to large strains, high strain rates and high temperatures. Proc. 7th Int. Symp. Ballist., vol. 21, The Netherlands; 1983, p. 541–7.
[3] Banerjee B. An evaluation of plastic flow stress models for the simulation of high-temperature and high-strain-rate deformation of metals. Science (80-) 2005:43.
[4] Rohr I, Nahme H, Thoma K. Material characterization and constitutive modelling of ductile high strength steel for a wide range of strain rates. Int J Impact Eng 2005;31:401–33.
[5] Bigger RP, Carpenter A, Scott N, Dannemann K, Chocron S, Williams C. Dynamic response of aluminum 5083 during Taylor impact using digital image correlation. Exp Mech 2018.
[6] Anderson Jr CE, Chocron IS, Nicholls AE. Damage modeling for Taylor impact simulations. J Phys IV 2006;134:331–7.
[7] Johnson GR, Cook WH. Fracture characteristics of three metals subjected to various strains, strain rates, temperatures and pressures. Eng Fract Mech 1985;21:31–48.
[8] Chocron S, Erice B, Anderson CE. A new plasticity and failure model for ballistic application. Int J Impact Eng 2011;38:755–64.
[9] Rakvåg KG, Børvik T, Hopperstad OS. A numerical study on the deformation and fracture modes of steel projectiles during Taylor bar impact tests. Int J Solids Struct 2014;51:808–21.
[10] Cockcroft MG, Latham DJ. Ductility and the workability of metals. J Inst Met 1968;96:33–9.
[11] McDonald B, Bornstein H, Ameri A, Daliri A, Orifici A. Plasticity and ductile fracture behaviour of four armour steels (In preparation) 2018.
[12] U.S Detail Specification. MIL-DTL-46100E(MR). Armor plate, steel, Wrought, high-hardness 2008.
[13] Herzig N, Abdel-Malek S, Meyer L, Cimpoeru SJ. Modeling of ductile failure in high strength steel. Procedia Eng 2017;197:285–93.
[14] Clos R, Schreppel U, Veit P. Temperature, microstructure and mechanical response during shear-band formation in different metallic materials. J. Phys. IV, vol. 110, 2003, p. 111–6.

[15] Roth C, Mohr D. Ductile fracture experiments with locally proportional loading histories. Int J Plast 2016;79:328–54.

[16] Meyer LW, Staskewitsch E, Burblies a. Adiabatic shear failure under biaxial dynamic compression/shear loading. Mech Mater 1994;17:203–14.

[17] Sengoz K. Development of a generalized isotropic yield surface for pressure insensitive metal plasticity considering yield strength differential effect in tension, compression and shear stress states. PhD Thesis, George Washington University, 2017.

[18] Ryan S, McDonald B. Modelling the deformation of high-hardness armour steel in Taylor ron-on-anvil experiments. Proc. Hypervelocity Impact Symp., Sandestin: 2019.

[19] Basaran M. Stress state dependent damage modeling with a focus on the lode angle influence. PhD Thesis, RWTH Aachen University, 2011.

[20] Børvik T, Hopperstad OS, Berstad T, Langseth M. A computational model of viscoplasticity and ductile damage for impact and penetration. Eur J Mech - A/Solids 2001;20:685–712.

[21] McDonald B, Bornstein H, Langdon GS, Ameri A, Orifici AC. Evaluation of constituitive models for predicting the deformation and rupture behaviour of armour steel under localised blast loading. Proc. 25th Mil. Asp. Blast Shock., The Hague: 2018.

[22] Teng X, Wierzbicki T, Couque H. On the transition from adiabatic shear banding to fracture. Mech Mater 2007;39:107–25.

[23] Hancock JW, Mackenzie AC. On the mechanisms of ductile failure in high-strength steels subjected to multi-axial stress-states. J Mech Phys Solids 1976;24:147–60.

[24] Strack OE, Leavy RB, Brannon RM. Aleatory uncertainty and scale effects in computational damage models for failure and fragmentation. Int J Numer Methods Eng 2015;102:468–95.

[25] Karlsen T, Kjølseth A. Fragmentation of metallic materials during impact. Norwegian University of Science and Technology, 2012.

[26] Edwards N, Song W, Cimpoeru SJ, Ruan D, Lu G, Herzig N. Mechanical and microstructural properties of 2024-T351 aluminium using a hat-shaped specimen at high strain rates. Mater Sci Eng A 2018;720:203–13.

[27] Fras T, Roth C, Mohr D. Fracture of high-strength armor steel under impact loading. Int J Impact Eng 2018;111:147–64.

[28] Klavzar A, Pursche F, Herzig N, Wolf T. Ballistic limit thickness and failure of steel targets with different strengths under EFP impact. Procedia Eng 2017;197:270–7.

Proceedings of the 2019 Hypervelocity Impact Symposium
HVIS2019
April 14-19, 2019, Destin, FL, USA
HVIS2019-051

Shear band characteristics in high strain rate naval applications

Michael J. P. Conway[a], James D. Hogan[a]*

[a]*Department of Mechanical Engineering, University of Alberta, Edmonton, AB T6G 2R3, Canada*

Abstract

This paper explores the dynamic behavior of HSLA 65 naval steels, specifically focusing on the initiation and growth of shear bands in quasi-static and dynamic compression experiments and how these bands affect stress-strain responses. The results indicate that the yield strength for this HSLA 65 increases from 541 ± 8 MPa for quasi-static (10^{-3} s^{-1}) to 1081 ± 48 MPa for dynamic rates 1853 ± 31 s^{-1}, and the hardening exponent increases from 0.376 ± 0.028 for quasi-static to 0.396 ± 0.006 for dynamic rates. Yield behavior was found to be associated with the onset of shear banding for both strain-rates, confirmed through visualization of the specimen surface using high-speed and ultra-high-speed cameras. For the quasi-static case, shear banding and yielding was observed to occur at 2.5% strain, and were observed to grow at speeds of upwards of 38 mm/s. For the dynamic experiments, the shear banding begins at approximately 1.18 ± 0.06% strain and these can grow upwards of 2122 ± 213 m/s during post-yield softening. Altogether, these measurements are some of the first of their kind in the open literature, and provide guidance on the critical time and length scales in shear banding. This information can be used in the future to design more failure-resistant steels, which has broader applications in construction, defense, and natural resource industries.

Keywords: shear band; dynamic failure; steel; split-hopkinson pressure bar; digital image correlation

---

Nomenclature

n	strain hardening coefficient
K	strength coefficient (MPa)
Greek symbols	
ε	strain (mm/mm)
σ	stress (MPa)
Subscripts	
f	flow stress (MPa)

1. Introduction

Shear bands are an important topic in understanding the mechanics of materials as it is a dominant failure mechanism and precursor for crack initiation and propagation at high strain rates in many materials [1]. For example, shear bands are commonly observed in manufacturing and operation in the ship building [2], rail transportation [3], and aerospace industries [4]. In these industries, the materials are exposed to a wide range of strain rates from quasi-static to explosive loading. With the ever-increasing expectations of steels being stronger and more cost effective, it becomes more important to develop a fundamental understanding of the failure of these materials. In these applications, the formation of shear bands plays a role in the failure and the stability of the material. Understanding these loading conditions and the behavior of shear bands (i.e. velocity, propagation direction and propensity) will improve the material selection process and the performance of newly developed materials for these industries.

---

* Corresponding author. Tel.: +1 780 492 7133.
E-mail address: jdhogan@ualberta.ca ; mjconway@ualberta.ca .

In the past, shear banding has been commonly observed in metals [1,5,6], polymers [7], and geologic [8] materials. Authors have studied shear banding through experiments [9,10], and through computational and analytical modelling [11,12]. Shear bands are commonly labelled as: 1. deformed shear bands, and 2. transformed shear bands, and these are mainly distinguished by whether a white etched hardened band is formed [13]. There has been several theories on the generation of shear bands in materials, including dynamic recrystallization (DRX) [14], grain deformation [13], rotational recrystallization [15], and phase transformation [16]. Regardless, shear bands are commonly believed to significantly affect material strength, stability, and have been noted to often be a precursor to cracking through the generation of voids [17]. Despite broad implications of shear banding to industry applications, the topic remains a very active area of research [1,12,18].

Motivated by these and many other past works, this paper focuses on studying the growth and evolution of shear bands in a high strength, low alloy steel, specifically HSLA 65. We study material behavior through mechanical testing in compressive stress-states and couple this testing to visualization of shear banding and failure using high-speed and ultra-high-speed cameras and Digital Image Correlation (DIC) techniques. Being able to pair visualization and quasi-static and dynamic loading events provides insight into the deformation of the material and, specifically, the propagation velocity, direction, and orientation of the shear bands on the surface of the specimen. The results are discussed in the context of our current understanding of shear banding in steels, especially discussing the strains at which they occur, the velocity they grow, and their effect on the stress-strain response.

2. Methodology

2.1. Material Tested

HSLA 65 (ASTM A945 [19]) was tested in compression, where specimens were removed from as-received steel plate that was 12" x 12" x 0.75" in size. This steel has been used on navy ships in Canada due to its cost effectiveness, weldability, and its high strength. The specimens were milled from the center of the thickness of the plate and were made in accordance with American Society for Testing and Materials (ASTM) standards following ASTM 09 for both specimen creation and testing. The HSLA 65 steel was specifically chosen due to the propagation of shear bands being clearly visible during the impact process. The chemical analysis of the HSLA 65 in this study is given in Table 1 and was provided from the manufacturer in the mill test certificate.

Table 1. Components of HSLA 65 (wt.% are listed)

C	Mn	P	S	Si	Cu	Ni	Cr	Mo	V	Ti	Al	B	Cb	N
0.08	1.43	0.020	0.004	0.299	0.020	0.01	0.04	0.006	0.067	0.016	0.030	0.0002	0.035	0.008

2.2. Quasi-static Testing

Compression tests were performed on a MTS 810 testing machine for rectangular specimens of size 5.00 mm x 4.00 mm x 3.50 mm using displacement control. The quasi-static tests were performed at a nominal strain rate of $10^{-3}\,s^{-1}$. This specimen shape was selected for the ability to visualize the deformation on the surface of the specimen during experiments, as well as to perform Digital Image Correlation measurements. The displacement and force were recorded through the MTS machine at a sampling rate of 100 Hz, while the surface and deformation features of the specimen were recorded through the use of a PROMON one-megapixel camera with a macro lens recording at 100 frames per second, which matched the sampling frequency of the MTS machine. The strains were measured during testing using DIC techniques with the VIC2D commercial software. This approach consisted of applying a randomly distributed pattern over the surface of the specimen using an airbrush and trying to minimize the overall particle size through atomization given the small specimen size. This was achieved through the use of an ultra-fine point airbrush to apply black paint that resulted in a speckle diameter between 50 and 300 microns. The quality of the speckle pattern was checked in the VIC2D software to minimize error before experimentation. During testing, an ultra-bright LED light equipped with a 7.6 mm diameter liquid light guide was used to illuminate the specimen in a highly contrasted and overexposed condition. Once the data was obtained, the DIC data was paired with the captured load data from the universal testing machine and matched to the DIC strain data through physical events captured in both measurement techniques. In this case, a sharp unload was induced which was matched in time between the camera and MTS measurements. Using DIC, both small strains and large plastic strains can be resolved, whereas conventional strain gauges cannot be used to provide similar measurements [20] without the need to apply compliance corrections for the universal testing machine. Additionally, a greater amount of data can be derived from the measurements given the ability to spatially resolve the strain across the surface of the specimen providing information such as local strains, displacement and ultimately verifying stress equilibrium.

2.3. Split-Hopkinson Pressure Bar Testing

A Split-Hopkinson Pressure Bar (SHPB) was used to perform dynamic compression measurements. The same specimen sizes were used in the dynamic experiments as the quasi-static experiments (5.00 mm x 4.00 mm x 3.50 mm), and the rectangular shapes were chosen for visualization purposes. The apparatus consisted of a 304.8 mm striker, 914.4 mm incident bar and a 1016 mm transmission bar that all have a diameter of 12.7 mm. The bars are made of Maraging C350 steel that has a minimum hardness of 52 HRC. The SHPB is an apparatus that takes an input from a striker that is fired from a gas gun. The striker then impacts a pulse shaper that is attached to the incident bar. This pulse shaper is used to shape the profile of the input pulse to achieve proper stress equilibrium and constant strain rate in the specimen [21]. In our experiments, mild steel with a diameter of 6.35 mm and a thickness of 0.635 mm was used, which follows from previous studies on improving stress equilibrium and reducing Pochhammer-Chree dispersion during SHPB testing [22,23]. The specimen and the pulse shaper are both placed with a small amount of high pressure grease to minimize friction between the interfaces [21]. Verification of stress equilibrium is performed through a force balance, which is found in reasonable agreement for all our tests. In these experiments, the surface of the specimen is recorded using a Kirana ultra-high-speed camera with a HBM GEN3i data acquisition system recording at 20 MHz. The data acquisition triggers the camera from the beginning rise of the incident pulse and the camera trigger pulse is recorded in the data set. The camera records 180 frames at 1 million frames per second with a resolution of 924 pixels x 768 pixels. At this frame rate and resolution, we can visualize deformation in materials at sufficient spatial and temporal resolution that are not widely reported in the literature.

3. Results

In this study we investigate the compressive behavior of HSLA 65 navy steel in quasi-static and dynamic loading conditions. To begin the analysis, we have shown three stress-strain curves for the quasi-static case and four stress-strain curves for dynamic case in Figure 1 to demonstrate the rate-effects and repeatability of the measurements. In quasi-static testing conditions at a strain rate of 0.001 s^{-1}, the yield strength of the HSLA 65 was found to be 541 ± 8 MPa, which are colored black in Figure 1. The yield stress found at a dynamic strain rate of 1853 ± 31 s^{-1} was 1081 ± 48 MPa (colored blue in Figure 1). These values are summarized in Table 2. The yield strengths and symmetrical limits are found from averaging the resulting values in each test and providing the standard deviation across the sample set. In quasi-static and dynamic loading of HSLA 65, there is a distinct yield point seen in the curves. For the quasi-static experiments, the yield strain is 0.93 ± 0.12 % and for the dynamic experiments, the yield strain is 1.05 ± 0.03 %. At the higher strain rates, this yield point is much more prominent, while the Pochhammer-Chree dispersion should be considered after the yielding of the material. Later, these yield features are linked to shear banding and failure in the material.

To further probe material behavior and generalize the results, the flow stress for this material was calculated using the true stress-true strain data for Equation (1).

$$\sigma_f = K\varepsilon^n \tag{1}$$

where the flow stress, σ_f, is the stress required to continue the plastic deformation of a metal. From Equation (1), K is the strength coefficient, ε is the plastic strain, and n is the hardening coefficient. Using a linear fit of the true strain-true stress data on a log-log scale, the strength coefficient for the quasi-static case is K = 2029 ± 172 MPa with a strain hardening coefficient of n = 0.376 ± 0.028. For the dynamic experiments, the strength coefficient is found to be K = 3098 ± 52 MPa and the hardening coefficient is n = 0.396 ± 0.006. These values are also summarized in Table 2 for both strain rates, along with the yield strength.

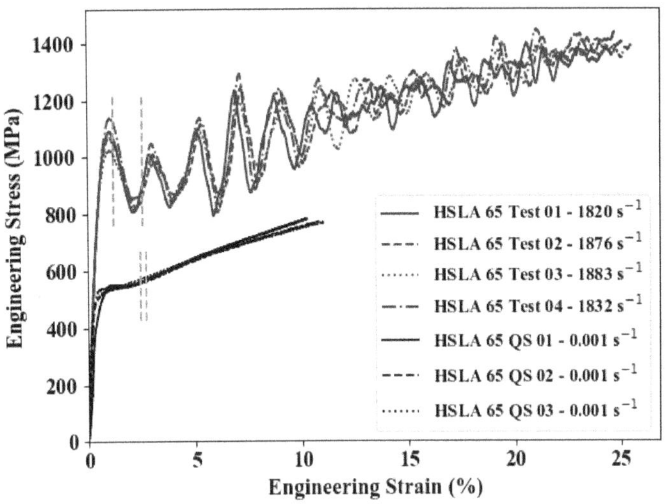

Figure 1. Engineering stress-strain curves of the tested HSLA 65 in quasi-static (black) and dynamic (blue) compression. The vertical lines indicate the region that shear band measurements were taken, which are linked with subsequent figures.

Table 2. Flow stress coefficients for HSLA 65 in quasi-static and dynamic compression.

Strain Rate (s^{-1})	Yield Strength (MPa)	K (MPa)	n
0.001	541 ± 8	2029 ± 172	0.376 ± 0.028
1853 ± 31	1081 ± 48	3098 ± 52	0.396 ± 0.006

Next, we investigate a series of high-speed camera images for the quasi-static experiments, which are shown in Figure 2, and dynamic experiments that are shown in Figure 3. In Figure 2, the collection of images were chosen that represents notable stages in shear band development in the quasi-static experiment: Figure 2a shows the intact surface before localizations occur; Figure2b shows localizations forming on the surface at 45° to the loading direction with the predominant bands developing from the edges and propagating inwards until they meet; Figure 2c shows the surface at which point the bands visually look consumed by the surface texturing that begins to develop at later stages of plastic deformation. The localizations began forming on the surface at approximately 2.5% strain. Next, shown in Figure 3 are a collection of high-speed camera images taken during a dynamic experiment that demonstrate the initiation and growth of shear banding. Figure 3a shows the specimen in pristine conditions. Figure 3b shows the surface at ~1% strain with the initial bands propagating across the surface. The initiation of the localizations occurs at the corners of the specimen and again propagate at approximately 45° angles. Across all the dynamic experiments, the initiation of these localizations on the surface begin at approximately 1.19 ± 0.06% strain. In comparison with the stress-strain curve (Figure 1), the localizations approximately begin around the yield of the material. Further, Figure 3c shows the specimen surface at ~2% strain, and this is when the bands begin to coalesce on the surface that is shown. The bands are the most visually distinct at this strain. At later strains (Figure 3d), we observe surface texturing occurring that consumes the shear bands on the surface and this continues until only the large and visually distinct localizations can be seen, which remain after the experiment. This occurs similarly to the quasi-static experiments (Figure 2c).

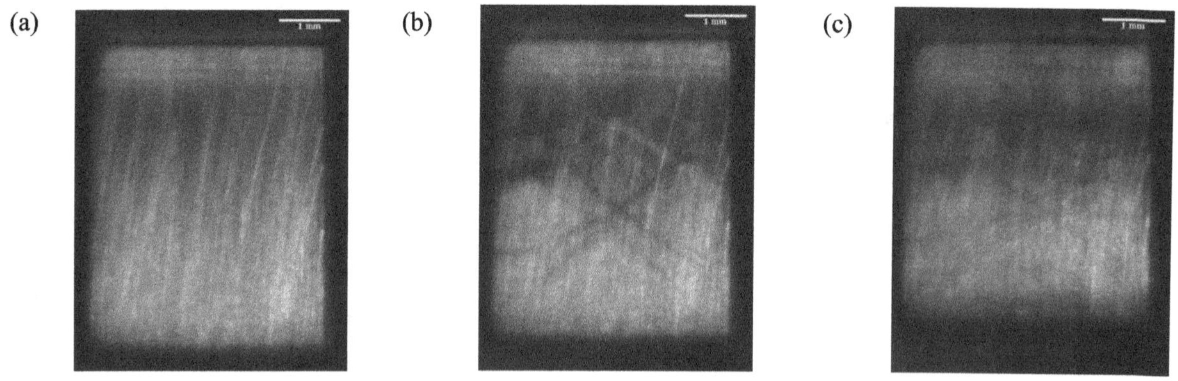

Figure 2. Surface of HSLA 65 specimen in quasi-static compression using a MTS 810 universal testing machine at a strain rate of 10^{-3} s^{-1}: Image a) surface of the specimen before deformation; b) surface of the specimen with localizations fully developed at ~4.6% strain; and c) surface of the specimen after formation of bands as surface texturing develops at ~11% strain demonstrating that the shear bands are consumed and are barely visible.

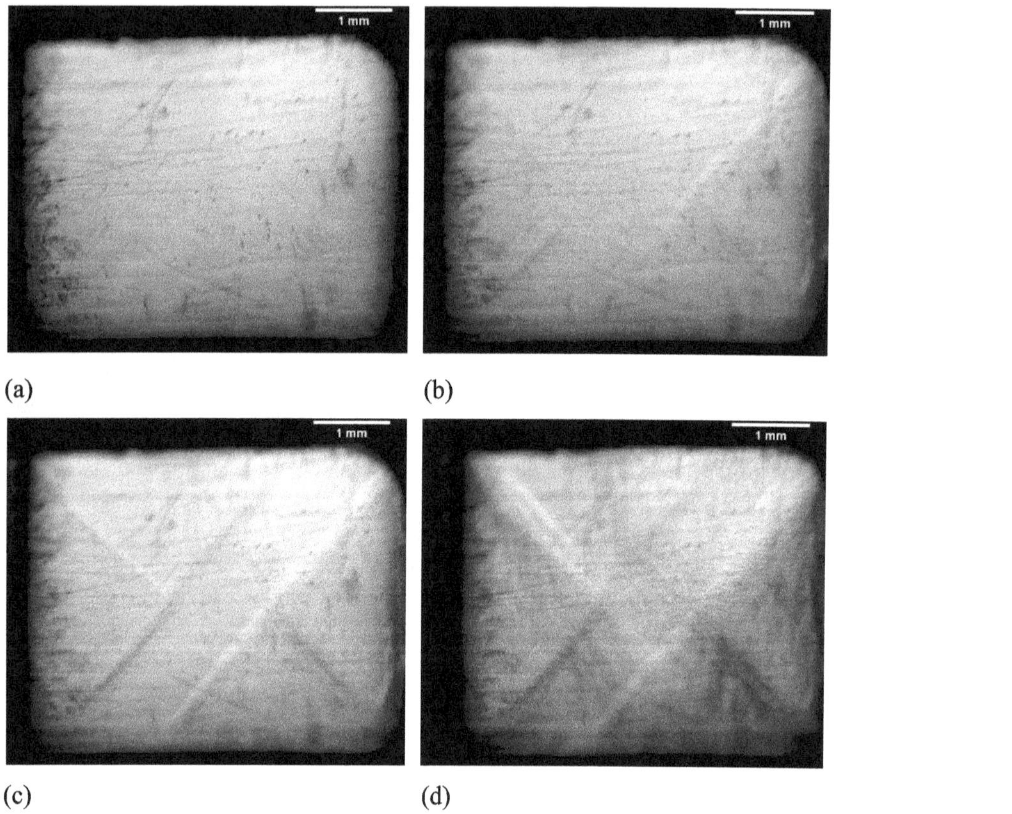

Figure 3. In-situ images of HSLA 65 using Kirana ultra-high-speed camera filming at 1 million frames per second in a Split Hopkinson pressure bar testing apparatus: a) surface of specimen prior to deformation; b) surface of specimen with bands beginning to form at ~1% strain; c) surface of specimen with bands formed before coalesce dominates at ~2% strain; d) surface of specimen with visible localizations and surface texturing at ~3.9% strain.

Using ImageJ [24,25], the initiation and growth of shear bands were tracked over multiple frames for both quasi-static and dynamic loading. The velocity of the localizations on the surface is plotted against the compressive strain for both strain rates in Figure 4, with the range of the strain they occur over denoted in Figure 1 as red vertical lines. The measurements were taken until the originally pristine surface was filled with the larger bands. At this point, coalescence is predominant and the resulting shear bands are not included in the analysis. In the quasi-static case, the shear bands begin to form on the surface at approximately 2.5% strain and accelerate from a velocity of 27 ± 5 mm/s to 38 ± 8 mm/s at a strain of 2.51%. Following this, the band growth slows to 19 ± 4 mm/s and stays approximately constant over the next 0.01% strain, at which it slows to 11 ± 2 mm/s before meeting other localizations seen on the surface at a strain of 2.71%. Again, these measurements are shown as black dots in Figure 4.

For the dynamic case (blue dots in Figure 4), the shear bands are tracked from ~1.24% to ~2.58% strain, and this occurs approximately over 4 μs during the experiment. The shear band velocity measurements for the dynamic cases are shown as black points in Figure 4. In this plot, the error bars correspond to the bias uncertainty in the calculation, including the camera resolution, variation in time for each frame, and in the dynamic case the random error in repeated measurements of the length of the band between experiments. These errors provide the variation in calculated shear band growth velocities given the short time scales over which they are calculated. The first measurement of velocity is 951 ± 96 m/s at approximately 1.24% strain, where the growth of the band accelerates for the next two frames to 1911 ± 192 m/s at 1.46% strain and 2122 ± 213 m/s at 1.92% strain. At the third frame, the band growth begins to decelerate, when it reaches approximately 2% strain and finally ends at 1464 ± 147 m/s. After this time, the bands may arrest after fully forming on the surface and the energy may be transformed into heat, further deformation and grain alterations [26]. Relating the strain measurement from the images to the stress-strain curves, the pace of growth begins and increases at the point of yield and slows near the trough of the softening (see Figure 1). Generally, the velocity first accelerates as new bands begin forming on the surface and decelerates until coalescence begins.

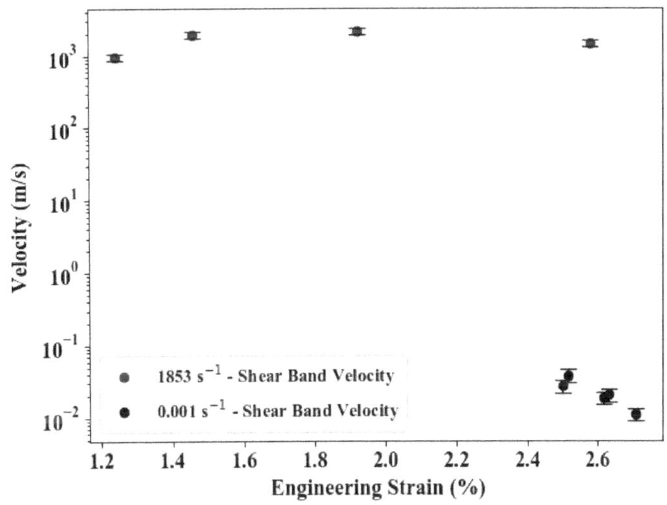

Figure 4. Velocity of shear bands on the surface of the specimens in quasi-static and dynamic compression where the first and last points for each respective case correspond to the lines on the stress-strain curve provided in Figure 1.

4. Discussion

In this paper, we explored the strain rate-dependent compression response of HSLA 65 naval steel, focusing on shear bands and failure. Consideration of the stress-strain response found that the yield strength is 541 ± 8 MPa for quasi-static and 1081 ± 48 MPa for dynamic rates, the K is 2029 ± 172 MPa for quasi-static and 3098 ± 52 MPa for dynamic rates, and the hardening exponent n is 0.376 ± 0.028 for quasi-static and 0.396 ± 0.006 for dynamic rates. These measurements are consistent with those reported in the literature for HSLA 65 [27], and they demonstrate that the HSLA 65 steel is comparable to common naval steels such as HY-80 [28] and DH36 [29].

To explore material responses further, mechanical testing was coupled to high-speed imaging of the specimen surface to investigate shear banding and failure in HSLA 65. To do this, the initiation and growth of the bands were tracked using ImageJ, and the resulting growth rates were plotted against compressive strain (Figure 4). In the quasi-static case, the shear bands begin to form on the surface of the specimen at approximately 2.5% strain and accelerate to a peak of 38 ± 8 mm/s at a strain of 2.51%. Following this, the band growth slows to 19 ± 4 mm/s and stays approximately constant over the next 0.01% strain, at which it slows 11 ± 2 mm/s before coalescing with other localizations observed on the surface at a strain of 2.71%. For dynamic strain rates of 1853 ± 31 s^{-1}, the bands begin to grow at approximately 1.2% strain, and they accelerate up to 2122 ± 213 m/s at 1.9% strain. Following this, the bands decelerate and then are consumed by deformation on the surface of the specimen. Relating the strain measurement from the images to the stress-strain curves, the pace of growth begins and increases at the point of yield and slows near the trough of the softening (see Figure 1 and Figure 4). Altogether, results from this paper will be important in the development of models, both for phenomenological [30] and for those models that capturing shear banding in steels [31,32].

The peak velocity of 2122 ± 213 m/s is 68% of the shear wave speed, which is acceptable to values of 40 to 67% that are commonly reported in the literature [9,10,33]. While numerical simulations show generally good agreement with experimental studies, shear band velocities in simulations have been shown to be higher using methods which do not require mesh generation [33]. In the study by Li et al. [33], a better curvature match is found in regards to Zhou et al. [9] experimental work, although the peak velocity was found to be 2005 m/s with an impact velocity of 35 m/s. Coupled with the current work, it may be plausible that the shear bands can meet or exceed previously found velocity values in the experimental literature. Namely, the shear band velocities found is this study are higher than those velocities found in studies relating to Maraging c300 steel by Zhou et al. [9], who found at a peak velocity of 1200 m/s with the same acceleration pattern that was found in this study even given the smaller time window our measurements exist within. In a separate study by Guduru et al. [10] also using Maraging c300 steel found the peak velocity to be 1100 m/s, although they found that the generated shear band momentarily slowed down and then proceeded to accelerate again. This behavior is not seen in our current work or in the work of Zhou et al. [9]. Our study also did not observe fracturing from the shear band, which was noted in the works by Zhou et al. [9] and Guduru et al. [10].

5. Concluding Remarks

In this paper, we investigate the dynamic failure and shear banding in HSLA 65 steel. To accomplish this, we performed strain-rate dependent compression measurements on the material and coupled high-speed and ultra-high-speed cameras to monitor the initiation and growth of shear bands during testing. These observations were linked to measurements of the stress-strain response of the materials. Results indicate that the yield strength increases from 541 ± 8 MPa for quasi-static (10^{-3} s^{-1}) to 1081 ± 48 MPa for dynamic rates (1853 ± 31 s^{-1}), and the hardening exponent increases from 0.376 ± 0.028 for quasi-static to 0.396 ± 0.006 for dynamic rates. The yield in the material was observed to be associated with the onset and evolution of shear banding in HSLA 65. For the quasi-static case, shear banding and yielding was observed to occur at 2.5% and 0.93 ± 0.12 % strain, respectively and these can reach speeds of upwards of 38 mm/s. For the dynamic experiments, the shear banding begins at approximately 1.18 ± 0.06% strain and these can grow upwards of 2122 ± 213 m/s. Altogether, this study provides new insights into shear banding and dynamic failure of steel materials, which is useful in manufacturing and modelling of these materials.

Acknowledgements

This work is supported by the Canadian Natural Sciences and Engineering Research Council, the Canadian Foundation for Innovation, and Defence Research and Development Canada. Calvin Lo and Haoyang Li are thanked for their aid while running the experiments.

References

[1] Mendoza, I., Villalobos, D., and Alexandrov, B. T., 2015, "Crack Propagation of Ti Alloy via Adiabatic Shear Bands," Materials Science and Engineering: A, **645**, pp. 306–310.

[2] Latourte, F., Feinberg, Z., Mori, L. F., Olson, G. B., and Espinosa, H. D., 2010, "Shear and Tensile Plastic Behavior of Austenitic Steel TRIP-120 Compared with Martensitic Steel HSLA-100," International Journal of Fracture, **162**(1–2), pp. 187–204.

[3] Nakkalil, R., Hornaday, J. R., and Bassim, M. N., 1991, "Characterization of the Compression Properties of Rail Steels at High Temperatures and Strain Rates," Materials Science and Engineering: A, **141**(2), pp. 247–260.

[4] Odeshi, A. G., Owolabi, G. M., and Nabil Bassim, M., 2007, "Effects of particulate reinforcement and strain-rates on deformation and fracture behavior of Aluminum 6061-T6 under high velocity impact," Materialwissenschaft und Werkstofftechnik, **38**(2), pp. 66–69.

[5] Odoh, D., Owolabi, G., Odeshi, A., and Whitworth, H., 2013, "Shear Band Formation in AISI 4340 Steel under Dynamic Impact Loads: Modeling and Experiment," Acta Metallurgica Sinica (English Letters), **26**(4), pp. 378–384.

[6] Nakkalil, R., 1991, "Formation of Adiabatic Shear Bands in Eutectoid Steels in High Strain Rate Compression," Acta Metallurgica et Materialia, **39**(11), pp. 2553–2563.

[7] Swallowe, G. M., 2012, "Adiabatic Shear Bands in Polymers," *Adiabatic Shear Localization*, Elsevier, pp. 363–398.
[8] Fakhimi A., Riedel J. J., and Labuz J. F., 2006, "Shear Banding in Sandstone: Physical and Numerical Studies," International Journal of Geomechanics, **6**(3), pp. 185–194.
[9] Zhou, M., Rosakis, A. J., and Ravichandran, G., 1996, "Dynamically Propagating Shear Bands in Impact-Loaded Prenotched Plates—I. Experimental Investigations of Temperature Signatures and Propagation Speed," Journal of the Mechanics and Physics of Solids, **44**(6), pp. 981–1006.
[10] Guduru, P. R., Rosakis, A. J., and Ravichandran, G., 2001, "Dynamic Shear Bands: An Investigation Using High Speed Optical and Infrared Diagnostics," Mechanics of Materials, **33**(7), pp. 371–402.
[11] Wright, T. W., and Batra, R. C., 1985, "The Initiation and Growth of Adiabatic Shear Bands," International Journal of Plasticity, **1**(3), pp. 205–212.
[12] Arriaga, M., McAuliffe, C., and Waisman, H., 2016, "Instability Analysis of Shear Bands Using the Instantaneous Growth-Rate Method," International Journal of Impact Engineering, **87**, pp. 156–168.
[13] Owolabi, G., Odoh, D., Odeshi, A., and Whitworth, H., 2013, "Occurrence of Dynamic Shear Bands in AISI 4340 Steel under Impact Loads," World Journal of Mechanics, **03**(02), pp. 139–145.
[14] Hines, J. A., and Vecchio, K. S., 1997, "Recrystallization Kinetics within Adiabatic Shear Bands," Acta Materialia, **45**(2), pp. 635–649.
[15] Meyers, M. A., Xu, Y. B., Xue, Q., Pérez-Prado, M. T., and McNelley, T. R., 2003, "Microstructural Evolution in Adiabatic Shear Localization in Stainless Steel," Acta Materialia, **51**(5), pp. 1307–1325.
[16] Syn, C. K., Lesuer, D. R., and Sherby, O. D., 2005, "Microstructure in Adiabatic Shear Bands in a Pearlitic Ultrahigh Carbon Steel," Materials science and technology, **21**(3), pp. 317–324.
[17] Xu, Y. B., Bai, Y. L., Xue, Q., and Shen, L. T., 1996, "Formation, Microstructure and Development of the Localized Shear Deformation in Low-Carbon Steels," Acta Materialia, **44**(5), pp. 1917–1926.
[18] Dolinski, M., Merzer, M., and Rittel, D., 2015, "Analytical Formulation of a Criterion for Adiabatic Shear Failure," International Journal of Impact Engineering, **85**(Supplement C), pp. 20–26.
[19] A01 Committee, *Specification for High-Strength Low-Alloy Structural Steel Plate with Low Carbon and Restricted Sulfur for Improved Weldability, Formability, and Toughness*, ASTM International.
[20] Štamborská, M., Kvíčala, M., and Losertová, M., 2014, "Identification of the Mechanical Properties of High-Strength Steel Using Digital Image Correlation," Advanced Materials Research, **980**, pp. 122–126.
[21] Chen, W. W., and Song, B., 2011, *Split Hopkinson (Kolsky) Bar: Design, Testing and Applications*, Springer, New York, NY.
[22] Panowicz, R., Janiszewski, J., and Kochanowski, K., 2018, "Influence of Pulse Shaper Geometry on Wave Pulses in SHPB Experiments," Journal of Theoretical and Applied Mechanics, **56**(4), pp. 1217–1221.
[23] ASM International, Kuhn, H., and ASM International, eds., 2000, *Mechanical Testing and Evaluation*, ASM International, Materials Park, Ohio.
[24] Schneider, C. A., Rasband, W. S., and Eliceiri, K. W., 2012, "NIH Image to ImageJ: 25 Years of Image Analysis," Nature Methods, **9**(7), pp. 671–675.
[25] Schindelin, J., Arganda-Carreras, I., Frise, E., Kaynig, V., Longair, M., Pietzsch, T., Preibisch, S., Rueden, C., Saalfeld, S., Schmid, B., Tinevez, J.-Y., White, D. J., Hartenstein, V., Eliceiri, K., Tomancak, P., and Cardona, A., 2012, "Fiji: An Open-Source Platform for Biological-Image Analysis," Nature Methods, **9**(7), pp. 676–682.
[26] Longère, P., 2017, "Respective/Combined Roles of Thermal Softening and Dynamic Recrystallization in Adiabatic Shear Banding Initiation," Mechanics of Materials.
[27] Nemat-Nasser, S., and Guo, W.-G., 2005, "Thermomechanical Response of HSLA-65 Steel Plates: Experiments and Modeling," Mechanics of Materials, **37**(2–3), pp. 379–405.
[28] Conn, A. F., Rudy, S. L., and Howard, S. C., 1974, *THE STRAIN RATE DEPENDENCE OF THREE HIGH STRENGTH NAVAL ALLOYS*, T. R. 7111-1, Office of Naval Research Department of the Navy.
[29] Nemat-Nasser, S., and Guo, W.-G., 2003, "Thermomechanical Response of DH-36 Structural Steel over a Wide Range of Strain Rates and Temperatures," Mechanics of Materials, **35**(11), pp. 1023–1047.
[30] Daridon, L., Oussouaddi, O., and Ahzi, S., 2004, "Influence of the Material Constitutive Models on the Adiabatic Shear Band Spacing: MTS, Power Law and Johnson–Cook Models," International Journal of Solids and Structures, **41**(11–12), pp. 3109–3124.
[31] Batra, R. C., 2012, "Analysis of Adiabatic Shear Bands by Numerical Methods," *Adiabatic Shear Localization*, Elsevier, pp. 173–214.
[32] Medyanik, S. N., Liu, W. K., and Li, S., 2007, "On Criteria for Dynamic Adiabatic Shear Band Propagation," Journal of the Mechanics and Physics of Solids, **55**(7), pp. 1439–1461.
[33] Li, S., Liu, W. K., Rosakis, A. J., Belytschko, T., and Hao, W., 2002, "Mesh-Free Galerkin Simulations of Dynamic Shear Band Propagation and Failure Mode Transition," International Journal of Solids and Structures, **39**(5), pp. 1213–1240.
[34] Hunter, J. D., 2007, "Matplotlib: A 2D Graphics Environment," Computing In Science & Engineering, **9**(3), pp. 90–95.

Proceedings of the 2019 Hypervelocity Impact Symposium
HVIS2019
April 14-19, 2019, Destin, FL, USA
HVIS2019-056

The Role of Inclusions in the Failure of Boron Carbide Subjected to Impact Loading

Andrew L. Tonge[a]* and Brian E. Schuster[b]

[a]*U.S. Army Research Laboratory, Building 390 Room 206, 321 Colleran Road Aberdeen Proving Ground MD 21001*
[b]*U.S. Army Research Laboratory, Building 309 Room 213, 321 Colleran Road Aberdeen Proving Ground MD 21001*

Abstract

This work investigates the importance of the microstructure of boron carbide for initiating inelastic deformation under impact conditions. Simple loading resulting from a flyer plate impact geometry is used to illustrate the importance of microstructure for the well-controlled and easily instrumented experimental geometry. A second set of simulations is performed on a miniaturized impact geometry to investigate the importance of the microstructure for the early stages of semi-infinite penetration for impact velocities between 0.9 km/s and 1.9 km/s. The effect of the microstructure is more pronounced for the flyer plate impact geometry.

Keywords: Boron carbide; microstructure; brittle failure; semi-infinite penetration

---

1. Introduction

Boron carbide is a hard, stiff, low density ceramic material that is attractive for impact applications where low weight is important. While boron carbide has a 1 kg Knoop Hardness of 25.5 GPa [1] and a Hugoniot Elastic Limit (HEL) between 15 and 18 GPa [2], no deformation mechanism that is not related to fracture has been observed. All reported experiments that observe the deformation process of boron carbide in real time suggest that the primary deformation mechanism in boron carbide is the growth and interaction of cracks leading to massive fragmentation, which allows subsequent motion of the fragmented material. There are numerous theoretical models that connect the distribution of cracks within a ceramic material to the volume averaged response [3] [4] [5]. All of these works focus on simplified defect structures like cracks and most make the argument that other types of defects can be mapped to cracks using a suitable scaling argument. In this work, we start with the defect structure from a commercially produced boron carbide and then look at how those defects affect the development of a 2 wave steady shock profile in simple planar shock loading and how that defect structure affects cracking and material failure at a slightly larger scale in miniaturized ballistic experiments between 1 and 2 km/s.

1.1. Boron carbide microstructure

Boron carbide plates are manufactured using a hot pressing technique where powder is consolidated under high temperature and moderate pressure to form fully dense parts. During processing second phases such as carbon are introduced into the microstructure. Typical grain sizes in boron carbide are on the order of 10 μm. The carbon inclusions range in size from less than 1 μm to over 100 μm. A more detailed discussion of the microstructure of boron carbide can be found in Moorhead et al. [6] and Hogan et al. [7].

---

* Corresponding author. Tel.:+1-410-278-1069;
E-mail address: Andrew.l.tonge.civ@mail.mil.

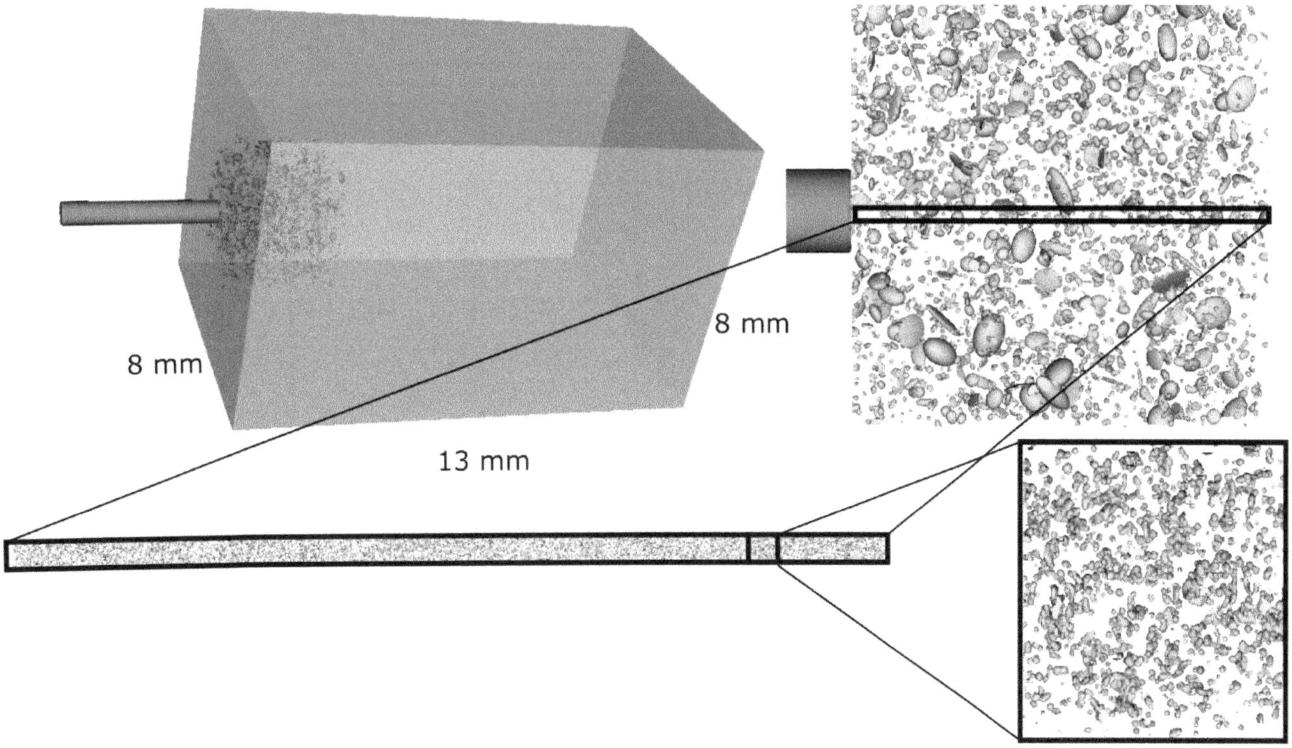

Figure 1: Images showing the small scale ballistic impact configuration and the simulated microstructures for both the ballistic simulations and the plate impact simulations. The graphite inclusions are shown in green.

This paper presents a technique for simulating the boron carbide microstructure followed by simulation results for planar shock propagation and small scale ballistic loading then concludes with a discussion of the results and how these observations could be extended to other systems or used to improve material models.

2. Microstructure simulation

The shock simulations used a microstructure derived from X-ray computed tomography (XCT) scans of CoorsTek PAD B4C. From the XCT image stack, a gray scale threshold was applied to extract the dark features, which represent carbon inclusions or voids. These features were then converted into a format that can be explicitly shaped into an ALE3D [8] simulation following the procedure discussed in Lloyd et al. [9]. To generate long thin simulated microstructures, the scanned volume was sliced then reassembled to produce a microstructure that was suitably long for the shock simulations. The shock simulations used a simulation cell size of 0.25 μm per cell. An example microstructure is shown in the bottom row of Figure 1. Such a fine mesh could not be used for the small scale ballistic simulations. The small scale ballistic simulations used a graded mesh with a minimum mesh size of 16 μm per cell. The high resolution region was a 3 mm cube centered below the impact site. This region is similar to the viewing area of interest for the corresponding experiments. Within this region a simulated microstructure was generated based on the statistical flaw population described in Moorhead et al. [6].

The simulations were conducted using ALE3D[8], which is an Arbitrary Lagrange Eulerian Finite element code developed by Lawrence Livermore National Laboratories (LLNL). For the plate impact simulations, the mesh was allowed to relax to minimize the gradient in the condition numbers of the elements while the nodes that represented the boundary between materials remained Lagrangian. The small scale ballistic simulations used an advecting region around the penetrator and the boron carbide material was kept Lagrangian as long as possible to maintain accuracy of the history variables during the formation and growth of crack features. Once material experienced significant inelastic deformation nodes connected to elements containing that material were allowed to relax.

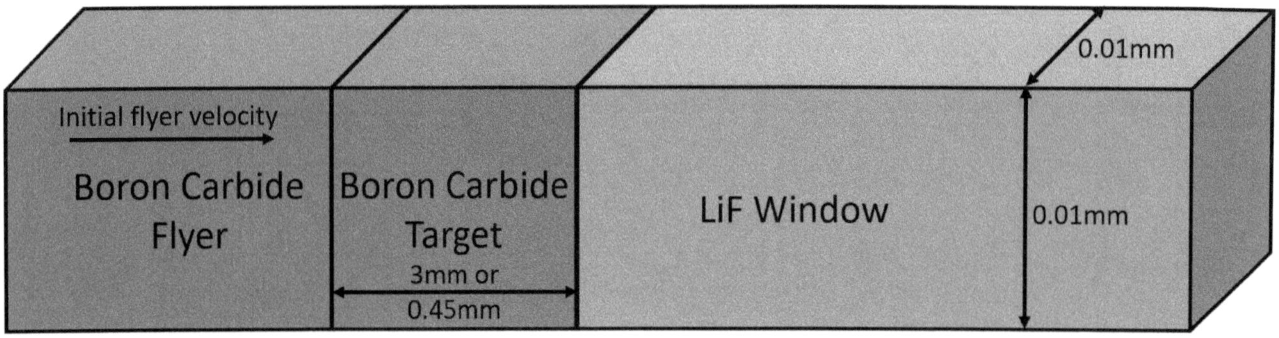

Figure 2: Planar shock loading geometry

The inclusions in the microstructure were approximated by ellipsoids that are described by a center location and the orientation and length of three principal axes. The orientation of the ellipsoids had a uniform random distribution. The major axes length was sampled from a bounded Pareto distribution with minimum size of 20 μm and a maximum size of 200 μm. Once the major axes length was generated, the next shorter axes length was uniformly distributed between the major axis length and one tenth of that length. The same procedure was repeated again for the smallest axis length. This procedure was used because the statistical characterization study did not find any significant trends in the ellipsoid orientation and there was only limited correlation between the major axis length and the other dimensions of the ellipses.

The inclusions are either porosity or residual carbon. They are low strength defects relative to a much stiffer and stronger matrix material. To represent this and maintain numerical stability, the inclusions are modeled as graphite particles with no strength and a Mie-Grüneisen equation of state. The matrix material is modeled as a brittle material with a pressure dependent strength that saturates at high pressure. Once the material is failed the fragmented material is treated as a granular material with associated flow, Lode angle and pressure dependent strength, and a porosity dependent crush strength. The failure strength of the material has superimposed variability [10] and the strain to failure is computed to enforce an energy dissipation per unit crack extension that is consistent with measured toughness values of boron carbide. The hydrostatic tensile strength variability in the matrix material represents the effect of microstructure that is not explicitly resolved in the simulation. For the planar shock simulations, these microstructural features are sub-micron porosity within the grains and residual stresses that result from anisotropic thermal expansion of the grains. In the small scale ballistic simulations, the strength variability represents all microstructural features below the simulation resolution of 16 μm. Both the local element strength variability and strain to failure for each element are sensitive to the element size. This model for the matrix material represents a necessary compromise between representing the fundamental processes in the material and being suitably robust to run in a large simulation with advection.

3. Results

3.1. Planar shock simulations

In this work three cases are investigated using the same set of material parameters. In the first case, there is no microstructure; it is just a homogeneous material. The simulated loading condition is a symmetric impact at 1.8 km/s. To save computational effort by reducing both the simulation domain and the required simulation time, the target thickness was reduced from 3.0 mm to 0.45 mm for simulations that showed the velocity history at the interface between the target and the LiF window. Figure 2 shows

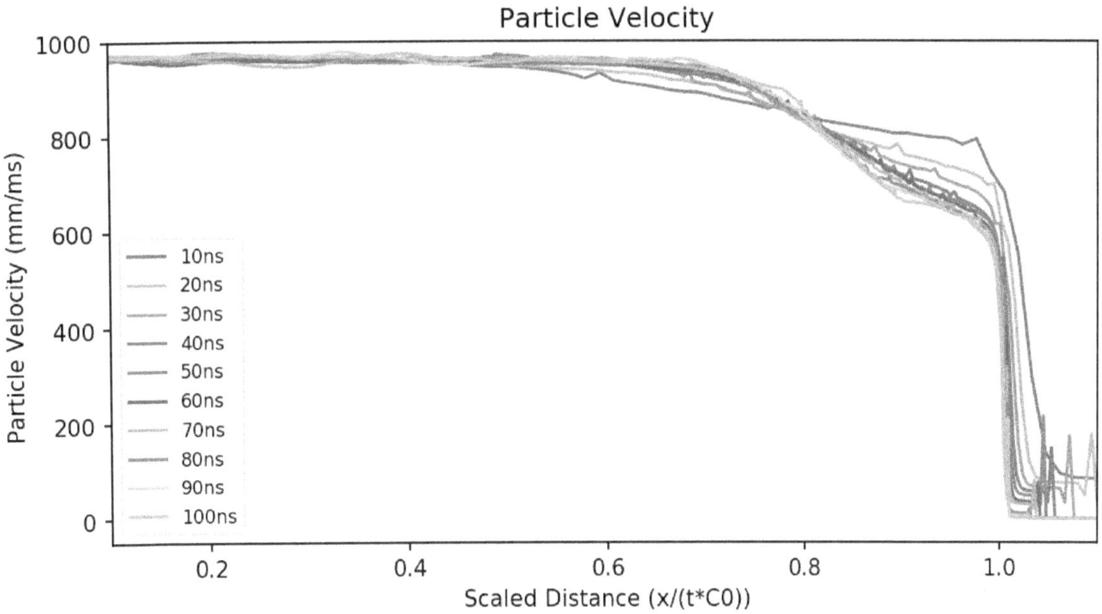

Figure 3: Shock profile evolution in long boron carbide sample with simulated microstructure.

a not to scale diagram of the loading configuration with the flyer plate, target, and the LiF window. Symmetric boundary conditions are used on the 4 transverse surfaces to reduce the size of the simulation.

To show the development of the steady shock wave the full 3.0 mm domain was simulated. In Figure 2 the axial velocity is averaged over bins that extend 2 μm in the axial direction and the entire simulation domain in the two transverse directions. This provides a reasonable reduction of the three dimensional velocity field data to a single dimension that is more easily interpreted. To illustrate the development of the shock wave, the spatial dimension is scaled by the distance that the elastic precursor has traveled. For a fully developed self-similar wave profile, the scaled velocity curves would lie on top of each other. Based on the results from the first 100 ns of the long domain simulation, the wave seems to be approaching the steady state after 100 ns, but has not reached that state yet. At 100 ns, the elastic precursor has traveled about 1.4 mm.

In experiments, the measured particle velocity comes from the back face of the target material where it is in contact with a transparent window material used for optical velocity measurements. In the experiments conducted by Vogler et al. [2], the window material was LiF. The simulated mean velocity and extreme velocities are shown at the LiF window interface in Figure 3. The distribution of velocities results from the material microstructure, which disturbs the planar shock wave.

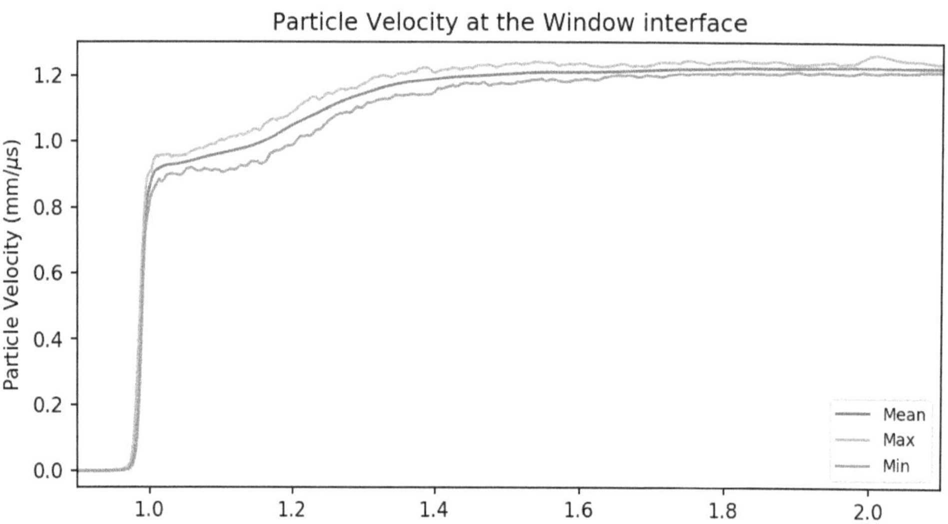

Figure 4: Particle velocity at the interface between the boron carbide and LiF window for simulations with a microstructure and a 0.45mm thick target.

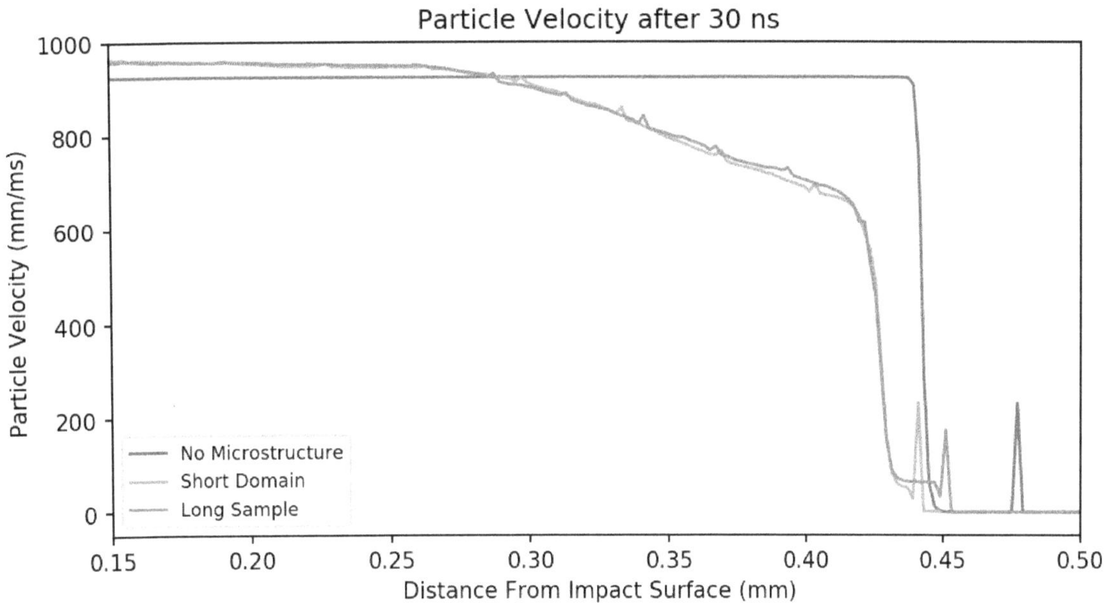

Figure 6: Particle velocity 30 ns after impact illustrating the importance of the material microstructure

Figure 5 shows the average velocity as a function of distance from the target-flyer interface 30 ns after impact for the simulation without any graphite inclusions and two simulations with graphite inclusions but different target plate thicknesses. These velocity profiles show the importance of the material microstructure for initiating inelastic deformation and introducing an elastic limit into the material behavior. This is the expected behavior because the material parameters were chosen so that pure uniaxial strain loading up to the shock state will not produce failure in the material. The local loading path for a material point is altered by the

Figure 5: Crack growth for simulations with and without microstructure

presence of the inclusions resulting in earlier failure of the material. The average velocity profile for the short and long domain are the same because the shock has only traveled 0.45mm and has not reached the end of the short domain.

3.2. Small scale rod impact

To determine the effect of the microstructure in a simplified ballistic loading configuration we simulate a copper rod with a diameter of 0.6 mm and a length of 4.0 mm impacting a block of boron carbide that is 8 mm on a side and 13.0 mm deep. The time of interest for the event is the first 500 ns of the event while the impact event is a reasonable approximation of a semi-infinite penetration problem. The simulated microstructure is confined to a cubic region extending 3 mm down from the impact surface and 1.5 mm transverse to the impact direction. The high resolution region is used to limit the computational cost because the resolution required to resolve the simulated microstructure for these impact simulations was 60 cells/mm. Outside of the high resolution region, a graded mesh was used. Simulations were performed with impact velocities of 0.9 km/s, 1.5 km/s and 1.9 km/s to determine the effect of impact velocity on the damage propagation away from the impact site and the importance of the microstructure at these impact velocities.

Figure 6 shows a volume rendering of the simulations at 1.5 km/s for simulations with and without the graphite inclusions. The graphite inclusions are shown as green ellipsoids. The copper penetrator is a solid orange and the remaining colors represent regions of material that have expanded beyond their original reference density. This expansion of the material in the target indicates failure and the opening of sub-scale porosity (or micro-cracks). Connected regions of expanded material are representative of macroscopic cracks in the structure. These are approximate cracks because they gradually introduce a zero stress region as the strain across the band increases. From this image it appears that the microstructure has some effect on the cracking pattern, but the effect is not as pronounced as originally expected. The major crack regions extending from the corners of the projectile have propagated about the same distance in both simulations.

The effect of the impact velocity is shown in Figure 6 for simulations that included the graphite microstructure. As expected increasing the impact velocity increases the density of the crack like features. At the 1.9 km/s impact velocity the crack patterns fill most of the field of view while at the lowest impact velocity of 0.9 km/s regions in the lower corners of the field of view have minimal cracking 200 ns after impact. At all of the impact velocities there is significant cracking near the surface of the ceramic.

To provide a better quantitative evaluation of the effect of impact velocity and microstructure, Figure 8 shows the penetration depth of the copper impactor as a function of time after impact. Increased impact velocity leads to an increase in the penetration velocity and greater total penetration after 200 ns. The 1.5 km/s simulation without the explicit microstructure has a slightly higher

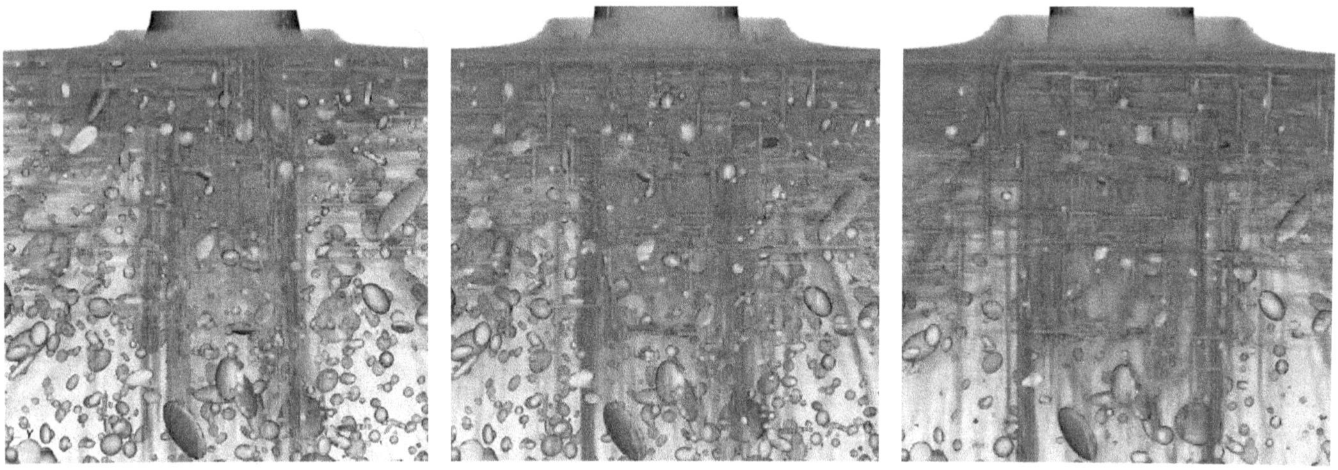

Figure 7: Cracking 200 ns after impact for impact at 0.9 km/s, 1.5 km/s, and 1.9 km/s. Velocity increases from left to right.

penetration velocity than the 1.5 km/s simulation that included the microstructure. The difference in penetration depth is less than 25 μm. In this region of the computation the cell size is 16.7 μm. The differences between these two runs is on the order of a computational cell supporting the conclusion that there is minimal effect of microstructural features over 20 μm in size for this impact geometry.

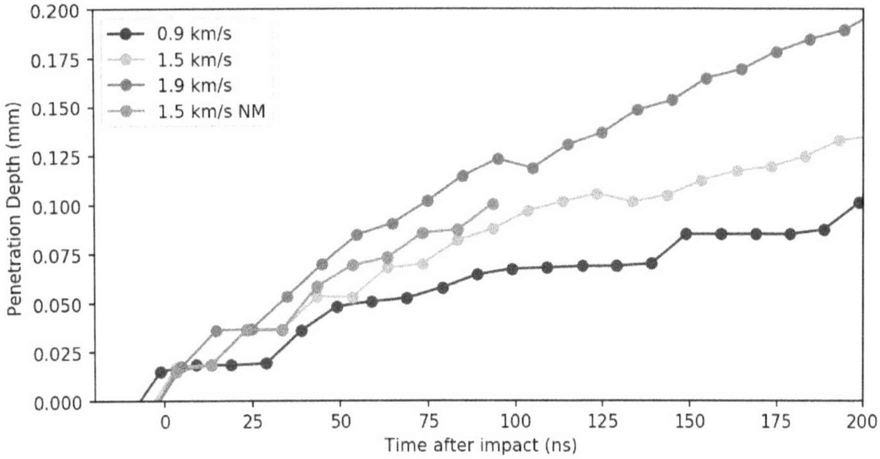

Figure 8: Penetration depth as a function of time and impact velocity for copper impactor

4. Discussion

For the planar shock simulations, the effect of the material microstructure was clearly visible in the averaged velocity profile resulting from the shock propagation. The interaction of the inclusions increased the local stress and caused local failure of the material. These local failures cooperatively form a process zone that accommodates the formation of an elastic precursor and a slower moving final rise to the shock state. In the simulations this behavior will be sensitive to the details of the material model including the exact shape of the yield surface, the direction of inelastic deformation, and the energy dissipated during formation of the localization region. In this work one set of material parameters was used that illustrated the effect of the microstructure. It is likely that many similar models or parameterizations would show similar effects. Simulations provide access to arbitrary loading paths that are not possible with experimental techniques. This should allow us to fill in some of the gaps in experimental loading configurations by using real material microstructures to perform mesoscale simulations. The information gained from these simulations will always be limited by our confidence in the physics that is captured in the component models.

The small scale ballistic simulations did not show as much sensitivity to the microstructure as the planar shock simulations. One possible explanation for this difference is the loading states interrogated by the small scale ballistic impact are very different than the loading states that occur in plate impact. The nominally uniaxial strain loading in the plate impact geometry suppresses tensile fracture for the homogeneous sample. In the semi-infinite penetration problem, there is much more shear and tensile loading. This could lead to failure of the material due to the stress transient associated with the projectile touchdown, which may not have time to be effected significantly by the microstructure. Understanding these details will require a closer analysis with multiple simulations looking at the very early times of the problem and using slightly different material models.

5. Summary

Two different impact problems were investigated to illustrate the effect of microstructure on the observed failure behavior of boron carbide. The plate impact simulations, which used a microstructure derived from XCT scans, showed a clear change in behavior when the microstructure was introduced. The effect of the microstructure was less clear in the small scale semi-infinite penetration simulations. In these simulations, there was not a clear change in the fracture pattern when the microstructure was introduced. Detailed simulations of the microstructure interaction with larger scale loading conditions are now possible, but estimation of the material behavior for both the matrix material and the inclusion materials remains a challenging problem. This is especially true of brittle materials that fail through a fracture process, which initiates at length scales much smaller than the resolved computational cell size then grows to span many computational cells.

Acknowledgements

The computational resources for this work were provided through the DOD High Performance Computing Modernization Program (HPCMP). The views and conclusions contained in this document are those of the authors and should not be interpreted as representing the official policies, either expressed or implied, of the Army Research Laboratory or the U.S. Government. The U.S. Government is authorized to reproduce and distribute reprints for Government purposes notwithstanding any copyright notation herein.

References

[1] J. J. Swab, C. S. Meredith, D. T. Casem and W. R. Gamble, "Static and dynamic compression strength of hot-pressed boron carbide using a dumbbell-shaped specimen," *Journal of Materials Science,* vol. 52, pp. 10073-10084, 5 2017.

[2] T. J. Vogler, W. D. Reinhart and L. C. Chhabildas, "Dynamic behavior of boron carbide," *Journal of Applied Physics,* vol. 95, pp. 4173-4183, 2004.

[3] B. Budiansky and R. J. O'Connell, "Elastic moduli of a cracked solid," *International Journal of Solids and Structures,* vol. 12, pp. 81-97, 1976.

[4] V. Grechka and M. Kachanov, "Effective elasticity of rocks with closely spaced and intersecting cracks," *Geophysics,* vol. 71, pp. D85-D91, 2006.

[5] M. F. Ashby and S. D. Hallam (Née Cooksley), "The failure of brittle solids containing small cracks under compressive stress states," *Acta Metallurgica,* vol. 34, pp. 497-510, 1986.

[6] C. A. Moorehead, J. M. Sietins and J. J. Swab, "Meso-scale microstructural agglomerate quantification in boron carbide using X-ray microcomputed tomography," *Materials Characterization,* vol. 141, pp. 177-185, 7 2018.

[7] J. D. Hogan, L. Farbaniec, M. Shaeffer and K. T. Ramesh, "The Effects of Microstructure and Confinement on the Compressive Fragmentation of an Advanced Ceramic," *Journal of the American Ceramic Society,* vol. 98, pp. 902-912, 2015.

[8] C. Noble, A. Anderson, N. Barton, J. Bramwell, A. Capps, M. Chang, J. Chou, D. Dawson, E. Diana, T. Dunn, D. Faux, A. Fisher, P. Greene, I. Heinz, Y. Kanarska, S. Khairallah, B. Liu, J. Margraf, A. Nichols, R. Nourgaliev, M. Puso, J. Reus, P. Robinson, A. Shestakov, J. Solberg, D. Taller, P. Tsuji, C. White and J. White, "ALE3D: An Arbitrary Lagrangian-Eulerian Multi-Physics Code".

[9] J. T. Lloyd, A. J. Matejunas, R. Becker, T. R. Walter, M. W. Priddy and J. Kimberley, "Dynamic tensile failure of rolled magnesium: Simulations and experiments quantifying the role of texture and second-phase particles," *International Journal of Plasticity,* 11 2018.

[10] O. E. Strack, R. B. Leavy and R. M. Brannon, "Aleatory uncertainty and scale effects in computational damage models for failure and fragmentation," *International Journal for Numerical Methods in Engineering,* vol. 102, no. 3-4, pp. 468-495, 2014.

Proceedings of the 2019 Hypervelocity Impact Symposium
HVIS2019
April 14-19, 2019, Destin, FL, USA
HVIS2019-061

Assessment and Validation of Collision "Consequence" Method of Assessing Orbital Regime Risk Posed by Potential Satellite Conjunctions

Travis F. Lechtenberg[a], and Matthew D. Hejduk[b]*

[a]*Conjunction Assessment Analyst, Omitron Inc., Colorado Springs, CO, USA*
[b]*Astrorum Consulting LLC, Woodway, TX USA*

Abstract

Collision risk management theory requires a thorough assessment of both the likelihood and consequence of potential collision events. Satellite conjunction risk assessment has produced a highly-developed theory for assessing the likelihood of collision but typically neglects to account for the consequences of a given collision. While any collision may compromise the operational survival of a space-craft, the amount of debris produced by the potential collision, and therefore the degree to which the orbital corridor may be compromised, can vary greatly among satellite conjunctions. Previous studies leveraged work on satellite collision modeling to develop a method to estimate whether a particular collision is likely to produce a relatively large or relatively small amount of resultant debris. The approximation of the number of debris pieces is dependent on a mass estimation process for the secondary objects utilizing the radar cross section of said object. This study examines the validity of the mass estimation process and establishes uncertainty bounds on the secondary object mass which will be used to best approximate the possible consequences of a prospective collision.

Keywords: COLLISIONS; RISK ASSESSMENT; PROBABILITY THEORY; SPACE DEBRIS; SATELLITE ORBITS; ATMOSPHERIC DENSITY; DRAG COEFFICIENTS; RISK MANAGEMENT

Nomenclature

A	Satellite Frontal Area (m^2)
BC	Inverse Ballistic Coefficient (m^2/kg)
C_d	Drag Coefficient (dimensionless)
CDM	Conjunction Data Message
L_c	Characteristic Length (m)
M_p	Mass of Primary Object (kg)
M_s	Mass of Secondary Object (kg)
P_c	Probability of Collision
RCS	Radar Cross Section (m)
SEM	Size Estimation Model
SNR	Signal to Noise Ratio (dimensionless)
$\vec{v}$	Satellite Velocity Vector (m/s)
V_{rel}	Relative Velocity (m/s)
x	Longest Dimension of Object (m)
y	Longest Dimension of Object in Plane Perpendicular to x (m)

* Corresponding author. Tel.: 1-785-766-3943.
E-mail address: travis.lechtenberg@omitron.com.

z	Longest Dimension of Object Perpendicular to both x and y (m)
ρ	Atmospheric Density (kg/m^3)

1. Introduction

In most current assessments of on orbit collision risk, operators typically are concerned only with the probability of collision (P_c). This approach has existed since the early 1980's with collision likelihood methodology initially developed by Kaplan and Garrick [1]. As such, operators tend to view all likely conjunctions as equally disastrous, with a single possible outcome from a collision, primary object inoperability. This approach may become untenable as potential conjunction events increase in rate as additional spacecraft tracking capabilities come online and additional debris creation events occur. The orbital debris environment has seen large increases in population recently due to several events, including the Chinese ASAT test in 2007 against the FY-1C Fengyun satellite, the Iridium-Cosmos collision in 2009, and the projected increase in tracked debris population due to Lockheed-Martin's Space Fence, S-Band radar capabilities coming online in the near future.

As the spacecraft operational environment becomes more densely populated, consideration should be given to the potential consequences of a prospective collision. Potential collisions may be characterized as either catastrophic or non-catastrophic depending on the relative velocities of two objects and their masses. A catastrophic collision is one in which the primary object and the secondary object are fully fragmented and hence contributing significantly to the spacecraft debris environment, while a non-catastrophic collision is one in which the primary object is largely intact, with minor additional generation of debris objects resulting from the collision. As such, consideration of the potential debris generation from a given event or the "consequence" of a prospective collision is an important addition to evaluating the risk a specific event poses to the orbital regime at which the potential collision may occur.

Considering the consequence of a prospective conjunction as it relates to the spacecraft operational environment is critical to maintaining a safe environment for current and future missions. This means that the risk posed by a potential conjunction to the space environment is a combination of two factors, the likelihood, or probability of collision, and the possible, additional debris production from a specific collision. Current practices in assessing collision risk are focused almost exclusively on the probability of collision aspect, whereas the use of an approach incorporating the consequences of a perspective collision allows for more informed decisions to be made with regard to the risk posed to the operational environment.

Taking into account the potential for debris object generation as a consequence of a prospective collision will allow operators to better tailor their collision risk management processes to preserve the space environment. If the potential for debris generation is not taken into consideration, future collisions may increase the debris object density in favorable satellite orbits and render them unusable because operators, faced with multiple serious conjunctions and able to remediate only a subset of them, may unwittingly choose to remediate only those with small debris production potential. Taking debris generation into account can also decelerate the onslaught of Kessler syndrome in the future [2]. This approach may be of considerable interest to satellite operators in geosynchronous orbit, where low velocity conjunctions are commonplace.

Previous work by Hejduk et al. [3] focused on developing methodologies to estimate debris generation for specific conjunctions and probability distribution functions of conjunction severity by orbit type based on historically observed conjunctions. These previous methods were dependent on a number of spacecraft estimation processes, and preliminary attempts at estimation were examined for accuracy against the RORSAT Coolant Spheres.

Current efforts contained in this paper are focused on refining the secondary object size estimation process using enhanced methodology, validating this process against known satellite objects, and presenting a readily readable method of interpreting these collision consequence results using this methodology. For validation purposes, a large number of NanoSats with known characteristics were compared to the secondary object size estimation process outlined in this paper.

2. Debris Generation Methodology

The NASA Orbital Debris Program Office (ODPO) has been studying the subject of collision and explosion-spawned space debris for the past several decades. Using information gleaned from known satellite collisions and staged hyperkinetic collisions, relationships have been developed to estimate the size and number of debris pieces that may be generated by a hyperkinetic impact. The relationships are contained as part of ODPO's EVOLVE 4 satellite break-up model [4].

The first consequence-based assessment of a potential conjunction is whether a conjunction, should it produce a collision, will result in a "catastrophic" collision, that is, a collision that produces widespread fragmentation of the primary object as well as the secondary; or "non-catastrophic," wherein only the secondary object is likely to fragment. This determination is made through a relationship based on the relative momentum of the two objects; a collision may be considered "catastrophic" if the relative momentum exceeds 40,000 J/kg.

$$\frac{M_S V_{rel}^2}{2M_p} > 40{,}000 \frac{J}{kg} \tag{1}$$

Once the "catastrophic" or "non-catastrophic" nature of the event has been determined, an additional relation is used to estimate the possible number of debris objects that may be generated by such an event.

$$N(L_c) = \begin{cases} 0.1(V_{rel} M_s)^{0.75} L_c^{-1.71}, & \frac{M_S V_{rel}^2}{2M_p} \leq 40{,}000 \\ 0.1\left(M_s + M_p\right)^{0.75} L_c^{-1.71}, & \frac{M_S V_{rel}^2}{2M_p} > 40{,}000 \end{cases} \tag{2}$$

In the previous equations, L_c refers to the characteristic length of the debris piece size threshold above which the operator is concerned, *i.e.*, how many pieces will be generated with a characteristic length larger than L_c. A reasonable value for this variable would be the minimum characteristic length for which an operator might reasonably expect tracking data, such as 5 cm for the perceived tracking fidelity of the Space Fence.

The previous equations are relatively straightforward to evaluate if all terms are known; however, for prospective conjunctions, often times the secondary mass values are not known as the secondary objects involved in conjunctions are frequently debris objects such as those from the Fengyun or Iridium-COSMOS debris generation events. There are many possible reasons why the secondary object characteristics may be unknown; this is simply the most dominant one. The relative velocity of the two objects is generally fairly well known due to orbit determination processes for the two objects, as is the primary object's mass which is known from the spacecraft's operator, but characteristic length and mass of the secondary object are still unknown.

3. Unknown Satellite Mass Estimation Process

A method presents itself as convenient for estimating the secondary object's mass at lower altitudes using parameters contained within the drag equation:

$$a_{drag} = -0.5 \frac{C_D A}{M} \rho \|\vec{v}\| \vec{v} \tag{3}$$

Estimation of an object's drag characteristics is of significant import in the orbit determination process, as this is the primary non-conservative force acting on objects in low earth orbit, where most spacecraft missions also operate. As such, the collective terms for satellite drag can be determined as part of the orbit determination process and collected as a term known as the ballistic coefficient and is typically included in reported spacecraft states.

$$BC = C_d A / M \tag{4}$$

Knowledge of the ballistic coefficient allows for estimation of the object mass, but presents a new problem, namely that the drag coefficient (C_d) and frontal area (A) of the spacecraft must then also be estimated or known. In this study methods to estimate reasonable drag coefficients and frontal areas from known satellite quantities are proposed and validated by examining the characteristics of known NanoSats, and as these objects are known, their drag coefficients can be estimated using a cuboid drag model as proposed by Walker et al. [5]. The cuboid drag model is broadly applicable regardless of object type for nanosatellites as they typically conform to the cuboid model, though there may be more profound variation due to nanosatellite orientation. This allows for a preliminary examination of model accuracy for when the object profile is relatively known, though operationally the most calamitous estimation of C_d may be used as a conservative estimate of the possible catastrophic nature of a collision. For this analysis the nanosatellite drag coefficients are generally presumed to be in line with the more oblong configurations, which show greater variation than strict cube-shaped approximations. Assuming an oblong cuboid satellite form, the drag values can be roughly approximated from Figure 8 of Walker et al. [5] as linearly varying based on exospheric temperature, which can be estimated by a given atmospheric model of choice.

$$C_d = 2.4 + \frac{0.6}{800}\left(T_{inf} - 200\right) \tag{5}$$

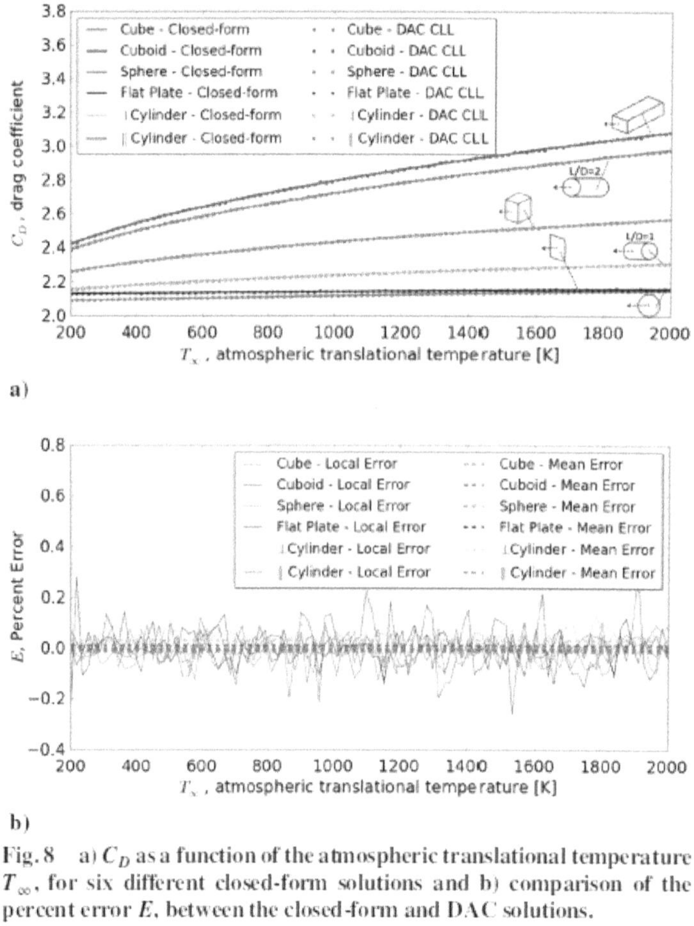

Fig. 1. Recreation of Figure 8 from Walker et al. [5] denoting drag coefficient dependence on exospheric temperature.

While there are certainly variations in the drag coefficient as estimated, these have been shown to often be overshadowed by frontal area variations and atmospheric density variations, particularly at lower altitudes as shown in Pilinski et al. [6], which estimated that C_d variations are typically only on the order of 2% except in extremely low altitude cases. For each determined best estimate of the drag coefficient, a number of samples are conservatively drawn using a normal distribution with a relative standard deviation of 5% (significantly larger than Pilinski's stated 2%) and a mean as determined above, using a sampling methodology allows for a predicted distribution of collision consequence. This larger sigma value is used so that a larger spread of C_d values may be examined for their effect on a collision.

4. Unknown Satellite Frontal Area Estimation Process

The frontal area of the secondary object may be estimated using a relationship between the presented area of the satellite and the object's Radar Cross Section (RCS). The intensity of the return of radar observations used in the orbit determination of an unknown object is characterized as the RCS and is a solved-for variable in the radar range equation if the Signal to Noise Ratio (SNR) and range are known. Typically this is reported as the median RCS value, as the RCS distribution does not conform to a normal distribution. If the wavelength of the observing radar station is known, the characteristic length of the object may be estimated using the median RCS (a value that is made available from the 18 SPCS, for example). This singular value is not necessarily an accurate representation of the spacecraft characteristic length and is better characterized as a distribution, the NASA ODPO model was developed in order to take a distribution of RCS values and produce a distribution of characteristic lengths. This study approximates the distribution of RCS values as a Swerling III, grounded by a single shape parameter that can be derived from the RCS median value. While this methodology is not an ideal manner of characterizing RCS distributions, it outperforms most other choices, as presented by Hejduk and DePalma [7].

This approach improves upon previous work, which used a less robust method of characterizing the RCS distribution that separated the RCS distribution into a sum of two distributions. Using this Swerling III distribution, a large set of samples representing the secondary object RCS are generated for use in the mass estimation process.

While this RCS value has units of area, it is not usually indicative of the satellite cross sectional area, and hence must be translated to an effective frontal area. The ODPO has examined this in prior research and developed a size estimation model (SEM), which was developed by fragmenting a satellite in a vacuum, measuring the characteristic dimensions of the resulting fragments, and then evaluating said pieces' radar returns when illuminated in all possible configurations [4]. Radar theory allows a dimensionless relationship to be established between an object's characteristic length (normalized by the radar wavelength) and the object's RCS (normalized by the square of the radar wavelength). This process provides a method by which the characteristic length may be determined if the RCS and the tracking radar's wavelength or frequency is known; if not, a good approximation of a generic radar may be established using radar frequencies of either UHF (~430MHz) or L-Band (~1200MHz), depending on object RCS value. The characteristic length distribution of the secondary object is then used to approximate a distribution of estimates of the satellite frontal area:

$$A = \frac{\pi L_c^2}{4} \tag{6}$$

5. Satellite Size Validation Method

At this point a method has been established by which an object's mass may be estimated using readily available data from the orbit determination process. To validate this approach a large set of NanoSats for a range of operational altitudes were examined. The initial set of examined satellites was comprised of roughly 1000 NanoSats. This set was then investigated to determine their prescribed masses, dimensions, satellite bus size, and Satellite ID numbers. The initial set of 1000 NanoSats was reduced to 530 satellites based on satellite specification availability and launch successes. As all of these spacecraft are cuboid in nature due to the nature of satellite buses available at that size, the expected frontal area for each spacecraft was estimated as the mean area presented by the spacecraft assuming a random tumbling nature. This estimation assumes a sinusoidal behavior as it relates to the presented satellite area, which makes the estimated area an average between the maximum and minimum presented areas. Using this approach, an independent data set of small object masses and frontal areas had been established. For an irregularly shaped object, a formula proposed in the NASA standard breakup model [4] and expounded upon by Hanada [8] for the average cross-sectional area is:

$$A_{total} = \frac{(A_{xy} + A_{yz} + A_{xz})}{3}$$

As the spacecraft in question are cuboid, the maximum triplet dimensions of x, y, and z can be determined in a relatively straightforward manner as the outer dimensions of the cuboid are known.

Next, orbit determination data was retrieved from relevant Conjunction Data Messages (CDMs) pertaining to these objects, including orbit regime, RCS, and ballistic coefficient. This exercise further reduced the available Nanosat population for validation to 371 based on CDM data availability.

6. Validation Results

Combining the previous sections' analyses and processes yields a distribution of object frontal areas and masses of the prescribed NanoSats. For each spacecraft 10,000 samples were derived, taking random draws of ballistic coefficient and frontal area as estimated from the RCS distribution. For validation purposes, the median values of frontal area and mass estimates are taken as the best estimate of the object's characteristics from this process and the quartile values are taken as the range of the uncertainty in this estimate. The estimated masses and frontal areas of these NanoSatellites were then compared with the independently determined values from the spacecraft specifications as researched, and compared in a normalized ratio sense. The distribution of these ratios is summarized below:

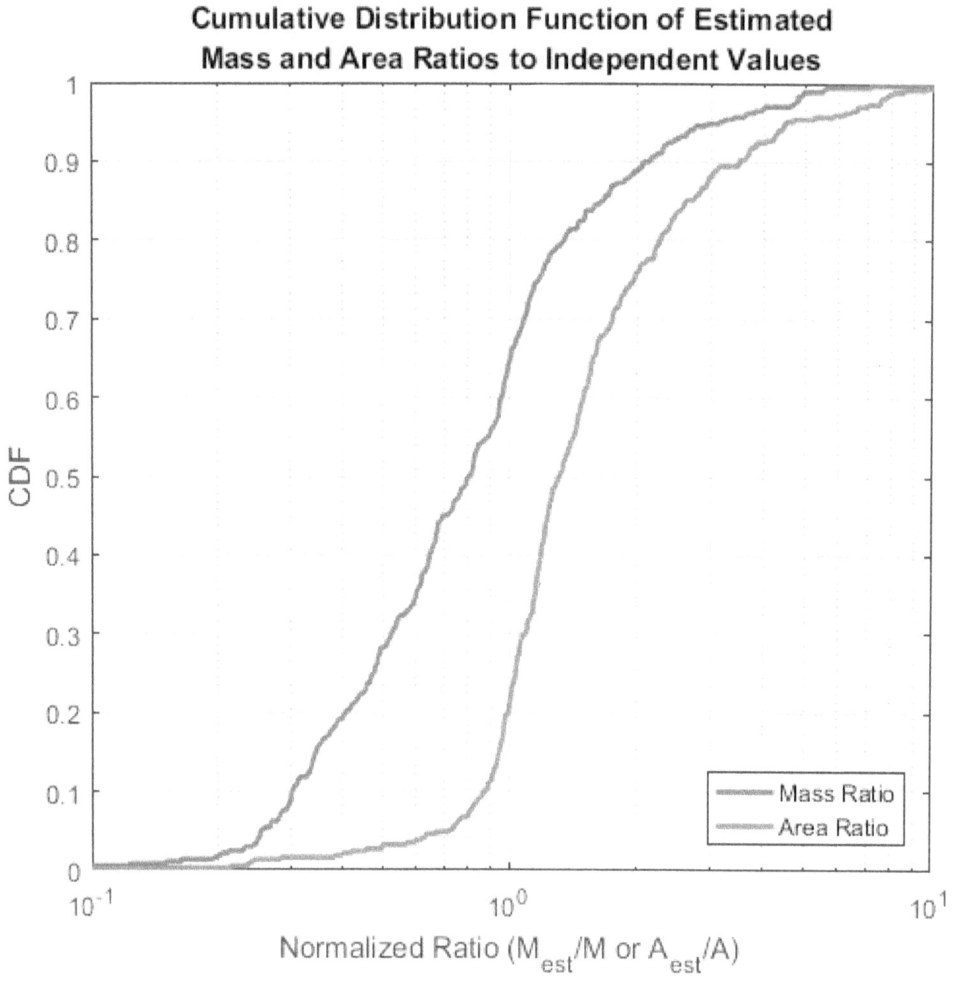

Fig. 2. Cumulative Distribution Function of Estimated Parameter Ratios to Researched Values for 371 NanoSats

Figure 2 shows the cumulative distribution of estimated parameter ratios and good general agreement for both area and mass ratios for described NanoSats. This figure shows that the frontal area estimates for the satellites in question typically overestimate the actual value, while the mass estimates tend to underestimate the actual vale. The mass estimates show the following ranges of accuracies for desired confidence intervals using this methodology:

- Mass Estimate Accuracy Ratios for NanoSatellites Show the Following Range at 68% Confidence: 0.353 - 1.574
- Mass Estimate Accuracy Ratios for NanoSatellites Show the Following Range at 95% Confidence: 0.227 - 4.709

These confidence intervals were selected to mirror the standard deviations of a normal distribution (1-sigma, 2-sigma), and the mass estimates show a 68% confidence that the mass estimates of the satellites will be roughly within 60% of the actual spacecraft mass, and a 95% confidence that the mass estimate will be within a factor of 5 of the actual mass estimate. In this process, the satellite masses were overestimated ~36% of the time. In order to err on the side of conservatism, users may opt in the future to artificially inflate the C_d values (or add a multiplicative biasing coefficient), which in turn will inflate the secondary object mass estimates, leading to a higher probability that the prospective collision may be "catastrophic".

7. Collision Consequence Assessment for Individual Events

The end goal of these sets of analysis is to provide satellite operators with a method by which to assess possible consequences of a prospective collision. Useful information to be provided to the operators would be the probability of a prospective conjunction being a "catastrophic" collision, the expected number of debris objects generated for both "catastrophic" and "non-catastrophic" events, and the distribution of the number of expected debris pieces based on a large number of trials. Using the validated methodology outlined above, a distribution of secondary object masses may be determined for a specific conjunction using a large sample number and the distributions outlined above. Using this distribution of secondary object masses as well as the characteristics of the prospective conjunction, a probability that the conjunction will be catastrophic may be assessed based on the number of mass estimates which result in a "catastrophic" collision assuming the two object's relative velocities and the primary object's mass are well known. By distinguishing which members of the secondary mass estimate distribution are catastrophic and which are not, the expected number of generated debris pieces for a given conjunction may be assessed for both the "catastrophic" and "non-catastrophic" cases. The two values for the expected number of debris pieces can then be reported to give operators a better perspective on the possible debris generation given a specific conjunction. Below follows two examples of conjunctions and the prospective consequences of a collision.

The following figures are examples of prospective outputs for an event which have very low and very high likelihood, respectively of being catastrophic. In these cases 10,000 samples of secondary mass estimate were used for assessment purposes.

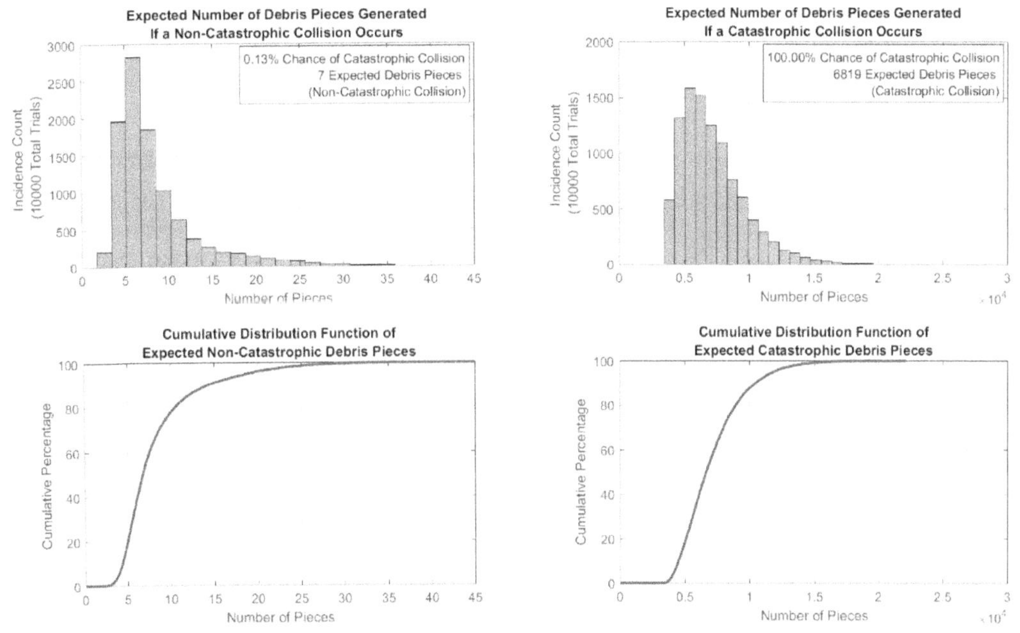

(a) (b)

Fig. 3. Collision Consequence Assessment for Independent Individual Events (a) Non-Catastrophic (b) Catastrophic

For many events, the probability that the prospective collision will be "catastrophic" or "non-catastrophic" is nearly certain. When this is the case, only the relevant chart is of interest. However, when this is not the case, the debris generation processes must be separated by trial on a "catastrophic"/"non-catastrophic" basis, and both sets of results presented. This is due to the radical difference in scale in prospective debris generation profiles for the two varieties of collision.

Above in Figure 3.a there is only a 0.13% chance the collision will be "catastrophic", with an expected number of generated debris pieces for the "non-catastrophic" case of: 7 debris pieces generated. However, for the few cases where the event was judged to be catastrophic, almost 2000 debris pieces were expected to be generated. In Figure 1.b, there is a >99.99% chance the prospective collision will be "catastrophic", as no trials resulted in "non-catastrophic" cases, with an expected number of debris pieces equal to 6819.

8. Conclusions and Future Work

This study presents steps forward in the estimation of the mass and size of secondary objects which pose a collision risk to operational spacecraft, but whose characteristics are not widely known. These estimation processes are validated against a set of known NanoSats, and show acceptable agreement, but are far from perfect in the estimation of these satellite states. These updated methodologies for assessing secondary object mass are used in a collision consequence assessment to provide operators with additional information regarding the potential for detrimental effects on the operational space environment through the likelihood of debris generation events, and the expected amount of debris articles generated from such an event. A method is then proposed for presenting relevant information to spacecraft operators. In future work, better drag coefficient estimation methodologies may be explored, mass estimates may be inflated to err on the side of conservatism, and additional characteristics of the predicted debris pieces may be assessed. One such additional characteristic may be the orbital half-life of the debris field based on drag modelling and atmospheric density predictions.

References

[1] Kaplan, S. and Garrick, B.: "On the Quantitative Definition of Risk." *Risk Analysis*, Vol. 1 No. 1 (1981), pp. 11-27.

[2] Kessler, D.J. and Cour-Palais, B.C.: "Collision Frequency of Artificial Satellites: The Creation of a Debris Belt." Journal of Geophysical Research 83 (1978), pp. 2637–2646.

[3] Hejduk, M., Laporte, F., Moury, M., Kelso, T.S., Newman, L., Shepperd, R. "Consideration of Collision "Consequence" in Satellite Conjunction Assessment and Risk Analysis", International Symposium on Space Flight Dynamics, Matsuyama, Japan, 2017.

[4] Johnson, N.L., Krisko, P.H., Liou, J.-C., and Anz-Meador, P.D.: "NASA's New Breakup Model of EVOLVE 4.0." *Advances in Space Research,* Vol. 28 No. 9 (2001), pp. 1377-1384.

[5] Walker, A., Mehta, P., Koller, J., "Drag Coefficient Model Using Cercignani-Lampis-Lord Gas-Surface Interaction Model", Journal of Spacecraft and Rockets, Vol. 51, No. 5 (2014), pp. 1544-1563.

[6] Pilinski, M., Argrow, B., Palo, S., "Semi-Empirical Satellite Accommodation Model for Spherical and Randomly Tumbling Objects", Journal of Spacecraft and Rockets, Vol. 50, No. 3 (2013), pp. 556-571.

[7] Hejduk, M., DePalma, D., "Comprehensive Radar Cross-Section "Target Typing" Investigation for Spacecraft", *Advances in the Astronautical Sciences*, Vol. 135, 2010, AAS 09-301.

[8] Hanada, T., Liou, J. C., Krisko, P., Nakajima, T., "For Better Calculation of the Average Cross-Sectional Area of Breakup Fragments", *Transactions of Space Technology Japan*, Vol. 7, Iss. 26, 2009, pp. 25-30.

Proceedings of the 2019 Hypervelocity Impact Symposium
HVIS2019
April 14-19, 2019, Destin, FL, USA
HVIS2019-085

Modeling Fragmentation within Pagosa Using Particle Methods

David B. Culp[a], Xia Ma[b]

[a]*XCP-2: Los Alamos National Laboratory, Los Alamos, New Mexico, USA*
[b]*T-3: Los Alamos National Laboratory, Los Alamos, New Mexico, USA*

Abstract

The mechanics involved in shock physics often involves materials undergoing large deformations being subjected to high strain rates and temperature variations. When considering high-velocity impacts and explosions, metals experience plastic flow, dynamic failures and fragmentation that are often too complex for a Lagrangian method, such as the finite element method, to properly resolve. Conversely, Eulerian methods are simple to setup, but often result in numerical diffusion errors [1]. These unpleasantries can be skirted by using an alternative technique that incorporates a blend of these aforementioned methods. FLIP+MPM (FLuid Implicit Particle + Material Point Method) employs Lagrangian points to track state quantities associated with materials as strength, as well as conserved quantities, such as mass. Concurrently, an Eulerian grid is used to calculate gradient fields and incorporate an algorithm that carries out the hydrodynamics [2]. By incorporating the FLIP+MPM method into Los Alamos National Laboratory's Pagosa hydrodynamics code, massively parallel architectures may be employed to solve such problems as those including fragmentation, plastic flow and fluid-structure interaction. This paper will begin with a mathematical description of the FLIP+MPM technique and describe how it fits into Pagosa. After a description of the implementation, the capabilities of this numerical technique are highlighted by simulating fragmentation as a result of high velocity impacts and explosions. Several strength and damage models will be exercised to demonstrate the code's flexibility. Comparison of the different models' fragment size distributions are given and discussed.
Keywords: MPM; Hydrodynamics; Fragmentation; Particles; Strength

1. Introduction

1.1. Motivation

Problems involving materials of strength undergoing fragmentation is an ongoing area of research among many disciplines. Two scenarios in which materials undergo fragmentation are explosions and high velocity impacts, both of which involve multiple materials that undergo high strain-rates, large deformations, heat transfer, phase change and multiphase (solid and fluid) flow. In order to capture all of these features in a computational environment that is capable of numerical modeling a wide variety of problems, the FLIP+MPM method is married to the hydrodynamics code Pagosa, both of which are tools developed at Los Alamos National Laboratory.

The FLIP+MPM method is the union of two separate, yet similar, particle methods. FLIP, which is short for FLuid Implicit Particle handles the calculation of thermodynamic variables by interpolating the Lagrangian marker, or particle, data to a uniform Eulerian mesh, calculating the changes to these variables during a timestep using continuum rules, and a remap of this data back onto the particles. MPM is similar except that rather than solving for updates to thermodynamic variables, the stress and history variables are updated using the pressure, temperature and velocity gradient data from the background Eulerian grid [4]. Together, the two methods are referred to as FLIP+MPM.

* Corresponding author. Tel.: +1-(505) 667-6826.
E-mail address: culp@lanl.gov.

Traditionally, problems in computational mechanics are set up in a Lagrangian or an Eulerian reference frame. The difficulty of using an Eulerian approach for these types of multiphase flow problems arise from the numerical diffusion errors that occur during the advection of materials across the computational grid, while the advantage lies in the simplicity of setting up a uniformgrid for the domain and avoiding mesh entanglement. Alternatively, Lagrangian problems are prone to mesh entanglement when materials flow and have difficulty handling interpenetration of materials with strength. However, using a Lagrangian approach is nice because it avoids the advection diffusion errors by allowing the markers to freely follow the material [7].

FLIP+MPM is considered a mesh-free method, despite having a background mesh. Because of this, its implementation avoids the disadvantages of both aforementioned approaches while capitalizing on their strengths. This hybrid method uses Lagrangian markers and a uniform, Eulerian background mesh with interpolation between the two domains to model the evolution of both conserved quantities, such as energy, mass and momentum, as well as the state quantities associated with materials of strength, such as stress and damage [8].

In the remainder of this chapter, a mathematical foundation of the recipe for the aforementioned algorithm will be shown, followed by an overview of the physics handled within Pagosa. The following chapter will demonstrate several applications of this suite of codes, namely demonstrations of fragmentations due high velocity impacts and explosion of a copper shell filled with high explosives. The final chapter will summarize the results and discuss how the codes will be validated with data.

1.2. The FLIP+MPM method

Over the past four decades, the T3 group at Los Alamos National Laboratory has developed the FLIP method, which is an augmented descendent of earlier Particle-in-Cell (PIC) methods, also developed in the group during the 1960's [1,2]. In the mid 1990's the PIC technique was extended to history-dependent variables, which ultimately became MPM [3]. Subsequently, literature began to emerge that combined the two methods in order to solve problems that involve materials with both thermodynamic variables as well as a stress history with rate and thermal effects [4].

$$\Delta \boldsymbol{x}_r = \Delta t \boldsymbol{u}_r \tag{1}$$

$$\rho_r \Delta \boldsymbol{u}_r = \Delta t[-\theta_r \boldsymbol{\nabla} \mathrm{p} + \textstyle\sum_s \mathcal{K}_{rs}(\boldsymbol{u}_s - \boldsymbol{u}_r)] + \rho_r \Delta_\sigma \boldsymbol{u}_r \tag{2}$$

$$\rho_r(\Delta e_r + p\Delta v_r) = \Delta t[\textstyle\sum_s \mathcal{R}_{rs}(T_s - T_r) - \boldsymbol{\nabla}_r \cdot \mathbf{j}_r + \boldsymbol{\sigma}_r : \dot{\boldsymbol{\epsilon}}_r] \tag{3}$$

$$\rho_r \Delta v_r = (\theta_r \beta_r \Delta T_r - f_r^\theta \textstyle\sum_s \theta_s \Delta T_s) + \Delta t[f_r^\theta \boldsymbol{\nabla} \cdot \boldsymbol{u}] \tag{4}$$

$$\sigma_i \longrightarrow \sigma_i + \Delta t \psi[\sigma_i, \boldsymbol{\nabla}_i \boldsymbol{u}_i, T_i, *_i] \tag{5}$$

Given a set of initial and boundary values, the state of the system at the markers (or particles) can be captured by the collection of $\mathbf{x}_i$, $\mathbf{u}_i$, e_i, v_i, m_i, σ_i and $*_I$, where $\mathbf{x}$ is position, $\mathbf{u}$ is velocity, e is internal energy, v is specific volume, m is mass, σ_i is stress. $\mathcal{K}_{rs}$ is the momentum exchange rate between materials on the *grid* whose subscripts are $_r$ and $_s$. $\mathcal{R}_{rs}$ is the r-s thermal energy exchange rate, and $*_i$ represents any number of additional history dependent variables. Similarly, θ_r is the volume fraction of material r, while β_r is its expansivity. f_r^θ is defined as the product of material r's compressibility and volume fraction divided by the total sum of each materials product of compressibility and volume fraction. Let the subscript $_i$ indicate that these variables are associated with a *marker*. These states may exist in a multi-field setting, with states for each desired material.

Along with equations of state, and rules associated with models for strength and damage, this thermodynamic state can be evolved using a finite difference scheme to compute the state after a timestep. The finite differences, for most state variables, are calculated not on the particles, but instead on the Eulerian mesh. In order for this to occur, the data on the particles are mapped to the grid using shape functions, where the gradient values are conveniently computed before remapping to the particles. They are convenient due to the uniform nature of the background mesh.

The finite differences calculations are shown above in (1) - (4), where ρ indicates the mass density, p is the pressure, T is the temperature, ϵ is the strain and ∇ is the gradient operator. These updates are remapped back onto the markers using shape functions. This process is analogous to material advection. Equation (5) shows the stress update, which does *not* involve a remapping of the marker data to the grid, but is rather updated as a function Ψ (which depends on the choice of constitutive model chosen) of p_i, T_i, and $\nabla \mathbf{u}_i$. The update algorithm developed in the previous paragraphs is sufficient for the scope of this paper. If a more rigorous, and exhaustive derivation of the method is desired, it can be found in [8].

1.3. Pagosa hydrodynamics code

At its core, the Pagosa code is a fully 3D, hydrodynamic, finite volume solver, which has the ability to calculate the evolution of thermodynamic properties on an Eulerian grid, based on conservation principles. It is implemented across a variety of architectures and is designed to use MPI for massive parallel processing. A wide variety of equations of state, strength models make its use ideal for modelling multiple material, multi-phase, high strain rate systems characterized by high-velocity impacts and explosions.

The governing equations consist of the continuity equation, which expresses the conservation of mass, the Euler equations, which express conservation of linear momentum, a constitutive relation, which expresses the conservation of internal energy, and an equation of state, which relates pressure, density, and internal energy. Together these equations form a closed system, which, along with initial and boundary conditions, give a full description of the system's motion and thermodynamic state.

Within Pagosa, the time integration is second-order accurate and fully explicit, having a timestep that is controlled by Courant condition. Because the code is designed to model shocks, artificial viscosity may be introduced, allowing a shock to be spread over several zones. The initial conditions, geometry and model options are user-driven through an input file. The integration takes the form of a Lagrangian operation at the beginning of a timestep, followed by advection phases, which remap the variables back to the Eulerian mesh [5] .

1.4. Pagosa with FLIP+MPM

Over the last several years, the FLIP+MPM technique has been implemented within the Pagosa hydrocode at Los Alamos National Laboratory, and has only recently become operational. Because Pagosa is an Eulerian code, the addition of FLIP+MPM took advantage of this for all of the mesh-based finite volume calculations, and needed only add on the particle functionality along with the proper bookkeeping and communication interface between the particles and grid. The remainder of this paper will focus on the verification of the particle methods by applying problems the pertain to high velocity impact of materials, and explosions, with a focus on the fragmentation of materials during these events.

2. Numerical Experiments

2.1. Exploding Cylinder

In this experiment, a cylindrical shell of copper, filled with C4 high explosive, is detonated. The high temperatures, strain-rates and stresses generated by the expanding gasses lead to failure and fragmentation of the copper shell. For this experiment, three different strength models are used for verification of their performance using Pagosa with FLIP+MPM.

The copper shell is modeled as a right cylinder with a height of 2.5 cm, an inner radius of 0.37 cm, an outer radius of 0.52 cm., while the wall thickness of the top and bottom caps is 0.175 cm. The entirety of the shell is filled with C4. A plane detonator is used at time t=0 to commence the explosion. The copper shell is modelled with physical parameters that are aligned with that of annealed copper.

The initial density is specified as 8.9673 grams/cm^3, and the initial internal energy is set to 7.11 E-6 Mbar-cm^3/g. A sesame equation of state is selected, which utilizes a tabular data set to retrieve the appropriate thermodynamic variables given a particular state in the problem. Table 1 in Appendix A shows the input parameters necessary for the Mechanical Threshold Stress (MTS), Johnson-Cook (JC), and Preston-Tonks-Wallace (PTW) implementations within Pagosa. Documentation on these plasticity models can be found in [6]. This table also includes the 20 input parameters for the Johnson-Cook style damage model used for fracture. It should be emphasized that this damage model is not fully documented yet – the scalar damage is defined as below, where d is a scalar damage variable, ϵ_p is the accumulated incremental plastic strain, and ϵ_f is the fracture flow strain, which is defined by parameters within the damage model.

$$d = \epsilon_p \,/\, \epsilon_f \qquad (6)$$

The uniform Eulerian mesh is set up with one zone every 0.01 cm in the X, Y and Z directions, with the initial particle layout having 64 uniformly distributed markers per zone. The initial timestep is set up to 1 E-3 microseconds, and is allowed to grow per the courant conditions of the problem. Data is output every 0.1 microsecond and the simulation runs until 140 microseconds. Figure 1 shows the evolution of the Johnson-Cook scalar damage over the course of the simulation using the Johnson-Cook strength model. It is worth noting how the top and bottom discs blow off as some of the initial fragments before the hoop stress begins making longitudinal tears in the cylinder walls.

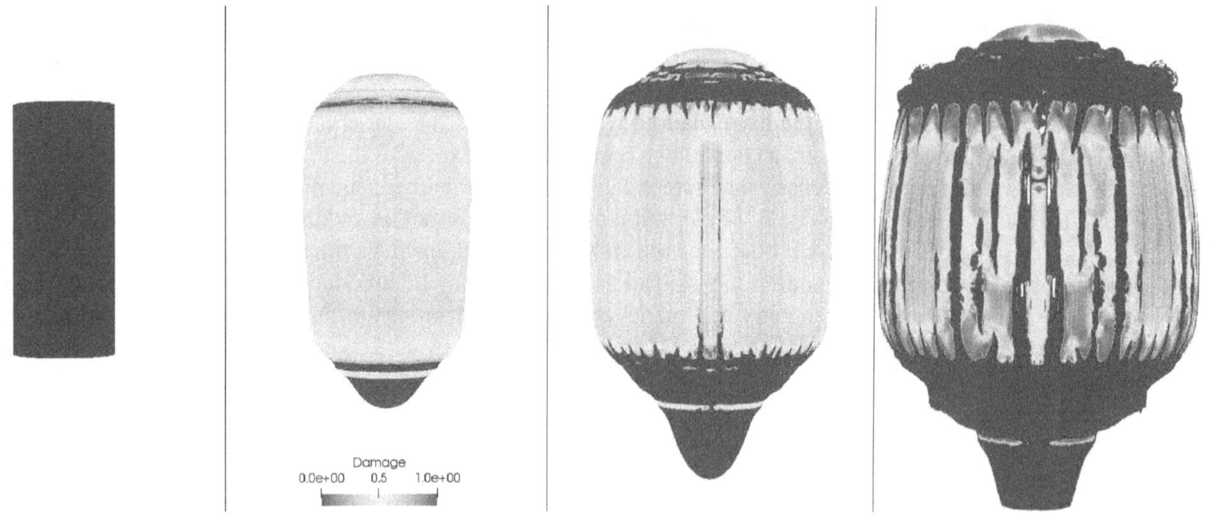

Figure 1: From left to right, the exploding cylinder using the Johnson-Cook strength model is shown at times 0, 4, 10, and 14 microseconds after detonation. The color represents a scalar damage variable where a value of 1, corresponding to red, indicates a fracture or failure.

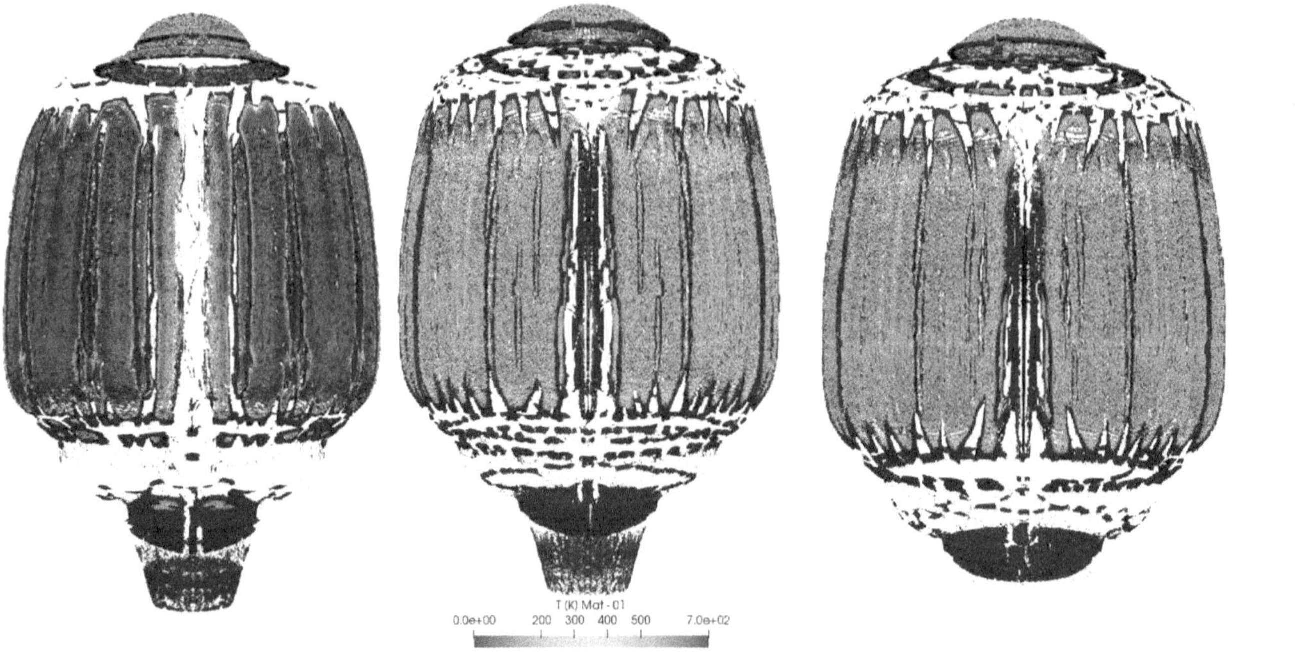

Fig 2: From left to right, the final fragmentation arrangement is shown for MTS, JC, and PTW strength models. The temperature distribution is also shown. MTS shows a significantly hotter end result, relative to the other two strength models.

Qualitatively, all three of the strength models predict very similar fragment and fracture patterns and can be seen at 140 microseconds in Figure 2. The MTS strength model appears to predict a substantially higher temperature in some portions of the failed shell, to the order of about 200 degrees Kelvin. Quantitively, the largest 20 fragments, measured by volume, from each of the three experiments are plotted in Figure 3. The JC and PTW strength models predict very similar fragment size distributions, with the MTS being the outlier.

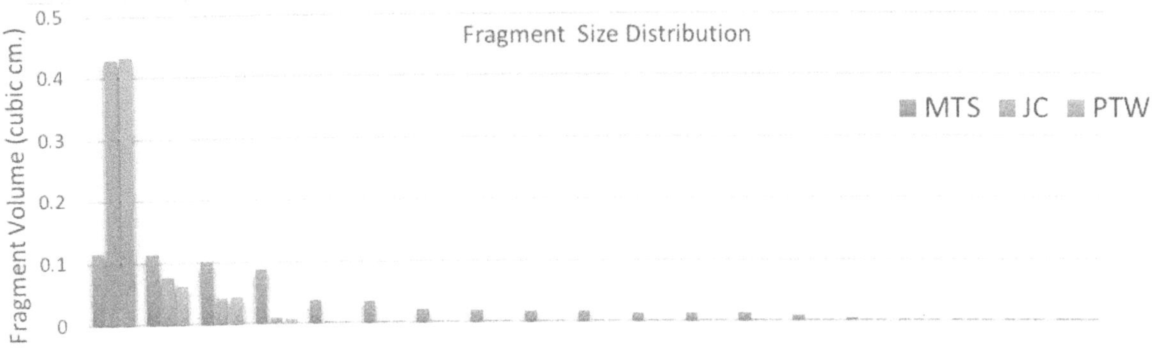

Fig 3: The 20 largest fragments are plotted by volume for the three strength models used in the exploding cylinder experiment.

2.2. High Velocity Impact

During this second experiment, a copper projectile collides with a stationary copper target, where damage ensues, and ultimately, fragmentation. The initial velocity of a copper projectile is set to be 2000 m/s, which is roughly mach 6, and the corresponding fragmentation distribution is noted while varying the strength models.

The projectile is a cylinder with a radius 0.15 cm and a height of 0.2 cm, while the target is a plate with length and width of 0.8 cm and a thickness, in the direction of the moving projectile, of 0.1 cm. The initial position of the projectile is 0.02 cm away from contacting the target plate. The projectile and the target are identical materials, with physical parameters that are aligned with that of annealed copper. The input parameters are the same as those used in the previous experiment, however, the Steinberg-Guinan plasticity is an additional strength model that is exercised.

After running the simulations at varying initial velocities, 2000 m/s was chosen because of the large number of fragments produced that were of an appreciable size. Figure 5 shows four different stages of the impact and fragmentation, with the color indicating the velocity in the x-direction. A simulated time of 15 microseconds was considered, with an initial timestep of 1E-3

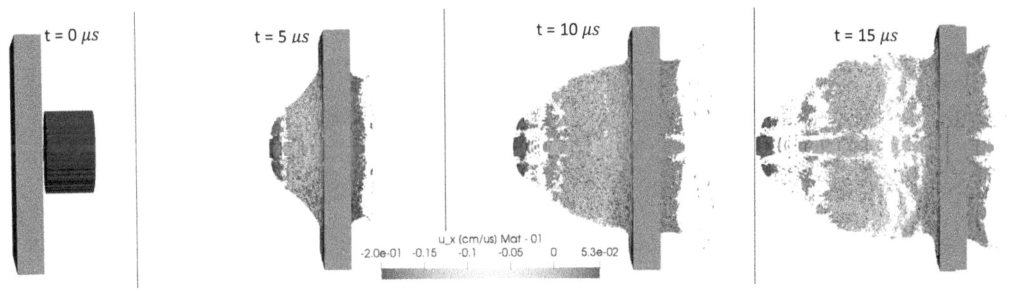

Fig. 4. The impact experiment using MTS is shown at four different timesteps with a plot of particle velocity in the x-direction. The projectile has an initial velocity of 2000 m/s in the negative x-direction.

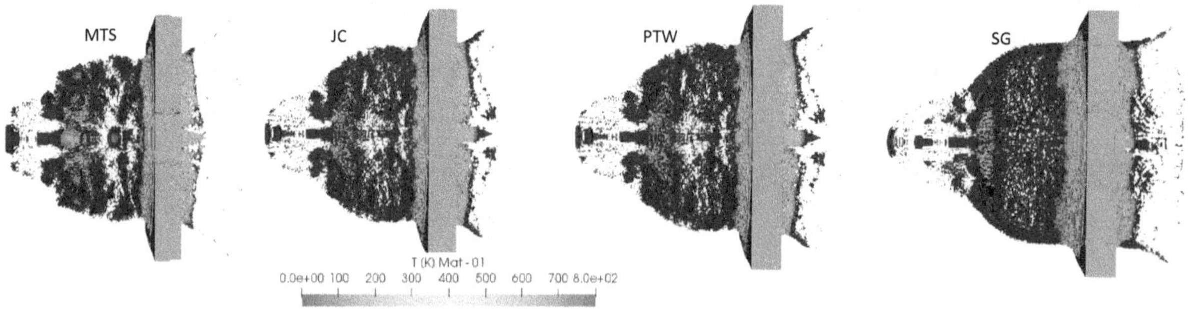

Fig 5. A plot of temperature at 15 microseconds after impact shows the outcomes using four strength models. MTS is an outlier in both temperature and shape.

microseconds, outputting data every 0.25 microseconds. The Eulerian mesh was discretized to 0.005 cm or 50 microns, which also was the initial spacing for the FLIP/MPM particles. 1024 processors were used to run each simulation.

Again, MTS appears to be the clear outlier, both qualitatively and quantitatively. Figure 6 shows a plot of temperature at 15 microseconds after impact, comparing all four of the strength models that were considered. The highest temperatures for MTS were around 80 degrees Kelvin higher than those predicted by the other three methods, which found good agreement. Furthermore, the particle size distribution was very tight between all but the MTS method, whose largest fragments were shown to be an order of magnitude greater than the variances between the other three methods. Figure 7 shows this distribution – the three largest fragment sizes were disregarded in this plot due to the large variance between the MTS fragment size and the those of the other three models.

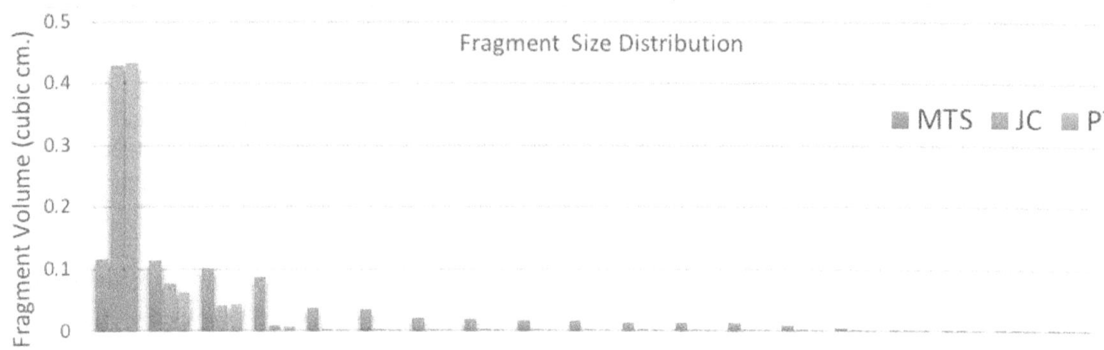

Fig 6. A histogram showing the 4th-24th largest fragments form the impact test, with the four strength models that were considered. The 3 largest fragment sizes were not plotted due to scale. Once again, MTS is the clear outlier.

2.3. Spall

A physical feature that we wish to demonstrate within this model is the process of spallation, which provides verification of the fragmentation process using FLIP+MPM. An analytical solution for the minimum velocity to predict the onset of spall in this situation can be written as

$$v_{spall} = \frac{2\sigma_u}{\sqrt{\rho E}}$$

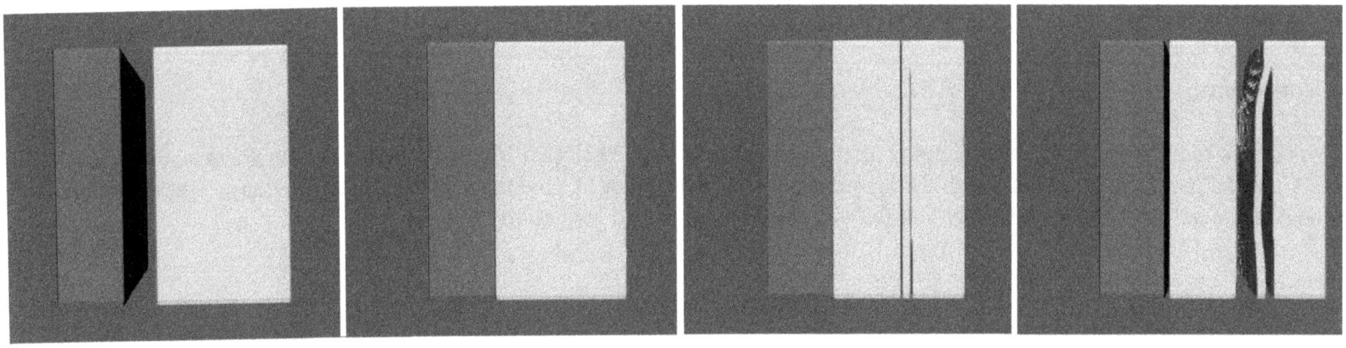

Fig. 7. The two-plate collision over time. From left to right, the plates are shown before impact, during impact, spall beginning to form, and full spall formation.

where v_{spall} is the minimum impact velocity required to produce the tensile stress, σ_u, which will induce failure, ρ is the material's mass density, and E is the material's young's modulus. A more in-depth calculation of this phenomenon can be found in [9].

To demonstrate this, we simulate two slabs of steel colliding, as shown in figure 7. One plane is modeled with thickness of 2 cm and is initially stationary, while a second slab of thickness 1 cm approaches with velocity of 360 m/s, which is greater than the calculated v_{spall} for steel. After the impact, compression waves travel in opposite directions in both of the plates, and rebound as tension waves at the free surfaces, where they meet at the half-thickness of the larger plate. The large amount of tensile stress experienced at this location causes a fracture.

Figure 7 shows the progress of the experiment, from the initial state on the left, to the final fractured state of the thick plate, due to spall. The far-right frame of the figure captures the spallation fracture nicely. This experiment verifies that the model does indeed capture this physical feature, and will be a cause of fracture leading to fragmentation. There is a need to focus more attention into this problem in order to address mesh refinement sensitivities associating with capturing spall, and also resolve the additional spall plane that is evident in the figure.

3. Conclusions

The preceding experiments were conducted to highlight the capabilities of FLIP+MPM when coupled with a powerful finite volume application. In particular, these capabilities working in conjunction with the hydrodynamics code Pagosa, offer a wide array of options for modeling materials that undergo high strain rate, multi-phase phenomenon associated with explosions and high velocity impacts. The two physical examples shown in this paper clearly demonstrate that the FLIP+MPM method is a promising and powerful method to couple the physics of conserved, thermodynamic quantities with those that have a stress history.

It was shown that there is good relative agreement between the fragmentation that is predicted from several plasticity models, specifically Johnson-Cook, Preston-Tonks-Wallace and the Steinberg-Guinan strength models, while it was noted that the Mechanical Threshold Stress model is somewhat of an outlier when compared to the other three. Due to its 34 input parameters, MTS is a complicated model that may need more tuning. Of course, there is the obvious need to validate these models with physical data, which will be considered in the immediate future. A continuation of this research will seek to validate the damage models, and tune parameters where needed.

Acknowledgements

This work was made possible by the ongoing implementation of this FLIP+MPM functionality within Pagosa by Bucky Kashiwa. A special thanks to Brandon Smith for his patience and investment of time in to sharing his knowledge weekly. Rob Kelsey and Xia Ma have also been extremely helpful with their guidance and availability to help.

References

[1] Harlow, F.H. 1964. The Particle-in-Cell Computing Method for Fluid Dynamics, Methods of Computational Physics Vol. 3, p. 319.

[2] Brackbill, J.U. Ruppel, H.M, 1986, FLIP: A Method for Adaptively Zoned Particle-in-Cell Calculations of Fluid Flows in Two Dimensions, Journal of Computational Physics, 65 p. 314.

[3] Sulsky, D., Chen, Z., Schreyer, H.L. A Particle Method for History-Dependent Materials, Computational Methods in Applied Mechanical Engineering. 118, p 179.

[4] Zou, Q., Zhang, D. Z., VanderHeyden, W. B., 2004. Application of a FLIP-MPM-MFM Method for simulating Weapon-Target Interaction. LA-UR-04-6592.

[5] Weseloh, W., Clancy, S., Painter, J. 2010. PAGOSA Physics Manual, LA-14425-M

[6] PAGOSA Input Reference Manual, 2017. Saftey Applications Project, LA-CP-17-20444

[7] Chen, Z., Brannon, R., 2002. An Evaluation of the Material Point Method. Sand Report SAND 2002-0482 UR

[8] Kashiwa, B., Rauenzahn, R., 2011. FLIP Recipes: An Evolving Notebook. LA-UR-11-01860.

[9] Ma, X., Culp, D., Smith, B., 2019. Pagosa Simulation of Hypervelocity Impact and Fragmentation from Hypersonic Explosions. High Velocity Impact Symposium Destin, FL, paper #089

Appendix A. Model Parameter Values

Table 1. Material model parameter values

Model Parameter No.	MTS Strength Model	JC Strength Model	PTW Strength Model	JC Damage Model	SG Strength Model
1	0.4578	0.00288	0.025	1.52	36.
2	0.03	0.025	3.	7.08	0.
3	180.	0.31	0.0092	-3.0	0.45
4	0.7	1.09	0.0022	0.	2.83
5	10.	4.2829E-3	0.17	0.	3.77E-4
6	1.6	7.112E-6	8.E-6	0.	1790.
7	0.	0.	1.E-4	0.	2.83
8	1.E-10	-	1.E-4	0.0022	3.77E-4
9	0.	-	0.094	0.	300.
10	1.E-10	-	0.575	0.	-
11	1.5	-	0.25	2.8E-6	-
12	1.	-	0.447	3.75E-6	-
13	0.	-	1356.	50.	-
14	0.	-	63.546	3.	-
15	0.	-	0.	0.5	-
16	0.	-	-	1.52	-
17	4.E-4	-	-	5.	-
18	0.	-	-	0.05	-
19	0.00745	-	-	0.5	-
20	2.56E-8	-	-	0.5	-
21	0.	-	-	-	-
22	1356	-	-	-	-
23	298	-	-	-	-
24	1.E-6	-	-	-	-
25	1.E-9	-	-	-	-
26	0.	-	-	-	-
27	0.2375	-	-	-	-
28	4.E-4	-	-	-	-
29	0.00775	-	-	-	-
30	2.	-	-	-	-
31	0.0239	-	-	-	-
32	1.2E-4	-	-	-	-
33	1.696E-5	-	-	-	-
34	6.2E4	-	-	-	-

Proceedings of the 2019 Hypervelocity Impact Symposium
HVIS2019
April 14-19, 2019, Destin, FL, USA
HVIS2019-089

Pagosa Simulation of Hypervelocity Impact and Fragmentation From Hypersonic Explosions

Xia Ma[a]*, David Culp[b], Brandon Smith[c]

[a]*Los Alamos National Laboratory, P. O. Box 1663, Mail Stop B216, Los Alamos, NM 87545-1663*
[b]*Los Alamos National Laboratory, P. O. Box 1663, Mail Stop K784, Los Alamos, NM 87545-1663*
[c]*Los Alamos National Laboratory, P. O. Box 1663, Mail Stop T082, Los Alamos, NM 87545-1663*

Abstract

We use PAGOSA's FLIP+MPM capability to simulate hypervelocity impact and fragmentation from hypersonic explosions. The scenario to be simulated involves a complex chain explosion from fragmentation impact which was caused by another explosion. The simulations also use the SURF model for shock to detonation transition (SDT) and the MATCH model for mechanical ignition and deflagration of high explosives. These models in PAGOSA working together are crucial for modeling complex system for real world applications. This shows the powerful modeling and predicting capability of PAGOSA that others cannot do. Since experimental data are not available for any complex scenario like this, we did verification and validation (V&V) in each separate steps, These include the fragmentation simulated by FLIP+MPM, the Shock to Detonation Transition (SDT) modeled by SURF and mechanical ignition and deflagration modeled by MATCH. PAGOSA is a shock hydrodynamics program developed at Los Alamos National Laboratory (LANL) for the study of high-speed compressible flow and high-rate material deformation. PAGOSA is a three-dimensional Eulerian finite difference code, solving problems with a wide variety of equations of state (EOSs), material strength, and explosive modeling options. It has high efficiency for simulations running on massively parallel supercomputers. It is a multi-material code using volume of fluid (VOF) interface reconstruction and second order fully explicit time integration. Standard von Neumann artificial viscosity is used. Newly added material point method (MPM) plus Fluid-Implicit Particle (FLIP) capability can simulate high-speed metal fragmentation.

Keywords: Hypervelocity Impact, MPM methods, Fragmentation, High Explosive

1. Introduction

Pagosa is an efficient Eulerian code for large-scale 3D simulations. FLIP+MPM has semi-Lagrangian feature like Arbitrary Lagrangian Eulerian (ALE) so it can simulate phenomena of Lagrangian nature, for example fragmentation. It has been recently incorporated into Pagosa [1]. This enables Pagosa for high-speed fragmentation applications. In this paper we compute a complicated simulation. The results show the great capability of Pagosa. And we then do the V&V on each of the critical parts of the simulation to make sure the simulation is accurate for the applications. In practice, we often meet problems of high speed

* Corresponding author. Tel.: 505-695-5703;
E-mail address: xia@lanl.gov.

projectile hitting a cased high explosive (HE) and causing a detonation of the HE. The detonation then fragments its case into pieces which can fly at hypervelocity to hit another nearby HE. We would like to know if this can cause a chain of explosions, and if so under what conditions. This is a very complex scenario to simulate. Here we only use simple geometry to demonstrate the capability of PAGOSA for modeling the complex processes. In the simulation, the mechanical ignition and deflagration is modeled by MATCH [2], which stands for Mechanically Activated Thermal CHemistry. The functional form reflects the deep nature of the process from mechanically activated ignition to deflagration with some adjustable parameters to be calibrated with experimental data for a specific HE considered. Damage and cracking are the main mechanism that convert mechanical energy to heat energy by producing inelastic work that acts to heat the HE and generate local regions of high temperature. So the pre-damage of HE is also a very sensitive factor for its initiation. MATCH was invented by Brad Clements at LANL [2] and together we implemented it into PAGOSA last year. SDT is modeled by the SURF [3] model which stands for the Scaled Uniform Reactive Front. In fragment impacts we need to be able to account for dead zones and explosive build up in the presence of release waves originating at boundaries (conditions off the pop plot). This is complicated by the fact that in any real application, two and three dimensional effects are always present. While important, these complications are usually ignored. SURF is suitable for problems, where SDT is likely. It captures the details of the build up to detonation, not just the pop plot. It is able to capture short shocks and dead zones, whereas other burn models, like Forest Fire, cannot. (This is critical for fragment impact simulations!) Forest Fire was the main model for SDT problems used in PAGOSA before we implemented the SURF model. SURF has been calibrated for PBX 9501, PBX 9502 and other explosives with different grid size when large scale 3D applications may require coarser grids. In this paper, we show our work on large 3D, complex (multi-material multi-physics) simulations involving hyper-velocity detonation and fragmentation. First, we start from governing equations.

2. Governing equations

Continuity equation:

$$\frac{D\rho}{dt} = -\rho\nabla\cdot\boldsymbol{u}\,, \tag{1}$$

where $\frac{D}{dt}$ is material derivative defined as

$$\frac{D}{dt} = \frac{\partial}{\partial t} + U\frac{\partial}{\partial x} + V\frac{\partial}{\partial y} + W\frac{\partial}{\partial z}\,, \tag{2}$$

and (U, V, W) are three component of velocity vector $\boldsymbol{u}$. The momentum equation is

$$\rho\frac{D\boldsymbol{u}}{dt} = \boldsymbol{F} - \nabla P + \nabla\cdot\boldsymbol{S}\,, \tag{3}$$

where $\boldsymbol{S}$ is the symmetric deviatoric stress tensor by subtracting the isotropic pressure P from the total stress tensor and

$$\nabla\cdot\boldsymbol{S} = \left[\left(\frac{\partial}{\partial x},\frac{\partial}{\partial y},\frac{\partial}{\partial z}\right)\cdot\begin{pmatrix} S_{xx} & S_{xy} & S_{xz} \\ S_{yx} & S_{yy} & S_{yz} \\ S_{zx} & S_{zy} & S_{zz}\end{pmatrix}\right]^T. \tag{4}$$

The volume of fluid (VOF) equation for multi-material is

$$\frac{\partial f_m}{\partial t} + \boldsymbol{u}\cdot\nabla f_m = 0\,, \tag{5}$$

where f_m is occupancy function of a certain material m (or fluid) and is defined as

$$f_m(x,y,z,t) \equiv \begin{cases} 1 & if\ point(x,y,z)\ has\ m\ material\ at\ time\ t \\ 0 & otherwise \end{cases} \tag{6}$$

The equations (1) to (6) are closed by equation of state (EOS) and stress strain relations of different model options. We used Mie-Gruneisen EOS for solid HE phase and the linear UsUp relation is

$$U_s = c_0 + sU_p\,, \tag{7}$$

where U_s is the shock wave speed, U_p is the particle speed behind the shock, c_0 is the sound speed of the material, and s is the slope of this linear UsUp relation; s is also a material property. We used Jones-Wilkins-Lee (JWL) EOS for the HE product (gas phrase)

$$P_m = A\left(1-\frac{\omega}{R_1V}\right)e^{-R_1V} + B\left(1-\frac{\omega}{R_2V}\right)e^{-R_2V} + \omega\rho(E_m+E_0), \tag{8}$$

where A, B, R_1, R_2, ω are coefficients calibrated with the experimental data. V is dimensionless volume defined as $V = \rho_0/\rho$. For mixing cells, pressure and temperature equilibrium are imposed, that is, they are equal between different phases.

2.1. Burn model equations

SURF model uses the Hugoniot function:

$$h = e - e_0 - 0.5(P + P_0)(V_0 - V)\,, \tag{9}$$

to detect the leading shock, where e is energy, $V = 1/\rho$ is the specific volume here, and P the pressure. The subscript zero denotes initial values. The burn rate λ is calculated as

$$\lambda = 1 - e^{-s^2}; \frac{ds}{dt} = f(P_s)\,, \tag{10}$$

where s is the ratio of the burn radius divided by the separation between hotspots, P_s is shock pressure, and f is relatively complicated, but essentially given by an exponential function of shock pressure:

$$f \sim e^{(A+B\cdot P_s)}\,, \tag{11}$$

where A and B are the two parameters to be calibrated for a specific HE.
MATCH [2] model solves the Frank-Kamenetskii equation for a sub-grid hot-spot model

$$\rho c_v \frac{\partial T}{\partial t} = \kappa \frac{\partial^2 T}{\partial y^2} + \rho Z \Delta H e^{-E/(RT)} + \mu P \dot{e}_{max}\,, \tag{12}$$

where thermal conductivity k is assumed to be spatially constant. Here $T = T(y,t)$ is the temperature field, ρ is the mass density, c_v is the solid specific heat, R is the universal gas constant, E is the Arrhenius activation energy, ΔH is the heat of combustion, Z is the Arrhenius pre-exponential factor, μ is the coefficient of friction, and $\dot{e}_{max}$ is the maximum principal deviatoric strain rate. The cracking provides the source of heat $\mu P \dot{e}_{max}$.

2.2. FLIP+MPM Lagrangian marker equations

The MPM Lagrangian marker carries its mass, its volume, its stress and stress history etc. Momentum equation (3) is solved using the Galerkin weak integral method. For conciseness we only show the stress term on (3)'s right hand-side. That is,

$$\rho \frac{D\boldsymbol{u}}{dt} = \nabla \cdot (\theta \boldsymbol{S})\,, \tag{13}$$

where θ is volume fraction of the considered species. We approximate the velocity $\boldsymbol{u} = \boldsymbol{u}(\boldsymbol{x}, t)$ as a linear combination of piecewise linear trial function $s_j(\boldsymbol{x})$ (also called shape function)

$$\boldsymbol{u}(\boldsymbol{x}, t) = \sum_{j=1}^{N} \boldsymbol{u}_j(t) s_j(\boldsymbol{x})\,, \tag{14}$$

where the coefficient of the trial function $\boldsymbol{u}_j(t)$ is also j particle's velocity with total of N number of particles. Then we project the differential equation to the test function, which is the same as the trial function (thus a Galerkin method), and set it to zero to seek a weak solution

$$\int_\Omega \left(\rho \frac{D\boldsymbol{u}}{Dt} - \nabla \cdot (\theta \boldsymbol{S})\right) s_i dv = 0\,, \tag{15}$$

so that the discretization error is perpendicular to the test space of the shape function. By plugging (14) into (15) we get the discrete system of equations

$$\sum_{j=1}^{N} m_{ij} \frac{D\boldsymbol{u}_j}{Dt} = -(\theta \boldsymbol{S}, \nabla s_i), \tag{16}$$

where $m_{ij} = \int_\Omega \rho s_i(\boldsymbol{x}) s_j(\boldsymbol{x}) dv$, and $(\theta \boldsymbol{S}, \nabla s_i)$ is the inner product. The minus sign in (16) comes from the integration by parts with boundary term being zero. Equation (16) is actually just one equation for i. Since the shape function $s_i(\boldsymbol{x})$ is local support, that is, it has value of one at grid i, and value of zero at all other grids. So i sweeps all grids and resulting equation (16) a system of equation of mesh size dimension of freedom. Giving a linear hat shape function $s_i(\boldsymbol{x})$, the integral for m_{ij} can be evaluated either analytically or numerically. The integral of the inner product $(\theta \boldsymbol{S}, \nabla s_i)$ is approximated as

$$(\theta \boldsymbol{S}, \nabla s_i) \equiv \int_{\Omega} \theta \boldsymbol{S} \nabla s_i(\boldsymbol{x}) dv \approx \sum_{j=1}^{N_p} v_j \nabla s_i(\boldsymbol{x}_j) \boldsymbol{S}_j \tag{17}$$

Because the localness of the shape function, N reduced to N_p, which is number of particles in a cell. Equation (17) is very much like the quadrature rule of finite element method, except the quantities at the actual particle marker position are used instead of that at the Gaussian quadrature points. Particle volume v_j and stress $\boldsymbol{S}_j$ are carried by particle j. With these approximations, the system of discrete ODE (16) can be numerically integrated and advance in time.

3. Numerical simulation and results

In this section, we show our work on PAGOSA simulation of hypervelocity impact and fragmentation from hypersonic explosions. Fig. 1 shows the target that will be showered by high-speed metal fragments. The target is also an HE covered by a metal shell, which is engraved the word "MATCH INSIDE". This demonstrates that PAGOSA can deal with complicated 3-D geometry for real applications. "MATCH INSIDE" really means that the HE inside (Comp B in this particular simulation) will be modeled by the MATCH model or SURF model.

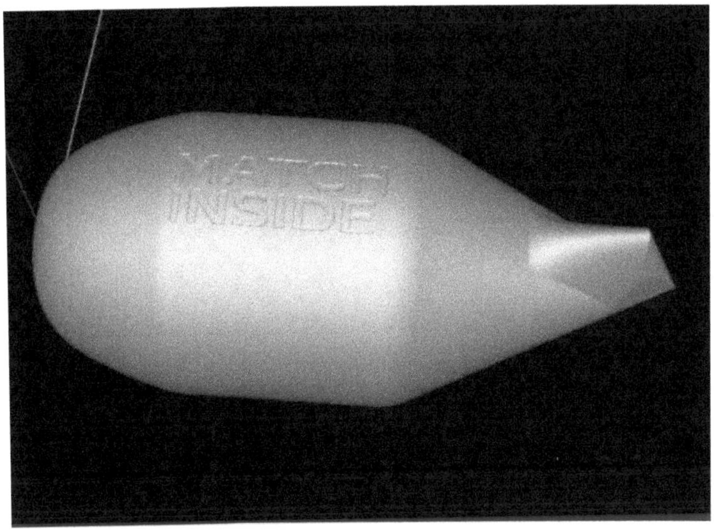

Fig. 1. HE covered by metal shell, which is the PAGOSA input for the target of impact in the simulation. The HE inside the shell is to be modeled by MATCH /SURF. The engraved word "MATCH INSIDE" shows that PAGOSA can simulate complicated geometry details.

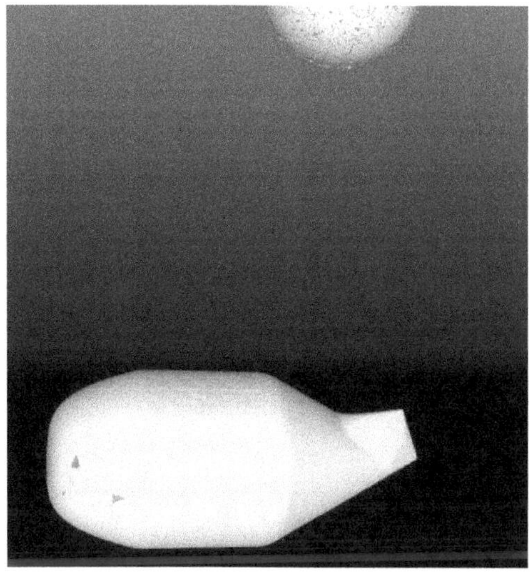

Fig.2. Graphic output of PAGOSA of simulation results at time=0 (so also the set-up of the simulation). The Hemisphere shell on top will be driven by HE to hyper-speed fragments then will impact the lower object, which is also an covered HE.

Fig. 2 shows the set-up of the simulation or the results at time equal zero. The top is a hemisphere of HE covered by a metal shell (Copper). There are experimental data for the fragmentation and separate works are performed for comparison with data for hemisphere fragmentation [4]. On next section, we do V&V for a slab spalling which has analytic solution to show that the FLIP+MPM fragmentation is accurate. Fig. 3 shows the simulation results at time= 30 μs. The hemi is fragmented, but not striking the target yet. We would think that the smaller fragments are light and thus would be accelerated faster by the driving HE (PBX 9501). Next, the air resistance would decelerate it more when the smaller pieces crossing the room to reach the target. However, this seems not the case as we observed here.

Fig. 3 shows that they are flying together. In a future simulation, we may neglect the air for higher computational efficiency. Fig. 4 shows the simulation at time=76 μs. The CompB is ignited by the MATCH model and exploding to the exterior through the holes penetrated by the fragments. The purple colored material are CompB. Fig. 5 of the cross section confirms that CompB did ignited at this moment. Fig. 5 shows a 2-D cross section of the simulation. Fig. 5a and Fig. 5c are at time = 60 μs, Fig. 5b and Fig.5d are at time = 76 μs. Fig. 5a and Fig. 5b are for pressure fields and Fig. 5c and Fig. 5d are for temperature fields. Comparing Fig. 5a and Fig. 5b, the pressure increases considerably and Fig. 5b shows the expansion of the metal shell due to the deflagration. Comparing Fig. 5c and Fig 5d, the temperature also increased considerably implying a full ignition and deflagration. The same simulation is performed (repeated) for SURF model (by replacing the MATCH model with the SURF model). Fig. 6a and Fig. 6b shows the SURF model's simulation. Fig. 6a compares with Fig. 4 and Fig. 6b compares with 5b. It is obvious that SURF did not predict a detonation. This is good, since no detonation is good for safety reasons. SURF can only be used for SDT problems. If no SDT, SURF cannot tell if there is HE ignition, deflagration, or explosion. (Explosion's pressure usually is one order of magnitude lower than detonation). However, deflagration

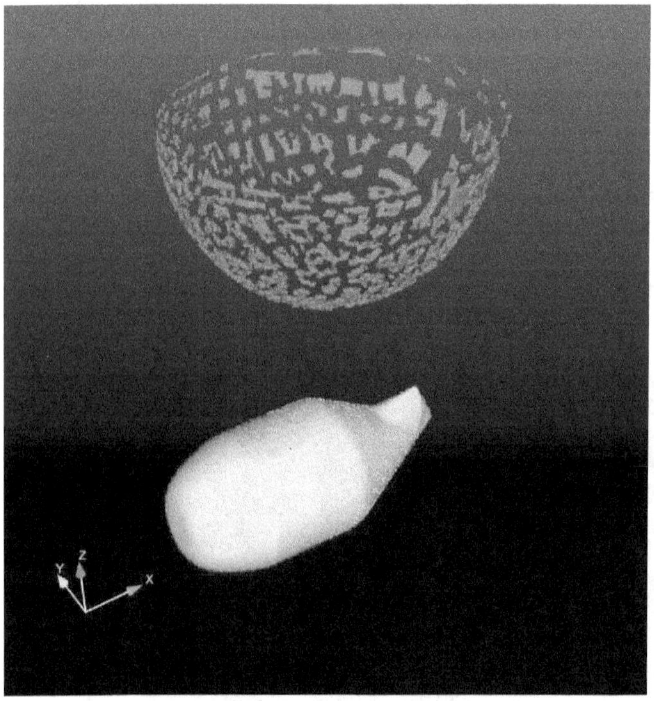

Fig 3. Simulation results: the hemispherical shell was fragmented to pieces, but not reach to the target yet (t=30 μs).

Fig 4. Fragments hit the target, causing an explosion. The purple colour represents HE (Comp B) exploding from the holes penetrated by the fragments. Time = 76 μs.

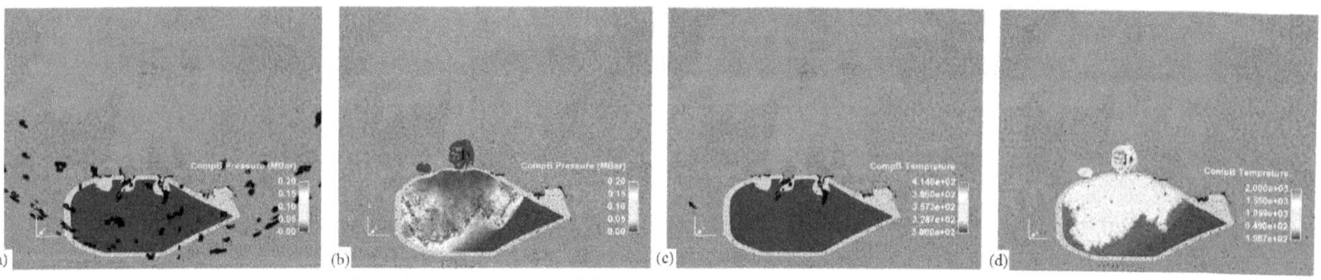

Fig.5. (a) CompB Pressure fields at time = 60 μs; (b) CompB Pressure fields at time = 76 μs; (c) CompB Temperature fields at time = 60 μs; (d) CompB Temperature fields at time = 76 μs.

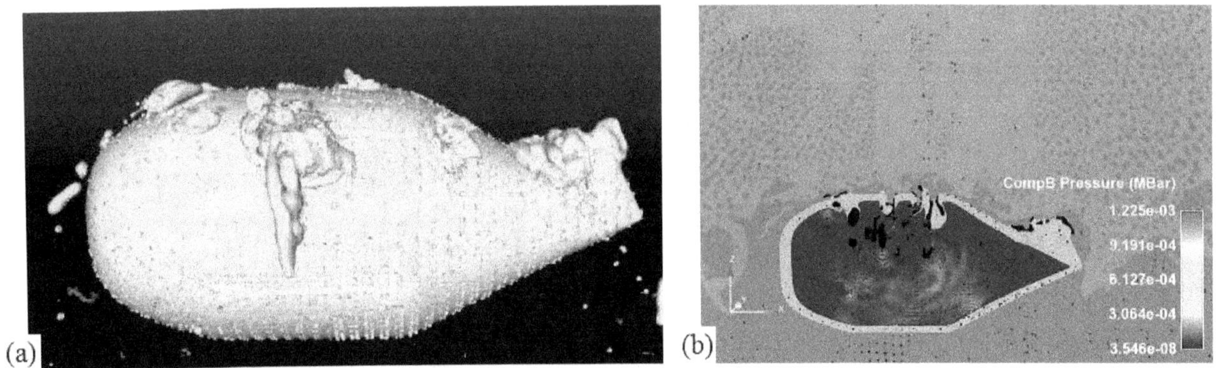

Fig. 6. SURF model prediction at t=76 µs, same time as Fig. 4 and Fig. 5b for MATCH model prediction. (a) Comparing with Fig. 4, there is no expansion of the metal shell and no explosion HE gas coming out of the shell; (b) This compares pressure at same time as Fig. 5b. The much lower pressure confirms that SURF did not predict a detonation.

and explosion can also be dangerous or a disaster by itself. So, MATCH is in handy for these situations. There is also a possibility for Deflagration to Detonation Transition (DDT). Initially there was no detonation, only deflagration, but later transitioned to detonation. Either SURF or MATCH can handle this condition. This will be our future work, that is, incorporate SURF into the MATCH model. Another potential effort is to incorporate the temperature based burn model AWSD [5], and its associate EOS [6] into MATCH.

The simulation was performed with 10 million of zones and two thousands of parallel processors. To make sure the simulation is dependable, as no direct experimental data are available for the whole system, we do V&V for each individual models involved.

4. V&V of separate parts (divide and concur)

On one hand, we need to do complicated large 3-D simulations for real world applications; on the hand, we do not have experimental data for the complicated scenario. Still we need it to be science based, accurate and dependable. We perform V&V for each component, which either have analytic solutions or experimental data to compare.

4.1. Slab impact spalling and fragmentation

To make sure the fragmentation of the FLIP+MPM method is accurate, we simulate a slab impact spalling experiment and compare with its analytic solution. There are four strength model in PAGOSA; a modified Johnson-Cook fracture model is used here. The basic form of the fracture model is define by a damage parameter D [7]

$$D = \sum \frac{\Delta\epsilon}{\epsilon^f}, \tag{18}$$

where the failure strain ϵ^f is defined as

$$\epsilon^f = \left[D_1 + D_2 e^{D_3\sigma^*}\right]\left[1 + D_4 \ln \dot{\epsilon}^*\right]\left[1 + D_5 T^*\right] \tag{19}$$

The sum in equation (18) is to accumulate all the incremental plastic strain $\Delta\epsilon$.

If D is greater than one, the material is considered fractured. The failure plastic strain ϵ^f versus dimensionless pressure-stress ratio σ^* is shown in Fig. 7 for four materials. σ^* is defined as $\sigma^* = \sigma_m/\sigma_{vms}$, where σ_m is the mean of the three normal stresses and σ_{vms} is the von Mises equivalent stress. σ^* is also known as the triaxial ratio. The stress and its history are carried by particles, thus FLIP+MPM can predict spall and fragmentation more accurately.

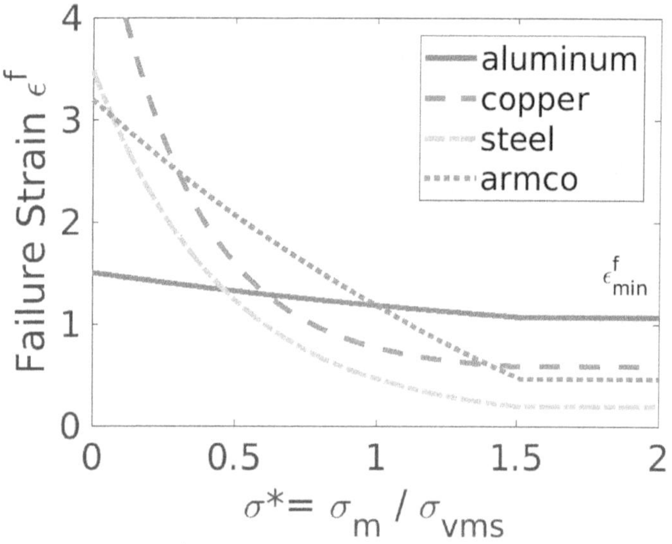

Fig. 7. Plastic failure strain ϵ^f as function of σ^*, the triaxial ratio.

Fig.8. Slab impact set-up, The red is the flyer with dimension 1cm * 5cm * 5cm. The spall slab has 2cm *5cm * 5cm.

The two-slab impact spall has an analytic solution, which may be used for validation. There are two pieces of same material with thickness of L and 2L as shown in Fig.8. The thinner one flies into the thicker one which is stationary. A compression

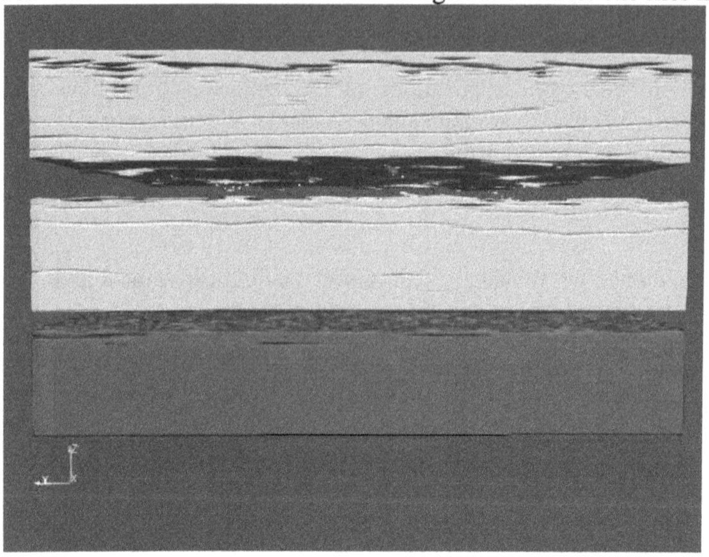

Fig. 9. Aluminium spall at velocity V=360m/s.

Fig. 10. Pressure field shows negative pressure near spalling location and uniform in the 2D plane.

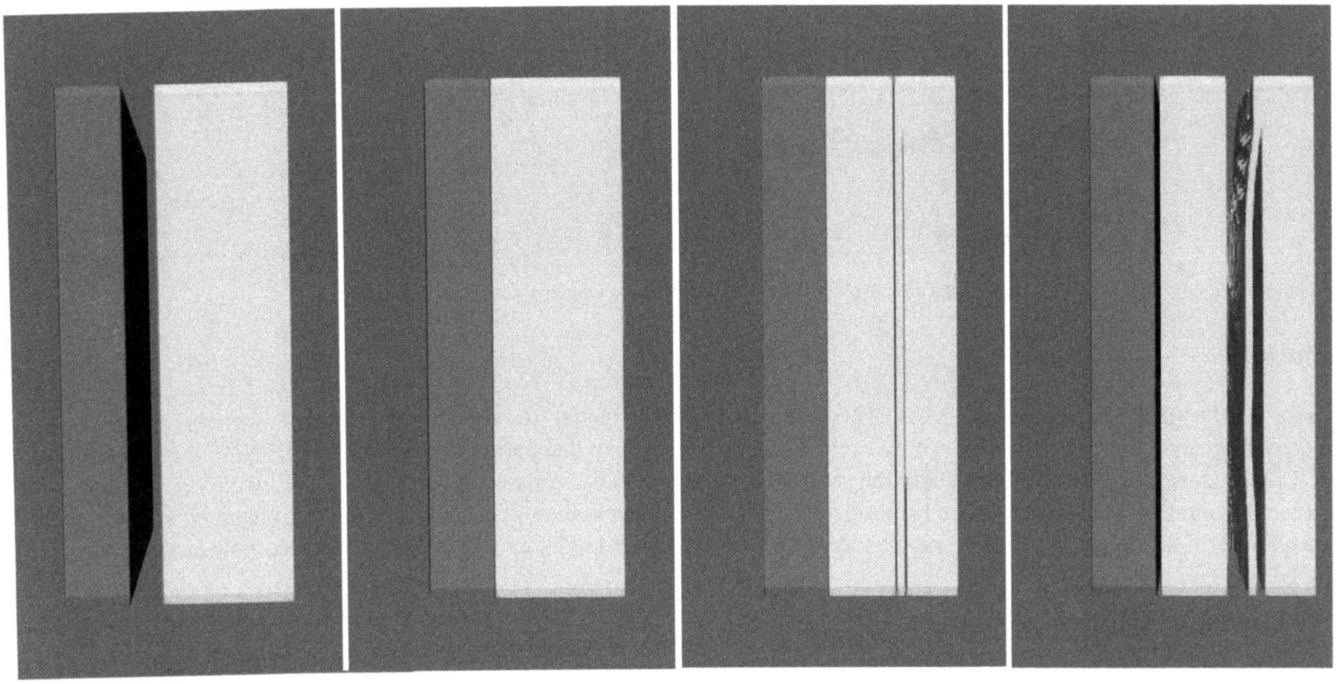

Fig. 11. Spall for steel. a) initial state, red flyer flies to the right with velocity v=190m/s; b) just touches the target; c) spall starts; d) big open spalls.

stress wave with speed c=$\sqrt{\rho E}$, where E is the Young's modulus of the slab material and ρ is the density of the elastic material. Since both slabs were made of the same material, the speed c of the compression stress waves propagating in both directions are the same. In the time interval *dt*, the waves travel distance $c \cdot dt$ and change the velocity in the material experiencing the stress waves in both directions. Using the principle of momentum conservation, we find the velocity in the region visited by the stress waves to be V/2 and the acquired momentum in the time interval to be

$$c \cdot dt \cdot A \cdot \rho \cdot \frac{V}{2} = \sigma \cdot A \cdot dt \tag{20}$$

where σ is the stress, A the section area of the plate. From this we can solve the stress as

$$\sigma = c \cdot \rho \cdot \frac{V}{2} = \sqrt{\rho E} \cdot \frac{V}{2} \tag{21}$$

When the compression wave reaches both ends of the plates, they reflect back as tension waves. When the tension waves meet, it produces an impulse of twice the tensile strength, that is 2*σ. If V is big enough such that 2*σ exceeds the tensile failure strength of the model, spall should occur. We set $\sigma_{spall} = 1.5\sigma$ and obtained spall for aluminium and steel as shown in Fig. 9. There seems some grid dependence, that is, the compaction direction need finer resolution. We need to study further the resolution effects. We used reflective boundary condition on plane edge to eliminate the boundary effects.

4.2. Burn models (MATCH and SURF)

We made comprehensive research and model V&V for both MATCH and SURF. Due to conciseness of the paper, we only show two results. Fig. 12 (a) and (b) shows agreement with experimental data [8] after calibration.

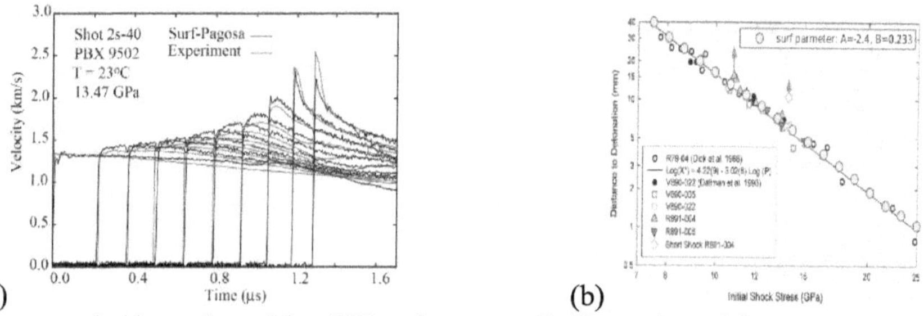

(a) (b)

Fig. 12. (a) Model predition compared with experimental data. (b) Pop plot agrees well with experimental data on a wide range.

5. Conclusions

We used FLIP+MPM fragmentation, MATCH model, and SURF model to simulate a complex scenario involving hypervelocity fragmentation and impact driven by explosion. SURF shows no detonation of the MATCH INSIDE target, which means that no chain reaction would occur under the set-up. However, MATCH predicts ignition of the HE, which means there will be some deflagrations and low-pressure explosion. The V&V of individual models suggests that the integrated simulation should be reliable. In the future, we need to combine the SURF and MATCH model to deal with even more complicated situation involving DDT.

Acknowledgements

We gratefully thank Mike Burkett, Daniel Garcia, Brian Key, Tariq Aslam and Dave Zerkle for their support of the project. We also thank Brad Clements as the present work is extensively based on his previous work..

References

[1] B. A. Kashiwa, etc., FLIP RECIPES: AN EVOLVING NOTEBOOK, LANL report, LA–UR–11–01860.
[2] B. Clements, etc. Mechanically Activated Thermal CHemistry (MATCH) Model for PAGOSA Users, LANL report
[3] R. Menikoff & M. S. Shaw (2012): The SURF model and the curvature effect for PBX 9502, Combustion Theory and Modeling, DOI:10.1080/13647830.2012.713994
[4] E.M.Mas, etc. "Application of Pagosa's Flip+MPM Simulation Capability to Fragmenting Metal Hemispherical Shells Driven by High Explosives". Presentation at 15th Hypervelocity Impact Symposium April 14-19, 2019 Destin, Florida, USA
[5] T.D. Aslam, "Shock temperature dependent rate law for plastic bonded explosives," J. Appl. Phys., 123, 145901 (2018); doi: 10.1063/1.5020172
[6] T.D. Aslam, "The reactants equation of state for the tri-amino-trinitro-benzene (TATB) based explosive PBX 9502," J. Appl. Phys., 122, 2017; doi: 10.1063/1.4989378
[7] G.R. Johnson and W.H. Cook., "Fracture Characteristics of Three Metals Subjectedto Various Strain, Strain Rates, Temperatures and Pressures", Engineering Fracture Mechanics, Vol. 21, No 1, pp. 31-48, 1985
[8] R. L. Gustavsen, etc "Measurements ofshock initiation in the tri-amino-tri-nitro-benzene based explosive PBX 9502: Wave forms from embedded gauges and comparison of four different material lots," J. Appl. Phys. 99(11), 114907 (2006)

Proceedings of the 2019 Hypervelocity Impact Symposium
HVIS2019
April 14-19, 2019, Destin, FL, USA
HVIS2019-097

Extended Abstract+

Characterization of the ballistic properties of ejecta from laser shock-loaded samples using high resolution picosecond imaging

Arnaud Sollier[a], Emilien Lescoute[a]

[a]*CEA, DAM, DIF, 91297 Arpajon,*

1. Introduction

Anticipating the generation of high velocity debris from shock-loaded specimens and the damage that their impacts may cause to nearby equipment is a major safety issue in applications involving shock waves, such as pyrotechnics [1] or inertial confinement fusion (ICF) experiments on large scale laser facilities [2]. Microjetting is one of the processes governing such debris generation. It is due to the interaction of a shock wave with a free surface presenting geometrical defects such as pits, cavities, scratches, or grooves, leading to material ejection from these defects, in the form of thin jets expanding ahead of the main surface and breaking up into small particles [3]. Over the last few years, we have used laser shock loading in order to expand microjetting investigations over ranges of small spatial scales (μm scale), extremely high loading rates (~ 10^7 s^{-1}) and very short pressure pulses (a few ns) [4-11]. Optical shadowgraphy and Photonic Doppler Velocimetry (PDV) have been used to measure both jet tip and planar surface velocities [4-6], while attempts to infer fragments size distributions, to be compared with model predictions, have been made using either fast transverse shadowgraphy [7] or ejecta recovery [8]. More recently, picosecond x-ray radiography has been used to provide estimates of the density gradients along the jets and of the total ejected mass at different times after shock breakout [9-11]. Here, we present the development of a new picosecond laser imaging diagnostic intended to overcome the limitations of our current transverse optical shadowgraphy setup. We describe our experimental setup and show the results of our first experiments performed using both visible (532 nm) and UV (355 nm) lightning of the sample. These results are compared to those obtained at LANL under high explosive loading using ultraviolet in-line Fraunhofer holography [12], and also to molecular dynamics (MD) simulations performed by our colleagues at lower space and time scales [15-18].

2. Experimental setup

Square samples of about 10 mm × 10 mm were cut from 100 μm-thick rolled foils of high purity (99.95 %) Sn purchased from GoodFellow. Their surface was first polished to mirror-like finish, and then a straight groove, about 40 μm deep, was imprinted into this smooth surface. For most of our experiments, we used samples with V-shaped grooves, produced using a steel blade that was polished into V-shaped edges with a controlled half-angle of 45°. For some experiments, we also used samples with U-shaped grooves produced by indentation of a 45 μm diameter tungsten wire into the target. For each sample, the spatial characteristics of the groove (width, depth and half angle) were quantified by optical surface metrology using both a Keyence VHX-5000 digital microscope and a Polytec TMS-1200 TopMap μ.Lab optical profilometer prior to the laser shot.

+ Full paper can found at https://www.sciencedirect.com/journal/international-journal-of-impact-engineering/special-issue/10MTGP5W4VJ

* Corresponding author. Tel.: +33-169264000; fax: +33-169267077.

E-mail address: arnaud.sollier@cea.fr.

The experiments were performed with the GCLT high power laser facility of CEA/DIF. The setup is schematically represented in Fig. 1. A high power laser pulse of λ=1053 nm wavelength, with a FWHM duration τ of 10 ns (square temporal profile), and an energy between 18 and 20 J, was focused on a 500 μm diameter spot opposite to the grooved surface. In order to insure a homogeneous illumination of the sample throughout the whole focal spot, we used a diffractive optical elements (DOE) homogenizer from Silios Technologies. The resulting peak power densities on the sample were in the range 0.9-1.3×10^{12} W/cm^2. The sample was set in primary vacuum (< 5×10^{-2} mbar) to avoid laser breakdown in air. At the back side of the sample, opposite to the irradiated spot, the grooved surface is monitored by picosecond laser imaging using a standard transverse optical shadowgraphy setup. This system provides quasi-instantaneous images of the debris ejection in vacuum at controlled delay times after the laser shot. Illumination is supplied by a custom SL234 EKSPLA ps laser system, optimized for single shot operation. This laser provides Gaussian pulses with a FWHM duration of 100±15 ps and a top hat spatial profile, with a maximum energy of 125 mJ at 532 nm and 80 mJ at 355 nm wavelength respectively. In order to break the speckle pattern of the laser beam, a 2 m length multimode fiber and several optical diffusers are used to smooth its spatial beam profile. A relay imaging system consisting of a Mitutoyo Infinity Corrected Long Working Distance Objective, with a 10× magnification, was used for both wavelengths. Indeed, this system allows reaching a better resolution than a standard afocal system, but at the expense of the depth of field which is reduced to less than 20 μm. The objective was placed closer from the target than the

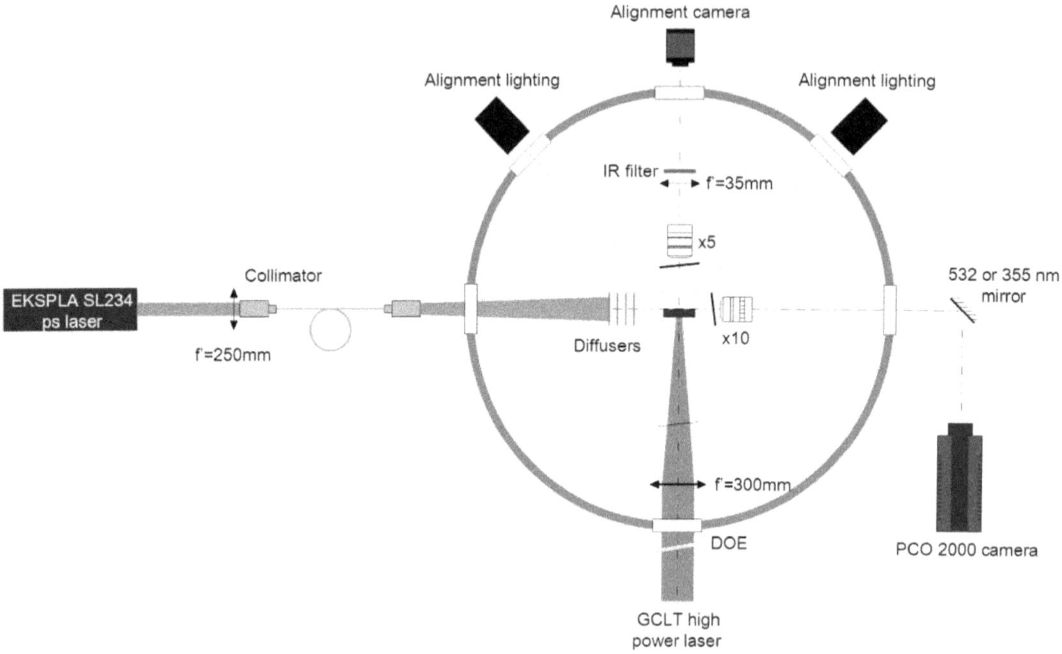

Fig. 1. Scheme of our experimental setup.

maximum working distance in order to reach a higher magnification. The resulting magnification (230 nm/pixel at 532 nm, and 260 nm/pixel at 355 nm) was measured using a 1951 USAF Resolution Test Targets placed at the location of the sheet of ejecta. A thin polycarbonate (PC) protecting shield was placed in front of the objective in order to protect it from the impact of debris. The images were recorded on a high resolution 14bit CCD cooled camera system (pco.2000 camera from PCO), consisting of an array of 2048×2048 pixels with a size of 7.4 μm×7.4 μm. In order to reduce the background noise on the recorded images, we used the shortest exposure time of the camera (500 ns). Moreover, a Rmax mirror was used in front of the camera in order to select only the light coming from the lighting laser.

3. Experimental results

Fig. 2 presents a typical sequence of six high-resolution shadowgraphs recorded behind a 100 μm-thick Sn sample subjected to a 25 GPa laser shock, showing the microsheet ejected from a triangular groove of 45° half-angle at successive times after the laser shot (100, 125, 150, 175, 200 and 225 ns). Each image originates from a distinct experiment. The laser drive is coming from the left, and the free surface is just to the left of the images. The V-shaped groove is oriented vertically so that we observe the whole microsheet from the side. To highlight the details of the breakup process of the sheet of ejecta, Fig. 3 presents crops of

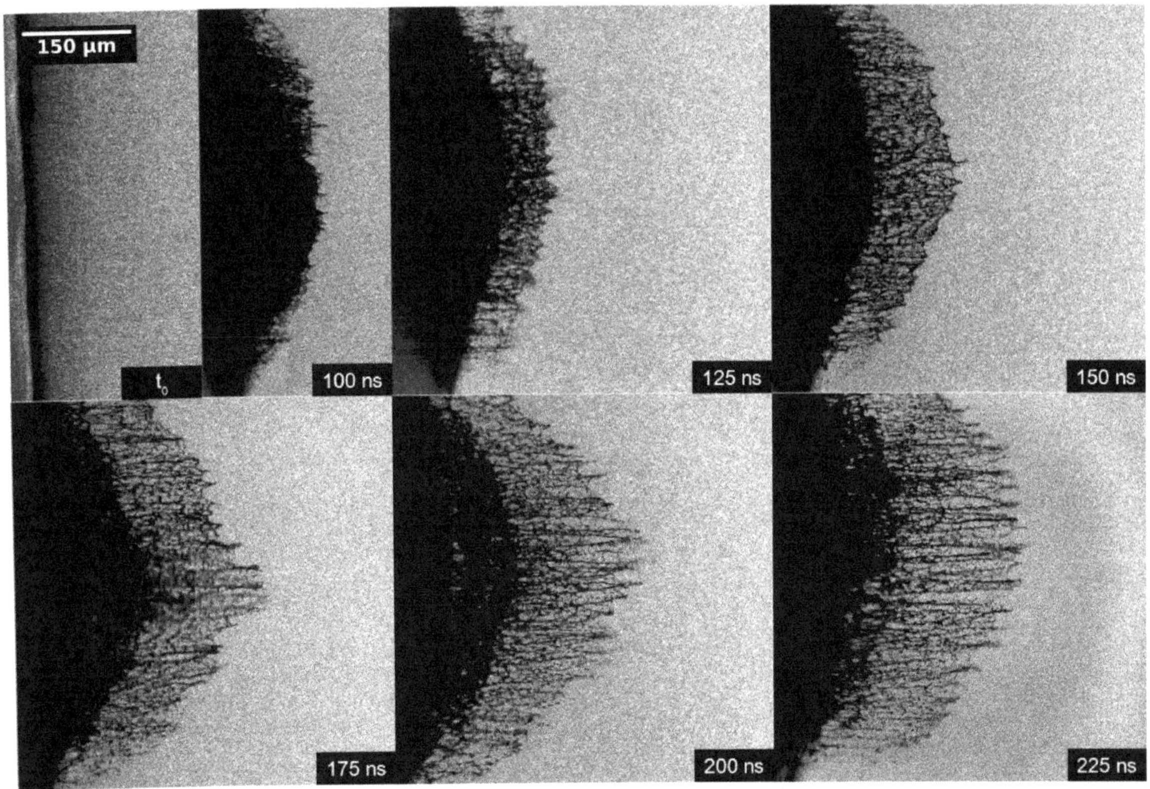

Fig. 2. Sequence of six high-resolution shadowgraphs recorded behind a 100 µm thick Sn targets, with a vertical triangular groove of 45° half-angle in its free surface. A 25 GPa laser shock is applied at time t_0 onto a 500 µm diameter spot on the front surface (to the left, not shown). Its reflection from the groove generates a microsheet of ejecta moving at high velocity from left to right. The six images originate from distinct experiments, and were recorded at different delays from t_0, using a 100 ps FWHM duration laser pulse at 355 nm wavelength. The free-surface is just to the left of the images, and the horizontal axis is the position above the static free-surface. The whole field of view of each image is 532.5×532.5 µm.

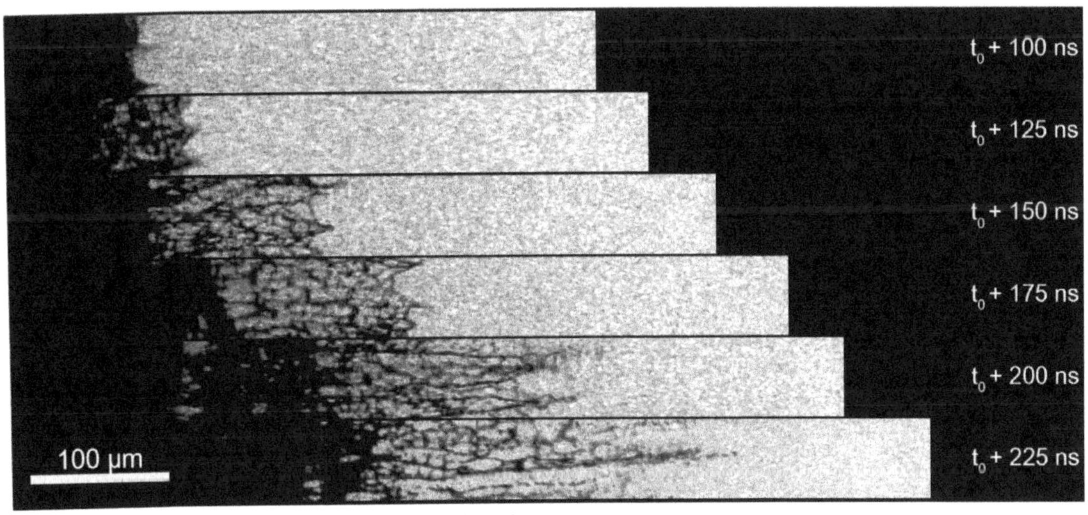

Fig. 3. Crops of the central part of the images of Fig. 2 displayed in false colors.

the central part of the images of Fig. 2 displayed in false colors. We can see on the first images, that the microsheet of ejecta consists of an array of quasi-parallel spikes, with transverse links between them. Near the free-surface, we can observe well-defined closed web-like structures that are still observed at longer delays. If we look further to the right on the two last images, we observe that these large web-like structures breakup into isolated tendrils, which in turn breakup into individual particles. These data seem to confirm the existence of three stages of breakup, which were first evidenced by Sorenson *et al.* at LANL [12-13], using ultraviolet in-line Fraunhofer holography. They are also consistent with the results derived from MD calculations and direct numerical simulations (DNS) of microsheet breakup as discussed in Ref. [14-17].

4. Conclusion

A high resolution picosecond laser imaging diagnostic has been developed for making high-resolution spatial measurements of ejecta particles moving at high velocities (a few km/s). Preliminary results obtained with both visible (532 nm) and UV (355nm) lighting were presented for laser shock-driven ejecta experiments performed on 100 μm thick Sn targets, with vertical grooves of different shape in their free surface. A comparison with ns laser imaging performed in similar conditions highlights the necessity of sub-nanosecond lighting if one wants to capture the movement of the fastest ejecta with typical velocities of a few $km.s^{-1}$ without any motion blur. The microsheet breakup process observed with both visible and ultraviolet lighting is in good agreement with the three stages breakup process first evidenced by Sorenson *et al.* at LANL [19-20], using ultraviolet in-line Fraunhofer holography, and also observed in MD simulations [21-24].

References

[1] Yeager, J. D., Bowden, P. R., Guildenbecher, D. R., Olles, J. D., 2017. Characterization of hypervelocity metal fragments for explosive initiation, Journal of Applied Physics 122, p. 035901.

[2] Tokheim, R. E., Curran, D. R., Seaman, L., Cooper, T., Anderson, A. T., Burnham, A. K., Scott, J. M., Latkowski, J., Schirmann, D., 1999. Hypervelocity shrapnel damage assessment in the NIF target chamber, International Journal of Impact Engineering 23, p. 933.

[3] Buttler, W. T., Williams, R. J. R., Najjar, F. M., 2017. Foreword to the Special Issue on Ejecta, Journal of Dynamic Behavior of Materials 3, p. 151.

[4] de Rességuier, T., Lescoute, E., Sollier, A., Prudhomme, G., Mercier, P., 2014. Microjetting from grooved surfaces in metallic samples subjected to laser driven shocks, Journal of Applied Physics 115, p. 043525.

[5] de Rességuier, T., Roland, C., Prudhomme, G., Lescoute, E., Loison, D., Mercier, P., 2016. Influence of edge conditions on material ejection from periodic grooves in laser shock loaded tin, Journal of Applied Physics 119, p. 185108.

[6] de Rességuier, T., Roland, C., Lescoute, E., Sollier, A., Loison, D., Berthe, L., Prudhomme, G., Mercier, P., 2017. Experimental study of microjetting from triangular grooves in laser shock-loaded samples, AIP Conference Proceeding 1793, p. 100025.

[7] Roland, C., de Rességuier, T., Sollier, A., Lescoute, E., Loison, D., Soulard, L., 2017. Ejection of Micron-Scale Fragments from Triangular Grooves in Laser Shock-Loaded Copper Samples, Journal of Dynamic Behavior of Materials 3, p. 156.

[8] Roland, C., de Rességuier, T., Sollier, A., Lescoute, E., Tandiang, D., Toulminet, M., Soulard, L., 2018. Ballistic properties of ejecta from a laser shock-loaded groove: SPH versus experiments, AIP Conference Proceeding 1979, p. 080012.

[9] de Rességuier, T., Prudhomme, G., Roland, C., Brambrink, E., Loison, D., Jodar, B., Lescoute, E., Sollier, A., 2018. Picosecond x-ray radiography of microjets expanding from laser shock-loaded grooves, Journal of Applied Physics 124, p. 065106.

[10] Prudhomme, G., Franzkowiak, J.-E., de Rességuier, T., Brambrink, E., Roland, C., Loison, D., Lescoute, E., Sollier, A., 2018. Ejecta from periodical grooves in tin foils under laser-driven shock loading, AIP Conference Proceeding 1979, p. 080010.

[11] de Rességuier, T., Roland, C., Prudhomme, G., Brambrink, E., Franzkowiak, J. E., Loison, D., Lescoute, E., Sollier, A., Berthe, L., 2018. Picosecond radiography combined with other techniques to investigate microjetting from laser shock-loaded grooves, AIP Conference Proceeding 1979, p. 080011.

[12] Sorenson, D. S., Pazuchanics, P. D., Johnson, R. P., Tunnell, T. W., Smalley, D. D., Malone, R. M., Kaufman, M. I., Marks, D. G., Capelle, G. A., Grover, M., Stevens, G. D., LaLone, B. M., Marshall, B. F., Turley, W. D., 2017. Ejecta particle size measurements from the break-up of micro-jets in vacuum and helium gas using ultraviolet in-line Fraunhofer holography, AIP Conference Proceeding 1793, p. 100026.

[13] Sorenson, D. S., Capelle, G. A., Grover, M., Johnson, R. P., Kaufman, M. I., LaLone, B. M., Malone, R. M., Marshall, B. F., R. W. Minich, R. W., Pazuchanics, P. D., Smalley, D. D., Stevens, G. D., Tunnell, T. W., Turley, W. D., 2017. Measurements of Sn Ejecta Particle-Size Distributions Using Ultraviolet In-line Fraunhofer Holography, Journal of Dynamic Behavior of Materials 3, p.233.

[14] Durand, O., Soulard, L., 2012. Large-scale molecular dynamics study of jet breakup and ejecta production from shock-loaded copper with a hybrid method, Journal of Applied Physics 111, p. 044901.

[15] Durand, O., Soulard, L., 2013. Power law and exponential ejecta size distributions from the dynamic fragmentation of shock-loaded Cu and Sn metals under melt conditions, Journal of Applied Physics 114, p. 194902.

[16] Durand, O., Soulard, L., 2015. Mass-velocity and size-velocity distributions of ejecta cloud from shock-loaded tin surface using atomistic simulations, Journal of Applied Physics 117, p. 165903.

[17] Durand, O., Jaouen, S., Soulard, L., Heuzé, O., Colombet, L., 2017. Comparative simulations of microjetting using atomistic and continuous approaches in the presence of viscosity and surface tension, Journal of Applied Physics 122, p. 135107.

Proceedings of the 2019 Hypervelocity Impact Symposium
HVIS2019
April 14-19, 2019, Destin, FL USA
HVIS2019-120

Extended Abstract+

Conical impact fragmentation test (CIFT)

Christopher Neel[a*], Peter Sable[b], Philip Flater[a], David Lacina[c]

[a] AFRL/RWMW, Eglin AFB 32542, FL USA
[b] Marquette University, Milwaukee, WI 53233 USA
[c] UDRI, Eglin AFB, FL 32542 USA

Abstract

Conical impact fragmentation tests (CIFT) were conducted to develop and demonstrate an experimental capability for uniformly inducing controlled fragmentation in structural metals. The setup involves a conical specimen impacting a mating conical target (similar in geometry to a funnel) at nominal velocities of 1 - 2 km/s. Three experiments were conducted as proof of concept to characterize the fragmentation behavior of 1018 steel. Photonic Doppler velocimetry probes on the free outer surface of the target cone allow for validating simulations that can indicate strain uniformity in the target cone. Overall, experimental results demonstrate CIFT to be a feasible method to evaluate fragmentation behavior. The conical geometry produces consistent and bounded strain rates that are maintained for at least 10 microseconds. Furthermore, when compared with other laboratory techniques, the CIFT technique is shown to be more ideal than sphere-on-plate impact (SPI) and more tunable than cylinder expansion (CYLEX), and so is a promising fragmentation characterization tool.

Keywords: fragmentation; CIFT; conical; fracture; CYLEX

+ Full paper can found at https://www.sciencedirect.com/journal/international-journal-of-impact-engineering/special-issue/10MTGP5W4VJ

* Corresponding author. Tel.: +1-850-882-7992.
E-mail address: christopher.neel.1@us.af.mil.

Proceedings of the 2019 Hypervelocity Impact Symposium
HVIS2019
April 14-19, 2019, Destin, FL, USA
HVIS2019-016

Extended Abstract+

Electrically-Launched mm-sized Hypervelocity Projectiles

W. Casey Uhlig*, Paul R. Berning, Peter T. Bartkowski, Matthew J. Coppinger

US Army Research Laboratory, 328 Hopkins Rd, Aberdeen Proving Ground, MD 21005

1. Introduction and Experimental Set-Up

A series of experiments and simulations were performed as proof of concept that an electrically powered research gun could propel small cylindrical projectiles to hypervelocities. Although small-caliber electrothermal accelerators and other electromagnetic launch systems have been utilized for some years for laboratory hypervelocity impact and other studies [1-5], we developed a simple, reproducible device that allows impact studies and direct comparison to magnetohydrodynamic (MHD) simulations for design considerations, efficiency improvements, and validation studies. This work focusses on 4.8 mm cylindrical 7075-T6 aluminum projectiles with a length to diameter ratio of one (nominally a mass of 0.24 grams) and a 150-mm long, 25-mm outer-diameter 4043 steel barrel with a 4.8-mm diameter bore and 9.5-mm chamber that acts as the electrical cathode. The anode consists of a 6.3-mm diameter copper rod that is reduced to 2.9 mm then tapered to a point with the tip length over diameter ratio (L/D) varying from 2 to 5. The tip is placed at the chamber/bore junction. The copper anode is insulated by a polyethylene sleeve and epoxy surrounding the electrode such that the arc initiates only at the very tip. A 191 μF capacitor was used as the power source for all experiments. The applied voltage was varied from 10 kV to 20 kV, and the resulting inductance of the system varied from approximately 320 nH to 450 nH (due to varying separation in the copper transmission lines). Fits to the current pulse using an LRC circuit resulted in resistance on the order of 10 mΩ to 15 mΩ. Typically the portion of the electrical pulse responsible for the bulk of the acceleration of the projectile occurs within the first 15 μs; however, the projectile is accelerated during the entirety of the time it remains in the barrel, which is on the order of 40 μs to 50 μs.

Projectile velocity was measured using several different techniques during the test series. The first technique used three make screens consisting of 0.025 mm aluminum foil along with photodiode light detectors that detected the flash from the impact of the aluminum projectile with the make screen. The second technique used high speed video to film the projectile flight. The third technique used a pair of magnetic pickup sensors to detect the passage of the projectile in time[6-7]. The timing between measurements along with the distances measured allowed the determination of an average projectile velocity for each of the first three techniques. The fourth technique used Photon Doppler Velocimetry (PDV) to record the projectile velocity[8]. The PDV system uses a laser to track the velocity of a surface. The Doppler shift imparted to the reflected laser beam is recorded and then analyzed to give a projectile velocity with time. One significant advantage of the PDV system is that the laser is capable of looking down the barrel of the gun, thus providing a complete velocity history of the projectile during acceleration in the barrel and free flight out of the barrel. In addition, electrical current in these experiments was measured using a calibrated Rogowski probe, and the voltage difference was measured using two 40 kV high voltage probes. Using the measured voltage and current, the energy delivered to the test stand was calculated.

Three-dimensional simulations of the electrothermal device were performed using the shock physics code ALEGRA with MHD capabilities [9] developed by Sandia National Laboratories. A cylindrical trisection mesh with ¼ symmetry and cell edges less than 0.1 mm in the region of the electrode tip was utilized. A material model for Lexan was used in place of the epoxy and polyethylene tube surrounding the electrode, and a simple elastic-plastic model in which the yield strength was varied was used for the steel barrel. An external circuit was applied to the mesh which included a capacitor, resistor, and inductor with electrical

+ Full paper can found at https://www.sciencedirect.com/journal/international-journal-of-impact-engineering/special-issue/10MTGP5W4VJ

* Corresponding author. Tel.: +1-410-278-3997
E-mail address: willard.c.uhlig.civ@mail.mil.

feeds going to the Cu electrode (hot lead) and the end of the barrel (zero potential). The capacitance was set to 191 μF as measured in the experimental apparatus, while an inductance of 340 nH and a resistance of 10.4 mΩ were selected as the other two external circuit parameters. The parameters were chosen to produce current output in the middle of the experimentally measured curves, which varied from shot to shot. The current generated by the simulation at 2.0 μs is shown in Fig. 1, just prior to initial tip vaporization. Electrical current initially flows from the Cu electrode through the projectile to the barrel.

The color map of the current density seen in Fig. 1 is for an L/D = 4 electrode. Copper vaporization and/or wire explosion typically requires current densities between 10^{10} A/m^2 and 10^{11} A/m^2. The current is initially isolated to the conductive surfaces

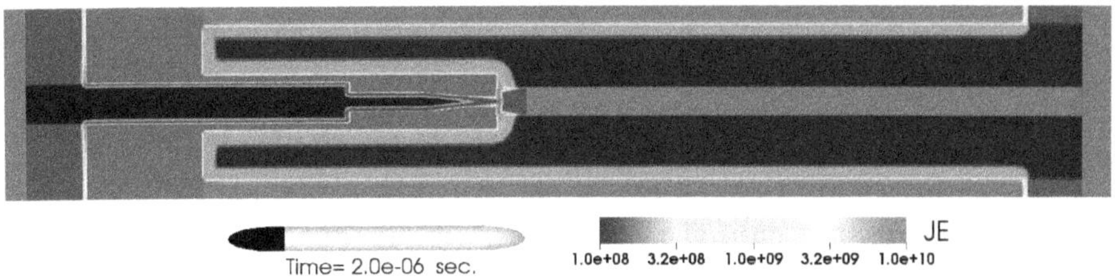

Fig. 1. ALEGRA simulation of the electrical current density at 2.0 μs. The color map is shown on a log scale overlaying the density profile of the gun.

flowing along the Cu electrode to the projectile and out along the surface of the barrel, but gradually soaks into the material. As the Cu electrode vaporizes (as well as a small crater at the back of the pellet), a plasma surrounding the tip is formed, and pressures over 5 GPa develop within the chamber, accelerating the projectile. Insulating material surrounding the tip is compressed and flows outward creating a cavity while some of the material is eroded and mixed into the plasma. Some current trails along with the projectile down the barrel during its early acceleration, but after separation from the initial contact with the anode, the current density decreases and is not sufficient to cause further damage or erosion to the pellet. Also note, there is a 1 mesh cell layer of air between the pellet and the sides of the barrel bore to allow projectile motion.

2. Experimental Results and Comparison to Simulation

Velocities in this test series ranged from 1740 m/s to 3410 m/s. Table 1 summarizes the experimental results. The column labeled Charge Voltage is the voltage to which the capacitor bank was charged before firing the gun, and Capacitor Energy is the calculated initial stored energy in the capacitor bank. Fixture Energy is the final energy delivered to the gun fixture, some of which is transmitted to the fixture after the projectile leaves the barrel. The velocity of the projectiles as measured using the 4 different techniques is also listed. In each experiment, two velocity measuring techniques were employed. In general, the high speed video velocity is higher than the make light velocity. This is due to the high speed video measurement being made closer to the muzzle than the make light screens. The velocity of the projectile decreases as it flies down range due to aerodynamic drag.

Table 1. Experimental data

Shot #	Tip Geometry	Charge Voltage (kV)	Capacitor Energy (kJ)	Fixture Energy (kJ)	Velocity (m/s) Make light	Video	Magnetic Pickup	PDV	Total Efficiency %
1	L/D=4	10	9.5	6.98	1703	1740	NA	NA	3.8
2	L/D=4	15	21.4	15.2	2519	2540	NA	NA	3.6
3	L/D=4	20	38.0	21.8	2953	2970	NA	NA	2.8
4	L/D=4	20	38.0	23.9	2971	2940	NA	NA	2.7
5	L/D=4	20	38.0	17.9	NA	NA	2920	2978	2.8
6	L/D=5	20	38.0	21.5	NA	NA	2865	2950	2.7
7	L/D=3	20	38.0	23.7	NA	NA	2985	3070	3.0
8	L/D=2	20	38.0	18.3	NA	NA	2500	2590	2.1
9	L/D=4	20	38.0	26.1	NA	NA	3305	3410	3.7

The same effect can be seen between the PDV projectile velocity and the magnetic pickup velocity. The PDV measurement yields the peak velocity of the projectile as it exits the muzzle, while the magnetic pickup sensors are located downrange and thus record a lower velocity. Total efficiency is the percent of stored capacitive energy imparted into kinetic energy of the pellet using the maximum recorded velocity.

For a given L/D, the velocity increased, but the efficiency of the gun decreased as the voltage on the capacitor bank increased. This efficiency was restored with higher velocity in shot 9 by modifications to the gun discussed below, while the fixture energy followed a similar trend, but varied significantly. The variation in fixture energy resulted from at least two sources: changes in the inductance due to transmission line shot to shot variation and the response of the measured current at late times. The late time current has no effect on projectile velocity and accounts for the largest fixture energy variation. Therefore, only total efficiency is tracked and compared in the table. For tips of L/D 3 to 5, no significant final velocity difference is noted. The efficiency and velocity drops significantly as the tip geometry is dropped to L/D of 2. This is simply due to a reduction in the quantity of working fluid available. The 2.9 mm diameter region of the Cu electrode is just on the cusp of vaporization and does not contribute to the high pressures produced for propulsion, which occurs in very early times. Thus, insufficient vaporized Cu is available for low L/D geometries, while a plateau region is achieved in the 3 to 5 range.

The simulation that generated Fig. 1 results in a projectile velocity of 3020 m/s. Fig. 2a shows the experimental electrical current curves of shot 5 and 7 in comparison to the current produced in the simulations, which were nearly identical for different tip geometries; L/D = 4 is shown in the figure. The simulations attain a peak current of 350 kA, which is in the approximate median range of the experimental peak currents. An LCR fit of the two experimental curves gives values of 340 nH and 14 mΩ, and 450 nH and 12.5 mΩ for shots 5 and 7, respectively. While the resistance chosen for the simulation is slightly lower than experiment, it is important to note that the experimental fit is the total resistance and the simulation circuit resistance is just external to the gun. While no attempt was made to identically fit the simulation current to each experiment, the simulation currents fit well within the experimental range. The resulting simulated velocity is directly compared in figure 2b to the PDV data captured in shot 5. The simulated velocity plot was produced by a Lagrangian tracer at the front center of the projectile. The correlation is actually exceptional both in final velocity and the acceleration profile.

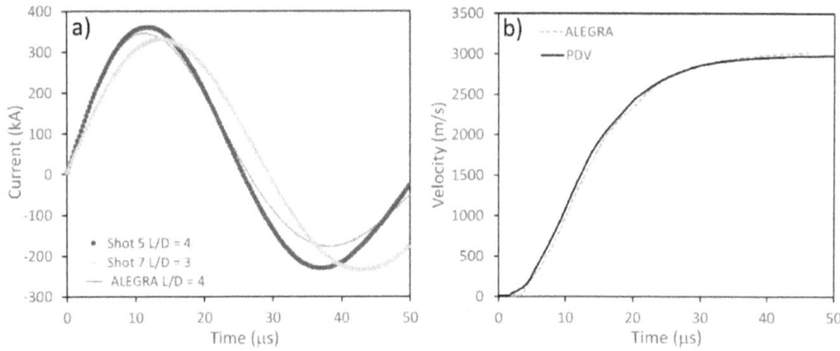

Fig. 2. ALEGRA simulations of the current (a) and resulting velocity (b) directly compared to the data from shot 5 and 7.

The barrel yield strength in the simulations seen above was set to 695 MPa which corresponds to the measured Brinnell hardness of 207 at the center of the barrel material as purchased. The barrel expansion produced in a simulation with a barrel strength of 695 MPa closely matched the expansion of the sectioned barrel from shot 3. The simulations, however, showed differences in the expansion of the barrel chamber with barrel yield strength. A simulation was run with a barrel yield strength of 1200 MPa, and the strengthened steel barrel resulted in a higher projectile velocity with minimal barrel deformation. Thus, the barrel used in shot 9 was heat treated for hardening after machining. A final reaming of the bore to 3/16 inch precision (4.8 mm) was then performed. A measurement of the hardness on the outer section of the barrel resulted in a Brinnell hardness of 387 corresponding to a yield strength of 1276 MPa. Simulated projectiles reached a velocity of 3575 m/s before exiting the 1200 MPa barrel; 4.6 % higher than shot 9 velocity of 3410 m/s. While the simulations predicted an 18% increase in velocity, the experiment yielded 14%.

The ALEGRA simulations proved to be a very useful design tool in furthering the performance of the gun. Additional work was performed on simulating the pellet / bore interface because excessive erosion of the pellet was observed in the simulations. Because the back of the pellet expands to the edge of the bore during acceleration and Eulerian simulations "weld" interfaces, excessive friction occurs along the pellet sides. Since the aluminum pellet is much softer than the barrel, it yields and the pellet accelerates smoothly. In order to experimentally quantify the amount of pellet mass lost more precisely, an aluminum pellet launched at 3100 m/s was soft captured in 2.0 meters of shaving cream with a density of 0.05 g/cm^3. Total travel distance in the 100 mm x 100 mm x 2 m block of shaving cream was approximately 1.92 m. Fig 3 shows an image of an unlaunched pellet alongside the captured pellet. The captured pellet was reduced in length from 4.8 mm to 4.5 mm and has a 1.4 mm deep crater at the back of the pellet. The back half of the pellet also had a slightly reduced diameter (less than 0.1 mm), while the top of the pellet still had machining marks visible from constructing the pellet. The mass decreased from 0.24 g to 0.19 g.

Fig. 3. Images of aluminum projectiles prior to launch and after launch (via soft capture) compared to a 3D isovolume plot of an ALEGRA simulation and 2D cross-section of the simulated pellet.

A feature of ALEGRA known as block suppression was applied to provide a friction free surface between the pellet and the mesh boundary. The mesh was split into two regions. One region from 7.5 mm beyond the electrode/pellet interface (i.e. the first 7.5 mm of the bore) to the back of the electrode and hot plate was left unchanged from the previous simulations, while the mesh in the second region (beyond the 7.5 mm of barrel bore) was reduced to the diameter of the bore. Thus, no steel would be present in the second region and the pellet would be free to accelerate without friction at the mesh interface. The yield strength and Young's modulus were then varied in an elastic-plastic model to produce pellet deformation that matched the experimentally captured pellet. A 3D isovolume of the simulated pellet and 2D slice are also shown in Fig 3 for comparison to the actual pellets. The simulated pellets shown have a yield strength and Young's modulus double that of the standard 7075-T6 aluminum used in the experiment, but the crater depth and the slight erosion of the back surface of the pellet match well. In addition, the velocities attained for the modified simulations are lower than the experimental results, 2800 m/s instead of 2970 m/s, a 5.7% difference.

3. Conclusion

A laboratory electrothermal gun has been demonstrated and notably improved using hydrocode simulations to guide experiment. Insights into barrel expansion and material strength determined from the simulations were highly beneficial. While acknowledging weaknesses in the simulations concerning erroneous erosion of the projectile inherent to Eulerian meshes, some potential avenues to address the issue were outlined and overall the simulations had a significant impact on the advancement of this work.

Acknowledgements

The authors would like to thank Aaron Bard for experimental insight, Mike Zellner for PDV guidance, and James Cazamias and John Niederhaus for simulation insights. This work was supported in part by a grant of computer time from the Department of Defense High Performance Computing Modernization Program at the United States Army Research Laboratory Department of Defense Supercomputing Resource Center and the Terminal Ballistics Frontier project.

References

[1] M. Rott, 1993, The LRT/TUM small caliber electrothermal accelerator, IEEE Transactions on Magnetics, vol. 29, no. 1
[2] H. Yoshida, Y. Hoshi, K. Uematsu, and Y. Kitazawa, 1997, A single, small particle launch system by electrothermal gun and microsabot, Rev Sci Inst **68**, 178
[3] Fair, H. D., 2007, Progress in Electromagnetic Launch Science and Technology, IEEE Transactions on Magnetics, vol. 43, no. 1
[4] Lee, T., Taber, G., Vivek, A., and Daehn, G. S., 2018, "Characterization of High-Speed Flyer Evolution by Multi-Probe Photon Doppler Velocimetry," 8th International Conference on High Speed Forming – 2018
[5] Pastore, R, Podlesak, T., Singh, H., 1993, Progress in 2N Megajoule Electrothermal-Chemical Gun Pulser. Proceedings of the 9th IEEE International Pulsed Power Conference, pp. 783-786. G. L. Katulka, M. DelGuercio, R. Pastore, D. Singh and R. J. Burdalski, 1995, A 4-MJ mobile pulse power facility for electrothermal-chemical gun research, Tenth IEEE International Pulsed Power Conference, Albuquerque, NM, USA, 1995, pp. 267-272 vol.1.
[6] Uhlig W.C., Heine A., 2015, Electromagnetic diagnostic techniques for hypervelocity projectile detection, velocity measurement, and size characterization: theoretical concept and first experimental test. Journal of Applied Physics. 2015;118:184901

[7] Uhlig W.C., Hummer C.R., 2013, In-flight conductivity and temperature measurements of hypervelocity projectiles. Procedia Engineering. 2013;58:48–57.
[8] Strand, O. T., Goosman, D. R., Martinez, C., Whitworth, T. L., and Kuhlow, W. W., 2006, Compact System for High-Speed Velocimetry Using Heterodyne Techniques., Review of Scientific Instruments 77 (8): 1-9
[9] Robinson AC, et al., 2008, ALEGRA: an arbitrary Lagrangian-Eulerian multimaterial, multiphysics code. Proceedings of the 46th AIAA Aerospace Sciences Meeting; Paper AIAA-2008-1235.

Proceedings of the 2019 Hypervelocity Impact Symposium
HVIS2019
April 14-19, 2019, Destin, FL, USA
HVIS2019-039

HyFIRE: Hypervelocity Facility for Impact Research Experiments at Johns Hopkins University

Gary Simpson[a*], Matthew Shaeffer[a], K.T. Ramesh[a]

[a]Hopkins Extreme Materials Institute, Johns Hopkins University, 3400 N. Charles St., Malone 140, Baltimore, MD, 21218, USA

Abstract

The Hopkins Extreme Materials Institute (HEMI) recently installed a hypervelocity impact facility (HyFIRE) including a two-stage light gas gun at Johns Hopkins University in Baltimore, MD. The HyFIRE launcher has a launch tube bore diameter of 7.62 mm and is designed to attain launch velocities up to 7 km/s. The enclosed ballistic range and terminal test chamber provide multiple axes with which to view both projectile free flight and terminal impact, maximizing diagnostic access to events of interest.
Initial test diagnostics include ultra-high-speed optical video and orthogonal 300 kV flash x-ray imaging. Photon doppler velocimetry for surface velocity measurement—currently used in HEMI's laser shock facility—as well as emission spectroscopy/pyrometry are planned, providing researchers across multiple disciplines with the ability to investigate the coupling of mechanics, physics and chemistry present in high energy density impact events. Initial experiments at the facility investigate the fragmentation of inert impactors on anvil targets, with an aim towards identifying the dominant mechanisms controlling the fragmentation characteristics, temperature distributions and trajectories of generated debris fields.

Keywords: Hypervelocity launchers, fragmentation, high speed diagnostics, dynamic behavior of materials

1. Introduction

The Hopkins Extreme Materials Institute (HEMI) at Johns Hopkins University (JHU) was founded to investigate material behavior in extreme environments, with impact phenomena and associated material response comprising a significant focus of the research. Recent work at the institute uses a combination of experiments and modeling to provide significant insight into the complex processes of impact on advanced armor ceramics, magnesium alloys and geologic materials [1-5]. As threat velocities increase and space impact problems become more prevalent, being able to accurately observe hypervelocity impact experiments has become crucial to informing computational models and developing the understanding of these impacts for relevant applications. The institute therefore decided to build a hypervelocity impact facility where academic research on these problems can take place and advanced diagnostics can be utilized and developed.

The Hypervelocity Facility for Impact Research (HyFIRE) was recently established on the JHU Homewood Campus. The two stage light gas gun and associated diagnostics expands the capabilities of HEMI researchers, adding hypervelocity impact experiments to existing plate impact, Kolsky bar and laser shock capabilities. Current goals are to implement a suite of diagnostics

* Gary Simpson. Tel: 1-480-266-4610
E-mail address:gsimps10@jhu.edu

including high speed imaging, flash X-ray imaging, spectroscopy and velocimetry techniques. The first application of the facility will be to investigate the impact fragmentation of materials.

2. Launcher and Range Facility

2.1 Two Stage Light Gas Gun Launcher

The two-stage light gas gun at the HyFIRE facility was fabricated by Physics Applications Inc. (PAI), and installed in December 2018. The HyFIRE launcher (Fig. 1) has a 1.5m long launch tube with a bore diameter of 7.62 mm to facilitate the launch of a variety of projectile shapes and configurations at laboratory scale. A 2.3 m long pump tube with a bore diameter of 30 mm is coupled to the launch tube by a hydraulically clamped central breech. The system utilizes a 10 ksi compressed gas breech to drive the first stage, which is brought to pressure by means of an electrically driven booster pump. A powder driven first stage was considered, but a compressed gas breech—hereafter referred to as the primary breech—was selected to maximize consistency, reduce fouling by combustion products, and eliminate powder handling and storage concerns in our student operated facility.

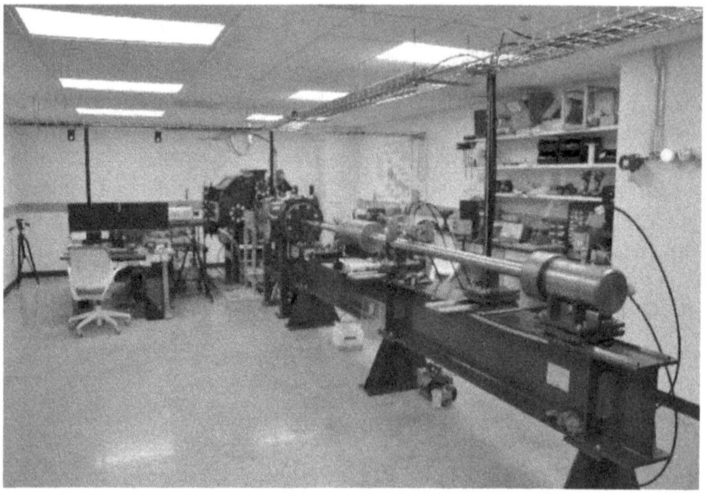

Fig. 1. HyFIRE facility at Johns Hopkins University

The primary breech is a wraparound design, using a two-piece piston assembly consisting of a reusable steel piston-weight intended to seal the breech prior to launch and a consumable polyethylene (HDPE) piston, which is deformed and lodged in the central breech during the shot. The primary breech uses N_2 or He drive gas depending upon desired launch velocity, and while this capability has not been implemented on the launcher, affords the opportunity for drive gas recapture from the breech/pump tube section post shot. The use of a compressed gas system also simplifies the adaptation of the enclosed range and reconfiguration of the launch system to a larger bore single stage gun, a future modification that with the addition of near-muzzle impact chamber will expand the capabilities of the facility to include more traditional shock experiments and experimental configurations. Maximum expected launch velocity for the system is approximately 7 km/s (Fig. 2) with launch masses below 1g, but the system will typically launch saboted projectiles at 1-5 km/s.

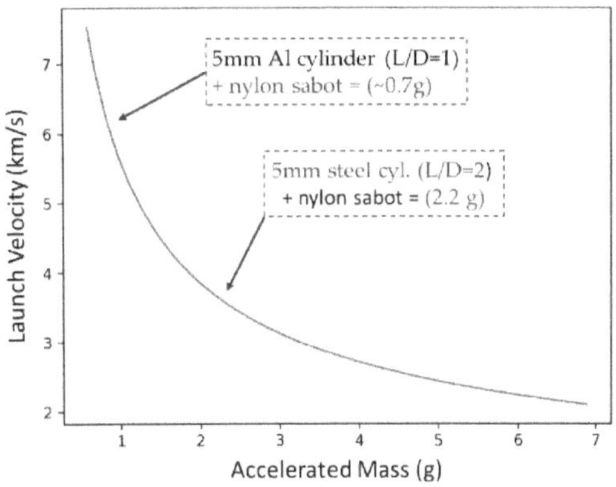

Fig. 2. Maximum expected performance curve, maximum achievable launch velocity vs. accelerated mass [6]. Marked on the curve are two representative launch packages which will be used in future experiments

Maximum expected launch velocities have not yet been achieved during initial testing. As of writing, only 8 test shots have been performed. A description of these test conditions and achieved launch velocities are given in Table 1. Additionally, the launch efficiency, calculated as the ratio of projectile kinetic energy to maximum available breech work assuming adiabatic expansion, is given. Poor launch efficiencies and velocity performance were observed with early test shots. We suspect that poor dimensional consistency with initial projectile packages may be partly to blame for the observed inconsistency. Additional shots under more controlled conditions are needed to fully vet the system.

Table 1. Initial Test Shots: Before (Top) and After (Bottom) Breech Redesign

Target/Projectile	Accelerated Mass (mg)	Breech Pressure (psi) / Gas	Pump Tube Pressure. (psi) / Gas	Velocity (km/s)	Launch Efficiency (%)
1/8" Al Plate / Polycarbonate Slug	375	4,000 / N_2	160 / He	1.8	2.1
1/8" Al Plate / Polycarbonate Slug	375	8,000 / N_2	160 / He	2.6	2.1
1/8" Al Plate / Polycarbonate Slug	375	6,000 / He	160 / He	2.8	3.5
3/8" AR500 Plate / Saboted 1/8" Al Sphere	500	6,000 / He	160 / He	3.6	7.8
3/8" AR500 Plate / Saboted 1/8" Al Sphere	500	10,000 / He	160 / He	4.2	6.3
(Breech/pump tube redesigned prior to last 3 shots, see section 2.3 for details)					
1/2" AR500 Plate / Polycarbonate Slug	375	3,000 / N_2	160 / He	2.5	5.3
2" Acrylic Cube / Polycarbonate Slug	375	3,000 / He	140 / H_2	3.3	9.8
1/2" AR500 Plate / Polycarbonate Slug	375	9,200 / He	140 / H_2	5.5	8.1

* The volume of the breech and pump tube are 2.5L and 1.5L respectively

2.2 *Enclosed Range and Test Chamber*

The available lab space restricts the maximum combined length of the range and launcher (including future modifications) to approximately 27'. To aid in sabot separation over the limited free flight distance available, an alternate rifled barrel providing centrifugal separation is available for test configurations for which aerodynamic separation is not achievable or smoothbore launch is impractical. The enclosed ballistic range (Fig. 3a) consists of a blast tank, drift tube, and terminal test chamber. While absent in the current configuration, accommodations have been made for the future installation of the aforementioned near muzzle test chamber, including additional uprange access in the blast tank and enlarged seal plate. The sabot stripper plate is located at the downrange wall of the cylindrical blast tank.

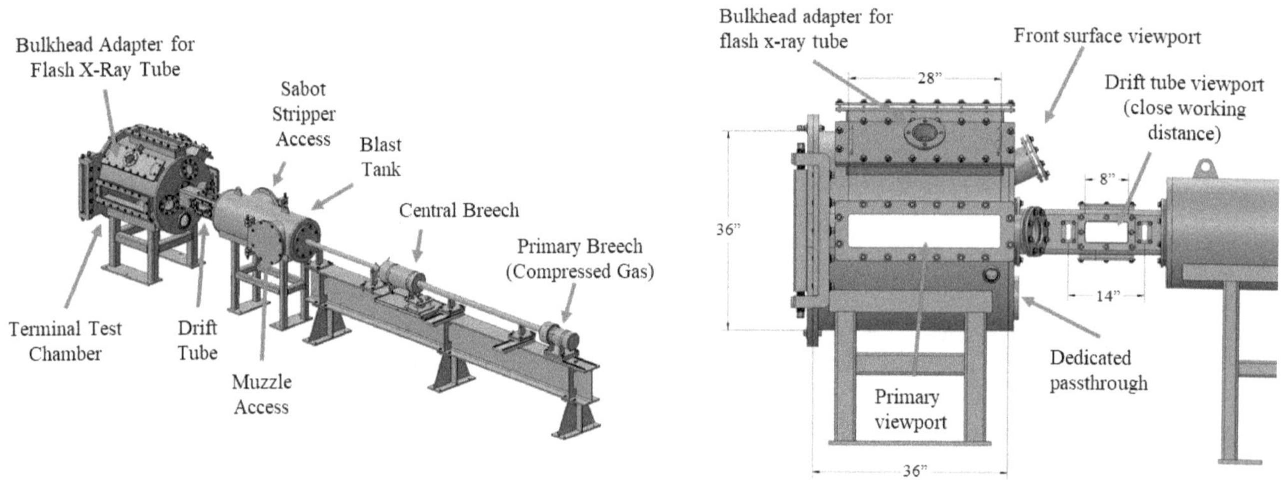

(a) HyFIRE Launcher (b) Range Test Section

Fig. 3. (a) HyFIRE launcher and current range configuration, (b) range test section including terminal test chamber and drift tube

The drift tube and terminal test chamber have been designed with the aim of maximizing simultaneous diagnostic access with emphasis on imaging. The drift tube, in which projectile free flight velocity is measured, contains four viewports arranged for orthogonal imaging of the projectile in flight as well as slit windows for the use of a laser break beam intervalometer and trigger. The terminal test chamber, seen in Fig. 3b, is fitted with five rectangular primary viewports along the cylindrical surface of the chamber. These large viewports also serve as removable bulkheads in which a variety of measurement equipment, imaging windows, or chamber accessories can be installed. A dedicated passthrough flange and three round 5" diameter viewports in the front surface of the test chamber have been installed to facilitate front surface viewing of a target assembly. In all cases, simple sealing geometries or off-the-shelf standardized flanges have been utilized to maximize the ease with which a researcher can fit diagnostic equipment and sensors to the test chamber. The high accessibility of the range, however, comes at the price of poor vacuum sealing. The current configuration of the system can provide range pressures down to approximately 2-3 Torr. Planned modifications are expected to reduce range atmospheres to at least 100 mTorr.

2.3 Compressed Gas Breech Redesign

For the benefit of current operators or future builders, a design issue associated with the compressed gas breech is discussed here. The wrap-around style primary breech consists of cup and insert construction. The downrange face of the breech assembly, breech sealing elements and gas channels for both drive gas and firing mechanism gas are contained in the insert, which is placed into the breech cup (Fig. 4). The breech storage reservoir is sealed when a piston weight is inserted into the wraparound cavity. The clamp force between the uprange face of the pump tube and downrange face of the breech insert maintains the joint between breech and pump tube, as well as seats and contains the breech insert inside the breech cup. An o-ring between the clamped faces of the breech insert and the pump tube seals the breech/pump tube joint.

A compressed gas breech is more conservative than a powder breech in that a compressed gas breech is loaded quasi-statically and unloaded dynamically, providing some measure of design safety. However, the breech/pump tube joint is loaded dynamically in the wraparound design and sees full application of breech pressure only after the initiation of piston movement when the breech is "uncorked". The initial design of the breech/pump tube joint distributed the clamp force between breech and pump tube over a relatively narrow boss. During initial high pressure shots (8000-10,000 psi initial breech pressure), it was observed that slight "mushrooming" deformation of this boss caused difficult separation of the breech and pump tube during breakdown.

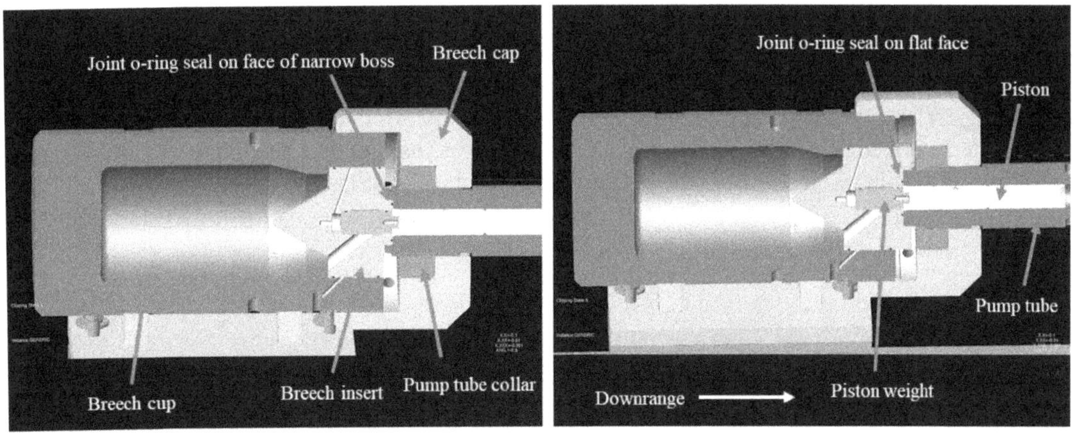

(a) Original Design (b) Revised Design

Fig. 4. Primary compressed gas breech cutaway with (a) initial design and (b) revised design. The extension of the revised breech insert allowed the elimination of the relativity narrow ring shaped boss over which the joint clamp force was exerted [6].

In consultation with PAI, it was suspected that insufficient clamping force at the joint was allowing some separation during the dynamic loading experienced during the shot, and the following springback and compression "hammering" on the boss were likely responsible for the observed deformation. Swift discusses this action and the associated forces [7]. The insufficient clamping force was likely due to a loose collar on the pump tube, against which the exterior breech cap is tightened to clamp the joint. A missing set screw in concert with the effects of a machining error in which synchronous threading was used on both the collar threads and the exterior breech cap, allowed slight synchronous rotation of the loose collar with the breech cap during joint assembly. To correct these issues, the breech/pump tube joint was redesigned (see Fig. 4b) and assembly procedure altered to ensure proper joint compression during shot preparation and gun assembly.

The pump tube was altered and a new breech insert was manufactured, distributing the clamp force and any resultant impulse over larger mating surfaces than the narrow boss present in the original design. High pressure test shots performed after installation of the redesigned components have revealed no further issues. It is recommended that breech or joint designs utilizing coaxial thread-tightening fixtures should take advantage of asynchronous threading in addition to proper anti-rotation set screws, and that operators should pay close attention to the positioning and potential rotation of load bearing elements to ensure proper clamp forces are achieved during gun assembly.

3. Facility Diagnostics and Initial Tests

3.1 Current and Planned Facility Diagnostics

The incident velocity of the projectile is measured in the drift tube (Fig. 3b), prior to entering the terminal test chamber, by a break-beam laser intervalometer. The terminal test chamber was designed to facilitate the use of multiple simultaneous imaging and diagnostic systems in order to capture the maximum amount of useful data available from a single impact event. Initial facility diagnostics will feature optical ultra-high-speed imaging—with Kirana and Shimadzu cameras—and 2-channel orthogonal 300 kV flash x-ray imaging. Up to four flash x-ray tubes can be utilized in a single experiment, providing two time-separated pairs of images if desired. The combination of optical ultra-high-speed video and flash x-ray absorption contrast imaging will provide both high time and spatial resolution imaging for the analysis of impact events and the evolution of generated fragmentation patterns, debris and ejecta clouds. In order to help measure and characterize the chemistry of energetic phenomena associated with kinetic impacts, such as impact generated plasmas or emitting products [8], the facility utilizes a Princeton Instruments Isoplane SCT320 imaging spectrograph in conjunction with a PiMAX4 1024i ICCD camera, which can acquire two images or spectra with 500 ns interframe time.The standard configuration of the spectrograph will utilize multimode fiber coupling to collect emissions from impact events in the terminal chamber. Photon doppler velocimetry (PDV), similar to that utilized for surface velocity measurement in HEMI's laser shock facility [9-10], is also available for use in the HyFIRE facility. Additionally, novel imaging and diagnostic systems in development by HEMI collaborators, including snapshot hyperspectral imaging and 3D particle tracking velocimetry (see Fig. 5), will be adapted for use in HyFIRE experiments. Both the design of the launcher and range, as well as the diagnostic package described here is intended to provide researchers across multiple disciplines the ability to investigate hypervelocity impact phenomena, with an emphasis on the coupling of mechanics, physics and chemistry.

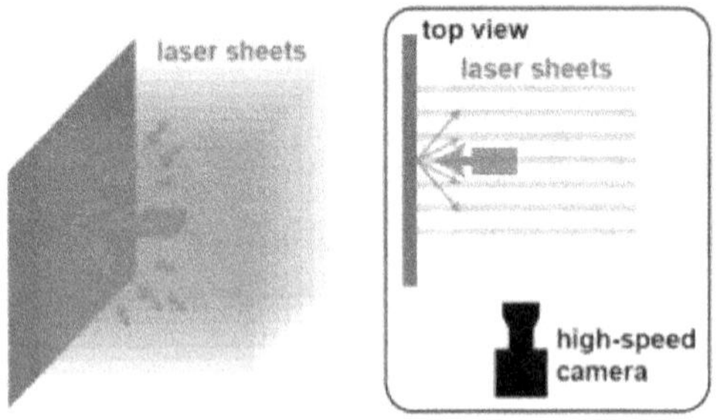

Fig. 5. Compressed sensing based 3D particle tracking velocimetry system uses multiple laser sheets and a single camera to track debris cloud trajectories. Multiple laser sheets allow tracking even in relatively dense particle fields.

3.2 Initial Diagnostic Tests

Following installation, initial test shots were performed to train HEMI researchers, validate launcher operation, and evaluate the use of ultra-high-speed imaging through the existing terminal impact chamber viewports. These test shots (Table 1) ranged in velocity from 1.8-5.5 km/s. Note, the most current three shots were performed after installation of the redesigned breech/pump tube, the last two using hydrogen for the pump tube media. Test shots were performed using 7mm polycarbonate slugs (L/D ~ 1) and 1/8" saboted aluminum spheres, impacting 1/8" thin aluminum plates or thick hardened armor steel (AR500) plates—with one exception, where a 2" acrylic cube was impacted by a polycarbonate slug. Images from two of these shakedown tests are shown below. Note, the intent of these experiments is only to evaluate the ability of our current imaging systems to capture a hypervelocity impact event.

Figure 6 shows the impact of a polycarbonate cylinder on a thin aluminum plate at 2.6 km/s. A Shimadzu HPV-X2 was used to collect images at 2 Mfps with 200 ns exposure. The target was backlit with diffused flashlamps. At initial impact (a), an impact flash can be seen which expands radially in (b). The previously mentioned spectroscopy system will be used to analyze such phenomena in future work. In (c), disrupted material from the rear surface of the projectile can be seen moving uprange. Additionally, the characteristic lobe shape of the debris cloud and ejecta cone is seen. By (d), a relatively sparse ejecta fragment field is seen, suitable for particle tracking using the 3D system in development. All images presented have been cropped from full available resolution.

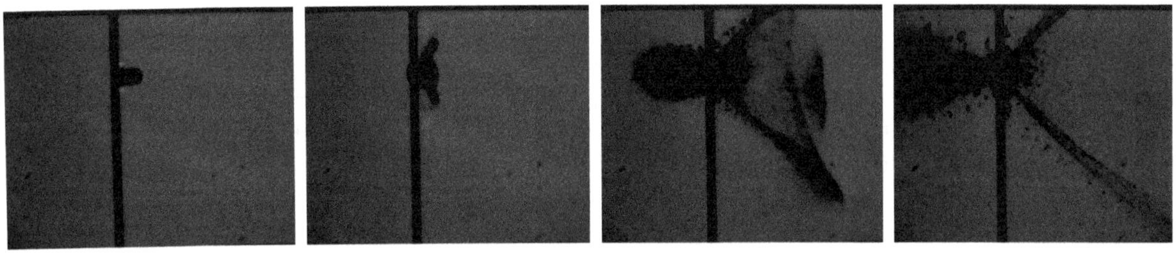

(a) Impact, t = ~0 μs (b) t = 2.5 μs (c) t = 20 μs (d) t = 40 μs

Fig. 6. Impact of polycarbonate cylinder on 1/8" aluminum plate at 2.6 km/s. An impact flash and its expanding front can be seen in (a) and (b). The characteristic lobed shape of the debris cloud and ejecta cone can be seen. An ejecta fragment field suitable for particle tracking develops in (d). A Shimadzu HPV-X2 ultra-high-speed camera was used at 2 Mfps with 200ns exposure. Images have been cropped for presentation.

Figure 7 shows the impact of a polycarbonate cylinder on a 2" acrylic cube at 3.3 km/s, using a Kirana ultra-high-speed camera at 2 Mfps with 200 ns exposure for image collection. Several ghost images and camera artifacts can be seen in (a) and (b). These ghost images complicate analysis as each frame seems to contain superimposed information from every tenth separated frame. Further evaluation is necessary to more accurately understand the phenomena. Careful distinction must be made to separate signal phenomena from potential artifacts, but several basic phenomena can be observed. An impact flash (a) and subsequent expansion of the plume toward the bottom edge of the target is seen (b). Shock and damage propagation included reflections at top and bottom boundaries are seen in (c), and a spall damage plane near the rear surface of the target is visible in (d). Surface velocimetry measurements in future experiments will be performed using photon doppler velocimetry systems developed by HEMI researchers [9-10]. Future tests will refine imaging capability, utilize front surface and pulsed laser illumination, as well as incorporate additional diagnostics including flash x-ray and spectroscopy for the purpose of capturing simultaneous information pertaining to multiple impact phenomena seen here.

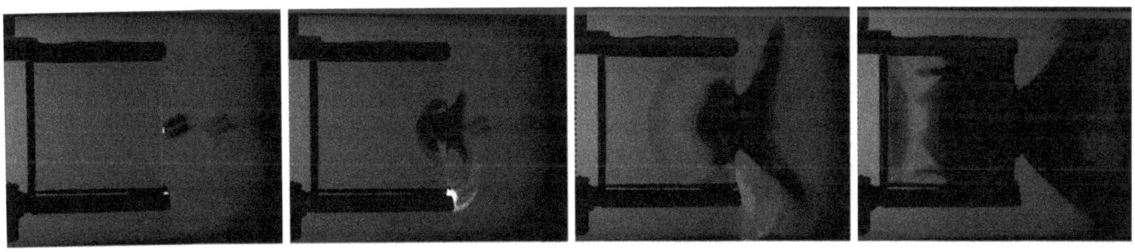

(a) Initial impact, t = 0 μs (b) t = 3 μs (c) t = 10 μs (d) t = 20 μs

Fig. 7. Impact of polycarbonate cylinder on acrylic cube at 3.3 km/s. Careful distinction must be made between camera artifacts and real signals, but flash phenomena, shock and damage propagation, and spall can be seen in the selected frames. A Kirana ultra-high-speed camera was used at 2 Mfps with 200ns shutter. Images have been cropped for presentation.

4. Planned Experiments and Initial Research Program

Apart from continuing shakedown tests to verify and improve the performance of the launcher and diagnostic systems, planned research investigates the fragmentation of impactors on anvil targets, and perhaps evaluates the effects of deformation and resulting fragmentation on the combustion of the generated debris cloud. Critical to the design of such experiments is an understanding of the deformation and failure mechanisms controlling the formation of the impact debris cloud, and a more complete understanding of fragmentation mechanics under the complicated states experienced during high velocity impact on finite structures. Fragmentation theories often assume brittle material behaviors in which dissipation due to fracture is minimal [11]. However, ductile deformation and the associated dissipation and heating may be the source of activation energy for the combustion of fragments. Additionally, while fragmentation is generally well posed in model problems involving tensile loading, such as an expanding rings or shells [11], impact fragmentation deals with complicated compression dominated loadings, non-homogeneous strain rates within the fragmenting body. The mechanisms dominating such a fragmentation event, which may additionally be influenced by the coupling of mechanics and chemistry, are poorly understood and require much more exploration.

5. Conclusions

The newly commissioned HyFIRE facility at Johns Hopkins provides researchers high experimental configurability, high accessibility to events of interest, and multiple diagnostic systems with which to analyze multiple facets of high velocity impact phenomena. To date, only a handful of test shots have been performed, and much continued effort is needed to reach the maximum potential of both the launcher and associated diagnostic and experimental systems, but initial research experiments will begin soon and attempt to expand the understanding of impact related fragmentation and coupling of mechanics and chemistry reactive materials under extreme loading.

Acknowledgements

We would like to thank Dave Strange of Physics Applications Inc. and his team for their work to satisfy the particular requirements of our facility and their continuing support. We would also like to thank the United States Congress, Department of the Army, and the Whiting School of Engineering for their support and assistance in building and commissioning this facility. Work was funded in part by the Defense Threat Reduction Agency under grant HDTRA1-18-1-0016. This research was sponsored by the Army Research Laboratory and accomplished under Cooperative Agreement Number W911NF-12-2-0022. The views and conclusions contained in this document are those of the authors and should not be interpreted as representing the official policies, either expressed or implied, of the Army Research Laboratory or the U.S. Government. The U.S. Government is authorized to reproduce and distribute reprints for Government purposes notwithstanding any copyright herein.

References

[1] Hogan, J.D., Farbaniec, L., Mallick, D., Dominich, V., Kuwelker, K., Sano, T., McCauley, J.W., Ramesh, K.T., 2017. "Fragmentation of an advanced ceramic under ballistic impact: Mechanisms and microstructure". International Journal of Impact Engineering, 102, pp. 47-54.

[2] Farbaniec, L., Hogan, J.D., Xie, K.Y., Shaeffer, M., Hemker, K.J., Ramesh, K.T., 2017. "Damage evolution of hot-pressed boron carbide under confined dynamic compression". International Journal of Impact Engineering, 99, pp. 75-84.

[3] Farbaniec, L., Williams, C.L., Kecskes, L., Ramesh, K.T., Becker, R., 2017. "Spall response and failure mechanisms associated with a hot-extruded AMX602 Mg alloy". Materials Science and Engineering, A 707, pp. 725-731.

[4] Mayercsik, N.P., Shaeffer, M., Graham-Brady, L., Kurtis, K.E., 2015. "Analysis of Portland cement mortar under impact: A combined material characterization, micromechanics modeling, and dynamic testing approach". Cement and Concrete Research, 73, pp. 190-206.

[5] Tonge, A.L., Ramesh, K.T., Barnouin, O., 2016. "A model for impact-induced lineament formation and porosity growth on Eros". Icarus, 266, pp. 76-87.

[6] Design documentation, David Strange, Physics Applications Inc., Dec. 2018.

[7] Swift, H.F., 2005. Light-Gas Gun Technology: A Historical Perspective in High-Pressure Shock Compression of Solids VIII. High-Pressure Shock Compression of Condensed Matter, Chhabildas L.C., Davison L., Horie Y., Editors, Springer, Berlin, Germany.

[8] Tandy, J.D., Mihaly, J.M., Rosakis, A.J., 2014. "Examining the temporal evolution of hypervelocity impact phenomena via high-speed imaging and ultraviolet-visible emission spectroscopy." Journal of Applied Physics, 116.

[9] Mallick, D., Shaeffer, M., Dean, S., Ramesh, K.T., 2017. "Investigating the velocity envelope of laser-driven micro-flyers for hypervelocity impact experiments." Proceedings of the 14th Hypervelocity Impact Symposium, pp. 215-222.

[10] Mallick, D. D., Zhao, M., Bosworth, B. T., Schuster, B. E., Foster, M. A., & Ramesh, K. T. ,2019. "A Simple Dual-Beam Time-Multiplexed Photon Doppler Velocimeter for Pressure-Shear Plate Impact Experiments". Experimental Mechanics, 59(1), pp. 41-49.

[11] Grady, D., 2006. Fragmentation of Rings and Shells: The Legacy of N.F. Mott, Springer, Berlin, Germany.

Proceedings of the 2019 Hypervelocity Impact Symposium
HVIS2019
April 14-19, 2019, Destin, FL, USA
HVIS2019-081

Extended Abstract[+]

Voitenko experiments with novel diagnostics detect velocities of 89 km/s

Douglas G. Tasker[a,*], Young K. Bae[b], Carl Johnson[a], Kevin Rainey[a], Christopher Campbell[a], Dave Oschwald[a], Cornelius Reed[a]†

[a]*Los Alamos National Laboratory, Los Alamos., NM 87545, USA*
[b]*Y.K. Bae Corporation, Tustin, CA 92780-7709, USA*

1. Introduction

Using a Voitenko accelerator [1-3], a series of experiments were performed with the goal of attaining shock velocities in gases approaching 90 km/s. Typically, the basic apparatus comprises a hemispherical bowl filled with a gas at atmospheric pressure; a metal piston across its diameter; and a small bore evacuated shock tube at its apex, Fig. 1. The evacuated shock tube is separated from the gas bowl by a thin diaphragm. A combination of a plane wave explosive lens and a high explosive pad accelerates the piston to a velocity of the order of 4 km/s and subsequently compresses the gas in the bowl. The thin diaphragm at the other end of the bowl then ruptures and the high pressure (shock compressed) gas escapes into the shock tube.

The purpose of the experiments was to demonstrate that gas shock velocities approaching 90 km/s could be attained with a Voitenko apparatus; to this end three experiments were performed. This goal was achieved with a measured velocity of ~89 km/s.

Methods of accelerating particles to high velocity and their impacts with solid targets have long been of interest to various communities including shock physics, high energy density physics and space travel. Recently a new avenue of interest is the impact of nanoparticles (1 to 10 nm in diameter) at velocities approaching 100 km/s. These particles undergo shock compression to stresses of ~10 TPa in times of the order of 10 fs to 100 fs. Bae demonstrated experimentally that as these excited particles relax they emit soft X-rays with high efficiency [4, 5].

Bae proposed that under such conditions, because electron-ion thermalization takes more than one picosecond while the compression occurs in femtoseconds, the initial compression energy is mainly mechanical, i.e., the compression is "cold." Under these unique conditions the nanoparticles are compressed into the "Metastable inner shell molecular state" (MIMS), which is bound by inner shell electrons with binding energies exceeding 100 eV. It is the decay of this MIMS state that is accompanied by X-ray emission.

+ Full paper can found at https://www.sciencedirect.com/journal/international-journal-of-impact-engineering/special-issue/10MTGP5W4VJ

* Corresponding author. Tel.: +01-505-665-2859; fax: +01-505-665-6774
E-mail address: tasker@lanl.gov.

2. Experiments

2.1. Large scale design and diaphragm

In a miniature system originally tested by the authors [6] the diameter of the shock tube was close to 1 mm, and at this scale it

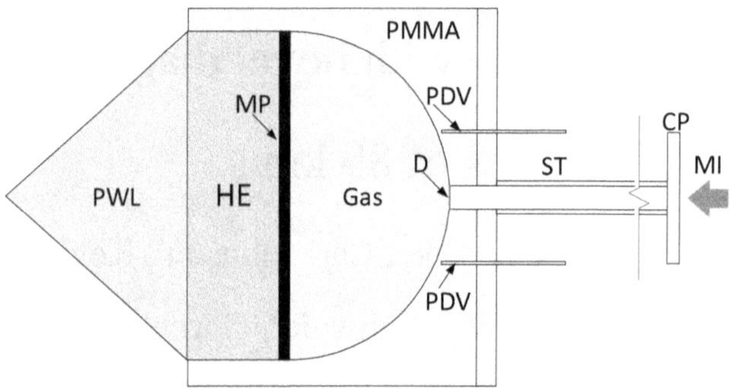

Figure 1. Voitenko apparatus. PWL = plane wave lens; HE = high explosive PBX-9501; MP = metal piston; CP = cap plate; D = diaphragm; ST = the evacuated shock-tube; PDV = photon Doppler velocimetry probes (2); MI = microwave interferometer.

was near impossible to obtain a diaphragm that was thin enough to function efficiently; peak jet velocities of only 42 km/s were attained. The shock tube bore was therefore increased in diameter to 12.7 mm which allowed the use of both a 100 mm diameter gas bowl diameter and microwave velocimetry, see below. With this diameter and the consequent increased mass of gas, an optimum thickness of 12-µm polyimide diaphragm could be used, which was readily available.

The maximum shock speed in the shock tube is primarily determined by the sound speed of the gas at the elevated pressure. So, for the highest escape velocity, hydrogen is preferred, followed by helium and so on. From a practical standpoint, hydrogen is highly flammable and helium was chosen.

2.2. Optical diagnostics

In the new 12.7-mm diameter experiments two conventional Photon Doppler Velocimetry (PDV) probes [7] were used to directly record the velocity of the metal piston and determine piston integrity. This was done to ensure uniformity of the drive and the absence of plate spall. A Shimadzu model HPV-X high speed digital framing camera, operating at 5 x 10^6 frames per second (110 ns exposure, 200 ns interframe time), was used to observe the luminous shock front as it travelled along the shock tube. In addition, a microwave interferometer was used for the first time in these experiments in an attempt to record the high velocity plasma.

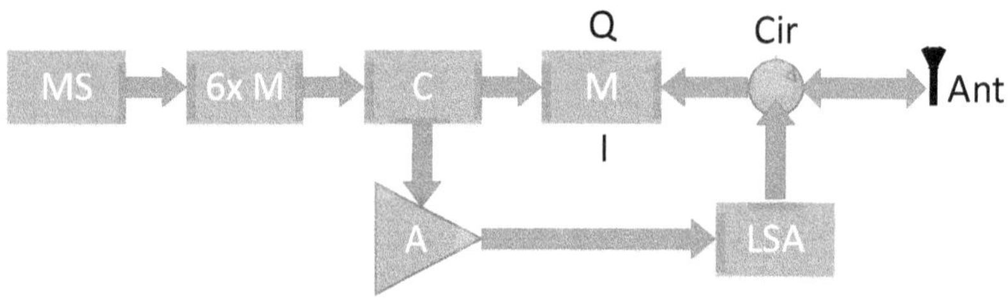

Figure 2. Millimeter-wave interferometer. Key: MS - microwave signal generator, Agilent E8247C; 6x M – six-fold Active Multiplier Chain, Millitech AMC-10-RFHB0; C – 20 dB Coupler, Millitech CL3-10-R2000; M – I/Q Mixer, Millitech MIQ-10-RLNA0; Cir – circulator, Quinstar QJY-94023W; A – 500 mW Power Amplifier, Quinstar QPN-94022730-P2; LSA – level-set attenuator, Quinstar QAL-W00000; Ant – Conical Horn Antenna, Sage Millimeter SAC-2309-10-S2; Q and I – quadrature and in-phase output signals.

2.3. Millimeter-wave interferometer

A circulator allowed the use of a single antenna for the transmission and receipt of data. The interferometer utilized a mixer to generate the in-phase and quadrature signals which were recorded by a digitizer. Conventional short-time Fourier transform The 94-GHz interferometer used a microwave source and six-fold active multiplier chain to generate the 94-GHz carrier, Fig. 2. analysis was used to convert the spectral data to velocities. The interferometer beam was projected into the shock tube through the cap plate, Fig. 1, towards the piston.

3. Results

The first large scale experiment was performed with no plastic attenuator between the explosive and the metal piston. The PDV velocity spectrogram showed clear evidence of spall and the ultimate piston velocity was just 2250 m/s. By adding PMMA attenuators between the piston and the explosive for the second and third experiment piston velocities of to 4200 m/s were attained.

3.1. Observations of experiment 3

In the experiment with the fastest jet velocity, experiment 3, the camera video showed a bright luminous front travelling down the tube at near 35 km/s. Then a bright flash appeared ahead of the luminous front at the bottom of the shock tube with a dark zone between the flash and the luminous front. This behavior was similar to that observed in experiment 2.

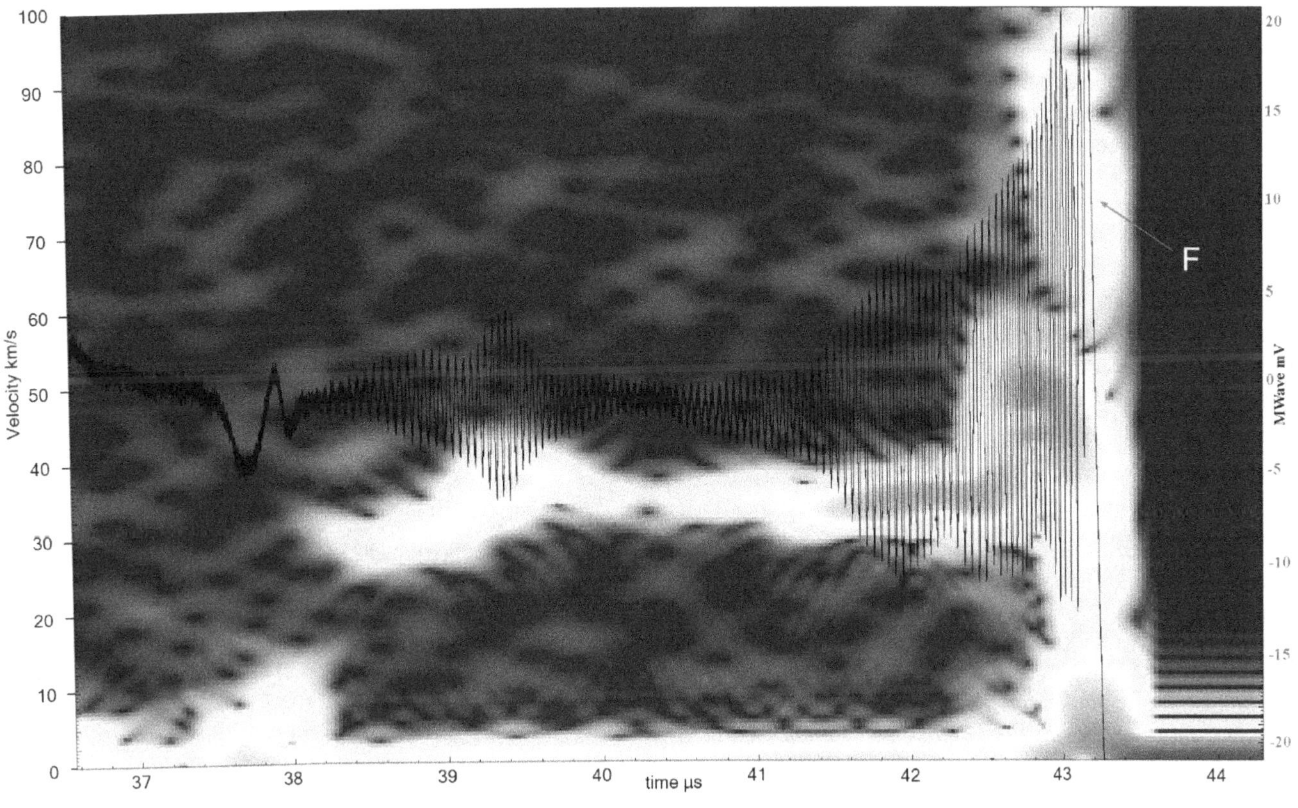

Figure 3. Experiment 3: Microwave spectrogram (colors) with velocities in km/s vs time in µs; raw voltage data (MWave) overlaid in black (mV), right–hand scale. There was a near-steady velocity of 35 km/s which continued until 42.5 µs when it jumped to 51 km/s. At 43.1 µs there was another jump to velocities of between 75 and 89 km/s (F) corresponding to the bright flash at the end of the tube.

The microwave data revealed a complex wave behavior, Fig. 3. The jet travel started with a relatively steady velocity of 35 km/s for the majority of the travel in the tube. Then, just before the end of the tube, the velocity stepped up to several simultaneousstates in a staircase formation with velocities ranging from 35 to 55 km/s. Finally, the wave apparently accelerated at the end of the tube to velocities spread between 70 and 90 km/s. Note again another weaker peak on the spectrogram near 100 km/s at 43.1 μs.

4. Discussion

4.1. Interpretation of imagery

In the video data from experiment 3, a cool luminous front was seen travelling well ahead of the brightly heated zone. This means the temperature of the jet front was probably below the ionization energy (24.6 eV) of helium; consequently it was neither visible nor reflective. The real shock front was probably not visible in the video data and could have been further advanced than the visible front. This is interpreted to mean that the camera and microwave data near the end of the tube are indeed real because the jet front (approaching 90 km/s) became detectable due to heating by the backward flowing gas.

In other words, there was a fast, near invisible portion of the jet flow travelling at near 90 km/s followed by a slower moving tail which was hotter and brighter. The fast portion became visible as it ploughed into the reflected gas flow at the end of the tube. In short, the wave did not accelerate towards the end of the tube, it merely becomes more visible. (Unfortunately, the time of shock arrival at the cap plate was not recorded.)

5. Conclusions

The combination of the photographic and microwave data provides strong evidence of the existence of wave speeds of 90 km/s or more in these Voitenko experiments. There are clearly many features of the experiments that are not fully understood and merit further study. Moreover, the explosive drive system should be optimized to get the highest plate velocity without spall.

The microwave interferometry data provided a rich suite of information; it is clearly a powerful diagnostic for such experiments. Additional diagnostics to record shock arrivals time and correlate them with the microwave signatures would be helpful in future experiments.

Acknowledgments

This collaborative study by the Los Alamos National Laboratory and the Y.K. Bae Corporation was supported by a contract from US Defense Threat Reduction Agency (DTRA). We are especially grateful to Drs. J. Davies and W. Wilson of DTRA for their generous support. 'LA-UR-19-20670' Version 4.

References

[1] Voitenko, A.E., *Generation of high speed gas jets.* Dokl. Akad. Nauk SSSR, 1964. **158**(6): p. 1278-1280.

[2] Voitenko, A.E., *Acceleration of a gas compressed under conditions of acuteangled geometry.* Journal of Applied Mechanics and Technical Physics, 1966. **7**(4): p. 79-82.

[3] Voitenko, A.E. and V. Kirko, *Efficiency of a high-explosive plasma compressor.* Combustion, Explosion, and Shock Waves, 1975. **11**(6): p. 813-815.

[4] Bae, Y., *Metastable innershell molecular state (MIMS).* Physics letters. A, 2008. **372**(29): p. 4865-4869.

[5] Bae, Y.K., *Table-top Generation and Spectroscopic Study of ~10 TPa High-Energy Density Materials with C+60 Hypervelocity (v ~ 100 km/s) Impact*, in *Shock Compression of Condensed Matter - 2013*. 2013, American Institute of Physics: Seattle, WA.

[6] Johnson, C.E., et al., *Experiments on a Miniature Hypervelocity Shock Tube*, in *APS Shock Conference, American Physical Society, Shock Compression of Condensed Matter, Topical Group*. 2013: Seattle, WA, USA.

[7] Strand, O.T., et al., *Compact system for high-speed velocimetry using heterodyne techniques.* Review of Scientific Instruments, 2006. **77**(8): p. 083108-8.

Proceedings of the 2019 Hypervelocity Impact Symposium
HVIS2019
April 14-19, 2019, Destin, FL, USA
HVIS2019-084

The JHUAPL Planetary Impact Lab (PIL): Capabilities and initial results

R. Terik Daly, Olivier S. Barnouin, Andrew M. Lennon, Angela M. Stickle, Emma S. Rainey, Carolyn M. Ernst, Andrew A. Knuth

Johns Hopkins University Applied Physics Laboratory, 11100 Johns Hopkins Road, Laurel, MD, 20723

Abstract

The Planetary Impact Lab (PIL) at the Johns Hopkins University Applied Physics Laboratory (JHUAPL) includes a single-stage, compressed inert gas gun that can be used for impact experiments. The impact angle can be varied from 15° to 90° with respect to horizontal, a capability which enables oblique impacts into unconsolidated or granular materials (e.g., regolith analogs). The gun currently achieves impact velocities up to 400 m/s, although future enhancements could increase the maximum projectile velocity. Experiments can be done with atmospheric pressures ranging from ambient pressure down to ~75 Pa. The gun uses sabots produced with state-of-the-art additive manufacturing techniques (AM). Several engineering challenges had to be overcome to create a reliable AM sabot; however, AM sabots are ~45% lighter than and provide substantial cost savings over machined sabots. The PIL gun is currently being used to investigate impact processes on sloped coarse-grained surfaces, with application to planetary science and, specifically, rubble-pile asteroids. In contrast to previous studies of impacts onto slopes, we kept the projectile trajectory perpendicular to the target surface, thereby disentangling the effects of oblique impacts from the effects caused by a sloped surface. Initial results show enhanced crater collapse in the sloped target, with most of the collapse occurring in the direction parallel to the surface gradient. Consequently, final craters on sloped targets have smaller volumes and reduced depth-to-diameter ratios.

Keywords: Impact cratering; impact experiments; planetary defense; asteroids; shock physics

1. Introduction

Experiments provide fundamental information about impact processes[e.g.,1]. They provide critical data needed to validate numerical models[2–4] and permit investigations of processes that shock physics codes have yet to fully capture. The advent of new experimental facilities has often led to breakthroughs in our understanding of impact processes. The development of the NASA Ames Vertical Gun Range[5] in the 1960s, for example, led to an array of groundbreaking insights into impact cratering on terrestrial planets[1,5,6]. Here we describe a new gun range developed for the Planetary Impact Lab (PIL) at the Johns Hopkins University Applied Physics Laboratory (APL). The PIL gun is currently being used as part of a study investigating crater formation and scaling during impacts into coarse, rubble-pile-like targets such as Itokawa[7], Ryugu[8] and Bennu[9].

2. Facility description as of January 2019

The PIL gun range consists of a single-stage, compressed inert gas gun and a large impact chamber (Fig. 1). The gun can be positioned over six different ports, which enables experiments at 15°, 30°, 45°, 60°, 75° and 90° with respect to horizontal without tilting the target. The impact chamber has five other viewports that can be used to mount diagnostics such as high-speed cameras or lasers. Three viewports are located on top of the chamber; the two others are on the chamber's sides. The largest viewport is

* Corresponding author. Tel.: +1-240- 592-1011.
E-mail address: Terik.Daly@jhuapl.edu.

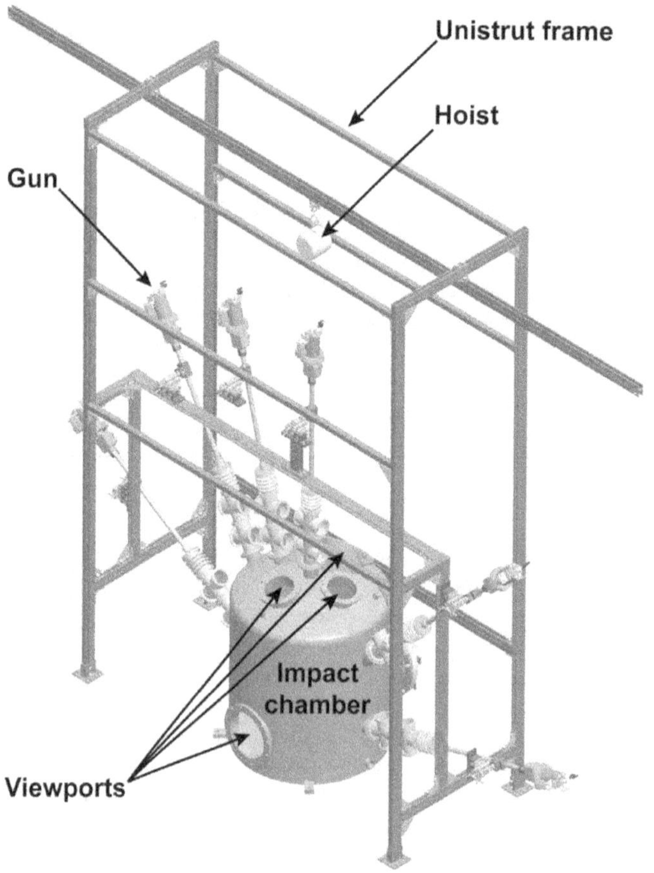

Fig. 1. Perspective drawing of the PIL gun. The facility consists of an impact chamber with several viewports over which cameras, lasers, or other diagnostics can be mounted; a single-stage compressed inert gas gun that can be positioned at six different locations using the indicated hoist; and a Unistrut superstructure for mounting and positioning equipment around the gun and impact chamber. The impact chamber is ~1.3 m in diameter.

50 cm across (located on the unseen side of the impact chamber in Figure 1) and is large enough to easily load, remove, and manipulate targets. The impact chamber is ~1.3 meters in diameter and ~2 meters tall. It can accommodate a wide range of targets (solid, liquid, particulate, granular, etc.). Due to the chamber's large size, most experiments can proceed with minimal interference from material bouncing off the chamber walls. The chamber can hold targets in many ways, including mounted on the floor, placed on sloped ramps, or hung from the ceiling of the chamber (e.g., for ballistic pendulum experiments relevant to planetary defense). The nominal target container for granular materials is 40 cm wide, 17 cm deep, and made of rubber. The container's large size relative to the sizes of craters formed by the PIL gun, and the container's stress-absorbing material properties, mitigate any interference to the experiment caused by rarefactions off the free surfaces of the container. The impact chamber is connected to a rotary vane pump that evacuates the entire system. Experiments can be done at any chamber pressure between atmospheric and ~75 Pa. This capability, in effect, allows experimenters to "turn on" or "turn off" atmospheric effects on the impact process at will. The lower end of the pressure range is sufficient for the excavation of fine-grained materials to proceed without atmospheric interference.

The gun system uses only compressed inert gasses, rather than gunpowder. This design choice that significantly reduces the hazards associated with the PIL gun. The gun currently uses pressured nitrogen gas to fill a reservoir with up to 6000 psi (~41 MPa). A projectile is launched on the front end of a sabot. Projectiles of any solid material up to 6.35 mm diameter can be accommodated. Once the launch system is triggered, a plug within the breech is rapidly displaced to release the pressurized gas into the smooth-bore gun barrel. The back of the sabot has a flange that allows pressure to build behind it. Once the force exerted by the gas exceeds the failure strength of the flange, the sabot is shot down the barrel. An internally tapered cylinder attached to the end of the gun barrel stops the sabot via radial compression, thereby separating the sabot from the projectile and allowing the projectile alone to enter the impact chamber. As the projectile travels from the gun barrel into the impact chamber, it interrupts two lasers (spaced a known distance apart) to measure projectile velocity. The system can launch a wide range of projectile shapes and materials in sizes up to 6.35 mm in diameter. To date, most of the experiments have used a gas pressure of 3000 psi (~21 MPa), which yields to velocities in the 310 – 330 m/s for launch packages that weigh ~1.8 g. Higher gas pressures provide faster launch velocities. The fastest shot to date had a velocity of 414 m/s.

The number of shots that can be performed per day depends on the complexity of the setup and the diagnostics required for a particular set of experiments. Thus far, the PIL gun has accommodated 6 to 8 shots per day for experiments that do not require vacuum conditions and are done at the same impact angle. Experiments in vacuum take longer due to the time required to evacuate and vent the chamber. Under these conditions, only 4 to 6 shots per day have been possible at the same impact angle. Moving the gun to a new position and re-aligning it takes additional time, so fewer shots are possible on days when the gun is moved to a new port.

The PIL has access to a range of instrumentation to support experiments, including two Phantom Miro high-speed cameras, several line lasers of various wavelengths, pressure sensors, oscilloscopes, and 3D scanners. Our baseline configuration places the Phantom cameras at two different ports, thereby providing complementary views of each experiment with frame rates of 2100 and 1900 frame per second. Higher frame rates are possible at the expense of spatial resolution. The cameras and line lasers can be configured to permit non-intrusive measurements of crater growth using the method reported by Barnouin et al.[10].

3. Additively manufactured sabots

Like many gun facilities, the PIL gun uses a sabot to carry the projectile. Sabot design is critical to gun performance. Although the PIL gun works reliably with sabots made of machined Delrin, these sabots are costly and heavy. Their high manufacturing costs increase the cost per shot; their weight reduces the velocities that can be achieved for a given gas pressure. Additive manufacturing (AM) techniques have the potential to produce lighter sabots at a lower cost. Therefore, we engineered sabots that can be produced using AM techniques. Specifically, we used fused filament fabrication (FFF). FFF is a commonly used printing process that uses a heated nozzle to extrude material and build up parts layer by layer.

Several challenges had to be resolved before honing in on the current—and highly successful—sabot design. As described in section 2, the PIL gun uses a smooth-bore barrel and relies on radial compression of the sabot to separate the sabot from the projectile. Early AM prototypes either broke up during launch or fragmented while being separated from the projectile, both of which resulted in sabot debris entering the impact chamber and interfering with the experiment. We tested AM sabot designs made using a range of polymer materials, printers, and print techniques. Parts produced using FFF are anisotropic due to the printing process and can be strong in one direction but be weak in another. A key breakthrough came from printing the sabot in the *longitudinal* direction, rather than the radial direction. Sabots printed in the direction parallel to the long axis of the sabot withstood the launch stresses and were stopped intact. A simple test proved that print direction was crucial: sabots printed in the *radial* direction using the same design and the same material broke apart while being stopped. The version printed in the longitudinal direction did not because the strong direction of the sabot was better aligned relative to the stresses that previously caused the sabot to fragment into several pieces.

The current version of the sabot (Fig. 2) is produced on a Markforged Onyx One or Mark Two printer using Markforged's proprietary Onyx material, which is a chopped-carbon-fiber-reinforced Nylon. The sabot is slightly longer than twice its diameter to stabilize it as it travels down the gun barrel. The sides are scalloped to save weight; three longitudinal ridges provide adequate stabilization. The back end has a groove for a Viton 75A O-ring. Earlier tests showed that gas escaping around the sabot was limiting the impact speeds achieved with the PIL gun; the O-ring reduces this problem by creating a seal between the sabot and the gun barrel. However, some gas leakage around the sabot is still suspected, and additional work is in progress to improve the seal.

As described in section 2, the thickened flange serves to increase the gas pressure required to initiate motion down the barrel, thereby preventing the sabot from traveling during the entire period of pressure build-up. The front is dimpled to provide a seat for the projectile, which is attached to the sabot using a small amount of vacuum grease. Compared to the previously used machined Delrin sabots, these AM sabots are less expensive to make and are ~45% lighter, which permits higher launch velocities for the same initial gas pressure.

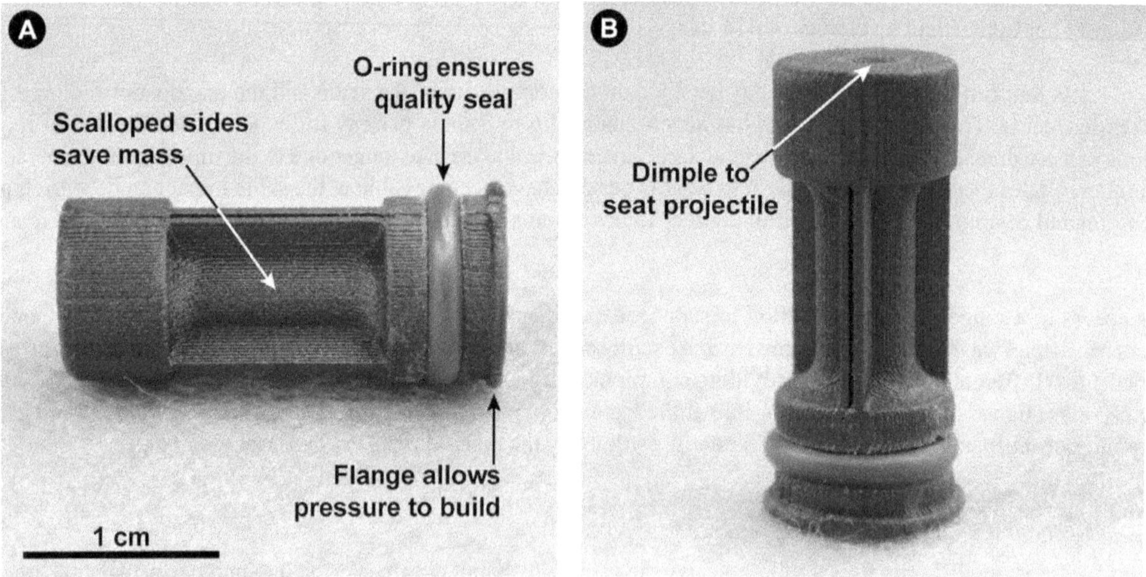

Fig. 2. Additively manufactured (AM) sabots used at the APL Vertical Gun Range. (A) A sabot lying on its side. (B) A sabot standing up on its rear flange.

4. Scientific investigations

Many impact craters on the Moon[11] and asteroids (e.g., Vesta[12], Lutetia[13], etc.) formed on slopes and have morphologic asymmetries that have been attributed to the sloped topography[11–13]. Some experiments have been done to investigate impacts into sloping targets[14,15]. However, these earlier studies convolved the effects of oblique impacts with those caused by a sloped surface. Oblique impacts affect crater formation and scaling[5,16–18], which makes it essential to isolate the role of a sloped surface from that of impact angle variations. In addition, many of the near-Earth asteroids visited by spacecraft, including Itokawa, Ryugu, and Bennu, have blocky, rubbly surfaces, in addition to an abundance of slopes[7–9]. Coarse-grained targets can affect crater formation and scaling[19–22]. The targets used in earlier studies[14,15] were composed of fine-grained materials. Hence, two competing effects—the presence of slopes and the coarse-grained character of the target—may influence crater scaling and morphology on many asteroids. We hypothesize that an interplay between these effects may explain some of the unusual, irregular morphologies of impact craters on rubble-rich asteroids.

To test that hypothesis, we are using the PIL gun to investigate the outcomes of impacts into sloped, coarse-grained targets. Table 1 summarizes the initial conditions and resultant craters from two experiments. The experiments use a target of uniform, 3-mm diameter alumina spheres. The shots used the sabot as the impactor (i.e., did not use the sabot stopper). The individual alumina grains are quite strong, but the bulk target is essentially strengthless. The spheres have an angle of repose of 28°. This is lower than the angle of repose of the sands used by Anderson et al.[15] (33°), but similar to the 29° for quartz sand and slightly higher than the 23° – 26° reported for glass spheres by Aschauer and Kenkmann[14] (in that study, the smaller beads had a lower angle of repose). The target lacks cohesion; hence, the internal friction angle is also ~28°. Impacts were done onto targets with surface slopes of 0° and 15° with respect to horizontal. In contrast to previous studies[14,15], we kept a constant ~90° angle between the target surface and the projectile trajectory. This angle was maintained by moving the gun to the 75° port for the experiments into the target on a 15° slope, with the target sloping in the opposite direction as the gun barrel. In other words, the gun barrel was at 75° with respect to horizontal (measured in the counterclockwise direction), and the target surface was at 15° with respect to horizontal (measured in the clockwise direction). This strategy ensured that the effects of sloped topography could be disentangled from the effects of an oblique impact. Experiments were recorded from two views using high-speed imaging. The final craters

Table 1. Comparison of craters formed on a flat surface vs. 15° slope.

Experiment ID	180222-06	180417-02
Impact velocity, U (m/s)	288	295
Impact angle (wrt. horizontal)	90°	75°
Target slope (wrt. horizontal)	0°	15°
Rim-to-rim major axis, a (cm)	18.4	28.6
Rim-to-rim minor axis, b (cm)	18.4	18.9
Rim-to-rim diameter, D (cm)	18.4	23.2*
Rim-to-floor depth, d (cm)	3.6	2.6
d/D	0.20	0.11
Displaced volume (cm^3)	181	135
Cratering efficiency[†]	296	221

*Diameter of a circle with the same area as the ellipse defined by the major and minor axes of the crater

[†]Defined as the ratio between the displaced mass and the projectile mass.

were captured using an Artec Eva using the scanner's proprietary Artec Studio software. Light et al.[23] demonstrated that crater morphometry can be accurately measured using these 3D scan products.

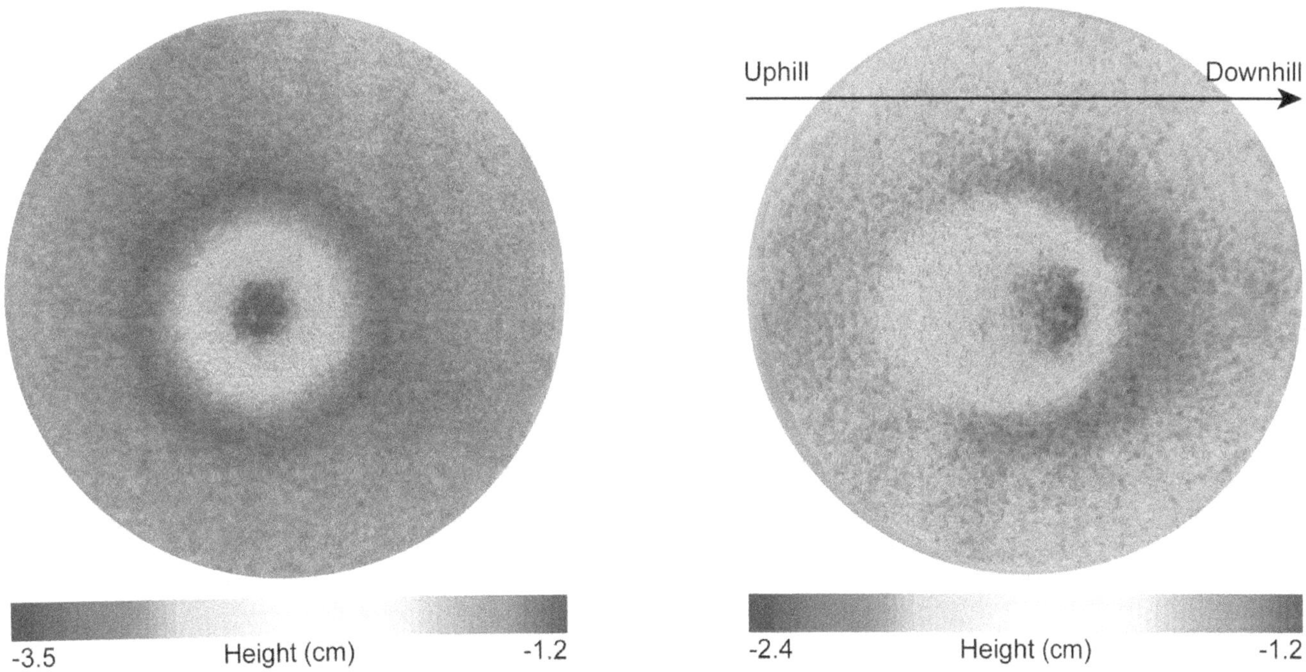

(A) Crater formed on flat surface **(B) Crater formed on 15° slope**

Fig. 3. Results of 3D scans of impact craters formed with the PIL gun viewed from a perspective normal to the pre-impact surface. Craters were formed in 3-mm alumina sphere targets on (A) a flat surface and (B) a 15° slope. The projectile trajectory was normal to the target surface. The scans have been colorized based on height relative to the pre-impact surface. This emphasizes crater morphology, but masks the slope of the pre-impact surface in (B). In (B), the slope dips down to the right. The crater formed on the flat surface is axisymmetric, as expected. The crater formed on the sloped surface is elliptical and shallower than the crater formed on the flat surface; note the difference in scale between the color bars in (A) and (B). Each scan is ~40 cm across.

The high-speed imaging data (including laser profiles visualized through the center of the crater in real-time) show that early-time crater growth is relatively unaffected by the sloped surface, but later-time collapse on the 15° slope is asymmetric in the slope dip direction. Enhanced collapse on the uphill side of the crater leads to a final crater with a steeper wall on the downhill side. Crater growth and collapse parallel to the strike of the slope appear similar that observed in flat targets.

3D scans reveal morphological differences between the final craters formed on the flat target (Fig. 3a) and on the 15° sloped target (Fig. 3b). The crater formed on the flat surface is circular, with a well-defined rim of uniform width. The crater formed on a 15° slope is elliptical with a long axis parallel to the dip direction. The uphill crater rim is absent; the downhill rim is wider than the rim of the crater formed on the flat target. The eccentricity of this crater planform is ~0.5. The effective diameter of the crater formed on the sloped target (i.e., the diameter of a circle with an area equal to the ellipse) is about 25% *larger* than the diameter of the crater formed on the flat slope (Table 1). In contrast, the displaced volume of the final crater formed on the 15° slope is about 25% *smaller* than that of the crater formed on the flat target, likely due to enhanced collapse in the sloped target. The depth-to-diameter (d/D) ratios of the craters also differ (Table 1). The crater formed on the flat target has a d/D of 0.20, comparable to the canonical d/D value of simple, fresh craters on the terrestrial planets[24,25]. The d/D of the crater formed on the 15° slope is 0.11, which indicates that the final crater is much shallower than expected. Oblique impacts can also lead to low d/D ratios[5,26,27]; this result shows that impacts on slopes, too, can cause small d/D ratios. Because this study kept the projectile trajectory perpendicular to the target surface, target slope, not impact angle, must be responsible for the shallow crater in this case. Enhanced crater collapse is likely the culprit.

Profiles extracted from the 3D scans shed further light on differences between the two craters (Fig. 4). In the crater formed on the 15° slope, the crater wall is much steeper on the downhill side (Fig. 4a). The profiles also show that the deepest part of the crater is offset downhill from the center of the elliptical crater planform. Profiles drawn parallel to strike are symmetric (Fig. 4b),

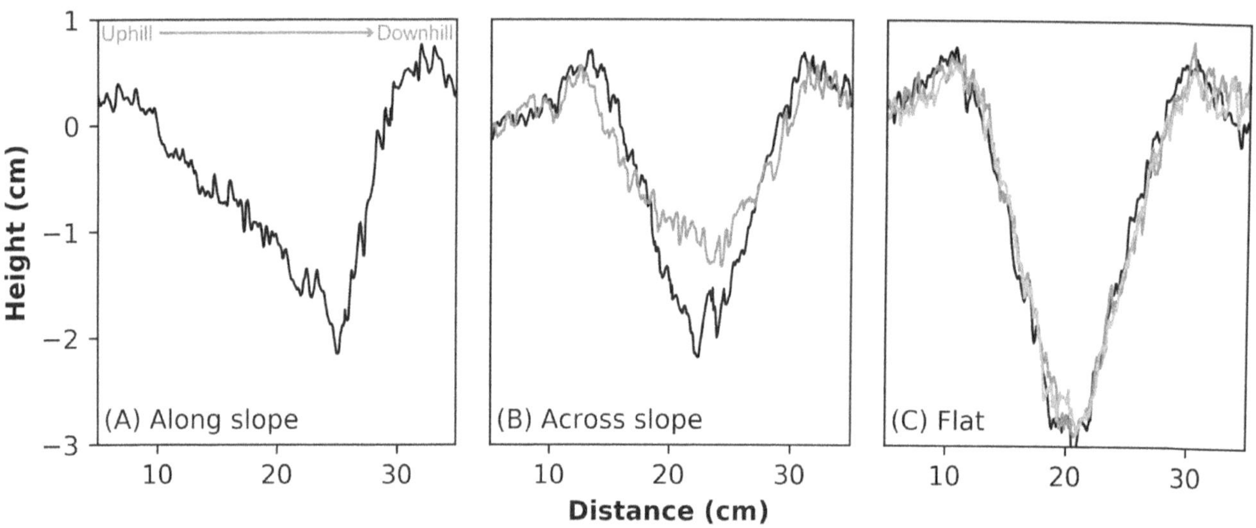

Fig. 4. Profiles across impact craters formed using the PIL gun. (A) and (B) show profiles across a crater formed on a 15° slope. (A) A profile in the dip direction that passes through the deepest point in the crater; the slope dips from right to left. (B) shows profiles of the same crater drawn parallel to strike. The black profile in (B) passes through the deepest point in the crater. The blue profile in (B) passes through the widest part of the crater. (C) Three profiles at different azimuths through the crater formed on the flat target are essentially identical.

but shallower than the profiles through the center of the crater formed on a flat surface (compare Figs. 4b and 4c). In the strike direction, the widest part of the crater is offset downhill from the deepest point. Because crater profiles parallel to strike are approximately symmetric, we infer that the processes that generate asymmetries in the dip direction (elongated planform; contrasting wall slopes) do not affect crater morphology parallel to strike.

Some aspects of these results are consistent with the findings of previous studies. Aschauer and Kenkmann[14] observed over-steepened material on the uphill wall slumping downward toward the center of the crater, a process observed here, too. The elliptical crater planform parallel to the dip direction, the offset of the deepest point downslope relative to the center of the elliptical crater planform, and the smaller d/D of the crater formed on the sloped surface are also similar to findings reported by Aschauer and Kenkmann[14]. Aschauer and Kenkmann[14] did experiments with projectiles traveling at ~180 m/s with a kinetic energy of 6.7 J. The impactors in this present study traveled somewhat faster (~1.6 time faster; see Table 1) and had much higher kinetic energies (~110 J).

Other aspects of the results, however, differ. Aschauer and Kenkmann[14] reported their data in terms of a normalized slope, which they defined as the ratio between the slope angle and the angle of repose. For the 3-mm diameter alumina spheres used here, a 15° slope equates to a normalized slope of 0.54. The crater formed on the 15° slope has an aspect ratio (ratio of major axis to minor axis) of ~1.5. Aschauer and Kenkmann[14] reported aspect ratios of ~1.1 for normalized slopes of near 0.5 in sand and small glass spheres. The crater formed in the alumina spheres is more elongated. Aschauer and Kenkmann[14] also reported that craters in their experiments had sharp uphill rims and smooth downhill rims. This differs from the morphology observed here: the crater formed on the 15° slope had no uphill raised rim and a well-defined, albeit broad, downhill rim. Although the uphill edge of the crater lacked a raised rim, a clear inflection in the crater profile (Fig. 4a) marks the uphill edge. However, our analysis of the figures in Aschauer and Kenkmann[14] indicates that not all of the craters in that study have a sharp uphill rim. For example, on a 15° slope, the glass sphere targets had no clear uphill rim, but the sand target did. Differences in rim morphologies may be due to several factors, including that the experiments reported here isolated the effect of a sloped target from the effects of an oblique impact, the difference in grain sizes (Aschauer and Kenkmann[14] used quartz sand and glass beads whose sizes ranged from 0.15 to 0.8 mm, much smaller than the 3-mm diameter spheres used here), differences in target strength properties (as manifested by the difference in the angles of repose), or differences in the grain size distribution of particles in the targets. Aschauer and Kenkmann[14] used sand and glass spheres with a range of particle sizes. The 3-mm diameter alumina spheres used here were monodisperse. The coarser grain size and monodisperse nature of the target may influence the crater collapse process because more work is required to move a particle over an adjacent one. Further experiments will elucidate the relative importance of these factors.

The sharp uphill rims observed by Aschauer and Kenkmann[14] are consistent with observations of craters formed on slopes on the Moon and Vesta, where such craters have sharp uphill rims but muted downhill ones[11,12]. Both the Moon and Vesta are laden with regolith rich in fine-grained material. On the block-rich asteroid Itokawa, however, craters lack well-defined rims[7]. Most of the craters on Itokawa formed on slopes. However, Itokawa is also orders of magnitude smaller than the Moon or Vesta, which means that gravity is much weaker there. Since gravity influences collapse, the very low g environment at Itokawa may also play a role. Data provided by the OSIRIS-REx and Hayabusa2 spacecraft at Bennu and Ryugu, respectively, will be essential to assessing the rim morphology of craters formed on slopes on small, rubble-rich bodies and will provide an additional basis for comparisons with the experiments reported here.

5. Conclusion

The gun at the Planetary Impact Lab (PIL) at the Johns Hopkins University Applied Physics Laboratory enables a range of impact experiments for planetary science, national security, and code validation. The PIL gun uses sabots produced using additive manufacturing, an innovation that provides significant cost and mass savings compared to machined sabots. Our preliminary investigation of impacts onto coarse-grained, sloping targets has yielded some similarities to results from earlier studies; the observed differences may stem from the coarse-grained nature of the targets used here, differences in target strength, or differences in target grain size distribution. The fact that the crater formed on a sloping target has a larger rim-to-rim diameter than the crater formed on the flat surface has implications for cratering statistics on rubble-rich bodies. Cratering statistics studies often focus on crater diameter[28]. The results of this study imply that the ages derived from crater counts on slope-rich areas of rubble-pile asteroids might yield artificially old ages due to the enlarged crater diameter which would, unless slope effects were accounted for, imply a larger projectile.

Acknowledgements

This work was supported by NASA grant NNX16AQ13G and the Johns Hopkins University Applied Physics Laboratory. It made use of the Small Body Mapping Tool (SBMT), a free, publicly available data visualization and analysis program developed at the Johns Hopkins University Applied Physics Laboratory (sbmt.jhuapl.edu).

References

[1] Gault, D., Quaide, W. & Oberbeck, V. "Impact cratering mechanics and structures" in *Shock metamorphism of natural materials* 87–99 (Mono Book Corp., 1968).
[2] McGlaun, J. M., Thompson, S. L. & Elrick, M. G. CTH: A three-dimensional shock wave physics code. *International Journal of Impact Engineering* 10, 351–360 (1990).
[3] Pierazzo, E. *et al.* Validation of numerical codes for impact and explosion cratering: Impacts on strengthless and metal targets. *Meteoritics & Planetary Science* 43, 1917–1938 (2008).
[4] Jutzi, M., Michel, P., Hiraoka, K., Nakamura, A. M. & Benz, W. Numerical simulations of impacts involving porous bodies: II. Comparison with laboratory experiments. *Icarus* 201, 802–813 (2009).
[5] Gault, D. E. & Wedekind, J. A. Experimental studies of oblique impact. *Proceedings of the Ninth Lunar and Planetary Science Conference,* 374–376 (1978).
[6] Gault, D. E. M. *Spray Ejected from the Lunar Surface by Meteoroid Impact.* NASA-TN-D-1767. (1963).
[7] Hirata, N. *et al.* A survey of possible impact structures on 25143 Itokawa. *Icarus* 200, 486–502 (2009).
[8] Hirata, N. *et al.* Shape and geophysical properties of the asteroid Ryugu from initial observations by Hayabusa2. *AGU Fall Meeting*, abstract P22A-06 (2018).
[9] Barnouin, O. S. *et al.* The shape of Bennu. *AGU Fall Meeting*, abstract P33C-3835 (2018).
[10] Barnouin-Jha, O. S., Yamamoto, S., Toriumi, T., Sugita, S. & Matsui, T. Non-intrusive measurements of crater growth. *Icarus* 188, 506–521 (2007).
[11] Plescia, J. B. Impacts on Sloping Surfaces: Lunar Examples. *Meteoritics and Planetary Science Supplement* 75, 5318 (2012).
[12] Krohn, K. *et al.* Asymmetric craters on Vesta: Impact on sloping surfaces. *Planetary and Space Science* 103, 36–56 (2014).
[13] Elbeshausen, D., Wünnemann, K., Sierks, H., Vincent, J. B. & Oklay, N. The Effect of Topography on the Impact Cratering Process on Lutetia. *43rd Lunar and Planetary Science Conference*, abstract 1867 (2012).
[14] Aschauer, J. & Kenkmann, T. Impact cratering on slopes. *Icarus* 290, 89–95 (2017).
[15] Anderson, J. L. B., Cintala, M. J., Dechant, L. E., Ebel, J. M. & Plescia, J. B. Ejecta Dynamics and Crater Topography During Experimental Impacts into Sloped Targets. *49th Lunar and Planetary Science Conference*, abstract 2646 (2018).
[16] Elbeshausen, D., Wünnemann, K. & Collins, G. S. Scaling of oblique impacts in frictional targets: Implications for crater size and formation mechanisms. *Icarus* 204, 716–731 (2009).
[17] Davison, T. M., Collins, G. S., Elbeshausen, D., Wünnemann, K. & Kearsley, A. Numerical modeling of oblique hypervelocity impacts on strong ductile targets. *Meteoritics & Planetary Science* 46, 1510–1524 (2011).
[18] Schultz, P. H. & Crawford, D. A. Origin and implications of non-radial Imbrium Sculpture on the Moon. *Nature* 535, 391 (2016).
[19] Tatsumi, E. & Sugita, S. Cratering efficiency on coarse-grain targets: implications for the dynamical evolution of asteroid 25143 Itokawa. *Icarus* 300, 227–248 (2018).
[20] Güttler, C., Hirata, N. & Nakamura, A. M. Cratering experiments on the self armoring of coarse-grained granular targets. *Icarus* 220, 1040–1049 (2012).
[21] Holsapple, K. A. & Housen, K. R. Cratering in Blocky Regoliths. *45th Lunar and Planetary Science Conference*, abstract` 2538 (2014).
[22] Susorney, H. C. M. et al. The role of target heterogeneity in impact crater formation: Numerical results. *Procedia Engineering* 204, 421–428 (2017).
[23] Light, S. L., Daly, R. T., Schultz, P. H. & Knuth, A. A. Cratering efficiency in metal targets: Results from impact experiments and 3D scanning. *50th Lunar and Planetary Science Conference.*
[24] Pike, R. J. Depth/diameter relations of fresh lunar craters: Revision from spacecraft data. *Geophys. Res. Lett.* 1, 291–294 (1974).
[25] Robbins, S. J. *et al.* Measuring impact crater depth throughout the solar system. *Meteoritics & Planetary Science* 53, 583–637
[26] Gault, D. E. Displaced mass, depth, diameter, and effects of oblique trajectories for impact craters formed in dense crystalline rocks. *Moon* 6, 32–44 (1973).
[27] Herrick, R. R. & Forsberg-Taylor, N. K. The shape and appearance of craters formed by oblique impact on the Moon and Venus. *Meteoritics and Planetary Science* 38, 1551–1578 (2003).
[28] Crater analysis techniques working group. Standard techniques for presentation and analysis of crater size-frequency data. *Icarus* 37, 467 – 474 (1975).

Proceedings of the 2019 Hypervelocity Impact Symposium
HVIS2019
April 14-19, 2019, Destin, FL, USA
HVIS2019-092

Transitional ballistics of electric high-velocity launchers

B. Reck*, S. Hundertmark, R. Hruschka, A. Zeiner, B. Sauerwein, M. Schneider

French-German Research Institute of Saint-Louis, 5 rue du General Cassagnou, 68301 Saint-Louis, France

---

Abstract

The high-velocity launch of a projectile is subjected to a number of disturbances which exert an influence on the flight trajectory. In the case of sub-caliber projectiles, sabot separation is one of the critical aspects. In this work, we focus on the projectiles and the launch package of an electric railgun launch, i.e. on the behavior of the launch-package, when transitioning from the gun barrel to free-flight. This work further addresses the use of a hydrocode for creating numerical models which are capable of predicting the motion and deflection of the sabot parts during their separation from the projectile after exiting the muzzle. An earlier study showed that the air flow around the projectile and the sabot can be modeled with sufficiently high accuracy by means of a simulation code that uses an Eulerian description of the gas flow. Within a time interval of several milliseconds, just the duration that a projectile needs to enter quasi-stationary flight, viscous effects of the air or gas flow have relatively little influence on the sabot discard process. If the Eulerian gas flow is coupled with the Lagrangian structural parts, the mechanical response of the latter to the gas pressure can be complex in terms of deformation and damage, and in that way, can affect the gas flow. In this study, the hydrocode model is applied to a medium caliber launch package concept for accelerating long rod projectiles. The computed results agree well with the corresponding experimental values obtained from a launch package model test in the shock tunnel at Mach 4.5. This demonstrates that the presented hydrocode model can be used for launch package design optimizations with high confidence.

Keywords: numerical simulation, railgun, launch package, transitional ballistics, sabot separation, shock tunnel.

---

1. Introduction

The ISL railgun activities are focused, among others, on applications for long-range artillery (Fig. 1) [1, 2] and multi-shot salvo applications [3] (RAFIRA, caliber 25 mm). Very high electric current on the order of Mega amps is required, as the resultant propulsive Lorentz force F is given by the well-known formula $F = 1/2\ L'\ I^2$ for a "simple" launcher without magnetic augmentation field (Fig. 2). The force is proportional to the square of the current I, since in many cases the inductance gradient of the rails L' can be considered as constant for a given barrel configuration.

The Lorentz force is a body force $\boldsymbol{f} = \boldsymbol{j} \times \boldsymbol{B}$, acting inside the current carrying parts, i.e. in the armature, where $\boldsymbol{j}$ is the vector of the current density and $\boldsymbol{B}$ the magnetic flux density vector. For the same reason, the rails experience a repulsive force lateral to the launch direction. Main challenges of the launcher design are therefore the mechanical load inside the barrel structure and inside the launch package, and the thermal load due to the high electric current. The combined investigation of these effects requires at least an electromagnetic/mechanical multi-physics approach as done in a recent study [4]. The thermal load will be considered in the near future.

The load transfer from the armature to the projectile can be done in different ways. At the ISL, the base push concept is usually chosen, where the load is introduced to the projectile base by compression (Fig. 1 b), while the sabot parts essentially work as bore rides ensuring lateral stability. This concept can be advantageous regarding lateral disturbances on the armature

* Corresponding author. Tel.: +33 389 69 5178; fax: +33 389 69 5193.
E-mail address: Bernhard.Reck@isl.eu.

when exiting the muzzle [5, 6]. An armature which is weakly coupled with the other parts of the launch package can only transmit limited undesired moment. However, the armature of this concept frequently consists of one solid metal block which, after sabot separation in the transitional ballistics stage, continues traveling behind the projectile. If a rapid separation of both, the armature and the sabot from the projectile is required, the side push concept provides a number of beneficial options, e.g. the accelerating load is transferred into the projectile by shear over a relatively large area [7]. Long and slender projectiles can be launched, and the projectile base does not need to be reinforced. This concept is basically identical to that of APFSDS (armor-piercing fin-stabilized discarding sabot) rounds for powder guns [8].

Asymmetric loading of the launch package at the muzzle generates a moment resulting in armature rotation. This is shown in the flash x-ray image (Fig. 1c) of a launch package in free flight a few meters after barrel exit. According to reference [9], the transition from metal-to-metal contact to a plasma contact at the rail/armature interfaces is supposed to be the principal source of symmetry breaking. If transition starts on one of the two rail/armature interfaces, plasma appears. It generates an asymmetric pressure component, which adds to the magnetic pressure, pushing the armature against the opposite rail. Besides a local redistribution of the electric current and mass-loss due to plasma heating and its associated material loss, at least two further asymmetry effects of the plasma contact may play a role at the muzzle. First, since one rail/armature interface is still a metal-to-metal contact, asymmetric friction drag between the two sides occurs. Secondly, the plasma attached to one of the rails exerts pressure over an area of the armature portion outside the barrel, which results in torque. The authors in [9] provided analytical formulas for getting order-of-magnitude estimates of these effects, which were then applied to the launch package presented in Fig. 1b [6]. However, the obtained results did not agree well enough with the experimental findings (e.g. flash x-ray images, Fig. 1c), and therefore the rough-estimate formulas could not be used for further work.

Our objective in this study is two-fold. On the one hand, the investigation focuses on the transitional ballistics, i.e. the sabot separation process of a small to medium caliber railgun launch package, capable of carrying a long rod projectile. A possible application for terminal ballistics studies in the laboratory scale is envisaged and therefore, all launch package parts except the projectile itself should be cleared rapidly from the launch trajectory. On the other hand, a hydrocode is used for the numerical modeling of the sabot separation. The aim is to verify its ability for launch package design optimization. To that end, a simplified version of the launch package is tested in the ISL shock tunnel to simulate sabot separation experimentally. The hydrocode results are then compared with the results obtained from the shock tunnel experiment.

Generally, prior to the numerical simulation of the sabot separation process, the launch package model is numerically tested for being able to withstand the axial force occurring at the launch inside the barrel. In the simplest case the lateral constraints due to the in-bore walls are applied by zero lateral motion boundary conditions. This analysis, e.g. performed in [6], is not the scope of this work.

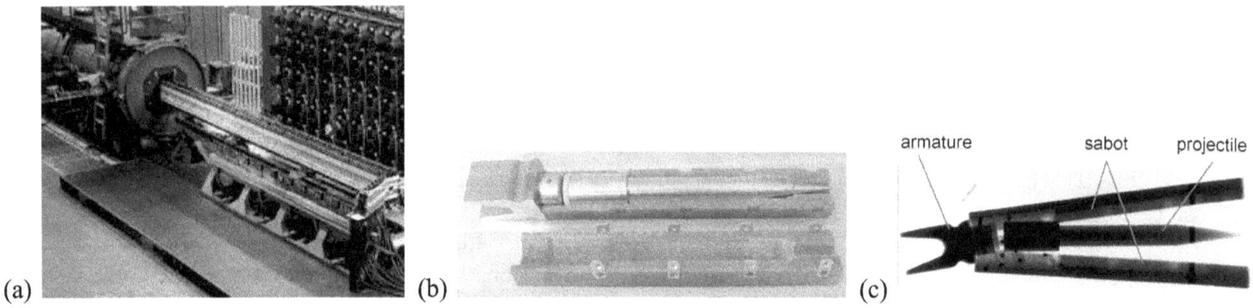

Fig. 1. (a) ISL-PEGASUS railgun facility with cal 60x60 mm², 6 m long NGL60 launcher for investigating long range artillery applications and connected 10 MJ capacitive energy storage. Launch package tested with former 40x40 mm² barrel [2], (b) photo before launch and (c) flash x-ray image after muzzle exit.

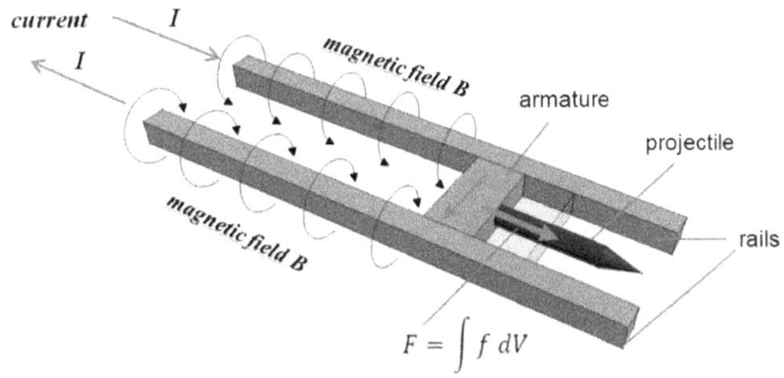

Fig. 2. Accelerating force F, electric current I and magnetic field $\boldsymbol{B}$ of a railgun (schematic view).

2. Launch package design

The launch package is intended for the use in a medium caliber railgun. It further should be able to accelerate slender long-rods to muzzle velocities close to Mach 4.5 or above. The side-push or lateral-push concept is suitable in that case, where a shear load is acting on a "threaded" area at the center part of the projectile, similar to that of kinetic energy APFSDS rounds. A possible application in the domain of terminal ballistics studies shall be considered as well, which means that all launch package components around the projectile should be rapidly removed from its trajectory. However, using a monolithic armature is an issue. It tends to travel closely behind the projectile and is difficult to stop or to deflect (see Fig. 1c). In this case, the idea is to use two symmetrical armature parts and to attach them to the two corresponding sabot parts, in order to obtain a common separation after muzzle exit. A CAD image of this launch package is presented in Fig. 3. Fig. 4 shows one of two sabot/armature parts. The rear C-shape part is wedged, in order to enable the rotation around the indicated pivoting axis. Fast lift-up is expected, because the moment of inertia is reduced in comparison with a geometry that rotates around the rear edge. A numerical analysis of this railgun/armature configuration was performed with the finite-element code LS-DYNA using mechanical/electromagnetic coupling. For that purpose, we adapted the FE-model presented in [4] to our needs, including a typical transient current pulse. Fig 5a shows the current distribution for a total armature current of 200 kA, a typical value of a 25-mm launch package at the muzzle. The upper rail is not shown for clarity. A necessary simplification was due to the fact, that numerical modeling did not allow the separation of the armature into two halves. Yet, it is possible to show the lateral compression stress in the armature in y-direction (Fig. 5b), indicating the attractive force. These values are quite moderate and differ from the higher axial stress values in the propulsion direction by two orders of magnitude. Attractive forces in the tapered rear end of the armature can have a beneficial effect for the sabot separation, i.e. contribute to sabot opening when the armature exits the muzzle, provided that the sabot halves are laterally well constrained inside the bore during launch, to prevent them from premature separation.

Armature and sabot should act like one stiff body during the separation process, but do not necessarily consist of two different materials. The body can alternatively be built from one single block of a current carrying material, e.g. aluminum, an option, which may be studied next as an intermediate and simplifying step toward a weight-saving composite solution. Making the sabot a conductor should not dramatically affect the electric performance of the railgun, since the electric current is concentrated in rear C-shape part, as a consequence of the velocity skin effect. However, depending on the shape and intensity of the driving current pulse, one cannot exclude the presence of magnetic fields in the sabot region, generating a potential risk to electronic devices inside the projectile. In our test case, the projectile is designed for kinetic energy impact without containing any electronic parts.

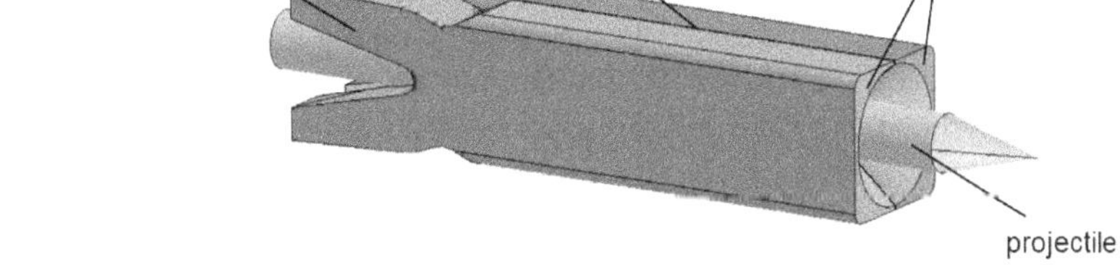

Fig. 3. Caliber 25 mm model of the railgun launch package consisting of the projectile and the two combined sabot/armature parts.

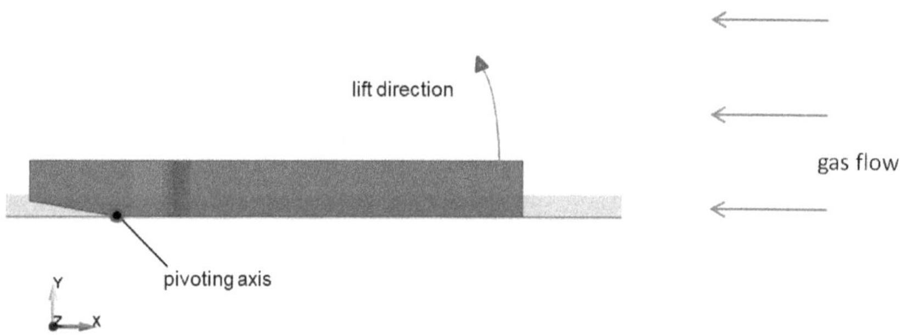

Fig. 4. Upper sabot/armature part. Wedged rear (left) and pivoting axis for rotation during initial separation stage.

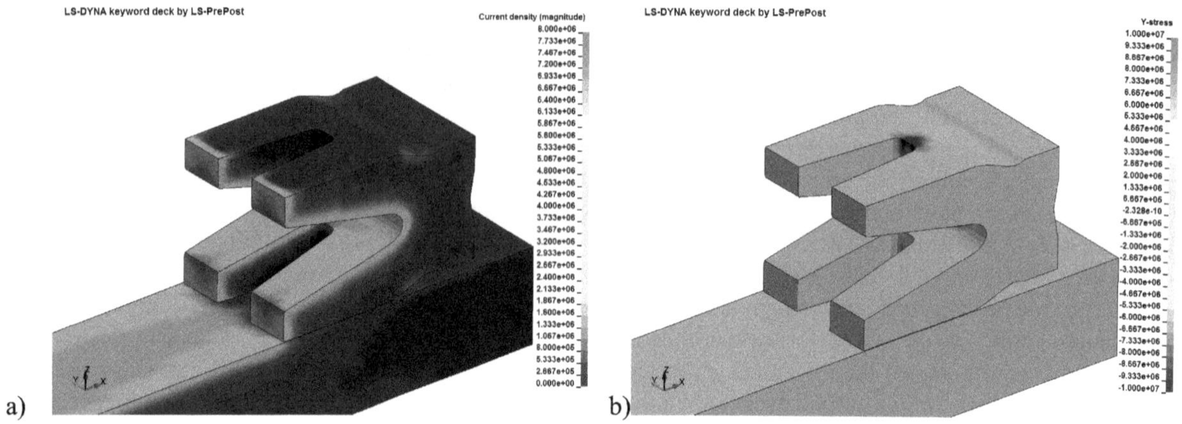

Fig 5. Numerical simulation (LS-DYNA) of the armature in a railgun (upper rail not shown). a) Current density [10^{-3} A/mm²] in the armature for I_{tot} = 200 kA. b) Corresponding lateral y-stress distribution [Pa] in the armature perpendicular to the separation plane.

3. Shock tunnel test

A free-flight test was performed in the ISL shock tunnel. Sabot-separation can be simulated experimentally thanks to the impulsively started flow which is typical for this type of wind tunnel. The sabot/armature model was built with a 3D printer using a suitable polymer material. The mass density is 1.156 g/cm³, which results in a total mass of 31 grams for each of the two parts. We have to keep in mind that the armature is usually made of aluminum whose density is significantly higher than that of the polymer material. Considering the test as a first proof of concept, it could be realized rapidly and furthermore, it serves for the validation of the numerical model. The simplified experimental model for the shock tunnel is shown in Fig. 6, left. The projectile is represented by a long cantilever steel sting located at the center of the test section, on which the polymer models of the sabot/armature components were loosely attached (see also [10] for further details). The principal diagnostics tool for this task is the differential interferometry, producing schlieren-like video images. This allows both the recording of the sabot trajectories as well as the visualization of the transient development of the flow structure (see section "Discussion of the shock tunnel results").

The gas flow parameters, shown in Table 1, were chosen as close as possible to sea level conditions, but nevertheless feasible for the shock tunnel facility in order to achieve a flow velocity of Mach 4.5. The simulation of the sabot separation in a shock tunnel considers neither the shock wave generated by the powder gases pushing the projectile, the so-called blow-out, nor the precursor shock. The former is not an issue for the railgun. However, the precursor shock wave in fact exists at the railgun muzzle. It shall be neglected here, because we estimate that the sabot discard process takes place after penetrating the corresponding shock front.

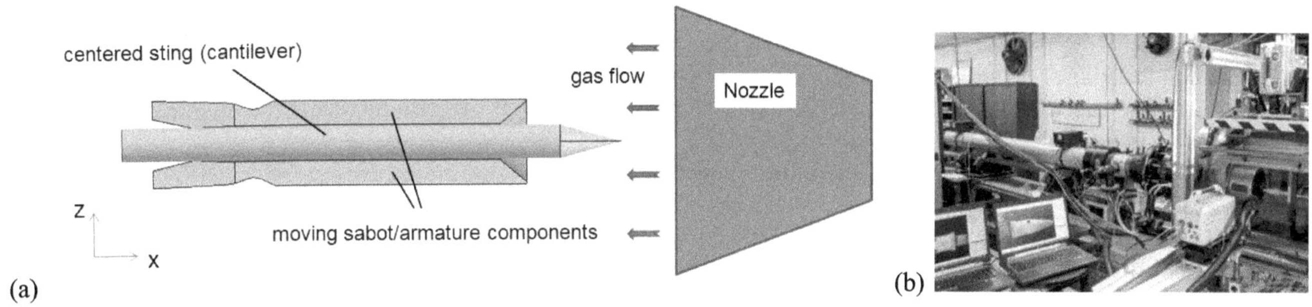

Fig. 6.(a) Schematic view of the shock tunnel test setup at the ISL. The sabot lift direction is normal to the sketch plane [10]. (b) Test facility .

Table 1. Gas flow conditions in the testing section (undisturbed)

Parameter	gas: nitrogen	
pressure	p [bar]	0.48
density	ρ [kg/m³]	0.7
temperature	T [K]	231
velocity	v [m/s]	1406
Mach number	Ma	4.54

4. Numerical simulation

The shock tunnel test was modeled and simulated numerically with the Autodyn hydrocode in parallel to the experimental work. The sabot part was imported in the Iges-format from the CAD software into LS-Prepost for meshing. The numerical mesh was then imported into the Autodyn-hydrocode for computation. The code was successfully used to investigate sabot separation in previous studies [6, 9]. Although the modeling of the gas volume is based on the Euler method, which cannot account for viscous effects in the flow, the duration of the phenomenon of some milliseconds is short enough that friction can be neglected when the focus is put on the sabot/armature motion (see also [10]). The structural parts, i.e. the sting and the sabot /armature component, are modeled as Lagrangian parts, interacting with the Eulerian volume by using the coupling option. One of the main advantages of the hydrocode is that the Lagrangian parts can deform under the gas flow pressure and additionally can interact with themselves. Depending on the chosen material models, it is therefore possible to consider damage and failure of these parts with subsequent feedback on the gas flow.

Inflow and outflow conditions were attributed to the front and to the rear side of the Euler volume respectively. The inflow parameters were basically taken from Table 1. In order to limit the computation time, the boundaries of the volume should not extend too far from the structural parts, but on the other hand, the gas flow around the parts should not be altered too much by the boundaries in order to avoid loss of accuracy.

5. Discussion of the shock tunnel results

The comparison of the experimental results with the numerical values was done by means of the differential interferometry images extracted from the high-speed video performing 40,000 fps. In a first step, the time-dependent displacement of the sabot/armature edges was measured. The edges are referred to as "leading edge" toward the arriving gas flow and "trailing edge" at the opposite side, as indicated in the sketch of Fig 7. Figures 8 and 9 show the results of displacement of both experimental measurement and the numerical computation. The numerical model allows to put "gauges" at the corresponding locations and to trace their displacements. The x-origin is located at the trailing edge position. For that reason, the x-displacement of the leading edge starts at x = 111 mm corresponding to the length of the whole part (see Fig. 8a). In this case, the displacement value decreases with increasing time, since the gas flow pushes the sabot/armature part in the negative x-direction. The vertical or y-displacement starts at y = 0, rising up to values greater than 100 mm within 3 milliseconds. The displacement of the trailing edge is presented in Fig. 8b. Again, the x-displacement shows decreasing values with time due to the gas flow in the negative x-direction. Concerning the y-displacement in the lift-direction, we observe a slight downward movement until approximately t= 1.7 ms, before the trailing edge lifts up. This initial down-move is the direct effect of the motion around the pivoting axis (Fig. 4).

By comparing the experimental with the numerical values presented in Fig. 8, a good agreement can be stated for all curves. In particular, the curve shapes are coherent over the whole considered time interval. The deviation between experiment and numerical computation increases to about 10% toward t= 3 ms for the y-displacement of both the leading and the trailing edge. The values of the numerical simulation are systematically higher than those of the experiment. One possible reason might be related to uncertainties in the experimental flow parameters of the undisturbed tunnel gas flow. In a previous case [10], the simulation in contrast underestimated the sabot velocity, which was explained by not taking into account the viscous effects in the gas flow when the Euler method is used.

Fig. 9 shows the time-dependent angle of attack of the lifting sabot/armature component. As in the case of the edge displacement, the comparison between the experimental and the numerical values show a good agreement, where again, a certain deviation is observed at later stages, i.e. at t ≥ 2.5 ms. Since the angle of attack is directly determined from the displacements of Fig. 8a and 8b, this result is not surprising. It should be further noted that the shapes of the experimental and the numerical curves are very similar.

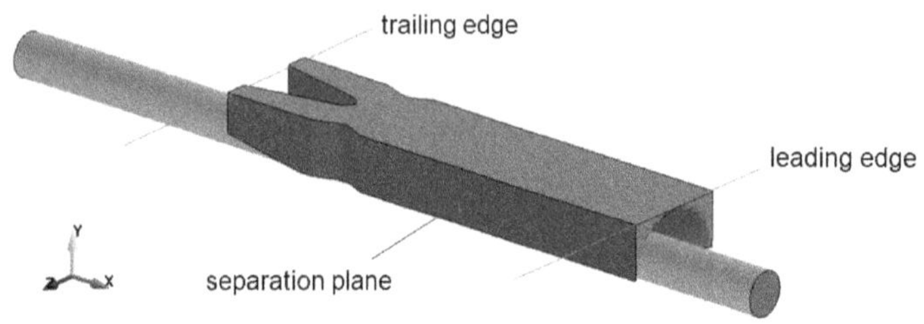

Fig. 7. Recorded edge positions on the upper sabot /armature part (see Fig. 8). Lower part not shown. Gas flow from right to left (negative x-direction).

The pictures presented in the Figures 10 to 12 show the comparison between experimental schlieren image (right-hand-side) and the numerically calculated density contour plot (left-hand-side) of the gas flow for three different instances in time.

Besides the fact that schlieren images show the density gradient, rather than the density itself, it should be emphasized that the numerical density contours are shown for the central plane which means that the view is not transparent. Therefore the aft portion of the c-shape is hidden. In order to compare the sabot/armature position of the experiment with that of the numerical simulation, its outline is indicated by a red frame showing the rear tapered part as well.

The shapes of the bow shock are mainly identical in both the experimental and the numerical images. However, viscous phenomena, such as the boundary layer which can be seen close to the sting surface region, cannot be simulated with a hydrocode based on the Euler method. Advanced CFD codes should be preferred for this purpose.

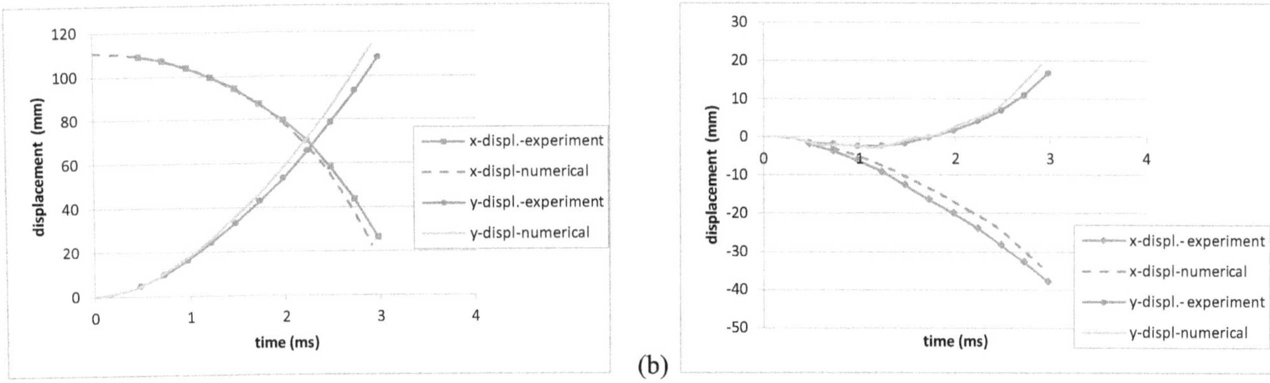

Fig. 8.Shock tube test and corresponding numerical simulation: Sabot/armature part separation from sting. Motion of (a) the leading edge and (b) the trailing edge of the upper part in the x-y-plane (see Fig. 7))

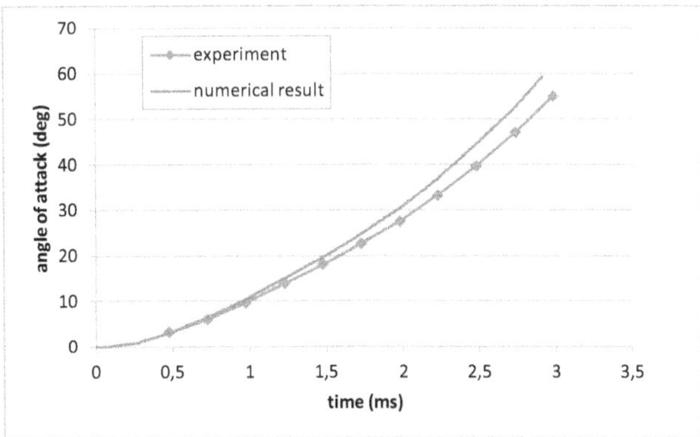

Fig. 9. Shock tube test and corresponding numerical simulation: Upper Sabot/armature part separation from sting. Angle of attack in the x-y-plane (see Fig. 7)).

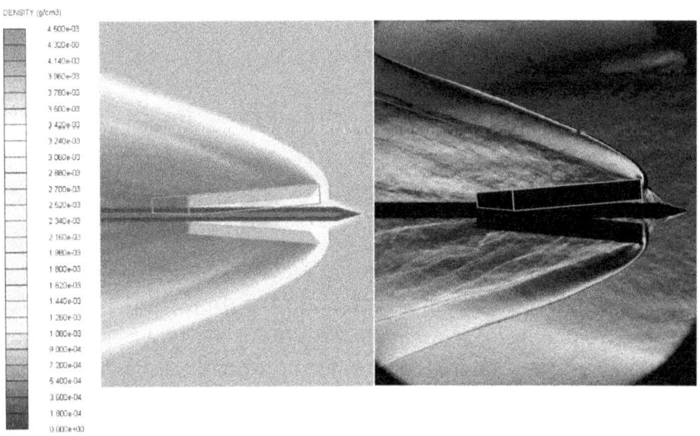

Fig. 10. Sabot/armature separation model in the shock tube at t=0.6 ms. Differential interferometry image (right) of the test and corresponding density contour plot of the numerical simulation in the central x-y-plane (left).

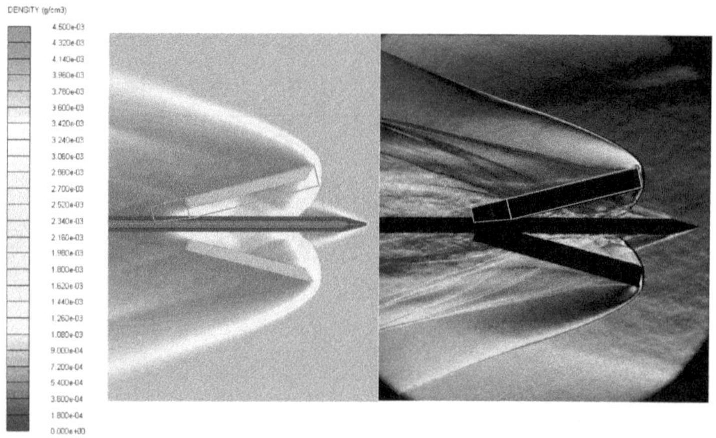

Fig. 11. Sabot/armature separation model in the shock tube at t=1.2 ms. Differential interferometry image (right) of the test and corresponding density contour plot of the numerical simulation in the central x-y-plane (left).

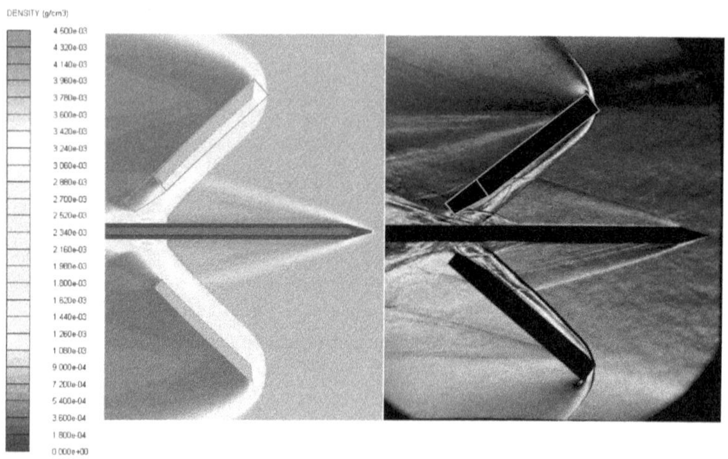

Fig. 12. Sabot/armature separation model in the shock tube at t=2.4 ms. Differential interferometry image (right) of the test and corresponding density contour plot of the numerical simulation in the central x-y-plane (left).

6. Conclusions

A railgun launch-package design is presented, which is suitable for medium-caliber rod-shaped projectiles. The integrated armature sabot design enables the lateral separation of the armature from the projectile. This makes the concept suitable for terminal ballistics testing in a laboratory environment. The C-shape armature geometry was adapted to the specific sabot discard requirements by using a wedge-shaped armature rear part. It facilitates lift up from the projectile surface after muzzle exit.

This design was experimentally tested in the shock tunnel and modeled by means of the Autodyn hydrocode. The numerical results of the hydrocode model agree well with the experimental results. However a slight deviation between the numerical calculations and the experiment can be observed at later time. The reason for that must be analyzed, since this deviation was not observed in earlier studies. It might be due to a small error of the gas flow velocity mesasurement.

The hydrocode model uses Euler-Lagrange interaction for the sabot separation process. It simulates the shock tunnel experiment quite accurately and shows that the code has proven to be suitable for further design optimization studies of railgun launch packages.

Originally developed for the sabot discard simulation, the hydrocode model can be easily adapted to simulate the dynamic loads acting on the launch package during the in-bore acceleration. It can be used to verify that it withstands the mechanical stresses occurring during that stage [6]. In the simplest case, lateral constraints are introduced which represent the lateral confinement due to the in-bore walls. The constraints are applied by zero lateral motion boundary conditions.

Acknowledgements

The authors wish to thank the ISL division "flight techniques for projectiles". In particular we wish to acknowledge the work of the corresponding groups, i.e. the "aerodynamics for measurement & simulations" group for performing the shock tunnel test and the "aerodynamics & exterior ballistics" group for manufacturing the tested sabot parts.

References

[1] Schneider M.,"Electromagnetic launch technology for hypervelocity applications - the multiple armature approach of ISL", Proceedings of the 11th Hypervelocity Impact Symposium (2011), pp. 653-665.

[2] Hundertmark S., Vincent G., Simicic D., "Developing a Launch Package for the PEGASUS Launcher", IEEE Transactions on Plasma Science, Vol. 45, Issue 7, July 2017.

[3] Schneider M., Schuppler C., Alouahabi F., "Single shot performance of RAFIRA", 17th International Symposium on Electromagnetic Launch Technology, 7-11 July 2014.

[4] Reck B., Hundertmark S., Vincent G., Schubert F., Schneider M., "Investigation of rail deformation and stress wave propagation in the ISL-NGL60 railgun", IEEE Transactions on Plasma Science, Vol. 47, Issue 5, May 2019.

[5] Satapathy S., McNab I.R., Erengil M., Lawhorn W.S., "Design of an 8-MJ Integrated Launch Package", IEEE Transactions on Magnetics, Vol. 41, No. 1, January 2005.

[6] Reck B., Hundertmark S., Simicic D, Hruschka R., Sauerwein B., Leopold F., Schneider M., " Behavior of a railgun launch package at the muzzle and during sabot discard", IEEE Transactions on Plasma Science, Vol. 47, Issue 5, May 2019.

[7] Erengil M.E., Zielinski A.E., "Effect of Electromagnetic Launch on Armature/Sabot Discard", IEEE Transactions on Magnetics, Vol. 37, No. 1, January 2001.

[8] Schmidt E M. Measurement of sabot discard and analysis of associated launch disturbances, Technical Report ARBRL-TR-02157, April 1979.

[9] Seiler S., Legner H.H., Miller M., Reinecke W.G., "Intermediate Ballistics Model for Hypervelocity Projectiles Launched from EM Guns", 15th International Symposium on Ballistics, Jerusalem, Israel, 21-24 May, 1995.

[10] Reck B., Hruschka R., Sauerwein B., " Flow Visualization of Projectile-Sabot Separation in a Pulsed Wind Tunnel in Comparison with Numerical Simulations ", 18th International Symposium on Flow Visualization, Zurich, Switzerland, 26-29 June, 2018.

Proceedings of the 2019 Hypervelocity Impact Symposium
HVIS2019
April 14-19, 2019, Destin, FL, USA
HVIS2019-117

Timing Delay Analysis of the Auxiliary Pump Technique to Improve the Performance of an Implosion-Driven Hypervelocity Launcher

Mafa Wang[a]*, Justin Huneault[b], Andrew J. Higgins[b], Sen Liu[a]

[a]*Hypervelocity Aerodynamics Institute, China Aerodynamics R&D Center, No.6, South Section, Second Ring Road, Mianyang, Sichuan, China*
[b]*Department of Mechanical Engineering, McGill University, Montreal, Canada*

Abstract

In order to understand the irreproducibility of the auxiliary pump technique, an interior ballistic solver taking into account reservoir collapse has been used to simulate the performance of launchers. Launchers with different detonation velocities, explosive lengths, and timing delays (the difference between the initiation time of the pump tube explosives and auxiliary pump explosives) of the auxiliary pump have been calculated. The effective timing delay region, which could achieve a velocity gain larger than 1.0 km/s, has been discussed. And its influence factors, such as the detonation velocity of auxiliary pump explosives and the inner-wall velocity of the reservoir, have been analyzed. Results show that the velocity gain decreases with an increase in the timing delay and increases with the increasing length of explosives. The effective timing delay region is about 2μs and depends weakly on the detonation velocity and the length of explosives when using the same explosives for the pump tube and the reservoir. Nevertheless, low detonation velocity of the reservoir explosives and high inner-wall velocity could improve the effective timing delay region, but the maximum effective timing delay region cannot exceed 10μs, which is not easily accomplished experimentally. Therefore, the auxiliary pump technique should not be a very reproducible technique.

Keywords:Hypervelocity launch; Implosion-Driven Launcher;Auxiliary pump; Timing delay; Velocity gain

1. Introduction

The Implosion-Driven Launcher concept (IDL) is shown in Fig. 1. It is a light-gas gun in which the driver gas is compressed by a linear explosive driver[1, 2]. The driver is composed of a thin-walled steel tube, a layer of explosives, a steel casing and a light gas (helium or hydrogen). As the explosive is ignited, the detonation drives the tube walls inward and forms an explosively driven piston that travels into the gas at the speed of the detonation. The principle of launching is similar to a Two-Stage Light Gas Gun (2SLGG). The implosion driven launcher concept was first developed by the Physics International Company (PI) in the late 1960's [3-6]. At that time, PI was able to launch a 2-g magnesium-lithium projectile with 15.9 mm caliber up to 12.2 km/s[5]. However, there have been few publications since then. McGill University has been redeveloping and improving the PI design over the last decades. They are now able to launch a 0.37 g magnesium projectile with 8.0 mm caliber up to 10.4 km/s with the single-stage IDL.

* Corresponding author.*E-mail address:*wangmafa@cardc.cn

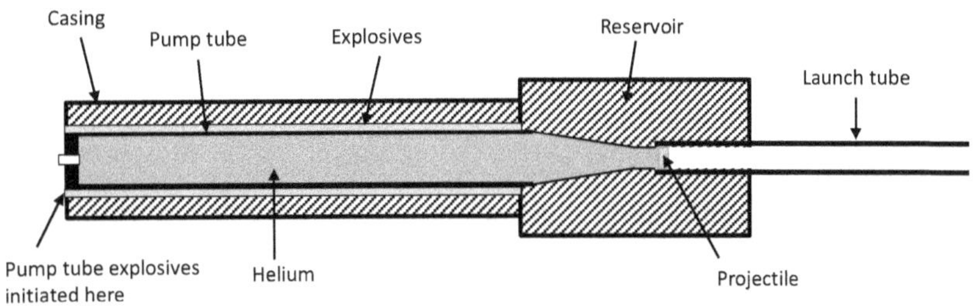

Fig. 1 The schematic of the implosion-driven launcher

There are many ways to improve the performance of the IDL such as using higher detonation velocity explosives, using a second stage augmentation technique, or using explosives with a gradient velocity to squeeze the pump tube. The auxiliary pump technique is another one of these techniques and was proven by PI in the 1960's. According to this technique, explosives are installed around the reservoir of the single-stage IDL in order to dynamically confine the reservoir, as shown in Fig. 2, and thus additional compression is provided to the driver gas[7]. It is simple to design the auxiliary pump geometry, but it's difficult to determine the initiation timing delay of explosives which is the difference between the initiation times of the pump tube explosives and the auxiliary pump explosives. On one hand, the auxiliary pump explosives must be ignited early enough to add the compression pressure effectively. On the other hand, the projectile can be damaged by the stress wave derived from the auxiliary pump detonation when the initiation time is too early. Therefore, the timing must be precisely designed to launch the projectile. The effectiveness of the auxiliary pump technique depends strongly on the timing delay of the explosive ignition.

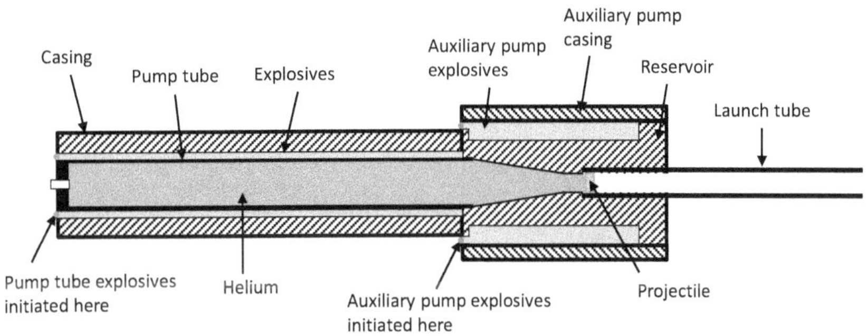

Fig. 2 The schematic of auxiliary pump launchers

The auxiliary pump technique has been tested, and velocity gains of 4 km/s and 1 km/s have been reported in the tests of the Physics International Company and McGill University, respectively. In our tests, a gain of about 1.5 km/s (from 10.28 km/s to 11.84 km/s) has been observed. Two repeated tests were carried out. However, the velocity gain of the muzzle velocity did not appear again. The results indicate that the auxiliary pump technique may not be reproducible.

In order to understand the irreproducibility problem, an interior ballistic solver taking into account reservoir collapse has been adopted to simulate the performance of launchers. Launchers with different detonation velocities, explosive lengths, and timing delays of the auxiliary pump have been calculated.

2. Interior ballistic solver

In order to simulate the internal ballistics of an IDL, a quasi-one-dimensional Lagrangian gas dynamics solver developed by Huneault[2] was used. A second-order accuracy (in space and time) numerical scheme, developed by von Neumann and Richtmyer[8] with a modified energy formulation developed by Noh[9]was adopted. The detailed finite difference equations for the scheme, test cases to validate the gas dynamics solver, and ideal IDL simulations were presented in Huneault's paper[2] and thesis[7]. Based on Huneault's work, a simple auxiliary pump model has been added to simulate the auxiliary pump technique in this paper.

2.1. Auxiliary pump model

A two-dimensional axisymmetric hydro-code is required to completely capture radial collapsing along the axial length of the reservoir. Besides a dynamic structural model that is capable of simulating the stress wave dynamics, an explosion is needed to determine the radial velocity and position history of the reservoir walls[7].

Nevertheless, the auxiliary pump effect has been simplified and added into the quasi-one-dimensional Lagrangian gas dynamics solver. The simplification contains two steps. In step one, the launcher reservoir is dissected into a series of axial slices, and the radial expansion of each slice is computed independently. In step two, a simple model is given to simulate the inner wall of the reservoir. It is assumed that the inner wall of the reservoir starts to collapse at a constant velocity when the stress waves generated by the auxiliary pump explosion arrive. Based on this assumption, only the velocity of the inner wall is needed as the boundary condition in the simulation.

2.2. Stress wave timing

The projectile could be damaged by the stress wave due to explosive detonation in the reservoir if the timing delay is not long enough. According to the length of the auxiliary pump explosive and the location of the projectile, there are two cases of note. The first one is when the projectile is surrounded by the explosive. The other one is when the projectile is in front of the explosive, as shown in Fig. 3. When the detonation velocity of the auxiliary pump explosive is faster than the speed of sound in steel, the speed of the stress wave along the axial direction of the launcher in the inner wall of the reservoir is equal to the detonation velocity of the explosives. Thus, one only needs to calculate whether the stress wave propagating at detonation velocity can catch up to the projectile in case one. In case two, the stress wave firstly propagates at the detonation velocity in the region surrounded by the explosives and catches up with the projectile at about 5 km/s ~5.5 km/s after the detonation finishes. A minimum timing delay exists when the stress wave effect is considered in simulation.

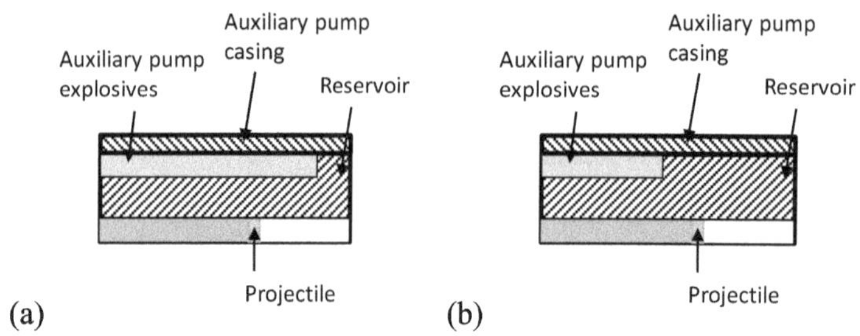

Fig. 3 Two cases of the projectile location (a) the projectile is surrounded by the explosive (b) the projectile is in front of the explosive

3. Simulation results and analysis

3.1. Results of the single-stage launcher

For convenience, a single-stage launcher with initial filling pressure of 5MPa is adopted. The parameters of the structure are shown in Fig. 4: the caliber is 8 mm, the length of the launch tube is 308 mm, the diameter of the pump tube is 16.5 mm, the length of the pump tube is 490 mm, and the length of gradual area change section is 49 mm. The velocity history and driving pressure history of the projectile in cases of different detonation velocities of the pump tube explosives are shown in Fig. 5, where D is the detonation velocity of the auxiliary pump explosive. As can be seen, the projectile begins to move earlier when the detonation velocity is higher, and the muzzle velocity of the projectile increases with the increasing detonation velocity, but the maximum driving pressure is higher too.

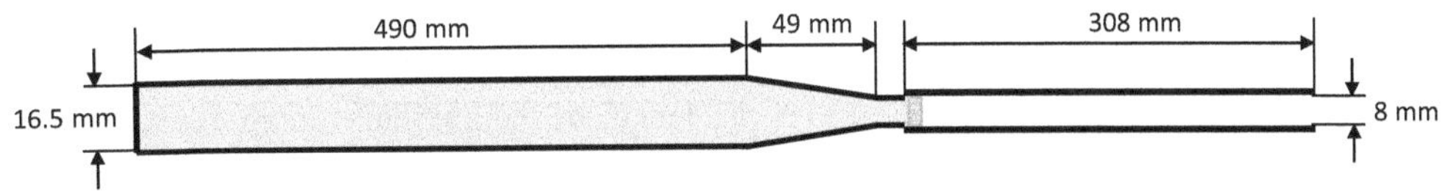

Fig. 4The structure of the single-stage launcher

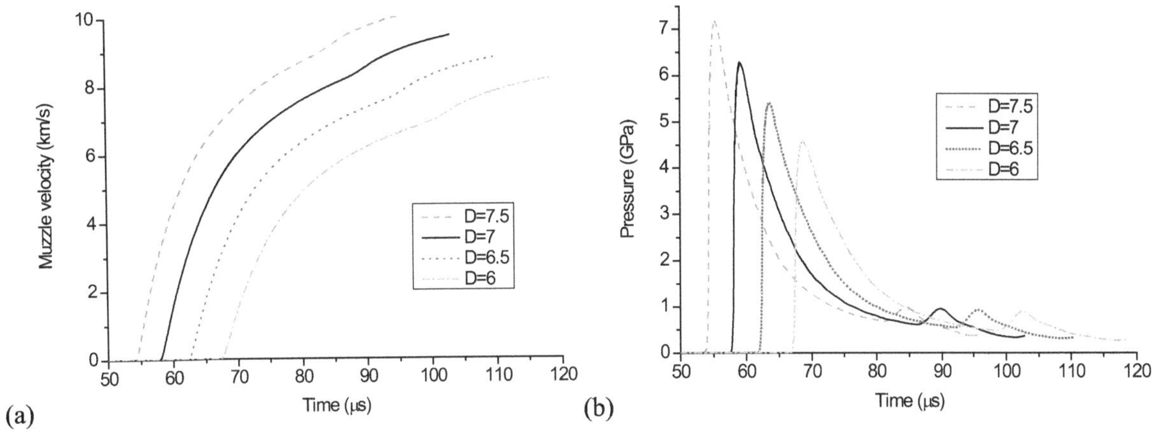

Fig. 5 Simulation results of the single-stage launcher (a) Velocity-time history of the projectile (b) Driving pressure-time history of the projectile

3.2. Results of the auxiliary pump technique launchers

A typical launcher with the auxiliary pump technique has been modeled. The length of the explosive that starts from the beginning of the gradual area change section is 10.8 cm, the detonation velocity is 7 km/s which is equal to that of the pump tube explosives, the timing delay is set at 55μs, and the inner wall velocity of the reservoir is 60m/s. The velocity and driving pressure history of the projectile are shown in Fig. 6. The velocity of the projectile is the same as that of the single-stage launcher at the beginning. The projectile starts to move faster at about 65μs and gets a higher muzzle velocity. The maximum pressures of the two cases are exactly the same. However, as can be seen in Fig. 6(b), the driving pressure of the launcher with the auxiliary pump technique is a little higher from 65μs to 90μs. This auxiliary pump technique improves the muzzle velocity from 9.46 km/s to 10.58 km/s.

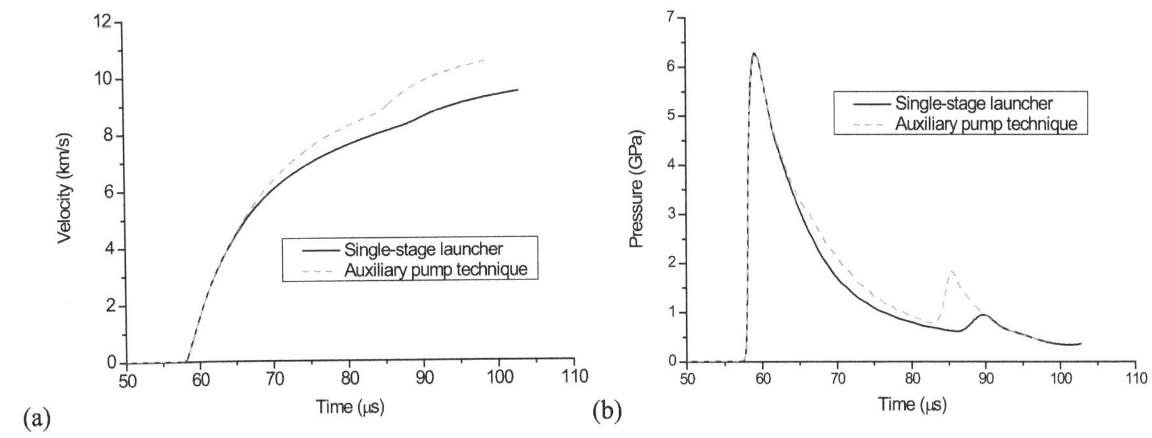

Fig. 6 Simulation results of a typical auxiliary pump technique launcher (a) Velocity-time history of the projectile (b) Driving pressure-time history of the projectile

Moreover, launchers with different detonation velocities, explosive lengths, and timing delays of the auxiliary pump explosives have been calculated. Simulations resulting in the projectile being crushed from the stress waves have been abandoned. The rest of the results are shown in Fig. 7, where L is the length of the auxiliary pump explosives starting at the beginning of the gradual area change section. The detonation velocity of the pump tube explosives is the same as that of the auxiliary pump explosives. The lengths of the auxiliary pump explosives are 10.8 cm for all launchers in Fig. 7 (a) and the detonation velocities of explosives is 7 km/s for all launchers in Fig. 7 (b). As can be seen, muzzle velocity decreases with increasing timing delay no matter what the detonation velocity and length of explosives are. The higher the detonation velocity is and the longer the explosive is, the higher the muzzle velocity is. The timing delay is larger when the detonation velocity decreases.

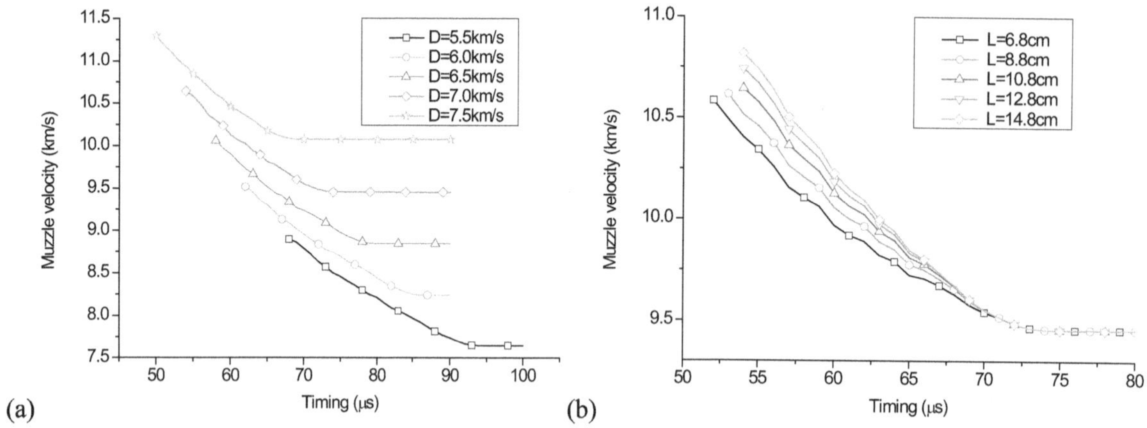

Fig. 7 Results of auxiliary pump technique (a) Velocity-timing relationship with different detonation velocities (b) Velocity-timing relationship with different explosive length

3.3. Velocity gain of the auxiliary pump technique

To obtain the velocity gain of the auxiliary pump technique, the muzzle velocities of launchers with different auxiliary pump parameters have been subtracted from the muzzle velocities of single-stage launchers. Figure 8 shows these velocity gains. We can see that the velocity gain decreases with an increase in the timing delay and increases with the increase of the length of explosives. However, the maximum gain is approximately the same for different detonation velocities and lengths of explosives.

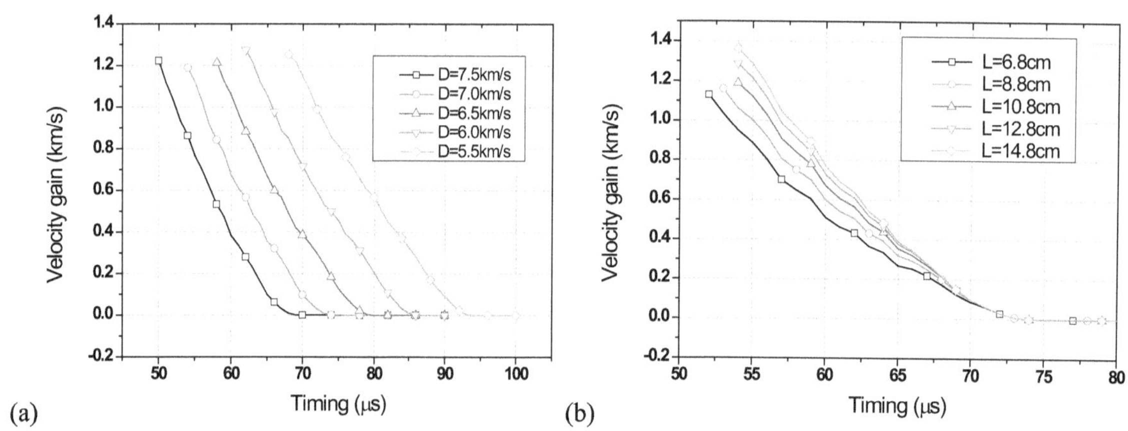

Fig. 8 Velocity gains of auxiliary pump launchers (a) launchers with different detonation velocities (b) launchers with different lengths of explosives

Further analysis shows that the timing delay region for obtaining a gain greater than 0.5 km/s is about 10μs. This region depends weakly on the detonation velocity and the charge length. In addition, the timing delay region for a gain larger than 1.0 km/s, named effective timing delay region in this paper, is about 2μs. This region is too small for the auxiliary pump technique to obtain a gain of 1.0 km/s in tests. However, there may be other parameters that could increase the effective timing delay region. Two parameters will be further discussed in the next section.

4. Discussion of the detonation velocity and the inner-wall velocity of the reservoir

4.1. The detonation velocities

Hereto, the detonation velocities of pump tube explosives and auxiliary pump explosives have been the same. In this section, the detonation velocity of pump tube explosives is set as 7 km/s and that of auxiliary pump explosives is changed from 5.5 km/s to 7.5 km/s. The length of the explosives is 10.8 cm. As can be seen from Fig. 9, the velocity gain is almost the same for different detonation velocities. However, a lower detonation velocity results in a smaller minimum timing delay. When reducing the detonation velocity from 7.5 km/s to 5.5 km/s, the effective timing delay region increases from 2μs to 6μs. It indicates that lower detonation velocity could make the auxiliary pump technique a little more reproducible.

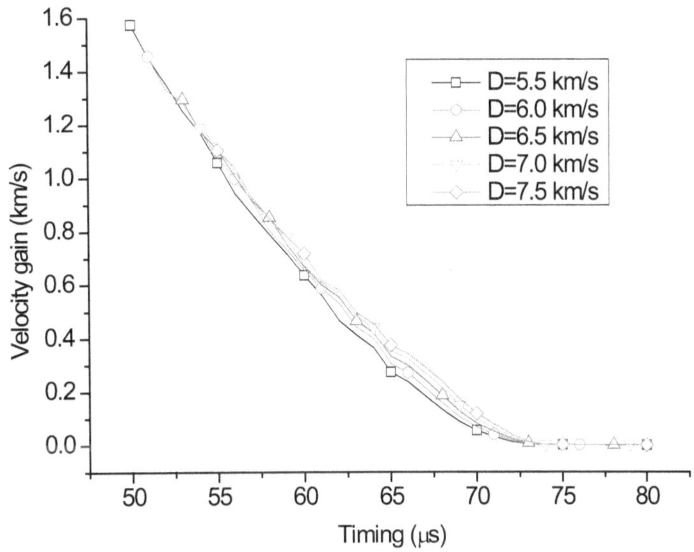

Fig. 9 Velocity gain results of launchers with different detonation velocities of pump tube explosives and auxiliary pump explosives

4.2. The inner-wall velocity of the reservoir

For simulation stability, the inner wall velocity of the reservoir, named U_{pump}, was set at 60 m/s in the above simulations. Obviously, the velocity should not be a constant velocity, and its average velocity is not equal to 60 m/s exactly. Therefore, launchers with different U_{pump} values are simulated here. Results are shown in Fig. 10. In these simulations, the lengths of the auxiliary pump explosives are 10.8 cm and the detonation velocities of explosives are 7 km/s. As can be seen, when U_{pump} increases, velocity gains increase rapidly, but the timing delay region that could improve the muzzle velocity is the same. However, the effective timing delay region could be enlarged by increasing the inner wall velocity of the reservoir. For instance, the effective timing delay region is enlarged from 3μs to 8μs by increasing U_{pump} from 60 m/s to 100 m/s.

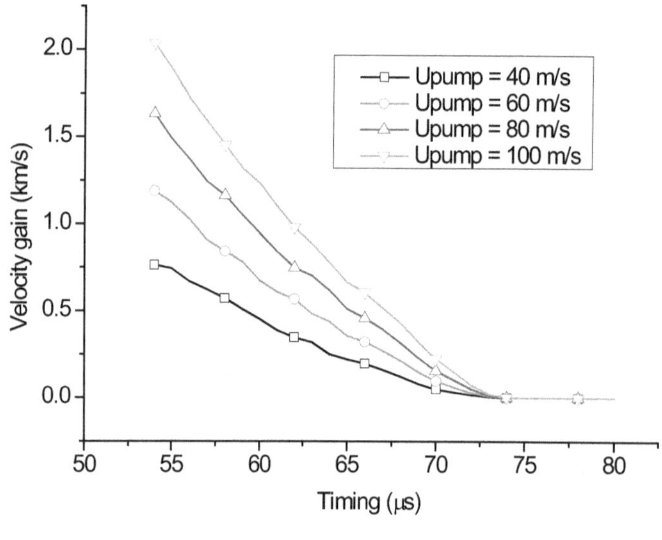

Fig. 10Velocity gain results of the launchers with different inner wall velocity

The inner wall velocity depends on the reservoir's mass and casing, the mass of the explosives, and the pressure inside the reservoir. According to the Gurney equations[7], the inner wall velocity strongly depends on the mass ratio of the reservoir to the explosives. It implies that increasing the mass of the explosives could improve the feasibility of the auxiliary pump technique to make this technique more reproducible. However, this method could not enhance the timing delay region for augmenting the muzzle velocity. Hence, the effectiveness from increasing explosive mass is limited.

5. Conclusion

An interior ballistic solver involving reservoir collapse has been used to simulate launchers utilizing the auxiliary pump technique. Launchers with different detonation velocities, lengths, and timing delays of the auxiliary pump explosives have been calculated. Results show that the auxiliary pump technique could improve the performance of launchers by initiating the auxiliary pump explosives at the appropriate time. However, the effective timing delay region is not easily managed in experiments.

In addition, the influences of the detonation velocity of auxiliary pump explosives and the inner-wall velocity of the reservoir have been further discussed. Results show that lower detonation velocity and higher inner-wall velocity could extend the effective timing delay region, but the effective timing delay region cannot exceed 10μs.

What's more, the reservoir expansion effect has not been considered in these simulations. This effect may reduce the muzzle velocity in the launchers without the auxiliary pump technique, and therefore, the velocity gain may increase and accordingly the effective timing delay region may be greater.

Acknowledgements

This work was supported by Nature Science Foundation of China (Grant 11802330).

References

[1] Loiseau, J., Huneault, J., Higgins, A. J., 2013. "Development of a Linear Implosion-driven Hypervelocity Launcher," The 12th Hypervelocity Impact Symposium. p. 77-87.

[2] Huneault, J., Loiseau, J., Higgins, A. J., 2013. "Coupled Lagrangian Gasdynamic and Structural Hydrocode Solvers for Simulating an Implosion-Driven Hypervelocity Launcher," 51st AIAA Aerospace Science Meeting including the New Horizons Forum and Aerospace Exposition. Grapevine, p. 1-21.

[3] Crosby, J. K., Gill, S. P. Feasibility study of an explosive gun. Physics Internatioanal Company, NASA CR-709, 1967.

[4] Moore, E. T. J. Explosive hypervelocity launchers. Physics Internatioanal Company, No. PIFR-051, 1968.
[5] Watson, J. D. High-velocity explosively driven guns. Phisics International Company, NASA CR-1533, 1970.
[6] Watson, J. D. A summary of the development of large explosive guns for reentry simulation. Physics International Company, AD712394, 1970.
[7] Huneault, J., 2013. Development of an Implosion-Driven Hypervelocity Launcher for Orbital Debris Impact Simulation. Montreal: McGill University.
[8] von Neumann, J., Richtmyer, R. D., 1950. A Method for the Numerical Calculation of Hydrodynamic Shocks, Journal of Applied Physics, 21, p. 232-237.
[9] Noh, W. F., 1987. Errors for Calculations of Strong Shocks Using an Artificial Viscosity and an Artificial Heat Flux, Journal of Computational Physics, 72, p. 78-120.

Proceedings of the 2019 Hypervelocity Impact Symposium
HVIS2019
April 14-19, 2019, Destin, FL, USA
HVIS2019-127

Extended Abstract+

Laser-Driven High-Velocity Microparticle Launcher In Atmosphere And Under Vacuum

David Veysset[a,*], Yuchen Sun[a,b], Steven E. Kooi[a], Keith A. Nelson[a,b]

[a]*Institute for Soldier Nanotechnologies, Massachusetts Institute of Technology, 77 Massachusetts Avenue, Cambridge, MA 02139, USA*
[b]*Department of Chemistry, Massachusetts Institute of Technology, 77 Massachusetts Avenue, Cambridge, MA 02139, USA*

1. Introduction

The study of high-velocity microparticles is important to a wide range of both space and terrestrial applications. In space, high- and hyper-velocity micro-debris and micrometeorites, while also a subject of study, pose a threat to equipment and personnel integrity [1–4]. On earth, high-velocity microparticle impact can be, for instance, utilized for therapeutic purposes in the field of biolistics [5] or to build metallic coatings via the cold spray method [6]. While macroscale projectile impacts have been studied using well established experimental tools, such as light-gas guns, optical methods are gaining interest in the field of micro-particle impacts.

The laser-induced particle impact test (LIPIT) is an all-optical, table-top platform that is used to accelerate microparticles to high velocities [7]. Particle trajectories and impact events can be visualized in real time using a microscope objective and an ultra-high-speed camera [8]. This experimental approach has been used to study the high-strain-rate behavior of a variety of materials including nanocomposites [7, 9], polymers [8, 10–12], gels [13], and metals [14–17]. Here, we present a modified version of the LIPIT that allows for experiments under vacuum conditions and higher velocities than previously reported through the design of a new launch pad configuration. We report velocities presently achievable for various micro-projectiles and discuss the performance of the modified platform.

2. Materials and methods

As shown in Fig. 1(a), micro-particles are sparsely deposited on a launch pad assembly, which is composed of a 210-μm-thick, 1-inch diameter, glass substrate (Corning No. 2 microscope cover slip) and a 670-nm-thick gold layer, fabricated using e-beam evaporation. A focused laser pulse (Nd:YAG, 532-nm wavelength, 10-ns duration, 50-μm spot diameter) ablates the gold film causing acceleration of the microparticle. The accelerated particle travels in free space toward a target and through a pinhole with 200-μm diameter that acts as a shield from ablation-generated debris. The pinhole is positioned about 1 mm away from the launch pad. Using an ultrahigh-frame-rate camera, which can yield up to 16 frames with nanosecond time resolution (SIMX16, Specialised Imaging), and a synchronized quasi-cw laser imaging pulse for illumination (Cavilux, Specialised Imaging 640-nm wavelength, 10-μs duration), the impact events can be recorded in real time (Fig. 1(b)). The camera provides a minimum exposure time of 3 ns, which is short enough to visualize microparticles travelling at velocities up to 50 km/s. Because of the localized laser ablation and the intentionally sparse distribution of the microparticles on the launch pad surface, a single particle is selected to be accelerated in each laser shot.

+ Full paper can found at https://www.sciencedirect.com/journal/international-journal-of-impact-engineering/special-issue/10MTGP5W4VJ

* Corresponding author. Tel.: +1-617-324-6425.
E-mail address: dveysset@mit.edu.

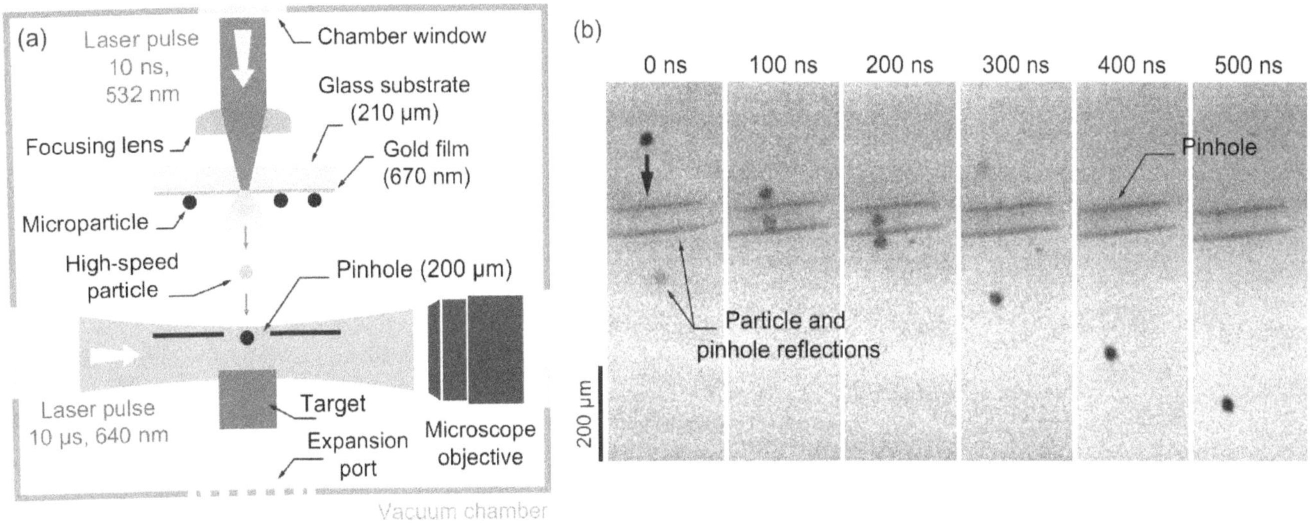

Fig. 1. (a) Microparticle launcher schematic. (b) Multi–frame sequence showing a 30-µm silica particle launched at 1085 m/s going through the pinhole. The reflections of the particle and the pinhole from the pinhole holder are visible in the images. 6 of 16 frame are presented here cropped from their initial size for ease of visualization. No target was present here. The time stamps, shown above the frames, indicate the delay relative to the first frame of the sequence.

The vacuum chamber was designed to fit the focusing lens for the excitation pulse, the launch pad, the target, the pinhole, and the microscope objective. A cubic chamber (16×16×16 inches internal dimensions) was purchased from Kurt J. Lesker Company. It is comprised of customized windows: two 2-inch windows with anti-reflection-coating for laser illumination and imaging optics and one 1-inch window (Brewster-angle cut) for the laser excitation pulse. An 8-inch flange at the bottom of the chamber, with its aperture in line with the impact direction, allows for potential expansion of the chamber. The minimum pressure achieved with the current design is 6×10^{-3} mbar. The target can be placed at a desired distance from the launch pad between <1 mm and ~30 cm.

3. Launcher performance

Figure 2(a) demonstrates how the laser pulse energy is adjusted to control particle velocity. Particle velocity increases monotonically with laser energy until ~8 mJ where saturation in optical absorption of the metallic film prevents further increase. Maximum velocities were measured for a wide range of particle masses both under atmospheric and vacuum conditions (Fig. 2(b)). With the current launch pad configuration, we note a power law relationship between the maximum particle velocity and the particle mass. This relationship can help predict the maximum velocity for a given particle type and size and suggests that submicron particles could be accelerated to tens of km/s, relevant to hypervelocity applications. However, such particles would fall below our optical resolution (~1 µm), and would be blurred at such velocities in a 3-ns measurement window, making direct optical observation difficult.

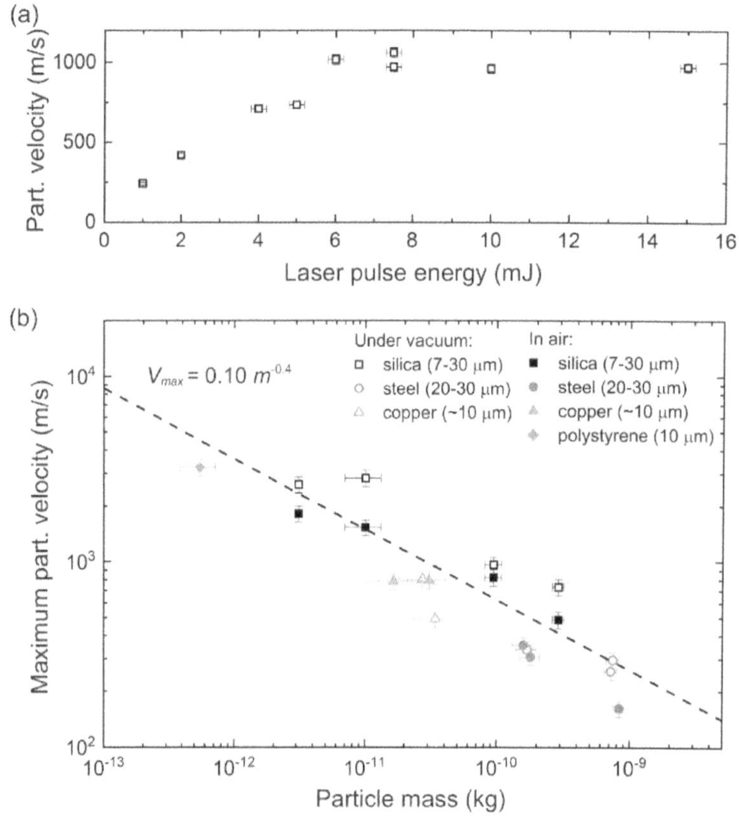

Fig. 2. (a) Particle velocity versus laser pulse energy for 21-μm silica particles under 0.3 mbar pressure. Particle velocity saturates at laser energies higher than 8 mJ. (b) Maximum particle velocity versus particle mass for a wide range of particle types and diameters under vacuum (0.3 mbar) and in atmospheric conditions.

As laser energy is further increased (typically above 20 mJ, 5% of total available laser pulse energy), particle damage signified by fragmentation for ceramic particles and melting for polymeric and metallic particles is often observed in real-time videos. This is detrimental when intact particles are desired though could be used to study fragmented or molten particle impacts. With further improvements of the launch pad assembly, we foresee substantially higher achievable velocities.

By performing experiments under vacuum, we suppress the undesirable blast wave that is generated upon laser ablation of the gold film. Indeed, the intense blast wave can damage fragile targets (such as nanometric films) when they are placed to close to the launch pad. Because the blast wave dissipates as it diverges spherically, one could place the target farther away from the launch pad to avoid blast effects. However, due to the air drag exerted on the projectile, the microparticle velocity rapidly decreases with distance so a compromise needs to be found between blast reduction and particle velocity. Under vacuum, both the drag and the blast are reduced allowing more flexibility in distance between the launch pad and target.

4. Conclusions

The modified LIPIT platform presented in this paper enables acceleration of a variety of microparticles to high velocities. We have measured the maximum velocities achievable with the method under air and vacuum and presented the particle mass dependence on the maximum velocity. With reduced blast and drag effects under vacuum, both fragile and atmosphere-sensitive samples can be tested. This platform will facilitate the study of a wide range of materials to be exposed to high-velocity microparticle impacts under both earth (atmospheric) and space conditions. This method will provide a better understanding of material responses under high strain rate and will guide future high-performance material design.

Acknowledgements

The authors would like to thank Prof. Hajime Yano for fruitful discussions and for providing silica and polymeric particles. This material is based upon work supported in part by the U. S. Army Research Office through the Institute for Soldier Nanotechnologies, under Cooperative Agreement Number W911NF-18-2-0048.

References

1. Berthoud, L., Mandeville, J.C., 1997. Material damage in space from microparticle impact, Journal of Materials Science 32, p. 3043.
2. Wyatt, M.C., 2008. Evolution of Debris Disks, Annual Review of Astronomy and Astrophysics 46, p. 339.
3. Westphal, A.J., Stroud, R.M., Bechtel, H.A., Brenker, F.E., Butterworth, A.L., Flynn, G.J., Frank, D.R., Gainsforth, Z., Hillier, J.K., Postberg, F., Simionovici, A.S., Sterken, V.J., Nittler, L.R., Allen, C., Anderson, D., Ansari, A., Bajt, S., Bastien, R.K., Bassim, N., Bridges, J., Brownlee, D.E., Burchell, M., Burghammer, M., Changela, H., Cloetens, P., Davis, A.M., Doll, R., Floss, C., Grun, E., Heck, P.R., Hoppe, P., Hudson, B., Huth, J., Kearsley, A., King, A.J., Lai, B., Leitner, J., Lemelle, L., Leonard, A., Leroux, H., Lettieri, R., Marchant, W., Ogliore, R., Ong, W.J., Price, M.C., Sandford, S.A., Tresseras, J.-A.S., Schmitz, S., Schoonjans, T., Schreiber, K., Silversmit, G., Sole, V.A., Srama, R., Stadermann, F., Stephan, T., Stodolna, J., Sutton, S., Trieloff, M., Tsou, P., Tyliszczak, T., Vekemans, B., Vincze, L., Von Korff, J., Wordsworth, N., Zevin, D., Zolensky, M.E., 2014. Evidence for interstellar origin of seven dust particles collected by the Stardust spacecraft, Science 345, p. 786.
4. Nna-Mvondo, D., Khare, B., Ishihara, T., McKay, C.P., 2008. Experimental impact shock chemistry on planetary icy satellites, Icarus 194, p. 822.
5. Kendall, M.A.F., 2002. The delivery of particulate vaccines and drugs to human skin with a practical, hand-held shock tube-based system, Shock Waves 12, p. 23.
6. Moridi, A., Hassani-Gangaraj, S.M., Guagliano, M., Dao, M., 2014. Cold spray coating: review of material systems and future perspectives, Surface Engineering 30, p. 369.
7. Lee, J.-H., Veysset, D., Singer, J.P., Retsch, M., Saini, G., Pezeril, T., Nelson, K.A., Thomas, E.L., 2012. High strain rate deformation of layered nanocomposites, Nature Communications 3, p. 1164.
8. Veysset, D., Hsieh, A.J., Kooi, S., Maznev, A.A., Masser, K.A., Nelson, K.A., 2016. Dynamics of supersonic microparticle impact on elastomers revealed by real-time multi-frame imaging, Scientific Reports 6, p. 46944.
9. Xie, W., Tadepalli, S., Park, S.H., Kazemi-Moridani, A., Jiang, Q., Singamaneni, S., Lee, J.H., 2018. Extreme Mechanical Behavior of Nacre-Mimetic Graphene-Oxide and Silk Nanocomposites, Nano Letters 18, p. 987.
10. Hsieh, A.J., Veysset, D., Miranda, D.F., Kooi, S.E., Runt, J., Nelson, K.A., 2018. Molecular influence in the glass/polymer interface design: The role of segmental dynamics, Polymer 146, p. 222.
11. Veysset, D., Hsieh, A.J., Kooi, S.E., Nelson, K.A., 2017. Molecular influence in high-strain-rate microparticle impact response of poly(urethane urea) elastomers, Polymer 123, p. 30.
12. Hyon, J., Lawal, O., Fried, O., Thevamaran, R., Yazdi, S., Zhou, M., Veysset, D., Kooi, S.E., Jiao, Y., Hsiao, M.-S., Streit, J., Vaia, R.A., Thomas, E.L., 2018. Extreme Energy Absorption in Glassy Polymer Thin Films by Supersonic Micro-projectile Impact, Materials Today 21, p. 817.
13. Veysset, D., Kooi, S.E., Maznev, A.A., Tang, S., Mijailovic, A.S., Yang, Y.J., Geiser, K., Van Vliet, K.J., Olsen, B.D., Nelson, K.A., 2018. High-velocity micro-particle impact on gelatin and synthetic hydrogel, Journal of the Mechanical Behavior of Biomedical Materials 86, p. 71.
14. Thevamaran, R., Lawal, O., Yazdi, S., Jeon, S.-J., Lee, J.-H., Thomas, E.L., 2016. Dynamic creation and evolution of gradient nanostructure in single-crystal metallic microcubes, Science 354, p. 312.
15. Hassani-Gangaraj, M., Veysset, D., Nelson, K.A., Schuh, C.A., 2018. In-situ observations of single micro-particle impact bonding, Scripta Materialia 145, p. 9.
16. Hassani-Gangaraj, M., Veysset, D., Nelson, K.A., Schuh, C.A., 2019. Impact-Bonding with Aluminum, Silver, and Gold Microparticles: Toward Understanding the Role of Native Oxide Layer, Applied Surface Science 476, p. 528.
17. Hassani-Gangaraj, M., Veysset, D., Nelson, K.A., Schuh, C.A., 2017. Melting Can Hinder Impact-Induced Adhesion, Physical Review Letters 119, p. 175701.

Proceedings of the 2019 Hypervelocity Impact Symposium
HVIS2019
April 14-19, 2019, Destin, FL, USA
HVIS2019-011

Extended Abstract+

In-situ observation of damage evolution in polycarbonate subjected to hypervelocity impact

Nobuaki Kawai[a]*, Mikio Nagano[b,c], Sunao Hasegawa[c], Eiichi Sato[c]

[a]Institute of Pulsed Power Science, Kumamoto University, 2-39-1 Kurokami, Chuo-ku, Kumamoto 840-8555, Japan
[b]Department of Aeronautics and Astronautics, Tokyo Metropolitan University, 6-6 Asahigaoka, Hino, Tokyo 191-0065, Japan
[c]Institute of Space and Astronautical Science, Japan Aerospace Exploration Agency, 3-1-1 Yoshinodai, Chuo-ku, Sagamihara 252-5210, Japan

1. Introduction

In the fields of space engineering and planetary science, hypervelocity impact phenomena have been studied as they relate to the space debris problem and planetary impact. With regard to hypervelocity-impact-induced damage, many studies focus on the evaluation of impact-damage geometry and morphology, for example, to construct the ballistic limit equations and/or penetrating equations for space structures, and to predict the size and shape of crater and fragments generated by planetary impact [1-4]. While the final state or late stage of an impact event are of primal interest, damage accumulation at early stages affect the overall outcome of the impact event. The understanding of hypervelocity-impact-damage processes lead to improvement of material-response models for hypervelocity impact and higher fidelity simulations of hypervelocity impact events. Under such a background, we have performed real-time imaging of hypervelocity-impact events on transparent materials to investigate the impact-damage formation and evolution processes [5-7]. In our previous work, the stress-wave-propagation behavior and damage evolution were observed by means of a transmitted light shadowgraph. In these measurements, the shape of the longitudinal-stress-wave front, crater and spall fracture were successfully visualized. On the other hand, these shadowgraph images provide little information about damage microstructure. The shadowgraph has difficulty in visualizing ramped waves, such as the release wave, and also for the shear wave which is not accompanied by the change of volumetric strain. Those play important role in initiating damage. This occurs because the intensity of the shadowgraph image depends on the second spatial derivative of the refractive index. In this study, we try two types of real-time imaging of impact events. One is imaging by using scattered light on the impacted target to visualize the microstructure of the impact-induced damage, the other is a shadowgraph using polarized light to visualize propagation of the impact-induced stress field.

2. Experiment

A 3.2-mm diameter spherical projectile was impacted onto a polycarbonate specimen by using a two-stage light-gas gun [8]. Spheres made of alumina (Al_2O_3) ceramic were used for impactors. Polycarbonate was chosen as the target material to compare these results to previously recorded shadowgraph images [7]. The size of the specimen was 80x80x30 mm. The projectile was normally impacted to an 80x80-mm plane. In the set-up for scattered-light imaging, two ultra-high-speed video cameras were used. One was arranged vertically to the ballistic axis to observe penetrating motion and the propagation of the stress-wave front and damage in the ballistic direction. The other was set behind the target on the ballistic axis to record the damage formation and propagation in-plane. Both cameras were triggered at the same time. Two flash lamps used as a light source were put off-axis to

+ Full paper can found at https://www.sciencedirect.com/journal/international-journal-of-impact-engineering/special-issue/10MTGP5W4VJ

* Corresponding author. Tel.:+81-96-342-3299; fax:+81-96-342-3193.
*E-mail address:*nkawai@kumamoto-u.ac.jp.

the shooting angles of cameras to get scattered light for imaging. In the polarized-shadowgraph set-up, two polarizers were installed in front of a video camera and a light source. The polarizing directions crossed each other to visualize stress fields using the photo-elastic effect [9, 10]. In this imaging, a pulsed laser (Cavitar Ltd., CAVILUX®) was used as a light source. Its pulse width and wave length were 150 ns and 640 nm, respectively. The timing of laser radiation was synchronized with the exposure of the high-speed camera. The ultra-high-speed video cameras used in this study were HPV-X (Shimadzu Corp.) and Kirana (Specialised Imaging Ltd.). The framing interval and exposure time of both cameras were set to 500 and 200 ns, respectively.

3. Results and Discussion

Figure 1 shows selected pictures captured by the scattered-light imaging obtained during the 4.99 km/s impact of an alumina ceramic sphere. The texture of damages is clearly visualized in all pictures, which is different from shadowgraph images. The side view images show stress-wave propagation, penetration-damage evolution and spall-fracture formation. The nose shape of penetration-damage is quite rugged. This feature is also observed in the rear view image at 13.0 μs after impact. This structure appears immediately after impact. These results indicate that the alumina impactor breaks up at impact and the lump of impactor-fragments penetrates into target with its collapse. The stress wave generated by the impact propagates into target and reflects on the rear surface as a rarefaction wave. This wave induces spall fracture inside target as shown in the side view image at 15.5 μs after impact. After some development of spall fracture, a stress wave appears to be propagating from the spall fracture at 20.5 μs. This suggests that a free-surface plane is formed by spallation and the released tensile stress propagates into the

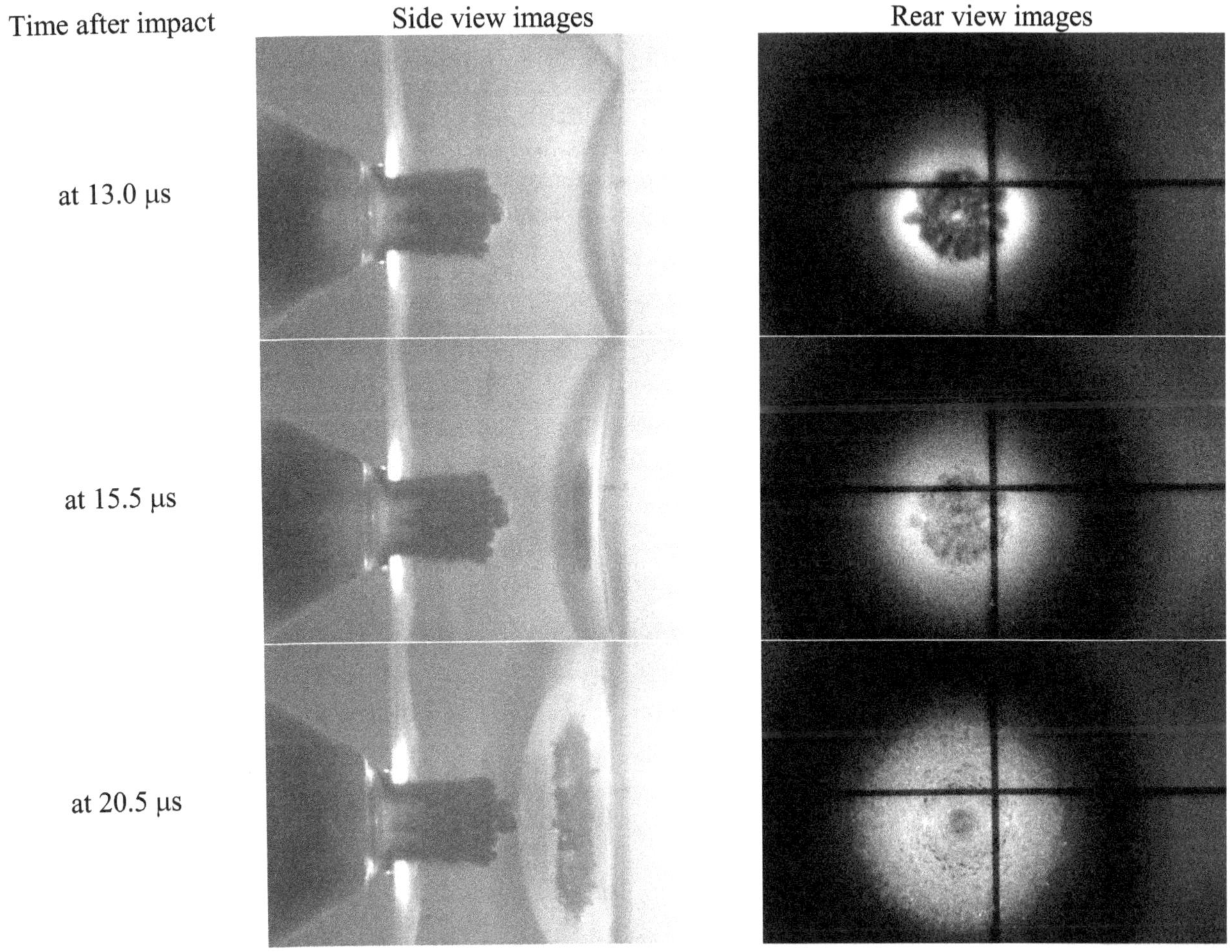

Fig. 1. Damage evolution in the polycarbonate impacted by a 3.2-mm alumina ceramic sphere at 4.99 km/s simultaneously obtained in side view and rear view by the scattered light imaging.

target. Information about the formation and development of the spall fracture in-plane can be obtained from the rear view images. At 15.5 μs when spall fracture is initiated, an apparent fracture surface is not observed while the image of the penetration damage is blurred by some changes on the spall plane. A scale-like structure is recognized on the spall plane at 20.5 μs. The timing of the formation of the scale-like texture corresponds to the timing of the stress wave generation at the spall plane, which is observed in side view images. These results suggest that the scale texture indicates the shape of the fracture surface generated by spallation.

The still pictures extracted from the polarized shadowgraph images at 3.66 km/s impact are shown in Fig. 2. The normal-shadowgraph images recorded under another impact test at approximately equivalent velocity which is 3.71 km/s are also shown in Fig. 2. The light intensity of the cross-polarized shadowgraph depends on stress birefringence. Therefore, the shape of stress field can be visualized in the cross-polarized-shadowgraph image while the shadowgraph image only shows the shape of stress-wave front. In the cross-polarized image at 11.5 μs after impact, the area where the stress wave has propagated show birefringence induced by compression stress, and the stress wave seems to propagate as a spherical stress pulse with a spatial width of about 6 mm. After passage of the stress pulse, some stress fields exist around penetration hole. This stress field is expected to be caused by the compression strain induced by plastic flow associated with impactor penetration. At 24.0 μs after impact, development of spall fracture is recognized, especially in the shadowgraph image. The image of the rarefaction front reflected from the rear surface becomes ambiguous in the shadowgraph picture because the rarefaction front ramps and the change rate of refractive index decreases with its propagation. In contrast, the rarefaction pulse is still observed in the cross-polarized picture as a negative image in contrast to a positive image for a compression wave. This image-inversion reveals the loading condition of stress waves, either loading or unloading. The picture also shows a birefringence change around the spall fracture. Some possible causes of this birefringence are expected, for example, the interaction between the residual stress field around penetration hole as shown in the cross-polarized image at 11.5 μs and the stress wave propagating from the spall plane as shown in Fig. 2, and the stress concentration on the spall fracture surface induced by the reverberated stress wave between the spall and rear surface. To reveal the relation between the spall fracture and stress field, it is planned to compare between results showing various spall-formation timings and locations, obtained by performing impact-tests on targets with different thickness.

As described above, we succeeded in visualizing propagation of the stress field induced by hypervelocity impact via cross-polarized shadowgraphy. However, the images are a two-dimensional projection of a three-dimensional stress field developed by propagation of a spherical stress wave in this study. It is important for precise analyses of images to understand how the three-

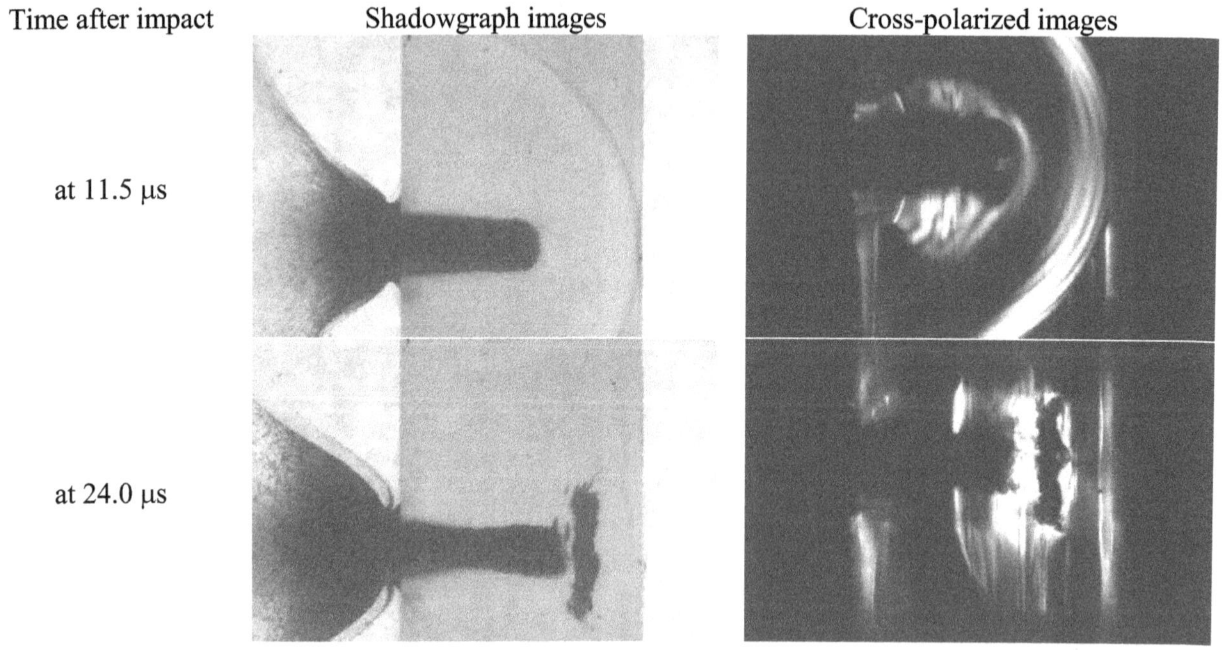

Fig. 3. Selected high-speed images captured .by means of normal shadowgraphy and cross-polarized shadowgraphy. Impact velocities are 3.71 km/s at the shadowgraph experiment and 3.66 km/s at the cross-polarized shadowgraph experiment.

dimensional stress-distribution affects the two-dimensional polarized shadowgraph images. To study this problem, we continue to extend this kind of experiment under various loading conditions such as uniaxial strain compression by plate impact experiments [11] and plane stress compression by edge-on-impact experiments [9, 10], in addition to the three-dimensional stress condition by sphere-on-block impact used in this study.

4. Conclusion

In this study, hypervelocity impact events on a polycarbonate target were visualized by means of scattered-light imaging and the cross-polarized shadowgraph. The pictures obtained by scattered-light imaging clearly show the evolution of impact-damage texture. The cross-polarized shadowgraph successfully visualized not only the propagation of stress-wave pulse but also formation of the stress field. Information obtained by both methods is difficult to approach by means of usual shadowgraph imaging. Combination of these various imaging methods is an effective way to investigate damage mechanisms under hypervelocity impact. These images also help with the evaluation of impact-damage modeling by a direct comparison to numerical experiments. Improvement of the imaging technique, experiments under various impact conditions and further analysis of the obtained images are still ongoing.

Acknowledgements

This study was supported by ISAS/JAXA as a collaborative program with the Hypervelocity Impact Facility (former facility name: Space Plasma Laboratory). This work was supported by JSPS Grant-in-Aid for Scientific Research (C) Grant Number JP18K04561.

References

[1] Cour-Palais, E. G., 1987. Hypervelocity impact in metals, glass and composites, International Journal of Impact Engineering 5, p. 221.

[2] Taylor, E. A., Tsembelis, K., Hayhurst, C. J., Kay, L., Burchell, M. J., 1999. Hydrocode modelling of hypervelocity impact on brittle materials: depth of penetration and conchoidal diameter, International Journal of Impact Engineering 23, p. 895.

[3] Christiansen, E. L., Kerr, J. H., 2001. Ballistic equations for spacecraft shielding, International Journal of Impact Engineering 14, p. 145.

[4] Kawai, N., Tsurui, K., Shindo, D., Motoyashiki, Y., Sato, E., 2011. Fracture behavior of silicon nitride ceramics subjected to hypervelocity impact, International Journal of Impact Engineering 38, p. 542.

[5] Kawai, N., Tsurui. K., Moriguchi, K., Hasegawa, S., Sato, E., 2013. Damage evolution in SiO_2 glass subjected to hypervelocity impact, Procedia Engineering 58, p. 702.

[6] Kawai, N., Zama, S., Takemoto, W., Moriguchi, K., Arai, K., Hasegawa, S., Sato, E., 2015. Stress wave and damage propagation in transparent materials subjected to hypervelocity impact, Procedia Engineering 103, p. 287.

[7] Kawai, N., Kuroda, Y., Nagano, M., Hasegawa, S., Sato, E., 2017. Stress-wave propagation and damage formation associated with hypaervelocity penetration into polycarbonate, Procedia Engineering 204, p. 255.

[8] Kawai, N., Tsurui, K., Hasegawa, S., Sato, E., 2010. Single microparticle launching method using two-stage light-gas gun for simulating hypervelocity impacts of micrometeoroids and space debris, Review of Scientific Instruments 81, 115105.

[9] Strassburger, E., Patel, P., McCauley, J. W., Templeton, D. W., 2005. "High-speed photographic study of wave and fracture propagation in fused silica," Proceedings of 22nd International Symposium on Ballistics, pp. 761-768.

[10] Strassburger, E., Patel, P., McCauley, J. W., Templeton, D. W., 2007. "Wave propagation and impact damage in transparent laminates," Proceedings of 23rd International Symposium on Ballistics, pp. 1381-1388.

[11] Bourne, N. K., Rosenberg, Z., Field, J. E., 1995. High-speed photography of compressive failure waves in glasses, Journal of Applied Physics 78, p. 3736.

Proceedings of the 2019 Hypervelocity Impact Symposium
HVIS2019
April 14-19, 2019, Destin, FL, USA
HVIS2019-014

Towards a Better Understanding of Shaped Charge Jet Formation and Penetration

David W. Price*, Ernest J. Harris and Frances G. Daykin

AWE Aldermaston, Reading, RG7 4PR,

Abstract

JeMMA, a set of relatively simple shaped-charge devices, has been designed in order to generate suitable data on jet formation, break-up and penetration for code validation purposes. The JeMMA Phase 1 device incorporated a copper liner and six of these shaped charges were manufactured as a technology demonstrator and fired in a special shaped charge facility in December 2016. The radiographic results obtained from the JeMMA Phase 1 and 2 devices, along with data reproducibility between trials, was excellent. This report gives an overview of the Phase 1 and 2 trials, including device design, the results of the firings conducted in Switzerland and details of the subsequent 2D and 3D hydrocode modelling carried out at AWE. The agreement between the data and both 2D and 3D modelling of the experiments is very pleasing, but highlights where further work is required. These JeMMA experiments will enhance the body of relevant data required to provide the validation of the hydrocode materials and modelling methodologies and enable us to better model the jetting threats of our experiments and have higher confidence in the results of the modelling.

Keywords: Eliptical Shaped Charge; Multi-material; Modeling

The Hydrodynamics Delivery Unit at the Atomic Weapons Establishment (AWE) at Aldermaston, UK, is required to be able to safely conduct trials with a wide variety of high explosive assemblies. These data are required to underwrite performance and safety under the Comprehensive Test Ban Treaty (CTBT) in support of the AWE core mission. The testing facilities need to be protected from the effects of the explosions, which can include blast and pressure loading, high velocity case fragments and potentially metallic jets. This is of concern in both the open firing chambers and containment vessel environments used within our testing facilities. Since the jetting threat is arguably the most damaging to the containment, the successful mitigation of any jet from our experiments is essential to enable the safe fielding of them within our facilities.

A number of trails have been previously carried out at the UK Ministry of Defence facility at dstl, Porton Down with K4 shaped charges [1], but due to the intrinsic variability of military-grade shaped charges a large number of expensive repeat experiments were required. Since the threats from experiments can vary greatly and safety justification trials are expensive, we need to be able to model the threats that any given experiment produces and that requires a good understanding of the processes that produce the threats. So, in order to better understand the physical processes involved in the formation and penetration of metallic jets, a series of simplified shaped charge devices (the Jetting Multi-Material Assembly, or JeMMA) have been designed to provide some of the required experimental validation data for a range of metallic liners. It is believed that relevant jet material data can be obtained by using these shaped charge devices, provided that they are manufactured to a high quality and specification. The shaped charges need to be manufactured to tight tolerances in order to get a good 2D-symmetric jet and ensure repeatability between experiments. The JeMMA devices are relatively simple and low cost shaped charges that incorporate readily available explosives and unclassified liner profiles. The design concept is considered to be suitable for a variety of single element and multi-material liners and will be carried out in phases.

* Corresponding author. Tel.: +44-118-985-58195; fax: +44-118-981-5320
E-mail address: david.price@awe.co.uk

1. JeMMA 1 Design and Fielding

The JeMMA Phase 1 device was designed as a technology demonstrator to give a coherent jet that was fast and penetrating to provide data in the most useful regime in jet-threat modelling. Six JeMMA 1 devices were manufactured to the AWE design for fielding in a specialised firing chamber. Fig. 1 a) shows the JeMMA 1 design, while Fig. 1 b) shows a photograph of one of the JeMMA Phase 1 shaped charges after assembly. The JeMMA 1 devices comprise elliptical copper liners embedded within cylindrical explosive charges. The LX-14 explosive is housed in a stainless steel cylindrical case with a centrally located rear detonator. A blast plate on the front of each device is designed to contain the explosive products and support the edge of the liner during early-time jet formation. The copper liners and explosive were precision machined to a tolerance of approximately 20 µm on diameter and the steel case and copper liner for each device were thermally fitted to the explosive to reduce gaps at the interfaces.

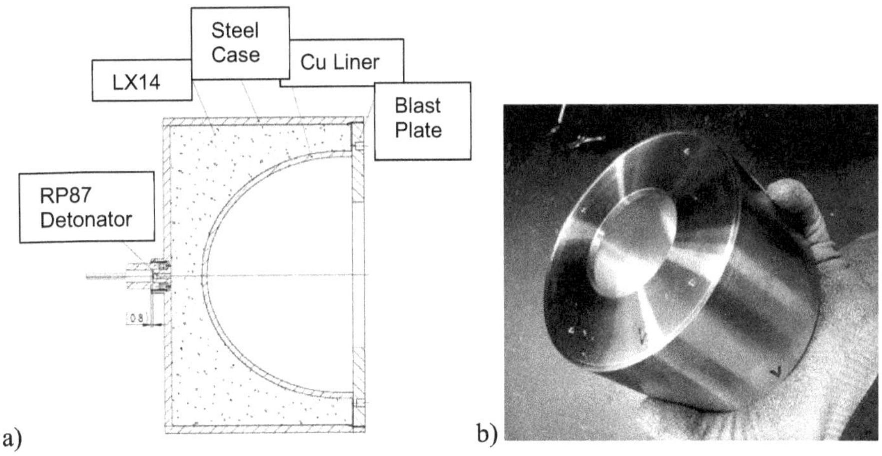

Fig. 1. JeMMA 1 shaped charge.

1.1. Experimental Setup

The JeMMA Phase 1 and 2 trials were conducted under contract in a shaped-charge firing facility. The specially designed facility consisted of firing chamber cut into the side of the mountain with three axis radiography capability. A Schematic of the experimental layout including the positions of the x-ray machines is given in Fig. 2. A permanent metal firing stand was used to support the wood and foam shaped charge stand, as shown schematically in Fig. 2, which gives a view of a longer stand-off setup looking from the x-ray heads towards the x-ray film position.

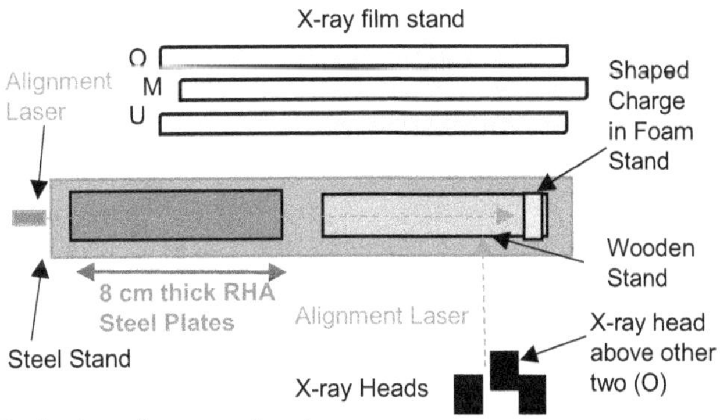

Fig. 2. Schematic of the experimental set-up for the shape charge experiments

Wooden planks were used to protect the x-ray film holders (labelled O, M and U for the upper, middle and lower) and the x-ray heads from the steel case fragments. The jet from the JeMMA device was arrested using a number of 8 cm thick armour steel plates which were moved to vary the stand-off of the device. A glass-steel interleaved target would be more efficient jet-stopper [1], but the armour steel was more suitable for a fast turn-around between firings. For all firings, the stand-off distance was measured from the front of the blast plate on the shaped charge to the front surface of the first steel target plate. In each case the x-ray field of view was approximately 750 mm, which was dictated by the x-ray film length and the x-ray source to jet / jet to x-ray film ratio.

For shaped charge firings, it is useful to add fiducial markers onto the x-ray film to determine jet length and tip position more accurately. This was achieved by taking an x-ray radiograph of a marker grid that was positioned in the field of view prior to installing the target plates, as shown in Fig. 3. Alignment lasers from the central (M) x-ray machine and along the horizontal axis of the jet were used to position the shaped charge and marker grid. Once the static exposure of the marker grid had been taken, this grid was removed and replaced with the target plates at the correct stand-off distance from the front of the shaped charge. Two of the x-ray machines, labelled "U" and "M", are approximately vertically aligned with the other machine (labelled "O") offset to one side by about 500 mm. For some of the firings, the static image of just the marker grid was developed in order to confirm that the field of view in the radiographs was adequate to capture the correct portion of jet for the chosen x-ray time.

Fig. 3. Photograph of the fiducial marker grid in place during a static x-ray radiograph

1.2. Pre-experiment modelling

Following the initial design phase for the JeMMA Phase 1, shaped charge using the Eulerian hydrocode, CTH [2], models were simulated to times up to approximately 200 μs in order to provide information on jet shape and possible jet break-up times as this would determine the firing plan for the seven shaped charges. A pseudo radiograph created from a CTH calculations at 31 μs after detonation is shown in Fig. 3 and indicated that this device would produce a jet with a relatively thin tip that widened to a much thicker jet at the rear.

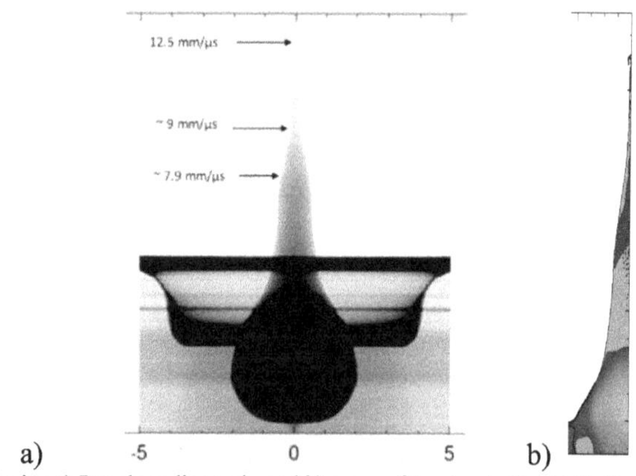

Fig. 3. Pre-trial modelling of JeMMA 1 device a) Pseudo radiograph, and b) areas of tension in the jet (indicated by blue colouration)

The predicted velocities of the jet material at particular locations along the jet are indicated in the Fig. as well. Using Chou's jet break-up theory [3], it was estimated that the thin jet tip would break-up at approximately 60 μs, and that the thicker rear jet would break-up much later, at approximately 200 μs after detonation. It important to note that, as indicated in Fig. 3a, the tip velocity is predicted to be approximately 12.5 $km.s^{-1}$ (≡12.5 $mm.\mu s^{-1}$). This is very fast for a copper jet and is predicted to be impossible according to some jet coherence theories, albeit for conical shaped charge liners [4]. It was also interesting to note that the CTH calculations showed voids appearing in the widening section of the jet at about 31 μs, as shown in Fig. 3b. It was uncertain prior to the trials how these voids would manifest themselves in reality or whether this was just an unphysical code artefact.

1.3. Radiographic results

The SBDS facility at Weisenbergburg was set up to record the x-ray times relative to direct trip (pressing the firing button). Since the detonator and booster pellet/PIC had not been modelled explicitly in the pre-trial calculations, the time zero in the calculation was taken as when the main charge was initiated. Interpretation of the simulation results required them to be shifted by approximately 4 μs in time for comparison with experiment.

Radiographs from the first experiment that was used to probe the jet-formation stage are shown in Fig. 4. It can be seen that the thick steel case on the device has obscured the slug at the rear of the jet but the shape of the front of the jet in the third radiograph is a good match to the pre-trial calculation (in Fig. 3), which gave confidence for the remainder of the trials. The effect of the offset in x-ray head position can be clearly seen in the third radiograph at 36.75 μs.

Following this firing, the jet tip location in the 'O' radiograph was measured as 70 mm from the fiducial origin (and hence front of the blast plate). The pre-trial simulation jet tip velocity of 12.5 km.s^{-1} was then used to determine the radiograph times requested for the next firing in order to achieve the required view of the jet to image the thin jet tip break-up (assuming good repeatability between the charges due to their precision build).

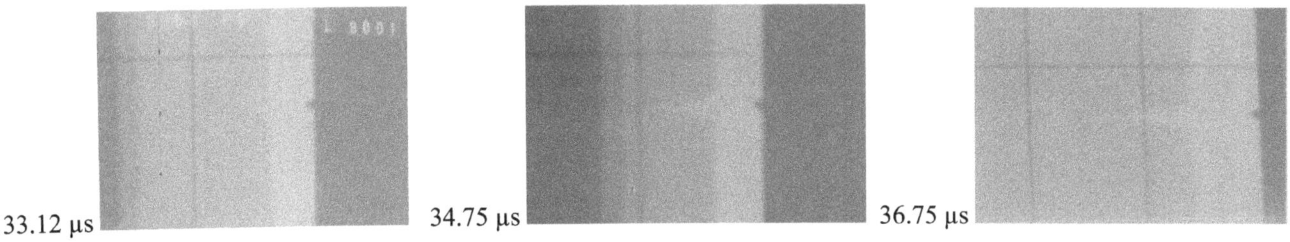

Fig. 4. Three early-time radiographs of the JeMMA 1 first shaped charge

Two radiographs obtained from second charge are given in Fig. 5. The aim of this firing was to investigate the break-up of the thin front jet tip. The first radiographic time was chosen to be the same as the last from the previous firing to check for repeatability between charges.

It can be seen that the second radiograph at 62.78 μs shows a particulated jet tip. It is also interesting to note that a section of the jet behind the thin tip has bulged (increased in diameter). This is in the approximate location where voids were appearing in the hydrocode model. A radiograph at 88.82 μs (not shown) indicates that both the thin jet tip and the bulge have particulated at this time but the rear of the jet is still fully continuous.

Fig. 5. Two radiographs of the JeMMA 1 second shaped charge

The fourth Charge was fielded with a 30 mm thick mild steel stripper plate positioned at 20 mm from the front of the blast plate on the shaped charge with the main jet-stopper plates at a stand-off distance of 250 mm, as shown in Fig. 6, so that there would be a direct comparison between the results from this firing and that of the first charge. Calculations performed using 2D CTH prior to the trials were used to inform the requested radiograph times for this trial (in addition to adding in the estimated timing delay offset from the firing of the first Charge). The radiographs from the fourth Charge are given in Fig. 7 along with a pseudo radiograph from the pre-trial calculation.

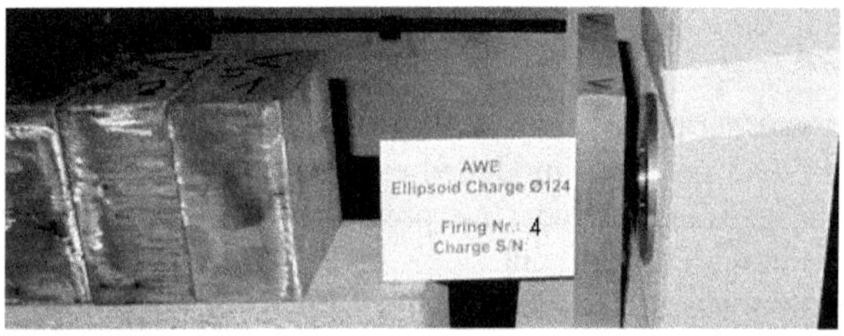

Fig 6. Firing setup with steel stripper plate for fourth charge firing

The stripper plate positioned close-in produces a significant amount of debris surrounding the jet as shown in the radiographs in Fig. 7 a and b. It is postulated that the debris could be either from the eroded jet material or spalled material from the rear of the steel stripper plate. Whilst the radiographs show a degree of asymmetry in the jet, the jet shape in the 2D calculation shown at a computational time of 52.5 μs is qualitatively similar to the shape of the jet in the experimental radiographs and this is shown in Fig. 7c.

Fig. 7. a) and b) Radiographs from the fourth JeMMA 1 Charge; c) CTH simulated radiograph of the fourth JeMMA 1 Charge at 52.5 μs

1.4. Target Penetration

As stated above, the six JeMMA 1 shaped charges were fielded at various stand-off distances from the jet-stopper and the depth of penetration into the jet-stack was measured; including the depth of penetration of the jet from the shaped charge fielded with a stripper plate. The 8 cm thick armour steel target plates were photographed, both grouped and individually, following each firing. Three whole jet-stoppers were later sectioned using water jet cutting along with the last penetrated plate from every firing and the mild steel stripper plate, as shown in Fig. 8 and 9. It can be seen that in all of the jet stoppers there is also residual Copper material from the jet both wedged into the front plate (most likely the slow moving slug) and in the bottom of the hole that the jet has drilled into the steel target plates. There is also a wider witness mark on the front surface of the first plate in the jet-stopper, presumably caused by the impact of the blast plate.

a) b)

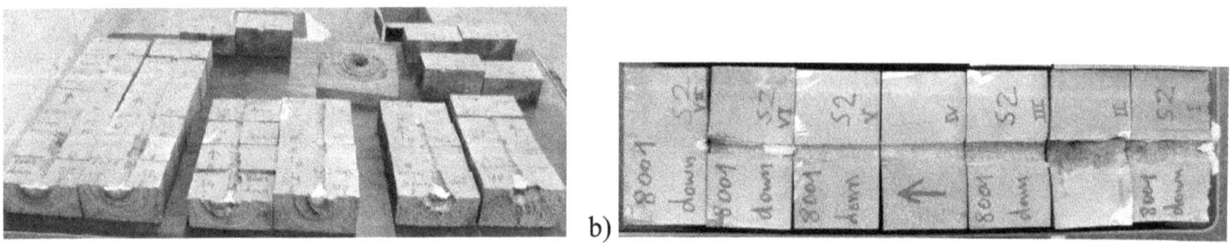

Fig. 8. Steel target plates following water jet cutting

The photographs shown in Fig. 8b and fig. 9 illustrate the effect of the mild steel stripper plate on the final penetration depth into a jet-stopper positioned at the same distance of 250 mm from the charge for each trial.

Modelling of the penetration of the JeMMA 1 devices has proven extremely challenging with simulation times well in excess of 1 ms.

Fig. 9. Steel target plates following water jet cutting for 4th JeMMA 1 Charge with a stripper plate

1.5. 2D CTH Modelling

Although Lagrangian simulations have been used in the modelling of shaped charge jet formation recently by Downes *et al* [5], the large distortions of material even shortly after the emergence of the jet-tip are more suited to be modelled using Eulerian hydrocodes, Smooth Particle Hydrodynamics [6] or Adaptive Lagrangian-Eulerian methods [7]. CTH has been shown to work well in the modelling of hyper-velocity jet by Helminiak *et al* [8] and Sable *et al* [9] and we have adopted this code. 2D CTH simulations were carried out prior to the JeMMA Phase 1 experiments and simulated radiographs were produced to aid comparison of the calculations with the experimental data for x-ray timing purposes during the trials. A Mie-Gruneisen EoS [10], was used for the copper with a Steinberg-Guinan-Lund [11] constitutive model with a fracture stress of -5 GPa. The JWL [12] EoS was used for LX14 from the CTH library and a model for the RP87 detonator was incorporated into some simulations.

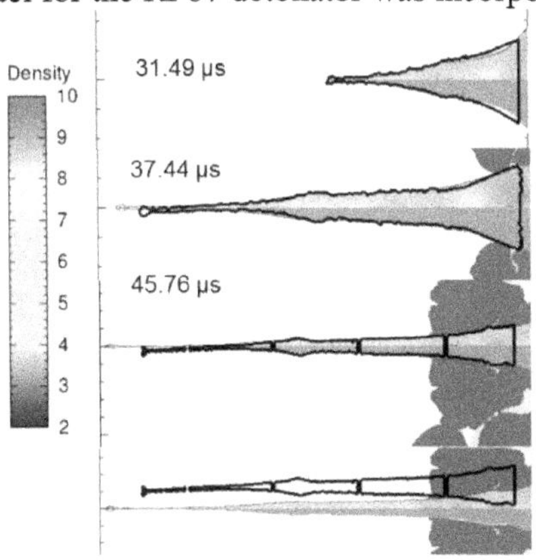

Fig. 10. Comparison of 2D CTH simulation with experimental edges (the bottom jet is a replication of the 45.76 µs jet with the edges of the radiograph shifted upwards)

A comparison of the CTH simulation with the edge analysis data from the radiographs is given in Fig. 10. The CTH output shows density contours at the top of each image and material location at the bottom. The third image has been repeated at the bottom of the Fig. but with the calculation and experimental edges deliberately misaligned so that it can be seen more easily that the bulge in the calculation is not as wide as in the experiment.

1.6. 3D CTH Modelling

In our 3D CTH simulations of the JeMMA 1 shaped charge, the same models were used for all materials except that the Johnson-Cook strength model [13] was used for the steel case along with the Grady-Kipp fragmentation model [14]. In the 3D CTH models, all space not occupied by components of the shaped charge was filled with air at 1 atm., with the air represented by a SESAME table, which was supplied with CTH. The foam support and all other external features of the experimental set-up were omitted from the model. The explosive was modelled using the default density of 1.835 g.cm^{-3}. Finally, most simulations were carried out in quarter symmetry except a small number were carried out with no symmetry as a comparison. As was observed in the 2D simulations and shown in Fig.11, a bulbous region is present behind the jet tip, however the full 3D simulation displays some break-up of this. This may be due to the lower mesh resolution that had to be used to complete the calculation since higher resolution simulations displayed less porosity of the bulbous section than the lower resolution simulations. Higher resolution simulations are planned for the future.

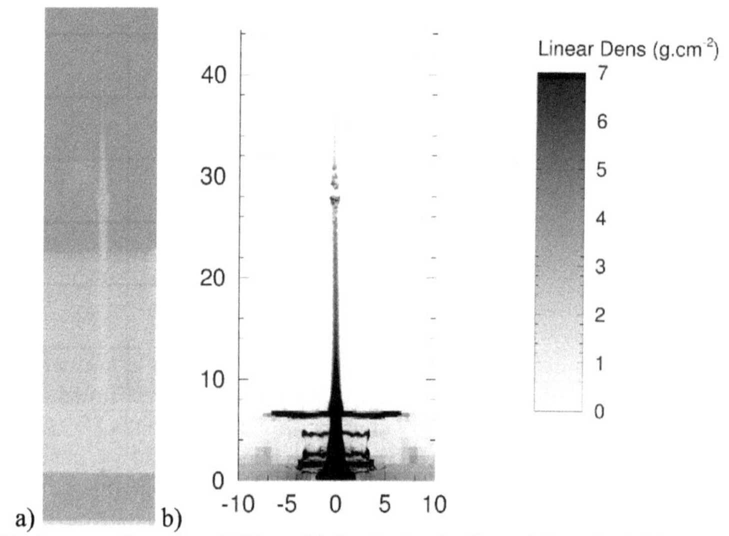

Fig. 11. Comparison between the JeMMA 1 x-ray radiograph at 62.78 μs with the simulated radiograph from the full 3D CTH calculation at 63.0 μs (length units = cm)

2. JeMMA Phase 2

2.1. JeMMA 2 Design

The JeMMA Phase 2 devices are extensions of the Phase 1 devices and comprise a dual layer elliptical liner of aluminium and tantalum embedded within profiled stainless steel case, as shown in Fig. 12 a and b. The LX-14 explosive is housed in the stainless steel case with a centrally located rear detonator as with JeMMA 1. A blast plate on the front of each device is designed to contain the explosive products and support the edge of the liner during early-time jet formation, again as in JeMMA 1. The JeMMA 2 device was designed to investigate the interaction between the liner materials and the effect on jet-formation and break-up in the regime of most significance to jet-threat modelling. The steel case and liner profile are loosely based on the shaped charge patent of Haselman [15], while the principle of including two layers in the liner is found in the patent of House [16].

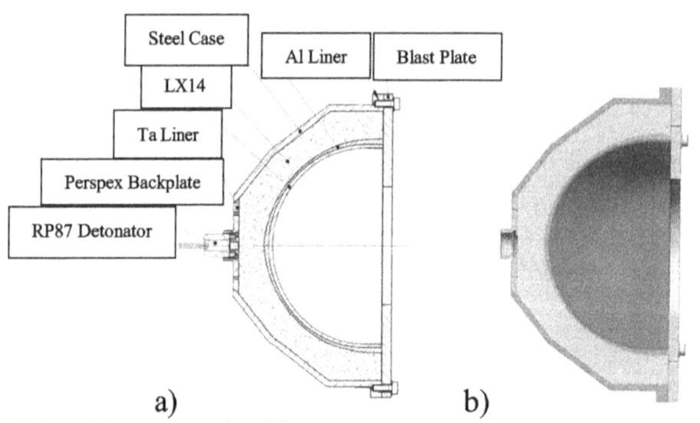

Fig. 12 – JeMMA 2 design, a), and a solid model representation, b)

2.2. Radiographic results and Computational Comparison

Early-time comparisons between the JeMMA Phase 2 radiographs and the simulated radiographs obtained from 3D quarter symmetry CTH calculations are very encouraging and an example of one such comparison is shown in Fig. 12, where the diffuse spray-like head of the jet jut behind the tip seen in the radiograph is reproduced. The x-rays were not penetrative enough in the experiment to see through the shaped charge case, whereas in the synthetic radiograph the slug of the shaped charge and the fracturing case can clearly be seen as strips of case material at the sides and bottom of the case region of the simulation.

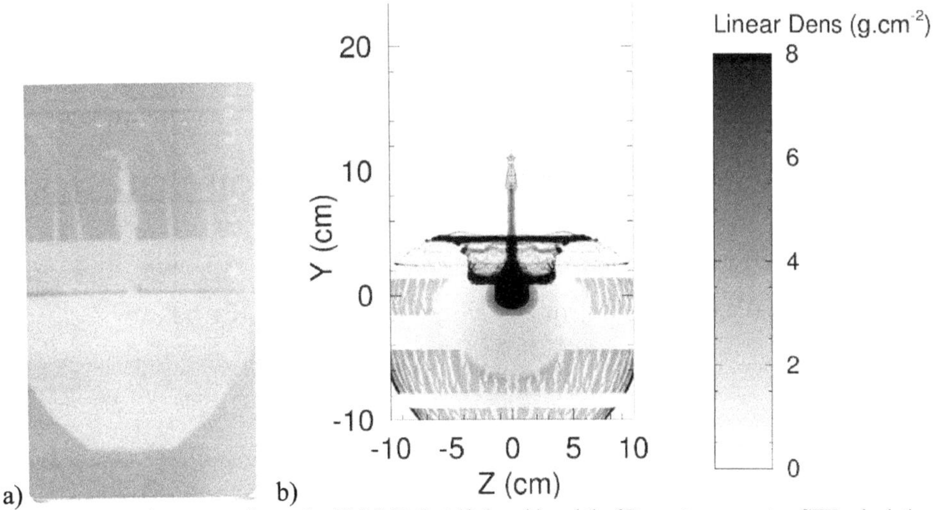

Fig. 13. Early-time comparison between the x-ray radiograph of JeMMA 2 at 40.6 μs (a) and the 3D quarter symmetry CTH calculation at 40 μs (b)

3. Summary and Conclusions

Two Simplified shaped charge devices, JeMMA 1 and JeMMA 2, have been designed and fielded by AWE. A wealth of data has been collected on them including target penetration and x-ray radiographs at multiple times during the jet-formation and break-up. CTH Eulerian calculations have been carried out in 2D and 3D and the synthetic radiographs generated by the program show remarkable similarity to the experimental ones and jet velocities are comparable. More work is required to simulate the penetration of the jets due to the long time-scales involved in stopping jets travelling at in excess of 8 km.s^{-1} and that is on-going.

Acknowledgements

The funding of this work by the UK MOD is gratefully acknowledged.

References

[1] Daykin, F.G., Harris, E.J., Price, D.W., 2012, 'Experiments to Determine the Relative Performance of Layered Jet-Stoppers', Proceedings of 27th International Symposium on Ballistics, p. 1639-1649

[2] McGlaun, J.M., Thompson, S.L., Elrick, M.G., 1990, 'CTH: A Three-Dimensional Shock Wave Physics Code', International Journal of Impact Engineering 10, p.351-360

[3] Chou, P.C. and Carleone, J., 1977, "The breakup of Shaped Charge Jets", Proceedings of the Third International Symposium on Ballistics, Karlsruhe, Germany

[4] Kelly, R.J., Curtis, J.P., Cowan, K.G., 1999, "An Analytical Model for the Prediction of Incoherent Shaped Charge Jets", Journal of Applied Physics 86, p. 1255-1265

[5] Downes, D., Bouamoul, A. and Ensan, M.N., 2014, Canadian Defense Research Reports, DRDC-RDDC-2014-P38

[6] Cheng, X, Huang, G, Lu, C and Feng, S, 2018, Appl. Sci. 8, p.1863

[7] W, Cheng, Ding, J, and Zhao H, 2015, Defence Science Journal, 65 (4), p. 279-286,

[8] Helminiak, N.S., Sable, P., Harstad, E., Gullerud, A., Hollenshead, J., and Hertel, E.S., 2017, Procedia Engineering 204, p. 178-185

[9] Sable, P., Helminiak, N.S., Gullerud, A., Harstad, E., Hollenshead, J., and Hertel, E.S., 2017, Procedia Engineering 204, p. 375-382

[10] Mie, G., 1903, Zur kinetischen Theorie der einatomigen Körper, Annalen der Physik 316, p. 657–697; Grüneisen, E., 1912, Theorie des festen Zustandes einatomiger Elemente, Annalen der Physik 344, p. 257–306

[11] Steinberg, D.J., Lund, C.M., 1989, Journal of Applied Physics 65, p.1528-1533

[12] Lee, E.L., Hornig, H.C., Kury, J.W., 1968, "Adiabatic expansion of high explosive detonation products", UCRL50422

[13] Johnson, G.R., Cook, W.H., 1983, Proceedings of the 7th International Symposium on Ballistics, pp. 541–547.

[14] Kipp, M.E., Grady, D.E., Swegle, J.W., 1993, "Numerical and experimental studies of high-velocity impact fragmentation", International Journal of Impact Engineering 14, p. 427-438

[15] Haselman Jr., L.C., 1996, 'Free Form Hemispherical Shaped Charge', US Patent 5,522,319 (6/1996)

[16] House, A.M., 1965, "Dual Liner Shaped Charge", US Patent 3,224,368 (12/1965)

Proceedings of the 2019 Hypervelocity Impact Symposium
HVIS2019
April 14-19, 2019, Destin, FL, USA
HVIS2019-023

Extended Abstract+

Laser-Driven Micro-Flyers for Dynamic Fragmentation Statistics of Boron Carbide

Debjoy D. Mallick[a,b,c]*, Daniel J. Magagnosc[c], KT Ramesh[a,b]

[a]*Department of Mechanical Engineering, Johns Hopkins University, 3400 N Charles St, Baltimore, MD 21218, USA*
[b]*Hopkins Extreme Materials Institute, Johns Hopkins University, Malone Hall 140, 3400 N Charles St, Baltimore, MD, 21218, USA*
[c]*U.S. Army Research Laboratory, Aberdeen Proving Ground, MD, 21005, USA*

1. Introduction

Dynamic fragmentation through high-rate impact generates large numbers of fragments with various shapes and sizes. The fragmentation failure mode is an important part of the protection capacity of advanced ceramics which typically feature high strength and low density but fail in brittle modes. The penetration resistance of these brittle materials has been linked to the fragment size and shape created through impact in the literature [1]. Such studies have shown that particular fragment size and shape combinations can more effectively erode incoming projectiles, presenting a possible route to improve penetration resistance. These results stand in contrast to other studies that examine links between penetration resistance and material properties (e.g. fracture toughness or stiffness) which have sometimes resulted in contradictory correlations. Boron carbide has received a strong focus in the literature in recent years as an advanced ceramic with one of the highest specific strengths and lowest densities [2]. Yet boron carbide exhibits poor penetration resistance at higher loads, a phenomenon that some researchers attribute to a phase transformation termed "amorphization" [2]. To better understand the protection capacity of boron carbide under high rate loading, we use a laser-driven micro-flyer apparatus to impact boron carbide specimens.

Laser-driven projectile launchers have demonstrated effectiveness as a way to probe novel micro-produced protection materials such as graphene sheets with high throughput on the lab-bench [3], but have not been utilized to evaluate mass-produced protection materials such as advanced ceramics. In this work, we explore the impact-induced fragmentation of a commercially available hot-pressed boron carbide by examining the fragmentation developed by a laser-driven flyer apparatus. Since the dynamic fragmentation failure mode is known to be crucial in the impact performance of protection ceramics [1], we assess the fragment geometries resulting from laser-driven impact and explore connections to the as-received microstructure.

2. Material

The as-received boron carbide material is a plate and is conceptualized in the schematic in Fig. 1 with the projectile impact direction along the hot-pressing direction. Fig. 1(a-b) shows optical micrographs of the material normal-to and transverse-to the hot-pressing direction, highlighting, (1) larger carbonaceous inclusions that present as circular in the hot-pressing direction and elliptical when transverse to the hot-pressing direction. These inclusions are disk-like in three dimensions with a preferred orientation that is normal to the hot-pressing direction (orange circles), (2) smaller spherical features that are primarily carbonaceous in composition, although sometimes they are pores (gray circles), and (3) brighter ceramic phases that have been identified as aluminum nitride and boron nitride (green circles) [4]. We perform the laser-driven impacts on 3mm diameter, 50μm thick disks manufactured from the as-received 8mm thick plate by first cutting a 1.2cm x 1.2cm sample that is 1mm thick (along

+ Full paper can found at https://www.sciencedirect.com/journal/international-journal-of-impact-engineering/special-issue/10MTGP5W4VJ
* Corresponding author. Tel.: +1-443-721-3006
E-mail address: dmallic1@jhu.edu

hot-pressing direction) and then mechanically polishing (Bomas Machine Specialties, Inc.) to a 50μm thick and 1cm x 1cm sample. We cut 3mm diameter disk-shaped specimens from the sample using a femto-second pulsed-laser based machining procedure outlined in Mallick et al. [5].

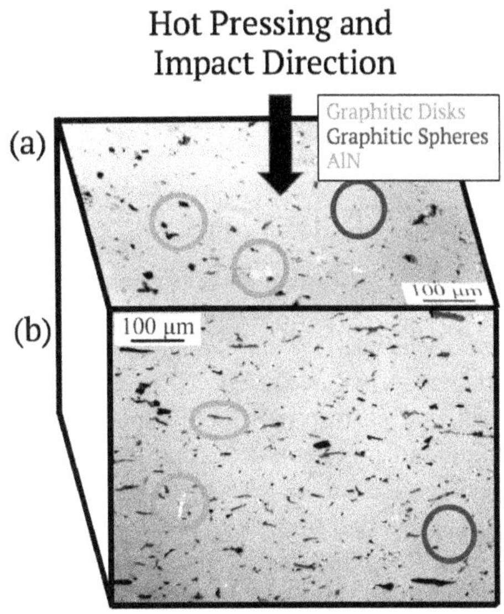

Fig. 1. Microstructure of the boron carbide target (reprinted from Hogan et al. [4]) with (a) an optical micrograph showing defects parallel with the direction of impact and (b) an optical micrograph showing microstructural defects transverse-to the loading/ hot-pressing direction. Orange circles denote carbonaceous disks. Gray circles denote graphitic or porous spheres. Green circles denote AlN inclusions.

3. Laser-Driven Micro-Flyer Experiments

In this study, we accelerate 1.5 mm diameter, 25 μm thick Al flyers to 1100 to 1200m/s impact velocities using a pulsed-laser -driven flyer launcher described in Mallick et al. [5]. The launched flyer velocity U_0, is related to flyer density ρ, and flyer thickness h_{flyer}, according to a conservation of energy launch model developed by Trott et al. and extended in our previous work [6,7]. These relationships are $U_0 \propto \rho^{-1/2}$, and $U_0 \propto h_{flyer}{}^{-1/2}$, so we select aluminum as the flyer foil material to optimize launch velocity and the resulting shock stress. The flyers become curved as time-of-flight increases (verified by high speed imaging [5]), so we perform impact experiments in two configurations: (1) a nominally flat flyer and (2) a highly curved flyer geometry. The flight distance for the flat flyer is nominally 125 μm, while the flight distance for the curved flyer is 700 μm. The experiments are performed so that all generated fragments are ejected into a chamber lined with PDMS to trap fragments and limit further deformation in a "soft-catch" configuration.

4. Results

The fragmentation resulting from the laser-driven impacts is characterized using a Zeiss optical microscope with an AxioCam MRC camera. We perform optical microscopy on the fragments captured in the "soft-catch" PDMS layer (example micrograph shown in Fig. 2) and then use MATLAB image thresholding techniques described in Hogan et al. [4] to obtain geometry statistics for the fragments. The primary statistic collected through this method is major axis size, the largest dimension through the fragment. Overall, 225 micrographs are captured for analysis through this technique, with a minimum measurable fragment size of ~6 μm. The resulting dataset contains ~7000 fragments, a number large enough where the distribution of fragments can then be examined.

The fragment major axis size is compared with the spacing of inclusions in the microstructure measured from the optical microscope images of the as-received plate shown in Fig. 1 through the use of the empirical cumulative distribution function φ,

$$\varphi(x) = \frac{1}{n}\sum_{i=1}^{n} I(x_i \leq x), \tag{1}$$

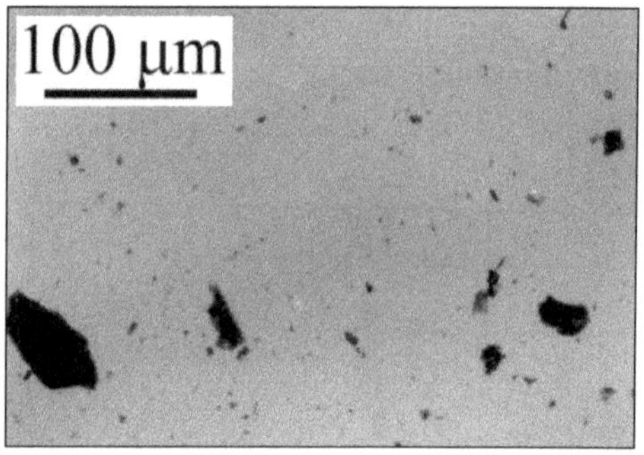

Fig. 2. Example micrograph that is analyzed in MATLAB to obtain fragmentation size and shape statistics.

with n fragments and indicator function I with value 1 when $x_i \leq x$ and 0 otherwise based on the variable of interest x (fragment size and inclusion spacing distributions are the variables shown in Fig. 3). In Fig. 3, the red line corresponds to the fragment size distribution for the curved flyer impact. The black line corresponds to the spacing between carbonaceous inclusions in the microstructure. The light blue line is for the distribution of fragments from the flat flyer impact. The cumulative distributions for the curved impact fragment size and the defect spacing have similar maximum values, but the cumulative distribution of the flat impact fragment size shows much larger fragments. These larger fragments exceed the specimen thickness, so after discriminating against fragments larger than the thickness (50 µm), the subset of flat flyer impact fragment sizes more closely follows the distribution of fragments for the curved flyer. That subset is shown as a dark blue line in Fig. 3. These results indicate that the majority of fragments from laser-driven impact follow the spacing between defects in the microstructure, suggesting that these defects control fragmentation of the boron carbide material. The curved flyer generates smaller fragments as the curvature generates a localized loading, while the flat flyer generates larger fragments from "structural fracture" away from the interface between the flyer and target during the impact [8].

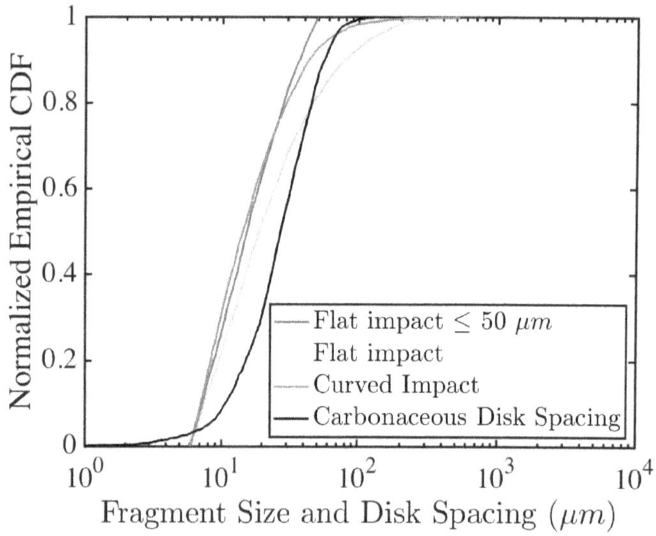

Fig. 3. Empirical cumulative distribution of fragment size for the two impact cases (blue and red) and carbonaceous inclusion spacing (black).

5. Concluding Remarks

The impact of a projectile into a target results in the rapid fragmentation of the target, where the resulting fragment size, shape and number strongly influence the ability of the target to resist the projectile. In this study, we examined the fragmentation of boron carbide using a laser-driven micro-flyer plate apparatus. The launched flyers have two geometries, as a flat flyer and as a curved flyer. In both cases, the fragmentation statistics suggest that the spacing between defects in the boron carbide microstructure may control size of generated fragments, but the flat flyer also creates larger fragments that are shown to be from regions of the specimen that are not in contact with the flyer during the impact experiment (in a mode termed "structural fragmentation"). Altogether, this study demonstrates the potential of laser-driven loading methods for use in the statistical analysis of dynamic fragmentation.

Acknowledgements

We thank Professor J. Hogan for his discussions on this work. We thank J. Parker for his assistance with fragment characterization. We thank Dr. B. Schuster for providing the target samples. Finally, we thank the Hopkins Extreme Materials Institute for their support, specifically Professor M. Foster, Professor T. Hufnagel, H. Sheng, and M. Shaeffer. This research was sponsored by the Army Research Laboratory and was accomplished under Cooperative Agreement Number W911NF-12-2-0022. The views and conclusions contained in this document are those of the authors and should not be interpreted as representing the official policies, either expressed or implied, of the Army Research Laboratory or the U.S. Government. The U.S. Government is authorized to reproduce and distribute reprints for Government purposes notwithstanding any copyright notation herein.

References

[1] Krell, A., & Strassburger, E. (2014). Order of influences on the ballistic resistance of armor ceramics and single crystals. Materials Science and Engineering: A, 597, 422-430.

[2] Chen, M., McCauley, J. W., & Hemker, K. J. (2003). Shock-induced localized amorphization in boron carbide. Science, 299(5612), 1563-1566.

[3] Lee, J. H., Loya, P. E., Lou, J., & Thomas, E. L. (2014). Dynamic mechanical behavior of multilayer graphene via supersonic projectile penetration. Science, 346(6213), 1092-1096. [4] Clark, T., Woodley, R., De Halas, D., 1962. Gas-Graphite Systems, in *"Nuclear Graphite"* R. Nightingale, Editor. Academic Press, New York, p. 387.

[4] Hogan, J. D., Farbaniec, L., Shaeffer, M., & Ramesh, K. T. (2015). The effects of microstructure and confinement on the compressive fragmentation of an advanced ceramic. Journal of the American Ceramic Society, 98(3), 902-912.

[5] Mallick, D. D., Zhao, M., Parker, J., Kannan, V., Bosworth, B.T., Sagapuram, D., Foster, M.A. & Ramesh, K. T. (2019). Laser-Driven Flyers and Nanosecond-Resolved Velocimetry for Spall Studies in Thin Films. Experimental Mechanics.

[6] R.J. Lawrence, W.M. Trott, Theoretical Analysis of a Pulsed-Laser-Driven Hypervelocity Flyer Launcher, Int. J Impact Engng, 14 (1993) 439-449.

[7] Mallick, D., Shaeffer, M., Dean, S., & Ramesh, K. T. (2017). Investigating the velocity envelope of laser-driven micro-flyers for hypervelocity impact experiments. Procedia engineering, 204, 215-222.

[8] Hogan, J. D., Farbaniec, L., Mallick, D., Domnich, V., Kuwelkar, K., Sano, T., ... & Ramesh, K. T. (2017). Fragmentation of an advanced ceramic under ballistic impact: mechanisms and microstructure. International Journal of Impact Engineering, 102, 47-54.

Proceedings of the 2019 Hypervelocity Impact Symposium
HVIS2019
April 14-19, 2019, Destin, FL, USA
HVIS2019-036

Extended Abstract+

Prediction of Micrometeoroid Damage to Lunar Construction Materials using Numerical Modeling of Hypervelocity Impact Events

Maria I. Allende[a*], Joshua E. Miller[b], B. Alan Davis[c], Eric L. Christiansen[d], Michael D. Lepech[a], and David J. Loftus[e]

[a]*Department of Civil and Environmental Engineering, Stanford University, Stanford, CA 94305, USA*
[b]*Department of Computer Science, University of Texas at El Paso, El Paso, TX 79968, USA*
[c]*Jacobs Technologies Engineering Science Contract Group, NASA Johnson Space Center, Houston, TX 77058, USA*
[d]*Astromaterials Research and Exploration Science, NASA Johnson Space Center, Houston, TX 77058, USA*
[e]*Division of Space Biosciences, NASA Ames Research Center, Moffett Field, CA 94035, USA*

1. Introduction

Constructing a human-occupied Lunar base presents a unique civil engineering challenge; the resources to make conventional construction materials are unavailable. One approach, known as *in-situ* resource utilization (ISRU), proposes transforming local resources into construction materials [1,2]. One of the Moon's most abundant resources is the unconsolidated surface "soil", known as *regolith*. Several methods for transforming regolith into useful engineering elements, known as *regolith stabilization*, have been proposed and are the subjects of ongoing research efforts [e.g., 3-9]. One class of stabilized regolith material, *Biopolymer-Bound Soil Composites (BSC)*, consists of regolith mixed with a small amount of biopolymer binding agent (10% w/w). BSC compares favorably to other stabilized regolith materials because it does not require high temperature or high energy input and uses a relatively small fraction of binder to achieve an average compressive strength of 20 MPa.

Validating the performance of BSC in Lunar-like conditions is necessary to ensure crewmember safety and surface system reliability for long-duration missions. One notable hazard of the Lunar environment is micrometeoroids. Micrometeoroids pose a difficult design problem because both particle size and impact velocity vary widely, with particle size most frequently on the order of microns to centimeters and impact velocities that range from 3 to 70 km/s, with an average of 20 km/s [10]. Further, it is estimated that 10 impact events of particles greater than 10^{-18} grams occur every second per square meter [11]. Thus, it is important to quantify the risk of damage to structures.

Micrometeoroids larger than 0.1milligrams can be approximated experimentally up to an impact velocity of approximately 8 km/s using a light-gas gun. Velocities in excess of 8 km/s must be simulated using numerical models. Numerical models are validated by hypervelocity impact (HVI) experiments conducted with projectile velocities between 3-8 km/s. Following validation, numerical models are used to predict craters caused by micrometeoroid impact by modeling velocities above 8 km/s. This method for velocity extrapolation is common practice for understanding cratering processes at speeds that are experimentally unachievable. [e.g., 12-14].

+ Full paper can found at https://www.sciencedirect.com/journal/international-journal-of-impact-engineering/special-issue/10MTGP5W4VJ
* Corresponding author. Tel.: +1-609-617-7935
E-mail address: mallende@stanford.edu

In this study, a numerical model is developed for hypervelocity impact simulations of BSC made of JSC-1A Lunar regolith simulant and AP920, a bovine blood protein powder [3, 15]. To validate the model, the simulations' crater dimensions are compared to those of the hypervelocity impact experiments. In the future, this model will be used to extrapolate micrometeoroid impact velocities for micrometeoroid risk assessments of lunar base concepts.

2. Hypervelocity Impact Experiments

Seventeen HVI tests were conducted at NASA's White Sands Testing Facility (WSTF) in Las Cruces, New Mexico using the 0.17-caliber two-stage light-gas gun. The BSC target samples were cylindrical in shape (101.6mm diameter by 50.8mm tall) and secured in an aluminum frame within the test chamber. Spherical projectiles made of aluminum (Al-2017-T4) and nylon with approximately 1, 2, and 3 mm diameters were accelerated to velocities ranging from 3-8 km/s. After the hypervelocity impact tests, three-dimensional (3D) surface scans were taken of each sample. To compare the experimental craters with two-dimensional, axisymmetric numerical results, two-dimensional (2D) crater profiles were constructed by taking 24 radially oriented cross sections from the crater center of each scan, as shown in Figure 1a. The cross sections were then rotated into the same plane, as shown illustratively in Figure 1b for Test 14.

The crater profiles (i.e., Figure 1b) of each test reveal a characteristic crater shape for BSC that is typical for craters formed in brittle materials. This characteristic crater shape is composed of two zones: a transient crater zone and a spall zone. The transient crater is the deep pit formed beneath the projectile's point of impact and the spall zone is the shallow ring around the transient crater. Formation of the spall zone is an inconsistent process that introduces significant variability into crater measurements. Transient crater dimensions, on the other hand, are much more consistent between similar impacts [e.g., 16, 17]. Thus, it can be insightful to consider these two zones separately [16]. The transient crater can be reconstructed from crater cross sections by identifying the inflection point and fitting a parabola to the data within the transient crater, as shown in Figure 1c [e.g., 16-19].

For each cross section, the transient crater diameter, D_{TC}, is two times the radius, r_{TC}, of the fitted parabola at the crater surface and transient crater volume, V_{TC}, is approximated as the volume of the paraboloid. The diameter of the spall zone, D_{SPALL}, is two times the radius, r_{SPALL}, and the volume of the spall zone, V_{SPALL}, is the total crater volume minus V_{TC}.

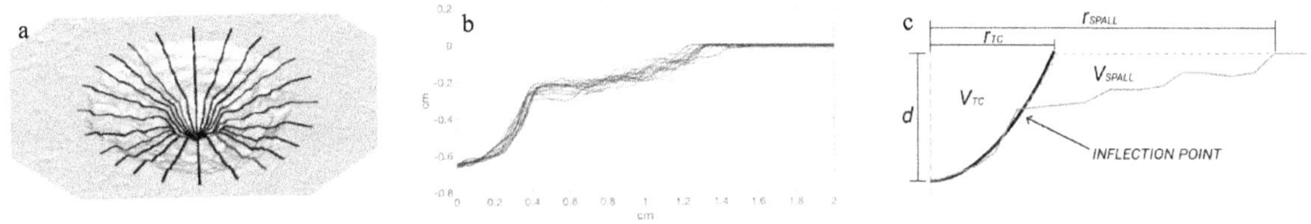

Fig. 1. Analysis of each experimental crater included (a) taking 24 radial cross sections from the crater center of a 3D scanned mesh, (b) rotating these cross-sections into the same plane to create a crater profile (HITF18056 shown here), and (c) for each cross section, separating the transient and spall regions and fitting a parabola to the transient crater to estimate crater dimensions.

3. Numerical Modeling

Numerical simulations of the HVI experiments were carried out using CTH, a shock physics code developed by Sandia National Laboratories [20]. Within CTH, a two-dimensional, axisymmetric geometry was defined to simplify the computational domain. To model the EOS of BSC, the EOS of JSC-1A's principal constituent, *fused quartz*, was selected from CTH's SESAME EOS package and slightly modified to account for known properties of BSC, such as initial density, and a P-alpha compaction model to capture pore collapse [21]. The *fused quartz* EOS was modified within each simulation for parameters that are known for the respective HVI samples, such as initial density (RP=1.98 g/cm^3). To characterize the strength of BSC, the elastic-perfectly plastic

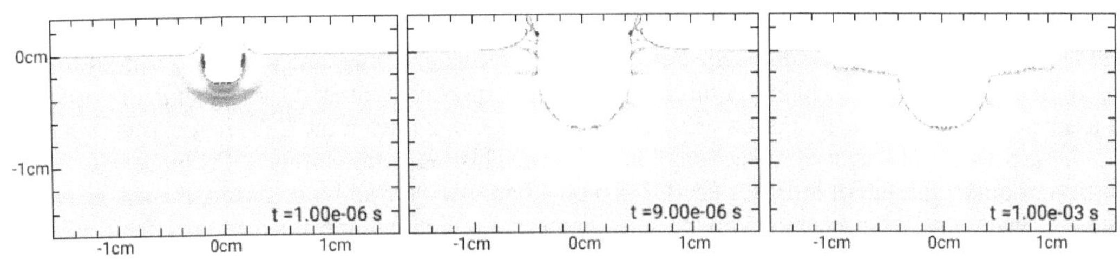

Von Mises (EPPVM) model within CTH has been selected. EPPVM requires a Poisson's ratio (POI=0.2) and a yield strength (Y=118 MPa) to define a constant yield surface. Lastly, a fracture model was included to introduce void into a cell after the fracture stress was reached (PFRAC=120 MPa). An example simulation is shown in Figure 2.

Fig. 2. An example CTH simulation of a hypervelocity impact experiment into BSC. Within the first microsecond, crater excavation begins. By nine microseconds, the transient crater has completed formation and the crater has reached its final depth. After nine microseconds, the spall zone forms.

4. Results

A visual comparison of the CTH model and the experiments can be made by overlaying the crater profile onto the CTH simulation for each test. A representative example is shown in Figure 3 for HITF18043. Visually, the CTH model captured the shape of the crater well for this particular test. A more in-depth evaluation of the CTH model can be made by comparing crater dimensions, such as crater depth, transient crater volume and diameter, and spall zone volume and diameter, between the simulations and the experiments. Because the CTH model is two-dimensional and axisymmetric, each simulation could represent any single cross section from the experimental crater. Thus, 24 cross sections were taken of each experimental crater (shown as the black lines in Figure 1a). This resulted in one crater depth measurement at the intersection of the cross sections and a range of 24 crater volume and diameter values. While a single value comparison is done for experimental and simulated crater depth, the single CTH volume and diameter values for both the transient crater and the spall zone are compared to the range of 24 diameter values calculated from the experimental cross sections.

The CTH model is able to predict crater depth within 0.5 millimeters of the experimental results, or a maximum percent difference of 6%, for all tests. In addition to accurate depth prediction, the CTH model was successful in predicting transient crater volume and diameter. All CTH values for transient crater volume and diameter fall within the range of experimental values. The CTH model had more difficulty in predicting the spall zone response seen in the experimental craters. In particular, the CTH model struggled with crater spall diameter. Only three simulations fell within the range of experimental values, while all others were less than the minimum diameter calculated from the experiments. Spall volume prediction was slightly better than spall diameter, with eight of the 17 simulations falling within the measured range of experimental values and most of the unsuccessful simulations being overestimations.

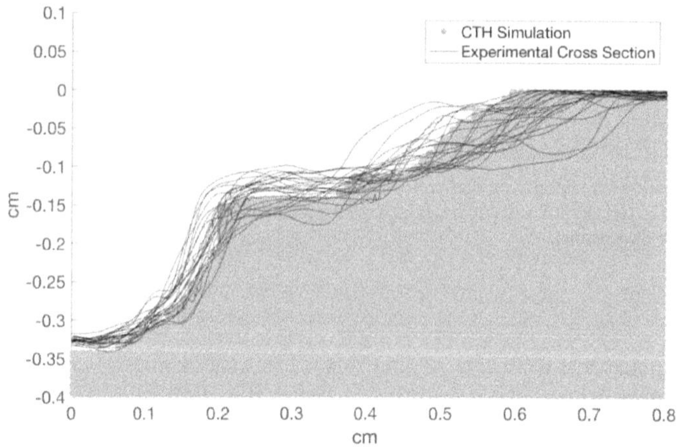

Fig. 3. The 24 cross sections taken of the HITF18043 target (black lines) are overlaid onto the corresponding CTH simulation (gray).

5. Discussion and Conclusions

The CTH model was used to simulate hypervelocity impact experiments conducted on BSC targets at NASA's WSTF. Comparisons of crater depth, volume, and diameter were made between the experimental crater and simulated crater for each test. The CTH model was able to predict crater depth within 0.5 millimeters of the experimental results, or a maximum difference of 6%, indicating successful predictions of crater depth. For transient crater volume and diameter, the CTH model is within the range of values determined from the experimental crater cross sections for each test. This indicates that the CTH model can accurately capture the significant portion of crater formation in BSC. The CTH model had more difficulty capturing the shape and size of the experimental spall region, leading to less accurate volume and diameter predictions and thus an area for further model refinement. Future work on the CTH model includes sensitivity analyses on crater dimensions as well as further material characterization experiments, such as triaxial and split-Hopkinson pressure bar tests on BSC, to better CTH input parameters and perhaps employ more complex material models.

The CTH model developed in this paper can be used as a material design tool that is capable of varying material parameters computationally to optimize BSC's performance under hypervelocity impact loading. This is especially useful in limiting the necessity of expensive light-gas gun experiments, as well as guiding the experimental design of these tests when they are required. The model also has applications as a damage prediction tool that allows for the necessary extrapolation of impact velocities past what is experimentally achievable. By simulating micrometeoroid impacts in the velocity regime that is relevant to micrometeoroids, this model could be used in a performance-based impact engineering framework that incorporates site-specific micrometeoroid hazard models to predict the damage a structure will incur over its lifetime and thus design the structure to achieve a certain degree of performance.

Acknowledgements

This work was supported by a NASA Space Technology Research Fellowship. We gratefully acknowledge the technical contributions from the teams at NASA White Sands Testing Facility and NASA Johnson Space Center's Astromaterials Research and Exploration Science. We also thank the Lepech Group, including Isamar Rosa, for their support and feedback.

References

[1] Khalili, E.N., 1989. Lunar structures generated and shielded with on-site materials. J. Aerosp. Eng., 1989, 2(3): 119-129.
[2] Bodiford, M.P., Burks, K.H., Perry, M.R., Cooper, R.W., Fiske, M.R., 2006. Lunar in situ materials-based habitat technology development efforts at NASA/MSFC. ASCE Earth Sp. Conf. 1–8. https://doi.org/https://doi.org/10.1061/40830(188)70
[3] Roedel, H., Lepech, M.D., Loftus, D.J., 2014. Protein-Regolith Composites for Space Construction. ASCE Earth and Space Conference 2014. pp. 291–300.
[4] Taylor, L.A., Meek, T.T. 2005. Microwave Sintering of Lunar Soil: Properties, Theory, and Practice. J. Aerosp. Eng. 18, 188–196.
[5] Hintze, P., Curran, J., Back, T. 2009. Lunar Surface Stabilization via Sintering or the Use of Heat Cured Polymers. 47th AIAA Aerosp. Sci. Meet. Incl. New Horizons Forum Aerosp. Expo. https://doi.org/10.2514/6.2009-1015
[6] Fateri, M., Gebhardt, A. 2015. Process parameters development of selective Laser Melting of lunar regolith for on-site manufacturing applications. Int. J. Appl. Ceram. Technol. 12, 46–52. https://doi.org/10.1111/ijac.12326
[7] White, C., Alvarez, F., Shafirovich, E. 2011. Combustible Mixtures of Lunar Regolith with Metals: Thermodynamic Analysis and Combustion Experiments. J. Thermophys. Heat Transf. 25, 620–625.
[8] Toutanji, H.A, Evans, S., Grugel, R.N. 2012. Performance of lunar sulfur concrete in lunar environments. Constr. Build. Mater. 29, 444–448.
[9] Faierson, E.J, Logan, K.V., Stewart, B.K., Hunt, M.P. 2010. Demonstration of concept for fabrication of lunar physical assets utilizing lunar regolith simulant and a geothermite reaction. Acta Astronaut. 67, 38–45.
[10] Moorhead, A.V., Koehler, H.M., Cooke, W. J., Space, M. 2015. NASA Meteoroid Engineering Model Release 2.0.
[11] Grun, E., Zook, H.A., Fechtig, H., Giese, R.H. 1985. Collisional Balance of the Meteoritic Complex. Icarus 272, 244–272.
[12] Pierazzo, E., Collins, G. 2004. A Brief Introduction to Hydrocode Modeling of Impact Cratering, in: Dypvik, H., Burchell, M.J., Claeys, P. (Eds.), Cratering in Marine Environments and on Ice. Springer Berlin Heidelberg, Berlin, Heidelberg, pp. 323–340.
[13] Holsapple, K.A., 1993. The Scaling of Impact Processes in Planetary Sciences. Annu. Rev. Earth Planet. Sci. 21, 333–373.
[14] Güldemeister, N., Wünnemann, K., Poelchau, M.H., 2015. Scaling impact crater dimensions in cohesive rock by numerical modeling and laboratory experiments. Geol. Soc. Am. Spec. Pap. 518, 17–29. https://doi.org/10.1130/2015.2518(02)
[15] Roedel, H. 2017. On the compressive strength and sustainability of protein bound concretes. PhD Thesis, Stanford University, Stanford, CA.
[16] Poelchau, M.H., Kenkmann, T., Hoerth, T., Schäfer, F., Rudolf, M., Thoma, K., 2014. Impact cratering experiments into quartzite, sandstone and tuff: The effects of projectile size and target properties on spallation. Icarus 242, 211–224. https://doi.org/10.1016/j.icarus.2014.08.018
[17] Dufresne, A., Poelchau, M.H., Kenkmann, T., Deutsch, A., Hoerth, T., Schäfer, F., Thoma, K., 2013. Crater morphology in sandstone targets: The MEMIN impact parameter study. Meteorit. Planet. Sci. 48, 50–70. https://doi.org/10.1111/maps.12024
[18] Kenkmann, T., Wunnemann, K., Deutsch, A., Poelchau, M.H., Schafer, F., and Thoma, K., 2011. Impact cratering in sandstone: The MEMIN pilot study and the effect of pore water. Meteorit. Planet. Sci. 36:890-902. https://doi.org/10.1111/j.1945-5100.2011.01200.
[19] Poelchau, M.H., Kenkmann, T., Thoma, K., Hoerth, T., Dufresne, A., Schäfer, F., 2013. The MEMIN research unit: Scaling impact cratering experiments in porous sandstones. Meteorit. Planet. Sci. 48, 8–22. https://doi.org/10.1111/maps.12016
[20] McGlaun, J.M., Thompson, S.L., Elrick, M.G. 1990. CTH: A three-dimensional shock wave physics code. Int. J. Impact Eng. 10, 351-360.
[21] Kerley, G.I. 1999. Equations of State for composite materials. Kerley Publishing Services KPS99-4, Albuquerque, NM.

Proceedings of the 2019 Hypervelocity Impact Symposium
HVIS2019
April 14-19, 2019, Destin, FL, USA
HVIS2019-046

Extended Abstract+

A penetration model for semi-infinite ultra-high molecular weight polyethylene composite

Long H. Nguyen[a]*, Shannon Ryan[a], Adrian C. Orifici[b] , Stephen J. Cimpoeru[a]

[a]*Defence Science and Technology Group, 506 Lorimer Street, Port Melbourne 3207, Australia*
[b]*School of Engineering, RMIT University, PO Box 71, Bundoora 3083, Australia*

1. Introduction

Ultra-high molecular weight polyethylene (UHMW-PE) composite has been shown to be an effective material for ballistic protection against blunt penetrators [1]. The material exhibits multiple stages of penetration, typically characterised by an initial local penetration phase followed by large bulge deformation of the back face [2]. The location at which transition occurs between the localised penetration stage and non-localised bulging stage is an important property of UHMW-PE composite armour. However, the conditions required to induce transition are poorly understood with a range of different mechanisms proposed to explain the behaviour [2,3], none of which can be used to predict the transition location within the target.

In this paper, the modelling strategy proposed in [4] is used to investigate the ballistic impact response of thick UHMW-PE composite. Analysis of the numerical simulation results highlight the mechanism for transition, the mode of penetration and the influence this has on the penetration rate. An analytical model is proposed based on axial momentum balance to describe the localised penetration phase for this class of composites.

2. Numerical model

A commercial hydrocode ANSYS® AUTODYN® is used to model the impact event as per the modelling approach developed by the authors in [4]. Modelling of UHMW-PE composite is accomplished using a non-linear orthotropic model that incorporates coupling of a shock equation of state and a non-linear orthotropic strength model with orthotropic strength and energy based failure. The model is discretised using lagrangian elements, with the composite discretised using a sub-laminate approach (multiple layers through the target thickness), to capture delamination failure. This modelling approach has been validated against fragment simulating projectiles (FSP) for impact velocities between 400-2000 m/s and targets between 10 mm to 100 mm thick [4].

+ Full paper can found at https://www.sciencedirect.com/journal/international-journal-of-impact-engineering/special-issue/10MTGP5W4VJ

* Corresponding author. Tel.:+61 (0) 3 9626 8622;
E-mail address: long.nguyen@dst.defence.gov.au

3. Penetration and transition mechanism

Fig. 1(a) shows numerical simulation results of a 20 mm FSP impacting UHMW-PE composite (Dyneema® HB26) at 2 km/s with two different target thickness and back face support conditions. The purpose is to investigate any changes in penetration mechanism when the back of the target is free (where the impact pressure is allowed to release), fixed (where compressive pressure increases) and where the target is semi-infinite (the projectile is slowed to rest before additional compressive or release waves have come into effect). These effects are studied with a 100 mm thick target unsupported or free on the back surface, 100 mm thick target rigidly fixed on the back surface and a 250 mm thick target free on the back surface respectively.

For the semi-infinite target ("250 mm free"), the projectile velocity profile is approximately linear for the duration of the penetration event. The velocity profile slope changes when the velocity drops below approximately 200 m/s. The stresses in the target material at the projectile-target interface is too low to cause any further penetration compared to the initial stage, so the projectile is further decelerated by compression of the material ahead of it. In the case where the target thickness is finite and the rear surface is fixed from any axial displacement, the initial velocity profile is the same as in the semi-finite case. At a later time, the slope steepens leading to an increase in the projectile deceleration rate. Here, the shock wave reflects on the back of the fixed support as a compressive wave. When this reflected compressive wave reaches the projectile, the pressure at the projectile-target interface increases. As a result, the local penetration mode persists, but the deceleration rate of the projectile increases. For the case where the back of the finite target is free to move, the shock wave reflects as a release fan or rarefaction, reducing the overall pressure in the material ahead of the projectile, reducing the penetration rate. Further, the release also leads to a velocity jump and bulging of the material on the back face, such that a larger portion of the target is now under strain and resisting penetration.

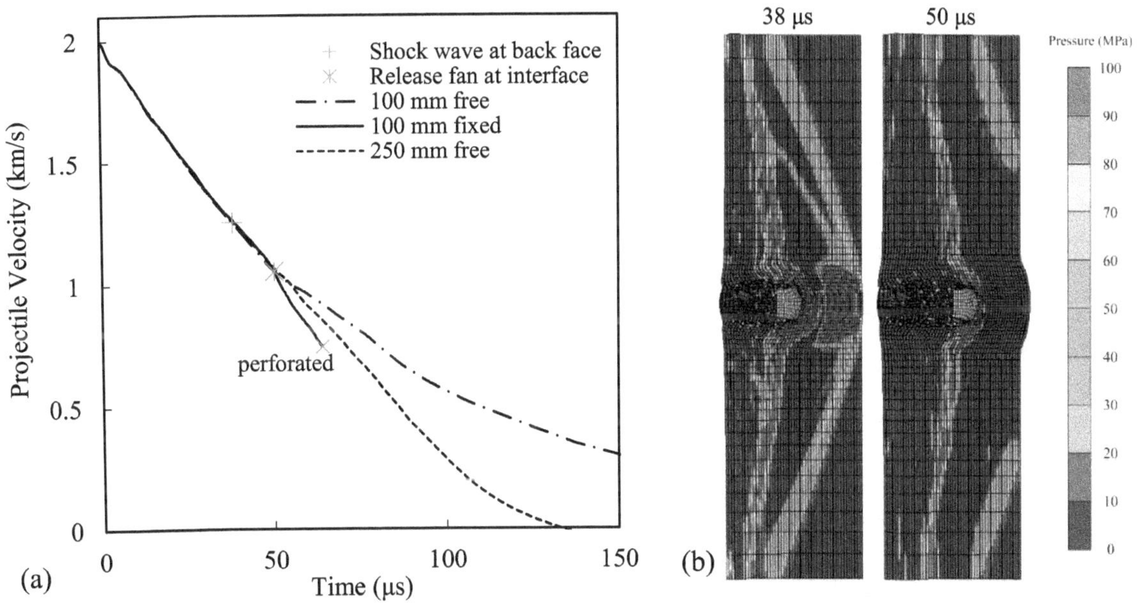

Fig. 1. (a) Projectile velocity from numerical simulation of 20 mm FSP impacting UHMW-PE composite at 2 km/s with different thickness and back support conditions. (b) Pressure contour at the mid-plane of a 100 mm unsupported target impacted by 20 mm FSP at 2 km/s. The left image is at an instant in time where the shock reaches the back (free surface) and is released, right image is time when the front of the release fan reaches the projectile.

Fig. 1(b) show images of the pressure wave propagating through the target at different points in time for the 100 mm thick unsupported target. In the first image, the pressure wave reaches the back of the target and the compressive stress is released. This point in time is marked in Fig. 1(a). From this point, a release fan travels back towards the projectile. When the front of the release fan reaches the projectile, the target material between the back of the target and the projectile transition from local penetration to bulging as shown in the second image. This instant in time is highlighted in Fig. 1(a) and corresponds to a change in projectile velocity slope. The transition occurs at 50 μs, and the slope of the velocity profile deviates from the curve of the semi-infinite target. It is clear that a change in the projectile velocity slope strongly indicates a change in the loading condition at the projectile–target interface and therefore the prevailing penetration mode.

In order to analytically describe the penetration of UHMW-PE composite, a model is required to describe the penetration of the material with time, the propagation of the shock wave on impact through the material and the propagation of the rarefaction from the free back surface. In the next section of the paper, penetration model is proposed to address the first of these challenges.

4. Penetration model

For penetration of UHMW-PE composite there are several general considerations and assumptions:

1. In experiments conducted with FSPs [2], negligible changes in mass of the recovered FSPs were observed and only at the highest impact velocities (~2 km/s) was a 2% reduction in projectile mass measured. Projectile erosion doesn't take place and the mass can be considered to be constant.
2. The penetration of UHMW-PE composite by hard metallic projectiles is generally by rigid body penetration, i.e no deformation of the projectile. Where the projectile is softer or impacted at high velocities (>1 km/s), plastic deformation of the nose typically occurs [4] and needs to be accounted for.

As a first step, the penetration is considered rigid body, such that the axial stress at the projectile-target interface, σ_z, is given by:

$$\sigma_z = \frac{m_p}{A}\frac{du}{dt} = -\left(\frac{1}{2}\rho_t u^2 + R_t\right) \tag{1}$$

where m_p and A are the mass and presented area of the projectile respectively, ρ_t is the target density, u the velocity of the projectile and R_t is the target resistance [5]. Where the projectile is plastically deforming, the target is providing greater resistance to penetration as a result of the increased presented area. The target resistance term needs to be modified to account for this. This is introduced in Eq. (2).

$$\frac{du}{dt} = -\frac{A_0}{m_p}\left(\frac{1}{2}\rho_t u^2 + S\left(\frac{d(x)}{d_0}\right)^2\right) \tag{2}$$

where $d(x)/d_0$ is the ratio of the current and initial projectile diameter as a function of the projectile position. The term is squared to account for the increase in the initial presented area, A_0. S is the strength of the material, for UHMW-PE it is the tensile strength of the cross-ply laminate, which is directly related and equal to its out-of-plane compressive strength [6]. For a rigid body penetrator, Eq. (2) can be solved for the velocity of the projectile with $d(x)/d_0 = 1$. For a plastically deforming projectile, a description for $d(x)/d_0$ is required, this is explained in the full paper.

The results of Eq. (2) is shown in Fig. 2 for impacts by a 20 mm FSP. The projectile mass, m_p is 53.8 grams, target density ρ_t is 980 kg/m^3. The tensile strength, S, of Dyneema® HB26 composite is 1.15 GPa [7], this value is consistent with the numerical model. Fig. 2(a) compares the velocity profiles predicted by the analytical and numerical model. The velocites are in reasonably good agreement with the semi-infinite target in the numerical model (250 mm thick free on the back surface). The analytical results differ as the projectile reaches velocites of about 200 m/s where it is too slow to be penetrated by the mechanisms described in this model. Fig. 2(b) compares the projectile axial stress with those predicted by the model. The numerical model predicts a cyclic load rupture behaviour, which has been observed experimentally when a blunt or hemispherical nose punch is quasi-statically or dynamically indented into UHMW-PE composite in the thickness direction [8]. The analytical model does not model this behaviour, instead its provides an approximation of the average stress (average stress peak and troughs) during the local penetration phase, for which agreement is good compared to the numerical simulation.

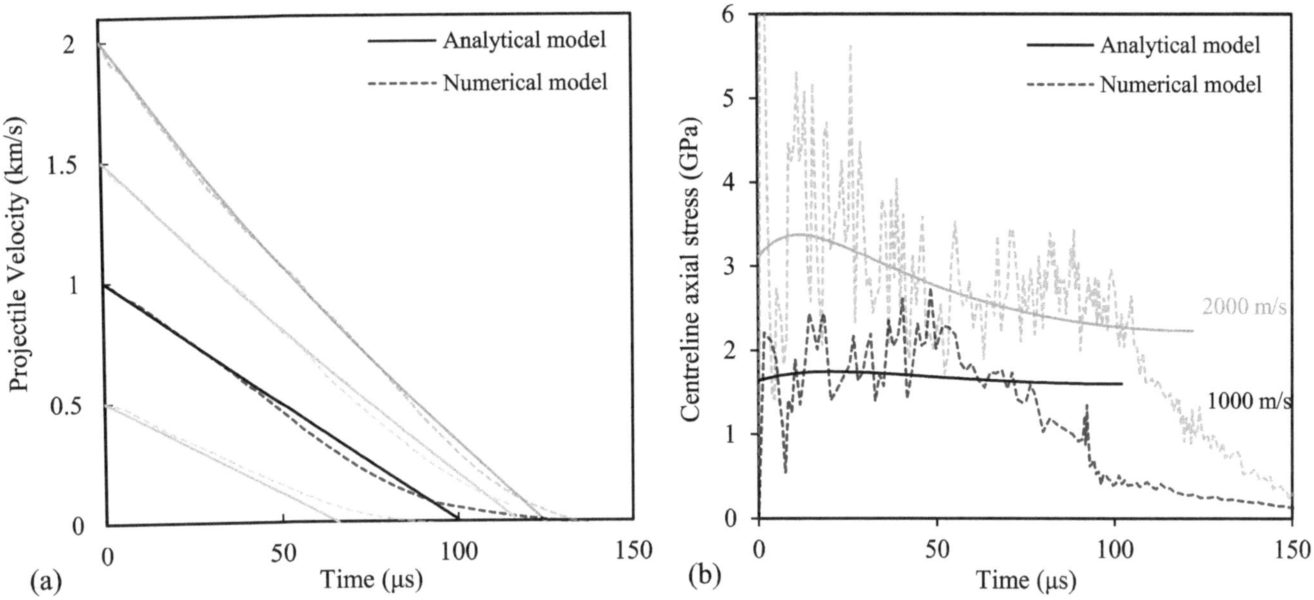

Fig. 2. Results from 20 mm FSP impacting UHMW-PE composite at various impact velocities (a) projectile velocity profiles and (b) projectile axial stresses (numerical results of a 250 mm unsupported target).

5. Conclusion

A validated numerical model was used to investigate the penetration of UHMW-PE composite under ballistic impact. For a target back face that is unsupported, a release fan is reflected from the free surface, initiating bulging. The intercept of the front of the release fan and the projectile-target interface reduces the interface stresses and changes the mode of penetration from local to increasingly global. For a semi-infinite target, local penetration persists until the projectile comes to rest. A penetration model based on axial momentum balance is proposed that describes the local phase of penetration. The analytical model shows good agreement with numerical simulations in terms of the projectile velocity profile and axial stresses at the projectile-target interface.

References

[1] Nguyen, L.H., Ryan, S., Cimpoeru, S.J., Mouritz, A.P., Orifici, A.C., 2016. The efficiency of ultra-high molecular weight polyethylene composite against fragment impact, Experimental Mechanics 56, p.595–605.

[2] Nguyen, L.H., Ryan, S., Cimpoeru, S.J., Mouritz A.P., Orifici, A.C., 2015. The effect of target thickness on the ballistic performance of ultra-high molecular weight polyethylene composite, International Journal of Impact Engineering 75, p.174–183.

[3] Greenhalgh, E.S., Bloodworth, V.M., Iannucci, L., Pope, D., 2013. Fractographic observations on Dyneema® composites under ballistic impact, Composites Part A: Applied Science and Manufacturing 44, p.51–62.

[4] Nguyen, L.H., Lässig, T.R., Ryan, S., Riedel, W., Mouritz, A.P., Orifici, A.C., 2016. A methodology for hydrocode analysis of ultra-high molecular weight polyethylene composite under ballistic impact. Composites Part A: Applied Science and Manufacturing 84, p.224–235.

[5] Tate, A., 1967. A theory for the deceleration of long rods after impact, Journal Mechanics and Physics of Solids 15, p.387–399.

[6] Attwood, J.P., Khaderi, S.N., Karthikeyan, K., Fleck, N.A., O'Masta, M.R., Wadley, H.N.G., Deshpande, V.S., 2014. The out-of-plane compressive response of Dyneema® composites, Journal of the Mechanics and Physics of Solids 70, p.200–226.

[7] Heisserer, U., Van der Werff, H., 2016. Strength matters: Which strength of Dyneema fiber composites to use in hydrocode models? A discussion, 29th International Symposium on Ballistics, Edinburg, Scotland.

[8] Woodward, R.L., Egglestone, G.T., Baxter, B.J., Challis, K., 1994. Resistance to penetration and compression of fibre-reinforced composite materials. Composites Engineering 4, p.329–341.

Proceedings of the 2019 Hypervelocity Impact Symposium
HVIS2019
April 14-19, 2019, Destin, FL, USA
HVIS2019-063

Modeling Hypervelocity Impact of Reinforced Carbon-Carbon Composite Thermal Protection System

Alexander J. Carpenter[a,*], Sidney Chocron[a], James D. Walker[a]

[a]*Engineering Dynamcs Department, Southwest Research Institute, 6220 Culebra Rd., San Antonio, TX 78238, U.S.A.*

Abstract

Reinforced carbon-carbon (RCC) composite is used in applications where structural stiffness and strength must be maintained at very high temperatures that may reach 2000°C or more. For example, it was used on both the Space Shuttle's nose cone and the leading edges of its wings. As exemplified by the Space Shuttle *Columbia* accident, the ability of these materials to survive impacts up to hypervelocity speeds can be critical for some applications. As computational modeling becomes an increasingly important component of the design process, the ability to accurately model RCC materials under impact conditions likewise becomes more and more important. This paper describes a computational model of the thermal protection used on the Space Shuttle orbiter. The model incorporates both the RCC comprising much of the protection system and its silicon carbide coating. The model was subjected to hypervelocity impacts with both steel and aluminum projectiles, and the results were compared to test data from the literature.

Keywords: Reinforced carbon-carbon, numerical simulations, orbital debris

---

Nomenclature

ρ	composite mass density
E_l	longitudinal Young's modulus for fibers
E_t	transverse Young's modulus for fibers
ν_{lt}	longitudinal Poisson's ratio for fibers
ν_{tt}	transverse Poisson's ratio for fibers
G_{lt}	shear modulus for fibers
E_m	Young's modulus for matrix
ν_m	Poisson's ratio for matrix
σ_f	fiber strength
$\sigma_{d,1}$	normal delamination strength
$\sigma_{d,2}$	shear delamination strength
V_f	volume fraction of fibers in composite

1. Introduction

Reinforced carbon-carbon (RCC) materials are an integral component of thermal protection systems used in spacecraft such as the Space Shuttle orbiter. High velocity impact by debris or other bodies during service can lead to damage of the RCC and failure of the thermal protection system. Failure of the RCC on the wing leading edge of the Space Shuttle *Columbia* led to loss of the

* Corresponding author. Tel.: +1-210-522-2621 ; fax: +1-210-522-6290 .
E-mail address: alexander.carpenter@swri.org .

orbiter during reentry from orbit in 2003. An understanding of the impact response of RCC may help to prevent similar incidents in the future.

A number of hypervelocity impact tests on specimens similar to the RCC utilized on the Space Shuttle orbiter were conducted by Christiansen *et al.* [1, 2] in 1993. In these tests, RCC panels with a silicon carbide coating on both sides were impacted by aluminum, steel, and nylon projectiles of various sizes at a variety of velocities and angles of obliquity. The responses of the panels ranged from mere surface coating damage to complete perforation.

In this study, a computational model of Christiansen *et al.*'s test samples was created. The model incorporates both the RCC substrate and the silicon carbide coating as separate parts with their own material models. The computational model was subjected to impact by aluminum and steel spherical particles, and the results were compared to Christiansen *et al.*'s data.

2. Material Models

2.1. Reinforced Carbon-Carbon Composite

In the tests, the RCC was formed from a 19-ply layup of phenolic-graphite subjected to pyrolysis and densification [1]. The RCC substrate utilizes a continuum composite model created by Carpenter *et al.* [3] that does not explicitly model individual layers within the composite. This model typically requires material properties for the composite's individual fiber and matrix constituents and the fiber volume fraction. These properties were not available in the literature, but typical values for RCC composites [4] suggest that the substrate has a stiffness of approximately 55 GPa and a tensile strength of approximately 120 MPa. One simple method of utilizing Carpenter *et al.*'s continuum composite model when only homogenized material properties are available is to attribute most of the longitudinal stiffness and strength to the fibers. Assuming a fiber volume fraction of 60% [4], the appropriate fiber longitudinal Young's modulus and strength are estimated to be 183 GPa and 400 MPa, respectively. For simplicity, all other elastic moduli were assumed to be 5 GPa (which is on the lower end of the range expected for graphite), while all Poisson's ratios were set to 0. In addition, the delamination strength for the composite is assumed to be 8 MPa [4]. Table 1 summarizes the RCC material properties used in the computations. RCC elements were eroded from the computations if the fibers in both directions failed or if the maximum principal strain exceeded 20%.

Table 1. RCC Material Properties Used in Computations

ρ (g/cm^3)	E_l (GPa)	E_t (GPa)	ν_{lt}	ν_{tt}	G_{lt} (GPa)	E_m (GPa)	ν_m	σ_f (MPa)	$\sigma_{d,1}$ (MPa)	$\sigma_{d,2}$ (MPa)	V_f (%)
1.52	183	5	0	0	5	5	0	400	8	8	60

2.2. Silicon Carbide Coating

The silicon carbide (SiC) coating, which is approximately 0.8-mm thick on both the front and back faces of the target, is approximately 25% of the target's 6.3-mm thickness [1]. Thus, the coating should be explicitly modeled in order to accurately reproduce the full target's response during hypervelocity impact. Johnson and Holmquist developed several models for simulating the dynamic behavior of ceramic materials. One of these, commonly referred to as the JH-1 model [5], was used in this study for computing the response of the silicon carbide coating. The silicon carbide constants used in this model were previously determined by Holmquist and Johnson [6].

2.3. Projectile Materials

The projectile materials in the computations utilized the Johnson-Cook strength and fracture models [7, 8]. Although the tests utilized a variety of projectile materials and sizes [1], those with spherical aluminum and steel projectiles approximately 3.17 mm in diameter were the focus of this investigation. The aluminum projectiles used in the tests were Al 2017-T4. Johnson-Cook constants were not available for this material, so constants for Al 6061-T6 [9] were used instead, as both alloys have similar strength and ductility. The steel projectiles were only listed as being "C-Steel," so constants for 1006 steel [7] were used, as this alloy was deemed a representative low-alloy carbon steel.

3. Problem Setup

Impact computations were conducted using LS-DYNA, a commercial dynamic explicit finite element code. A close-up image of the meshes used in the computations is shown in Figure 1. A 5-cm × 5-cm portion of the target was explicitly modeled using constant stress solid elements. The thickness of the RCC substrate is 4.7 mm, while the thickness of each SiC layer is 0.8 mm. The RCC composite shares nodes with the SiC layers, i.e. the SiC is assumed to be perfectly bonded to the RCC substrate. An average element size of 0.4 mm was used, and all elements were constant stress elements with a single integration point. The tests were generally conducted with large angles of obliquity, so the initial position of the spherical projectile, which also utilizes constant stress solid elements, was offset from the center of the target, and the non-normal velocity component is toward the target center. The target was assumed to be free-floating in space, i.e. no nodal constraints were added to the target boundaries in the computations. A contact condition was used that allows for the contact surfaces to update when elements were eroded from the computation.

A small number of Christiansen *et al.*'s RCC impact tests [1] were simulated during this study. Generally, multiple tests were performed under very similar conditions (e.g. projectile, impact velocity), so the simulations utilized average values from similar tests. The average values used in each simulation are summarized in Table 2.

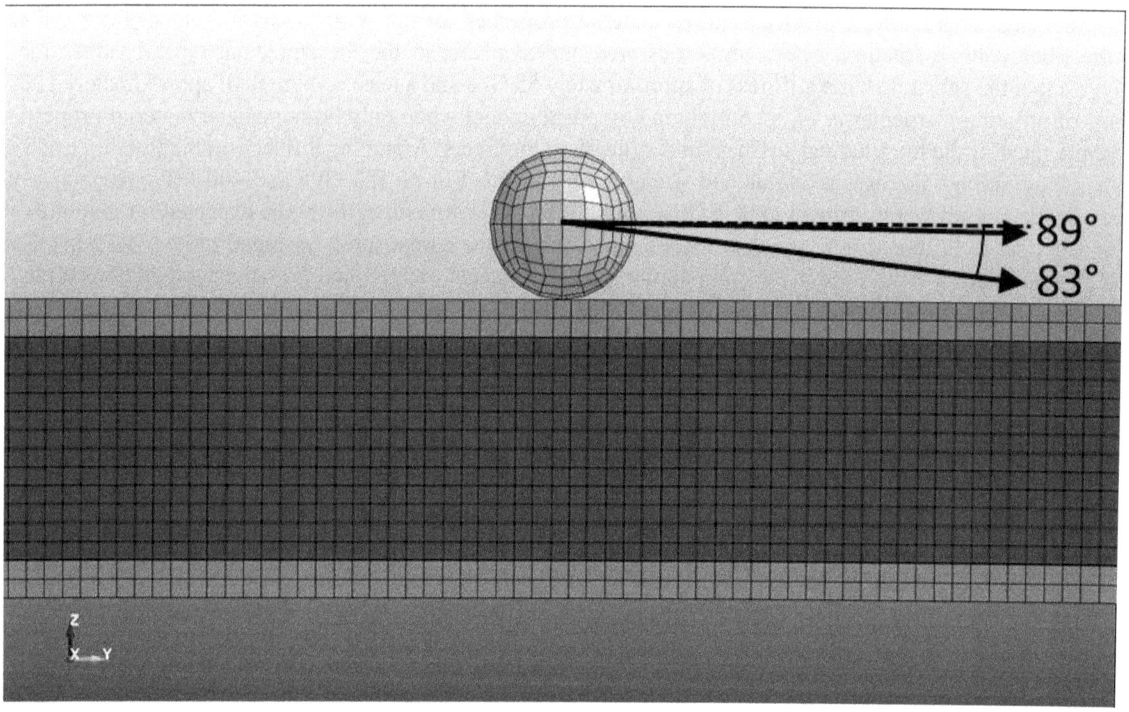

Fig. 1. The finite element mesh used in the impact computations of the SiC-covered RCC targets is shown along with the investigated impact angles. The RCC substrate is shown in red, the SiC coatings on the front and back of the target are shown in blue, and the spherical projectile is shown in yellow.

Table 2. Summary of Simulated RCC Impact Tests [1]

Projectile Material	Projectile Diameter (mm)	Avg. Impact Velocity (km/s)	Avg. Impact Angle (deg)	Damage Mode Observed in Tests
Aluminum	3.17	6.15	89	Surface coating damage
Aluminum	3.17	6.30	83	Rear-side spall
Steel	3.17	2.44	83	Complete perforation

4. Results/Discussion

Due to the high degree of obliquity, the projectiles in all three computations exhibited a ricochet, in which the projectile slid across the surface for a time before rebounding off the target, as illustrated in Figure 2. However, the resulting effects on the target were very different in each of the three cases. Figures 3-5 present images of the top/bottom surfaces after impact from each of the computations. Generally, SiC elements were not automatically eroded from the computations by the solver. However, several of the elements experienced significant, i.e. nonzero, plastic strains. Because ceramics typically do not admit significant plastic deformation before fracture occurs, elements exhibiting effective plastic strains larger than 0.1% were assumed to have failed, and therefore are not included in the visualizations even though they were still present in the calculations, i.e. the elements were removed in post-processing. Future references to "eroded" elements in this paper include these ceramic elements. The images demonstrate the ability of the model to match the damage modes observed from the tests.

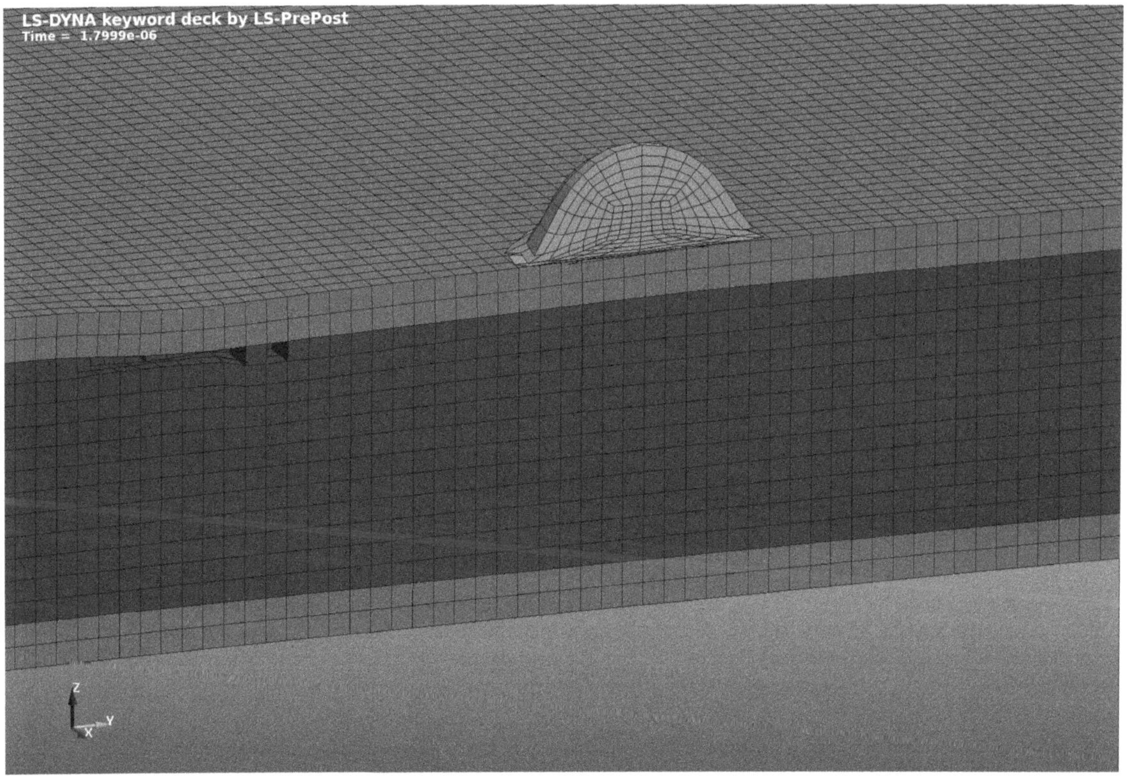

Fig. 2. A cross-section of the projectile and target meshes from the aluminum impact at 83° is presented midway through the computation. The RCC substrate is shown in red, the SiC coatings on the front and back of the target are shown in blue, and the spherical projectile is shown in yellow. Projectile is moving from left to right in the computation. Note that damage within the target continues to evolve beyond this point.

In the simulation where the aluminum projectile impacts at 89°, the only eroded elements were located within the front SiC coating. This agrees well with the tests, which only exhibited surface coating damage. The size of the damaged region in the computation (i.e., where plastic strain is greater than zero) was 17-mm × 2-mm. This is somewhat smaller than the average reported coating damage area of 27-mm × 4-mm from the tests [1], but the overall agreement in the size and shape of the coating damage is still somewhat reasonable given the uncertainty in many of the appropriate material properties.

When the aluminum projectile's angle of incidence was changed to 83° (and impact velocity slightly increased), damage was observed in both the front and rear SiC coatings. This also agrees well with the observation of rear-side spall damage from the corresponding tests. Some RCC elements were also eroded, but complete perforation of the RCC was not observed in this computation. This behavior was also observed from the tests, i.e. the rear SiC can be damaged without completely perforating the RCC. The size of the large damaged area in the front SiC mesh was approximately 18-mm × 6-mm. The length agrees well with the corresponding value from the tests, 16 mm, although the simulation underpredicts the average front coating damaged width of 11 mm [1]. The RCC substrate was damaged to a depth of approximately 1 mm in the computation. When this number is added to the 0.8-mm thickness from the front SiC coating, the value agrees well with the average reported penetration depth from the corresponding tests, which was 2.3 mm [1].

For the final simulation, the projectile material was changed to steel, but the impact velocity was significantly reduced. Despite the large reduction in velocity, the steel projectile is more damaging to the RCC target due to its significantly higher density. This is true even if the velocity of the previous aluminum computation is increased slightly so that both projectiles have the same momentum. RCC elements were eroded through the entire thickness of the target, forming a hole. Likewise, the reported damage mode from the corresponding tests was complete perforation of the RCC. The size of the hole is approximately 4.5 mm, which is only slightly smaller than the average reported hole diameter of 5.4 mm from the corresponding tests. The front coating damaged area from the computation is approximately 21-mm × 8-mm. Like the previous computation, the damaged area's length agrees well with the average test value, 17.5 mm, but the width, which was 16 mm on average, was underpredicted [1]. The hole in the computations is formed despite the fact that the steel projectile ricochets off without any part passing through the target. It is unclear from the description of the tests as to how much of the steel projectile, if any, passed through the target in the tests.

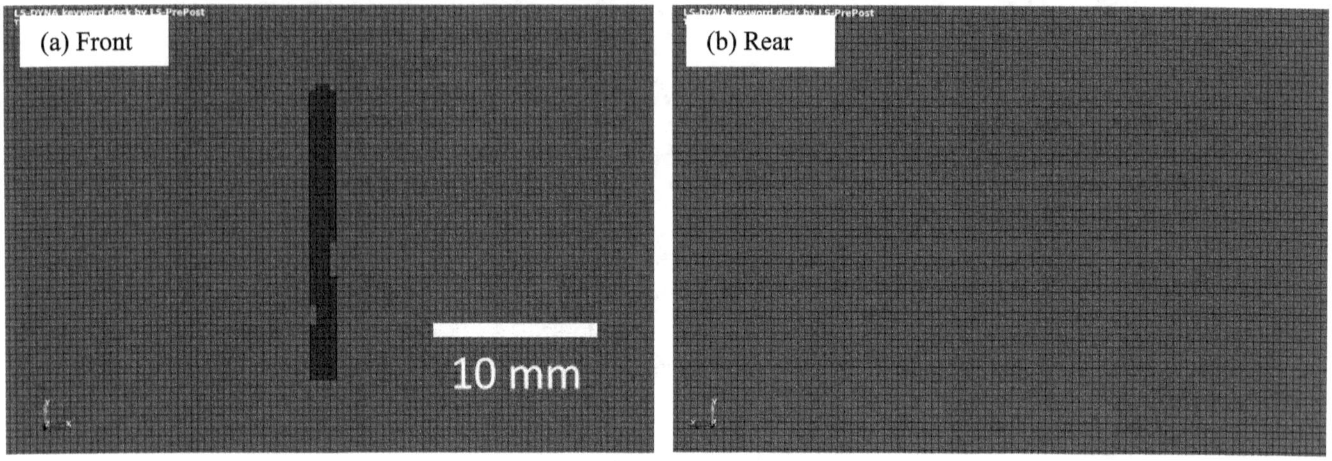

Fig. 3. The (a) front and (b) rear surfaces of the SiC-coated RCC target mesh are shown after impact by an aluminum projectile with a velocity of 6.15 km/s and obliquity of 89°. Red elements represent the RCC substrate, while blue elements represent the SiC coating. The projectile moved from the bottom to the top of the page during the computation. The initial position of the projectile as it came into contact with the front SiC coating was approximately 7.7 mm below the bottom of the damaged region in (a), which is below the region shown in the figure. Damage is limited to the front SiC coating.

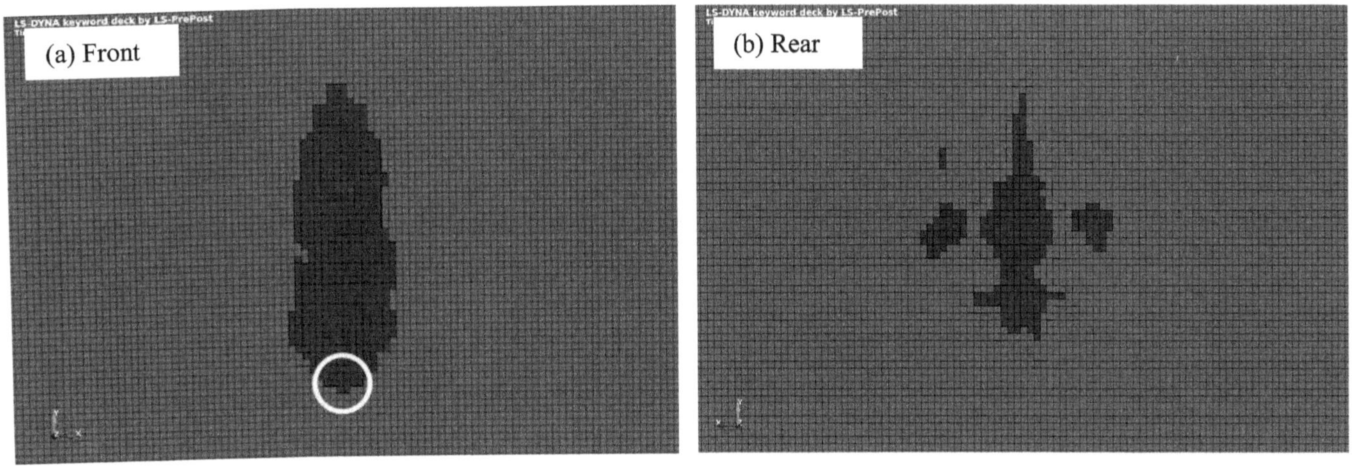

Fig. 4. The (a) front and (b) rear surfaces of the SiC-coated RCC target mesh are shown after impact by an aluminum projectile with a velocity of 6.30 km/s and obliquity of 83°. Red elements represent the RCC substrate, while blue elements represent the SiC coating. The projectile moved from the bottom to the top of the page during the computation. The initial position of the projectile as it comes into contact with the front SiC coating is shown with a yellow circle. Both the front and rear SiC coatings have been damaged, but the RCC has not been completely perforated. The scale is the same as in Figure 3.

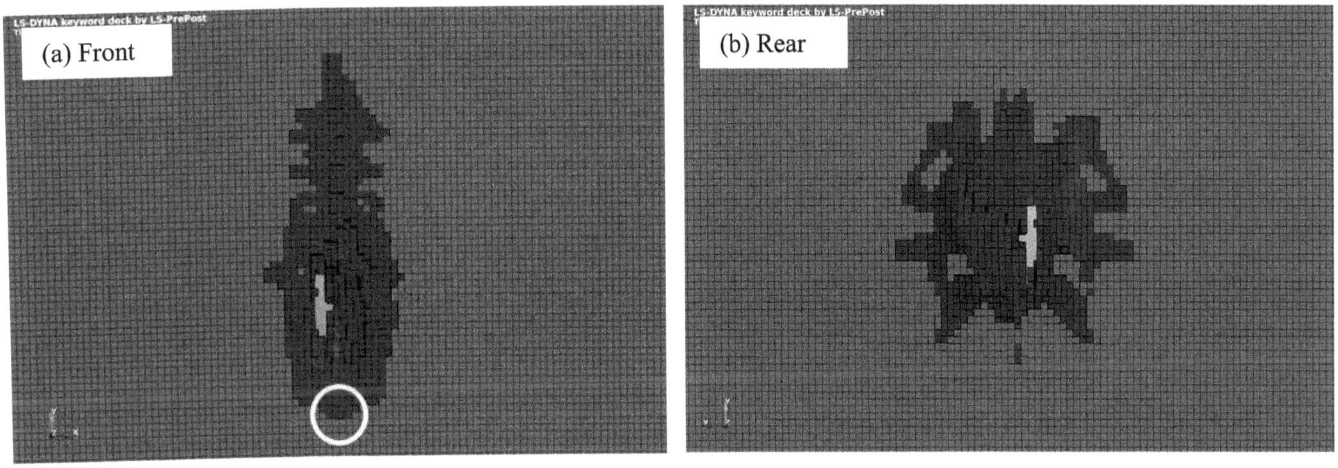

Fig. 5. The (a) front and (b) rear surfaces of the SiC-coated RCC target mesh are shown after impact by a steel projectile with a velocity of 2.44 km/s and obliquity of 83°. Red elements represent the RCC substrate, while blue elements represent the SiC coating. The projectile moved from the bottom to the top of the page during the computation. The initial position of the projectile as it comes into contact with the front SiC coating is shown with a yellow circle. In addition to damaging the front and rear SiC coatings, the impact has caused a hole to form through the full thickness of the RCC substrate. Note that the projectile ricochets off the target, i.e. it does not pass through the hole. The scale is the same as in Figure 3.

5. Summary and Conclusions

Simulations of the hypervelocity impact of SiC-coated RCC composites were conducted. Material properties for the projectiles and target constituents were derived from literature data. Simulation results for three different projectile impacts were compared to corresponding experimental observations of damage mode. The results demonstrate that the model can match the differing damage modes from these tests and reproduce the length of the damaged areas that occur during these impacts. The model is also able to reproduce the fact that the steel projectile is more damaging than the aluminum projectile even though it impacts the target at a very similar angle and kinetic energy. For the 83° impacts, the simulated damage length was accurate to within 20%. However, the simulated damage width was always approximately half of what was observed in each of the tests, indicating that there is room for improvement. It is likely that the accuracy of the model could be improved further if additional characterization data from the individual fiber and matrix components of the RCC composite were available.

References

[1] Christiansen, E., Curry, D., Kerr, J., Cykowski, E., Crews, J., 1993. "Evaluation of the impact resistance of reinforced carbon-carbon," Metal Matrix Composites: Proceedings of the Ninth International Conference on Composite Materials, pp. 498-507.

[2] Christiansen, E., Friesen, L., 1997. Penetration equations for thermal protection materials, International Journal of Impact Engineering 20, pp. 153-164.

[3] Carpenter, A., Chocron, S., Anderson, C. 2018. Bridging the scales: Continuum-based material constitutive modeling of mechanical and ballistic test data from composites and fabrics, International Journal of Impact Engineering 120, pp. 31-45.

[4] CeraMaterials. "Carbon-Carbon Composites," <ceramaterials.com/carboncarboncomposites.html>.

[5] Johnson, G., Holmquist, T., 1990. "A computational constitutive model for brittle materials subjected to large strains, high strain rates and high pressures," Shock Waves and High-Strain-Rate Phenomena in Materials, pp. 1075-1081.

[6] Holmquist, T., Johnson, G. 2002. Response of silicon carbide to high velocity impact, Journal of Applied Physics 91, pp. 5858-5866.

[7] Johnson, G., Cook, W., 1983. "A constitutive model and data for metals subjected to large strains, high strain rates, and high temperatures," Proceedings of the 7th International Symposium on Ballistics, pp. 541-547.

[8] Johnson, G., Cook, W., 1985. Fracture characteristics of three metals subjected to various strains, strain rates, temperatures, and pressures, Engineering Fracture Mechanics 21, pp. 31-48.

[9] Beissel, S., Gerlach, C., Holmquist, T., Walker, J., 2010, "A comparison of numerical methods in the simulation of hypervelocity impact," Proceedings of the 11th Hypervelocity Impact Symposium.

Proceedings of the 2019 Hypervelocity Impact Symposium
HVIS2019
April 14-19, 2019, Destin, FL, USA
HVIS2019-068

Extended Abstract⁺

Comparison of the Hydrodynamic Ram Caused by Single and Two Projectiles Impacting Water-filled Containers

Yangziyi Ji, Xiangdong Li*, Lanwei Zhou, Xiaoying Lan

School of Mechanical Engineering, Nanjing University of Science and Technology, 200 XiaoLingWei, Nanjing, 210094, China.

1. Introduction

When high-velocity projectiles penetrate an enclosed fluid-filled structure, the projectiles transfer the kinetic energy to the surrounding structure through the fluid and cause excessive structural damage. This process is referred to as a hydrodynamic ram (HRAM) [1]. The HRAM can be divided into five phases[1]: penetration, shock, drag, exit, and cavity oscillation. The HRAM phenomenon has been studied analytically [2,3], experimentally [4–7] and numerically [8–10]. However, few studies have been performed on the hydrodynamic ram caused by multiple penetrators. In the present work, a previously validated numerical model was used to analyze the differences of hydrodynamic rams caused by one and two projectiles impacting water-filled containers.

2. Comparison results and analyze

2.1. Simulation model

In our previous work [11,12], experimental and numerical studies on the cumulative damage of water-filled containers impacted by two projectiles were conducted. The model shown in Fig. 1 is comprised of projectiles, flanges, plates, a cylinder, water, and air. Both the plate deformation along the centerline and the pressure histories inside the fluid of the experiment and simulation

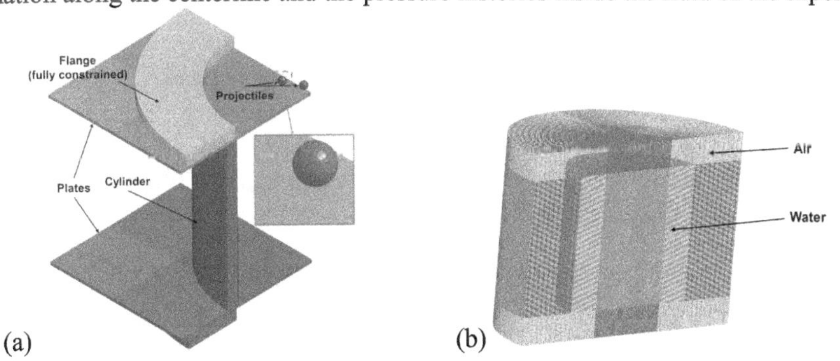

(a) (b)

Fig. 1. Mesh of the model: (a) Solid and (b) fluid

+ Full paper can found at https://www.sciencedirect.com/journal/international-journal-of-impact-engineering/special-issue/10MTGP5W4VJ

*Corresponding author. Tel.: +86 02584317634
E-mail address: lixiangd@njust.edu.cn (X. Li)

were compared to verify the model. More details about the experimental setup, simulation model and validation can be found in [11,12].

2.2. Comparison of the pressure profiles

The pressure contour in the water at different times in two cases are shown in Fig. 2 and Fig. 3, respectively. The shock front and the cavity in the single-projectile case are clearly seen in Fig. 2. The average velocity of the shock front was about 1600 m/s. In the two-projectiles case, two hemispherical shock waves were generated in the water at the shock phase and collided with each other when they propagated outward (Fig.3 (a) and Fig.3(b)). These two shock waves then interacted in two triple points and formed a new shock wave, which is so-called "Mach disk" [13] (Fig. 3(c)). The propagation velocity of the superimposed shock wave was faster than that of the initial shock wave (Fig. 3(b) and Fig. 3 (c)), and the distance between the wavefront of the superimposed region and the initial shock wave was gradually decreased. As the propagation distance increased, the superimposed shock wave merged with the initial shock wave and formed a nearly spherical wavefront with a slightly flat head, as shown in Fig. 3(d). The average velocity of the wavefront was about 1481 m/s, slightly higher than the sound speed in the water.

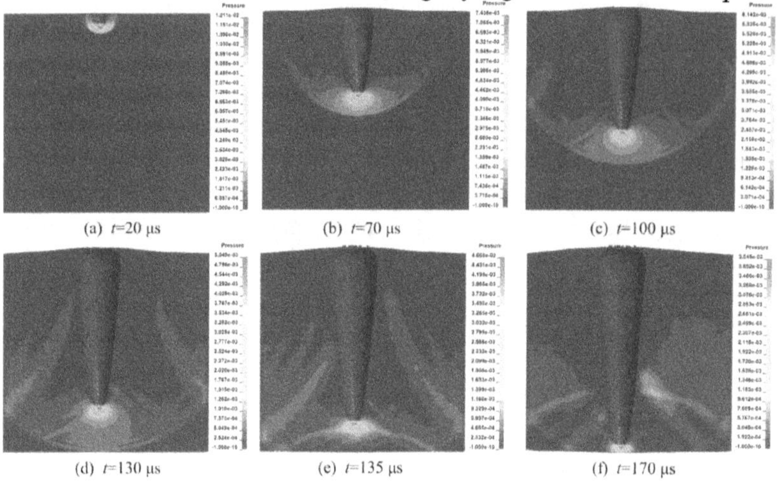

Fig. 2. Pressure contours in the water, pressure units in GPa (Single-projectile case, E_{k0}=11.16 kJ, v_{s0}=1697 m/s.)

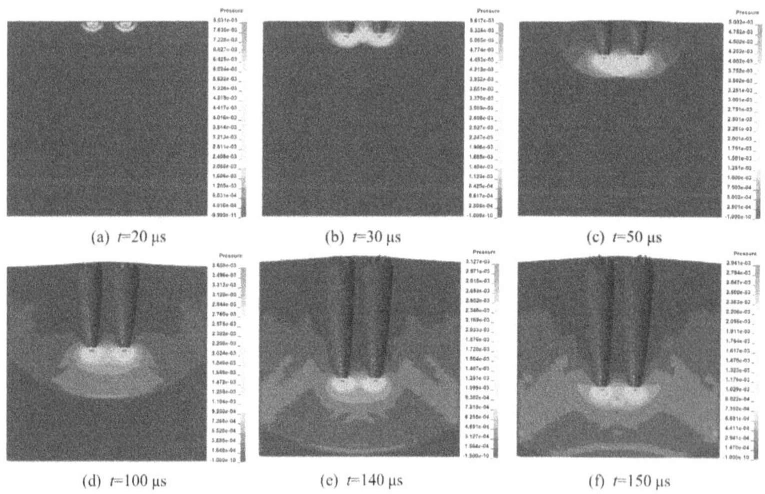

Fig. 3. Pressure contours in the water, pressure units in GPa (Two-projectile case, E_{k0}=11.16 kJ, v_{d0}=1200 m/s)

2.3. Comparison of the energy-time histories

Fig. 4 shows the overall energy-time histories of various parts (projectiles, water, plates) in two cases with three different projectile impact energies. It can be observed that the energy projectiles dissipated during the whole process, as well as the peak energy of the water in the single-projectile case, is larger than that of the two-projectiles cases. Furthermore, the difference of the residual projectile energy between the two cases increases with E_{k0}. It is interesting to note that the energy change of the entry plate was almost identical in both cases, which is consistent with our previous finding that the residual deformation of the entry plate is not influenced by the number of projectiles [11]. The energy of the exit plate began to rise before the exit phase and it always rose slightly earlier and more rapidly in the single-projectile case at the drag phase. The rise of the exit plate energy was caused by the arrival of the initial shock front[8], while the magnitude of the plate energy was related to the shock wave intensity. Therefore, the velocity and intensity of the shock wave in the single-projectile cases were higher than that of the two-projectile cases, as shown in Fig. 2 and Fig. 3. It can also be observed that this difference became smaller as the E_{k0} increased. At the early stage of the cavity oscillation phase, the energy of the exit plate in the single-projectile cases was larger than that of the two-projectile cases. However, at the late cavity oscillation phase, when the energy started to stabilize, the latter one overtook the former one. Contrary to the drag phase, the difference of the stabilized energy between these two cases became larger as E_{k0} increased.

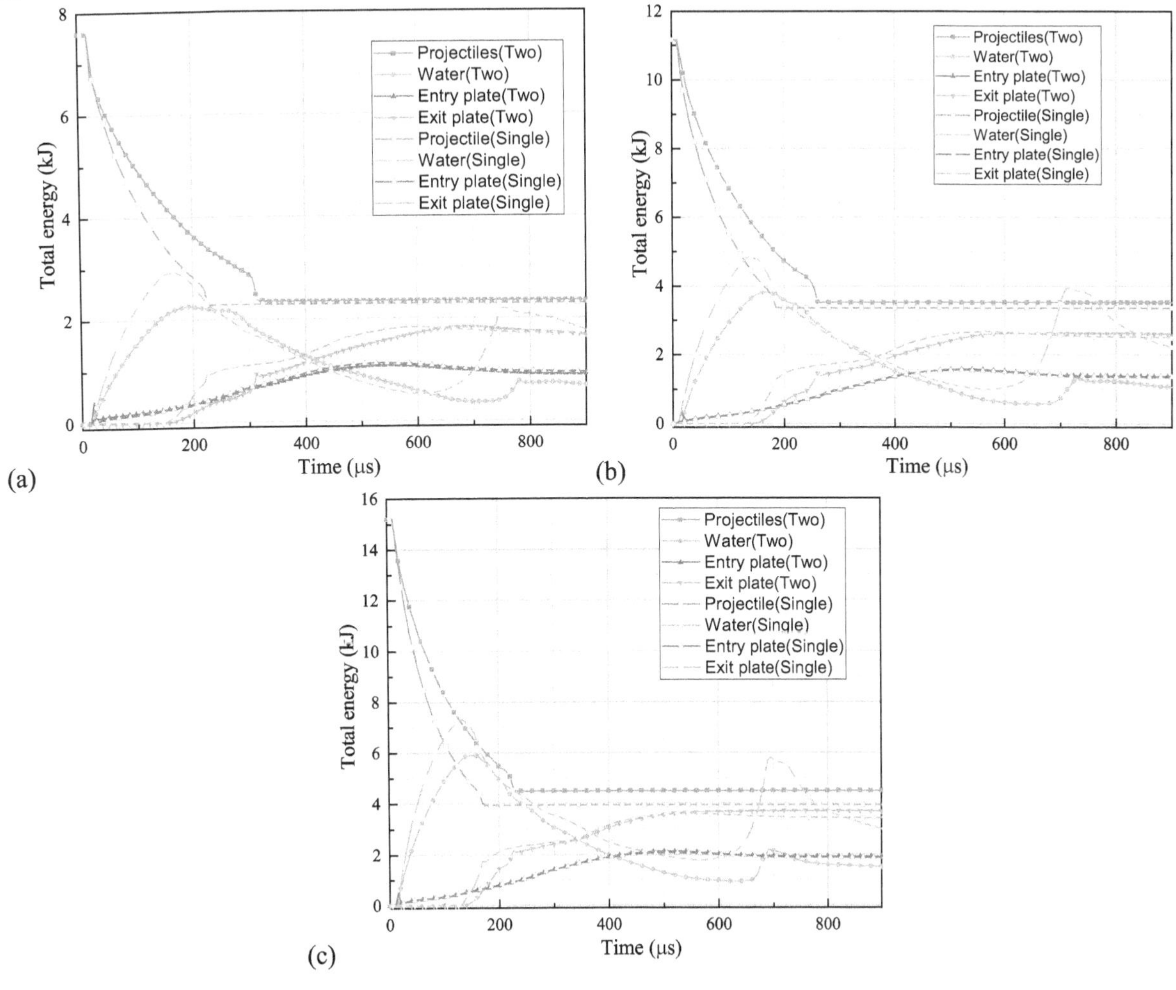

Fig. 4. The energy-time histories of the projectile(s), water and plates. (a) E_{k0}=7.60 kJ; (b) E_{k0}=11.16 kJ; (c) E_{k0}=15.20 kJ

The energy loss of the projectiles at different HRAM phases is shown in Fig. 5. As the figure shows, the energy loss during the shock and drag phase increase approximately linearly as the impact kinetic energy increased, while remaining quite stable during the penetration and exit phases. For the penetration phase, energy losses in the two-projectile case are larger than that of the single-projectile case when E_{k0} is 7.60 kJ, while the difference becomes less noticeable with the increasing of E_{k0}. The projectile(s) dissipated more energy in the penetration phase than the exit phase, mainly owing to that the exit plate was already pre-stressed and it took less energy to penetrate it. Besides, the energy loss during the exit phase in the two-projectiles case is larger than that of the single-projectile cases.

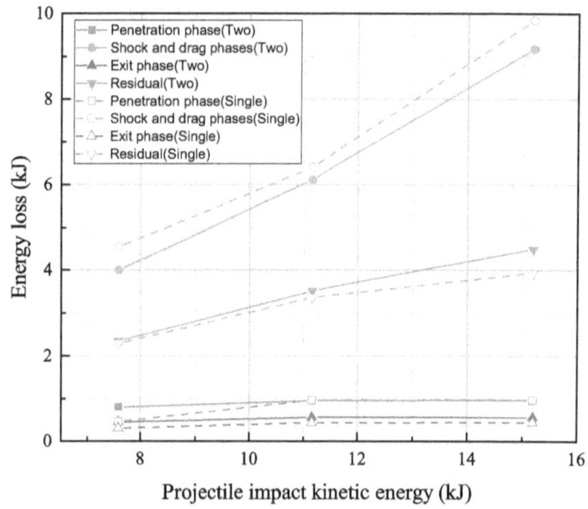

Fig. 5. The energy loss of the projectiles during different HRAM phase.

Acknowledgments

We would like to thank the National Natural Science Foundation of China (Grant No.11572159) and Postgraduate Research & Practice Innovation Program of Jiangsu Province (Grant No. SJKY19_0360) for funding this project.

References

[1] Ball, R. E., 2003. The Fundamentals of Aircraft Combat Survivability: Analysis and Design, 2nd Edition. American Institute of Aeronautics and Astronautics, New York, U. S.

[2] Lee, M., Longoria, R. G., Wilson, D. E., 1997. Ballistic waves in high-speed water entry, Journal of Fluids and Structures 11, p. 819.

[3] Fourest, T., Laurens, J.-M., Deletombe, E., Dupas, J., Arrigoni, M., 2014. Analysis of bubbles dynamics created by Hydrodynamic Ram in confined geometries using the Rayleigh–Plesset equation. International Journal of Impact Engineering 73, p. 66.

[4] Townsend, D., Park, N., Devall, P. M., 2003. Failure of fluid filled structures due to high velocity fragment impact, International Journal of Impact Engineering 29, p. 723.

[5]Disimile, P. J, Swanson, L. A., Toy, N., 2009. The hydrodynamic ram pressure generated by spherical projectiles, International Journal of Impact Engineering 36, p. 821.

[6] Varas, D., López-Puente, J., Zaera, R., 2009. Experimental analysis of fluid-filled aluminium tubes subjected to high-velocity impact, International Journal of Impact Engineering 36, p. 81.

[7] Ren, P., Zhou, J., Tian, A., Ye, R., Shi, L., Zhang, W., 2018. Experimental investigation on dynamic failure of water-filled vessel subjected to projectile impact, International Journal of Impact Engineering 117, p. 153.

[8] Varas, D., López-Puente, J., Zaera, R., 2012. Numerical Analysis of the Hydrodynamic Ram Phenomenon in Aircraft Fuel Tanks, AIAA Journal 50, p. 1621.

[9] Kwon, Y. W., Yang, K., Adams, C., 2016. Modeling and Simulation of High-Velocity Projectile Impact on Storage Tank, Journal of Pressure Vessel Technology 138, p. 041303.

[10] Liu, F., Kong, X., Zheng, C., Xu, S., Wu, W., Chen, P., 2017. The Influence of Rubber Layer on the Response of Fluid-Filled Container Due to High-Velocity Impact. Composite Structures 183, p. 671.

[11] Ji, Y., Li, X., Zhou, L., Lan, X., 2019. Experimental and numerical study on the cumulative damage of water-filled containers impacted by two projectiles, Thin-Walled Structures 135, p. 45.

[12] Lan, X., Li, X., Zhou, L., Ji, Y., 2019. Research on Damage of Vessel Walls Caused by Double Fragments Penetration and Coupling Load, Acta Armamentarii 40, p. 159. (in Chinese)

[13] Krehl, P. O., 2008. History of shock waves, explosions and impact: a chronological and biographical reference. Springer Science & Business Media.

Proceedings of the 2019 Hypervelocity Impact Symposium
HVIS2019
April 14-19, 2019, Destin, FL, USA
HVIS2019-070

Study on Shaped Charge Jet Initiation of Solid Rocket Motors

Hao CUI*[1], Rui GUO[2], Pu SONG[3], Jinsheng XU[4],Xiaohui GU[5] ,Rongzhong LIU[6]

[1]School of Mechanical Engineering, Nanjing University of Science and Technology, Nanjing 210094, Jiangsu, China. e-mail address:cuihao_93@126.com
[2] Corresponding Author, School of Mechanical Engineering, Nanjing University of Science and Technology, Nanjing 210094, Jiangsu, China. e-mail address: guorui@njust.edu.cn
[3]Science and technology on combustion and explosion laboratory, Xi'an modern chemistry research institute, Xi'an 710065, shanxi, China. e-mail address:songpu73@163.com
[4] School of Mechanical Engineering, Nanjing University of Science and Technology, Nanjing 210094, Jiangsu, China. e-mail address: xujinsheng@njust.edu.cn
[5] School of Mechanical Engineering, Nanjing University of Science and Technology, Nanjing 210094, Jiangsu, China. e-mail address: gxiaohui@mail.njust.edu.cn
[6]School of Mechanical Engineering, Nanjing University of Science and Technology, Nanjing 210094, Jiangsu, China. e-mail address: liurongz116@163.com

Abstract
In order to study the mechanism of initiation of solid rocket motors under the impact of shaped charge jets, a shaped charge jet initiation test was experimentally studied to evaluate the safety of the motor under attack in the battlefield environment. The experimental results indicated that the motor had a detonation reaction under the shaped charge jet impact. The response of the motor was recorded by a high-speed camera. In addition, the mechanism of initiation of the propellant charge was evaluated using by numerical simulations. Pressure-time and reaction-time curves of propellants were analyzed in this paper.
Keywords: Rocket motors; solid propellant; shaped charge jets; initiation

---

Nomenclature	
P	pressure (Mbar)
A、B、R_1、R_2、ω	constants
E_0	initial volume energy (J/m³ or Pa)
V	relative specific volume of detonation products
I, b, a, x, G_1, c, d, y, G_2, e, g, z	constants
Greek symbols	
λ	degree of reaction
ρ	density (g/cm³)
Subscripts	
0	initial state

1. INTRODUCTION

High energy solid propellant contains a large number of high energy components such as HMX, nitrate ester and aluminum powder. The composition of the high energy solid propellant is similar to that of mixed explosives and has the initiation and detonation characteristics of explosives.[1-2]

The concern about safety of tactical solid rockets is increasing with the use of high energy solid propellants. Several specific tests, including but not limited to, bullet impact and jet initiation tests, have been performed to evaluate the safety of explosives and propellants. Hamaide et al[3] reported a new scenario that a delayed detonation occurred in bullets against true rocket motors. Numerical modeling was used to analyze the mechanism of that, indicating both the sensitivity of propellants and the general design of rocket motors caused the detonation of propellants. Eishu et al[4] conducted a series of experiments including fast cook-off, card gap and bullet impact tests to evaluate

* Corresponding author E-mail address: cuihao_93@126.com.

the sensitivities of GAP and BAMO/NMMO propellants. All the samples tested showed no reaction other than burning. Overall, AN-based composite propellants had a relatively lower sensitivity. Hao et al[5] numerically simulated the falling process of solid motors. They concluded that the critical detonation pressure of the propellant was 2.675GPa and the critical detonation velocity range of the motor was ~150-200m/s.

The attack of shaped charge jets is one of the most common threats in the battlefield for the safety of rockets. Many scholars have studied the mechanism of jet initiation of explosives. In 1968, Held[6] found that the critical velocity (v) of the jet particles was related to the jet diameter (d) by the relationship v^2d = constant. The v^2d-criterion for detonation has been generally confirmed in other work on blunt-projectile impact tests against various high explosives [7]. Chink and Hatt[8]studied the initiation and build-up to detonation of Composition B impacted by a shaped charge jet. They found that the strong precursor shock from the thin barrier system produced initiation close to the explosive surface while for the weak precursor shock from the thick barrier system there was a long run to detonation. Held[9-11] found that easier initiation of a high explosive charge in the case of a wider air gap could only in part be attributed to the lack of pre-compression by an elastic precursor wave or of strong compression by plastic deformation of the target material, resulting in desensitization of the high explosive.

In this article, a solid rocket motor (ϕ170×200mm) is selected to experimentally study the response of motors under shaped charge impact. Numerical simulations were conducted to study the initiation mechanism and the shock wave evolution of propellants.

2. Experimental Procedure and Results

The motor was filled with high energy solid propellant (36%HMX,20%AP,18%～20%AL, et.al), of which the density was 1.836g/cm^3. The two end covers of the motor were made of aluminum, and the shell was made of carbon fiber. As is shown in Figure 1, a metal jet generated from a conventional 56mm diameter shaped charge with a copper liner horizontally impacted the center of the motor. The shaped charge was fixed on the surface of a table which was as high as the center of the motor and ignited from a standard standoff of 80mm. At this standoff, the head velocity of the jet is about 6600 m/s.

The response of the motor and firelight of the propellant were recorded by a high-speed camera. In addition, two velocity-measuring screens were placed 2m away from the motor. A bracket with a limiting device was designed to fasten the motor. The high-speed camera was located inside a protective cover with an observation window.

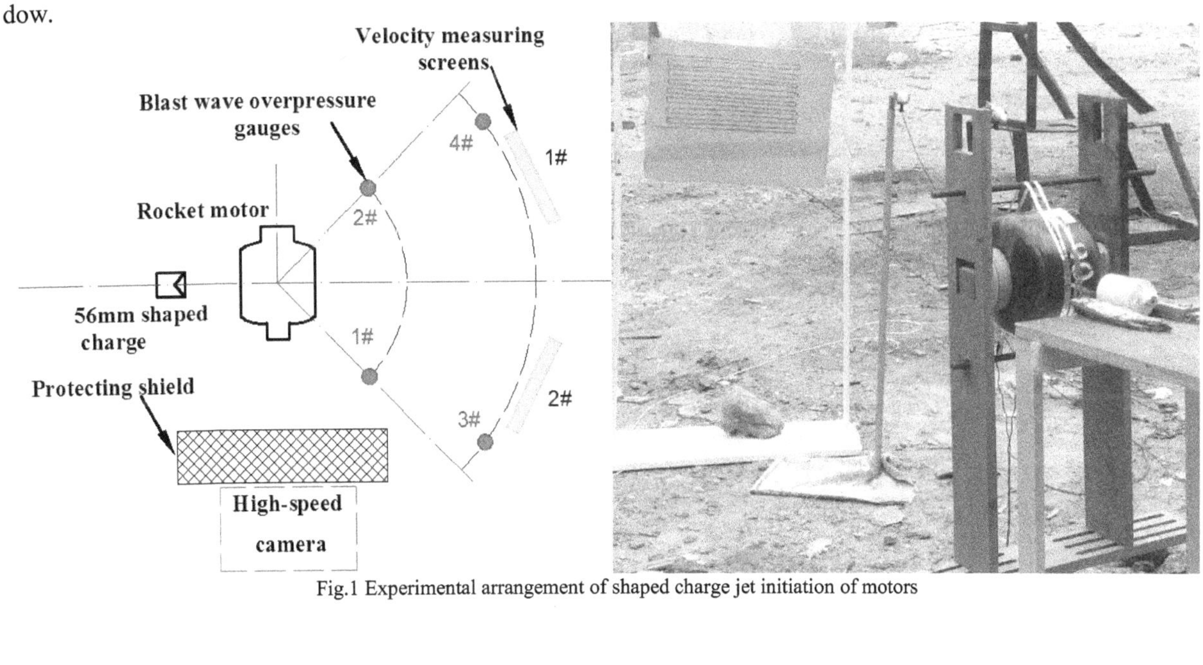

Fig.1 Experimental arrangement of shaped charge jet initiation of motors

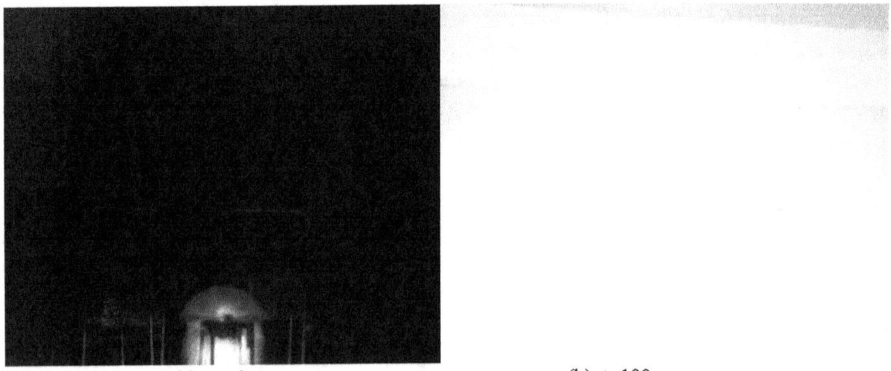

(a) t=0μs (b) t=100μs

Fig.2 High-speed images of the motor initiation

The sampling frequency of the high-speed camera was 10000 Hz. Figure 2 shows ignition and detonation response processes of the motor in the test. As shown in figure 2(a), firelight immediately emerged when the jet impacted the motor. In the next image shown in figure 2(b) at 100μs, expansion of detonation products was recorded. Since the detonation speed of this propellant is 9130.4 m/s[12], compared with the detonation growth rate and expansion speed of detonation products, the sampling frequency of the high-speed camera used in the test was relatively low. Speeds of fragments or shock wave measured by the two velocity measuring screens are shown in Table 1.

Tab. 1 speeds of fragments or shock wave

Screens	1	2
Velocity m/s	4790.42	4752.48

3. Numerical Modeling and Discussion

Figure 3 shows the forming process of the jet used in the test, which was simulated by AUTODYN-2D. After the main charge detonated at about 5μs, the top of liner was first subjected to the impulse action of detonation pressure, which led to the collapse, closure and collision of the liner. Convergence was almost completed by 10μs. Due to the speed differences in the head, the metal penetrator was gradually stretched to a high-speed jet and slug with lower velocity. The standoff of the 56mm shaped charge is 80mm and the ideal standoff was formed at 23.7μs in the simulation.

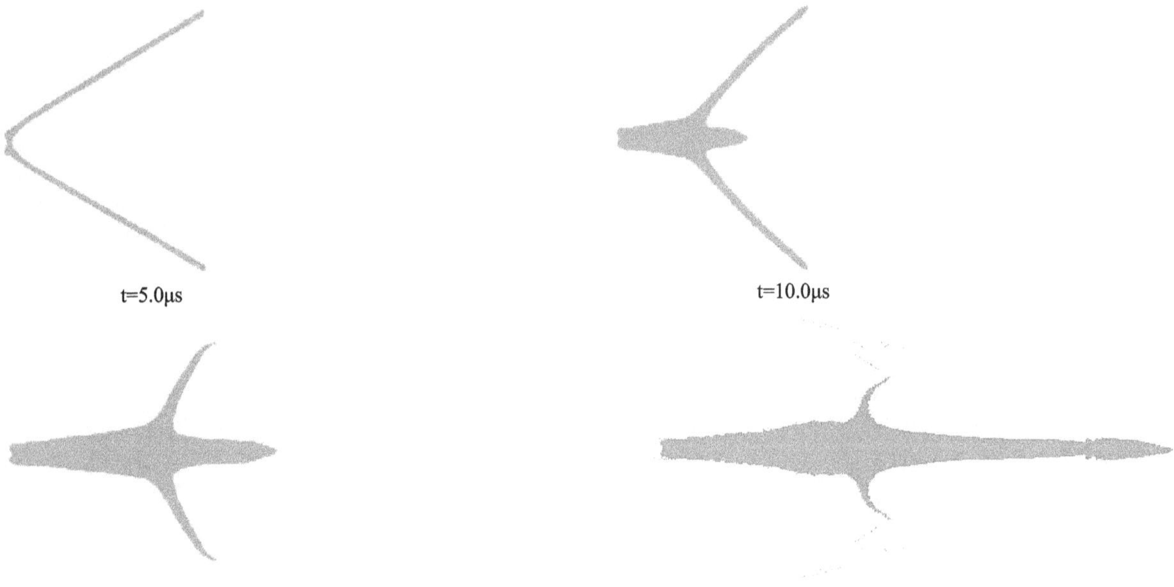

Fig.3 The forming process of the jet

The formed jet model then was transported from AUTODYN-2D to AUTODYN-3D, and the algorithm of the jet was converted from Euler to Lagrange. So there was no need to establish an air fluid domain in the simulation of the jet impacting the rocket motor. Using this method the calculation efficiency was significantly improved.

The Lee-Tarver model was applied to the initiation and detonation process of the propellant. This model includes two JWL state equations: the JWL equation of state of the unreacted explosive and the JWL equation of state of the detonation products. The pressure form of the JWL state equation for explosives is shown in equation 1. A new type cylinder test [5,12] was designed to calibrate the parameters (as shown in table 2) of the JWL equation of state of the detonation products for this high energy solid propellant. As for the parameters of the unreacted explosive, parameters for PBX whose density is similar to this propellant were used.[13]

$$p = A\left(1-\frac{\omega}{R_1V}\right)e^{-R_1V} + B\left(1-\frac{\omega}{R_2V}\right)e^{-R_2V} + \frac{\omega E_0}{V} \tag{1}$$

Tab. 2 JWL equation parameters for the propellant

ρ_0/ (g·cm^{-3})	D/ (m·s^{-1})	R_1	R_2	□ω	A/ GPa	B/ GPa
1.836	9 130.4	5	1.82	0.2	909.59	62.05

The rate equation for the conversion of the unreacted explosive to reaction products for the propellant is shown in equation 2. Parameters of the reaction rate equation for the propellant are shown in table 3. [5]

$$\frac{d\lambda}{dt} = I(1-\lambda)^b\left(\frac{\rho}{\rho_0}-1-a\right)^x + G_1(1-\lambda)^c\lambda^d P^y + G_2(1-\lambda)^e\lambda^g P^z \tag{2}$$

Tab. 3 parameters of the reaction rate equation for propellant

I	B	a	x	G_1	c
44	0.222	0.01	4	111	0.222
D	y	G_2	e	g	z
0.667	1.66	200	0.333	0.667	2

Coarsening the mesh for the end cover and refining the propellant mesh was performed to reduce the overall grid size and improve the calculation efficiency. As shown in Figure 4, a ½ symmetry three-dimensional finite element model was established. Four gauges were set on the propellant trauma channel to record the time-pressure and reaction-time curves of the propellant. The distance between gauges was 5mm.

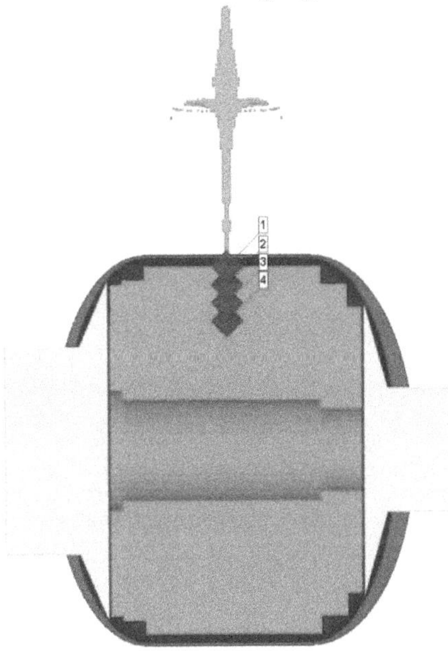

Fig.4 Finite element model of jet impacting solid motor

For shock initiation of the explosive charge within the shell, when the mass and velocity of fragments are constant, the mechanism of shock initiation is related to the thickness of the shell. It is generally acknowledged that the initiation mechanism is shock initiation when the shell is relatively thin, and it is both shock and shear initiation when the shell is relatively thick.[14] Considering the fact that the thickness of the shell where the metal jet impacted the motor in both the test and the simulation was no more than 3mm, shock initiation is regarded as the dominant initiation mechanism in this study. Figure 5 shows the pressure-time diagram of the motor under the impact of a jet whose head velocity is 6600m/s. When t=4μs, the solid motor is horizontally impacted by the jet, resulting in rapid formation of hot spots inside the propellant. When t=6μs, the shock wave basically grows into a detonation wave, and after that the detonation wave propagates steadily.

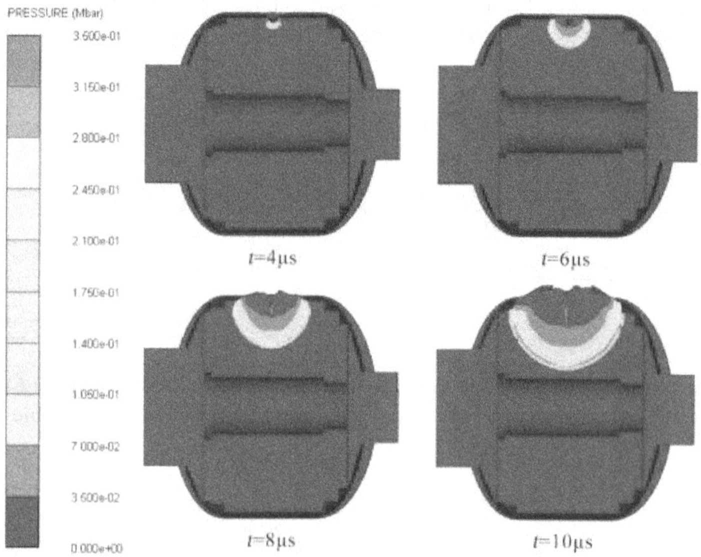

Fig.5 Time-pressure diagram of jet impacts solid motor

Figure 6 shows the pressure-time and reaction-time curves of the 4 gauges respectively. The incident pressure reaches 0.159Mbar (1Mbar=100GPa), and the reaction rate of the propellant is fast at this moment. When the shock wave arrives at gauge 2, the peak pressure reaches about 0.28Mbar, while the reaction rate increases further. It can be inferred that the detonation run distance is less than 5mm. When detonation wave arrives at gauge 3, steady detonation in propellant is formed. Due to the hypervelocity propagation of the jet and large incident pressure, no detonation is observed in the simulation.

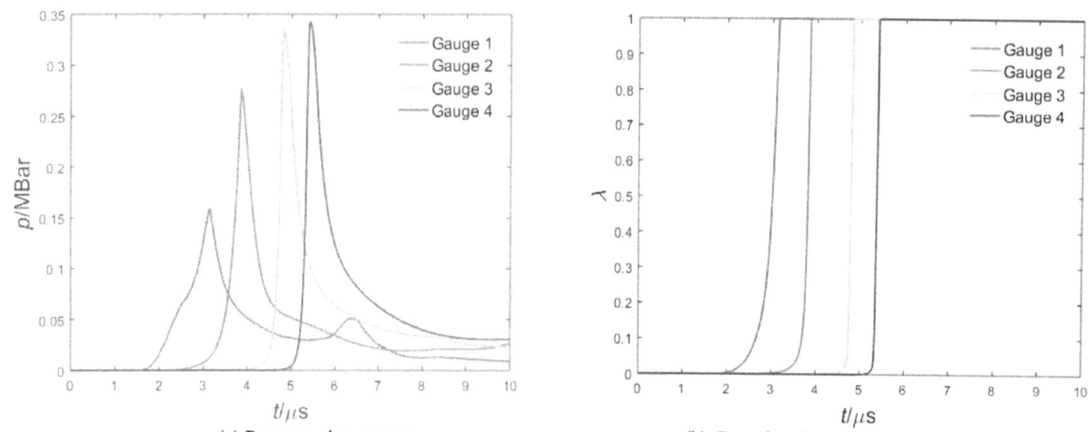

(a).Pressure-time curves (b) .Reaction-time curves

Fig.6 Pressure-time and Reaction-time curves of 4 gauges in propellant

4. Summary

In this work, a shaped charge jet initiation test and the corresponding numerical simulation are carried out to study the mechanism of initiation of the solid rocket motor. The experimental results show that the motor has a detonation reaction under impact of a shaped charge jet. The simulation process indicates that the incident-pressure reaches 0.159Mbar and there is a short run to detonation (about 5mm,4μs).

The aim of this work is to test the safety of the motor under the impact of 56mm shaped charge. A preliminary qualitative result for the detonation of motors was obtained. Different size designs of motors and more testing devices, such as air overpressure sensors, will be used in future tests.

References

[1] WANG Yong-jie, LU Jian-ying, CHEN Lin-quan, et al, 2009. Numerical analysis for safety of solid rocket motor impacting target, Journal of Solid Rocket Technology 32, pp.274-278. (in Chinese)

[2] WU Junying, CHEN Lang, LU Jianying, et al, 2008. Research on Shock Initiation of the High Energy Solid Propellants, Acta Armamentarii 29,pp.315—1319. (in Chinese)

[3] S.Hamaide, M.Quidot, J.Brunet, 1992. Tactical Solid Rocket Motors Response to Bullet Impact, Propellant, Explosives, Pyrotechnics 17, pp.120-125.

[4] Eishu Kimura, Yoshio Oyumi, 1997. Sensitivities of azide polymer propellants in fast cook-off, card gap and bullet impact tests. Journal of Energetic Materials 15, pp.163-178.

[5] CUI Hao, GUO Rui, SONG Pu, 2018. Numerical Analysis of Safety for Solid Motor During Falling Process. Acta Armamentarii. (accepted) (in Chinese)

[6]M. Held, 1968. Initiierung von Sprengstoffen, ein vielschichtiges Problem der Detonationsphysik, Explosivstoffe 5, pp.2-17,

[7] M. Held, 1983. Critical Area for the Initiation of High Explosive Charges, Shock Waves in Condensed Matter, Santa Fe, NM, [Proc.] Chapter XII, pp. 555-557.

[8] M. C . Chick and D. J. Hatt, 1983. The Initiation of Covered Composition B by a Metal Jet", Propellants, Explosives, Pyrotechnics 8, pp.121-126..

[9] M. Held 1987. Experiments of Initiation of Covered but Unconfined High Explosive Charges by means of Shaped Charge Jets, Propellants, Explosives, Pyrotechnics 12, pp.35-40.

[10] M. Held, 1987. Experiments of Initiation of Covered, but Unconfined HE Charges under Different Test Conditions by Shaped Charge Jets, Propellants, Explosives, Pyrotechnics 12, pp.97-100.

[11] M. Held, 1987. Discussion of the Experimental Findings from the Initiation of Covered, but Unconfined High Explosive Charges with Shaped Charge Jets, Propellants, Explosives, Pyrotechnics 12, pp.167-174.

[12] CUI Hao, GUO Rui, XU Jinsheng. Study on the State Equation of Detonation Products of High Energy Solid Propellant. Journal of National University of Defense Technology.(under review)(in Chinese)

[13] Zhengdi Pi, Lang Chen, Danyang Liu,et al,2017. Shock initiation of CL-20 based explosive. Acta Armamentarii 37, pp. 915-923. (in Chinese)

[14] Xianzhen Jia, Song Chen, Jian Yang, et al, 2011. Numerical study of explosive initiation by simultaneous impact from two fragments. Chinese Journal of High Pressure Physics25, pp.469-474.(in Chinese)

Proceedings of the 2019 Hypervelocity Impact Symposium
HVIS2019
April 14-19, 2019, Destin, FL, USA
HVIS2019-073

Extended Abstract+

Filling the Gap between Hypervelocity and Low Velocity Impacts

Werner Arnold[a,]*, Thomas Hartmann[b], Ernst Rottenkolber[b]

[a] *MBDA-TDW Gesellschaft für verteidigungstechnische Wirksysteme mbH, Hagenauer Forst, D-86529 Schrobenhausen, Germany*
[b] *NUMERICS GmbH, Mozartring 6, D-85238 Petershausen, Germany*

1. Introduction

During more than one decade of studying initiation phenomenology numerous papers at the previous HVIS and other symposia ([1] - [12]) were published. Most of them dealt with the hypervelocity impact initiation of plastic bonded high explosive charges by shaped charge jets (SCJ) and a few ones reported results in the ordnance velocity impact regime with STANAG projectiles and explosively formed projectiles (EFP) ([2] & [11]). A recent finding of our investigations of shaped charge jet (SCJ) attacks suggests that the critical stimulus $S = v^2 \cdot d$ (v = SCJ / projectile velocity; d = SCJ / projectile diameter) for the initiation of a munition can no longer be seen as a constant ($S \neq$ const.) ([11] & [10]). Also, known equations, e.g. Jacobs-Roslund [13], are not capable to describe low velocity and hypervelocity impacts with the same parameter set.

Therefore a new initiation model is needed taking these findings into account [10]. On the way to such a new model further investigations was necessary trying to make a link between the initiation phenomenology of shaped charge jet impacts in the hypervelocity regime and of projectile impacts in the lower velocity regime. Instead of using STANAG projectiles [16] to further bridge this gap a ***Simplified Shaped Charge (SSC)*** was introduced in [12], which generates a slower and thicker jet thus allowing more flexibility in the jet / projectile parameters. The new SSC charge used a caliber of 145 mm with a copper liner of 90° angle. In this study an SSC with a wider liner angle of 110° will be applied to generate even thicker and slower jets.

2. Hypervelocity and Low Velocity Regimes

2.1. Hypervelocity Regime (SCJ)

The hypervelocity regime with jet tip velocities causing increasing reaction inside the explosive from 2200 m/s up to 5800 m/s and generated by high performance military shaped charges was already thoroughly investigated and published in [10]. Nevertheless, for the reason of completeness these tests and their results are briefly summarized in the following. Thereby, the maximum caliber of the tested (COTS = components off-the-shelf) shaped charges was 200 mm.

In all conducted trials the so-called standard charge filled with the TDW high explosive KS32 (HMX/PB 85/15, $\rho = 1.64$ g/cm^3) was used in the investigations. The charge consists of a high explosive cylinder with 100 mm in diameter and 200 mm in length and a mild steel casing with 10 mm thickness and screwed lids on both sides (standard threads).

The experiments were assessed according to STANAG 4439 [14] introducing six Explosive Reaction Levels (ERL) from "Propulsion" (ERL = VI) to "Detonation" (ERL = I). Lower level reactions were estimated from the high explosive remnants of the charge and from broken pieces / fragments of the casing. The high-level reaction modes (ERL = I & II = (partial) detonation) were determined from the fragment pattern caused in the event, i.e. by evaluation of a 4 mm thick mild steel witness plate.

+ Full paper can found at https://www.sciencedirect.com/journal/international-journal-of-impact-engineering/special-issue/10MTGP5W4VJ

* Corresponding author. Tel.: +49-8252-99-6267; fax: Tel.: +49-8252-99-6114.
E-mail address: werner.arnold@mbda-systems.de.

Figure 1 shows the results obtained with five (in service) military SC charges (taken form [10]) with increasing SC caliber and SCJ diameter. The ERL-curves for the varying stimulus $S = v^2 \cdot d$ clearly illustrate the above mentioned new finding: the so far accepted assumption that the critical stimulus S = const. - as e.g. described in [15] - is clearly not valid. Further, the "strange" behavior with the SC-200 crossing three other ERL-curves indicates that the stimulus S is not an appropriate parameter to compare the ERL caused by different shaped charges. When plotted over the velocity v (as will be shown later) the data becomes much more consistent.

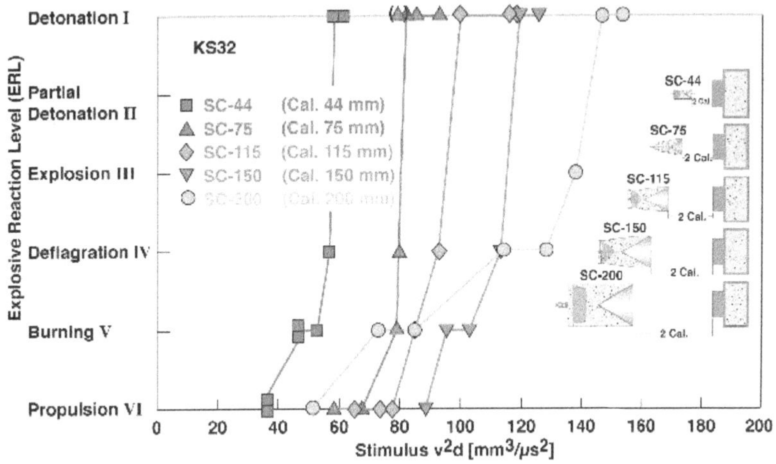

Fig. 1. ERL-curves obtained with five military SCs (taken from [10]).

2.2. Lower Velocity Regime (STANAG Projectile)

There were no (COTS) high performance military shaped charges available generating jets with a diameter of more than about 8 mm (SC-200). Besides, threats from fast fragments are typically represented by STANAG projectiles (projectile diameter: 14.3 mm) [16]. Therefore, these projectiles were used for the lower velocity regime. Further experimental trials with original STANAG projectiles (L/D = 1) and elongated ones made out of steel and copper were conducted [11].

The diagram in Figure 2 combines the data of the SCJ trials (see above) with those from the STANAG projectile trials. Different from Figure 1, the jet / projectile velocities (instead of $S = v^2 \cdot d$) were used as x-axis. The dashed lines from ERL = III to ERL = I indicate that an extrapolation was necessary since the velocity of the powder gun was limited to about 2600 m/s. The STANAG projectile data fit well into the overall ERL trends and the data generally look consistent. It should be noted that this is not at all the case when the ERL is plotted over the stimulus $S = v^2 \cdot d$. Hence, also this result confirms that the stimulus $S = v^2 \cdot d$ is not an appropriate parameter for describing the initiation behavior.

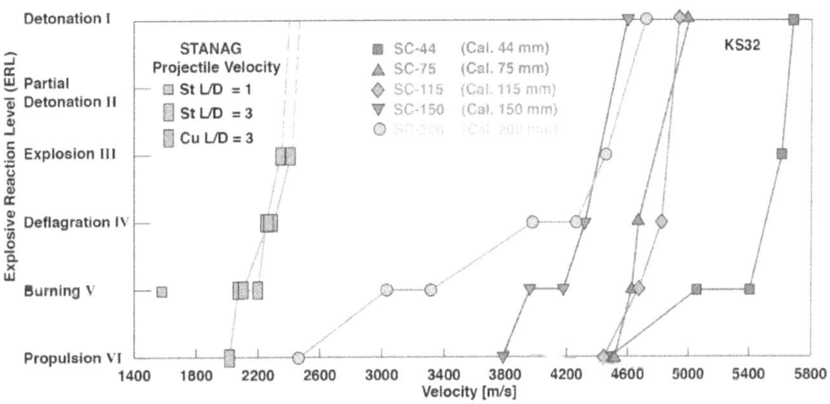

Fig. 2. Comparison of the SCJ results (from [10]) with the STANAG projectile results (taken from [11]).

Despite the fact that the STANAG data fit in very well, there are still a couple of questions that must remain unanswered for the moment, e.g.:

- Why are ERL-slopes different between SCJ and STANAG projectile results?
- What is the reason for the ramp-like ERL-slope especially for the SC-200?
- How does the gap between SCJ and STANAG projectile results look like?
- What is the influence on the transition between impact mode and penetration mode initiation?

In order to answer these questions and to take the next step towards a new unified initiation model [10] further trials were necessary. Instead of using larger caliber shaped charges (> 200 mm, COTS were not available) for getting thicker jet diameters (d > 8 mm) and lower jet velocities a new ***"Simplified Shaped Charge"*** (SSC) with a thicker jet and jet tip velocities in the intermediate regime (between high performance / hypervelocity SCs and STANAG projectiles in the ordnance velocity regime) was designed.

3. Intermediate Velocity Regime

The SSC design and jet characterization was already presented in [12]. The charge is based on an SC-145 with a liner angle of 90°, a design intentionally being very close to the SC-200 to make the results comparable afterwards. In this work a wider liner angle of 110° was selected to further increasing the SCC-jet diameter while reducing the jet velocity and jet length.

Like in the earlier tests the standard charge filled with KS32 was used (see [1] – [11]). Again, a mild steel barrier with varying thickness P was applied to tune the SCCJ-stimulus $S = v^2 \cdot d$. Thereby, the stand-off to the steel barrier P was kept constant at two calibers (here 290 mm).

The initiation test results were again assessed according to STANAG 4439 [14] determining the respective ERLs. A complete ERL-curve (from ERL = VI/V to ERL = I) was determined by incrementally varying the barrier thickness. The initiation ERL-curves with all SSC-145 (90° & 110°) results were plotted over the velocity in Figure 3.

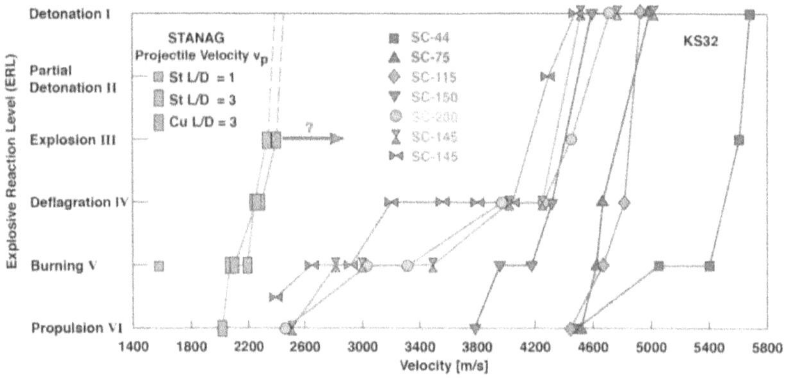

Fig. 3. Comparison of all ERL-curves achieved so far with the SCs, SSCs and the STANAG projectiles.

Comparing the SSC-145 (90°) with the SC-200 curves it can be seen that both curves are nearly identical. The initiation behavior obtained with the high performance SC-200 could thus be reproduced with the simplified design (despite the shorter and slower jet) requiring significantly thinner barriers P to tune the jet. The results of the new ERL-curve achieved with the SC-145 (110°) shaped charge 6 show several interesting features:

- The ERL-curve shows almost the same flat slope and plateau as the curves for SC-145 (90°) and SC-200,
- The ERL-curve is shifted to the left towards lower velocities getting closer to the STANAG results,
- The distinctive plateau of the SC-145 (110°) seems to be one ERL-level higher than for the SC-145 (90°): ERL = V -> IV,
- At the high level reactions (ERL = I/II) it seems that a partial detonation took place (less holes in the witness plate),
- At the low level reactions (ERL = VI/V) it seems that the lowest level reaction (VI) could not be achieved (instead VI/V).

4. Unified Initiation Model

It was already realized in [11] that the stimulus $S = v^2 \cdot d$ is not an appropriate parameter for the description of the observed initiation phenomena. Neither the so far assumed "v^2d-rule" ($S = v^2 \cdot d = const.$) nor another conclusive relation to the parameter S can be observed for the different SC and projectile initiation tests.

Figure 4 therefore exhibits the correlation of jet / projectile diameter and the velocities leading to higher and lower order reactions (from [11]) including the new results. As already proposed earlier [11], a more linear relationship, $v = A - B \cdot d$, is more appropriate to describe the initiation phenomena correctly.

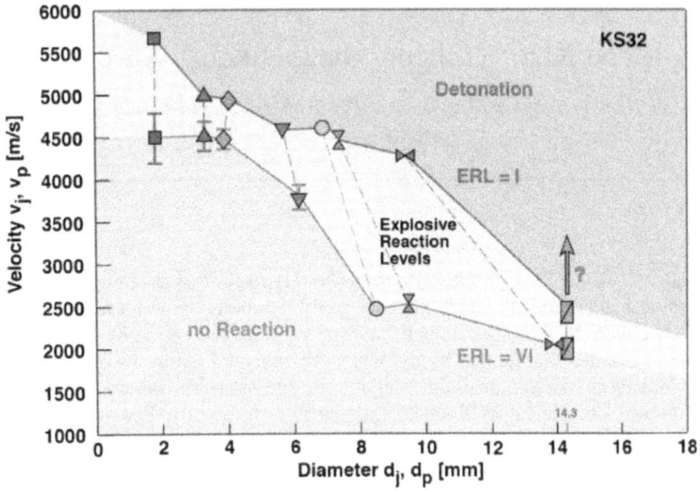

Fig. 4. Approximately linear initiation behavior of the data for ERL = I and ERL = VI.

Acknowledgements

The authors would like to thank the BAAINBw Team K 1.2 at Koblenz for the funding of this study.

References

[1] W. Arnold, E. Rottenkolber, "Sensitivity of High Explosives against Shaped Charge Jets", Insensitive Munitions & Energetic Materials (IMEMTS) 2007, Miami, USA

[2] W. Arnold, "High Explosive Initiation by High Velocity Projectile Impact", HVIS 2010, Freiburg, GE

[3] W. Arnold, M. Graswald, "Shaped Charge Jet Initiation of High Explosives equipped with an Explosive Train", IMEMTS 2010, Munich, GE

[4] W. Arnold, E. Rottenkolber, "Shaped Charge Jet Initiation Phenomena of Plastic Bonded High Ex-plosives", IMEMTS 2012, Las Vegas, USA

[5] W. Arnold, E. Rottenkolber, "High Explosive Initiation Behavior by Shaped Charge Jet Impacts", Hypervelocity Impact Symposium HVIS 2012, Baltimore, USA

[6] W. Arnold, E. Rottenkolber, "Significant Charge Parameters influencing the Shaped Charge Jet Initiation", IMEMTS 2013, San Diego, USA

[7] W. Arnold, E. Rottenkolber, T. Hartmann, "Analysis of Shock and Jet Initiation Tests of High Explosives", 19th Int. Symposium on Detonation ISD 2014, San Francisco, USA

[8] W. Arnold, E. Rottenkolber, T. Hartmann, "Challenging v^2d ", IMEMTS 2015, Rome, IT

[9] W. Arnold, E. Rottenkolber, T. Hartmann, "Testing and Modeling the Initiation of Insensitive Explosives by Projectile Impact", 11th EUROPYRO International Seminar 2015, Toulouse, France

[10] W. Arnold, T. Hartmann, E. Rottenkolber, "Towards a Unified Initiation Model", IMEMTS 2016, Nashville, TN, USA

[11] W. Arnold, T. Hartmann, E. Rottenkolber, "Filling the Gap between the Initiation Behavior of Shaped Charge Jets and Fragments", IMEMTS 2018, Portland, OR, USA

[12] W. Arnold, T. Hartmann, E. Rottenkolber, "Initiation Phenomenology from Hypervelocity to Low Velocity Impacts", 16th Detonation Symposium 2018, Cambridge, MD, USA

[13] T.P. Liddiard, L.A. Roslund, "Projectile/Fragment Impact Sensitivity of Explosives" NSWC TR 89-184, Naval Surface Warfare Center, Dahlgren, VA, June 1993

[14] STANAG 4439 (Edition 3), "Policy for Introduction and Assessment of Insensitive Munitions (MURAT)", 2006

[15] STANAG 4526 (Edition 2), "Shaped Charge Jet Munitions Test Procedure", 2004

[16] STANAG 4496 (Edition 1), "Fragment Impact, Munitions Test Procedure", 2006

Proceedings of the 2019 Hypervelocity Impact Symposium
HVIS2019
April 14-19, 2019, Destin, FL, USA
HVIS2019-076

Investigation on the Impact and Penetration Performance of the Particulated Jet into Concrete Targets

Qi-feng Zhu[a], Qiang-qiang Xiao[a]*, Zheng-xiang Huang[a], Xu-dong Zu[a], Xin Jia[a]

[a] *School of Mechanical Engineering, Nanjing University of Science and Technology, Xiao lingwei 200, Nanjing 210094, PR China*

Abstract

In this study, the performance of titanium alloys (TC21, TC1), nickel-titanium (Ni-Ti) alloy, and zirconium-niobium (Zr-Nb) alloy lined shaped charge impact and penetration into concrete targets are investigated experimentally. Shaped charge jet radiographs reveal that the resulting jets of titanium alloys and Ni-Ti alloy exhibit particulate, radially dispersed behaviors, whereas that of the Zr-Nb alloy is coherent. Cavity diameters, penetration depths and parameters of the impact craters generated by the jets were analyzed using the depth of penetration (DOP) experiment method. Data indicate that the particulated jet causes more extensive damage to the surface of the concrete targets compared to the coherent jet. The penetration depth decreases to some degree, but the cavity diameter increases significantly. Penetration efficiency varies with degree of dispersion of the particulated jet and, as such, is also sensitive to stand-off distance.

Keywords: particulated jet, concrete, cavity diameter, impact crater, penetration depth

---

Nomenclature

CD	charge diameter
d	diameter of the impact crater (mm)
$\bar{d}$	average diameter of the impact crater (mm)
h	depth of the impact crater (mm)
$\bar{h}$	average depth of the impact crater (mm)
D	diameter of the entrance hole (mm)
$\bar{D}$	average diameter of the entrance hole (mm)
P	penetration depth (mm)
$\bar{P}$	average penetration depth (mm)
ρ_j	density of the selected alloy (g/cm^3)
v_j	jet tip velocity (m/s)
d_j	jet tip diameter (mm)

1. Introduce

Concrete has been extensively used as a construction material for the civil engineering and military field. To destroy these hard targets, a two-stage tandem warhead consisting of a precursor shaped charge and a following kinetic energy projectile was developed [1]. The cavity diameter generated by the precursor shaped charge jet, as a key index of the damage effectiveness, has been the focus of scholars' research over the past years [2-3].

---

* Corresponding author. Tel.: +86-18901583363.
E-mail address: xiao_wawj@njust.edu.cn.

Wang et al. [4] experimentally studied the penetration of shaped charges with different liner materials, cone angles and liner thicknesses at different stand-off distances into concrete targets. Their results validated that the impact crater created by the aluminum liner was significantly greater than that of copper and steel liners. Zhu et al. [5] conducted similar research using Autodyn computational software. Kang et al. [6] investigated the penetration performance of titanium alloy, low carbon steel and copper liners against a multi-layer target. Results showed that the kinetic energy obtained by the titanium alloy liner was significantly improved compared with that of the low carbon steel and copper liners under the same structure. Moreover, the shape of the jet was relatively short and thick, and the jet velocity gradient was small, leading to the penetration hole being greatly increased and the impact crater substantially enlarged. Guo et al. [7] analyzed the effect of the yield strength of a titanium alloy liner on the penetration aperture diameter and found that the aperture diameter decreased with increasing liner strength. Murphy et al. [8-9] and Pincosy et al. [10] used aluminum as a liner material to evaluate the effect of single and multiple jets penetration into geologic materials to enhance the target borehole. Edwards et al. [11] selected low-density materials, including polypropylene, nylon and other polymers as the liner materials to obtain a larger entrance diameter for concrete targets. Xiao et al. [12] analyzed the penetration of a jacketed jet into concrete, and analytically calculated the cavity diameters. The study inferred that the final cavity diameter of the jacketed jet formed by a double-layered liner was larger than that in the single material jet when "co-erosion" occurred. Moreover, an average diameter larger than the diameter of the single material jet was obtained.

In the past, design of the shaped charge for creating a large hole was mainly concentrated on optimizing the charge structure, changing the liner material, etc, which always formed a coherent jet with a minor diameter increase but shorter length and lower velocity. On the other hand, the particulated jet, which consists of a series of high-speed moving dispersed particles, having both larger diameter and higher velocity, may be a possible solution to create a large penetration hole, and has not undergone significant investigation, with only a limited number papers on this topic. Baker et al. [13] reported the use of traditional silica-based glasses as liner materials. They found that the resulting jet behaviors with extreme particulate, ductile and radially dispersed behaviors. Kecskes et al. [14] experimentally determined that the diameter of the broken hole resulting from the particulated jet increased significantly under the appropriate stand-off distance. Other researchers such as Saran et al. [15] studied the penetration performance of the particulated jet formed by AA5000, AA6000 and AA7000 series aluminum alloys into steel and sand targets. Experimental phenomenon revealed that shaped charges having trumpet liners penetrated deeper into the target. Moreover, decreasing liner thickness had an increasing effect on penetration depth with a minor decrease in bore hole diameter.

In consideration of this limited research on the impact and penetration performance of the particulated jet into concrete targets, TC21, TC1, Ni-Ti alloy, and Zr705 were selected as the liner materials in the current study. X-ray experiments and DOP experiments were conducted. The present study attempted to provide a feasible further research direction for the design of the precursor shaped charge of the two-stage tandem warhead.

2. Experimental program

2.1. The shaped charges and the concrete targets

The nominal components and densities of the TC21, TC1, Ni-Ti alloy, and Zr705 are shown in Table 1. The designed structures used for the shaped charges are the same and the schematic is shown in Fig 1. The explosives are JH-2 (95% RDX and 5% TNT) with a density of 1.713g/cm^3 and a detonation velocity of 7980m/s. The uniaxial compressive strength of the concrete targets is 35MPa and the density is 2.4g/cm^3. The thickness and diameter of the concrete are 600mm and 1000mm, respectively. To prevent scabbing at the lateral face and diminish the effect of the rarefaction wave to the experimental results, steel hoops are used with a thickness of 6mm to confine the concrete targets.

Table 1. Nominal components and densities of the TC21, TC1, Ni-Ti alloy and Zr705.

Alloy	Nominal components	ρ_a
TC21	Ti-6Al-2Zr-2Sn-2Mo-1.5Cr-2Nb	4.75
TC1	Ti-2Al-1.5Mn	4.55
Ni-Ti	Ti-10Ni	4.51
Zr705	Zr-2.2Nb	6.71

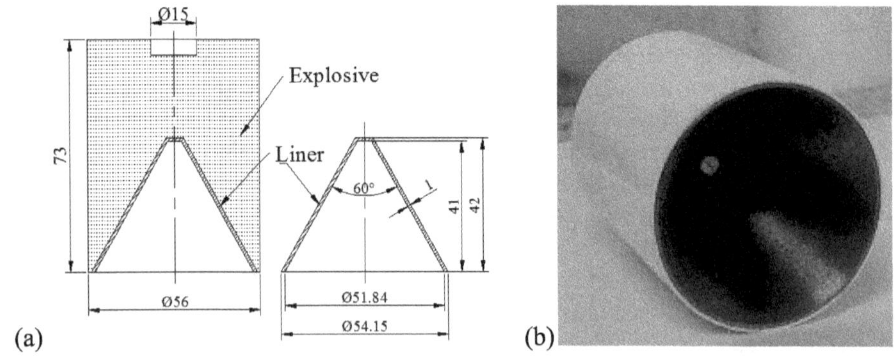

Fig. 1 Shaped charges used in the experiments (a) configuration and sizes (b) photograph.

2.2. Experimental setup

Two 450KV flash X-ray devices were used to visualize the shapes and parameters of the formed jets. Fig 2 shows the schematic and image of the X-ray experiment setup of the shaped charge in front of the double flash X-ray tube. Two protected cassettes containing film were positioned to capture the images of the resulting jets. Aluminum protective cases for the cassettes were attached with two marker lines. In the experiments, two X-ray images were captured at different times for each shot to find the corresponding jet space position. According to the jet displacement difference Δz, trigger time difference Δt, and the amplification coefficient k, the velocity and diameter of jet tip and tail were calculated. As shown in Fig 2, the amplification coefficients were $k_1 = (d_1 + d_1')/d_1$ and $k_2 = (d_2 + d_2')/d_2$, respectively.

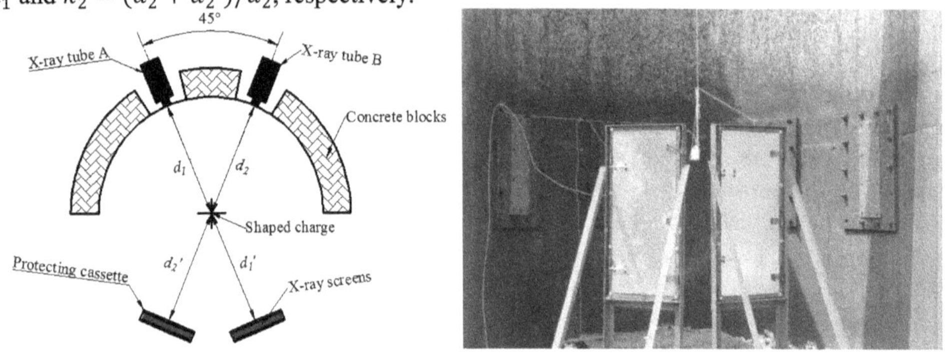

Fig. 2 Schematic and image of the X-ray experiment setup of the shaped charge in front of the double flash X-ray tube

The schematic of the DOP experiment is shown in Fig 3. A smaller stand-off distance of 80mm (roughly 1.43CD) was chosen. The initiation was at a single point by a #8 flash detonator.

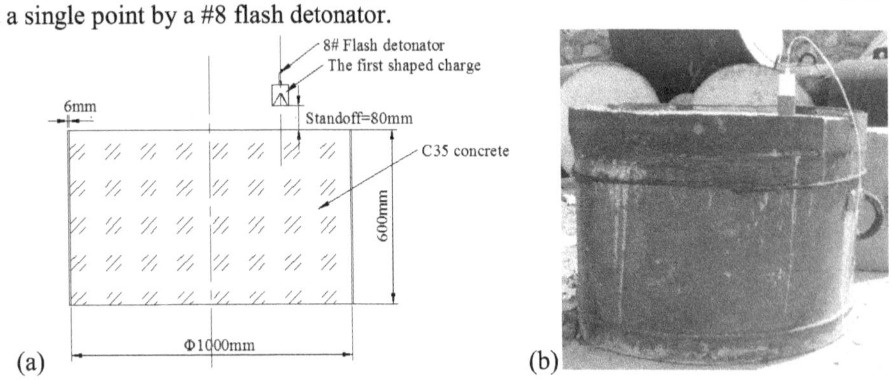

Fig. 3 DOP experiment setup for (a) schematic diagram (b) site layout

3. Results and discussions

3.1. Analysis of the formed jets

Fig 4 shows the double flash X-ray images of the shaped charge jets. The jets travel from left to right. Unfortunately, the radiograph of TC1 was not captured, perhaps due to the set-up problem of the trigger time. It can be seen from Fig 4 that the jet behaviour of TC21 appears to be incoherent and similar to the form of fog. The jet particles are discontinuous and disperse in the radial direction. The image to the right of the "gap between negatives" shows nothing and the jet tip has moved out of the left negative. Thus, the jet tip may be somewhere in the gap between the two negatives. Herein, the jet tip is assumed to sit in the middle of the gap, and its velocity is calculated to be 8265m/s. The diameter of the jet tip is about 11.5mm at 25μs and about 15.7mm at 45μs. The jet formed by the Ni-Ti alloy is found more severely dispersed in the radial direction than the TC21 and appears to be a hollow two-part structure. The calculated jet tip velocity is 6928m/s. The diameter of the jet tip is about 19.8mm at 45μs and about 37.6mm at 55μs, which are larger than that of the TC21. The jet formed by Zr705 does not exhibit radial dispersion, but it fractures earlier and the jet tip is skewed. The jet tip velocity is measured to be 7079m/s. The jet tip diameters at 35μs and 50μs are basically the same; both are about 4.1mm. The diameter in a direction perpendicular to the motion axis is about 14.6 mm. The parameters of the jets are listed in Table 2.

Table 2. Parameters of the jets

Alloy	v_j	d_j	
		first exposure	second exposure
TC1	—	—	—
TC21	8265	11.5	15.7
Ni-Ti	6928	19.8	37.6
Zr705	7079	4.1 (14.6)	4.1 (14.6)

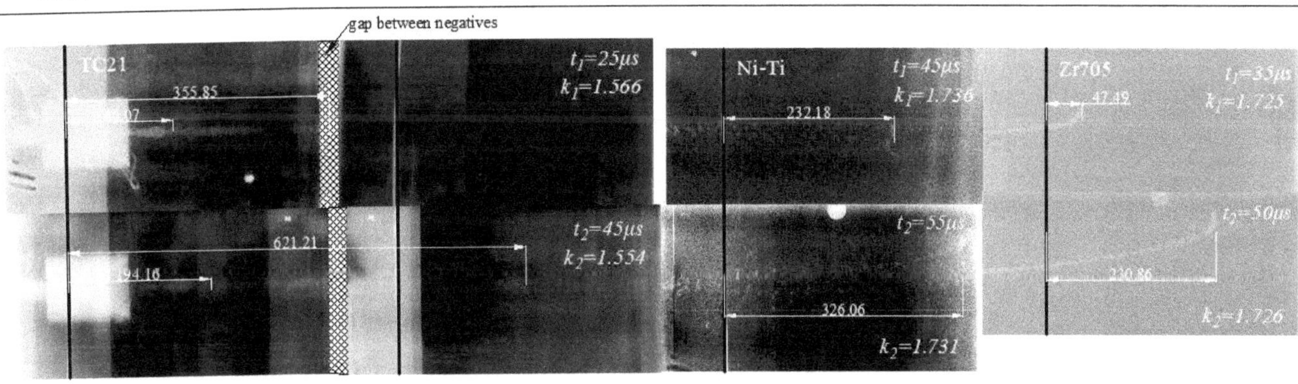

Fig. 4 Flash X-ray images of the shaped charge jets

3.2. Analysis of the DOP experiments

In order to reduce errors, the DOP experiments for each alloy were conducted twice. Fig 5 shows the spalling and scabbing in the front surface of the concrete. When the specimens were subjected to the impact loads by the particulated jet, a typical crater similar to that created by the condensed jet was formed. The formation of the extensive cracking region can be primarily attributed to the propagation of elastic compressive longitudinal waves. When the wave reaches the rear surface, it is reflected as a tensile wave. Superposition of the original compressive wave and the reflected tensile wave resulted in a rapidly decreasing compressive wave and then increasing tensile wave. As the amplitude of the resulting tensile stress wave exceeds the dynamic tensile strength at any point in the concrete matrix, a crack is observed. Compared to the one-time and continued impact of the condensed jet, each dispersed particle of the particulated jet tip hits the concrete surface discontinuously. It is obvious that the larger the radial velocities of the jet particles, the larger the diameter of the subsequent particulated jet when it impacts the concrete. As the tip diameter expands to a diameter larger than the entrance hole, some of the jet particles will strike the surface of the impact crater and further destroy the target surface. Thus, the concrete matrix around the impact crater exhibits a more pronounced failure and is substantially comminuted.

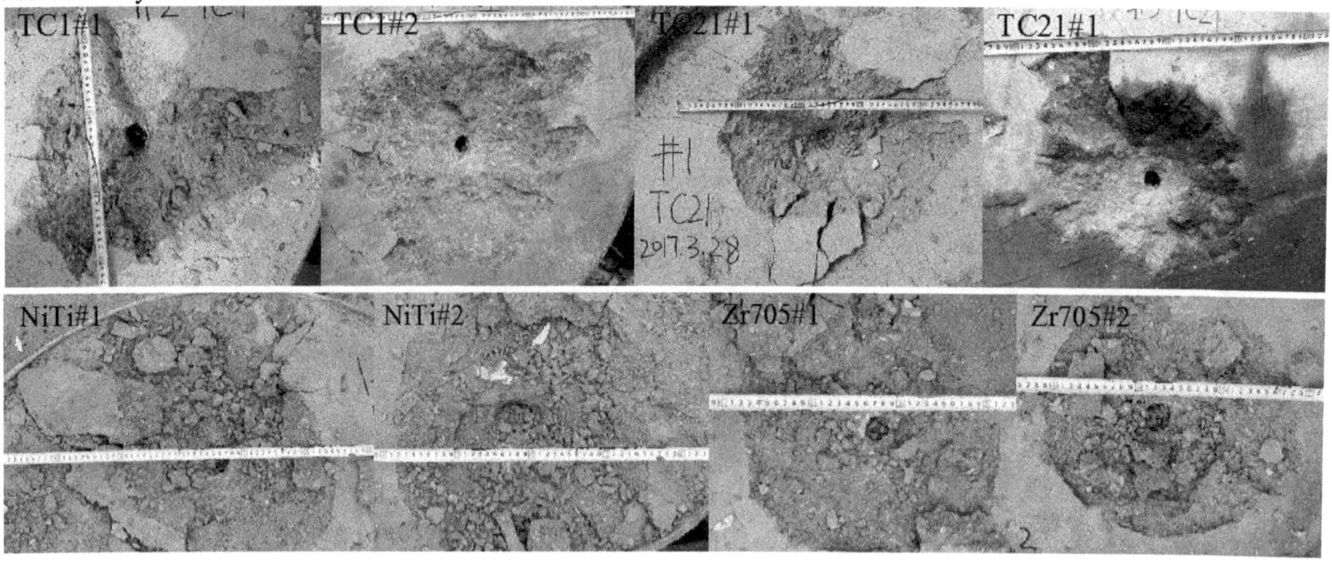

Fig. 5. Damage views of the front faces for the concrete targets after experiment.

Table 3. Experimental data of the impact craters

Alloy	d	$\bar{d}$	h	$\bar{h}$
TC1#1	290	326.5	75	76.5
TC1#2	363		78	
TC21#1	333	311.5	80	78
TC21#2	290		76	
Ni-Ti#1	450	400	60	76
Ni-Ti#2	350		92	
Zr705#1	270	280	70	70
Zr705#2	290		70	

To further analyze the impact performance of the particulated jets into the concrete targets, the diameters and depths of the impact craters were measured. The equivalent diameter d was obtained by averaging the maximal dimensions of the crater from four directions: 45°, 135°, vertical and horizontal, respectively. The crater equivalent diameter and depth are listed in Table 3. It is seen that the average diameters of the impact craters for TC1 and TC21 are basically the same, which are 5.83CD and 5.56CD, respectively. The diameter of the impact crater generated by Zr705 is the smallest, which is 5CD. It should be noted that the impact crater for Ni-Ti as shown in Fig 5 extends all the way to the confining tube on one side, which may actually skew the measurements. But beyond question, the crater diameter for Ni-Ti is the largest, which is measured to be 7.14CD herein, increasing by 23%, 28%, and 43%, respectively, compared to TC1, TC21, and Zr705. The crater depths created by the four kinds of alloy materials range from 1.25 to 1.39CD, and the smallest is still the Zr705. Based on the formed jet shape as shown in Fig 4, it can be inferred that the jet diameter is a key factor affecting the degree of damage on the surface of the concrete targets. For the two-stage tandem warhead, the larger impact crater generated by the precursor shaped charge will contribute to the subsequent damage of the following kinetic energy projectile. Thus, the damage ability of a particulated jet to the concrete surface is significantly improved compared to a condensed jet.

Table 4. Experimental data of the entrance hole diameters and penetration depths

Alloy	D	$\bar{D}$	P	$\bar{P}$
TC1#1	34	32	242	240
TC1#2	30		238	
TC21#1	35	32.5	242	242
TC21#2	30		242	
Ni-Ti#1	60	63.5	210	215
Ni-Ti#2	67		220	
Zr705#1	35	37.5	410	408
Zr705#2	40		406	

Table 4 shows the experimental data of the entrance hole diameters and penetration depths of the formed jets. The average entrance hole diameter and penetration depth for TC1 are 0.57CD and 4.29CD, respectively, while that for TC21 are 0.58CD and 4.32CD, respectively. Therefore, the penetration performance of the particulated jets formed by TC1 and TC21 into C35 concrete targets are basically the same. The average entrance hole diameter for Ni-Ti is up to 1.13CD, and the penetration depth is 3.84CD. Comparing with the two titanium alloy materials, the entrance diameter increases by 95.4% but the penetration depth decreases only 10.8%. The experimental results show that at a proper stand-off distance, the larger the radial dispersion of the particulated jet when hits the target, the stronger the perforation ability. Simultaneously, the penetration depth declines slightly. As can be seen from the X-ray images in Fig 4, the tip of the condensed jet formed by Zr705 is deflective, which increases the jet diameter as it hits the target, thus, the diameter of the entrance hole is enlarged. The diameter of the entrance hole and the penetration depth for Zr705 are measured to be 0.67CD and 7.29CD, respectively, which is an increase of 17.5% and 69.9%, respectively, compared with the titanium alloy. Nevertheless, the diameter of the entrance hole is still smaller than that for Ni-Ti. These results show that the dispersion morphology of the particulated jet has a great influence on the penetration results, so it is necessary to reasonably select the liner material and design the charge structure.

The experimental results in Ref. [6] show that the maximum entrance hole diameter and the penetration depth of an EFP (explosive formed projectile) formed by the titanium alloy penetration into a C35 concrete are 0.60CD and 3.10CD, respectively. Therefore, it can be speculated that the particulated jet has as good performance of creating a large hole into concrete as an EFP, but without the obvious decrease of penetration depth.

4. Summary

Shaped charge penetration into C35 concrete targets were studied by the DOP experiment method. TC21, TC1, Ni-Ti and Zr705 alloys were selected as liner materials. The jet behaviours were obtained by X-ray experiments. The impact and penetration performance of the particulated jet into concrete were analyzed and compared with the condensed jet. The following conclusions were drawn:

(1) TC1, TC21, and Ni-Ti alloys can form particulated jets under certain kinds of charge structures (i.e., the structure in this paper). It may be related to the material mechanical properties. The behaviour of the particulated jet varies with liner material, and radial velocities of the jets are different.

(2) The impact and penetration performance of the particulated jet into the concrete targets are greatly improved compared with that of the condensed jet. The concrete matrix material is more severely broken under the discontinuous impact of the jet particles, and the particulated jet opens a large hole to the concrete target while guaranteeing a certain penetration depth. The performance is sensitive to the stand-off distance due to the radial movement of the particles. By optimizing the charge structure, it is of great significance to the design of the precursor shaped charge in the two-stage tandem warhead to study the penetration performance of particulated jets formed by different liner materials when impacting concrete targets.

Acknowledgments

This study is supported by the National Natural Science Foundation of China (Grant Nos. 11702144).

References

[1] Wang C, Ma T, Ning J. 2008. Experimental investigation of penetration performance of shaped charge into concrete targets[J]. Acta Mechanica Sinica, 24(3): 345.

[2] Zhu, Qi-feng, et al. 2018. "Theoretical and Experimental Study of Shaped Charge Jet Penetration into High and Ultra-high Strength Concrete Targets. *International Journal of Impact Engineering* 122:431-438

[3] Gerami N D, Liaghat G H, Moghadas G H R S, et al. 2017. Analysis of liner effect on shaped charge penetration into thick concrete targets[J]. Journal of the Brazilian Society of Mechanical Sciences and Engineering, 39(8): 3189-3201.

[4] WANG Cheng, WANG Wan-Jun, NING Jian-Guo. 2015. Investigation of shaped charge penetrating into concrete targets[J]. Chinese Journal of Theoretical and Applied Mechanics. 47(4):672-686.

[5] QI-FENG, Z. H. U., et al. 2017. "Numerical Study of High-Density Titanium Alloy Shaped Charge Jet Penetration into Concrete. *30th International Symposium on Ballistics*.

[6] KANG Yan-Long, JIANG Jian-Wei, WANG Shu-You, MEN Jian-Bing. 2011. Experimental and Numerical Simulation Study of Penetration into Multi-Layer Target by Shaped Charge with Different Liner Material[J]. CHINESE JOURNAL OF HIGH PRESSURE PHYSICS. 26.5: 487-493.

[7] J. Guo et al. 2017. "Influence of Yield Strength of Ti-Alloy Liner on Crater Diameter Formed by Shaped Charge Jet Penetration", Solid State Phenomena, Vol. 263, pp. 30-37.

[8] M. J. Murphy, D. W. Baum, R. M. Kuklo, S. C. Simonson. 2001. Effect of Multiple and Delayed Jet Impact and Penetration on Concrete Target Borehole Diameter[C]. 19th International Symposium of Ballistics, Interlaken, Switzerland, 1553-1559.

[9] M. J. Murphy, R. M. Kuklo, T. A. Rambur, L. L. Switzer, M. A. Summers. 2004. Single and Multiple Jet Penetration Experiments into Geologic Materials[C]. 21st International Symposium on Ballistics, Adelaide, South Australia, 41-48.

[10] P. Pincosy, M. J. Murphy. 2007. Calculated Concrete Target Damage by Multiple Rod Impact and Penetration. 23rd International Symposium on Ballistics, Tarragona, Spain, 1453-1460.

[11] M. R. Edwards, R. Arulanandam, S. M. Hille. 2011. Polymers as Potential Shaped Charge Liner Materials[C]. 26th International Symposium on Ballistics, Miami, U.S.A. 1512-1521.

[12] Xiao Q Q, Huang Z X, Zu X D, et al. 2013. Penetration research of jacketed jet into concrete[J]. International Journal of Impact Engineering, 54: 246-253.

[13] Baker, Ernest L., et al. 2011. "Glass as a shaped charge liner material. *Proc. 26th Int. Symp. on Ballistics, E. Baker and D. Templeton (eds). Destech Publications, Lancaster, PA.*

[14] Kecskes L J, Walters W P. 2007. Investigation of a bulk metallic glass as a shaped charge liner material[C]//Proc. of the 23th Int. Symp. on Ballistics, Tarragona, Spain. 31-38.

[15] Saran S, Ayısıt O, Yavuz M S. 2013. Experimental investigations on aluminum shaped charge liners[J]. Procedia Engineering, 58: 479-486.

Proceedings of the 2019 Hypervelocity Impact Symposium
HVIS2019
April 14-19, 2019, Destin, FL, USA
HVIS2019-080

Automatic Mesh-Generation (FEM/SPH) for HVI-Simulations of Arbitrary Rotational Symmetric Impactors

Marvin Becker[a,1], Marina Seidl[a], Miriam Mehl[b], Mhamed Souli[c]

a Institute Saint Louis, 5 Rue de Général Cassagnou, 68301 Saint-Louis, France
b University of Stuttgart, Universitätsstr. 38, 70569 Stuttgart, Germany
c Université de Lille, 42 rue Paul Duez, 59000 Lille, France

Abstract

For the numerical description of high velocity impact, Smooth-Particle-Hydrodynamics (SPH) has gained more and more interest. The standard Lagrangian Finite-Element (FE) approach has difficulties in describing large deformations and fracture. However, a simulation based on SPH only is very expensive due to the small size of the particles. A well adopted solution to this is to couple both methods, using SPH only where it is necessary, and capturing the outer boundary conditions with a bias FE-mesh correctly - without considerable extra computational cost.
We apply such a hybrid approach in LS-DYNA® for the characterization of threats in terminal ballistics. Different meshing approaches for the projectile and target were implemented to guarantee an optimal initial condition. The particle size and the required size of the SPH-region were studied to exclude discretization effects. Exemplarily, a projectile surrogate with simplified geometry is investigated for a fixed impact velocity and two different angles of obliquity. A qualitative comparison between experiments, observed with X-ray cinematography, reveals a good potential of this approach towards predicting fracture and ricochet during high velocity impact events.

Keywords SPH, Fracture, HVI, Ricochet, LS-DYNA®

Nomenclature

Latin symbols

e	energy
f	force
k	kernel width
u	function
v	velocity
N	shape function
W	kernel function

Greek symbols

ρ	density
ω	weight
σ	Cauchy stress tensor

Abbreviations

DOF	Degrees of Freedom
EOS	Equation of State
FEM	Finite-Element Method
HVI	High Velocity Impact
SPH	Smooth-Particle-Hydrodynamics

[1] * Marvin Becker. Tel.: +33 (0)3 8969 5754 *E-mail address:* Marvin.BECKER@isl.eu

1. Introduction

Being able to predict the behavior of a projectile during a High Velocity Impact (HVI) event is of high interest for the design of protective structures and projectiles. The Finite-Element Method (FEM) has been the standard approach used to investigate the projectile impact (1; 2). While FEM-based approaches are able to accurately compute the exit velocity and ricochet behavior (3; 1), their capabilities of predicting fracture–inherited from their numerical description –are limited. In order to address the deficiencies of FEM, several numerical methods has been developed in the recent years. Of prime interest for this paper are the particle methods, such as SPH, that allow to reliable description of the evolution of cracks. However, the family of methods capable of describing fractures consists also of: a phase-field model proposed by Brodon (4), a pre-fragmented FE-mesh of Brescani (5), Peridynamics of Silling (6) (7), and a moving least squares method of Dilts (8).

This paper focuses on SPH. It follows the Johnson-Cook model description (9) (10) combined with the Grüneisen EOS, considered as the standard material description used for this kind of application. The influence of material models is neglected hereinafter.

The paper discusses an approach to predict the behavior of a projectile during a high velocity impact event, which can be divided into two separate tasks: the calculation of forces acting on the projectile, and the simulation of the target response. A full SPH discretization of the target would exceed the computational possibilities, however. Thus, a hybrid discretization of FE-mesh and SPH particle distribution is suggested in this paper. A similar setup has already been compared by Aktay et al. with PAM-CRASH and the contact forces and were in good agreement to the experiment (11). The LS-DYNA® input deck is generated based on an *xml*-input file allowing the user to easily access different geometries of impactors or targets, and adjusting the resolution of the model. This allows running and comparing various simulation scenarios with little effort, which enables grid convergence or parameter studies as well as geometry optimization.

2. SPH and FEM

2.1. Models

This section briefly recapitulates FEM and SPH approaches with respect to the formulation used in the applied multi-physics code, LS-DYNA® (11), and highlights the major differences between the two. A more detailed description for FEM can be found in the book of Zienkiewicz et al. (12) and for SPH one can refer to the original paper of Gingold (13), the improvements proposed by Libersky, Randles and Yreux (14; 15; 16), and the book of Filho (17). Both FEM and SPH solve the conservative equations of continuum mechanics:

$$\frac{d}{dt}\rho_i = -\rho_i \nabla \cdot \vec{v_i}, \qquad \rho_i \frac{d}{dt}\vec{v_i} = -\nabla \cdot \sigma_i + f_{ext}, \qquad \frac{d}{dt}e_i = -\frac{1}{\rho_i}\sigma_i \cdot \nabla \cdot v_i,$$

where the density is ρ, velocity is v, and energy is e with stresses σ and external forces f_{ext}. While FEM uses a weak formulation with shape functions, SPH utilizes a collocation method. The function approximation for FEM is defined by

$$u(x_i) = \sum_j u_j N_j(x_i),$$

where the function u is represented by coefficients u_j and shape functions N_j for each neighbor node j. For SPH, the influence of the neighbors is captured by a kernel function W that integrates to one and has a finite support domain:

$$u(x_i) = \sum_j \omega_j u_j W(x_i - x_j, h).$$

Here, ω_j notes the volume of the particle and u_j are the coefficients. W is the kernel function evaluated at the distance between the nodes $x_i - x_j$ for the kernel width h. The standard collocation method is unable to reproduce neither linear nor constant functions, and is, thus, not consistent. Such a problem can be circumvented using a renormalization approach, where a corrected kernel that satisfies the interpolation condition is used. Hence, higher accuracy is obtained, however for the price of lower numerical stability. Nonetheless, instability problems are not encountered, even if a good initial setup is applied. Thus, in this paper the renormalized approach is applied, for which an appropriate setup has been selected beforehand.

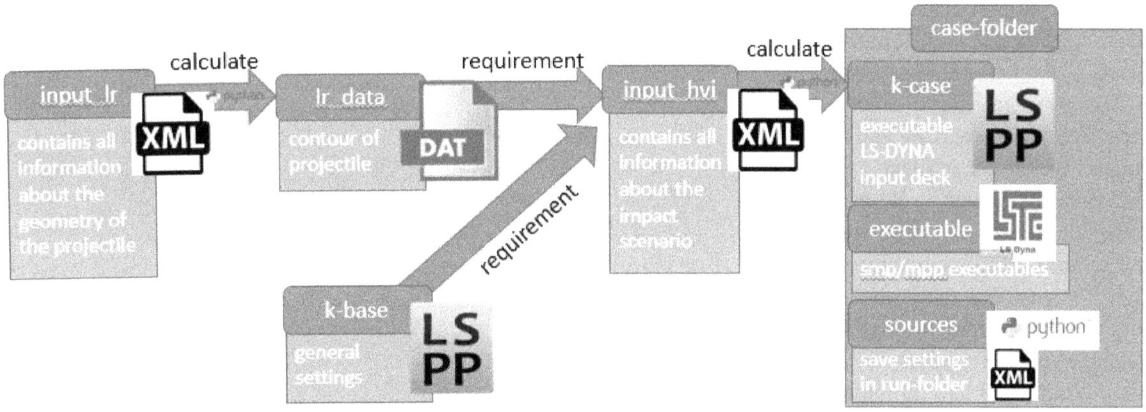

Figure 1: Workflow for the HVI-SPH preprocessing tool: First, the *lr-file* for the projectile geometry has to be generated. This can be done with a small *xml*-file or the contour data itself. Next, a k-file referred to as *k-base* and containing all case specific settings (*CONTROL*-flags, *DATABASE*-flags and material data) is required. Combining the general setup, the *lr-data* of all projectile parts, and the *k-base* settings, a k-file is generated. It considers all requirements of LS-DYNA®. Finally, a run-script is generated, and the sources are copied into the case folder.

2.2. Preprocessing Tool

The capabilities using the LS-DYNA® preprocessor for SPH are limited–in particular concerning the generation of a particle distribution (18). In addition to being more accurate, a self-generated particle distribution simplifies the automation of an entire LS-DYNA® keyword deck. This allows the engineer to compare different applications with little effort.

For setting up a case study, the following tasks are facilitated by the proposed tool:

- Mesh generation of all parts of a desired resolution
- Contact definition for a hybrid mesh in order to improve the runtime–especially the selection of all segments and nodes for the contact; that is time consuming if done manually
- Definition of a simulation box and cutoff velocity for the SPH computation based on impact velocity, oblique angle, and projectile and target size
- Auto-generation of a title and simulation folder for the purpose of tracking, evaluation, and reproducibility
- Consideration of the unit system in the material definition
- Accessing materials by name instead of IDs
- Accessing the most important settings easily

The requirements of the input-deck demand for an automated workflow to cover a wide range of applications and to avoid input mistakes.

Without repeating the single steps already highlighted in the list above, the general workflow of preprocessing with the tool is depicted in Figure 1. First, a file describing the ratio-over-length (lr-data) distribution of the impactor is needed, which can be selected out of two options: either an existing set of lr-data, or by generating a new one based on an xml-description (compare Figure 2). The second required input is a reduced keyword deck that contains all case-specific settings used throughout all cases.

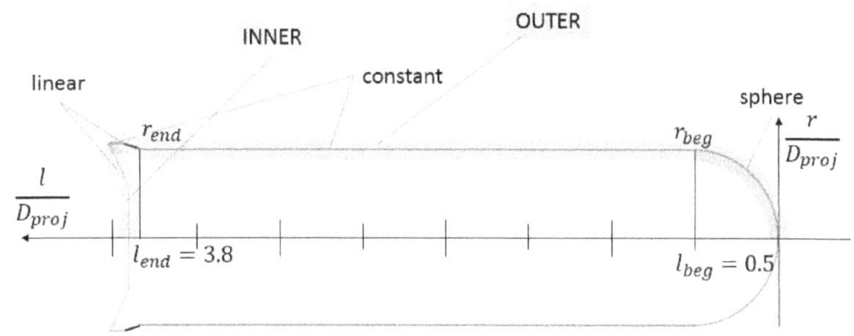

Figure 2: Projectile description based on *xml*-input shown in Figure 10. Each *xml*-element describes one segment of the outline.

The size of the target and boundary conditions can be defined in the xml-input. This allows investigation of the influence of different boundary conditions, and comparison of different materials, thicknesses, or sandwich combinations for advanced protective structures.

In Figure 3, the free parameters for defining the target are explained for an exemplary resolution (reduced for the sake of clarity). Some of the parameters, such as the dimension of the plate, are prescribed by the experiment. For the FE-part, the three parameters N_ϕ, N_R, N_Z can be chosen independently, but it is advised to choose them such that the target shape is as regular as possible. Therefore, in the circumferential direction, three times as many elements are chosen compared to the radial direction. The tool will not only distribute the circumferential points with constant angle, but also accounts for the targets' aspect ratio. The ratio for the bias mesh in radial direction is noted as $ratio_{FEM}$. A value of ten produces a regular mesh. Another free parameter is the size of the SPH region R_{SPH}. The larger the SPH domain, the more expensive the computation becomes. Hence, the question is: what is the minimum size for the SPH region required for accurate results? It has been observed that choosing R_{SPH} such that the whole contact of the projectile is captured has sufficient accuracy. To improve the results, the particle size should be reduced instead.

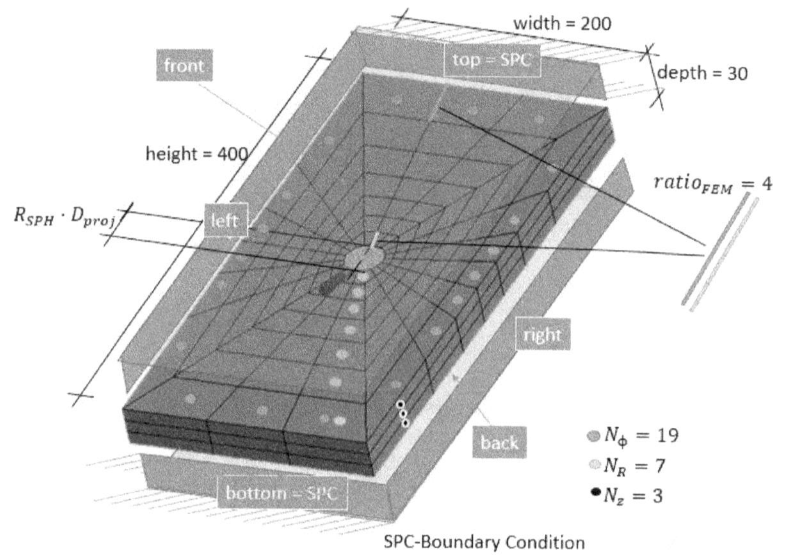

Figure 3: Definition of the target for auto-generating the hybrid FEM/SPH mesh. The boundary conditions are described in blue boxes. Dimensions and the resolution have black font.

2.3. Versatility of the Tool

This paper investigates an exemplary impactor for which all the experimental validation data is available. However, this type of analysis could be applied for any rotationally symmetric impactor and even for multi-material, such as small caliber projectiles (Figure 4), kinetic energy penetrators, or any other kind of threat.

Especially for more complex geometries, the SPH automatic particle generator has a considerable advantage over FEM in generating the mesh. Concerning multi-material, it was observed that instead of populating each part separately, one obtains better results when populating the whole geometry and mapping each particle to the correct part as this results in a homogeneous particle distribution without initial stresses.

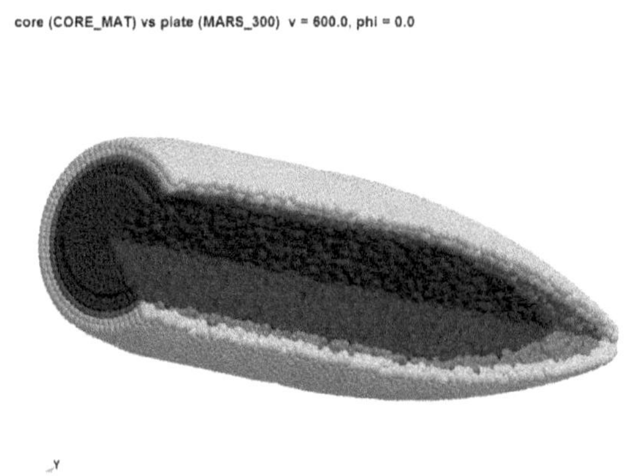

Figure 4: Particle population for an exemplary small caliber ammunition consisting of three different material layers.

3. Setup of the Numerical Experiment

3.1 Geometry

The projectile surrogate investigated in this paper is designed for the validation of numerical simulations, particularly with regard to fracture mechanisms during the impact. It represents the dimensions and impact velocity of a small caliber projectile with a simplified geometry, shown in Figure 2. The surrogate is eroded from a well-characterized armor steel (MARS®300). The target plate is made from the same material as the impactor and is 10 mm thick to avoid perforation.

3.2 Model parameters

The numerical model is described by its physical and numerical parameter sets, and by the discretization itself. The SPH solver uses the renormalized Eulerian formulation. Eroded particles are not considered in the computation, and particles which exceed the initial velocity by more than ten percent are disabled to stabilize the simulation and maintain the time step. Additionally, a box is defined around the impact area, such that the particles that are not of interest are not calculated. The model calculates the default FEM constant stress solid element, and includes hourglass energy and dissipation in the energy balance. The time step is manually reduced, compared to the numerically estimated time step limitation for the explicit solver, by a factor of two, in order to assert that the contact is correctly detected. The contact is computed between all SPH parts by the standard interpolation method, instead of the penalty based contact algorithm SPH_TO_SPH_COUPLING. The latter is case-sensitive and requires calibration. The material parameters for the Johnson-Cook material model and the Grüneisen EOS are taken from the literature (19). However, as the projectile and target consist of similar material, the accuracy of these parameters is not as important for this study. The boundary conditions of the experiment are simplified in the numerical setting by fixing all DOF of the nodes at the top and bottom of the target, whereas four screws fix the steel plate in the experiment.

3.3 Coupling

In LS-DYNA® coupling between different numerical methods is still under active development (20), (21). For the coupling of this hybrid approach, the TIED_NODES_TO_SURFACE_CONSTRAINED_OFFSET keyword is used. In general, the tied contacts can be categorized as constraint-based and penalty-based. The algorithm applied hereinafter is penalty-based. The reason for not using the constraint-based algorithm- where the nodes are translated onto the surface if the contact criteria are fulfilled,-- is that it destroys the particle distribution, and thus artifacts may be observed. This makes it not applicable for a wide range of applications, unless a set of requirements is applied to the FEM and SPH discretization. Furthermore, the penalty based algorithm does not affect the time step size. The option "CONSTRAINED" in the keyword means that the contact allows transferring momentum and forces between slave and master parts.

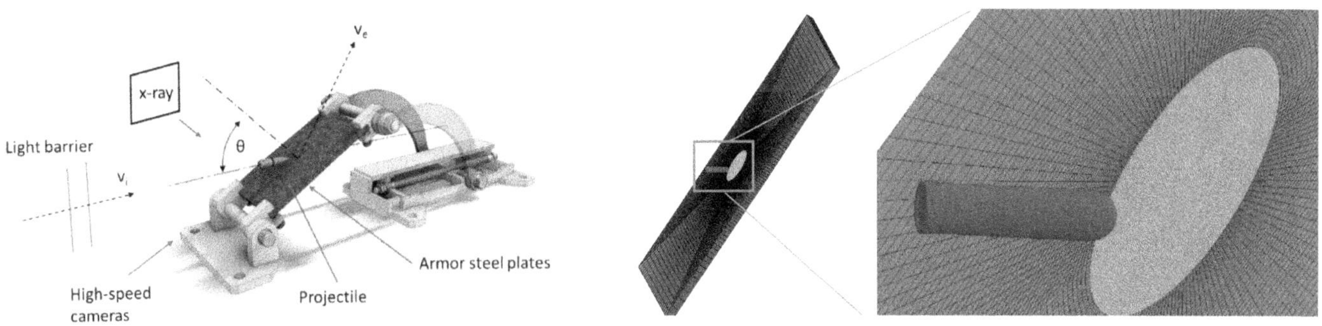

Figure 5: Setup of the real-world experiment (left) and the numerical experiment (right). The projectile is positioned not exactly in the center, such that all interaction between impactor and threat happen inside the SPH region. The SPH particles of the projectile and the target have the same particle diameter.

4. Real World Experiment

The real world experiments are conducted in-house for validation of the numerical model.

A multi-anode X-ray cinematography setup, explained in Figure 7, captures six separate images for each experiment. The raw images contain noise and parallax error, which demand for image processing and alignment. The processed images can be used for direct comparison to numerical results, for tracking of the projectile, and for extracting the physical properties. This paper does a qualitative comparison between numerically and experimentally observed behaviors. Thus, the mass and length of the remainder, which were measured after the impact, are compared in a future report.

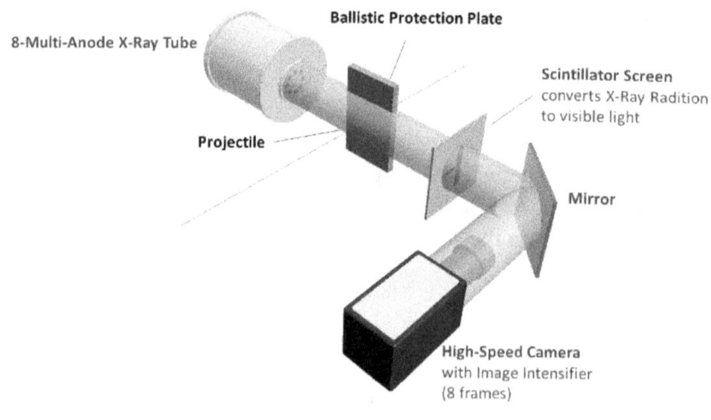

Figure 6: Schematic procedure of multi-anode X-ray cinematography: The impact is X-rayed six times and each time the resulting X-ray distribution is converted to visible light on the scintillator screen and captured with a high-speed camera. With this approach separate images of a single experiment can be captured. (22)

The geometry of the surrogate consists of a cylinder with a spherical nose and a small groove at the back (compare Figure 2). The groove transfers the rotational momentum, during the acceleration phase, from the barrel rifling to the impactor. This is needed to achieve a spin-stabilized trajectory. However, aerodynamic forces on the nose cause disturbance and thus the yaw angle was measured for all experiments. Those test cases with more than two degrees of yaw were not considered in the evaluation. All tests are conducted at an impact speed of $800 \pm 20\ m/s$. The differences in size of the impactor are within the production tolerance of $0.01\ mm$ and thus negligible. The target is a milled armor steel plate of the same material like the surrogate, with a thickness of $10\ mm$, a width of $150\ mm$, and a length of $300\ mm$. The decreased width was chosen to exclude shadows on the X-ray images.

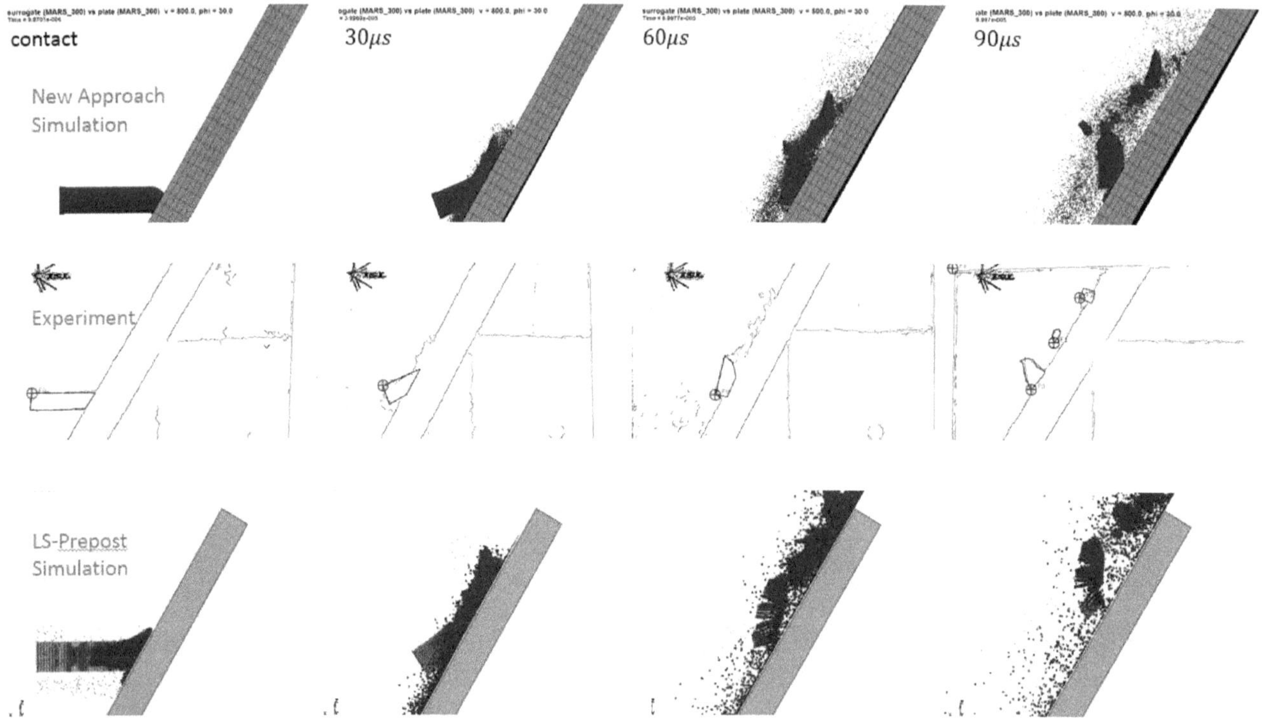

Figure 7: Comparison: Experimental (second row of images) and numerical results (first row new approach, third row original particle distribution) for an impact of the surrogate with initial velocity of 800 m/s and obliquity angle of 30 degrees (NATO). Fracture in general, and rotation of the remainder, is well predicted by the new approach, but the scattering of the front part predict larger fragments. Results of the original approach have large dependency on the discretization.

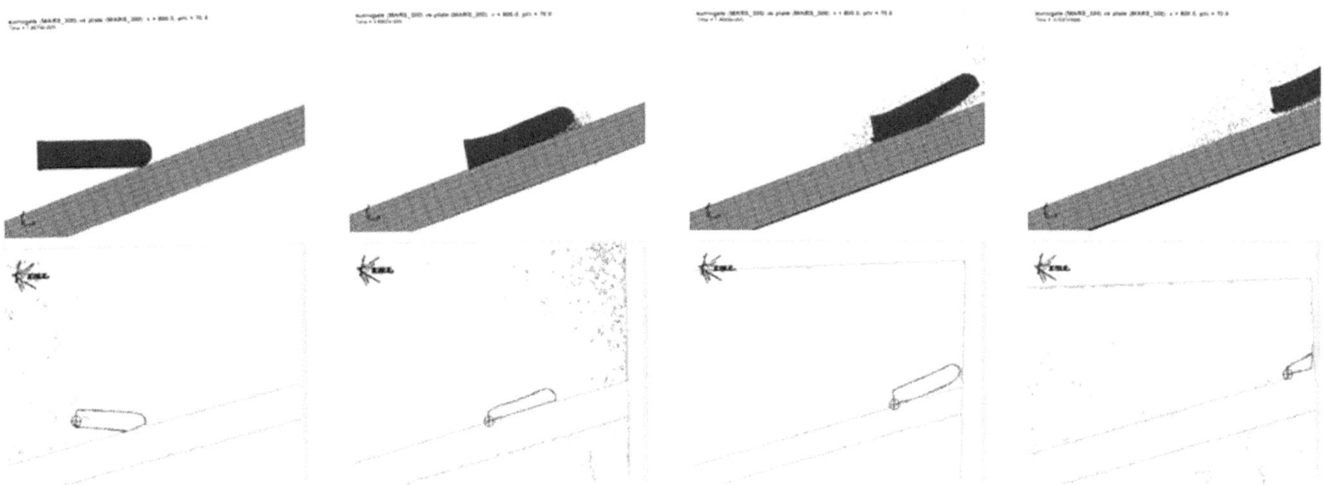

Figure 8: Comparison SPH (upper images) and experiment (lower images) for an impact of the surrogate at 75 degrees obliquity and an impact speed of 800 m/s. The SPH method is able to predict the experimentally observed behavior well.

5. Results

This section presents results of the case study with the surrogate impacting at 800 m/s, for obliquity angles of 30 and 75 degrees. The SPH particle size and the grid spacing of the FE-mesh are set in order to assure convergence. This extract of results focuses on the qualitative behavior, which has not been correctly described with the pure FE-approach used in the past (3). Figure 7 shows comparison for an angle of 30 degrees. The experimental data; the proposed hybrid discretization with an improved particle distribution; and the original approach, with the particle distribution generated from the FEM mesh in LS-PREPOST® are shown. The fragmentation of the surrogate is well-predicted by the simulation obtained using our approach, as opposed to the original one. Although the fracture has been previously already observed using the standard SPH particle distribution, these initial results suffered from discretization-induced errors: not only physical but also numerical cracks were visible. This physically incorrect behavior could be improved with the approach suggested in this paper. The counter-clockwise rotation of the remainder which was observed in the experiment can be seen in the simulation results of the new approach as well. Small differences were observed in the remainder size, which is slightly over predicted. Furthermore, the fragmentation at the front yields lerger fragments than in the experiment.

Although the surrogate fragments at smaller oblique angles, an intact ricochet can be observed in the experiment for 75 degrees obliquity or higher (Figure 8). This behavior can be reproduced with the traditional FE-analysis, since much smaller forces and deformations occur for this case. The question was whether it is also possible to predict this behavior with the SPH method. It definitely is: the simulation and experiment data show the same qualitative result-- the intact ricochet. Moreover, the quantitative results concerning the shape, ricochet angle, and velocity agree as well, and thus SPH can be considered as a capable method of predicting the intact ricochet.

6. Conclusions

A comprehensive tool for simulating high velocity impact has been developed and applied to different applications. It is written as an addition to the LS-DYNA® preprocessor LS-PREPOST®. The tool generates the SPH particles for any desired rotational symmetric geometry, and is able to use different materials. The method utilizes a hybrid FEM/SPH mesh for the target description. For the validation of the numerical data, experiments were conducted using multi-anode X-ray cinematography. The numerical results show good agreement with the experimental data, and can produce results without the unphysical numerical cracks, which are observed if the particle distribution is generated from the traditional FE-mesh. To exclude any influence from the mesh and particle size, a mesh sensitivity study has been conducted previously, but is not presented in this paper. A small caliber projectile has been meshed with the tool to demonstrate its versatility, and showed possible use cases of the SPH-HVI add-on for LS-PREPOST. Results of these simulations are not presented in this paper.

Having designed an automated way for preprocessing simulations, the authors aim at applying the tool to a wide variety of applications and assessing both the possibilities and limitations of the approach. Being able to auto-generate a FE-mesh of the impactor would be a very valuable option for direct comparison with the FEM, but is also difficult to implement for a broad range of applications. A few preliminary steps towards an automated post-processing have already been done, e.g. the auto detecting of fragments, which is required to extract and compare the physical values of the fragments. This approach will be applied, and results will be compared in the future.

Acknowledgements

The author acknowledges ISL to finance the project, and the two universities, University of Stuttgart and Lille University, for supervision.

References

1. **Seidl, Marina, Wolf, Thomas und Nuesing, Rainer.** *Numerical Investigations on Ricochet of a Spin-Stabalised Projectile on Armour Steel Plates.* Salzburg, Austria, 11th European LS-Dyna Conference 2017.
2. **Kilic, Namik and Bedir, Said.** *Ballistic behavior of high hardness perforated armor plates against 7.62 mm armor piercing projectile.* Materials and Design. 2014.
3. **Borden, Michael J., et al.** *A phase-field description of dynamic brittle fracture.* Elsevier, 2012, Computer Methods in Applied Mechanics and Engineering, Vol. 217, pp. 77-95.
4. **Bresciani, L. M., et al.** *Numerical modelling to reproduce fragmentation of a tungsten heavy alloy projectile impacting a ceramic tile: Adaptive solid mesh to the SPH technique and the cohesive law.* 2016, International Journal of Impact Engineering, Vol. 87, pp. 3-13. SI: Experimental Testing and Computational Modeling of Dynamic Fracture. ISSN: 0734-743X.
5. **Silling, S. A.,** *Reformulation of elasticity theory for discontinuities and long-range forces.* 2000, Journal of the Mechanics and Physics of Solids, Vol. 48, pp. 175-209. ISSN: 0022-5096.
6. **Diehl, Patrick.** *Modeling and Simulation of cracks and fractures with peridynamics in brittle materials.* Rheinische Friedrich-Wilhelms-Universität Bonn. 2017. Ph.D. dissertation.
7. **Dilts, Gary A.** *Moving least-squares particle hydrodynamics II: conservation and boundaries.* : Wiley Online Library, 2000, International Journal for numerical methods in engineering, Vol. 48, pp. 1503-1524.
8. **Johnson, G.R., Cook, W.H.** *A constitutive model and data for metals subjected to large strains, high strain rates and high temperatures.* Proceedings of the Seventh International Symposium on Ballistics. 1983.
9. **Johnson, G.R., Cook, W.H.** *Fracture characteristics of three metals subjected tovarious strains, strain rates, temperatures and pressures.* Engineering Fracture Mechanics. 1985, Vol. 21, 1.
10. **Hallquist, J.O.** *LS-DYNA Theory Manual.* 2006.
11. **Aktay L. et al.** *Combined FEM/Meshfree SPH Method for Impact Damage Prediction of Composite Sandwich Panels.* ECCOMAS, 2005.
12. **Olek C. Zienkiewicz, Robert L. Taylor, J.Z Zhu.** *The Finite Element Method: Its Basis and Fundamentals.* s.l. : Butterworth-Heinemann, 2013.
13. **Gingold, Robert A. and Monaghan, Joseph J.** *Smoothed particle hydrodynamics: theory and application to non-spherical stars.* Oxford University Press Oxford, UK, 1977, Monthly notices of the royal astronomical society, Vol. 181, pp. 375-389.
14. **Libersky, Larry D. and Petschek, A. G.** *Smooth particle hydrodynamics with strength of materials.* Berlin : Springer Berlin Heidelberg, 1991. Advances in the Free-Lagrange Method Including Contributions on Adaptive Gridding and the Smooth Particle Hydrodynamics Method. pp. 248-257.
15. **Yreux, Edouard and Chen, Jiun-Shyan.** *A quasi-linear reproducing kernel particle method.* Wiley Online Library, 2017, International Journal for Numerical Methods in Engineering, Vol. 109, pp. 1045-1064.
16. **Randles, P. W. and Libersky, L. D.** *Smoothed Particle Hydrodynamics: Some recent improvements and applications.* 1996, Computer Methods in Applied Mechanics and Engineering, Vol. 139, pp. 375-408. ISSN: 0045-7825.
17. **Fraga Filho, Carlos Alberto Dutra.** *Smoothed Particle Hydrodynamics.* Springer, 2018.
18. **Becker, Marvin.** *A Study on Initial Conditions in Smooth-Particle-Hydrodynamics Modelling of High-Velocity-Impact.* Sweden, Gothenburg : s.n., 2018. 2018 Nordic LS-DYNA® Users Conference.
19. **Fras, Teresa, Roth, Christian and Mohr, Dirk.** *Fracture of high-strength armor steel under impact loading.* International Journal of Impact Engineering. 111, 2017.
20. **Ramzi, Messahel and Mhamed, Souli.** *SPH and ALE Formulations for Fluid Structure Coupling.* Computer Modeling in Engineering & Sciences. 2013, Vol. 96, 6.
21. **J Xu, J Wang, M Souli.** *SPH and ALE formulations for sloshing tank analysis.* International Journal of Multiphysics. 2015, Vol. 9, 3.
22. **Becker, Marvin.** *Numerical Ricochet Model of a 7.62 mm Projectile Penetrating an Armor Steel Plate.* 2018. 15th International LS-DYNA Users Conference.
23. **Faderl, Norbert and Becker, Marvin.** *High-Speed 300keV flash X-ray cinematography for investigations of the ballistic impact behavior of target materials and structures.* Institute of Saint-Louis. 2018. Tech. rep.

Proceedings of the 2019 Hypervelocity Impact Symposium
HVIS2019
April 14-19, 2019, Destin, FL, USA
HVIS2019-095

Study on the explosion-proof performance of polyurea-reinforced masonry walls with different spraying methods

Wei Shang[a], Xu-dong Zu[a], Zheng-xiang Huang[a,*], Wen-ni Shen[a]

[a]*School of Mechanical Engineering, Nanjing University of Science and Technology, Xiao lingwei 200, Nanjing 210094, PR China*

Abstract

Based on the propagation theory of blast waves and the strain rate effect of polyurea, the explosion-proof performance of polyurea-reinforced masonry walls with different spraying methods is discussed in this paper. The impact fracture of masonry walls after contact explosion was analyzed, and the fracture results of a blast wave on polyurea-reinforced masonry walls with different spraying methods were predicted. Furthermore, explosion-proof experiments of a standard masonry wall (2m×1.2m×0.37m) under three conditions including non-sprayed, back surface sprayed polyurea and double-sided sprayed polyurea were carried out to verify the theoretical predictions. Finally, the impact fracture results of standard masonry walls after a 1 kg TNT contact explosion under the three conditions were obtained. The test results were in good agreement with the theoretical predictions. It clearly demonstrated that polyurea coating can significantly improve the explosion-proof performance of masonry walls, and double-sided sprayed showed better explosion-proof performance than back surface sprayed at the same coating thickness.

Keywords: Polyurea; Masonry walls; Impact fracture; Explosion-proof performance

Nomenclature

p_c	compressive stress of the compressional wave (Mpa)
p_t	Tensile stress of the expansion wave (Mpa)
σ_{cmax}	ultimate compressive strength of the wall (Mpa)
σ_{tmax}	ultimate tensile strength of the wall (Mpa)
ρ_0	initial density of the wall (Kg/m^3)
ρ_1	density of the wall after compressional wave (Kg/m^3)
ρ_2	density of the wall after expansion wave (Kg/m^3)
vs, vp	shock-velocity and particle-velocity (m/s)
C	the intercept of the (vp) curve (m/s)
E_0	initial internal energy of the wall (KJ)
E_1	internal energy of the wall after compressional wave (KJ)
γ_0	the unit less Gruneisen gamma
a	first order volume correction to γ_0
S_1, S_1, S_1	unit less coefficients of the slope of the $vs(vp)$ curve
σ_0	tensile yield stress of polyurea under quasi-static condition (Mpa)
σ_y	dynamic tensile yield stress of polyurea (Mpa)
$\dot{\varepsilon}$	strain rate (s^{-1})
DIF	stress enhancement factor
A,B	fitting parameters of the experiment

* Corresponding author. Tel.: +86-18061888202
E-mail address: huangyu@mail.njust.edu.cn.

1. Introduction

In the contemporary world, the frequent occurrence of terrorist incidents has brought a great challenge to building protection. At present, most walls of existing buildings were designed as a masonry structure or reinforced concrete structure. A blast wave will cause damage to the walls and form a large spalling area on the back of the walls which seriously threatens the safety of personnel and equipment inside the walls. Therefore, it becomes an urgent matter to carry out research on enhancing explosion-proof performance of walls under the premise of guaranteeing the existing building structure unchanged. Experimental and simulation studies found that [1] polymer coating can enhance the explosion-proof performance of existing walls, which made the explosion shock response characteristics of sprayed polymer-reinforced walls a research hotspot.

Polyurea is a green polymer coating with excellent impact and tear resistance. It has great potential in the field of anti-terrorism and explosion protection. After spraying it on the surface of walls, polyurea can absorb blast wave energy and reduce secondary injury caused by wall spalling during the explosion process [2-3]. Over the past decade, much research has been done on the dynamic mechanical properties of polyurea and the impact response characteristics of walls. Jongmin Shim [4] and S.N. Raman [5] tested the tensile and compression dynamic performance of polyurea over the strain rate region from 10^{-2} to 10^4/s. Results showed that the stress–strain behavior of polyurea at high strain rates is considerably non-linear and exhibits significant strain rate sensitivity. Hui Guo et al. [6] developed a three-dimensional visco-hyperelastic constitutive model to describe the finite deformation mechanical behavior of polyurea materials at different strain rates, and carried out nonlinear stress-strain behavior experiments of polyurea under uniaxial tension and compression to validate the effectiveness of the constitutive model. Jun Li et al. [7] investigated spall damage of generic reinforced concrete columns subjected to blast loads numerically. Kai X et al. [8] carried out a numerical simulation study on the concrete spallation under various blast loading and structural conditions and established the spallation criteria for different levels of spallation. Lee G. Moradi et al. [9] studied the resistance of membrane retrofitted concrete masonry walls to lateral pressure. Sanam Aghdamy et al. [10] developed numerical models to simulate the response of concrete masonry unit walls retrofitted with polyurea coating and proved that the sprayed polyurea improved the explosion resistance of walls. James S. Davidson et al. [11-12] carried out failure mechanisms and experimental testing investigations of polyurea-reinforced concrete masonry walls subjected to blast loading. Results showed that the explosion-proof performance of a masonry wall can be significantly enhanced by spraying polymer on the walls. The above research only showed that the sprayed polyurea wall has a good explosion-proof performance, however, there is a lack of in-depth analysis on its explosion-proof mechanism and spraying method.

Based on the propagation theory of blast waves and the strain rate effect of polyurea, this paper analyzes the impact fracture of masonry walls under the blast wave of a contact explosion and predicts the impact fracture results of a blast wave on polyurea-reinforced masonry walls with different spraying methods. At the same time, explosion-proof experiments of standard masonry walls (2m×1.2m×0.37m) under three conditions were carried out to verify the theoretical predictions.

2. Impact fracture analysis of walls under blast loading

For a non-sprayed wall, detonation products with high temperature and high pressure characteristics have a strong impact on the wall. Because the rate of heat conduction is much less than that of a blast wave, the effect of temperature increase on the explosion-proof performance of the wall is ignored [13]. Therefore, research on the explosion-proof mechanism of a polyurea-reinforced wall should start from propagation of the shock wave. According to the impedance effect of the medium, a blast wave will exhibit multiple reflections at the interface of the wall. At the initial moment of detonation, due to the fact that the impact impedance of the wall is greater than that of the explosive, compressive stress generates a compression wave under blast loading. A pit will appear on the surface of the wall. The Gruneisen equation of state with cubic shock-velocity as a function of particle-velocity (vp) defines pressure p_c for compressed materials as [14]

$$p_c = \frac{\rho_0 C^2 \mu_1 \left[1+\left(1-\frac{\gamma_0}{2}\right)\mu_1-\frac{a}{2}{\mu_1}^2\right]}{\left[1-(S_1-1)\mu_1-S_2\frac{{\mu_1}^2}{\mu_1+1}-S_3\frac{{\mu_1}^3}{(\mu_1+1)^2}\right]}+(\gamma_0+a\mu_1)E_0 \tag{1}$$

where ρ_0 is the initial density of the wall, ρ_1 is the density of the wall after compression, C is the intercept of the (vp) curve (in velocity units), S_1, S_2, and S_3 are the unit-less coefficients of the slope of the $vs(vp)$ curve, γ_0 is the unit-less Gruneisen gamma, a is the unit-less first order volume correction to γ_0, $\mu_1 = \rho_1/\rho_0 - 1$, and E_0 is the initial internal energy of the wall.

If the shock wave strength is greater than the ultimate compressive strength of the wall, the wall will be crushed. σ_{cmax} is the ultimate compressive strength of the wall.

$$\frac{p_c}{\sigma_{cmax}} \geq 1 \tag{2}$$

When the shock wave propagates to the back of the wall due to the impact impedance of air being much smaller than that of the wall, a tensile stress is generated due to the influence of the expansion wave, and a bulge appears on the wall. The tensile stress p_t is defined as

$$p_t = \rho_1 C^2 \mu_2 + (\gamma_0 + a\mu_2)E_1 \tag{3}$$

In equation (3), $\mu_2 = \rho_2/\rho_1 - 1$, ρ_2 is the density of the wall after the expansion wave, and E_1 is the internal energy of the wall after the compressional wave.

When the reflected expansion wave propagates to the front of the wall, the crushed wall will be thrown outward which will further enlarge the crater of the wall.

If the shock wave strength is greater than the ultimate tensile strength of the wall during this process, it will generate spallation where σ_{tmax} is the ultimate tensile strength of the wall.

$$\frac{p_t}{\sigma_{tmax}} \geq 1 \tag{4}$$

The expansion wave is reflected back and forth at the interface between the wall and air, until the wave energy inside the wall is completely attenuated. At this moment, maximum fracture occurs [15].

From the above analysis, when the blast wave impacts on the non-sprayed wall, obvious fractures appear or even cause collapsing on both sides of the wall. Fig. 1 shows the process of a blast wave impacting on the non-sprayed wall.

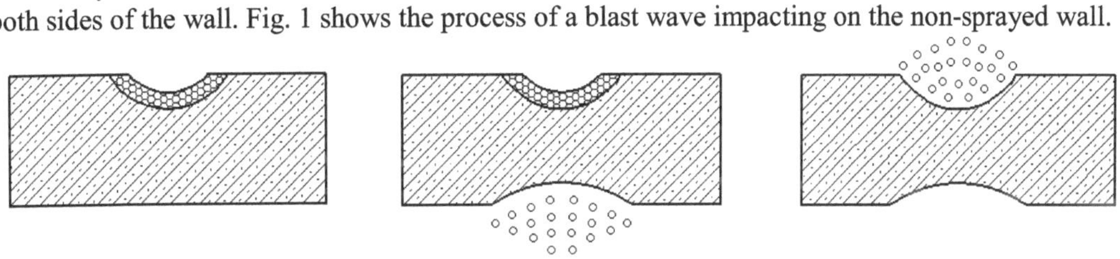

Fig.1. Impact process of a blast wave on the non-sprayed wall (a) Front surface of the wall is crushed after the compressional wave. (b) Back surface of the wall generates spallation after the expansion wave. (c) The crushed wall is thrown outward after the reflected wave.

For composite walls, the ultimate tensile and compressive strengths are determined by the component of the maximum ultimate tensile and compressive strengths [16]. Polyurea is a strain rate sensitive material. S.N. Raman et al. [5] found that the relationship between tensile yield stress of polyurea σ_y , the stress enhancement factor DIF, and the strain rate $\dot{\varepsilon}$ can be well described by equations (5) and (6).

$$\mathrm{DIF} = \mathrm{A}\ln(\dot{\varepsilon}) + B \tag{5}$$

$$\sigma_y = DIF \times \sigma_0 \tag{6}$$

where σ_0 is the tensile yield stress of polyurea under quasi-static conditions, and A and B are fitting parameters of the experiment. S.N. Raman et al. gave the fitting results shown in Table 1. Research has shown [17] the strain rate of solid materials under explosive load is about 10^4/s~10^6/s. Therefore, the dynamic tensile yield stress of polyurea is up to 3.5~5.0 times larger than that under quasi-static conditions.

According to strain rate effect analysis, the dynamic tensile strength of polyurea under explosive load is much larger than the dynamic tensile strength of the wall [18]. It predicts that the back surface of a sprayed wall will not exhibit a large area collapse under the protection of the polyurea coating, while the front side of the wall will be seriously damaged. For the double-sided sprayed wall, both the front and the back surfaces are protected well by a polyurea coating. The overall fracture area is much smaller.

Table 1. Coefficients A and B for DIF of yield stress at different ranges of strain rates

Strain rate range	A	B
$\dot{\varepsilon} < 10^{-2}$	0.0068	1.0572
$10^{-2} \leq \dot{\varepsilon} < 10^{2}$	0.1043	1.5065
$\dot{\varepsilon} \geq 10^{2}$	0.3347	0.4422

3. Explosion-proof experiments of walls

In order to verify the impact fracture results of a blast wave on a polyurea-reinforced wall with different spraying methods, explosion-proof experiments of standard masonry walls were carried out to study the deformation and fracture of the walls. The thickness of the masonry walls is 0.37m, which is a widely used thickness in Chinese buildings. Figs. 2 and 3 respectively show the structural dimensions of standard masonry and the experimental setup. The explosives used in these experiments were TNT columns (see Fig.4) detonated with an 8# fire detonator. The back-surface sprayed wall had uniformly sprayed polyurea on the back surface of the masonry wall, and the thickness of the coating was 10mm. The double-sided sprayed wall had polyurea sprayed on both sides of the masonry wall. The thicknesses of the coatings were 3mm on the front surface and 7mm on the back surface. The numbers of non-sprayed, back-surface polyurea sprayed and double-sided polyurea sprayed masonry walls were # 1, # 2 and # 3.

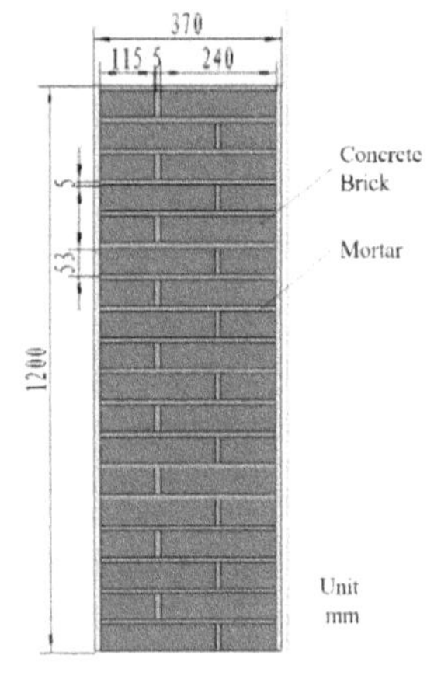

Fig 2. Structural dimensions of standard masonry walls

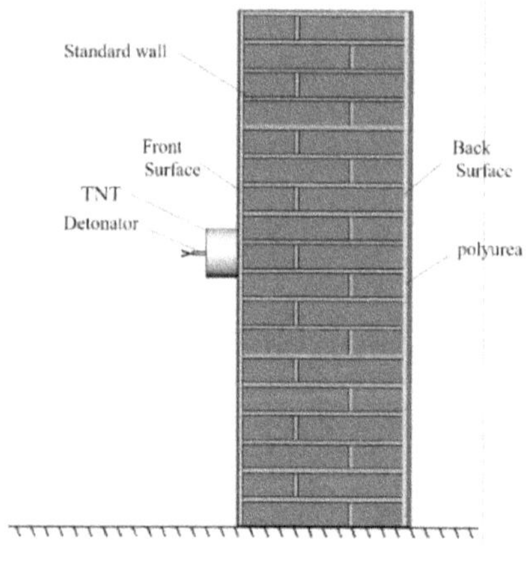

Fig 3. Experimental setup

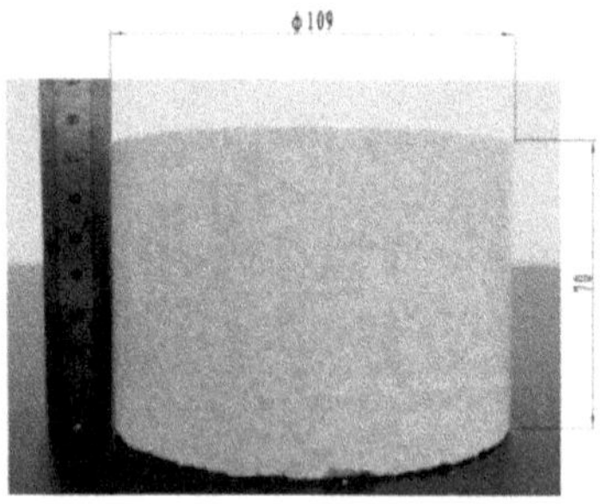

Fig 4. Physical map of TNT column

4. Experimental results and analysis

Fig. 5 shows the impact fracture results of a non-sprayed masonry wall after a 1 Kg TNT contact explosion. A large hole with a diameter of 900mm appeared and large cracks diffused around the explosion center. Because there were no constraints, the top side of the wall formed a gap 1400 mm in length. The masonry wall is left perforated and the overall damage is significant.

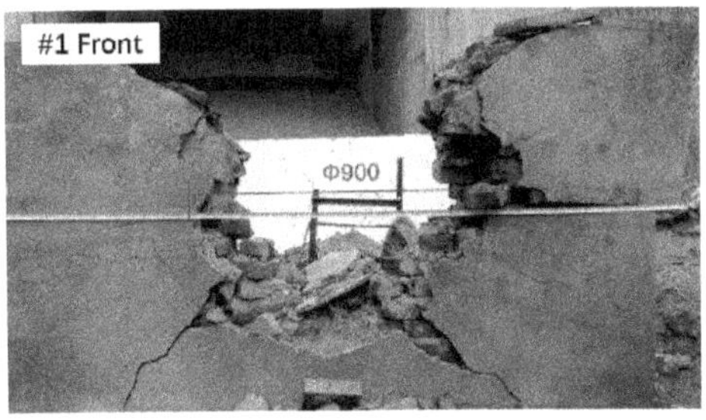

Fig 5. Impact fracture of non-sprayed masonry walls on the front surface

Fig. 6 shows the impact fracture of the masonry wall sprayed with 10mm polyurea on the back surface after a 1 Kg TNT contact explosion. It can be seen that the fracture of the front surface of the wall is similar to that of the non-sprayed masonry wall with large holes and cracks. But the aperture and cracks are greatly reduced compared with those of the non-sprayed masonry wall. The aperture is 520 mm, which is 42.2% less than that of the non-sprayed masonry wall. Moreover, there is only a 68 mm thick bulge appearing on the back surface of the wall and no collapsing or cracks occurred on the back surface due to the protection of the polyurea coating. The wall is not completely perforated, and the depth of the hole is 245 mm which is 33.8% less than that of perforation. There is no obvious gap appearing on the top side of the wall. Experimental results show that the impact fracture under a blast wave can be effectively reduced by spraying polyurea on the back surface of the wall.

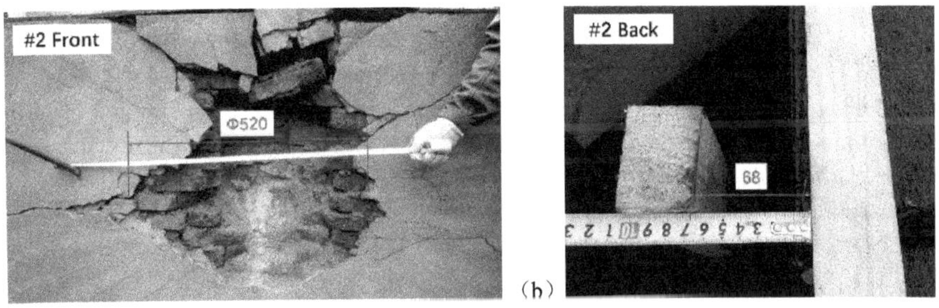

Fig 6. Impact fracture of back surface sprayed polyurea masonry walls (a) Front surface ; (b) Back surface

Fig. 7 shows the impact fracture of the masonry wall with 3 mm of polyurea sprayed on the front surface and 7 mm of polyurea sprayed on the back surface after a 1 Kg TNT contact explosion. On the front surface of the masonry wall, there is a hole of only 196 mm in diameter. The aperture is 78.2% smaller than that of the non-sprayed cases. There is also no obvious fracture on the back surface of the wall, which only exhibits a bulge with diameter of 800 mm and thickness of 38 mm. The thickness of the bulge decreases by 44.1% compared with the back-surface polyurea sprayed masonry wall. There are no obvious cracks on both sides of the wall. In addition, the wall is not completely perforated, and the depth of the hole is 72 mm which is 80.5% less than that of perforation. There is almost no fracturing on the top side of the wall. However, it is noteworthy that there are obvious bulges on the front surface of the wall, and the inner wall of the bulging area is severely broken. Experimental results show that when the blast wave is reflected back and forth in the wall, the bricks inside the wall are crushed in a large area and the energy of the shock wave is consumed, while the surface of the wall is damaged less due to the protection of the polyurea coating. Therefore, the explosion-proof performance of a double-sided sprayed masonry wall is better than that of a back-side sprayed wall when the thickness of the polyurea coating is kept unchanged.

According to the analysis of the experimental results, the deformation and fracture of the wall under three different working conditions are shown in Table 2. It can be concluded that a polyurea coating can greatly improve the explosion-proof performance of masonry walls, and when the thickness of the coating is a constant, the explosion-proof effect of a double-sided sprayed wall is better than that of a back-sprayed wall.

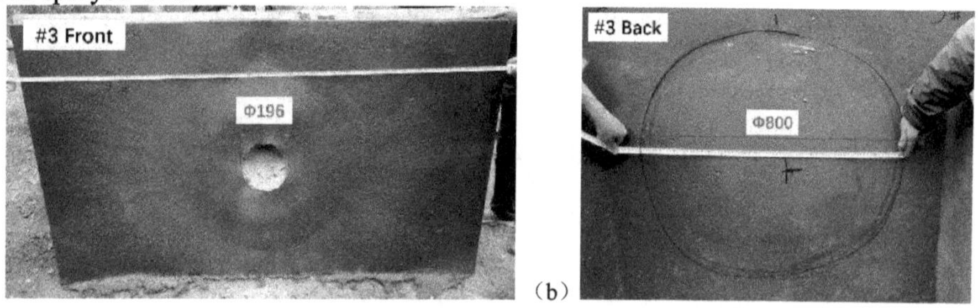

Fig 7. Impact fracture of double-sided polyurea sprayed masonry wall (a) Front surface ; (b) Back surface

Table 2 Anti-explosion experimental results of concrete masonry walls

Wall	Component description	Failure characterization
#1	Standard masonry wall	Control wall severely fragments and collapses. Large holes and cracks appear in the wall. The diameter of centre hole is 900mm. The length of the gap on the top side of the wall is 1400 mm. The wall is perforated.
#2	Standard masonry wall+ 10 mm thick polyurea coating on the back surface	The front surface of the wall is severely overloaded. Holes and cracks appear on the front surface. The diameter and depth of the hole are 520mm and 245mm. A bulge but no cracks appear on the back surface. The thickness of the bulge is 68mm. The wall is not completely perforated.
#3	3 mm thick polyurea coating on the back surface +Standard masonry wall+ 7 mm thick polyurea coating on the back surface	A small hole and no cracks appear on the front surface. The diameter and depth of the hole are 196mm and 72mm . A bulge and no cracks appear on the back surface of the wall. The diameter and thickness of bulge are 80mm and 38mm. The wall is completely perforated.

5. Conclusion

Based on the propagation theory of blast waves and the strain rate effect of polyurea, the impact fracture of sprayed polyurea reinforced walls under blast loading are predicted. Explosion-proof experiments of masonry walls were carried out to verify the predicted results. The following conclusions are obtained:

(1) Sprayed polyurea reduces the tensile fracture of masonry walls caused by an expansion wave. Thus, it reduces the gaps and cracks of the wall under a blast wave, significantly improving the explosion-proof performance of the wall.

(2) When the thickness of the polyurea coating is the same, the deformation and damage of a double-sided sprayed masonry wall is much smaller than that of a back-surface sprayed wall. Therefore the explosion-proof performance of a double-sided sprayed wall is better.

Acknowledgements

This research was supported by the National Natural Science Foundation of China (Grant No. 11872214).

References

[1] Buchan P A, Chen J F. 2007, Blast resistance of FRP composites and polymer strengthened concrete and masonry structures – A state-of-the-art review. Composites Part B: Engineering. 38(5-6): 509-522.

[2] Y.A. Bahei El Din, G.J. Dvorak, O.J. Fredricksen. 2006, A blast-tolerant sandwich plate design with a polyurea interlayer. Int. J. Solids Struct. 43:7644–7658.

[3] S.A. Tekalur, A. Shukla, K. Shivakumar. 2008, Blast resistance of polyurea based layered composite materials. Compos. Struct. 84:271–281.

[4] Shim J, Mohr D. 2009, Using split Hopkinson pressure bars to perform large strain compression tests on polyurea at low, intermediate and high strain rates. International Journal of Impact Engineering. 36(9): 1116-1127.
[5] Raman S N, Ngo T, Lu J, et al. 2013, Experimental investigation on the tensile behavior of polyurea at high strain rates. Materials & Design. 50: 124-129.
[6] Guo H, Guo W, Amirkhizi A V. 2017, Constitutive modeling of the tensile and compressive deformation behavior of polyurea over a wide range of strain rates. Construction and Building Materials. 150: 851-859.
[7] Li J, Hao H. 2014, Numerical study of concrete spall damage to blast loads. International Journal of Impact Engineering. 68: 41-55.
[8] Xu K, Lu Y. 2006, Numerical simulation study of spallation in reinforced concrete plates subjected to blast loading. Computers & Structures. 84(5-6): 431-438.
[9] L.G. Moradi, J.S. Davidson, R.J. Dinan. 2008, Resistance of membrane retrofit concrete masonry walls to lateral pressure. Perform Constr Facil, 22 (3): 131-142.
[10] Aghdamy S, Wu C, Griffith M. 2013, Simulation of Retrofitted Unreinforced Concrete Masonry Unit Walls under Blast Loading. International Journal of Protective Structures. 4(1): 21-44.
[11] Dinan R J, Porter J R, Davidson J S, et al. 2005, Failure Mechanisms of Polymer-Reinforced Concrete Masonry Walls Subjected to Blast. Journal of Structural Engineering. 131(8): 1194-1205.
[12] Davidson J S, Porter J R, Dinan R J, et al. 2004, Explosive Testing of Polymer Retrofit Masonry Walls. Journal of Performance of Constructed Facilities. 18(2): 100-106.
[13] Ruan Z, Chen L, Fang Q. 2015, Numerical investigation into dynamic responses of RC columns subjected for fire and blast. Journal of Loss Prevention in the Process Industries. 34: 10-21.
[14] LS-DYNA. 2015, LS-DYNA KRYWORDS USER'S MANUAL, LS-DYNA R8.0.
[15] Cullis I G. 2001, Blast waves and how they interact with structures. J R Army Med Corps. 147(1): 16-26.
[16] Sitnikova E, Guan Z W, Schleyer G K, et al. 2014, Modelling of perforation failure in fiber metal laminates subjected to high impulsive blast loading. International Journal of Solids and Structures. 51(18): 3135-3146.
[17] Marc Andr Meyer. 1994. Dynamic behaviors of materials[M].
[18] Habel K, Gauvreau P. 2008, Response of ultra-high performance fiber reinforced concrete (UHPFRC) to impact and static loading. Cement and Concrete Composites. 30(10): 938-946.

Proceedings of the 2019 Hypervelocity Impact Symposium
HVIS2019
April 14-19, 2019, Destin, FL, USA
HVIS2019-104

Extended Abstract+
High Velocity Impact Testing for Evaluation of Intermetallic Projectiles

Colton B. Cagle[a], Kevin J. Hill[a], Connor Woodruff[a], Michelle L. Pantoya[a]*,
Joseph Abraham[b], Casey Meakin[b]

[a]*Texas Tech University, Mechanical Engineering Department, Lubbock, TX 79409 USA*
[b]*Karagozian and Case, Inc.,700 N Brand Blvd, Suite 700, Glendale, CA 91203 USA*

Abstract

Experiments were performed to study penetration through multiple aluminum plates followed by impact into an inert steel anvil using a High-velocity Impact-ignition Testing System (HITS). The projectiles are intermetallic pellets launched from a propellant driven gun into a catch chamber equipped with view ports and imaging diagnostics. Penetration, impact and reaction are monitored using high-speed cameras that provide local and macroscopic perspectives of projectile and target interaction as well as overall reactivity. Results demonstrate the range of visual data that can be captured by a non-gas generating intermetallic projectile that fragments and reacts upon penetration and impact. Results show that higher velocity projectiles (~ 1300 and 800 m/s) produce smaller fragments upon target penetration that result in flame spreading through the chamber upon impact while lower velocity projectiles (~ 500 m/s) negligibly fragment upon target penetration and produce no flames even upon anvil impact.

Keywords: Penetration, Fragmentation, Intermetallic Projectile, Flame Spread, Impact Ignition

1. Introduction

Intermetallic compounds consist of two or more metallic elements that are combined in order to exploit their mechanical, chemical, electrical or physical properties toward an application. Some examples of intermetallic reactions include shape memory alloys, corrosion and microbial resistant coatings, or high temperature structural materials. Varma [1] presents an overview of various intermetallic reactions and the properties that make the product alloys useful for various applications. From a chemical perspective, intermetallic reactions are interesting to study because they undergo various stages of reaction such as phase change and sintering, and alloying, followed by reactions dependent upon metalloid or alloy phase, as well as oxidation reactions with the surrounding environment [2]. Most stages of reaction are exothermic to varying degrees and offer the potential to harness chemical energy for localized heating.

Intermetallic projectiles offer the potential to increase energy deposition into a target through exothermic reactions triggered by fragmentation from the penetration or impact event. Many studies have examined intermetallic projectiles that include high density metals such as aluminum (Al) and tungsten (W) [3-7]. Previous research laid the foundation for understanding that monitoring the chamber quasi-static pressure upon penetration and reaction provides little insight into energy deposition because intermetallic reactions are generally not gas generating, so pressure variations are minimal. This study extends previous work with

+ Full paper can found at https://www.sciencedirect.com/journal/international-journal-of-impact-engineering/special-issue/10MTGP5W4VJ

* Corresponding author. Tel.: 806-834-3733.
E-mail address: michelle.pantoya@ttu.edu.

qualitative observations of the penetration and impact event. The objective of this study is to examine the effect of impact velocity on projectile penetration, fragmentation and reaction through two aluminum plates followed by impact onto an inert steel anvil.

2. Experimental

The High-velocity Impact-initiation Testing System (HITS) is a lab-scale diagnostic to examine penetration and impact of projectiles at velocities that can range from 150-1500 m/s. The projectiles penetrate through 0.32 cm thick 6061-T6 aluminum plates before impacting a modular, hardened steel anvil. Tests were designed to allow penetration through two plates separated by about 10 cm. All data captured is visual using two high-speed cameras: a Phantom v710 color camera and a Phantom v2512 monochrome camera, that record both a macroscopic view of the entire catch chamber (at 49,000 fps) and a detailed view of the penetrating plate (at 460,000 fps), respectively. The cameras are positioned perpendicular to the direction of projectile motion and aligned with a side viewing window in the catch chamber. The window is made of 1.5 cm polycarbonate to prevent blowout. A white sheet of paper is placed along the back of the chamber to assist in projectile visualization and flame tracking during post processing. The high-speed cameras enable measurements of fragmentation and flame spreading as well as impact velocity.

3. Results and Discussion

Table 1 summarizes the projectile impact velocities at each target. In comparison, there is less of a reduction in velocity through the first penetration plate than the second, and as the initial projectile velocity increases, the overall velocity losses decrease. Using the velocities reported in Table 1, the relative specific kinetic energy (*KE*) is calculated as ½ multiplied by the velocity (V) squared ($KE = \frac{1}{2}V^2$). The *KE* of the system is measured from the leading edge of the fragments after penetration through the Al plates. Because of this, the energy measurements are only valid for the projectile masses that fully and completely penetrate through the plates.

An inverse trend is seen when comparing the energy lost upon penetration. Energy losses of 38, 121, and 171 *kJ/kg* are observed for the first Al plate at the low, mid-range, and high velocities respectively. Energy losses of 71, 178, and 385 *kJ/kg* are observed for the second Al plate at the low, mid-range, and high velocities respectively. These results indicate that more energy is lost overall at the higher velocity ranges. Comparing the two observed trends, even though more energy is lost during penetration at higher velocities, there is less of a percentage reduction in energy lost. The differences in energy losses are possibly due to different penetration mechanics associated with different impact velocity regimes. For example, the ratio of the wave speed through aluminum to the impact velocity could be critical in restricting damage toward cold-forming the aluminum and limiting plastic deformation. Also, the sound speed through aluminum is estimated as 6320 m/s, such that the ratio of impact velocity to sound speed is 0.076, 0.134, 0.206 for the first plate (using Table 1 impact velocities). At a certain threshold (i.e., > 0.10) the plate may fracture more readily, preserving more of the projectile's initial energy. The increased overall energy loss can be explained due to the increased severity of fragmentation, dissipating larger amounts of energy into the pellet and throughout the chamber rather than into the plate.

Table 1. Impact velocity and specific kinetic energy for various locations throughout the test chamber for both low and high impact velocity cases.

Location of measurement	Impact Velocity (*m/s*)	Kinetic Energy (*kJ/kg*)	Impact Velocity (*m/s*)	Kinetic Energy (*kJ/kg*)	Impact Velocity (*m/s*)	Kinetic Energy (kJ/kg)
First Al plate	481	116	850	361	1300	845
Second Al plate	396	78	693	240	1161	674
Steel Anvil	121	7	352	62	760	289

Figure 1 shows the Al plates from both the front and back perspectives after penetration. There is about a 20 % increase in size when the second Al plate is penetrated for the lower impact velocity case. The size increase could be attributed to many different phenomena. Most likely the reduced velocity and specific kinetic energy upon impact is low enough that the second aluminum plate experiences extreme plastic deformation before fracturing. Figure 1 indicates extreme plastic deformation at lower impact velocity by the hinged piece of aluminum protruding out from behind the second plate. The hinged piece of aluminum is also consistent with theory that there are different penetration mechanics associated with different velocity regimes. At the lower impact velocity regime, most of the kinetic energy may be directed toward cold-forming the aluminum rather than projectile fragmentation. It is also possible that effects such as increased cross-sectional area of the pellet due to plastic deformation after penetration through the first plate might contribute to the size increase. The fragmentation field after the first penetration event is almost non-existent, and any fragmentation produced is extremely fine.

For the mid-range velocity there is about a 34% increase in size after penetration through the second Al plate. The size increase implies that the fragmentation field after the first penetration event widens. Cratering and pitting is observed around the peripheral of the hole in the second Al plate indicating that some debris does not contribute towards the second penetration event.
For the high velocity impact case there is about a 45 % increase in size when the second Al plate is penetrated. Similar phenomena are observed in the mid-range velocity case; however, the high velocity case exhibits more extreme behavior. Figure 1 observations are consistent with the conclusions from Table 1 that show more of the overall kinetic energy of the projectile goes into deforming the Al plate for the lower impact velocity case and there is more visual deformation of the plate in Fig. 1a for the lower impact velocity case. Figure 1b and 1c also suggests that some kinetic energy in Table 1 goes toward fragmenting the projectile for the higher impact velocity case.

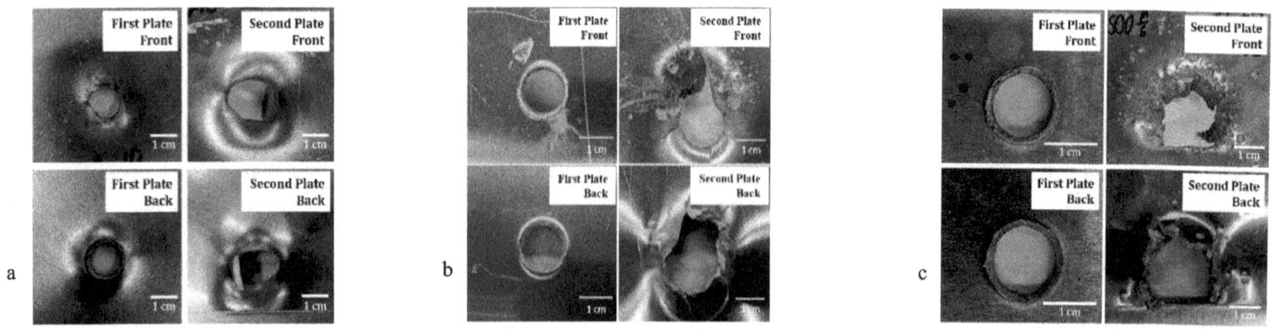

Fig. 1. Photographs of plates (both front and back views) after projectile penetration at initial impact velocities of (a) 500 m/s (b) 850 m/s and (c) 1300 m/s.

4. Conclusions

Flame spreading is observed for intermetallic projectile impact at 1300 and 850 m/s and appears to result from small particle debris within the fragmentation field. The rates of flame spread for the 1300 m/s shot range from 500 down to 115 m/s, corresponding to the first and second penetration events. At higher impact velocities less kinetic energy is consumed through plate fracture and deformation and more energy goes into fragmenting the intermetallic projectile. At the lower impact velocity of 500 m/s impact velocity, no flame spreading is observed, and the debris field appears negligible with the projectile relatively intact even upon penetration of the second aluminum plate. For this lower velocity impact case, the plate appears to be cold-formed with the puncture remaining hinged to the plate. These results show that impact velocity has a strong effect on projectile fragmentation that further influences reactivity as observed through flame spreading. The dramatic differences in plate morphology after penetration indicate impact velocity may induce different penetration mechanics.

Acknowledgements

MP is grateful for partial support from the Office of Naval Research (ONR) and the Army Research Office (ARO). The authors also acknowledge Matsys, Inc. for fabricating all the projectiles and test specimens used in this study.

References

[1] Varma, A., 2000. Form from Fire, Scientific American 283(2), p. 58-61.
[2] Fredrickson, D.C., Miller, G.J., 2018. Intermetallic Chemistry: New Advances in Humanity's Age-Old Exploration of Metals, Acc. Chem. Res. 51(2), p. 213.
[3] Ames RG. Vented Chamber Caloriemtry for Impact-Initiated Energetic Materials. Rd AIAA Aerosp Sci Meet Exhib 2005:1–14.
[4] Ames R. Energy Release Characteristics of Impact-Initiated Energetic Materials. Multifunct Energ Mater MRS Proc Vol 896 2006;896:1–10.
[5] Wang H, Zheng Y, Yu Q, Liu Z, Yu W. Impact-induced initiation and energy release behavior of reactive materials. J Appl Phys 2011.
[6] Zhang XF, Shi AS, Qiao L, Zhang J, Guan ZW. Experimental study on impact-initiated of multifunctional energetic structural materials. J Appl Phys 2013;113.
[7] Lee RJ. Reactive Materials Studies. AIP Conf. Proc., vol. 845, AIP; 2006, p. 169–74.

Proceedings of the 2019 Hypervelocity Impact Symposium
HVIS2019
April 14-19, 2019, Destin, FL, USA
HVIS2019-105

Optimized Impact Mitigation Barriers for Insensitive Munitions Compliance of a 120mm Warhead

Kevin T. Miers*, Daniel L. Prillaman, Nausheen M. Al-Shehab

U.S. Army Combat Capabilities Development Command Armaments Center, Picatinny Arsenal, NJ 07806-5000

Abstract

The U.S. Army Combat Capabilities Development Command (CCDC) Armaments Center at Picatinny Arsenal, NJ is working to develop technologies to mitigate the violent reaction of a 120 mm warhead, loaded with an aluminized HMX-based enhanced blast explosive, when subjected to the NATO Insensitive Munitions (IM) Fragment Impact (FI) test. As per NATO STANAG 4496, FI testing is conducted at 8300±300 ft/s with a 0.563" diameter, L/D~1, 160° conical nosed mild steel fragment. Reaction violence resulting from FI can be mitigated by the use of liners or barriers applied to the munition itself or its packaging, commonly referred to as a Particle Impact Mitigation Sleeves (PIMS). Previous development efforts for this item focused on a lightweight plastic warhead support which was able to reduce the severity of the input shock sufficiently to prevent high order detonation. However, violent sub-detonative responses were still observed which occurred over several hundred microseconds, consumed part of the explosive charge, and ejected hazardous debris over large distances. These responses are driven by rapid combustion coupled with damage to the explosive as well as mechanical confinement. Quantitative modeling of these scenarios is a challenging active research area. Prior experimental results and modeling guidance have shown that mitigation of these reactions requires a more substantial reduction in the overall mechanical insult to the explosive. In particular, steel and aluminum PIMS have been able to efficiently provide the necessary fragment velocity reduction, breakup and dispersion in typical packaging applications. Packaged warheads were tested at the GD-OTS Rock Hill facility with several PIMS designs incorporated into the ammunition containers. Several designs were demonstrated to provide benign reactions with minimal added weight. Future iterations will attempt to further improve the design using advanced lightweight barrier materials.

Keywords: Insensitive Munitions; Fragment Impact; Explosives; Deflagration; Hydrocode Modeling

1. Introduction

All U.S. munitions are required by international agreement to be made Insensitive Munitions (IM) compliant to the extent practical. This entails that the munition in both its operational and logistical configurations must react nonviolently when subjected to a suite of tests intended to simulate hazards commonly encountered on the modern battlefield. These include Fragment Impact (FI), Bullet Impact (BI), Fast Cookoff (FCO), Slow Cookoff (SCO), Sympathetic Reaction (SR), and Shaped Charge Jet Impact (SCJI). Reaction severity is categorized into detonation (Type I), partial detonation (Type II), explosion (Type III), deflagration (Type IV), and burn (Type V) responses. The determination of reaction severity is made based on photographic evidence, witness plate damage, fragment size and throw distance, and blast gauge pressure readings. Currently, work funded by the Joint Program

* Tel.: 973-724-1180
E-mail address: kevin.t.miers.civ@mail.mil

Executive Office for Armaments and Ammunition is being performed to improve the IM response of a 120 mm warhead packaged in its logistical configuration. This paper documents several modeling and experimental results for this configuration subjected to the NATO standard fragment impact test.

2. NATO Fragment Impact Testing

The NATO Standardization Agreement (STANAG) 4496 Ed. 1 [1] describes the NATO standard FI test of interest in this work. It specifies a 287 grain, 0.563" diameter, L/D ~ 1, 160° conical-nosed mild steel cylindrical fragment with a Brinell hardness of less than 270, which impacts the item under test at a velocity of 8300±300 ft/s. Two aim points are specified in STANAG 4496: the center of the largest presented area of explosive, and the most shock-sensitive location. In the U.S., a smooth bore 40 mm powder gun is typically used to propel the fragment, which is mounted in a plastic sabot. Velocity is measured using time-of-arrival gauges and/or by high speed camera. Fig. 1 shows the fragment, sabot, and a typical testing setup [2].

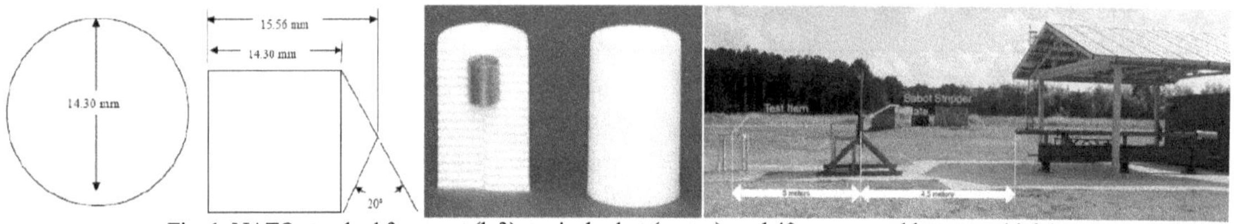

Fig. 1. NATO standard fragment (left), typical sabot (center), and 40 mm smoothbore gun (right)

3. General Phenomenology and Previous Work

The 8300 ft/s NATO FI test is intended to simulate a heavy bomb or artillery fragment. At this impact velocity, extremely strong shock waves and hydrodynamic deformation due to pressures well above material flow strength are commonly observed. The typical strategy for preventing violent response of high explosive warheads under such conditions is to buffer out the initial impact shock, and secondarily to reduce the mechanical insult to the explosive by providing sufficient fragment velocity reduction, breakup and dispersion. To achieve this, Particle Impact Mitigation Sleeve (PIMS) technology is normally tailored to the munition of interest. PIMS are specialized mechanical barriers designed to mitigate the FI threat by addressing the aforementioned physical phenomena. The use of a thin plastic PIMS internal or external to the warhead body has been observed to prevent shock initiation of high order detonation in a variety of munition items [3]. However mitigation of the violent sub-detonative responses which usually still occur typically requires substantially heavier barriers.

The phenomenology of sub-detonative response due to high velocity impact is extremely complicated, and outcomes are challenging to quantitatively predict. These responses are driven by rapid combustion coupled with damage to and fracture of the explosive, conductive and convective heat transfer, as well as mechanical confinement, and can occur over relatively long timescales, e.g., several milliseconds. Typically some of the explosive charge is left unconsumed, large fragments are projected over long distances, and pressures which develop are relatively low. Modeling challenges include (but are not limited to) production of damage and porosity changes, ignition criteria, reaction rates, flame propagation in damaged reactant, equations of state and constitutive models for the solid explosive reactant and gaseous products (as well as mixtures of both), interaction with confinement, and possible transition to detonation, shown in Fig. 2 [4, 5]. All of these phenomena need to be experimentally quantified in a consistent theoretical framework which must have a robust numerical implementation that can be used for practical length and time scales, and this remains a challenging active research area. Some material models which have been developed for this purpose include Coupled Damage and Reaction with Kinetics (CDAR-K) [5] and High Explosive Response to Mechanical Stimulus (HERMES) [6]. In any case, no such models have been parameterized for the explosive of interest or anything similar, to the authors' knowledge. Additionally, other modeling challenges such as prediction of dynamic fracture for the inert materials might limit the utility of such an explosive model in many applications. As a result, while computer modeling can currently provide some useful qualitative and quantitative information, testing is still relied upon for IM design.

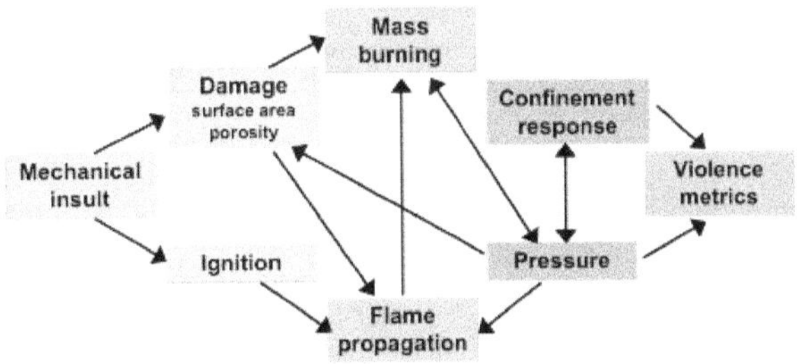

Fig. 2. Sub-detonative response phenomenology (from [4, 5])

Much work has been performed in the past to mitigate the 8300 ft/s NATO FI threat for a variety of larger caliber explosive warheads. Several experimental programs have found PIMS consisting of steel and/or aluminum layers several millimeters thick to provide the necessary fragment velocity reduction and dispersion to pass the FI test criteria, although this obviously varies to some extent with the munition case and explosive fill. Examples include recent IM programs for the TOW2B missile [7, 8] and the M72 LAW [9]. Testing of the 8300 ft/s NATO fragment against inert flat plates of these materials has also been performed. A recent CCDC Armaments Center experimental program [10, 11] obtained residual velocities for FI against a variety of steel plates using flash radiography. Some results from this study are shown in Fig. 3. This work demonstrated that hardness of steel plates at the thicknesses of interest had a minor effect on the residual velocity but a significant effect on the fragment debris mass distribution. Similarly, high speed video (HSV) has been used to obtain residual velocities for a variety of protection schemes under the aforementioned TOW2B program [12]. These tests have shown that successful designs provide velocity reductions of 1000-2000 ft/s and result in significant fragment breakup, which is clearly an essential feature of the problem. Other more exotic materials have been able to achieve the same velocity reduction for somewhat less areal density, e.g., [7-8, 12]. However these may not make good packaging materials, which must generally be inexpensive and endure extreme environmental and rough handling conditions. Additionally, the extent to which the PIMS thermally insulates the warhead is an important consideration from the standpoint of mitigating cookoff response.

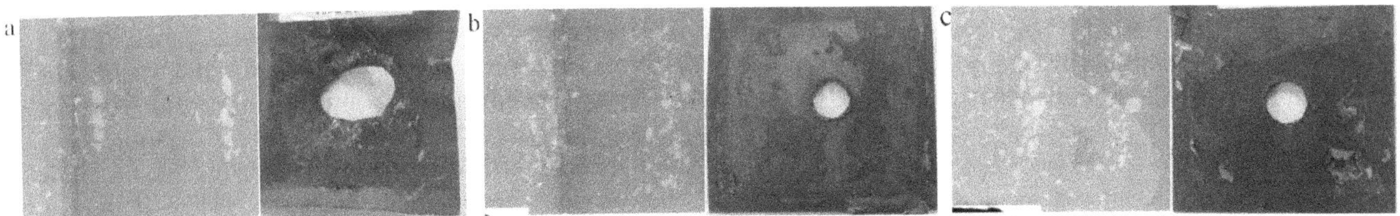

Fig. 3. 8300 ft/s NATO FI test radiographs for (a) ~0.04" thick steel ammunition container (velocity reduction of 500 ft/s); (b) 0.118" thick steel plate (velocity reduction of 950-1200 ft/s); (c) 0.236" thick steel plate (velocity reduction of 1975 ft/s)

4. Item and Configuration Description

The configuration of interest in this work, shown in Fig. 4, is a 120 mm warhead packaged in a ~0.04" thick sheet steel ammunition container. The steel warhead was designed by the CCDC Armaments Center and is loaded with ~4.5 lb of an aluminized HMX-based enhanced blast explosive (65% HMX, 20% Al, 15% binder). Its wall thickness is approximately 0.4", and it is mounted in a 0.35" thick high density polyethylene (HDPE) warhead support which slides over a recess on the warhead body, introducing a small air gap. In previous work, this configuration was demonstrated to prevent high order detonation of the warhead when subjected to FI. However, it was not sufficient to2 address the violent sub-detonative responses which still occurred. It should be noted that the container is expected to break up the incoming fragment (see Fig. 3a). This test series experimentally evaluated several PIMS designs which were downselected with the aid of high rate continuum modeling, past experience and engineering judgment. Approximately 5/8" of space is available between the OD of the warhead support and the ID of the container to incorporate PIMS. These could be attached to the container, the warhead support, or both to achieve a multi-layer design. Cylindrical steel nose and aft simulants were threaded onto the warhead. The PIMS and HDPE support were placed over the warhead, and the entire assembly was centered in the container by laser cut acrylic discs, shown in Fig. 5.

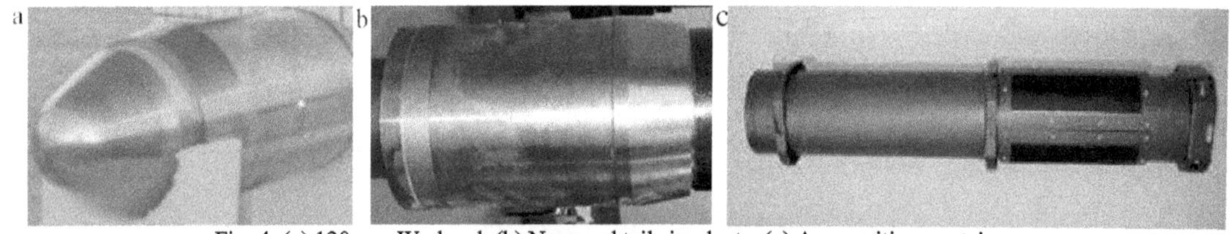

Fig. 4. (a) 120 mm Warhead; (b) Nose and tail simulants; (c) Ammunition container

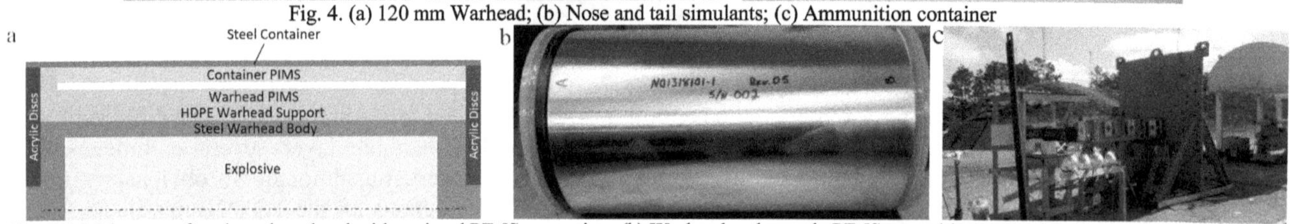

Fig. 5. (a) Diagram of packaged warhead with notional PIMS protection; (b) Warhead and sample PIMS assembly prepared for insertion; (c) Test setup

5. High Rate Continuum Modeling

To investigate what kinds of protection designs might be required to fully mitigate FI, several calculations were performed in the multi-material Arbitrary Lagrangian-Eulerian (ALE) hydrocode ALE3D [13]. The ALE formulation allows for severe distortions to arise, while better preserving material interfaces and retaining some of the natural zoning adaptivity a pure Lagrangian calculation would provide. ALE3D advances the solution in time with a Lagrangian step followed by a remap/advection step. A modified equipotential mesh relaxation scheme is used to dynamically move mesh into regions of interest. A second order monotonic advection algorithm is utilized. Standard artificial viscosity is used to spread shock fronts over several zones. The Mie-Gruneisen equation of state (EOS) was used for all of the solid materials. Standard Steinberg-Guinan [14] forms for the flow stress as a function of equivalent plastic strain, strain rate, and temperature (defined using the zero-Kelvin isotherm) with a von Mises yield surface were utilized. Spall failure was approximated using a maximum tensile hydrostatic stress criterion. Failed material is unable to withstand shear or hydrostatic tensile stress, and voids are inserted upon failed material reaching an excessive relative volume. The explosive was modeled as an inert Mie-Gruneisen solid using the unreacted Hugoniot, and was assumed to be elastic perfectly plastic with a small amount of strength; no attempt was made to model its energetic response.

The ballistic limit of the warhead body has been suggested as an approximate threshold for violent sub-detonative munition response in heavily confined charges [15], although such a criterion is clearly independent of the particular explosive. It has also been observed for several common explosives that the transition from essentially no reaction to violent explosion occurs over a fairly narrow range of impact velocities, e.g., less than ~650 ft/s [16]. As such the plan was to computationally identify designs which resulted in relatively minor deformation of the warhead body as an upper bound on the required protection, then hone in on the lightest successful configuration experimentally. Based on experience and logistical considerations, steel and aluminum PIMS were investigated under this effort. Warhead and container PIMS configurations were modeled, and some sample calculations are shown in Fig. 6. The calculations indicated that the HDPE warhead support, while not providing a significant amount of penetration resistance, would still easily buffer out the impact shock from the incoming fragment debris. It also appeared that a container PIMS provided noticeably less penetration of the warhead body than a warhead PIMS for the same thickness. Clearly the code does not explicitly model fragment breakup, but the distances in question are so short that the effect is well approximated by the loss of strength due to extensive spall failure. It appeared that a 0.25" thick steel container PIMS would likely be an upper bound on the thickness required to prevent severe penetration of the warhead body. A 0.375" thick aluminum warhead PIMS also appeared viable, with a substantial weight reduction compared to the steel design. As a result PIMS in several appropriate increments up to these thicknesses were downselected for testing.

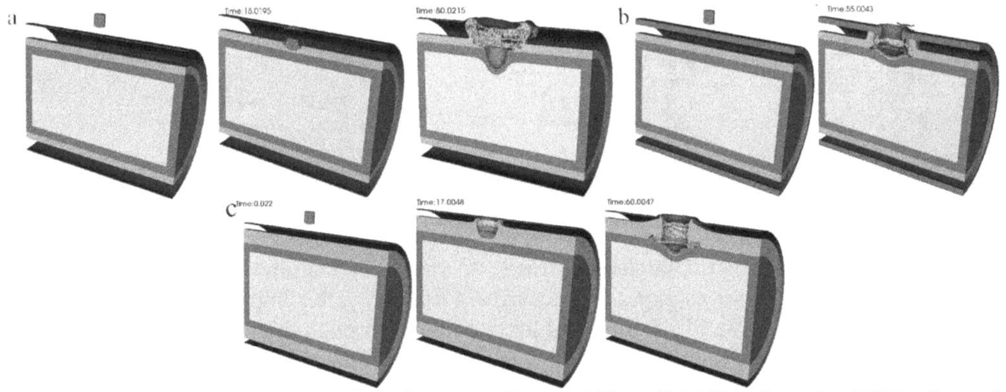

Fig. 6. Sample ALE3D FI calculations for (a) baseline configuration at 0 μs, 15 μs, and 80 μs; (b) 0.25" steel container PIMS at 0 μs and 55 μs; (c) 0.375" aluminum warhead PIMS at 0 μs, 17 μs, and 60 μs

6. Experimental Methodology and Results

FI testing was performed at the General Dynamics Ordnance and Tactical Systems (GD-OTS) Rock Hill test facility. Ten warheads were available for this test series. The warheads were to be tested in a configuration most closely representing a potential fielded configuration: in thin steel ammunition containers, inside a plastic warhead support, with full cylindrical tube PIMS applied to either the outside of the warhead support or the inside of the container providing full 360° protection (see Fig. 5). This ensures that the confinement from the PIMS and ammunition container is taken into account in the test results. Because it was not practical to test against the entire munition, the warheads and the rest of the hardware had to be mounted concentrically within the containers, which were not modified. The PIMS can be split into arcs if more practical from a manufacturing standpoint, but the test results from the tubes should be conservative as the strength of the metal will add confinement. To conserve test assets, a significant amount of extra hardware for different PIMS designs was obtained to provide flexibility in terms of testing configurations which it seemed logical to evaulate, or repeating a configuration for any no-tests. Several thicknesses of 4140 steel and 6061-T6 aluminum warhead and container PIMS were obtained, as these materials were readily available in the geometries of interest. Additionally, it was hypothesized that strength of the PIMS might appreciably affect fragment breakup, as well as penetration since full deceleration of the fragment debris is desired. To evaluate whether this would provide any improvements, some of the steel PIMS were heat treated to as high of a hardness as was practical. In addition, a limited quantity of 7075-T6 aluminum PIMS were fabricated. The hardware selection turned out to be very good, with most of the PIMS being used and hardware availability only slightly constraining testing decisions. Additionally, very good aimpoint accuracy and impact conditions [2] were achieved throughout this test series, resulting in consistent and valid tests which allowed for optimization of the design to progress quite satisfactorily. Accordingly, this data is tabulated in Table 1, along with peak blast overpressures (BOP) which were comparable to those produced by the gun, even for violent sub-detonative responses. Comparisons of modeling predictions with the results are also shown for several of the configurations.

A baseline shot (test number RT18592) was conducted first, shown in Fig. 7. A violent deflagration was observed. This reaction launched the aft simulant into the berm surrounding the test arena, and it was found buried several feet into the soil. The warhead body broke up into several large pieces, most of which were recovered. The remaining debris was likely launched outside the test facility, at least several hundred yards. During several test series with this item where such reactions were observed, isolated chunks of explosive were found at such distances and further. Gouging was also noted on the witness plate, which occurred for all of the deflagrations.

Fig. 7. Baseline (RT18592) (a) HSV at 310 μs, 895 μs, 1517 μs from impact; (b) Recovered debris; (c) Witness plate gouging

The next test (RT18591) was a 0.125" steel container PIMS. A violent deflagration was observed, shown in Fig. 8. Interestingly, little reaction was observed until an explosion occurred some 2ms after impact. What appeared to be a plug of warhead body material was also recovered. The PIMS fractured longitudinally through the impact point and was blown through the container in one piece, straightened into a flat plate and thrown ~20 yards. Thus one disadvantage of using a metal PIMS is that if it is overmatched, it can itself become hazardous debris, although one large hazardous fragment is always better than several. In IM testing, fragments over 0.66 lb found past 50 feet are generally considered hazardous [17]. The next test (RT18593) was a 0.1875" steel container PIMS. The packaged warhead was knocked off the test stand onto the ground. The warhead was pulled from the packaging intact with a large indent in the warhead body that contained some of the residual fragment, shown in Fig. 9.

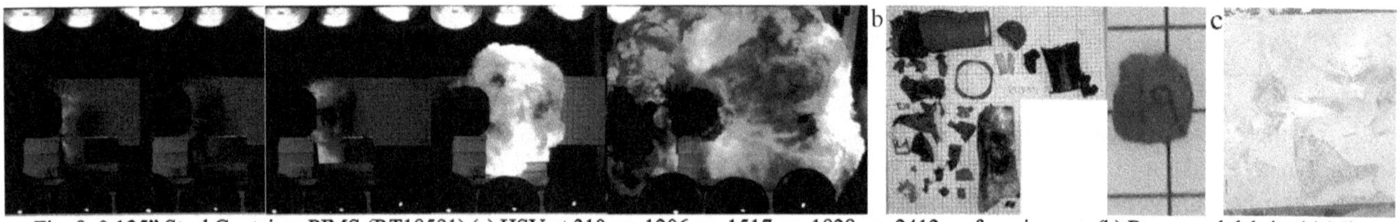

Fig. 8. 0.125" Steel Container PIMS (RT18591) (a) HSV at 310 μs, 1206 μs, 1517 μs, 1828 μs, 2412 μs from impact; (b) Recovered debris; (c) Minor witness plate gouging

Fig. 9. 0.1875" Steel Container PIMS (RT18593) (a) HSV at 310 μs, 2723 μs from impact; (b) Recovered debris, intact warhead, model comparisons

At this point a mitigation solution was found, eliminating a fraction of the available hardware from consideration. As a result all of the following tests attempted to further reduce the weight per unit length of the PIMS with the remaining hardware, in light of currently available test data. Since a hardened steel 0.125" container PIMS was not available, it was decided to test a 0.1875" steel warhead PIMS (RT18594). A violent deflagration was observed, shown in Fig. 10. Again, the PIMS fractured longitudinally and straightened into a flat plate, which was found ~100 yards away. The warhead body split into several pieces, all of which were recovered. The next design tested was a 0.25" 6061 aluminum container PIMS (RT18595). This was much lighter than the steel designs, but significant penetration of the explosive charge was observed in the models. A deflagration was observed, shown in Fig. 11. The PIMS fractured longitudinally along the impact point and peeled open into a flat plate, which was thrown ~30 yards. The warhead body was split into two halves, both of which were recovered ~50 yards from the test stand. The next configuration tested was a 0.1875" thick hardened heat treated steel container PIMS (RT18597), which was intended to investigate whether the choice of steel for a practical fielded design would result in a significantly different outcome. The warhead was pulled from the packaging intact with a large dent in the warhead body that contained some of the residual fragment, shown in Fig. 12.

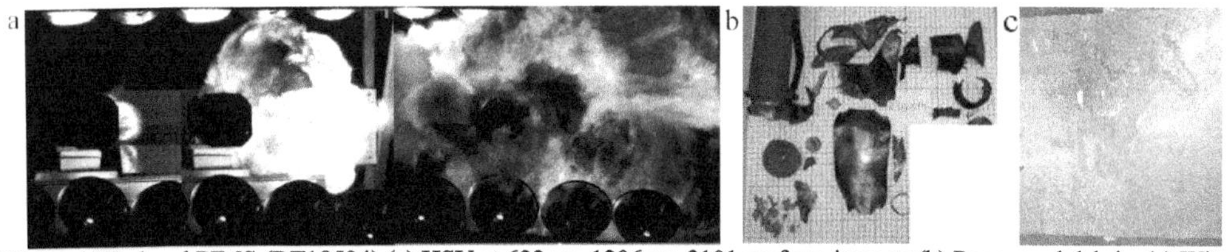

Fig. 10. 0.1875" Steel Warhead PIMS (RT18594) (a) HSV at 622 μs, 1206 μs, 2101 μs from impact; (b) Recovered debris; (c) Witness plate gouging

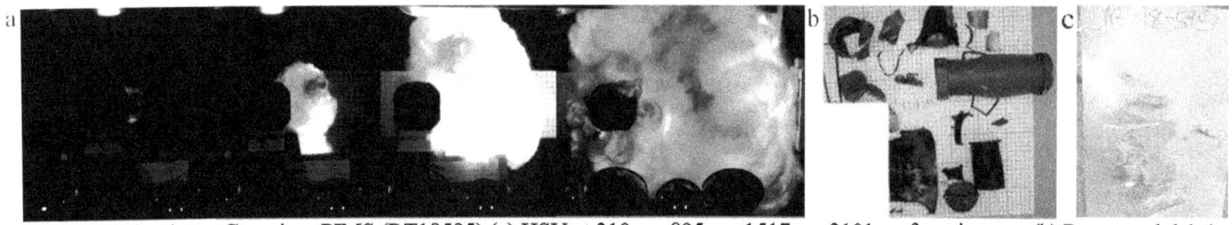

Fig. 11. 0.25" 6061 Aluminum Container PIMS (RT18595) (a) HSV at 310 μs, 895 μs, 1517 μs, 2101 μs from impact; (b) Recovered debris; (c) Witness plate gouging

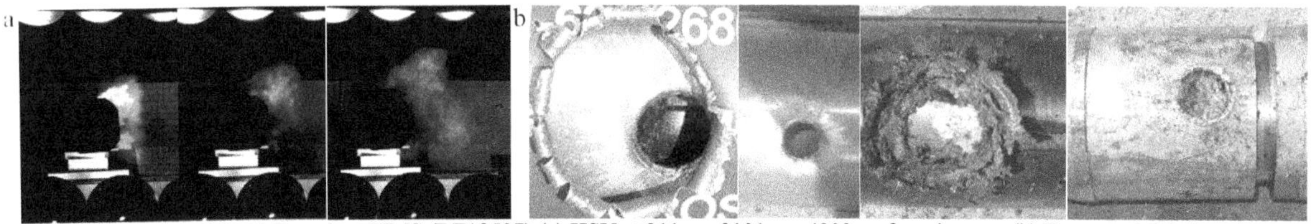

Fig. 12. 0.1875" Hardened Steel Container PIMS (RT18597) (a) HSV at 311 μs, 2101 μs, 4202 μs from impact; (b) Recovered debris and intact warhead

The next configuration tested was a spaced plate design: a 0.125" 7075 aluminum warhead PIMS inside of a 0.125" steel container PIMS (RT18598). This produced about as much of a weight reduction as using the 0.1875" steel warhead PIMS. The warhead was pulled from the packaging intact with a large dent in the warhead body that contained some of the residual fragment, shown in Fig. 13. The next configuration tested (RT18599) was a spaced plate design: a 0.25" 7075 aluminum warhead PIMS inside of a 0.125" 6061 aluminum container PIMS. This would provide a higher areal density than the 0.25" aluminum container PIMS, and a substantially higher flow strength for the aluminum. The warhead was pulled from the packaging intact with a smaller dent in the warhead body that contained some of the residual fragment, shown in Fig. 14. The observed deformation compared well with modeling predictions. This spaced aluminum PIMS design produced a 33% weight reduction compared to the 0.1875" steel container PIMS demonstrated to produce a benign reaction. The next configuration tested was a 0.25" 6061 aluminum warhead PIMS inside a 0.125" 6061 aluminum container PIMS (RT18620). This was done to determine whether the grade of aluminum made a difference, and also since the 0.25" 6061 aluminum container PIMS was unsuccessful. A completely benign reaction was observed, shown in Fig. 15. Unfortunately the warhead could not safely be removed for visual inspection.

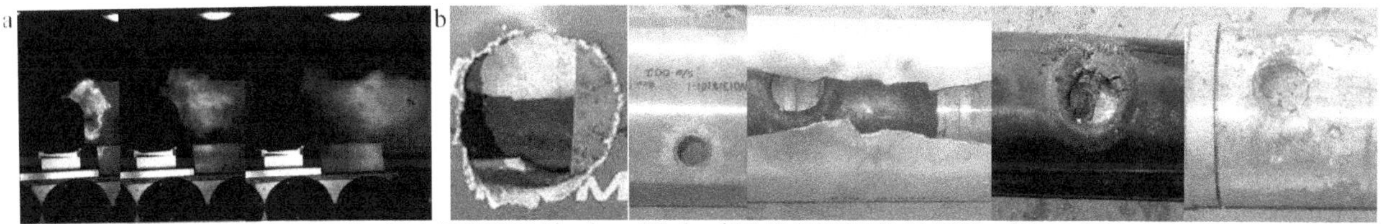

Fig. 13. 0.125" Steel Container PIMS with 0.125" 7075 Aluminum Warhead PIMS (RT18598) (a) HSV at 622 μs, 2101 μs, 4202 μs from impact; (b) Recovered debris and intact warhead

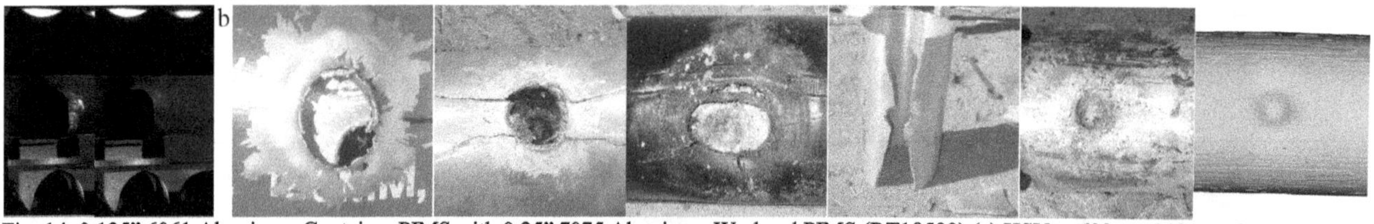

Fig. 14. 0.125" 6061 Aluminum Container PIMS with 0.25" 7075 Aluminum Warhead PIMS (RT18599) (a) HSV at 622 μs, 2101 μs from impact; (b) Recovered debris and intact warhead, modeling comparison

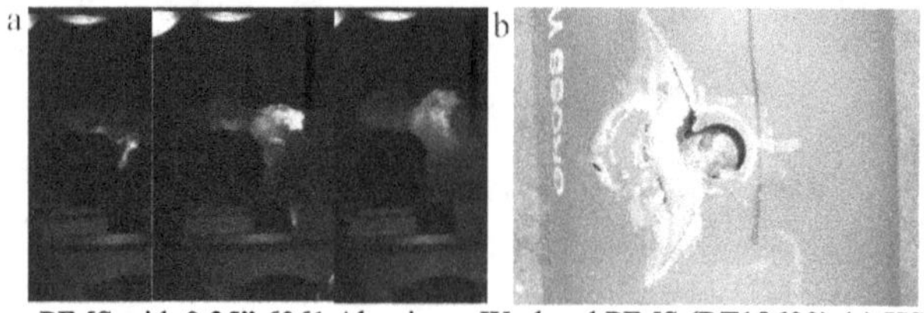

Fig. 15. 0.125" 6061 Aluminum Container PIMS with 0.25" 6061 Aluminum Warhead PIMS (RT18620) (a) HSV at 622 μs, 1206 μs, 2101 μs from impact; (b) Recovered debris

The final test was a 0.125" hardened steel warhead PIMS (RT18621). This would achieve another 5% weight reduction over the previous best design, and would confine the warhead less. Unfortunately a violent deflagration was observed. Both halves of the warhead, as well as the PIMS, were recovered ~100 yards away. The PIMS still fractured longitudinally through the impact point. This concluded the test series; all results are listed in Table 1.

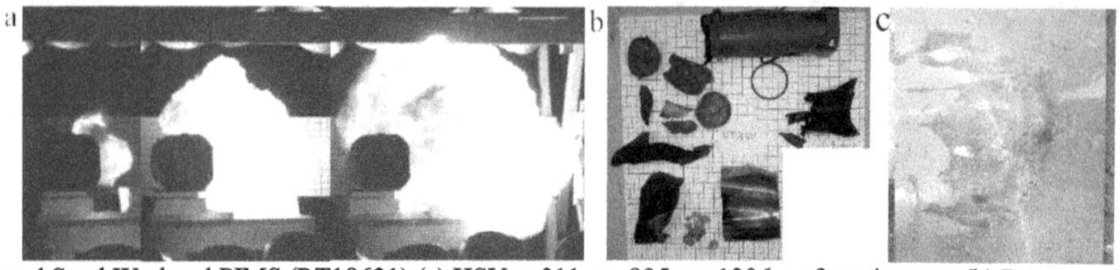

Fig. 16. 0.125" Hardened Steel Warhead PIMS (RT18621) (a) HSV at 311 μs, 895 μs, 1206 μs from impact; (b) Recovered debris; (c) Witness plate gouging

Table 1: Summary of Results

Serial No.	Container PIMS Material	Container PIMS Thickness (in)	Warhead PIMS Material	Warhead PIMS Thickness (in)	PIMS Weight per Length (lb/in)	Velocity (fps)	BOP at 10ft Peak 1 (psi)	BOP at 10 ft Peak 2 (psi)	Witness Gouge	Type
RT18591	Steel	0.125	---	---	0.729	8391	3.9	3	Y	III
RT18592	---	---	---	---	---	8253	---	1.9 at 20ft	Y	III
RT18593	Steel	0.1875	---	---	1.072	8345	1.75	---	N	V
RT18594	---	---	Steel	0.1875	0.959	8219	2.3	1.35	Y	IV
RT18595	6061 Al	0.25	---	---	0.490	8273	2.8	1.75	Y	IV
RT18597	Hardened Steel	0.1875	---	---	1.074	8287	2.2	---	N	V
RT18598	Steel	0.125	7075 Aluminum	0.125	0.956	8275	1.75	---	N	V
RT18599	6061 Aluminum	0.125	7075 Aluminum	0.25	0.716	8416	2.25	---	N	V
RT18620	6061 Aluminum	0.125	6061 Aluminum	0.25	0.700	8511	1.5	---	N	V
RT18621	---	---	Hardened Steel	0.125	0.628	8463	2.9	2	Y	IV

7. Summary and Conclusions

Several FI mitigation schemes for a 120 mm warhead were modeled and tested. Steel and aluminum PIMS were chosen based on past experience and engineering judgment. High rate continuum modeling was utilized to qualitatively identify designs which produced minimal damage to the warhead body. This provided an upper bound on the required protection, as the ballistic limit of the warhead body has been suggested as an approximate threshold for violent sub-detonative response. A variety of warhead and container PIMS were fabricated from steel at two different hardnesses as well as two grades of aluminum. A viable solution was found very early on in the test series, allowing for progressive reduction in PIMS weight over the remaining tests while maintaining a Type V reaction. It was observed that a steel PIMS of a given thickness was significantly more effective when applied to the container rather than to the warhead support. It also appeared that impacts sufficient to perforate the warhead body always resulted in ignition of the explosive charge and violent deflagration shortly thereafter, and that the transition from violent to almost no reaction occurred over a very narrow range of mechanical insults. This observation might be used to further optimize the design computationally. Overall, an aluminum spaced plate design with a 0.25" warhead PIMS inside a 0.125" container PIMS was identified to fully mitigate FI with the least amount of weight per unit length. This is a ~35% improvement over the thinnest steel PIMS found to work, which might be expected since the aluminum design better utilizes the available space in the container. The steel hardness or aluminum grade did not appear to have an appreciable effect on test outcomes. Future work will cover integration of these designs into the packaging for production purposes, as well as attempts to further reduce the weight using different materials.

Acknowledgements

This work was supported by the Joint Program Executive Office for Armaments and Ammunition. The authors would also like to thank Sean Glass, Karen Huddleston, and Todd Gordon of GD-OTS for procuring the PIMS hardware and arranging and conducting these experiments. Dr. V. M. Gold of the CCDC Armaments Center is thanked for reviewing the manuscript and providing helpful comments.

References

[1] Fragment Impact, Munitions Test Procedure, STANAG 4496 Ed. 1, NATO Standardization Agency, Brussels, Belgium, 2006.

[2] Baker, E. L., Al-Shehab, N., Miers, K. T., Pudlak, D. J. "Insensitive Munitions Fragment Impact Gun Testing Technology Challenges". Propellants, Explosives, Pyrotechnics **41**, 572-579 (2016).

[3] Daniels, A., Pham, J., Ng, K., Pfau, D. "Development of Particle Impact Mitigation Sleeves to Reduced IM Response". 2007 Insensitive Munitions & Energetic Materials Technology Symposium, October 15-17, Miami, FL, USA, 2007.

[4] Reaugh, J. E. "Progress in Model Development to Quantify High Explosive Violent Response (HEVR) to Mechanical Insult". LLNL-TR-405903 (2008).

[5] Hertel, E. S., Springer, H. K., Milne, A. "Multiphase and Reactive Flow Modeling". SAND2008-6386C, US UK PA Meeting, Huntsville, AL (2008).

[6] Reaugh, J. E. "HERMES: A Model to Describe Deformation, Burning, Explosion, and Detonation". LLNL-TR-516119 (2011).

[7] Peterson, N. R., Sweitzer, J. C. "Composite Material Particle Impact Mitigation Sleeve Testing". Procedia Engineering **103** (2015) 475-481.

[8] Sweitzer, J. C., Peterson, N. R. "Method for Prediction of Fragment Impact Response Using Physics Based Modeling and Statistical Analysis". Procedia Engineering **103** (2015) 601-609.

[9] Pudlak, D. J. "Development & Demonstration of a Lightweight Particle Impact Mitigation Sleeve (PIMS) with Specified Hardness and Perforation Features". 2018 Insensitive Munitions and Energetic Materials Technology Symposium, Portland, OR, USA, April 2018.

[10] Pudlak, D. J., Al-Shehab, N. M., Miers, K. T., Daniels, A. S., Chang, D. M., Baker, E. L. "Fragment Breakup Investigation Analysis of X-Ray and Computed Tomography Data". 29th JANNAF Propulsion Systems Hazards Joint Subcommittee Meeting, May 2016.

[11] Pudlak, D. J. "Characterization of Fragment Break-up as a Result of Ballistic Impact with Munitions and Associated Materiel". 2016 Insensitive Munitions and Energetic Materials Technology Symposium, Nashville, TN, USA, September 2016.

[12] Peterson, N. R., Al-Shehab, N. M., Sweitzer, J. C., Miller, J. F., Wortman, C. A. "High-Velocity Impact Fragmentation of Projectiles Experimental Results". AMRDEC Technical Report RDMR-WD-16-47 (2016).

[13] Nichols, A. L. III, Editor. "ALE3D User's Manual". LLNL-SM-681737, Lawrence Livermore National Laboratory, November 2015.

[14] Steinberg, D. J. "Equation of State and Strength Properties of Selected Materials". UCRL-MA-106439, Lawrence Livermore National Laboratory, February 1996.

[15] Howe, P. M., Watson, J. L., Frey, R. B. "The Response of Confined Explosive Charges to Fragment Attack". Seventh Symposium (International) on Detonation, June 1981, pg. 1048-1054.

[16] Frey, R., Melani, G., Chawla, M., Trimble, J. "Initiation of Violent Reaction by Projectile Impact". Sixth Symposium (International) on Detonation, August 1976, pg. 325-334.

[17] Miers, K. T., Pudlak, D. J., Fuchs, B. E. "Ballistic Trajectory Modeling for the Insensitive Munitions Type IV/V Hazardous Fragment Threshold". 2018 International Explosives Safety Symposium and Exposition, San Diego, CA, USA, August 2018.

Proceedings of the 2019 Hypervelocity Impact Symposium
HVIS2019
April 14-19, 2019, Destin, FL, USA
HVIS2019-108

Experiment guided simulation of multi-fragment impact into PBXs

Andrew M. Schmalzer[a]*, John D. Yeager[a], Patrick R. Bowden[a], Daniel R. Guildenbecher[b], and Joseph D. Olles[b]

[a]*Los Alamos National Laboratory, P.O. Box 1663, Los Alamos, NM 87545, USA*
[b]*Sandia National Laboratory, P.O. Box 5800, Albuquerque, NM 87185, USA*

Abstract

Multi-fragment impact of energetic materials can provide the impetus initiation and growth to detonation when shockwaves from these discrete fragments collide. The Sandia hydrocode CTH is used with reactive burn modeling to identify relationships between spherical fragment separation distances, variable fragment arrival timing, and initiability in energetic materials. This work demonstrates that detonation is most likely to occur is when multiple fragments collide with a surface simultaneously, because of the cumulative pressure rise of two equal colliding waves compared to the colliding waves generated by fragment impacts offset in time.

Keywords: Energetic Material Initiation; Hydrodynamic Simulation; Multiple Fragment Impact; Reactive Burn Modeling

Nomenclature

D_F	Fragment diameter, (mm)
D_S	Fragment separation distance, (mm)
P_i	Pressure at the leading edge of the shockwave for the 1st (P_1) and 2nd fragments (P_1).
D_O	Offset for the second fragment.
V_F	Fragment Velocity, (mm/μs)

1. Introduction

In the assembly and disassembly of munitions, metal encapsulated detonators are placed in contact with or removed from bare explosive charges. This creates the potential for a scenario where unintentional detonator functioning could lead to multiple small fragments hurtling towards exposed energetic material (EM). Insult of EM by highspeed impact has long been an interest to the explosives community to help establish safety metrics and procedures for worker safety and accident prevention. Both experimental work and simulation of explosive initiation by high speed impact have been investigated to determine what might cause a prompt shock to detonation transition (SDT) [1-3]. However, these efforts have looked at simple systems that only involve one insult at a time.

This present work studies a more complex system where multiple highspeed fragments that are below the threshold for initiating a PBX on their own, whether by size, shape, or velocity can synergistically provide enough pressure for SDT to occur. To this

* Corresponding author. Tel.: +1-505-665-6918.
E-mail address: aschmalz@lanl.gov.

extent, a variety of generic detonator experiments previously studied [4] are coupled to several PBXs to provide data to compare against simulation.

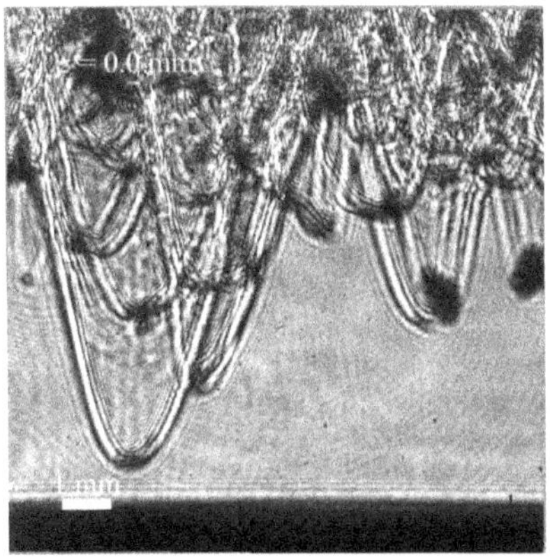

Fig. 1. Holographic image of detonator fragments from a generic detonator [4].

When a spherical particle impacts an explosive surface normally, a nearly spherical compression wave begins to propagate through the target body. When the pressure rise resulting from such compression waves is high enough, and with a long enough duration, detonation will occur. Previous studies on shock-generated impact phenomena have noted some of the complexities relating to solutions for single particle impacts into semi-infinite media, but these are limited to assumptions such as neglecting material strength and assuming that the size and speed of the spherical impactor generates an inviscid flow that can be approximated by an ideal gas [5,6]. While these assumptions may be valid or at least instructive for certain scenarios, they are limited in applicability and do not work out the details for colliding waves.

2. Simulation approach

2.1. Modeling with CTH and the HVRB model

Simulation of explosive initiation via multi-fragment impact is accomplished here by using the Sandia hydrocode CTH (**C**hart-D to the **T**hree **H**alves) to demonstrate the effects of Mach stem interactions from multiple fragments. The fragment material properties are provided by the Mie-Grueniesen equation of state (EOS) for 1100-0 series aluminum, while the EM properties are provided by the HVRB (History-Variable Reactive Burn) model for the plastic-bonded explosive PBX 9501 [7]. The HVRB model provides the impetus for initiation, and is coupled to the PBX 9501 Mie-Gruieneisen parameters for unreacted material or the SESAME tabular EOS for the detonation products for the reactants [8,9].

2.2. Simultaneous fragment impact

When multiple spherical fragments collide with a surface, the resulting compression waves from each fragment collide and produce "Mach stems", or regions where the pressure is greatly amplified. A simple system containing two fragments of equal sizes and velocities impacting a semi-infinite explosive charge simultaneously provides the first step in understanding this pressure amplification (see Fig. 2. below). Here, the fragments were chosen to have a Fragment Diameter, D_F, of 2mm and velocity, V_F, of 3 mm/μs, which is a condition that each fragment imparts an energy insufficient to initiate the explosive alone. If the separation distance, D_s, between the particles is >2.8 mm, the resulting pressure rise is not significant enough to lead to detonation [see Fig. 3. a)]; however, when the separation distance is reduced below 2.8 mm the Mach stem pressure rise is enough to initiate the EM [see Fig. 3. b)].

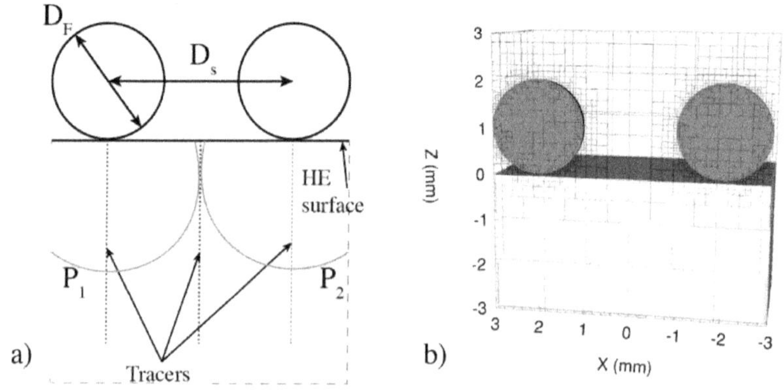

Fig. 2. a) a schematic of simultaneous multi-fragment impact resulting in colliding shockwaves is drawn. D_F represents the fragment diameter, D_s is the separation distance and P_1 and P_2, being equal in this case, represent the pressure of the shockwave's leading edge generated by the impact of the spherical fragments. The black and blue dashed lines represent where the tracers for collecting data are located for this problem. b) The initial condition for the materials in the multi-fragment impact scenario is shown with the mesh defined in CTH.

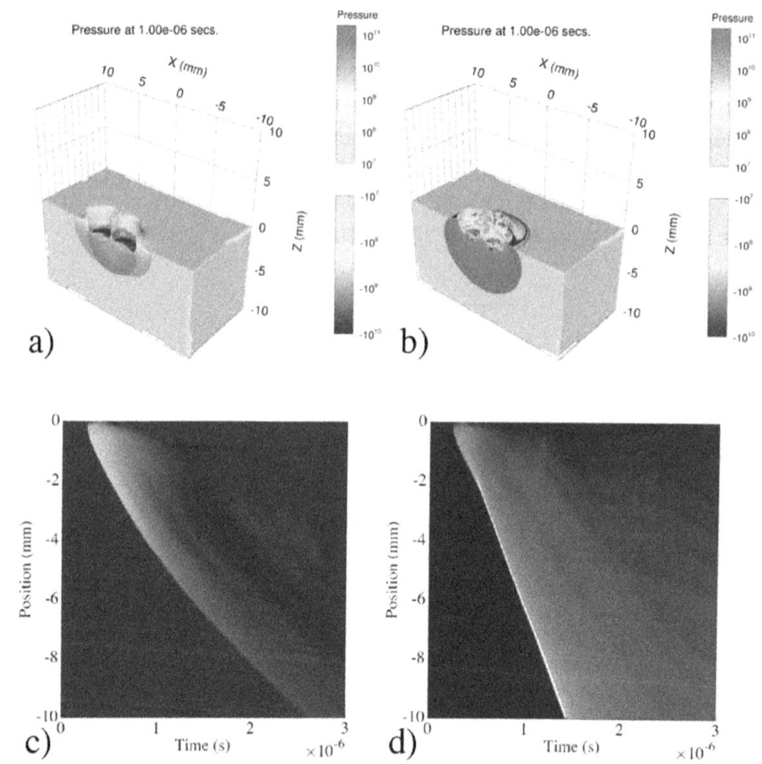

Fig. 3. Material pressure and pressure streak of colliding shockwaves in a non-detonating case [a) and c)], and in a detonating case [b) and d)]. The pressure streaks are generated from data taken at the tracer locations corresponding to the blue dashed line in Fig. 1.

In the case of simultaneous fragment impact, the pressure where the compression waves first meet follows a linear trend, where the maximum pressure recorded at the interface of the colliding waves is ~4.16x the arriving pressure. When the combined pressure of the colliding waves exceeds 22 GPa, the pressure rise results in a turnover to detonation as shown in Figure 3 above, where the pressure oscillates between 48 and 59 GPa for separation distances less than 2.8 mm. Cases where detonation occurs can be identified by several different presentation methods. Shown in Figure 3 a) and b), the graphics output from CTH demonstrates distinct differences in the pressure fields for cases that turn over to detonation. Figure 3 a) and c) shows a case where the compression waves impact but dissipate over time, however Figure 3 b) and d) shows a transition to detonation. Figure 3 c) and d) can be thought of as a streak image of the pressure along the tracers directly between the impactors. In the non-initiating case, the pressure wave begins to slow, and the pressure decays toward ambient conditions. In the initiating case, the streak image shows a maximum in pressure that proceeds through the material with a slope of 8.85 mm/μs. This velocity matches the known detonation velocity for PBX 9501, demonstrating that the simulation is self-consistent. The relationship between the arrival pressure, and the subsequent combined pressure is also given in Figure 4 below.

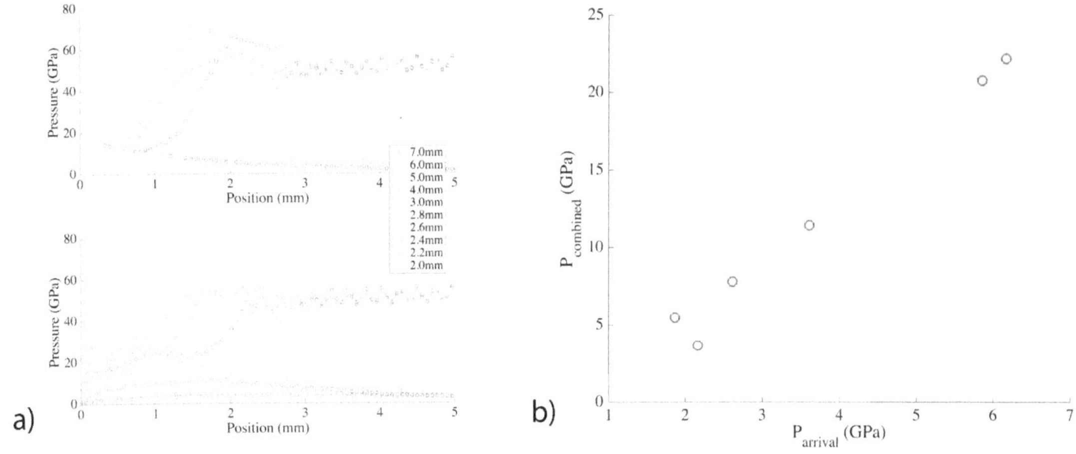

Fig. 4. The maximum pressure recorded at each tracer during simulations of simultaneous impacts at varying separation distances in line with each fragment (top left), and along the center-line of the two fragments (bottom left). For non-initiating cases, the resulting pressure from symmetric colliding waves is plotted against the pressure of the wave immediately before they contact.

2.3. *Asynchronous fragment impact*

Experimental results (e.g. Figure 1.) show that the arrival time of fragments may not be simultaneous, but in fact occurs over the span of several microseconds. As such, a simulation was designed to model a scenario where the spheres impact an EM surface asynchronously. Using timing data for the position of the shockwave from the previous set of simulations, and with the assumption that the wave expands nearly spherically (except in close proximity to the EM surface), the fragments can be given initial conditions such that a lagging compression wave (P_2 in Figure 5) from a late-time impactor can arrive and collide with a leading compression wave (P_1 in Figure 5) spatially centred below the late-time impactor, at a predetermined depth into the material.

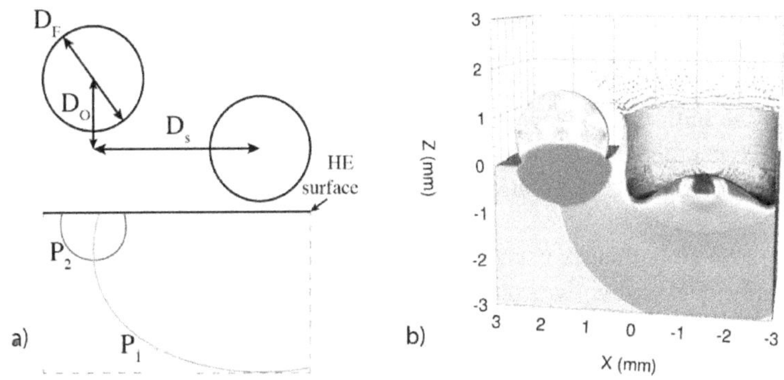

Fig. 5. a) a schematic of an asynchronous impact scenario. The added complexity for this problem arises from determination of what offset, D_O, should be given to the lagging impactor. b) demonstrates the location of the pressure fronts, and their relative magnitude at 0.7μs after impact of the first fragment.

We chose to compare the case where fragments separated by 3 mm impact simultaneously and asynchronously. In the latter case, an intersection 500 μm below the surface was desired because at this point the front of the second compression wave would still be at a high pressure when it intersects the wave from the first impactor. In order for the pressure wave intersection to occur at the right time, the second impactor was given an additional offset, D_O, of 1.79 mm from the surface of the EM.

By comparison, the case of the simultaneous impact provides a much higher pressure rise compared to the pressure rise in an asynchronous system. In Figure 6 below, the plot represents the series of tracers directly below the lagging fragment at x=1.5mm. The maximum pressure for the asynchronous case of 18 GPa recorded here is at 537 μm in depth. The lower plot shows the series of tracers in between the fragments at x=0 mm. The maximum pressure for the simultaneously impacting fragments is recorded as 21 GPa, which is 13.5% greater than in the asynchronous case.

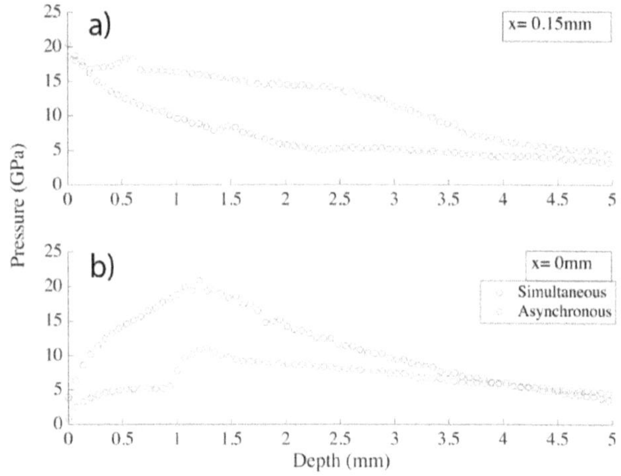

Fig. 6. Maximum pressures observed along the tracers in-line with the lagging impactor [a)] and in between the impactors [b)].

3. Discussion

These initial simulations are encouraging and demonstrate that CTH can be used to help understand fragment impact scenarios of relevance to detonation train assembly and disassembly. Using experimentally determined parameters as a starting point, the simulations demonstrate that small differences in fragment arrival time and spacing can result in drastically different EM response. While this work will ultimately need to be verified by experiment, the use of reactive burn models such as the HVRB model used in coordination with hydrocodes such as CTH provide for inexpensive opportunities to probe what constitute important characteristics of hypervelocity multi-fragment impact of EM. One aspect of this work that might be considered for future improvement is the incorporation of a reactive burn model that considers shock desensitization, such as XHVRB or SURF (Scaled Uniform Reactive Front)[10,11]. This will probe whether colliding shockwaves from an asynchronous event might be more or less "live" depending on how the initial shockwave compresses hotspots prior to the arrival and resulting pressure rise from colliding waves. In the case of asynchronous colliding waves such as the scenario considered in section 2.3 above, the colliding waves generate the greatest pressure rise at the leading edge of the shockwaves in virgin EM, so the HVRB model should be sufficient.

Perhaps the most interesting result from this study was that the pressure rise from simultaneous impact was higher than asynchronous impact. This was initially counterintuitive, because it was initially thought that a lower pressure shockwave interacting with a higher pressure shockwave would more easily lead to detonation. However, this is advantageous to real world assembly/disassembly scenarios where detonator fragmentation generates an asynchronous fragmentation field.

In the experimental scenarios referenced earlier, a field of fragments was seen to impinge on the HE surface. Detailed holography studies [4] showed that the fragment shape for a certain type of detonator was decidedly non-spherical, and undoubtedly the fragment morphology has some effect on the resulting shock pressure and duration. Both of these experimental factors will need to be considered in future work. While incorporating more fragments into the simulation is relatively easy, generating the exact fragment morphology may be more difficult. CTH does allow for user-defined particle shape, so the existing holography data could be used to generate more realistic fragment morphology in future studies. Of course, if this approach is taken then a series of other types of simulation studies (e.g. mesh refinement around the sharp fragment edges) may need to also be undertaken.

4. Conclusions

Spherical multi-fragment impact has been simulated using CTH coupled with the HVRB model for PBX 9501 to show initiation of EM via colliding shockwaves. Spherical 2 mm aluminum fragments, traveling at 3 mm/μs, that impact a PBX 9501 surface simultaneously can lead to detonation if they impact less than 2.8 mm from each other. If the fragments impact asynchronously such that the lagging compression wave impacts the edge of the leading wave directly below the late-time impactor, the maximum pressure reached is significantly lower than if the particles had impacted simultaneously. Future work will investigate systems with N>2 particles, and particles of different morphologies to match fragment shapes seen in experiment.

Acknowledgements

The authors would like to thank Eric Mas for helpful discussions relating to colliding shockwaves. Funding was provided jointly by Nuclear Safety Research and Development (via Paul Peterson) and the Joint Munitions Program (via Thomas Mason). Los Alamos National Laboratory, an affirmative action equal opportunity employer, is operated by TRIAD National Security, LLC, for the National Nuclear Security Administration of the U.S. Department of Energy under Contract No. 89233218-NCNA000001. Sandia National Laboratories is a multi-mission laboratory managed and operated by National Technology and Engineering Solutions of Sandia, a wholly owned subsidiary of Honeywell International Inc., for the U.S. Department of Energy's National Nuclear Security Administration under Contract No. DE-NA0003525.

References

[1] Deal, B., Grove, A., 1965. General Relationship for the Thermal Oxidation of Silicon, Journal of Applied Physics 36, p. 3770.

[1] Chen, J., Ching, H., Allahdadi, F., 2007. Shock-induced detonation of high explosives by high velocity impact, Journal of Mechanics of Materials and Structures, 2, p. 1701.

[2] Clements, B., 2011. Temperature Calculation Using Viscoscram with Discussion of Supporting Experiments, (No. LA-UR-11-04989; LA-UR-11-4989). Los Alamos National Lab.(LANL), Los Alamos, NM (United States).

[3] Pfeil, M., Stanfield, J., Neidert, J., Harstad, E., White, B., and Springer, H., 2016. Parameters Influencing the Response of MSP-1 Propellant Subject to Fragment Impact, (No. SAND2016-8207C). Sandia National Lab.(SNL-NM), Albuquerque, NM (United States).

[4] Yeager, J., Bowden, P., Guildenbecher, D., and Olles, J.D., 2017. Characterization of Hypervelocity Metal Fragments for Explosive Initiation, Journal of Applied Physics, 122, p. 035901.

[5] Rae, W., 1970. Analytical Studies of Impact-Generated Shock Propagation: Survey and New Results, High-Velocity Impact Phenomena, p. 214.

[6] Holsapple, K. 1987. The Scaling of Impact Phenomena, International Journal of Impact Engineering, 5, p. 343.

[7] Lam, K., Foley, T., Novak, A., Dickson, P. and Parker, G., 2010. Testing and modeling of PBX-9501 Shock Initiation (No. LA-UR-10-01496; LA-UR-10-1496). Los Alamos National Lab.(LANL), Los Alamos, NM (United States).

[8] Starkenberg, J., 2002, August. Modeling Detonation Propagation and Failure Using Explosive Initiation Models in a Conventional Hydrocode. In The Twelfth Symposium (International) on Detonation, p. 1001.

[9] Starkenberg, J., Dorsey, T. M. 1998. An Assessment of the Performance of the History Variable Reactive Bum Explosive Initiation Model in the CTH Code (No. ARL-TR-1667). Army Research Lab Aberdeen Proving Ground MD Weapons Technology Directorate.

[10] Tuttle, L. Schmitt, R., Kittell, and Harstad, E., 2018 Evaluation of XHVRB for capturing shock desensitization, AIP Conference Proceedings, vol. 1979, no. 1, p. 100044.

[11] Shaw, M. S., & Menikoff, R.,2010 A reactive burn model for shock initiation in a PBX: Scaling and separability based on the hot spot concept, Proceedings of the 14th Detonation Symposium, Coeur d'Alene, ID.

Proceedings of the 2019 Hypervelocity Impact Symposium
HVIS2019
April 14-19, 2019, Destin, FL, USA
HVIS2019-113

Extended Abstract+

Investigating Pressure Wave Impact on a Surrogate Head Model Using Numerical Simulation Techniques

Rohan Banton [a,*], Thuvan Piehler [a], Nicole Zander [a], Richard Benjamin[a], Josh Duckworth [b], Oren Petel [c]

[a]*U.S. Army Research Laboaratory, Aberdeen Proving Ground,MD , 21005-5066*
[b]*Uniformed Services University of the Health Sciences, 4301 Jones Bridge Rd, Bethesda, MD 20814*
[c]*Carleton University, 1125 Colonel By Dr, Ottawa, ON K1S 5B6, Canada*

1. Introduction

There is an urgent need to understand the mechanism leading to mild traumatic brain injury (mTBI) resulting from blast wave impact to the head. The recent conflicts in Iraq and Afghanistan have heightened the awareness of head impact injuries to military personnel resulting from exposure to blast waves [1, 2]. A blast wave generated in air is a by-product of the detonation of an explosive [3]. To date the mechanism resulting in mTBI from primary blast insult is still unclear.

In the present study, we are building on our previous work [4] through a systematic approach to improve and develop our numerical model. We used a simple skull/brain surrogate assembly to examine the early time primary blast-wave interaction on the surrogate head. The numerical model results of blast wave head impact and intracranial pressure wave transmission into surrogate brain were examined at strategic locations at the point of impact as well as in the surrogate brain. Specifically, the intracranial pressure loading in the form of both positive (compressive) and negative (tension) stress waves were monitored. We are building our model to explore the spatial regions of the brain for maximum pressure distributions. We have also built our model to account for the biomechanical shear response of the brain simulant material based on research discussed in the literature [2, 5, 6, 7]. The following section discusses the numerical methodology utilized in the present work.

2. Methodology and Results

2.1. Computational approach

The simulations performed in this work were carried out to investigate blast wave impact approaching a surrogate head from the front or anterior direction. A meshed head model assembly was inserted in an Eulerian air domain in the CTH code with standoff distance of 180 mm from a modeled RDX charge. Figure 1 shows the numerical problem setup for investigation of the blast wave/head interaction. The simulations were started after time zero with point initiation of the RDX charge. This subsequently produced detonation of the explosive followed by the blast wave propagation towards the surrogate head. Adaptive mesh refinement techniques were used to track the blast propagation waves as well as refine the shock front. The blast waves were allowed to transition across the surrogate head with hydrodynamic outflow boundary conditions imposed to allow mass flow out of the computational domain. All simulations were carried out on the U.S. Army Research Laboratory DoD Supercomputer

---

+ Full paper published in another journal. Please contact author.

* Corresponding author. Tel.:410-278-6042.
E-mail address: rohan.j,banton.civ@mail.mil.

Computing Research Center facilities using 64 processor for 24 hours with calculation integrated out to 1 ms involving 32.7 X 10^6 cells.

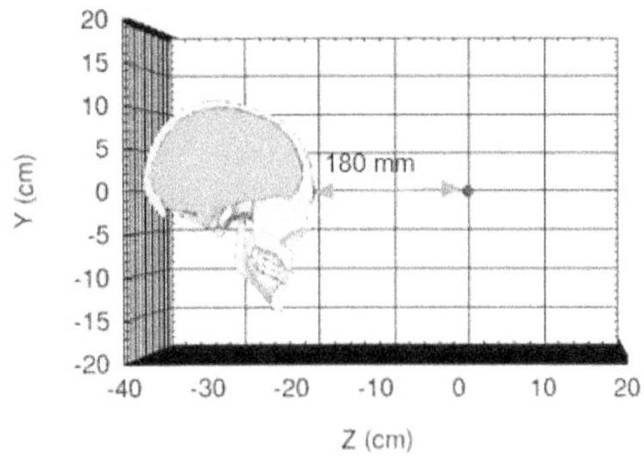

Fig. 1. Numerical setup of surrogate head model showing 180 mm standoff distance from RDX charge.

2.2. Results and Discussion

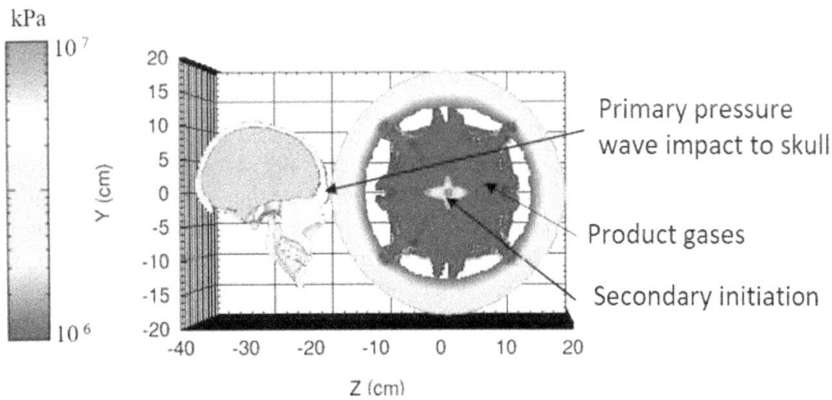

Fig. 2. Simulated blast wave showing impact to the surrogate head after 140 μs after initiation from RDX charge.

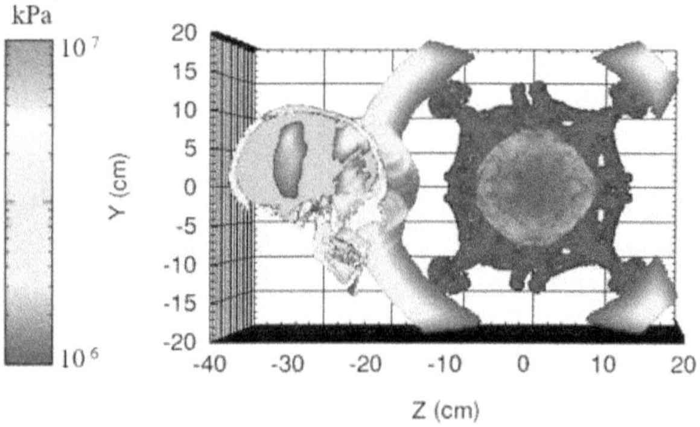

Fig. 3. Numerical simulation showing blast wave transition into the surrogate skull, reflection off the skull front, and traversing the top of the skull.

Figure 2 shows the simulated results at time 140 μs after initiation from a RDX charge. The simulation results captured the primary blast wave, the secondary initiation and the product gases. The expansion gases, after the detonation, compressed the surrounding air into the pressure wave (blast wave) that propagated towards the skull. Figure 3 shows the interaction of the blast wave with the surrogate head at time, 240 μs, after the explosive initiation. As the incident blast wave encountered the higher density skull, it reflected off it, and continued traversing the upper part of the skull and the lower part of the skull face. The midsagittal view of the surrogate head shows the transmitted compressive waves moving through the skull and into the brain simulant. The compressive wave appeared to move faster through the skull and the brain simulant compared to the wave in the free air. A secondary initiation also produced a second blast wave, although at a much more weakened strength compared to the primary wave. The simulated results of our numerical models showed good qualitative agreement with experiments (not shown here) based on the blast wave impact, reflection and the external flow across the front surface of the skull [6]. However, further developments are still needed to improve the deviatoric shear response under blast loading conditions.

2.3. Conclusion

In summary, this work presented the development of a numerical model based on comparisons and validations against experiments to investigate frontal blast impact on a surrogate head form. The results revealed strong compression waves in the anterior region of the brain simulant with the pressure waves propagated faster inside the skull and brain regions compared to the waves in the free air. This impedance mismatch has the potential to create tearing and shearing of brain tissues as the pressure waves transition from the external air, through the skull, to the brain. Unfortunately, these shearing phenomena were not verified against experimental data as of the writing of this manuscript. We will continue to systematically develop our model to investigate brain tissue response under blast loading scenarios and to provide information to mitigate brain injury to this research community.

Acknowledgements

The authors would like to thank Drs. Robert Schmidt and Aaron Brundage of Sandia National Laboratory for many helpful suggestions in using CTH models. Special thanks is also extended to Mr. Steve Schraml and Mr. Dwight Hofstetter in the use of CTH AMR techniques and the use of numerical pre and post processing modeling tools.

References

[1] Moore, D.F., Jérusalem, A., Nyein, M., Noels, L., Jaffee, M.S., Radovitzky, R.A. 2009. Computational biology—Modeling of primary blast effects on the central nervous system. NeuroImage. **47**, T10–T20

[2] Taylor, P.A., Ford, C.C.2009. Simulation of blast-induced early-time intracranial wave physics leading to traumatic brain injury. J. Biomech. Eng. **131**(6), p. 61007

[3] Kinney G.F. Graham K.J., 1985. Explosive Shocks in Air 2nd edition, Springer-Verlag, New York

[4] Banton, R., et al. "Comparison of Numerical Simulations with Experiments of Blast-Induced Pressure Wave Impact on a Surrogate Head Model," Dynamic Behavior of Materials, Volume 1 - Proceedings of the 2017 Annual Conference on Experimental and Applied Mechanics, pp. 181-187.

[5] Wang, C., Pahk, J.B., Balaban, C.D., Miller, M.C., Wood, A.R., Vipperman, J.S. 2014. Computational study of human head response to primary blast waves of five levels from three directions. PLoS One. **9**(11), p. 113264

[6] Zhang, L., Yang, K.H., King, A.I., 2004. A proposed injury threshold for mild traumatic brain injury. J. Biomech. Eng. **126**(2), p. 226

[7] Zhu, F., et al., 2012. Using a gel/plastic surrogate to study the biomechanical response of the head under air shock loading: a combined experimental and numerical investigation. Biomech. Model. Mechanobiol. **11**(3–4), p. 341–353

Proceedings of the 2019 Hypervelocity Impact Symposium
HVIS2019
April 14-19, 2019, Destin, FL, USA
HVIS2019-119

A Predictive Non-Dimensional Scaling Law for the Plate Perforation of Several Aluminum Alloys by Fragment-Simulating Projectiles

Weinong Chen[a*], Zherui Guo[a*]

[a]*Purdue University, School of Aeronautics & Astronautics, Neil Armstrong Hall of Engineering, 701 W Stadium Ave, West Lafayette, IN 47906, USA*

Abstract

An equation was previously-presented to predict the ballistic-limit velocity for the perforation of aluminum armor plates by fragment-simulating projectiles (FSP). The ballistic-limit equation was presented in terms of dimensionless parameters so that the geometric and material problem scales are identified. Previously published predictions and data for two different FSP projectile calibers (12.7 mm and 20 mm) and two different strength aluminum alloys show the scaling law to be accurate. In this paper we extend the same concept to several other alloys and show that this scaling law is predictive.

Keywords: Aluminum plates; fragment-simulating projectiles; scaling law.

Nomenclature

D	projectile diameter [mm]
L	projectile length [mm]
M	projectile mass (g)
m	shear plug mass (g)
T	plate thickness [mm]
V_{bl}	ballistic limit velocity (m/s)
V_r	residual velocity (m/s)
V_s	striking velocity (m/s)
W	work done by shear stress
Greek symbols	
ε	true strain [mm/mm]
ρ	density [kg/m^3]
σ_0	yield strength (MPa)
τ	shear strength (MPa)

1. Introduction

A scaling law was previously formulated [1] for the impact of fragment-simulating projectiles (FSPs) on aluminum plates. By modifying the Recht-Ipson model [2] and deriving an expression for the work done during shear plug formation in the energy balance equation, non-dimensional predictive equations were given for the ballistic limit velocity for three different aluminum alloys, namely 6055-T651 [3], 2139-T8 [4], and 6061-T651 [5] for 12.7-mm (0.50-in) and 20-mm caliber FSPs. Data obtained

* Corresponding author. Tel.: +1-765-494-1788.
E-mail address: wchen@purdue.edu

from experiments proved the effectiveness of the scaling law in predicting the ballistic limit velocities for untested aluminum plates.

In this study, we extend the scaling law to several other different strength alloys where the dynamic stress-strain curves may not be readily available. We use a combination of a modified Ludwik model and Hollomon model to estimate dynamic compressive strengths of alloys and use these estimated values to further predict the ballistic performance of these aluminum alloy plates.

2. Analytical Model

2.1. Modified Recht-Ipson Model

Recht and Ipson used an energy balance analysis for the problem of a cylindrical projectile perforating a thin plate. In their analysis, the work done due to shear stresses during plug formation is derived as a function of the ballistic limit velocity V_{bl}, which is obtained empirically from experimental data. In contrast, in our previous study [1], the work done is derived as a function of the shearing mechanism, and predictive equations are provided for V_{bl} and residual velocity V_r. As a cylindrical projectile of mass M impacts a plate at striking velocity V_s, conservation of energy gives the equation

$$MV_s^2/2 = (M+m)V_r^2/2 + E_i + W \tag{1}$$

During this process, a target shear plug of mass m is formed, and from inelastic collision with the projectile, they both exit the plate with residual velocity V_r. We also further assume that the plate does not deflect significantly, and so the only resistive force acting on the projectile is via the shear strength of the target. The energy lost during inelastic collision E_i is given as

$$E_i = \left(\frac{m}{M+m}\right)\frac{MV_s^2}{2} \tag{2}$$

The overall energy balance equation is then given as

$$\frac{MV_s^2}{2} = \left(\frac{m}{M+m}\right)\frac{MV_s^2}{2} + \frac{(M+m)}{2}V_r^2 + W \tag{3}$$

At the ballistic limit velocity V_{bl}, $V_r = 0$, and rearranging Equation3, the work done by shear stress is then given by

$$W = \left(\frac{M}{M+m}\right)\frac{MV_{bl}^2}{2} \tag{4}$$

The work done during plug formation is explicitly calculated by integrating the incremental shear plug movement and is currently assumed to be rate-independent. Subsequently, using the von Mises yield criterion to relate the shear stress τ to the uniaxial compressive yield stress σ_0, the energy balance equation is then given as

$$\frac{MV_s^2}{2} = \left(\frac{m}{M+m}\right)\frac{MV_s^2}{2} + \frac{(M+m)}{2}V_r^2 + \frac{\pi D T^2 \sigma_0}{2\sqrt{3}} \tag{5}$$

Further expressing the projectile and shear plug masses as

$$M = \rho_p\left(\frac{D}{2}\right)^2 \pi L, \qquad m = \rho_t\left(\frac{D}{2}\right)^2 \pi T \tag{6}$$

Rearranging Equation 5, the equation is then given in a non-dimensional form as

$$\frac{V_{bl}}{(\sigma_0/\rho_p)^{1/2}} = \left\{\frac{4}{\sqrt{3}}\left(\frac{L}{D}\right)\left(\frac{T}{L}\right)^2\left[1+\left(\frac{T}{L}\right)\left(\frac{\rho_t}{\rho_p}\right)\right]\right\}^{1/2} \tag{7}$$

The compressive true stress σ_0 values have been previously estimated as 400 MPa for 6055-T651 and 6061-T651, and 600 MPa for 2139-T8, which gives reasonably good estimates for the scaling law. An extension of this scaling law requires σ_0 for the other alloys, for which the dynamic stress-strain curves may not be readily available. We thus provide a method of estimation for σ_0 values for other aluminum alloys based on the modified Ludwik and Hollomon equations for quasi-static compression tests. In this formulation, the T/L ratios are assumed to be within the range of 0.5-2.5. Thinner targets involve more energy dissipation via bending mechanisms, but when the targets are too thick with respect to the projectile, the scenario becomes a deep penetration rather than a plate perforation problem.

2.2. Estimated true compressive stress values

In a previous study on APM2 bullets perforating aluminum plates, Ryan *et al.* [6] provided quasi-static compressive stress-strain curves for a large number of aluminum alloys. For this comprehensive list of estimated σ_0 values, the modified Ludwik and Hollomon power law fits were used to calculate the true stresses. Both power law fits tend to overestimate maximum true stress values at large strains above 0.5 when compared to the actual dynamic experimental stress-strain curves given in Ref. [7] obtained using a compression Kolsky bar. Actual experimental stress values tend to level out as the strains increase (Figure 1), and at large strains, the quasi-static and dynamic stress values do not appear to deviate significantly. The dynamic compressive strength values were therefore calculated at a true strain value of 0.5 using both fits and averaged to provide an estimation for strengths using quasi-static power-law hardening curves, using data from existing curves as a reference.

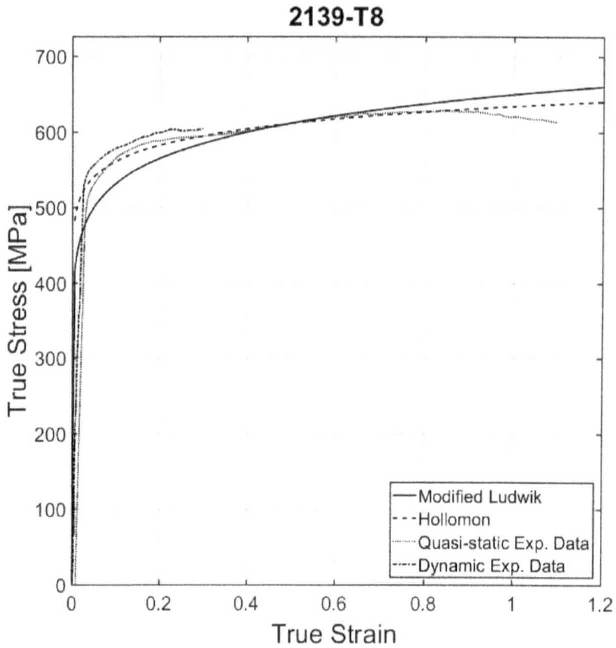

Fig. 1. Modified Ludwik and Hollomon models for 2139-T8 alloy showing overestimation of true compressive strengths for $\varepsilon > 0.7$ in comparison with digitized quasi-static data from Ryan *et al.* [6] and dynamic data from Cao et al. [7].

Table I. Compressive stress values for several aluminum alloys estimated at true strain value $\varepsilon = 0.5$.

Alloy	Modified Ludwik (MPa)	Hollomon (MPa)	Ave. quasi-static σ_0 (MPa)	Dynamic σ_0 from curves (MPa)	No. of V_{bl} data points	Ref.
AA2060-T86	661	666	664		14	[8]
AA2139-T8	612	611	611	610 [4]	6	[4]
AA5059-H131	497	500	498		5	[5]
AA5083-H131	449	450	449	440 [9]	1	[5]
AA6055-T651	445	448	446		17	[3]
AA6061-T651	389	392	390	400 [10]	3	[5]
AA7017-T6	500	500	500		12	[11]
AA7017-T7	485	484	485		2	[11]
AA7056-T721/T751	504	502	503		10	[12]
AA7056-T761	622	616	619		6	[12]
AA7085-T7E01	655	659	657	650 [13]	12	[14]
AA7085-T7E02	533	530	531		12	[14]

Comparison of the averaged values to the available curves shows reasonably close approximations for 2139-T8 [7], 5083-H131 [9], 6061-T651 [10], and 7085-T711 [13]. Where the dynamic compressive strengths are available, these values are used in Equation 7; otherwise, the averaged values estimated from quasi-static compression tests was used.

2.3. Non-dimensional predictions

The predicted true compressive stress values are subsequently plotted using Equation 7 for ballistic limit velocities of the 10 aluminum alloys given in Table I. Data points for ballistic limit velocities were collected from existing literature for 12.7-mm (0.50-in.) and 20-mm caliber FSPs impacting aluminum plates of different thicknesses between 12.7-50.8 mm (0.5-2.0 inches). Target and FSP projectile material densities are calculated as 2710 and 7830 kg/m^3 respectively. FSP *L/D* aspect ratios for both 12.7-mm and 20-mm calibers are calculated as 1.17 and 1.19 respectively, and so a nominal value of 1.18 was used for the model. A total of 96 data points for all 12 aluminum alloys are plotted in Figure 2.

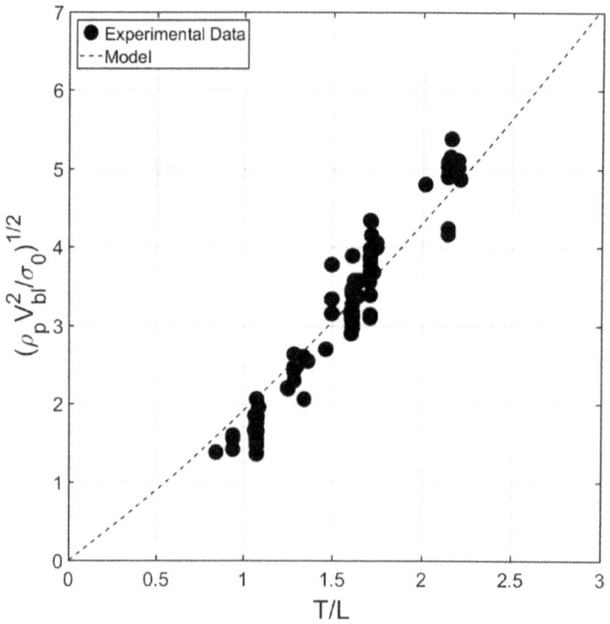

Fig. 2. Dimensionless model predictions using Equation 7 and experimental data for 12 different aluminum alloys.

The non-dimensional model predictions show the scaling law to be accurate. It is noted that the model starts to deviate from experimental data at lower *T/L* ranges where bending mechanisms come into play, since the current model only assumes failure via shear plugging under normal projectile impact. The average percentage error across all data points is about 5%. These errors are likely due to approximations of the unknown compressive strengths of some aluminum alloys.

3. Conclusions

A previously-formulated scaling law was further extended to predict the ballistic limit velocities of fragment-simulating projectiles perforating various aluminum alloys where the stress-strain responses may not be readily available. The FSPs were modeled as an ideal cylinder impacting a flat plate, and the ballistic limit velocities were predicted as a function of the work done during shearing. Using a combination of the modified Ludwik and Hollomon curve-fits for high strain values, the alloys' dynamic compressive strengths may be approximated at a true-strain value of 0.5 from power-law hardening curves, and shear stresses were subsequently estimated using the von Mises yield criterion. This dimensionless scaling law shows the geometric and material scales of the impact problem and aligns well with experimental data for 12 different alloys and 96 different tests.

Acknowledgements

The authors would like to thank Dr. Michael J. Forrestal for his help and guidance on the topics covered in this work. We would also like to thank Stephenie Martinez-Morales for her help with digitizing the extensive amount of ballistic data presented.

References

[1] Guo, Z., Forrestal, M.J., Chen, W. 2019. A Non-Dimensional Scaling Law for Aluminum Plate Perforation by Fragment-Simulating Projectiles. Journal of Dynamic Behavior of Materials 5, pp. 409-415.

[2] Recht, R.F., Ipson., T.W., 1963. Ballistic Perforation Dynamics, Journal of Applied Mechanics 30, pp. 384–390.

[3] ARL-MR-0904: Ballistic Evaluation of 6055 Aluminum, Army Research Laboratory, 2015.

[4] Cheeseman, A., Gooch, W., Burkins, M., 2008. Ballistic evaluation of aluminum 2139-T8, Proceedings of the 24th International Symposium on Ballistics, pp. 651.

[5] Gooch, W.A., Burkins, M.S., Squillacioti, R.J., 2007. "Ballistic Testing of Commercial Aluminum Alloys and Alternate Processing Techniques to Increase the Availability of Aluminum Armor," Proceedings of the 23rd International Symposium on Ballistics. Tarragona, Spain, pp. 2007: pp. 981–988.

[6] Ryan, S., Nguyen, L.H., Gallardy, D., Cimpoeru, S.J., 2018. A scaling law for predicting the ballistic limit of aluminium alloy targets perforated in ductile hole formation, Int. J. Impact Eng. 116, pp. 34–50.

[7] Cao, B., Shaeffer, M., Cadel, D., Ramesh, K.T., Prasad, S., 2018. An Analysis of Strengthening Mechanisms and Rate-Dependence in a High Strength Aluminum Alloy, J. Dyn. Behav. Mater. 4, pp. 6–17.

[8] ARL-MR-0930: Ballistic Evaluation of 2060 Aluminum, Army Research Laboratory, May 2016.

[9] Huskins, E.L., Cao, B., Ramesh, K.T., 2010. Strengthening mechanisms in an Al–Mg alloy, Mater. Sci. Eng. A. 527, pp. 1292–1298.

[10] Ramesh, K.T., 2008. High Strain Rate and Impact Experiments, in: Handb. Exp. Solid Mech., Springer, pp. 874.

[11] ARL-TR-7727: The Examination of the Aluminum Alloy 7017 as a Replacement for the Aluminum Alloy 7039 in lightweight Armor Systems, July 2016.

[12] ARL-TR-7947: Ballistic Evaluation of 7056 Aluminum, Army Research Laboratory, February 2017.

[13] Bhattacharyya, J.J., Agnew, S.R., Lee, M.M., Whittington, W.R., El Kadiri, H., 2017. Measuring and modeling the anisotropic, high strain rate deformation of Al alloy, 7085, plate in T711 temper, Int. J. Plast. 93, pp. 46–63.

[14] ARL-TR-5952: Ballistic Evaluation of 7085 Aluminum, Army Research Laboratory, 2012.

Proceedings of the 2019 Hypervelocity Impact Symposium
HVIS2019
April 14-19, 2019, Destin, FL, USA
HVIS2019-015

DEVELOPMENT OF ARL'S MULTI-ENERGY FLASH COMPUTED TOMOGRAPHY DIAGNOSTIC: CAPABILITY TO TRACK MASS-FLUX THROUGH A RECONSTRUCTION VOLUME

Michael B. Zellner[a,*], Melissa S. Love[a], Kyle Champley[b]

[a]*CCDC Army Research Laboratory, 321 Colleran Rd, Aberdeen, Md 21005*
[b]*Lawrence Livermore National Laboratory, 7000 East Ave., Livermore, CA 94550*

Abstract

The U.S. Combat Capabilities Development Command Army Research Laboratory and Lawrence Livermore National Laboratory are currently developing a Multi-Energy Flash Computed Tomography (MEFCT) diagnostic for multi-frame, *in situ,* three-dimensional radiographic assessment of ballistic impact phenomena. To accomplish this, we combine the capabilities of medical X-ray computed tomography and high-speed computed tomography, to produce a system that captures three independent, time-sequenced volume reconstructions throughout the timespan of a typical dynamic ballistic event. Because this system has the capability to image an event across three spatial dimensions and time, it is the first of its kind to track mass/material-flux of an un-bounded system through a volume at ballistic timescales.

To demonstrate the diagnostic's capabilities, an assessment of a bullet penetrating an aluminum plate is performed. A compilation of the three volume reconstructions were computed to describe the event. The results were compared to a state-of-the-art simulation of the event using EPIC, a Lagrangian hydrocode with penetration applications. This comparison shows how using a four-dimensional computed tomography system can benefit the validation of physical failure and mass/material-flow models.

Keywords: Computed Tomography; Flash Radiography Ballistic Penetration; EPIC; Hydrocode Simulation

The U.S. Combat Capabilities Development Command Army Research Laboratory and Lawrence Livermore National Laboratory are currently developing a Multi-Energy Flash Computed Tomography (MEFCT) diagnostic for assessing ballistic impact phenomena [1-4]. In theory, x-ray computed tomography is well suited for ballistic event assessment because it uses penetrating radiography to visualize features beyond the outermost material layers, dust clouds, fire-balls, etc. and is capable of imaging three-dimensional features commonly attributed with fragmentation processes that occur during ballistic impact events. This work pushes towards turning that theory into reality by extending the capabilities of medical X-ray computed tomography [5-7] and high-speed computed tomography [8], to produce a system that captures three independent, time-sequenced volume reconstructions throughout the timespan of a typical dynamic ballistic event. This enhancement was accomplished by incorporating fifteen individual flash x-ray systems into a single unit that acquires fifteen unique image projections of a scattering body with sub-microsecond temporal resolution. Each of three computed tomography volume reconstructions is then created independently using five of the projections. Because this system has the capability to image an event across three spatial dimensions and time, it is the first of its kind at being able to track mass/material-flux of an un-bounded system through the reconstruction volume [4].

* Corresponding author. Tel.: +1-410-278-1183.
E-mail address: michael.b.zellner.civ@mail.mil.

To demonstrate the diagnostic's capabilities, an assessment of a bullet penetrating an aluminum plate is performed. This relatively simple experiment has many aspects that inform about the MEFCT's complexities, while serving to demonstrate its benefits in expanding visualization of high-rate phenomena to four-dimensions. The primary focus of this work is to demonstrate how the data collected can be compared to a state-of-the-art simulation and how the comparison using a four-dimensional computed tomography system can benefit the validation of physical failure and mass/material-flow models. For this comparison, simulations of the event were computed using EPIC, which is a Lagrangian hydrocode with penetration applications [9].

1. Results

2.1. Experiment

Details pertaining to the design and functionality of the MEFCT device are published elsewhere [1-4]. Generally, the system uses fifteen flash X-ray source/detector pairs to collect 15 radiographic projections during an event. Of the fifteen projections, five are collected simultaneously* using L-3 Applied Technology flash sources that have a maximum X-ray energy of 150 keV, five are collected simultaneously at a later time using flash sources that have a maximum X-ray energy of 300 keV, and five are collected simultaneously at yet a later time using flash sources that have a maximum X-ray energy of 450 keV. The arrangement is such that three independent volumetric reconstructions can be computed from the off-axis projections at three specified times throughout a dynamic event.

To demonstrate the diagnostic's capabilities, an assessment of a 0.30-caliber jacked bullet penetrating an aluminum plate was performed. The 7059 aluminum target, measuring 50.8 x 101.6 x 9.5 mm, was placed on a suspended platform in the center of the MEFCT device. The target was impacted through the thin dimension and had relatively free boundary conditions (a small strip of tape held the target upright on the platform). A polyvinylidene fluoride piezoelectric thin film transducer was adhered to the rear surface of the target for diagnostic triggering purposes. Figure 1 shows a photograph of the target prior to the impact event.

Figure 1: Photograph of the aluminum plate target suspended in the center of the MEFCT device.

The bullet was launched out of a percussion-initiated cartridge along the Y-axis of the MEFCT device using a solenoid-driven 0.30-caliber gun. The gun was located approximately 3 meters up range from the target. Two paper-board based wire-printed break-screens of known separation distance were placed between the gun and the target to acquire projectile impact velocity. Measurements indicated that projectile was traveling 892±5 m/s prior to impact.

The MEFCT device collected independent data sets at 31, 73, and 115 μs after target impact. The projections were collected using Carestream GP digital imaging plates, which were subsequently scanned using a Carestream HPX-1 scanner at a 50 μm

* For the demonstrated experiment, inter-system flash times occurred within ~500 ns. At this separation, motion blur could induce a spatial error of up to 0.44 mm. For hypervelocity experiments (velocity ≥ 3000 m/s), the system can be set to acquire inter-system projections with ~280 ns separation, resulting in minimum motion blur of 0.84 mm.

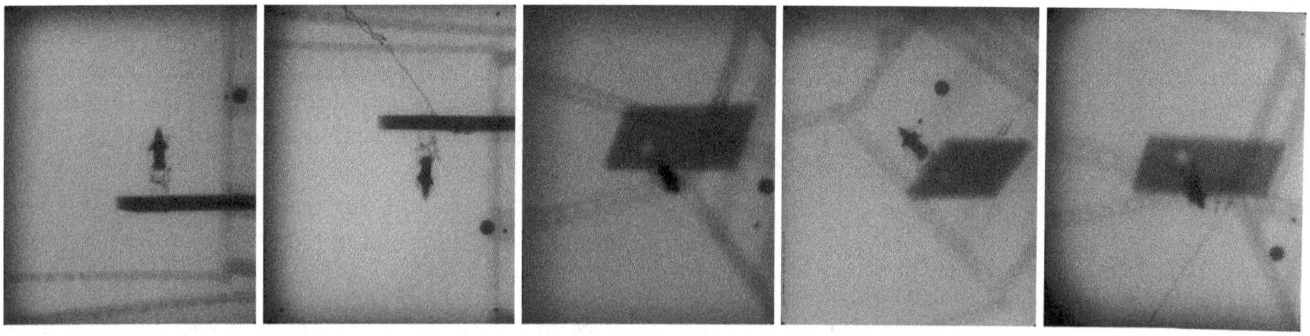

Figure 2: Representative radiographic projections collected using the MEFCT device (the 300 keV set collected at t=73 μs).

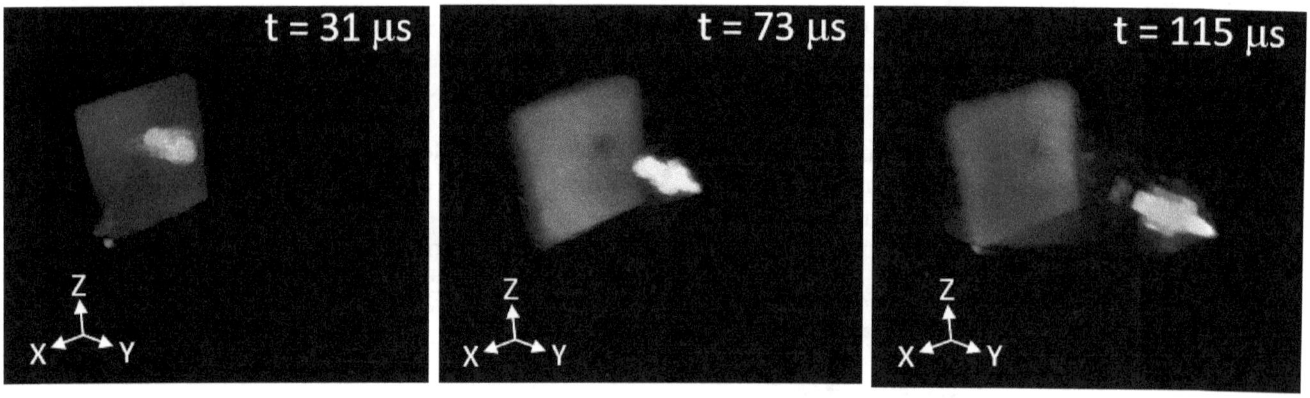

Figure 3: Maximum intensity projections extracted from each of the three-dimensional volume reconstructions.

pixel resolution. The projections were processed using custom MATLAB® routines that orientated and calibrated intensity to a reference [4] before being passed into Livermore Tomography Tools (LTT) software suite [10], which perform further image enhancement and the volumetric reconstructions. For this work, LTT utilized a proprietary reconstruction algorithm [11], however slightly reduced resolution reconstructions were also achieved using the ASDPOCS algorithm [12]. Figure 2 shows representative projections collected during the experiment, and Figure 3 shows three maximum intensity projections of the event extracted from the volume reconstructions. These projections are cropped to the upper half of the plate to focus on the region of interest where the dynamics occur.

2.2. Simulation

A simulation of the event was computed using EPIC, which is a Lagrangian hydrocode with penetration applications [9]. The bullet was simulated using dimensions from its engineering drawings at a velocity of 890 m/s. The target was simulated using truly free boundaries, i.e. floating in a vacuum. The dimensions of the target during initial setup were those measured post-experiment, which are assumed to be the same as pre-experiment as the damage was very local to the impact region. The simulation also utilized the actual hit-point as measured from the post-experiment target in an effort to produce the most quantitative comparison. All meshing utilized hexahedral elements with a 0.26 mm element length, which resulted in 36 elements across the smallest dimension of the aluminum plate at the point of impact and 3 elements through the jacket thickness. Each element was then converted to 24 tetrahedral elements. Simulations were fully three-dimensional with no symmetry planes. Johnson-Cook strength models [13], Johnson-Cook failure models [14], and EPIC library parameters were used for all materials. The simulation initiated just prior to impact and continued for 120 μs. When computed on 160 processors (4 nodes, 40 cores), the simulation required a wall-clock time of approximately 28 hours to complete.

Figure 4 shows material plots of the simulation. Unlike the computed tomography analysis, in these plots, the entire target plate is shown. For visualization purposes, the bullet jacket, core, and filler are shown in blue, red, and green, respectively, and the aluminum plate is shown in light gray for all simulation images.

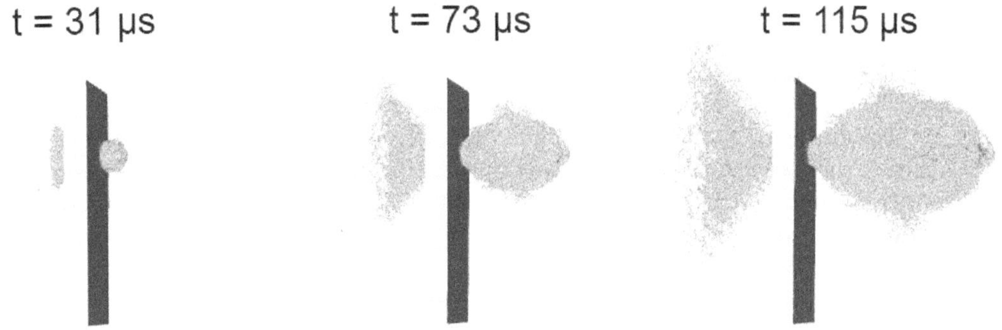

Figure 4: Simulation of the ballistic impact event depicting materials and their evolution. The bullet jacket, core, and filler are shown in blue, red, and green, respectively, and the aluminum plate is shown in light gray for all simulation images

2. Discussion

Comparison of the material position from the computed tomography analysis to simulations reveals many correlations and discrepancies that could possibly be overlooked if only captured in two dimensions. The comparisons which we will discuss are drawn from figures 2, 3, & 4, as well as from projections, isosurfaces, and cut/slice visualizations that were not displayed in this text. Here we would like to stress that these comparisons are derived from a single experiment that includes stoichastic material failures, and many variables. These comparisons are made primarily to demonstrate the newly developed technique and highlight the possible insights to which it can inform. More experiments would be needed to determine if the comparisons are statistically significant.

Focusing first on the fragment cloud of aluminum, EPIC appears to do a reasonable job of predicting the outermost envelope within which the particles extend. This is confirmed through comparison with optical photography. The radiography has a deficiency in this area as the contrast from what are assumed to be very tiny particles falls below the detection noise floor. This can be confirmed from the projections in figures 2 and 3. However, the simulations appear to fail at predicting many larger fragments created during the penetration process. These fragments are apparent within the radiographic projections, and further

Figure 5: Three-dimensional perspective of bullet, target, and fragments at t=115 µs from impact. This rendering was created by compositing numerous isosurfaces extracted from the computed tomography reconstructed volume. The perspective was rotated to easily visualize the doughnut-shaped fragment, so the bullet is now traveling from right to left within the image. False coloring was applied.

enlightened upon via the computed tomography analysis. Of particular note is a doughnut-shaped fragment that appears to ride along the case of the bullet through t=73 μs, and then dislodge itself by t=115 μs. Figure 5 shows a composite of four three-dimensional isosurfaces extracted from the volume rendering at t=115 μs computed using a combination of MATLAB® for threshold processing and Google's Unity® software for visualization. The individual isosurfaces were false-colored to represent the authors' hypothesis of correlating attenuation to known materials within the volume space†. The color "amber" was applied to the isosurface that was hypothesized to correlate with bullet jacket material, the color "gray" was applied to the isosurfaces hypothesized to correlate with bullet core material, the color "light gray" was applied to the isosurfaces hypothesized to correlate with aluminum target material, and the color "cyan" was applied to the isosurfaces of unknown particles and fragments (i.e. one cannot easily discern from which material these particles originated). The view within figure 5 was selected to most easily visualize that the central particle is of doughnut shape with a hollow center (in which the bullet travels from right to left).

The simulation's inability to capture these larger fragments is likely linked to deficiences in the material failure model. These features, however, may be important when ballistic interactions are more complex, as they could propagate and accumulate to cause non-predictive or catastrophic results. This may be applicable to cases such as that when a multi-layer target is impacted. Simulations and experiments also display some discrepencies when focusing on the bullet and its jacket erosion and core exposure dynamics. The post-penetration velocity and trajectory of the simulation and experiment agree well, as would be expected for a normal impact. The experiments measure a residual velocity of 877±9 m/s from t = 31 to t = 73 μs and 849±9 m/s from t = 73 to t = 115 μs, and the simulation predicts a residual velocity of 856 m/s at 31 μs, 845 m/s at 73 μs and 839 m/s at 115 μs. However, comparisons of the amount of jacket that is eroded (or equivelently how much of the core is exposed) appear to disagree. Figure 6 shows a depiction of the simulated results compared to maximum intensity projections and volume rendered isosurfaces extracted from the reconstructed volume. Because the bullet is penetrating the plate at t=31 μs, large attenuation gradients inhibit discerning features in the radiographic analysis. At t=73 and 115 μs, the bullet/jacket have sufficiently separated from the particle cloud allowing for comparisons. Comparing the t=73 μs experimental results to the simulation indicate qualitative agreement of the shape through which the jacket erodes and peels to expose the core. However, by this time, experiments indicate that a significantly larger portion of the core (approximately two times the amount) has become exposed than that in the simulations. At t=115 μs, results are similar to those at t=73 μs, with the simulated length of predicted core exposure being half that imaged by the experiment. At this time, however, the experiment and simulation also deviate from qualitative agreement of the eroding jacket shape. In the experiments, the jacket begins to erode in an asymetric fashion. In the simulations, small asymetries are noticed, but the magnitude and overall shape of these significantly deviate from that in the experiment. In addition to the asymetries in the jacket erosion, the experiments indicate that the core becomes slightly misaligned with the jacket (axially). This claim is supported when viewed from multiple angular perspectives around the bullet axis [4]. This misalignment, however, is not captured in the simulations.

An extended simulaton investigation was performed to probe if either the jacket erosion or bullet/jacket alignment differences between experiment and simulation arise from potential pitch and/or yaw of the bullet just prior to target impact (not measured in the experiments). These investigations indicated that up to two degrees of total pitch and yaw did not influence either phenomena in the simulated results.

If three-dimensional phemonon such as the jacket erosion asymmetry, bullet/jacket misalignment, and material breakup/fragmentation processes are found to be statistically relevant, then adjustments to the simulation methods may be warranted to obtain more predictive results. The ability to collect experimental data in three spatial dimensions as time evolves is vital to this understanding and will help inform how/if simulation methods should be modified to capture stochastic phenomena that ultimately define the material dynamics. This will be increasingly important as engagements become more complex, such as that within a multi-layer target, where effects could accumulate or deviate to engage different penetration mechanisms.

† The assignment of this isosurface is subjective and user-specific. Upgrades to the diagnostic and software to include a spectral-based assignment of isosurfaces based on densitometry and material specificity are being developed/incorporated.

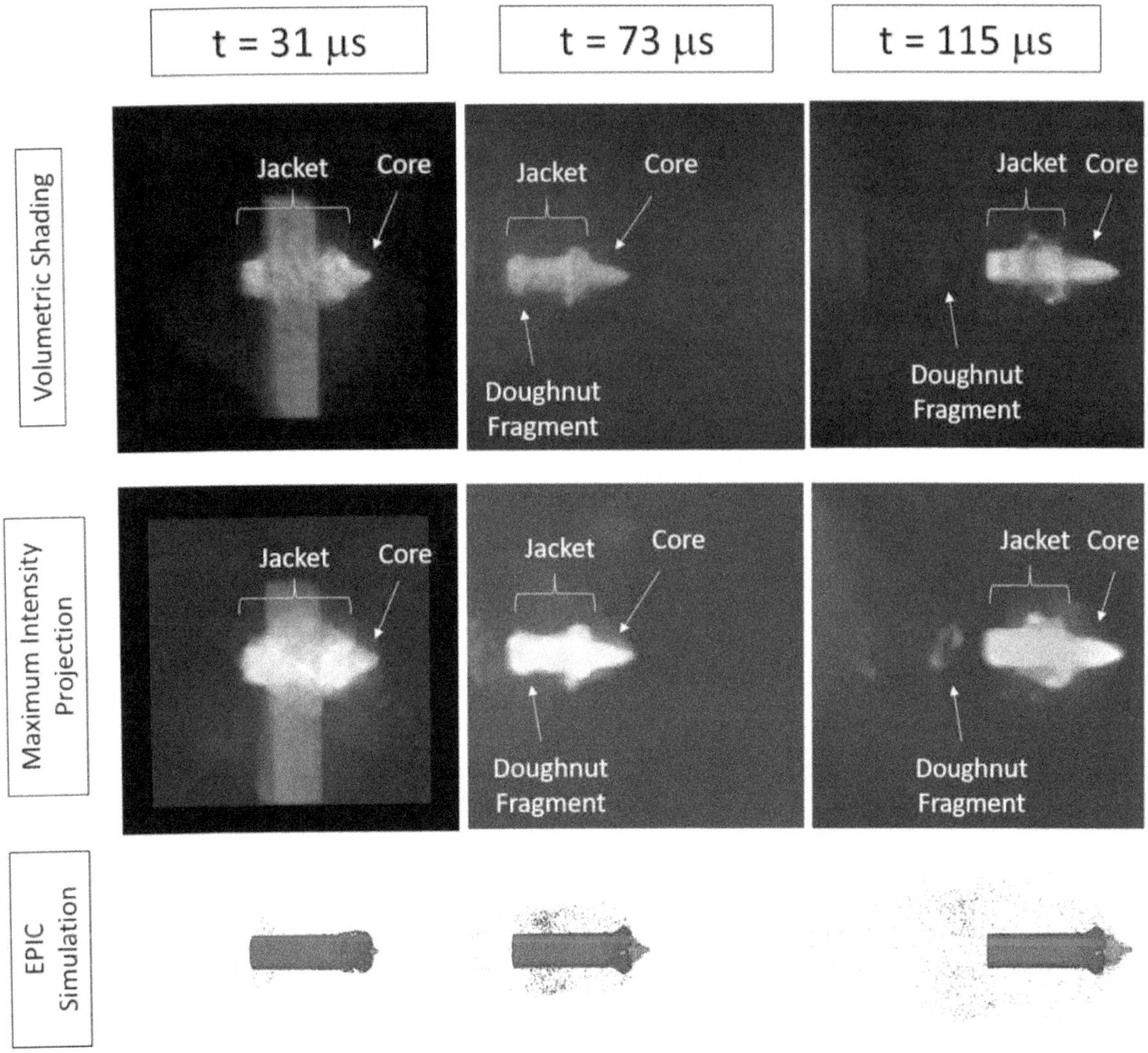

Figure 6: Comparison of bullet/jacket erosion process at t=31, 73, and 115 μs after impact. This comparison depicts simulated results compared to extracted maximum intensity projections and volume rendered isosurfaces from the reconstructed volume. The relative trend of bullet positions within the individual images are not correlated to a standard, as the window presented was arbitararly chosen. Therefore assumptions about the bullet velocity should not be made from this image compilation.

3. Conclusions

ARL and LLNL have developed a flash computed tomography capable of measuring mass/material-flux through a reconstruction volume space at ballistic timescales. This diagnostic has the potential to provide new insight into ballistic-timescale dynamic events. To demonstrate the diagnostic's usefulness, a comparison was made between experimental results and an EPIC simulation of a bullet impacting an aluminum plate. The diagnostic was capable of measuring the attenuation within a three-dimensional volume at three independent time steps, *in situ* to the penetration process. The comparison of experimental measurements and simulation reveled numerous regions where the two techniques agree and disagree. In general, the techniques agreed well when collecting/predicting post impact bullet velocities and relative trajectories. Because of the radiographic scattering cross-section, the experimental measurements lacked resolution of the small particles within the behind armor debris cloud. However, the experimental technique enlightened upon large volume/mass fragments which the simulations did not predict.

The computed tomography results also highlighted some differences in the erosion/peeling process of the bullet jacket. If these observations are statistically relevant, it points to a deficiency in the material failure model and/or the method used in simulations.

Acknowledgements

The authors would like to acknowledge R. Cantrell, K. Dudeck, J. Perrella, personnel of the ARL Wood/Model Shop, and personnel of the ARL Metal Fabrication Shop for assistance in design and fabrication of the many parts that comprise the MEFCT diagnostic. We would also like to thank D. Schall, A. Ducote, T. Nellenbach, T. O'Connor, T. Quigg, N. Sturgill, B. Huntzinger, and J. Benjamin for their expertise in operating and fielding the experimental test described within this work. We would like to thank the Livermore Tomography Tools team of Trevor Willey, Hyojin Kim, and Karina Bond, who have supported development of the computational software and routines used to iteratively reconstruct the XCT volume. We would like to thank S. Moser and S. Nau of Ernst Mach Institute, for their early input on development of the MEFCT diagnostic. The simulation work was supported in part by a grant of computer time from the DOD High Performance Computing Modernization Program at ARL. Finally, funding was provided by ARL and LLNL mission.

References

[1] Michael B. Zellner, Gerard T. Chaney, Chester A. Benjamin, Ronald Cantrell, Robert W. Borys, Robin D. Strickland, Joshua M. Sturgill, James A. Perrella, Gerald L. Schafer, Corey E. Yonce, Martin L. Potter, Stephen Alleyne, and John M. Zajicek. Considerations for the Design of a Multi-Color High-Speed X-ray Computed Tomography Diagnostic. ARL-TR-6969, US Army Research Laboratory, Aberdeen Proving Ground, MD (June 2014)

[2] Michael B Zellner, Charles L Randow, Ronald Cantrell, and Corey E Yonce. Blast-Loading Assessment of Multi-Energy Flash Computed Tomography (MEFCT) Diagnostic ARL-TR-7741, US Army Research Laboratory, Aberdeen Proving Ground, MD (August 2016)

[3] Michael B. Zellner, Kyle Champley, Larry McMichael, Harry Martz, Ronald Cantrell, Corey E. Yonce, Kenneth W. Dudeck, Chester A. Benjamin, Robert W. Borys, David R. Schall, Allen P Ducote III, Thomas J. O'Connor, Thomas E. Nellenbach, Nathan J. Sturgill, Thomas L. Quigg, Seth T. Halsey, Jennifer A. Benjamin, and Benjamin P. Huntzinger, Development of a Multi-Energy Flash Computed Tomography Diagnostic for Three Dimensional Imaging of Ballistic Experiments, Proceedings of the Shock Compression of Condensed Matter Topical Group Meeting of the American Physical Society, St. Louis, Mo, (July 9-14, 2017)

[4] Zellner, M.B.; Champley, K; "Development of a Computed Tomography System Capable of Tracking High-Velocity Unbounded Material through a Reconstruction Volume" Int. J. Impact Eng. Vol 129, July 2019, Pages 26-35

[5] Gordon, Richard; Bender, Robert; Herman, Gabor (1970), "Algebraic reconstruction techniques (ART) for three dimensional electron microscopy and x-ray photography", Journal of Theoretical Biology, 29 (3): 471–481, doi:10.1016/0022-5193(70)90109-8, PMID 5492997

[6] Radon J (1917). "Uber die Bestimmung von Funktionen durch ihre Integralwerte Langs Gewisser Mannigfaltigkeiten" [On the determination of functions from their integrals along certain manifolds]. Ber. Saechsische Akad. Wiss. 29: 262. Reprint: Radon J (1 December 1986). "On the determination of functions from their integral values along certain manifolds". IEEE Transactions on Medical Imaging. 5 (4): 170–176. doi:10.1109/TMI.1986.4307775. PMID 18244009.

[7] Kaczmarz, Stefan (1937), "Angenäherte Auflösung von Systemen linearer Gleichungen" (PDF), Bulletin International de l'Académie Polonaise des Sciences et des Lettres. Classe des Sciences Mathématiques et Naturelles. Série A, Sciences Mathématiques, 35, pp. 355–357

[8] Moser, S.; Nau, S.; Manfred, S.; Klaus, T. In Situ Flash X-Ray High-Speed Computed Tomogra-phy for the Quantitative Analysis of Highly Dynamic Processes. Meas. Sci. Technol. 2014, 25 025009

[9] Johnson, G. R.; Stryk, R. A.; Holmquist, T. J.; Beissel, S. R. Numerical Algorithms in a Lagrangian Hydrocode. WL-TR-1997-7039, Wright Laboratory, Eglin AFB, FL (June 1997)

[10]Kyle Champley, Livermore Tomography Tools (LTT) Technical Manual LLNL-SM-687016 (March 2016).

[11]Kyle Champley, "Few-view image reconstruction" https://patents.google.com/patent/US20180182128A1/en (12/28/2016).

[12]Emil Y Sidky and Xiaochuan Pan, "Image reconstruction in circular cone-beam computed tomography by constrained, total-variation minimization." Phys. Med. Biol. 53 (2008) 4777–4807

[13]Johnson, G.R.; Cook, W.H. "A constitutive model and data for metals subjected to large strains, high strain rates, and high temperatures." Proceedings of the 7th International Symposium on Ballistics, The Hague, the Netherlands, April 1983.

[14]Johnson, G.R.; Cook, W.H. "Fracture characteristics of three metals subjected to various strains, strain rates, temperatures, and pressures." Eng. Frac. Mech., 1985, 21, 31-48.

Proceedings of the 2019 Hypervelocity Impact Symposium
HVIS2019
April 14-19, 2019, Destin, FL, USA
HVIS2019-024

Extended Abstract+

Simulating Lateral Drift of a Shaped Charge Jet in ALEGRA

Matthew J. Coppinger[a]*, W. Casey Uhlig[a], John H. J. Niederhaus[b]

[a]*US Army Research Laboratory, 328 Hopkins Rd, Aberdeen Proving Ground, MD 21005*
[b]*Sandia National Laboratories, P.O. Box 5800 MS-1323, Albuquerque, NM 87185*

1. Introduction

Shaped charge jet (SCJ) research has long been an active area for industrial, academic, and defense organizations. Traditionally, the depth of penetration (DOP) has been one of the most important metrics for the evaluation of shaped charge jet performance, and early 1D analytical penetration models based on hydrodynamic penetration were created with this metric in mind [1]. As the standoff of a shaped charge jet increases, the DOP reaches a maximum and then begins to decrease. A simple 1D hydrodynamic penetration model must account for the totality of the jet material on axis penetrating, and as a result experimental DOP at longer standoffs is lower than the analytical models predicted. Some analytical models reasoned that since a velocity gradient evolves as a SCJ forms, contributions to penetration from jet material below a minimum jet or penetration velocity should be eliminated. These were better able to account for the difference between analytical hydrodynamic and experimental DOPs [2]. The actual difference between analytical hydrodynamic penetration theory and experimentally recorded values is now regarded to be a result of 3D phenomena including particle tumbling and motion transverse to the jet axis known as lateral drift [2]. The origins of these 3D phenomena have been attributed to sources including variability in the uniformity of the explosive charge or the microstructure of the liner [2,3].

The effect of explosive non-uniformity or liner microstructural variations on particle tumbling or lateral drift in a SCJ should be observable using a continuum level shock hydrocode simulation if the code is capable of modeling this variability. In this study, the manifestation of non-uniformity or microstructure is approximated through stochastic variations in material parameters at the start of a continuum level simulation. Sandia National Laboratories' multiphysics shock hydrocode, ALEGRA [4], is used to computationally examine the extent to which stochastic variations in the explosive charge and liner material parameters contribute to lateral drift in a SCJ.

This study builds upon previous computational work for 2D copper SCJs that was successful in modeling key experimentally measured physical metrics including jet core temperature and jet tip velocity [5,6]. A 2D framework is not, however, sufficient to simulate fundamentally 3D lateral drift; thus, the simulations for this work are conducted using a 3D Cartesian mesh for the full domain. Correctly capturing lateral drift in 3D could alleviate the need for corrective analysis that might be required in 1D or 2D when considering the DOP. ALEGRA offers significant versatility when accommodating stochastic variations in material parameters, and it has been used successfully to model aleatory uncertainty in ceramic materials [7]. In this study we apply aleatory uncertainty to the explosive and the metallic liner of a SCJ. To limit the scope of this study to a manageable size, four simulation parameters were chosen to apply variability. Since the yield strength of copper has a dependency on grain size [8], the first parameter of interest selected to be varied was the initial yield strength of the liner material. The second material parameter was the initial fracture pressure of the copper liner material. Given the physical processes involved in liner collapse and jet extrusion and the fact that a wide range of values for the fracture pressure of copper exists in the literature [9], it is postulated that liner material fracture pressure could play a role in lateral drift. Since the Weibull distribution has been shown to approximate variations in material parameters [10], this distribution was applied to the initial yield strength and the initial fracture pressure. The densities

+ Full paper can found at https://www.sciencedirect.com/journal/international-journal-of-impact-engineering/special-issue/10MTGP5W4VJ

* Corresponding author.
E-mail address: matthew.j.coppinger.civ@mail.mil

of the liner and the explosive were the final two parameters that included variability. The density variations for the liner and the explosive employed a uniform distribution ranging from 1% below the density at ambient temperature to 1% above. Since materials used for shaped charges are highly uniform, these variations are likely an exaggeration of actual physical features; however, they were chosen to provide insight into the magnitude of drift possible with material variations.

2. Configuration and Diagnostics

To capture the fundamentally 3D nature of a SCJ event, an Eulerian mesh consisting of a Cartesian grid with cubic elements was implemented in ALEGRA. Edge sizes for mesh elements were 1/3 mm and the simulation duration was 300 μs. A shaped charge with a 65-mm charge diameter was situated so that the jet axis coincided with the *Z*-axis of the Cartesian coordinate system and the jet traveled in the positive *Z*-direction. The copper liner was modeled using tabular SESAME EOS 3336 and the Johnson Cook yield model [11], while the explosive used the analytical JWL EOS for LX-14. 60 cm of air was included in the mesh ahead of the mouth of the liner to provide travel space for SCJ analysis.

Lagrangian tracers were used to evaluate the lateral drift of the jet as it traversed the mesh. ALEGRA has a feature that allows tracers to be activated in a simulation at a predetermined time. This functionality was employed to enter Lagrangian tracers into a shaped charge jet as it was extruding. Beginning at 20 μs four tracers were placed annularly in the jet at an equal spacing; two in the *XZ*-plane and two in the *YZ*-plane. Each tracer was located in the center of the third cell from the jet axis 15 mm downstream (positive *Z*-direction) of the stagnation point of the emerging jet. Additional groups of Lagrangian tracers were activated in the jet in this manner at ten-microsecond intervals until a total of six groups of four tracers was available for analysis. Allowing an interval between the activation times of the tracers provided a sampling of data with differing axial velocities. For an axisymmetric jet, displacement normal to the plane in which they were placed is not expected to occur. Thus, the *XZ* and *YZ*-planes are used as reference planes in this study, and the displacement of the tracers normal to the reference planes in which they were placed is used to approximate the magnitude of the lateral drift. One can imagine that each group of four tracers is contained in a cross-sectional slice of a single jet particle moving through the mesh. The average displacement of each pair of coplanar tracers normal to their reference plane provides *X* and *Y* coordinates that can be used to give an approximate location of the center of a particle in a particular *Z*-plane as it moves through the mesh. These *X* and *Y* coordinates can then be used to calculate the magnitude of the radial displacement from the jet axis, in this case the *Z*-axis.

Material parameter variations were included during initialization through the use of space-filling curves. The initial yield strength and fracture pressure of the copper were varied using the modified Weibull distribution included in ALEGRA with a Weibull modulus of 4 and median values of 89.7 MPa and -2.3 GPa, respectively. Lagrangian material tracers were used in these simulations to track material properties at a subgrid resolution. When applying parameter variations in ALEGRA the aggregate size of the variation may be selected. In this study three aggregate sizes ranging from 0.5 to 3 mm were chosen for comparison. Figure 1 shows pseudocolor plots depicting the stochastic variations applied to the initial liner yield strength and the density of

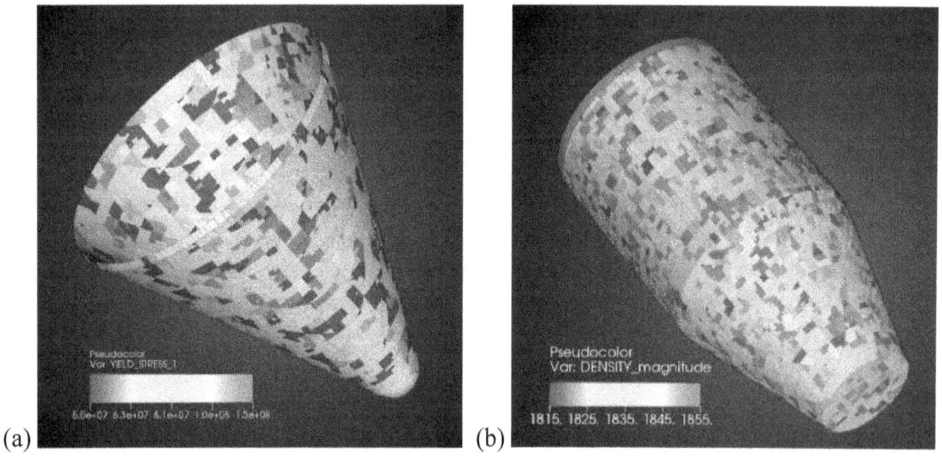

Fig. 1. Pseudocolor plots of (a) variations in the initial yield stress of the copper liner with a 3 mm aggregate size and (b) variations in the density of the explosive with a 3 mm aggregate size.

the explosive.

3. Results and Analysis

Figure 2 compares the magnitudes of the radial displacements for a uniform baseline simulation in which no material variations were included in either the liner or explosive with simulations in which variations were included in the initial yield strength of the Johnson Cook yield model for the copper liner, the initial fracture pressure of the copper liner, and the densities of the liner and the explosive. The magnitude of the radial displacement from the jet axis shown in Figure 2 indicates that some lateral drift is present even for the baseline case in a 3D Eulerian simulation. If the jet tip radius is approximated to be 2.5 mm, the largest particle drift from the jet axis for the baseline simulation was 11.4% of the jet tip radius at 300 μs, which is not insignificant. Also note that the magnitude of the displacement from the jet axis can be smaller than the size of a cell edge. These values are suitable as a subgrid measurement because of the finite element basis functions that support the nodal displacements in ALEGRA. In this study, the magnitude of the radial displacement from the jet axis at the final time available for each tracer group was used as a comparative measure for the lateral jet drift. In effect, the motion of six arbitrary jet particles with a range of initial velocities are directly compared between simulations to determine the effects of various stochastic variations on lateral drift.

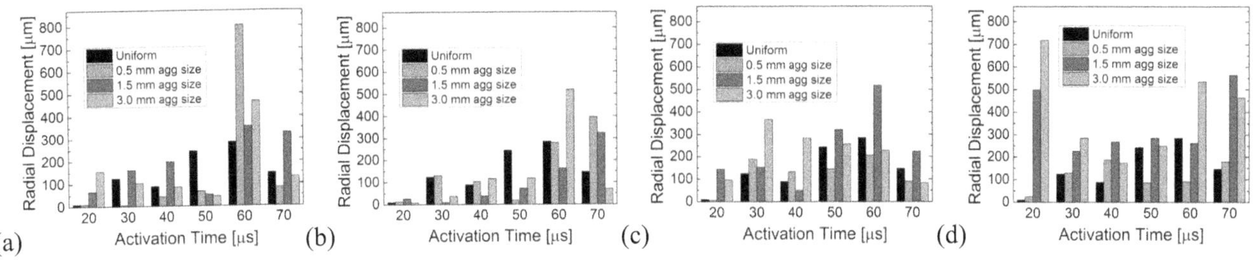

Fig. 2. Comparison of the simulated magnitudes of the radial displacements for each group of Lagrangian tracers in a uniform baseline simulation and simulations with variations at each aggregate size included in (a) the initial yield strength of the Johnson Cook yield model for the copper liner, (b) the initial fracture pressure of the copper liner, (c) the density of the explosive, and (d) the density of the liner.

With variations included in the initial yield strength of the liner, each aggregate size exhibited at least one lateral displacement that was greater than the maximum of baseline case; however, the majority of the displacements were comparable to the baseline displacements. When stochastic variations were applied to the initial fracture pressure, the largest aggregate variation size of 3 mm resulted in a lateral displacement of 520 μm which is nearly 21% of the jet tip radius (2.5 mm) while the remainder of the displacements were comparable to the baseline displacements. With stochastic variations applied to the explosive density, the maximum lateral drift occurred again in the tracer group that was activated at 60 μs. This time, however, the maximum occurred for the second largest aggregate size and the magnitude of the lateral displacement was again about 20% of the jet tip radius. For the liner density variations, the lateral drift of the jet appears to increase noticeably for the two largest aggregate sizes. Nearly all of the displacements for the two largest aggregate sizes are comparable to the baseline maximum and several exceed it significantly. For the largest aggregate variation size of 3 mm the magnitude of the lateral displacement increased to over 700

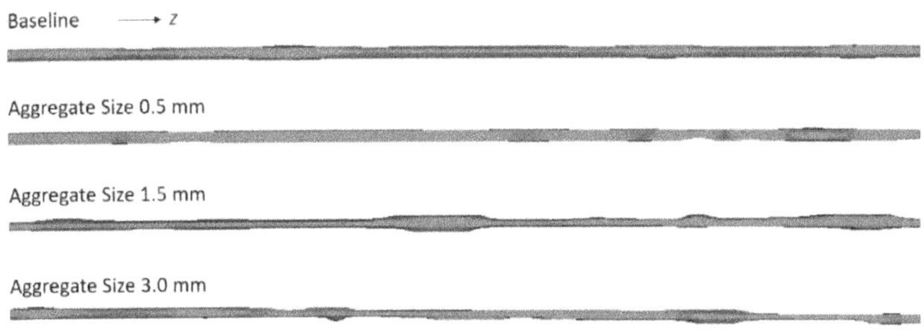

Fig. 3. Profiles of a 6 cm portion of the jet taken 100 μs after detonation showing the effects of increasing a variation's aggregate size on the symmetry of the jet.

μm which is nearly 28% of the jet radius. Notably, the maximum value occurs in the first group. This is an indication that density variations are capable of inducing lateral drift in any portion of the jet.

Figure 3 shows a comparison of jet profiles for the baseline simulation and the simulations with liner density variations. The profiles are approximately 6 cm in length and were generated 100 μs after detonation. The asymmetric effects of incorporating variability into the liner is visually evident from the profiles in Figure 3. Asymmetry in the jet is an indication of lateral drift since material must be moving non-uniformly with respect to the jet axis. The significant increases in lateral drift for the two largest aggregate sizes appears to be evident from Figure 3 as well. In each jet, there are portions that appear symmetric with respect to the jet axis. If the Lagrangian tracers used for evaluating the magnitude of the lateral drift were initialized in one of these symmetric jet particles, it will not appear to have drifted. This gives insight into why some of the simulations that included stochastic variations appeared similar to the baseline case for some tracers while other tracers showed significant differences.

4. Conclusion

This study demonstrated that stochastic variations in initial material parameters are capable of affecting lateral drift in a SCJ. A measurable displacement normal to the jet axis, a maximum of 11% of the jet tip radius, was observed even in a baseline Eulerian simulation without any stochastic variation. For the cases of variations in initial liner yield strength, liner fracture pressure, and explosive density the highest increases in the magnitude of the lateral drift were measured to be approximately 30%, 21%, and 20% of the jet tip radius, respectively. With liner density variations of ± 1%, the overall lateral drift increased significantly for the variations with the two largest aggregate sizes, and visible differences in the jet profiles were apparent. Ultimately, simulations that involve penetration will be necessary to determine full extent of how changes in lateral drift though stochastic parameter variation affect the final DOP.

Acknowledgements

This work was supported in part by a grant of computer time from the Department of Defense High Performance Computing Modernization Program at the United States Army Research Laboratory Department of Defense Supercomputing Resource Center and the Terminal Ballistics Frontier project. The authors would like to thank Andrew Porwitzky and Robert Doney for meaningful discussions.

Sandia National Laboratories is a multimission laboratory managed and operated by National Technology and Engineering Solutions of Sandia, LLC., a wholly owned subsidiary of Honeywell International, Inc., for the U.S. Department of Energy's National Nuclear Security Administration under contract DE-NA0003525.

References

[1] Birkhoff, G., MacDougall, D. P., Pugh, E. M., Taylor, S. G., 1948. Explosives with Lined Cavities, Journal of Applied Physics 19, p. 563.

[2] Walters, William P. and Zukas, Jonas A., 1989. Fundamentals of Shaped Charges, John Wiley & Sons, Inc, New York, NY.

[3] Aseltine, C.L., "Analytical predictions of the effect of warhead asymmetries on shaped charge jets," Aberdeen Proving Ground (MD): Ballistic Research Laboratories, ARBRL-TR-02214 1980.

[4] Robinson A.C., Brunner T.A., Carroll S., Drake R., Garasi C.J., Gardiner T., Haill T., Hanshaw H., Hensinger D., Labreche D., Lemke R., Love E., Luchini C., Mosso S., Niederhaus J., Ober C.C., Petney S., Rider W.J., Scovazzi G., Strack O.E., Summers R., Trucano T., Weirs V.G., Wong M., 2008. "ALEGRA: an arbitrary Lagrangian-Eulerian multimaterial, multiphysics code," Proceedings of the 46th AIAA Aerospace Sciences Meeting, paper AIAA-2008-1235.

[5] Uhlig W.C., Hummer C.R., 2013. "In-flight conductivity and temperature measurements of hypervelocity projectiles," Procedia Engineering (Proceedings of the 12th HVIS) 58, p. 48–57.

[6] Zellner MB, Vunni GB, "Photon Doppler velocimetry characterization of shaped charge jet formation," Aberdeen Proving Ground (MD): Army Research Laboratory, ARL-TR-5961 2012.

[7] Strack, O.E., Leavy, R. B., Brannon R.M., 2015. Aleatory Uncertainty and Scale Effects in Computational Damage Models for Failure and Fragmentation, International Journal for Numerical Methods in Engineering 102, p. 468-495.

[8] Yuejian, W., Jianzhong Z., Qiang W., and Yusheng Z., 2013. Grain Size Effects on the Compressibility and Yield Strength of Copper. Journal of Physics and Chemistry of Solids 74, p. 75–79.

[9] Razorenov S.V., Zaretsky E.B., Savinykh A.S., 2014. "The spall strength and Hugoniot elastic limit of mono-crystalline and polycrystalline copper near melting temperature." Journal of Physics: Conference Series 500.

[10] Weibull, W.A., 1939. A Statistical Theory of the Strength of Materials, Ingeniors Ventenskaps Akademien Handlinger 151, p. 5-45.

[11] Johnson, G.R., Cook, W. H., 1983. "A Constitutive Model and Data for Metals Subjected to Large Strains, High Strain Rates, and High Temperatures" in Proceedings of the 7th International Symposium on Ballistics, Hague, Netherlands.

Proceedings of the 2019 Hypervelocity Impact Symposium
HVIS2019
April 14-19, 2019, Destin, FL, USA
HVIS2019-030

Experiments using a light gas gun to investigate the impact melting of gunshot residue analogues

V. Spathis[a]* and M. C. Price[a]

[a]*School of Physical Sciences, University of Kent, Canterbury, Kent, CT2 7NH, UK.*

Abstract

In this set of experiments, the versatility of the University of Kent's light gas gun was utilised to obtain a selection of corroborative data regarding the formation and impact of metallic gunshot residues onto high purity silicon wafers. The results from the two experiments are presented. The first experiment investigated how the formation of metallic residues varied as gunshot residue analogues traversed through air under a range of pressures from 0.056 millibar (5.6 Pa) to 1 bar (100 kPa), using solely the energy released during primer ignition; the second involved firing a metallic powder mix of pre-determined composition (via a split-sabot) under vacuum at two velocities- 500 ms^{-1} and 2000 ms^{-1}. This ensured that there was no ignition or heating of the powders, unlike the first experiment, and so the morphology of the particles collected would be solely due to impact. The residues on the substrates were then analysed using a cold Field Emission Gun Scanning Electron Microscope (FEG) and Energy Dispersive X-ray (EDX) detector. By separating the ignition process of the primers from the residue impacts, it allows for a closer look into the formation of these particles and helps determine whether their varied morphologies are due to the heating caused during the activation and combustion of the primer or whether its due to impact melting. This information can aid in the understanding of metallic particle formation in different pressure environments and give insight into the physical state of firearm residues when they impact a surface. Hydrocode modelling was also incorporated to corroborate the results observed during these experiments and gave results which mimicked the experimental data.

Keywords: Gunshot residue; Light Gas Gun; Impact melting; EDX-SEM; Hydrocode modelling.

---

1. Introduction

The development of light gas guns (LGG) commenced after the Second World War in order to study terminal ballistics [1]. However, over the years, due to their unique abilities to create various shock environments and fire projectiles into the hypervelocity range, their utility has evolved and as a result, they are now used in a range of diverse fields such as material science- studying materials undergoing dynamic shock compression; planetary science- studying the process of cratering on solar system bodies; as well as the highly topical field of astronautics and evaluating the hazard of orbital debris to space assets. In this set of experiments, the versatility of the University of Kent's LGG [2] was utilised to obtain a selection of corroborative data regarding the formation and impact of metallic gunshot residues (GSR). When a cartridge-based weapon is fired, both gaseous and solid residues are expelled from the barrel. These consist of both organic and inorganic particles from the burnt and un-burnt primer, propellant, and can even contain traces from the bullet, cartridge or the firearm barrel itself [3] [4] [5]. As these residues rapidly cool and condense in-flight, they form into inorganic, spheroidal particles of a characteristic elemental composition that is used to identify them as GSR [6]. For a decade, the analysis of these residues has played an integral part in forensic investigations with regards to shooting cases. Prior investigation of GSR using subsonic rim-fire ammunition [7] led to the discovery that at certain distances from the firearm the residues exhibit irregular, typically molten, morphologies as they hit the target, rather than the traditionally acknowledged spheroidal shape [8] [9]. The morphologies of these impact-disrupted GSR were classified according to the degree of melt exhibited by the particles [7] and investigations commenced into determining whether their morphology was temperature-related, due to their molten appearance, or due to their impact onto the substrate.

* Corresponding author. Tel.: +44 1227 823266.
E-mail address: v.spathi@kent.ac.uk.

1.1. Materials and Method

Both of the experiments were carried out using the University of Kent's two-stage LGG. The substrates for the experiments were also kept consistent - 600 micrometre thick optically polished, high purity silicon wafers. This substrate was originally chosen as a way of determining the velocity at which the residues hit the target- the premise being that the stress induced on the silicon upon impact would cause a shift in the silicon's distinctive Raman peak (at 520 cm^{-1}) due to the strains induced within the crystal lattice [10]. Prior to use, the silicon substrates were cleaned using 99.5% isopropyl alcohol (IPA) and cotton buds to remove any surface contaminants. Once dried, they were each fixed onto an aluminium SEM stub with double sided adhesive carbon tape and placed in an enclosed container. They were only taken out for the test firing in the lab and gun room using gloves to avoid any external contamination. This ensured that any residue-like particles found on them during the analytical process were a product of the firing itself and not contamination.

1.2. Experimental Setup

1.2.1 Variable Pressure Experiments

In order to gain a deeper insight into the formation of these metallic particles as they move through the air, the first experiment involved firing primer under different pressures. The primer was placed in an empty shotgun cartridge and loaded in the same way as when doing a standard high speed shot on the Kent LGG. The pump-tube (the first stage of the light gas gun) was then used as a sealed gas container, allowing for the internal gas pressure to be accurately controlled. The gas that was used was air, and the pressures investigated were 2 bar (200 kPa), 1 bar (100 kPa), 200 millibar (20 kPa), 100 millibar (10 kPa), 50 millibar (5 kPa), 1.8 millibar (180 Pa), 1.5 millibar (150 Pa), 0.52 millibar (52 Pa), 0.33 millibar (33 Pa) and 0.056 millibar (5.6 Pa). The target was then fixed at the entry to the launch-tube, thereby enabling particles to land on the substrate. The primer was initiated using an external striker and the residues were allowed to move down the pump-tube based solely on the energy released during the ignition of the primer

1.2.2 Impact Melting Experiments

To examine the extent of impact melting of the particles, the LGG was used in a more conventional way. After having examined a number of particles of all morphologies from the previously acquired rim-fire ammunition data, an average relative elemental composition was obtained for the particles that had in some way not fully formed into a sphere. This consisted of the 'characteristic' GSR elements: lead (Pb), antimony (Sb) and barium (Ba), as well as copper (Cu). The corresponding metallic powders were obtained (with the exception of barium which, due to its reactivity, was replaced by barium nitrate) and sieved within a range of 20 micrometer - 50 micrometer. Two powder mixes were then created using the pre-determined relative elemental composition (75% Pb, 15% Ba, 7% Sb, 3% Cu by wt%) considered to represent the typical composition of a GSR particle. The first projectile powder mix was used after simply mixing the powders, whereas the second projectile powder mix was heated to 913 K and allowed to cool, exceeding the melting point of three out of the four components to create a more intimate mix and be representative of GSR that was molten (or partially molten) prior to impact due to the heat of primer and powder ignition. The rest of the experiment was identical for the two powders. The powders were loaded in a split-sabot and two shots were fired under vacuum at two velocities- 500 ms^{-1} and 2000 ms^{-1} using the same silicon target substrates. This ensured that there was no ignition or heating of the powders, unlike the first experiment, and so the morphology of the particles collected would be due solely to their acceleration and impact.

1.3. Analytical Method

The analyses of all the samples obtained were carried out using two different Scanning Electron Microscopes (SEM) equipped with Energy Dispersive X-Ray detectors (EDX). Firstly, a Hitachi S-3400 SEM with an Oxford Instruments *X-Max* 80 mm^2 EDX detector was used to acquire low magnification topographical maps of the substrates, showing the abundance and dispersion of particles on each sample. Following that, a Hitachi S-4700 cold Field Emission SEM (FEG-SEM) with a Bruker *X-flash* 5060f flat-quad EDX detector was used for the individual morphological and elemental particle analysis due to its high spatial resolution allowing for elemental discrimination down to the sub-micron level. An accelerating voltage of 10 kV was used for all analyses to ensure sufficient X-ray emission stimulation from the metallic elements present, thereby providing discriminatory X-ray peaks for each, while still retaining topographical detail.

2. Results

2.1 Particle formation

Some typical maps from the particle formation samples acquired using the SEM are presented in Figures 1 and 2 below. Fig.

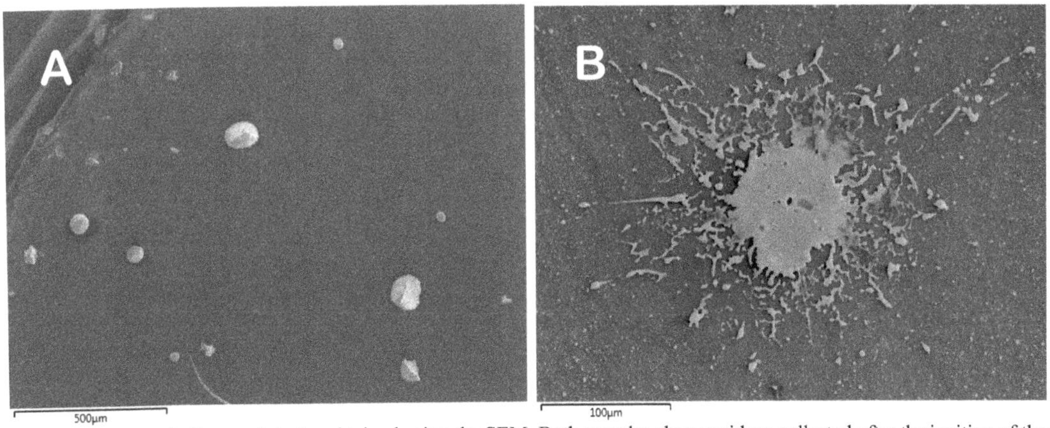

Fig. 1: Secondary electron images of silicon substrates obtained using the SEM. Both samples show residues collected after the ignition of the primer at different pressures. **A**: Image showing the particle distribution and morphology after particle launch under atmospheric pressure (1 bar). Beam accelerating voltage = 10 kV. **B**: Image of an impacted particle ejected during the ignition of the primer under a pressure of 1.5 mbar. Beam accelerating voltage = 10kV. Note the much more molten appearance indicating significantly less in-flight cooling of the particle during its passage down the pump tube.

1 displays residues found after firing the primer under atmospheric pressure (Fig. 1A) and an example of the residues seen at

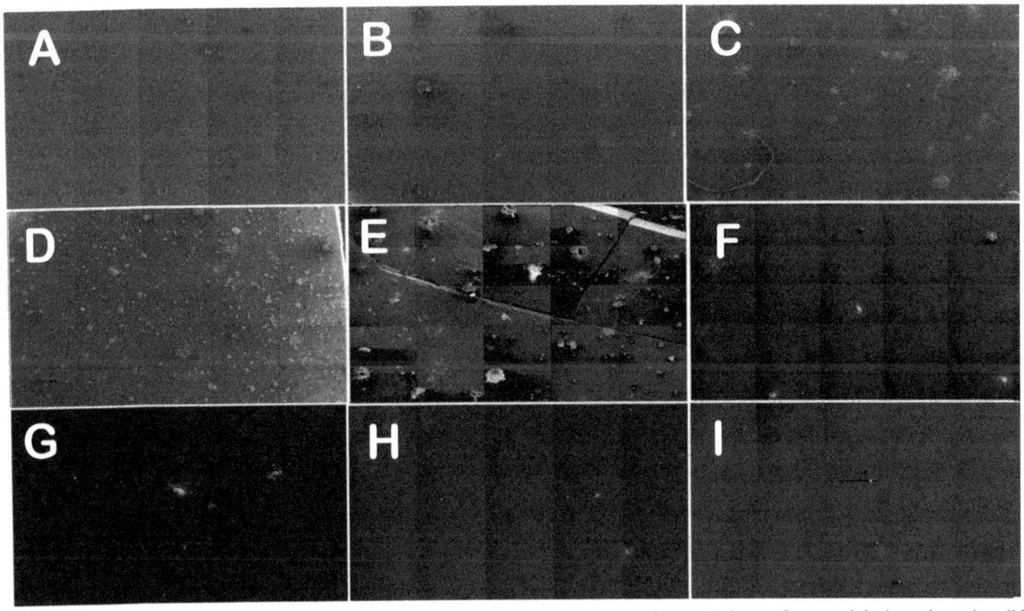

Fig 2: Secondary electron maps of each silicon substrate showing the particle distribution and morphology after particle launch under different air pressure environments. **A:** 0.056 millibar, **B:** 0.33 millibar, **C:** 0.52 millibar, **D:** 1.5 millibar, **E:** 1.8 millibar, **F:** 50 millibar, **G:** 100 millibar, **H:** 200 millibar, **I:** 2000 millibar. Beam accelerating voltage = 10 kV.

1.5mbar (Fig. 1B), which is also reproduced by the AUTODYN hydrocode modelling discussed in Section 3. Fig. 2 shows the results over a range of pressures from near-vacuum to two times atmospheric pressure. From these data, it is apparent that as the pressure changes, so does the particle morphology and spatial density frequency on the silicon substrate.

2.2 *Particle Impacts*

The results indicate that the substrates used when firing at 2000 ms^{-1} display a significant amount of cratering and surface damage, whereas the 500 ms^{-1} target substrates barely show any residues on them at all. An elemental and morphological comparison of two impacted particles of different velocities is shown in Fig. 3 below. Although both particles appear to be molten to some extent, only the particle launched at 2000 ms^{-1} exhibits visible, structural damage to the silicon substrate. At 500 ms^{-1}, there are three distinct particle populations present: molten (Fig. 3B), semi-solid, structured with a number of nodules attached to the main body of the particle, both of which are displayed in Fig. 4 and Fig. 5 respectively.

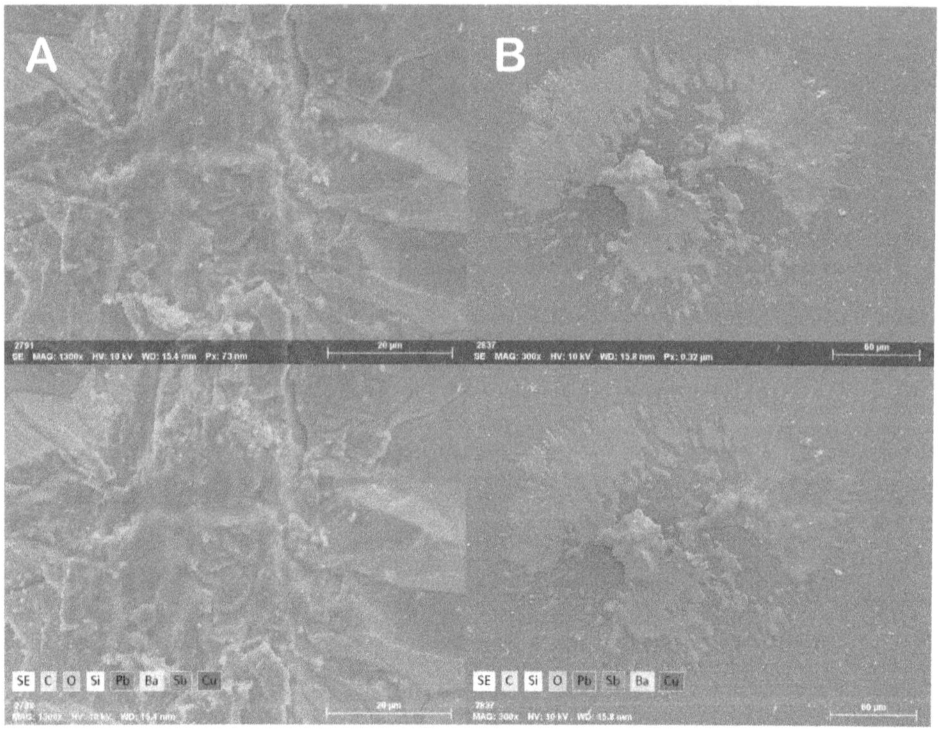

Fig. 3: FEG SEM images (top panels) and overlaid EDX images (bottom panels) showing molten impact Pb residues formed at different velocities. **A**: 2000 ms^{-1} residue inside a silicon crater with some antimony on the outskirts of the crater. *Magnification 1300×, beam accelerating voltage= 10kV*. **B**: 500 ms^{-1} residue with a relatively structured particle centre and no evidence of damage to the silicon substrate. *Magnification 3000×, beam accelerating voltage= 10 kV.*

3 Hydrocode modelling

The experimental data described above show predominantly that Pb-impactors were either molten on impact, or melted during the impact due to the liquid appearance of the residue. To gain insight into the processes going on during impact, and the initial launch of the projectile, hydrocode modelling was performed using Ansys' AUTODYN [11] software. The goal was to investigate the temperatures experienced during the launch, and the subsequent impact, to investigate the point at which significant impact melting of the projectile material begun. 3-D simulations (with a resolution of 20 cells across the radius of the projectile) were ran using 100% Pb (melting point = 760 K) spherical projectiles 50 microns in diameter impacting silicon at a range of velocities (100 ms^{-1} – 2000 ms^{-1}) and at a range of initial temperatures (from 300 K – 900 K). The Pb incorporated a Steinberg-Guinan [12] constitutive model which incorporates both strain-hardening and thermal softening. Additionally the projectile was modelled using Smoothed Particle Hydrodynamics (SPH) impacting a lagrangian target mesh. The rationale for this was that

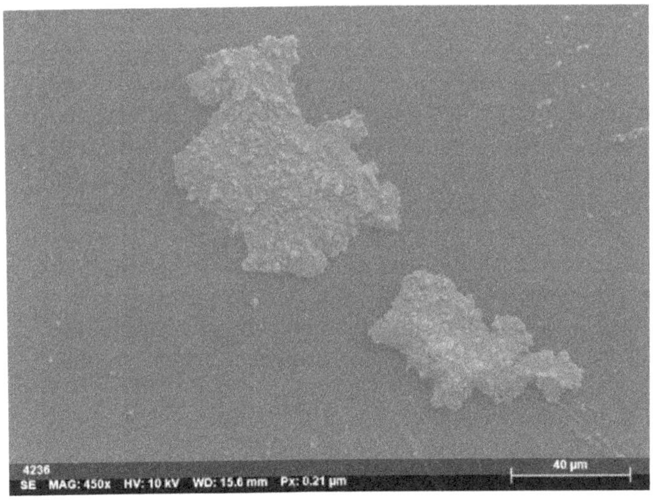

Fig. 4: FEG/SEM image of semi-solid particles at 500ms^{-1} using the shaken, unheated powder. The particles appear to be on the precipice of impact melting due to its flattened appearance and homogenous matrix. Magnification = 450×

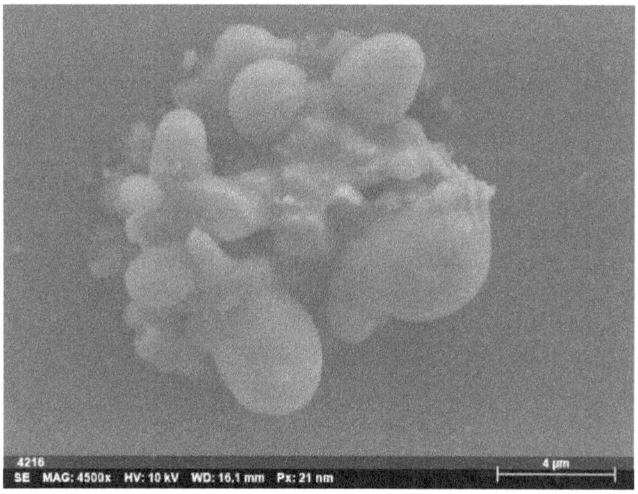

Fig. 5: FEG/SEM image of a nodular particle at 500ms^{-1} using the intimately mixed, heated powder. The seemingly condensed particles appear to be coalesced via a molten particle segment in the centre. This population of particles is also observed at the same velocity using both powders. Magnification = 4500×

initial modelling carried out using a lagrangian mesh for the projectile resulted in numerical erosion of highly distorted cells, which also were the cells with the highest temperature, which led to the model "running cold". By changing to an SPH solver the problem of numerical erosion is not encountered and it is much more suitable for modelling the disruption of small impactors travelling at high speed. Similar models using a range of solvers have been used successfully to model the splat formation of metal aerosols during thermal and cold spraying coating processes (i.e. [14] – [17]).

3.1 Compressional heating due to projectile launch process

One hypothesis was that during the launching process (where the projectile can experience accelerations of ~1×10^6 ms^{-2}) the projectile was subjected to sufficient force to experience partial or full melting, and thus was exiting the sabot in an already-molten state. To test this hypothesis a 2-D multi-material Euler model was setup to simulate the acceleration by a driver gas at an initial pressure of 20 kbar of a Pb-loaded sabot down the 0.7 metre long launch-tube of the Kent LGG. Gauges were placed within the Pb to measure the velocity, pressure (and thus temperature) of the projectile during the acceleration process. The ambient (pre-launch) temperature was set to 300 K.

Figure 6 below shows the results of the simulation that gave a final projectile velocity of 1600 ms^{-1} (it should be noted that the sabot was still being accelerated at the point it exited the launch tube). The results of the simulation suggest that a solid

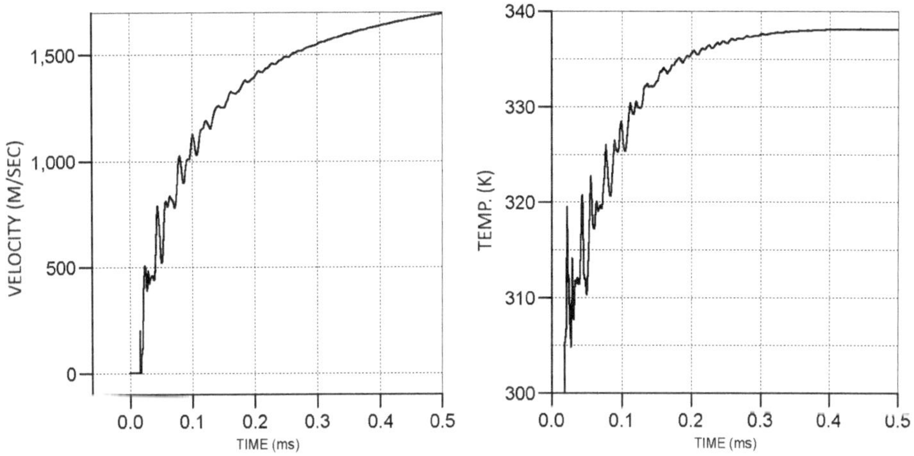

Fig 6: Graphs showing the modelled velocity (left) and temperature (right) of a solid Pb-projectile during acceleration down the launch tube of the Kent LGG.

Pb projectile would undergo some heating due to the launch acceleration, but that the amount of heating would be modest (~40 K) which would be insufficient to cause any appreciable softening or melting. However, in the experiments described above, the metal powder had been sieved and was loosely packed into the sabot forming a pseudo-porous projectile. Modelling of impacts into porous bodies (see, for example, [13]) has shown that the temperature experienced by a porous target can be significantly higher than for impacts into solid targets. Therefore, we ran a second, fully 3-D, simulation (with the same spatial resolution as the 2-D simulation) which modelled the projectile as a discrete set of Pb particles (with an overall porosity of ~50% - based on the void spacing inside an actual sabot filled with metallic powder) confined within the sabot. The results of this simulation show a much wider range of temperatures within the projectile material, with a maximum of ~750 K, and significant compaction of the projectile particles during the launch acceleration. It is therefore likely that during launch the individual particles making up the projectile experienced differing degrees of heating and that some of the particles could be significantly softened, or even totally molten, at the time of impact.

Experimental evidence for this 'flattening' of the projectile during the launch phase is given by the output trace from the oscilloscopes mounted on the LGG to measure the projectile's speed. The output is from a pair of laser curtains situated in the gun's range. As a projectile passes through the laser curtain it interrupts the laser beam and a pulse is seen on the oscilloscopes. The height (and width) of the pulse gives an indication of the size of the projectile. As the projectile material was packed into the sabot as a finely sieved powder is was not expected to give a trace (as an individual metallic particle is too small to be detected). However, what was observed was a well-defined, symmetrical peak indicating a mm-sized consolidated projectile. These experimental data therefore support the modelling result that the loose powder gets compacted into a 'slug' and that the temperature of the grains within this slug have a distribution of temperatures due to the compaction during the launch phase.

3.2 Results from modelling of Pb impactors as a function of impact velocity

Table 1 below gives the modelled temperature rise of a 50 micron diameter sphere impacting a silicon target at a range of impact velocities. The quoted "ΔT" gives an indication of the amount of heating experienced averaged across the whole projectile, however in order to investigate localised heating and melting a range of 3-D models were ran to capture the changes in splat morphology: these are presented in Fig. 7 below.

Table 1: Results from 2D simulations of impacts of a solid 50-micron diameter Pb lead sphere onto a silicon target. ΔT represents the temperature increase of the whole sphere if the temperatures of all nodes are averaged and the third column is the percentage of the kinetic energy that is needed to produce the temperature increase.

Impact Velocity ($m\ s^{-1}$)	**ΔT (K)**	**Kinetic Energy Converted to Heat (%)**	**Impact Velocity ($m\ s^{-1}$)**	**ΔT (K)**	**Kinetic Energy Converted to Heat (%)**
100	38	95	750	218	10
200	106	66	1000	245	6
400	200	31	1500	356	4
500	212	21	2000	495	3

Interestingly there is a distinct change in the splat morphology from an impact speed of 500 – 1000 ms^{-1} which is due to crater formation in the target material. As the crater forms, the projectile ejecta is forced back along the line-of-flight and no longer lands on the target surface, additionally the crater contains the projectile material so it remains localised.

3.3 Results from modelling of Pb impactors as a function of temperature with a fixed impact speed (200 ms^{-1}).

Additional models were ran to look at the splat formation as a function of impactor temperature (Fig. 8). These results are fascinating and show a definite progression from a solid splat state, to a very liquid state where the projectile material is totally molten and shows a fine filigree patination, which has been observed experimentally with nylon and glass projectiles [18], [19].

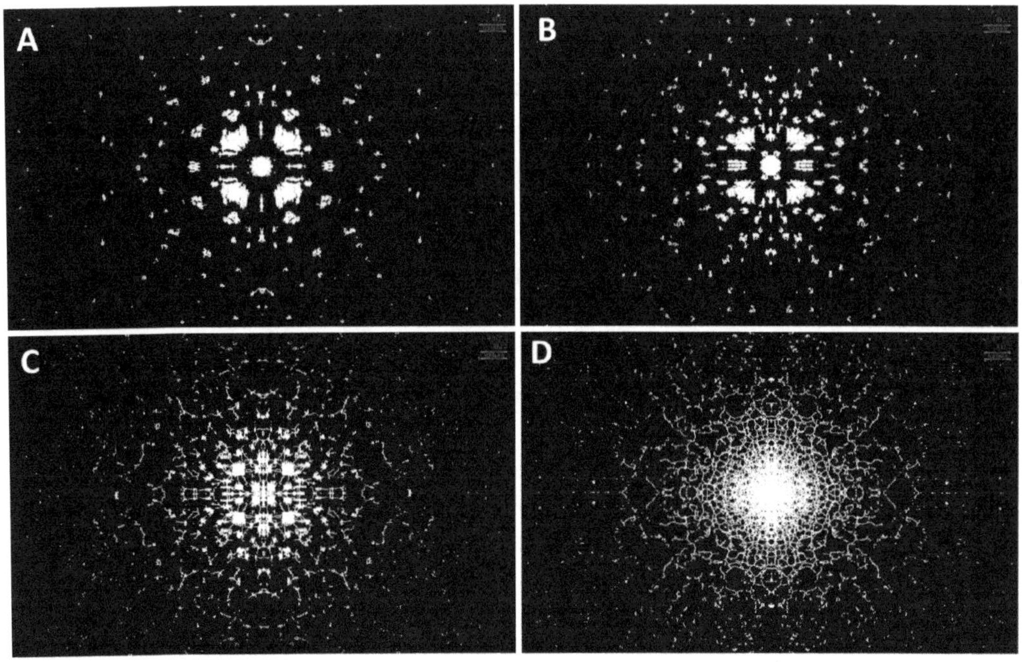

Fig. 7: Splat morphology (Pb is white, black is the silicon substrate) as a function of impactor velocity for an initial fixed impactor temperature of 300 K. **A:** 100 ms^{-1}, **B**: 200 ms^{-1}, **C:** 500 ms^{-1}, **D:** 1000 ms^{-1}. Images are to the same scale: each panel is 400 μm across.

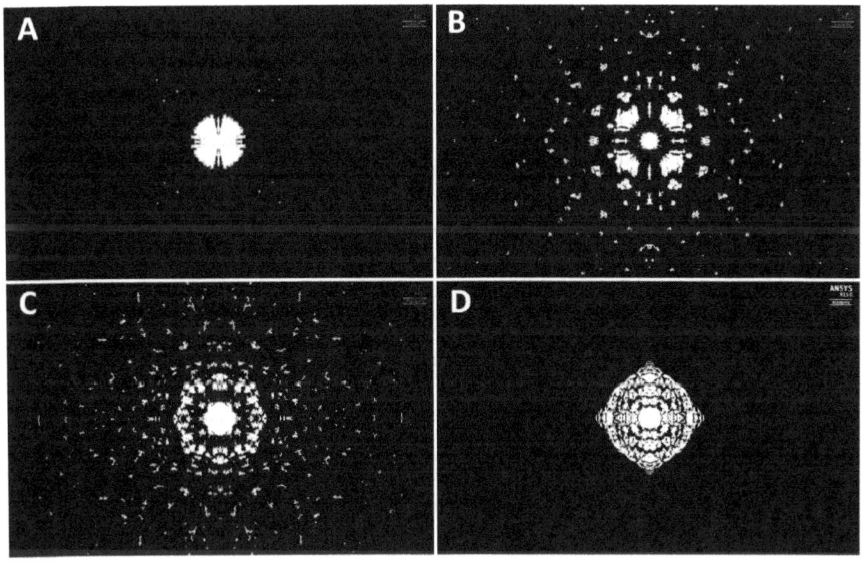

Fig. 8: Splat morphology (Pb is white) as a function of impactor temperature for a fixed impact velocity of 200 ms^{-1}. **A:** 300 K, **B**: 500 K, **C:** 700 K, **D:** 900 K. Images are to the same scale: each panel is 400 μm across.

3.4 Hydrocode modelling: discussion.

The results of the hydrocode modelling seem to capture the observed evolutionary morphologies of Pb splats. Most importantly the results help to interpret the range of morphologies seen for the fixed impact velocity experiments: i.e. that a mix of solid, semi-solid and molten Pb residues were observed which may be due to the inhomogeneous compressional heating experienced by the projectile during the LGG launch process. They also reveal that the size of the splat seems to *decrease* as the impact speed is increased. This seems to be due to the projectile material being ejected back along the impact direction and not landing on the target material and/or the formation of a crater in the silicon which confines the projectile material.

Additional work is planned to refine the hydrocode modelling so that a solid (or porous) projectile can be modelled which includes a mix of metals (i.e. Ba, Sb and Cu).

4 In-flight cooling of particles

A full mathematical model describing the cooling of a particle in-flight is complex (see for example [20], [21]) and beyond the scope of this paper. However, a first-order approximation of the final temperature, T_f, of a sphere radiating into a 0 K radiation sink (i.e. with no re-absorption of radiation from the surroundings) can be found from:

$$T_f = \left(\frac{\alpha T_h^3}{t T_h^3 + \alpha} \right)^{\frac{1}{3}}$$

where

$$\alpha = \left(\frac{m N_A k}{2 M \varepsilon \sigma A} \right)$$

and T_h = the initial (hot) temperature (K), t is the time (s), N_A = is Avagadro's number, M is the molar mass (207 for Pb), m is the mass of the sphere (kg), ε is the emissivity (here assumed to be 1), σ is Stefan's constant, k is Boltzman's constant and A is the area of the radiating surface (m^2). Substituting values for a 50-micron diameter Pb sphere with an initial temperature of 750 K and a velocity of 500 ms^{-1} travelling four metres (i.e. the distance along the LGG) suggests that it would have radiatively cooled to approximately 725 K on impact with the target and we suggest that radiative cooling during flight is not a significant effect.

5 Discussion and Conclusions

Upon examination of the results, it is clear that there are general trends followed in each of the experiments. Firstly, as can be seen in Fig. 1 and Fig. 3, the frequency of the particles present varies with pressure. The relation appears to be that as the pressure decreases, the number of particles increases, suggesting an inverse relationship. At a pressure of 1.5 mbar and 1.8 mbar, displayed in Fig. 3D and Fig 3E respectively, it appears as though the observable particle frequency is at its maximum, despite firing primer at even lower pressures. The particles also exhibit a molten morphology at these pressures, as shown in Fig. 2, which is corroborated by the results obtained during the hydrocode modelling (Fig. 8). However, what is not immediately observable in the SEM images in Fig. 3 is that the samples collected using a pressure of 0.056 mbar and 0.33 mbar (Fig. 3A and Fig 3B respectively) were completely coated in a dark residue that could be seen under ambient light, obscuring smaller particles. This meant that trying to get overall topographical detail with the SEM was challenging due to this additional layer of residue. Conversely, as the pressure is increased to atmospheric pressure and beyond, the particle frequency per substrate decreases, with only a few particles being visible. This is most likely attributed to the densely populated pump tube as a result of the increased pressure. By increasing the air pressure in the pump tube, a larger number of particles are present so there are too many gas particles packed within the tube that they inhibit the movement of the residues coming down towards the substrate.

With respect to the particle morphologies, it is evident that the general pattern followed by the residues is that the lower the pressure, the more molten their appearance. When looking at the individual particles on the samples below atmospheric pressure, it is clear that there is a significant amount of progressive melting. At the lowest pressures, it appears as though the particles were either substantially molten, and/or hit the substrate with great speed as there is a thin, residual, 'splatted' outline around each particle core, which is the only part of the particles that is clearly discernible. As the pressure is increased, the 'splats' become more structured, until they reach a semi-molten, coagulated shape at 1 bar (Fig. 1A). However, their structured morphology also appears to be size-dependent as the smaller particles seem to have solidified into spheres whereas the larger ones still retain a molten centre. This is expected as, to first order, the relative particle cooling time is a function of the volume of a particle, and thus is eight times faster for a particle with half the diameter. On the other hand, the larger particles would not have had the chance to cool because of their added size. Adding to that the lack of aerodynamic drag and friction at the lower pressures, there is insufficient time for the particles to cool by the time they reach the substrates.

Comparing the samples from the LGG in Fig. 3, it can be seen that the impacted residues appear to be molten at both velocities examined. Despite the main purpose of this experiment being to determine whether the melting previously seen on the rim-fire samples was due to the heat of combustion or impact melting, there is no clear difference discerned. As it can be seen in Figs. 3, 4 and 5, despite particles being molten even at 500 ms^{-1} they still retain some of their structure. In Fig. 3, the impacted side of the particle is molten, however the remaining part of the particle appears to retain its structure as the impact energy dissipates. However, Fig. 4 shows a lead particle that has collapsed in on itself. The powders all contained spheroidal particles of a diameter between 20 - 50 μm, and so despite the lack of obvious melting, it is evident that the shape of the particle has been deformed either in-flight or due to the force of the impact, which could only occur if the particles were in a semi-molten state when they hit the target. Therefore, despite their solid appearance, it would appear as though at 500 ms^{-1}, the shaken powder residues are on the verge of melting due to impacting the target. The range of molten states observed (from solid to completely liquid) at nominally the same impact velocity can, potentially, be explained by the heterogeneous compressional heating of the individual particles in the projectile mix during the launch process.

Taking into account the results from both of these experiments, it can be established that no firm conclusions regarding the discrimination between the two effects can be made at this stage. Although the impact melting morphologies resemble the impact-disrupted GSR particles more closely, there are still a number of particles that resemble those seen during the combustion heat tests, particularly as the pressure is varied. In order to limit the extraneous effects that may be causing these findings, a range of additional velocities will be investigated in order to establish where there is little/no particle melting on impact, as well as to try and mimic the impact velocity the particles experienced when using the rim-fire ammunition. Both of these will help determine whether there is a singular cause for the molten morphologies observed, or a range of different effects experienced by the particles in-flight. In doing so, this will allow for an accurate determination of the physical properties and metallurgy of GSR.

Acknowledgements

V. Spathis would like to thank the HVIS committee for the award of an Alex Charters scholarship to present these data at HVIS 2019, and Prof. President for his continued encouragement. The authors thank Mike Cole for operating the Kent LGG.

References

[1] Crozier W. D. & Hume W., 1957. "High-velocity, Light-gas gun", Journal of Applied Physics, 28, 8, 892.
[2] Burchell M. J. et al., 1999. "Hypervelocity impact studies using the 2 MV Van de Graaff accelerator and two-stage light gas gun at the University of Kent at Canterbury", Measurement Science and Technology, 10, 4.
[3] Schwoeble A.J. & Exline D.L., 2000, "Current methods in forensic gunshot residue analysis", CRC Press
[4] Wallace J. S., 2008, "Chemical analysis of firearms, ammunition and gunshot residue", CRC Press
[5] Dalby O., Butler D., Birkett J. W., 2010, "Analysis of gunshot residue and associated materials – a review", Journal of Forensic Sciences, 55, 924-943
[6] Wolten G. M. et al., 1979. "Particle analysis for the detection of gunshot residue. I: scanning electron microscopy/energy dispersive X-ray characterization of hand deposits from firing", Journal of Forensic Sciences, 24, 409.
[7] Spathis V., 2017. "Impact-disrupted gunshot residue: A sub-micron analysis using a novel collection protocol", Journal of Defence Technology, 13, 3, 143
[8] Wolten G. M., Nesbitt R. S., Calloway A. R., Loper G. L., 1979, "Particle analysis for the detection of gunshot residue II: occupational and environmental particles", Journal of Forensic Sciences, 24, 2, 423-430
[9] Kara I., Sarikavakc Y., Lisesivdinb S. B., Kasapb M., 2016, "Evaluation of morphological and chemical differences of gunshot residues in different ammunitions using SEM/EDS technique", Environ Forensics, 17, 1, 68-79
[10] Price M. C. et al., 2014. "Using Raman mapping to determine hypervelocity impact peak shock pressures in high purity silicon, germanium, gallium arsenide and peridot targets", European Planetary Science Congress, Abstract EPSC2014-91.
[11] Hayhurst C. J. & Clegg R. A., 1997. "Cylindrically symmetric SPH simulations of hypervelocity impacts on thin plates", International Journal of Impact Engineering, 20, 337.
[12] Steinberg D. J. et al., 1980. "A constitutive model for metals applicable at high-strain rate", Journal of Applied Physics, 51, 1498.
[13] Davison T. M. et al., 2010. "Numerical modelling of heating in porous planetesimal collisions", Icarus, 208, 1, 468.
[14] Fauchais P. et al., 2004. "Knowledge concerning splat formation: An invited review", Journal of Thermal Spray Technology, 13.3, 337.
[15] Zhang M. Y. et al., 2008. "Simulation of droplet spreading, splashing and solidification using smoothed particle hydrodynamics method", Journal of Heat and Mass Transfer, 51, 3410.
[16] Yildrim B. & Müftü S, 2012. "Simulation and analysis of the impact of micron-scale particles onto rough surface", International Journal of Solids and Structures, 49, 1375.
[17] Kumar S. et al., 2017. "Correlation of splat state with deposition characteristics of cold sprayed niobium coatings", Acta Materialia, 130, 177.
[18] Loft K et al., 2013. "Impacts into metal targets at velocities greater than 1 km s-1: A new online resource for the hypervelocity impact community and an illustration of the geometric change of debris cloud patterns with impact velocity", International Journal of Impact Engineering, 56, 47.
[19] Horz F et al., 1994. "Dimensionally scaled penetration experiments: aluminum targets and glass projectiles 50 μm to 3.2 mm in diameter", International

Journal of Impact Engineering, 15, 257.
[20] Planche M. P. et al., 2001. "In-flight characteristics of plasma sprayed alumina particles: measurements modeling, and comparison", Journal of Thermal Spray Technology, 12, 1, 101.
[21] Cizek J. et al., 2013. "In-flight temperature and velocity of powder particles of plasma-sprayed TiO_2", Journal of Thermal Spray Technology, 22, 8, 1320.

Proceedings of the 2019 Hypervelocity Impact Symposium
HVIS2019
April 14-19, 2019, Destin, FL, USA
HVIS2019-035

Extended Abstract+

Four-View Split-Image Fragment Tracking in Hypervelocity Impact Experiments

Erkai Watson[a], Nico Kunert[a], Robin Putzar[a], Hans-Gerd Maas[b], Stefan Hiermaier[a]

[a] *Fraunhofer Institute for High-Speed Dynamics, Ernst Mach-Institut, EMI, 79104 Freiburg, Germany*
[b] *Technische Universität Dresden, Institute of Photogrammetry and Remote Sensing, 01062 Dresden, Germany*

1. Introduction

Hypervelocity impacts (HVI) often cause significant fragmentation to occur in both target and projectile materials, and is often encountered in space debris and planetary impact applications [1]–[5]. In this paper, we focus on determining the individual velocities and sizes of fragments tracked in high-speed images. Inspired by velocimetry methods such as Particle Image Velocimetry (PIV) [6] and Particle Tracking Velocimetry (PTV) [7] and building on past work [8], we describe the setup and algorithm used for measuring fragmentation data.

2. Experimental Setup

The experimental setup for multi-view fragment tracking involves high-speed cameras and backlit flash lighting to create shadowgraph images of the HVI fragments. An experimental setup with two high-speed cameras is shown in Figure 1A. Two high-speed cameras, capable of micro- to nanosecond exposure times and framerates of up to 10 million frames per second are synchronized and used to record the ejected fragments. In this paper, we will use one specific experiment to describe a method for measuring individual fragment velocities in a HVI experiment. The experiment involved the impact of a 3mm aluminum sphere on a 5mm thick CFRP plate at 7.2 km/s. The image sequence was taken at 250 000 fps with a field of view of approximately 70 x 100 mm in each view. The target chamber was evacuated to 120 mbar. The goal of this experiment was to measure the properties of the backward ejecta that is accelerated up-range from the CFRP target plate shown in Figure 1B.

Although 3D particle measurements with two views is possible, using three or four views greatly decreases the number of ambiguities that arise in stereo image matching of detected particles [9]. In order to increase the number of perspectives with a limited number of cameras, the images of each high-speed cameras is split using a custom build mirror rig shown in Figure 1C, yielding four views from two cameras. The split-image setup requires that the camera have a large Depth of Field, which ultimately limits the magnification achievable. Intrinsic and extrinsic camera parameters are determined with a 2D camera calibration routine [10]. Radial and tangential distortion are corrected for with Brown's distortion model [11].

+ Full paper can found at https://www.sciencedirect.com/journal/international-journal-of-impact-engineering/special-issue/10MTGP5W4VJ

* Corresponding author. Tel.: +497612714499.
E-mail address: Erkai.Watson@emi.fraunhofer.de.

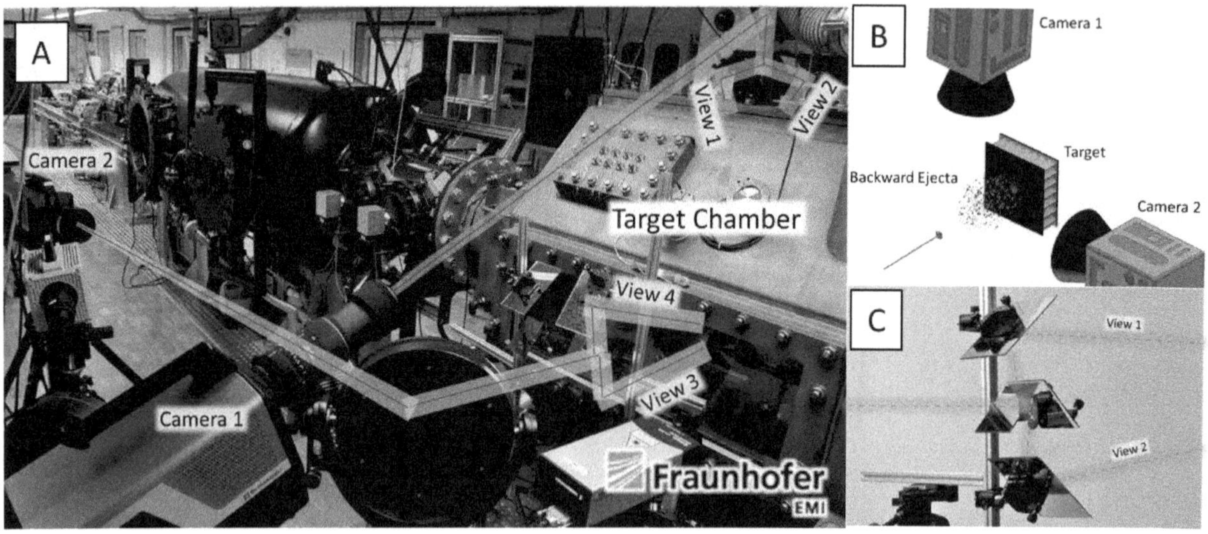

Figure 1: The experimental setup consists of two synchronized high-speed cameras. Both cameras have their images split into two views with the help of a mirror rig shown in C. This results in four separate views of the backward ejecta emanating from the target.

3. Method

The analysis of the multi-view high-speed images can be separated into three main parts: object detection, object tracking and multi-view correspondence establishment. Figure 2 shows a set of raw images from the high-speed image sequences. These images will serves as inputs to the algorithm.

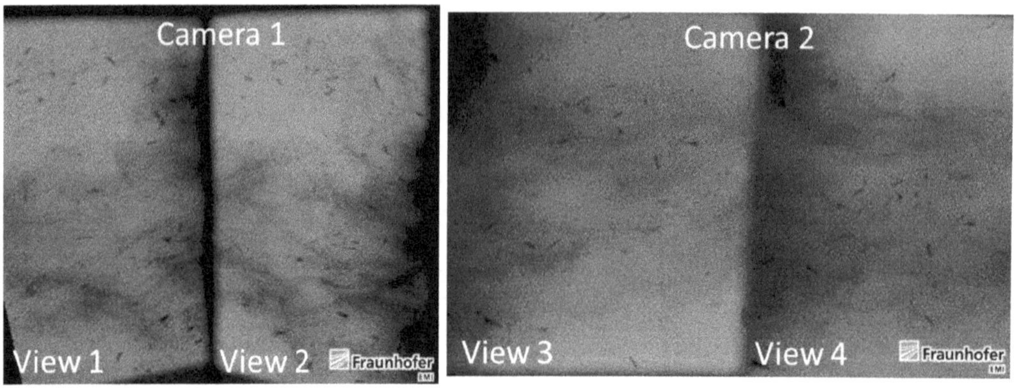

Figure 2: High-speed images of the ejected fragments seen from two synchronized cameras and four different viewing angles. These images sequences are used to measure the velocity of the recorded fragments.

3.1. Object Detection

To identify the fragments in each image, we apply the following steps:

- Apply a top-hat filter to preserve intensity peaks and eliminate other features
- Remove background pixels below given threshold
- Apply Gaussian blur to remove small variations in intensity
- Find all local maxima in the image
- Apply Gaussian fit for subpixel accuracy

These steps are applied to every frame of each image sequence, which results in a list of detected fragments with their 2D locations and times.

3.2. Object Tracking

Once fragments have been identified in each frame of the image sequence, they need to be tracked from frame to frame. We base our fragment tracking on the assumption that each fragment travels in a straight line and at a constant velocity, due to the evacuated target chamber and short time scales of the experiment. Fragments are tracked by plotting all identified points from each of the four image sequences into a 3D-XYT space representing 2D space and time. Points in this XYT space that lie on a line represent the same fragment in different images traveling in a straight line and at constant velocity. These lines are identified with a random sample consensus algorithm (RANSAC) [12].

3.3. Trajectory Correspondence

In photogrammetry, correspondence refers to identifying parts of a 3D scene in one image that correspond to parts in another image. Once such a correspondence is made in at least two views, then the 3D location of that point can be determined via triangulation. In this paper, we determine the correspondences of the fragments based not simply on individual fragments, but also consider the fragment's trajectory. Searching for the correspondences in this way greatly decreases the likelihood of ambiguities.

We use the constraints provided by epipolar geometry and epipolar point transfer to determine the correspondences across the four views [7]. Ambiguities are dealt with by using a scoring system, where distance from the epipolar transfer, fragment size measured in each image, and number of occurrences are evaluated. The correspondence search is performed over sets of three views and then sets of two views. Figure 3 shows some of the corresponded trajectories overlaid on the high-speed images.

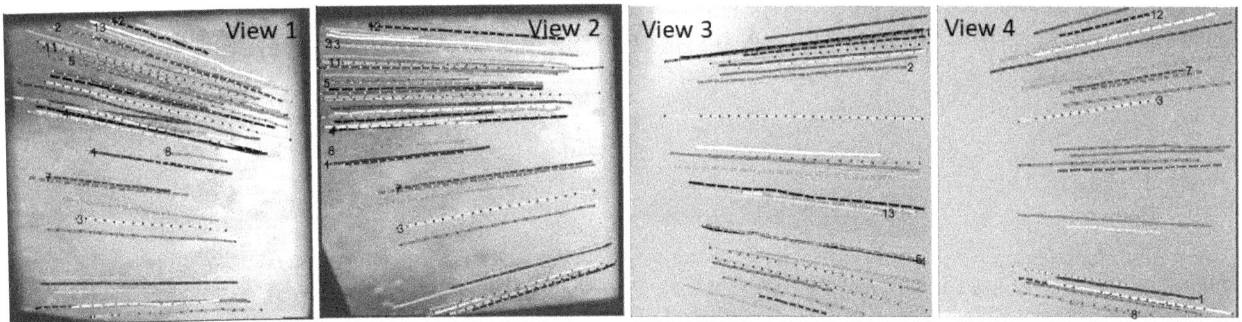

Figure 3: Corresponded trajectories in all four views are overlaid on the high-speed images. Each trajectory's colors are the same in all four views and the first few are numbered to aid the reader in finding the same trajectory in different views.

4. Results and Measurement Accuracies

Figure 4a shows the fragments triangulated into their 3D locations. The coordinate system is centered about Camera 1, View 1. With the known 3D locations and the known frame rate of the synchronized cameras, an individual velocity vector for each fragment can be calculated. Figure 4b shows a histogram of the velocities of the fragments from the experiment.

We determine the measurement accuracies of this method by evaluating control points with know positions. 144 control points are images by the four views and their measured positions compared. The analysis shows a 2σ error of no more than 300 μm, which represents a 0.4% error over images. The error analysis did not accounted for errors due to camera synchronization and the effects of slightly shifted fragment centers. The error in the velocity can be derived from the position and time errors. If we assume that the timing uncertainties from the high-speed cameras (± 10 ns) are negligible compared to the position uncertainties of the fragments, then the velocity uncertainty becomes

$$\epsilon_v = \frac{\epsilon_x}{t} = \frac{\pm 1\, mm}{16\ \mu s} = \pm\, 63\ \mathrm{m/s}\,.$$

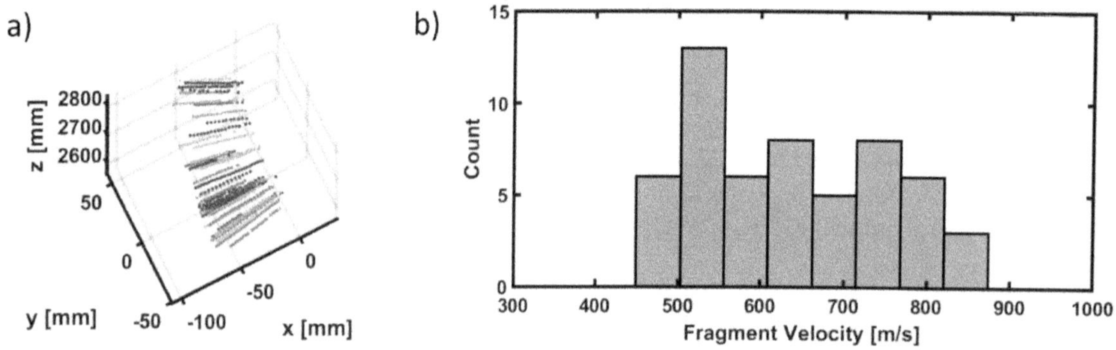

Figure 4: a) The 3D position of fragments. Points of the same color belong to one trajectory. b) Histogram showing the velocity of the fragments measured.

5. Conclusion

This paper describes the setup and analysis method for measuring individual fragment velocities in 3D from HVI experiments. The setup makes use of two synchronized high-speed cameras whose images are further split in half yielding four completely separate views. Having these views allows the vast majority of the identified fragment positions to be determined in 3D without ambiguity. The steps taken to identify, link, correspond, and triangulate each fragment are described in detail. The final result is an approach that is able to measure fragment positions in 3D with sub-millimeter accuracy and 3D velocities with an accuracy of ± 63 m/s. The approach can be applied to study a variety of HVI scenarios ranging from satellite breakup, to momentum transfer, to crater formation.

Acknowledgements

Part of the work described herein was carried out under a program funded by, the European Space Agency under the project "Microparticle impact related attitude disturbances", ESA Contract No. 4000124111/18/NL/GLC.

References

[1] E. Ausay, A. Cornejo, A. Horn, K. Palma, T. Sato, B. Blake, F. Pistella, C. Boyle, and N. Todd, "A Comparison of the SOCIT and DebriSat Experiments," no. April, pp. 18–21, 2017.

[2] S. Flegel, J. Gelhaus, C. Wiedemann, P. Vörsmann, M. Oswald, S. Stabroth, H. Klinkrad, and H. Krag, "The master-2009 space debris environment model," *Eur. Sp. Agency, (Special Publ. ESA SP*, vol. 672 SP, no. April, 2009.

[3] L. Chappaz, H. J. Melosh, M. Vaquero, and K. C. Howell, "Transfer of impact ejecta material from the surface of Mars to Phobos and Deimos," *Adv. Astronaut. Sci.*, vol. 143, no. 10, pp. 1627–1646, 2012.

[4] J. N. Head and H. J. Melosh, "Launch velocity distribution of the martian clan meteorites," in *Lunar and Planetary Science XXXI*, 2000.

[5] E. Buhl, F. Sommer, M. H. Poelchau, G. Dresen, and T. Kenkmann, "Ejecta from experimental impact craters: Particle size distribution and fragmentation energy," *Icarus*, vol. 237, pp. 131–142, 2014.

[6] M. Raffel, C. E. Willert, S. T. Wereley, and J. Kompenhans, *Particle Image Velocimetry: A Practical Guide*. Springer Berlin Heidelberg, 2007.

[7] H. G. Maas, A. Gruen, and D. Papantoniou, "Particle tracking velocimetry in 3-dimensional flows. 1. Photogrammetric determination of partacle coordinates," *Exp. Fluids*, vol. 15, no. 2, pp. 133–146, 1993.

[8] E. Watson, M. Gulde, L. Kortmann, M. Higashide, F. Schaefer, and S. Hiermaier, "Optical fragment tracking in hypervelocity impact experiments," *Acta Astronaut.*, vol. 155, pp. 111–117, 2019.

[9] H. Maas, "Complexity analysis for the establishment of image correspondences of dense spatial target fields," *Int. Arch. Photogramm. Remote Sens.*, vol. 29, pp. 102–107, 1992.

[10] Zhengyou Zhang, "Flexible camera calibration by viewing a plane from unknown orientations," *Proc. Seventh IEEE Int. Conf. Comput. Vis.*, no. c, pp. 666–673 vol.1, 1999.

[11] D. Brown, "Decentering Distortion of Lenses," *Photom. Eng.*, vol. 32, no. 3, pp. 444–462, 1966.

[12] M. a Fischler and R. C. Bolles, "Random Sample Consensus: A Paradigm for Model Fitting with Applicatlons to Image Analysis and Automated Cartography," in *Communications of the ACM*, 1981, vol. 24, pp. 381–395.

Proceedings of the 2019 Hypervelocity Impact Symposium
HVIS2019
April 14-19, 2019, Destin, FL, USA
HVIS2019-112

X-ray diffraction diagnostic paired with gas gun driven compression of polyethylene

Rachel C. Huber[a,]*, Erik B. Watkins[a], Dana M. Dattelbaum[a] and Richard L. Gustavsen[a]

[a]*Los Alamos National Laboratory, P.O. Box 1663, Los Alamos NM 87544, USA*

Abstract

Understanding the kinetics of phase transitions, including decomposition from reactants to products under extreme condition events is challenging. Capturing these processes require: 1) diagnostics that probe on the timescales and at energies capable of interacting with the dynamically evolving products, penetrating the opaqueness of the changing system; and 2) detectors sensitive enough to observe these events. Synchrotrons and free electron lasers provide ke-V-energy x-ray beams capable of penetrating the optical-opaqueness of the temporally evolving products. At the Dynamic Compression Sector at the Advanced Photon Source, the x-ray beam is coupled to single and two-stage gas guns capable of producing planar shocks at a range of projectile velocities while capturing *in situ* x-ray diffraction/scattering of the evolving material under dynamic compression. In this work, we demonstrate the utility of this approach in measuring the evolution of crystalline domains in shocked high-density polyethylene to P = 7.45 GPa, and have observed the compression and orientation of the polymer chains in real time.

Keywords: Shock phyics, x-ray diffraction, polyethylene, HDPE

Nomenclature

u_p	particle velocity (km/s)
U_S	shock velocity (km/s)
u_{proj}	projectile velocity (km/s)
P	pressure (GPa)
u_p	free surface velocity (km/s)
ρ_0	initial density (g/cm^3)
Greek symbols	
λ	wavelength (Å)
θ	diffraction angle
ρ	density (g/cm^3)

1. Introduction

The degree of polymer crystallinity is well known to affect a polymer's mechanical and electrical properties.(1-5) For example polyethylene (PE, $(\text{-}C_2H_4\text{-})_n$), can be obtained in a variety of forms/% crystallinity as low density PE (LDPE, 35-55% crystallinity), high density PE (HDPE, 78.5% crystallinity) and ultra-high molecular weight PE (UHMWPE, 70-80% crystallinity). PE's strength is derived from both intra- and intermolecular forces including C-C and C-H covalent bonds, and van der Waals forces. As with most polymers, compression along the chain axis is typically stiff, where compression between polymer chains is

* Corresponding author. *E-mail address:* rhuber@lanl.gov
LA-UR-19-20281

relatively soft. Polymer crystallinity is defined by the unit cell, similar to metals, however this repeat unit is composed of the polymer chains and multiple monomer units instead of single atoms. The PE unit cell under ambient conditions is orthorhombic. Low pressure dynamic experiments, if carefully designed, can access the hexagonal PE crystal structure; however these experiments can easily bypass the hexagonal phase and go directly to monoclinic.(6) Furthermore, Fontana *et al* has shown that at 2.5 GPa PE is still orthorhombic during diamond anvil cell (DAC) experiments, as pressure is increased to 15.5 GPa PE transitions to monoclinic.(7) With gas gun driven dynamic compression paired with x-ray diffraction we can directly probe polymer response through crystallinity at a higher strain rate than accessible with DAC experiments, further contributing to the understanding of how strain rate drives material response.

Dynamic loading of polymer samples will change the diffraction patterns as compression forces the polymer chains closer together, squeezing the unit cell thereby resulting in peak shifts to higher q (smaller spacing). With respect to polymer crystallinity, if the polymer is truly amorphous no diffraction peaks or a broad amorphous carbon peak will be observed. Bragg peaks will be present if the polymer is crystalline, where in semi-crystalline polymers the amorphous and crystalline peaks are both observed providing a ratio of amorphous to crystalline regions. Crystalline polymers often show isotropic diffraction, meaning there is no preferred orientation of those crystalline regions.(8) However, by synthesis techniques, molten compression or by applying shear, anisotropic orientation can be obtained.(4, 9, 10) X-ray diffraction as a diagnostic to study dynamic loading can extract information about the changing crystallinity and/or phase.

Carter and Marsh published a large subset of polymers that exhibited a reactants and products Hugoniot during dynamic compression, with a transition region between these two branches.(11) This report suggested that the transition region, where the polymer is undergoing decomposition, was reflective of polymer chain compression with ultimately a phase change from sp^2 (graphite-like) to sp^3 hybridization (diamond-like). Diagnostics traditionally paired with planar shock fronts are Photon Doppler Velocimetry (PDV) and Velocimetry Interferometry System for Any Reflector (VISAR) that capture the particle velocity (u_p) at an interface.(12, 13) Phase transitions may be observed in the u_p profile if there is a large volume change and the shock velocity (U_S) for the second wave is slower than that of the first wave.(14, 15) However these multi-wave structures are often difficult to observe with velocimetry; organic materials where multi-wave structures have been captured include benzene, phenylacetylene and cyanate ester.(16-18) By pairing gas gun driven shock waves to x-ray diffraction we no longer require large volume changes or U_S variations, phase changes may now be observed *in situ* with x-ray diffraction.

At the Dynamic Compression Sector (DCS, Advanced Photon Source, at Argonne National Laboratory) x-ray diffraction has been paired with gas guns capable of projectile velocities upwards of 5.5 km/s (setup in Figure 1). This is the first time, to our knowledge, that polymer diffraction captured during gas gun driven planar shocks has been reported. Specifically, HDPE was dynamically loaded to 7.45 GPa by impaction with a LiF projectile at 2.116 km/s on the single stage powder gun. Herein, we will discuss the change in HDPE polymer crystallinity and polymer chain orientation due to dynamic shock.

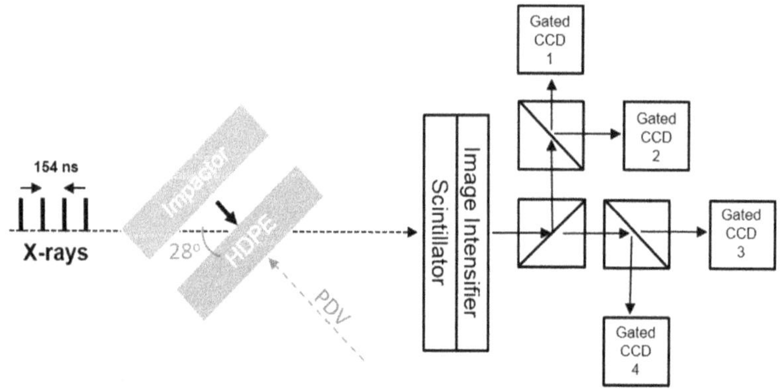

Figure 1: X-ray diffraction setup. X-ray bunches separated by 154 ns pass through the dynamically evolving system to the scintillator and image intensifier. The now visible light (converted by the scintillator) is directed to the four CCD cameras via beam splitters. Each camera is capable of capturing one image per gas gun shot, resulting in a total of four diffractograms total for each experiment.

2. Experimental

HDPE was purchased from Polymer Industries (Densetec grade), with a density of 0.957 g/cm^3 and percent crystallinity of 78.5 ± 2.0.(2) The HDPE was machined into a disk that measured 3 mm in thickness and 10 mm in diameter. The rear surface of the HDPE was vapour coated with an Al mirror for PDV. The three PDV probes that were focused on the back surface captured the arrival time of the shock at the back surface, tilt and the free surface particle velocity, u_{fs} (sample holder in Figure 2A-C).

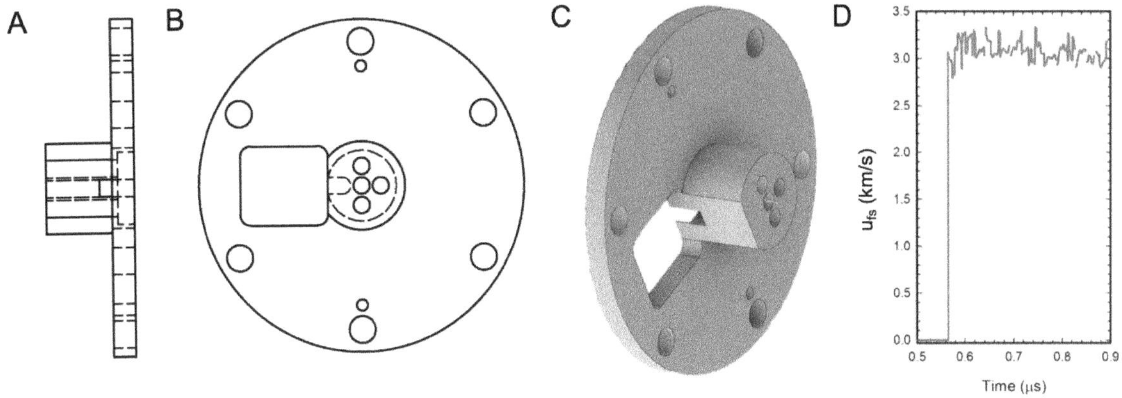

Figure 2: (A-C) Sample Holder for HDPE samples. (D) PDV trace for HDPE sample at the free surface

These x-ray experiments were performed using the Advanced Photon Source's 24-bunch mode, which consists of 80 ps x-ray bunches spaced 154 ns apart with a wavelength of 0.5276 Å (24 keV). The x-ray beam (800 μm^2 at the sample impact face) was directed through a 2.7 cm period undulator at DCS, along with two horizontal and two vertical KB mirrors to focus the x-ray beam at the sample. The experimental setup can be found in Figure 1. These measurements were performed with the x-ray beam on the center of the sample, with the incident beam 28° relative to the sample surface as shown in Figure 1. Due to the angle of the x-ray beam in relation to the shock direction, the x-rays penetrated through the LiF/Lexan impactor, as well as the HDPE during the shock event. A 75-mm scintillator (lutetium oxyorthosilicate:Ce3+ (LSO:Ce)) was used with a pixel size of 43.5 μm x 43.5 μm. The detector plane was oriented perpendicular to the x-ray beam. Once the x-ray beam diffracted off the sample it was directed at the scintillator and image intensifier (Photek MCP140), then to the beam splitters towards four CCD cameras (Princeton Instruments PiMax-4). The CCD cameras were gated to capture sequential bunches.

The sample-to-detector (Sa-Det) distance was calibrated with a polycrystalline silicon (Si) standard. Figure 3A shows the diffractogram with the first two Si diffraction rings. This diffraction pattern was calibrated in Dioptas(19), providing a Sa-Det distance of 134 mm and a q-range of 0.5-3.0 Å^{-1} ($q=4\pi sin\theta/\lambda$). This Sa-Det distance was chosen because the first two orthorhombic peaks of HDPE are the (110) at 1.5 Å^{-1} and the (200) at 1.7 Å^{-1}, as shown in Figure 3B, along with the molecular structure in Figure 3C (created in VESTA).(20-22)

Planar shocks were imparted to the samples using projectile impact from a single stage powder gun located in hutch 35-ID-E at DCS capable of velocities from 0.5-2.5 km/s. In these experiments, lithium fluoride (LiF) impactors were moulded into a Lexan projectile. The projectile velocity (u_{proj}) of the impactor was measured with PDV located in the barrel of the gun. The PDV scopes were triggered by an optical beam block (OBB) placed ~1.2 mm from the front surface of the sample, which initiates the scopes when the projectile has blocked ~33% of the PDV beam. With the known distance from OBB to the front of the sample, u_{proj}, and cable lengths for signal delays, the cameras are gated to capture four frames of dynamically shocked sample. Since 24-bunch mode was used, we collected one frame on each camera, where the time between cameras was 154 ns. Figure 2D shows the PDV u_{fs} trace captured at the mirrored surface on the back of the sample. To calculate u_p the free surface approximation was used ($2u_p=u_{fs}$), resulting in u_p = 1.5 km/s.

3. Results and Discussion

Pairing x-ray diffraction with shockwave compression allows for observation of *in situ* compression of polymer chains, and potentially, phase transformations. Figure 3B shows the static diffraction peaks for HDPE. The two most intense peaks corresponding to (110) and (200) were measured at d = 1.5 and 1.7 Å^{-1}. The (110) peak maintained enough intensity under shock loading to observe the polymer chain compression along the a and b-axis and assuming retention of the the *Pnma* space group. As mentioned above compression along the polymer chain is stiff (c-axis) therefore we assumed that the compression was mainly along the a and b-axis. This static diffraction pattern provided a baseline for analyzing the dynamic frames; as the shock wave moved through the 3 mm sample the static/unshocked diffraction peaks reduced in intensity until no longer observed in the diffractograms (Figure 4B). This disappearance of the static diffraction peaks also provided redundancy to the calculated timing for each camera frame.

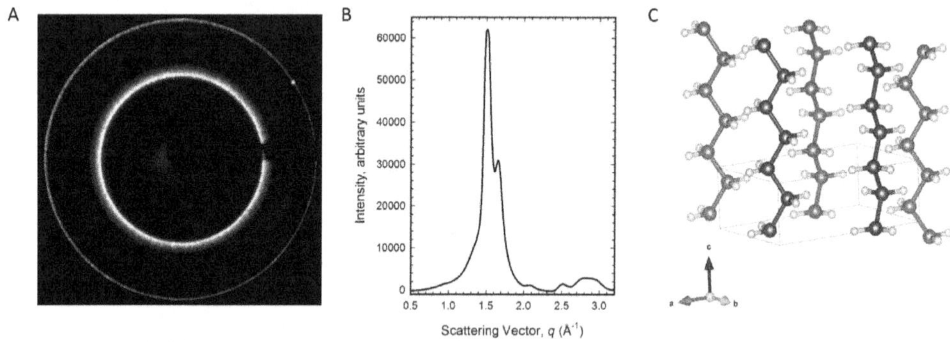

Figure 3: (A) Silicon calibration standard showing the first two diffraction rings, used to determine the sample-to-detector distance. (B) Static diffraction pattern of HDPE, the first two diffraction peaks are at 1.5 Å^{-1} (110) and 1.7 Å^{-1} (200). (C) Crystal structure of HDPE created in VESTA.

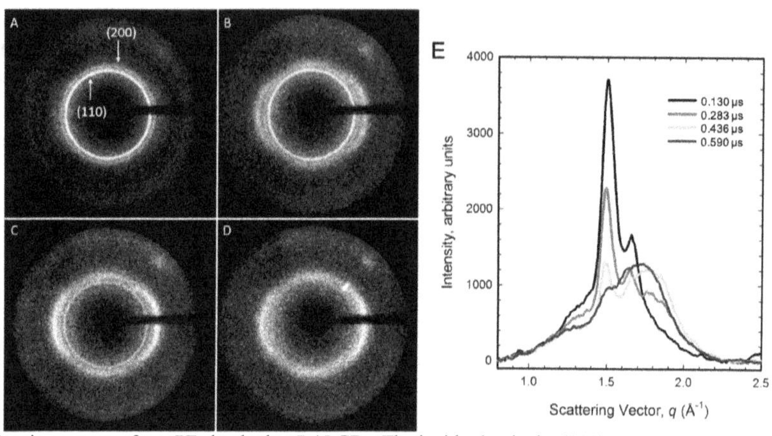

Figure 4: (A-D) Dynamic diffraction patterns from PE shocked to 7.45 GPa. The inside ring is the (110) peak, the outside ring is the (200) peak. By radially integrating the diffractograms we produced graph (E). (E) Scattering pattern from the frames A-D, as time advanced and the shock wave progressed through the HDPE sample the un-shock HDPE peaks decrease and the shifted (110) peak at ~1.8 Å^{-1} increases in signal. By the final frame the shifted (110) peak relaxes back to ~1.75 Å^{-1}, as the PE was released.

To determine the shock locus, PDV probes were placed in the barrel of the gun and facing the mirrored surface at the back of the sample. The PDV probes in the gun barrel were to capture the u_{proj}, which was 2.116 km/s. Since the cameras were triggered from the optical beam block placed in front of the sample, a good approximation of the U_S through the sample is not achievable because we did not have a PDV probe at the impactor/HDPE surface. Therefore we employed the linear Rankine-Hugoniot for U_S-u_p to determine U_S, which relies on the parameters C=2.655 km/s and s=1.692.(11) The U_S from this relationship was calculated as 5.19 km/s. From here we used the momentum jump equation, $P=\rho_0 U_S u_p$, to calculated the pressure (7.45 GPa). Finally, the volume compression was determined from the mass jump equation, $V/V_0=1-u_p/U_S$, which was 0.711.

The diffractograms for the 7.45 GPa shot can be found in Figure 4A-D. The first frame at 0.130 µs was shortly after impact, and still contains both the (110) and (200) peaks. As mentioned above, each x-ray pulse is separated by 154 ns, therefore the next three frames were 0.283, 0.436 and 0.590 µs. Initially the HDPE diffraction rings were isotropic, the crystalline regions had no preferred orientation. However, as the shock progressed through the sample the (110) diffraction ring became anisotropic. The peaks or partial rings in the horizontal plane are referred to as in-plane diffraction, often observed from static grazing incidence diffraction of polymer samples whose backbone is oriented perpendicular to the surface it was deposited on.(4) The anisotropic diffraction rings imply partial orientation of the crystalline regions, further experiments will be performed to determine polymer orientation.

By radially integrating the diffractograms, less intense peaks are observable; these 1D diffraction patterns are found in Fig. 4E. The first frame was captured at 0.130 μs, 0.67 mm into the HDPE sample. This 1D pattern appears similar to the static diffraction in Fig. 3B. As the shock wave progressed further into the sample (0.283 μs, 1.45 mm) the intensity of the un-shocked HDPE decreased and a peak at ~1.8 $Å^{-1}$ appeared. At 0.436 μs, 2.24 mm into the HDPE, the un-shocked HDPE peaks have almost completely been shifted and/or compressed and the ~1.8 $Å^{-1}$ peak has grown in intensity. Finally, at 0.590 μs, 3.03 mm, the shock wave has progressed through the entire thickness of the 3 mm sample. Almost no un-shocked material is observed and the ~1.8 $Å^{-1}$ peak has relaxed to ~1.75 $Å^{-1}$. This new peak at ~1.8 $Å^{-1}$ was the shifted (110). As the polymer chains compressed the diffraction peak shifted to higher q, indicative of smaller chain spacing. The shift of the final frame to low q is due to the relaxation of the polymer once it had been released. Additionally, the broadness of the shifted (110) is likely due to the (200) peak not retaining enough intensity to break through as an individual peak. The smaller diffraction peaks that were present in the static scattering in Fig. 3B are too small to observe under shock compression.

As explained above, the shock parameters were calculated with the jump equations and the measured u_{fs}. Those entities are related to the bulk compression of the polymer, similar calculations starting with the crystalline volume can also be performed. For an orthorhombic crystal structure, the volume is $V=abc$, where a, b and c are the lengths of the unit cell in Ås. We calculated the initial (before shock) and shocked volume to be 93.8 $Å^3$ and 73.7 $Å^3$. Therefore V/V_0 for this 7.45 GPa shot was 0.786, which is larger than what was calculated for the bulk compressed volume of 0.711. The crystalline sections of the polymer are less compressible than the amorphous sections due to van der Waals forces, hydrogen bonding, or other intra-polymer chain interactions. The compression of the crystalline domains is therefore less than the bulk compression; this is the first time that the partitioning of compression energy has been measured in a two-phase semicrystalline polymer under shockwave loading. Untangling how the amorphous and crystalline polymeric regions are affected by shock wave propagation will require further study through systematic manipulation of polymeric crystallinity in a controlled and deliberate fashion.

4. Conclusions

In this paper, HDPE was shock compressed to 7.45 GPa using a LiF impactor traveling at 2.116 km/s. The HDPE polymer chains showed both compression and anisotropy. Compression was observed by the (110) peak shifting to higher q (smaller d spacing) at ~1.8 $Å^{-1}$. Polymer chain orientation was captured by the 2D diffractograms, with the diffraction rings transforming from isotropic rings to anisotropic rings.

Dynamic compression of polymers is not a new area of study, however the ability to observe crystal structure changes, *in situ*, during dynamic loading presents an exciting new diagnostic. X-ray diffraction affords the pathway to observe crystal structure changes in complicated polymeric systems. HDPE is a semi-crystalline polymer that has multiple phase changes below the decomposition pressure (24.7 GPa), phase changes that have not been captured in the u_p profiles. With the addition of x-ray diffraction as a diagnostic for shock compression it presents the ability to observe compression, phase change, melt and decomposition for the first time in these carbon-based, low scattering systems.

Acknowledgements

Funding for this work was provided by DOE/NNSA from Science Campaign 2. We thank Lee Gibson and Brian Bartram for experimental target assembly. Los Alamos National Laboratory is operated by LANS LLC for the Department of Energy. This publication is based upon work performed at the Dynamic Compression Sector, which is operated by Washington State University under the U.S. Department of Energy (DOE)/National Nuclear Security Administration award no. DE-NA0002442. This research used resources of the Advanced Photon Source, a DOE Office of Science User Facility operated for the DOE Office of Science by Argonne National Laboratory under contract no. DE-AC02-06CH11357

References

1. Brown EN, Willms RB, Gray GT, Rae PJ, Cady CM, Vecchio KS, et al. Influence of Molecular Conformation on the Constitutive Response of Polyethylene: A Comparison of HDPE, UHMWPE, and PEX. Experimental Mechanics. 2007;47(3):381-93.
2. Coe JD, Brown E, Cady CM, Carlson CA, Clements BE, Dattelbaum DM, et al. Equation of State and Damage in Polyethylene. LANL; 2017. Report No.: LA-UR-17-29234.
3. Haro EE, Odeshi AG, Szpunar JA. The Effects of Micro- and Nano-Fillers' Additions on the Dynamic Impact Response of Hybrid Composite Armors

Made of HDPE Reinforced with Kevlar Short Fibers. Polymer-Plastics Technology and Engineering. 2017;57(7):609-24.
4. Li H, Hwang Y-J, Earmme T, Huber RC, Courtright BAE, O'Brien C, et al. Polymer/Polymer Blend Solar Cells Using Tetraazabenzodifluoranthene Diimide Conjugated Polymers as Electron Acceptors. Macromolecules. 2015;48(6):1759-66.
5. Zhang G, Huber RC, Ferreira AS, Boyd SD, Luscombe CK, Tolbert SH, et al. Crystallinity Effects in Sequentially Processed and Blend-Cast Bulk-Heterojunction Polymer/Fullerene Photovoltaics. The Journal of Physical Chemistry C. 2014;118(32):18424-35.
6. Priest RG. Calculation of the High-Pressure Phase Diagram of Polyethylene. Macromolecules. 1985;18:1504-8.
7. Fontana L, Vinh DQ, Santoro M, Scandolo S, Gorelli FA, Bini R, et al. High-pressure crystalline polyethylene studied by x-ray diffraction and ab initio simulations. Physical Review B. 2007;75(17).
8. Russel KE, Hunter BK, Heyding RD. Monoclinic polyethylene revisted. Polymer. 1997;38(6):1409-14.
9. Pujari S, Dougherty L, Mobuchon C, Carreau PJ, Heuzey M-C, Burghardt WR. X-ray scattering measurements of particle orientation in a sheared polymer/clay dispersion. Rheologica Acta. 2010;50(1):3-16.
10. Hyon S-H, Taniuchi H, Kitamaru R. The orientation of crystal planes in polyethylene crystallized under compression. Bull Inst Cehm Res. 1973;51(2):91-103.
11. Carter WJ, Marsh SP. Hugoniot Equation of State of Polymers. Los Alamos, NM: Los Alamos National Laboratory; 1995.
12. Hemsing WF. Velocity sensing interferometer (VISAR) modification. Rev Sci Instrum. 1979;50(1):73.
13. Strand OT, Goosman DR, Martinez C, Whitworth TL, Kuhlow WW. Compact system for high-speed velocimetry using heterodyne techniques. Review of Scientific Instruments. 2006;77(8):083108.
14. Hayes DB. Polymorphic phase transformation rates in shock - loaded potassium chloride. Journal of Applied Physics. 1974;45(3):1208-17.
15. Erskine DJ, Nellis WJ. Shock - induced martensitic transformation of highly oriented graphite to diamond. Journal of Applied Physics. 1992;71(10):4882-6.
16. Dattelbaum DM, Sheffield SA, Coe JD. Shock-driven chemistry and reactive wave dynamics in liquid benzene. 2017;1793:040020.
17. Dattelbaum DM, Sheffield SA, Coe JD, Margevicius MA. Shock-induced chemistry of phenylacetylene. Journal of Physics: Conference Series. 2014;500(2):022004.
18. Dattelbaum DM, Coe JD, Rigg PA, Scharff RJ, Gammel JT. Shockwave response of two carbon fiber-polymer composites to 50 GPa. Journal of Applied Physics. 2014;116(19):194308.
19. Prescher C, Prakapenka VB. DIOPTAS: a program for reduction of two-dimensional X-ray diffraction data and data exploration. High Pressure Research. 2015;35(3):223-30.
20. Momma K, Izumi F. VESTA 3for three-dimensional visualization of crystal, volumetric and morphology data. Journal of Applied Crystallography. 2011;44(6):1272-6.
21. Yamamoto T, Miyaji H, Asai K. Structure and Properties of High Pressure Phase of Polyethylene. Japanese Journal of Applied Physics. 1977;16(11):1891-8.
22. Thomas EL, Sass SL. On the Orthorhombic to Hexagonal Phase Transformation in Polyethylene Single Crystals. Die Makromolekulare Chemie. 1973;164:333-41.

Proceedings of the 2019 Hypervelocity Impact Symposium
HVIS2019
April 14-19, 2019, Destin, FL, USA
HVIS2019-007

Extended Abstract+

Wave speeds in single-crystal and polycrystalline copper

Sarah A. Thomas[a,*], Robert S. Hixson[a], M. Cameron Hawkins[b], Oliver T. Strand[a]

[a]*Mission Support and Test Services, LLC, New Mexico Operations, Los Alamos, New Mexico 87544, USA*
bMission Support and Test Services, LLC, Nevada Operations, Las Vegas, Nevada 89030, USA

1. Introduction

While the equation of state for copper has been fairly well studied, wave speeds at low stress are not as well known. Systematic errors may be present in the lowest stress data presented in the Marsh [1] compendium due to the use of the flash gap method to collect the data. Additionally, little data has been gathered on the wave speeds in single-crystal copper, which may vary from polycrystalline due to the different longitudinal and shear sound speeds. Hugoniot information at low pressures is useful in constraining and improving predictive hydrodynamic codes. Knowledge of single-crystal behavior provides input for mesoscale computer models that use tens-of-micron-sized grains of single crystals to build a model of polycrystalline systems. We undertook experiments to measure wave speeds in polycrystalline and single-crystal copper at low pressures using a novel technique to limit error, and to determine if single-crystal shock velocities are systematically different than polycrystalline shock velocities at the same stress. The best previous research on this topic is from Chau et al. [2] at relatively high shock stress; they reported no observed difference between orientations. It is of interest to do careful measurements at low stress, and that is the principal goal of this work.

2. Experimental methods

A copper flyer plate impacted the base of a copper "top hat," with a larger diameter, thinner brim, and a smaller diameter, thicker top. The polycrystalline top hat was cut from a single piece of material. However, the single-crystal top hat was made from two crystals, the smaller-diameter piece glued to the center of the larger-diameter piece. We then used five channels of photonic Doppler velocimetry (PDV) [3], four around the brim, and one in the center of the top, to measure the free surface velocity at these points. Data from the four channels around the brim helped us determine the time that the elastic and plastic waves entered the base of the top part of the top hat. The center probe recorded the times the elastic and plastic waves exited the back of the top hat. The five channels of PDV allowed accurate cross timing.

Using timing data and thickness parameters of the top part of the top hat, we determined the speeds of the elastic and plastic waves. The wave profiles in Figure 1 show data taken from each of the five probes of a typical shot, with elastic and plastic waves labeled as E1 and P1, respectively. Having measured elastic wave speed, we compared it to the ultrasonically measured longitudinal sound speed (C_L) of the material and found it to be in fairly good agreement, lending credence to our subsequent shock speed measurements. We calculated the elastic wave speed by recording the time the elastic wave entered the center of the top section of the top hat (the same as the time the wave exited the back of the brim) and subtracted from it the time it exited the top, then divided the time difference by the top thickness. A similar calculation determined the plastic shock speed. Data were

+ Full paper can found at https://www.sciencedirect.com/journal/international-journal-of-impact-engineering/special-issue/10MTGP5W4VJ

* Sarah Thomas. Tel.: +1-505-663-2068; fax: +1-505-663-2003.
E-mail address: thomassa@nv.doe.gov.

collected for polycrystalline and single-crystal [111], [100], and [110] copper at nominally 300 m/s. Two additional shots provided data on polycrystalline and single-crystal [110] copper at nominally 600 m/s.

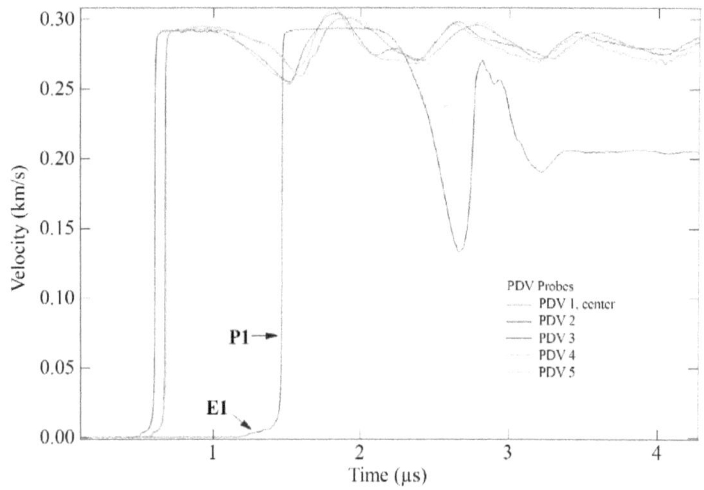

Fig. 1. Wave profile extracted from PDV data taken at the brim of the target.

3. Experimental results

Values for shock speed vs. particle speed are plotted in Figure 2 along with data taken from the Marsh [1] compendium. After careful error analysis, we believe the error in these data is about 2%, as concluded through a comparison of the elastic wave speeds with our ultrasonically measured ambient sound speeds. Note that one source of uncertainty is the perturbation caused by the elastic wave reflecting from the free surface as a small release and interacting with the oncoming plastic wave. We have quantified this effect in other work, and we find that in a material with a small elastic precursor such as copper, it is negligible with respect to other sources of error [4]. Figure 2 shows both the polycrystalline and single-crystal Hugoniot data plotted alongside archival data from Marsh. Taking an average of the shock velocities for each orientation of single-crystal as well as the polycrystalline velocities, we see that the differences between our results and the Marsh data is within around 2%.

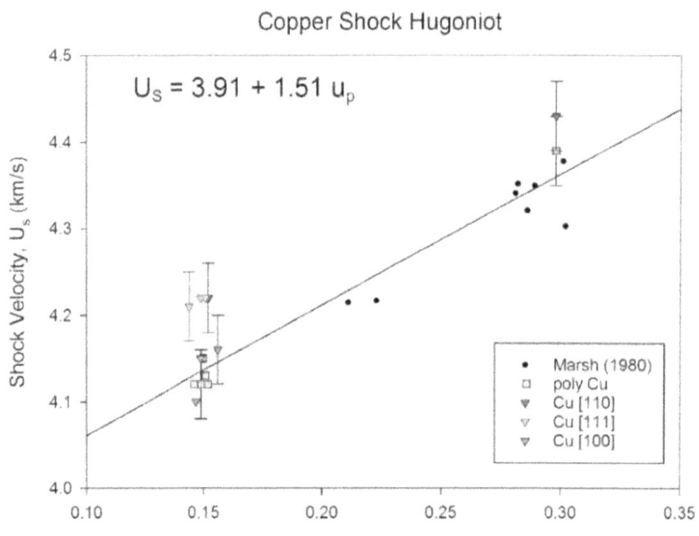

Fig. 2. Final results of shock Hugoniot of copper, including points from literature in black.

4. Conclusions

Careful measurements have been made of shock wave speeds in the three principal orientations of single-crystal and polycrystalline copper. Using a top hat target configuration, we observe differences in shock speeds between the different single-crystal orientations and polycrystalline copper, but those differences appear to be within the uncertainty from the experimental technique as determined from the elastic wave analysis. We stress that these are preliminary results, and believe that the technique is capable of making wave speed measurements with less uncertainty than we report here. Work will continue on analysis of this dataset. Extracting longitudinal sound speeds from copper wave profiles is difficult because the precursor wave is small and does not have a fast initial rise. We are working on a more comprehensive uncertainty analysis to better understand the inherent uncertainty budget. Because material anisotropy is an important technical question in the dynamic materials community, we will work to improve this technique.

Acknowledgments

This manuscript has been authored by Mission Support and Test Services, LLC, under Contract No. DE-NA0003624 with the U.S. Department of Energy and supported by the Site-Directed Research and Development Program, National Nuclear Security Administration, NA-10 USDOE NA Office of Defense Programs (NA-10). The United States Government retains and the publisher, by accepting the article for publication, acknowledges that the United States Government retains a non-exclusive, paid-up, irrevocable, worldwide license to publish or reproduce the published form of this manuscript, or allow others to do so, for United States Government purposes. The U.S. Department of Energy will provide public access to these results of federally sponsored research in accordance with the DOE Public Access Plan (http://energy.gov/downloads/doe-public-access-plan). The views expressed in the article do not necessarily represent the views of the U.S. Department of Energy or the United States Government. DOE/NV/03624--0488.

References

[1] Marsh, Stanley P., ed., 1980. LASL Shock Hugoniot Data, University of California Press, Berkeley, California.

[2] Chau, R., Stölken, J., Asoka-Kumar, P., Kumar, M., 2010. Shock Hugoniot of Single Crystal Copper, Journal of Applied Physics 107, p. 023506.

[3] Strand, O. T., Goosman, D. R., Martinez, C., Whitworth, T. L., Kuhlow, W. W., 2006. Compact system for high-speed velocimetry using heterodyne techniques, Review of Scientific Instruments 77, p. 083108.

[4] Thomas, S. A., Hawkins, M. C., Matthes, M. K., Gray III, G. T., Hixson, R. S., 2018. Dynamic Strength Properties and Alpha-Phase Shock Hugoniot of Iron and Steel, Journal of Applied Physics 123, p. 175902.

Proceedings of the 2019 Hypervelocity Impact Symposium
HVIS2019
April 14-19, 2019, Destin, FL, USA
HVIS2019-008

High velocity impact of an Fe/Cr/Mn/Ni high entropy alloy

M. Cameron Hawkins[a], Sarah Thomas[b], Robert Hixson[b], Nan Li[c] and Saryu Fensin[c]

[a] *Mission Support and Test Services, LLC, Las Vegas Operations, Las Vegas, NV 89030, USA*
[b] *Mission Support and Test Services, LLC, New Mexico Operations, Los Alamos, NM, 87544, USA*
[c] *Los Alamos National Laboratory, Los Alamos, NM, 87545, USA*

Abstract

A series of shock loading experiments were conducted on High Entropy Alloy (HEA) samples consisting of Fe/Cr/Mn/Ni in weight percentages of 25.2/23.5/24.8/26.5. Characterizing this transition metal alloy is a first step in understanding the shock compression response of this relatively new class of alloy. A single stage light gas gun was used to conduct a series of flyer plate symmetric impact experiments to obtain fundamental dynamic properties. Photonic Doppler Velocimetry (PDV) diagnostics were employed to measure the free surface velocity on the back of each target during dynamic compression. These experiments yielded four data points that are in reasonable agreement with an estimated Hugoniot for the material.

Keywords: PDV, Hugoniot, High Velocity Impact, High Entropy Alloys, transisiton metal alloy

---

1. Introduction

High entropy alloys (HEA) are a relatively new class of materials with high compressive strength coupled with ductility. These materials have many potential dynamic applications, but first they must be characterized for relevant properties. Ultimately these properties can be used to build physics models for use in numerical simulations. HEAs have been studied for several years, but mostly metallurgical studies have been done, as well as testing for hardness, fracture toughness, fatigue, quasi-static compression, and tensile experiments [1-4]. Very little research on dynamic properties studies were found in the literature [5-7].

The HEA alloy for this study was characterized beforehand to investigate its microstructure and relevant mechanical properties under quasi-static loading. Specifically, uniaxial tension specimens were tested quasi-statically at ambient temperature and shown to exhibit large ductility [8].

It was of interest to measure the dynamic properties of this alloy using shock compression techniques, and this study was a first step in characterizing this HEA material for shock compression response. Four HEA target experiments were designed and tested in a symmetric impact configuration to specifically study spall response. This work was intended to be a short study to determine if this alloy was of interest; if so, more experiments would be performed.

2. Experimental Procedure

The high entropy alloy used in this study, FeCrNiMn, is a single phase alloy with a face centered cubic structure. This particular HEA is a transition metal alloy, which consists of Fe/Cr/Mn/Ni in weight percentages of 25.2/23.5/24.8/26.5. Figure 1 shows an inverse pole figure map from EBSD for the HEA used in this study.

---

* Corresponding author. Tel.: +1-702-295-9807.
E-mail address: HAWKINMC@nv.doe.gov.

Dynamic experiments were performed using a single stage large bore (76.2 mm) light gas gun capable of planar impacts (tilts typically range from 1-2 mrad). Impact tilts were measured using four electrical shorting pins. Photonic Doppler Velocimetry (PDV) [9] was used during the shock experiments to record the velocity of the back surface of the target with a two-fiber PDV probe. The velocity of the projectile was recorded using a collimating PDV probe pointed down the barrel at the front surface of the projectile. The center of the back surface of the target was used as the location for collecting PDV data. The probe location and cross-sectional view of the experiment are both illustrated in Figure 2.

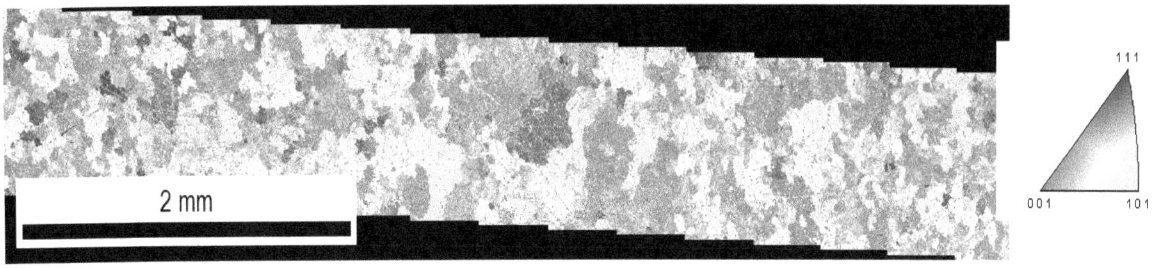

Fig. 1. Inverse pole figure map from electron back scatter diffraction (EBSD) for the high entropy alloy used in this study.

Since this material has not been previously characterized for its shock compression response, the Hugoniot was unknown and we developed an estimated shock Hugoniot to design initial experiments. One goal of this research was to measure shock (U_s) and particle velocities (u_p). Additionally we wished to determine compressive and tensile strength using dynamic measurement techniques.

Targets of the HEA material were of nominal 22 mm diameter x 3.5 mm thick; impactors of the HEA material had thicknesses that were approximately half that of the target thickness, chosen to induce spall close to the middle of the target. Since the sample diameters were small, this limited the observation time in uniaxial strain, so timing, including that for lateral releases, was carefully considered in the design of the experiment. A piece of syntactic foam (nominal 22 mm diameter x 4.00 mm thick) was used to insulate the HEA impactor from the aluminum projectile, and allow for a deep release from that interface.

The material was first characterized for its density, longitudinal sound speed, and shear sound speed. Density was measured using the Archimedes Principle with a Mettler Toledo analytical balance and ultrasonic measurements were conducted with an Olympus ultrasonic pulser/receiver and transducers to determine the longitudinal and shear sound speeds. A Heidenhain gauge was used to measure the targets and impactors to within 5-10 microns of flatness and parallelism. All targets and impactors were taken from the same batch of material, however the density and sound speed measurements showed some small differences between samples. Values for the density of the targets averaged 7.73 g/cc, with variations up to 0.05 g/cc from sample to sample, as shown in Table 1. Longitudinal sound speeds are also shown in Table 1, and values ranged from 5.40 km/s to 5.49 km/s. This variation (about 3%) is well outside the uncertainty for this technique, which for longitudinal waves is typically less than 1%.

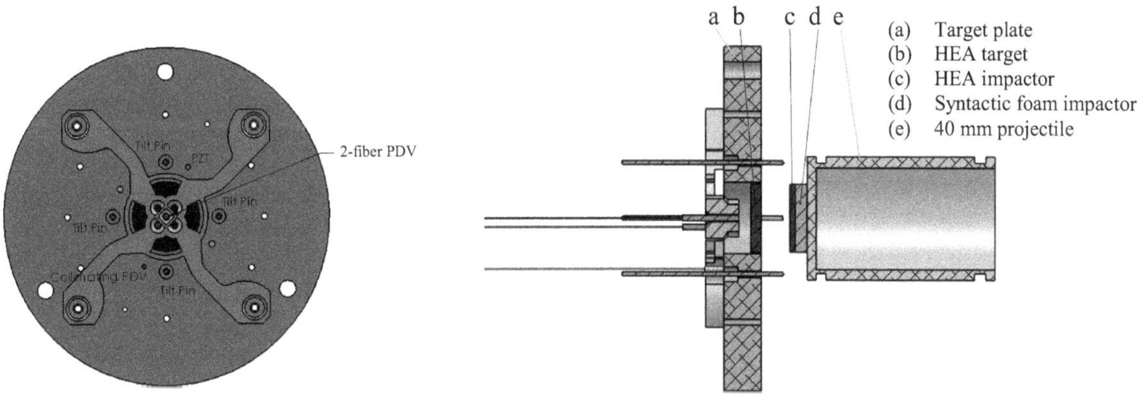

Fig. 2. Experimental setup (left) PDV probe placement on back side of the target, (right) cross-sectional view target plate

Shear sound speed was particularly difficult to measure because of large signal attenuation, and a very small number of detectable pulse echoes. We note that we used the pulse-echo method to determine sound speeds, and that this metal is very dissipative. Echoes were strongly attenuated, which lead to relatively large uncertainties in sound speed (about 1.0 % for longitudinal, and 5.8 % for shear). Because of the large scatter we observed in these measurements, we used average values of 5.453 km/s for C_L and 3.101 km/s for C_S. The bulk sound speed was calculated as ~4.115 km/s from the longitudinal and shear sound speed values.

3. Experimental Results and Discussion

Data were obtained from the four experiments with a nominal projectile velocity of 0.300 km/s. Details of each shot are shown in Table 1, with the impactor thickness defined as x_I and the target thickness defined as x_T. The initial density (ρ_o), longitudinal sound speed (C_L), and shear sound speed (C_S) are listed for the targets. Shown in Figure 3 are the velocity vs. time data obtained from the four experiments from the center of the target. All four velocimetry curves showed clear evidence of spall by the ringing in the data following the velocity pullback from the peak. The first shot was the only curve to show a well-defined elastic wave and also had a higher spall amplitude. Shown in Figure 4 are the velocimetry curves for all four shots shown on the same graph, providing a close up of the Hugoniot elastic limit (HEL).

Table 1. Metrology and shot parameters.

Shot Number	Projectile velocity (km/s) (+/-0.002)	Tilt (mrad)	x_I (mm) (+/-0.01)	x_T (mm) (+/-0.01)	ρ_o (g/cc) (+/-0.05)	C_L (km/s) (+/- 0.05)	C_S (km/s) (+/-0.18)
1-97	0.300	0.67	1.770	3.527	*	5.400	2.920
2-98	0.297	*	1.753	3.499	7.682	5.457	3.284
3-142	0.304	0.74	1.749	3.498	7.719	5.466	2.970
4-143	0.308	0.71	1.748	3.499	7.782	5.490	3.230

*data not collected

Quantities derived from the velocimetry curve were the shock velocity (U_s), particle velocity (u_p), HEL, spall strength (σ_{spall}), and spall scab thickness (h) using methods as described by Thomas, et al. [10] and are shown in Table 2. The data revealed that this alloy has a relatively low compressive strength. Interestingly, the HEL was lower for the first two shots (0.22 GPa and 0.21 GPa, respectively) than for the latter two shots (0.67 GPa and 0.51 GPa, respectively). Because homodyne velocimetry is known to have difficulty resolving small elastic waves, we used heterodyne (up-shifted) PDV. But even with upshifted PDV, the details of small elastic precursors can be difficult to resolve, and can possibly depend upon details of data analysis. Despite this, we believe these differences are real, and may be another indication that the HEA plate, where samples were extracted, was not isotropic.

The approximate spall strength was calculated using simple assumptions. We use the momentum jump condition, as modified to calculate approximate spall stress.

$$\sigma_{spall} = \frac{1}{2}\rho_0 C_b \Delta u_{fs} \tag{1}$$

Where ρ_0 is the initial density, C_b is the bulk sound velocity, and Δu_{fs} is the particle velocity change in the spall signature. We make the assumption that this alloy is ductile enough to justify use of the bulk sound speed in this equation. As shown in Table 2, the spall strength results for shots 2-98, 3-142, and 4-143 are similar, where the value for shot 1-97 is considerably different. The spall plane is estimated to have been located at approximately the midline (1.75 mm) of the target thickness (3.50 mm), as calculated from the PDV velocimetry curve to be between 1.77 mm and 1.88 mm, depending on the shot analyzed.

The experiments were designed using an approximate Hugoniot, constructed by assuming that the intercept in the shock velocity vs. particle velocity plot equals the bulk sound velocity discussed above (C_b=4.115 km/s). This is well known to be approximately true for many metals. For the approximate Hugoniot slope, we estimated the Gruneisen parameter to be 1.6, since the range for metals is around 1.5 to 2.0 [11], as well as the well-known estimate [12] that:

$$S = \frac{(\gamma+1)}{2} \quad (2)$$

This value for γ_G yields an approximate Hugoniot slope of 1.3, which is typical for many metals. Note that all this is just a rough estimation, done simply to allow initial design of experiments.

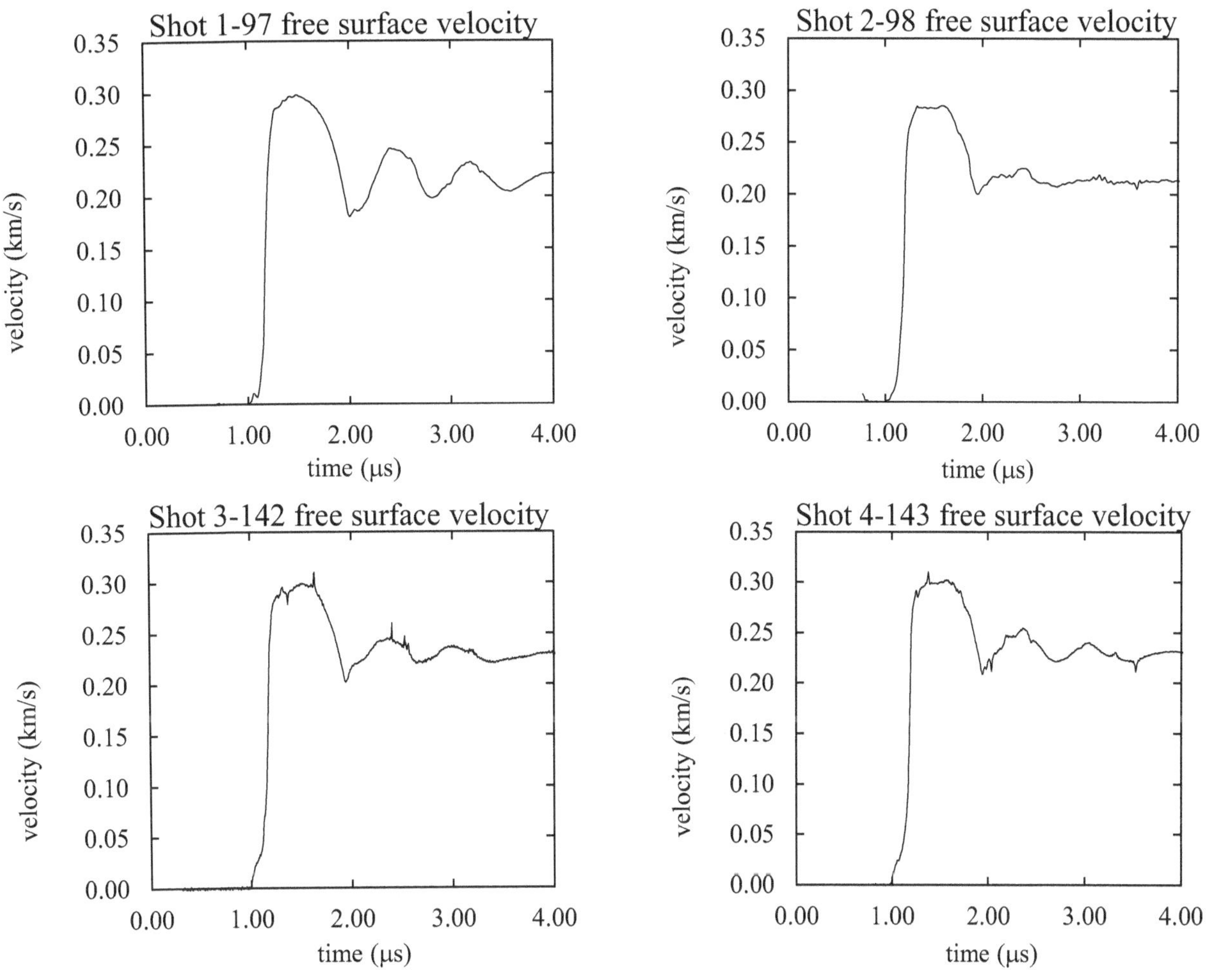

Fig. 3. Velocimetry curves for each experiment showing data from the center probe.

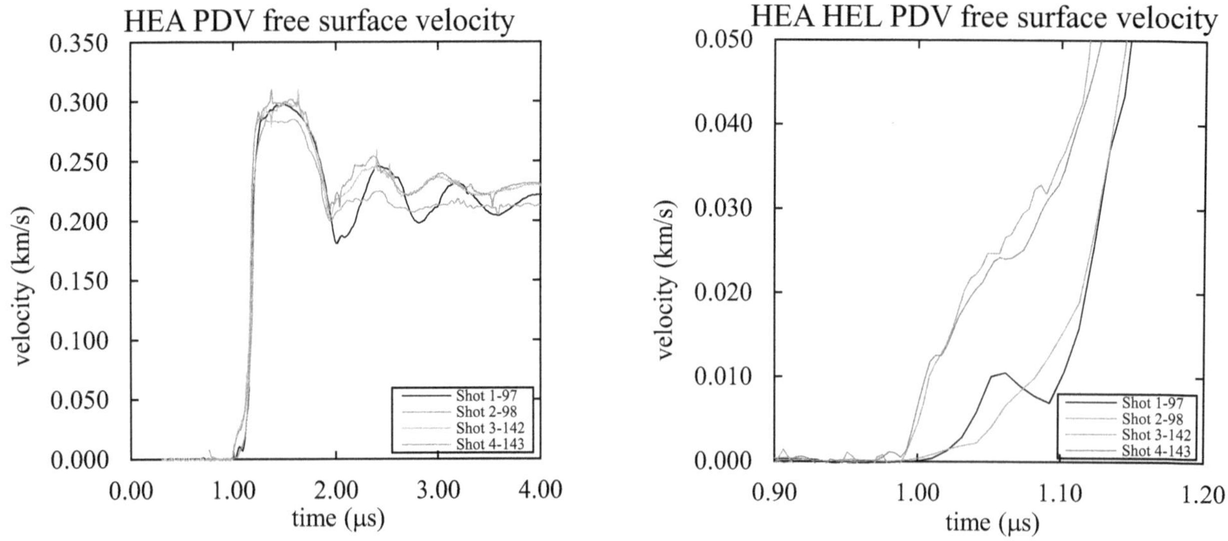

Fig. 4. Velocimetry curves of the four experiments on this HEA alloy.

Table 2. Derived quantities from velocimetry.

Shot Number	U_D (km/s) (+/-0.002)	U_s (km/s) (+/-0.050)	u_p (km/s) (+/-0.002)	HEL (GPa) (+/-0.003)	σ_{spall} (GPa) (+/-0.03)	h (mm) (+/-.02)
1-97	0.300	4.26	0.150	0.22	1.90	1.88
2-98	0.298	4.31	0.149	0.21	1.28	1.85
3-142	0.304	4.25	0.152	0.67	1.58	1.77
4-143	0.308	4.28	0.154	0.51	1.46	1.84

Time resolved data was used to estimate the shock velocity. This was done by making the assumption that the elastic wave moves at the ultrasonic longitudinal sound speed. We then use the arrival time of the elastic wave to find the impact time. The arrival time of the plastic (shock) wave was found by measuring the time at the midpoint in the shock rise, as discussed previously by Thomas et. al. [10]. This method allows a shock velocity to be determined to about the 1 % level. To increase accuracy requires a different experimental technique.

The experiments yielded four data points that are in reasonable agreement with an estimated Hugoniot for the material. Measured wave speeds were close to the estimated Hugoniot ($U_s = 4.115 + 1.3\ u_p$), as shown in Figure 5. The particle velocity, u_p, is taken as half the flyer velocity. Also of note is the long rise time of the plastic wave (~110-150 ± 13 ns), which leads to a relatively large uncertainty in the shock velocity. As can be seen from the plot, there is some scatter in U_s due to this large uncertainty as well as inherent error in metrology, data analysis, and extracting the data from the velocimetry curve. One representative error bar (+/-1%) is shown on the graph for shot 4-143.

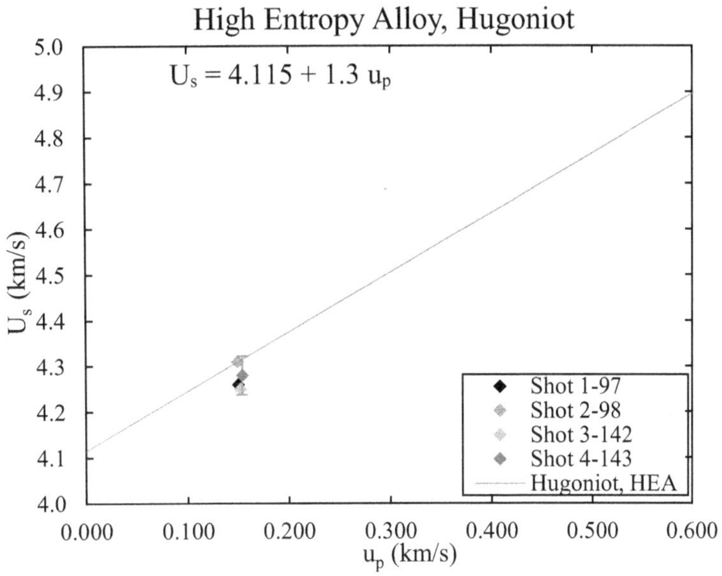

Fig. 5. Approximate Hugoniot line shown with experimental data.

4. Conclusions

Initial scoping experiments at a single shock stress have been done on a particular type of HEA, to begin the process of obtaining a better understanding of the dynamic properties of this class of metal. Experiments were done at a single stress state to check for reproducibility in derived quantities. The initial characterization of this particular HEA alloy revealed relatively large deviations in density and sound speeds at ambient conditions, as compared to elemental metals. These sample-to-sample differences may be caused by local differences in the plate stock from which our samples were taken. The root causes for what we observed are not clear at this point, but more studies would possibly shed light on this issue. We also observed scatter in the observed compressive strength as deviations in amplitude and shape of the elastic precursor wave, as well as in approximate spall strength

However, what is clear from these results is that this alloy does not possess a very high compressive strength. In addition, the approximate spall (tensile) strength for this alloy was relatively high (much higher than the compressive strength), and showed a large variation in value for our experiments. More experiments are warranted (on other HEA's) to investigate if the observed scatter in results is typical for this kind of alloy. Because this alloy has low dynamic compressive strength, it is not a good candidate for some potential applications (such as armor), and we will move on to study another HEA, probably a cobalt-based alloy.

The quasi-static analysis performed on samples from the same batch of HEA material showed relatively similar properties until the very end of the experiment after peak load was achieved. Perhaps this HEA material is not as uniform as, for example, elemental metals, which creates differences in the measured and calculated material properties. It is possible that our observed ultrasonic attenuations are a characteristic of this kind of multi-component alloy. A more detailed analysis of the microstructure coupled with recovery of the targets may provide insights into the scatter in the dynamic properties of this material.

Acknowledgements

This manuscript has been authored by Mission Support and Test Services, LLC, under Contract No. DE-NA0003624 with the U.S. Department of Energy, National Nuclear Security Administration, Office of Defense Programs, and supported by the Site-Directed Research and Development Program. The United States Government retains and the publisher, by accepting the article for publication, acknowledges that the United States Government retains a non-exclusive, paid-up, irrevocable, worldwide license to publish or reproduce the published form of this manuscript, or allow others to do so, for United States Government purposes. The U.S. Department of Energy will provide public access to these results of federally sponsored research in accordance with the

DOE Public Access Plan (http://energy.gov/downloads/doe-public-access-plan). The views expressed in the article do not necessarily represent the views of the U.S. Department of Energy or the United States Government. DOE/NV/03624—0354.

References

[1] Miracle, D.B., Senkov, O.N., 2017. "A critical review of high entropy alloys and related concepts", Acta Materialia 122, p. 448-511.

[2] Gludovatz, B., Hohenwarter, A., Catoor, D., Chang, E.H., George, E.P., Ritchie, R.O., 2014. "A fracture-resistant high-entropy alloy for cryogenic applications", Science 345.

[3] Stepanov, N.D., Shaysultanov, D.G., Salishchev, G.A., Tikhonovsky, M.A., Oleynik, E.E., Tortika, A.S., Senkov, O.N., 2015. "Effect of V content on microstructure and mechanical properties of the CoCrFeMnNiVx high entropy alloys", J. Alloys Compd. 628.

[4] Hemphill, M.A., Yuan, T., Wang, A.M., Yeh, J.W., Tsai, C.W., Chuang, A., Liaw, P.K., 2012. "Fatigue behavior of Al0.5CoCrCuFeNi high entropy alloys", Acta Mater 60.

[5] Geanta, V. Voiculescu, I., Stefanoiu, R., Chereches, T., Zecheru, T., Matache, L., Rotariu, A., 2018. "Dynamic Impact Behaviour of High Entropy Alloys Used in the Military Domain". IOP Conf. Ser.: Mater. Sci. Eng. 374 012041.

[6] Li, Z., Zhao, S., Diao, H., Liaw, P.K., Meyers, M.A., 2017. "High-velocity deformation of Al0.3CoCrFeNi high-entropy alloy: Remarkable resistance to shear failure". Scientific reports (7), 42742. doi: 10.1038/srep42742.

[7] Geanta, V., Chereches, T., Lixandru, P., Voiculescu, I., Stefanoiu, R., Dragnea, D., Zecheru, T., Matache, L., 2017. "Virtual testing of composite structures made of high entropy alloys and steel". Metals, 7(11), 496. doi:10.3390.

[8] Liu, C., Valdez, J.A., 2018. HEA Uniaxial Tension Tests. MST-8 Materials Science and Technology Division, Los Alamos National Laboratory. Internal report.

[9] Strand O.T., Goosman, D.R., Martinez, C., Whitworth, T.L., and Kuhlow, W.W., 2006. "Compact system for high-speed velocimetry using heterodyne techniques. Review of Scientific Instruments". 77, 083108.

[10] Thomas, S.A, Hawkins, M.C., Matthes, M.K., Gray, G.T. III, Hixson, R.S., 2018. "Dynamic Strength Properties and Alpha-Phase Shock Hugoniot of Iron and Steel", Journal of Applied Physics 123 (17).

[11] Selected Hugoniots, 1969. Prepared by Group GMX-6, Los Alamos Scientific Laboratory, University of California and Los Alamos, New Mexico. LA-4167-MS.

[12] Deal, W.E. Jr., 1962. Dynamic High Pressure Techniques, Modern Very High Pressure Techniques, ed. R.H. Wentorf, Jr., Washington, DC. pp. 204-207.

Proceedings of the 2019 Hypervelocity Impact Symposium
HVIS2019
April 14-19, 2019, Destin, FL, USA
HVIS2019-010

Extended Abstract+

Effects of EOS and constitutive models on simulating copper shaped charge jets in ALEGRA

Robert L. Doney[a,*], John H. J. Niederhaus[b], Timothy J. Fuller[b], Matthew J. Coppinger[a]

[a]*U.S. Army Research Laboratory, 328 Hopkins Rd, Aberdeen Proving Ground, MD, 21005, USA*
[b]*Sandia National Laboratories, P.O. Box 5800 MS-1323, Albuquerque, NM, 87185, USA*

1. Introduction

Metallic shaped charge jets (SCJ) have been studied for many decades across multiple communities for applications ranging from military warheads to earth penetrators for accessing oil-rich areas [1]. Researchers have had varied success in modeling these jets using simulation codes such as CTH, ALEGRA, and ALE3D. Recently, a large amount of work has been performed at the US Army Research Lab investigating the behavior of jets with increasingly sophisticated experimental diagnostics. Advances in computational resources, code enhancements, and material models have allowed us to model jets and probe uncertainties caused by algorithms, equations of state (EOS), constitutive models, and any of the available parameters each one provides. In this work we explore the effects that various EOS and constitutive models have on the development and characteristics of a 65-mm diameter, 2D copper SCJ using the Sandia National Laboratories' multiphysics hydrocode, ALEGRA [2]. Specifically, we evaluate the tabular SESAME 3320 [3], 3325 [4-5], and 3337 [6] EOS models, analytic EOS (ANEOS) 3331 [7], as well as the Johnson-Cook (JC) [8], Zerilli-Armstrong (ZA) [9], Preston-Tonks-Wallace (PTW) [10], Steinberg-Guinan-Lund (SGL) [11-12], and Mechanical Threshold Stress (MTS) [13] constitutive models. Note that while the SGL model supports rate-dependence, there is no current characterization for copper, thus we are using rate-independent version. We do not consider the MieGrüneisen equation of state here as we expect parts of the jet to be near or cross into melt.

Nikkel and Lassila [15] investigated the effect of the SGL and MTS constitutive models on temperature and stress in jets. More recently Uhlig and Hummer [16] measured the temperature of a separated jet tip (the technique being sensitive to the outer 50% of the jet by volume), finding its bulk value to be roughly 1190 K. They further found that temperature distributions produced from the SGL model using ALEGRA were nearly identical to experimental data, with hot axial temperatures (~1400 K) and a cooler surface with an average temperature of ~1140 K. The very high axial temperature is consistent with the emergence of a "probiscus" or protruding region—presumably liquid—ahead of the jet tip. Consequently, we will use the SGL model as a baseline in most cases. CTH simulations of temperature were also performed by Sable et al. [17], agreeing with the work of Uhlig and Hummer.

+ Full paper can found at https://www.sciencedirect.com/journal/international-journal-of-impact-engineering/special-issue/10MTGP5W4VJ

* Corresponding author. Tel.: +1-410-278-7309; fax: +1-410-278-6061
E-mail address: robert.L.doney4.civ@mail.mil

2. Main text

2.1. Computational Setup

In ALEGRA, we created an ensemble of 2D axisymmetric SCJ simulations formed by a 65-mm copper liner and LX-14 explosive in an air environment. A programmed burn capability detonated the explosive symmetrically at its base. Multiple diagnostics were used for visualization and to quantify material properties. An array of Lagrangian tracer particles were inserted near the liner's apex to try and capture a particle that traverses the compression region during liner collapse and then flow along with the jet tip. In all cases, most of the tracers drifted back into the slug; usually only one or two tracers would flow towards the jet tip.

All of the EOS models investigated contain solid and liquid phases with melt and vaporization [18]. The most recent SESAME EOS for copper, 3337, is of particular interest. According to the authors [4], "the new [3337] EOS has been compared to numerous data sets examining the behavior of the EOS both in expansion and compression and found to be in excellent agreement. Additionally, the new equation of state is superior to the previous SESAME equation of state 3336, in thermal expansion, heat capacity, and treatment of the melt. In all other comparisons, the two equations of state are comparable."

2.2. Effects of strength models

Material strength is occasionally dismissed when modeling jets since pressures generated during liner collapse greatly exceed the yield strength—thus placing the system in the hydrodynamic regime. However, strength *is* important as we find measurable variation in the thermal properties of jets with differing strength models. Even though jet profiles and tip speeds are fairly similar, the various strength models generate substantial differences in the jets' internal energy after about 15 μs. Fig. 1 illustrates this for the 3337 EOS with MTS and JC models reporting similar results and the SGL and PTW models responding with higher energies.

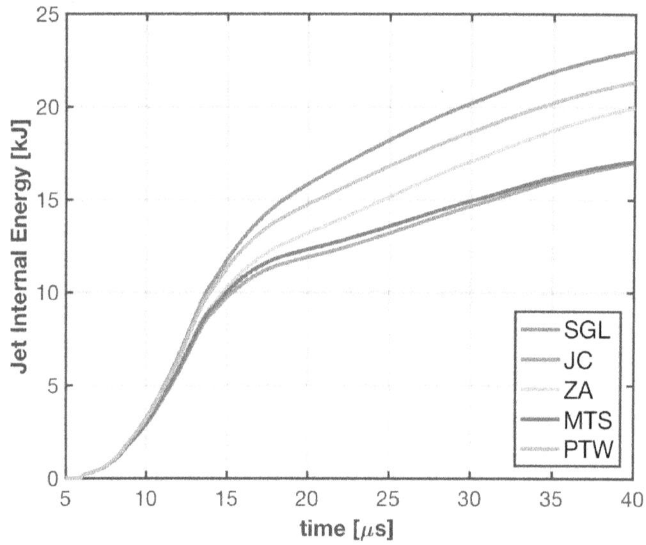

Fig. 1. Jet internal energy for the various strength models using the 3337 EOS.

To investigate the effect of strength models on SCJ behavior in more detail, we monitor the trajectory of a Lagrangian tracer particle immersed in the copper liner near the initial apex: from the ambient state through the initial moments of jet formation (i.e., during liner collapse). Density, pressure, and temperature data from the tracer are overlaid on (grey) isotherms extracted from the SESAME 3337 EOS table. The result is a simulated state-space trajectory for material participating in jet formation, shown in Fig. 2 with density plotted logarithmically. In density-pressure space (top panel), the melt transition region is visible as inflections that appear on several of the higher-temperature isotherms near the center of the plot. It can be seen that the trajectories do not traverse this region until the pressure oscillations have subsided. The state-space trajectories overlay nearly perfectly through the first series of pressure oscillations. However, during the strongest compression, the trajectories diverge noticeably: with SGL migrating to higher isotherms during compression, but particularly during release, while JC remains at lower isotherms.

Superimposed over temperature space (lower panel) is the density-dependent Lindemann melt curve [19] as well as a static copper melt temperature of 1381 K. For the SGL, PTW, and ZA models, the material coincident with the tracer appears to become liquid. Interestingly, this happens very close to the intersection of the two melt curves. In the JC and MTS models, the material stays solid as the static melt temperature is not exceeded. In pressure space, the trajectories show ringing as the point oscillates in density. The SGL and PTW models achieve the highest values (40-45 GPa) as the jet tip forms at 8 us. Both JC and MTS models follow similar trajectories and report higher peak densities.

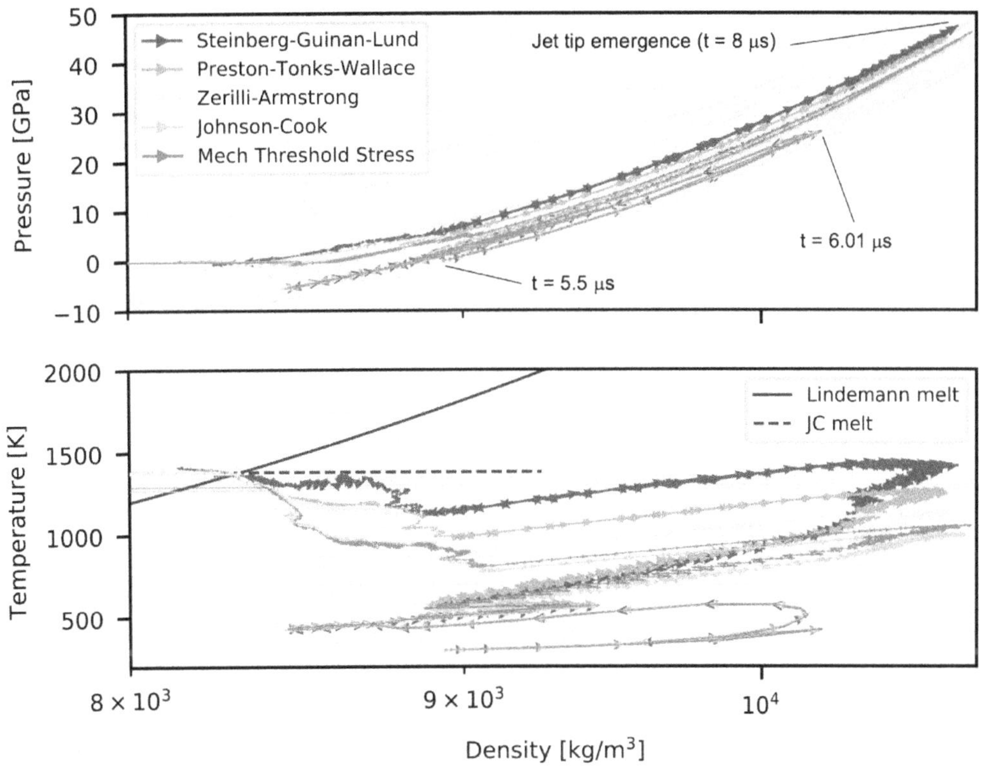

Fig. 2. Trajectories of a tracer in state space for the 3337 EOS and various strength models.

We can look more closely at the presence of liquid since we track the material phase through ALEGRA's phase control capability, which turns off the strength model when melt or vaporization is flagged in a cell. When using the JC model, the melt region and probiscus are not present. This presence of liquid copper in the SGL model is consistent with experimental findings by Lassila, *et al.* [20]: "the metallurgical analyses (optical light microscopy, SEM) performed on the recovered slug and jet particles indicate that the regions in the vicinity of the centerline of slug and jet particles were melted." None of the other yield models generated a melt region of this extent. Further, when the pressure-dependent term in SGL is turned off, the melt region—and experimentally observed probiscus—vanishes.

2.3. Summary

In this work we evaluated the effects that equations of state and strength models have on SCJ development using the Sandia National Laboratories multiphysics shock code, ALEGRA. Results were quantified using a Lagrangian tracer particle following liner collapse, passing through the compression zone, and flowing into the jet tip. We found consistent results among several EOS: 3320, 3331, and 3337. The 3325 EOS generated a measurable low density and hollow region near the jet tip which appears to be reflected in a lower internal energy of the jet. At this time, we cannot tell, experimentally, if such a hollow region exists. The 3337 EOS is recent, well documented [6], and produces results similar to 3320 [3]. The various strength models produced more noticeable differences. In terms of internal energy and temperature, SGL had the largest values followed by PTW, ZA, and finally JC and MTS, which were quite similar to each other. We looked at melt conditions in the SGL and JC models using the 3337 EOS. The SGL model reported a liquid region along the jet axis all the way to the tip—seemingly consistent with experiment—while the JC model does not indicate any phase transition. None of the other yield models indicated melt along the jet axis. For all EOS and strength models, we found similar results for the velocity history of the jet tip as measured against experiment using photon Dopper velocimetry.

Acknowledgements

This work was supported in part by a grant of computer time from the DoD High Performance Computing Modernization Program at the Army Research Laboratory DoD Supercomputing Resource Center and the Terminal Ballistics Frontier project. We also thank Dr. Michael Zellner for providing the PVD data and Mr. Bill Clark and Dr. Bill Walters for helpful discussions regarding tracer and jet flow during liner collapse. We thank Kyle Cochrane and John Kaushagen (Sandia) for the "sesutils" tools for processing EOS data. Mr. Cochrane was also helpful on discussions regarding SESAME isotherms. We also thank Dr. Dan Hornbaker for reviewing this work.

Sandia National Laboratories is a multimission laboratory managed and operated by National Technology and Engineering Solutions of Sandia, LLC., a wholly owned subsidiary of Honeywell International, Inc., for the U.S. Department of Energy's National Nuclear Security Administration under contract DE-NA0003525.

References

[1] Walters, W. "The shaped charge concept. Part III: applications of shaped charges." BRL-TR-3168, Ballistics Research Laboratory, Oct. 1990.

[2] Robinson, A., Brunner, T., Carroll, S., and coauthors, 2008. ALEGRA: An arbitrary Lagrangian-Eulerian multimaterial, multi-physics code. In: Proc. 46th AIAA Aerospace Sciences Meeting: AIAA-2008-1235.

[3] Kerley, G., "Equations of state for Copper and Lead." KTS02-1, Kerley Technical Services Report, March 2002.

[4] Cochrane, K., Lemke, R., Riford, Z., Carpenter, J., 2016. "Magnetically launched flyer plate technique for probing electrical conductivity of compressed copper." Journal of Applied Physics, 119, 105902.

[5] Kraus, R., Davis, J.-P., Seagle, C., Fratanduono, D., Swift, D., Brown, J., Eggert, J., 2016. "Dynamic compression of copper to over 450 GPa: a high-pressure standard." Physical Review B, 93, 134105.

[6] Peterson, J., Honnell, K., Greeff, C., Johnson, J., Boettger, J., Crockett, S. Global equation of state for copper. LA-UR-11-04443, Los Alamos National Laboratory, 2011. Also in: AIP Conference Proceedings (Shock Compression of Condensed Matter – 2011), 1426, 763 (2012); doi: 10.1063/1.3686390.

[7] Thompson, S., "ANEOS: Analytic equations of state for shock physics codes input manual." SAND89-2951, Sandia Report, March 1990.

[8] Johnson, G., Cook, W., "A constitutive model and data for metals subjected to large strains, high strain rates and high temperatures," (source unknown).

[9] Zerilli, F., Armstrong, R., 1987. "Dislocation-mechanics-based constitutive relations for material dynamics calculations." Journal of Applied Physics, 61, 1816.

[10] Preston, D., Tonks, D., and Wallace, D., 2003. "Model of plastic deformation for extreme loading conditions." Journal of Applied Physics, 93, 211.

[11] Steinberg, D., Cochran, S., Guinan, M., 1980. "A constitutive model for metals applicable at high-strain rate." Journal of Applied Physics, 51, 1498.

[12] Steinberg, D., Lund, C., 1998. "A constitutive model for strain rates from 10^{-4} to 10^{6} s^{-1} " Journal of Applied Physics, 65, 1528.

[13] Follansbee, P., Kocks, U., 1988. "A constitutive description of the deformation of copper based on the use of the mechanical threshold stress as an internal state variable." Acta Metallurgica, 36 (1), pp. 81-93.

[14] Zellner, M., Vunni, G., Photon Doppler velocimetry characterization of shaped charge jet formation. ARL-TR-5961, Army Research Laboratory, April 2012.

[15] Nikkel D., Lassila, D., 1994. "The effect of the constitutive response on the predicted temperature in copper jets." AIP Conf. Proc., 309, pp. 1857–1860.

[16] Uhlig, W., Hummer, C., 2013. "In-flight conductivity and temperature measurements of hypervelocity projectiles." Procedia Engineering (Proceedings of the 12th HVIS), 58, pp. 48–57.

[17] Sable, P., Helminiak, N., Gullerud, A., Harstad, E., Hollenshead, J., Hertel, E., 2017. "Characterizing in-flight temperature of shaped charge penetrators in CTH." Procedia Engineering (Proceedings of the 17th HVIS), 204, pp. 375–382.

[18] Carpenter, J., "What's in an Equation of State?" SAND2016-3982PE, 2016.

[19] Stacey, F.D., Irvine, R.D., 1977. "Theory of melting: thermodynamic basis of Lindemann's Law." Australian Journal of Physics, 30, pp. 631-640.

[20] Lassila, D., Walters, W., Nikkel, D., Kershaw, D. "Evidence of melt in 'soft' recovered copper jets." UCRL-JC-120717. Lawrence Livermore National Laboratories, August 1995.

Proceedings of the 2019 Hypervelocity Impact Symposium
HVIS2019
April 14-19, 2019, Destin, FL, USA
HVIS2019-012

Extended Abstract+

Bayesian model selection for metal yield models in high-velocity impact

Teresa Portone[a]*, John Niederhaus[b], Jason Sanchez[b], Laura Swiler[b]

[a]*Institute for Computational Engineering and Sciences, UT Austin, 201 E 24th St, Stop C0200, Austin, Texas 78731-1229, USA*
[b]*Sandia National Laboratories, P.O. Box 5800 MS-1323, Albuquerque NM 87185 USA*

1. Introduction

There are multiple candidate models for the von Mises yield stress in elastoplastic models that are used to simulate material deformation in high-velocity impact. Previous work has studied the effects of such models on quantities of interest in high-strain-rate deformation, in order to select the most appropriate model in comparison to experimental data [1-2]. This work focuses on selecting between three such models, based on their ability to reproduce time-varying depth of penetration data of a tungsten-alloy rod impacting a hardened steel plate at high velocity, measured by Anderson, Hohler *et al.* [3]. Novel in the present study is the systematic treatment of uncertainty in the process, and automation of the process. The three models considered are the Johnson-Cook (JC), Zerilli-Armstrong (ZA), and Steinberg-Guinan-Lund (SGL) flow stress models [4-6].

2. Bayesian uncertainty quantification and calibration

In developing predictive models for high-velocity impact based on empirical material data, sources of uncertainty include measurement uncertainty, incomplete knowledge of model parameters, and model uncertainty. Uncertainty quantification (UQ) is the science of mathematically representing these uncertainties, typically using probabilities, and propagating them to a modeled quantity of interest (QOI). First, consider measurement uncertainty. Typically, it can be assumed that the measurement error is additive, independent, and normally distributed. Based on these properties, a probability density function can be developed to answer the question, "what is the probability of the observed data $\mathbf{d}$ arising from the model M, given parameter values $\mathbf{p}$?", called the likelihood density, denoted as $\pi(\mathbf{d}|\mathbf{p})$. Next, consider parameter uncertainty. There is limited knowledge about the correct value of the parameters $\mathbf{p}$ for the model M. Prior knowledge about the parameter and the physical system can be used to represent the uncertainty in the parameters. For instance, uniform or normal probability density may be appropriate. Such probability densities describe the prior parameter information available and are called prior probability densities, denoted here as $\pi(\mathbf{p})$.

Bayesian calibration allows for uncertainty representations, for instance of the uncertain parameters $\mathbf{p}$, to be updated by comparing them to data. This is done using Bayes' theorem, also called Bayes' rule, which combines information from the prior and likelihood densities to give a *new* distribution for the parameters, called the posterior density:

$$\pi(\mathbf{p}|\mathbf{d}) = \frac{\pi(\mathbf{d}|\mathbf{p})\pi(\mathbf{p})}{\int \pi(\mathbf{d}|\mathbf{p})\pi(\mathbf{p})d\mathbf{p}}. \tag{1}$$

+ Full paper can found at https://www.sciencedirect.com/journal/international-journal-of-impact-engineering/special-issue/10MTGP5W4VJ

* Corresponding author E-mail address: tnportone@gmail.com.

Bayesian calibration not only provides updated "most-likely" parameters to have produced the data, but also updated information about how sure one can be of those "most-likely" values. An analytical expression for the posterior distribution is unavailable for models that are nonlinear in the parameters **p**, because the integral in the denominator cannot be computed directly. Thus, the most common approach to Bayesian calibration is to use Markov Chain Monte Carlo (MCMC) methods, which produce samples from the posterior distribution, which are used to compute summary statistics. One or more model evaluations is required per sample of the posterior, which quickly becomes costly. Surrogate models are commonly used to reduce this cost. They approximate model responses by fitting to a small number of model evaluations. Most popular are Gaussian processes (GPs) and Polynomial Chaos Expansions (PCEs), both used as part of this work. Another way to reduce cost is to only update the uncertainty representations of the parameters that significantly affect the modeled quantity of interest. This was done using global variance-based sensitivity analysis.

3. Bayesian model selection

Bayesian model selection provides a systematic way to select between multiple candidate models to represent a phenomenon. It selects the model that best fits observational data of the phenomenon with the least amount of tuning of parameters. This is done in the UQ context using Bayes' theorem. The question is, "which of these models is most likely to have produced this data?" Prior to comparing to the data, it is not clear which model is most likely, so it is typical to assign equal probability to each model. If there are N_m models, indexed by i, the prior probability that the model M_i produced the data is $P(M_i) = 1/N_m$.

The likelihood density, also called the model evidence, is obtained by asking, "what is the likelihood the data arose from the model M_i, given all possible parameter values?" It is computed by averaging the likelihood density for the model's parameters over all possible parameter values:

$$L(M_i) \equiv P(d|M_i) \equiv \int \pi(\mathbf{d}|\mathbf{p}_i)\pi(\mathbf{p}_i)d\mathbf{p}_i \tag{2}$$

where $\mathbf{p}_i$ are the parameters associated with model M_i. Combining these probabilities using Bayes' theorem, and noting that the prior probabilities are the same for all of the models, the posterior probability of each model having produced the data is

$$P(M_i|\mathbf{d}) = \frac{P(\mathbf{d}|M_i)}{\sum_{j=1}^{N_m} P(\mathbf{d}|M_j)}. \tag{3}$$

The model with the highest posterior probability is the one that is selected in the Bayesian model selection paradigm.

Computing the model evidence is the most important aspect of Bayesian model selection. Because it requires an integration over all possible parameter values for each model, it is also the most computationally intensive part. Different Bayesian model selection methods approximate Equation 2 in different ways. Three different techniques were used for this work, varying in accuracy and computational cost from most to least: Monte Carlo (MC) integration, a Kernel Density Estimate (KDE) approximation, and a Laplace approximation.

4. Problem description and simulation results

For this study, we applied these ideas to a simple model selection problem between material models, evaluating their ability to reproduce time-dependent depth of penetration (DOP) data in high-velocity impact. The DOP data are obtained from repeated-shot x-ray observations of a 4-mm-diameter, 50-mm-long tungsten-alloy rod impacting a hardened steel plate at approximately 1250 m/s, reported by Anderson, Hohler, *et al.* in 1995 [3]. These data are the basis for our Bayesian calibration and model selection process. The plate material was HzB steel, which is a hardened steel found in some ballistic protection applications. The DOP history appears as red points below in Figure 1.

The observational standard deviation for the DOP data is reported as 0.01 cm by Anderson, Hohler, *et al.* [3] However, there is a significant inflection in the DOP history at $t = 40$ μs. Because this inflection was not observed in simulations, we postulate that there were uncertainties not fully accounted for in the estimate. This could be problematic for our analysis, because the relative effect of the likelihood density on the posterior distribution increases with decreasing measurement error. This can cause an overfitting to the data that could lead to unreliable calibration and model selection outcomes. To address this missing uncertainty, a polynomial regressive fit was performed, and a standard deviation of approximately 0.06 cm was estimated using the data's deviation from that experimental fit. This inferred uncertainty was used in the UQ analysis and appears as red bars in Figure 1.

As mentioned above, we considered three well-known strength models used for metals subject to large strains, high strain rates, and high temperatures: JC, ZA, and SGL. All models are implemented in the finite-element software package ALEGRA [7], which was used here to generate model evaluations for the analysis by simulating the impact event in steel in two-dimensional axisymmetry. Management of the UQ study based on these model evaluations was done with Dakota [8].

A sample ALEGRA simulation for this system using the JC model is shown in Figure 1. The DOP as a function of time was extracted from the simulation for comparison to data. The mesh was refined until the difference in the time history of the tracer was smaller than the data's measurement error and the behaviour of the tracer had effectively converged. This cell size was 0.1 mm, or 20 elements spanning the radius of the tungsten-alloy rod. The modeled DOP history after Bayesian calibration ("posterior") appears as a blue line in Figure 1.

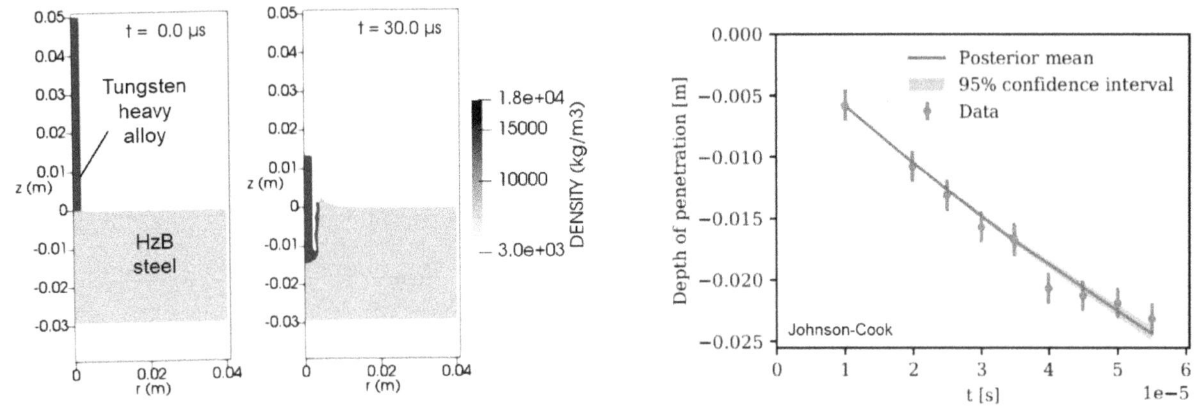

Figure 1: (Left) Sample ALEGRA simulation results for 1250-m/s impact using the JC model for HzB steel and a cell size of 0.1 mm. (Right) Comparison of the DOP histories computed by ALEGRA using the nominal "starting" JC parameterization, and the final "maximum *a posteriori* probability" (MAP) configuration after Bayesian calibration (same as the posterior mean).

For the UQ studies, surrogate models were used. Model evaluations were collected using Latin Hypercube Sampling (LHS), and both Gaussian Processes and Polynomial Chaos Expansions were employed. A Bayesian calibration using the impact data was performed for each of the models using MCMC, implemented in Dakota. The length-10^6 Markov chain produced from the Bayesian calibration for each model was used in the Kernel Density Estimate (KDE) approximation of its model evidence. Furthermore, the evidence was approximated using MC integration with 10^6 samples, as well as using the Laplace approximation.

5. Bayesian calibration results.

The Bayesian calibration and model selection processes were performed on the JC, ZA, and SGL flow stress models, with parameters assumed to fall into normal distributions. Mean values were assumed to be ALEGRA material library values for HzB or comparable steels, and standard deviations were assumed to be 5%. These priors are shown as blue curves in Figure 2.

A set of 256 LHS samples was generated for each model using ALEGRA's embedded Dakota interface. These samples were used to generate a GP for sensitivity analysis, and the most significant model parameter(s) for the DOP were determined. For both JC and SGL, the initial/ambient yield stress dominated the sensitivity (A for JC, Y_0 for SGL). For ZA, the sensitivity analysis indicated that c_1, c_3 and c_5 are all significant. The c_1 and c_3 coefficients appear in a term related to initial yield strength, and c_5 appears in the ZA work-hardening term. Calibrated posterior distributions for some of the JC and ZA parameters are shown in Figure 2, with the priors. The posterior mean and 95% confidence interval of the depth of penetration as a function of time were also generated in this analysis. For the JC model, these are plotted against the data with their estimated uncertainty in Figure 1.

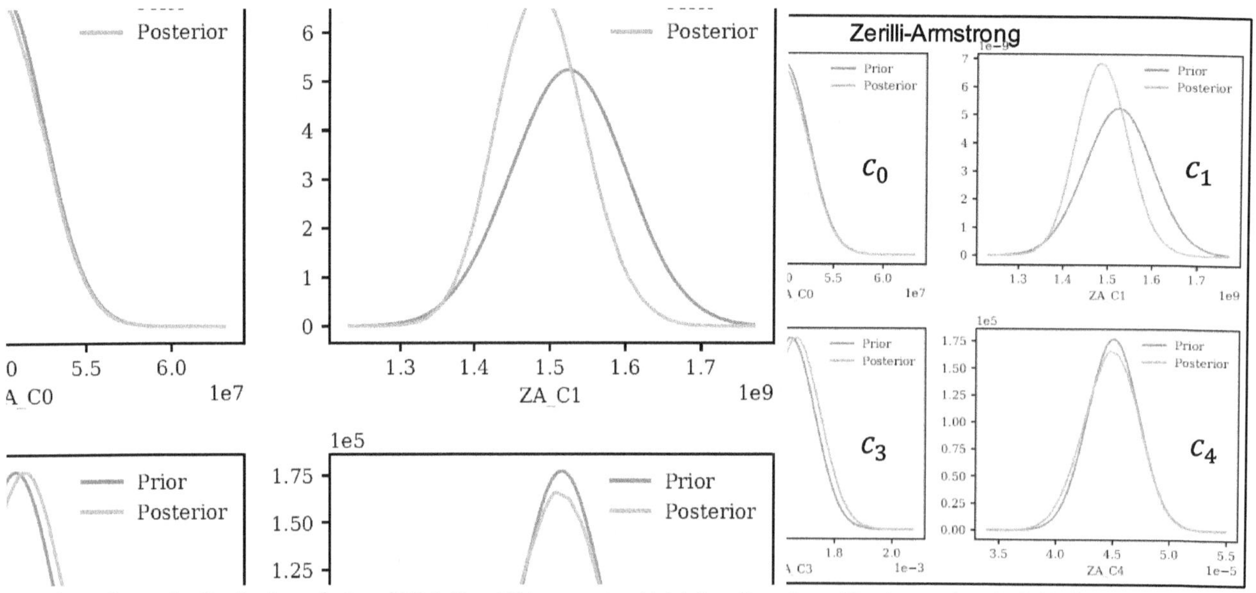

Figure 2: Prior and posterior distributions of selected JC (left) and ZA parameters (right) from Bayesian calibrations against the DOP data.

5. Bayesian model selection results

As described in Section 3, the model evidence (the likelihood for each model) was approximated using the MC, KDE, and Laplace methods. The results of the model selection are shown in Figure 3. All the models were assumed to be equally likely to start. All three methods of approximating the model evidence chose the ZA model as most likely to have produced the observational data. This result may motivate further exploration of ZA as a suitable model for representing DOP at high velocities.

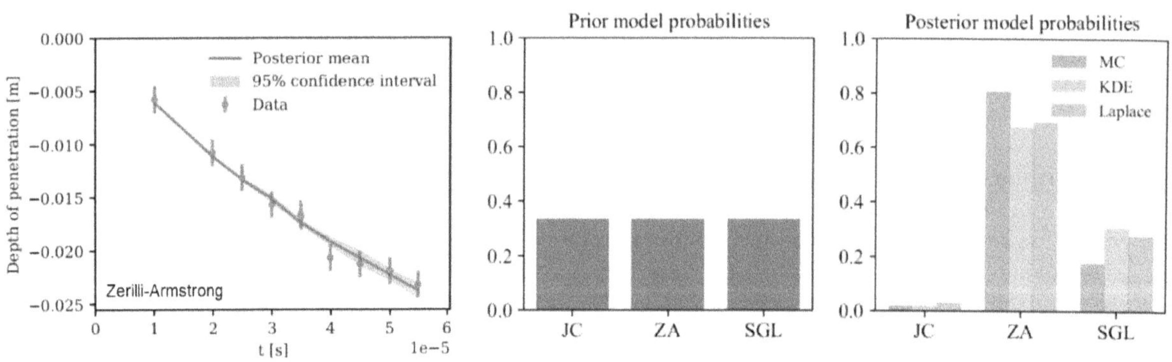

Figure 3: (Left) DOP history computed by ALEGRA for the posterior mean ZA material model configuration after Bayesian calibration, with 95% confidence interval. (Middle and right) Bayesian model selection results, showing prior (middle) and posterior (right) probabilities for the three flow stress models. For posterior probabilities, each color indicates a different method of approximating the model evidence. The study shows ZA has the highest probability.

Acknowledgements

This work was supported by the U.S. Army Research Laboratory. Sandia National Laboratories is a multimission laboratory managed and operated by National Technology and Engineering Solutions of Sandia, LLC., a wholly owned subsidiary of Honeywell International, Inc., for the U.S. Department of Energy's National Nuclear Security Administration under contract DE-NA0003525. This paper describes objective technical results and analysis. Any subjective views or opinions that might be expressed in the paper do not necessarily represent the views of the U.S. Department of Energy or the United States Government.

References

[1] Uhlig, W. C., Hummer, C. R., 2013. In-Flight Conductivity and Temperature Measurements of Hypervelocity Projectiles. Procedia Engineering, 58, p. 48.

[2] Helminiak, H. S., Sable, P., Harstad, E., Gullerud, E., Hollenshead, J., Hertel, E. S., 2017. Characterizing in-flight temperature of explosively formed projectiles in CTH. Procedia Engineering 204, p. 178.

[3] Anderson, C. E., Hohler, V., Walker, J. D., Stilp, A. J., 1995. Time-resolved penetration of long rods into steel targets. International Journal of Impact Engineering, 16(1), pp. 1-18

[4] Johnson, G. R., Cook, W. H., 1983. A constitutive model and data for metals subjected to large strains, high strain rates and high temperatures. In: Proc. 7th International Symposium on Ballistics: pp. 541-547.

[5] Steinberg, D. J., Cochran, S. G., Guinan, M. W., 1980. A constitutive model for metals applicable at high strain rate. J. Appl. Phys. 51 (3), 1498-1504.

[6] Zerilli, F. J., Armstrong, R. W., 1987. Dislocation-mechanics-based constitutive relations for material dynamics calculations. Journal of Applied Physics, 61(5), pp. 1816-1825.

[7] Robinson, A. C., Brunner, T. A., Carroll, S., and coauthors, 2008. ALEGRA: An arbitrary Lagrangian-Eulerian multimaterial, multi-physics code. In: Proc. 46th AIAA Aerospace Sciences Meeting: AIAA-2008-1235.

[8] Adams, B. M., Ebeida, M. S., Eldred, M. S., and coauthors, 2018. Dakota, a Multilevel Parallel Object-Oriented Framework for Design Optimization, Parameter Estimation, Uncertainty Quantification, and Sensitivity Analysis: Version 6.9 User's Manual. Technical report SAND2014-4633, Sandia National Laboratories. (https://dakota.sandia.gov/)

Proceedings of the 2019 Hypervelocity Impact Symposium
HVIS2019
April 14-19, 2019, Destin, FL, USA
HVIS2019-020

Extended Abstract+

Statistics of energy dissipation in the hypervelocity impact shock failure transition

Dennis Grady[a]*

[a]*Applied Research Associates, 4300 San Mateo Blvd NE, Albuquerque, NM, 87110, USA*

1. Introduction

In the hypervelocity impact event, shock waves subject material to failure transitions with the attendant dissipation of the imparted energy. Under shock compression, failure and dissipation entail intense compression, inelastic shear and compaction. Through shock interactions, states of dynamic tension are achieved and further failure dissipation involves fracture and fragmentation. The nature of failure of solids in the shock environment has encouraged considerable experimental effort through the past several decades. Such efforts have yielded results that suggest universality in the shock failure response over significant spans of shock intensity. Examples include the fourth-power relation between pressure and strain rate in solid-material compressive shock waves, and power-law relations capturing spall fracture strength and fragmentation size scale in dynamic tensile failure. Comparable power-laws also describe the shock compaction of distended solids. The present paper explores a statistical perspective of the underlying micro failure dynamics for the purpose of achieving better understanding of the macro failure trends noted above. A statistical correlation function description of the random micro velocity field is introduced. Through the attendant kinetic dissipation, the statistical fluctuation-dissipation principle is applied to the shock failure transition. From this statistical approach, power-law relations for compressive and tensile shock failure emerge that replicate the reported experimental behaviors.

The hypervelocity impact event entails the transfer of kinetic energy of the impacting body into internal energy and kinetic energy of the impacted material. Energies and energy rates are extreme. Subsequent to the impact shock event, the energy not remaining as kinetic energy resides as forms of thermal and athermal energy, a consequence of energy dissipation in the shock wave. The nature of energy dissipation in the shock wave is complex, and resides at the heart of much of shock wave physics.

Energy dissipation is heterogeneous, consummated through the collective dynamics of localized dissipative structures. Dissipative structures are unique to the material of concern and individually self-organize on the local scale in response to the dissipation and dissipation rates dictated by the encompassing shock wave. Dissipative structures can be, for example, thermal adiabatic shear bands in metals, shear and tensile fractures in brittle compounds, or planar compaction zones in porous matter. The formation and dynamics of dissipative structures are integral to material transitions caused by the shock, including onset of crystallographic and thermodynamic phase transformation, initiation of reaction in energetic materials, and the fracture and fragmentation of solid matter.

The dissipation of energy by the elemental dissipative structure is dichotomous. First, dissipation is frictional, occurring interior to the dissipative structure, contributing to the thermal phonon field and local temperature rise, as well as local athermal dissipation. Second, dissipation is kinetic, proceeding through acoustic phonon radiation from the dissipative structure.

+ Full paper can found at https://www.sciencedirect.com/journal/international-journal-of-impact-engineering/special-issue/10MTGP5W4VJ

* Corresponding author. Tel.: 505-883-3636; fax: 505-872-0794.
E-mail address: dgrady@ara.com.

Radiation dissipation and frictional dissipation are comparable in magnitude, and governed by the same failure and dissipation properties of the material.

Within the framework of Boltzmann statistics, the statistics-based fluctuation-dissipation theorem has application to the shock-wave failure transition. Through application of the fluctuation-dissipation relation, the kinetic dissipation, namely the acoustic phonon field, is identified with the viscosity responsible for structuring the shock wave. Through application of the fluctuation-dissipation relation, the correlated kinetic dissipation necessary to achieve the shock failure transition is determined. From application of the fluctuation-dissipation relation emerge relations for the dependence of spall stress and fragment size on the intensity of the imparting shock wave, and the fourth-power relation for steady structured shock waves.

2. Dissipation dynamics and the shock failure transition

Through the use of statistical principles that are central to other areas of physics, a statistical framework for failure dynamics is constructed within the structured shock wave. Through this framework, expressions are developed for the viscosity appropriate to the spatial structuring of the shock wave and for critical state relations determining the shock-induced failure transition. Out of the development emerge relations for the experimental fourth-power behavior observed in structure shock measurements in solids and the rate dependence of spall strength and fragment size in the dynamic tensile failure of materials.

2.1. The Variant Velocity Field

At any point on the microscale within the material, as shock passage proceeds, particle velocity within a planar shock wave can be decomposed into the vector sum $\mu+\vec{u}$, with $\mu(t)$ the expected value of the velocity at any time over a sufficient span of a plane normal to the shock direction. The expected value $\mu(t)$ is the continuum transport particle velocity in the structure shock wave. The expected value of the variant velocity over a comparable span of the plane is necessarily zero. The variance of the particle velocity, the expected value of the square of the variant velocity, $\langle \vec{u}^2 \rangle = \sigma^2(t)$, is nonzero.

2.2. Dissipation in the shock transition

Kinetic dissipation, $E_k(t)=\langle \vec{u}^2 \rangle /2=\sigma^2/2$, associated with the variant velocity field, is the random acoustic phonon energy that is a consequence of the collective dissipative structure activity underway as failure proceeds through the shock transition. This kinetic dissipation will ultimately thermalize. Within the interim, however, the kinetic dissipation has an integral role to play in the correlation dynamics of the shock failure transition.

Concurrently, frictional deformation dissipation E_f occurs within the interior of elemental active dissipative structures. Depending on the material and mode of deformation in the shock transition, frictional dissipation is a collective term that can include fracture bond breakage, dislocation plasticity and distended matter compaction among others. Frictional dissipation within the elemental dissipative structure is characterized by a dissipation constant Γ with dimensions of energy per unit area. Familiar is that of fracture, where Γ is a measure of strain energy release as commonly specified through measurements of the material fracture toughness. Less familiar is application of the dissipation constant Γ to other forms of failure dissipation including plastic shear slip or distended matter compaction.

Dissipation dichotomy is central to the statistical physics governing the shock failure transition. Frictional dissipation and kinetic dissipation are both a consequence of the same underlying microstructural failure mechanics [1,2]. Both frictional dissipation and kinetic dissipation scale with the dissipation constant Γ. Dissipation dichotomy was, perhaps, first noted in an analysis of dynamic fracture by Mott [1].

2.3. Correlation dynamics in the shock transition

Within the shock transition the variant velocity $\vec{u}$ is a statistically random function of time and position. On any plane within the shock, with normal collinear to the planar shock direction, the variant velocity is statistically stationary; that is, no point on the plane is statistically preferred. At any time t within the structured shock wave, a temporal variant velocity correlation function on that plane, referenced to an arbitrary position $x=0$ within that plane, is the expected value expression $K(s,t)=\langle \vec{u}(0,t)\vec{u}(s,t) \rangle$. The temporal coordinate s is related to position coordinate x within the plane through the acoustic impedance $Z = \rho c$ through $s = \rho x / Z$. The variant velocity correlation function $K(s,t)$ is a measure of the temporal correlation of kinetic dissipation as dynamic failure ensues through the shock transition. Note that $K(0,t) = \sigma^2(t)$.

Here, a powerful result emerging from the applications of Boltzmann statistics to complex systems [3] is applied to the shock failure transition. Namely, the integral of the variant velocity correlation function equates to a continuum self-diffusion property of the material,

$$D(t) = \int_0^\infty \langle \vec{u}(0,t)\vec{u}(s,t) \rangle ds = \int_0^\infty K(s,t) ds . \tag{1}$$

Equation (1) is identified as the fluctuation-dissipation theorem [4] and, in this form, more commonly identified with the Green-Kubo equation [5]. The fluctuation-dissipation relation has wide application in relating continuum transport properties of a system to the underlying micro dynamics [4]. In the present treatment this expression relates the correlated kinetic acoustic dissipation fluctuations to a continuum phonon momentum diffusivity property $D(t)$ within the advancing shock wave. As such, $D(t)$ is identified with the kinematic acoustic phonon viscosity within the shock transition. The property $D(t)$ is integral to the structuring of the time history of the wave through the shock transition.

The variant velocity correlation function plays a further role in the shock failure transition. The differential expression $dm = Zds$ is an element of areal mass. The failure transition within the shock, within a temporal correlation period τ, is the critical state at which the correlated kinetic dissipation $E_k(\tau)dm$ achieves the requisite shock transition dissipation as expressed through the dissipation constant Γ. Dissipation dichotomy through the shock transition is realized by equating the frictional dissipation constant Γ to the correlated kinetic dissipation through the shock transition period τ,

$$\Gamma = \frac{1}{2}\int_0^\infty \langle \vec{u}(0,\tau)\vec{u}(s,\tau) \rangle Zds = \frac{1}{2}\int_0^\infty K(s,\tau) Zds . \tag{2}$$

Equation (2) yields an energy-time criterion for the shock transition.

A functional form for the velocity correlation function is not known. Selected statistical processes (e.g., Gaussian, Markovian), common to other random physical applications, dictate an exponential decay for the temporal correlation span [5]. Application of an exponential variant velocity correlation function, with temporal correlation span τ, is accepted here,

$$K(s,\tau) = \left\langle \vec{u}(0,\tau)^2 \right\rangle e^{-s/\tau} = \sigma^2(\tau) e^{-s/\tau} . \tag{3}$$

Integration of Equation (2) then yields,

$$\Gamma = \frac{1}{2}\sigma^2 Z\tau . \tag{4}$$

As expressed, $E_k(\tau) = \sigma^2(\tau)/2$ is the specific kinetic dissipation necessary to the shock failure transition. As such, Equation (4) relating frictional and kinetic dissipation takes the elemental energy-time form,

$$\Gamma/Z = E_k\tau = \sigma^2\tau/2 . \tag{5}$$

Equation (5), barring uncertainties of a proportionality constant of order unity, is an energy-time criterion descriptive of the shock failure transition. Equation (5) follows from considerations of the underlying micro-statistics failure and correlation dynamics that bring about the macro shock-failure properties observed in the shock-wave experiment. Such observations, among others, include the rated dependence of shock-induced spall strength, and the fourth-power dependence of steady structured shock waves in solids.

A comparison of Equation (2) with Equation (1) reveals the ratio Γ/Z equates to a critical state amplitude of the self-diffusion constant $D(\tau)$ within the temporal span of the shock transition. This critical state constant is identified with the kinematic viscosity and dissipative action A associated with the shock failure transition [6]; namely, $D(\tau) = A = \Gamma/Z$.

3. Closure

Page constraints subjected to the present extended abstract prohibits a fair discussion of the application of Equation (5) to the span of experimental shock transition data noted earlier. Briefly, dissipation dichotomy dictates that the kinetic dissipation

$E_k(\tau)$ equate to one-half the total shock dissipation in a steady structured shock wave transition. For shock transitions sensibly described by linear shock versus particle velocity constants C_o and S, shock dissipation to Hugoniot strain ε is to first order $SC_o^2\varepsilon^3/3$. Shock transition time is related to strain and strain rate through $\tau = \varepsilon/\dot{\varepsilon}$. Constancy of energy-time in Equation (5) yields the universal fourth-power dependence of strain rate on shock amplitude observed in many solid matter materials [6]. Similarly, application of Equation (5) to steady structured shock waves in porous solids, accounting for the differences in shock dissipation, yield the more diverse power-law relations between strain rate and shock amplitudes observed in these materials [6].

Further, Equation (5) has application to fracture and fragmentation in the tensile spall failure of condensed matter. Relating kinetic dissipation $E_k(\tau)$ to one-half the strain energy at the tensile strain amplitude of spall failure, and again relating the spall transition time to strain rate through $\tau = \varepsilon/\dot{\varepsilon}$, expressions for the dependence on strain rate of the tensile spall strength and nominal fragment size are arrived at. Relations mirror those derived through alternative arguments in earlier studies, *e.g.*, [11]. Here, the dissipation constant Γ can be identified with liquid surface energy, solid fracture toughness, or other energy properties characterizing spall cavitation of the material under dynamic tension.

Representative applications to compressive shock transition data are illustrated in Fig. (1). Pressure step versus strain rate structured shock wave data for a chemical compound Al2O3, an unalloyed uranium metal and an HMX-binder mixture exhibit the fourth-power nature emerging from application of Equation (5). Alternatively, pressure step versus shock width plot of the same data are compared with polycrystal grain size. Further, structured wave measurements for sapphire and for single crystal HMX overlay the polycrystal data. These observations collectively suggest underlying dynamic plasticity accounts for viscous structuring of the shock waves through the shock transition. Dissipation constants, assessed for the three materials from Equation (5) and plotted against respective experimental HEL values, illustrate a systematic with this dynamic strength measure.

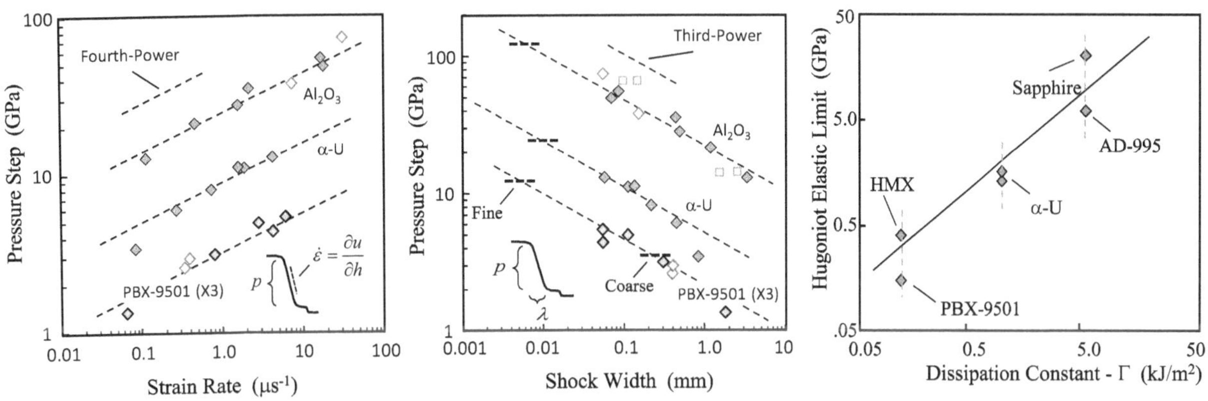

Fig. 1. Steady-wave strain rates (a) and shock widths (b) versus Hugoniot pressure step for structured shock-wave measurements on aluminum oxide, uranium metal [7] and unreacting HMX based energetic material [8]. Open symbols are for single crystal HMX [8] and aluminum oxide (sapphire) [9,10]. Note times three vertical scaling for PBX-9501 and single crystal HMX. Horizontal dashed lines (b) identify nominal grain size. Hugoniot elastic limit versus the dissipation constant (c), as assessed from Eq. (5), and the fourth-power presentation (a) of the structured wave experimental data.

References

[1] Mott, N., 1948, Brittle Fracture in Mild Steel Plates, Engineering, 165, p. 16.

[2] Grady, D., 2018, Principles Underlying the Fourth Power Nature of Structured Shock Waves, Proc. APS Topical Group on Shock Compression of Condensed Matter, St. Louis, MO, July 10-14, 2017; AIP Conf. Proc., 1979, 070014.

[3] Reif, F., 1965, Fundamentals of statistical and thermal physics, McGraw Hill.

[4] Kubo, R., 1966, The Fluctuation-Dissipation Theorem, Rep. Prog. Phys., 29, p. 255-284.

[5] Resibois, P. and de Leener, M., 1977, Classical Kinetic Theory of Fluids, Wiley InterScience Publ.

[6] Grady, D., 2015, Unifying Role of Dissipative Action in the Dynamic Failure of Solids, J. Appl. Phys., 117, 165905.

[7] Grady, D., 1986, Steady-Wave Rise-Time and Spall Measurements on Uranium (3-15 GPa), in Metallurgical Applications of Shock-Wave and High-Strain-Rate Phenomena, Murr LE, Staudhammer KP, Meyers MA, eds., Marcel Dekker, Inc., p. 763.

[8] Dick, J,, Martinez, A., Hixson R., 1998, Plate Impact Response of PBX 9501 and its Components Below Two GPa, Los Alamos National Laboratory Rept. LA-13426-MS, April.

[9] Kanel, G., Nellis, W., Savinykh, A., Razorenov, S., Rajendran, A., 2009, Effect of Crystalline Anisotropy on Shock Propagation in Sapphire, *Shock Compression of Condensed Matter,* American Institute of Physics, p. 851-854.

[10] Reinhart, W., Chhabildas, C., Vogler. J., 2006, Investigating Phase Transitions and Strength in Single-Crystal Sapphire using Shock–Reshock Loading Techniques, Int. J. Impact Engng., 33, p. 655-669.
[11] Grady, D., 1982, Local Inertial Effects in Dynamic Fragmentation, J. Appl. Phys., 53, p. 322-325.

Proceedings of the 2019 Hypervelocity Impact Symposium
HVIS2019
April 14-19, 2019, Destin, FL, USA
HVIS2019-027

Study on phase transformation in Tin under dynamic compression

Camille Chauvin[a], Frédéric Zucchini[a], David Palma de Barros[a]

[a]*CEA, DAM, GRAMAT, F-46500 GRAMAT, FRANCE*

Abstract

We propose to study experimentally the polymorphic transition of Tin under dynamic compression. These transformations have been investigated for a long time through usual velocity measurements under shock from ambient condition. At CEA Gramat we have improved our understanding of such phase transformations through both experimental and theoretical means.

Experimental velocity measurements have long suggested that non equilibrium behavior and kinetics is an important part of the dynamic compression response of materials undergoing phase transformations. Empirical kinetic models can in many cases reproduce the experimental velocity profiles, but without clearly identifying the nature of the transition.

For nearly two decades, the CEA Gramat operates several gas guns for shock loading and high pulsed power (HPP) drivers dedicated to Isentropic Compression Experiments (ICE) up to several GPa. These experimental devices and associated diagnostics (velocimetry and temperature measurements and x-ray diffraction experiments) help to begin to study kinetics under dynamic transition in a more rigorous manner. We have used these experiments to examine various compression paths and have used the results to improve equation of state (EOS) models incorporated in our numerical codes. The latter can be used to run simulations starting with ambient initial conditions, then load metallic materials from various non ambient initial temperatures. This can significantly extend the range of our studies into previously unexplored thermodynamic paths.

We propose to describe our preheating devices for gas gun experiments and our HPP driver, and to present our preliminary results on shock loading and on isentropic compression at various initial temperatures, to explore the phase diagram of Tin.

In addition, we present the design of promising testing on X-ray diffraction under shock to help to develop a more physical kinetic model relying on nucleation and growth mechanisms, which are implemented in our continuum level codes.

Keywords: phase transition, tin, isentropic compression, shock compression, X-ray diffraction

---

1. Introduction

Using synchrotron X-ray diffraction techniques with Diamond anvil cells, the crystallographic structures of Tin have been probed under static high pressures. Tin has been exhaustively studied over a wide range of the P-T space so its phase diagram has been determined under static conditions. In particular, these techniques reveal a crystal structure transition from a 2-lattice body-centered-tetragonal (bct) ambient β-Sn phase to a single bct γ-Sn phase which was first observed at 9.8 GPa [1] with a volume change up to 5%. This β–γ phase boundary presents a negative and steep dP/dT up to the triple β–γ–liquid point at about 3.3 GPa and 581 K [2]. So, at static high pressure and high temperature, the determination of phases is relatively well known, but under rapid dynamic compression, the determination of the transition boundaries and of the existing phases has been an important challenge. Dynamic loading studies have been a priority research activity at CEA Gramat for several decades, using planar impact experiments and isentropic compression techniques. The objective of this study was to acquire data to study kinetics in our code under dynamic transition in a more rigorous manner.

---

* Corresponding author. Tel.: +33 5 65 10 55 68.
E-mail address: Camille.CHAUVIN@CEA.FR.

2. Exploring new thermodynamic paths for Tin with preheating devices on gas guns and on magnetically–driven compression devices

Dynamic loading studies have been a priority research activity at CEA Gramat for several decades [3,4]. Plate impact experiments with gas guns and isentropic compression experiments with High Pulsed Power (HPP) drivers are used to investigate the dynamic behavior of materials and, in particular, we focus on phase transitions under dynamic compression.

2.1. Shock experiments

Plate impact experiments are commonly used to investigate the dynamic behavior of materials and, in particular, phase transitions under dynamic compression. This method is useful to detect polymorphic transitions, indicated by two-wave compression data. Shock phase boundaries can be established by preheating samples to different initial temperatures; this opens new promising avenues for exploring structural changes by starting from non-ambient conditions.

2.1.1. Gas gun heating device

We have integrated a device capable of pre-heating metallic samples up to 200°C (Fig 1) prior to loading in a gas gun.

The sample temperature was controlled by a K-type thermocouple. The heating system was designed in order to be non-intrusive:

- for the projectile to impact the preheated sample, so a spiral resistive heater transfers heat from the outer annulus of 40-mm diameter tin samples,
- for the measurements at the sample free surface (velocimetry or temperature measurements).

To avoid thermal loss and to concentrate thermal conduction to the sample, thermal flat and circular shields are used. The sample heater assembly was placed in a vacuum chamber attached to the muzzle of the gas gun.

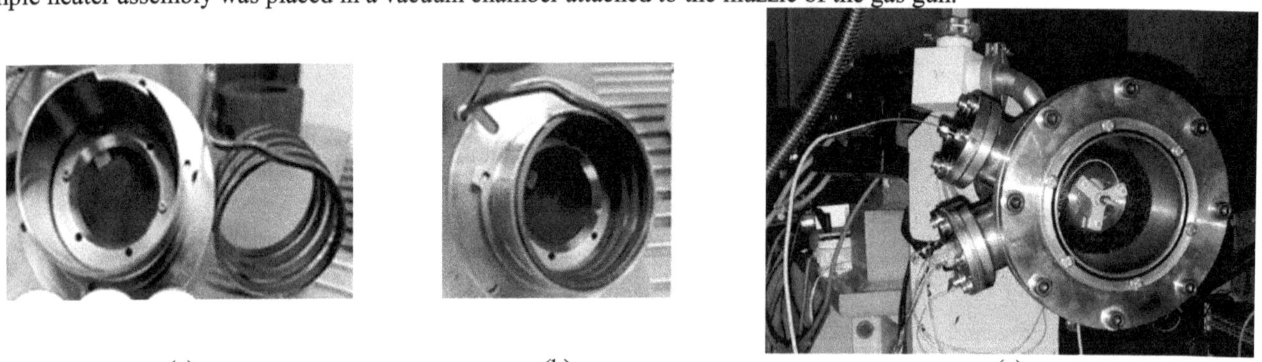

(a) (b) (c)

Fig. 1. (a) Sample holder and spiral resistive heater (b) and their assembly (c) Pre-heating system on gas gun.

2.1.2. Results

Fig 2 (a) presents typical free surface velocity histories obtained with high purity (99.999%) polycrystalline tin (grain from 300-500 µm) at different initial temperatures (293K, 418K and 468 K) and at different shock pressures. The graphs display the β–γ transition; the compression shock waves are composed of the elastic precursor wave (first peak) followed by the plastic shock waves of the β phase (second peak) and then of the γ phase (plateau). As already noticed [5], the Hugoniot elastic limit (HEL) grows with temperature. From each wave in the front from the different shots, the pressure corresponding to the direct polymorphic transition ($\beta \rightarrow \gamma$) was determined by using the elastic precursor wave as a timing fiducial, and the Hugoniot relations, and plotted in the pressure-temperature (P,T) diagram (see Fig 2 (b)). The corresponding dynamic boundary comes directly from the β-phase EOS of Tin indicated by the Hugoniot paths which is assumed to be well known. These data indicate a hysteretic overshoot under shock loading which is closest to the static boundary as the initial temperature increases.

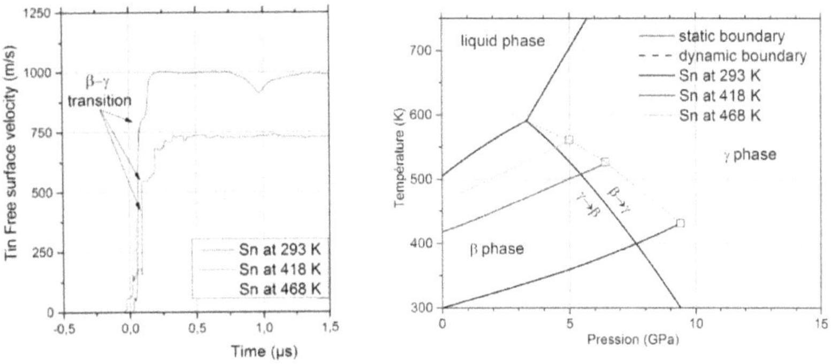

(a) (b)

Fig. 2. (a) Free surface velocity measurements obtained at different initial temperature and (b) Phase diagram for Tin: static boundary from static data (—), present shock results (□).

2.2. Ramp experiments

Within the framework of studies on dynamically-compressed materials, High-Pulsed Power drivers are operated at the CEA. These facilities enable various types of samples to be magnetically ramp-compressed up to Mbar stress levels to carry out Isentropic Compression Experiments (ICE).

In the central part of the generator GEPI, two plane electrodes are separated by an insulator composed of dielectric foils. This configuration is called the strip-line configuration (Fig 3.) [6]. Additional peaking capacitors have also been employed to smooth the temporal current profile and to push away the formation of a shock front when high pressures are generated.

By varying the strip-line width and the intensity of the current (the intensity is controlled by choosing the charging voltage and the number of stages used during a test up to 3.3 MA, rising time of 500 ns with a square sinus shape), we can master the magnetic pressure level that is applied to the electrodes over a wide range from several hundred MPa up to 80 GPa over durations of several hundred nanoseconds. We can observe that the pulse shape is progressive, unlike a discontinuous pulse that is produced by the front of a shock wave.

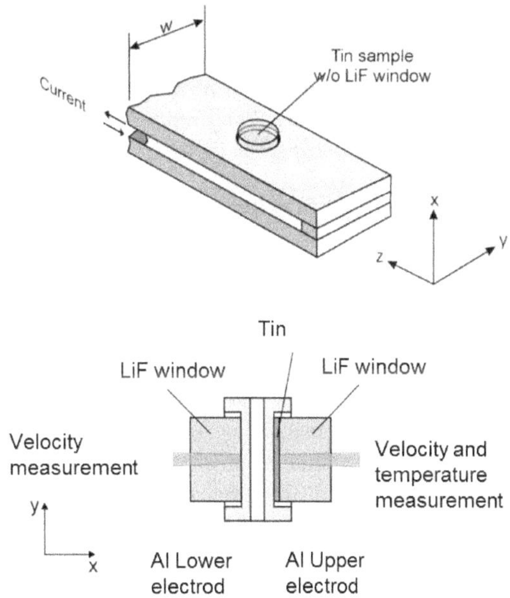

Fig. 3. Schematic of the experimental configuration on ICE.

Several diagnostics (velocity and temperature measurements), are fielded during these quasi-isentropic experiments in order to characterize structural changes occurring within these materials under dynamic loading. In addition, the dynamic phase boundary can be compared to that obtained under shock loading at high strain rate. The latter is usually produced starting from ambient conditions.

Fig. 3 shows the experimental configuration for ICE. The left part is used to obtain the pressure history in electrodes by comparing velocity measurements and simulations obtained using a Lagrangian code with magneto-hydrodynamics considerations. Different tin samples with high purity (99.999%) with fine-grain (<100μm) and large-grain (300-500 μm) are tested and velocity profiles in Fig 4. (a) are presented, revealing no significant influence on transition. As displayed in the shock experiments, the singularity in the ramp wave indicates the phase transition. The temperature measurement achieved with a pyrometer [4] allows the mapping of the β–γ boundary under this loading, complementary to shock, and reveals that the ICE data are closer to the static boundary than the shock data (Fig. 4 (c)). The rise of temperature obtained by measurement is close to the rise in temperature of β-phase tin along the supposedly-known isentrope.

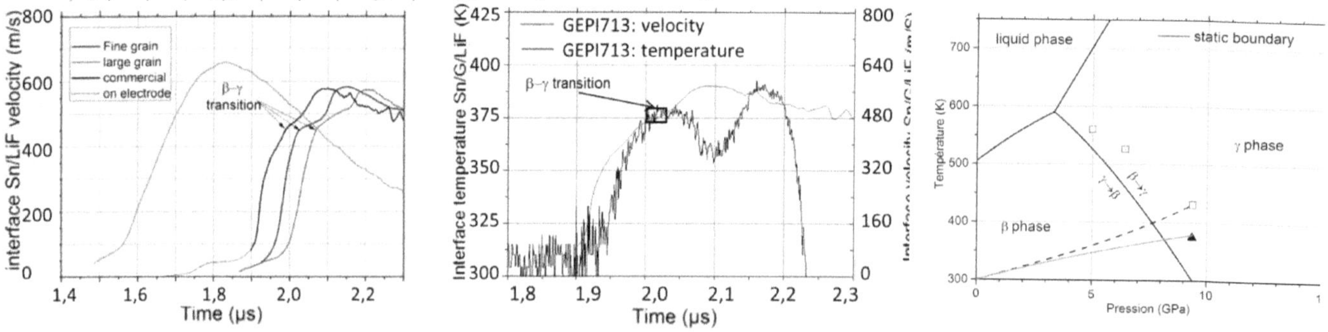

(a) (b) (c)

Fig. 4. Interface Sn/LiF (a) velocity and (b) temperature measurements obtained under isentropic paths.
Phase diagram (c) for Tin: static boundary from static data (—), present shock results (□), present ICE results (▲).

Ramp compressing samples from various non ambient initial temperatures can significantly extend the range of our studies into previously unexplored thermodynamic paths [7] and help constrain Equation Of State (EOS) models incorporated in numerical codes. However, attempts to couple reliable, high performance pre-heating devices with the operation of HPP drivers have encountered a number of technical limitations due to the severe electromagnetic environment and nature of the components associated with this type of experimental platform. We have recently developed a novel configuration which allows metallic samples to be heated to several hundred degrees prior to their loading. With this new development, promising avenues will be opened for exploring phase changes starting from non-ambient conditions.

3. Development of single X-ray diffraction on gas-gun experiments

3.1. Analysis

We have developed a multiphase equation of state capable of reproducing our experimental velocity and temperature measurements under shock and ICE, revealing kinetic effects in phase transition mechanisms. Shock and ramp compressed materials may undergo physical changes including solid-solid or solid-liquid phase transformations. At CEA Gramat, a three-phase equation of state (EOS) has been developed [3]. This model, integrated into a 1D Lagrangian numerical code Unidim, is able to satisfactory reproduce our experimental results from common velocity and temperature measurements. The phase boundaries, dP/dT & Δvij are known from static conditions (Diamond Anvil Cell). An EOS formalism with a Mie-Gruneisen with thermal isotherm reference is used. Along the phase boundaries between 2 phases i and j, some properties are verified:

- the equality of Gibbs free energies Gi(P,T) = Gj(P,T),
- the Clausius-Clapeyron law.

Parameters of the 3 phases are optimized to reproduce the phase boundaries as close as possible. This model describes the β, γ solid phases and the liquid phase of tin (Fig. 5).

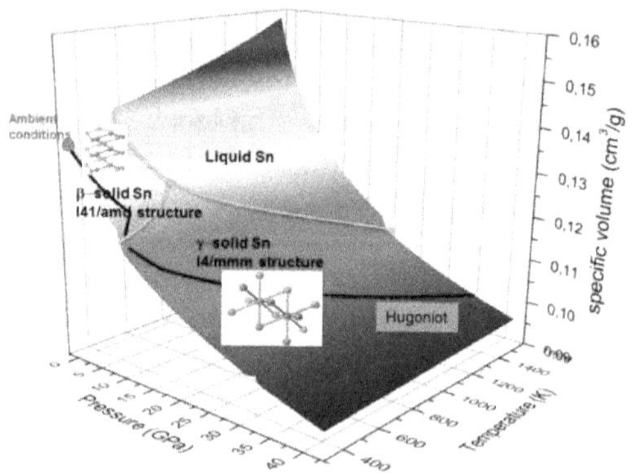

Fig. 5. Phase diagram for Tin.

Data obtained with velocity and temperature measurements around the polymorphic β–γ transition of Tin on gas gun and isentropic compression experiments have displayed the importance of kinetics of the transition. Two models of kinetics can be used in our multiphase EOS [8,9]. Each model gives the same velocity profile reproducing the experimental profile displayed on

Fig. 6 (a), but shows discrepancies between the transformation rate and the final proportion of β and γ phase inside the material visible on the plateau in Fig. 6 (b).

The model of Mabire [8] uses a homogeneous mixture model described by Eq. (1) in which τ, θ and A_R are specific parameters. A_R allows the delay of the transition from the equilibrium phase boundary.

$$\dot{\varphi} = \frac{1 - e^{\vartheta \Delta G^*} - \varphi}{\tau} \tag{1}$$

A phenomenological model, developed by Greeff [9], gives an expression of the transformation rate described in the Eq. (2), where ν and B are specific parameters.

$$\dot{\varphi} = \upsilon(1-\varphi)\frac{\Delta G^*}{B} e^{\left(\frac{\Delta G^*}{B}\right)^2} \tag{2}$$

All parameters of the models were fitted to reproduce the experimental velocity profiles.

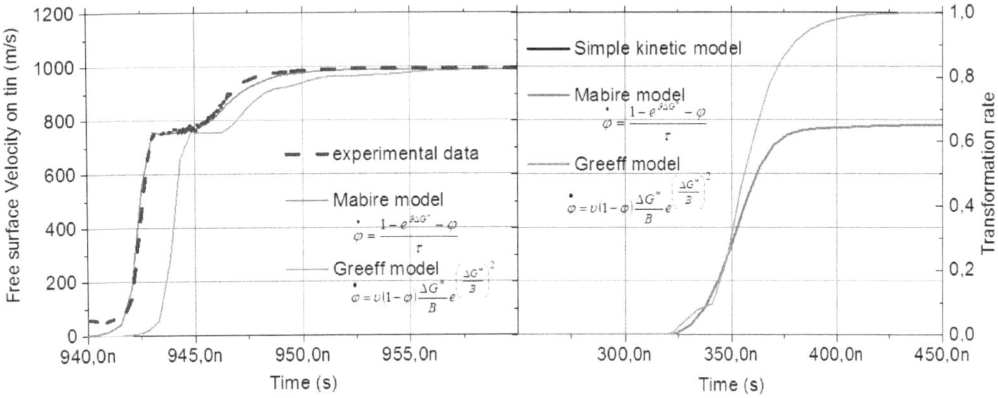

Fig. 6. Comparison between the velocity profiles at Tin free surface (a) obtained by experimental data and numerical calculation and (b) the rate of γ phase inside the material.

These results underline the limitation of empirical kinetic models and of the macroscopic measurements; the macroscopic results are not sufficient to understand mechanisms governing shock-induced structural changes. The Mabire model presents an incomplete transition because of the use of a parameter which delays the transition, and the Greeff model describes a transformation rate equal to zero at the equilibrium phase boundary. No model seems better than the other and no in situ crystallographic information was measured to confirm the crystal structure of the high-pressure phases during dynamic loading. Direct insight about the crystallographic structure under shock would be extremely valuable to elucidate the time scale of a transition or the presence of a mixture phase. Likewise, a more physical kinetic model relying on nucleation and growth mechanisms could be implemented into our continuum level codes.

3.2. Experimental

X-ray diffraction measurements are needed to obtain direct insight about the crystallographic structure of the phase under shock and, as a result, deducing the beginning and the end of the transition under dynamic loading. These microscopic data would help to constrain or choose the best kinetic model in the multiphase EOS, and to understand mechanisms governing shock-induced structural changes. Previous complex experiments [10,11] allow the inference of phase change in the material.

X-ray based techniques have proved their unique potential in diagnosing dynamic compression experiments at the atomic level. Therefore, developing flexible, high flux, high brilliance, compact X-ray sources is of considerable interest, as complementary devices to synchrotron facilities [11,12], for exploring new ideas in the laboratory prior to implementation in large-facility experimental endeavours.

The current work is focused on using the X-ray diffraction method to determine the change in the interplanar spacing of a crystal which presents a phase change during shock compression. The principle relies on the optimization of the X-ray beam to diffract from the crystal according to the Bragg condition (Fig. 7).

The experimental configuration for the X-ray diffraction measurement is classical (Fig. 7) and consists of a flyer plate impacting a sample to be studied backed with an X-ray window. This is used to maintain a uniform state of pressure within the X-ray probe emission during the diffraction measurements. The X-ray source is aligned to satisfy the Bragg condition. The wave profile measurements are obtained with PDV, a common diagnostic in shock experiments to know the pressure in the material and to obtain the chronometry of the experiments.

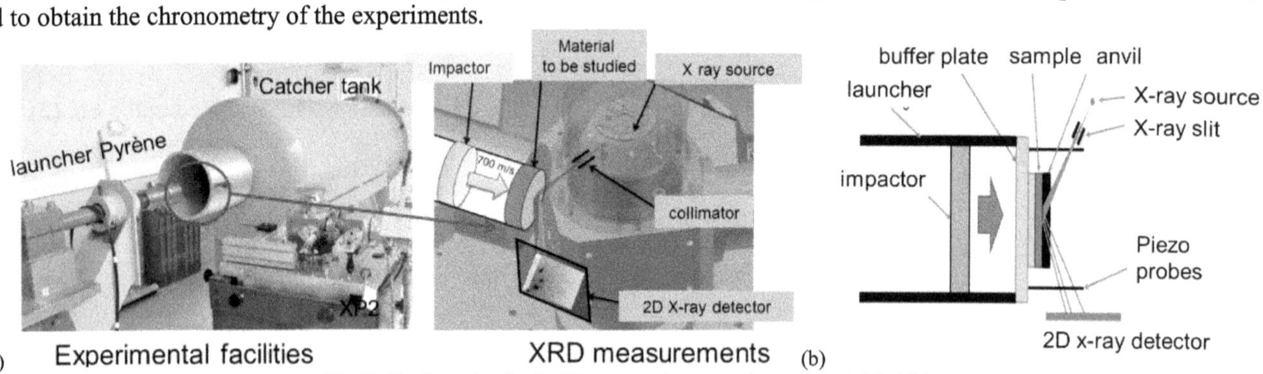

Fig. 7. Configuration for the X-ray experiments and experimental facilities

Presently, the goal of this study is only to observe a change in the pattern of a shocked material in comparison with that of the initial phase under shock loading at different pressure. We choose a polychromatic X-ray source to examine the existence of the initial phase under dynamic loading and its disappearance. A high pulsed power X-Pinch generator satisfies several requirements to be an X-ray probe source for diffraction under shock [13]. A fast electrical discharge (1kA/ns) in a short circuit load under vacuum which is made of fine 20-25-μm diameter crossed metallic wires (silver, gold, molybdenum) generates X-ray from 10 to 50 keV with a duration from 10 to 50 ns with a 400 μm-diameter.

The experimental configuration is shown in Fig. 8. Because between the β and γ phases of tin, no crystallographic mechanism is known and no evidence of conservation of monocrystalline structure is proved, it is necessary to guarantee the spectral emission (continuum) of the source on a reference sample, and to ensure the beginning of the transition, in spite of the apparition of a

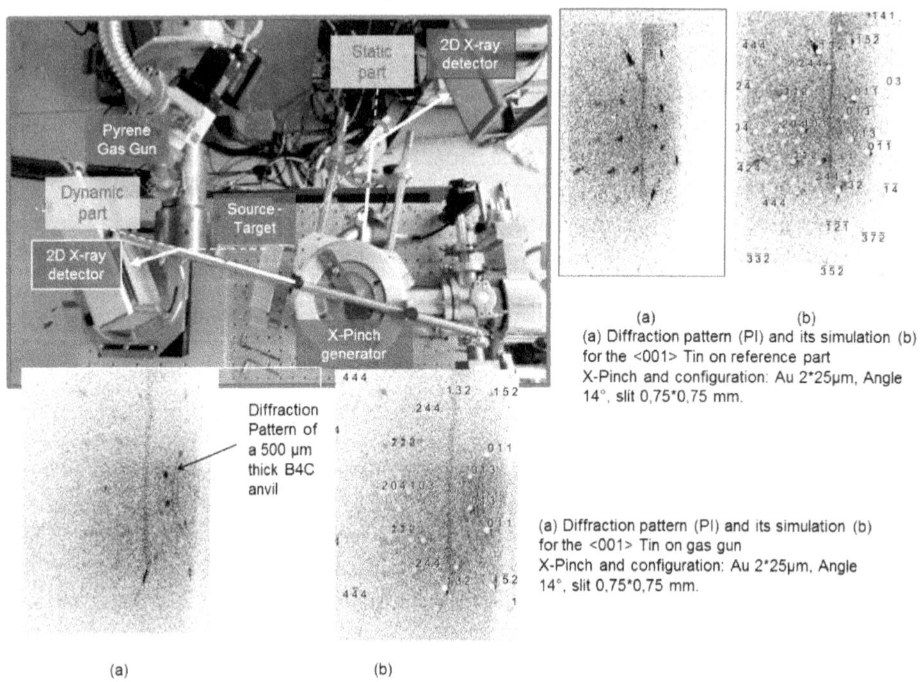

Fig. 8. Sample and experimental geometry.

polycrystalline γ–Tin and the possibility of loss of the diffraction pattern. A static part with a reference sample on a static part has been developed in parallel with the dynamic part made of the sample on the gas gun, using the original 4π steradian polychromatic X-ray source. Fig. 8a describes the diffraction pattern obtained on <001> single crystal Tin placed on the muzzle of the gas gun on the left bottom part, and that obtained on the static part on the right upper part. Each point of the pattern can be identified using a simulation with CrystalMaker software (Fig. 8b).

We have developed a flexible and compact X-ray source for diagnosing material under dynamic compression in the laboratory. We succeed in designing specific shock experiments with X-ray diffraction measurements by synchronizing X-Pinch emission and by minimizing the debris projection. To validate our experimental methods, we have performed experiments on KCl because its phase change has been examined extensively (under shock loading) and it has already been studied with X-ray diffraction under shock loading [14,15]. Similar experiments at different stresses in Tin will be conducted, the beginning and the end of the shock induced solid-solid phase transitions can be identified at the macroscopic level in order to discriminate our model.

4. Conclusion

In this paper, we present pressure and temperature on the β–γ phase boundary determined on shock and isentropic compression experiments compared the results directly to the equilibrium boundary. We observed a hysteresis under all dynamic loading conditions considered. Further experiments are needed to explore new thermodynamic paths with ramp or shock compression of samples from various non ambient initial temperatures. These can help constrain Equation of State (EOS) models incorporated in numerical codes.

But, these macroscopic results are not sufficient to understand mechanisms governing shock-induced structural changes. Direct insight about crystallographic structure under shock is extremely valuable to appreciate the time scale of a transition or the presence of a mixture phase. We are developing a novel experimental setup based on a compact High Pulsed Power generator capable of producing intense X radiation through X-pinch. This laboratory source is specifically designed for X-ray diffraction in the Bragg geometry on gas gun experiments. Promising preliminary X-ray diffraction images are obtained.

Acknowledgements

This work was supported by the Délégation Générale de l'Armement (DGA) of the French Ministry of Defense.

References

[1] Barnett, J. D. et al., Science, **141**, 1041, (1963)
[2] Cannon, J.F., J. Phys Chem. Ref. Data **3**, 781 (1974)
[3] Mabire, C. and Hereil, P.-L., J. Phys. IV France **10**, pp. 9-749-9-754, (2000)
[4] Chauvin C., AIP Conference Proceedings 1793, 060013 (2017)
[5] Zaretsky E. B. and Kanel G.I., DYMAT 2009, pp 27-33, EDP Sciences, (2009)
[6] Mangeant C., et al., AIP Conference Proceedings, Shock Compression of Condensed Matter 2001, pp. 1173–1176, (2002)
[7] Davis J.P. and Hayes D., CP955, Shock Compression of Condensed Matter-2007, pp. 159-162, (2007)
[8] Hayes D.B., *J. Appl. Phys.* **46**, 8 (1975)
[9] Greeff, C. W., Rigg, P.A., Knudson, M.D., Hixson, R. S. and Gray, III, G.T., CP706, Shock Compression of Condensed Matter-2003, pp. 209-212, (2003)
[10] Kalantar D. H. and al, Phys. Rev. Lett., PRL **95** (7), 075502 (2005)
[11] Gupta Y. M, Turneaure S. J., et al., Rev. Of scientific instrum., **83**, 123905 (2012)
[12] Branch B. and Jensen B., AIP Conf. Proc. 1979, Shock Compression of Condensed Matter-2017, 040001-1, (2017)
[13] Zucchini F., Bland S.N., Chauvin C., Rev. Sci. Instrum. **86**, 033507 (2015)
[14] d'Almeida T. and Gupta Y.M., Phys. Rev. Lett, Vol, 85, 330-333, (1999)
[15] Turneaure S.J., Y. M. Gupta Y.M., Rigg P., Jour. Of Applied Phys., **105**, 013544 (2009)

Proceedings of the 2019 Hypervelocity Impact Symposium
HVIS2019
April 14-19, 2019, Destin, FL, USA
HVIS2019-031

Corning0120 High-Lead Glass Subject to Shock Loading

Bernardo Farfan[a] William Reinhart[a] and Scott Alexander[a]

[a]*Sandia National Laboratories, Albuquerque, NM, 87185*

Abstract

Equation of state properties were studied for the high-lead glass Corning 0120, which is a potash-soda-lead glass also referred to as G12. This glass, which contains approximately 30% PbO by weight and has a density, ρ_o, of 3.034 g/cm^3 possesses properties suitable for many applications in industry such as optical components for space exploration instrumentation. Further understanding of its mechanical properties is desired for more complex applications in various fields, including applications where the glass may experience high-pressure shock loading. In this work plate impact experiments were conducted to determine the dynamic response of Corning 0120 at high stress levels. Tests were conducted over the pressure range from approximately 5 to 24 GPa utilizing the 90 mm bore single-stage powder driven gas gun at the Sandia National Laboratories STAR Facility. For this study, we used one-inch diameter Corning 0120 glass samples of two different thicknesses (3 mm and 7 mm) to use the evolution of the shock wave propagation through the material for analysis. The time-resolved material response was measured by means of a Velocity Interferometer System for Any Reflector system (VISAR). Results will be presented detailing the high-pressure shock loading response characteristics of the high-lead glass Corning 0120. Comparisons are made with similar results for lead free glass to assess the most prominent changes compared to lower density glasses and other lead filled glasses.

Keywords: High-lead glass, Equation of State, Hugoniot states.

Nomenclature

ρ	Density (g/m^3)
σ_x	Uniaxial stress (GPa)
σ_x^{HEL}	Hugoniot Elastic Limit (GPa)
U_s	Shock velocity (mm/us)
u	Particle velocity (mm/us)
C_L	Longitudinal speed of sound (mm/us)
C_o, S	Coefficients of linear relation $U_s = C_0 + S\ u_p$
Z	Shock impedances of the lithium-fluoride window
ν	Poisson ration
E	Young's modulus

1. Introduction

High-lead glass Corning 0120, also known in the industry as G12, is a material of interest for applications where the glass may experience extreme shock loading conditions such as optical components for space exploration instrumentation [1]. High-lead glasses exhibit lower Young's modulus and slow crack growth exponent as compared to lead free soda-lime silica glass or high purity glasses such as fused silica [1] [2] and are an ideal candidate for mission critical applications. To adequately implement lead filled glasses for high stress applications designers need to understand the response of Corning 0120 glass under shock

* Corresponding author. Tel.: +1- 505-844-5426.
E-mail address: bgfarfa@sandia.gov.

loading. Computational models are usually the tool of choice to aid the design process for a rapid turnaround of simulated shock conditions. However, numerical simulations rely on accurate experimental data to inform high fidelity material models to have the ability to precisely represent the material under shock conditions. The study presented in this work is intended to support future numerical models and general understanding of Corning 0120 glass with experimental data in a region of interest for the scientific community working in the field.

Corning 0120 is a potash-soda-lead glass with the chemical composition $55SiO_2$-$3B_2O_3$-$4Na_2O$-$9K_2O$-$27PbO$-$2Al_2O_3$ and contains approximately 30% PbO by weight and has a density, ρ_o, of 3.034 g/cm^3 [3] [4]. Glasses modified with oxides such as PbO exhibit lower fracture toughness and are more susceptible to stress corrosion. Like many glasses, silicon dioxide, SiO_2, forms the basis of Corning 0120. By employing the random network model [5] for amorphous solids, the glass structure can have ions that play the role of either network former or modifiers. Weaker ionic bonding leads to network modification while stronger orbital bonding leads to formation. The role of Pb as a modifier or former is still debated [6] but the different glass compositions and filler types can be associated to changes in the material response to shock loading. This paper considers the high-pressure shock loading response of Corning 0120 glass and comparisons are made with similar results for lead free glass [7] and other lead filled glasses [8] to assess its loading paths and mechanical deformation regions characteristics. Incremental and two-wave analysis were performed to study the elastic region and Hugoniot states respectively. We investigated the pressure range from approximately 5 to 24 GPa which covers most useful applications.

2. Experimental details

Plate impact experiments were conducted utilizing the 90 mm bore powder driven gas gun at the Sandia National Laboratories Shock Thermodynamics Applied Research (STAR) Facility. The experimental configuration, shown in Fig. 1, is designed to measure the shock wave transmitted through the sample material at two different distances to study the shock response as well as the wave evolution through the material. The target plate includes two 1 inch in diameter Corning 0120 glass samples of different thicknesses each backed with a Lithium Fluoride (LiF) window coated by a thin layer of vapor deposited aluminum on the sample interface. The velocity of the sample–window interface is monitored with a VISAR [9]. Adjacent and flush to each sample we installed smaller LiF windows, vapor deposited in the same manner as described above, to monitor shock time of arrival (TOA) to calculate closing impact times. These TOA windows were monitored with the same VISAR system. The target plate is impacted with a phenolic projectile faced with a thick copper impactor traveling at a range of velocities. The thickness of the impactor is designed to avoid early rarefactions that could possibly disrupt the plane shock wave formation and propagation through the sample. The velocity and tilt of the impactor/target impact was recorded with a set of electric trigger pins.

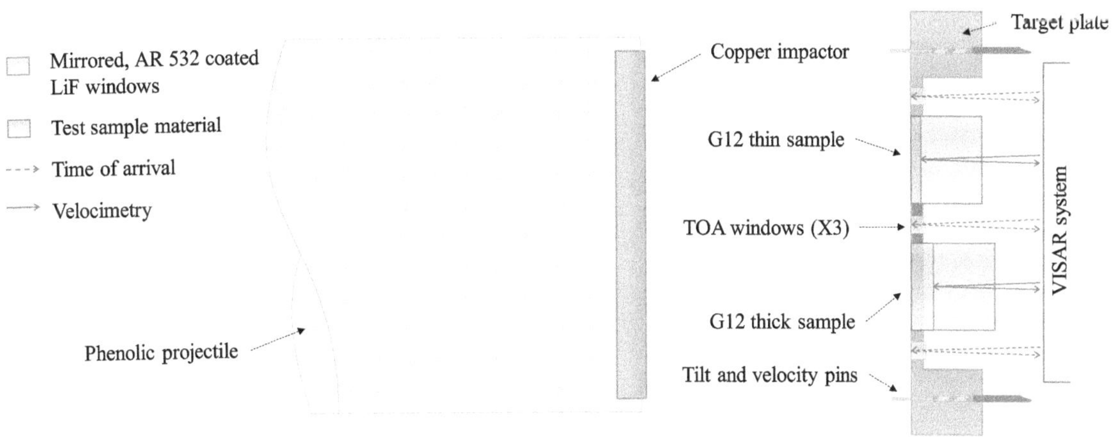

Fig. 1. Illustration of double sample plate impact experimental configuration for Corning 0120 glass at high velocity impact.

Table 1 shows the details of the experiments performed for this work. Five independent experiments were performed. Each experiment had two Corning 0120 samples from the same lot of approximately 3 mm and 7 mm thick. Ultrasonic measurement of the samples was used to determine the longitudinal elastic wave velocity (C_L) in the material at zero stress as well as Young's modulus (E) and Poisson ratios (v). The samples were impacted with a 12.7 mm copper projectile at a range of velocities from 0.52 km/s to 2.00 km/s, these impact velocities generated stress levels from 4.79 GPa to 23.46 GPa in the sample interface.

Table 1. Experimental details

Shot ID	Sample thickness (mm)		C_L (mm/us)	E (GPa)	v	Impact Velocity (km/s)	Stress (GPa)
SHOT - 1	A: 7.015	B: 3.005	4.834	60.08	0.240	0.52	4.79
SHOT - 2	A: 7.015	B: 3.010	4.808	59.77	0.237	0.91	8.57
SHOT - 3	A: 7.023	B: 3.007	4.759	59.44	0.228	1.31	13.39
SHOT - 4	A: 7.018	B: 3.015	4.800	59.77	0.235	1.86	20.95
SHOT - 5	A: 7.025	B: 3.007	4.782	59.80	0.237	2.00	23.46

3. Results and discussions

3.1. Wave profiles

Wave profiles were obtained at two different propagation distances using the two sample thicknesses in the experimental setup described above. The data collected for the time-resolved velocity at the sample/window interface is plotted in Fig. 2, where the wave profile formation and propagation are observed, the thick sample data was shifted by one unit for visualization purposes. By looking at the wave profiles we can see that the two lowest pressure shots (SHOT 1 and 2) are mostly in the elastic region where deformation is reversible. Experiment SHOT-3 exhibits what it seems like the initial formation of a dispersed plastic shock somewhere after the Hugoniot Elastic Limit (HEL) where densification begins to occur resulting in irreversible deformation of the material. Subsequent experiments (SHOT 4 and 5) show a more definite and steeper formation of a plastic shock wave in the material at the higher stress levels. It has been reported in the literature that glass structures such as fused quartz and borosilicate glass exhibit a complex response, as seen in this data, when subjected to plane shock wave dynamic loading [7] [10]. Lower density glass response usually starts with an initial elastic ramp of the stress, or particle velocity, followed by a steep rise to the HEL. This, unusual behavior has been attributed to the decreasing elastic moduli of these glasses with pressure [11]. Highly modified glasses such as soda-lime lack the initial elastic ramp region due to densification [7] [8].

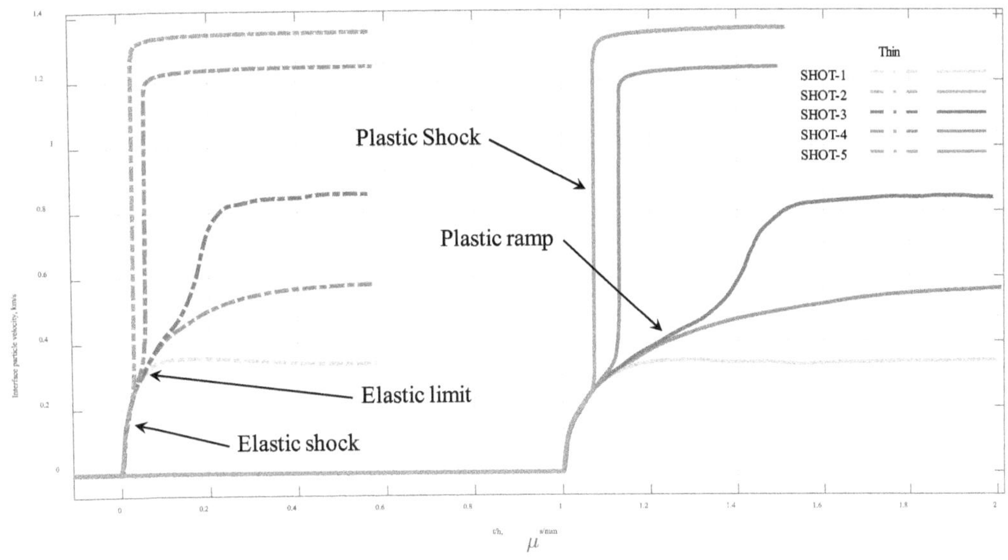

Fig. 2. Time resolved wave profiles of high-lead Corning 0120 glass at different propagation distances (thick sample data shifted by one unit).

By comparing the dynamic shock loading response of the potash-soda-lead Corning 0120 glass to low density glasses we can say that its response is more like the one of soda-lime glass [7]. The wave profiles show, in order of arrival, an elastic shock, a plastic ramp, and a plastic shock as labeled in Fig 2. Unlike open structure glasses, like fused quartz, Corning 0120 does not show the nearly lineal response before the steeper rise to the HEL [7] [10]. Other higher density lead-filled glasses such as type D, extra dense flint glass (DEDF), ρ_o, of 5.18 g/cm^3 show an extremely fast rise response and a breakage in its slope [10] characteristic of ceramics not seen in the Corning 0120 glass studied here. It can also be observed that the experimental series have not overdriven the HEL level by the still evident elastic region at the highest stress level of 23.46 GPa in experiment SHOT-5. The experimental series has covered the transition from the elastic to the plastic region pressure range. The loading paths and mechanical deformation regions will be discussed in the following sections.

3.2. Hugoniot characterization

Hugoniot states were measured by conducting a series of experiments varying the impact velocity on the Corning 0120/LiF samples from 0.5-2 km/s which resolved the wave profiles shown in Fig 2. The foot of the elastic wave will travel at the wave velocity at zero-stress, thus, the time of arrival at the measurement interface relative to impact can be calculated from the thickness of the sample divided by the zero-stress elastic wave speed. This data is used as a fiducial to determine the time of impact. The zero-stress longitudinal wave velocities C_L were obtained by ultrasonic measurement of each sample before testing. By this absolute time reference, we can determine the wave speed of each component of the wave profile by Lagrangian analysis. The Rankine–Hugoniot equations describe the relationship between the shock wave speed, the particle velocity in the material, the stress and the density of the material. Impedance matching is used to determine the *in-situ* particle velocity from that measured at the sample/window interface and to uniquely define the Hugoniot state.

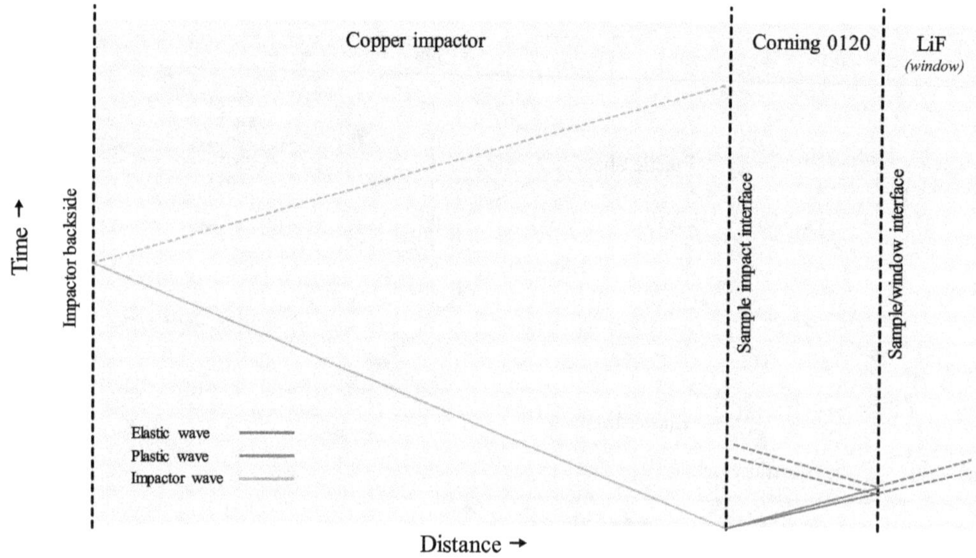

Fig. 3. x-t diagram of plate impact experiments on Corning 0120.

Fig 3. Shows a general X-t diagram of the plate impact experiments executed in this work. A shock is generated at the sample impact interface between impactor and sample producing an elastic and plastic wave traveling through the sample material (the experimental series did not overdrive the elastic wave). The traveling waves reach the window interface where the in-material particle velocities are measured. Incremental and two-wave analysis were performed to study the elastic region and Hugoniot states respectively. The Hugoniot elastic limit stress, (σ_x^{HEL}), is determined using the relation:

$$\sigma_x^{HEL} = \rho_o C_L u_e$$

where ρ_o is the initial density of the glass, and u_e is the in-material particle velocity measurement prior to transition to a plastic wave. The in-material particle velocity is determined through the impedance matching relation:

$$u_m = \frac{u_w(Z_w + Z_m)}{2Z_m}$$

where u_w, Z_w, and Z_m are the measured velocity in the window material, and the shock impedances of the lithium-fluoride window and the glass material, respectively. The loading path of the elastic region can be determined from transmitted wave experimental data. Utilizing the method of characteristics [12], an incremental analysis of the differential form of the Rankine–Hugoniot equations can be found from the time resolved velocity history data.

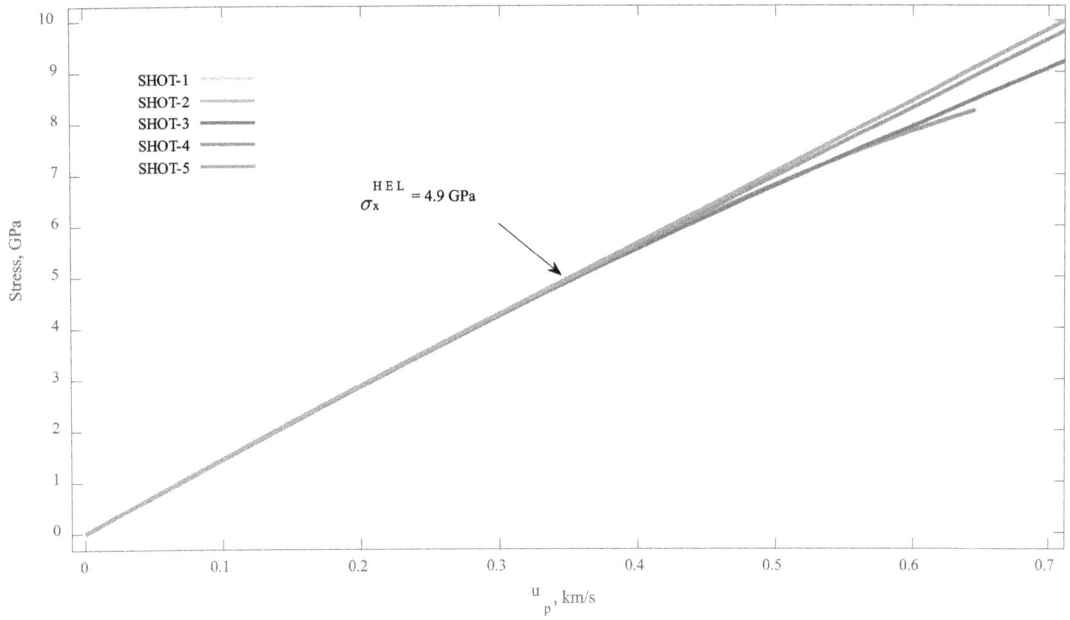

Fig. 4. Loading paths for Corning 0120 glass. Arrow indicate the point where divergence occurs representing stresses above the HEL level.

Applying the Rankine–Hugoniot equations by an incremental analysis, the pressure–particle velocity of the target can be found from the time resolved velocity history data obtained. Fig.4 shows the resulting loading paths for Corning 0120 glass from the data obtained in the experimental series. When a material is shocked at a stress level where the elastic deformation has not been overdriven, the material will exhibit an elastic response which is independent from the ultimate stress level the material is shocked to. Therefore, when elastic, the loading paths will be identical regardless of the final stress state. The point where the loading paths begin to deviate from each other is indicative of the HEL. The incremental analysis was applied through the plastic shock region to be able to distinguish the divergence where HEL occurs. By analyzing the loading paths in Fig 4 we can identify that the HEL for Corning 0120 is around 4.9 GPa. The identification of this limit was obtained by evaluating the stress respect to particle velocity of the data collected and defining the divergence point when all loading paths have deviated from each other by at least 20 KPa.

The planar impact produces a compressive wave of uniaxial strain, which propagates across the target specimen and into the lithium-fluoride window. The measured velocity exhibits a two-wave structure. The subsequent structure following the elastic precursor represents pressure hardening of the material and this two-wave structure is the result of a transition from elastic to plastic deformation. As compression within the shock increases during the shock loading process, shear stresses will exceed the critical strength of the material (HEL) and plastic deformation occurs in the observed second wave. Because finite rise times are measured for the plastic wave, the plastic-wave velocity, U_{sp}, is taken at the center of the wave profiles of the different known thicknesses. For the Hugoniot states, or plastic region, a two-wave analysis was used by the following relation.

$$\sigma_{ph} = \sigma_e + \rho_0 U_{sp}(u_{ph} - u_e)$$

Where σ_{ph} , u_{ph} are the plastic wave stress and particle velocity, and σ_e, u_e, ρ_e are the elastic wave stress, particle velocity and density behind the wave. Table 2 shows the Hugoniot parameters of the experimental series where a plastic shock wave formation has been observed overcoming the HEL of Corning 0120.

Table 2. Hugoniot summary

Shot ID	Impact Velocity (km/s)	U_{sp} (mm/us)	u_p (mm/us)	Strain	σ_{ph} (GPa)
SHOT - 1	0.52	-	-	-	-
SHOT - 2	0.91	-	-	-	-
SHOT - 3	1.31	3.712	1.041	0.260	12.682
SHOT - 4	1.86	4.406	1.453	0.323	19.639
SHOT - 5	2.00	4.563	1.582	0.343	21.959

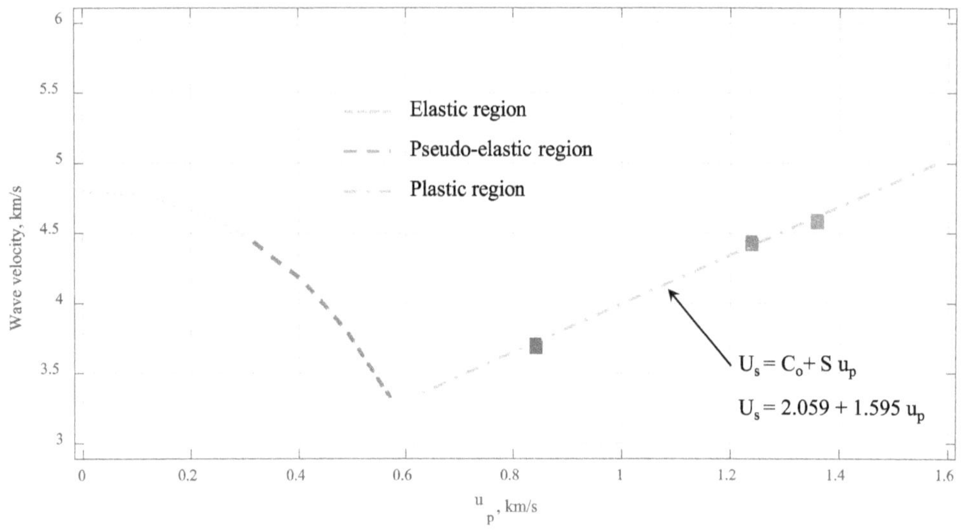

Fig. 5. Wave velocity/particle velocity relationship from experimental data illustrating the different regions of Corning 0120 glass under shock conditions. The initial elastic region at low pressures, the pseudo-elastic region, and the plastic fitted Us–up relation for the plastic region at higher pressures.

3.3. Wave velocities

The wave velocity to particle velocity relationship can be used to describe the material response to dynamic shock loading. Fig 4 shows this relation for the Corning 0120 glass studied in this work. The solid yellow curve indicates the elastic region and is the differential analysis of the elastic portion of the wave profiles (SHOT 1-5) that overlap before they start diverging. The green dashed curve represents the pseudo-elastic region where plastic deformation has begun to occur, as evidenced by loading paths dependent on the final states, but no clear plastic shock wave is yet observed. Experiment SHOT-2 is a good representation of this state and defines the green dashed curve for the pseudo-elastic region. The rest of the experiments (SHOT 3-5), where a plastic shock wave is readily apparent, define the plastic region by the linear U_s-u_p relation represented by the gray dashed line.

The relationship in Fig 5 shows the initial negative dependence of wave velocity on particle velocity typically observed in glass materials [7]. It is important to notice that after passing the HEL of 4.9 GPa (end of yellow line) determined in Fig 3, the loading path continues through the negative pseudo-elastic region (green dashed line), so the material responds pseudo-elastically up to 7.5 GPa before the formation of a plastic shock wave. Above this limit, the shock wave velocity is linear with respect to particle velocity and is well described by the general form of the relation $U_s=C_o+Su_p$ where C_o is the bulk wave speed and S is a dimensionless parameter describing the slope of the line.

4. Conclusions

High-lead glass Corning 0120 was studied under high-pressure dynamic shock loading to determine its response in the pressure range of approximately 5 to 24 Gpa which is relevant to many applications. The shock loading response of this potash-soda-lead Corning 0120 glass is more similar to the lead-free soda-lime glass [7] than a higher density lead filled glasses such as DEDF [10]. The wave profiles show, in order of arrival, an elastic shock, a plastic ramp, and a plastic shock. The elastic region prolongs to its HEL and continues through a negative dependent pseudo-elastic response to higher pressures before entering the plastic region. The HEL determined for Corning 0120 is higher than soda-lime and DEDF. DEFE glass with a density of 5.18 g/cm^3 has a HEL of 4.3 GPa, and soda-lime with 2.49 g/cm^3 has a HEL of 4 GPa compared to Corning 0120 with a density of 3.03 g/cm^3 which has a HEL of 4.9 GPa. The linear U_s-u_p relation that represent experimental Hugoniot data is defined by the coefficients C_o =2.059, S=1.595.

Acknowledgements

This paper describes objective technical results and analysis. Any subjective views or opinions that might be expressed in the paper do not necessarily represent the views of the U.S. Department of Energy or the United States Government.

Sandia National Laboratories is a multimission laboratory managed and operated by National Technology & Engineering Solutions of Sandia, LLC, a wholly owned subsidiary of Honeywell International Inc., for the U.S. Department of Energy's National Nuclear Security Administration under contract DE-NA0003525.

References

[1] J. A. Salem, N. A. Smith and A. Ersahim, "Mechanical Properties of a High Lead Glass Used in the Mars Organic Molecule Analyzer," NASA STI Program, Cleveland, Ohio, 2015.
[2] J. A. Salem and R. Tandon, "Test Method Variability in Slow Crack Growth Properties of Sealing Glasses," NASA STI Program, Cleveland, Ohio, 2010.
[3] J. Sanghera and I. D. Aggarwal, Infrared Fiber Optics, Washington DC.: Naval Research Laboratories, 1998.
[4] C. A. Harper, Handbook of Ceramics, Glasses, and Diamonds, McGraw-Hill, 2001.
[5] W. D. Kingery, H. K. Bowen and D. R. Uhlmann, "Chapter 3," in Introduction to Ceramics, New York, John Wiley & Sons, 1972, pp. 91-124.
[6] I. B. Kacem, L. Gautron, D. Coillot and D. R. Neuville, "Structure and Properties of Lead Silicate Glasses and Melts," Chemical Geology, Elsevier, vol. 461, pp. 104-114, 2017.
[7] C. S. Alexander, L. C. Chhabildas, W. D. Reinhart and D. W. Templeton, "Changes to the shock response of fused quartz due to glass modification," International Journal of Impact Engineering, pp. 1376-1385, 2008.
[8] J. C. F. Millett and N. K. Bourne, "Direct measurements of strain in a shock-loaded, lead filled glass," Journal of Applied Physics, vol. 87, no. 8457, 2000.
[9] L. M. Barker, "Laser Interferometry in Shock-wave Research," Experimental Mechanics, vol. 12, 1972.
[10] N. K. Bourne, Z. Rosenberg and J. C. F. Millett, "The Plate Impact Response of Three Glasses," Transactions on the Built Environment, vol. 22, pp. 553-562, 1996.
[11] K. Kondo, S. Iio and A. Sawaoka, "Nonlinear pressure dependence of the elastic moduli of fused quartz up to 3 GPa," Journal of Applied Physics, vol. 52, no. 2826, 1981.
[12] D. E. Grady and E. G. Young, "Evaluation of Consitutive Properties From Velocity Interferometer Data," SAND 75-0650, Albuquerque, 1975.
[13] J. Wackerle, "Shock-wave Compression of Quartz," Journal of Applied Physiscs, vol. 33, no. 3, pp. 922-936, 1962.

Proceedings of the 2019 Hypervelocity Impact Symposium
HVIS2019
April 14-19, 2019, Destin, FL, USA
HVIS2019-040

Modelling the deformation of a high-hardness armour steel in Taylor rod-on-anvil experiments

Shannon Ryan[a]*, Brodie McDonald[a,b], Nikki Scott[c], Rory Bigger[d], Sidney Chocron[c]

[a]*Defence Science and Technology Group, 506 Lorimer Street, Fishermans Bend, VIC 3027, Australia*
[b]*School of Engineering, RMIT University, PO Box 71, Bundoora, VIC 3083, Australia*
[c]*Southwest Research Institute, 6220 Culebra Road, San Antonio, TX 78238, USA*
[d]*Los Alamos National Laboratory, Los Alamos, NM 87545, USA*

Abstract

A high hardness armour steel (HHA) has been subjected to mechanical characterization under tension, compression, and shear loading at quasi-static and dynamic rates incorporating ambient and elevated temperatures. The resulting data has been used to derive constants for four plasticity constitutive models: Johnson-Cook (JC), Zerilli-Armstrong (ZA), modified Johnson-Cook (MJC), and a generalized J_2-J_3 yield surface (GYS). The resulting models have been used to predict the response of the HHA material during Taylor rod-on-anvil experiments. High speed photography and digital image correlation was used during the rod-on-anvil experiments to capture both transient deformation profiles and maximum principal strain along the surface of the rod (i.e. compression along the length of the rod). The JC, MJC, and GYS models were found to provide the best prediction of the shape of the rod (nose diameter and length), within 2% of the experimental measurement in all four rod-on-anvil experiments which did not result in fracture. The JC and GYS models, furthermore, were found to provide the best agreement with the measured transient surface strain profiles, predicting the experimental measurement to within 10% at all measurement locations and time steps for the experiment resulting in maximum deformation (impact velocity = 208 m/s). The results suggest that the added complexity of models such as the MJC and GYS, which incorporate strain hardening saturation, two-part strain rate dependency, and J_3 plasticity effects, are unnecessary for HHA under the loading conditions experienced during rod-on-anvil experiments.

Keywords: Plasticity; Numerical modelling; Armour steel.

1. Introduction

The rod-on-anvil impact experiment, based on the work of Taylor [1], involves firing a flat-nosed cylindrical projectile into a rigid target. The experiment was initially proposed to provide a simple means of measuring material yield strength at high strain rates, but more recently is used as a validation experiment for constitutive models implemented in computational codes due to the wide range of strains, strain rates, and temperatures generated in the projectile material. Classically, validation was performed via comparison of the final projectile shape predicted by the model with that measured in the experiment, i.e. length, nose diameter, and profile, e.g. [2]. More recently, the deceleration profile measured at the rear of the projectile has be used [3], or the deformation-time history for a 2D projectile profile (e.g. [4]). Digital image correlation (DIC) techniques now allow the additional measurement of time-dependent strain and strain-rate on the surface of the deforming projectile (e.g. [5]). Rod-on-anvil experiments have been used to validate constitutive models of copper [4], aluminium alloy [5], tantalum [6], and others.

In this paper four plasticity models will be assessed for modelling the deformation of a high hardness armour steel during rod-on-anvil experiments via comparison with experimental measurements of deformation and surface strain.

* Corresponding author. Tel.: +61 (0) 3 96267706
E-mail address: shannon.ryan@dst.defence.gov.au

2. Testing

2.1. Mechanical characterisation

A quenched and tempered high hardness armour steel (HHA), with a nominal hardness of 500 HB and conforming to MIL-DTL-46100 Class 1, was subject to an extensive mechanical characterization program at the University of New South Wales, Canberra's Impact Dynamics Laboratory. Details of the tensile characterization testing, covering both quasi-static and dynamic tests on cylindrical, notched cylindrical, flat grooved, flat notched, and flat dogbone specimens, are presented in [8]. Additional shear and compressive testing was performed on inclined cylindrical and modified hat specimens at quasi-static and dynamic rates to generate data at zero and negative stress triaxialities, the details of which are reported in [9]. The characterization testing generated data at a range of strain rates, temperatures, stress triaxiliaties, and Lode angles necessary for deriving the constants for a wide range of plasticity and fracture models.

2.2. Taylor rod-on-anvil

A series of eight rod-on-anvil impact experiments were conducted at Southwest Research Institute using the setup described in [5]. The steel rods were machined into cylinders with 6.35 mm diameter and 31.75 mm length and subject to impact velocities ranging from 148 m/s to 236 m/s. Four of the impact experiments resulted in deformed but intact rods, while four resulted in fractured rods. A stochastic speckle pattern was applied to the rods to enable digital image correlation (DIC), from which transient strain and strain rate fields can be recorded during the impact event. Images were captured by a stereo pair of Phantom v711 cameras, generating 12880 pixel images at an interval of 3.125 μs (i.e. 320,000 Hz). The DIC facet size was 10×10 pixels, which is small than ideal for measuring small strains (<1%) with DIC, but works well with the large strains observed in these tests. The transient profile of the rod was also extracted from the high speed imagery, although as the cameras were setup to capture 3D DIC and were not positioned normal to the shot line there is some uncertainty in the measured profile due to parallax. Successful DIC measurements were achieved for 6 of the 8 tests.

3. Plasticity Constitutive Modelling

3.1. Johnson-Cook

The Johnson-Cook (JC) model is an empirical phenomenological model which defines the von Mises flow stress as a function of strain hardening, strain-rate hardening, and thermal softening [6],

$$\sigma = (A + B\varepsilon^n)(1 + C\ln\dot{\varepsilon}^*)(1 - T^{*m}) \tag{1}$$

where ε is the equivalent plastic strain; $\dot{\varepsilon}^* = \dot{\varepsilon}/\dot{\varepsilon}_0$ is the dimensionless equivalent strain rate; $\dot{\varepsilon}_0$ is the reference strain rate; $T^* = (T - T_R)/(T_M - T_R)$ is the homologous temperature; T is the temperature of the material at the corresponding stress and strain; T_R is the ambient (or reference) temperature; T_M is the melt temperature, and; A, B, C, m, and n are the empirical model coefficients. The constant A represents the material yield stress at the reference strain rate $\dot{\varepsilon}_0$ and temperature T_R.

3.2. Modified Johnson-Cook

McDonald et al. [8] apply a modified Johnson-Cook (MJC) plasticity model to simulate the deformation of high strength steel plates under localized blast loading. The Ludwik-type strain hardening formulation of the JC plasticity model (i.e. the first bracketed term) prescribes a monotonically increasing flow stress with increasing plastic strain. A number of modelling studies using high strength steel materials, e.g. [10] demonstrate near-linear isotropic strain hardening, which, when fitted to a Ludwik power law, results in a hardening exponent $n \cong 1$. Incorporation of a Voce-type saturation expression in the strain hardening formulation can provide a more suitable fit to mechanical characterization data by capturing both the initial strain hardening while ensuring saturation of the strain hardening effect at high plastic strains. Additionally, a two-stage strain rate hardening expression is used, incorporating the modified form of Johnson and Cook's relationship from [7], to enable improved fitting of high-rate phenomena with significant strain rate enhancements,

$$\sigma = \left[A + B\varepsilon^n + \sum_{i=1}^{2} Q_i(1 - \exp(-c_i\varepsilon))\right][(1 + \dot{\varepsilon}_1{}^*)^{C_1}(1 + \dot{\varepsilon}_2{}^*)^{C_2}](1 - T^{*m}) \tag{2}$$

where Q_i and c_i are empirical constants that describe saturation of the strain hardening, and $\dot{\varepsilon}_1{}^* = \dot{\varepsilon}/\dot{\varepsilon}_{0,1}$ and $\dot{\varepsilon}_2{}^* = \dot{\varepsilon}/\dot{\varepsilon}_{0,2}$.

3.3. Zerilli-Armstrong

The Zerilli-Armstrong (ZA) model [2] is a semi-empirical physically-based model that uses dislocation mechanics to define a material's response to plastic flow. Von Mises flow stress formulations are specified for body centred cubic (bcc) and face centred cubic (fcc) materials, however, HHA is bcc through the temperature range investigated, hence

$$\sigma_{bcc} = \sigma_0 + C_1 \exp(-C_3 T + C_4 T \ln\dot{\varepsilon}) + C_5\varepsilon^n \tag{3}$$

where σ_0 is the sum of the athermal, strain rate and temperature independent yield strength, σ'_G, and the Hall-Petch grain strengthening relationship, $kd^{-1/2}$; k is the microstructural stress intensity (MPa mm$^{1/2}$); d is the average grain diameter (mm), and; C_1, C_3, C_4, C_5 and n are the empirical model coefficients.

3.4. Generalised J_2-J_3 Plasticity

Plasticity formulations based on the J_2 invariant of the deviatoric stress tensor such as the JC and ZA models are unable to reproduce deviations between tensile and compressive loading conditions, i.e. the strength differential (SD) effect [11] and shear. For materials exhibiting such behavior, inclusion of the third stress deviator invariant, J_3, becomes important. Although not having a physical representation, J_3 is effectively a weighting parameter that induces asymmetry in the yield and plastic flow behavior of a material under compressive and tensile loading. The SD effect is commonly observed in high strength metals and, as such, a number of J_3-dependent plasticity and fracture models have been proposed for use with such materials, e.g. [11], [12], [13]. The generalized yield surface (GYS) proposed in [13] is effectively an empirical extension of any J_2 plasticity model,

$$\sigma = \sigma_{\mathrm{vm}}[g_1 + g_2\cos(3\xi) + g_3\cos(3\xi)] \tag{4}$$

where σ_{vm} is the von Mises flow stress defined by a J_2 plasticity formulation, e.g. Eq. (1), (2), (3), etc, g_1, g_2, g_3, are empirical constants that define the asymmetry of the yield surface in the deviatoric plane, and ξ is the third deviatoric invariant, related to the Lode angle, ϑ, as

$$\xi = \cos(3\vartheta) = \frac{27J_3}{2\sigma_{vm}{}^3} \tag{5}$$

For this study the GYS model is used with the MJC formulation for J2 plasticity (Eq. (2)), i.e.

$$\sigma = \left[\left(A + B\varepsilon^n + \sum_{i=1}^{2} Q_i(1 - \exp(-c_i\varepsilon))\right)((1 + \dot{\varepsilon}_1{}^*)^{C_1}(1 + \dot{\varepsilon}_2{}^*)^{C_2})(1 - T^{*m})\right][g_1 + g_2\xi + g_3\xi^2] \tag{6}$$

In LS-DYNA, Eq. (6) is not explicitly defined (MAT_224). Rather the data is entered in tabular form for tension, compression, and shear from which the yield surface is interpolated.

3.5. Overview of the Plasticity Constitutive Models

The constants used in the four introduced plasticity constitutive models are summarized in Table 1. For the JC and ZA models the constants were determined simultaneously via least-squares fitting of the experimental stress-strain measurements. Due to this approach, instead of the conventional step-by-step fitting of each bracketed term in series, the strain hardening terms of the JC and MJC models vary (i.e. A, B, n). The MJC and GYS constants were derived from an inverse modelling approach, in which numerical simulations of the mechanical test are performed and constitutive model constants are optimized to provide simulated output signals (e.g. deformation-time history) that most closely match those recorded experimentally. A detailed description of the inverse modelling approach used in this study is provided in [8]. The high hardness steel under investigation exhibits a large strength differential effect, as shown in Fig. 1a. As such, for constitutive models which do not incorporate third deviatoric invariant effects (here: JC, ZA, MJC) a choice must be made to either fit the compression data, tension data, shear data, or some combination/average of the three. Clearly, this choice can have a significant effect on the derived constitutive model constants. For a Taylor rod-on-anvil test, compression data is the most suitable, and has thus been used to derive the JC, ZA, and MJC constants. The experimental shear curve is not shown in Fig. 1a as the hat specimen generates a simple shear condition, not a pure shear condition as per a torsion test. Thus, conversion to effective stress is not as straight forward as multiplying the shear stress by $\sqrt{3}$. Simulations of the shear test were used to derive the GYS shear parameters via an optimisation of the plasticity model constants – the experimental data was not used to derive constants in a conventional manner.

It should be noted that derivation of the plasticity model parameters was conducted prior to simulation of the rod-on-anvil experiments, i.e. derivation of the JC and ZA constants and inverse modelling-optimisation of the MJC and GYS constants was performed on the mechanical characterisation data only. No manipulation or tuning of the plasticity model parameters was performed to improve numerical agreement with the rod-on-anvil experiment.

Table 1. Constitutive model parameters derived for the four plasticity constitutive models from mechanical characterisation data.

JC (Compression only)	
Constant	**Value**
A (MPa)	1500
B (MPa)	810
n	0.2787
C	0.0089
m	0.9627
$\dot{\varepsilon}_0$ (/s)	0.001
T_{ref} (K)	293
T_{melt} (K)	1793

ZA (Compression only)	
Constant	**Value**
σ_0 (MPa)	1500
C_1	19409
C_3	0.297
C_4	0.0022
C_5	810
n	0.2787

MJC (Compression only)	
Constant	**Value**
A (MPa)	1137
B (MPa)	420
n	0.38
Q_1 (MPa)	158.6
c_1	594.2
Q_2 (MPa)	315
c_2	95.6
C_1	0.0055
$\dot{\varepsilon}_{0,1}$	0.001
C_2	0.015
$\dot{\varepsilon}_{0,2}$	800
m	1.2
T_{ref} (K)	293
T_{melt} (K)	1793

GYS	**Compression**	**Tension**	**Shear**
Constant	**Value**	**Value**	**Value**
A (MPa)	1137	1154	568.5
B (MPa)	420	406.1	210
n	0.38	0.3026	0.38
Q_1 (MPa)	158.6	420.4	79.3
c_1	594.2	240.8	594.2
Q_2 (MPa)	315	177.7	157.5
c_2	95.6	47.46	95.6
C_1	0.0055	0.001	0.001
$\dot{\varepsilon}_{0,1}$	0.001	0.001	0.001
C_2	0.015	0.015	0.015
$\dot{\varepsilon}_{0,2}$	800	800	800
m	1.2	0.8	0.8
T_{ref} (K)	293	293	293
T_{melt} (K)	1793	1793	1793

Fig. 1 provides some examples of experimental stress-strain curves compared with the constitutive model fits. Additionally, the strain rate and thermal dependency of the different models are shown to provide an indication of the prescribed model trends. With regards to strain rate hardening in compression, the ZA, MJC, and GYS models are shown to provide the best agreement with the experimental data. For this condition the MJC and GYS models are the same. Under tensile loading, the JC, ZA, and MJC curves are identical to that shown for the compression loading as they are independent of third deviatoric invariant effects. The GYS model, however, incorporates these effects and is shown to provide a much better level of agreement with the experimental data, as expected. For the thermal dependency in tension, the JC model provides the best agreement with the experimental data, followed by the GYS model. Interestingly, the ZA model defines very minimal thermal softening (effectively zero at 5% strain and quasi-static rates as shown in Fig. 1, however at higher strain rates there is some apparent thermal softening). In deriving the ZA model constants a conflict was apparent between the rate dependency and thermal dependency characteristics. In other words, constants could be selected which provided a good level of agreement with the elevated temperature tests but a poor level of agreement with the elevated strain rate tests, or vice-versa. This is a characteristic of the coupled rate and thermal effects of the ZA model, which effectively consist of a negative exponential log dependency on temperature, and a coupled linear rate and thermal dependency.

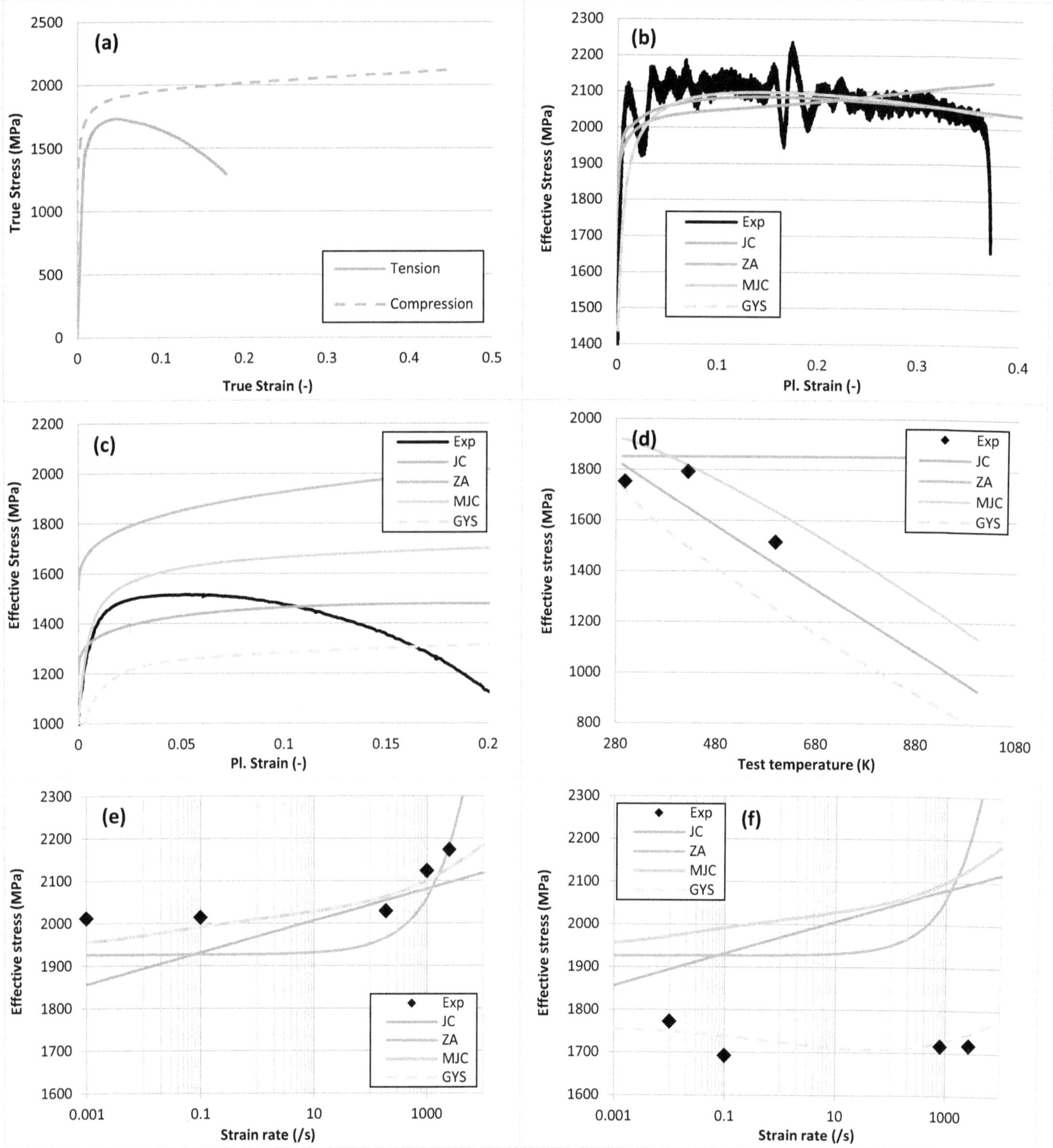

Fig. 1. (a) Experimental quasi-static effective stress vs. plastic strain curves for compression and tension loading conditions highlighting the strength differential effect, (b) effective stress vs. plastic strain curves for compression loading at 1000 /s and 298 K, (c) effective stress vs. plastic strain curves for tensile loading at 0.001 /s and 598 K, (d) temperature dependency of the four plasticity models at 5% plastic strain and 0.001 /s (tension loading), (e) strain rate dependency of the four plasticity models at 10% plastic strain and 298 K start temperature (compression loading), (f) strain rate dependency of the four plasticity models at 5% plastic strain and 298 K start temperature (tension loading).

4. Numerical Modelling

The Taylor rod-on-anvil simulations were simulated in LS-DYNA with full 3D geometry, discretized using hexahedral elements. The mesh used for the rod had an edge length ranging from 0.1 mm at the centre of the sample to 0.2 mm at the outer surface, which was sufficient for mesh-independent results. The anvil was modelled as an elastic material with a mesh size of 0.25 mm and a yield-limited friction coefficient of 0.1 as suggested in [14]. No failure model was used in any of the numerical simulations. A mesh sensitivity study was conducted to ensure that the results were mesh-independent. The computational cost of the simulations was negligible – less than four hours on a single compute node.

A comparison between the experimentally measured and numerically predicated maximum diameter and rod length is given in Fig. 2. In Fig. 3 the final rod profiles are plotted for the maximum deformation case: 208 m/s impact velocity. Up to the fracture velocity, the maximum diameter and length is predicted within 2% of experiments by the JC, MJC, and GYS models. Beyond the fracture velocity, the final rod diameter is underpredicted due to the neglect of fracture in the numerical model. The ZA model is shown to underpredict deformation for all impact velocities, the final rod diameter is predicted to within 7% and the final length is predicted to within 2% of the unfractured experimental measurements. For the conditions evaluated there is effectively no difference in the MJC and GYS simulation results, confirming the dominance of the compressive response of the material (i.e. the third invariant effects are shown to have effectively no influence on the simulation result). Furthermore, the JC predictions are as good as the MJC predictions, suggesting that the modifications to the strain and strain rate hardening terms incorporated in the MJC model are unnecessary for this material under these loading conditions.

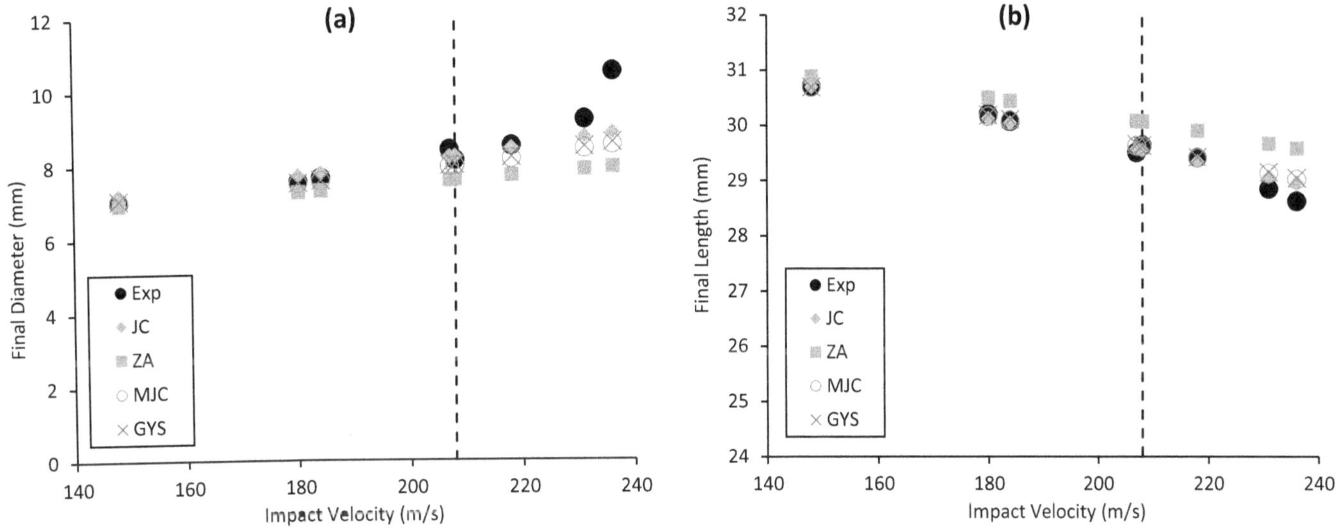

Fig. 2. Comparison of the final experimental (Exp) and numerical dimensions of Taylor rod samples: (a) maximum rod diameter (b) rod length. The dashed line indicated the fracture threshold, above which the numerical predictions are not expected to be accurate (no fracture model was used).

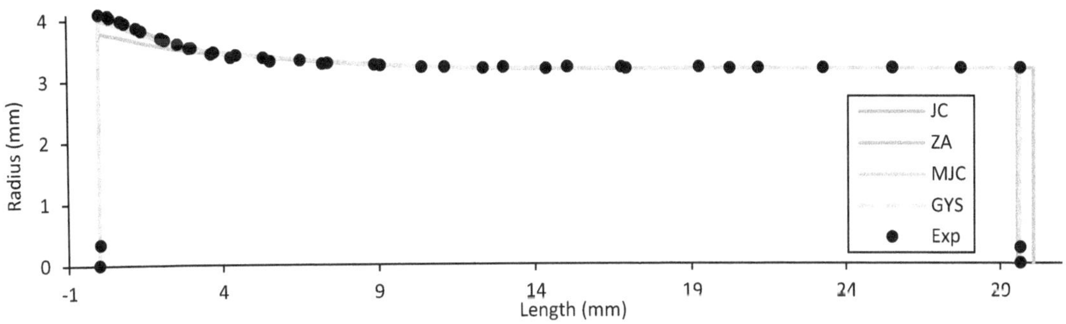

Fig. 3. Final rod profile for the 208 m/s impact experiment compared with numerical results.

The strain distribution throughout the impact experiment is recorded via DIC at a series of five evenly spaced locations along the first 10 mm of the rod, marked in Fig. 4. The equivalent strain, measured in the direction of the velocity vector and with negative values representing compression, is plotted in Fig. 4 for the 208 m/s impact experiment. The JC and GYS models are shown to provide the best agreement with the experimental measurements, followed by the MJC model and then the ZA that provides the worst agreement. The agreement between the JC and GYS models is excellent, within 10% of the measured value over the range of gauge locations and measurement times. Interestingly, the MJC model performed worse than the JC and GYS model, suggesting that there are two effects which need to be further examined: (a) there is some J3 effect, more clearly observable as we move towards the nose of the rod, which is not reproducible by the MJC model, and; (b) the JC model, which also uses J2 plasticity and thus would also be unable to reproduce third invariant effects, nonetheless provides a more accurate prediction of strain – perhaps due to the difference in strain and strain rate formulation in the equations. The ZA model is shown to significantly underpredict the measured strain level, which corresponds to the underprediction of deformation in Fig. 3.

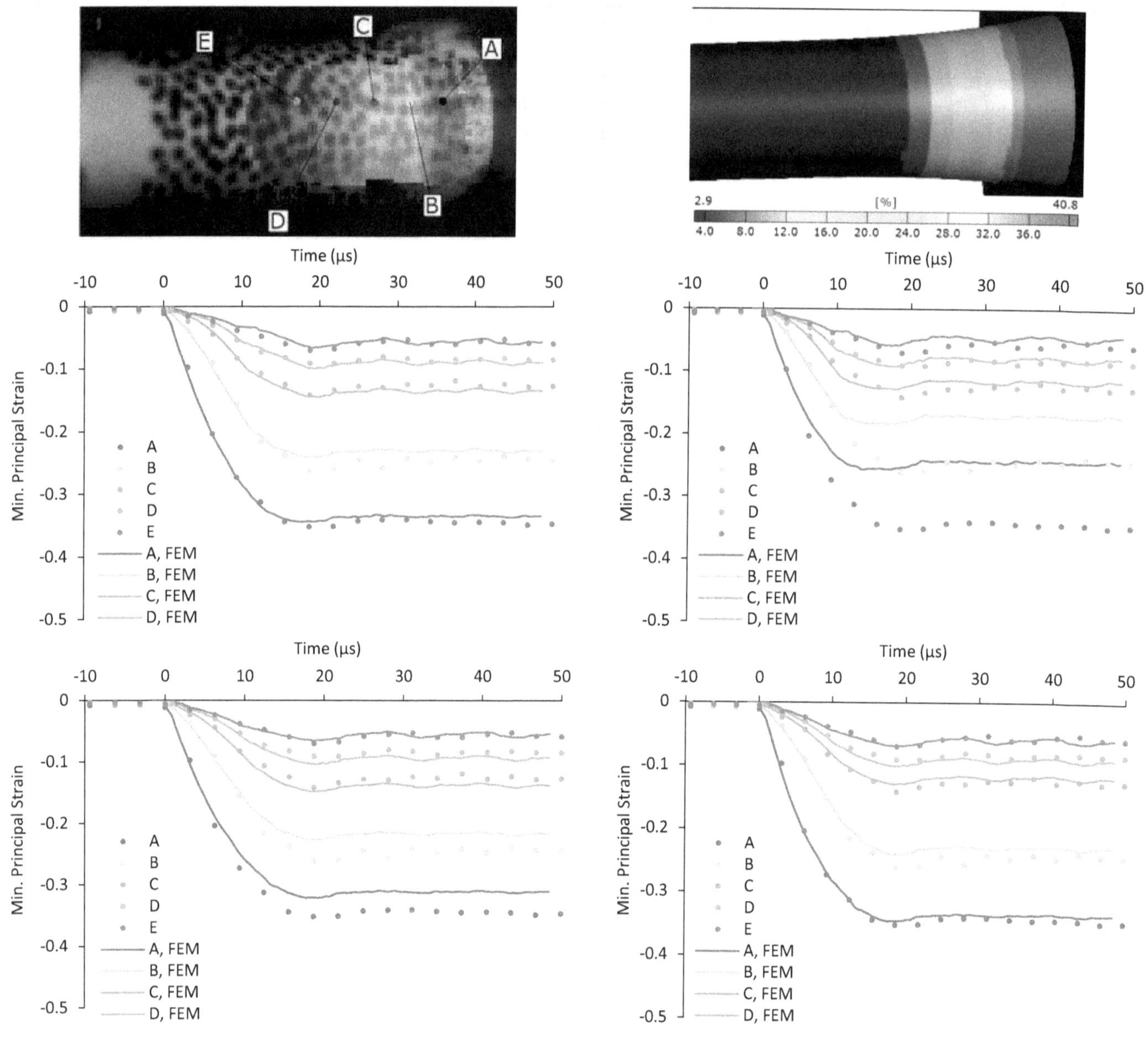

Fig. 4. Comparison of experimental and numerical strain measured in the direction of the rod velocity vector during the 208 m/s impact experiment. (a) measurement gauge locations and strain contours; (b) representative simulation result showing principal strain contours; (c) JC; (d) ZA; (e) MJC, and; (d) GYS.

5. Conclusions

An extensive mechanical characterization campaign has been performed on a high hardness armour steel (HHA), covering quasi-static and dynamic loading rates at ambient and elevated temperatures under a combination of tensile, compressive, and shear loading. The mechanical data has been used to derive constants for four plasticity constitutive models: Johnson-Cook (JC), Zerilli-Armstrong (ZA), modified Johnson-Cook (MJC) incorporating alternate strain and strain rate hardening formulations, and a Generalised Yield Surface (GYS) which extended the MJC expression to incorporate third deviatoric invariant effects (i.e. J_3 plasticity). A series of Taylor rod-on-anvil impact experiments have also been conducted, during which high speed photography and digital image correlation was used to capture both transient deformation profiles and maximum principal strain measurements (i.e. compression along the length of the rod) on the surface of the rod. The captured data is used to assess the accuracy of the plasticity constitutive models via comparison of the experimental measurements and numerical predictions.

The JC, MJC, and GYS models were found to accurately predict the post-test profile of the rods to a high degree of accuracy (within 2% of the experimental measurements) for impacts which did not result in rod fracture. These three plasticity models are built on the same phenomenological basis, albeit with modified expressions for strain, strain rate, and Lode angle dependencies. In predicting the maximum principal surface strain, the JC and GYS models were found to have the best agreement with experimental measurements. For the impact experiment with the greatest deformation and no fracture (208 m/s) the GYS and JC model predictions were within 10% of the experimental value across all gauge locations and measurement times. Interestingly, the GYS and MJC models used in this study are identical for uniaxial compression loading. Thus, minor differences in the surface strain predictions indicate small deviatoric third invariant effects. The comparison of surface strain measurements also identifies a decrease in the predictive accuracy from the modified strain and strain rate hardening expressions of the MJC model.

This work has shown that for numerical simulation of Taylor rod-on-anvil impacts, during which material response is dominated by the prescribed behaviour under uniaxial compression, the additional complexity of models such as GYS which incorporates strain hardening saturation, two-part strain rate hardening, and deviatoric third invariant effects, is unnecessary for the high hardness armour steel investigated.

References

[1] Taylor, G.I., 1948. The Use of Flat-Ended Projectiles for Determining Dynamic Yield Stress, I. Theoretical Considerations, Proceedings of the Royal Society of London, Series A: Mathematical and Physical Sciences 194, p. 289-299.

[2] Zerilli, F.J., Armstrong, R.W., 1987. Dislocation-Mechanics-Based Constitutive Relations for Material Dynamics Calculations, J. of App Phys 61, p. 1816.

[3] Rohr, I., Nahme, H., Thoma, K., 2005. Material Characterization and Constitutive Modelling of Ductile High Strength Steel for a Wide Range of Strain Rates, International Journal of Impact Engineering 31, p. 401.

[4] Eakins, D.E., Thadhani, D.D., 2006. Instrumented Taylor Anvil-on-Rod Impact Tests for Validating Applicability of Standard Strength Models to Transient Deformation States, Journal of Applied Physics 100, p. 1.

[5] Bigger, R., Carpenter, A., Scott, N., Dannemann, K., Chocron, I.S., Williams, C., 2018. Dynamic Response of Aluminum 5083 During Taylor Impact Using Digital Image Correlation, Experimental Mechanics 58, p. 951.

[6] Maudlin, P.J., Bingert, J.F., House, J.W., 1999. On the Modeling of the Taylor Cylinder Impact Test for Orthotropic Textured Materials: Experiments and Simulations, International Journal of Plasticity 15, p. 139.

[7] Johnson, G.R., Cook, W.H., 1983. A Constitutive Model and Data for Metals Subject to Large Strains, High Strain Rates and High Temperatures, Proceedings of the 7th International Symposium on Ballistics, The Hague, 19-21 April.

[8] Børvik, T., Hopperstad, O.S., Berstad, T., Langseth, M., 2001. A Computational Model of Viscoplasticity and Ductile Damage for Impact and Penetration, European Journal of Mechanics – A/Solids 20, p. 685.

[9] McDonald, B., Bornstein, H., Ameri, A., Daliri, A., Orifici, A., 2019. Plasticity and Ductile Fracture Behaviour of Four Armour Steels, submitted to International Journal of Solids and Structures.

[10] McDonald, B., Ryan, S., Edwards, N., Orifici, O., 2019, Modelling the Fracture of High-Hardness Armour Steel in Taylor Rod-on-Anvil Experiments, Proceedings of the Hypervelocity Impact Symposium, Destin, 14-19 April.

[11] Børvik, T., Dey, S., Clausen, A.H.., 2009. Perforation Resistance of Five Different High-Strength Steel Plates Subjected to Small-Arms Projectiles, International Journal of Impact Engineering 36, p. 948.

[12] Hirth, J.P., Cohen, M., 1970. On the Strength-Differential Phenomenon in Hardened Steel, Metallurgical Transactions 1, p. 3.

[13] Bai, Y., Wierzbicki, T., 2008. A New Model of Metal Plasticity and Fracture with Pressure and Lode Dependence, Int. Journal of Plasticity 24, p. 1071.

[14] Chocron, S., Erice, B., Anderson, C.E., 2011. A New Plasticity and Failure Model for Ballistic Application, Int. Journal of Impact Engineering 38, p. 755.

[15] Sengoz, K., 2017. Development of a Generalized Isotropic Yield Surface for Pressure Insensitive Metal Plasticity Considering Yield Strength Differential Effect in Tension, Compression, and Shear Stress States, PhD Thesis, George Washington University.

[16] Rakvåg, K.G., Børvik, T., Hopperstad, O.S., 2014. A Numerical Study on the Deformation and Fracture Models of Steel Projectiles During Taylor Bar Impacts, International Journal of Solids and Structures 51, p. 808.

Proceedings of the 2019 Hypervelocity Impact Symposium
HVIS2019
April 14-19, 2019, Destin, FL, USA
HVIS2019-042

Dynamic response of graphene and yttria-stabilized zirconia (YSZ) composites

Christopher R. Johnson[a*], John P. Borg[a]

[a]*Marquette University – Department of Mechanical Engineering, 1515 W. Wisconsin Ave., Milwaukee, WI 53233, United States*

Abstract

A series of dynamic compaction studies were performed on yttria-stabilized zirconia (YSZ) and graphene composites using uniaxial flyer plate impact experiments. Studies aimed to characterize variation in dynamic behavior with respect to morphological differences for eight powdered YSZ and graphene compositions. Parameters of interest included YSZ particle size (nanometer or micrometer) and added graphene content (graphene weight percentage: 0%, 1%, 3%, 5%). Experiments were performed over impact velocities ranging between 315 and 586 m/s, resulting in pressures between 0.8 and 2.8 GPa. Hugoniot states measured appear to exhibit dependence on particle size and graphene content. Shock velocities tended to increase with graphene content and were generally larger in magnitude for the micrometer particle size YSZ. Compacted densities tended to increase as graphene content was increased and were generally larger in magnitude for the micrometer particle size YSZ samples. Resulting Hugoniot curves are compared and summarized to convey the dynamic behavior of the specimens.

Keywords: Dynamic Behavior, Shock Compaction, Porous Media, Yttria Stabilized Zirconia, Graphene

1. Introduction

Processing of nano-particle materials using non-equilibrium techniques is a current thrust area in material design. Such approach may enable the fabrication of metastable materials with desirable and tailorable properties. Using experimental techniques, such as; explosives, light-gas guns, or laser shock processing, thermodynamic behavior and resulting material properties may be quantified. Characterization provides telling information regarding dynamic behavior and the densified state, thus allowing fabrication techniques to be customized.

Yttria stabilized zirconia (YSZ) is a ceramic of interest in the scientific community due to its wide use in engineering applications [1]–[5]. Although relatively susceptible to fracture, previous research has indicated that shock compaction of YSZ and graphene composites may exhibit increased fracture toughness [6], [7]. This approach may be viewed as analogous to rebar in concrete, increasing its affinity for use in many different applications [8], [9]. Previous investigations have successfully quantified the dynamic behavior of porous and nonporous cubic YSZ media as well as porous YSZ and carbon nanotube (CNT) mixtures [10], [11]. These works set the foundation of this research, shedding light on the expected dynamic behavior of the porous YSZ and graphene compositions.

This study aims to develop Hugoniot relations for various graphene and YSZ mixtures. Specimens will be subjected to high velocity impacts (315-586 m/s) to analyze the dynamic behavior of the powdered graphene and YSZ media. Results will illustrate the significance of two variables: YSZ particle size and added graphene content. It was hypothesized that the influence of added graphene may increase the relative shock velocity observed. Additionally, smaller magnitude particle sizes may also exhibit larger

* Corresponding author. Tel.: 414-288-3522.
E-mail address: christopher.r.johnson@marquette.edu.

shock velocities when compared to their larger magnitude counterparts. These hypothesis stem from observations stated in previous works.

2. Materials

Specimens studied were composed of yttria-stabilized zirconia (YSZ) powder with added graphene. YSZ was sourced from Skyspring Nanomaterials Inc. (8532QI) and was fabricated by stabilizing zirconium dioxide at room temperature with yttrium oxide. The resulting particulate exhibits a cubic fluorite structure, and has a theoretical density of 5.89 g/cc. Additive mixtures were then formed with graphene nanomaterials and the YSZ specimens by Oceanit Laboratories. In total, eight different YSZ and graphene compositions were fabricated. As previously mentioned, the two control variables are YSZ particle size and graphene weight percentage. It was determined that the average nanometer particle YSZ and micrometer particle YSZ sizes were ~20 nm and ~700 nm. For this work, "nano-YSZ" and "micro-YSZ" will be used to differentiate between the two particle sizes. Various weight percentages of graphene were also added (0%, 1%, 3%, 5%) to the YSZ. Table 1 contains a breakdown of the weight percentage and mole percentage of the various mixtures.

Samples were prepared by compacting the particulate into four designated cavities within the experimental targets. This configuration facilitates direct comparison of the specimens, and subjects the four samples to identical conditions. To apply similar loading conditions to the eight specimen types, two experiments were performed at similar impact velocities. Each experiment contained four samples of the same particle size YSZ while varying graphene compositions. Preparation of each sample was performed by pouring ~1.8 grams of the powdered sample into a vial and measuring the weight. Each sample was then poured into designated cavities in the target plate. A PMMA window with vapor deposited gold palladium was then placed downrange of the samples, and driven into the samples with a hydraulic press. A pressure of 4,000 psi (27 MPa) was applied to each PMMA window, where the surface area of the window was 1.77 in^2. Thicknesses of the PMMA windows and depths of the cavities were determined using a coordinate measuring machine (CMM), and were used to calculate sample thicknesses. Diameters of the cavities were measured using a diameter micrometer, and volumes and initial bulk densities of the samples were calculated with the discussed dimensions and sample weights. Initial bulk densities have been tabulated in Table 2, where it can be noticed that grain size and weight percentage of the graphene effected the bulk density.

Table 1. Weight percentage and mole percentage breakdown of the mixed power specimens.

Weight Percentage (Wt. %)		Mole Percentage (Mol %)	
YSZ	Graphene	YSZ	Graphene
100	0	100	0
99	1	77	23
97	3	53	47
95	5	39	61

3. Methods

The experimental series performed in this work utilized Marquette University's 50.8 mm bore dual diaphragm light-gas gun (Figure 1). Multiple uniaxial impact experiments were performed over varied impact velocities. In the experiments, a polycarbonate sabot embedded with a 1045 steel impactor struck a 1045 steel target containing the four samples. A momentum trap and recovery system were implemented to aid in the retrieval of samples. With the system, samples remained contained in the target throughout the duration of the experiment.

Diagnostics used included a laser diode system, piezoelectric pins (PZT), and photon Doppler velocimetry (PDV). The laser

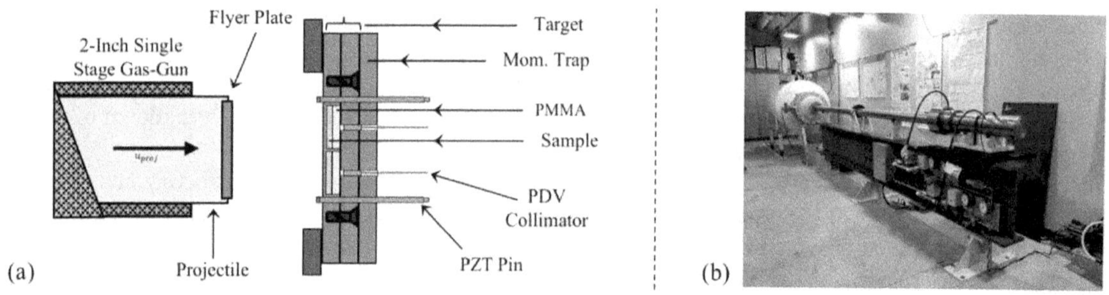

Fig. 1: (a) Uniaxial experimental configuration. (b) Marquette University's 2-inch bore diameter light gas-gun.

diode system was used to determine projectile velocity, where the projectile would obstruct a series of light beams just prior to impact. Planarity of impact was determined using PZT pins to calculate the tilt angle. Shock and particle velocities were determined using PDV. Time corrections were performed for PZT and PDV signals using the tilt angle. Hugoniot states were found through traditional impedance matching with the acquired PDV signals, and mapped into shock velocity / particle velocity space ($U - u$).

4. Results

Eight experiments were performed on 32 samples at impact velocities between 315 m/s and 586 m/s. Collected PDV signals are illustrated in Figure 2. Particle velocity measurements were made at the interface between the sample and PMMA window. Traces are grouped by graphene weight percentage, while coloration indicates YSZ particle size. Many observations can be made from the particle velocity traces. Higher particle velocities were generally measured in the micro-YSZ samples (black/darker shade). It was observed that as graphene content was increased the particle velocity increased. The exception was the 1% wt. graphene samples which observed generally higher particle velocities then the other samples. Additional features in the PDV traces may also indicate certain compaction phenomena in the sample, such as the Hugoniot state achieved or pullback caused from fracture.

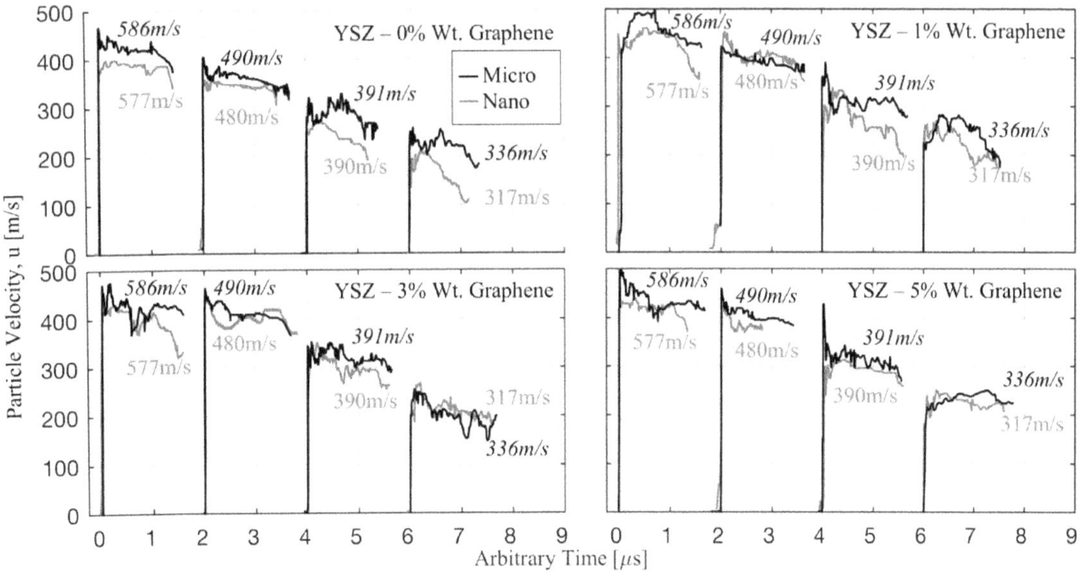

Fig. 2: Reduced PDV traces for the various samples. Black (darker) and red (lighter) colors indicate micro-YSZ and nano-YSZ particle sizes. Impact velocity from the experiments have also been cited above and below the traces.

With the exhibited particle velocities, data was impedance matched and mapped onto shock and particle velocity ($U - u$) space. The relationships were then converted into pressure-density ($P - \rho$) space, based on the Rankine-Hugoniot equations, to compare densification of the porous media. Figure 3 illustrates these relations, which are again grouped based on graphene weight percentage. It can be noticed that higher shock velocities were generally observed in the micro-YSZ samples. Additionally, as the graphene content was increased, shock velocity typically increased. To summarize these relations, linear regressions were fit to the Hugoniot states, and are recorded in Table 3. Phenomena seen in $P - \rho$ space illustrate the densified state of the media. It can be observed that higher density states were typically exhibited for the micro-YSZ particle size samples. Additionally, as the graphene content was increased, the densified states typically increased. These trends, however, should be taken with caution considering the high variability observed in the densified states.

Variation in results suggest that bulk density of the samples affected dynamic behavior. Particle size and structure also likely attribute to this variation, though the addition of graphene appeared to allow for slight increase in the bulk density state. As would be expected, increased shock velocities are generally exhibited regardless of the particle sizes and structures. Such statements appear to align with previous work done studying the compaction of YSZ and carbon nanotube (CNT) composites [11]. Previous findings indicated that as CNT content was increased, the bulk density and shock velocity increased. It is important to keep in mind, however, that this work did have slight deviation in sample thickness and density, and therefore may have caused different dynamic behavior based on wave propagation through the porous media.

Table 2: Sample measurements for the various compositions. Initial bulk densities, compacted bulk densities, particle velocities, and shock velocities are tabulated for each sample.

Powder Composition	Particle Size	Initial Density (g/cc)	Particle Velocity (m/s)	Shock Velocity (m/s)	Compacted Density (g/cc)
0% wt. graphene/YSZ	Nano	2.31	302	933	3.41
		2.36	371	978	3.79
		2.46	473	1439	3.66
		2.49	569	1202	4.72
0% wt. graphene/YSZ	Micro	2.71	312	1020	3.90
		2.83	363	1109	4.20
		2.90	446	1233	4.54

		2.85	532	1316	4.79
1% wt. graphene/YSZ	Nano	2.95	295	1035	4.13
		2.90	367	1121	4.31
		3.03	455	1686	4.15
		3.06	538	1502	4.78
1% wt. graphene/YSZ	Micro	3.29	310	1007	4.75
		3.28	351	1409	4.37
		3.35	429	1539	4.65
		3.28	517	1492	5.01
3% wt. graphene/YSZ	Nano	2.85	297	989	4.07
		2.83	362	1160	4.11
		3.07	456	1624	4.26
		3.17	549	1387	5.24
3% wt. graphene/YSZ	Micro	3.40	313	1017	4.91
		3.44	347	1510	4.46
		3.31	430	1531	4.60
		3.31	520	1420	5.22
5% wt. graphene/YSZ	Nano	2.82	296	1080	3.88
		2.73	364	1150	3.99
		2.91	463	1523	4.19
		2.94	545	1595	4.47
5% wt. graphene/YSZ	Micro	3.48	301	1201	4.65
		3.27	355	1261	4.55
		3.46	431	1654	4.68
		3.41	517	1439	5.33

Table 3. Summary of the $U-u$ Hugoniot parameters for the powdered mixtures. The initial density for each sample type was an average from the specimens listed in Table 2.

Material	Micro-YSZ			Nano-YSZ		
	$\rho_{00}\ (g/cc)$	$C_0(km/s)$	s	$\rho_{00}(g/cc)$	$C_0(km/s)$	s
0% wt. Graphene	2.82	0.626	1.31	2.40	0.538	1.40
1% wt. Graphene	3.30	0.558	1.99	2.98	0.336	2.42
3% wt. Graphene	3.36	0.834	1.33	2.98	0.499	1.90
5% wt. Graphene	3.40	0.831	1.38	2.85	0.381	2.29

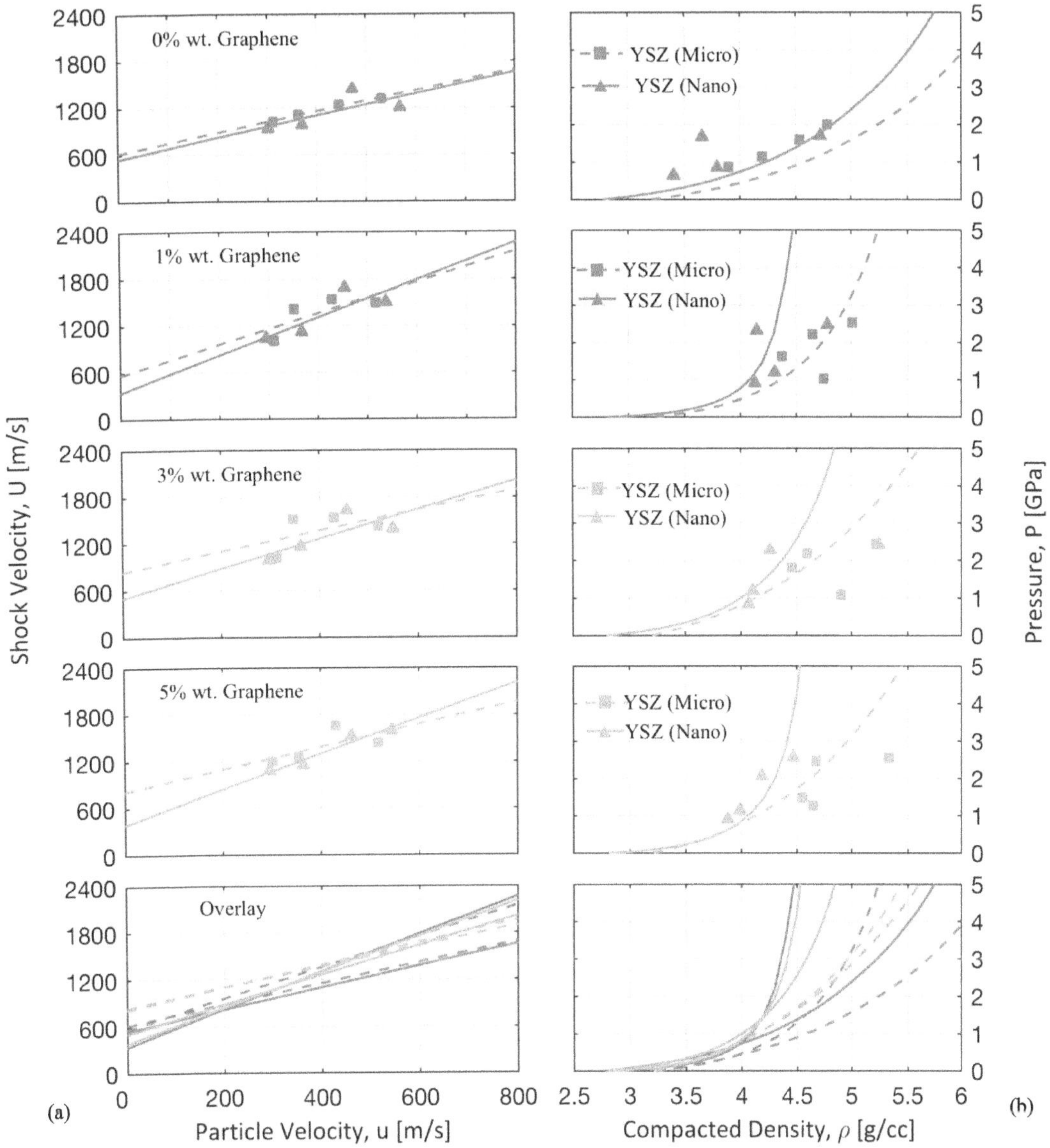

Fig. 3: (a) Shock-particle velocity relations. (b) Pressure-density relations. Plots are organized by graphene weight percentage, and each illustrates the relationship between YSZ particle size.

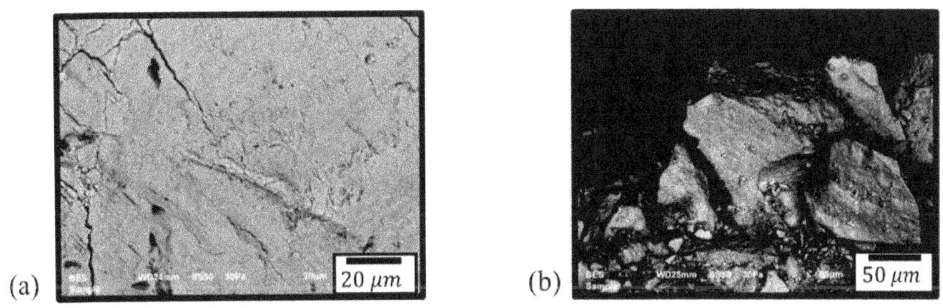

Fig. 4: Recovered 1% wt. graphene and YSZ samples from an impact velocity of 586 m/s: (a) top down view (b) side view of a small fractured specimen.

Figure 4 illustrates two compacted specimens. Note the large degree of densification observable in the images, considering the previous powdered state. Figure 4a illustrates the top down view of the specimen, while Figure 4b illustrates a side view of a small fractured specimen. Without further post-mortem mechanical property characterization, the mechanical effect cannot be commented on. However, overall results demonstrate how the mesoscale and microscale features appear to influence the overall bulk response of the specimens with regards to shock compaction.

5. Conclusion & Future Work

Eight uniaxial impact experiments were performed on the YSZ and graphene composites. Experiments were conducted over a spectrum of impact velocities (315-586 m/s). Configurations featured a 1045 steel impactor striking a 1045 steel target containing four YSZ/graphene samples of varying graphene content. The results suggest significant variations in dynamic behavior are caused by differing particulate size and graphene content. Measured shock velocities increased with graphene content, and were typically larger in magnitude for the micro-YSZ particle sizes. Compacted densities observed tended to increase with graphene content, and were generally larger in magnitude for the micro-YSZ particle sizes. Future directions should consider isolating the compacted bulk densities, which may yield more interesting results regarding wave propagation and the Hugoniot states achieved in the specimens. Additional studies may also investigate shock wave development throughout varied thicknesses of the heterogeneous materials.

Acknowledgments

The authors gratefully acknowledge the financial support of the United States Air Force (under Grant No. AF10-BT26), and collaboration with Oceanit Laboratories. The authors would also like to acknowledge and thank Dr. Jeff LaJeunesse, Peter Sable, Nathaniel Helminiak, and Sebastian Konewko for the many joyful conversations and discussions.

References

[1] G. Stapper, M. Bernasconi, N. Nicoloso, and M. Parrinello, "Ab initio study of structural and electronic properties of yttria-stabilized cubic zirconia," Phys. Rev. B, vol. 59, no. 2, pp. 797–810, 1999.

[2] W. J. Fleming, "Physical Principles Governing Nonideal Behavior of the Zirconia Oxygen Sensor," J. Electrochem. Soc., vol. 124, no. 1, p. 21, 1977.

[3] M. A. Parkes, "ab initio Modeling of Yttria Stabilised Zirconia for Solid Oxide Fuel Cells," Imperial College London, 2015.

[4] S. P. S. Badwal and I. Introduction, "Zirconia-based solid electrolytes: microstructure, stability and ionic conductivity," Solid State Ionics, vol. 52, no. 1–3, pp. 23–32, 1992.

[5] C. Piconi and G. Maccauro, "Zirconia as a ceramic biomaterial," Biomaterials, vol. 20, no. 1, pp. 1–25, 1999.

[6] M. Mazaheri, D. Mari, R. Schaller, G. Bonnefont, and G. Fantozzi, "Processing of yttria stabilized zirconia reinforced with multi-walled carbon nanotubes with attractive mechanical properties," J. Eur. Ceram. Soc., vol. 31, no. 14, pp. 2691–2698, 2011.

[7] M. Mazaheri, D. Mari, Z. R. Hesabi, R. Schaller, and G. Fantozzi, "Multi-walled carbon nanotube/nanostructured zirconia composites: Outstanding mechanical properties in a wide range of temperature," Compos. Sci. Technol., vol. 71, no. 7, pp. 939–945, 2011.

[8] A. Gallardo-lópez, I. Márquez-abril, A. Morales-rodríguez, and A. Muñoz, "Dense graphene nanoplatelet / yttria tetragonal zirconia composites : Processing , hardness and electrical conductivity," Ceram. Int., vol. 43, no. 15, pp. 11743–11752, 2017.

[9] J. H. Shin and S. H. Hong, "Fabrication and properties of reduced graphene oxide reinforced yttria-stabilized zirconia composite ceramics," J. Eur. Ceram. Soc., vol. 34, no. 5, pp. 1297–1302, 2014.

[10] T. Mashimo, A. Nakamura, M. Kodama, K. Kusaba, K. Fukuoka, and Y. Syono, "Yielding and Phase-Transition Under Shock Compression of Yttria-Doped Cubic Zirconia Single-Crystal and Polycrystal," J. Appl. Phys., vol. 77, no. 10, pp. 5060–5068, 1995.

[11] P. A. Sable, J. LaJeunesse, C. Sullivan, V. Kamavaram, and J. P. Borg, "Dynamic compaction of yttria-stabilized zirconia with the addition of carbon-nanotubes," in AIP Conference Proceedings, 2017, p. 120004.

Proceedings of the 2019 Hypervelocity Impact Symposium
HVIS2019
April 14-19, 2019, Destin, FL, USA
HVIS2019-047

Extended Abstract+

High Strain-rate Shear and Friction Characterization of Fully-Dense Polyurethane and Epoxy

Peter A. Sable[a,*], Christopher H. Neel[b], John P. Borg[a]

[a]*Dept. of Mechanical Engineering, Marquette University, Milwaukee WI 53233 USA*
[b]*Damage Mechanisms Branch, Air Force Research Laboratory Munitions Directorate, Eglin AFB FL 32542 USA*

1. Introduction

The characterization of polymer behavior at high strain-rates is a critical area of research driven by the use of polymers as adhesives, structural components, or as binders for energetic systems. Previous investigations have successfully defined Hugoniot equations-of-state for many relevant polymers such as epoxy, polycarbonate, and polyether ether ketone [1]–[4].Several studies have also explored the ramifications of polymer viscoelasticity at high strain-rates and pressures. This has been observed as nonlinearities in both propagating stress wave structure and Hugoniot state spaces [5]–[7]. The mechanisms driving such behavior is hypothesized as the breakdown and extended time interaction of polymer-chains but is yet to be fully understood.

The majority of high strain-rate polymer characterizations were achieved through uniaxial (plane strain) flyer plate impact experiments using light-gas guns. Such testing is well suited for Hugoniot development, however is limited as a method for evaluating strength. In many cases, strength may be quantified through the Hugoniot elastic limit (HEL) observable in the elastic wave preceding a shock. Polymers lack an observable elastic precursor, and so strength can only be inferred from the material shear stress response. To observe shear in uniaxial tests embedded transverse stress gages must be implemented [4]. More so, the loading-path in this configuration is fixed (plane strain) and so implications of combined loading on material strength are unobservable.

Given these limitations, it is of interest to consider alternative experimental methodologies which expand high strain-rate characterizations to include other phenomena. Oblique impact experiments consist of an angled projectile propelled toward a target set at the same angle via gas gun. Upon impact, both normal (pressure) and shear stress waves are generated and propagate away from the impact surface. Often term pressure shear plate impact experiments (PSPI), targets are subjected to combined loading and analysis results in direct measurement of material strength [8]. The technique has been used to successfully evaluate the yield strength behavior of a metals, granular materials, and a few polymers [9]–[12]. It has additionally been used to observe dynamic friction behavior between tribological pairs [13]–[15].

In this study, two fully dense thermosetting polymers, polyurethane (PUR) and epoxy, were subjected to rapid (high strain-rate) pressure-shear stress loading via oblique impact and the material responses were observed. Two experimental variations were utilized to observe both strength and friction phenomena. Impact speeds were varied to consider the effects of increasing pressure on polymer strength, and the roles of surface roughness and polymer-adhesion were considered. Constitutive and equation-of-state models, informed by experiment, were implemented into the CTH hydrocode [16] to demonstrate the applicability of such a characterization.

+ Full paper can found at https://www.sciencedirect.com/journal/international-journal-of-impact-engineering/special-issue/10MTGP5W4VJ

* Corresponding author. Tel.: +1-417-818-1483;
E-mail address: peter.sable@marquette.edu

2. Methodology

2.1. Polymer Description

Polyurethane and epoxy were the chosen materials of interest. This was due to their prevalence in most polymer applications, ease of acquisition, and the availability other dynamic characterization data with which to compare. The polyurethane was 1512 two-part liquid plastic purchased from *Polytek*. Relevant mechanical properties include a mix viscosity of 400 cP, a shrinkage of about 5×10^{-3} in/in, a quasi-static yield strength of approximately 46 MPa, and a 20-minute pot life available for mixing and molding. Post-mold densities were on average 1.135 g/cc. The epoxy-of-interest was the F110 formulation from *Atom Adhesives* and was also two-part resin/hardener system. Specifications include a mix viscosity of 310 cP, negligible shrinkage, a yield strength of 36 MPa, and a 30-minute pot life. Density was nominally 1.155 g/cc. Each had degradation temperatures around 100 deg-C. The presence of air bubbles post-molding was a significant concern, and so samples were degassed during pot life with a vacuum less than or equal to 29 inHg. This was sufficient to remove most bubbles from the mixed polymer with the remainder being crushed by the return of ambient pressure, any remaining microbubbles were considered negligible.

Mechanical testing was done measure relevant properties and establish a baseline for polymer behavior. Sound speed measurements using *Olympus* ultrasound through transmission equipment resulted in nominal longitudinal and shear wave speeds of 2.30 and 0.99 mm/us respectively for polyurethane, with 2.64 and 1.24 mm/us for epoxy (±0.04 mm/us). Quasi-static/low strain-rate shear strength was measured conducting the ASTM-D143 standard test with an Instron machine. Results were consistent with manufacturer specifications and confirmed strain-rate dependence – a few megapascals increased strength for an order of magnitude increasing rate. Thin-film specimens exhibited very brittle behavior compared to larger counterparts. From this comparison, toughness was observed to be very dependent on specimen geometry. Uniaxial plane-strain impact experiments were conducted at the High-Pressure Particulate Physics (HP3) facility at Eglin AFB. Results demonstrate that the selected polyurethane and epoxy have similar Hugoniot behavior to others observed in literature [6][7].

2.2. Oblique Impact Experiments

Experiments were conducted using a two slotted-bore light-gas guns, including a 2.5" bore 6 ft barrel gun at Eglin AFB and 2" bore 12 ft barrel gun at Marquette University. Two configurations of oblique impact were utilized referred to as constant-pressure pressure shear plate impact (CPPSPI) experiments and dynamic friction impact experiments respectively. Figure 1 presents a schematic of both tests and labels relevant components and their orientation before impact. In both cases, upon impact, a normal (pressure) and shear stress wave are produced and propagate into the target package and impactor respectively.

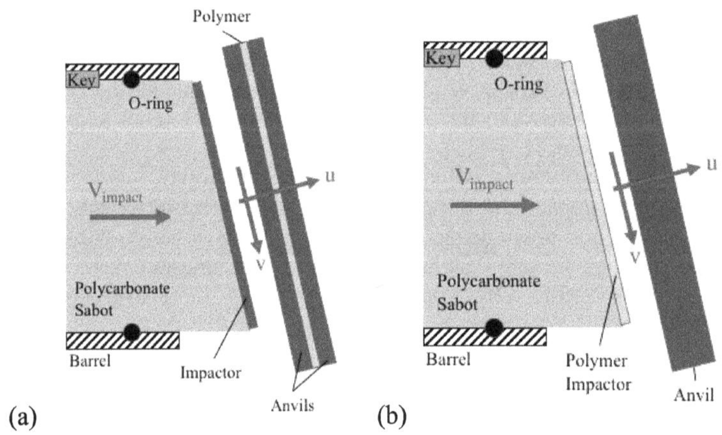

Fig. 1. Oblique impact experiment configurations (a) A CPPSPI experiment which consists of an angled projectile planarly impacting a target package set at an angle of obliquity. The target package consists of a front anvil, polymer sample, and rear anvil respectively. Note a key is shown which maintains projectile orientation down the slotted-barrel. (b) A dynamic friction impact experiment, wherein the same angled planar impact is facilitated, however the polymer of interest serves as the impactor.

For the CPPSPI experiment, the target package consisted of a front and rear anvil confining the polymer of interest. The front anvil thickness was set such that the normal stress wave arrives at the polymer sample and reverberates between anvils bringing the polymer to an equilibrium (constant) pressure state prior to the arrival of the slower moving shear wave. This ensures the effect of shear stress is temporally distinct from that of normal stress and is more readily distinguished within observed wave profiles. Samples were kept thin, nominally 0.5 millimeters, so reverberation did not take more than approximately a microsecond. After propagating through the polymer, waves continued through the rear anvil and their behavior was observed via free surface particle velocity traces measured through transverse photon Doppler velocimetry (PDV) [17][18]. Anvil materials and impact conditions were chosen such that the yield strength of all components, except the sample, was never exceeded. This was necessary to ensure only anvils could sustain shear loading and enables the use of one-dimensional elastic wave theory for analysis, wherein particle velocity is directly proportional to stress [19]. Impactors and anvils consisted of 7075-T6 aluminum with the polymer sample being either the PUR or epoxy of interest. The measurement of both normal and shear stress states resulted in the characterization of the polymer stress response to the imparted loading conditions. If no yielding occurs, the expected observable shear stress magnitude is known based on impact conditions and material properties. As such, any deficiency in shear stress is assumed to be representative of polymer yield – i.e. shear strength.

For dynamic friction experiments, the polymer of interest served as the angled projectile impacting a 7075-T6 aluminum anvil target, the yield strength of which was again not exceeded. Given its role as impactor, the polymer simultaneously experiences both normal (pressure) and shear stress unlike the CPPSPI configuration. Stress waves propagate elastically through the anvil and were observed using the same PDV methods. A coefficient of friction (COF) for the anvil/polymer interface was then quantified through the ratio of shear to normal stress [13][15]. Slip occurring at the interface was indicative of polymer yield, on the level of surface roughness asperities, and so shear stress in these experiments was also assumed to be representative of shear strength. Notably, loading in this configuration is pressure-*and*-shear as opposed to pressure-*then*-shear and the polymer is not confined so strength behaviors are not directly comparable to those obtained in CPPSPI testing [20].

3. Results & Discussion

Six CPPSPI and two dynamic friction experiments were conducted on fully dense polyurethane, while seven CPPSPI and one dynamic friction experiment were done for epoxy. Impact velocities were varied allowing for different pressures. Anvil surface roughness Ra (average roughness) was tracked using a *Pocket Surf* ® sapphire stylus profilometer. Roughness was allowed to vary between approximately 0.2 and 10 μm, in an effort to identify possible ramifications. Notably, over said range roughness had no observable effect on bulk stress wave profiles -which was further corroborated by mesoscale simulations. Coefficient of friction values were successfully calculated from dynamic friction experiments, with values of 0.18 and 0.42 for PUR and epoxy respectively each with a 7075-T6 aluminum tribological partner.

Normal and shear stress values were measured for all experiments (both configurations). These shear strengths can then be plotted against pressure to evaluate any dependence, which has been seen at lower strain-rates [21]. Figure 2a shows these results plotted alongside literature data [11][15] and the Mulliken-Boyce viscoelastic model [22].

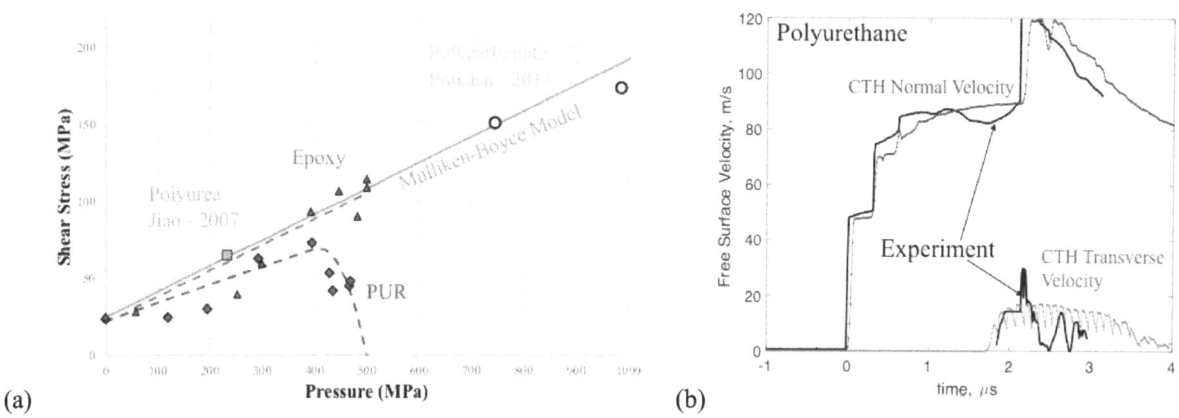

Fig. 2. (a) Shear strength as a function of pressure. The tested PUR and epoxy are shown alongside other literature data for different materials as well as the theoretical Mulliken-Boyce viscoelastic model. (b) CTH simulations, with data-based constitutive models, compared against experimentally observed PDV.

Results show shear strength to have a strong dependence on pressure for both polyurethane and epoxy. This persists even with respect to the dynamic friction case, which is under different loading conditions. For polyurethane specifically, this trend eventually decays reminiscent of a Drucker-Prager cap [23]. Observed epoxy data does not suggest the same behavior, however it is hypothesized a similar trend would be evident if the tested pressure regime were extended. Strength data is used to formulate a *Geo-Yield* constitutive model for each polymer in CTH. Along with Hugoniot information, CTH is able to recreate experimental wave profiles quite well for both normal and shear stress waves, shown as particle velocities in Figure 2b. This suggests pressure-dependence to be a dominate mechanism governing polymer strength. Viscoelastic effects are known to be another crucial factor and, despite the good agreement between simulation and data, are notably not implemented – nor readily isolated in this methodology. Temperature degradation was a concern, and while not measured directly, values were inferred from CTH simulations. CPPSPI experiments did not appear to induce temperatures exceeding degradation, however localized surface asperities in simulated dynamic friction experiments did, providing another reason for which the two configurations should be compared with caution.

As a last consideration, polymers in CPPSPI were adhered to confining anvils, as such it was initially unknown whether strength being measured was that of the polymer or that of interface-bond. However, post-mortem of CPPSPI targets revealed that adhesion survived the impact event suggesting that the observations were indeed of the polymer itself. A more precise evaluation of the adhesive bond is of particular interest for future work.

References

[1] J. C. F. Millett, N. K. Bourne, and N. R. Barnes, "The behavior of an epoxy resin under one-dimensional shock loading," *J. Appl. Phys.*, vol. 92, no. 11, pp. 6590–6594, 2002.

[2] J. C. F. Millett and N. K. Bourne, "Shock and release of polycarbonate under one-dimensional strain," *J. Mater. Sci.*, vol. 41, no. 6, pp. 1683–1690, 2006.

[3] J. C. F. Millett, N. K. Bourne, and G. T. Gray, "The response of polyether ether ketone to one-dimensional shock loading," *J. Phys. D. Appl. Phys.*, vol. 37, no. 6, pp. 942–947, 2004.

[4] N. K. Bourne, "On the Shock Response of Polymers to Extreme Loading," *J. Dyn. Behav. Mater.*, vol. 2, no. 1, pp. 33–42, 2016.

[5] L. M. Barker and R. E. Hollenbach, "Shock-wave studies of PMMA, fused silica, and sapphire," *J. Appl. Phys.*, vol. 41, no. 10, pp. 4208–4226, 1970.

[6] G. J. Appleby-Thomas, P. J. Hazell, and C. Stennett, "The variation in lateral and longitudinal stress gauge response within an RTM 6 epoxy resin under One-dimensional shock loading," *J. Mater. Sci.*, vol. 44, no. 22, pp. 6187–6198, 2009.

[7] G. J. Appleby-Thomas, P. J. Hazell, C. Stennett, G. Cooper, and R. Cleave, "The dynamic behaviour of a modified polyurethane resin," *DYMAT 2009 - 9th Int. Conf. Mech. Phys. Behav. Mater. under Dyn. Load.*, vol. 2, pp. 1081–1087, 2009.

[8] A. S. Abou-Sayed and R. J. Clifton, "The oblique-plate impact experiment," *Exp. Mech.*, vol. 16, no. 4, pp. 127–132, 1976.

[9] K. J. Frutschy and R. J. Clifton, "High-temperature pressure-shear plate impact experiments on ofhc copper," *J. Mech. Phys. Solids*, vol. 46, no. 10, pp. 1723–1744, 1998.

[10] W. D. Reinhart, T. F. Thornhill III, L. C. Chhabilidas, T. J. Vogler, and J. L. Brown, "Shock response of dry sand," vol. SAND2007-3, pp. 1–28, 2007.

[11] T. Jiao, R. J. Clifton, and S. E. Grunschel, "Pressure-sensitivity and tensile strength of an elastomer at high strain rates," *AIP Conf. Proc.*, vol. 955, pp. 707–710, 2007.

[12] V. Prakash and N. Mehta, "Uniaxial compression and combined compression-and-shear response of amorphous polycarbonate at high loading rates," *Polym. Eng. Sci.*, vol. 52, no. 6, pp. 1217–1231, 2012.

[13] M. A. Irfan and V. Prakash, "Time resolved friction during dry sliding of metal on metal," *Int. J. Solids Struct.*, vol. 37, no. 20, pp. 2859–2882, 2000.

[14] F. Yuan, N. S. Liou, and V. Prakash, "High-speed frictional slip at metal-on-metal interfaces," *Int. J. Plast.*, vol. 25, no. 4, pp. 612–634, 2009.

[15] P. Loading and V. Prakash, "Investigations of Material Strength and Interface Friction Using," 2014.

[16] J. M. McGlaun, S. L. Thompson, and M. G. Elrick, "CTH: A three-dimensional shock wave physics code," *Int. J. Impact Eng.*, vol. 10, no. 1–4, pp. 351–360, 1990.

[17] O. T. Strand, D. R. Goosman, C. Martinez, T. L. Whitworth, and W. W. Kuhlow, "Compact system for high-speed velocimetry using heterodyne techniques," *Rev. Sci. Instrum.*, vol. 77, no. 8, 2006.

[18] C. R. Johnson, P. A. Sable, J. W. Lajeunesse, A. Dawson, A. Hatzenbihler, and J. P. Borg, "Photon Doppler velocimetry measurements of transverse surface velocities," *AIP Conf. Proc.*, vol. 1979, 2018.

[19] H. Kolsky, "Stress Waves in Solids," *J. Sound Vib.*, vol. 1, pp. 88–110, 1964.

[20] K. T. Ramesh and R. J. Clifton, "Finite Deformation Analysis of Pressure-Shear Plate Impact Experiments on an Elastohydrodynamic Lubricant," vol. 59, no. 92, 2015.

[21] R. C. Bowers and W. A. Zisman, "Pressure effects on the friction coefficient of thin-film solid lubricants," *J. Appl. Phys.*, vol. 39, no. 12, pp. 5385–5395, 1968.

[22] A. D. Mulliken and M. C. Boyce, "Mechanics of the rate-dependent elastic-plastic deformation of glassy polymers from low to high strain rates," *Int. J. Solids Struct.*, vol. 43, no. 5, pp. 1331–1356, 2006.

[23] B. L. Resende, J. B. Martin, and M. Asce, "Formulation of drucker-prager c a p m o d e l," vol. I, no. 7, pp. 855–881, 1985.

[24] S. Measurements and L. M. Barker, "Shock-Wave Studies of PMMA, Fused Silica, and Sapphire," *J. Appl. Phys.*, vol. 41, no. 10, p. 4208, 1970.

Proceedings of the 2019 Hypervelocity Impact Symposium
HVIS2019
April 14-19, 2019, Destin, FL, USA
HVIS2019-057

Eulerian Hydrocode Modeling of a Dynamic Tensile Extrusion Experiment

M. W. Burkett*

Los Alamos National Laboratory, P. O. Box 1663, Mail Stop T085, Los Alamos, NM 87545, USA

Abstract

Eulerian hydrocode simulations using the Mechanical Threshold Stress (MTS), Zerilli-Armstrong (Z-A), and Johnson Cook (J-C) flow stress models were performed to provide insights into dynamic tensile extrusion (DTE) experiments with copper (Cu) and tantalum (Ta). The extrusion of Cu and Ta projectiles was simulated with an explicit, two-dimensional Eulerian continuum mechanics hydrocode and compared with data to determine if this extrusion concept is a useful indirect hydrocode material strength model evaluation experiment. The data consisted of high-speed images of the extrusion process, photon Doppler velocimetry (PDV) to measure the projectile velocity history and die transit time, dynamic temperature measurements of the extruded material, recovered extruded samples, and post-test metallography. The hydrocode was developed to predict large-strain and high-strain-rate loading problems. The code features a high-order advection algorithm, material interface tracking scheme, and van Leer monotonic advection-limiting algorithm. The strength models were utilized to evolve the flow stress (σ) as a function of strain, strain rate, and temperature. Average strain rates on the order of 10^4 s^{-1} and plastic strains exceeding 300% were predicted in the extrusion of copper at impact velocities between 400–450 m/s, while plastic strains exceeding 800% were predicted for Ta. The predicted and measured deformation topologies, projectile velocity profiles and die transits times, plastic strains, and temperatures were qualitatively compared. The flow stress distributions predicted by the three strength models were also compared for one experiment. Finally, the feasibility of using DTE to evaluate hydrocode strength models will be discussed.

Keywords: Hydrocode, Mechanical Threshold Stress, Zerilli-Armstrong, Johnson Cook, strength model, Eulerian, dynamic tensile loading, anisotropy, flow stress.

1. Introduction and motivation

Eulerian hydrocode [1] simulations utilizing the Mechanical Threshold Stress (MTS) flow stress (σ) [2], Zerilli-Armstrong (Z-A) [3], and Johnson Cook (J-C) [4] flow stress models were performed to provide insight into a dynamic extrusion experiment. The experimental concept [3] was developed by George T. Gray, III, of Los Alamos National Laboratory to evaluate the behavior of selected materials (Cu and Ta will be discussed) when subjected to large plastic strain deformation at relatively high strain rates without the need for a high-explosive launch system. The objectives of the computational study were to perform quantitative and qualitative comparisons between the hydrocode simulation output parameters and the available experimental data to determine whether this experiment is a useful for evaluating indirect hydrocode constitutive models in which material strength is an important component.

2. Extrusion experiment description and data summary

Spherical and bullet-shaped Cu and Ta projectiles (7.62 mm) were accelerated in a high-pressure helium (He) gas gun to velocities ranging between 330 and 700 m/s and extruded through a high-strength steel die with a tapering angle of ~10° (Fig. 1).

* M. W. Burkett. Tel.: +1 505-665-6336
E-mail address: mwb@lanl.gov

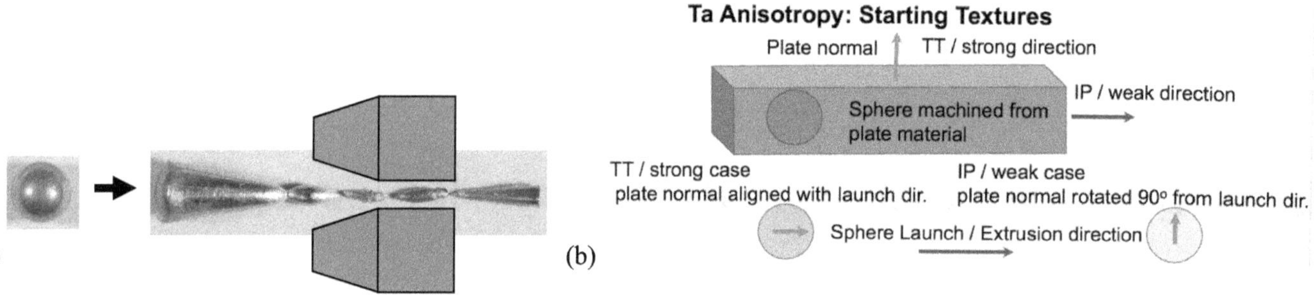

Fig. 1. (a) 7.62-mm spherical Cu spherical projectiles of three different grain sizes were accelerated in a He-gas launcher to velocities of ~400 m/s and extruded through a steel die (Image courtesy G. T. Gray, III). (b) Tantalum anisotropy initial conditions.

High-speed optical imaging revealed the extruded topologies, formation of plastic instabilities, and leading-edge velocities of the extruded materials (Figs. 2–3). Metallographic analyses were performed on the soft-recovered extruded particles. In the high-speed images, a spray, or "ejecta cloud," of the lubricant used in the die was observed preceding the exit of the sphere's leading edge (Fig. 2). Grain size sensitivity to extrusion was investigated for Cu, while the effect of initial texture in Ta (material anisotropy) was investigated by controlling the sphere's orientation during launch and subsequent extrusion. Through-thickness (TT) and in-plane (IP) orientations were studied (Fig. 1).

The texture evolution in Cu deformed during dynamic tensile extrusion loading displayed a strong grain-size dependency [5] (Fig. 2). Additionally, as the heat treatment temperature was increased (yielding a larger grain size) the mass (amount) of extruded material was found to decrease (projectile velocity held constant). Extruded lengths and masses were found to range from 34 to 39 mm and ~0.22 to ~0.39 g (10%–13% of the initial sphere mass) over all heat treatments and projectile velocities.

After the initial Cu DTE series was completed, a single PDV probe was added to the experiment to provide additional quantitative data for modeling comparisons. Also, the projectile shape was modified to a bullet shape with the same caliber to control the projectile orientation and initial texture. Figure 3 shows PDV data and imaging for the DTE of Cu (Follansbee Cu heat treated for 1 h at 600°C) at 340 m/s. At this velocity, the leading edge of the Cu projectile did not separate from the main portion of the bullet and was arrested in the die. The PDV and imaging data show the projectile's leading edge exiting the die but rapidly decelerating without particulation (separation). A neck or plastic instability does develop (Fig. 7) but particulation does not occur. As the projectile velocity is increased, the projectile's leading edge exits the die at a high velocity, plastic instabilities develop due to the spatial velocity gradient, a reduction in the cross-sectional area of the extruded material (necking process) begins and particulation eventually occurs on a microsecond time scale.

Analysis of the Ta data [6] revealed that particle elongation is mainly affected by velocity (Fig. 4). Larger elongations and propensity for instability (more particles) were achieved at higher velocities (>500 m/s). The extrusion velocity was found to vary approximately linearly with the impact velocity. The effect of starting texture of the spheres is exhibited at velocities lower than 400 m/s and becomes insignificant at higher velocities. The dynamic extrusion process significantly reoriented the texture. Similar shear textures and hardness increments were developed in segments recovered at all velocities, regardless of the starting textures. No shear bands were observed in dynamically extruded Ta.

3. Eulerian hydrocode and modeling details

MESA is a serial two-dimensional Eulerian hydrocode. The code has several key features that minimize advection errors including a Young's interface reconstruction algorithm and a Van Leer advection limiter. Several common hydrocode strength and damage models and standard equation of state options are available. Commercially available post-processing software can be used to perform analyses of simulations. The MTS, Z-A, and J-C strength models were used to approximate the plastic deformation behavior of Cu and Ta. The parameters for Cu and Ta were taken from [7] and [8], respectively. A tabular SESAME [9] equation of state was used with all yield models. A von Mises yield criterion was used that results in a "radial return" to the yield surface [10]. A uniform zoning of ~40 computational cells was employed to discretize the extrusion exit diameter (the reduced end of the die) region. Also, a thin layer was defined along the inner surface of the die to allow the projectile to "slip" through the computational mesh at the projectile/die interface with minimal artificial drag induced by the hydrocode's vertex centered velocity scheme. If a coarse mesh is used, the projectile can realize numerical drag at this interface, which can significantly affect the predicted extrusion rate, extruded material exit velocity, and σ evolution. This slip condition enabled the predicted projectile die transit time to be in very close agreement with the transit time measured by PDV while, allowing the projectile to deform in a manner that closely approximates the actual extrusion conditions. The high-strength stainless steel die performed as a rigid structure because of its high yield strength and was modeled as an elastic-perfectly plastic material.

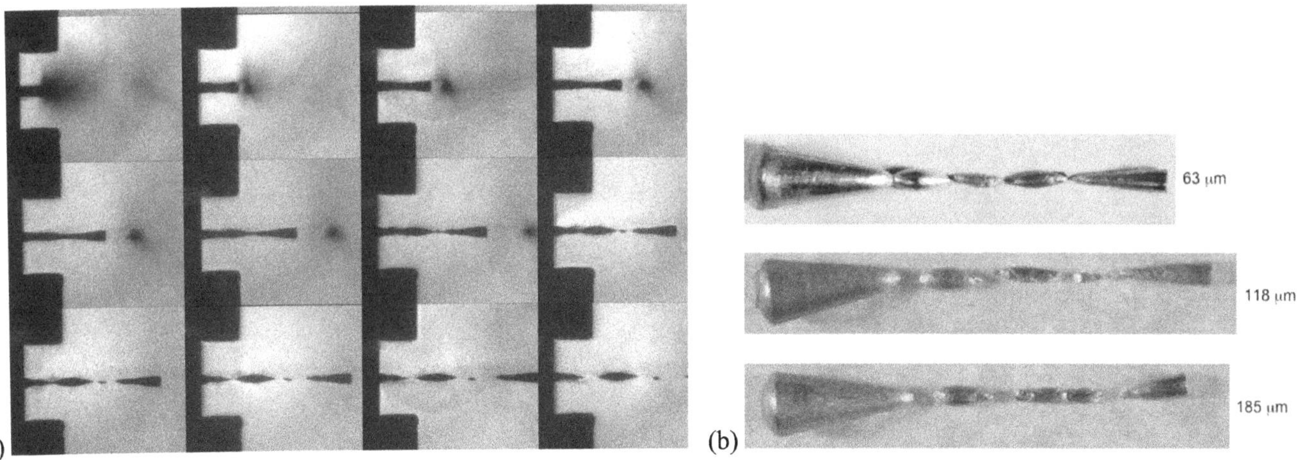

Fig. 2. High-speed imaging (a) showing the deformation topology of an extruded Cu sphere which entered the die at ~400 m/s. The images advance in time from the upper left to lower right at 4 μs intervals. The propensity for shear, strain, and localizations in Cu is seen (b) to increase with increasing grain size thereby reducing the overall dynamic ductility in the recovered samples [5]. (Images courtesy G. T. Gray, III.)

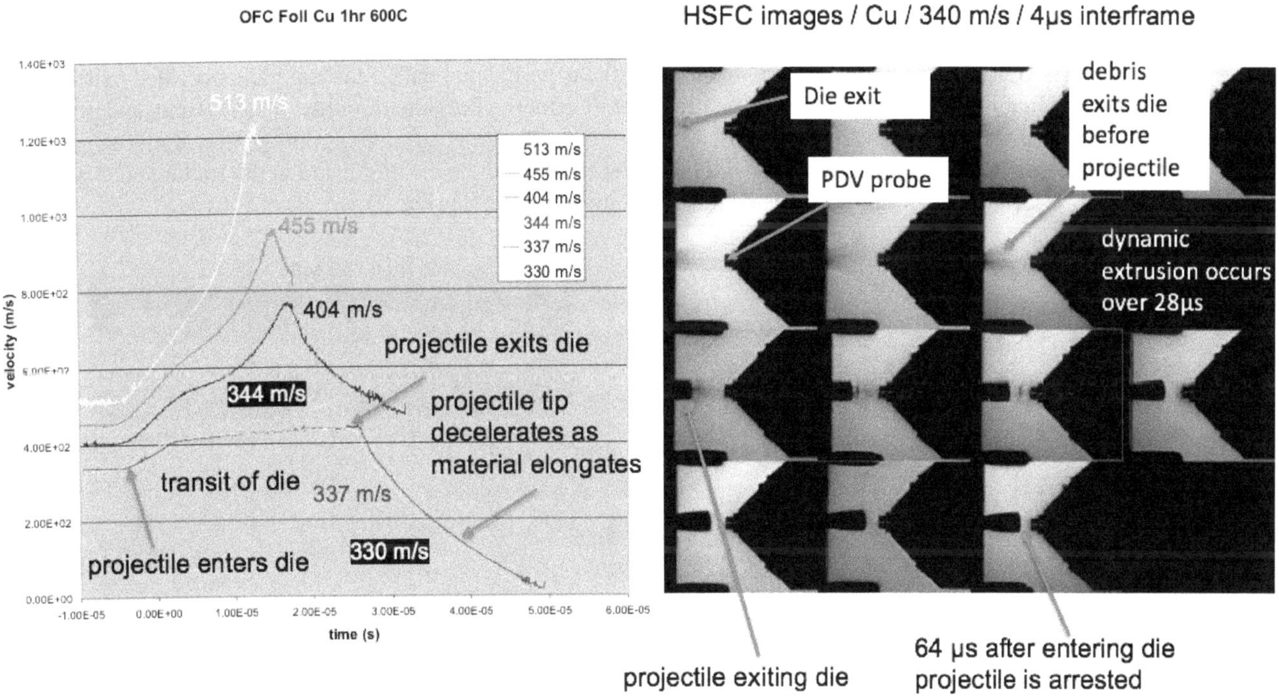

Fig. 3. Cu PDV data (left) and high-speed imaging (right). The Cu projectile was arrested in the die at an initial velocity of 340 m/s. (Images and PDV data courtesy C. P. Trujillo.)

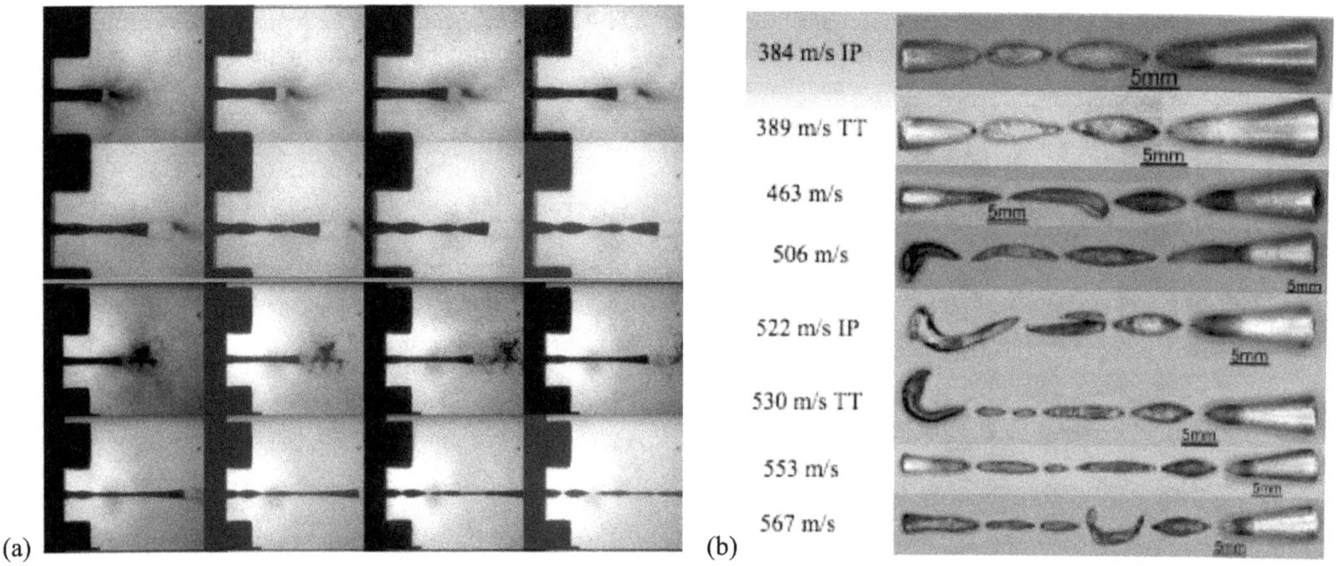

Fig. 4. (a) Ta sphere extrusion at ~390 m/s (top) with as-received material in the TT orientation (4 μs/frame) and at ~530 m/s (bottom) with as-received material in the TT orientation (5 μs/frame). The images advance in time from the upper left to lower right, and a red line separates images of the two expriments. (Images courtesy G. T. Gray, III, and C. P. Trujillo.) (b) Topology comparison of recovered extruded Ta spheres for several initial textures and velocities. Material instabilities increased with increasing velocity. (Images courtesy F. Cao.)

4. Computational results

Hydrocode output variables were used to provide comparisons with the experimental data. Material plots provided extrusion trajectories and shape(s), and Lagrangian tracers provided the extrusion exit velocity. Pressure response, plastic strain, and σ were extracted from the computational cell-centered variables.

In all the simulations and experiments, the extruded projectile material that exited the die originated in the leading edge of the sphere and experienced minimal rigid body rotations induced the by geometry of the die (Fig. 5).

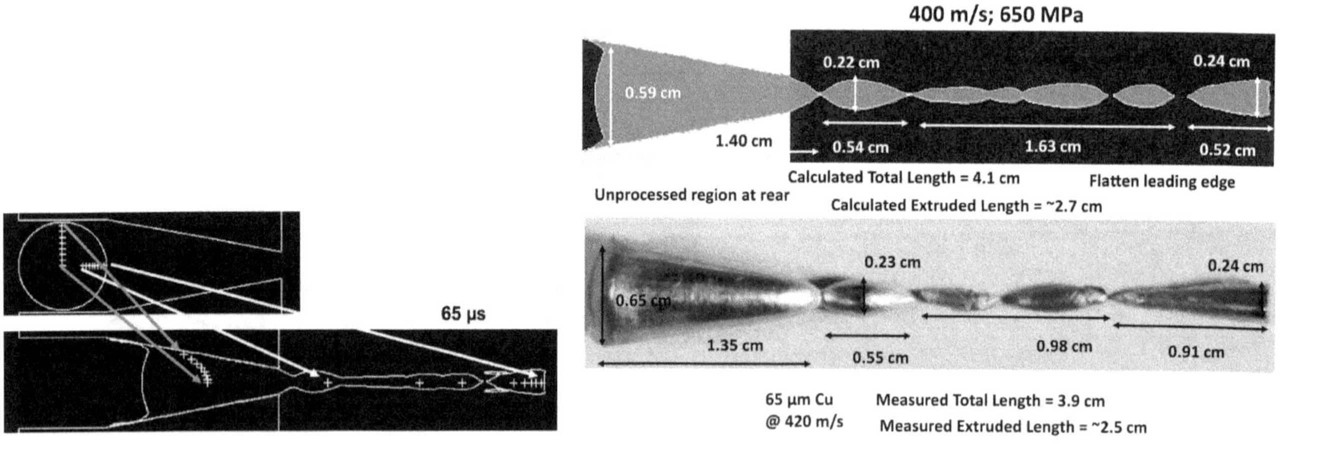

Fig. 5. Material trajectories and topologies predicted for Cu at 400 m/s. Only material at the leading edge of the spherical projectile extruded through the die.

Peak pressures of ~1200 MPa (~12 kbar) were realized in Cu (Fig. 6), well below pressures experienced in an explosively driven extrusion-like configuration. However, the plastic strains and strain rates realized during DTE of Cu are comparable to explosively driven systems. Strain rates on the order of 10^4 s^{-1} and plastic strains exceeding 300% were predicted and observed from postmortem analysis of the recovered Cu projectiles.

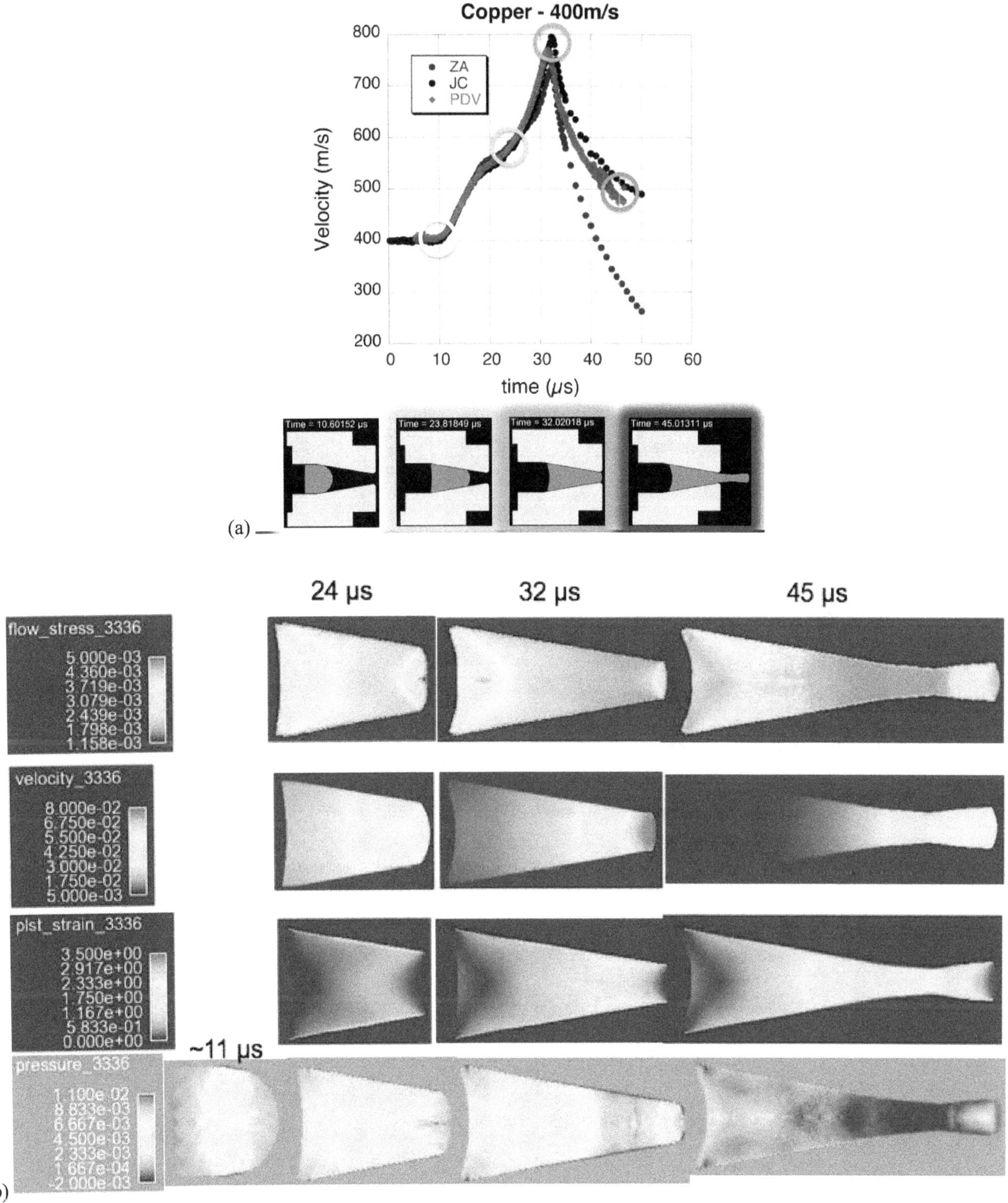

Fig. 6. Simulation using the J-C model for Cu at 400 m/s: (a) predictions for the projectile's leading-edge velocity using the Z-A and J-C models compared with PDV data and (b) projectile flow stress (Mbar), velocity (cm/μs), plastic strain, and pressure (Mbar) as the projectile enters, transits, and exits the die.

The hydrocode predicted that the leading edge of spheres would be accelerated significantly when exiting the die and that the exit velocity can be nearly double the entrance velocity (Fig. 6). Additionally, the exit velocity was found to be sensitive to the predicted σ magnitude and profile. Fig. 6 shows that the J-C model was in close agreement with the measured exit velocity, while Z-A underpredicted the exit velocity. The J-C model also predicted the deceleration of the projectile's leading edge quite well, while Z-A significantly overpredicted the deceleration. Finally, the use of J-C led to good agreement between the measured and predicted projectile die transit time. Fig. 7 compares predicted and actual Cu topologies for a low velocity extrusion at 340 m/s. In both cases, the bullet-shaped projectile did not fully extrude but was arrested in the die. The J-C model gave the best agreement with the measured topology and PDV data, while Z-A predicted less plastic deformation. Electron backscatter diffraction (EBSD) results indicated severe plastic deformation in the necking region (region 3 in Fig. 7). The J-C simulation predicted plastic strains exceeding 150%. The data suggests that J-C slightly underpredicted the σ, leading to a slightly higher exit velocity, while Z-A overpredicted σ which resulted in less deformation and more deceleration in the projectile's leading edge when exiting the die.

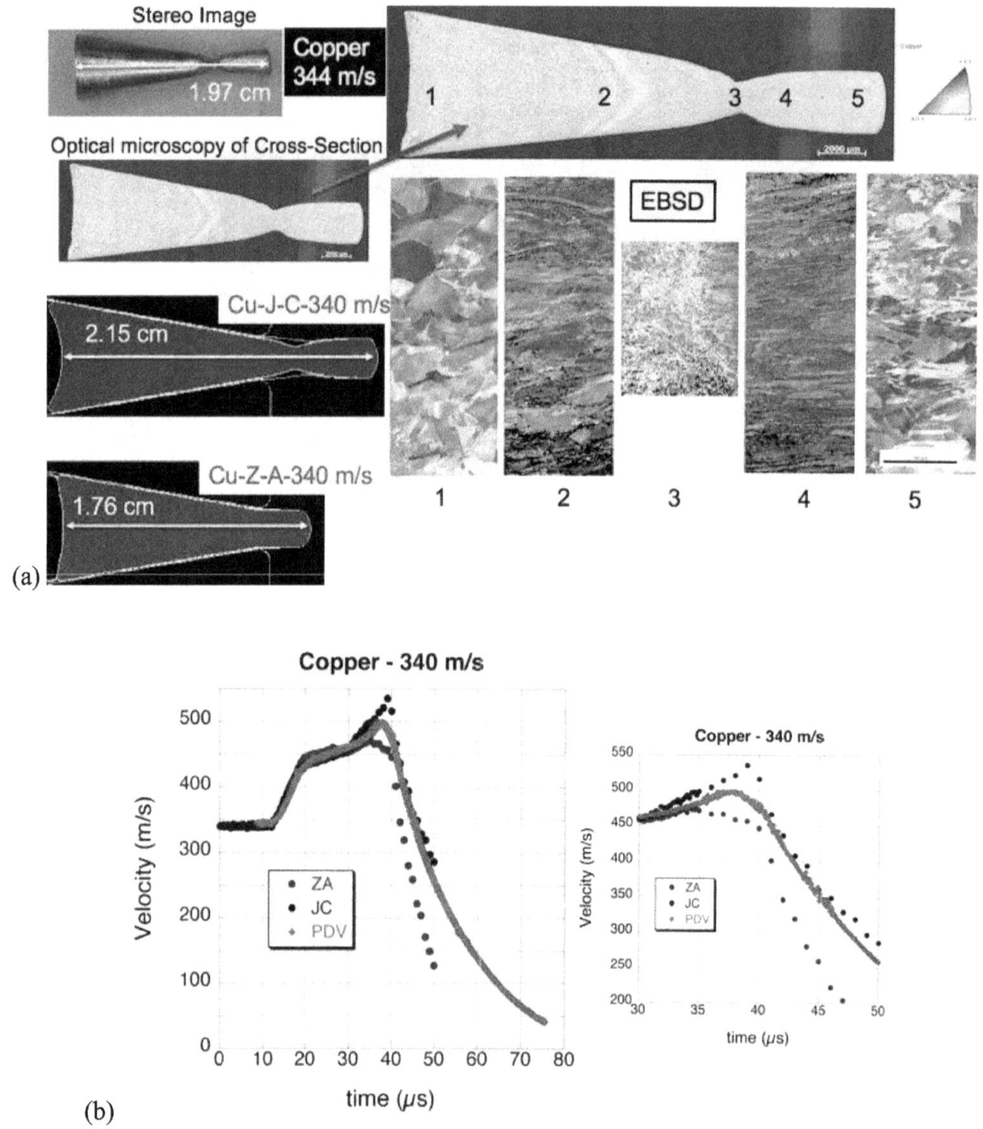

Fig. 7. Comparisons between simulations using J-C and Z-A and the data for Cu DTE at 330 m/s: (a) final topologies and grain texture and (b) leading-edge velocities. EBSD analysis of a recovered Cu bullet extruded at 340 m/s showed initial microstructure in region 1, elongated grain development with increasing plastic deformation along the extrusion direction in regions 2 and 4, significant plastic deformation in region 3, and orientations similar to the initial microstructure with minimal plastic deformation in region 5.

Ta spheres were launched with TT and IP texture orientations (see Fig. 1). The TT orientation is referred to as the "strong" direction while, the IP orientation is referred to as the "weak" direction. The TT is defined as "strong" because the σ magnitude in this direction is significantly higher than the value in the IP direction. In a previous unpublished study, anisotropic plasticity simulations under a 2D axisymmetric approximation were performed to investigate the effect of the initial texture on the predicted extruded topologies. The modeling predicted little if any sensitivity to the initial texture for velocities of 384 and 553 m/s even though a sensitivity was observed at velocities below 400 m/s. In general, the predicted topologies and number of particles were in poor agreement with the experimental data (Fig. 8). However, as shown in Fig. 8, the MTS simulation was found to be in good agreement with the measured particulation time, particle length, and the velocity of the leading particle for the DTE at 384 m/s. Fig. 9 compares the σ distributions for all strength models at an extrusion velocity of 487 m/s. The distributions predicted by MTS and Z-A are similar (σ varies between 3–8 kbar) while, the J-C evolved σ varies between 3–4 kbar. Plastic strains in the range 500%–800% were calculated in Ta, while strains up to 800% were measured during the metallographic analyses. Currently, PDV data is not available for the DTE of Ta but will be collected in an ongoing test series. A preliminary temperature measurement of Ta at ~410 m/s with an infrared camera was found to be in qualitative agreement with the calculated temperature profile (Fig. 10). Simulation showed MTS predicted more σ variations along the extrusion direction that produce regions of necking and elongation that are in better agreement with the imaging data. The low σ predicted by J-C suppresses the development of plastic instabilities and promotes excessive elongation.

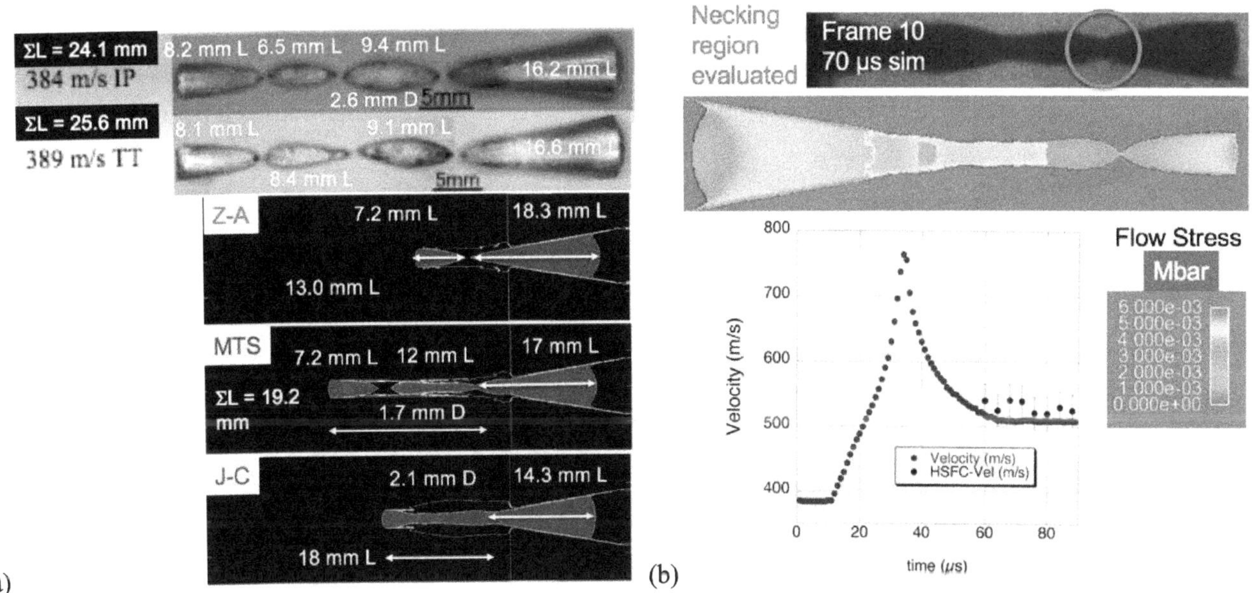

(a)

Fig. 8. (a) Topology comparisons between simulations and data at ~81 µs for Ta at 384 m/s: (b) data and MTS simulation showing flow stress distribution, late time leading-edge velocity comparison, and the predicted particulation time occurs slightly earlier than measured. The good agreement with the late-time velocity profile is a result of evolving the flow stress magnitude accurately in the necking region of interest.

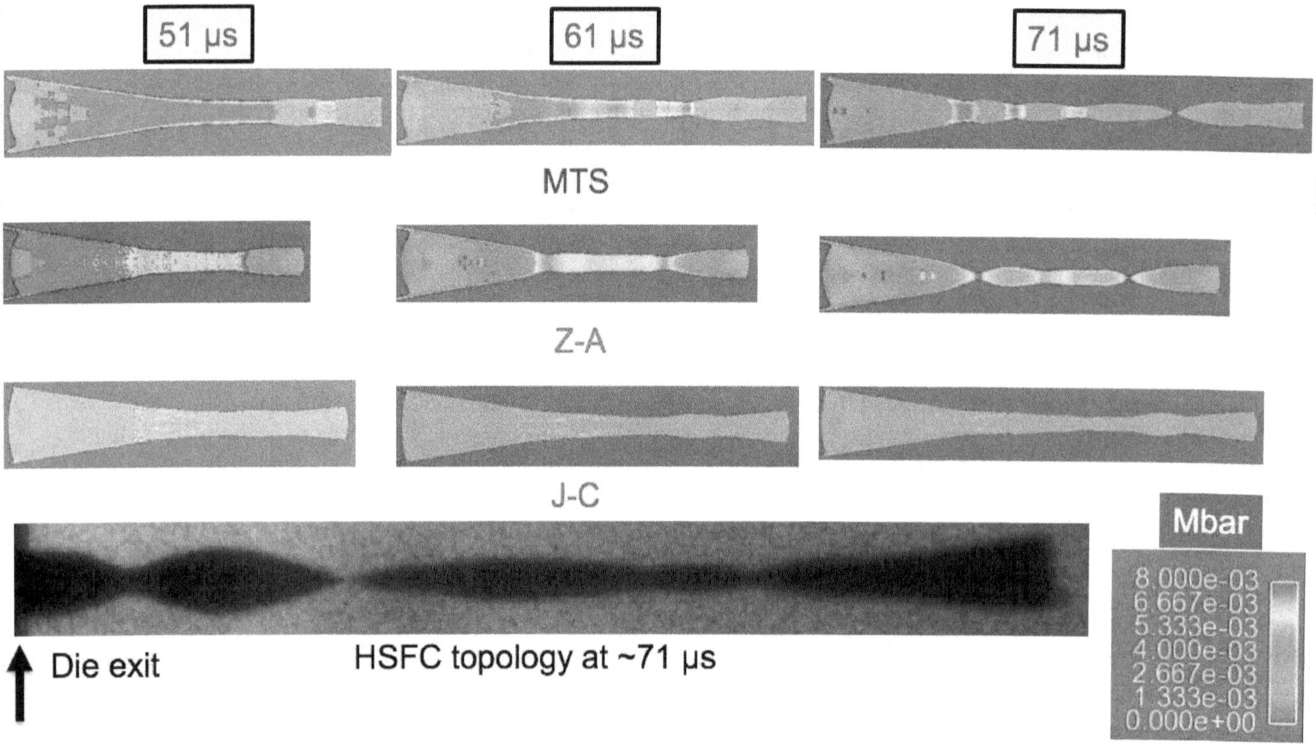

Fig. 9. Predicted flow stress profiles and topologies for Ta at ~487 m/s.

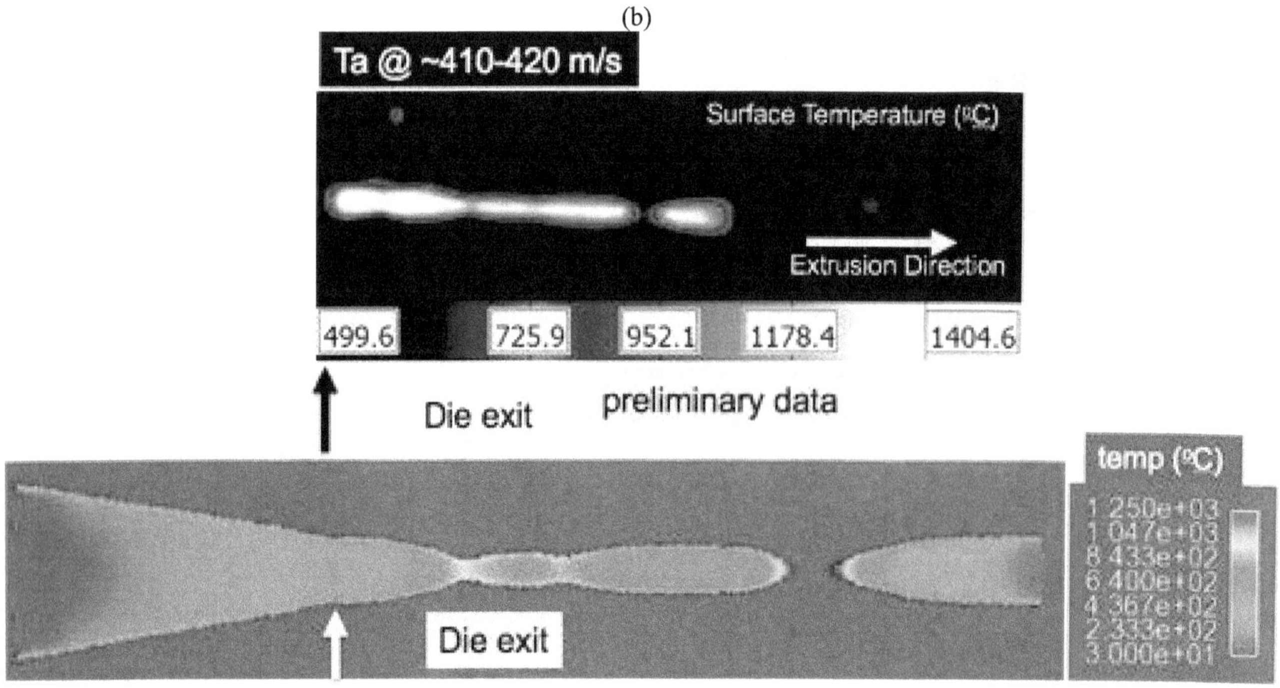

Fig. 10. Preliminary temperature (ºC) measurement of an extruded Ta projectile (~410 m/s) with Telops infrared camera. Preliminary results indicate calculated temperature ranges are in qualitative agreement with infrared camera data.

5. Findings, discussion, conclusions, and future studies

The DTE is a conceptually simple and cost-effective laboratory experiment for evaluating and comparing the deformation behavior of materials undergoing large plastic strains at elevated strain rates while controlling the initial conditions of the material prior to the dynamic tensile loading condition. The Eulerian hydrocode simulations have been able to provide insights into the material processing during DTE. Since the DTE experiment can provide relevant data, the experiment should be considered to evaluate strength models when applications to large-plastic-strain and high-strain-rate problems are anticipated. However, from a modeling perspective, approximating the projectile "slip" through the die is an important computational detail that can affect the predicted extrusion process characteristics and introduce additional complexity for the evaluation. The addition of PDV quantified the projectile die transient time (which typically ranges between ~15 and 20 μs) and greatly assisted the DTE modeling efforts by providing additional quantitative data to evaluate the strength models. Closely approximating the measured transit time in the simulation enables a more accurate assessment of the strength models. For Cu DTE, there was reasonable agreement between the predicted and measured total extruded length, characteristics of the leading particle's front surface contour, and characteristics of the rear contour of the sphere segment remaining in the die (Fig. 5). In general, for Cu and Ta, the qualitative agreement found between the predicted and measured particulation (material necking) as a function of launch velocity was reasonable. More particles and smaller diameter particles (stretched or elongated particles) were predicted and observed at higher launch velocities. Comparisons between actual and predicted extruded masses would be beneficial. Finally, the low-velocity extrusions of Cu modeled with J-C were found to be in good agreement with the data. The extruded particle topologies and number of particles (especially for Ta) were found to be in poor agreement with the data for all strength models. Additionally, the texture evolution and grain size sensitivity observed for Cu and the anisotropic effects observed in Ta at lower velocities were not addressed in the modelling assumptions. In general, Z-A was found to evolve a higher value of σ compared to MTS and J-C. Follow-on computational studies are required to investigate the effects of mesh resolution, possible three-dimensional effects in the deformation topologies induced by texture evolution and damage evolution (porosity and shear localization development by direct application of the TEPLA model [11]). Additional simulations of Ta at higher velocities and comparisons with other strength models like PTW [12] and SCG [13] would be beneficial. Further work is required to determine if the surface temperature measurements can be used as an additional quantitative metric.

Future studies applying the tensile stability criteria developed by Romero [14] and Walsh [15] to the DTE would be useful. Higher strain rates, a larger radius of extruded material, and a lower σ promote stability (reduce the propensity for necking during elongation). The high-speed imaging indicated that the extruded material appears to be unstable (necking observed) immediately after exiting the die. Additionally, the local σ at necking regions can be estimated from the local strain rate and necking region geometry (provided by the camera images) in a uniaxially stretching rod hydrocode simulation [16]. The estimated σ can be compared with the evolved σ from the integrated hydrocode simulation.

In summary, the DTE experiments have demonstrated that a variety of high-quality data can be obtained that can be used for qualitative and quantitative comparisons with simulations. We concluded that this experiment is very useful for hydrocode strength model evaluation when applications at large plastic strain and elevated strain rates are being considered.

Acknowledgements

The author would like to acknowledge the following individuals for contributions to this computational study: Sean P. Clancy, MESA code developer, George T. Gray, III, experimental concept development and data, C. P. Trujillo, experimental configuration and data reduction, Ellen K. Cerreta, F. Cao, L. B. Addessio, B. L. Henrie, metallography and experimental data reduction, and S. R. Chen, strength model parameters. Los Alamos National Laboratory is managed by Triad National Security, LLC for the U.S. Department of Energy's NNSA. The LANL reference number for this publication is LA-UR-18-31889.

References

[1] Holian KS, Mandell D, Adams TF, Addessio FL, Baumgardner JR, Mosso SJ. MESA: computer code for armor/anti-armor applications. Supercomputing World Conference, San Francisco, CA, Oct. 17-20,1997.

[2] Follansbee PS, Kochs UF. A constitutive description of copper based on the use of the mechanical threshold stress as an internal state variable. *Acta Metall.* 1988; **36**(1): 81-93.

[3] Zerilli, FJ, and Armstrong, RW. J. Appl. Phys., 61(5) 1 March 1987, pg. 1816.

[4] Johnson GR, Cook WH. A constitutive model and data for metals subjected to large strains, high strain rates, and temperatures. *Proceedings of the seventh International Symposium on Ballistics*. The Hague, The Netherlands, 541-548 (2003).

[5] Gray III GT, Maudlin PJ, Burkett MW, Cerreta EK, Yablinsky CA, Henrie, BL, Trujillo, CP. Influence of shock prestraining and grain size of the dynamic tensile extrusion response of copper: experiments and modeling. 14th American Physical Society Topical Conference on Shock Compression of Condensed Matter, Baltimore, MD, 2005.
[6] Cao F, Cerreta, EK, Trujillo CP, Gray III GT. Dynamic tensile extrusion of tantalum, *Acta Metall.* 56 (2008) 5804-5817.
[7] Chen, SR, Los Alamos National Laboratory, internal communication (2018).
[8] Chen, SR and Gray, GT, "Constitutive Behavior of Tantalum and Tantalum-Tungsten Alloys," Metall. and Matl. Trans. A Vol. **27A** 2994 (1996).
[9] Holian KS (Ed.), T-4 Handbook of material databases, vol. 1c: equation of state, Los Alamos National Laboratory, Report LA-UR-10160-MS, Los Alamos, NM 1984.
[10] Clancy SP, Burkett MW, Maudlin PJ. Modeling anisotropic plasticity: eulerian hydrocode applications of high strain-rate deformation processes. J. Phys. IV France 7 (1997). DYMAT.
[11] Maudlin PJ, Harstad EN, Mason TA, Zuo QH, Addessio FL. TEPLA: an elastoplasticity, damage, and failure approach for ductile materials. Los Alamos National Laboratory technical document (unpublished), 2003.
[12] Preston DL, Tonks DL, Wallace DC. Model of plastic deformation for extreme loading conditions. *J. Appl. Phys.* **93** (1) 211 (2003).
[13] Steinberg, DJ, Cochran, SG, and Guinan, MW, "A Constitutive Model for Metals Applicable at High=Strain Rate," *J. Appl. Phys.* **51** (3) 1498-1504 (1980).
[14] Romero LA. The instability of rapidly stretching plastic jets. *J. Appl. Phys.* **65**, 3006 (1989).
[15] Walsh JM. Plastic instability and particulation in stretching metal jets. *J. Appl. Phys.* **56** (7) 1(1984).
[16] Karpp, RR, Hull, LM and Price, ML. A technique for estimating the strength of materials in stretching shaped-charge jets. Fifteenth International Symposium on Ballistics, Jerusalem, Israel, May 21-24, 1995

Proceedings of the 2019 Hypervelocity Impact Symposium
HVIS2019
April 14-19, 2019, Destin, FL, USA
HVIS2019-072

Extended Abstract+

Numerical and Experimental Evaluations of a Glass-Epoxy Composite Material Under High Velocity Oblique Impacts

Christopher T. Key[a,*], C. Scott Alexander[b]

[a]*Applied Physical Sciences Corp., 475 Bridge Street Suite 100, Groton CT, 06340, USA*
[b]*Sandia National Laboratories, PO Box 5800, Albuquerque NM, 87185, USA*

1. Introduction

Composite materials are widely used in both the defense and commercial sectors for various applications. Many of these applications present the possibility of the composite material being subjected to high-velocity impacts, where the strain rates can be on the order of 10^6 s^{-1} or more and the pressures are often in the tens of GPa. The work presented in this paper expands upon previous work for the same materials where the impacts were normal to the composite material [1] by considering the non-ideal condition of oblique impacts that are more likely to occur in regular use.

In order to capture the appropriate physics responses for these high-velocity and large strain impact problems, the multi-constituent composite model (MCM) [2] within the shock physics hydrocode CTH [3] was used to model the impact event and subsequent response of the composite material. The MCM model was chosen as it provides an advanced damage model which allows for the prediction and degradation of the composite material response according to the mode of damage achieved. This paper details both the oblique impact experimental testing and the associated numerical modeling efforts which evaluate the oblique impact response and damage evolution of an S-2 glass / SC15 epoxy woven fabric composite material.

2. MCM Model Overview

The multiple constituent model (MCM) provides the micro-scale constituent (i.e. fiber and matrix) level responses of a composite microstructure without the need to model the micromechanical details explicitly. This means that one can model macroscopic objects and still capture the constituent level details without explicitly resolving the microscopic details.

In order to accomplish this, the MCM model leverages micromechanics simulations run *a priori* to extract volume averaged constituent stress and strain fields. Specifically, the MCM model assumes that at each continuum point within a macro-scale structure, there exists a representative volume element (RVE) or repeating microstructure. Equation (1) is the fundamental relationship for the MCM model from which the volume averaged constituent strain fields are extracted from the individual constituent volume fractions (ϕ), composite strain fields (ε) and a concentration tensor (A) which is a function of the composite and constituent stiffness matrices.

$$\{\varepsilon_{fiber}\} = \left(\phi_{fiber}[I] - \phi_{resin}[A]\right)^{-1}(\{\varepsilon\}) \tag{1}$$

+ Full paper can found at https://www.sciencedirect.com/journal/international-journal-of-impact-engineering/special-issue/10MTGP5W4VJ

* Corresponding author. Tel.: +1-307-760-8799; fax: +1-(860) 772-1876.
E-mail address: ckey@aphysci.com

The volume averaged constituent stress fields can similarly be determined using their respective linear elastic constitutive relationship.

It is important to note here that Equation (1) decomposes the composite strain field for a two constituent system (i.e. fiber and resin). Therefore, for the woven fabric microstructure where there are assumed to be three constituents (warp bundle, weft bundle and pure resin pockets), the decomposition relationships presented in Equation (1) must be applied successively to obtain the stress and strain fields for each of the 3 constituents in the microstructure. The reader is referenced to Hill [4], Garnich and Hansen [5] and Key [6] for a detailed derivation and description of the MCM modeling approach.

Additionally, the damage and failure criterion within the MCM model treats each of the fiber bundles of the woven fabric microstructure as individual unidirectional composites such that the transversely isotropic failure criterion developed by Hashin [7] can be applied to the bundles individually. The detailed development of these criterion are provided in [8, 9] and not restated here in the interest of brevity. Finally, the MCM model contains an anisotropic coupling algorithm which accounts for the coupling between the pressure (EOS) and deviatoric stress components which is necessary in a conventional hydrocode formulation. For this work, a coupling approach developed and detailed by Lukyanov [10] was used.

3. Materials and Experimental Setup

The target materials for all testing performed in this study were comprised of S-2 glass / SC15 epoxy composite. Samples tested were all 6.35 mm (¼") thick and were manufactured with a [0/90] layup. All material stiffness, strength and equation of state parameters used for this material are detailed in [1]. The projectiles were 6.35 mm (¼") diameter spheres composed of soft carbon steel (type 1018).

The high velocity oblique impact experiments for this work were performed at Sandia National Laboratories using single and two stage light gas guns. The spheres were launched at velocities from 300 to 1000 m/s with obliquity angles of 22.5°, 45°, 56.2° and 67.5° as defined in Figure 1.

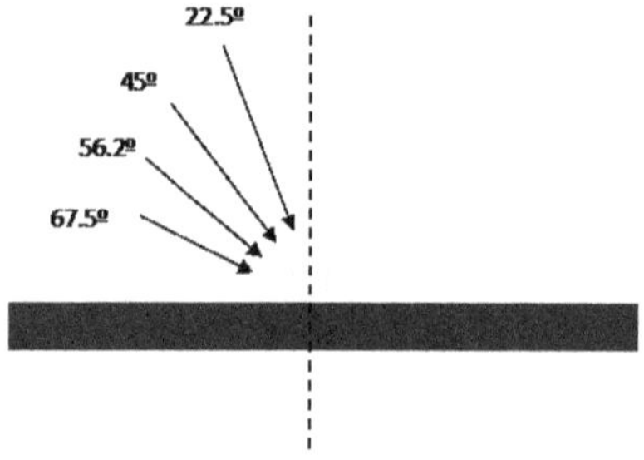

Figure 1. Obliquity Angles for Experimental Testing.

The sphere passes through a Magnetic Velocity Induction System (MAVIS), an external magnetic field and a pair of pickup coils, prior to impact to determine the impact velocity. The impact event was also monitored with three Phantom V2511 high speed video cameras running at 200,000 frames per second.

4. Results

4.1 Residual Velocity

The MCM predicted residual velocity and the residual velocity measured during the experimental testing are shown below in Table 1, where green represents conditions where the projectile was defeated and orange represents conditions where the projectile defeated the target plate. The simulations were conducted *a priori* and as a result there are minor differences between the simulated and tested impact velocities. Untested velocities where defeat could not be inferred are left uncolored.

Table 1. Simulation results for oblique impacts.

Impact Velocity (m/s)	22.5° Residual Velocity (m/s)	
	CTH – MCM Simulation	Experimental
300	DNP	-
345	-	DNP
400	DNP	DNP
455	-	61
500	130	-
517	-	220
600	260	-
700	370	-
800	475	-
900	575	-

Impact Velocity (m/s)	45° Residual Velocity (m/s)	
	CTH – MCM Simulation	Experimental
300	DNP	-
400	DNP	-
417	-	DNP
500	20	-
530	-	DNP
600	60	-
609	-	234
700	230	-
705	-	428
800	350	-
1000	525	-

Impact Velocity (m/s)	56.2° Residual Velocity (m/s)	
	CTH – MCM Simulation	Experimental
400	DNP	-
419	-	DNP
447	-	DNP
500	DNP	-
600	DNP	-
647	-	238
700	DNP	
777	-	465
800	50	-
900	250	-
1000	350	-

Impact Velocity (m/s)	67.5° Residual Velocity (m/s)	
	CTH – MCM Simulation	Experimental
500	DNP	-
593	-	DNP
600	DNP	-
700	DNP	-
717	-	DNP
777	-	279
800	DNP	-
855	-	438
900	DNP	-
1000	DNP	-

The results summarized in Table 1 show that the MCM model predictions are in good agreement with the experimental results for the 22.5° impact condition, where both penetrate/no-penetrate conditions are between 400 and 500 m/s. For this obliquity, the residual velocity for the MCM model was calculated to be 130 m/s at an impact velocity of 500 m/s while the experimental residual velocity was measured to be 220 m/s at an impact velocity of 517 m/s. For the 45° configuration, the MCM model predicts penetration at approximately 600 m/s while the experimental results indicate the penetrate/no-penetrate impact condition to be between 530 and 609 m/s. A calculation at 500 m/s impact velocity indicated penetration with a residual velocity of 20 m/s. However, under these conditions, penetration occurred slowly over a much longer time scale than other conditions making it unclear if this result clearly indicates a prediction of penetration. At obliquity angles greater than 45° an increasing angle results in a divergence of the MCM model predictions from the experimental results. For the 56.2° configuration the simulations predicted a penetrate/no-penetrate impact condition between 700 – 800 m/s, while the experimental data resulted in penetration occurring between 447 and 647 m/s. Finally, for the 67.5° configuration the MCM model simulation were run up to an impact velocity of 1000 m/s under which each simulation predicted that the projectile would not penetrate the panel and would instead deflect. However, under the experimental testing efforts a penetration event for this configuration was observed to occur between 717 and 777 m/s with a residual velocity of 279 m/s for the 777 m/s impact condition.

4.2 Damage Pattern and Extent

Damage is quantified for the 22.5° and 67.5° specimens in the following by measuring the horizontal extent of damage, referred to as width, and the vertical extent, measured above and below the impact point. Additionally, the bias ratio, or the ratio of the vertical extent of damage above to below the impact point is used.

Comparison of the predicted and measured damage patterns and extent for the 22.5° specimens at ~500 m/s are shown below in Figure 2. This image shows that the extent of damage is similar between the simulations and experiment. Specifically, the MCM model predicted a damage width of 6.35 cm (2.5") while the measured width from the experiment was 7.87 cm (3.1"). Similarly, the predicted damage above and below the impact point for the simulations was 3.0 cm (1.18") and 4.09 cm (1.61"), respectively; while the experimental testing resulted in damage extents of 3.63 cm (1.43") above the impact point and 4.42 cm (1.74") below the impact point. These measured damage extents correspond to bias ratio defined as the above impact extent to the below impact extent of 0.73 for the simulations and 0.82 for the experiment which are considered to be in good agreement. One point of note is that the damage pattern predicted by the MCM model is more of a "+" type pattern while the experimental backlit images reveal more of an oval pattern. As detailed in the normal impact results [1], this difference is likely due to the lack of an interstitial failure mode in the current MCM failure criteria and the inability to capture delamination planes in the Eulerian framework of CTH.

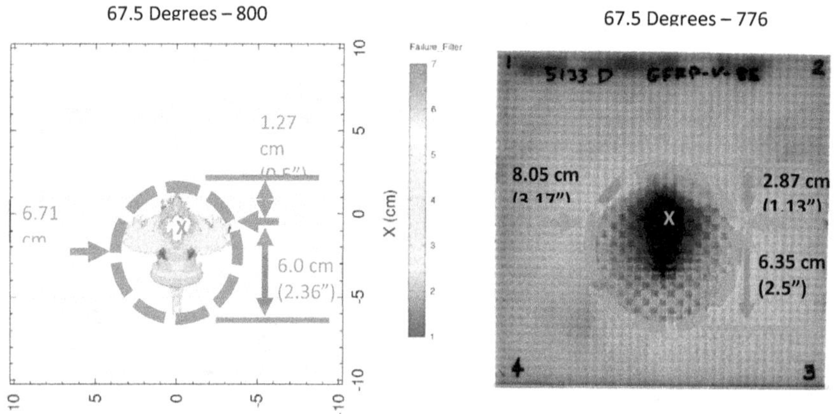

Figure 2. Predicted and experimentally observed damage pattern and extent for 22.5° obliquity at ~500 m/s impact.

For the 67.5° impact condition shown in Figure 3, the model predicted a damage width of 6.71 cm (2.64") with an extent of 1.27 cm (0.5") above the impact point and 6.0 cm (2.36") below the impact point (bias ratio = 0.21), while the experimental test resulted in a damage width of 8.05 cm (3.17") with a 2.87 cm (1.13") extent above the point of impact and 6.35 cm (2.5") below the point of impact (bias ratio = 0.45). Although the bias ratio is larger for the experimental testing, this is likely due to the fact that the simulations did not predict complete penetration of the panel at 800 m/s which would lead to a lesser damage extent on the exit face. Again, the model predictions result in a cross or "t" pattern which is different than the experimental results. However, it is noted that the damage pattern predicted by the model does correspond fairly well with the black regions of the test specimen photo corresponding to highly damaged material.

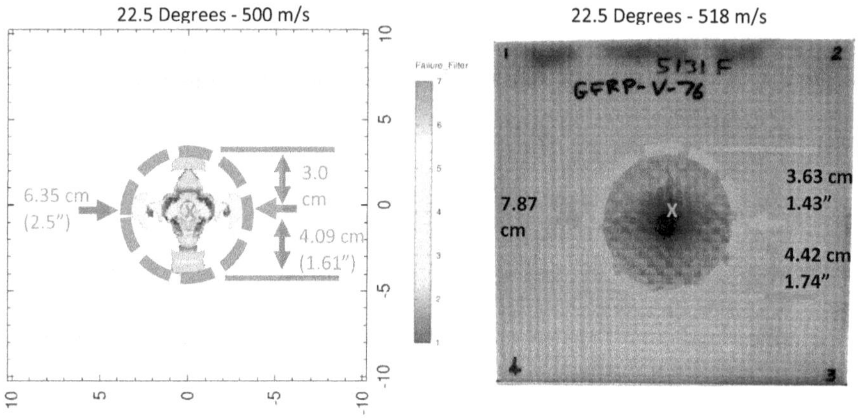

Figure 3. Predicted and experimentally observed damage patterns and extent for 67.5° obliquity at ~800 m/s impact.

5. Conclusions

Overall, the results shown in this paper demonstrate that the MCM model provides good correlation in residual velocity for obliquity angles up to 45°. The model also shows good agreement in damage extent with the experimental results throughout all angles of obliquity studied, with a limited set of comparisons shown herein. The results show that the MCM model predicted penetrate/no-penetrate level begins to diverge from the experimental results as the obliquity angle increases with divergence becoming apparent for angles over 45°. This divergence is believed to be a function of both the longitudinal equation of state parameters and the fiber dominated strength coefficients of the model which are exercised under the increased obliquity angles.

Acknowledgements

The views expressed in the article do not necessarily represent the views of the U.S. Department of Energy or the United States

Government. Sandia National Laboratories is a multimission laboratory managed and operated by National Technology & Engineering Solutions of Sandia, LLC, a wholly owned subsidiary of Honeywell International Inc., for the U.S. Department of Energy's National Nuclear Security Administration under contract DE-NA0003525.

References

[1] Key, C.T. Alexander, C.S., 2018. Experimental Testing and Numerical Modeling of Ballistic Impact on S-2 Glass/SC15 Composites, Journal of Dynamic Behavior of Materials, 4(3): 373-386.
[2] Schumacher, S.C., Key C.T. 2012. CTH Reference Manual: Composite Capabilities, Sandia National Laboratories Report SAND2012-7714.
[3] McGlaun, J.M., Thompson, S.L., Kmetyk, L.N. and M. G. Elrick, 1990. A brief description of the three- dimensional shock wave physics code CTH, Sandia National Laboratories report SAND89-0607.
[4] Hill, R. 1964. Theory of Mechanical Properties of Fibre-Strengthened Materials: I. Elastic Behavior, Journal of Mechanics and Physics of Solids 12, 199-212.
[5] Garnich, M.R., Hansen, A.C. 1997. A multicontinuum theory for thermal-elastic finite element analysis of composite materials, Journal of Composite Materials 31(1): 71-86.
[6] Key, C.T., Six R.W., Hansen, A.C., 2003. A three-constituent multicontinuum theory for woven fabric composite materials, Composites Science and Technology 63(13): 1857-1864.
[7] Hashin, Z. 1980. Failure criteria for unidirectional fiber composites, Journal of Applied Mechanics 47(6): 329-349.
[8] Key, C.T., Schumacher, S.C., Alexander, C.S., 2015. Evaluation of a strain based failure criterion for the multi-constituent composite model under shock loading, DYMAT 2015 Conference on Mechanical and Physical Behaviour of Materials under Dynamic Loading, Lugano, Switzerland, 7-11 September.
[9] Mayes J.S., Hansen, A.C., 2004. Composite laminate failure analysis using multicontinuum theory, Composites Science and Technology 64(3): 379-394.
[10] Lukyanov, A.A., 2008. Constitutive behavior of anisotropic materials under shock loading, International Journal of Plasticity 24: 140-176.

Proceedings of the 2019 Hypervelocity Impact Symposium
HVIS2019
April 14-19, 2019, Destin, FL, USA
HVIS2019-083

Thin film graded density impactors for high rate off-Hugoniot loading: Application to Ta strength

J. L. Brown, D. P. Adams, C. S. Alexander, J. L. Wise, W. D. Reinhart

Sandia National Laboratories, Albuquerque, New Mexico, 87185, USA

Abstract

Graded density impactors (GDIs) have long been of interest to provide off-Hugoniot loading capabilities for impact systems. We describe a new technique which utilizes sputter deposition to produce an approximately 40 μm-thick film containing alternating layers of Al and Cu. The thicknesses of the respective layers are adjusted to give an effective density gradient through the film. The GDIs were launched into samples of interest with a 2-stage light gas gun, and the resulting shock-ramp-release velocity profiles were measured over timescales of ~10 ns with a new velocimetry probe. Results are shown for the direct impact of the film onto a LiF window, which allows for the dynamic characterization of the GDI, as well as from impact onto a thin (~40 μm) sputtered Ta sample backed by a LiF window. These measurements were coupled into mesoscale numerical simulations to infer the strength of Ta at the high rate (10^7 s^{-1}), and high pressure (1 MBar) conditions this unique capability provides. Initial results suggest this is a viable strength platform which fills a critical gap and aids in cross-platform comparisons with other high-pressure strength platforms.

Keywords: Graded density impactor; ramp compression, strength, tantalum, mesoscale simulation

1. Introduction

Graded density impactors (GDIs) have long been recognized as a viable technique to produce ramp compression loading using plate impact gun systems. The fundamental concept requires a gradient in the shock impedance, from low at the impact surface to high at the rear of the GDI, such that the impact results in an initial compressive wave with a rise time which can be orders of magnitude slower than the nearly instantaneous shock produced using the traditional impact of a single material. Initial work in this area utilized cold compression of a powder blanket with a smoothly varying distribution of materials [1]. More recently, hot pressing of various powder mixtures to produce tapes of varying impedance and then a subsequent pressing of a stack of these tapes has produced impressive results [2]. In both cases, the GDI thicknesses are several mm thick and result in loading wave rise times on the order of ~ 1 μs.

Ramp compression techniques offer the ability to perform dynamic loading without the entropy generated by shocks. This expands the thermodynamic space which can be explored to off-Hugoniot loading paths, where significantly cooler temperatures can be achieved. Additionally, the ability to tailor the loading rise-time is of interest to the dynamic strength community. Traditionally, quasi-static and Split-Hopkinson pressure bar data are used to parameterize dynamic strength models [3], which only provide constraints for low pressures (< 1 GPa) and moderate strain rates (up to ~1000 /s). Since most high-pressure high-rate applications contain a massive extrapolation of strength models beyond their calibration range, there has been interest in developing experimental techniques to validate strength models under more extreme conditions. Modern high strain rate strength experiments largely fall into two categories: perturbation growth and release wave. Ramp wave loading of sinusoidal perturbations using either

* Corresponding author. Tel.: +1-505-844-4705.
E-mail address: jlbrown@sandia.gov.

explosives [4] or lasers [5] exhibit Rayleigh-Taylor (RT) instability growth which is sensitive to the material strength. Alternatively, the elastic-plastic transition observed through velocity measurements of release profiles from a shocked state [6] or ramp compressed using magnetic loading [7] have also been used to estimate strength. Unfortunately, there is little overlap in the available data to provide comparisons between these techniques. Tantalum, for example, is one of the most well studied metals, but the data for the laser RT is for strain rates of $\sim 10^6 - 10^8$ /s [8] while the magnetic release data is for strain rates of $\sim 10^5$ /s [7]. Thus, there is a need to establish a capability to provide flexibility in the ramp loading rate for the purposes of cross-platform validation and to isolate physical effects related to rate, pressure and temperature. In this work, we present a GDI capability which can provide a well-defined shock-ramp-release loading path at a characteristic strain rate of 10^7 /s to provide comparisons with the magnetic loading experiments which follow a similar thermodynamic trajectory [7].

2. GDI fabrication and characterization

The key idea in this approach is to deposit hundreds of individual nanometer-scale layers of different materials to establish an effective density gradient through the thickness of the film. We chose a bimetal system of Al and Cu for this initial work because this system contains a relatively large difference in shock impedance, and these materials can be deposited at high rates while maintaining low residual film stresses (important for film adhesion and integrity). The GDI fabrication begins with the choice of a substrate. A low impedance material is required to provide a release from the peak state, and we chose Al because it is easily diamond machined to provide an optimal coating surface. The substrates were 4 mm thick and 12.7 mm in diameter and a custom platen was designed such that tens of GDIs could be produced in a single deposition run. The deposition begins with a 25 nm thick layer of Ti for improved substrate adhesion along with a 35 nm thick layer of Ni for use as a diffusion barrier. It is desirable to have a thick high impedance base layer to provide a steady peak state in the loading condition, so 5.5 μm of Cu was then deposited on the diffusion barrier for this purpose. Next, the key feature of the GDI was applied: alternating layers of Cu and Al were deposited where each layer thickness was varied to give an effective gradient in density. This multilayer portion of the GDI can be envisioned as having 100 bilayers, where a bilayer is defined to be 1 Al layer and 1 Cu layer. The bilayer thickness is fixed at 340 nm throughout the multilayer deposition, and the thickness of the Al and the Cu within each bilayer was adjusted to give a 1% change in molar composition from the previous bilayer. Thus, the 100 bilayers (200 individual layers) result in a 34 μm deposition representing an effective change in composition from pure Cu to pure Al as the film growth proceeds. The GDI was then finalized by depositing a 2 μm Al cap, which is intended to produce a steady post-shock state prior to the arrival of the ramp wave induced by the multilayer. A cryopumped, KDF 744i pulsed DC magnetron sputter deposition system having a base pressure of 8e-8 Torr was utilized for GDI film deposition.

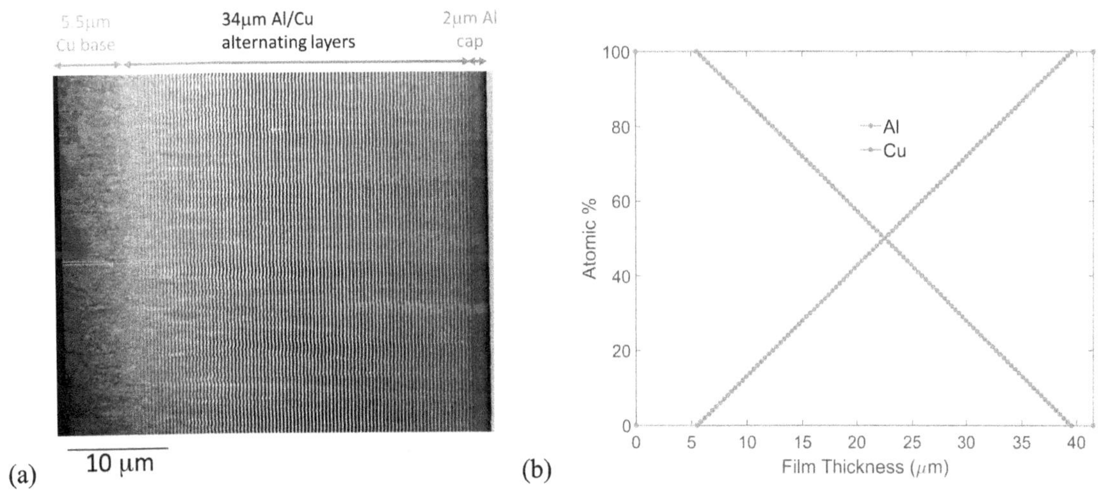

Fig. 1. (a) SEM cross section view of the Cu/Al GDI, where the Al layers are dark while the copper layers are light. Film composition is plotted versus thickness in (b). Note, the net composition of each bilayer is indicated in the multilayered section (from 5.5 to 39.5 μm).

To characterize the films, the GDIs were sectioned by focused Xe ion beam sputtering and imaged with a Scanning Electron Microscope (SEM). The resulting micrograph is shown in Fig. 1 (a). There is no detectible delamination and the film as well as the substrate-film interface appear to be free of contamination and particles. Individual layer thicknesses have been measured and are within 5%, confirming the composition profile shown in Fig. 1 (b). In addition, the measurements confirm the bilayer thickness (combined thickness of a single Al and single Cu layer) is constant and equal to 340 nm, so the sputter process did not drift out of calibration during the deposition. Additional images obtained from different substrates and locations suggest no differences and show excellent film thickness uniformity. Specifically, surface roughness measurements using a white light interferometer reveal a root mean square deviation of 0.11 μm and average peak to valley of 0.5 μm. These remarkably low values are comparable to the surface roughness of the diamond turned Al suggesting the film surface roughness is likely a result of conformal deposition onto the substrate.

3. Experimental technique

3.1. Experimental configuration

Experiments were performed using the two-stage light gas gun at the Shock Thermodynamics Applied Research (STAR) facility at Sandia National Laboratories [9]. This gun launches 25.4 mm diameter projectiles at velocities of 2-7 km/s. The target configuration utilized a standard transmitted wave plate impact orientation, a cartoon of which is shown in Fig 2 (a). The 12.7 mm diameter GDIs were glued into the projectiles such that ~ 1 mm was protruding from the surface and retroreflective tape was applied to the remaining surface of the projectile. A 19 mm diameter and 4 mm thick [100] LiF window was coated with a sample material of interest and potted in an aluminum target plate. The sample materials were deposited directly onto the window and consisted of either 5 μm of Al or 41 μm of Ta. The choice of these materials and thicknesses is discussed further in Section 5.

(a) 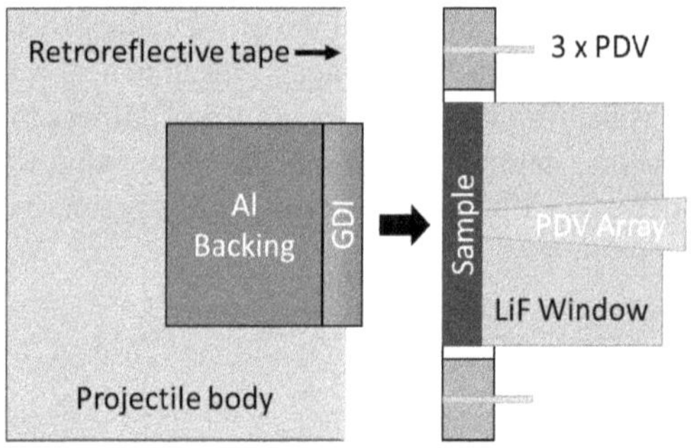

(b) 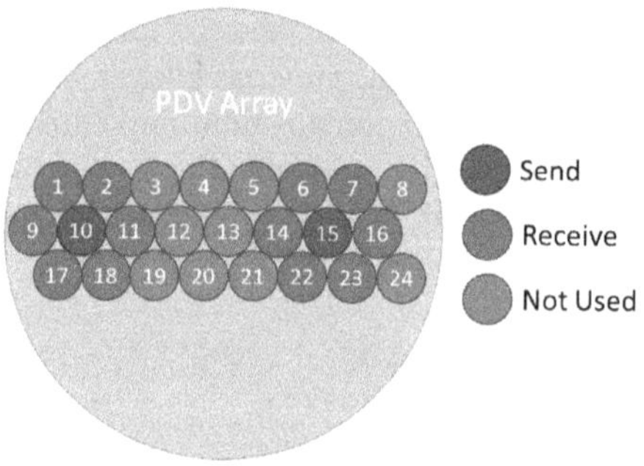

Fig. 2. (a) Plate impact experiment configuration. The GDI impacts a sample of interest (either 5 μm of Al or 41 μm of Ta) and PDV is used to measure both the impact velocity and the velocity of the sample/LiF window interface. The PDV array shown in (b) was used to measure the velocity of 10 distinct spots to provide the statistics necessary to improve the precision in the velocity.

3.2. Velocimetry diagnostic

Photonic Doppler Velocimetry (PDV) [10] was utilized during testing. Three standard collimating PDV probes (AC photonics) were located 120 degrees apart along a 21 mm diameter. These probes were used to accurately measure the velocity of the projectile by using 1550 nm laser light returned from the retroreflective tape. The primary measurement of interest is the velocity of the sample/LiF interface. The challenge in this measurement is the timescale associated with the thin film GDIs. As will be shown in Sections 4 and 5, the timescale of the ramp generated with these films is on the order of 5 ns. This is comparable to dynamic experiments performed on laser platforms [11, 12] where, traditionally, a line-imaging velocity interferometer using a streak camera is used to obtain the necessary temporal resolution [13]. As this type of line-imaging system was not available for these experiments, we pursued PDV measurements for this purpose. We utilized a new probe obtained from Berkshire photonics which incorporates a tight packing of an 8x3 array of SMF-28 fibers into a 2.1 mm diameter steel tube. The fiber core diameters are 9 μm while the outer cladding diameter is 125 μm, giving a spacing of ~1 mm across the probes. The philosophy with this probe was to make as many redundant measurements as possible and use statistical averaging to provide the required precision. Due to limitations in the availability of oscilloscope channels, only 12 PDV measurements were taken; a schematic of the arrangement is shown in Fig. 2 (b). The probe was glued directly to the back of the LiF window, and probes numbered 10 and 15 were used to illuminate the Al or Ta reflecting surface with 1550 nm laser light. The 6 probes surrounding the point of illumination were used to collect the reflected light before being coupled (3 GHz upshift) into the PDV system.

4. Numerical simulations

Predictive simulations of the experiments were performed using the ALEGRA hydrocode [14]. A significant advantage of this GDI approach is the use of full density layers of well-known materials. The GDI profile shown in Fig. 1 (a) was imported directly into ALEGRA to provide a mesoscale simulation capability. The simulations were configured using a 1D Lagrangian mesh in which 10 elements were constructed over the smallest layer thickness and then mass matched across the domain, resulting in cell sizes of 0.47 nm or 1.56 nm across the Al or Cu elements, respectively. The equations of state were modeled with Sesame tables 3700 and 3325 and the Steinberg-Cochran-Guinan form was used to model the strength [15]. The simulation results for the GDI impacting an Al sample at 6.2 km/s is shown in Fig. 3. A plot of the initial (time = 0) spatial configuration is shown in Fig. 3 (a) which illustrates the density differences across the GDI (discrete layers of Al and Cu) and into the Al sample and LiF window. A low pass filter was applied to give a sense for the average density gradient through the alternating layers of the GDI. The time history of the velocity at the impact (Al cap/Al sample) and window interfaces is shown in Fig. 3 (b) which contains the desired features of an initial shock with a steady hold, ramp to the peak state with a steady hold, and release back down to the close to the initial shock state. Given the nearly identical shock impedances between Al and LiF at these pressures, this configuration can be approximated as a direct impact on the LiF. We note the timescale for this entire experiment is only ~10 ns which is about 2 orders of magnitude faster than conventional GDI techniques [1, 2].

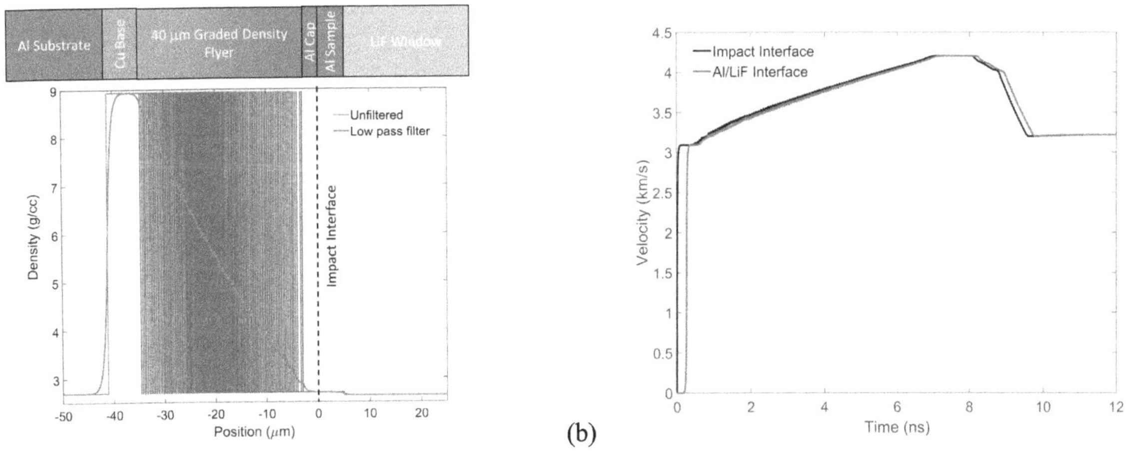

Fig. 3. (a) Initial (time = 0) configuration for the direct numerical simulation of a GDI impact. The low pass filter was applied as a visual aid representing the average density gradient through the GDI. The simulated velocities in (b) represent the predicted experimental observable. The Al is shocked to 80 GPa followed by a ramp to 120 GPa before release down to 80 GPa over ~10 ns.

5. Experimental results and analysis

5.1. Summary of experimental results

Table 1. Summary of the GDI experiments.

Shot Number	Sample Material	Impact velocity (km/s)	Impact Stress (GPa)	Peak Sample Stress (GPa)
GDI-1	5 μm Al	6.179	80	122
GDI-2	41 μm Ta	4.055	78	151
GDI-3	5 μm Al	4.608	53	80
GDI-4	41 μm Ta	3.010	52	99

Four experiments were conducted which are summarized in Table 1. The experiments were performed in pairs where a thin Al sample experiment was followed by a separate thick Ta experiment and the impact velocities were coordinated such that initial shock stresses were as close as possible. This pair of experiments was chosen such that the first (Al) provides a drive characterization while the second is used to infer the Ta strength. The drive characterization would ideally consist of an impact directly on the LiF window, but the thin Al coating was used to provide a reflecting surface and shield the measurement from any gas traveling ahead of the projectile.

Spectrograms along with 95% confidence intervals on the extracted velocities (shown as the black regions) obtained from 4 of the 12 fibers (labeled in Fig. 2 (b)) in the PDV array from experiment GDI-1 are shown in Fig. 4 (a). This subset was chosen to represent the best and the worst measurements and illustrates some qualitative trends common across the entire experimental series. First, the signal quality appears to be completely random and can vary drastically between different fiber locations. Signal dropouts also seem to be completely uncorrelated and occur at random times. However, by making 12 redundant measurements we consistently obtain several high-quality signals.

(a)
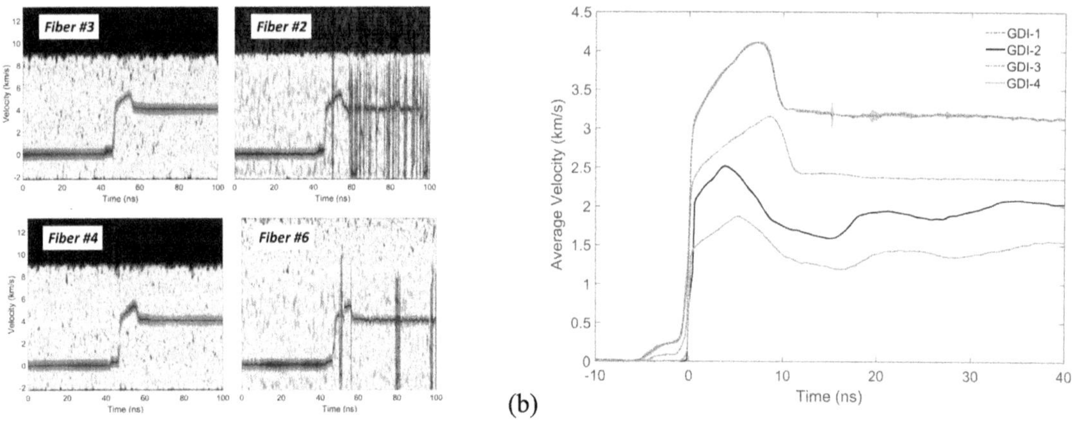

Fig. 4. The subset of the 12 measured PDV spectrograms in (a) represents the range of signal quality. The spectrum represents signal contrast and the black regions represent the 95% credible intervals on the velocity extraction. After aligning the initial shocks to time = 0, the estimated variance for each signal is used to produce (b), the weighted averaged velocity for each experiment. The shaded regions in (b) represent the estimated 95% credible intervals; except for GD1-1, these intervals are indistinguishable from the line thickness.

Rather than selecting the "best" signal, we chose to statistically average all 12 signals to provide the best estimate of the velocity. This was done by first shifting each signal such that the initial shocks overlap. The shock arrivals have a spread of ~5 ns in the raw data due to impactor tilt and slight differences in the fiber lengths, so using the shock as a fiducial was preferable to a detailed accounting of these sources of timing differences. This assumption is justified by the excellent spatial uniformity of the GDI described in Section 2. Next, a weighted average of the signals was taken where the weights were given by the reciprocal of the known variance in each measurement. In practice, this weighting results in regions of the signals with high noise (such as dropouts) contributing very little to the average, and so the highest quality signals naturally dominate the averaging. The averaged velocities after additionally correcting for the index of refraction of LiF [16] are given in Fig. 4 (b). The velocity detected before the shock arrival on GDI-1 and GDI-3 is believed to be gas blow-by traveling ahead of the projectile prior to impact and does significantly influence the interpretation of the data.

5.2. Drive determination

The Al sample experiments (GDI-1 and GDI-3) can be used to perform a dynamic characterization of the loading history provided by GDI. This was done by coupling the numerical simulations described in Section 2 to the measured experimental velocities. The idea, illustrated in Fig. 5 (a) is to use the simulations to describe the thermodynamic differences between the drive characterization impact and the Ta sample impact. The experiments were designed such that the initial impact pressures were nearly the same; as shown in Fig 5 (a), the state of each GDI is similar at early times. At later times, the wave interactions within the GDI reflect off the higher impedance Ta sample, causing higher pressures and a divergence between the two loading conditions. If we assume that the simulations accurately calculate this divergence, then a mapping can be constructed to describe the difference between the loading on the thin Al sample and the loading on the Ta sample. We use the transfer function methodology described in reference [17]. Briefly, this is a linear perturbation theory approach in which a transfer function, TF, can be constructed in the frequency domain using a fast Fourier transform (FFT) and the simulated velocities, u_{Al} and u_{Ta}:

$$TF(\omega) = \frac{FFT[u_{Al}(t)]}{FFT[u_{Ta}(t)]} \tag{1}$$

Once the TF has been constructed, it can then be applied (through the inverse of Eqn. 1) to the measured velocity in the Al experiment to generate an estimate of the time history of the velocity at the impact interface in the Ta experiment. This mapping is illustrated in Fig. 5 (b) where the orange and blue curves are used to generate the TF. Subsequently, the TF is applied to the black curve to generate the mapped purple profile.

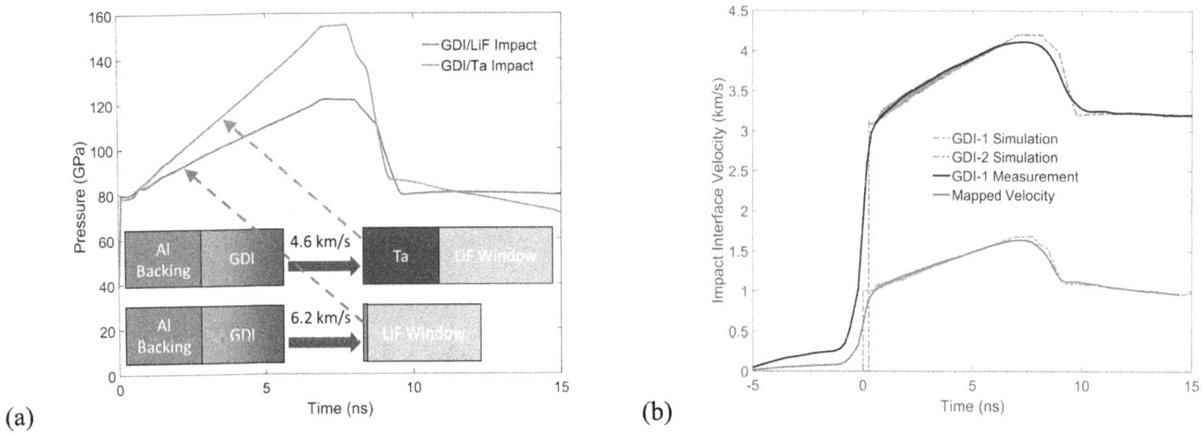

Fig. 5. Drive calculation for experiments GDI-1 and GDI-2. The simulated pressure-time histories at the impact interface in (a) describes the loading differences between the drive characterization and Ta sample experimental configurations. The differences between the GDI-1 simulation and experiment are linearly mapped in (b) to the simulated loading velocity in the GDI-2 simulation.

5.3. Wavespeed analysis

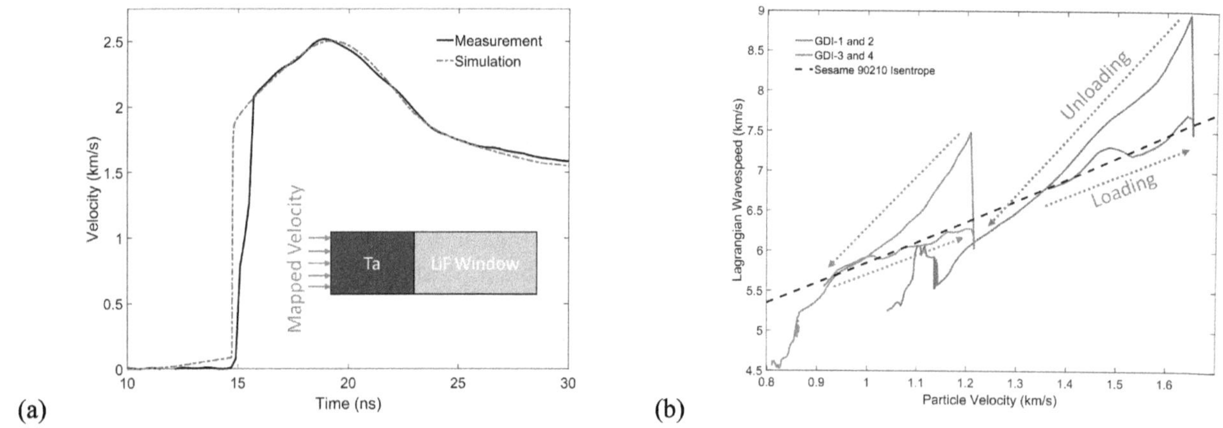

Fig. 6. A comparison between a best effort simulation and the measurement for GDI-2 is shown in (a). The subsequent wavespeed analysis gives the Lagrangian wave velocity profiles in (b) which can be used to extract properties related to the Ta strength.

Once the velocity-time history at the impact interface in the Ta sample experimental configuration is known, it can then be used as a boundary condition to drive 1D simulations to aid in the analysis of the Ta experiment. For example, consider Fig. 6 (a) in which the entire GDI in the simulations described in Section 4 is replaced by the mapped velocity which is applied directly as the input to the Ta sample. Using the Sesame 90210 equation of state (EOS) [18] coupled to the Preston-Tonks-Wallace strength model [3] and a quasi-elastic release model [19] results in excellent agreement between simulation and experiment. We note that a timeshift has been applied to the experimental measurement to obtain agreement with the initial ramp wave. We believe this is a reasonable approximation because the EOS has been well studied [7, 20] over these conditions; by forcing agreement in the initial ramp we are assuming the EOS is known. Once this level of agreement between simulation and experiment has been obtained, the methodology described in [7, 17] can be used to determine the wavespeed response shown in Fig. 6 (b). The loading wavespeed, by construction, is in good agreement with the estimated post-shock ramp compression path. The unloading portion of the wavespeed represents the transition from elastic to plastic deformation and agrees well with the magnetically driven ramp-release experiments performed under similar thermodynamic paths at the Z-machine [7].

5.4. Ta strength estimates

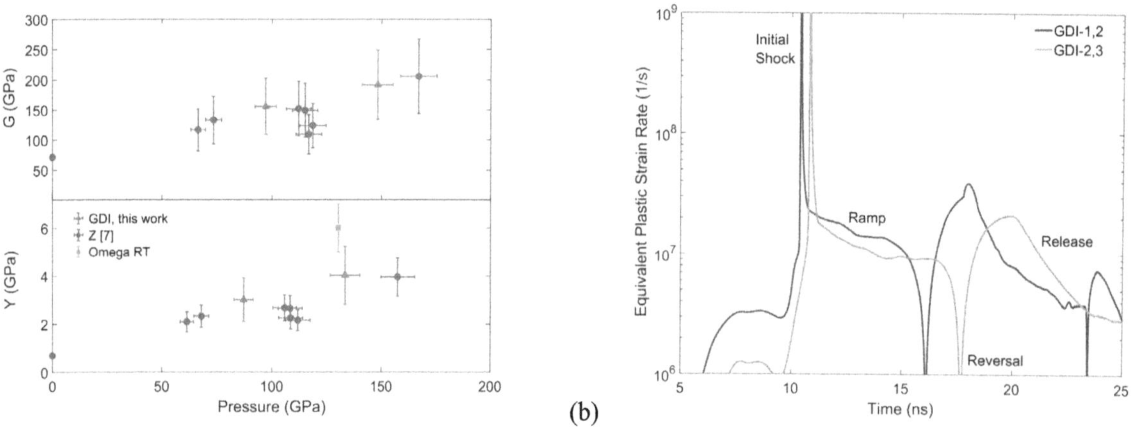

Fig. 7. The Ta strength properties in (a) suggest good agreement with this GDI work and the ramp-release experiments performed at the Z-machine. The simulated equivalent plastic strain rates halfway into the Ta sample in (b) suggest a characteristic rate of $\sim 10^7$ s^{-1} can be assigned to the GDI experiments, which is roughly two orders of magnitude faster than the Z experiments.

The wavespeed profiles in Fig. 6 (b) can be used to obtain estimates of the shear modulus, G, at peak compression as well as the average change in shear stress, Y, during the elastic-plastic unloading transition [7]. The shear modulus is calculated with the peak longitudinal elastic and bulk wavespeeds (height of the "triangle") while the shear stress is proportional to the area of the "triangle". The results of this analysis are shown in Fig. 7 (a) overlaid with an updated (use of a non-linear LiF index of refraction [21]) analysis of the experiments from [7]. As shown, the results are in excellent agreement with previous Z experiments. The shear modulus is consistent to within the error bars, while the strength is slightly higher. The higher strength is consistent with expectations based on both the higher loading rate [3] and the unique microstructure produced from the sputter deposition process [7], but these effects appear to be small relative to the error bars in the experiments. The GDI data do not compare as well with the strength inferred from Omega laser RT experiments [22]. This suggests the differences between Z and Omega could be related to platform dependencies and how strength is inferred from these experiments instead of rate effects.

A plot of the simulated equivalent plastic strain rate halfway into the Ta sample is given in Fig. 7 (b) for both experiments. The initial shock is followed by the ramp at a rate of ~ 10^7 s^{-1}. Upon reversal of the loading (peak of the velocity profile), the rate drops to zero and then increases again as the plastic strain begins to accumulate on release. The unloading rate is remarkably consistent with the loading rate, which is an ideal scenario in these types of experiments.

6. Conclusions and future work

A new GDI capability has been developed using a sputter deposition methodology to perform dynamic materials experiments at extreme loading rates. By depositing alternating layers of Al and Cu an effective density gradient smoothly transitioning between these two materials was established. The GDIs were used to load Ta samples, and experimental velocity measurements recorded with a new PDV probe were coupled to mesoscale numerical simulations to infer the strength of Ta. The dynamic loading path induced by the GDI resulted in initial shock states in the Ta of 50-80 GPa, a shockless ramp up to 100-150 GPa, and a release back to the initial shock pressure. The characteristic rates of the ramp and release were ~10^7 s^{-1}. The resulting shear modulus and shear stress associated with these experiments are consistent with expectations based on work on other platforms and suggests this is a viable methodology for performing strength experiments at these extreme conditions.

Future work will incorporate improvements in both the GDI and experimental configurations. The original intent of the GDI design was to incorporate steady holds after the initial shock and in the peak state. The experiments suggest the Cu base layer and Al cap thicknesses should be increased to achieve this objective. Further, moving to a higher impedance mismatch multilayer system (such as Al and Au) would provide higher peak pressures. Finally, improvements can be made to incorporate rigorous timing corrections. The diameter of the GDI is over 300 times larger than the thickness, so there is a massive amount of space to incorporate multiple measurements without the influence of edge waves. By multiplexing the PDV system, it should be possible to incorporate both the drive characterization configuration and the sample configuration into a single experiment. Three drive configuration measurements where we are currently making the projectile velocity measurements, for example, could provide a simultaneous estimate of the drive as well as absolute timing information.

Acknowledgements

The authors would like to thank C. Sobczak for Ta thin film deposition, J. Michael for ion sectioning, and B. Farfan for fielding the PDV. We also appreciate the insightful discussions with T. Mattsson. Sandia National Laboratories is a multimission laboratory managed and operated by National Technology & Engineering Solutions of Sandia, LLC, a wholly owned subsidiary of Honeywell International Inc., for the U.S. Department of Energy's National Nuclear Security Administration under contract DE-NA0003525. This paper describes objective technical results and analysis. Any subjective views or opinions that might be expressed in the paper do not necessarily represent the views of the U.S. Department of Energy or the United States Government.

References

[1] Barker, L. M., 1984, "High-pressure quasi-isentropic impact experiments," Proc. Shock Compression of Condensed Matter, J. R. Asay, R. A. Graham, and G. K. Straub, eds., Elsevier Science Publishers, p. 217.

[2] Yep, S. J., et al., 2013, "Fabrication and application of high impedance graded density impactors in light gas gun experiments," Rev. Sci. Instrum., 84(10), p. 103909.

[3] Preston, D. L., et al., 2003, "Model of plastic deformation for extreme loading conditions," J. Appl. Phys., 93(1), pp. 211-220.

[4] Barnes, J. F., et al., 1974, "Taylor instability in solids," J. Appl. Phys., 45(2), pp. 727-732.

[5] Park, H. S., et al., 2010, "Strong stabilization of the Rayleigh-Taylor instability by material strength at megabar pressures," Phys. Plasmas, 17, p. 056314.

[6] Asay, J. R., and Lipkin, J., 1978, "A self-consistent technique for estimating the dynamic yield strength of a shock-loaded material," J. Appl. Phys., 49(7), pp. 4242-4247.
[7] Brown, J. L., et al., 2014, "Flow strength of tantalum under ramp compression to 250 GPa," J. Appl. Phys., 115(4), p. 043530.
[8] Park, H. S., et al., 2012, "Experimental results of tantalum material strength at high pressure and high strain rate," Shock Compression of Condensed Matter - 2011, M. L. Elert, W. T. Buttler, J. P. Borg, J. L. Jordan, and T. J. Vogler, eds., Amer Inst Physics, Melville, p. 1371.
[9] Asay, J. R., 1981, "The Sandia National Laboratories Shock Thermodynamics Applied Research (STAR) Facility," Sandia National Laboratories.
[10] Strand, O. T., et al., 2006, "Compact system for high-speed velocimetry using heterodyne techniques," Rev. Sci. Instrum., 77(8).
[11] Smith, R. F., et al., 2011, "High strain-rate plastic flow in Al and Fe," J. Appl. Phys., 110(12), p. 123515.
[12] Swift, D. C., and Johnson, R. P., 2005, "Quasi-isentropic compression by ablative laser loading: Response of materials to dynamic loading on nanosecond time scales," Physical Review E, 71(6), p. 066401.
[13] Celliers, P. M., et al., 2004, "Line-imaging velocimeter for shock diagnostics at the OMEGA laser facility," Rev. Sci. Instrum., 75(11), pp. 4916-4929.
[14] Robinson, A. C., et al., 2008, "ALEGRA: An arbitrary Lagrangian-Eulerian multimaterial, multiphysics code," Proceedings of the 46th AIAA Aerospace Sciences Meeting and Exhibit, American Institute of Aeronautics and Astronautics, Inc, Reston, VA.
[15] Steinberg, D. J., et al., 1980, "A constitutive model for metals applicable at high-strain rate," J. Appl. Phys., 51(3), pp. 1498-1504.
[16] Rigg, P. A., et al., 2014, "Determining the refractive index of shocked [100] lithium fluoride to the limit of transmissibility," J. Appl. Phys., 116(3), p. 033515.
[17] Brown, J. L., et al., 2013, "Extracting strength from high pressure ramp-release experiments," J. Appl. Phys., 114(2), p. 223518.
[18] Greeff, C. W., et al., 2009, "THE COLD EQUATION OF STATE OF TANTALUM," Shock Compression of Condensed Matter - 2009, M. L. Elert, W. T. Buttler, M. D. Furnish, W. W. Anderson, and W. G. Proud, eds., Amer Inst Physics, Melville, pp. 681-684.
[19] Johnson, J. N., et al., 1992, "Quasielastic release in shock-compressed solids," J. Appl. Phys., 72(2), pp. 429-441.
[20] Davis, J.-P., et al., 2014, "Analysis of shockless dynamic compression data on solids to multi-megabar pressures: Application to tantalum," J. Appl. Phys., 116(20), p. 204903.
[21] Davis, J.-P., et al., 2016, "Mechanical and optical response of [100] lithium fluoride to multi-megabar dynamic pressures," J. Appl. Phys., 120(16), p. 165901.
[22] Park, H. S., et al., 2015, "Grain-Size-Independent Plastic Flow at Ultrahigh Pressures and Strain Rates," Physical Review Letters, 114(6), p. 065502.

Proceedings of the 2019 Hypervelocity Impact Symposium
HVIS2019
April 14-19, 2019, Destin, FL, USA
HVIS2019-102

Validating Ice Impacts Using Adaptive Smoothed Particle Hydrodynamics for Planetary Defense

Dawn Graninger*, Megan Bruck Syal, J. Michael Owen, Paul Miller

Lawrence Livremore National Laboratory, PO Box 808, Livermore CA 94551

Abstract

Understanding how a potentially hazardous object (PHO) responds to a kinetic impactor is of great interest to the planetary defense community. Target response depends upon the detailed material properties of the PHO, which may not be well constrained ahead of time. Hence, it is useful to explore a variety of target compositions for kinetic impact deflection. Previous validation efforts have focused primarily on understanding the behavior of common rocky materials, though PHOs are not exclusively composed of such material. Water ice is one material for which there has been only limited code validation against cratering experiments. It is known that comets consist of primarily icy material and some asteroids likely contain some amount of ice. Therefore, it is useful to understand the model sensitivities for ice in deflection simulations. Here we present Adaptive Smoothed Particle Hydrodynamics simulations of impacts into water ice by an aluminum projectile. We explore the sensitivities to the damage model within our code and find that the best-fit simulations of ice occur with a Weibull modulus of 12, though results can be obtained with values of the Weibull modulus near the published value of 9.59. This work demonstrates the efficacy of using an adaptive smoothed particle hydrodynamics code to simulate impacts into ice.

Keywords: Numerical Simulations, Adaptive Smoothed Particle Hydrodynamics, Cratering, Code Validation, Planetary Defense

Nomenclature

c_l	longitudinal sound speed
k	Weibull scale parameter
m	Weibull modulus
ASPH	Adaptive Smoothed Particle Hydrodynamics
BA	Benz-Asphaug Strain Algorithm
EOS	Equation of State
PPS	Pseudo-Plastic Strain Algorithm
SPH	Smoothed Particle Hydrodynamics
Greek symbols	
ϵ	strain

1. Introduction

Should an asteroid be heading towards Earth, a kinetic impactor is one of the most well-developed methods for deflection. A kinetic impactor works through impacting and transferring its momentum to a target. During this process, a crater will be formed, and ejecta will act as an additional momentum source, hopefully averting any disaster [1]. Testing the feasibility of this deflection

* Corresponding author. Tel.: +1-925-422-6210.
E-mail address: graninger1@llnl.gov.

technique is a central motivation for the Double Asteroid Redirection Test (DART), which will be the first full-scale demonstration of a kinetic impactor. In the lead up to this demonstration, it is necessary that we understand in detail one of the key unknowns to the success of a kinetic impactor strategy: formation of the crater.

Crater formation is something that has been studied in detail for many years [2-4 and references therein]. One interesting question is how does cratering occur in different materials, such as rock and ice. Experimental studies have looked into these questions in detail for both rock and water ice, however simulations are still needed to fully understand the processes occurring. Specifically, many cratering simulations focus on rocky material for the planetary science community; ice cratering simulations are much less common [5-8]. Ice is an important material to study for cratering in planetary defense as it is the primary constituent of comets and it is believed that some asteroids may contain ice as well.

In this work, we aim to validate our code `Spheral++` for hypervelocity ice impacts, based on the experimental data presented in Harriss & Burchell [9]. Section 2 presents a brief overview of `Spheral++`. Section 3 discusses the validation experiments and criteria. Sections 4 presents the Results and Discussion and Section 5 lists the Conclusions and Caveats of the work.

2. Numerical Methods

Simulations were performed using the Adaptive Smoothed Particle Hydrodynamics (ASPH) code `Spheral++` [10-12]. ASPH builds upon the Smoothed Particle Hydrodynamics (SPH) technique, generalizing the scalar smoothing length to a tensor formalism, allowing for non-spherical smoothing scales. The ASPH technique lends itself naturally to the capture of failure and dynamic fragmentation in the materials.

In `Spheral++`, the dynamic fragmentation model used is the Grady-Kipp formalism [13] which was generalized by [14]. This fracture and fragmentation model begins with the assumption that the materials of interest are brittle. As materials in nature are inherently imperfect, to realistically simulate a brittle material and its fracture, the simulated material is assumed to be flawed such that flaws can be activated, and damage can then be accumulated. The number density of flaws in the solid with a certain threshold activation strain is described by a Weibull distribution, which is based on weakest link theory, and is defined as:

$$n(\epsilon) = k\epsilon^{m} \tag{1}$$

where ϵ is the strain, k is the Weibull scale parameter, and m is the Weibull modulus [14, 15] The Weibull scale parameter and modulus are values obtained from experiments and are assumed to be characteristic of a given material. Seeding of these flaws is done randomly, using the volume of each particle within the simulation to determine the number of flaws assigned. This is particularly important in the simulations presented in this work as the mass resolution in the simulation is not constant. In order to activate a flaw, the strain must exceed the threshold set by Equation (1), at which point damage can accumulate. It is worth noting that once a strain exceeds the threshold strain prescribed by Equation (1), the particle within the simulation will not fail instantaneously. Damage will only be able to accumulate up to a certain point, with catastrophic failure exclusively occurring once every flaw within the particle has been activated.

To understand when these flaws will be activated, the stress-strain response of each particle must be calculated. Within `Spheral++`, two different models for calculating the strain are available. The first choice is the Benz-Asphaug strain (BA), based on the strain evolution prescribed in [16]. Here, the strain is given by:

$$\epsilon^{\alpha\beta} = E^{-1}\left(S^{\alpha\beta} - P\delta^{\alpha\beta}\right) \tag{2}$$

where $\epsilon^{\alpha\beta}$ is the strain tensor, E is the Young's modulus, $\boldsymbol{S^{\alpha\beta}}$ is the deviatoric stress tensor, P is the pressure and $\boldsymbol{\delta^{\alpha\beta}}$ is the Kronecker delta function. Equation (2) alone describes the strain in a perfectly elastic regime. To capture the plastic behavior intrinsic to brittle materials, a von Mises yielding criterion is applied such that the deviatoric stress tensor is limited based on the material yield strength [16]. The second model that can be used in `Spheral++` is known as Pseudo-Plastic Strain (PPS). The time evolution of the strain in this model is given by:

$$\frac{D\epsilon^{\alpha\beta}}{Dt} = \mu^{-1}\frac{DS^{\alpha\beta}}{Dt} \tag{3}$$

where μ is the shear modulus. Here we again are modeling the elastic region of the stress-strain curve, however we do not place any limits on the deviatoric stress tensor and fully follow its evolution in time. To create a "pseudo-plastic" regime, damage begins to accumulate once the yield strength is exceeded. If the stress exceeds the yield strength, the difference between the yield strength and the stress is logged as the plastic stress and is accessible for use in calculations should it be necessary.

The extent to which the damage evolved is based on the number of flaws activated in a given particle, but it is also determined by the material's strength. Within the simulations, the strength of the target is the Pressure-Dependent yield prescription from [17], given by:

$$Y_i = Y_0 + \frac{\mu_i P}{1 + \mu_i P / (Y_m - Y_0)} \tag{4}$$

where Y_i is the yield strength, Y_0 is shear strength at zero pressure, μ_i is the coefficient of internal friction, P is the pressure, and Y_m is the von Mises plastic limit. Once a particle is fully damaged, it is assumed that it is strength-less.

Combining the strength, stress-strain relationships, and the flaw distribution allows for the damage in a particular particle to be determined. The use of tensor notation in the damage allows for description of damage along a certain direction to be captured accurately and result in cracks opening along certain planes. The value for any one damage eigenvalue ranges between 0 and 1, with 0 being undamaged and 1 being completely damaged. Once the damage in each direction is calculated, it is evolved in time via:

$$\frac{D(\Delta_i^v)^{1/3}}{Dt} = \frac{0.4 c_l}{W_L h_i} \tag{5}$$

where Δ_i^v is the damage scalar, c_l is the longitudinal sound speed, W_L is the normalized extent of the ASPH kernel, and h_i is the smoothing scale. Damage is then applied to the particles in a pair-wise manner. In addition to not overestimating the damage artificially, the pair wise application of damage allows us to use a method called `useDamageGradient` which adds a further constraint on the damage application across nodes. When `useDamageGradient` is applied to the damage propagation, before determining the damage to apply across nodes, the gradient of the damage of the two nodes is calculated. If the gradients are aligned, damage is not applied between the nodes but if they are misaligned, damage is applied. This aims to aid in opening cracks in the simulation as those particles that are in alignment move together and those that are not, form a crack between them.

3. Validation Methodology: Experiments, Material Properties, and Validation Criteria

The experimental hypervelocity impacts that are our source of verification are presented in Harriss & Burchell [9]. Impacts were performed by firing a 1.5 mm aluminum sphere impactor with a gas gun into an optically clear cylinder of ice (r=10.5cm, h=8cm). The impactor struck the ice at an angle of 45°. Following the impacts, measurements of the crater diameter and depth were obtained. In these experiments, a range of impact velocities were investigated however for the purposes of this paper, we are looking only at the 4.96 km/s shot, as it is the most relevant to both planetary scale cratering events and planetary defense scenarios.

To minimize the computational resources required for our simulations, we approximate the ice cylinder using a hemisphere (r=10.5cm). Resolution of the simulation is based on the impactor, which has 22 particles across the 1.5 mm diameter of the aluminum impactor. Further, the resolution of the simulation is not constant; the impactor sets the smallest resolution element in the target. The resolution in the target then ratios out in a parabolic manner out to a radius of 7 cm, after which the resolution is constant. This results in a simulation with 2.2 million particles, requiring 612 processors to run. To determine at which point the simulation had completed, we waited until the damage stopped accumulating.

In an effort to minimize the parameter space, we relied on several published standards leaving only three free parameters in the simulation. The first parameter we fixed was the choice of the equation of state (EOS) for both ice and aluminum. We chose to use the Tillotson EOS [18], which is analytic in nature. The necessary parameters for the Tillotson EOS can be found in Table 1. Although the Tillotson EOS lacks detailed phase transition information, it is sufficient for the metrics of interest in this particular study: assessing crater size and damage extent at relatively modest impact velocities. The second piece of information that we fix is the strength of the materials. For aluminum, we use a standard Steinberg-Guinan Strength and for ice we use the pressure dependent strength as described in Section 2, with value obtained from [7]. The final material input that we fix is the Weibull scale parameter, k, in Equation (1) for ice. We set this value to be $\mathbf{1.4\times10^{32}}\text{cm}^{-3}$ from [6]. Previous validation results from Remington *et al.* [19] of the Nakamura & Fujiwara experiments [20] for aluminum and basalt sphere impacts have shown that damage is not necessary for aluminum impactors, thus the damage model parameters, such as the Weibull distribution [Equation (1)], are not explored for aluminum. Given these above fixed parameters, we are left to explore the strain algorithm, `useDamageGradient, and` the Weibull modulus; a brief discussion of each will be given in later sections.

Table 1: Tillotson EOS Parameters

	ρ_0 (g/cm³)	A (erg/ cm³)	B (erg/ cm³)	E_0 (erg/g)	$E_{liq\text{-}ice}$ (erg/g)	$E_{vap\text{-}ice}$ (erg/g)	a	b	α	β
Ice[a]	0.917	$\mathbf{9.47\times10^{10}}$	$\mathbf{9.47\times10^{10}}$	$\mathbf{1.00\times10^{11}}$	$\mathbf{7.73\times10^{9}}$	$\mathbf{3.04\times10^{10}}$	0.3	0.1	10.0	5.0
Aluminum	2.700	$\mathbf{7.52\times10^{11}}$	$\mathbf{6.50\times10^{11}}$	$\mathbf{5.00\times10^{10}}$	$\mathbf{2.70\times10^{10}}$	$\mathbf{1.41\times10^{11}}$	0.5	1.63	5.0	5.0

[a] Data for ice obtained from [6].

To compare against the results of the experiments presented in Harriss & Burchell [9], it is necessary to come up with a set of criteria which we deem most important. These criteria may be different for different validation efforts, but for the purposes of planetary defense we are interested in:

- Crater depth
- Crater width
- General fracture patterns – specifically, radial cracks from the crater to the edge of the target

These also coincide with the measurements listed in the work of [9] and therefore, in addition to relevance, allow us to compare quantitatively to the results of numerical simulations.

4. Results and Discussion

To benchmark the code against experimental results, we focus only on the 4.96 km/s shot into pure ice. We looked primarily at three different free parameters: the strain algorithm, the `useDamageGradient` method within `Spheral++`, and the Weibull modulus of ice. Initial analysis was done by eye, focusing on the extent of crack propagation, using VisIt [21] to visualize the post-impact targets. Further analysis was performed using more advanced techniques (discussed in detail below).

Beginning with the first free parameter, the strain algorithm, we ran two full-scale simulations: one for the BA strain and another for PPS using the default behavior for `useDamageGradient` (which is set to false by default) and a Weibull modulus of 10, which is near the published value for the Weibull modulus of ice of 9.59 from [22,23]. For the BA strain, we find that a crater is formed, however, it is significantly smaller in both width and depth than the published experiments. Furthermore, the resulting fracture pattern is not comparable to what is described in the experimental results. In the BA simulations, we find that cracks begin to form from the edges of the crater, but these cracks only extend about a centimeter and do not reach the edge of the target. For PPS, we find that the crater size is not quite right, but the radial crack pattern is a much better fit to the experiments. Hence, for the remaining validation efforts, PPS is the strain algorithm used.

This is different from previous validation results of basalt from [19] with `Spheral++`. There they find that the BA strain algorithm is able to reproduce the fragment population more accurately than PPS. There may be several reasons behind this. First, we are not looking at fragment distributions. It may be the fact that PPS is better at capturing general fracture behavior if the body is remaining competent and BA strain better captures complete dynamic fracture. This would not be surprising as BA strain was developed to do such a problem [16] The second, and more likely, explanation is that the validation efforts published in [19] are for basalt and we are studying ice. As the stress-strain response is related to the material properties, it is possible that BA strain is simply not the best strain algorithm for describing ice.

The second free parameter is the `useDamageGradient` option, a method that is unique to `Spheral++` to promote the opening of cracks. As mentioned above, the default behavior of the code is to not use this method for opening up cracks. However, as the investigation into strain algorithm revealed, the `useDamageGradient` method may be needed to allow for efficient crack opening to the edge of the target. Indeed, when simulations are run using this method, we find that the combination of PPS and the `useDamageGradient` method produces cracks that reach the edge and thus, we use the `useDamageGradient` method for the remaining of the validation efforts presented in this study. By applying this `useDamageGradient` method, we are attempting mimic crack opening and propagation dynamics that are not included in `Spheral++`. It should be noted that an investigation of the useDamageGradient method was done with the BA strain algorithm and no appreciable crack formation was observed, which agrees with the work of [19] that the useDamageGradient method coupled to BA strain does not result in large changes.

The final and most detailed analysis we did was studying the effect that the Weibull modulus had on the formation of the cracks and crater in the ice. In order to see how the Weibull modulus affected the crack propagation and the crater properties, we developed a new way of analyzing the simulation data that relied on some of the built-in functionalities of VisIt [21] Before presenting the results, we will walk through the analysis technique to see how the data for comparison to the experiments were generated and why each step in the analysis was performed.

The first step in the analysis is to convert the 3D simulation data to something that is akin to the images presented in [9] of the impacts. In order to do this, we need to convert the point particle visualization output by Spheral++ into a 2d image. We accomplish this by resampling the points onto a regular mesh in VisIt. To appropriately account for the varying mass resolution of the particles in the current simulations, it is necessary to use the "SPH Resample" function in VisIt, which will perform the resampling by the mass (or your choice of weighting parameter) correctly. We do the resampling over the trace of the damage tensor and, for the purposes of this paper, we will refer to this in later sections as the effective damage trace or simply as the damage. The trace of the damage tensor informs us of how "damaged" something may be and is similar to the eigenvalues of the damage tensor: a value of 0 signifies that the material is undamaged and a value of 3 (for the 3 dimensional nature of the presented simulations) means the material is completely damaged or destroyed. Since cratering and cracks are, in the simplest of terms, elements of damage, we feel that the use of damage as a metric to be appropriate.

The resampling allows us to visualize the damage in a more intuitive manner, but in order to quantitatively determine the crater depth and width, another step is necessary. The second step in the analysis is to perform a lineout of the data, either radially for the crater width or down from the surface for the crater depth. This allows us to sample the crater width and radius with a higher frequency and obtain an "average" width or depth. This analysis method was performed for all the figures presented in this paper and is the basis of our validation efforts.

In order to constrain the effect of the Weibull modulus on the damage, we looked at values for m between 10 and 12.5. Figures 1, 2, and 3 display the results for $m = 10$, 11, and 12, respectively. The first thing to note in Figures 1-3, are the upper panels which display the results for the crater width. Looking at the upper right panels of Figures 1 – 3, there are radial cracks which extend out to the edge of the targets. Differences arise however in Figures 1 – 3 when you look at the lineouts of the damage from the center of the target. Moving from $m = 10$ to $m = 12$, the lineouts all appear to have a knee, however the sharpness of the knee degrades as m increases. This is suggesting that increasing the Weibull modulus will work to "smooth out" the damage, allowing it to be more far-reaching and less sharply localized. Specifically looking at Figure 3, we can also note that the crater appears to the eye to be slightly more asymmetric, though we do not constrain that quantitatively. The crater depth is displayed in the bottom panels of Figure 1 – 3. Again, to the eye, it appears as though there are more cracks present in the simulations which use an $m = 12$ (Figure 3). This would support the idea that the increase in the Weibull modulus increases how far damage accrues. Again, we observe a knee-like shape in the lineouts.

From these results we can see that the Weibull modulus is important in determining the simulation results, but it is non-trivial to determine which simulation results are the most like the experiments. In order to do this, the meaning of damage in our simulations and the assumptions made about it must be taken into consideration. When damage is used as our criterion for determining the crater depth and width, we are looking only at what particles in the simulation have accumulated damage, not whether or not those particles have emptied the crater. Due to the numerical cost of these simulations, it is extremely difficult to run the simulation to a point where the crater is fully excavated. A velocity threshold could be used to determine which particles will escape the crater, however, it becomes uncertain as to which value one would want to use in these situations, as there is not a good "escape velocity" proxy in an experimental cratering test. Therefore, we choose to use the damage results as an upper limit – where it is unlikely that all of the damaged material will exit the crater, but no more will exit than what is damaged.

With this caveat, we note that the red line, which indicates the experimental results from [9], is located at the point before the knee where there is a slope change of both the crater width and depth lineouts in Figure 1 ($m = 10$) and at the knee of the lineout in crater depth in Figure 2 ($m = 11$). Using the assumption that not all of the material will exit the crater that is damaged, the simulations with a value of the Weibull modulus of $m = 10$ and 11 are underpredicting the crater width and depth. Looking at Figure 3, we see that the empirical value of the crater width and depth is less than the results from the simulation and so we believe that a Weibull modulus of 12 represents the best-fit model for our simulations. This best-fit value for the Weibull modulus is larger than the previously published value of 9.59 [22, 23], although not so high that it seems unphysical. It is important to note that published value was derived using experiments on crushed ice, which would have a different flaw distribution than a cylinder of optically clear, competent ice. Since the flaw distribution is a material property, one could assume this value to be similar between the crushed and competent ices, as we did by keeping the Weibull scale parameter constant. However, it is likely that the exact values will differ.

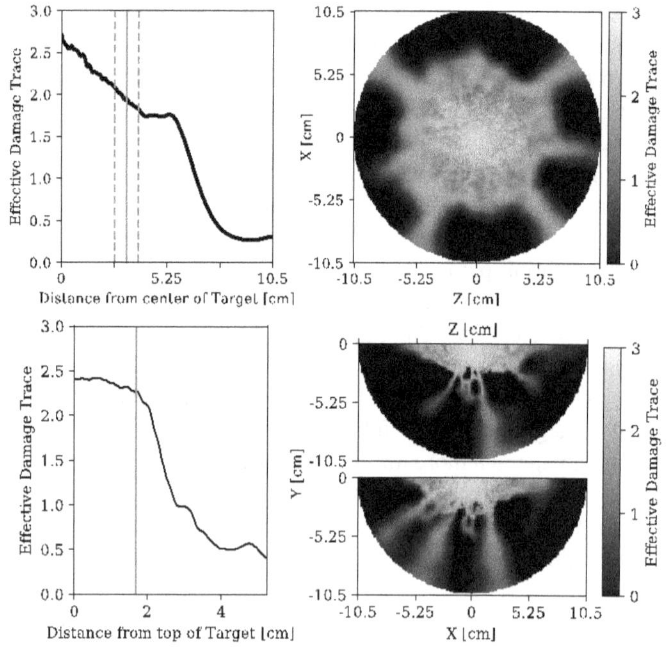

Figure 1: Results from the simulation of an impact into water ice with a Weibull modulus of $m = 10$. The top right panel displays the surface of the ice target and the top left panel displays lineouts sampled radially from the surface. The red solid line marks the measured crater surface and the red dashed lines mark 1σ from Harriss & Burchell [9]. The bottom panels are the same but instead of the top surface, it shows slices through the target. Again the red line is the measurement from Harriss & Burchell [9]

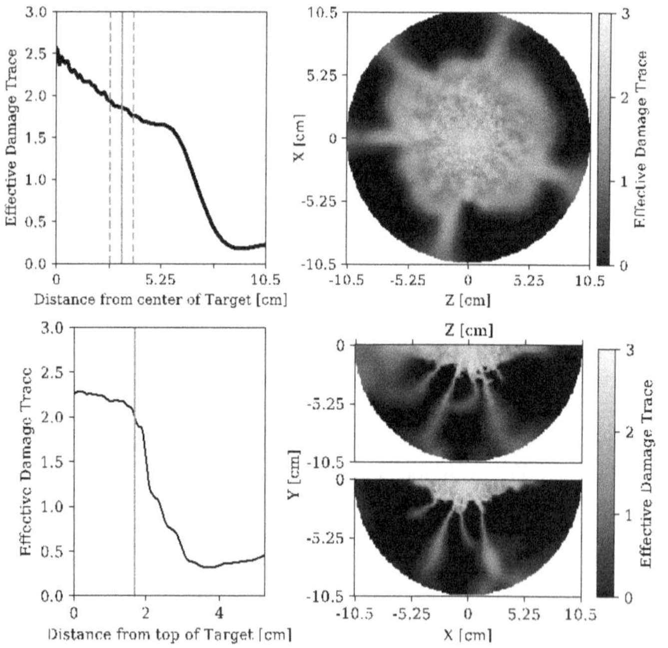

Figure 2: Same as Figure 1 but for a Weibull Modulus of $m = 11$.

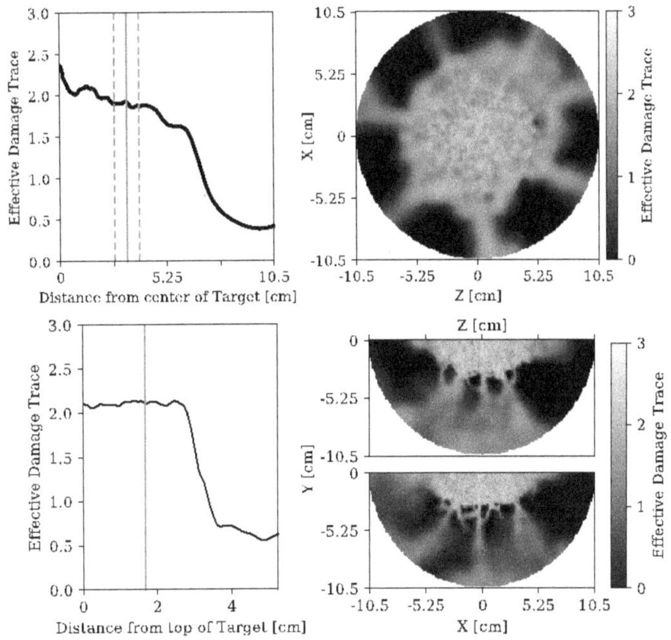

Figure 3: Same as Figure 1 but for a Weibull Modulus of $m = 12$.

5. Conclusions and Caveats

In this work, we aimed to validate our ASPH code, `Spheral++`, for impacts into ice. We found that using a Tillotson EOS, the pseudo-plastic strain algorithm, the `useDamageGradient` method, and a Weibull modulus of 12 to be the best fit solution. This validation allows for simulations of planetary defense scenarios and impact cratering events to be carried out using the best parameters that replicate experimental results. Further, this work demonstrates the ability of ASPH, and potentially other SPH techniques, to simulate ice in a realistic manner.

It is important to mention the caveats of this work when thinking about the conclusions. First, we interpret the damage value as a way of measuring the upper-limit of any parameter of interest. This method of looking at impact craters in simulations is relatively new and removes the "by eye" technique and as such, has not been used for any other validation experiments. Second, our validation solutions may not be unique solutions as it would be too numerically expensive to run every combination of parameters to get the best experimental match. With that, it should be noted that these best-fit simulations provide the input parameters for `Spheral++` simulations and may not be universally applicable to simulations with other codes or techniques. Finally, these results are only valid for hypervelocity impacts. Although not presented here, these best-fit simulation parameters do not extend to lower velocities. That suggests that they may only be valid in the hypervelocity regime and not in the high-speed regime where different strains may be observed. However, this work presents the first validation of code for ice impacts and therefore provides important data for the simulations of icy impacts.

Acknowledgements

VisIt is supported by the Department of Energy with funding from the Advanced Simulation and Computing Program and the Scientific Discovery through Advanced Computing Program. This work was performed under the auspices of the U.S. Department of Energy by Lawrence Livermore National Laboratory under Contract DE-AC52-07NA27344.
LLNL-PROC-766438

References

[1] Stickle, A.M.; Atchison, J.A.; Barnouin, O.S.; Cheng, A.F.; Crawford, D.A.; Ernst, C.M.; Fletcher, Z.; Rivkin, A.S. Modeling Momentum Transfer from Kinetic Impacts: Implications for Redirecting Asteroids, Procedia Engineering, 103, 2015, p. 577

[2] Melosh, H. J. (1989). Impact Cratering: A Geologic Process. Oxford Monographs on Geology and Geophysics. 11. Oxford University Press.

[3] Gault, D. E.; Wedekind, J. A., 1978, Experimental studies of oblique impact, Lunar and Planetary Science Conference, 9th, Houston, Tex., March 13-17, 1978, Proceedings. Volume 3. (A79-39253 16-91) New York, Pergamon Press, Inc., 1978, p. 3843-3875.

[4] Maurer, W. C.; Rinehart, J. S. Impact Crater Formation in Rock. Journal of Applied Physics, 1960, 31, 1247

[5] Asphaug, E., & Melosh, H. J. The Stickney Impact of Phobos: A Dynamical Model. Icarus, 1993, 101, 144

[6] Benz, W., & Asphaug, E. 1999, Icarus, 142, 5,doi:

[7] Senft, L. E., & Stewart, S. T. 2008, Meteoritics and Planetary Science, 43, 1993, doi:

[8] Pernas-Sánchez, J., Pedroche, D., Varas, D., López-Puente, J., & Zaera, R. 2012, International Journal of Solids and Structures, 49, 1919 ,

[9] Harriss, K. H., & Burchell, M. J. 2017, Meteoritics & PlanetaryScience, 52, 1505, doi:

[10] Owen, J. M., Villumsen, J. V., Shapiro, P. R., & Martel, H. 1998, ApJS, 116, 155, doi:

[11] Owen, J. M. 2010, in 5th International SPHERIC SPH Workshop, Manchester, United Kingdom

[12] Owen, J. M. 2014, International Journal for Numerical Methods in Fluids, 75, 749, doi:

[13] Grady, D., & Kipp, M. 1980, International Journal of Rock Mechanics and Mining Sciences & Geomechanics Abstracts, 17, 147

[14] Benz, W., & Asphaug, E. 1994, Icarus, 107, 98,doi:

[15] Jaeger, J. C., & Cook, N. G. W. 1969, Fundamentals of rock mechanics / J.C. Jaeger N.G.W. Cook (Methuen London), xiv, 513

[16] Benz, W., & Asphaug, E. 1995, Computer PhysicsCommunications, 87, 253 ,doi:

[17] Collins, G. S., Melosh, H. J., & Ivanov, B. A. 2004, Meteoriticsand Planetary Science, 39, 217,doi:

[18] H. Tillotson, J. (1962). Metallic Equations of State for Hypervelocity Impact. General Atomic Report. GA-3216. 141.

[19] Remington, T.; Owen, J. M.; Nakamura, A.; Miller, P. L.; Bruck Syal, M. Benchmarking Asteroid-Deflection Simulations: Basalt Spheres, 2019, in prep

[20] Nakamura, A. Fujiwara, A. Velocity distribution of fragments formed in a simulated collisional disruption. Icarus, 1991, 92, 132

[21] Childs, H.; Brugger, E.; Whitlock, B.; Meredith, J.; Ahern, S.; Pugmire, D.; Biagas, K.; Miller, M.; Harrison, C.; Weber, G. H.; Krishnan, H.; Fogal, T.; Sanderson, A.; Garth, C.; Bethel, E. W.; Camp, D.; Rübel, O.; Durant, M.; Favre, J. M.; Navràtil, P. VisIt: An End-User Tool For Visualizing and Analyzing Very Large Data. High Performance Visualization--Enabling Extreme-Scale Scientific Insight. 2012, pp. 357-372

[22] Lange, M. A., & Ahrens, T. J. 1983, Journal of Geophysical Research: Solid Earth, 88, 1197,doi:

[23] Stewart, S. T.; Ahrens, T. J., Correction to the Dynamic Tensile Strength of Ice and Ice-Silicate Mixtures (Lange & Ahrens 1983), 1999. 30th Annual Lunar and Planetary Science Conference, March 15-29, 1999, Houston, TX, abstract no. 2037

Proceedings of the 2019 Hypervelocity Impact Symposium
HVIS2019
April 14-19, 2019, Destin, FL, USA
HVIS2019-018

Consequences of micrometeoroid/orbital debris penetrations on the International Space Station

Hilary J. Evans[a], James L. Hyde[b]*, Eric L. Christansen[c], Dana M. Lear[c]

[a] *Independent Contractor – Vendor for Jacobs/LZ Technology, Inc., Athens, AL 35613, United States*
[b]*Barrios Technology—Jacobs JETS, 2101 NASA Parkway, Code XI4, Houston, TX 77058, United States*
[c]*NASA/Johnson Space Center, 2101 NASA Parkway, Code XI411, Houston, TX 77058, United States*

Abstract

Risk from micrometeoroid and orbital debris (MMOD) impacts on space vehicles is often quantified in terms of the probability of no penetration (PNP). However, for large spacecraft, especially those with multiple compartments, a penetration may have a number of possible outcomes. The extent of the damage (diameter of hole, crack length or penetration depth), the location of the damage relative to critical equipment or crew, crew response, and even the time of day of the penetration are among the many factors that can affect the outcome. For the International Space Station (ISS), a Monte-Carlo style software code called Manned Spacecraft Crew Survivability (MSCSurv) is used to predict the probability of several outcomes of an MMOD penetration—broadly classified as loss of crew (LOC), crew evacuation (EVAC), loss of escape vehicle (LEV), and nominal end of mission (NEOM). By generating large numbers of MMOD impacts (typically in the hundreds of billions) and tracking the consequences, MSCSurv allows for the inclusion of a large number of parameters and models as well as enabling the consideration of uncertainties in these models and parameters. MSCSurv builds upon the results from NASA's Bumper software (which provides the probability of penetration and critical input data to MSCSurv) to allow analysts to estimate the probability of LOC, EVAC, LEV, and NEOM. This paper provides an overview of the methodology used by NASA to quantify LOC, EVAC, LEV, and NEOM with particular emphasis on describing in broad terms how MSCSurv works and its capabilities and most significant models.

Keywords: micrometeoroid; orbital debris; MSCSurv; risk; survivability

---

Nomenclature

N	expected number of penetrations
R-factor	number of penetrations resulting in outcome/N
K	stress intensity factor
σ	stress
c	half crack length
R	radius of pressure vessel
h	pressure vessel wall thickness
E	modulus of elasticity (Young's Modulus)
TUC	Time of Useful Consciousness from Wolf, 2014 [11]

* Corresponding author. Tel.: +1-281-244-5068; fax: +1-281-483-3908.
E-mail address: james.l.hyde@nasa.gov

A	Uncertainty factor for Wolf equation
B	Adjustment factor for key differences between an on-orbit penetration and the data sets used to develop Wolf's equation
Greek symbols	
ν	Poisson's ratio
η	dimensionless constant
β	nonlinear bulging factor developed by Jeong
α	empirical constant (0.671 determined by Jeong);
Subscripts	
h	hoop

1. Introduction

The threat from micrometeoroid and orbital debris (MMOD) impacts on space vehicles is often quantified in terms of the risk of penetration and vehicle requirements are often put in terms of probability of no penetration (PNP). However, for large spacecraft, especially those with multiple compartments such as the International Space Station (ISS), an MMOD penetration may have a number of possible outcomes of interest. This paper will briefly describe the overall process of estimating the probability of loss of crew (LOC), crew evacuation (EVAC), loss of escape vehicle (LEV), and nominal end of mission (NEOM) for the ISS as a result of MMOD penetration with emphasis on NASA's software tool, Manned Spacecraft Crew Survivability (MSCSurv).

2. Overall methodology

The overall goal of predicting probability of LOC, EVAC, and LEV is accomplished using two distinct software codes, Bumper and MSCSurv, and combining the results. Bumper is NASA's accepted tool for predicting the probability of a penetration by MMOD for the ISS (and many other space assets). The probability of penetration (PP) can be calculated from the predicted number of penetrations (N) as follows:

$$PP = 1 - e^{-N} \tag{1}$$

MSCSurv produces ratios called R-factors. An R-factor is:

$$R - factor = \frac{Number\ of\ penetrations\ resulting\ in\ outcome}{Total\ number\ of\ penetrations\ (N)} \tag{2}$$

It follows that the probability of a particular outcome can be found by:

$$Probability\ of\ Outcome = 1 - e^{-N*R-factor} \tag{3}$$

Detailed descriptions of the Bumper software tool and applications may be found in [1].

The remainder of this paper will focus on MSCSurv and the process for predicting the R-factor in Equations (2) and (3) as well as the variety of outcomes for which R-factors are generated.

3. MSCSurv overview

MSCSurv is a Monte-Carlo style code designed to simulate hundreds of billions of MMOD impacts to the ISS and for each impact that results in a penetration determine the outcome. Figure 1 presents a flowchart for MSCSurv v11.1.

For each impact generated, MSCSurv must determine the location of the impact and the characteristics of the projectile (size, impact velocity, angle of impact, and density). MSCSurv then compares the size of the impacting projectile with the minimum projectile size that will penetrate the ISS in that location. The minimum projectile sizes are produced by Bumper. The vast majority of impacts will not result in a penetration, so MSCSurv spends most of its time generating impacts. When a projectile is generated that does penetrate the ISS, MSCSurv enters into the portion of the code that evaluates the damage in terms of hole and crack size as well as the projectile's penetration depth into the interior. After that, as shown in Figure 1, MSCSurv tracks the following six outcomes which result in LOC:

A. Critical external equipment failure;
B. Crew compartment critical cracking (unzipping);
C. Internal hazard penetration;
D. Hypoxia;
E. Fatal fragmentation injury to crew; and
F. Loss of control of ISS (Damage to key guidance, navigation, & control (GN&C) equipment combined with excessive thrust prevent safe emergency vehicle departure).

MSCSurv also tracks the following four outcomes which result in at least one crew evacuation:

G. Nonfatal crew injury;
H. Late loss of control of ISS;
I. Critical compartment depressurization; and
J. Loss of access to critical compartment(s)

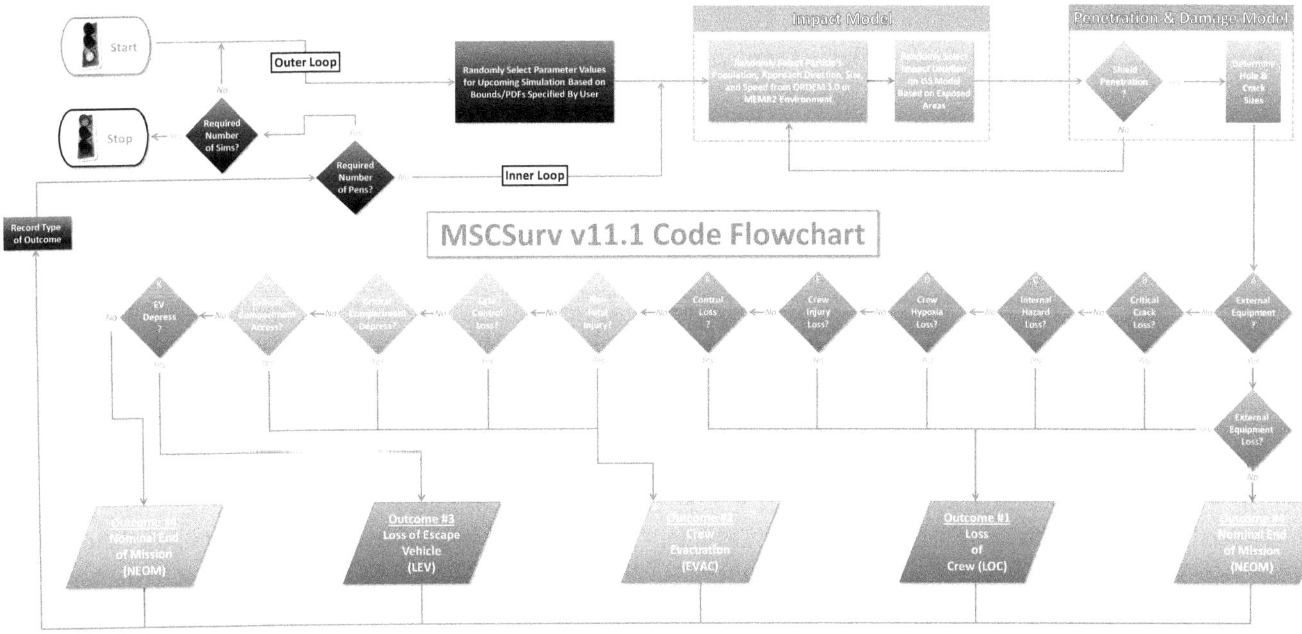

Figure 1. MSCSurv flowchart

If none of the LOC or EVAC outcomes occur, MSCSurv checks for LEV (K on Figure 1)--if no LEV occurs, MSCSurv tallies a NEOM. MSCSurv tallies a LOC if at least one crewmember is lost. MSCSurv does not distinguish between a penetration resulting in the loss of one crewmember and a penetration resulting in the loss of all crewmembers or any number in between.

The flowchart shows that MSCSurv is structured with an outer loop and an inner loop. The outer loop sets numerous key parameters in the analysis, each time varying them by drawing random numbers and applying those numbers to pre-defined probability distributions intended to represent the uncertainty in each parameter. These probability distributions are specified by the user via a user-input file. Once generated, these parameters will be held constant for that simulation of a user-specified number of penetrations. Some parameters are thought to be correlated. These parameters are linked within MSCSurv, which means that while the probability distributions may be different for the different parameters, the same random variable will be used to establish the values of the linked parameters. Since MSCSurv deals with the consequences of a penetration, the user is asked to specify the number of penetrations per simulation (and the number of simulations) as opposed to the number of impacts. Depending on the environment files and the difficulty or ease with which the compartments may be penetrated, the number of impacts required to achieve a specified number of penetrations can vary widely. The inner loop is repeated until the specified number of penetrations is achieved.

For each penetration, MSCSurv establishes the location of each crewmember in terms of compartment and crew station (longitudinally defined segment of about 1 meter in length). During the day, the crew locations are determined by drawing random variables against a distribution developed from specified crew frequencies based largely upon the information included in the 2012 Probabilistic Risk Assessment (PRA) report, *ISS Crew Traffic Patterns* [2]. At night (an eight-hour period), the crewmembers are assigned to the crew quarters.

The following subsections briefly describe each of the major blocks of the flowchart (Figure 1): impact model, penetration and damage model, and outcomes.

3.1. Impact Model

MSCSurv must generate hundreds of billions of impacts to model the one billion penetrations of a typical run used for an ISS MSCSurv analysis report. For each impact, the key characteristics of the projectile and the location of the impact are determined in the following order:

1. Population: MSCSurv is set up to handle data files in the proper format for two NASA environment models: ORDEM 3.0 and MEMR2. Population selection applies only to ORDEM 3.0 which has five populations defined (NaK, low density, medium density, high density, and intacts).
2. Threat direction: The threat may approach the ISS from many different directions (614 for ORDEM 3.0 and 1652 for MEMR2).
3. Diameter: MSCSurv assumes a spherical projectile, so this is the diameter of the sphere.
4. Velocity: Impact velocity of projectile.
5. Compartment impacted by projectile.
6. Element impacted by projectile: Each compartment usually has thousands of elements. The specific element impacted is determined so that the various pertinent data (shielding, equipment flags, internal shielding, and so forth) can be properly assigned.

3.2. Penetration and Damage Model

Penetration Ballistic Limit Equations—Using generated impact characteristics, MSCSurv looks up the critical diameter (minimum projectile size [at the determined density, velocity, and impact angle] at the impact location resulting in penetration) from a table generated by Bumper. To determine the critical diameters, Bumper utilizes numerous ballistic limit equations (BLEs), developed over years with hyper-velocity impact test data and hydrocode analyses. [3] [4]

The finite element model (FEM) used for Bumper runs and the geometry file it creates have regions defined in them, called property identifiers (PIDs). PIDs allow each shield configuration to be defined in terms of bumper and rear wall materials, thicknesses, spacing, and other parameters. These PIDs are maintained in the MSCSurv data files.

If the diameter of the projectile does not exceed the critical diameter, a penetration does not occur, so MSCSurv generates another impact and checks to see if this one penetrates. When a projectile does penetrate, the next step is to determine the extent of the damage. To this end, MSCSurv uses models to estimate the hole size, crack length, and, if applicable, the penetration depth of the projectile in the interior of the crew volume.

Hole and Crack Size Models—The holes created in the rear wall of the compartment by a projectile are sized by one of two models included in MSCSurv. The standard baseline model currently is the Williamsen/Schonberg Generic Hole-Size model. [5] The Burch model is included for comparison purposes but is rarely used. [6]

Similarly, the Williamsen/Schonberg Generic Crack-Length model is the standard baseline model for predicting crack lengths but a Burch model is included though rarely used.

Penetration Depth Model—The Burch report, issued in the 1960s, documented work done to develop several semi-empirical damage equations—one of which was a penetration depth equation for multiple aluminum plates penetrated by aluminum projectiles. A term to modify the equation based on the density of the projectile was developed with more limited data. The penetration depth equations used in MSCSurv must be applied in situations outside those for which the Burch equations were developed, including velocities outside the range of the data used to formulate the Burch equations, a wider range of materials, and irregular shielding configurations. Burch assumes multiple identical plates equally spaced and predicts how many plates the projectile will penetrate. The shielding (including internal shielding) present on the ISS does not fit into this specific configuration.

Burch was developed from testing performed in the transition region and as such the velocity is restricted to this region in MSCSurv when using the Burch penetration depth equation.

In order to improve upon the Burch equation predictions when applied to MSCSurv's broader application, for each penetration, Burch is "calibrated" by determining the penetration depth predicted by Burch for the critical diameter from Bumper (as used to determine whether or not a penetration occurred). Since extensive hypervelocity impact testing has been performed for ISS shielding configurations and this data is used to develop the BLEs within Bumper, these BLEs are more specific to the shielding configurations to which they are applied and so are considered of higher fidelity for this application than the Burch equations. So the Burch results are calibrated to match the Bumper BLE results at threshold penetration of the rear pressure wall. Then Burch is used to predict the amount of further penetration by the actual projectile into internal material and adjusted by the same amount as was necessary at the threshold penetration point.

3.3. Outcomes tracked by MSCSurv

MSCSurv checks for outcomes following a penetration in a cascading nature, moving onto the next impact if an outcome occurs, so no single penetration can result in more than one outcome. In reality, the same penetration may result in more than one outcome, but MSCSurv does not attempt to quantify that. Figure 1 shows the order of this "cascade". All outcomes fall into one of four major categories: LOC, EVAC, LEV, and NEOM.

Within LOC, six different outcomes resulting in LOC are tracked. Within EVAC, four different outcomes resulting in EVAC are tracked.

Outcomes Resulting in Loss of Crew—The following outcomes result in loss of crew (LOC): External Equipment, Unzipping, Internal Hazard, Hypoxia, Fatal Injury from Fragments, and Loss of Control.

<u>Critical External Equipment Failure.</u> The first possible outcome resulting in LOC examined by MSCSurv is due to external equipment penetration. This outcome is unique in that whether or not it happens, if an area of the model is identified as external, a penetration in this area is assumed to preclude a penetration to the internal crew compartment. Thus, the outcomes past external equipment shown on Figure 1 are not considered for external penetrations.

Each element on MSCSurv's geometry model analysis has a flag identified with it that tells MSCSurv whether or not the element is external and if so, identifies what type of external component is located there. Most external elements are in tank regions and are generally highly pressurized or contain hypergolic fuel. In terms of failure criteria, tanks generally fall into two categories—those where a partial penetration to the tank wall will result in LOC failure and those where a full penetration which results in damage sufficient to cause one or more of three loss modes is required. These loss modes are detonation, unzipping, and thrust-induced structural failure. For other external components, such as compressors and batteries, a probability of LOC in the case of penetration is provided.

The thresholds for failure from detonation and thrust-induced structural failure are set in a data file. The tank unzipping requires MSCSurv to do more of the work. A data file provides key tank characteristics, such as the expected pressure profile, the fracture toughness (onset of unstable propagation), diameter and shape of the tank, and certain material properties of the tank wall.

When a tank is penetrated, MSCSurv draws a random variable and, using the pressure profile defined, establishes the tank pressure at the moment of impact. Using this pressure and the other tank characteristics provided by the data file, MSCSurv calculates a critical crack length (CCL). The CCL is the minimum crack length that will initiate unstable crack growth (unzipping).

MSCSurv uses two methods to determine the CCL and selects the minimum CCL from the two. Both methods are closed form, isotropic shell equations. One, the Folias method, uses the Folias Factor and simple non-bending assumption as presented at Dr. Folias' website [7]:

$$K = \sigma_h(\pi c)^{1/2}(1 + 0.317\lambda^2)^{1/2} \tag{4}$$

$$\lambda^2 = \left(12(1 - \nu^2)\right)^{1/2}[c^2/(Rh)] \tag{5}$$

Where:

K = stress intensity factor; σ_h = hoop stress; c = half crack length; ν = Poisson's ratio; R = radius of pressure vessel; and h = pressure vessel wall thickness.

The second method was used by Lutz [8] to determine critical crack lengths for two ISS tanks and adopts a nonlinear bulging factor (β) developed by Jeong [9]:

$$K = \sigma_h(\pi c\eta)^{1/2}\beta \tag{6}$$

$$\beta = \left(1 + \alpha[(E/\sigma_h)(c/R)^2]^{2/3}\right)^{1/2} \tag{7}$$

Where:

K = stress intensity factor; σ_h = hoop stress; c = half crack length; η = dimensionless constant; α = empirical constant (0.671 determined by Jeong); E = modulus of elasticity (Young's Modulus); R = radius of pressure vessel.

For each penetration, MSCSurv starts with a very small half-crack length; and using the equations above, incrementally increases the half-crack length until the stress intensity factor exceeds the fracture toughness for unstable crack growth. The last incremental change is subtracted, and that is the CCL used as the failure threshold. This results in the smaller of the two CCLs determined by the two methods described above being used as the CCL.

Using the crack models discussed earlier, the crack length is determined and compared to the CCL. If the crack length exceeds the CCL, a LOC is tallied due to tank unzipping.

Crew Compartment Critical Cracking (Unzipping). For penetrations into the crew cabins, MSCSurv checks first for unzipping. As with external tanks, this occurs when the crack length determined with the crack model discussed earlier exceeds the critical crack length for that region. Since the pressure is fairly close to constant in the crew cabin (prior to penetration), the CCLs for the crew cabins are not calculated by MSCSurv but rather are specified in a data file by PID.

Internal Hazard Penetration. The next outcome considered is LOC due to an internal hazard. An internal hazard is anything internal to a crew cabin that poses a potentially fatal hazard to the crew if struck by debris. Examples include high-pressure tanks, toxic materials, and fire hazards. This may be equipment, payload, or stowage items. The analyst must identify internal equipment by providing on an element-by-element basis a flag which when set to 1 or 2 lets MSCSurv know there is an internal hazard there. Just because a penetration occurs to an element where an internal hazard is located doesn't mean it actually reaches the hazard. There may be additional protection, such as the back of an equipment rack, between the pressure wall and the hazard. This protection is specified as an areal density (g/cm^2) in the same data file as the flag. The Burch penetration depth model discussed earlier is used to determine if the projectile makes it through the protective areal density. Finally, a criticality factor probability (CRFAC) is assigned to each element where there is an internal hazard flag. This may be used to model the probability that a rack contains an internal hazard. For the last baseline analysis performed on the ISS [10], system racks in the United States On-Orbit Segment (USOS) were identified by subsystem such as Thermal Control System (TCS). For each subsystem type, an estimate was made as to how likely it was to contain an internal hazard. For a TCS rack, for example, a 0.10 CRFAC was assigned. Figure 2 provides an example of how internal hazards might be distributed on the USOS. If a penetration occurs to an element flagged as having internal equipment, two criteria must be met for a LOC to occur: 1) the penetration depth must exceed the internal protection and 2) a random number generated by MSCSurv must be less than or equal to the CRFAC probability assigned.

Figure 2. Example of internal and external hazard locations on the USOS

Hypoxia. Hypoxia is the next outcome in the cascade. As the cabin air pressure drops following a penetration, there comes a point at which the crew will become incapacitated. The time between the impact and crew incapacitation is called the time of useful consciousness (TUC). If the crew cannot safely isolate itself from the leak prior to crew incapacitation, hypoxia loss will be tallied. MSCSurv includes a model of the crew response given different circumstances. Figure 3 shows a very high-level overview. For a hole large enough to constitute a hypoxia threat (Emergency Scenario), the code utilizes a streamlined crew protocol; whereas for smaller holes (Non-Emergency Scenario), the ISS-Program's developed leak protocol is utilized. For each penetration, MSCSurv models the crew response (in much more detail than shown in Figure 3) by modeling events and crew actions and developing a timeline while keeping track of the pressure and available volume throughout.

Here is a notional example. MSCSurv generates an impact that results in the penetration of the US Lab. Using the previously discussed hole-size model, a large 11-cm hole is predicted. No outcomes prior to hypoxia occur, so the crew action subroutine is called. This subroutine models the crew's reaction to the penetration event, keeping track of the elapsed time along the way. There are many parameters that affect the timeline generated. What time of day or night did the impact occur (i.e. was the crew awake or asleep)? Where was the crew when the impact occurred? Were any crewmembers injured? In this case, the impact occurred while the crew was awake. When awake, the crew locations are determined by drawing random variables against a distribution developed from specified crew frequencies based largely upon the information included in the 2012 Probabilistic Risk

Assessment (PRA) report, *ISS Crew Traffic Patterns* [2]. The crew executes the Emergency protocol. MSCSurv keeps a running total of the time since impact and checks after the completion of most tasks to determine if certain conditions may have developed

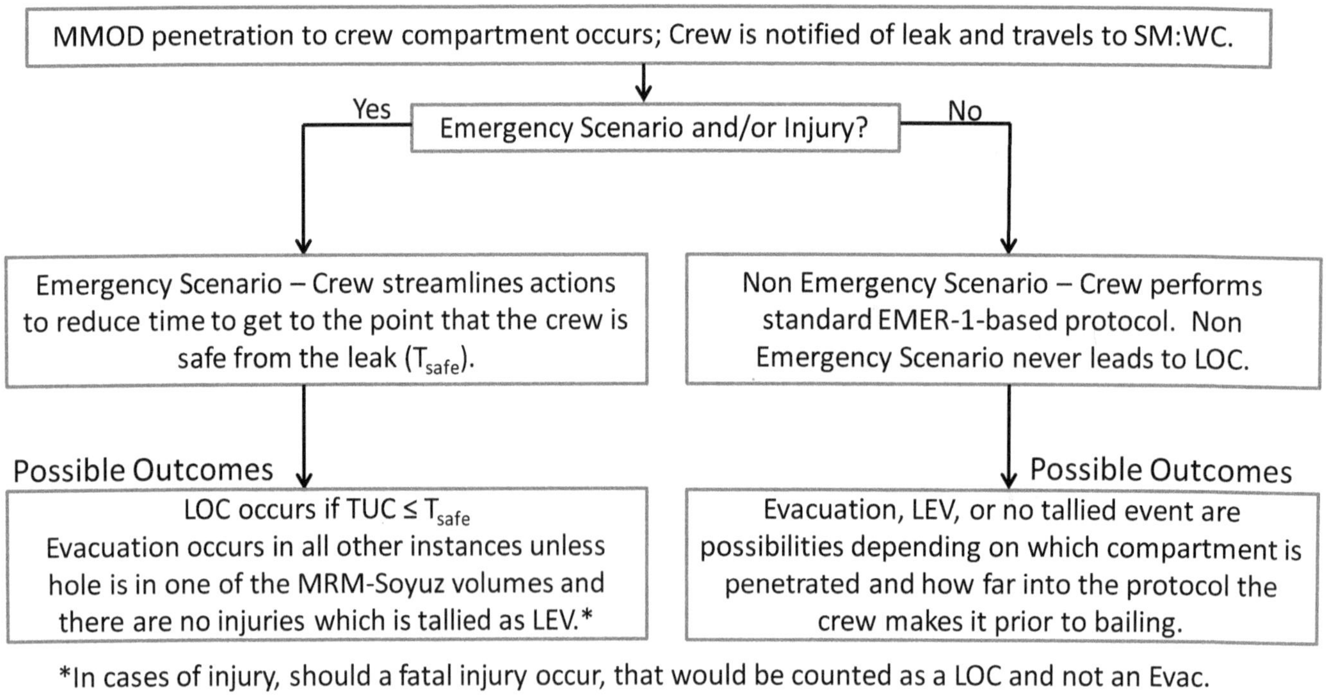

Figure 3. Rough flowchart of crew protocol coding

that will affect the time required to complete the next step in the protocol. Performance degradation due to the onset of hypoxia symptoms increases the time it takes for the crew to perform any task, and the formation of fog in the crew cabins decreases travel rates. After notification of the depress event (either from the ground or by alarm), all crewmembers first meet at the Russian Service Module Working Compartment. At this point if any injuries occurred, the crew will have either encountered any injured crewmember or the injured crewmember will not show up at the Service and the crew will go to look for the injured. For the sake of this example, there are no crew injuries. Travel times are calculated by taking the distance from each crewmember's initial location to the destination divided by the appropriate rate of travel.

A data file provides MSCSurv with the distances from each crew station on each compartment to every other crew station in all of the compartments. Crew stations are longitudinal sections about one meter in width. They are used to locate crew and to determine distances. The first crewmember to arrive at the Service Module obtains the manometer. In the example, the crewmember sees quickly that the pressure is dropping fast. When the rest of the crew arrives, the crewmembers split into two crews according to their designated Soyuz escape vehicles. Each crew travels to its designated Russian Mini-Research Module (MRM)-Soyuz volume and clears the MRM hatchway before determining which side of the hatch the hole is on. In the example, the hole is not in either MRM-Soyuz volume; so after the hatch is tapped closed to check which side the hole is on, the crew is safe. MSCSurv checks the timeline at that point. If the time since impact exceeds the TUC at the current pressure (which it did not in this notional case), a hypoxia LOC is counted.

To calculate TUC, MSCSurv uses the following equation:

$$TUC = A * B * TUC(Wolf) \tag{8}$$

Where:

TUC(Wolf) = Wolf equation extracted from Wolf's 2014 paper [11]. The pressure at the time the crew would be safe from the leak is used to calculate the H (altitude) term in Wolf's equation.

A = Uncertainty factor for Wolf equation which will be set in MSCSurv's outer loop based on the user-defined uncertainty distribution.

B = Adjustment factor for five potential key differences between an on-orbit penetration and the data sets used to develop Wolf's equation. The factor's uncertainty distribution is defined by the user. This will be set in the outer loop and is independent of the uncertainty factor A.

MSCSurv uses the simplifying assumption that all crewmembers have the same loss of useful consciousness (LUC) threshold. Likewise, the same onset of hypoxia symptoms is assumed to apply to all six crewmembers.

Had the hole been smaller, the crews might have been able to use the non-emergency (standard) protocol and after checking the MRM-Soyuz volumes, begun the leak isolation protocol. For our example, the crews do not have time to attempt the isolation protocol, so both crews evacuate the ISS.

Fatal Fragmentation Injury to Crew. Fragmentation injury is next in the cascade. In order for a fatal fragmentation injury to be counted, any internal shielding (such as equipment racks) must be penetrated. The areal density for crew protection specified for the penetrated element is converted to equivalent plates and compared to the penetration depth (calculated as discussed previously). Two areal densities are specified in the data file for each element, one for internal hazard shielding and the other for crew internal shielding. The crew areal density is usually greater than that specified for the internal hazard in the same location. If the penetration depth is greater than that required to penetrate completely through the crew internal shielding and there is a crewmember in the penetrated crew station or the one immediately adjacent to it (and in the same compartment), a fatal fragmentation will be tallied. The crew locations at the moment of impact (determined as discussed previously) are used for this check.

Loss of Control. When key Guidance, Navigation, and Control (GN&C) equipment is impacted and the hole size is above a user-specified size, a relatively rapid loss of ISS control can occur which will not allow the crew to evacuate the ISS safely. Two conditions must be met for this loss to be tallied: 1) GN&C equipment must be impacted (so the projectile must penetrate through any internal shielding there may be) and 2) the hole in the crew cabin must exceed a user-specified size.

Outcomes Resulting in Crew Evacuation—If no LOC occurs, penetrations to crew cabin areas are evaluated to see if crew evacuation occurs. Depending on the situation, one crew or both crews may evacuate. MSCSurv keeps track of the proportion of evacuations of one crew versus both crews. The following outcomes result in at least one crew evacuating the ISS: non-fatal crew injury, late loss of control of ISS, critical compartment depressurization, and loss of access to critical compartment(s).

Non-Fatal Crew Injury. Two types of non-fatal injuries are considered in MSCSurv—those due to overpressure and those due to light flash. For either of these types of injury to occur, the crewmember must be inside the penetrated compartment at the time of the MMOD penetration.

The equation used to determine the overpressure was developed based on testing performed at the University of Alabama in Huntsville (UAH). [12] Overpressure is highly dependent on the distance between the penetration and the crewmember.

The UAH test setup also allowed for the measurement of the intensity of the light flash resulting from the impact. The conclusions regarding light flash were that the flash can be intense enough to cause visual impairment. However, a crewmember would have to be looking in the direction of the flash at the time of penetration. The probability that the crew is looking in that direction would be about 1 in 6 (this probability is provided in the user-input file). In addition, for light flash injury to occur, the projectile would have to completely penetrate the crew internal shielding.

Late loss of control of ISS. Under LOC, the control loss was due to venting from a hole of a certain minimum size and damage to key GN&C equipment. If there is damage to key GN&C equipment and the hole size is below the minimum size or if certain coolant loops which cool GN&C equipment are damaged, loss of ISS control may occur, but there will be adequate time for both crews to safely undock and evacuate the ISS.

Critical Compartment Depressurization. A compartment is considered critical when, should it depressurize, the ISS can no longer support two crews. In the latest ISS MSCSurv assessment, two critical compartments were identified, the Service Module Working Compartment and the US Node 3. If one or both critical compartments depressurize, crew evacuation is tallied. MSCSurv keeps track of whether one or both critical compartments depressurize.

The non-emergency protocol (Figure 3) is used to determine how far along the crew makes it into the protocol. This can dictate whether or not there is critical compartment depressurization. For example, if there is a penetration to the US Node 2 and the crew has time to successfully isolate the hole by closing the US Lab hatch between the US Lab and the Node 2, no critical compartment depress occurs. However, if the crew runs out of time before completing the protocol, there are three possible outcomes (assuming no LOC or non-fatal injuries): both critical compartments depress (and both crews evacuate), one critical compartment depresses (and one crew evacuates), or no critical compartment depresses and no crew evacuates. Which of the three occurs depends on how far into the protocol the crew was when "bail" time was reached causing the crew to isolate itself from the hole using the knowledge available at the time.

Loss of Access to Critical Compartment(s). Depending on where the leak is and how far along in the crew protocol the crew gets before it must isolate itself from the hole, one of the two crews may find itself without access to either critical compartment or both crews may lose access to the same critical compartment yet neither critical compartment depressurizes. Either of these situations requires at least one crew to evacuate.

Loss of Escape Vehicle—If neither LOC nor EVAC occurs, the LEV is considered. An LEV is tallied when either crew loses its escape vehicle. The escape vehicles are the Soyuz in all analyses to date. If the Soyuz Descent Module depressurizes, an LEV will be counted. This may occur due to a hole in the Descent Module. It may also occur when the hole is in the Soyuz Orbital Module or the MRM and there is not time to further isolate the hole and evacuate.

A hole in the Soyuz Orbital Module or the MRM may result in a LOC, single-crew evacuation, or an LEV. Assuming no LOC, when the crew reaches the point of knowing the hole is somewhere in the MRM-Soyuz volume, time permitting, it will attempt to narrow down where the leak is. If there isn't time to further isolate the hole, the crew will close off the MRM-Soyuz volume. They will then lose access to the Soyuz and an LEV will be tallied. Smaller holes in the MRM may result in time to isolate the MRM and evacuate in the Soyuz, thus resulting in a single-crew evacuation. If the hole is in the Soyuz Orbital Module, the crew will attempt to don the Russian Sokol space suits and leave in the Soyuz. However, this takes quite a bit of time; and if the crew hits "bail" time before getting the suits on, the crew will isolate itself from the damaged Soyuz and remain on the ISS, and an LEV will be tallied. Any hole in the Soyuz Descent Module crew cabin (where no LOC or injury occurs) results in an LEV.

4. MSCSurv output

Figure 4 shows MSCSurv output in broad terms. The output of the outcomes discussed earlier is in the form of tables of R-factors such as that shown in Table 1. As shown in Eq. (2), an R-factor is the ratio of the number of penetrations resulting in the outcome of interest to the total number of penetrations. This output is provided for each compartment in the model for each outcome discussed in Section 3. This allows an analyst to combine MSCSurv results (R-factors) with Bumper predictions of the number of penetrations (N) to arrive at the number of occurrences of that particular outcome (N*R-factor). The analyst must be mindful of each outcome's place in the cascade, understanding that the overall occurrences for an outcome will not include occurrences from penetrations that also resulted in an outcome earlier in the cascade.

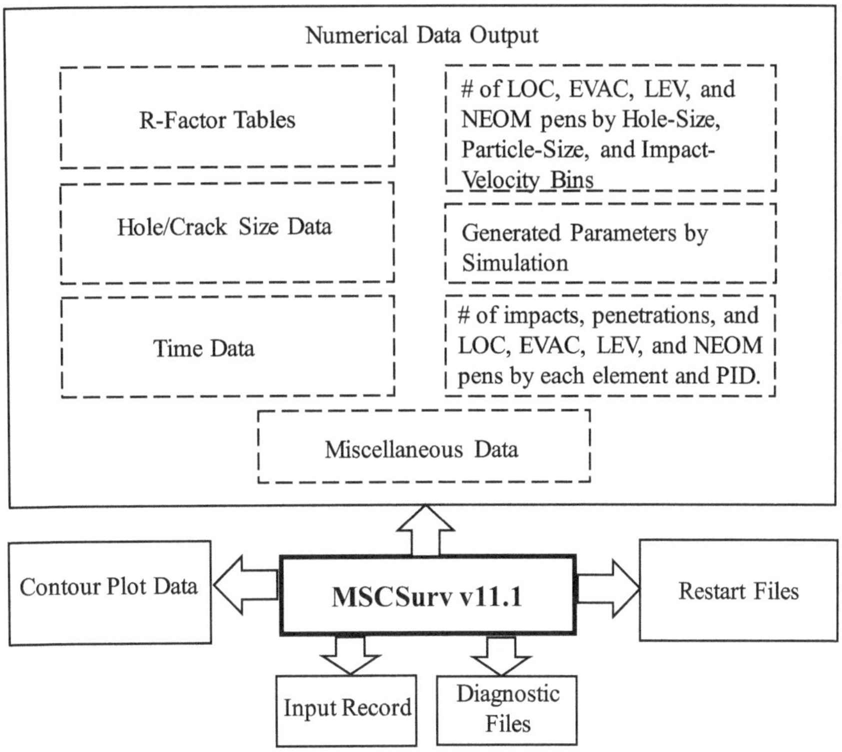

Figure 4. MSCSurv data output

Table 1. An example of a table of R-factors (where column heading letters correspond to the letters in Figure 1)

COMPARTMENT	% PENS	R-Factor (LOC)	A+	B+	C+	D+	E+	F	G	H	I	J	K
NODE 2	0.15	0.499	0.000	0.415	0.002	0.059	0.024	0.000	0.055	0.000	0.178	0.000	0.000
COLUMBUS	0.48	0.308	0.000	0.215	0.037	0.052	0.004	0.000	0.005	0.000	0.162	0.000	0.000
JPM	0.43	0.337	0.000	0.267	0.021	0.044	0.005	0.000	0.009	0.000	0.185	0.000	0.000
JLP	0.74	0.081	0.000	0.067	0.001	0.012	0.000	0.000	0.001	0.000	0.068	0.000	0.000
US LAB	0.08	0.537	0.000	0.215	0.183	0.115	0.024	0.000	0.039	0.000	0.201	0.000	0.000
NODE 1	0.04	0.185	0.000	0.103	0.008	0.038	0.036	0.000	0.041	0.000	0.340	0.434	0.000
CUPOLA	0.61	0.081	0.000	0.000	0.000	0.080	0.001	0.000	0.008	0.000	0.080	0.000	0.000
AIRLOCK:Crew	0.02	0.424	0.000	0.312	0.000	0.110	0.002	0.000	0.001	0.000	0.345	0.012	0.000
AIRLOCK:Equip	0.02	0.472	0.000	0.311	0.000	0.154	0.007	0.000	0.001	0.000	0.406	0.013	0.000
PMM	1.20	0.223	0.000	0.186	0.009	0.028	0.001	0.000	0.001	0.000	0.192	0.000	0.000
PMA1	0.01	0.057	0.000	0.012	0.000	0.011	0.034	0.000	0.008	0.000	0.341	0.594	0.000
PMA2	0.07	0.229	0.000	0.229	0.000	0.000	0.000	0.000	0.000	0.000	0.000	0.000	0.000
PMA3	1.11	0.006	0.000	0.006	0.000	0.000	0.000	0.000	0.000	0.000	0.000	0.000	0.000
NODE 3	0.13	0.409	0.000	0.283	0.000	0.096	0.030	0.000	0.033	0.000	0.558	0.000	0.000
TCS-ATA	0.09	0.159	0.159	0.000	0.000	0.000	0.000	0.000	0.000	0.000	0.000	0.000	0.000
TCS-NTA	0.09	0.331	0.331	0.000	0.000	0.000	0.000	0.000	0.000	0.000	0.000	0.000	0.000
TCS-PM	0.04	0.173	0.173	0.000	0.000	0.000	0.000	0.000	0.000	0.000	0.000	0.000	0.000
CMG	0.90	1.000	1.000	0.000	0.000	0.000	0.000	0.000	0.000	0.000	0.000	0.000	0.000
PCU	0.00	0.616	0.616	0.000	0.000	0.000	0.000	0.000	0.000	0.000	0.000	0.000	0.000
AMS	0.07	0.041	0.041	0.000	0.000	0.000	0.000	0.000	0.000	0.000	0.000	0.000	0.000
HPGT	0.10	0.371	0.371	0.000	0.000	0.000	0.000	0.000	0.000	0.000	0.000	0.000	0.000
USOS	**6.38**	**0.295**	**0.156**	**0.101**	**0.008**	**0.027**	**0.003**	**0.000**	**0.005**	**0.000**	**0.095**	**0.004**	**0.000**

R-factors have uncertainties associated with them. MSCSurv produces tables of R-factors for a specified upper and lower bound. However, these are of limited practical use as the results do not combine realistically. They do give an analyst an idea of the uncertainties involved in the results, however. A table of median values is also provided but is likewise of limited quantitative value. The values can give an analyst a qualitative idea of the spread of results and the general shape of the distribution of R-factors over all simulations.

MSCSurv provides more tables of R-factors including one specifying whether, in the case of EVAC, one or both crews depart and one which breaks the external component LOC penetrations down further by detonation, unzipping, and joint failure due to thrust. Some tanks may fail from partial penetration of the tank wall. These are broken out in the table as well.

The distribution of hole and crack sizes may also be of interest. MSCSurv writes out such data in both an easy-to-read tabular form and in a format easily imported into Excel. Several files are written out which contain the distributions of hole and crack sizes of certain sets of penetrations that may be of interest such as all penetrations, only penetrations resulting in LOC, only penetrations resulting in EVAC, only penetrations resulting in LEV, only those penetrations to crew compartments (excluding external component penetrations), only penetrations resulting in hypoxia loss, and only penetrations successfully repaired (repair is not considered a part of the baseline analyses).

Similar distributions are written out to represent certain times which include the time it would have taken to complete the emergency crew protocol in cases where hypoxia LOC is tallied and there were no injuries, time it takes to complete the emergency crew protocol when there is a crew injury, and the time available from alarm to bail time for a constant available volume (the initial volume).

4. Summary

In order to understand the risk from MMOD penetration to the ISS, Bumper is used to predict the probability of penetration. MSCSurv is then used to predict the likelihood of certain key outcomes in the event of a penetration. Combining results from both codes provides estimates as to the probability of LOC, probability of EVAC, and the probability of LEV.

Since the first version of MSCSurv was developed in the 1990's by Dr. Joel Williamsen [13], the code has been further developed and enhanced. Over the years, it has been used to provide input into the ISS PRA process and to assist in making Program design and operational decisions.

MSCSurv v12.0 is currently being developed and will be applicable to the permanent configuration which will have one Soyuz and one US Crew Vehicle (USCV) docked to ISS.

References

[1] Bjorkman, M., Christiansen, E., and Lear, D., *Bumper 3 Software User Manual*, NASA/TM-2014-218559, October 2014.

[2] DeSimpelaere, E.C., Grant, W., Brown, P., *ISS Crew Traffic Patterns*, PRA Report No: ISSPRA-12-42, Prepared for JSC, July 2012.

[3] Christiansen, E., *Meteoroid/Debris Shielding*, NASA/TP–2003-210788, August 2003.

[4] Christiansen, E., et.al, *Handbook for Designing MMOD Protection*, NASA/TM–2009–214785, June 2009.

[5] Williamsen, J.E., Schonberg, W.P., Evans, H.J., 2012. "Generic module wall damage prediction equations for habitable spacecraft crew survivability evaluations", *International Journal of Impact Engineering*, http://dx.doi.org/10.1016 /j.ijimpeng.2012.07.011.

[6] Burch, G. T., *Multiplate-Damage Study*, Air Force Armament Laboratory, AFATLTR-67-116, September 1967.

[7] Folias, E.S., http://www.math.utah.edu/~folias/esfolias _foliasfactor.html.

[8] Lutz, B.E.P., Goodwin, C.J., *Catastrophic Failure Modes Assessment of the International Space Station Alpha*; NASA-CR-4720, Prepared for MSFC, February 1996.

[9] Jeong, D.Y., Tong, P., "Nonlinear Bulging Factor Based on R-Curve Data", U.S. Department of Transportation, N95-14476, 1995.

[10] Evans, H.J., Hyde, J.L., Hoffman, K.D., *Hazard Assessment Following Micrometeoroid/Orbital Debris Penetration of International Space Station: ITA-15 Stage 1*, JSC-66916, March 2016.

[11] Wolf, Matthew, "Physiological Consequences of Rapid or Prolonged Aircraft Decompression: Evaluation Using a Human Respiratory Model", *Aviation, Space, and Environmental Medicine*, Vol. 85, No.4, April 2014.

[12] Serrano, J., Liquornik, D., and Schonberg, W.P., *Vulnerability of Space Station Freedom Modules: A Study of the Effects of Module Perforation on Crew and Equipment*, NASA CR 4716, February 1996.

[13] Williamsen, J.E., *Vulnerability of Manned Spacecraft to Crew Loss from Orbital Debris Penetration*, NASA Technical Memorandum TM-108452, April 1994.

Proceedings of the 2019 Hypervelocity Impact Symposium
HVIS2019
April 14-19, 2019, Destin, FL, USA
HVIS2019-019

Investigation on Sustained Discharge of Satellite's Power Harness Due to Plasma from Space Debris Impact

Yuki Mando[a]*, Koji Tanaka[a,b], Takayuki Hirai[c], Shirou Kawakita[d], Masumi Higashide[e], Hirohisa Kurosaki[e], Sunao Hasegawa[b], and Kumi Nitta[f]

[a] *The Graduate University for Advanced Studies, SOKENDAI, 3-1-1, Yoshinodai, Chuo, Sagamihara, Kanagawa, 252-5210, Japan*
[b] *Institute of Space and Astronautical Science/ Japan Aerospace Exploration Agency, 3-1-1, Yoshinodai, Chuo, Sagamihara, Kanagawa, 252-5210, Japan*
[c] *Planetary Exploration Research Center, Chiba Institute of Technology, 2-17-1, Tsudanuma, Narashino, Chiba, 275-0016, Japan*
[d] *Japan Aerospace Exploration Agency, 2-1-1, Sengen, Tsukuba, Ibaraki, 305-8505, Japan*
[e] *Japan Aerospace Exploration Agency, 7-44-1, Higashimachi , Jindaiji, Chofu, Tokyo, 182-8522, Japan*
[f] *Japan Aerospace Exploration Agency, Ochanomizu sola city ,4-6 Kandasurugadai, Chiyoda-ku, Tokyo, 101-8008, Japan*

Abstract

Space debris travels at a velocity of 7-8 km/s in low Earth orbit (LEO) and at 3 km/s in geostationary Earth orbit (GEO). An impact between space debris and spacecraft will result in tremendous damage. In particular, particles less than 1mm in diameter pose a risk of causing permanent sustained discharge (PSD). PSD may affect a satellite's power system. The effect on solar arrays has been well-studied given their large area, but the effect on the bundle of a satellite's wire harness (called the power harness) has yet to be clarified, even though the power harness is usually exposed to the space environment without protection. We conducted hypervelocity impact experiments using a two-stage light gas gun, and investigated the risk resulting in PSD from hypervelocity impacts of particles less than 1mm in size. In addition, we compared two kinds of circuit configurations: a more realistic circuit configuration with internal resistance and a circuit configuration without it, so as to investigate whether internal resistance affects the occurrence of PSD. Stainless steel and aluminum oxide projectiles measuring from 0.3 to 1 mm in diameter were gun-accelerated up to 7.16 km/s. Targets entailed a three-layered power harness under a simulated power condition of typical satellites operating in LEO or GEO. As a result, 11 of 28 shots resulted in PSD. With the more realistic circuit configuration we could not confirm any results regarding PSD. We thus found that PSD is less likely to occur in a more realistic circuit configuration.

Keywords: Sustained discharge; Impact-induced plasma; Power harness; Space debris

Nomenclature

V_{SAS1}	source voltage on the side of solar array simulator 1 (V)
I_{SAS1}	source current on the side of solar array simulator 1 (A)
V_{SAS2}	source voltage on the side of solar array simulator 2 (V)
I_{SAS2}	source current on the side of solar array simulator 2 (A)
I_p	plasma current (A)
N_e	electron density (m^{-3})
T_e	electron temperature (K)
S	area of the double probe (m^2)
e	elementary charge (C)

* Corresponding author. Tel.: +81-50-3362-7559
E-mail address: mando.yuki@ac.jaxa.jp

k	Boltzmann constant (J/K)
m_i	mass of ion (kg)
V_p	plasma voltage (V)
R	detected resistance in the double probe circuit (Ω)
C	capacitance of bypass condenser in the double probe circuit (F)

1. Introduction

Space debris such as abandoned artificial satellites, rocket bodies, and the fragments thereof poses an urgent issue relative to human activity in space. Space debris travels at a velocity of 7-8 km/s in low Earth orbit (LEO) at an altitude of around 160 to 500 km above the surface of Earth and at 3 km/s in geostationary Earth orbit (GEO) at an altitude of about 36,000 km above the surface of Earth [1]. An impact between space debris and spacecraft will result in tremendous damage. Space debris impact (i.e. hypervelocity impact) on spacecraft not only causes mechanical damage such as destruction of the main body or solar panel, but also causes electrical damage due to the impact-induced plasma. The effect of electrical damage on solar arrays has been well-studied given their large area [2-4]. An impact-induced plasma leads to arcing between the solar cells or between a cell and substrate on the solar array. At worst, Joule heating from the discharge carbonizes an insulation layer and creates a permanent short-circuit path. This is called permanent sustained discharge (PSD). Past studies have suggested that PSD would occur due to impact of a projectile less than a few millimeters in diameter. However, the effect of electrical damage on the bundle of a satellite's wire harness (called the power harness) has not been well-studied, even though the power harness is usually exposed to the space environment without protection. The sustained arc between the primary power cables of a satellite was investigated by Kawakita et al. [5]. Nitta et al. analyzed the response of the cable harness to hypervelocity impacts by comparing experimental results with the numerical simulation of AUTODYN-3D [6]. Hirai et al. conducted hypervelocity impact experiments using the two-stage light gas gun of ISAS/JAXA and examined the risk of electrical damage on the power harness [7]. They suggested that impacts from a stainless steel projectile less than 1mm in diameter may result in PSD in a power configuration of 100V/3A. However, a physical interpretation of the occurrence of PSD has yet to be reached.

In this paper, we conduct hypervelocity impact experiments using the two-stage light gas gun and investigate the of PSD resulting from hypervelocity impacts of particles less than 1mm in diameter. In addition, we compare two kinds of circuit configurations: a more realistic circuit configuration with internal resistance and a circuit configuration without it, so as to investigate whether internal resistance and capacitance affect the occurrence of PSD.

2. Hypervelocity Impact Experiments

Hypervelocity impact experiments were conducted using the two-stage light gas gun installed at the Institute of Space and Astronautical Science (ISAS)/ Japan Aerospace Exploration Agency (JAXA). Stainless steel and aluminum oxide projectiles measuring less than 1mm in diameter were accelerated up to 7.16 km/s. The air pressure of a target chamber was evacuated to 1-10 Pa (mean free path: 0.7-6.7mm). Figure 1 shows a photo of the power harness specimen. The specimen consists of the power harness and an A2024 substrate plate with a thickness of 3 mm to simulate the single line of the power harness behind the solar array panel. For the power harness, we selected the SPEC 55 ® wire of TE connectivity (single wall, AWG22) that is commonly used for Japanese spacecraft.

Figure 2 shows two kinds of circuit configurations used for different purposes. For both types, a three-layered power harness is connected and two solar array simulators (SAS) supply the voltage and current. One is the E4362A (Keysight Technologies), and the other is the E4360A (Agilent Technologies). The applied voltage and current values were simulated in the spacecraft configuration in LEO and GEO, and then recorded using an oscilloscope. For type 1, SAS1 is connected to the power harness bundle between the first and second layers with the circuit open. SAS2 is connected to the power harness bundle and the aluminum plate through bleeder resistance with the circuit also opened. For type 2, variable resistance and capacitance were added to simulate load resistance of the satellite bus power.

In order to investigate the effect of impact-induced plasma, arrays of double probes were arranged near the impact point as shown in Fig. 1. Unlike the conventional double probe method [8], a constant voltage of 9.6 V was applied between the electrodes to get the ion-saturated current due to a short-time phenomenon lasting dozens of microseconds [9]. Figure 3 shows the circuit diagram of the double probe. A plasma current I_p flows in accordance with electron density N_e or electron temperature T_e of the impact-induced plasma. I_p is obtained by calculating the potential difference V_p between the detected resistance R. The bypass condenser C was inserted parallel to the applied voltage to prevent the voltage from fluctuating. The electron density N_e is calculated by the following equation [9]:

$$I_p = 0.61 N_e e S \, (kT_e/m_i) \quad (1)$$

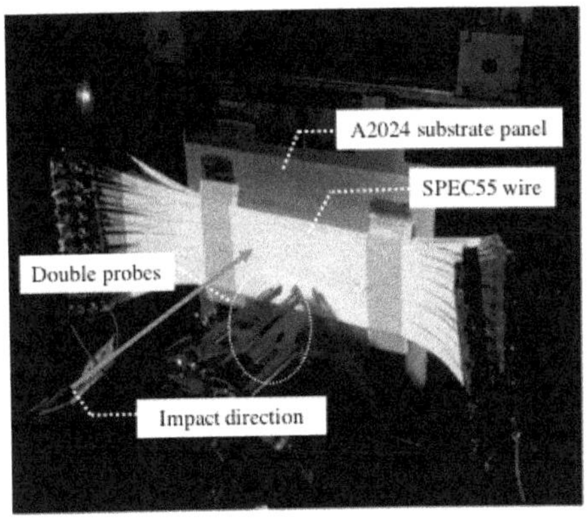

Fig. 1. Photo of the power harness specimen.

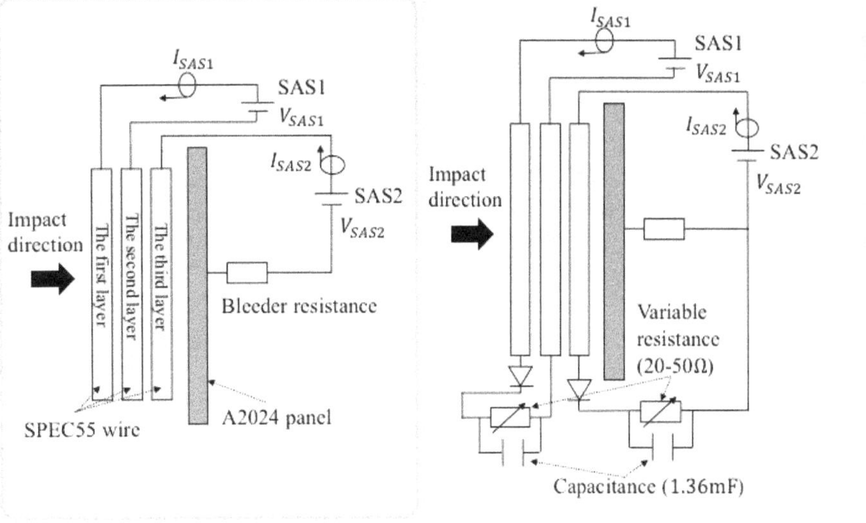

(a) Type 1 (b) Type 2

Fig.2. Circuit diagrams of two types of configurations.

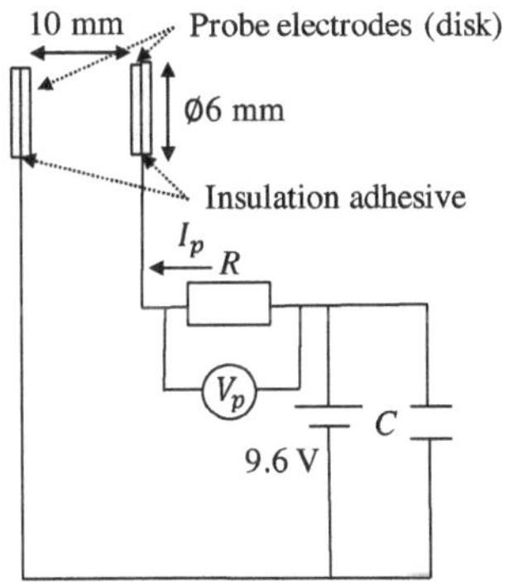

Fig.3. Circuit diagram of the double probe.

3. Results and Discussion

Table 1 lists the results of the hypervelocity impact experiments in which 11 of 28 shots resulted in PSD that lasted for more than 180 ms. The type 1 circuit configuration resulted in PSD, but the type 2 configuration has yet to be confirmed. For type 1, we can categorize three patterns of the power harness specimen's appearance as shown in Fig. 4. One is the case where the projectile penetrated through the first and second layers of the power harness as shown in Fig. 4 (a). The scars on the surface of the power harness can be seen on the two layers. Secondly, Fig.4 (b) shows the case where PSD occurred on the first and second layers. The surface appearance near the scar looks burnt in color on the first and the second layers. This means that the insulation of the power harness was molten and carbonized on the first and second layers, but we could not confirm any scar on the third layer. Lastly, Fig. 4 (c) shows the case where PSD occurs on all three layers. The large areas of the power harness on all three layers were molten and carbonized.

In Fig. 5, we also categorized three patterns of voltage and current waveforms. In Fig. 5 (a), the voltage V_{SAS1} surges to 188.2 V and returns to 100 V after approximately 234 μs. In this case, it did not result in PSD and the molten and carbonized power harness could not be seen. In Fig. 5 (b), the voltage V_{SAS1} drops and the current I_{SAS1} increases after the impact, but PSD from the waveforms V_{SAS2} and I_{SAS2} has not occurred. In Fig. 5 (c), both V_{SAS1} and V_{SAS2} drop, and both I_{SAS1} and I_{SAS2} increase after the impact. We obviously confirmed that both circuits with SAS1 and SAS2 have reached PSD. For the results of type 1, PSD is likely to occur regardless of the diameter and velocity of the projectile.

For the type 2 circuit configuration, we have not confirmed the results regarding PSD. Figure 6 shows the typical results in the case of the type 2 circuit configuration. In this case, six projectiles collided with the power harness specimen. Figure 6 (a) shows the appearance of the specimen. A certain portion of the power harness was damaged, but we could not confirm that the insulator was molten and carbonized. Figure 6 (b) shows the voltage and current waveforms. Waveforms V_{SAS1} and I_{SAS1} drop for approximately 40 ms, but the waveform V_{SAS1} gradually returns and the waveform I_{SAS1} suddenly returns. For the results of type 2, a temporary discharge lasting less than approximately 40.9 ms occurred, but did not lead to PSD. These results suggest that the difference between the type 1 and type 2 circuit configurations is whether internal resistance is present or not. Because the primary discharge is mainly triggered by dense plasma around the crater and the size of the crater is smaller than the mean free path (0.7-6.7mm) of this experimental condition (1-10Pa), we can confirm that occurrence of PSD is successfully simulated in this experimental condition.

Table 1. Experimental results

Shot ID	Specimen Type	Projectile material	Projectile diameter (mm)	Impact number	Impact velocity (km/s)	Source current (A)	Source voltage (V)	Discharge duration (ms)	PSD occurrence
170905-7	Type 1	SUS304	0.3	15	3.85	3.0/3.0	100/80	>180	Yes (on both circuits with SAS1 and SAS2)
170906-2	Type 1	SUS304	0.3	6	3.82	3.0/3.0	100/80	>180	Yes (on the circuit with SAS1)
170906-1	Type 1	SUS304	0.3	3	3.88	3.0/3.0	100/80	0	No
171017-1	Type 1	SUS304	1.0	1	4.12	3.0/3.0	100/80	>180	Yes (on both circuits with SAS1 and SAS2)
170906-3	Type 1	SUS304	0.3	1	4.16	3.0/3.0	100/80	>180	Yes (on the circuit with SAS1)
170907-1	Type 1	SUS304	0.5	1	4.25	3.0/3.0	100/80	>180	Yes (on the circuit with SAS1 and SAS2)
171016-3	Type 1	SUS304	0.3	1	6.87	3.0/3.0	100/80	>180	Yes (on the circuit with SAS1)
171016-2	Type 1	SUS304	0.8	1	6.95	3.0/3.0	100/80	>180	Yes (on the circuit with SAS1 and SAS2)
171016-1	Type 1	SUS304	1.0	1	7.03	3.0/3.0	100/80	13.0+22.5	No
171016-4	Type 1	SUS304	0.5	1	7.07	3.0/3.0	100/80	>180	Yes (on the circuit with SAS1 and SAS2)
170906-5	Type 1	SUS304	0.3	1	7.07	3.0/3.0	100/80	>180	Yes (on the circuit with SAS1)
171017-2	Type 1	SUS304	0.5	1	7.10	5.5/5.5	50/50	>180	Yes (on the circuit with SAS1 and SAS2)
171016-6	Type 1	Al_2O_3	0.3	1	7.16	3.0/3.0	100/80	<0.1	No
171013-1	Type 1	SUS304	0.5	1	7.17	3.0/3.0	100/80	>180	Yes (on the circuit with SAS1 and SAS2)
170627-4	Type2	SUS304	0.3	6	4.01	3.0/3.0	100/80	41.2	No
170626-5	Type2	SUS304	0.3	1	4.06	5.5/5.5	50/50	0	No
170626-7	Type2	SUS304	0.5	1	4.05	5.5/5.5	50/50	0	No
170626-1	Type2	SUS304	0.3	1	4.09	5.5/5.5	50/50	0	No
170626-4	Type2	SUS304	0.3	12	4.11	5.5/5.5	50/50	1.2	No
170622-1	Type2	SUS304	0.3	1	4.11	5.5/5.5	50/50	<0.1	No
170627-3	Type2	SUS304	0.3	1	4.11	3.0/3.0	100/80	<0.1	No
170623-1	Type2	SUS304	0.3	1	4.12	5.5/5.5	50/50	0	No
170627-1	Type2	SUS304	0.3	1	4.12	3.0/3.0	100/80	<0.1	No
170907-5	Type2	SUS304	0.3	~20	6.80	3.0/3.0	100/80	3.5	No
170907-4	Type2	SUS304	0.3	1	6.84	3.0/3.0	100/80	<0.1	No
171017-3	Type2	SUS304	0.5	1	7.17	5.5/5.5	50/50	<0.1	No
171017-4	Type2	SUS304	0.3	1	7.16	5.5/5.5	50/50	<0.1	No
171017-5	Type2	SUS304	0.3	1	7.16	5.5/5.5	50/50	0	No

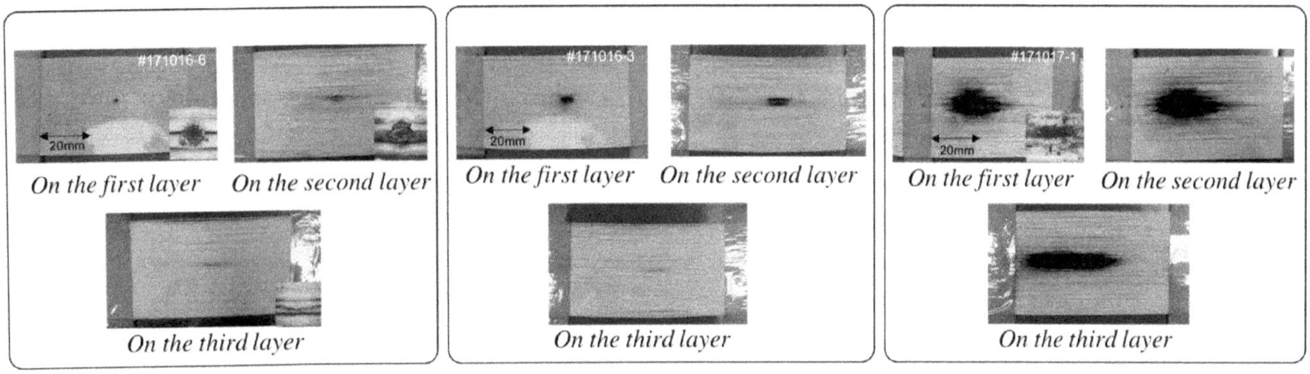

(a) Non-molten and carbonized specimen. (b) The molten and carbonized specimen on the first and the second layers. (c) The molten and carbonized specimen on all layers.

Fig. 4. Three types of cataloged power harness specimens: (a) non-molten and carbonized specimen of shot ID 171016-6, (b) molten and carbonized specimen on the first and second layers of shot ID 171016-3, and (c) molten and carbonized specimen on three layers of shot ID 171017-1.

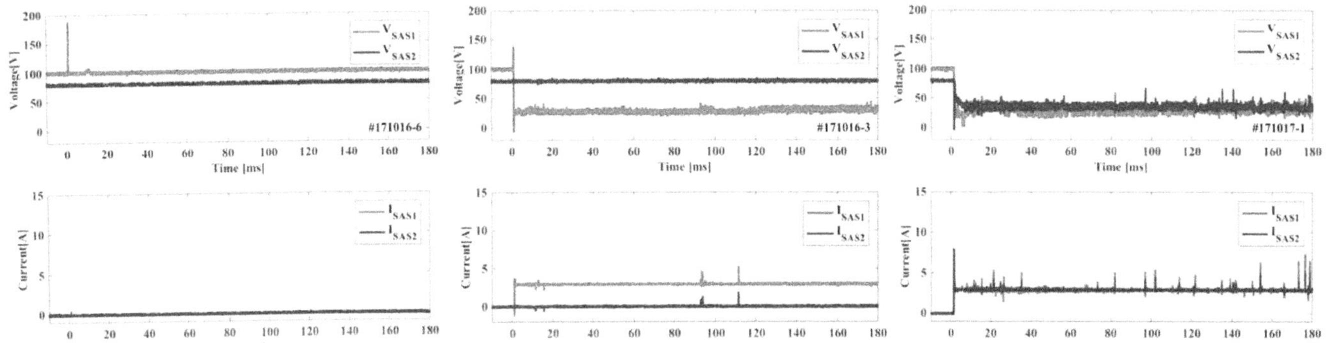

(a) Non-PSD occurrence. (b) PSD occurrence on the circuit with SAS1. (c) PSD occurrence on both circuits with SAS1 and SAS2.

Fig. 5. Three patterns of voltage and current waveforms of type 1 in case of (a) Non-PSD occurrence of shot ID 171016-6, (b) PSD occurrence on the circuit with SAS1 of shot ID 171016-3, and (c) PSD occurrence on both circuits with SAS1 and SAS2 of shot ID 171017-1.

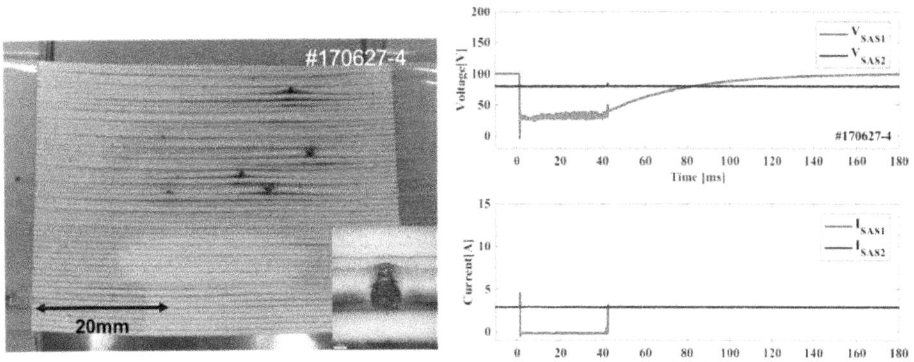

(a) An appearance of power harness specimen. (b) Voltage and current waveforms.

Fig. 6. Typical results in case of the type 2 circuit configuration (shot ID: 170627-4).

4. Conclusions

We conducted hypervelocity impact experiments using a two-stage light gas gun and confirmed that 11 of 28 shots resulted in PSD. In these experiments, we prepared two kinds of circuit configurations to investigate whether internal resistance affects the occurrence of PSD. For type 1, 11 of 14 shots resulted in PSD. There are three patterns: non-PSD occurrence, PSD occurrence on the circuit with only SAS1, and PSD occurrence of both circuits with SAS1 and SAS2. When a projectile less than 1 mm in diameter collides with the power harness, PSD may occur on either circuit. For type 2, which is the more realistic circuit configuration, we have not confirmed that PSD occurs on any circuit. Our results suggest that internal resistance prevents PSD from occurring. We therefore conclude that PSD is less likely to occur in a more realistic circuit configuration.

Acknowledgements

The hypervelocity impact experiments were supported by ISAS/JAXA.

References

[1] Anil K. Maini, and V. Agrawal, 2010. Satellite Technology Principles and Applications, A John Wiley and Sons, Ltd., Publication, Second Edition, pp. 66-68.

[2] S. Fukushige, Y. Akahoshi, K. Watanabe, T. Nagasaki, K. Sugawara, T Koura, and M. Cho, 2008. Solar-Array Arcing Due to Plasma Created by Space-Debris Impact, IEEE Transactions on Plasma Science, Vol. 36, No. 5, pp.2434-2439.

[3] Y. Akahoshi, T. Nakamura, S. Fukushige, N. Fukusawa, S. Kusunoki, Y. Machida, T. Koura, K. Watanabe, S. Hosoda, T. Fujita, and M. Cho, 2008. Influence of space debris impact on solar array under power generation, International Journal of Impact Engeering, Vol. 35, Issue 12, pp.1678-1682.

[4] S. Kawakita, G. Segami, K. Nitta, H. Kusawake, M. Takahashi, H. Matsumoto, and S. Hasegawa, 2009. Discharge of Spacecraft Solar Panels by Hypervelocity Impact, Trans. JSASS Space Technology of Japan, Vol. 7, No. ists26, pp. Tr_2_53-Tr_2_56.

[5] S. Kawakita, H. Kusawake, M. Takahashi, H. Maejima, J. Kim, S. Hosoda, M.Cho, K. Toyoda, and Y. Nozaki, 2004. "Sustained Arc between Primary Power Cables of a Satellite," 2nd International Energy Conversion Engineering Conference.

[6] K. Nitta, S. Kawakita, A. Takeda, M. Katayama, 2010. Response of Cable Harnesses Subjected to High-Velocity Impacts, Trans. JSASS Aerospace Technology of Japan, Vol. 8, No. ists 27, pp. Pr_2_55-Pr_2_62.

[7] T. Hirai, M. Higashide, H. Kurosaki, S. Kawakita, S. Hasegawa, Y. Mando, S. Yamaguchi, and K. Tanaka, 2017. Re-examination of electrical failure risk on satellite's power harnesses caused by space debris impacts: simultaneous measurements of sustained discharge and plasma density, Procedia Engineering, Vol. 204, pp.445-451.

[8] E. O. Johnson and L. Malter, 1950. A Floating Double Probe Method for Measurements in Gas Discharges, Phys. Rev. 80, 58.

[9] Y. Nagaoka, K. Tanaka, and S. Sasaki, 2012. Propagation of Plasma Generated by Hyper-Velocity Impact on Thin Plate Structure, Journal of JSASS, Vol.60, Issue 2, pp. 89-95.

Proceedings of the 2019 Hypervelocity Impact Symposium
HVIS2019
April 14-19, 2019, Destin, FL, USA
HVIS2019-026

A rupture limit equation for COPVs following a perforating MMOD impact

William P. Schonberg[a]*

[a]*Civil Engineering Department, Missouri University of Science & Technology, Rolla, MO 65409*

Abstract

Most spacecraft have at least one pressurized vessel on board. One of the primary design considerations for earth-orbiting spacecraft is the anticipation and mitigation of the possible damage that might occur in the event of a micrometeoroid or orbital debris (MMOD) particle impact. To prevent mission failure and possibly loss of life, protection against perforation by such high-speed impacts must be included. In addition to a hole, it is possible that, for certain pressure vessel designs, materials, impact parameters, and operating conditions, a pressure vessel may experience catastrophic failure (i.e. rupture) as a result of a hypervelocity impact. If such a tank rupture were to occur on-orbit following an MMOD impact, not only could it lead to loss of spacecraft, but quite possibly, for human missions, it could also result in loss of life. In this paper we present an update to a Rupture Limit Equation, or RLE, for composite overwrapped pressure vessels (COPVs) that was presented previously. The update consists of modified RLE parameters and coefficients that were obtained after the RLE was re-derived using new / additional data. The updated RLE functions in a manner similar to that of a ballistic limit equation, or BLE, that is, it differentiates between regions of operating and impact conditions that, given a tank wall perforation, would result in either tank rupture or only a relatively small hole or crack. This is an important consideration in the design of a COPV pressurized tank – if possible, design parameters and operating conditions should be chosen such that additional sizable debris (such as that which would be created in the event of tank rupture or catastrophic failure) is not created as a result of an on-orbit MMOD particle impact.

Keywords: space debris; COPV; MMOD; tank rupture; hypervelocity impact; composite material; pressure vessel

1. Introduction

Most spacecraft have at least one pressurized vessel on board. For robotic spacecraft, it is usually a liquid propellant tank, while for human spacecraft, there are also pressurized living quarters (or modules). Because of the potential of serious mission- and life-threatening damage that might result following an on-orbit micro-meteoroid or orbital debris (MMOD) particle impact, one of the primary design considerations of such spacecraft is the anticipation and mitigation of the possible damage that might occur in the event of such an impact. To prevent mission failure and possibly loss of life, protection against perforation by high-speed impacts must be included.

Considerable energy and effort has been expended in the study of the response of flat unstressed (i.e. non-pressurized) spacecraft components under hypervelocity impact conditions intended to simulate those of an on-orbit MMOD impact. However, fuel tanks are pressurized internally, and their main walls will develop bi-axial stress fields because of that internal pressurization. Until recently, technical and safety challenges have limited the number of tests that have been conducted on the pressurized elements of such spacecraft. Especially limited has been the number of hypervelocity impact tests performed on composite

* Corresponding author. Tel.: +1-573-341-4787.
E-mail address: wschon@mst.edu.

overwrapped pressure vessels (COPVs). In an effort to address this issue, a testing and analysis program has been undertaken to characterize the hypervelocity impact response of COPVs.

In addition to a hole, it is possible that, for certain COPV designs, materials, impact parameters, and operating conditions, a COPV may experience catastrophic failure (i.e. rupture) as a result of a hypervelocity impact. If such a tank rupture were to occur on-orbit following an MMOD impact, not only could it lead to loss of spacecraft, but quite possibly, for human missions, it could also result in loss of life.

In this paper we present an update to a Rupture Limit Equation, or RLE, for COPVs that was presented previously [1]. The update consists of modified RLE parameters and coefficients that were obtained after the RLE was re-derived using new / additional data. The updated RLE to be presented functions in a manner similar to that of a ballistic limit equation, or BLE. Namely, it differentiates between regions of operating and impact conditions that, given a tank wall perforation, would result in either tank rupture or only a relatively small hole or crack.

This is an important consideration in the design of a COPV pressurized tank – if possible, design parameters and operating conditions should be chosen such that additional sizable debris (such as that which would be created in the event of tank rupture or catastrophic failure) is not created as a result of an on-orbit MMOD particle impact.

Based on a comparison of the updated RLE with experimental results, the equation developed herein is seen to be able to make a clean separation between experimentally-obtained non-rupture and rupture data points. As such, the equation presented is highly accurate in predicting the rupture / non-rupture response of COPVs for the impact conditions considered.

2. Test Data Overview

Data from approximately 50 high speed impact tests on several different types of COPVs are pooled together and used in the development of the RLE presented herein. Tables 1 and 2 provide a summary of the impact test conditions and the geometric parameters and material properties of the COPVs used in the test program that generated the data that were subsequently used in the development of the RLE. Additional information regarding these tests can be found in References [2-4].

Table 1 Overview of Impact Test Conditions

Parameter		**Reference [2]**				**Refs. [3,4]**	**Units**
		Type I COPV		**Type II COPV**			
Projectile	**Material**	Al 2017-T4	440C SS	Al 2017-T4	440C SS	Aluminum	
	Density	2.79[a]	7.63[a]	2.79[a]	7.63	2.8[b]	gm/cm^3
	Diameter	1.0 – 2.5	1.09, 1.29	2.71, 3.4, 5.01	1.70 – 1.99	8.0	mm
Trajectory	**Obliquity**	0, 45, 60	45	0, 45	45	0	deg
	Velocity	~ 4.09, ~7.05	~7.06	~7.05	~7.00	~4.35 – ~4.46	km/s
Impact Location		Cylinder		Cylinder		Cylinder	
COPV Contents		Air		Air		Water, GN2	
COPV Pressure		0, ~29		~31		~0, 20-50	MPa

[a]Nominal values from www.matweb.com, [b]Assumed average nominal value

Table 2 Geometric Parameters and Material Properties of COPVs

Parameter		**Reference [2]**		**Refs. [3,4]**	**Units**
		Type I COPV	**Type II COPV**		
COPV Diameter (O.D.)		16	42.5	10	cm
Composite	**Fiber**	T1000	T1000	CFRP	
	Matrix	LRF-092	HARF-53		
	Tensile Strength	3040[a]	3040[a]	1515[e]	MPa
	Density	1.56[a]	1.56[a]	1.65[e]	gm/cm^3
	Thickness[b]	2.54	6.35	4.8	mm
Liner	**Material**	Al 6061-T6	Inconel	Aluminum	
	Tensile Strength	455[c]	1282[c]	455[e]	MPa
	Shear Strength	273[c]	882[c]	273[e]	MPa
	Yield Stress	285[c]	1041[c]	285[e]	MPa
	Density	2.70[c]	8.19[c]	2.8[e]	gm/cm^3
	Thickness[d]	1.98	0.81	2.3	mm

[a]Nominal value from manufacturer's literature, [b]Nominal composite thickness in cylindrical portion of COPV, [c]Nominal values from www.matweb.com, [d]Nominal liner thickness in cylindrical portion of COPV, [e]Assumed average nominal values

3. Rupture Limit Equation Development

Following the approach used in several previous studies (see, e.g. [5-7]), a simple power law was chosen as the form for the updated RLE to be developed. To render the equation as broadly applicable as possible, the operating conditions (x-axis) were parameterized as the hoop stress in the tank (non-dimensionalized by the uni-directional ultimate stress of the tank wall composite material), and the impact conditions (y-axis) were parameterized as impact momentum (non-dimensionalized by a number of appropriate tank wall material properties). Specifically, the power law for the curve that separates regions of rupture and non-rupture was chosen as follows:

$$\textit{Non-dimensional Projectile Momentum} = A\left(\frac{\sigma_{hoop}}{\sigma_{ult}^{comp}}\right)^{B} \tag{1}$$

where σ_{hoop} and σ_{ult}^{comp} are the COPV hoop stress and uni-directional ultimate stress of the COPV composite material, respectively. COPV hoop stresses were calculated using the equation

$$\sigma_{hoop} = \frac{p_{int} r_{OD}}{t_{tot}} \tag{2}$$

where $t_{tot} = t_{comp} + t_{liner}$ is the total nominal thickness of the cylindrical portion of the composite material overwrap, p_{int} is the internal pressure in the COPV, and r_{OD} is the COPV outer radius. The non-dimensional form of projectile momentum was taken initially to be given as follows:

$$\textit{Non-dimensional Projectile Momentum} = \frac{m_{proj} V_{proj}^{norm}}{(\rho_{comp} t_{tot}^{3})\sqrt{\frac{\sigma_{ult}^{comp}}{\rho_{comp}}}} \tag{3}$$

where V_{proj}^{norm} is the normal component of the impact velocity. The first term in the denominator in Eqn (3) has units of mass while the second has units of velocity, thereby rendering the right-hand-side of Eqn (3) unitless, or non-dimensional, so long as there is consistency in the units of mass and velocity used in its numerator and denominator. To include the effects of other material parameters, Eqn (3) was modified through the addition of a number of other unitless terms, with the following result:

$$\textit{Non-dimensional Projectile Momentum} = \frac{m_{proj} V_{proj}^{norm}}{(\rho_{comp} t_{tot}^{3})\sqrt{\frac{\sigma_{ult}^{comp}}{\rho_{comp}}}}\left(\frac{\rho_{proj}}{\rho_{comp}}\right)^{Q}\left(1+\frac{\sigma_{ult}^{liner}}{100}\right)^{S}\left(1+\frac{\sigma_{yld}^{liner}}{100}\right)^{T} \tag{4}$$

where the ultimate and yield stresses of the liner material, σ_{ult}^{liner} and σ_{yld}^{liner}, respectively, are in MPa. The liner material yield and ultimate stress values were divided by 100 so the terms involving these quantities can have a meaningful effect without being excessively large compared to the values of the other terms in the equation. Combining Eqns (1) and (4) yields the final form of the RLE as follows:

$$\frac{m_{proj} V_{proj}^{norm}}{(\rho_{comp} t_{tot}^{3})\sqrt{\frac{\sigma_{ult}^{comp}}{\rho_{comp}}}}\left(\frac{\rho_{proj}}{\rho_{comp}}\right)^{Q}\left(1+\frac{\sigma_{ult}^{liner}}{100}\right)^{S}\left(1+\frac{\sigma_{yld}^{liner}}{100}\right)^{T} = A\left(\frac{\sigma_{hoop}}{\sigma_{ult}^{comp}}\right)^{B} \tag{5}$$

The particular forms of the terms in Eqns (1)–(5) were motivated by the desire to include terms that not only characterized the impact loads and internal pressurization conditions, but at the same time included material parameters that were seen to differ among the various test programs considered herein. As such, the terms used in Eqns (1)–(5) are consistent with the identifying test specimen and material parameters shown in Table 1. Furthermore, the ability to use these equations does not rely on having to know the exact lay-up or stacking sequence of the composite laminae in the COPV – all that is needed is the uni-axial ultimate strength of the composite material. This serves to increase the ease of use of these equations and their potential applicability to a wider family of COPV constructions.

The exponents Q,S,T in Eqn (4) were selected so as to allow, as much as possible, a natural separation between the ruptured and non-ruptured data points. This would, in turn, facilitate the development of a RLE that would, again, as much as possible, lie between those two regions. The attractiveness and benefit of this approach is that if additional tests results were to become available, the exponents Q,S,T in Eqn (3) could be adjusted to allow the incorporation of whatever new rupture / non-rupture data becomes available and the subsequent development of a new RLE. To this end, the values of the exponents Q,S,T used in the non-dimensionalization scheme defined in Eqn (4) are $Q = 0.6$, $S = -1.0$, and $T = -1.0$.

The constants A and B in Eqn (1) are first determined through a linear regression of the rupture / non-rupture data using the Levenberg–Marquardt algorithm [8]. This is a common algorithm for minimizing a function over the space of parameters of the function (i.e. the coefficients in a user-defined function). In addition to solving for the unknown coefficients, the algorithm also provides statistical information that can be used to assess the "goodness of fit" of the regression results and to develop standard deviation curves for the regression model.

This initial regression exercise was actually performed by first creating a function Z that would take on values of +1 or -1 depending on whether a particular test resulted in either a rupture or a non-rupture. This process is described in detail in References [6,7], and yields $A = 2.961x10^{-3}$, and $B = -0.9912$ with a correlation coefficient of 72.7%. This completes the development of the updated RLE for the test data considered in this study.

4. Comparison with Empirical Data

Figures 1 and 2 show plots of the RLE developed using the process outlined above compared to experimental rupture / non-rupture results. These figures also show plots of $\pm$ one standard deviation curves and $\pm$ two standard deviation curves about the RLE curve. Figure 1 is a plot of all the data, while Figure 2 is a "zoomed-in" version of Figure 1 to allow closer inspection of the cluster of data at lower momentum values. In Figures 1 and 2, the orange data points represent those tests that resulted in tank rupture, while the green data points show those that did not rupture, but did sustain a perforation through the composite overwrap and internal metallic liner. The blue data points correspond to those tests in which perforation of the COPV liner did not occur, although in such tests the composite overwrap did sustain damage, even down to the metallic liner in some cases.

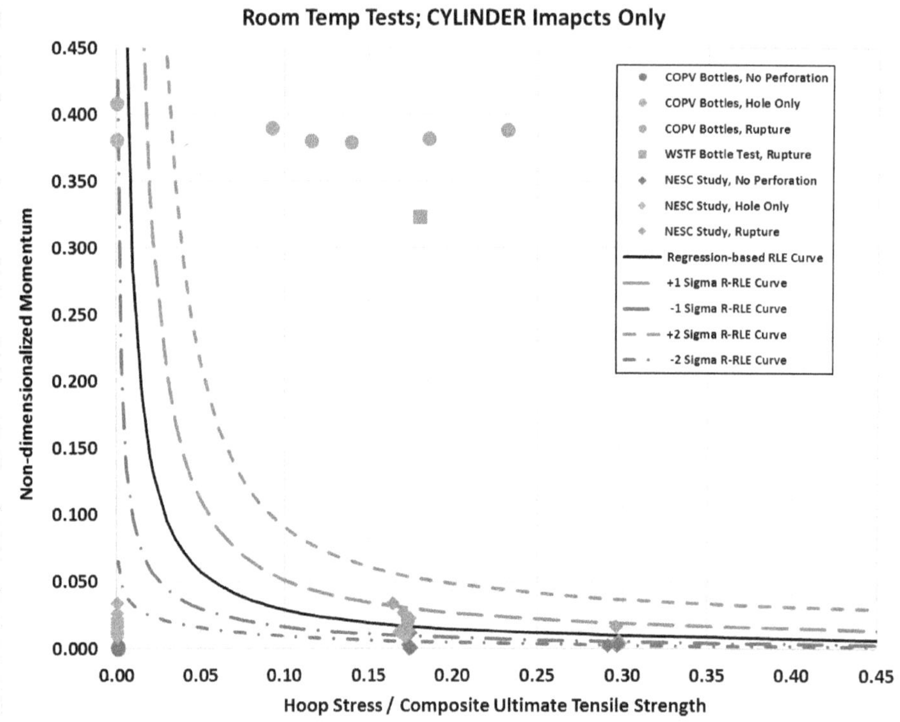

Figure 1. Plots of Regression-based RLE Curve, Standard Deviation Curves, and Experimental Rupture/No-rupture Data

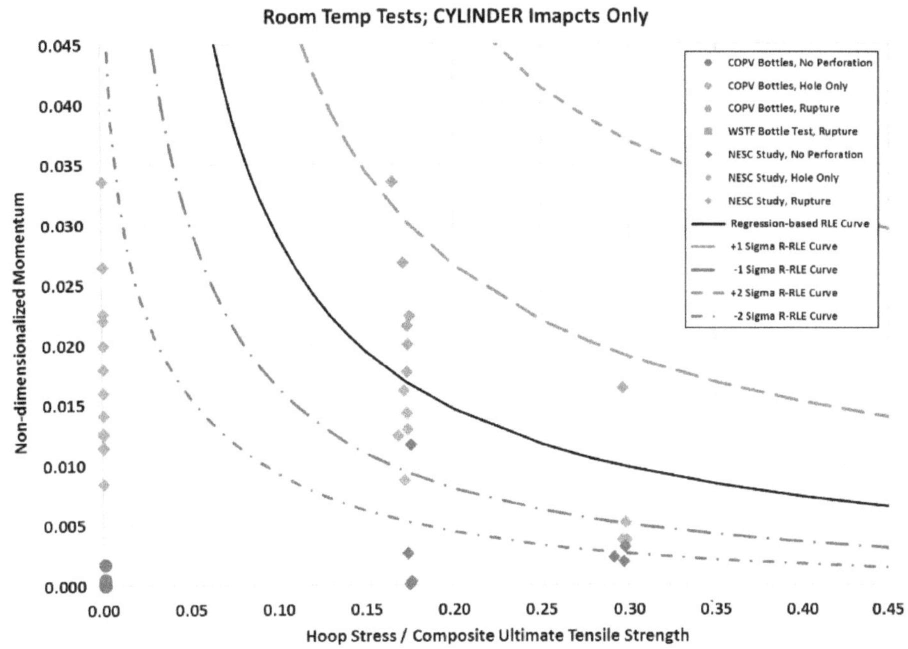

Figure 2. Zoomed in Plots of Regression-based RLE Curve, Standard Deviation Curves, and Experimental Rupture/No-rupture Data

A closer look the curves in Figures 1 and 2 reveals the following points of interest.

1. The RLE shown in Figure 1 as well as in Figure 2 appears to make a clean cut between the green (non-rupture) and orange (rupture) data points. Thus, the RLE curve developed using this approach is 100% accurate in discriminating between COPV ruptures and non-ruptures as indicated by the data obtained thus far.
2. The RLE curve appears to be "skewed" to lie closer to the NR (green) data points near ordinate values of 0.40-0.45. However, this is merely an artifact of the regression process which was performed using log values of actual non-dimensionalized hoop stress and projectile momentum.
3. The confidence bounds on the RLE shown in Figures 1 and 2 appear to be a bit wide. However, this is not unexpected for the following reasons.
    a. As can be seen in these figures, the data are grouped in four clusters: horizontally near a non-dimensional momentum value of approx. 0.4 (i.e. the test data from Refs. [3,4]), vertically about non-dimensional hoop stress values of ~0 (i.e. the unpressurized tests from Ref. [2] and the unpressurized Phase A tests), at approx. 0.173 (i.e. the remainder of the Phase A tests, which were pressurized), and at approx. 0.297 (i.e. the pressurized Phase B tests). Although the data from unpressurized tests (i.e. those with hoop stress values of ~0 from Ref. [2] and some Phase A tests) were not used in the development of the RLE, wide confidence bounds such as those shown in Figures 1 and 2 are to be expected considering the distribution of the test data that were used in the development of this updated RLE.
    b. The validity of any regression analysis is predicated on the assumption of relatively continuous changes in dependent variable values with respect to corresponding changes in the values of the independent variables. In this exercise, values of +1 or -1 were assigned to the dependent variable function, depending on whether a particular test resulted in either a rupture or a not-rupture. This approach, while certainly capable of producing a curve fit for this type of non-continuous dependent variable data, is known to produce highly conservative error bounds, which are clearly evident in Figures 1 and 2.
4. While this RLE is certainly satisfactory from the perspective of being able to separate the rupture data points from the non-rupture data points, the wide confidence bounds in Figures 1 and 2 indicate that the RLE could be anywhere in the confidence region, and that it contains sensitive parameters whose slight variation could possibly significantly change the location and shape of the curve. Until and unless additional tests are performed to generate data points in other regions of the plots shown

in these figures, it is unlikely that best fit curves with "tighter" confidence bounds can be developed using this approach. If a reduction in the uncertainty bounds shown in Figures 1 and 2 is desired, a more sophisticated or robust approach for determining alternative curves that discriminate between rupture and non-rupture data points is required.

5. The decided lack of overlapping of the rupture/non-rupture test data is primarily due to the judicious selection of the values of the exponents *Q,S,* and *T*. However, this lack of overlap is admittedly rather unusual, especially in hypervelocity impact test programs (see, e.g. the overlap and scatter in such testing of metallic dual-wall systems [9]). It is indeed possible that additional testing of similar COPVs could yield test results that cannot be so cleanly separated. In this case, there might be a bit of overlap and engineering judgement will be needed regarding where to best place the RLE so as to best meet program needs and requirements. For example, in a program where a more conservative curve is required (e.g. one in which crew safety is paramount), the RLE can be shifted "downward" amongst any overlapping data. However, if a higher element of risk is acceptable in a particular program, then it can be shifted "upward" amongst the overlapping R/NR data.

5. Conclusions

A study was performed to generate data to support the development of a new response predictor equation for COPVs. The purpose of this equation is to predict which combinations of operating and impact conditions, given a tank wall perforation by a high-speed MMOD particle, would result in either tank rupture or only a relatively small hole or crack. This is an important consideration in the design of a COPV tank – if possible, design parameters and operating conditions should be chosen such that additional sizable debris is not created as a result of an on-orbit MMOD particle impact. Based on a comparison with experimental results, the equation developed herein is able to make a clean separation between experimentally-obtained non-rupture and rupture data points. This is primarily due to a decided lack of overlapping in the rupture/non-rupture test data, which is itself due to the judicious selection of the values of the some of the exponents in the equation. As such, the equation presented is highly accurate in predicting the rupture/non-rupture response of the COPVs and impact conditions considered in this study.

Acknowledgments

The author would like to gratefully acknowledge the support by the NASA Engineering Safety Center which made this study possible. The author also gratefully acknowledges the NASA White Sands Test Facility staff who performed the high speed impact testing that generated the test data used in this study. The guidance and helpful suggestions of Dr. V.A. Samaranayake, Curator's Teaching Professor of Mathematics and Statistics at Missouri S&T, are also gratefully acknowledged.

References

[1] Schonberg, W.P., "Rupture of Composite Pressure Vessels (COPVs) Following a Hypervelocity MMOD Particle Impact", Proceedings of the 2018 AIAA Sci-Tech Forum, Paper No. AIAA-2018-0231, Kissimmee, Florida, January, 2018.

[2] Garcia, M.A, Davis, B.A., and Miller, J.E, "Hypervelocity Impact Testing of a Pressurized Composite Overwrapped Pressure Vessel and Comparison to Numerical Analysis", *Procedia Engineering*, Vol. 204, 2017, pp. 476–483.

[3] Hereil, P.L., Mespoulet, J., and Plassard, F., "Hypervelocity Impact of Aluminum Projectiles against Pressurized Aluminum-Composite Vessel", *Procedia Engineering*, Vol. 103, 2015, pp. 181-188.

[4] Hassin, J., Pfaab, K., Puillet, C., Limido, J., Lacome, J.L., and Hereil, P.L., "Hypervelocity Impacts on Composite Over-wrapped Pressurized Vessels", Proceedings of DYNCOMP 2015, Arles, France, June 2015.

[5] Schonberg, W.P., and Ratliff, M., "Hypervelocity Impact of a Pressurized Vessel: Comparison of Ballistic Limit Equation Predictions with Test Data and Rupture Limit Equation Development", *Acta Astronautica*, Vol. 115, 2015, pp. 400-406.

[6] Schonberg, W.P., "Rupture of a Cryogenic Composite Overwrapped Pressure Vessel Following a High-Speed Particle Impact", *Aerospace*, Vol. 5, 2018, Paper No. 20.

[7] Schonberg, W.P., "A Rupture Limit Equation for Pre-Loaded Composite Plates", *Journal of Composites Science*, Vol. 2, 2018, Paper No. 3.

[8] PSIPlot: Scientific Spreadsheet and Technical Plotting User's Guide, Poly Software International, Pearl River, NY, 2010.

[9] Schonberg, W.P., and Compton, L.E., "Application of the NASA/JSC Whipple Shield Ballistic Limit Equations to Dual-Wall Targets under Hypervelocity Impact", *International Journal of Impact Engineering*, Vol. 35, 2008, pp. 1792-1798.

Proceedings of the 2019 Hypervelocity Impact Symposium
HVIS2019
April 14-19, 2019, Destin, FL, USA
HVIS2019-029

Predicting orbital debris-induced failure risk of wire harnesses using SPH hydrocode modeling

Joel Williamsen[a], Michael Squire[b], Steven Evans[a]*

[a]*Institute for Defense Analyses, 4850 Mark Center Drive , Alexandria, VA 22311, USA*
[b]*NASA Langley Research Center, 11 Langley Boulevard, Hampton, VA 23681, USA*

Abstract

This paper describes a method derived to assess the probability of two types of complex cable failures (partial and full wire breaks), considering their location with respect to the debris spray from penetration of multi-layer insulation (MLI) suspended over them, and the likelihood of impacting particle sizes and velocities as predicted by NASA's model for predicting orbital debris impact size and velocity distributions for satellites in low earth orbit, ORDEM. The smooth particle hydrodynamics (SPH) code was used to determine the onset of these two failure types following hypervelocity impact for different orbital debris velocities, sizes and orientations relative to four different wire locations for a prototypical satellite in a 98-degree polar orbit at an altitude of approximately 750 km (i.e., a typical weather satellite). Interpolations between hydrocode results, combined with ORDEM predictions of orbital debris likelihoods, were used to predict overall risk of each failure type. Adding a few layers of beta cloth over the wires cut the risk of each failure type in half.

Keywords: Orbital debris, spacecraft wiring failure, risk prediction, spacecraft failure prediction, hypervelocity impact.

Nomenclature

cm	centimeters
km	kilometers
km/s	kilometers per second
EMI	Electromagnetic Interference

1. Background and Approach

The National Aeronautics and Space Administration (NASA) recently released a new orbital debris environment (ORDEM 3.0) that included an increase in the orbital debris population for many satellites in low earth orbit [1]. In the light of this increased environment, the NASA Engineering and Safety Center (NESC) has sponsored several recent assessments of the risk of orbital debris impact-induced failures of satellites in low Earth orbit [2]. Part of this assessment was an effort to predict failure risk of wire harnesses, usually located in cable trays under thin layers of multi-layer insulation (MLI). This paper describes a method derived to assess the probability of two types of complex cable failures, considering their location with respect to the debris spray from penetration of MLI suspended over them, and the likelihood of impacting particle sizes and velocities as predicted by ORDEM. The Smooth Particle Hydrodynamics (SPH) code was used to determine the onset of these two failure types following hypervelocity impact for different orbital debris velocities, sizes and orientations, relative to four different wire locations for a prototypical satellite in a 98-degree polar orbit at an altitude of approximately 750 km (i.e., a typical weather satellite).

---

* Corresponding author. Tel.: 1-703-994-6551
E-mail address: jwilliam@ida.org

Interpolations between hydrocode results, combined with ORDEM predictions of orbital debris likelihoods, were used to predict overall risk of two failure types (partial and full wire breaks).

1.1. *Objectives*

The two objectives of this task were to estimate the number of failures for the cables for primarily aluminum projectiles, and identify potential risk reductions by modifying the design.

1.2. Approach

As shown in Figure 1, the wires were assumed to be located in a tray inclined "back" at a 45-degree angle to the spacecraft velocity vector (or "ram direction"). An MLI blanket is positioned parallel to the cable mounting surface at a distance of 2.54 cm from the surface (this MLI layer is referred to as the "bumper"). Any impacting particle will break up on contact with the MLI bumper and produce a spray of particles onto the cables. Because of this effect, cables that are not directly in the path of the particle can still be affected. The analyses considered four cable positions relative to the impact point: on the shot line (P1), 2.54 cm below the shot line (P2, normal to the impact point), 1 cm below the shot line (P3), and 1 cm above the shot line (P4). Each cable is oriented so that the long side faces the impact point on the MLI, which is consistent with most cable runs (i.e., wires run along, not perpendicular to a cable tray).

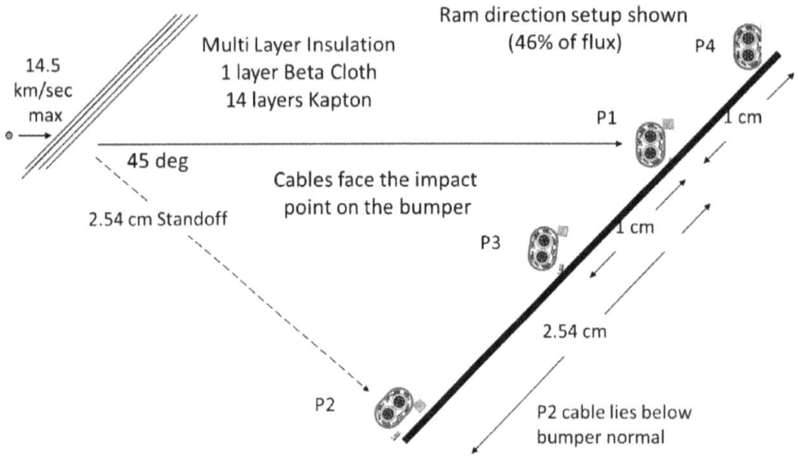

Fig. 1. Wire positions relative to impact point considered in analyses

Per ORDEM3.0, the maximum orbital debris impact velocity in a 90-degree polar orbit will be 14.52 km/s and be from the ram direction (i.e., 0° impact angle). The impact velocity decreases as the impact angle (from the ram direction along the orbital plane) increases, as shown in Table 1.

Table 1. Orbital debris impactor velocity as a function of impact angle derived from ORDEM.

Velocity (km/s)	Angle (degrees) from Ram	Cumulative Percentile Flux by Impact Direction
14.52	**0 (5)**	**.46**
14.52	15	.62
13.58	25	.71
12.30	35	.78
10.76	45	.84
8.76	55	.89
6.40	65	.94
3.96	75	.98
2.16	85	1.0

As shown in Figure 2, the cable was assumed to comprise two 19-strand copper wires. Each of the two wires was encased in a Teflon® insulator, and the pair was wrapped in copper, Kapton®, and electromagnetic interference (EMI) tape (metal and nylon).

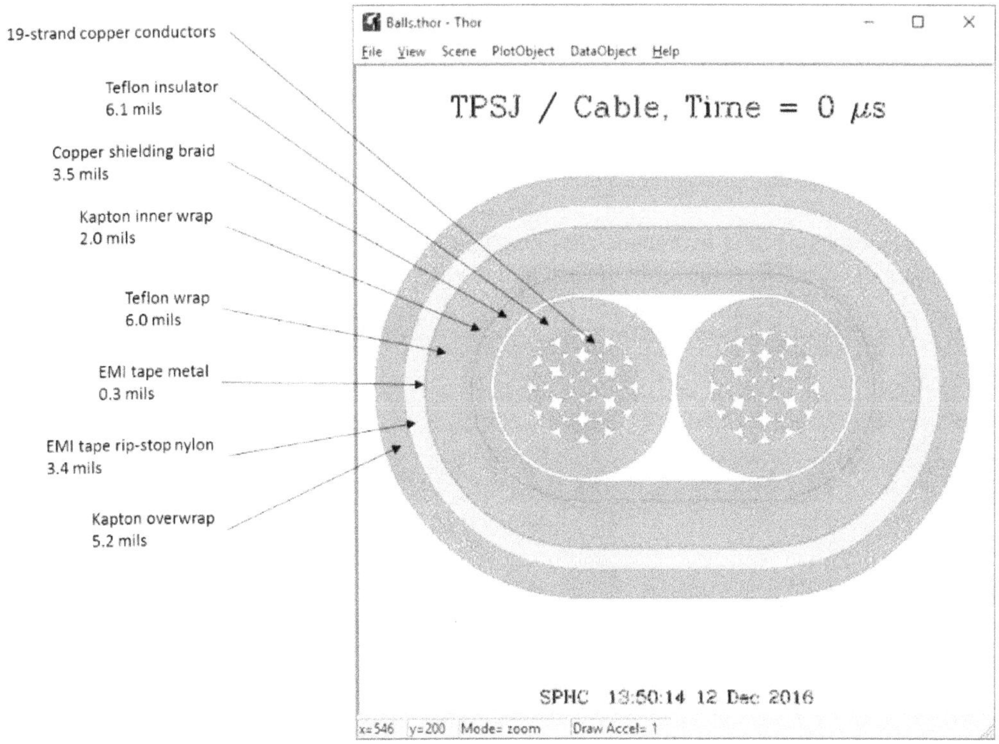

Fig. 2. Modelled characteristics of wire cable

In this example, the dimensions of the region with the cables was taken as 0.10 m by 2.31 m. The percentage of the cabled area at risk to MMOD damage for each of the positions was estimated by taking a scaled drawing of one-third of the region and overlaying a grid. The number of grid elements covered by possible impact points relative to each of the four cable positions defined (P1 through P4) was counted to estimate the total area at risk for each of the cable positions. Using this method, for any given impact, there was a 38 percent probability that there was an affected cable in the P1 position (that is, closer to P1 than any of the other three postulated wire positions). The probabilities for the four cable positions, plus the regions where no wires are affected, are shown in Table 2.

Table 2. Cable impact position coverage following examination of overall cable runs

Cable Position	Coverage Percentage
P1 (on shot line)	38%
P2 (normal to impact/2.54 cm below shot line)	4%
P3 (1 cm below shot line)	20%
P4 (1 cm above shot line)	24%
No wires affected	14%

All hydrocode runs in this study were made considering two velocities: 14.52 km/s (i.e., from the ram direction), resulting in a 45-degree obliquity to the MLI, and at 12.30 km/s (35 degrees off the ram direction), resulting in a 67-degree obliquity to the MLI, representing a range of ~78 percent of the expected flux velocities. Type 1 failures (potentially leading to shorts) were those hydrocode runs that exposed the wire conductors, while type 2 failure (potential wire breaks) were taken in hydrocode runs to show more than 50% conductor penetration. According to Miller, et al, [3] type 1 failures typically manifest as a short, and type 2 failures manifest as an open circuit failure in the power or data cable.

Figures 3 and 4 show examples of simulations of Type 1 and Type 2 failures, respectively. The simulation critical diameter results were interpolated for approach angles between 0 degrees and 35 degrees. Above 35 degrees (below 12.30 km/sec), the results were assumed to be the same as the 35-degree results (considered a conservative assumption, since these impact velocities are lower, and generally less energetic/damaging). Once critical diameters were obtained, ORDEM3.0 determined the flux for particles of that size. That flux was then multiplied by the effective area for each approach angle from 5 degrees to 85 degrees in five-degree increments, and by the coverage percentage for the cable position (from Table 2) to obtain the total predicted number of failures for a 3.5-year mission. The use of 5 degrees for the direct shot line was to avoid singularities that were associated with using 0 degrees. The predicted number of values were then summed to calculate a total number of predicted failures of that failure type and wire position.

Figures 3 and 4 show typical results of SPH hydrocode runs resulting in Type 1 and Type 2 failures. By rotating the 3D hydrocode outputs, one can more easily interpret the onset of each failure type.

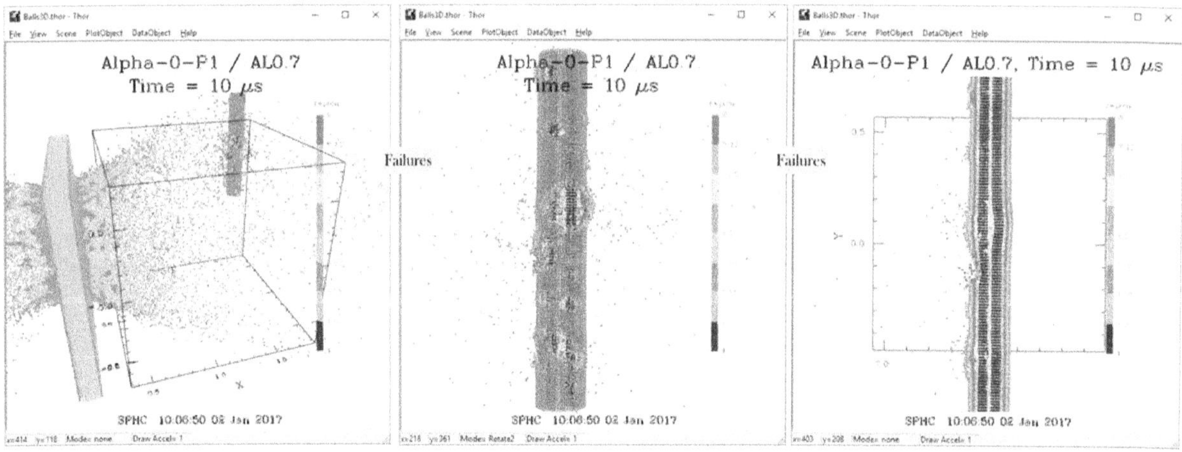

Fig. 3. Example of SPH simulation of a Type 1 failure (exposed conductor)

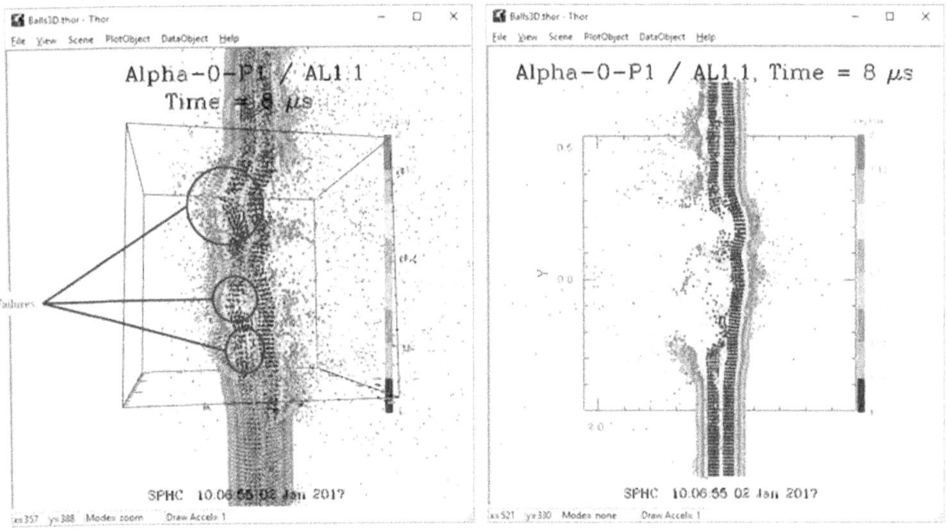

Figure 4. Example of SPH simulation of a Type 2 failure (>50% conductor penetration)

2. Baseline Results

SPH runs were performed for aluminum particle sizes from 0.5 mm up to 1.7 mm, in increments of 0.2 mm for each of the four cable positions. Damage to the cable after each run was evaluated and judged to be either a Type 1 or Type 2 failure, or no failure at all. The smallest particle size to cause damage defined the critical diameter for that damage type. The results of the simulations are summarized in Table 3 (0-degree approach angle) and Table 4 (35-degree approach angle). For these tables and all others like them, T1 = Type 1 failure, T2 = Type 2 failure, NF = no failure, red text = inferred value (i.e., no simulation was run at these conditions).

Table 3. SPH Results for Aluminum Particles Approaching at 0°/14.5 km/s

	Particle Size (mm)					
Position	**0.5**	**0.7**	**0.9**	**1.1**	**1.3**	**1.5**
P4	NF	T1	T1	T1	T2	T2
P1	NF	T1	T1	T2	T2	T2
P3	NF	T1	T1	T1	T1	T2
P2	NF	T1	T1	T1	T1	T1

Table 4. SPH Results for Aluminum Particles Approaching at 35°/12.3 km/s

	Particle Size (mm)						
Position	**0.5**	**0.7**	**0.9**	**1.1**	**1.3**	**1.5**	**1.7**
P4	NF	NF	T1	T1	T1	T1	T1
P1	NF	T1	T1	T1	T2	T2	T2
P3	NF	T1	T1	T1	T1	T2	T2
P2	NF	NF	T1	T1	T1	T1	T1

Using the values in Tables 3 and 4 for critical diameter, the exposed area of the cable tray (0.10 m by 2.31 m), and the probabilities of impacting each of the wire positions P1-P4 from Table 2 as inputs, the ORDEM3.0 analysis was applied to derive the number of failures (for 2 failure types) as shown in Table 5 over 3.25 years starting in 2018 considering only medium density (aluminum) particles (at the request of the sponsor).

Table 5. Predicted Number of Failures

Failure Type	Calculated Number of Failures P1	P2	P3	P4	Total
Type 1	1.001	0.082	0.527	0.497	**2.107**
Type 2	0.128	0.002	0.0206	0.038	**0.194**

Recall that the critical diameter for any impact angle greater than 35 degrees was assumed to be equal to that for the 35-degree conditions. This is a conservative assumption, so a sensitivity study was performed to assume perfect shields for impact angles above 35 degrees, meaning damage would not be possible for any particle size (this would be an extreme non-conservative assumption). This assumption had a small effect on the results, where the Type 1 damage dropped from 2.1 to 1.9 impacts, and the Type 2 dropped from 0.19 to 0.18 impacts. Therefore, the assumption stated at the beginning of this paragraph was appropriate for this exercise.

3. Improved Design Results

Next, two layers of beta cloth were simulated adjacent to the cable (see Figure 5). The critical diameter results for these SPH runs are shown in Table 5.

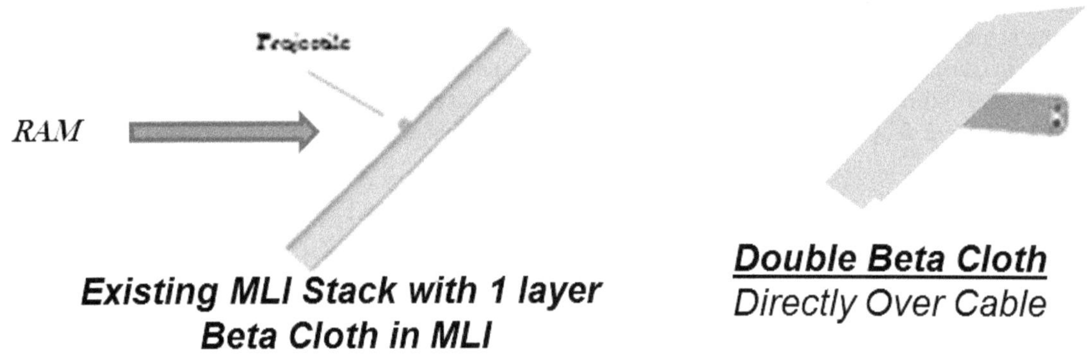

Figure 5. Improved design, with double Beta cloth placed directly over cables

Table 5. SPH Results for aluminum particles at 0°/14.5 km/s for baseline configuration plus two layers of Beta cloth adjacent to cables

Position	Particle Size (mm) 0.5	0.7	0.9	1.1	1.3	1.5
P4	NF	NF	T1	T1	T1	T1
P1	NF	NF	T1	T1	T1/T2	T2
P3	NF	NF	T1	T1	T1	T1
P2	NF	NF	T1	T1	T1	T1

Both Type 1 and Type 2 failures were improved with the additional beta cloth layer. The adjusted predicted numbers of impacts for the new critical diameters are shown in Table 6.

Table 6. Predicted number of impacts for the dual layer beta cloth (BC) adjacent to the cables configuration (changes from baseline are shown in red)

	Calculated Number of Failures					
	Failure Type	**P1**	**P2**	**P3**	**P4**	**Total**
Baseline	Type 1	1.001	0.082	0.527	0.497	**2.107**
	Type 2	0.128	0.002	0.0206	0.038	**0.194**
Dual BC Adjacent	Type 1	0.386	0.032	0.203	0.191	**0.812**
	Type 2	0.056	0.002	0.014	0.018	**0.090**

In this case, adding the two layers of beta cloth resulted in a significant (over 50 percent) reduction in risk for both Type 1 and Type 2 failures. Type 1 failures decreased from 2.107 to 0.812 and Type 2 failures from 0.194 to 0.090. All of the SPH runs for the risk reduction were at 14.5 km/s impact velocity, and the results were extrapolated to the other velocities. The intent of this activity was to produce a first-order approximation of the impact risk to the cables and show how it can be reduced. More detailed models and runs using the Bumper software, higher fidelity cable models, and inclusion of steel particles would be necessary to produce results that are more accurate.

Acknowledgements

The authors wish to thank the NASA NESC for its sponsorship of this research.

References

[1] Matney, et al., 2014. Orbital Debris Environment Model, ORDEM 3.0, NASA Johnson Space Center Presentation to NESC, March 2014.
[2] Squire, M., et al., 2014. Joint Polar Satellite System Micrometeoroid and Orbital Debris (MMOD) Assessment, NASA Engineering and Safety Council (NESC) Study TI-14-00948, April 2015.
[3] Miller, J.E., et al., 2019. Cumulative distribution of ballistic impact failures of common twisted-pair data cables at orbital speeds," Int. J. Imp. Engng. 124 (2019) 61-66.

Proceedings of the 2019 Hypervelocity Impact Symposium
HVIS2019
April 14-19, 2019, Destin, FL, USA
HVIS2019-037

Depth of penetration criteria on metallic surfaces for use in MMOD risk assessment

Henry Nahra[a]*, Louis Ghosn[a], Eric Christiansen[b], Joshua Miller[c,d], Bruce A. Davis[d]†

[a]NASA/GRC, 21000 Brookpark Rd., Cleveland, OH 44135, USA
[b]NASA/JSC, 2101 NASA Parkway, Houston, TX 77058, USA
[c]University of Texas at El Paso, 500 W. University Ave., El Paso, TX 79968, USA
[d]Jacobs NASA/JSC, 2224 Bay Area Blvd., Houston, TX 77058, USA

Abstract

System level assessment of hypervelocity impacts by micrometeoroids and orbital debris (MMOD) relies on the definition of the spacecraft geometry and trajectory, the natural environment of the micrometeoroids and induced environment of the orbital space debris, ballistic limit equations and the failure criteria. The definition of the MMOD environments provides the particles flux and when is combined with the ballistic limit equations will determine the number of the critical penetrating particles that could result in the failure of the underlying component is calculated and is used to calculate the risk based on some failure criterion. Spacecraft geometry provides the shielding configuration over the spacecraft critical body which defines the selection of the ballistic limit equations to be used in the risk assessment. The definition of the failure criterion for metallic pressure systems involves the definition of the allowable depth of penetration that could result in leakage or burst of the component. This paper addresses the definition of the allowable depth of penetration of generic metallic tanks from MMOD impacts. The allowable penetration depth of metal tanks is based on a fracture mechanics approach calibrated using biaxially stressed coupons tests subjected to Hypervelocity Impacts (HVI). The planar crack-crack spacing was based on the craters spacing distribution of the HVI coupon tests. The Stress Intensity Factor (SIF) as a function of crater depths and crater spacing and applied remote stress is calculated using NASGRO®, a linear fracture mechanics software. The calculated SIF is compared with the material fracture toughness to determine if the craters result in a failure of the coupons under biaxial stress. This work resulted in a recommended allowable depth of penetration of 20% on the surfaces of metallic pressure vessels on spacecraft.

Keywords: Micrometeoroids, Orbital Debris, Depth of penetration criterion, Fracture Mechanics, Shielding, HVI coupon tests

---

Nomenclature

a	Crack depth
Al	Aluminum
c	Crack half length
FS	Factor of Safety
$K(a)$	Stress Intensity Factor based on crack depth
K_0	Calculated stress intensity factor

* Corresponding author. Tel.: 216 433 5385; fax: 216 433 8050.
E-mail address: henry.k.nahra@nasa.gov.

K_{IC}	Fracture toughness
ksi	1000 pounds per square inch
DOP	Depth of penetration
DOPC	Depth of penetration criterion
h	Distance between cracks
HVI	Hypervelocity Impact
MCF	Multiple crack factor
MMOD	Micrometeoroids and Orbital Debris
MS	Margin of Safety
s	Distance between cracks tips
SIF	Stress Intensity Factor
STD	Standard
t	Plate thickness
Ti	Titanium
W	Plate width
Greek symbols	
σ_0	Applied tensile stress
ϕ	Angle, 10° or 90

1. Introduction

Propellant tanks in spacecraft have the potential of being vulnerable to Micrometeoroids and Orbital Debris (MMOD) impacts. The damage on the propellant tanks can vary from simple cratering to full perforation and even burst depending on the particle diameter and velocity, and material properties and geometrical configuration of the pressure tank. Micrometeoroids and Orbital Debris system level risk assessments rely on the definition of the spacecraft geometry, definition of the trajectory and the environment, ballistic limit equations and the failure criteria. The failure criterion for a propellant tank needs to be defined in terms of maximum allowable damage level above which leak or rupture could occur for the propellant tanks over the life of the mission, which contributes to the failure risk assessment for the whole spacecraft from MMOD impacts. By defining a depth of penetration criterion (DOPC) on the metallic tanks surfaces, a reduction in the shielding mass without compromising the safety and success of the spacecraft mission can be established.

1.1. Objective and motivation

This work is motivated by the need to determine a depth of penetration criterion that is supported by tests and analysis for realistic propellant tank conditions. The objective of this work is to define a depth of penetration criterion that can be used by MMOD system level risk assessment. Hypervelocity testing and fracture analysis will guide the definition of such criterion for metallic propellant tanks of spacecraft.

1.2. Background

During the Apollo era, a definition of the allowable damage was used for metallic tanks. Based on HVI tests, where glass particles of 3.2 mm were launched at a shield tank assembly, a criterion of depth of penetration not greater than 25% was adopted [1]. This definition of the depth of penetration criterion was revisited [2] because the environment changed by adding the orbital debris contribution to the particle flux. This revisited definition which was based on limited tests called for a depth of penetration of 10% on the surfaces of metallic tanks [2]. Two unpressurized Ti-6Al-4V tanks were impacted by a plume of ejecta generated by impacting an aluminum shield as shown in Figure. 1. Unexpectedly high damage to one tank was induced from an aluminum 1.6 mm projectile launched at 7 km/s. This was due to the late separation of the projectile carrier "Sabot". The late separation resulted in pieces of the sabot impacting the 0.04" thick aluminum shield, and consequently generating more fragmentation that impacted the titanium tank and resulted in craters depths that were over 50% of the tank wall thickness. Burst pressure testing was conducted on both tanks post impact. One tank failed at the expected design pressure on the opposite end of the impacted area while the other failed by crack propagation at a lower pressure (73% of the burst pressure seen on other tanks) and the crack was shown to initiate at the impact site. Measurements of the initial crack dimensions were estimated using microscopy and a linear fracture analysis based on the crack dimensions was performed. The results showed that the stress intensity factor exceeded the material toughness, and such a condition resulted in unstable crack propagation below the design burst pressure. Finite element stress analysis on the tanks showed that the von Mises stresses did not exceed the material yield strength at the lower burst pressure whereas the stresses from the shell theory exceeded the materials strength at the higher pressure. The stress analysis corroborated the argument that the failure of the tank at lower pressure was driven by crack propagation and fracture and not by exceedance of the material strength [2].

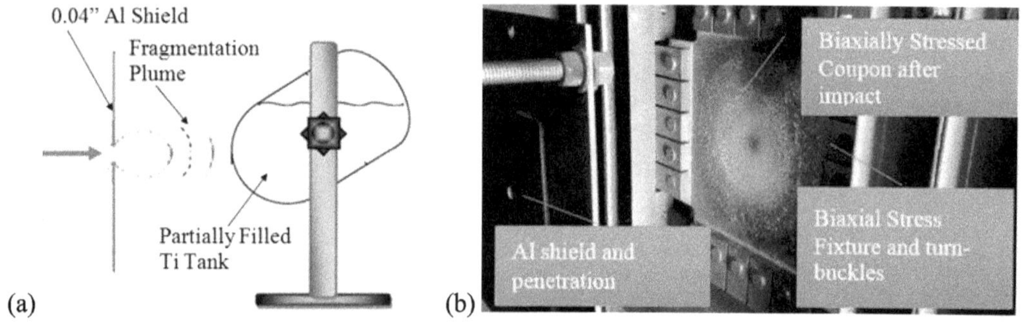

Fig. 1. Pressure tank HVI test set up.

The DOPC of 10% was deemed to be too conservative by the MMOD community given that the observed reduction in the burst pressure occurs at much greater depth of penetration. An effort was undertaken to revisit the DOPC for metallic tanks. The proposed effort was based on new coupon tests under biaxial stress loading conditions. The HVI coupon testing approach was chosen due to cost. HVI testing on pressurized propellant metallic tanks is costly due to the safety measures that need to be taken into consideration which raises the cost of performing such tests.

2. Materials & Methods

In an effort to define a sound failure criterion for impacts on metallic pressure tanks, hypervelocity impact (HVI) tests were performed on coupons of aluminum and titanium plates under a biaxial stress state. Two aluminum and two titanium coupons were indirectly impacted using different projectile diameter and materials. The biaxial stress levels of the 2219-285 aluminum and Ti-6Al-4V coupons used in the HVI coupon tests were based on the typical stress state of pressurized aluminum and titanium propellant tanks. The loads on the coupon samples were applied using a biaxial loading test fixture built at the University of Dayton Research Institute (UDRI). The determination of the biaxial stress state was based on the readings of the biaxial strain gauges that were placed on the backside of the test coupons. The biaxial stresses were calculated from the two dimensional stress-strain relation using the appropriate Poisson ratio for each of the metals. The HVI test matrix that was performed at UDRI is shown in Table 1 and the achieved impact parameters are shown in Table 2. The 2219-T85 aluminum and Ti-6Al-4V coupons were not directly impacted by the hypervelocity projectiles. The coupons were shielded by a 1mm aluminum bumper plate and the resulting ejected plume of debris from the shielding plate impacted the test coupons.

Table 1. An example of a table

Test No.	Target Material	Target Characteristics	Shield Materials	Target Material Thickness (cm)	Al Shield Thickness (cm)	Projectile Material	Projectile Diameter (mm)	Projectile Speeds (km/s)	Expected Penetration
1	Al 2219-T62	Whipple Shield	Al 606-T6	0.2	0.1	Al	4.8	7	Spall
2	Al 2219-T62	Whipple Shield	Al 606-T6	0.3	0.1	Al	5	7	Spall
3	Ti-6Al-4V	Whipple Shield	Al 606-T6	0.2	0.1	Al	5	7	Spall
4	Ti-6Al-4V	Whipple Shield	Al 606-T6	0.1	0.1	Al	4.4	7	Spall

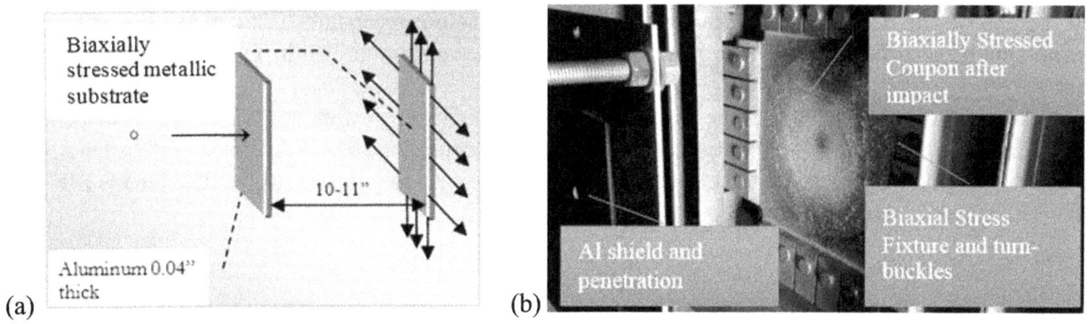

Fig. 2. Biaxial stress test a) schematic of experimental configuration and b) biaxial stressed target impacted through an aluminum shield.

3. Results

Results of the HVI tests are reported below in Table 2. Case run 4-2201 was repeated because a fragment of the steel stripper plate 15 time more massive than the projectile penetrated the Al shield and impacted the Ti-6Al-4V coupon causing it to break in two pieces. The results of the repeated test is designated as case 4-2202 and is also shown in Table 2.

Table 2. Summary of results of the HVI testing

Shot No.	Target Materials and thickness (cm)	Projectile Diameter (mm)	Projectile Materials	Projectile mass (g)	Impact Velocity (km/s)	Results
4-2198	0.2 cm 2219-T85 Al	4.76	Al 2017-T4	0.1587	6.98	1.1 cm diameter hole in bumper. Panel peppered with fine craters. Back of panel had 0.1" diameter attached spall. All strain gauges spalled from the back surface of the panel.
4-2199	0.3 cm 2219-T85 Al	5.54	Al 2017-T4	0.25	7.12	1.064 cm diameter hole in bumper. Panel peppered with fine craters. Panel not spalled. All strain gauges spalled from the back surface of the panel.
4-2200	0.2 cm Ti-6Al-4V	4.76	Al 2017-T4	0.1587	6.99	0.94 cm diameter hole in bumper. Panel peppered with fine craters. Panel not spalled. All strain gauges spalled from the back surface of the panel.
4-2201	0.1 cm Ti-6Al-4V	4.39	Al 2017-T4	0.1587	6.85	1.061 cm diameter hole in bumper. Panel peppered with fine craters. 0.51 by 0.76 by 1.5 cm steel stripper plate fragment weighing 2.46 g penetrated bumper and struck panel causing it to break in two. All strain gauges spalled from the back surface of the panel.
4-2202	0.1 cm Ti-6Al-4V	4.76	Al 2017-T4	0.1587	7.22	1.1 cm diameter hole in bumper. Panel peppered with fine craters. Panel not spalled. All strain gauges spalled from the back surface of the panel.

4. Discussion

4.1. Fracture analysis-single surface crack

A linear fracture analysis was performed on test run sample 2198 and 2200 (0.2 cm 2219-T85 Al coupon and 0.2 cm Ti-6Al-4V) using NASGRO®. NASGRO® is computer software developed to provide easy-to-use analytical tools to calculate the stress intensity factor and the fatigue life of space flight hardware using basic fracture mechanics principles. For this analysis, the material properties and applied stresses are similar to the test conditions and are summarized in Table 3 for 2219-T85 Al and in Table 4 for Ti-6AL-4V.

The stress intensity factor (SIF) *K(a)* based on the crack depth was calculated as a function of the depth of penetration for different values of *a/c* where *a* is the crack depth (corresponding to the crater depth) and *c* is ½ the crack surface length. The *K(a)* determination was based on the Newman and Raju methodology [4] as implemented in NASGRO®. The tensile SIF calculation in [4] is based on the following relation,

$$K_0 = M_0 g_1 f_\phi f_W f_x \sqrt{\pi a} \tag{1}$$

After a direct substitution of these individual terms, the SIF expression becomes a function of crack depth, *a*, the crack half width, *c*, the applied tensile stress, σ_0, the plate width, *W*, and the plate thickness, *t*, and the angle, ϕ, for the SIF at the crack depth ϕ=90 degrees,

$$K_0 = f(a, c, \sigma_0, W, \phi, t) =$$

$$\left(\left(1.13 - \frac{0.09`a}{c} + \frac{a^4\left(0.5 + 14\left(1-\frac{a}{c}\right)^{24} - \frac{1}{0.65+\frac{a}{c}}\right)}{t^4} + \frac{a^2\left(-0.5 + \frac{0.89}{0.2+\frac{a}{c}}\right)}{t^2}\right)\right.$$

$$\sqrt{\pi a}\sqrt{Sec\left[\frac{c\pi\sqrt{\frac{a}{t}}}{W}\right]}\left(1+\left(0.1+\frac{0.35a^2}{t^2}\right)(1-Sin[\phi])^2\right)$$

$$\left.\left(\frac{a^2\,Cos[\phi]^2}{c^2} + Sin[\phi]^2\right)^{1/4}\sigma_0\right)\Bigg/\left(\sqrt{1+1.464\left(\frac{a}{c}\right)^{1.65}}\right) \tag{2}$$

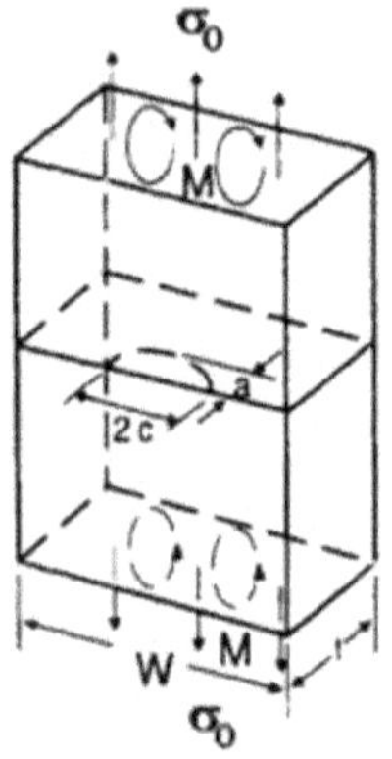

Definition of the parameters used in the Newman-Raju SIF formulation

The results of the calculations of *K(a)* SIF are shown in Figure 3 which also shows the maximum depth of penetration (DOP) measured on this particular sample and the Aluminum 2219-T85 fracture toughness, K_{IC}. It can be seen from Figure 3 that there is a significant margin except for the red curve which corresponds to an *a/c* of 0.1 which at a depth of penetration of about 0.066" (1.68 mm), the SIF exceeds the toughness.

Table 3. Assumed parameters for 2219-T85 Al for fracture analysis

Parameter	Value
Material	2219-T85 Al
Plate thickness	0.079" (2.0 mm)
Plate geometry	10"x10" (254x254mm)
Fracture toughness	30 ksi-in$^{1/2}$ (33 MPa. m$^{1/2}$)
Applied stress	30.6 ksi (211MPa)
Yield stress	57 ksi (393 MPa)
Multiple crack factor	1.2

Table 4. Assumed parameters for Ti-6AL-4V for fracture analysis

Parameter	Value
Material	Ti-6AL-4V
Plate thickness	0.079" (2.0 mm)
Plate geometry	10"x10" (254x254mm)
Fracture toughness	40 ksi-in$^{1/2}$(44 MPa. m$^{1/2}$)
Applied stress	88.47 ksi (610 MPa)
Yield stress	160 ksi (1103 MPa)
Multiple crack factor	1.2

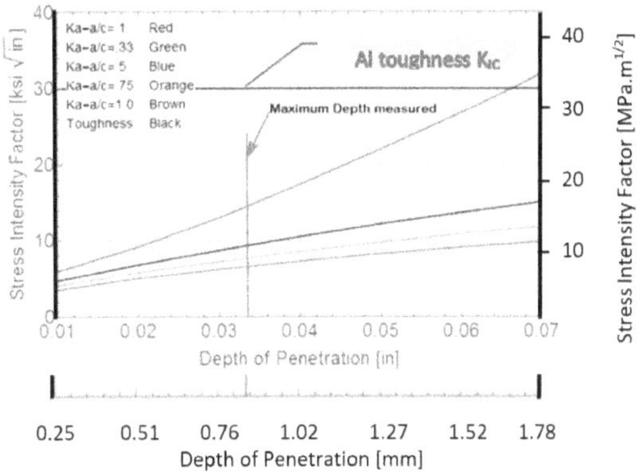

Fig. 3. Calculated stress intensity factor (SIF) as a function of crater depth for 0.079" (2.0 mm) thick 2219-T85 Aluminum material.

4.2. Multiple cracks interaction and multiple crack factor

The calculation of the SIF based on one crater is not however representative of the physical situation of the cratered substrate. Crater-crater interaction and contribution to unstable crack propagation must be taken into consideration in order to represent realistically the failure mode of a tank caused by the interactions with nearby cracks and in turn resulting in the failure of a tank. A multiple Crack Factor (MCF) must be considered in the analysis to account for multiple craters aiding in the failure of the cratered surface.

Interaction of surface cracks depends on the distance between tips of cracks, s, and on the distance between the middle of two crack widths, h, as shown in Figure 4a. The decrease in SIF is computed in terms of these two parameters for various combinations of s and h [5, 6]. In order to determine the multiple crack factor to be used in the analysis, experimental measurements were conducted using a photo of the cratered surface of an impacted Al coupon shown in Figure 4b. Results from the measurements made using Figure 4b were used to determine a conservative Multiple Crack Factor of 1.2 from the data provided in Reference [5, 6].

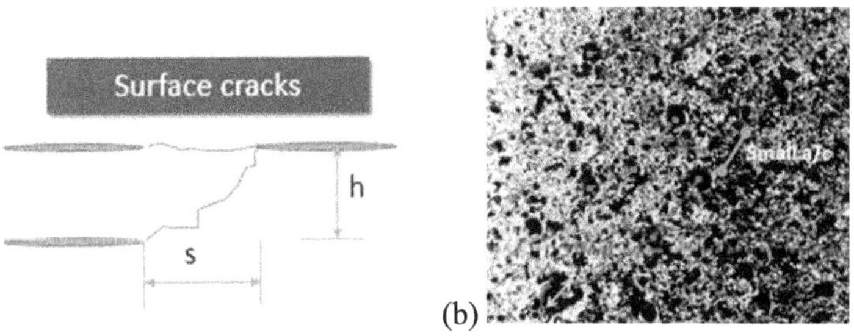

(a) (b)

Fig. 4. Multiple crack model a) parameters influencing multiple surface crack effects on the stress intensity factor and b) multiple cracks parameter measurements.

4.3. Margin of safety for unstable fracture

Based on NASA-STD-5019 "Fracture Control Requirements for Spaceflight Hardware" [7], a factor of safety (FS) of 1.4 must be considered in the calculation of the margin of safety for metallic pressure vessels. The margin of safety on unstable fracture (MS) is defined as,

$$MS = \frac{K_{IC}}{1.4 \times MCF \times K_{Applied}} - 1 \tag{3}$$

Where K_{IC} in equation 3 is the fracture toughness of the test article.

4.4. Margin of safety for aluminum and titanium impacted samples

Margin of safety analysis for sample 2198-Al 2219 T85 shows that there is a difference in the margin of safety between including and not including the multiple crack factor. Margin for unstable fracture is plotted as a function of depth of penetration for different values of *a/c* as shown in Figure 5a and 5b. For an *a/c* ratio of 0.1, and without consideration of crack-crack interaction (Figure 5a), the zero margin occurs at a depth of penetration of little over 60% which is much greater than the maximum depth of penetration measured on the sample. The zero margin of safety however, and for the same case (*K(a)*, and *a/c*=0.1) shifts to 50%.

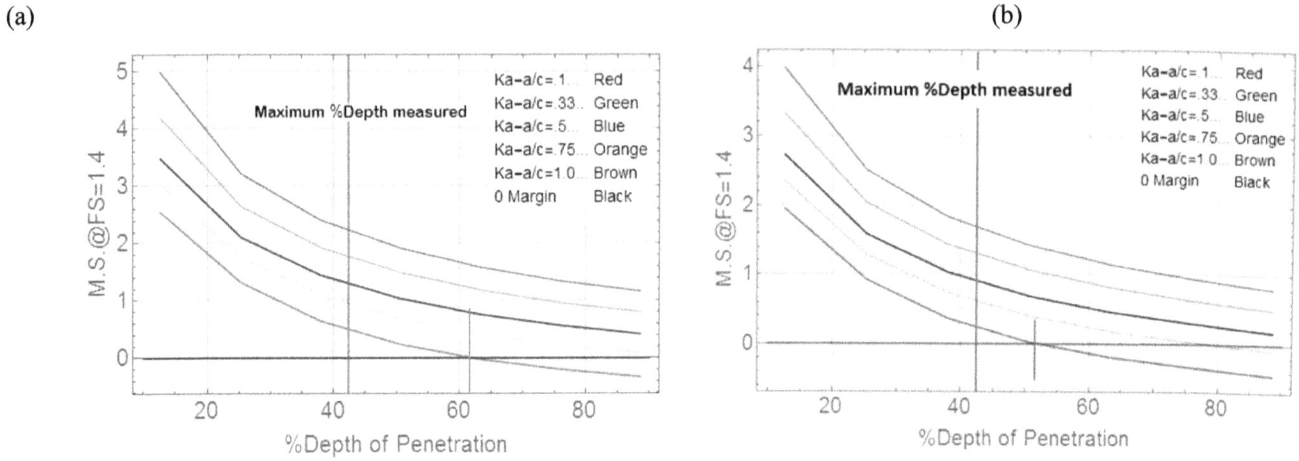

Fig. 5. Margin of safety based on a factor of safety as a function of (a) % depth of penetration without the multiple crack factor and (b) % depth of penetration with the multiple crack factor (note the shift in the zero margin of safety).

Margin of safety for unstable fracture was also assessed for the Ti-6Al-4V coupon samples. Figure 6 shows the SIF calculated based on the parameters summarized in table 4. For the titanium sample, the maximum depth of penetration observed is much less than 0.034" (0.864 mm) measured for the aluminum although the projectile mass, diameter and density and the bumper are similar between the two runs. At the maximum measured depth of penetration of the Titanium sample, the calculated stress intensity factor is much less than the Titanium toughness of 40 ksi-in$^{1/2}$ (44 MPa m$^{1/2}$). Data in Figure 6a were used to calculate the margin of safety as prescribed in Equation 3, and the results are shown in Figure 6b. For the worst case crack geometry of *a/c*=0.1, and at the maximum 20% DOP, the MS is still positive.

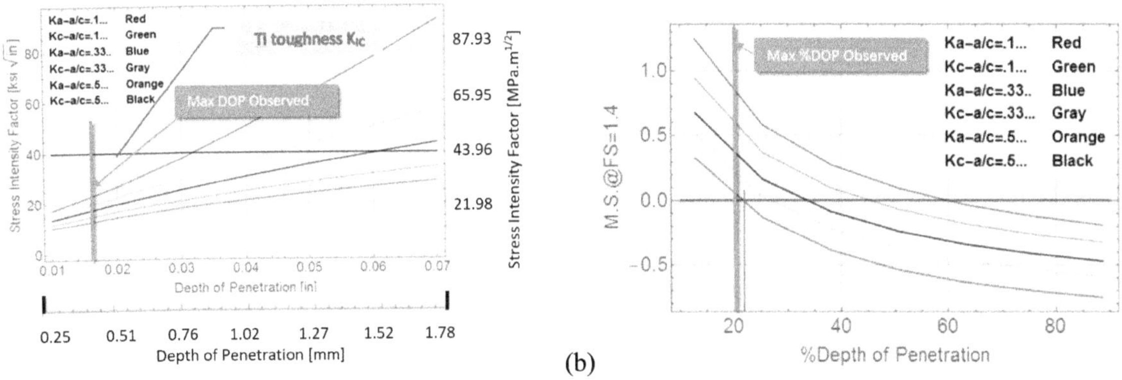

(a) (b)

Fig. 6. Results for the Ti-6Al-4V coupons (a) stress intensity factor as a function of depth of penetration (note at maximum depth of penetration, SIF is than Ti-6Al-4V toughness) and (b) margin of safety for unstable fracture as a function of % depth of penetration for Ti-6Al-4V sample (note positive margin at the maximum % depth of penetration).

5. Conclusions

Based on the Margin of Safety analysis for the Aluminum and Titanium coupon samples, a maximum depth of penetration of 20% is recommended for use in the MMOD system level risk analyses for metallic pressure vessels used in spacecraft. The use of this allowable depth of penetration offer an advantage in shield mass savings without compromising the overall failure risk of spacecraft from damaging hypervelocity impacts of orbital debris and micrometeoroids. The recommendation is based on the fact that no observed unstable crack growth after impact of preloaded plates were observed and this was validated with a fracture mechanics approach. The DOP criterion is adequate for typical metallic propellant tanks under internal pressure. Allowable depth of penetration for composite overwrapped tanks is subject to further studies due to the complexity of the composite overwraps.

References

[1] Cour-Palais BG., "Meteoroid protection by multiwall structures," AIAA 69-372, 1969.

[2] Nahra, H., Ghosn, L., Christiansen, E., Davis, B., Keddy, C., Rodriguez, K., Miller, J., Bohl, W., "Burst Pressure Failure Of Titanium Tanks Damaged by Secondary Plumes From Hypervelocity Impacts on Aluminum Shields, American Physical Society Conference, 2010.

[3] NASGRO®, Fracture Mechanics and Fatigue Crack Growth Analysis Software, User Manual.

[4] Newman, J. C., and Raju, I. S., "Stress Intensity Factor Equations in Three Dimensional Finite Bodies Subjected to Tension and Bending Loads," Computational Methods in the Mechanics of Fracture, Elsevier Science Publishing Company, New York, NY 1986.

[5] G.C. Shih, Proceeding of the 2nd Conference Theoretical Applied Mechanics, 1964, p. 547.

[6] W. Hu, Q Liu and S. Barter, "A Study of Interaction and Coalescence of Microsurface Fatigue Cracks in Aluminum 7050", 25th ICAF Symposium, Rotterdam, 27-29, May 2009

[7] NASA Technical Standard, "Fracture Control Requirements For Spaceflight Hardware", 2016.

Proceedings of the 2019 Hypervelocity Impact Symposium
HVIS2019
April 14-19, 2019, Destin, FL, USA
HVIS2019-043

Numerical modeling of impacts of twisted-pair data cables

Joshua E. Miller[a,b]*

[a]*University of Texas at El Paso, El Paso, TX 79968, United States of America*
[b]*Jacobs, NASA Johnson Space Center, Houston, TX 77058, United States of America*

Abstract

Data wire cable runs are a significant presence on the exterior of the International Space Station (ISS), and continued ISS mission support requires detailed assessment of cables due to micrometeoroid and orbit debris (MMOD) impact. These data wire cables are twisted-pair cables consisting of two 22 gauge stranded conductors inside a tight-fitting, braided-copper shield and jacket having a nominal outer diameter of 3.76 mm. Previous work has documented a total of 97 impact experiments that were performed into these cables to develop an empirical, statistical model for the failure of these cables in reliability studies; however, the experimental work left open the internal behaviors that contribute to the probabilistic findings. To address this shortcoming, numerical impact simulations have been performed to expand the understanding of the acquired dataset. This paper summarizes the dependence of impact location and speed to the penetration of wire jackets based upon particle size and provides an empirical ballistic limit equation based on the assumption that exposed conductors may lead to a short circuit. This work is consolidated with the previous experimental work for design and reliability assessments to cover projectile types, speeds and obliquities.

Keywords: Cables; MMOD; Numerical Simulation; Probabilistic Failure

Nomenclature

C&DH	command and data handling
d	diameter (mm)
$\bar{d}$	mean diameter (mm)
δd	diameter scale length (mm)
MMOD	meteoroids and orbital debris
O	offset (mm)
δO	offset scale length (mm)
θ	angle (°)
ρ	density (g/cm^3)
V	speed (km/s)
$\bar{V}$	mean speed (km/s)
Subscripts	
c	critical
Cu	copper
i	impact
o	initial
p	projectile

* Corresponding author. Tel.: 01-281-244-8093.
E-mail address: joshua.e.miller@nasa.gov.

1. Introduction

The ISS command and data handling (C&DH) system utilizes the MIL-STD-1553 data bus system [1,2] to interconnect its various subsystems such as the multiplexer/de-multiplexer [3]. The 1553 data bus system is composed of shielded, twisted-pair cables and is doubly redundant. A previous test conducted in 1995 (Test #A2372) resulted in a shorted conductor [4] under an impact from a 0.4 mm Al 2017-T4 projectile travelling approximately 7km/s and has since dominated reliability calculations for the system [5]. To improve the understanding of the ballistic performance of these systems, an experimental effort that consisted of 97 impact evaluations were performed by the ISS program office [6]. This in-depth campaign led to the development of a probabilistic failure description of impacts into these cables based on particle size that showed that the mean critical particle size is higher; in other words, failure of the cable cannot be described by a step-function at a single particle size and instead failure probability gradually increases with increasing particle size [7].

While reexamining the early design test results were a major part of the experimental campaign, the study also identified additional probability function parameters for other impact conditions like varied impact obliquities and materials; albeit, all at the same impact speed. Together the collection of the identified parameters represented a significant advancement in the modeling approach of this common flight hardware system; however, the models leave open important considerations like what causes the probability of failure and how to extend beyond the impact conditions attainable by the experimental campaign. To address these short comings, a numerical analysis has been performed to study the behavior of impact debris movement within the cable. The use of numerical analysis allowed the study of how the precise location of the impact affects the end result of an impact event. This paper presents the numerical model used in the analysis, the validation of the numerical model based off a subset of the experimental data, and the results of numerical simulations over a range of precisely controlled impact locations across the cable. These findings have been summarized into empirical models for use in reliability assessments of the ISS and other orbital architectures that use similar hardware.

2. Methods

The MIL-STD-1553 twisted-pair data cable by necessity of function has a complex internal architecture as compared to monolithic materials. A brief discussion of the previous experimental observations into this cable, the general architecture of these cables, the numerical model and the validation of the numerical model are provided here.

2.1. Experimental observations

A total of ninety-seven impact experiments were performed by the NASA White Sands Test Facility's Remote Hypervelocity Test Laboratory into the MIL-STD-1553 twisted-pair data cable [7]. A variability for the outcome was observed with the same particle size resulting in both intact cables and failed cables depending on the point-of-impact. An illustration of this variability is reproduced in Fig. 1 from Ref. 7. In the figure, the stacked count of experiments are shown where the intact cables are shown as an orange bar and the failed cables are shown as a blue bar. These results are shown for both aluminum (Al2017) and steel (SS440C) and for obliquities of 0° and 45° relative to the normal in the plane of the length of the cable. As can be seen, the count of intact results generally decreases with particle size, and the count of failed results generally increases with particle size.

The experimental work led to the development of a failure probability function, P, based on the diameter, d, given by

$$P[d] = \frac{1}{Exp\left[-\left(\frac{d-\bar{d}}{\delta d}\right)\right]+1} \tag{1}$$

where the two fitting parameters are the diameter at 50% probability of failure, $\bar{d}$ and a diameter scale length, δd. The values for the coefficients that produced the models shown in Fig. 1 are reproduced in Table 1. This model is useful for anchoring a ballistic limit model; however, little insight is given as to what resulted in the variability and how it might affect extrapolations. Addressing this limitation requires the development of numerical simulation models to study the internal geometry of the cables.

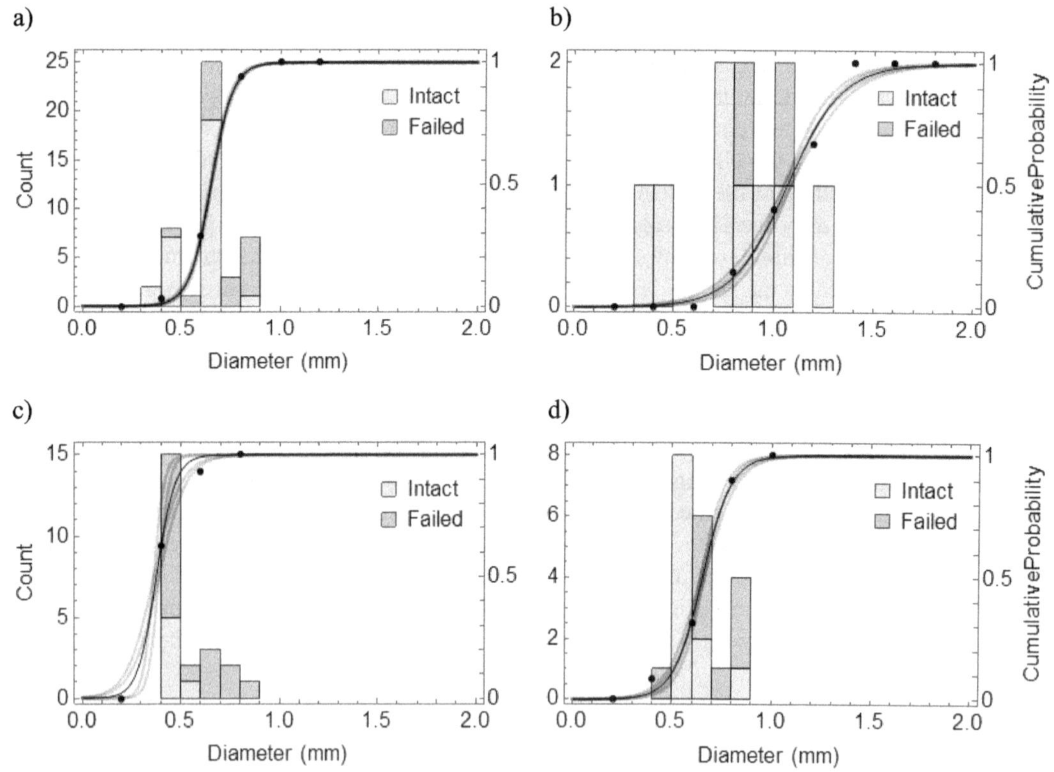

Fig. 1. The ISS C&DH cable statistical analysis for 7 km/s impacts of spherical Al2017 a) at normal incidence and b) 45° to normal, and spherical SS440C c) at normal and d) 45° to normal. The plots show the cumulative count and probability of failure as a function of projectile diameter [7].

Table 1. Failure probability function coefficients [7]

Material	Obliquity (°)	$\bar{d}$ (mm)	δd (mm)
Al2017	0	0.6494±0.0037	0.0546±0.0032
Al2017	45	1.067±0.017	0.135±0.015
SS440C	0	0.377±0.013	0.047±0.021
SS440C	45	0.652±0.010	0.0720±0.0088

2.2. Cable architecture

A CT-radiograph reconstruction of an actual experimental specimen from Ref. 7 is shown in Fig. 2a with the major components of the cable labeled. The cable consists of two 22 gauge stranded conductors with nominal diameters of 0.76 mm that are covered with polytetrafluoroethylene (PTFE) insulation such that the nominal diameter of the finished basic wire is 1.32 mm. The wires are twisted, along with two 0.889 mm diameter fillers inside a tight fitting braided, copper-shield (through-thickness, areal-density of 0.034 g/cm^2) and covered with a fluorinated ethylene propylene (FEP) jacket having a nominal outer diameter of 3.76 mm. To match the on-orbit configuration the experimental cable was wrapped with 25.4 mm wide strips of beta-cloth tape (areal-density of 0.029 g/cm^2) using an approximate 50% overwrap. The 50% overwrap means that the cable has two layers of beta-cloth tape over its length; although, the tape occasionally exceeds the desired overwrap along the run of a cable in both flight and experiment.

A numerical analysis model has been developed using the CTH hydrodynamic simulation tool [8] to explicitly advance impacts at 3, 7 and 11 km/s into the cable design under a variety of impact locations. A density contour of the modeled cable is shown in

Fig. 2b. Progressing from the outside towards the cable's central-axis are three concentric cylinders of 0.27 mm, 0.53 mm and 0.038 mm thickness to represent the beta-cloth tape, the cable jacket and the braided-copper shield, respectively. A 0.050 mm braided-copper shield is modeled as a single copper shell of the same mass as that considered in Ref. 6. Both the beta-cloth tape and the cable jacket are modeled with a density of 2.15 g/cm^3 yielding the through-thickness, areal-densities of 0.058 g/cm^2 and 0.114 g/cm^2, respectively, and the braided-copper shield is modeled with a density of 8.93 g/cm^3 yielding the through-thickness, areal-density of 0.034 g/cm^2. Internal to the copper-shielding are two cylinders of diameter 0.89 mm at densities of 0.915 g/cm^3 to represent the fillers, and cylinders to represent the wire at 0.76 mm diameter and 8.93 g/cm^3 and the wire jacket at 0.28 mm thickness at 2.15 g/cm^3 with an areal density of 0.060 g/cm^2.

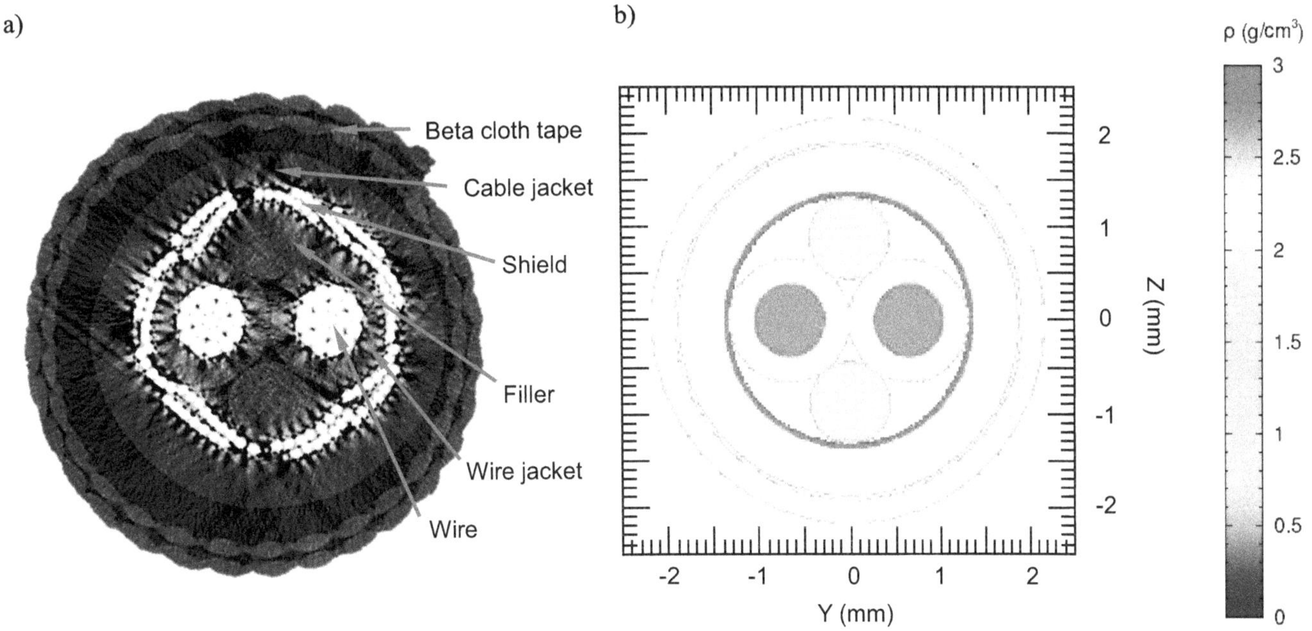

Fig. 2. The MIL-STD-1553 cable under consideration a) cross-section image from Ref. 7 and b) a corresponding density contour from the numerical model.

2.3. *Material Models*

The constitutive models, equation-of-state (bulk properties) and strength (deviatoric properties), have been selected using native CTH data inputs as much as possible. The aluminum projectile model used in the simulations is the Al2024 Mie-Gruneissen equation-of-state referenced to the principal Hugoniot for aluminum [8]. The bulk modulus, as approximated from the zero intercept of the principal Hugoniot, is 76.2 GPa, and all other moduli are determined by assuming a homogeneous, isotropic material using the Poisson ratio of 0.33. The strength model is the native CTH Steinberg-Guinan-Lund model for aluminum [9].

The copper model uses the native copper Mie-Gruneissen equation-of-state referenced to the principal Hugoniot for copper [10]. The bulk modulus, approximated from the zero intercept of the principal Hugoniot, is 138.6 GPa. The strength and fracture models used are the native CTH Johnson-Cook models for copper [11,12].

The plastic components except the filler, including the beta-cloth tape, used CTH native PTFE models for equation-of-state [10]. The beta-cloth tape is a composite of PTFE and fiberglass fiber, and while the cable jacket is FEP, it is similar to PTFE. The equation-of-state is of the quadratic form to account for the dissociation of the PTFE at high pressures. The Poisson's ratio for PTFE used is 0.46 [13]. Similarly, the filler material in the simulations used polyethylene model that is native to CTH [10].

2.4. *Numerical Model Validation*

The simulations have considered aluminum impacts normal to the length of the cable from direct impacts to grazing hits. The velocity vector has been rotated about the central axis of the cable at varied directions relative to the internal components. The simulations use a fixed three-dimensional cubic mesh with spatial resolution of 50 μm, and the model extends 3 cm along the length of the cable and 3.75 mm from the center of the cable in the orthogonal directions.

Using the numerical model, simulations have been performed to compare the output density contours with experimental results from Ref. 7 as shown in Fig. 3. In these simulations, it is assumed that HITF15035 resulted from an impact approximately normal to a plane defined by the conductors and offset by about 1.5 mm from the center of the fillers, and it is assumed that HITF15040 resulted from an impact approximately normal to a plane defined by a fillers and conductor pair with an offset of 0.5 mm toward the filler. It should be noted that the experimental impact vector is not known with certainty due to twisting of the conductors within the cable and potential shifting of the conductors and fillers due to the impact event.

Both experiments shown resulted in a short of the conductor with the shielding. Simulations have been carried out to 20 μs, which is beyond the time necessary for crater formation as determined by examination of time lapse output. The images from left to right are the initial cross-sectional material configuration and the cross-section after the 20 μs simulated damage evolution. In these simulations, the jacket (~ 2g/cm^3) surrounding the copper wire (> 3g/cm^3) is perforated and the grounded shield material (> 3g/cm^3) is in contact with the conductor. Examination of the wire jackets in the simulations showed perforation, and as can be seen in the provided density contours, copper from the shielding is touching or very near the copper of the conductors.

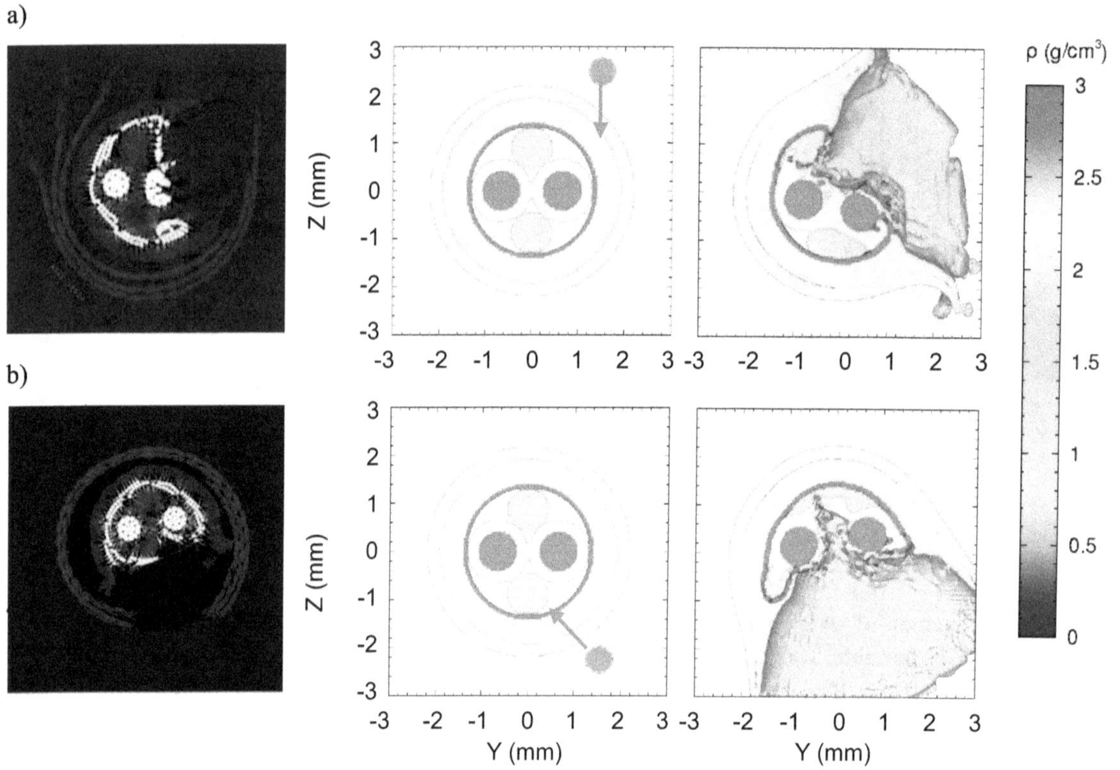

Fig. 3. X-ray CT images from two different impact experiments a) HITF15035 and b) HITF15040 from Ref. 7. Along with the two cross-sectional images are density contour plots from comparative simulations showing the assumed initial condition of the impact (orthogonal to the plane with the conductors) and after 20 μs of impact evolution.

The agreement between the experiment and the comparative simulations is good; although, the braided shielding tended to show less damage in experiment than the simulated shielding. This is likely due to the use of a mass averaged shielding representation rather than the more realistic bundled wire configuration. While for comparison to single tests it might be possible to better represent this performance, in general it cannot be known the true mass of braided shielding encountered during an impact; therefore, the averaged approach has been used which might yield somewhat conservative estimates. This model and this assumption of failure has been used to study the effect of both impact location and impact speed for aluminum projectiles.

3. Results

A series of over 250 simulations have been performed to study the effect on critical particle size to perforate the wire jacket as related to the impact location and speed. Three general impact vectors have been studied as shown in Fig. 4 at three impact speeds: 3, 7 and 11 km/s. The velocity vectors in Fig. 4 are shown as red arrows. The first velocity vector corresponds to the normal of the plane defined by the fillers, Fig. 4a; the second velocity vector is normal to the plane defined by the conductors, Fig. 4b; and, the third velocity vector is normal to the plane defined by a conductor and filler pair, Fig. 4c. The origin of the velocity vector has been varied in the respective parallel planes as illustrated by the gray arrows in the density contours of Fig. 4.

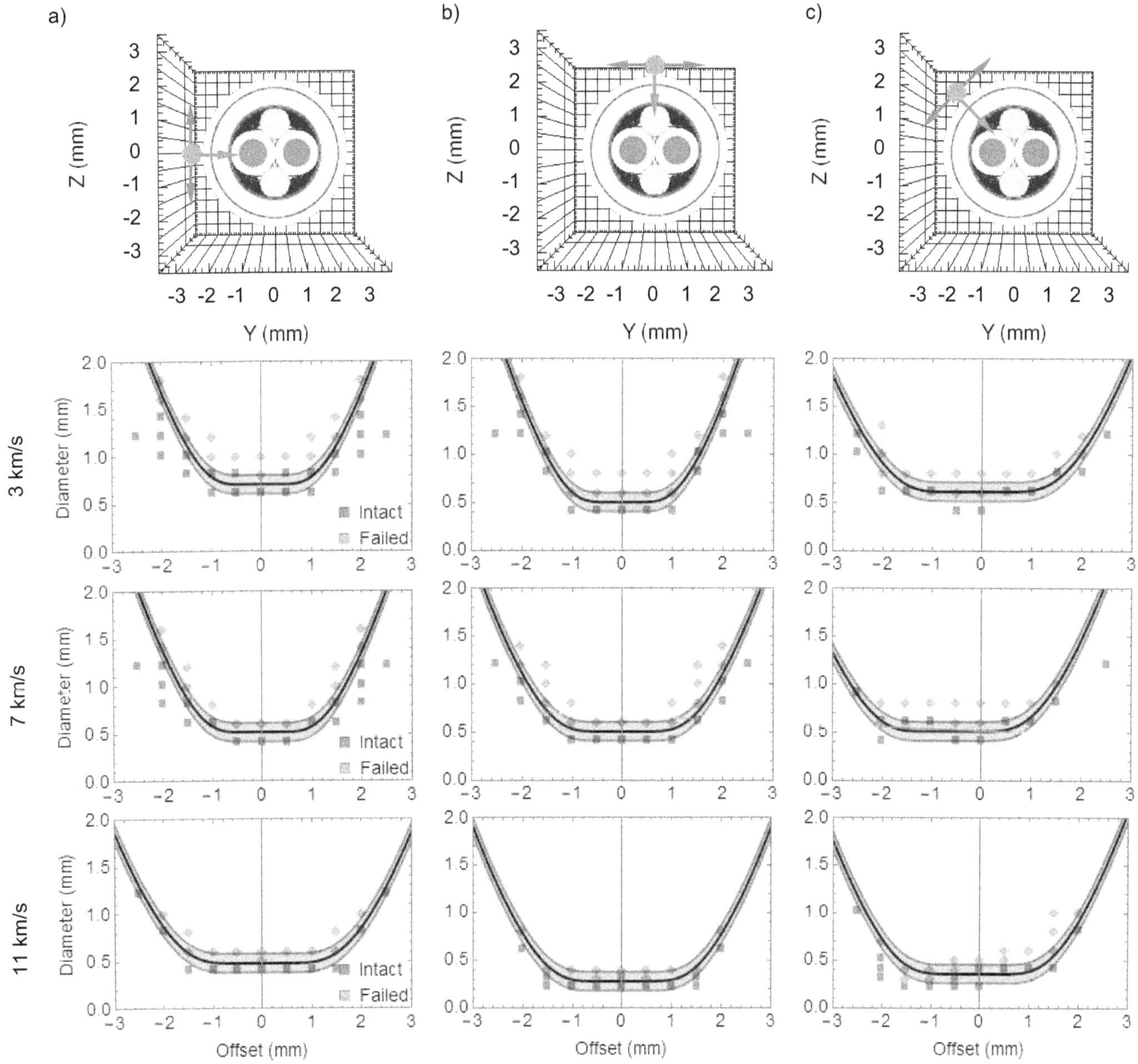

Fig. 4. Simulation results for impacts a) normal to the plane defined by the fillers, b) normal to the plane defined by the conductors and c) normal to the plane defined by the conductor and filler pair. Projectile impact offset is determined from the projectile vector that goes through the center of the cable. The blue squares are conditions where the wire jacket is intact and the orange diamonds are conditions resulting in wire jacket perforation.

For each of these impact vectors, the projectile is initiated outside the cable. The offset amount relative to the impact vector that goes through the center of the cable has been systematically varied on 0.5 mm increments, and the impact event has been simulated through time to model the crater formation. In Fig. 4, the final state of the wire jacket as determined from the numerical simulations is shown as a function of offset distance for the ballistic variables considered. Presence of continuous insulating material over the conductor is counted as intact (blue), and to the limits of the mesh, complete erosion of the jacket material down to the conductor as failed (orange). The positive offset is defined as vectors originating to the right of the reference vector through the center of the cable in the direction of travel. The transition between the blue and the orange data in the figure represents the failure threshold for the shot line assuming that perforation of the wire jacket is a necessary condition for the short of a conductor.

As can be seen in the figure, the onset of observed diameter necessary for failure generally goes up as the impact origin moves away from a shot line through the center of the cable. The gap impact vector results in a small asymmetry from the cable centerline shifted toward the shot line through the wire and away from the filler. To model these observations, the mid-point of the simulated transition has been fit to a reflected nonlinear function where the critical diameter is given by

$$d_c[O] = \frac{\delta O}{Exp\left[\frac{\delta O}{2\,(O-O_o)}\right]-1} + d_o. \tag{2}$$

The critical diameter is a function of the offset, O, referenced from the center of the cable. The three fitting parameters: δO, O_o, and d_o are the offset scale length parameter, an initial offset, and an initial critical diameter that is the function's minimum, respectively. This function is approximately flat as it approaches the symmetry plane in the simulations and has an asymptote of the critical radius when the offset is large. The parameters for the three impact vectors are summarized in Table 2 where the uncertainty estimates have been determined from the standard error associated with the regression. The regressed function along with the uncertainty bands for the simulated conditions are shown with the simulation points in Fig. 4. As can be seen, the smallest function minimums are from shot lines that are orthogonal to the conductor pair; in other words, impacts through the filler, as illustrated in Fig. 4b, actually result in smaller critical particles than those going through the conductor.

Table 2. Critical diameter coefficients for impact offset (vector is orthogonal to the plane defined)

Speed (km/s)	Orthogonal Plane	O_o (mm)	δO (mm)	d_o (mm)
3	Filler Pair	0	9.56±1.01	0.707±0.055
	Filler/Conductor	-0.096±0.059	15.14±0.69	0.614±0.026
	Conductor Pair	0	9.01±0.27	0.503±0.017
7	Filler Pair	0	10.30±0.56	0.519±0.024
	Filler/Conductor	-0.57±0.11	14.13±1.38	0.512±0.041
	Conductor Pair	0	12.19±0.67	0.506±0.027
11	Filler Pair	0	14.76±1.13	0.480±0.025
	Filler/Conductor	-0.117±0.096	13.73±1.11	0.357±0.044
	Conductor Pair	0	13.28±0.57	0.276±0.016

To relate these numerical results to the results found under impact experiments, the model given by Eqn. 2 along with parameters given in Table 2 have been averaged using the equation

$$\overline{d_c} = \frac{\int_{-O_B}^{O_B} d_c[O]\, dO}{2\, O_B}, \tag{3}$$

where O_B is the integration limit. To find the cable average critical diameter, this averaging technique is performed for each of the three directions considered (normal to the filler, the conductor and the filler/conductor pair). For the cable average, the filler/conductor pair is double weighted as the occurrence of this impact vector is twice as likely as the other planes. The integration limit has been determined to be 2.28 mm such that the numerical model matches the experimental value for a normal impact of an aluminum projectile at 7 km/s. The cable average critical diameters is shown for all three speeds in Fig. 5a. These critical diameters have been regressed to a power function with an exponent of -0.34±0.12, which is shown along with the average values.

The mean impact speed where the uncertainty bounds intersect is at 5.7 km/s, which is the normalization speed that achieves the shown error bands. The diameter at this point is 0.696 mm. Figure 5 also shows the residuals from the simulated data to the power function in Fig. 5b. While it is possible that more than the monotonically decreasing treatment is needed to completely model the simulated data, the uncertainty associated with the simulations and model regression do not justify a more complex speed description.

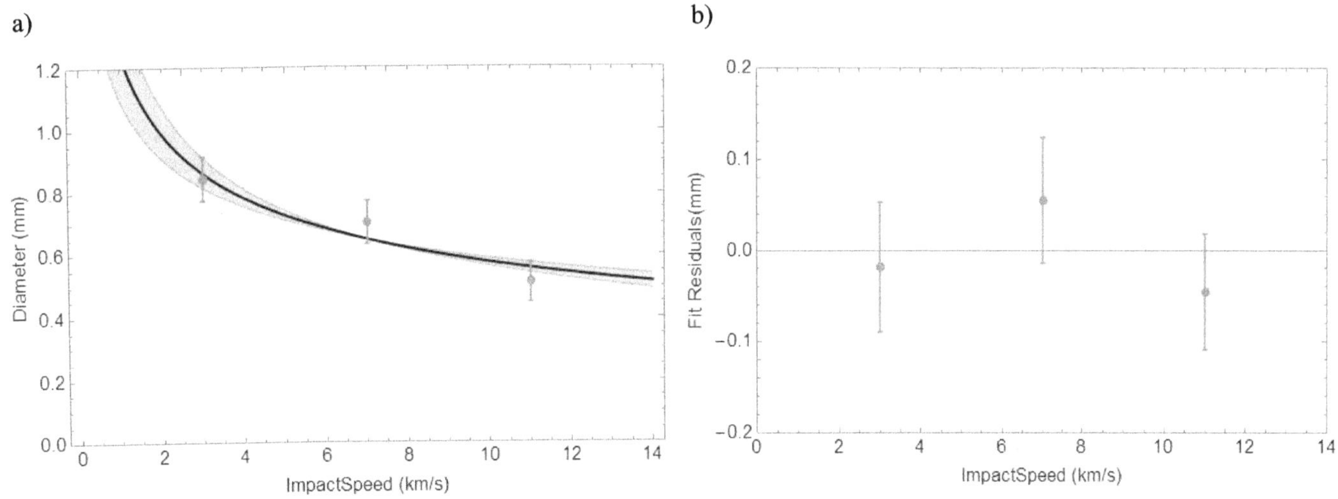

Fig. 5. Cable ballistic performance from numerical simulation for aluminum projectiles impacting orthogonal to the cable's central axis a) as a monotonic fit of impact speed shown along with the b) residuals from the fit compared to the numerical simulations.

To translate these observations into a ballistic limit equation, the determined mean diameters, $\bar{d}$ from Table 1 have been fit to a power function given by

$$d_c[\rho_p, V_i, \theta_i] = \bar{d}_o \left(\frac{\rho_p}{\rho_{Cu}}\right)^j \left(\frac{V_i}{\bar{V}}\right)^k Cos^l[\theta_i]\,. \tag{4}$$

The density of the projectiles have been normalized by ρ_{Cu}, an assumed density of 8.96 g/cm^3, and the impact speed has been normalized by $\bar{V}$, which is set to the previously defined mean impact speed of 5.7 km/s. The coefficients $\bar{d}_o$, j and l have been determined simultaneously using variance weighting of $\bar{d}$ for the four impact conditions studied yielding 0.361±0.010, -0.503±0.023 and 1.456±0.056, respectively. The exponent k is -0.34±0.12 from the numerical simulations as discussed. A comparison of the model performance and the values of Table 1 is shown in Table 3.

Table 3. Calculated critical diameter comparison with experimental observation

Material	ρ_p (g/cm^3)	θ_i (mm)	$\bar{d}$ (mm)	d_c (mm)
Al2017	2.796	0	0.6494±0.0037	0.649
Al2017	2.796	45	1.067±0.017	1.075
SS440C	7.667	0	0.377±0.013	0.391
SS440C	7.667	45	0.652±0.010	0.647

4. Conclusions

The ISS C&DH cable runs are a significant presence on the exterior of the International Space Station (ISS), thus these components are affected by the external environments including MMOD. Furthermore, as the cables are of a universal architecture, they have a broader presence in many other satellites and vehicles. As this work is part of a goal to extend the ISS mission beyond the initial life expectancy for the laboratory, it is necessary to revisit some of the early reliability engineering assessments for the vehicle. To this end, a total of ninety-seven impact experiments were performed to develop a statistical model for future analysis of this component [7].

Using the data Ref. 7, cumulative distribution functions failure were found for these cables for steel and aluminum components of the orbital debris environment at a representative impact speed and obliquities for ISS operation. While the scope of the experimental work is significant and touched many areas of concern for the cable's continued operation, some impact conditions have been left unexplored. Most notable among the dataset insufficiencies is the lack of observations to check the impact speed dependence on the performance of the cable. To partially fill this void, numerical simulation models have been developed of the cable to more deeply probe the effects of internal geometry of the cable.

It has been found that as the impact shot line moves away from passing through the center of the cable, that the minimum required particle for failure goes up; however, it has also been found that passing through various components of the cable other than the conductor doesn't improve performance. This behavior has been summarized by the development of nonlinear functions that describes the critical diameter for failure as a function of particle origin, and this function allows for the development of an average particle critical diameter for a general cable impact. These average critical diameters from the numerical model have been used to expand to a broader range of impact speeds. Using the numerical model along with the observations of Ref. 7, the ballistic performance of a MIL-STD-1553 cable has been summarized into an empirical, ballistic limit equation of a power function form for typical environment variables of MMOD reliability assessments.

In spite of the large set of impact experiments and numerical analysis, the breadth of variability in a cable and the many factors in the MMOD environment definition leaves room for exploration. The use of numerical analysis of the impact has allowed extrapolation for impact speed with higher confidence; however, uncertainties in material models and higher refinement of the model may necessitate future verification for impact speed dependence. The use of other projectile materials could affect the speed parameters used to extrapolate from the existing experimental data. Additionally, as it is likely that other impacting materials and shapes are needed for full reliability assessments, these parameters have received limited variation here and may need to be explored in more depth numerically and verified through experiment.

Acknowledgements

The author wishes to gratefully acknowledge the Jacobs JETS contract NNJ13HA01 in its support to the Hypervelocity Impact Technology group at NASA Johnson Space Center and its sponsorship to University of Texas at El Paso under EN41520TMS.

References

[1] Military Standard, "Aircraft Internal Time Division Command/Response Multiplex Data Bus". MIL-STD-1553b, April 1975.

[2] Crossgrove, A., McCall, M., Smith, L., Turner, R., and Ellis, D. "Multiplex Applications Handbook MIL-STD-1553", ADA-119529, May 1980.

[3] MSFC Technical Standard, "Electrical, Electronic, and Electromechanical (EEE) Parts Management and Control Requirements for MSFC Spaceflight Hardware". MSFC-STD-3012, February 2012.

[4] Westbury, R., Kerr, J., and Christiansen, E. L., "Hypervelocity Impact Test for McDonnell Douglas International Space Station Avionics Wire Harness, Phase I", JSC-27285, October 1995.

[5] Cano, T and Pan A., "Meteoroid/Orbital Debris Functional Failures Assessment of USOS Segments Z1, P6, S0, S1 P1, P3, P4, S3, S4, and S6", MDC-00H2387, July 2001.

[6] Read, J. "C&DH Wire Harness Hypervelocity Test Results," JSC-66971, September 2016.

[7] Miller, J. E., et al., "Cumulative distribution of ballistic impact failures of common twisted-pair data cables at orbital speeds," Int. J. Imp. Engng. 124 (2019) 61-66.

[8] McGlaun, J., et al., CTH: a three-dimensional shock wave physics code, Int. J. Imp. Eng. 10 (1990) 351-360.

[9] D. J. Steinberg, "Equation-of-state and Strength Properties of Selected Materials", Lawrence Livermore National Laboratory report UCRL-MA-106439, 1991.

[10] Group GMX-6, "Selected Hugoniots", Los Alamos Scientific Laboratory report LA-4167-MS, 1994.

[11] G. T. Grey, et al. "Constitutive equations for annealed metals under compression at high strain rates and high temperatures", Los Alamos National Laboratory report LA-12669-MS, 1994.

[12] Johnson, G. R. and Cook, W. H., "Fracture characteristics of three metals subjected to various strains, strain rates, temperatures, and pressures," Engrg. Frac. Mech. 21 (1985) 31-48.

[13] DuPont™ Flouroproducts. "DuPont™ Teflon PTFE Properties Handbook". Online report. http://www.ptfeparts.com/_resources/common/userfiles/file/Dupont%20Teflon%20PTFE%20Properties%20Handbook.pdf

Proceedings of the 2019 Hypervelocity Impact Symposium
HVIS2019
April 14-19, 2019, Destin, FL, USA
HVIS2019-044

Simulation study of non-spherical, graphite-epoxy projectiles

Joshua E. Miller[a,b]*

[a]*University of Texas at El Paso, El Paso, TX 79968, United States of America*
[b]*Jacobs, NASA Johnson Space Center, Houston, TX 77058, United States of America*

Abstract

The DebriSat hypervelocity impact experiment, performed at the Arnold Engineering Development Center, is intended to update the catastrophic break-up models for modern satellites. To this end, the DebrisSat was built with many modern materials including structural panels of carbon-fiber, reinforced-polymer (CFRP). Subsequent to the experiment, fragments of the DebrisSat have been extracted from porous, catcher panels used to gather the debris from the impact event. Thus far, one of the key observations from the collected fragments is that CFRP represents a large fraction of the fragments and that these fragments tend to be thin, flake-like structures or long, needle-like structures; whereas, debris with nearly equal dimensions is less prevalent. As current ballistic limit models are all developed based upon spherical impacting particles, the experiment has pointed to a missing component in the current approach that must be considered. To begin to understand the implications of this observation, simulations have been performed using cylindrical structures at a representative orbital speed into an externally-insulated, double-wall shield that is representative of shielding on the current International Space Station crew transport vehicle, the Soyuz. These simulations have been performed for normal impacts to the surface with three different impact angles-of-attack to capture the effect on the shield performance. This paper documents the simulated shield and the models developed to study the effect of fragments and derives the critical characteristics of CFRP impacting particles for the selected shield. This work gives a deployable form of a critical, non-spherical projectile ballistic limit equation for evaluating non-spherical space debris for orbital debris environment modeling.

Keywords: Non-spherical projectiles; CFRP; Numerical simulation

Nomenclature

AOA	angle-of-attack
CFRP	carbon-fiber, reinforced-polymer
D	diameter (mm)
EOS	equation-of-state
L	length (mm)
MLI	multilayer insulation
MMOD	meteoroids and orbital debris
Symbols	
α	angle-of-attack (rad)
Subscripts	
Cyl	cylinder
Sph	sphere

* Corresponding author. Tel.: 01-281-244-8093.
E-mail address: joshua.e.miller@nasa.gov.

1. Introduction

The DebriSat hypervelocity impact experiment, performed at the Arnold Engineering Development Center, is intended to update the catastrophic break-up models for modern satellites [1]. To this end, the DebriSat was built with many modern materials including structural panels of carbon-fiber reinforced polymer (CFRP). Subsequent to the experiment, fragments of the DebriSat have been extracted from porous catcher panels that were used to gather the debris from the impact event [2]. Thus far, one of the key observations from the collected fragments is that CFRP fragments represent a large fraction of the collected debris and that these fragments tend to be thin, flake-like structures or long, needle-like structures; whereas, debris with nearly equal dimensions is less prevalent. As current ballistic-limit models for shields are generally based upon spherical impacting particles [3], the experiment has pointed to a missing component in the current approach to ballistic modeling and may be beneficial [4].

While numerous shield types are currently in use for impact mitigation from orbital debris and meteoroids, the double-wall or Whipple shield [5] is among the most prevalent. This shield achieves a high level of ballistic performance for minimal weight because the stresses induced in a projectile during impact are far above the stresses the solid particle can withstand resulting in a break-up of the particle. In the double-wall approach, an empty volume between the two walls of the shield gives a space for the debris to expand resulting in a distributed impact on the second, shield-wall. Even with the increased performance of this design, the shield-wall reaches a limit, which is the ballistic-limit of the shield-system [3]. Commonly, the ballistic-limit is characterized through experiment to a characteristic dimension of the projectile, which has largely been based on the diameter of a sphere [3].

Spheres have beneficial qualities for shield characterization as there is no angle-of-attack (AOA) dependence for the shape (i.e. a sphere looks the same regardless of the orientation of the projectile on impact); whereas, non-spherical projectiles impacting normal to a shield add at least two degrees of freedom. Spheres also have some ballistic benefits creating a relatively uniform stress-state across the projectile [6], and the shock waves originating from the sphere push material approximately radially outward improving material spread [6]. It has been experimentally observed that departing from spheres diminishes these benefits, at least, for particles without an AOA impacting normal to the surface [7-10].

Complicating these experimental observations is that the pure, all-metal, Whipple shield is less prevalent in deployment due to thermal considerations. The presence of thermal shielding materials are known to affect the performance of double-wall shields depending on the construction and the placement of the thermal blanket, thus thermal materials should be included in an analysis to be representative of actual spacecraft [3]. NASA and other agencies have performed numerous experimental campaigns on flight configurations with thermal materials, and among those, a Whipple shield with a thermal blanket attached to the front of the outer bumper has been selected for this numerical study [11-12]. This shield has the material configuration shown in Fig. 1.

a) b)

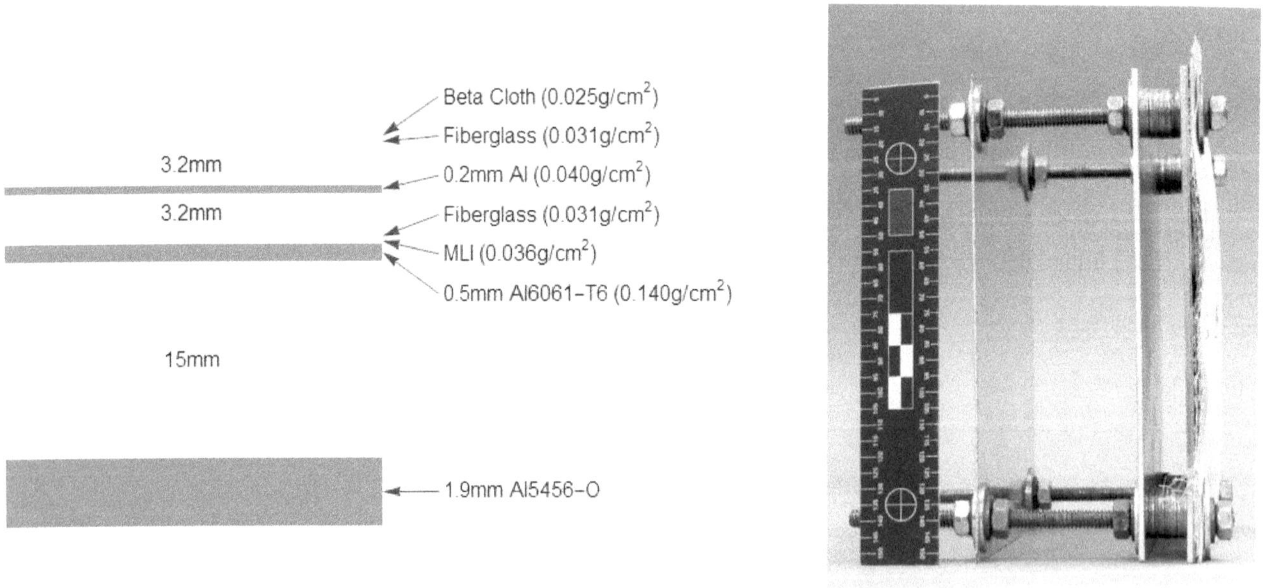

Fig. 1 Shield a) schematic for numerical simulation (layers scaled by mass; separations to scale) which represents b) a previously considered shield [11-12].

2. Methods

This section documents the shield design considered in simulation, the numerical solution approach, the rationale for the material models used, and a comparison of validation simulations with previously obtained impact experiment results [11-12].

2.1. Shield Description

The shield considered here is fundamentally an aluminum, Whipple shield; however, the shield also includes an enhanced, thermal-blanket over the top as shown in Fig. 1a. This shield is representative of an operational shield at the ISS, but has some subtle differences based on material model availability. The outermost aluminum wall, the "bumper", is made of 0.5 mm Al6061-T6, and the rear wall is made of a 1.9 mm Al5456-0. The two walls are separated from each other by 15 mm. Over the top of the outer wall is the enhanced, thermal-blanket. The outermost material is a combined Beta cloth and fiberglass layer that weighs 0.056 g/cm^2. Beyond those materials is an enhancement intended to boost protection from meteoroids and orbital debris (MMOD), which includes a bent 0.2mm thick aluminum plate weighing 0.040 g/cm^2. The plate that separates the layers of 0.031 g/cm^2 fiberglass cloth is bent in opposite directions on each side by 3.2 mm [11-12]. Beneath the MMOD enhancement is a 0.036 g/cm^2 multilayer insulation (MLI) blanket. The total mass of the enhanced, thermal-blanket is 0.163 g/cm^2.

Fig. 1b is an image of the actual experimental article with a millimeter scale reference considered from a previous study [9-10]. In the experiments, a projectile impacts the target setup from the right. The schematic in Fig. 1a is of the two, right-most walls in Fig. 1b. The left-most plate is an article structural witness and is not necessarily representative of any operational structure.

2.2. Numerical Model

The simulations have been performed using the multi-dimensional, nonlinear, structural-dynamics, continuum-analysis suite of codes, CTH [13]. The simulations used a three-dimensional, rectangular mesh with an assumed symmetry plane orthogonal to the direction of motion. This symmetry plane allows complete rotational adjustment while cutting the computational space in half.

The mesh is illustrated by the contour of the normalized mesh area in Fig. 2a. The mesh uses transverse rectangular cells measuring 0.3 mm in each direction (0.3 mm x 0.3 mm) with an area of 0.09 mm^2 in the central region that extends 1.5 cm from the impact location. Outside the central region, the mesh has been extended an additional 4.5 cm with a mesh size that increases linearly to 1.5 mm. The entire mesh in the through-thickness direction, as illustrated by a density contour in Fig. 2b, is discretized at a constant 0.3 mm such that the central region of the mesh is a cube of 0.3 mm dimensions. The mesh in the through-thickness direction extends 9 mm below the rear-wall and 5 mm or more above the outermost fabric to completely fit the projectile.

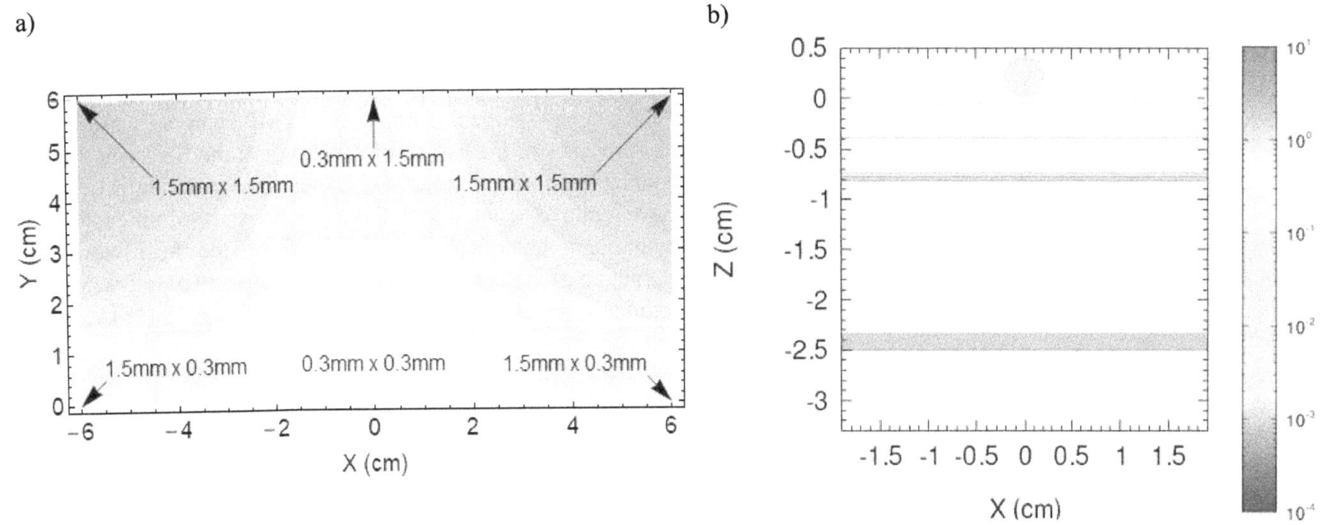

Fig. 2 Simulation space description with contours of the a) transverse mesh area with a 0.09 mm^2 minimum in the central region and 2.25 mm^2 maximum farthest from the center, and a representative b) density contour prior to impact of a spherical projectile. Contour in central region (0.3 mm x 0.3 mm x 0.3 mm).

All together the simulation space extends to a minimum of 6 cm from the impact location. Material is allowed to exit the mesh through all of the boundaries except the reflective boundary, and material is removed if its density falls below 10^{-5} g/cm^3 (a medium vacuum). In time, the simulations are performed to at least 35 μs after impact. If the rear wall continues to evolve, simulations are extended until the peak rear wall speed in the impact direction is less than a meter per second.

2.3. Material Models

Limited availability of material models required simplifications from the shield configuration considered in Ref. 11 and 12. Given the location of the blanket materials the shock wave strengths are expected to greatly exceed the limits of the component materials; therefore, the simplest deviatoric models available along with properties from material datasheets have been used for blanket materials. There are no models specifically developed for MLI; however, an analytical model [14-15] for Kapton® has been used to represent these materials [16]. The fiberglass uses style 7781 E-glass properties for both the beta-cloth and the fiberglass [17]. The breaking strength in the warp direction has been converted to an approximate failure strength using the fabrics thickness to arrive at a failure strength of 270MPa. Each fabric layer has been assumed to be isotropic and to remain elastic until reaching a yield surface that is 85% of the failure strength, and the material has also been assumed to have a Poisson ratio of 0.125. No work hardening or strain-rate hardening is assumed for this material.

The fiberglass fabric has no developed EOS; therefore, the fabrics have been treated as porous-layers of their chief constituent. For the purpose of the fiberglass blankets, the tabulated SESAME EOS for fused quartz [18] has been used with a porous density of 1.25 g/cm^3 and a fully crushed state achieved at 100MPa. The EOS includes a fused quartz to stishovite phase transition, and a liquid phase that includes dissociation to silicon and oxygen, electronic excitation and ionization.

While parts of the aluminum layers also see high stress states, the aluminum walls are responsible for the final arresting of the debris, and as such, stress states necessarily drop below the strength of the material. For this reason, a more-advanced, Johnson-Cook deviatoric model that accounts for work hardening, strain-rate dependence and thermal softening of the yield surface has been selected for the aluminum [19]. This yield surface along with the Poisson ratio of 0.33 describes the deviatoric behavior of the aluminum walls. For the volumetric behavior, the tabulated SESAME EOS for aluminum [20] has been used with energy and pressure scaling from molecular density of pure aluminum to that of the alloys in the experimental article. This EOS includes aluminum melting and vaporization as well as electronic excitation and ionization.

The steel, for the projectile material used in target validation, also undergoes high stresses, but because of its high impedance, the ability of steel to support shear stresses even at orbital impact speed is notable. As a consequence the steel has been modeled using the same, more advanced, deviatoric model as the aluminum [19]. The tabulated SESAME EOS for iron has been density scaled to SS440C [21]. This EOS includes four solid phases (alpha, gamma, epsilon, delta), melting, vaporization and electronic excitation and ionization.

As mentioned, projectile undergo strong, stress-waves during impacts at orbital speeds, and in the case of the CFRP, the material does not have a high ability to support shear at these conditions. The CFRP has been assumed to have a reference density 1.61 g/cm^3 with a slight porosity yielding a final density of 1.46 g/cm^3. The failure strength used for the CFRP is 2.01 GPa, which corresponds to M55J fibers in epoxy [22]. The von Mises yield surface has been assumed isotropic at 140 MPa with a Poisson's ratio of 0.2. The EOS, on the other hand, is more complicated due to the presence of the two chief constituents: the fiber and the matrix. The epoxy binder in graphite-epoxy composites dissociates at shock pressures above 20 GPa; therefore, material models must account for this energy sink [18]. While all chemical processes have a rate associated with them, the presence of unreacted materials results in a chemical decomposition that occurs at a rate comparable to the time of equilibration across the shock-wave; therefore, an Arrhenius burn model [14] has been used to determine the fraction of material in an equilibrium mixture of unreacted and reacted products. Both the unreacted material and the reacted material are modeled with the tabular SESAME EOS [18].

2.4. Numerical Model Validation

The results from the experimental efforts for this shield using SS440C projectiles hitting at 30° impact obliquity are summarized in Fig. 3 [11-12]. The impact speeds were varied from just under 3 km/s to just over 9 km/s, and due to that range of conditions, this work represents a relatively broad dataset for checking material models for the shield. These results are shown as squares in Fig. 3b where unfilled markers represent experiments where the rear, shield-wall remained intact, and filled markers represent the conditions where the rear, shield-wall was penetrated.

a)

HITF Number	Diameter (mm)	Impact Speed (km/s)	Result
12257	1.79	7.18	Fail
12258	1.59	4.99	Fail
12259	1.72	5.78	Fail
12273	1.29	2.87	Fail
12262	1.49	7.14	Pass
12263	1.49	6.23	Pass
12264	1.29	4.90	Pass
12266	1.39	3.19	Fail
12276	1.60	7.00	Fail
12277	1.50	4.83	Fail
12412	1.40	9.27	Pass
12413	1.60	9.01	Fail

b)

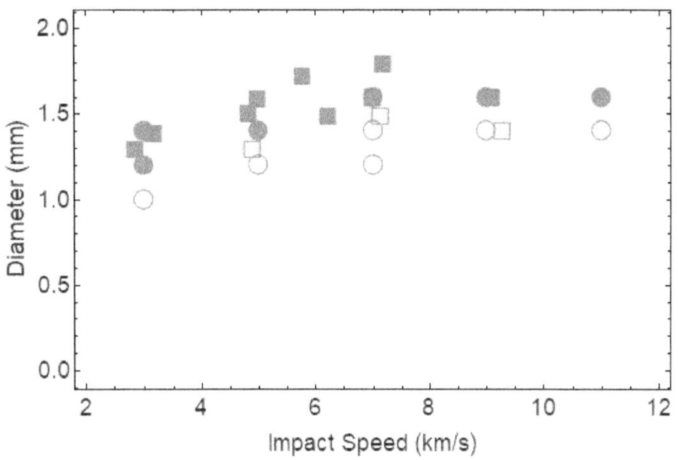

Fig. 3 Shield validation results from a) impact experiments summarized from Ref. 11 and Ref. 12 and b) the comparison of these experiments (squares) and the results from simulations (circles) where defeated shield markers are filled and intact shield markers are unfilled.

Similarly, simulations using the shield model are also shown in Fig. 3b as circles following the same filling procedures. The simulations have been performed for SS440C projectiles impacting at speeds between 3 km/s to 11 km/s in 2 km/s increments with a 30° impact obliquity relative to the target. The region between the filled and unfilled square markers is the ballistic limit of the shield, and as can be seen, the simulated ballistic limit from the numerical model is similar to that found experimentally, in other words, the transition diameter from intact to defeated shields occurs at a similar value. As no shaped or CFRP experimental evidence has been developed thus far, this comparison is the limit of model validation that can occur at this time; in other words, at this time only the shield material model is validated and future effort is required to validate the CFRP projectile.

3. Results

With the numerical model developed and validated to the limits of current capability, simulations using the assumed characteristics of CFRP projectiles impacting the shield normal to its surface have been performed. The simulations have considered the spherical ballistic limit at a relative impact speed of 7 km/s and evaluated the effect of cylindrical CFRP projectiles at the same speed. At these impact speeds, the typical peak stress in the CFRP projectiles has been found to be between 20 and 40 GPa. As the cylindrical impact condition does not necessarily have rotational symmetry about the axis orthogonal to the impact vector, the evaluation has considered three different AOA between the impact vector and the cylinder central-axis.

3.1. Spherical CFRP Projectiles

An illustration of a pair of representative simulations is shown in Fig 4 for spherical impactors. In this graphic, logarithmic density contours of the mid-plane of the simulation space are shown. The density contours range over five decades from 10^{-4} g/cm^3 to 10 g/cm^3; although, in the simulations the material is tracked until 10^{-5} g/cm^3. These density contours are the evolved result from the impact of a 4.2 mm (left) and a 4.4 mm (right) projectile. The left density contour, for the time-evolved 4.2 mm impact, is the same type and scale as the initial condition plot of Fig. 2b. The contours are the results after the simulation have been carried out for 35 μs. As can be seen, the 4.2 mm projectile dented in the rear and the 4.4 mm projectile bursted the rear-wall.

By identifying a projectile dimension that does not defeat a shield and then a slightly larger one that defeats the shield, the arithmetic mean between the projectile parameters can then be used as the critical projectile dimension. As the 4.2 mm diameter sphere left the shield in-tact but the 4.4 mm diameter sphere defeated the shield, the critical spherical diameter, D_{Sph}, of CFRP is then determined to be 4.3±0.1 mm for a normal impact at 7 km/s into the shield. This approach is similar to that required for cylindrical CFRP projectiles, but the cylinder has a larger parameter space to fully identify the critical projectile characteristic.

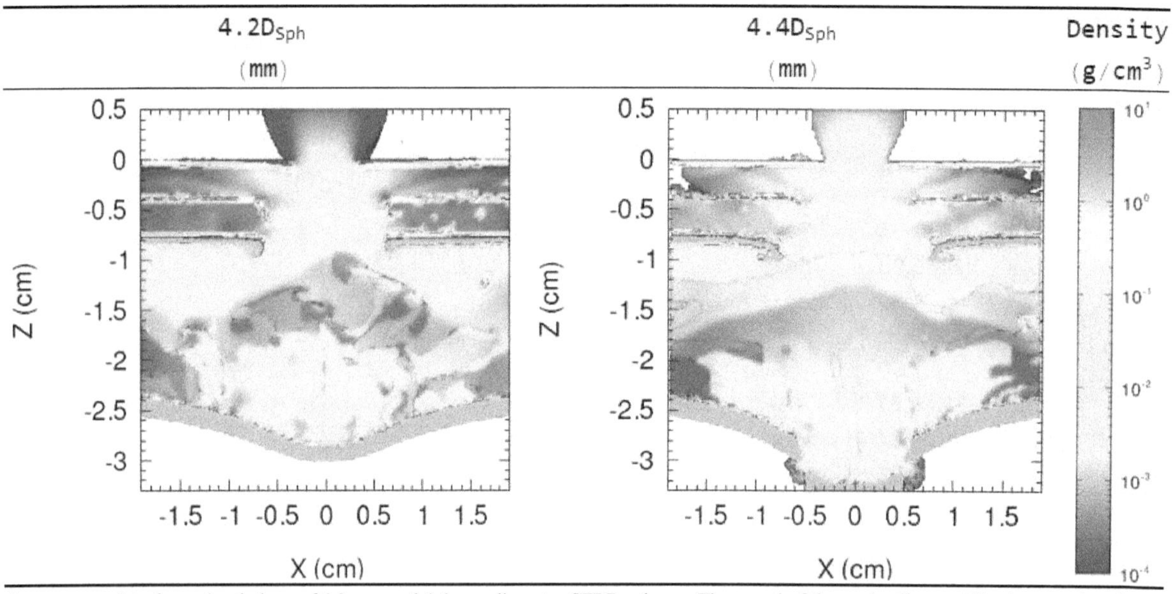

Fig. 4 Density contour plots from simulations of 4.2 mm and 4.4 mm diameter CFRP spheres. The speed of the projectiles are 7 km/s normal to the surface. The contours show the density contours after 35 μs. The critical diameter for this simulated condition would be 4.3±0.1 mm.

3.2. *Cylindrical CFRP Projectiles*

A similar analysis has been performed for critical cylinder lengths for a range of cylinder diameters as illustrated in Fig. 5. Different AOA have been considered for cylinders. Fig. 5a is central axis aligned to the velocity vector ($\alpha = 0°$), Fig. 5b is central axis rotated 45° from the velocity vector ($\alpha = 45°$), and Fig. 5c is central axis orthogonal to the velocity vector ($\alpha = 90°$).

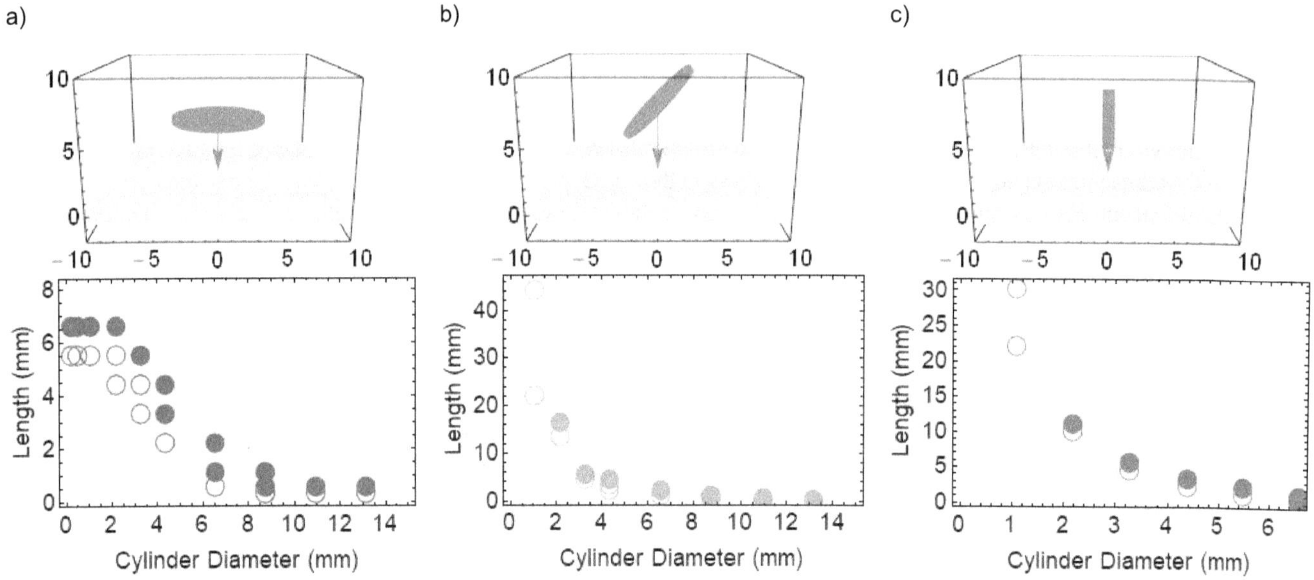

Fig. 5 Simulated results for ranges of cylinder diameters when the cylinder's symmetry axis is a) aligned with the velocity vector, b) rotated 45° from the velocity vector and c) rotated 90° from the velocity vector. The orientations are illustrated by L/D < 1 "disks" relative to the velocity vector (red arrow). In the plots, open circles indicate a cylinder length where the shield defeated the projectile, and filled circles indicate a cylinder length where the shield is defeated.

In Figure 5a, the simulation results are shown for cylinder diameters ranging from 0.275 mm to 13.2 mm. In this figure, the markers for the lengths that did not penetrate the shield are unfilled, and the lengths that did penetrate are filled. These two lengths identify the critical length as a function of cylinder diameter for the 7 km/s impact condition, which can be seen to plateau at small diameters near 6 mm then decrease to another plateau at less than a 0.5 mm for large diameters. In addition to these three orientations, AOA of 22.5° and 67.5° have also been simulated with similar findings.

As noted before, a cylinder does not look the same if it impacts at an attack-angle relative to its velocity vector as illustrated in Fig. 5, so similar simulations have been performed with the cylinders' central-axis rotated 45° and 90° to the velocity vector as shown in Figures 5b and 5c, respectively. Like before, the lengths that do not penetrate the rear wall of the shield are left open, and the lengths that do penetrate the rear wall are filled. Comparing the three figures side-by-side, it can be seen that while the 0° case can reach a length that no cylinder fails the shield (i.e. a cylinder less than 0.5 mm is stopped regardless of diameter when hitting flat), impacts of cylinders more through the radius of the cylinder pierce the shield even when they are very thin for a sufficiently large diameter. Similarly, when the diameter of the cylinder is small (D < 2 mm), impacts rotated away from aligned (i.e. AOA > 22.5°) can have critical lengths that are asymptotically long.

4. Discussion

As has been identified in the previous section, patterns of critical length-to-diameter are apparent for cylinder orientations relative to the velocity vector. A model for the critical length, L_{Cyl}, of an arbitrarily oriented cylinders has been fit to the function:

$$L_{Cyl}[D_{Cyl},\alpha] = 3.04\ Coth\left[\frac{D_{Cyl}-(-9.26+26.8\ Sin[\alpha]-16.3\ Sin[\alpha]^2)}{1.94}\right]\left(1 - Tanh\left[\frac{D_{Cyl}-4.13}{2.68}\right]\right) + 0.496\ Cos[\alpha]^2 \qquad (1)$$

where α is the AOA and D_{Cyl} is the cylinder diameter in millimeters. This critical length model is for a normal impact at 7 km/s. This fit has been accomplished using the arithmetic mean of the nearest intact and penetrated lengths from the simulations. The applicable fit parameters and their error estimates are shown in the Fig. 6a. This function is shown as a function of the AOA (abscissa) and cylinder diameter (curve color) in Fig. 6b. The cylinder diameters ranged from 0.55 mm to 13.2 mm, and the AOA ranges from 0° to 90°. In addition to the model, the full set of simulated critical length data (circles) developed in this study is shown for all the orientations that have been considered (0°, 22.5°, 45°, 67.5° and 90°). The curves and the simulated data are color synchronized. As previously done, the markers for the penetrated shield are shown filled, and the markers for intact shields are left empty. As can be seen, this empirical model captures much of the data well over the broad range of conditions considered.

a)

Estimate	Standard Error
3.04	0.4
-9.26	6.6
26.8	15.5
-16.3	9.45
1.94	1.15
4.13	0.554
2.68	0.717
0.496	0.296

b)

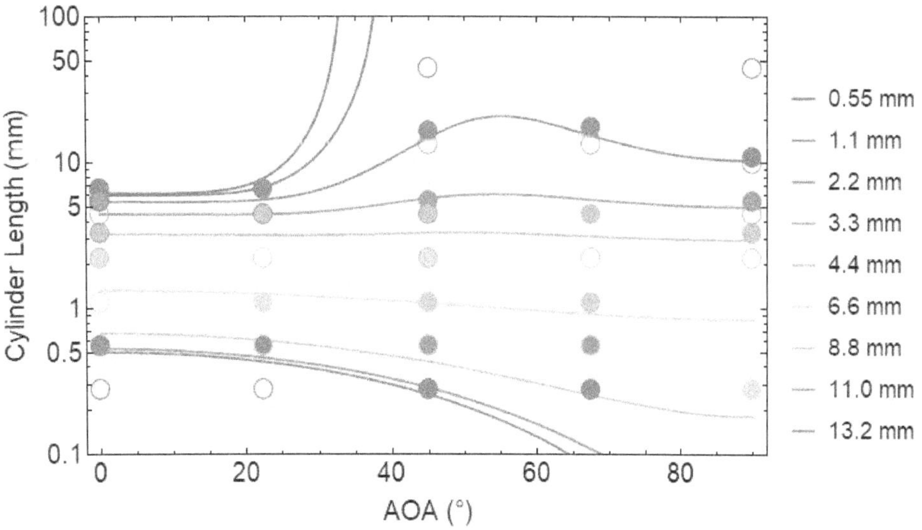

Fig. 6 Critical length model for cylindrical projectiles impacting normal to the shield surface at 7 km/s a) model coefficients and b) plot showing the model and data (circles). Defeated shield markers are filled, and intact shield markers are unfilled. Each curve represents the critical length for specified cylinder diameters.

The critical length model has been chosen to exploit trigonometric and hyperbolic properties to achieve the observed features of Figures 5. The parenthetical term that contains the hyperbolic tangent and the trailing term reproduce the shape of Fig. 5a for $\alpha = 0°$. The cosine term goes from 0 to 1 as α goes from 0° to 90°, thus the trailing term vanishes with increasing AOA. The hyperbolic cotangent produces the asymptotic behavior of the $\alpha = 45°$ and $\alpha = 90°$ with the onset of asymptotic length behavior for small diameter cylinders defined by the parenthetical term in the hyperbolic cotangent.

5. Conclusions

A numerical model of an operational shield with thermal blankets has been studied against varying types of CFRP projectiles at a single impact speed of 7 km/s and a single, normal impact obliquity. The types of CFRP projectiles evaluated consisted of spherical and cylindrical shapes. The simulations of the spherical shape has determined that the critical diameter is 4.3±0.1 mm for the shield. A similar series of simulations identified the critical length of a cylindrical projectile for various cylinder diameters. These simulations have been repeated for three different impact orientations. The collection of the simulations has been consolidated to a single critical length function, which can be subsequently converted to critical areal density or critical mass.

In all, it required almost eighty simulations to characterize a shield to the level that has been outlined here for this single impact condition. Extending to additional projectile AOA beyond the five that have been considered for more refined results would increase this requirement further. Future simulation work is underway to expand on impact speed and obliquity of the cylindrical projectiles into the subject shield along with other shields. In addition to the simulation efforts, hypervelocity impact experiments are also in progress to develop the means to validate simulations for the variety of projectile shapes that are to be evaluated.

Acknowledgements

The author wishes to gratefully acknowledge the Jacobs JETS contract NNJ13HA01 in its support to the Hypervelocity Impact Technology group at NASA Johnson Space Center and its sponsorship to University of Texas at El Paso under EN41520TMS.

References

[1] Liou, J.-C., et. al., "DebriSat-A planned laboratory-based satellite impact experiment for breakup fragment characterization", 6th European Conference on Space Debris at Darmstadt, Germany, ESA SP-723, 41 (2013).
[2] Fitz-Coy, N., et. al., "Characterization of debris from the Debrisat hypervelocity test", 66th International Astronautical Congress at Jerusalem, Israel, IAC-15-A6.2.9x30343, pp. 1-10 (2015).
[3] Christiansen, E.L., "Meteoroid/debris shielding", NASA/TP-2003-210788 (2003).
[4] Schonberg, W.P. and Williamsen, J.E., "RCS-based limit curves for non-spherical projectiles impact dual-wall spacecraft systems", Int. J. Impact Eng., 33, pp. 763-770 (2006).
[5] Whipple, F.L., "Meteorites and space travel", Astronomical Journal, 52:1161, 131 (1947).
[6] Miller, J.E., Bjorkman, M.D., Christiansen, E.L., and Ryan, S.J., "Analytic ballistic performance model of Whipple shields", Procedia Eng., 103, pp. 389-397 (2015).
[7] Christiansen, E.L. and Kerr, J.H., "Projectile shape effects on shielding performance at 7 km/s and 11 km/s", Int. J. Impact Eng., 20:1-5, pp. 165-172 (1997).
[8] Piekutowski, A.J., "Hypervelocity impact of non-spherical projectiles: observations and lessons learned from impact tests", Proceedings of the Hypervelocity Shielding Workshop, ed. H. Fair, Institute of Advanced Technology: Texas (1998).
[9] Schafer, F., Hiermaier, S. and Schneider, E., "Ballistic limit equation for the normal impact of unyawed ellipsoid-shaped projectiles on aluminium Whipple shields", 54th International Astronautical Congress at Bremen, Germany, IAC-03-IAA.5.3.06, pp. 1-10 (2003).
[10] Carrasquilla, M.J. and Miller, J.E., "Shape effect analysis of aluminum projectile impact on Whipple shields", Procedia Eng., 204, pp. 308-314 (2017).
[11] Lyons, F., "Determine ISS Soyuz Orbital Module ballistic limits for steel projectiles hypervelocity impact testing", NASA Johnson Space Center report 66540 (2013).
[12] Davis, B.A., "NASA NESC high-velocity impact (HVI) study at UDRI: Phase-1", NASA Johnson Space Center report 66578 (2013).
[13] McGlaun, J., et. al., "CTH: a three-dimensional shock wave physics code", Int. J. Impact Eng., 10, pp. 351-360 (1990).
[14] Hertel, E.S. and Kerley, G.I., "CTH reference manual: the equation-of-state package", Sandia report SAND98-0947 (1998).
[15] Kerley, G.I., "Kapton ANEOS model", unpublished distributed with CTH (1992).
[16] Thompson, S.L., "ANEOS analytic equations of state for shock physics codes input manual", Sandia report SAND89-2951 (1989).
[17] Kerley, G.I., "Equations-of-state for composite materials", Kerley Publishing Services report KPS99-4 (1999).
[18] "HexForce™7781:Fiber Glass Fabric", Hexcel® Corporation (Accessed: November 2017).
[19] Steinberg, D.J., "Equation-of-state and strength properties of selected materials", Lawrence Livermore National Laboratory report UCRL-MA-106439 (1991).
[20] Kerley, G.I., "Theoretical equation-of-state for aluminum", Int. J. Impact Eng., 5, pp. 441-449 (1987).
[21] Kerley, G.I., "Multiphase equation-of-state for iron", Sandia National Laboratories report SAND93-0227 (1993).
[22] "TORAYCA® M55J data sheet", TORAYCA® Toray Caron Fibers America, Inc. (Accessed: November 2017).

Proceedings of the 2019 Hypervelocity Impact Symposium
HVIS2019
April 14-19, 2019, Destin, FL, USA
HVIS2019-048

Investigation on Response of an Aluminum Honeycomb Subjected to Hypervelocity Impacts using Lagrange and SPH for Numerical Modeling

Kumi Nitta[a)], Masumi Higashide[b)], Mirai Sueki[c)], Atushi Takeba[c)]

[a] *New Enterprise Promotion Department Strategic Planning and Industrial Promotion Division, Japan Aerospace Exploration Agency (JAXA)*
Ochanomizu Sola City, 4-6 Kandasurugadai Chiyoda-ku, Tokyo 101-8008, Japan
[b] *Research and Development Directorate, JAXA*
7-44-1, Higashimachi, Jindaiji, Chofu, Tokyo, 182-8522, Japan
[c] *Science and Engineering Systems Division, ITOCHU Techno-Solutions, Kasumigaseki Bldg., 3-2-5, Kasumigaseki, Chiyoda-ku, Tokyo, Japan*

Abstract

Numerical modeling has been conducted with the commercial code AUTODYN 2D, using the Lagrange and Smooth Particle Hydrodynamics (SPH) processors. The numerical results are compared and discussed with the corresponding experimental results from the standpoint of assessing the protection of satellites against M/OD hypervelocity impacts. The material models used in the numerical simulation are also discussed, as well as a wide range of impact velocities, including shock-induced vaporization. The projectiles used to simulate M/OD consist of 100 μm to 1 mm diameter alumina with impact velocities of 2–15 km/s.

In order to assess the structural integrity of unmanned spacecraft subjected to the threat of hypervelocity impact by space debris, the numerical method was proposed mainly from the standpoint of material modeling suitable for extremely severe physical conditions such as high pressure, high temperature, high strain, and high strain rate, sometimes accompanied by shock-induced vaporization.

The numerical results adopting these material models were compared with the corresponding hypervelocity impact tests by using the two-stage light-gas gun at ISAS/JAXA. Although examples of the impacts on the aluminum honeycomb can be shown, it has been demonstrated that the numerical analysis can effectively simulate the overall corresponding experimental results.

We show the response of an aluminum honeycomb as derived from analysis of hypervelocity impact at 2 km/s to 15 km/s using the Lagrange and SPH processors. We also verified that the ballistic limit curve of an aluminum honeycomb panel is shown as a downward line using both processors, which is unlike the up and down ballistic limit curve of a Whipple shield.

Keywords: Orbital debris, hydrocode AUTODYN, two-stage light gas gun, numerical simulation, spacecraft

Nomenclature

u_0	*Impact velocity (km/s)*
t	*Thickness of plate (mm)*
D	*Diameter of projectile (mm)*

* Corresponding author. Tel.: +81-50-3362-7996.
E-mail address: nitta.kumi@jaxa.jp.

1. Introduction

We have been conducting hypervelocity impact tests and numerical simulations in order to create Japanese design guidelines that protect satellites against a certain degree of impact by micrometeoroids and orbital debris (M/OD). Certain members of JAXA, university researchers, computer analysts, and spacecraft manufacturers formed a working group to investigate the impact effect of OD on critical parts and bumpers, by conducting hypervelocity impact (HVI) tests and analysis. The knowledge acquired is now being reflected in spacecraft design [1].

We plan to continue and extend our experimental and numerical investigations covering a wider and more diverse range of conditions, in order to complete the guidelines at JAXA for the purpose of protecting unmanned spacecraft against impacts from space debris and micrometeoroids.

In line with those plans, we compared the numerical simulation results with the experimental results and investigated the response of an aluminum honeycomb subjected to high-velocity impacts.

In previous work, we analyzed the response of an aluminum honeycomb sandwich panel to hypervelocity impacts ranging from 2 to 15 km/s using Lagrange models. We verified that the ballistic limit of an aluminum honeycomb panel becomes progressively smaller as the impact velocity increases [2]. On the other hand, SPH is a relatively gridless Lagrangian numerical technique that offers advantages over the more established grid-based Lagrangian and Euler methods [3]. The paper discusses the simulation of hypervelocity impacts on an aluminum honeycomb, particularly in comparison with two numerical techniques [4][5].

2. Numerical analysis method – material model

2.1. Hydrocode

Numerical simulations were performed using the ANSYS AUTODYN hydrocode. As in almost all hydrocodes, the material model in ANSYS AUTODYN consists of three parts: 1) an equation of state (E.O.S.) describing the relation between pressure, density, and internal energy (not temperature); 2) a material strength model that expresses the constitutive relation of solid materials; and 3) a failure or fracture model. While a failure model is typically used for solid materials, it can also incorporate the spalling (fracture by negative pressure) of liquid materials. Despite their name, hydrocodes are formulated to simulate the highly dynamic and nonlinear behavior of materials not only in the liquid phase but also in the solid and gas phases, and sometimes even the plasma state can be approximately simulated by using an appropriate E.O.S.

2.2. Projectiles and targets

Orbital space debris consists of artificial objects that have been created through human space activities in near-Earth space ever since the launch of Sputnik I. The average density of space debris is considered to be 2.8–4.7 g/cm^3 for objects smaller than 10 mm, whereas the relative impact velocity of orbital space debris in low Earth orbit (LEO) is theoretically limited to approximately 16 km/s. In this assessment, the space debris material is assumed to be aluminum oxide (alumina). The propellant used in solid rocket motors (SRM) contains about 18% aluminum by weight, and these motors eject alumina particles from μm-order dust up to cm-order slugs into Earth orbit.

In our study, the Johnson-Holmquist II model [6] is adopted to simulate alumina projectiles. The material model includes an equation of state ("pressure"), a constitutive ("strength") model, and a fracture or failure ("damage") model. The Steinberg-Guinan strength model [7] was applied to stainless steel when we compared and discussed the simulation results with the corresponding experiments.

The targets tested and simulated numerically in this assessment are aluminum honeycomb panels (5056 aluminum core, 25.4-mm thick; 2024-T4/2024-T3 aluminum skins, 0.254-mm thick). The Tillotson equation of state [8] was applied to the aluminum skin materials subjected to extremely severe physical conditions, because it considers shock-induced vaporization. The Tillotson model [8] was applied to the ductile materials (aluminum alloys in case of projectile velocities under 10 km/s). When we used 2024-T3 with SPH, we did not use an erosion rule for the adapted SPH element.

3. Experimental facilities

A series of experiments was conducted using a two-stage light-gas gun installed at ISAS/JAXA, as shown in Fig. 1. This gas gun can accelerate a 0.2 g spherical projectile up to 7 km/s. The projectiles impact a target in a test chamber evacuated to a pressure of around 0.1 Pa. The spherical projectile is initially covered with a cylindrical polycarbonate sabot (measuring 7.1 mm in diameter and 10.5 mm in length) that guides the spherical projectile in the helium driving gas during acceleration, and then detaches from the projectile after sufficient velocity is attained.

The simulation results are compared with the corresponding test results and discussed from the standpoint of assessing the protection of satellites against M/OD hypervelocity impacts. The material models used in the numerical simulation to enable the assessment of phenomena corresponding to a wide range of impact velocities including shock-induced vaporization are also discussed. To simulate M/OD, we used spherical alumina projectiles ranging in diameter from 0.1 mm to 1 mm, with impact velocities of 2–15 km/s.

Fig. 1. Two-stage light-gas gun at ISAS/JAXA

4. Results and discussion

4.1. Experimental and computational configuration

We developed two-dimensional axial symmetry analytical models. Projectiles impact the target of an aluminium honeycomb sandwich typically used in the construction of satellite skin panels. The projectiles are assumed to be spherical with a diameter from φ0.3 mm to 1mm and have the properties of alumina in Fig.2.

Modelled as a disc of sufficient size, the target is an aluminium honeycomb core composed of cells measuring 6.35 mm (1/4 inch) in height and arranged in a regular hexagonal grid as shown in Fig. 3, sandwiched between aluminium skin sheets. In the numerical study, each cell is modelled as a circle bounded by the corresponding hexagon. This numerical model requires a large computational effort because the skin sheet thickness is much smaller than the height of the honeycomb core.

Accuracy of the numerical simulation was validated by a comparison of results computed from the gas gun experiments. In order to confirm the reliability of the numerical model, the projectiles are assumed to be spherical with a diameter of φ0.3 mm and have the properties of SUS304 stainless steel to simulate the kinetic energy of 6 km/s velocity impacts. Projectiles are modelled on the Lagrange and SPH solvers. Figure 4 shows an image of the experimental setup and Fig. 5 shows the experimental results. In the experiments using projectiles over 0.3 mm in diameter, damage produced by impact of the fragment cloud was observed on the back face sheet and surface of the aluminium alloy plate, as shown in Fig. 5. The projectile was considered to have been changed into the fragment cloud by the impact on the front face sheet. The honeycomb core acted similarly to the standoff of a double-wall bumper shield and dispersed the impact energy of the projectiles. Figure 6 depicts the numerical results using Lagrange that simulate the test case shown in Fig. 4. The particle diameter was 0.3 mm of SUS304 and particle velocity was 6 km/s. Figure 6 shows the damage at 50 μs, which is the end time. In both cases, the projectile penetrated the second (interior) skin. From the results of repeated experiments [9] and a comparison between the simulations and experimental data, we confirmed the numerical modelling.

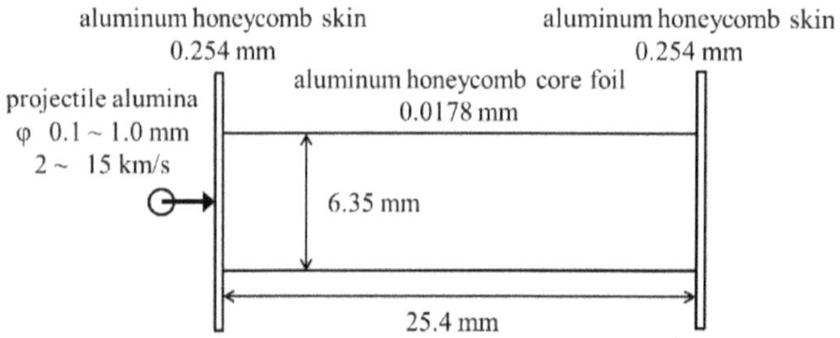

Fig. 2. Two-dimensional axial symmetry analytical model

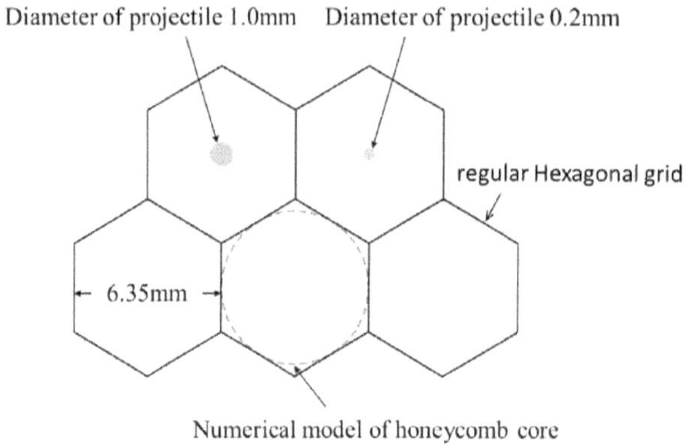

Fig. 3. Numerical model of honeycomb core

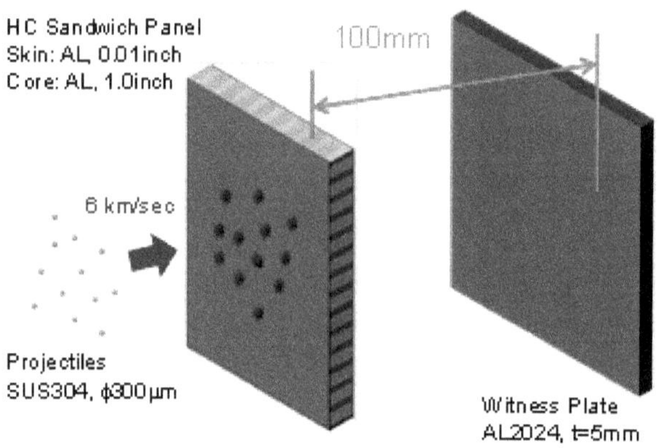

Fig. 4. Image of the experimental setup [9]

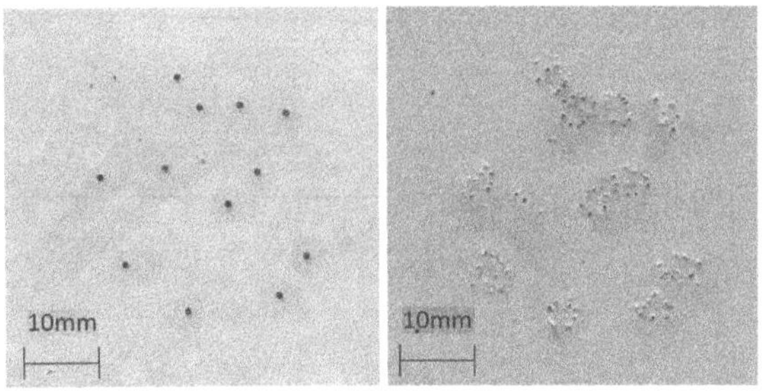

Fig. 5. Experimental results of Honeycomb sandwich panel impacted by $D = 0.3$ mm projectiles at 5.92 km/s: (left) front face sheet, (right) back face sheet [9]

Fig. 6. Deformation diagram of calculation results using Lagrange and fragment phenomena at 50 μs, impact velocity of 6 km/s

4.2. Hypervelocity impact results

We investigated the hypervelocity impact of space debris (M/OD) at 2-15 km/s by numerical simulation. The material models used in the simulation to cover a wide range of impact velocities were investigated and are discussed below. Alumina projectiles measuring between 0.1 mm and 1 mm in diameter were used to simulate M/OD.

Before presenting the results of our parametric study, we discuss the effect of the honeycomb core. In the case of no honeycomb core, after a particle penetrates the first (exterior) skin, there is a great expansion of fragments over a wide area, which consequently reduces damage to the second (interior) skin (Fig. 7). With a honeycomb core between the skins, however, the honeycomb cell walls serve to contain the expansion so that the fragments are concentrated in a small area of the second skin (Fig. 8). Our simulations confirm that the honeycomb core thus increases the damage to the inner skin.

Figure 9 shows the ballistic limit of the aluminum honeycomb panel. NO GO, Threshold, and GO denote no penetration to the second skin, a borderline score for penetration, and penetration to the second skin, respectively. As the impact velocity increases, the ballistic limit of the aluminum honeycomb sandwich panel becomes progressively smaller. This behavior differs from the ballistic limit curve of a Whipple shield. The ballistic limit of the aluminum honeycomb sandwich panel has a flexion point at an impact velocity of approximately 10 km/s. Our comparison shows that the aluminum honeycomb core negates the beneficial "bumper" effect of two skins. Usually, the projectiles perforated the front face sheet of the honeycomb sandwich panel. When only one projectile impacted a cell, there was no perforation hole on the back face sheet. Perforated holes on the back face sheets were only generated in the area under the impacted cells. The foils of cells adjacent to those impacted were damaged, but the fragments did not perforate the back face sheet under the adjacent cell. This result shows that, in only the impacted cell, the fragments kept sufficient energy to perforate the face sheet and that the honeycomb foils acted as bumpers for the fragment cloud.

The SPH and Lagrange results are quite similar, and the curves are almost the same at impact velocities above 10 km/s. At impact velocities under 10 km/s, however, the SPH results are smaller than the Lagrange results. This is caused mainly due to the difference in break time when projectiles perforated the panel. In the case of SPH, the projectile continues to penetrate the target at post impact.

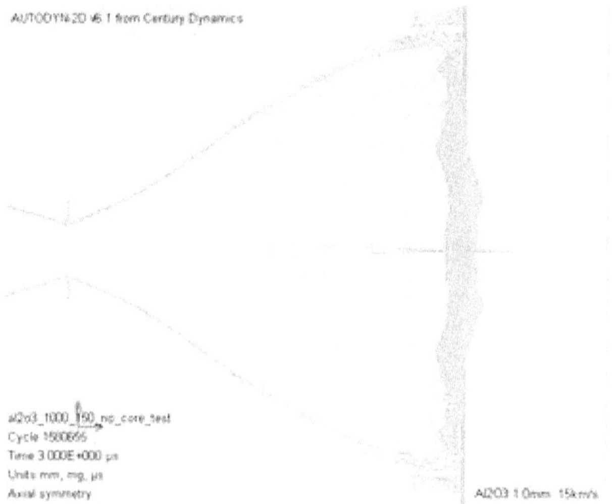

Fig. 7. Calculation results and fragment phenomena in case of no honeycomb core

(a) SPH

(b) Lagrange

Fig. 8. Calculation results and fragment phenomena in case of a honeycomb core at 30 μs, impact velocity of 2 km/s, projectile size of ϕ 0.7 mm

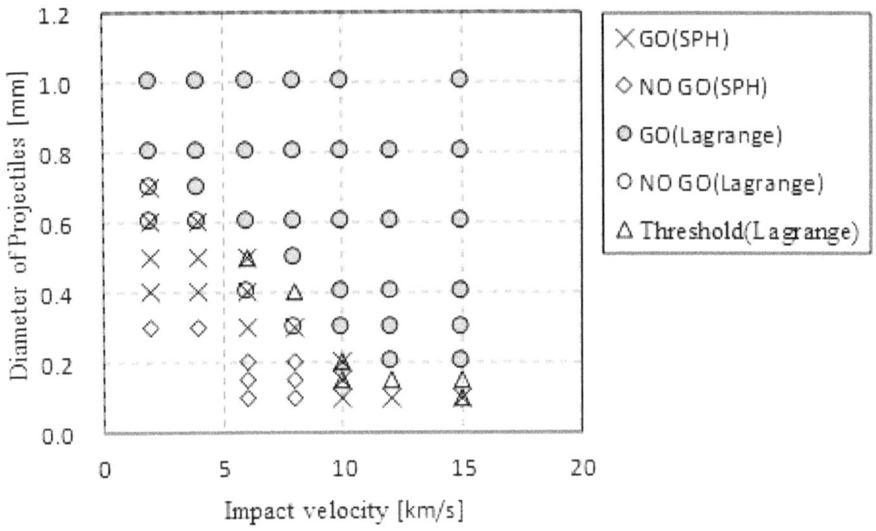

Fig. 9. Summary of the numerical simulation results

5. Conclusions

The results of numerical simulations adopting these material models were compared with hypervelocity impact tests using a two-stage light-gas gun, and examples of the impacts on an aluminum honeycomb panel were shown. The results obtained using the Lagrangian and SPH methods were discussed in this paper, and the numerical analysis was shown to effectively simulate the overall corresponding impact test results.

We analyzed the response of an aluminum honeycomb sandwich panel to hypervelocity impacts ranging from 2 km/s to 15 km/s and presented the results. We also verified that the ballistic limit of an aluminum honeycomb panel becomes progressively smaller as impact velocity increases. This differs from the ballistic limit of a Whipple shield. In the future, we will consider the effects and uses of different types of solvers after impact, and will continue to extend our experimental and numerical investigations to cover a wider range of conditions, in order to complete the guidelines for the protection of unmanned spacecraft against space debris and micrometeoroid impacts.

Acknowledgement

The experiments were conducted and supported by the Space Plasma Laboratory, ISAS, and JAXA. This research was also supported by the JAXA design guideline Working Group 3.

References

[1] JAXA: Space Debris Protection Design Manual, JERG-2-144-HB001, 2009.
[2] Kumi Nitta, Masumi Higashide, Yukito Kitazawa, Atushi Takeba, Masahide Katayama, and Haruhisa Matsumoto, "Response of an Aluminum Honeycomb Subjected to Hypervelocity Impacts," Procedia Engineering, Vol. 58, 2013, pp. 709-714.
[3] Colin J. Hayhurst and Richard A. Clegg, "Cylindrically Symmetric SPH Simulations of Hypervelocity Impacts on Thin Plates," *Int. J. Impact Eng.*, Vol. 20, (1997), pp. 337-348.
[4] S.Ryan F.Schaeter R.Destefanis M.Lambert," A ballistic limit equation for hypervelocity impacts on composite honeycomb sandwich panel satellite structures", Advances in Space Research, Volume 41, Issue 7, 2008, Pages 1152-1166
[5] J.-M.Sibeaud L.Thamié C.Puilletc," Hypervelocity impact on honeycomb target structures: Experiments and modeling", International Journal of Impact Engineering, Volume 35, Issue 12, December 2008, Pages 1799-1807

[6] G. R. Johnson and T. J. Holmquist, in: S. C. Schmidt et al. (eds.), Proceedings of the High-Pressure Science and Technology-1993, *AIP Conf. Proc.*, No. 309, AIP, New York (1994) pp. 981-984.
[7] D. J. Steinberg, S. G. Cochran, and M. W. Guinan, A constitutive model for metals applicable at high-strain rate, *J. Appl. Phys.*, 51, (1980), pp. 1498–1504.
[8] J. H. Tillotson, Metallic equations of state for hypervelocity impact, GA-3216, General Atomic, CA, July 1962.
[9] M. Higashide, N. Onose, and S. Hasegawa, Evaluation of Space Debris Impact on Spacecraft Panels, 2011-r-28, ISTS28th, 2011.

Proceedings of the 2019 Hypervelocity Impact Symposium
HVIS2019
April 14-19, 2019, Destin, FL, USA
HVIS2019-054

Debris Risk Evolution And Dispersal (DREAD) for post-fragmentation modeling

Daniel L. Oltrogge[a]* and David A. Vallado[a]

[a]*Center for Space Standards and Innovation (CSSI), Analytical Graphics, Inc., 7150 Campus Drive, Suite 260, Colorado Springs, CO 80920*

Abstract

The Debris Risk Evolution And Dispersal (DREAD) tool facilitates the 3D modeling and risk analysis of the fragmentation cloud after a collision or explosion. This tool uses the NASA Standard Breakup Model and other breakup models "under the hood" that are capable of estimating the Probability Density Function (PDF) of induced relative velocity, mass and area of fragments as a function of object size. DREAD can be further enhanced by incorporation of alternate, more detailed hypervelocity simulations that enforce conservation laws (conservation of mass, angular and linear momentum and kinetic energy). We also discuss our recent incorporation of an improved technique to normalize risk by the expansion volume occupied by debris fragments. DREAD is then used to examine the likely debris fragmentation cloud created by the Fengyun 1C (FY1C) antisatellite (ASAT) intercept test conducted by the Chinese in 2007 and the risk it subsequently posed to other spacecraft and the cloud's evolution and dispersal.

Keywords: Fragmentation, dispersal, hypervelocity, collision probability, Fengyun 1C

Nomenclature

$\mathcal{L}(\Delta v)$ Likelihood of obtaining a specified imparted velocity magnitude (Δv) from a fragmentation event.
a_f and a_g Equinoctial elements of primary and secondary conjuncting objects
$\overline{w_p}$ and $\overline{w_s}$ Angular momentum vectors of primary and secondary conjuncting objects

1. Introduction

Space debris analysts commonly employ two types of simulation tools to assess post-fragmentation event debris fields and their evolution: (1) simulation tools that create a notionally-representative depiction or scenario of what the individual debris fragments and relative velocities might look like [1, 2], and (2) simulation tools which assess the overall spatial volume of where such fragments may go in the ideal case that the imparted fragmentation relative velocity directions adhere to preconceived assumptions or distributions of fragmentation relative velocities [3, 4, 5, 6, 7, 8, 9, 10, 11, 12]. To date, researchers typically assume a specific initial or maximum relative fragmentation velocity (e.g. a fixed value or at best an adopted percentile of the Δv Probability Density Function, or PDF).

* Corresponding author. Tel.: +1-719-482-4552.
E-mail address: oltrogge@agi.com.

1.1. DREAD tool

While the authors have developed tools over the years that fall into both categories, the focus of this paper is on extending and applying the Debris Risk Evolution And Dispersal (DREAD) tool [13] to a high-interest space collision event, the kinetic anti-satellite (ASAT) intercept of the Fengyun spacecraft by the Chinese in 2007. The overall DREAD analysis process is shown in **Figure 1**. Such work is important because of the need for actionable and operationally realistic fragmentation event Space Situational Awareness (SSA) and associated risk assessments. This is especially relevant in the increasingly crowded Low Earth Orbit (LEO) and Geosynchronous Earth Orbit (GEO) orbit regimes, where analyses have shown that LEO and GEO collision risks can be up to four orders of magnitude higher than previously envisioned.

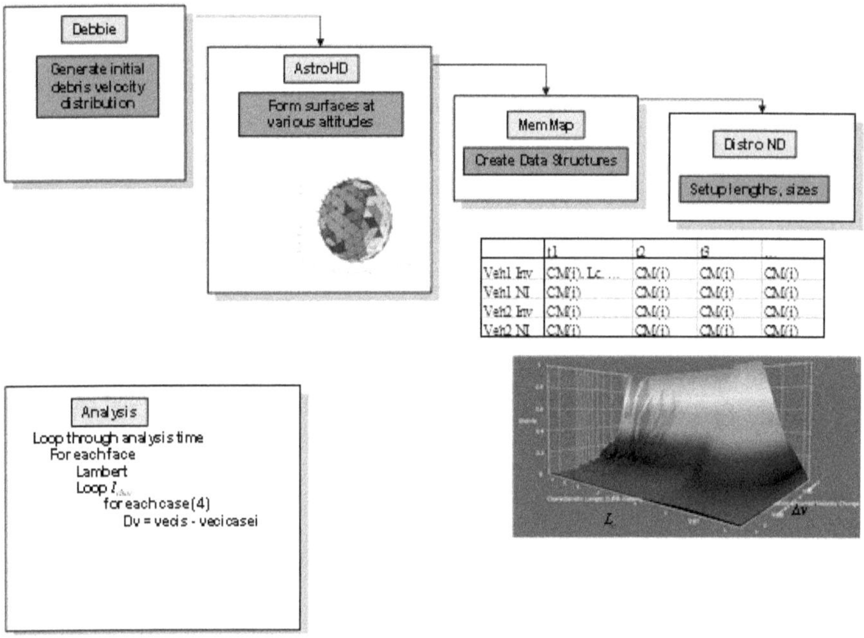

Figure 1. Overall DREAD analysis process. This figure shows the overall process flow in the DREAD tool. Several computer programs are combined to arrive at the final solution, accomplishing (a) estimation of the initial fragmentation imparted velocity PDF, (b) establishment of an equal angles grid, (c) data structures to store time-varying 3D data, and (d) accumulation of the Δv PDF.

2. Primary improvement to DREAD tool analysis process

As detailed in [9 and 13], the DREAD tool uses a breakup fragmentation model to obtain the imparted fragmentation velocity Δv PDF. It is assumed that fragment spread velocity directions are equally likely about the center-of-mass velocity vectors of both colliding objects. As a function of imparted velocity change, DREAD estimates the likelihood of a fragment's presence at a later time and place, spanning spacecraft 1 and 2's involved and non-involved [9 and 13] fragment sources, via:

$$\mathcal{L}(\Delta v) = \left\{1 - \left(1 - \mathcal{L}_{\Delta v L_{c_{Sat\,\#1_{Inv}}}}\right)\left(1 - \mathcal{L}_{\Delta v L_{c_{Sat\,\#1_{Noninv}}}}\right)\left(1 - \mathcal{L}_{\Delta v L_{c_{Sat\,\#2_{Inv}}}}\right)\left(-\mathcal{L}_{\Delta v L_{c_{Sat\,\#2_{Noninv}}}}\right)\right\} \tag{1}$$

Previously, this likelihood was then normalized by the digitized volume of space that fragments occupy at any time of interest as shown in Equation (2). But this approximation was flawed in that the volume of the collision event is very small and typically not available. In our current work, we replaced this by the method originally presented in [3] as shown in Equation (3). While other Jacobian formulations exist [3, 14], we selected the formulation of [15], consistent with the formulation selected in [8].

$$\mathcal{L}(\Delta v,\ volumetric\ presence) = [\mathcal{L}(\Delta v)]\left[\frac{Vol_{Collision\_Event\_Cell}}{Vol_{Total\_Dispersal}}\right] \tag{2}$$

$$\mathcal{L}(\Delta v,\ volumetric\ presence) = \left[\frac{[\mathcal{L}(\Delta v)]}{|\det Jacobian\ (\bar{p}_2\ w.r.t.\bar{v}_1\ |}\right] \tag{3}$$

3. Fengyun ASAT Intercept Fragmentation Event

The People's Republic of China (PRC) conducted its first successful direct-ascent ASAT weapons test in 2007, launching a ballistic missile armed with a kinetic kill vehicle (not an exploding conventional or nuclear warhead) to destroy the PRC's Fengyun-1C (FY1C) weather satellite [16]. The Fengyun satellite, shown in **Figure 2**, had a mass of 950 kg [17]. The Chinese ASAT intercept of Fengyun 3C occurred on 11 Jan 2007 at an estimated time of 22:26:10 UTC [18], in contrast to the original report in AW&ST suggested an 11 Jan 2007 22:28:00.000 time. While the time difference is small, the geometry of the Fengyun satellite over the ASAT launch site changes modestly, and the impact goes from a mostly head-on case to that of a hit from below trajectory. The ASAT launcher was reported to have been launched from the Xichang Space Center in China [19].

Figure 2. Picture of Fengyun 1C (FY1C) spacecraft [20].

For the purpose of this characterization we assumed the interceptor's mass to be 600 kg, based upon the ASAT launcher presumably used [20]. We also chose, to assume a slightly post-apogee ballistic trajectory for the ASAT, intercepting in a slightly downward, direction, in the roughly opposite direction of the Fengyun spacecraft velocity vector as shown in **Figure 3**. This assumption causes the ASAT-derived fragments to reenter shortly after this relatively high-altitude intercept as shown in **Figure 4**. In stark contrast, the debris fragments originating from the FY1C spacecraft's mass are not be reduced by such a ballistic trajectory and will remain orbital for quite some time.

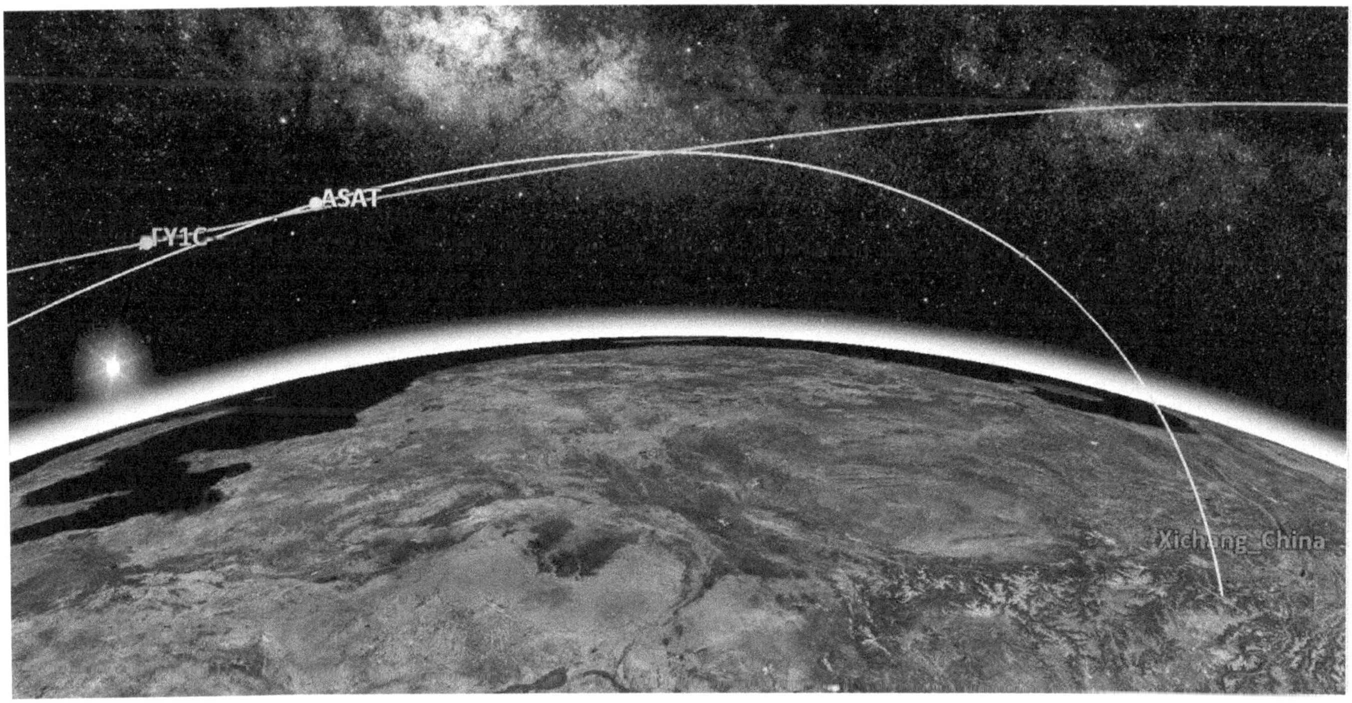

Figure 3. Pre-engagement scenario. This shows the as-modeled ASAT pre-engagement scenario based upon the final TLE published for Fengyun prior to ASAT intercept, and our assumed ASAT trajectory.

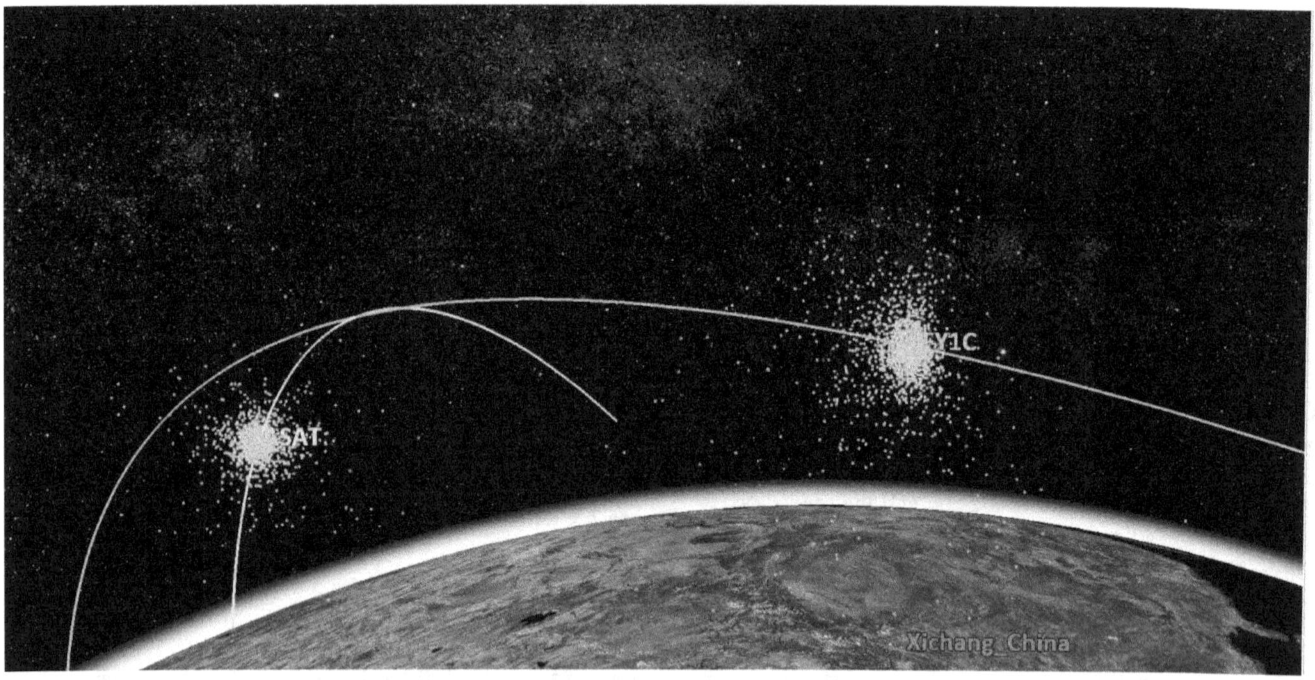

Figure 4. Post kinetic ASAT intercept discrete fragmentation. This shows the results of applying the DEBBIE instantiation of the NASA Standard Breakup Model to our assumed intercept conditions. Note that the ASAT-derived fragments are headed on a short ballistic trajectory to their reentry.

It is estimated [21] that the satellite body (excluding solar panels) is approximately 1.4 x 1.4 x 1.8 m, or ~2 m^2 as viewed from the top or bottom. The solar panels are about 3.5 m long, giving around a ~12.5 m^2 area for the satellite front or back.

3.1. Aftermath of the Fengyun ASAT intercept event

The official debris count from China's anti-satellite missile test has reached 3440 pieces big enough to be tracked by the 18th Space Control Squadron (18SPCS), and NASA's Orbital Debris Program Office is estimating more than 35,000 pieces larger than 1 cm [22]. By December 2018, there were about 2590 fragments still in orbit. The debris fragments associated with this event have gradually been released by the United States as shown in **Figure 5**, which reflects all newly tracked debris and/or newly associated objects stemming from the event up to 2019. Note that these are only the fragments which have been catalogued AND associated with the event; many thousands of additional fragments were likely produced that were either not observed or not properly associated with the event.

3.2. Fengyun ASAT fragmentation velocity characterization

Our first step is to infer intercept conditions and imparted velocities based upon post-intercept objects tracked by the Space Surveillance Network (SSN). At the Time of Closest Approach (TCA), all the objects should in theory be located at the same point and not be spread around the globe. We used the earliest Two-Line Element (TLE) orbit states released, backwards propagated to the TCA time, and evaluated the state differences. With the large number of fragments catalogued sometimes years after the intercept, we wondered if it was possible to generate the original state vectors of the two objects. The initial process of simply propagating the earliest TLE back to the TCA time resulted in a population of satellites spread literally around the Earth from the Fengyun/ASAT assumed impact location. Experimenting with true anomaly initial states seemed the good choice to modify as the orbital plane "should" remain relatively constant through time, once the secular J_2 drift is included via the Semi-analytic General Perturbations (SGP4) orbit propagator. The SGP4 propagator's B-Star drag term (which predominantly tries to account for atmospheric drag effects) can in turn also affect both the node and the argument of latitude. Unfortunately, we were unable to adjust these two parameters to get all elements to converge at TCA.

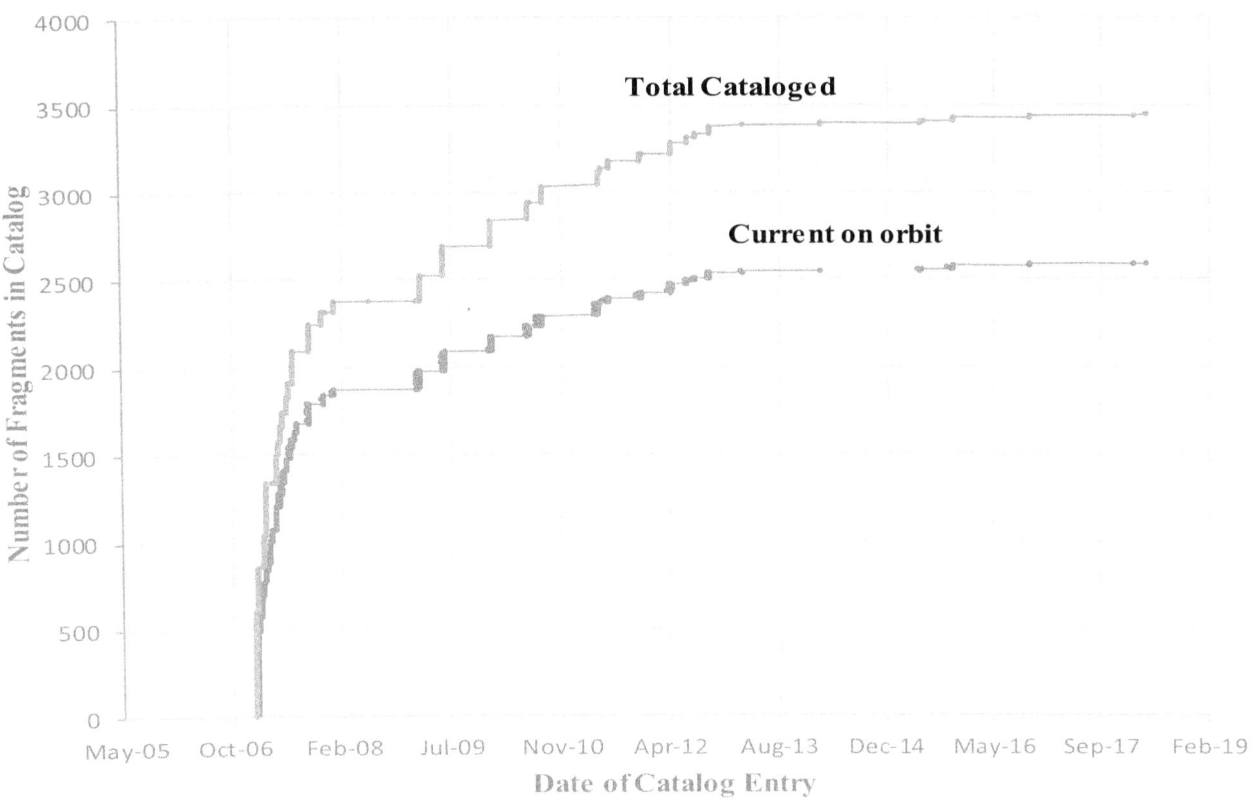

Figure 5. Debris objects catalogued by 18SPCS vs time

This made us question how the post-intercept fragments were originally associated with the event. From Air Force Space Command (AFSPC) document (TP-008 Sec 8.3 [23]), we observed that (a) the planar rate expressions contained higher-order terms than SGP4; and (b) the specific rudimentary process used to associate objects with known events was approximately described. It mentions the J_2 secular effects on the right ascension of the node and the argument of perigee, and mentions matching the refined orbits using equinoctial elements (a_f and a_g), evaluating parameters a and β shown in Equations (3) and (4), respectively. The β value compares the normal angular momentum component (*W*-axis in the EQW system). The values indicate proximity when they are *small*, however, the quantity of "small" is unspecified. The latter test seems a little more sensitive in that when values were large, the two orbits are likely unconnected.

$$\bar{a} = \sqrt{\left(a_{fp} - a_{fs}\right)^2 + \left(a_{gp} - a_{gs}\right)^2} \tag{3}$$

$$\beta = \cos^{-1}\left(\hat{w}_p \cdot \hat{w}_s\right) \tag{4}$$

AFSPC also transitioned from using a dedicated SGP4 Orbit Determination to an extrapolated Differential Correction (EGP) in late 2012. We observed this in the behavior of the argument of latitude of the debris fragments when viewed over time. This makes sense because the numerical OD would pick up additional frequencies in the observations not discernable in a SGP4 OD.

Ultimately, a combination of adjustments to in-track position and planar orientation were required in order to obtain FY1C fragments in proximity when back-propagating post-intercept debris states back to TCA. This allowed us to produce the accompanying distribution of imparted velocity changes (both magnitudes and directions) based upon JSpOC's SSN-tracked fragments as shown in **Figure 6**.

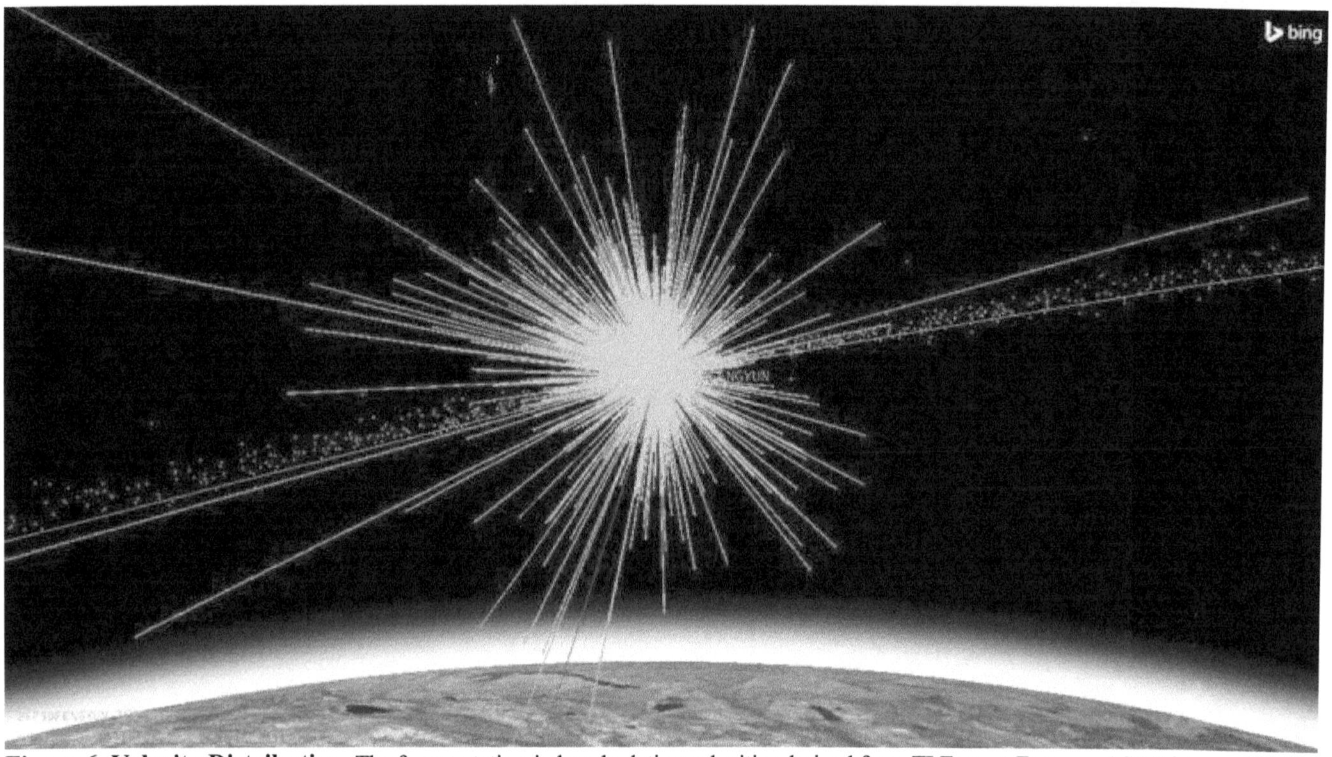

Figure 6. Velocity Distribution. The fragmentation-induced relative velocities derived from TLEs post-Fengyun ASAT intercept

4. Fengyun intercept fragmentation cloud evolution and dispersal

As previously discussed, for this representative scenario we assumed that the ASAT would have flown an interceptor trajectory which intercepts the FY1C a short while after apogee passage. Based upon these assumptions and an assumption that thirty percent of FY1C mass and one hundred percent of the interceptor's mass were directly involved in the collision, the 2D and 3D evolution of the predicted debris fragmentation were assessed via a DREAD simulation tool for the first two orbit revolutions after intercept as shown in **Appendix A**.

5. Assessing satellites at risk from fragmentation debris

As presented in Ref. 9, a primary concern that satellite operators have is whether collision- or explosion-induced debris fragments may pose a risk to their satellite, and if so, how that risk evolves over time. Because of its "digitized" nature, the DREAD tool is well suited to identify active satellites placed in harm's way. The risk can be assessed by integrating the satellite orbits as they pass through the DREAD three-dimensional PDF cloud. The risk is assessed in a relative sense, allowing rank-ordering of risk, notification to the operators flying those satellites, and identification of when and where the greatest relative risk will occur. As an example, the top 25 satellites at greatest risk for this Fengyun 1C ASAT kinetic intercept fragmentation scenario are provided below. While the absolute risk is quite low, one can see that certain satellites (e.g., COROT) were orders of magnitude at greater risk than others.

Table 2: Fengyun 1C ASAT intercept "Top 25" active satellites at high risk.

SSC	International Designator	Risk posed to S/C
29678	COROT	1.8E-09
2874	OPS 5712 (P/L 153)	6.7E-10
11416	NOAA 6	3.2E-10
26034	CELESTIS-03 & TAURUS R/B	2.7E-10
25980	ORBCOMM FM30	2.7E-10
28737	SJ-7	2.3E-10
25118	ORBCOMM FM06	2.2E-10
29052	FORMOSAT-3 FM2	2.0E-10
27004	MAROC-TUBSAT	1.9E-10
28493	NANOSAT-1	1.7E-10
29658	SAR-LUPE 1	1.7E-10
27375	IRIDIUM 95	1.6E-10
27606	LATINSAT B	1.5E-10
25338	NOAA 15	1.4E-10
26102	MTI	1.4E-10
28810	REIMEI (INDEX)	1.4E-10
25982	ORBCOMM FM32	1.3E-10
2872	SURCAL 159	1.3E-10
25635	ORSTED	1.2E-10
28372	APRIZESAT 1	1.2E-10
26931	PCSAT (NO-44)	1.0E-10
25420	ORBCOMM FM13	9.9E-11
1512	TEMPSAT 1	8.8E-11
28413	SJ-6A	8.7E-11
28376	AURA	8.6E-11

Figure 7. DREAD-estimated time history of debris risk to active satellite w/greatest risk (COROT) caused by Fengyun 1C ASAT intercept.

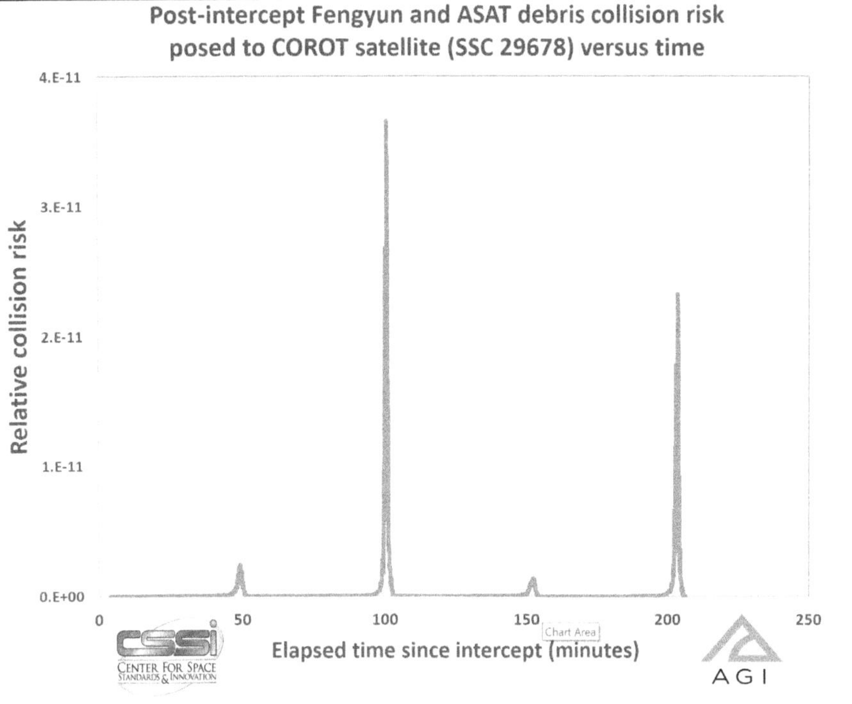

6. Conclusion

As evidenced by the population of actual space objects detected and tracked by the 18SPCS, the Fengyun 1C ASAT kinetic intercept fragmentation event generated a large amount of orbital debris that adversely increased the space population in a wide range of orbital regimes. The generation of a representative post-fragmentation debris scenario using the DREAD tool that is consistent with the observed imparted relative velocities allows us in turn to analyze the resulting fragmentation cloud's aggregate distribution volume and the risk posed to other spacecraft in the immediate aftermath of the event. These in turn demonstrate the adverse nature of any fragmentation event and the problems it poses to the long-term sustainability of space activities.

7. References

[1] Oltrogge, Dan. 2008. Breakup Modeling with 1Earth's DEBBIE Module and STK. Paper AGI-UTH-Oltrogge presented at the AGI Users Conference, October 7-9. Chicago, IL.

[2] Carrico, Timothy and John Carrico. 2008. Investigating Orbital Debris Events using Numerical Methods with Full Force Model Orbit Propagation. Paper AAS 08-126 presented at the AAS/AIAA Spaceflight Mechanics Conference, January 28-31. Galveston, TX.

[3] Housen, K.R., "An Analytical Model of Debris Evolution for Fragmented Artificial Satellites," Boeing Document D180-32710-t prepared for and released by N. JohnsonNASA Orbital Debris Office, 22 May 1991.

[4] Hujsak, Richard S. 1991. Nonlinear dynamical model of relative motion for the orbiting debris problem. Journal of Guidance, Control, and Dynamics. 14.2, pp. 460-465.

[5] McKennon-Kelly , R. and Hoots, F. 2013. Space Debris Visualization, Characterization, and Volume Modeling. Paper AAS 13-846 presented at the AAS/AIAA Astrodynamics Specialist Conference, August 11-15. Hilton Head, SC.

[6] Hansen, B. and Hoots, F. 2015. Containment of Moderate-Eccentricity Breakup Debris Clouds within a Maximum Isotropic Spreading Speed Boundary. Paper AAS 15-534 presented at the AAS/AIAA Astrodynamics Specialist Conference, August 10-14. Vail, CO.

[7] Hoots, F. and Hansen, B. 2016. Satellite Breakup Debris Cloud Model. Paper AAS 16-474 presented at the AAS/AIAA Spaceflight Mechanics Conference, February 14-18. Napa, CA.

[8] Healy, Liam, et al. 2016. Spatial Density maps from a Debris Cloud. Paper presented at the AMOS Advanced Maui Optical and Space Surveillance Technologies Conference. Maui, HI. September 20-23.

[9] Vallado, D.A. and Oltrogge, D.L. 2017. Fragmentation Event Debris Field Evolution Using 3D Volumetric Risk Assessment. Paper presented at the 7th European Conference on Space Debris, April 18-21. Darmstadt, Germany.

[10] Healy, L. Kindl, S., Rolfe, E. and Binz, C. 2016. Structure and evolution of a debris cloud in the early phases. Paper AAS 16-430 presented at the AAS/AIAA Spaceflight Mechanics Conference, February 14-18. Napa, CA.

[11] Healy, Liam, Scott Kindl, and Christopher Binz. 2017. Spatial Density Maps from a Debris Cloud. Paper presented at the 7th European Conference on Space Debris, April 18-21. Darmstadt, Germany.

[12] McKennon-Kelly, Ryan et al. 2015. Seeing Through the Clutter: The Power of the Torus. Aerospace Crosslink 16:1. Pp. 40-45.

[13] Oltrogge, D.L. and Vallado, D.A., "Application of New Debris Risk Evolution And Dispersal (DREAD) Tool to Characterize Post-Fragmentation Risk," AAS 17-600, Astrodynamics Specialist Conference, Stevenson,, Washington, USA.

[14] Gooyear, W.H., "A General Method for the Computation of Cartesian Coordinates and Partial Derivatives of the Two-Body Problem," NASA CR-522, 1972.

[15] Arora, N. et al, "Partial Derivatives of the Solution to the Lambert Boundary Value Problem". In: Journal of Guidance, Control, and Dynamics 38.9, pp. 1563–1572. ISSN: 0731-5090. DOI: 10.2514/1.G001030. URL: https://arc.aiaa.org/doi/pdf/10.2514/1.G001030 (visited on 24 January 2019).

[16] Kan, Shirley, "China's Anti-Satellite Weapon Test," CRS Report to Congress, Order Code RS22652, 23 Apr 2007.

[17] "Feng Yun FY-1 Earth Observation System," https://www.globalsecurity.org/space/world/china/fy-1.htm#ref622, (visited on 14 Feb 2019).

[18] Kelso, T.S., "Chinese ASAT Test, http://www.centerforspace.com/asat/ (visited on 26 Dec 2018).

[19] "SC-19 ASAT," https://www.globalsecurity.org/space/world/china/sc-19-asat.htm, (visited on 7 Apr 2019).

[20] Forden, G., "A Preliminary Analysis of the Chinese ASAT Test," MIT, 2007.

[21] Skinner, M.A., "Time-Resolved Infrared Spectrophotometric Observations of Fengyun 1C And Associated Resident Space Objects," IAC-09-A6.1.17, 2009.

[22] "China's Anti-Satellite Test: Worrisome Debris Cloud Circles Earth," https://www.space.com/3415-china-anti-satellite-test-worrisome-debris-cloud-circles-earth.html (visited on 2 February 2007).

[23] NORAD Technical Publication - SPADOC Computation Center Computer Program Product Specification for SCC Astrodynamic Theory. TP-SCC-008. April 6, 1982. Headquarters North American Aerospace Defense Command.

Appendix A. Post-fragmentation graphics

These sequences show 2D and 3D evolution and dispersal of Fengyun 1C and ASAT interceptor fragments over three hours.

A.1. Two-dimensional depiction at one thousand second step size post-impact

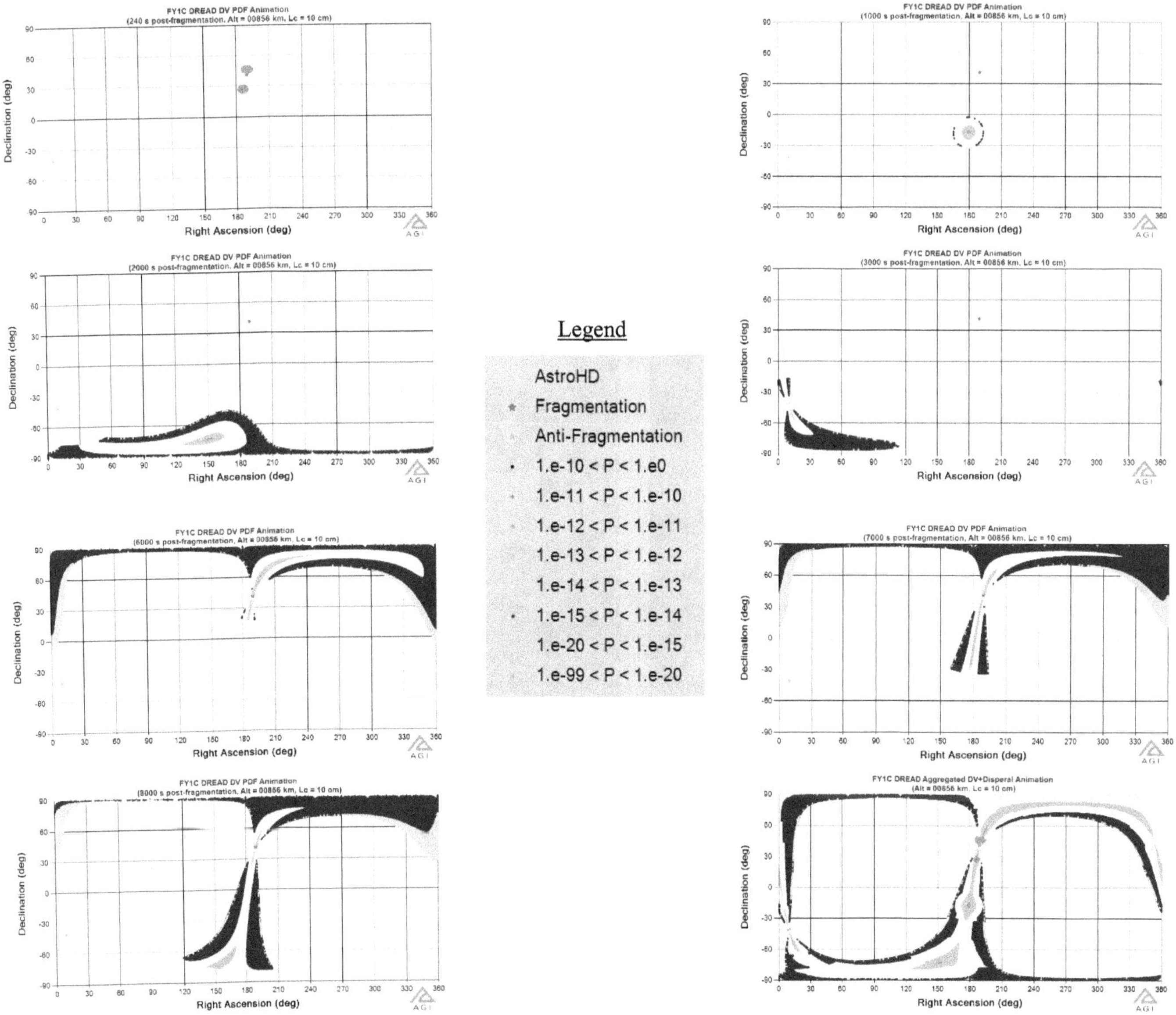

Figure 8. **LEO collision 2D slice animation.** Likelihood a Feng Yun or ASAT interceptor fragment will be present at a location and time. Aggregate in final picture.

A.2. Three-dimensional depiction at one thousand second step size post-impact

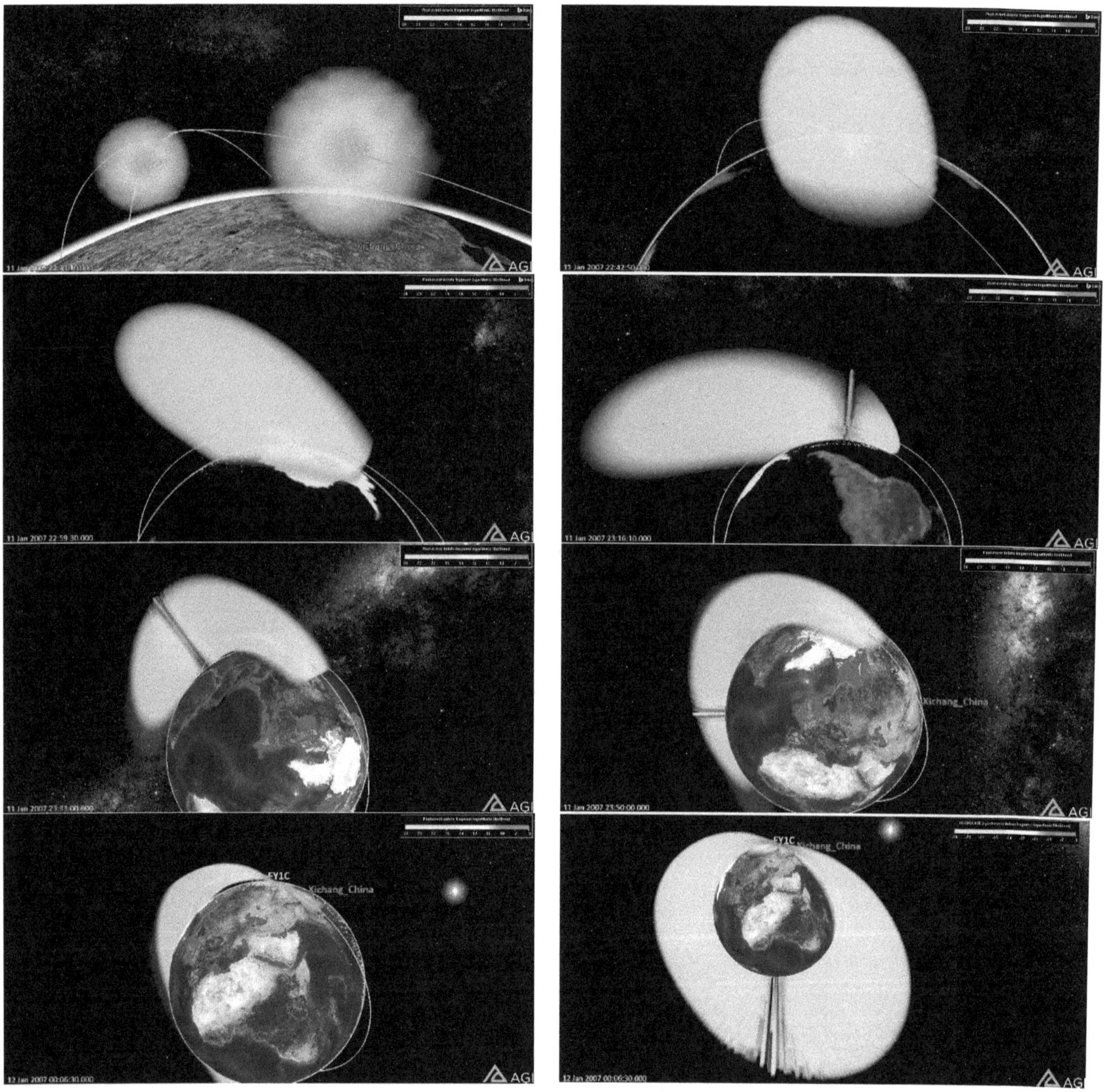

Figure 9. **LEO collision 3D "volumetric" animation.** These 3D snapshots from System Tool Kit's volumetric portrayal generated by the DREAD tool show the evolution of a fragment's presence as shown in the previous 2D pictorial sequence. The likelihood scale of the volumetric color is as shown in the legend in the upper right. Aggregate depiction in final picture; note the ASAT-originated debris fragment ballistic reentry, the intercept's "pinch point," and the red "pinch line" on the opposite side.

Proceedings of the 2019 Hypervelocity Impact Symposium
HVIS2019
April 14-19, 2019, Destin, FL, USA
HVIS2019-055

Hypervelocity impact performance of 3D printed aluminum panels

Bruce A. Davis[a]*, Richard A. Hagen[b], Robert J. McCandless[a] , Eric L. Christiansen[b] and Dana M. Lear[b]

[a]*Jacobs, NASA Johnson Space Center, Houston, TX 77058, USA*
[b]*NASA Johnson Space Center, Houston, TX 77058, USA*

Abstract

NASA, JSC has been developing a light-weight, multi-functional sandwich core for habitable structure over the last several years. Typically honeycomb-based structures have been and still are a common structural component for many applications in the aerospace industry, unfortunately, honeycomb structures with an ordered, open path through the thickness have served to channel the micro-meteoroid or orbital debris into the pressure wall (instead of disassociating and decelerating). The development of a metallic open cell foam core has been explored to enhance the micro-meteoroid or orbital debris protection, which is heavier than comparable honeycomb-based structures when non-structural requirements for deep space environments (vacuum, micro-meteoroids/orbital debris, and radiation) have not been considered. While the metallic foam core represents a notable improvement in this area, there is an overwhelming need to further reduce the weight of space vehicles; especially when deep space (beyond low earth orbit, or LEO) is considered. NASA, JSC is currently developing a multi-functional sandwich panel using additive machining (3D printing), this effort evaluated the material response of a limited amount of 3D printed aluminum panels under hypervelocity impact conditions. The four 3D printed aluminum panels provided for this effort consisted of three body centric cubic lattice structure core and one kelvin cell structure core. Each panel was impacted once with nominally the same impact conditions (0.34cm diameter aluminum sphere impacting at 6.8 km/s at 0 degrees to surface normal). All tests were impacted successfully, with the aforementioned impact conditions. Each of the test panels maintained their structural integrity from the hypervelocity impact event with no damage present on the back side of the panel for any of the tests. These tests and future tests will be used to enhance development of 3D printed structural panels.

Keywords: hypervelocity impact; body centric cubic; additive; protection; 3D

1. Introduction

Deep space exploration will require large habitats for both orbital and surface missions [1]. The weight of a habitat is driven largely by its structure such that future vehicles require lighter weight structures to minimize launch mass and cost. NASA, JSC structural engineering has been working on developing a multi-functional core material that can provide structural reliability, Micrometeoroid and orbital debris (MMOD) and radiation protection, over the last several years. Additive machining (3D printing) is being evaluated as an option to print core materials that can integrate features into the primary structure that traditionally add mass. This project seeks to evaluate a 3D printed structural core that can be optimized for flights loads and provide integrated MMOD protection.

With the continued development of additive manufacturing methods, control over the shape of ligaments, cell regularity, and macroscopic shape can be easily tuned. The ability to tailor these components can promote performance optimization and allow for potential mass savings in a habitat and space vehicle structure. These tests will be used to further enhance development of 3D printed core structures.

* Corresponding author. Tel.: +1-281-483-0861.
E-mail address: bruce.davis-1@nasa.gov.

2. Materials and Methods

For this study, four different types of aluminum (AlSi10Mg) panels have been 3D printed for testing where three being of a body centric cubic (BCC) lattice structural core design (see Figure 1) and one of a kelvin cell structural core design (see Figure 2). The BCC panels have a similar architecture, but each were printed with different lattice diameters. All of the panels have a 0.13cm (0.05") thick printed face plate on front and back of each panel. The panels for this testing were provided by JSC Structural Engineering, all panels were generated using Altair Optistruct Software and printed by Volunteer Aerospace in Knoxville, Tennessee.

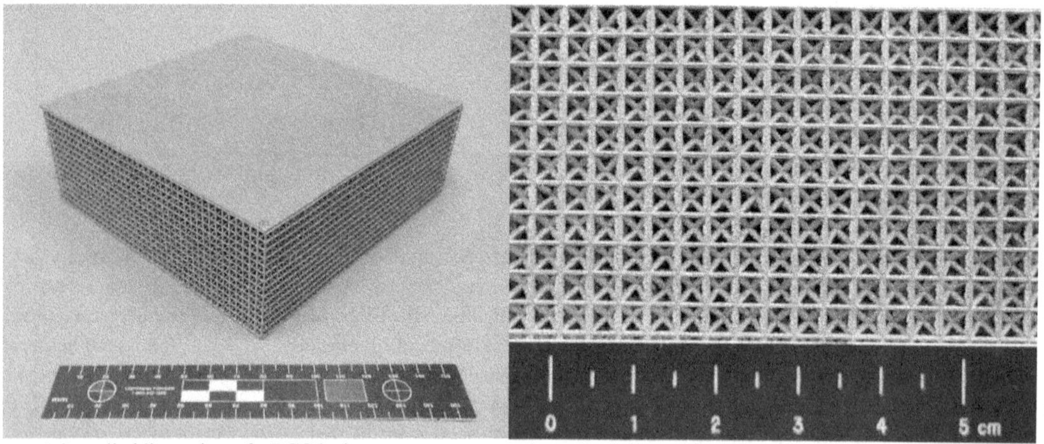

Figure 1 Overall oblique view of a BCC lattice structure core panel sample (left) and close-up view of lattice core structure (right).

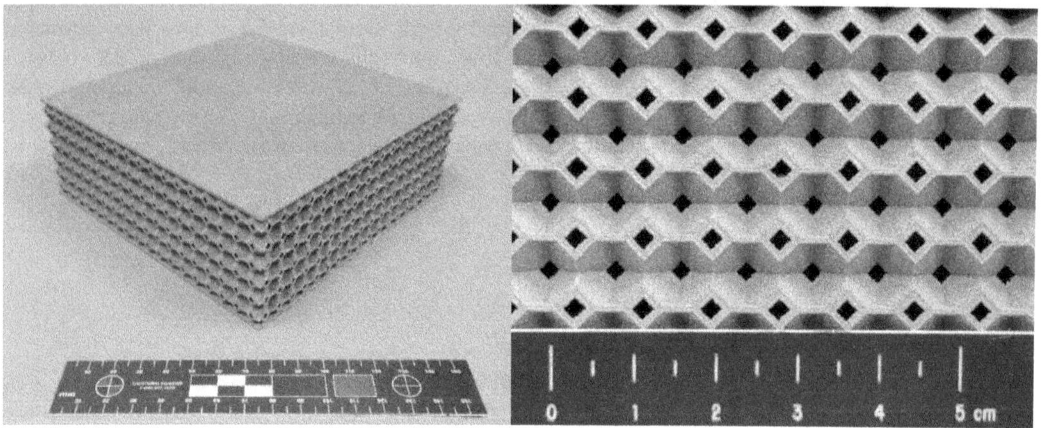

Figure 2 Overall oblique view of a Kelvin cell structure core panel sample (left) and close-up view of Kelvin cell core structure (right).

Table 1. The table below provides a summary of each 3D printed aluminum panel sample being used in this hypervelocity impact (HVI) test program. Each sample has its own unique serial number (S/N) to maintain a record of what test was impacted on each sample. The mass per unit area was calculated using the dimensions and mass from each sample. The sample S/N's in this table are also reflected within the test results summary in Table 2.

Panel Serial Number	Panel Core Type	Core Dimensional Information (cm)	Panel Mass (g)	Overall Sample Nominal Dimensions (cm)	Mass Per Unit Area (g/cm²)
1	BCC	0.076 nominal lattice diameter	1150.0	15.24 L x 15.24 W x 5.33 H	4.951
2	Kelvin	0.076 nominal wall thickness	977.0	15.24 L x 15.24 W x 5.33 H	4.207
3	BCC	0.051 nominal lattice diameter	720.9	15.24 L x 15.24 W x 5.33 H	3.104
4	BCC	0.064 nominal lattice diameter	922.9	15.24 L x 15.24 W x 5.33 H	3.974

The panels for each test were trapped between two aluminum frames, with rubber gasket material residing between the aluminum frame and panel to promote minimal compression of the panel. An aluminum witness plate was staged behind each test panel, to capture signature of debris if the back side of the panel were to perforation. Each test configuration was secured by means of all-thread, Belleville spring washers and nuts. See Figure 3 below for illustrations of a typical test configuration.

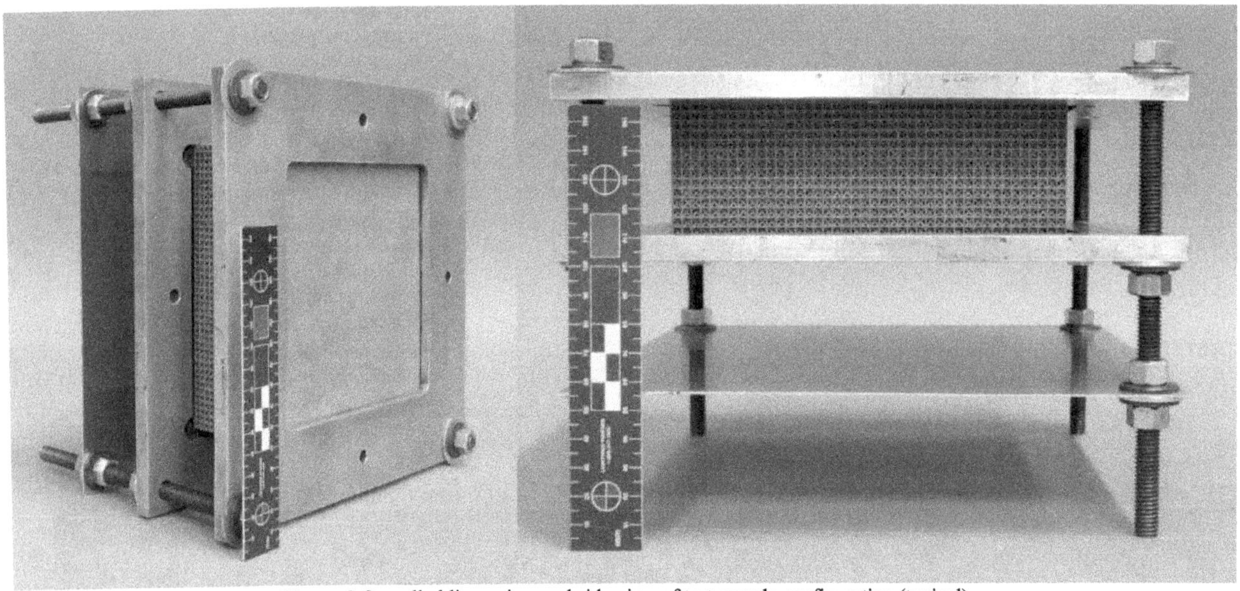

Figure 3 Overall oblique view and side view of test sample configuration (typical).

These HVI tests were conducted in support of NASA JSC Structural Engineering by the NASA JSC Hypervelocity Impact Technology (HVIT) group [2]. The tests were conducted at the NASA White Sands Test Facility (WSTF) Remote Hypervelocity Test Facility (RHTL) in Las Cruces, New Mexico using a 0.17-caliber two-stage light gas gun (2SLGG). All tests were conducted nominally at 6.8 km/s with 0.34cm diameter aluminum spheres impacting at 0° normal to the surface.

3. Results

The results of the tests are summarized in Table 2. The entry damage measurements were taken with the largest measurement being the first and the second measurement being taken perpendicular to the first measurement, all other damage measurements are as specified within the damage measurements column in Table 2. The internal damage illustrations for these four tests can be found in Figures 5 – 7. All of the internal damage measurements (primary cavity depth, maximum cavity depth and maximum cavity width) were obtained by cross-sectioning the impacted region of each test panel. All of the printed panels were impacted within a quadrant on each panel, the impacted quadrant of each panel was cut away from the sample and cross-sectioned through the impacted region to reveal the internal impact damage. See illustrations below in Figure 4 for typical cross-sectioning process in this effort. The measurements were captured using a Keyence VHX-5000 microscope.

a) b) c)

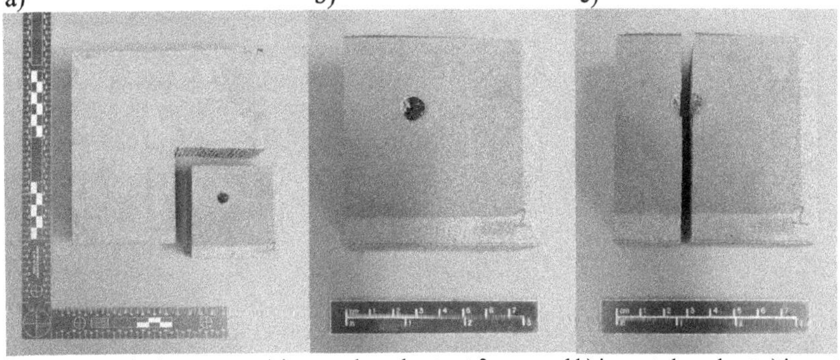

Figure 4 Overall view of typical cross-sectioning process a) impacted quadrant cut from panel b) impacted quadrant c) impacted quadrant cross-sectioned.

Table 2. The results of the tests are summarized in table below. All projectile diameters were calculated from the mass with a given projectile density of 2.796 g/cm³ (Al 2017-T4).

Test ID Number Panel Serial Number	Panel Core Type	Projectile Diameter (cm)	Velocity (km/s)	Damage Measurements (cm)
HITF18375 1	BCC	0.341	6.73	Entry damage = 0.86 x 0.83 entry hole Primary cavity depth =1.72 Maximum cavity depth = 2.17 Maximum cavity width = 1.56
HITF18376 2	Kelvin	0.341	6.72	Entry damage = 0.86 x 0.81 entry hole Primary cavity depth = 2.97 Maximum cavity depth = 2.97 Maximum cavity width = 1.71
HITF18377 3	BCC	0.341	6.77	Entry damage = 0.83 x 0.81 entry hole Primary cavity depth = 2.55 Maximum cavity depth = 2.90 Maximum cavity width = 2.12
HITF18378 4	BCC	0.341	6.87	Entry damage = 0.86 x 0.82 entry hole Primary cavity depth = 2.70 Maximum cavity depth = 2.70 Maximum cavity width = 2.31

a) HITF18375 (all internal damages) b) HITF18377 (all internal damages)

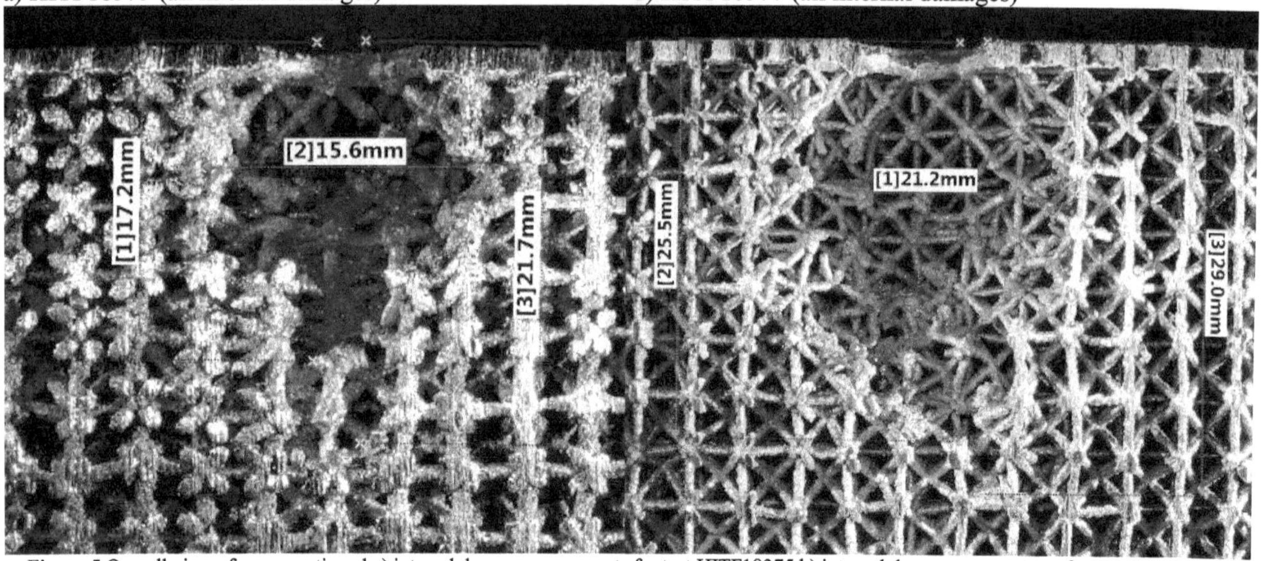

Figure 5 Overall view of cross-sectioned a) internal damage assessments for test HITF18375 b) internal damage assessments for test HITF18377.

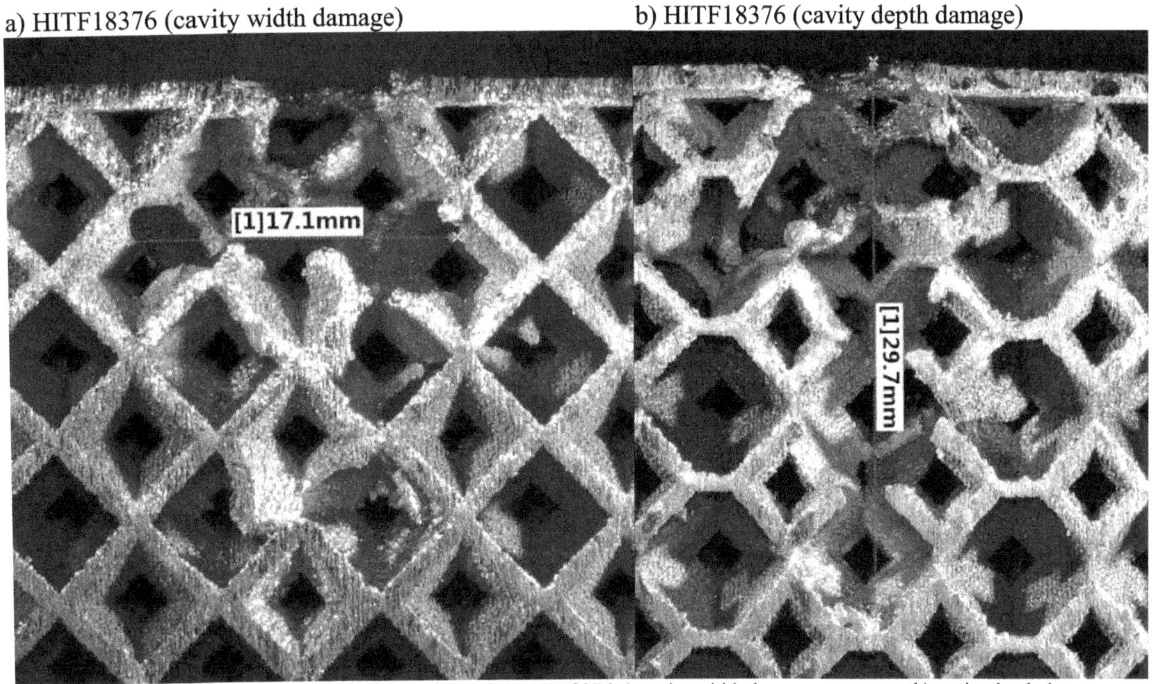

Figure 6 Overall view of cross-sectioned damage assessments for HITF18376 a) cavity width damage assessment b) cavity depth damage assessment.

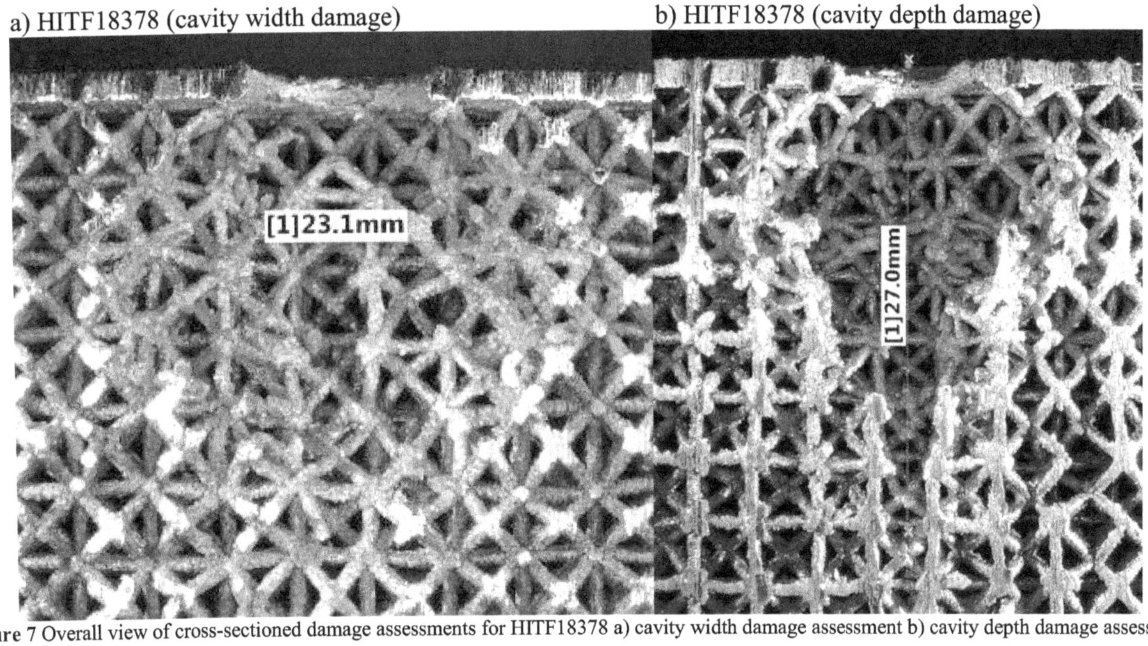

Figure 7 Overall view of cross-sectioned damage assessments for HITF18378 a) cavity width damage assessment b) cavity depth damage assessment.

4. Conclusions

The primary objective of these tests was to understand how 3D printed panels perform under HVI conditions and capture the damages for each test. All panels maintained their structural integrity during the course of testing and no tests yielded any damage to the back side of any of the panels. With the limited amount of tests, the BCC core type material does perform better that the Kelvin core material, from a penetration depth perspective. These tests and future tests will be used to refine development of 3D printed structural panels.

5. Acknowledgements

The authors wish to gratefully acknowledge the NASA JSC Structural Engineering for developing the 3D printed panels and the NASA White Sands Test Facility's Remote Hypervelocity Test Laboratory for executing these HVI tests.

6. References

[1] Deep Space Habitation Overview, https://www.nasa.gov/deep-space-habitation/overview/, April 2016.
[2] E.L. Christiansen and D.M. Lear, the NASA JSC Hypervelocity Impact Technology Group, https://hvit.jsc.nasa.gov/.

Proceedings of the 2019 Hypervelocity Impact Symposium
HVIS2019
April 14-19, 2019, Destin, FL, USA
HVIS2019-058

Extravehicular Activity Micrometeoroid and Orbital Debris Risk Assessment Methodology

Kevin D. Hoffman[a], James L. Hyde[a], Eric L. Christiansen[b], and Dana M. Lear[b]

[a]*Jacobs, NASA Johnson Space Center, Mail Code XI, Houston, TX 77058, USA*
[b]*NASA Johnson Space Center, Mail Code XI, Houston, TX 77058, USA*

Abstract

A well-known hazard associated with exposure to the space environment is the risk of vehicle failure due to an impact from a micrometeoroid and orbital debris (MMOD) particle. Among the vehicles of importance to NASA is the extravehicular mobility unit (EMU) "spacesuit" used while performing a US extravehicular activity (EVA). An EMU impact is of great concern as a large leak could prevent an astronaut from safely reaching the airlock in time resulting in a loss of life. For this reason, a risk assessment is provided to the EVA office at the Johnson Space Center (JSC) prior to certification of readiness for each US EVA. This paper will detail the methodology for an ISS EVA risk assessment. The soft goods regions (multilayer fabric over a pressurized bladder) are the highest contributors of risk for an ISS EVA. The gloves, due to reduced fabric layers to allow for improved dexterity, carry the highest risk per area. ISS EVA risk can be reduced by minimizing the exposure of the front of the suit and gloves to the orbital debris flux.

Keywords: Extravehicular activity; Micormeteoroid; Orbital debris

1. EVA Assessment Overview

EVA assessments are performed using standard risk assessment methodology illustrated in Figure 1. Central to the process is the Bumper risk assessment code (Figure 2), which calculates the critical penetration risk based on geometry, shielding configurations and flight parameters.

The assessment process begins by building a Finite Element Model (FEM) of the EMU, which defines the size and shape of the EMU as well as the locations of the various shielding configurations. Each region of the EMU having a different shield configuration and/or failure criteria are given a unique property identifier (PID) number in the FEM. The model was built using the NX I-deas software package from Siemens PLM Software. The FEM was constructed using triangular and quadrilateral elements that define the outer shell of the EMU. Bumper-3 uses the model file to determine the geometry of the EMU for the analysis.

The next step of the process is to identify the failure criteria and appropriate ballistic limit characteristics for the various regions of the FEM. The two failure criteria for the EMU are perforation threshold failure and a critical hole size threshold failure of the EMU bladder. Perforation threshold failure would result in any size leak in the EMU bladder while the critical hole size threshold failure (>4mm hole in the bladder) would result in an uncontrolled leak. The ballistic limit equations define the critical size particle

* Corresponding author. Tel.: +1-281-483-2174 ; fax: +1-281-483-3908 .
E-mail address: Kevin.Hoffman-1@nasa.gov

that will cause failure (as defined by the failure criteria) at a given impact angle and impact velocity. When the finite element model is built, each individual element is assigned a PID to act as an index for its shielding properties. Ballistic limit equations are based on results from impact tests on samples that are representative of the vehicle shields. The appropriate ballistic limit equations (BLEs) are added to the Bumper-3 code as needed for each and every PID in the model.

The final stage of the analysis is to determine the predicted number of failures for the EMU. This is calculated using the micrometeoroid and orbital debris (MMOD) environment definitions that are specified by the spacecraft design. The MMOD environment models predict an impacting particle flux based on inputs for orbital parameters, spacecraft attitude and analysis date.

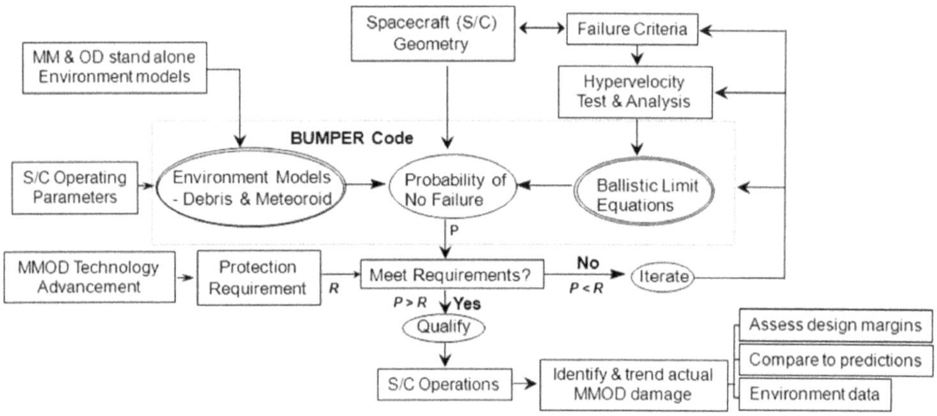

Fig. 1. EVA MMOD Risk Assessment Flowchart

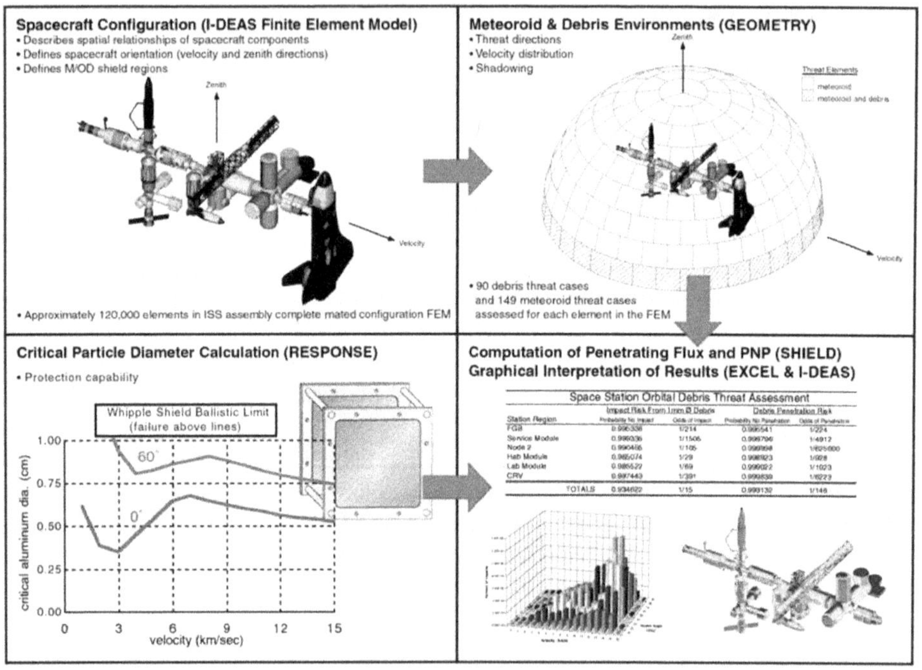

Fig. 2. Bumper-3 MMOD Risk Assessment Code

Using Poisson statistics, and combining the previously defined geometry, shielding characteristics and ballistic limit equations with the MMOD environment model calculations, the Bumper-3 code determines a predicted number of failures for the EMU.

1.1. US EVA Timeline Analysis

For every US EVA, a timeline MMOD risk analysis is provided to the EVA office at JSC prior to the certification of EVA readiness. With assistance from the EVA office, a summary of EMU positions (including body orientation) for the specific EVA worksite locations on the International Space Station (ISS) is produced using detailed EVA summaries/presentations and/or EVA training run videos from the Neutral Buoyancy Lab (NBL.) From this information, the analysis FEMs are built by orienting an EMU FEM at each worksite location on a simplified ISS FEM. When one or more worksites require multiple body orientations, additional analysis FEMs are built. The analysis FEM used for the US 52 and 53 EVAs is shown in Figure 3.

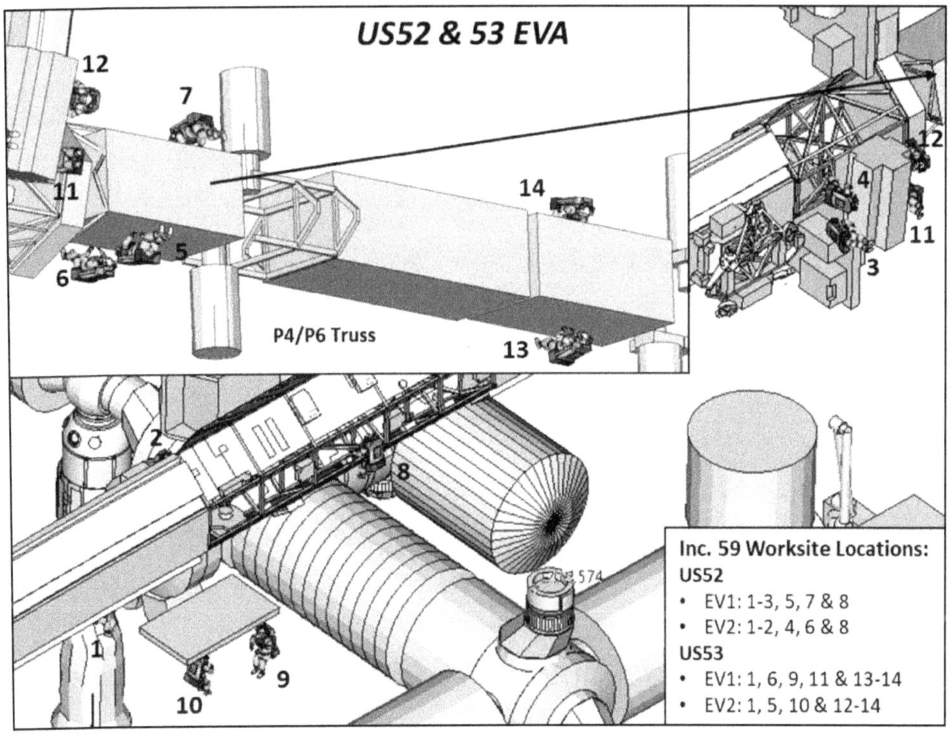

Fig. 3. Increment 57 (US EVAs 52 & 53) Analysis Model including ISS Shadowing

The analysis FEM(s), environment files and BLE parameters are inputted into the Bumper-3 code to obtain the expected number of failures at each worksite for a one year exposure. These results are scaled based on actual worksite exposure duration to generate total EVA risk and individual EMU risk for two failure criteria: any size leak risk of the bladder and an uncontrolled leak risk (> 4mm hole in the bladder.) Meteoroid and orbital debris enhancement factors (see Section 1.2) are applied to obtain the failure odds for the planned EVA date.

1.2. Meteoroid and Orbital Debris Enhancement Factors

The meteoroid and orbital debris environment files used for the timeline analysis provide the average flux for a single year exposure duration. Due to the short exposure duration of 6 to 7 hours, EVAs are more sensitive to sporadic events outside of the single year background environment (i.e. orbital debris breakups and meteor showers.) Environment enhancement factors are used to account for these events. For every US EVA, the Meteoroid Environment Office (MEO) at Marshall Space Flight Center (MSFC) provide 6-hour meteoroid enhancement factors for each day of the EVA window. The fluence factor is for particles with a kinetic energy of at least 6.7 J (the kinetic energy of a 0.04 cm diameter meteoroid traveling at 20 km/s.) For example, meteoroid enhancement factors provided by the MEO for the December 2018 window are shown in figure 4 and indicate a peak fluence enhancement factor of nearly 40% for an EVA performed during the Geminids meteor shower on December 14th, 2018. Likewise, the Orbital Debris Program Office (ODPO) at JSC provide an orbital debris enhancement factor to account for any recent orbital debris breakups that might increase the flux at ISS orbit.

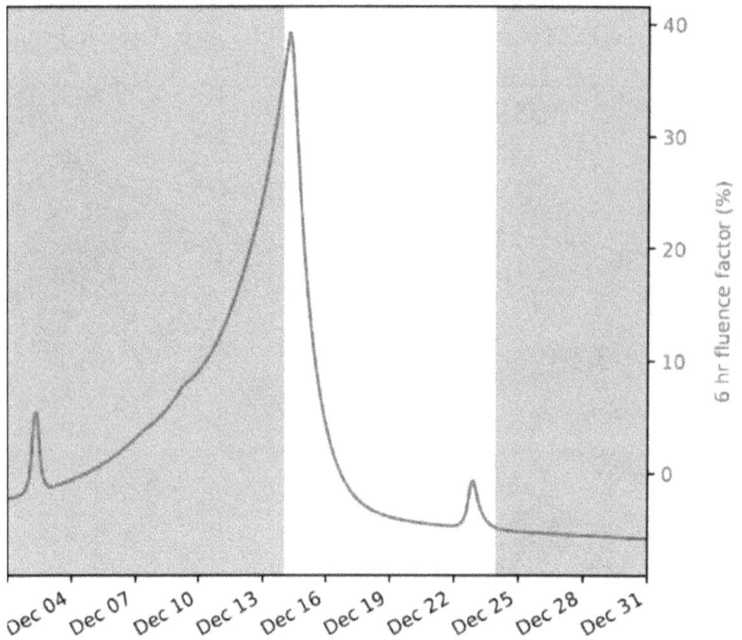

Fig. 4. Increment 57 (US EVAs 52 & 53) Meteoroid Environment Enhancement Factors

2. EMU Finite Element Model

A detailed finite element model (FEM) of the EMU (shown in Figure 5) has been created which has regions for the various shielding configurations. Each shielding configuration is based on the layers and materials over the innermost bladder layer that maintains the acceptable atmospheric environment for the astronaut. The EMU outer mold line is defined by over 14,500 quadrilateral and triangular elements. The color changes in the model views shown in Figure 5 illustrate the 42 different surface property ID (PID) types.

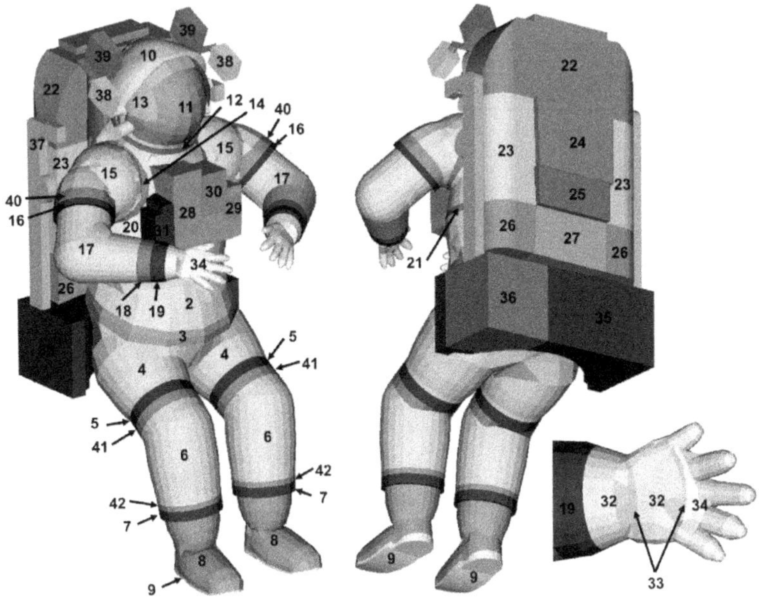

Extravehicular Mobility Unit (EMU) FEM					
Region *	PID	Code	Region	Color	BLE
Lower Torso Assembly	2	L1	Waist TMG	Green	soft goods, 1
	3	L2	Waist Bearings/Sizing Rings	Light Magenta	KE, 39 J
	4	L3	Thigh TMG	Green	soft goods, 1
	5	L4	Thigh Bearings/Sizing Rings	Blue	KE, 39 J
	6	L5	Lower Leg TMG	Green	soft goods, 1
	7	L6	Ankle Bearings/Sizing Rings	Blue	KE, 39 J
	8	L7	Boot TMG	Light Blue	soft goods, 1
	9	L8	Boot Heel & Sole	White	softgoods, 4, AD 5.860
Helmet	10	H1	EVVA Helmet Shell	White	KE, 71 J
	11	H2	Face Plate	Light Blue	KE, 71 J
	12	H3	Helmet Bearings/Sizing Rings	Red	KE, 39 J
	13	H4	Helmet Eyeshades	Pink	KE, 71 J
	38	H1	Helmet Lights	Yellow	shadow
	39	H2	Light Base & Head Mount	Dark Green	shadow
Arms	14	A1	Shoulder Bearings/Sizing Rings	Red	KE, 39 J
	15	A2	Shoulder TMG	Green	soft goods, 1
	16	A3	Upper Arm Bearings/Sizing Rings	Blue	KE, 39 J
	17	A4	Forearm TMG	Green	soft goods, 1
	18	A5	Gauntlet/TMG	Red	softgoods, 3
	19	A6	Gauntlet/Wrist Connect	Blue	KE, 39 J
Hard Upper Torso	20	T1	Hard Upper Torso (HUT)	Cyan	KE, 44 J
	21	T2	HUT Bearings/Sizing Rings	Red	KE, 39 J
Primary Life Support System	22	P1	PLSS - Other Equipment (upper)	Orange	KE, 15 J
	23	P2	PLSS - Primary O2	Cyan	KE, 60 J
	24	P3	PLSS METOX CCC	Dark Green	KE, 21.4 J
	25	P4	PLSS Advanced Battery	Red	KE, 3.5 J
	26	P5	Secondary O2	Light Blue	KE, 13.4 J
	27	P6	PLSS - Other Equipment (lower)	Golden Orange	KE, 15 J
Display and Control module	28	D1	DCM	Light Magenta	shadow
	29	D2	DCM Controls & Display	Orange	shadow
	30	D3	DCM Multiple Connector	Dark Green	shadow
	31	D4	DCM Critical Region	Blue	KE, 10 J
Gloves	32	G1	Glove Palm	Cyan	softgoods, 4, AD 1.285
	33	G2	Glove Gimbals & Palm Bar	Pink	KE, 39 J
	34	G3	Glove Back & Fingers	Yellow	softgoods, 2
SAFER Assembly	35	S1	SAFER Base (non-tank)	Blue	shadow
	36	S2	SAFER Base - N2 tank	Red	KE, 20 J
	37	S3	SAFER Thrusters & Supports	Pink	shadow
Overlap Regions	40	O1	Upper Arm SSA TMG Overlap	Orange	KE, 20 J
	41	O2	Thigh SSA TMG Overlap	Orange	KE, 20 J
	42	O3	Ankle SSA TMG Overlap	Orange	KE, 20 J

* Color corresponds to the regional risk breakdown bar charts

Fig. 5. Extravehicular Mobility Unit (EMU) Finite Element Model (FEM)

3. Hypervelocity Impact Testing

The EMU can be subdivided into two major shielding configurations: soft goods and hard goods. The soft goods consists of fabric layers called the thermal meteoroid garment (TMG) over an inner bladder and includes the arms, legs, lower torso, gloves and boots overlap regions. The hard goods (TMG over metallic, composite and/or plastic components) include the helmet, face plate, sizing rings, hard upper torso (HUT), portable life-support system (PLSS), and the display and control module (DCM.)

Testing of the EMU TMG has shown that it performs as a mini bumper shield. The ortho-fabric layers induce a shock pulse that breaks up the projectile and creates an expanding debris cloud. The inner layers of the TMG (MLI and ripstop) as well as the pressure garment restraint layer help with further particle breakup and create spacing for the debris cloud to expand before reaching the bladder or underlying critical component. Testing of the soft goods layup included both pressurized and unpressurized bladders with no difference found in the ballistic limit with pressurization. Testing results were used to set the ballistic limit equations (BLE) for each shielding configuration. Figure 6 shows the ballistic limit equations for the basic EMU soft goods layup (TMG over bladder) for both aluminum and steel debris.

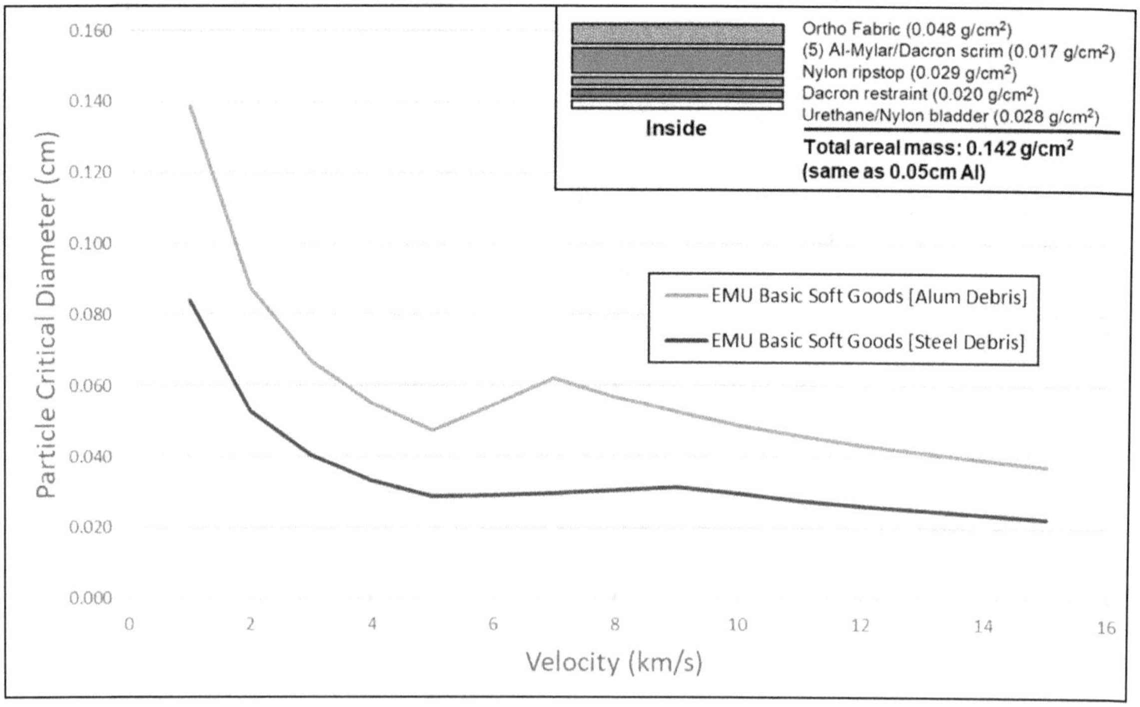

Fig. 6. Extravehicular Mobility Unit (EMU) Soft Goods Ballistic Limit Equation (Normal, 0° Impacts)

4. EMU Random Orientation Risk Study

A risk assessment of the EMU was performed using the ORDEM 3.0 orbital debris environment and the MEM R2 meteoroid environment. Twenty-four distinct attitude (body orientation) runs of the unshadowed EMU FEM were made for the exposure period of 2018 to 2024. Figure 7 shows the percentage of total risk contribution for the main regions of the EMU, with 75% of the total risk from the soft goods regions of the lower torso assembly, arms and gloves. The risk to area ratio for each of these sections is shown in Figure 8 and highlights that the gloves carry the highest risk per area for the EMU. Figure 9 is a risk contour (red areas are higher risk, blue areas are lower risk) and indicates the soft good regions of the gloves, arms, legs and torso carry the most risk.

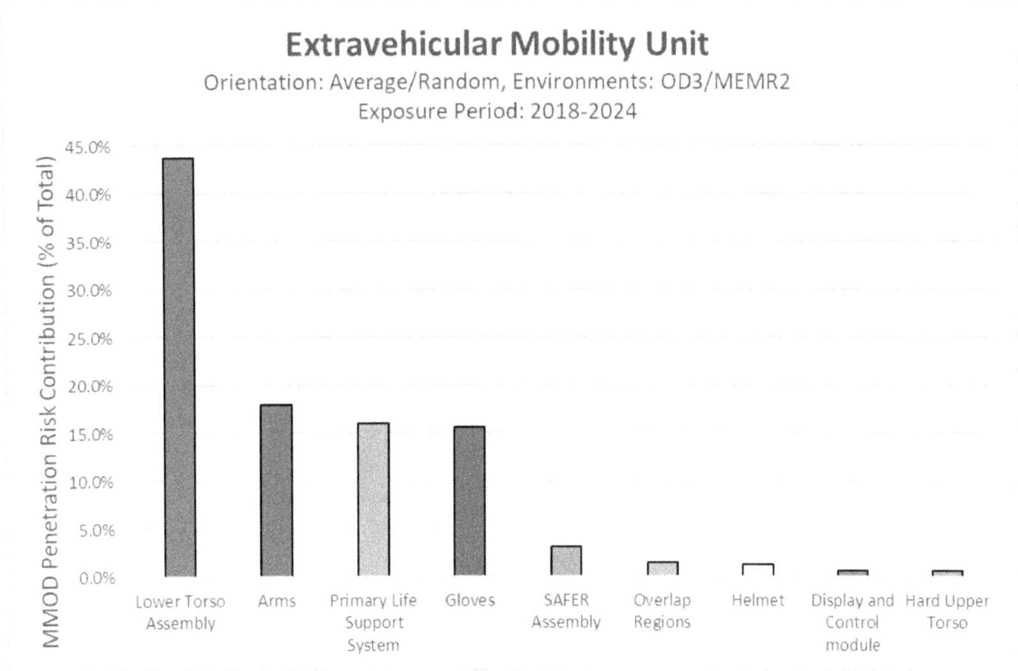

Fig. 7. Extravehicular Mobility Unit (EMU) Unshadowed Penetration Risk Summary (ORDEM3 and MEM R2 environments)

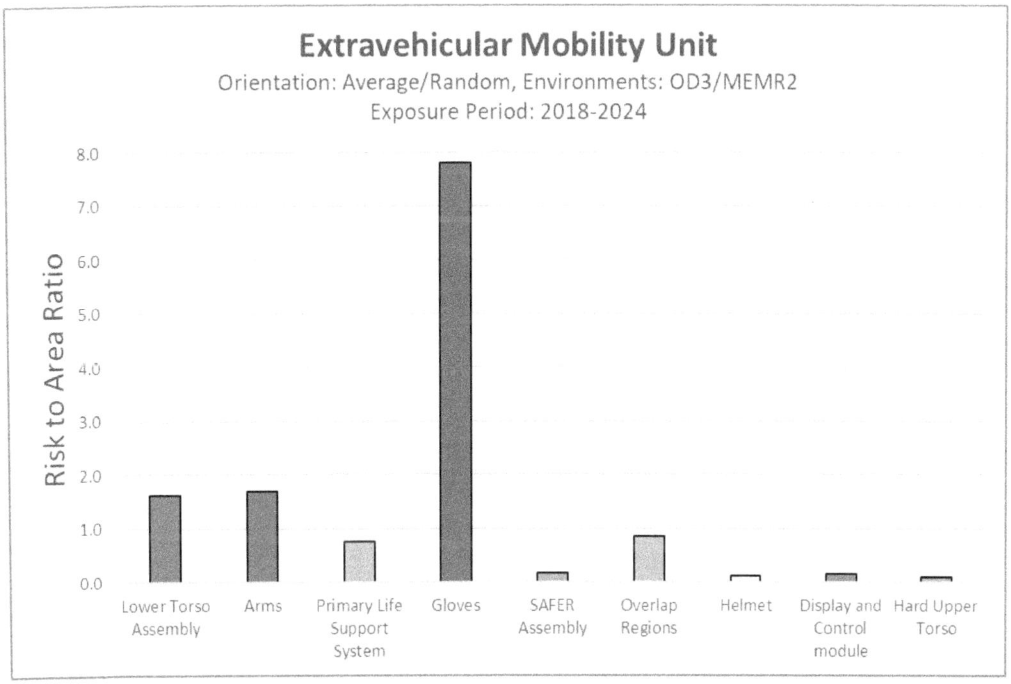

Fig. 8. Extravehicular Mobility Unit (EMU) Unshadowed Penetration Risk to Region Area Summary (ORDEM3 and MEM R2 environments)

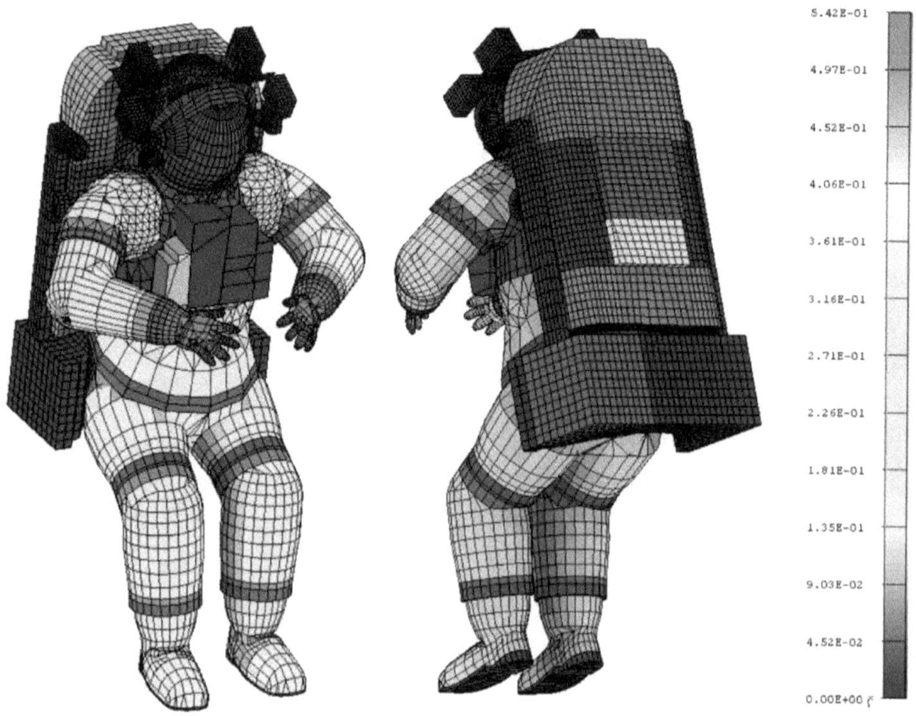

Fig. 9. Extravehicular Mobility Unit (EMU) Unshadowed Average Risk Contour Plot (red areas = higher risk, blue areas = lower risk)

5. Conclusion

The soft goods regions of the EMU contribute up to 75% of the risk (depending on body orientation at the worksites) for an ISS EVA. The gloves, due to reduced TMG layers to allow for improved dexterity, carry the highest risk per area. ISS EVA risk can be reduced by positioning the body at worksites to minimize the exposure of the front of the suit and gloves to the orbital debris flux (away from the velocity vector) and also to utilize ISS structure as "shadowing".

Acknowledgements

In support of the Jacobs JSC Engineering, Technology and Science (JETS) contract (NNJ13HA01C) for the Astromaterials Research and Exploration Science (ARES) division at the NASA Johnson Space Center (JSC.)

References

[1] Christiansen, Eric L., 2003. Meteoroid/Debris Shielding, NASA TP-2003-210788 . Houston: NASA.
[2] Friesen, L.J. and Christiansen, E.L., 1997. Hypervelocity Impact Tests of Extravehicular Mobility Unit (EMU) (Space Suit) Material Samples, Part 1, JSC 27856, Houston: NASA
[3] Lyons, Frankel, 1998. Hypervelocity Impact Testing of Extravehicular Mobility Unit (EMU) Alternative Material Samples, Part 1, JSC 28445, Houston: NASA

[4] Christiansen, E.L., Cour-Palais, B.G., and Friesen, L.J., 1998. "Extravehicular Activity Suit Penetration Resistance" Proceedings of the 1998 Hypervelocity Impact Symposium, International Journal of Impact Engineering, Vol. 23.
[5] Chase, T.D., Splawn, K., and Christiansen, E.L., 1998. "Extravehicular Mobility Unit Penetration Probability from Micrometeoroids and Orbital Debris – Revised Analytical Model and Potential Space Suit Improvements" Proceedings of the 2007 Hypervelocity Impact Symposium, International Journal of Impact Engineering, Vol. 35.

Proceedings of the 2019 Hypervelocity Impact Symposium
HVIS2019
April 14-19, 2019, Destin, FL, USA
HVIS2019-066

Updates to the DebriSat Project in Support of Improving Breakup Models and Orbital Debris Risk Assessments

Heather Cowardin[(1)] *, Phillip Anz-Meador[(1)], James Murray[(1)], J.-C. Liou[(2)], Eric Christiansen[(2)], Marlon Sorge[(3)], Norman Fitz-Coy[(4)], Tom Huynh[(5)]

[(1)] *Jacobs, NASA Johnson Space Center, 2224 Bay Area Blvd, Houston, TX 77058, USA*
[(2)] *NASA Orbital Debris Program Office, NASA Johnson Space Center, 2101 NASA Parkway, Houston TX 77058, USA,*
[(3)] *The Aerospace Corporation, 2155 Louisiana Blvd. NE #5000, Albuquerque, NM 87110, USA*
[(4)] *University of Florida,*
[(5)] *Air Force Space and Missile Systems Center 483 North Aviation Blvd., El Segundo, CA 90245, USA*

Abstract

Existing DOD and NASA satellite breakup models are based on a key laboratory test, the 1992 Satellite Orbital debris Characterization Impact Test (SOCIT), which has supported many applications and matched on-orbit events involving older satellite designs reasonably well over the years. To update and improve these models, the NASA Orbital Debris Program Office, in collaboration with the Air Force Space and Missile Systems Center, The Aerospace Corporation, and the University of Florida, conducted a hypervelocity impact test using a high-fidelity mock-up satellite, DebriSat, in controlled and instrumented laboratory conditions. DebriSat is representative of present-day LEO satellites, having been constructed with modern spacecraft materials and techniques. The DebriSat fragment ensemble provided a variety of shapes, bulk densities, and dimensions. Fragments down to 2 mm in size are being characterized by their physical and derived properties. A subset of fragments will be analyzed further in laboratory radar and optical facilities to update the existing radar-based NASA Size Estimation Model (SEM) and develop a comparable optical-based SEM. Thoroughly understanding size estimates from ground-based optical and radar sensors is one of the key parameters used in assessing the environment and the risks that debris present to operational spacecraft. The data will inform updates to the current NASA Standard Satellite Breakup Model (SSBM);, which was formulated using laboratory and ground-based measurements of on-orbit fragmentation events to describe an average breakup for spacecraft and upper stage collisions and explosions. DebriSat will extend the laboratory data ensemble. The DebriSat shape and density categories provide a baseline for non-spherical projectile hypervelocity impact testing for damage assessment. The data from these tests, simulations, and analyses will be used to update the NASA Orbital Debris Engineering Model (ORDEM) with more realistic simulations of catastrophic fragmentation events for modern satellites and to assess the risk posed by the orbital debris environment. This paper provides an overview of the project, updates on the characterization process, and the NASA analysis status.
Keywords: orbital debris; hypervelocity laboratory test;environmental models; non-spherical projectiles; DebriSat

Nomenclature

L_c	*Characteristic Length*
AMR	*Area-to-mass ratio*
Δv	Delta velocity
Subscripts	
SB	*solid body*
DIM	*characteristic length dimension*

* Corresponding author. Tel.: +1-281-483-2938.
E-mail address: Heather.Cowardin@nasa.gov.

1. Introduction

The DebriSat project is a collaboration of the NASA Orbital Debris Program Office (ODPO), the Air Force Space and Missile Systems Center (SMC), The Aerospace Corporation (Aerospace), and the University of Florida (UF). The project has four primary goals: 1) design and fabricate a 56-kg class spacecraft ("DebriSat") representative of modern spacecraft in the low Earth orbit (LEO) environment; 2) conduct a hypervelocity laboratory impact test to simulate a catastrophic fragmentation event of DebriSat; 3) collect, measure, and characterize all fragments down to 2 mm in size; and 4) use the data to improve space situational awareness applications and satellite breakup models for better orbital debris environment definition [1].

Existing DOD and NASA satellite breakup models are based on a key laboratory test, Satellite Orbital debris Characterization Impact Test (SOCIT), which was conducted by the Department of Defense (DOD) and NASA at the Air Force Arnold Engineering Development Complex (AEDC) in 1992 to support the development of satellite breakup models. The main target for SOCIT was a fully functional U.S. Navy Transit 1960's era satellite. The DOD and NASA breakup models based on the SOCIT data have supported many applications and matched on-orbit events reasonably well over the years [1].

As new materials and construction techniques are developed for modern satellites; however, there is a need for new laboratory-based tests to acquire data to improve the existing DOD and NASA breakup models. The need for such tests also is supported by discrepancies between model predictions and observations of fragments generated from the breakup of modern satellites, including the Iridium 33 and Fengyun 1-C [2]. The DebriSat project is an outgrowth of this need.

The design of DebriSat was based on a survey of modern satellites in LEO [3, 4]. All major design decisions, including the selection of components, subsystems, mass fractions, structure, and construction methods were reviewed and approved by subject matter experts from Aerospace, [5]. In addition, the DebriSat body was covered with multi-layer insulation (MLI) and three solar panels were attached to one side of the main body (Fig. 1).

To reduce the project cost, a decision was made to emulate the majority of components. The emulated components were based on existing designs of flight hardware, including structure, dimensions, materials, avionics components, and connection mechanisms. At the end of the assembly, DebriSat was subjected to a NASA General Environmental Verification Specification qualification vibration test to ensure the integrity of the fabrication/assembly process.

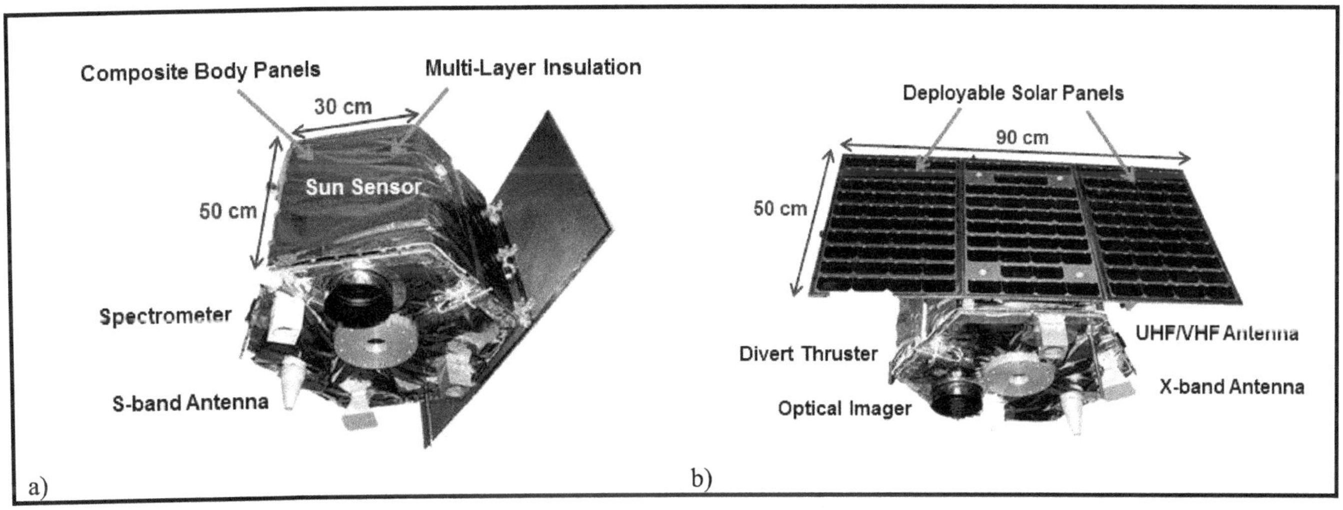

Figure 1. Illustration of the DebriSat satellite, a) highlighting various components and b) showing components and solar panels

To increase the project's benefits further, Aerospace designed and built a target resembling a launch vehicle upper stage ("DebrisLV") for the pre-test shot; details of the two targets are listed in Table. 1. The pre-test shot was conducted two weeks prior to the DebriSat impact test with identical conditions to DebriSat (facility set-up, projectile, and impact speed). The resulting fragments were collected in boxes and will be characterized after DebriSat characterization is complete.

Another key element implemented in DebriSat and DebrisLV hypervelocity impact tests was the use of a soft catch system inside the chamber. In the SOCIT test series the foam catch system only installed panels in the downrange and sideways directions, but to minimize secondary damage based on lessons learned from SOCIT the chamber was fully encompassed with various density panels for the DebriSat and DebrisLV tests. Secondly, the first layer of the soft catch panel in DebriSat and DebrisLV tests were slightly lower densities than that used in SOCIT to also minimize secondary damage to the fragments. Lastly, all foam panels were thicker in the DebriSat and DebrisLV tests in comparison to SOCIT, required to minimize damage to the chamber walls. Table. 1 shows a comparison of some SOCIT, DebriSat, and DebrisLV test conditions.

2. Experiment Test Background

From 2009-2014 the planning and fabrication for the DebriSat and DebrisLV were in process. On 1 April and 15 April 2014, respectively, the DebrisLV and DebriSat impacts were successfully carried out at AEDC Range G. To maximize the projectile mass at the 7 km/sec impact speed without a sabot, the AEDC team developed a special projectile design featuring a hollow aluminum cylinder embedded in a nylon cap. The nylon cap served as a bore rider for the aluminum cylinder to prevent hydrogen leakage and to protect the two-stage light gas gun barrel [1]. Details and images from the test at AEDC can be found in Ref. [6]

Table 1. SOCIT, DebriSat, & DebrisLV Test Comparison

	SOCIT/ Transit	DebriSat	DebrisLV
Target body dimensions	46 cm (dia) × 30 cm (ht)	60 cm (dia) × 50 cm (ht)	35 cm (dia) × 88 cm (ht)
Target mass	34.5 kg	56 kg	17.1 kg
MLI and solar panel	No	Yes	No
Projectile material	Al sphere	Hollow Al cylinder with attached nylon bore-rider	Hollow Al cylinder with attached nylon bore-rider
Projectile dimension/mass	4.7 cm diameter, 150 g	8.6 cm × 9 cm, 570 g	8.6 cm × 9 cm, 598 g
Impact speed	6.1 km/sec	6.8 km/sec	6.9 km/sec
Impact Energy to Target Mass ratio (EMR)	81 J/g (2.8 MJ total)	235 J/g (13.2 MJ total)	832 J/g (14.2 MJ total)
Soft-Catch System: Polyurethane foam stacks	3 densities: 0.06, 0.096, and 0.192 g/cm3; 25 cm thick	3 densities: 0.048, 0.096, and 0.192 g/cm3; up to 61 cm thick	3 densities: 0.048, 0.096, and 0.192 g/cm3; up to 51 cm thick

After the impacts of DebrisLV and DebriSat, all intact foam panels, broken foam pieces, loose fragments, and dust were carefully collected, processed, documented, and placed in bags or plastic containers for shipping to a storage facility. The initial estimates using the NASA Standard Satellite Breakup Model (SSBM) indicated the number of 2 mm (and larger) fragments from DebriSat and DebrisLV were approximately 85,000 and 35,000, respectively. Since August 2014, the UF team has been working to extract fragments from foam panels, characterize each fragment down to 2 mm in size, and upload the data into a database.

3. Fragment Characterization

The project's primary goal is to recover at least 90% of the total DebriSat mass from the fragments and to measure the physical properties of all recovered fragments. To achieve this goal, all fragments with at least one dimension ≥ 2 mm are carefully collected and/or extracted from the foam panels/pieces and assigned unique identification numbers before their physical characteristics are determined. A priority of the characterization process is to develop and implement methodologies to minimize biases and/or errors associated with "human-in-the-loop" activities during measurements.

During the characterization process, each fragment's physical (observed and derived) parameters are archived in the DebriSat Categorization System (DCS). The DCS is a database solution designed and developed specifically to manage the large amounts of data generated by the DebriSat project. The list of parameters is shown in Table. 2. In addition to the information shown in Table. 2, associated metadata (e.g., location the fragment was found within the chamber, images of the fragments, etc.) are also archived in the DCS.

The characterization process is broken into three major categories: Assessment, Measurement, and Calculation. In this process, the initial step involves assessing each uniquely identified fragment in terms of material, shape, color, and what imaging system the fragment qualifies for: two dimensional (2D) or three dimensional (3D) imaging. To expedite the large number of fragments, duplicate 2D imagers and one 3D imager are available. Smaller, flat fragments are measured in the 2D imaging systems, while larger objects up to 55 × 55 × 55 mm in volume are processed using the 3D system. For larger objects, an alternative scanning system will be utilized. Details on the imaging systems can be found in Ref [7]. After assessments are completed, the measurement process acquires the mass and physical size based on the solid body (SB) dimensions (X_{SB}, Y_{SB}, and Z_{SB}). The last step in the characterization process involves calculating the characteristic length (L_c), the average of the object's three maximum orthogonal projected dimensions (X_{DIM}, Y_{DIM}, and Z_{DIM}); average cross-sectional area; and volume. Details on how the volume dimensions are calculated can be found in Ref [8]. The process undergoes thorough verification checks to ensure repeatability and reproducibility on a recurring basis.

Table 2. Fragment parameters captured in database

Parameter	Definition/Notes
Unique ID with associated barcode	Earlier studies grouped multiple, similar small fragments to get average characteristics
Material	Predefined categories based on as-built design; material density is auto-populated once material is selected.
Shape	Flat Plate Bent Plate Straight Needle/Rod/Cylinder Bent Needle/Rod/Cylinder Parallelepiped/Nugget/Spheroid Flexible/MLI
Color	Predefined categories based on as-built design
Principle dimensions	x, y, z (mm) {relative to solid body coordinates}
Characteristic length	$L\_c = (X\_DIM + Y\_DIM + Z\_DIM)/3\ (mm)$
Average cross-sectional area	Weighted average of projected areas visible in multiple 2-D images; pixel-to-length (and area) relationship calculated from hardware characteristics and scene geometry (mm^2).
Mass	Fragment mass (g)
Area-to-mass ratio (AMR)	Calculated average cross-sectional area divided by measured mass (mm^2/g).
Volume	Calculated from point cloud using convex hull and alpha-shape subtraction (mm^3).
Bulk Density	Measured mass divided by calculated volume (g/mm^3).

Figures 2-5 provide an update of the status of the data collection process. Based on the mass of the target, projectile, and impact velocity, the expected number of fragments generated using the NASA SSBM was estimated to be close to 85,000, shown in Figures 2 and 3. After one year of collection, the number of collected fragments surpassed the amount projected by the SSBM and continues to increase as the project continues. In Figure 2, the recorded fragments are total counts for unique objects who have been entered into the database, versus collected which are estimates on how many fragments have been collected from the fragment collection (either loose fragments or extracted from the soft catch panels). In Figure 3, the status of fragments characterized using the 2D imager, 3D imager, and passed the verification process as on September 2018 is shown. Once the fragments are characterized, each fragment goes through a verification process where the technicians confirm the data are archived in the database. The fragment verification ensures the integrity of the fragment data generated throughout the post-impact process. In addition, a Gage repeatability and reproducibility (Gage R&R) tests are conducted every 1,000 verified fragments to measure the variations in equipment and procedure. The Gage R&R tests will help maintain high levels of confidence in the fragment characterization data.

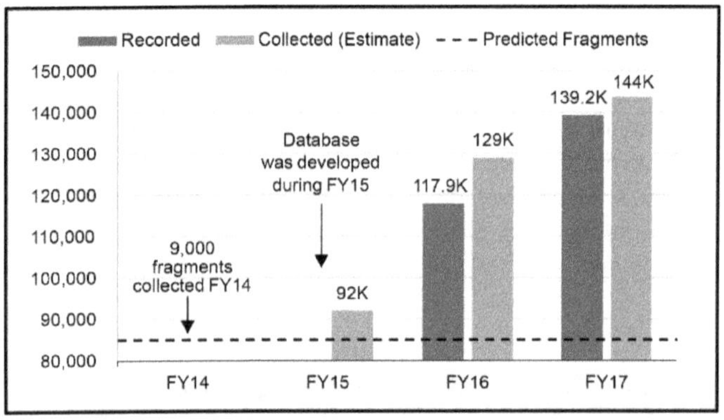

Figure 2. Collected and recorded fragment trend since the initial test in FY14.

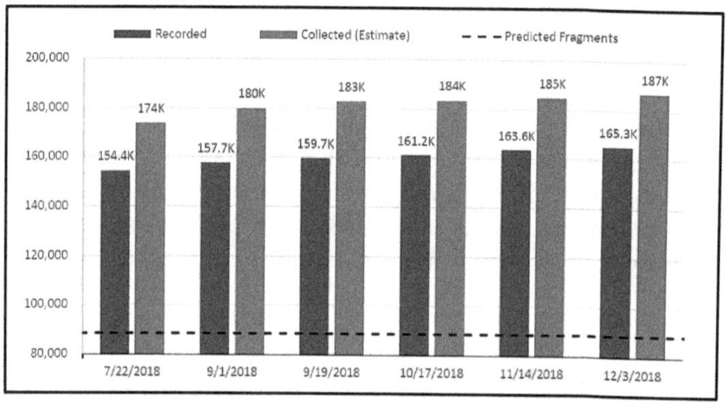

Figure 3. Recorded/Collected fragments, last updated December 2018

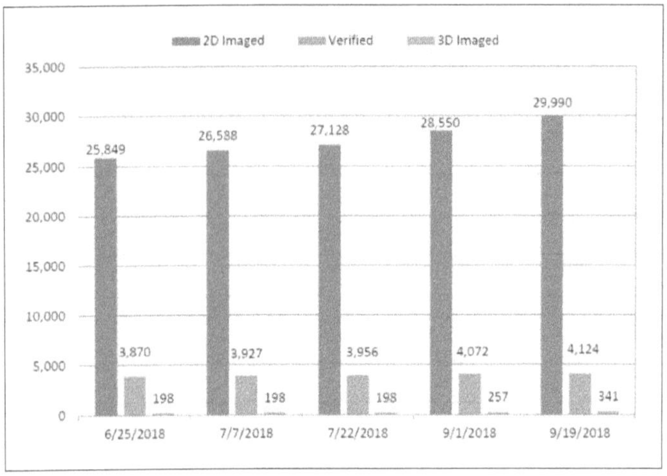

Figure 4. Historical record of collected/recorded DebriSat fragments

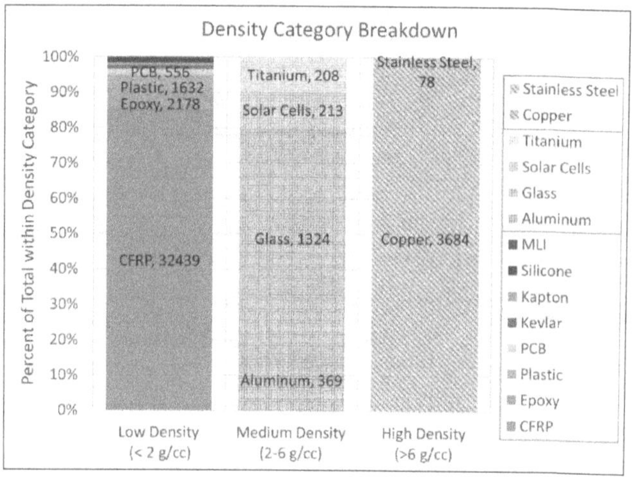

Figure 5. Fragments breakdown by material/density type

4. Shape Categorization

As summarized in Table. 2, six shape categories have been defined that allow DebriSat fragment's shapes to be uniquely described. A rubric developed by the project team allowed workers to unambiguously categorize what may otherwise have been a process prone to subjective assessment when measuring the solid body dimensions (SB) X_{SB}, Y_{SB}, and Z_{SB}. The SB of a fragment is the box that best represents the shape of the fragment and contains at least 75% of the fragment volume (better described as fragment area for 2D objects rather than SB volume). SB dimensions are used in qualitative shape assessment and will provide a baseline to determine the X_{SB}, Y_{SB}, and Z_{SB} dimensions of each fragment. The X_{SB} dimension will be the longest length of the fragment's SB, Y_{SB} dimension is the second longest dimension of the fragment's SB that best represents the fragment's SB width, orthogonal to the X_{SB} dimension, and the Z_{SB} dimension is the last longest dimension orthogonal to X_{SB} and Y_{SB} that best represents the fragment's SB height The definition of SB and the corresponding shape categories are illustrated in Fig. 6. Fig. 7 reflects the actual rubric used to determine what shape category each fragment should be binned into.

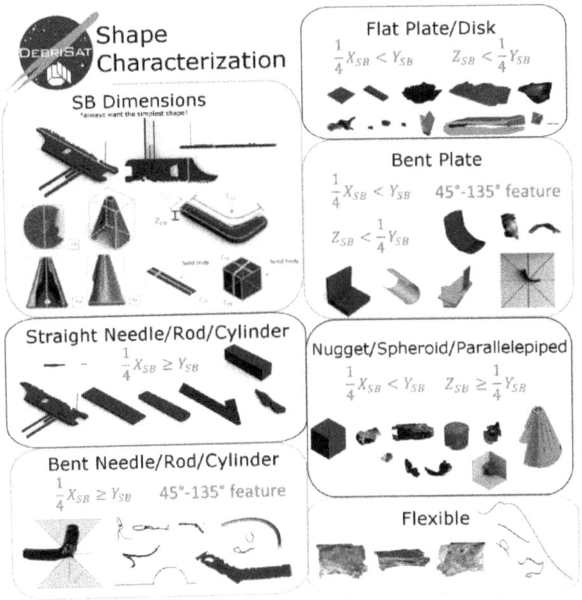

Fig. 6: Definition of solid body (SB) and six fragment shape categories

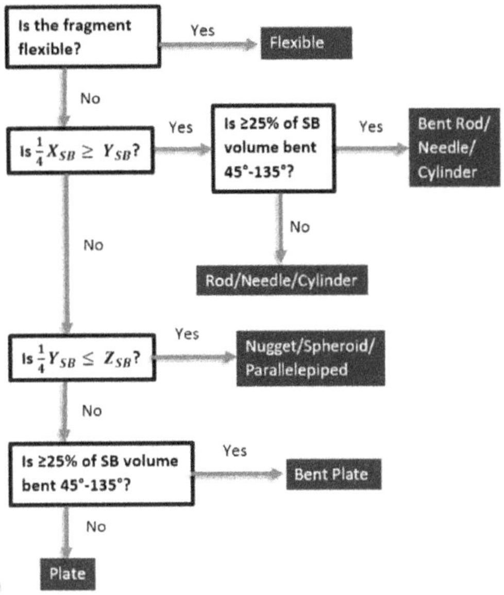

Fig. 7: Illustrated shape categories, in general order of rubric "Yes"

5. NASA Standard Satellite Breakup Model

The NASA SSBM was formulated using laboratory tests and ground-based remote measurements of on-orbit fragmentation events to provide an average breakup ensemble for spacecraft and upper stage collisions and explosions. Following significant validation efforts against observable data, including the development of a radar Size Estimation Model (SEM) based on SOCIT fragments [9], the SSBM was deployed in EVOLVE 4.0, NASA's long-term Monte Carlo environment simulation computer program [10]. The SSBM is described in detail in [11], which also describes other U.S. Government debris models.

The SSBM uses L_c as the fundamental independent variable in lieu of mass. This choice of L_c in preference to mass was driven by the on-orbit observables, namely Two Line Element (TLE) orbital data sets and radar cross section (RCS) time series. The TLE sets provide sufficient information to determine or infer number of fragments and their post-event orbits, the area-to-mass ratio (AMR), and separation velocities for members of specific breakup event ensembles. The RCS time series provide sufficient information to infer L_c and area. Orbital data was used to derive and validate distributions for $L_c > 11$ cm. Laboratory data, including the SOCIT [12] and the ESA Ariane subscale model explosion series, were essential in defining the distributions below approximately 8 cm, and a bridging function was applied to the 8-11 cm L_c gap. Distributions in cumulative number, area, and AMR for collision and explosion fragments are provided by the SSBM; the cumulative number distribution for collisions additionally incorporates mass characteristics of the event to model catastrophic and non-catastrophic collisions. The separation velocity (Δv) distribution uses AMR as its independent variable.

The NASA Engineering Safety Center (NESC) provided guidance for the Orbital Debris Engineering Model (ORDEM) update that resulted in the deployment of ORDEM 3.0, specifically including mass density as a parameter, which was accomplished in ORDEM 3.0. To accommodate this new parameter, the SSBM ensemble functions were partitioned by mass density. This was done by integrating SOCIT density data into three major density categories [13] for fragments between 1 mm and 10 cm in L_c (objects larger than 10 cm are assigned the nominal density of 2.8 g/cm^3). The density categories are shown in Table. 3

Table 3. SSBM Density Bins

Low Density	< 2.0 g/cm^3	Nominal Low Density	1.4 g/cm^3 (includes 0.9 g/cm^3 sodium-potassium reactor coolant droplets)
Medium Density	2.0 – 6.0 g/cm^3	Nominal Medium Density	2.8 g/cm^3
High Density	> 6.0 g/cm^3	Nominal High Density	7.9 g/cm^3

The AMR is monitored to ensure that a given statistical sampling from the distribution does not result in an object more dense than is reasonable in a given category—the consequence of which would be unrealistically long orbital lifetimes. Based on the ORDEM 3.0 and NASA BUMPER II impact risk models, the high- and medium-density categories pose the principal risk to resident space objects. The NASA Debris Assessment Software (DAS) v. 2.1.1 [14] also incorporates ORDEM 3.0 and the effects of mass density on probabilistic risk assessment for compliance with NASA Technical Standard 8719.14A, the Process for Limiting Orbital Debris.

The NESC also recommended that shape be included in the breakup ensemble, which will be included in the ORDEM 4.0 development. Analysis of the DebriSat and DebrisLV datasets will allow for the current three category density distribution to be further refined into an effective density distribution and, for the first time, shape effects. Incorporation of debris fragment shape into ORDEM 4.0 requires characterization of observed space object signatures as observed by ground-based assets. In order to apply shape characteristics to remote data used for ORDEM 4.0, a subset of DebriSat fragments will be further analyzed in laboratory radar and optical facilities to update the existing radar-based NASA SEM and develop a comparable optical-based SEM. Planned activities focused on assessing fragment characterization via laboratory tests (optical and radar) are discussed in Ref [11]. The goal is to have a more refined model showing cumulative number of fragments as a function of size and/or mass to analyze and compare with the current SSBM provided the larger number of fragments from new materials that were not included in previous laboratory tests (i.e., MLI and polymers). The SSBM utilizes a power law to fit the empirical based model for fragments of size (L_c) or larger, thus the distribution from the DebriSat analyses may require a new fit based on the final results.

6. Current Analytical Results

Much of the current fragment distribution analysis has focused on the cumulative number distributions as a function of L_c. Figure 8 shows the cumulative number distribution of fragments as a function of L_c broken down into the DebriSat material categories. Additionally, the SSBM prediction for the cumulative number distribution for DebriSat is shown, scaled to the total mass collected to date. Each point on the curves represents the total number of fragments of size L_c or larger. To date, the majority of fully characterized fragments have been carbon-fiber reinforced polymer (CFRP); 22,266 of the 31,225 fully characterized fragments, or approximately 71%, are CFRP. This is due to an initial focus on the characterization of the CFRP fragment over the other material categories, a sampling effect which explains the current predominance of CFRP in the cumulative number distributions. An inflection point in the distribution of all fragments can be seen at approximately 10 mm. This is another effect of the sampling of the fragments for characterization, where objects smaller than 10 mm are mostly processed using a 2D imaging system and objects larger than 10 mm are mostly processed using a 3D imaging system that has lower measurement throughput. The inflection point is expected to flatten as data become more evenly sampled in the measurement transition region. Current fragment characterization efforts focus on filling in the higher L_c part of the distribution by characterizing more 3D fragments and characterizing more fragments of all material types to reduce the skew in our current material distribution.

Figure 9 shows the cumulative number distribution of fragments as a function of L_c split into the DebriSat shape categories. Since the majority of fragments smaller than 10 mm are CFRP, they dominate the shape breakdown in that size regime, and consist mostly of flat plates until about 1 mm when straight rods become a significant portion of the population. Objects greater than 10 mm, or 3D fragments, are primarily classified as Parallelepiped/Nugget/Spheroid. Work on the material and shape distributions is ongoing and will evolve as the number of fully characterized fragments grows.

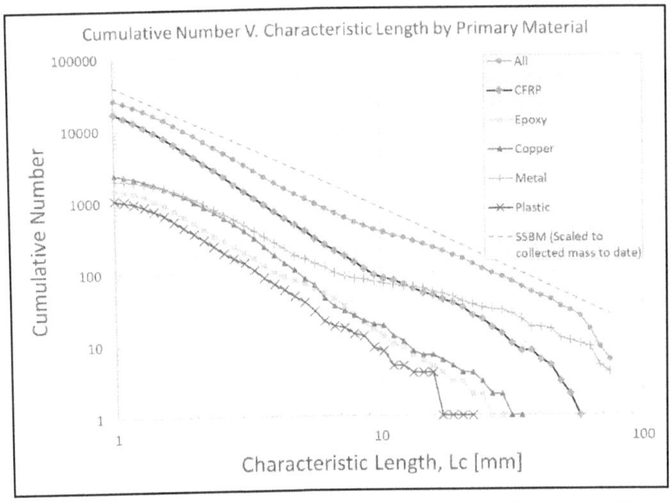

Fig. 8. Cumulative number vs L_c by material type, as of December 2018

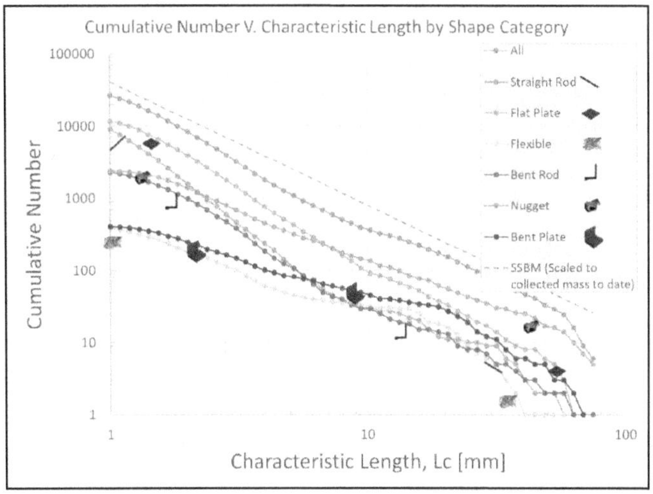

Fig. 9. Cumulative number vs L_c by shape type, as of December 2018

7. Shape Effects on Ballistic Limit Equations

Fragment physical characteristics, including shape and mass density, play a significant role in determining the damage caused by these fragments when they impact satellites and other resident space objects and hence the risk to a given surface. Shape-dependent effects are recognizable when examining the in situ record of impacts on returned surfaces and, by inference, even a cursory examination of the shapes of fragments generated in ground-based laboratories. These effects are notoriously difficult to replicate in the hypervelocity impact (HVI) laboratory. However, the need to better estimate damage and risk and the quest for improved shielding provides the motivation for incorporation of shape effects into an evolved SSBM. The distributions of fragment dimensions, shapes, and bulk densities as a function of material will enable identification of the most common density and shape characteristics for different classes of materials. These most common fragment characteristics can be used to focus damage assessment testing and modeling on the most likely types of debris to be encountered by satellites in on-orbit collisions with small fragments. Given the added difficulty of testing and modeling shape effects on collision damage, DebriSat data-based choices on where to focus the efforts can maximize the correlation between test parameters and what is encountered on-orbit.

To date, the most prominent material recorded from the DebriSat fragment catalog is CFRP, and the most likely shapes are nuggets (L_c > 10 mm) and flat plates (L_c < 1 mm). Numerical impact simulation studies on CFRP cylinders of varying length:diameter ratios [15] and non-spherical CFRP projectiles [16] have been conducted.

The ODPO also is initiating studies on the number of parameters needed to characterize the six DebriSat shape categories and to describe their impact attitude with respect to a target surface. In addition, the manner by which these factors or functions are to be included in ballistic limit equations (BLEs) for all categories and all attack attitudes requires attention, as only simple shapes at 0° attack attitude and normal incidence have been validated in a conservative sense. For example, Schäfer, et al. [17] describe a course of research using ellipsoids described by a single shape parameter, *f*, the parameter's incorporation into a Cour-Palais/Christiansen BLE, and programs of HVI testing and numerical simulations to validate the modified BLE form. In this relatively simple case, the BLE required not only the shape parameter *f*, but also five additional fit parameters to yield a reasonable fit. Significant work, theoretical, computational, and in the HVI laboratory, remains to be done.

8. Summary

This paper has reviewed the DebriSat project's motivation; the data collection plan and its current status; fragment characterization and categorization by shape; and current modeling and risk assessment activities. Currently, the DebriSat ensemble is dominated by CFRP (mass density) and nuggets and plates (shape); this may segue to other densities and shapes as the data collection incorporates larger fragments and discrete components. Fundamental measurement and analytical work will continue and will be incorporated into next-generation models for space object characterization, the orbital debris environment model, and risk assessment.

References

[1] Liou, J.-C. *et. al.*, 2014. Successful Hypervelocity Impacts of DebrisLV and DebriSat, Orbital Debris Quarterly News 18(3), pp. 3-6.

[2] Liou, J.-C., 2009. An Update on Recent Major Breakup Fragments, Orbital Debris Quarterly News 13(3), pp. 5-6.

[3] Werremeyer, M., 2013. Design of Sub-systems for a Representative Modern LEO Satellite, Dissertation, University of Florida.

[4] Clark, S., 2013. Design of a Representative LEO Satellite and Hypervelocity Impact Test to Improve the NASA Standard Breakup Model, Dissertation, University of Florida.

[5] DebriSat Design Review, 2013, Internal University of Florida report: unpublished.

[6] Polk, M., *et. al.*, 2015, "Orbital Debris Assessment Testing in the AEDC Range G Procedia Engineering," 103, 490– 498, 2015.

[7] Shiotani, B., 2017. "Characterizing DebriSat Fragments: So Many Fragments, So Much Data, and So Little Time," 68th International Astronautical Congress; 25-29 Sep. 2017, Adelaide; Australia

[8] Moraguez, M., *et al*, 2015. "An imaging System for Automated Characteristic Length Measurement of DebriSat Fragments," 66th International Astronautical Congress; 12-16 Oct. 2015, Jerusalem; Israel

[9] Xu, Y., *et al.*, 2005. A Statistical Size Estimation Model for Haystack and HAX Radar Detections, IAC 2005.

[10] Johnson, N.L., *et al*. 2001. "NASA's New Breakup Model of EVOLVE 4.0,"Adv. Space Res. 28(9), pp. 1377-84.

[11] Cowardin, H., *et al.*, 2017. "Characterization of Orbital Debris via Hyper-velocity Laboratory-Based Tests" Proc., 7th European Conf. Space Debris, ESA, 2017.

[12] Krisko, P.H., Horstman, M., Fudge, M.L., 2008. "SOCIT4 collisional-breakup test data analysis: with shape and material characterization." Adv. Space Res. 41(7), pp. 1138-46.

[13] Krisko, P.H., *et al.*, 2008. "Material Density Distribution of Small Debris in Earth Orbit," IAC-08.A6.2.1.

[14] Opiela, J., Vavrin, A., 2017. "New Version of DAS Now Available," Orbital Debris Quarterly News 21(1), pp. 4-7.

[15] Miller, J., 2018. "Implication of Debris Fragment Shape on Shield Protection Capability," Orbital Debris Quarterly News 22(4), pp. 2-4.

[16] Miller, J., 2019. "Simulation study of non-spherical, graphite-epoxy projectiles," Proceedings of the 2019 Hypervelocity Impact Symposium, *in work.*

[17] Schäfer, F., Hiermaier,S., Schneider, E., 2003. "Ballistic Limit Equation for the Normal Impact of Unyawed Ellipsoid-shaped Projectiles on Aluminum Whipple Shields," IAC-03-IAA.5.3.06.

Proceedings of the 2019 Hypervelocity Impact Symposium
HVIS2019
April 14-19, 2019, Destin, FL, USA
HVIS2019-071

Numerical study of dynamic behavior of foams subjected to high- to hyper-velocity impact

Xiaotian Zhang[a,*], Ruiqing Wang[a], Q.M. Li[b]

[a]*Beihang University, Beijing 100191, China*

[b]*The University of Manchester, Manchester M13 9PL, United Kingdom*

Abstract

Hypervelocity tests and numerical studies have been reported in the literature for aluminum foam to show its potential applications in spacecraft shielding against space debris based on "shielding set-up". Meanwhile the "forward impact" set-up has been widely reported in the literature to study the dynamic behavior of the foam materials in the range of low to intermediate impact velocities. This paper extends the forward impact to high- and hyper-velocity impacts to understand the dynamic deformation and failure mechanisms based on numerical simulation. The focused impact velocity range is from about 1km/s to 6km/s. The cell-based numerical model of the foam material is used along with the Smoothed Particle Hydrodynamics (SPH) method to simulate the deformation and the failure process. The failure of the foam materials in the range of intermediate to high impact velocities is related to the plastic yielding and crushing of the foam cell, while that in the hypervelocity impact regime is related to the cell material erosion. Dynamic effects in different impact velocity ranges also lead to shock and strain-rate effects. Understanding of the dependence of the deformation/failure mechanisms on the impact velocity helps to determine the application of foam materials in the relevant range of impact velocities.

Keywords: aluminum foam, hypervelocity impact, stress wave, SPH, space debris

1. Introduction

Aluminum foam has cellular structure consisting of a solid aluminum matrix and gas-filled pores that occupy a large portion of the volume. The pores can be sealed (closed-cell foam) or interconnected (open-cell foam). Reticulated aluminum foam[1] is a very porous, low density, extremely open-cell foam. Only the lineal boundaries where the bubbles meet remain in the reticulated foam. For commercial reticulated foams, the porosity is typically 95%, but can be as high as 98% making this material ultralight.

Open celled aluminum foam is a potential structural and functional material in heat exchanger or energy absorber applications in aerospace. NASA[2, 3], ESA[4-6] and China[7] have already tested the feasibility of using the aluminum foam material in space debris shielding applications. Representative shielding configurations from the ISS (International Space Station) and ATV (Automated Transfer Vehicle) were modified by replacing the aluminum bumper or the honeycomb sandwiched panel into the aluminum foam panel while maintaining the unified aerial density. Using the open celled aluminum foam material, an increase in the ballistic limit is observed with the impact velocity ranging from 3 to 7 km/s. The ultralight reticulated aluminum foam material is used in most of these studies.

The impact compression properties of the aluminum foam have been widely studied and revealed at the low and medium impact velocities[8, 9]. However, the hypervelocity impact response of aluminum foam material, especially the mechanism of mitigating impact energy, is hardly reported. Neither the uniform deformation occurring in the quasi-static load nor the cell crush and compaction occurring in the low to medium velocity impact can be observed in the hypervelocity impact process. Hypervelocity impact generates such a high pressure that even dense metals cannot withstand the load and the material becomes distorted, spalled and phase-changed due to the high pressure shock.

The authors have presented a preliminary numerical study about aluminum foam under hypervelocity impact in HVIS-2017[10]. The results showed that the failure pattern of aluminum foam material was more appropriately described as erosion

* Corresponding author. Tel.: +86-18601938269; fax: +86-10-82338404.
E-mail address: zhangxiaotian@buaa.edu.cn.

rather than compaction or crush[11]. In the non-perforation case, it was shown that the material of the spherical projectile and the material of the target foam panel in the impact area are all spalled and vaporized after a hypervelocity impact, and only the residual of the target panel was left with a crater in it. The crater size was much larger than the original projectile[10].

The different failure phenomenon indicates that the aluminum foam material must behave differently at hypervelocity from that at lower velocities. This paper extends the direct impact to high- and hyper-velocity impacts to understand the dynamic deformation and failure mechanisms based on numerical simulation. The impact velocity range evaluated in this study ranges from about 1km/s to 6km/s.

2. Numerical modeling of the foam material failure test

2.1. Impact configuration

The impact configuration is shown in Figure 1. The "forward impact" setup is used here, i.e. the aluminum foam material is fixed on the rigid supporting base, and the rigid striker is moving towards the foam material at hypervelocity. The velocity of the striker is prescribed so that the complicated effect due to reduction of the impact velocity can be removed.

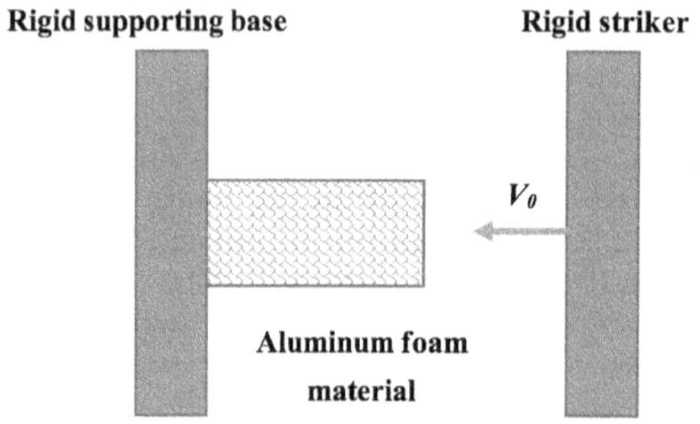

Figure 1 Impact configuration to generate failure in the foam material

The "direct impact" setup or the usual Taylor test setup are not chosen because the configuration is designed to be consistent with the hypervelocity test in future study. The foam rod is easily torn apart in the launching process when being used as the projectile, and thus, it is more convenient to use the foam as the impact target.

The aluminum foam sample here is a cylinder with a diameter of 1.27 cm and height of 2.54 cm. The porosity of the foam material is chosen to be 40 PPI (pore per inch), which is currently the largest one in general commercial production.

2.2. Cell-based modeling of the foam material

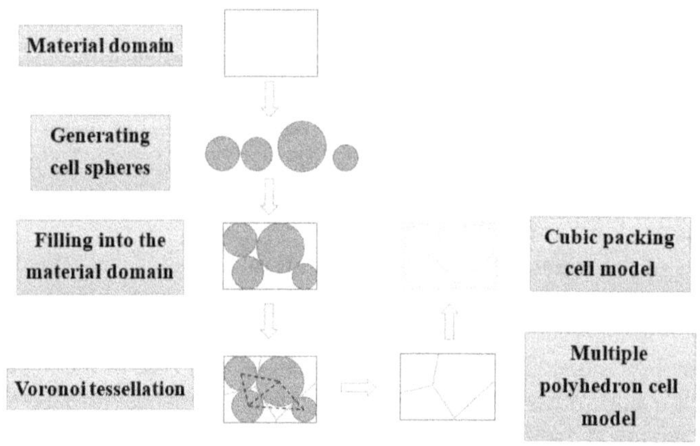

Figure 2 2D illustration of the cell-based modeling method of the foam material

A 2D illustration of the cell-based modeling method of the foam material is given in Figure 2. The material is set according to the designed sample dimensions given above. After that, the cell spheres, which represent the size of the foam cells, are generated by a Monte-Carlo method with the given mean and variance. The probability density function used here is the normal distribution function. The spheres are filled into the material domain with a random close packing algorithm. The centers and radius of the spheres are extracted after the packing, with which the weighted Voronoi tessellation is applied on the material domain and the multiple polyhedron model is exported. Finally, the regular arrayed cubic elements are packed into the adjacent domains of the edges in the polyhedrons to form the ligaments in the foam material with the specified width.

The generated foam model is shown in Figure 3.

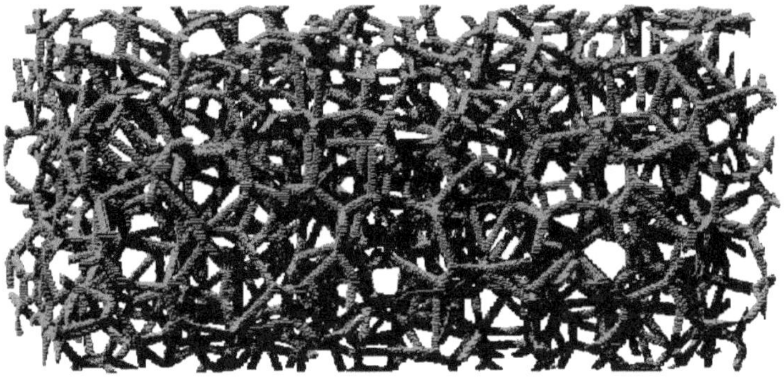

Figure 3 Cell-based foam model

2.3. SPH model in LS-DYNA

The SPH model for hypervelocity impact simulation is generated from the cubic packing model by creating a SPH particle at the centroid of each cubic element. The unified distance between the neighbor particles is set to 0.00625 cm. The total amount of the particles in this model is 632279. The SPH particle inherits the mass of the cubic element from which it is generated (Figure 4). Two rigid walls are defined at both ends of the foam cylinder with the keyword *RIGIDWALL_GEOMETRIC_FLAT. The rigid wall on the top is moving towards the foam cylinder with a prescribed velocity.

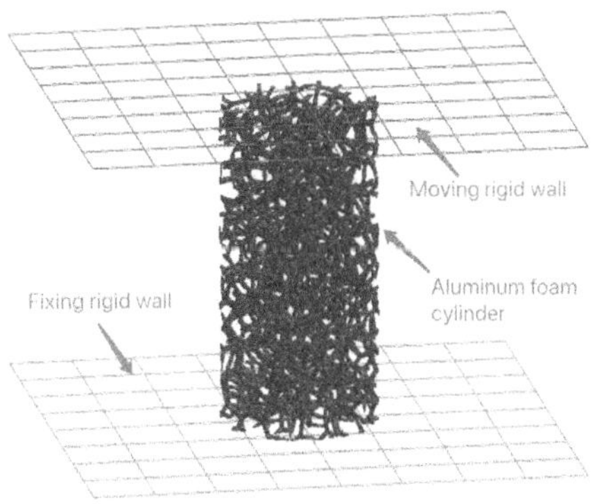

Figure 4 SPH model of the foam cylinder impact in LS-DYNA

The material model[12] used in this simulation for the base material of the foam cylinder is the Johnson-Cook strength model and the Gruneisen equation of state. Since the porosity of the foam is taken into consideration in the cell-based geometry, the base material model and parameters used here are exactly the same as those for dense aluminum. The material parameters used in this simulation are shown in Tables 1 and 2[12].

Table 1 Johnson-Cook material model parameters

Parameters	Value	Parameters	Value
Shear modulus (kPa)	2.76E+7	Pressure limit (kPa)	-1.2E+6
Yield stress (kPa)	2.65E+5	D1	0.112
Harding constant (kPa)	4.26E+5	D2	0.123
Harding exponent	0.34	D3	1.5
Strain rate constant	0.015	D4	0.007
Thermal softening exponent	1	D5	-1
melting temperature (K)	775		

Table 2 Gruneisen equation of state parameters

Parameters	Value
Gruneisen coefficient	1.97
C1 (m/s)	5386
S1	1.339
reference temperature (K)	300
Specific Heat (J/kg/K)	884

Both the spall and shear failure of the base material are taken into consideration in this simulation. The hydrostatic tension spall model detects spall if the pressure becomes greater than the specified tensile pressure limit, i.e. $P<P_{cutoff}$. Once detected, the deviatoric stresses are set to zero and the pressure is required to be compressive. D1~D4 are damage parameters of the Johnson-Cook model[13].

3. Numerical results

3.1. Overall failure pattern

A typical simulation output is shown in Figure 5 where the impact velocity is 6 km/s. The finite element reconstruction technique[14] is used to distinguish the spalled debris from the intact residual. The failure pattern is very different from that observed in the low velocity impact.

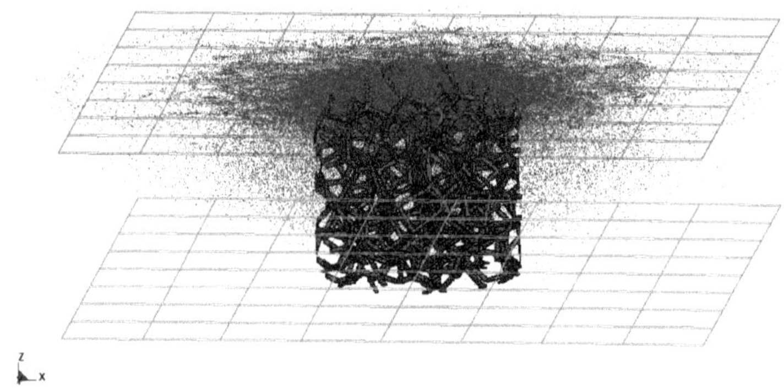

Figure 5 Failure of the foam material (v_0 = 6 km/s)

The material at the contact interface between the foam material and the above moving rigid wall is failed and spread laterally. But the deformation and failure can hardly be viewed in the intact residual. Crushing does not occur for the low-to-high velocity impact.

Failure patterns are compared with the unified displacement of the moving rigid wall at different lower impact velocities. The failure pattern does not change much until the impact velocity was reduced to 0.5 km/s. Figure 6 shows the intact residuals of the foam material at 0.75 km/s and 0.5 km/s. It indicates that the hypervelocity failure pattern for this specified foam sample only exists above 0.5 km/s, below which plastic deformation and compaction dominate.

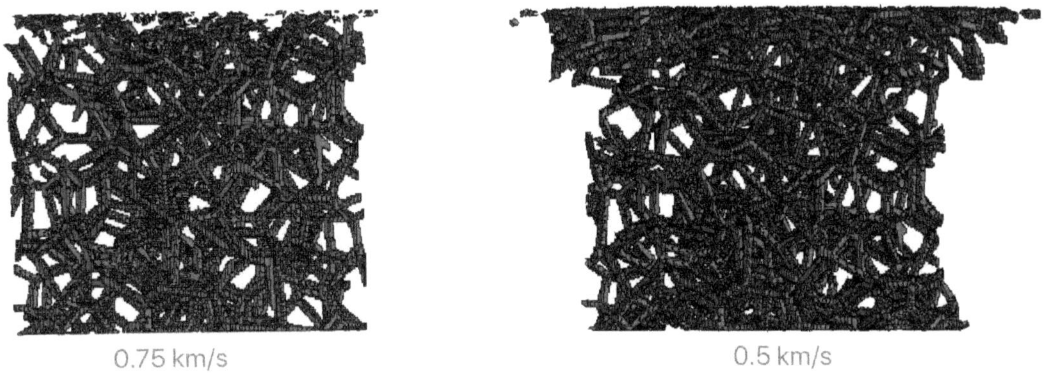

Figure 6 Intact residuals of the foam material at 0.75 km/s and 0.5 km/s

3.2. Base material failure mechanism

As introduced in section 2.3, the material failure criteria used in this numerical study contain the hydrostatic tension spall model (bulk failure) and the Johnson-Cook damage model (shear failure). The bulk failure and shear failure are treated individually, which means the material particle failure state can be bulk failure, shear failure or both. Figure 7 shows the contours of the 1 km/s case; the pressure and temperature levels are higher in the 2-6 km/s cases.

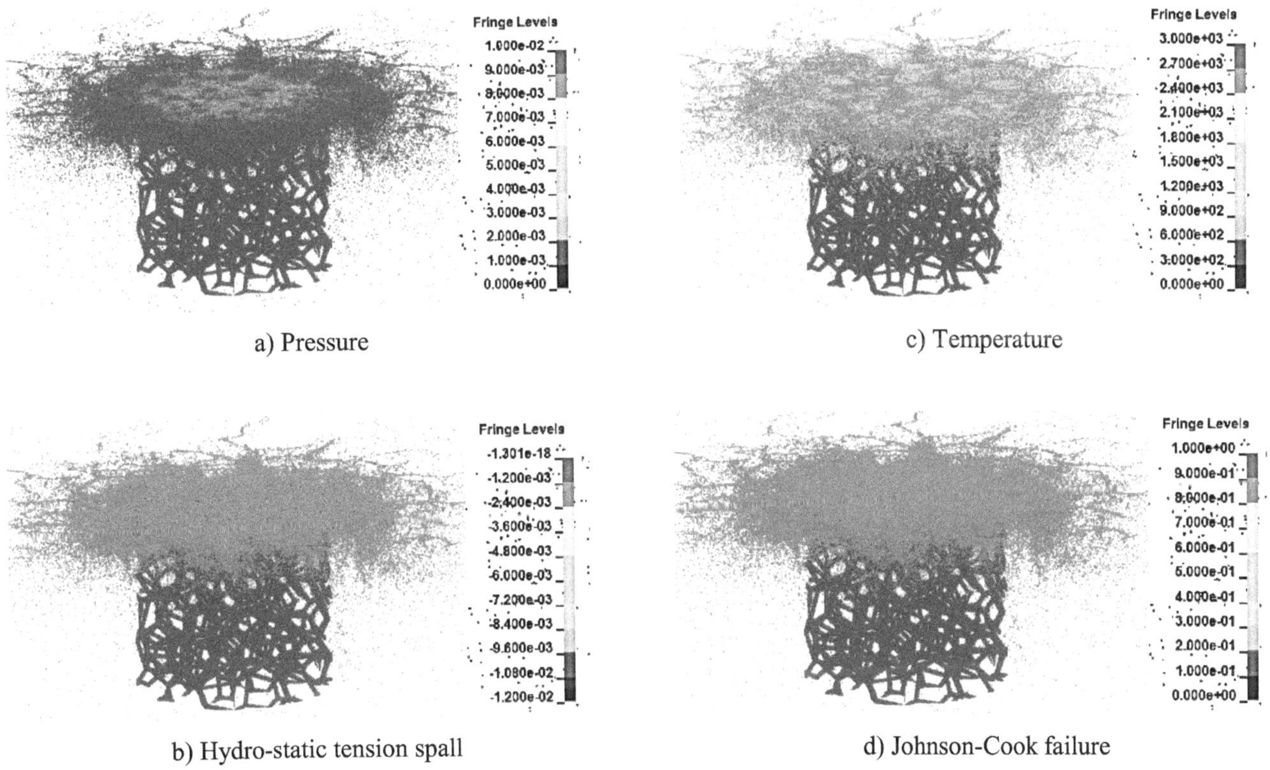

Figure 7 Contours of the 1 km/s case

Very high pressure is generated in the base material near the center of the impact interface, and then the material flows outwards and generates the hydro-static tension (Figure 7.a). Spall occurs when the tensile pressure exceeds the pressure cutoff (Figure 7.b). Moreover, the rapidly increasing pressure leads a rise in temperature (Figure 7.c). This process is so fast that there is not enough time for heat exchange, and therefore, it can be treated as adiabatic. The Johnson-Cook failure equation is:

$$\varepsilon^f = \left(D_1 + D_2 \exp(D_3\sigma^*)\right)\left(1 + D_4 \ln \dot{\varepsilon}^*\right)\left(1 + D_5 T^*\right) \quad (1)$$

$$T^* = \frac{T - T_{room}}{T_{melt} - T_{room}} \quad (2)$$

It indicates that when the temperature reaches the melting temperature, the failure strain would be 0, i.e. the material would fail immediately (Figure 7.d).

3.3. Reaction force on the rigid walls

The reaction force curves of the moving end and the fixed end of the foam material at an impact velocity of 1~6 km/s are given in Figure 8. The unit system used here is cm-g-μs.

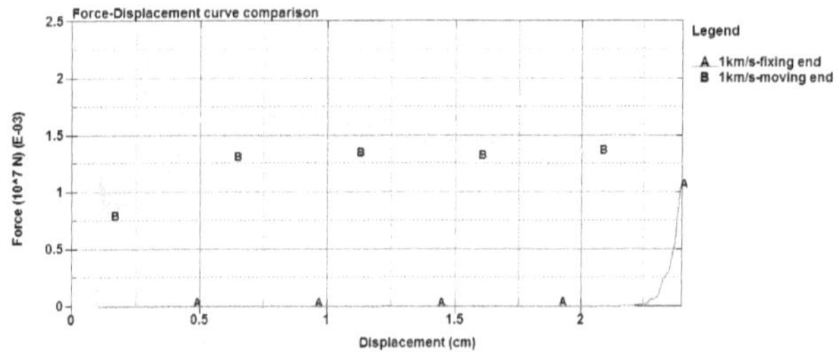

a) 1 km/s

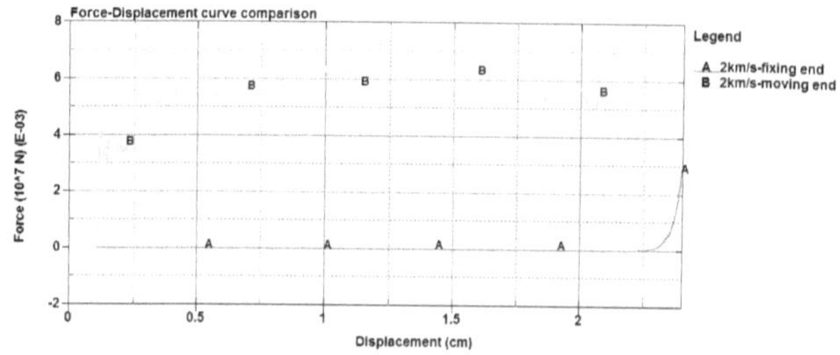

b) 2 km/s

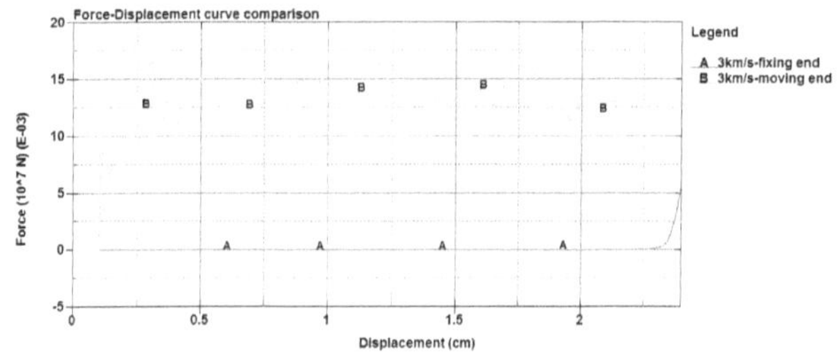

c) 3 km/s

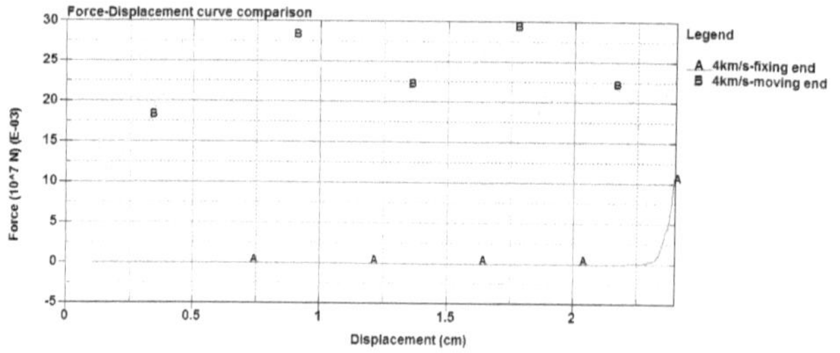

d) 4 km/s

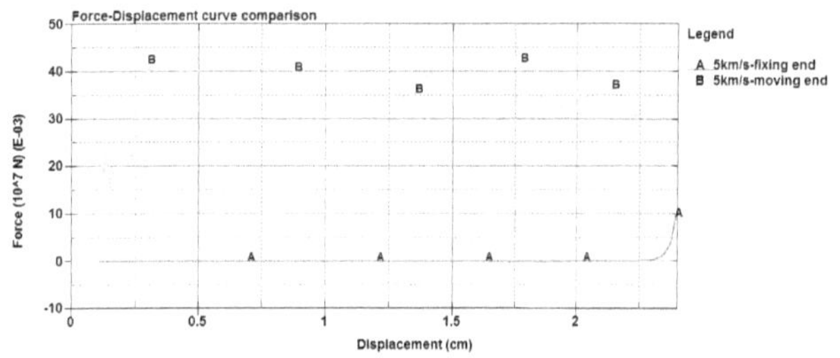

e) 5 km/s

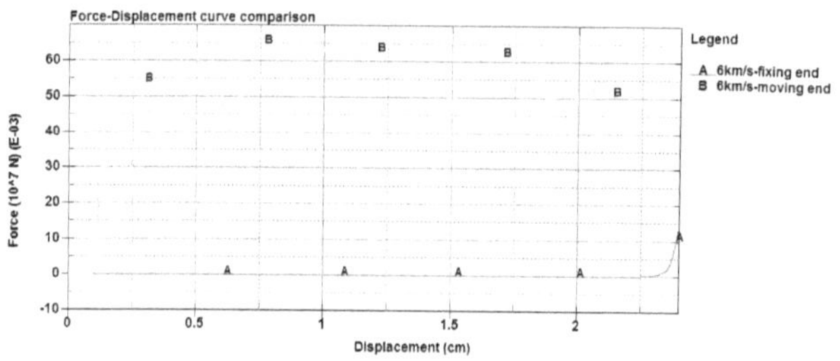

f) 6 km/s

Figure 8 Reaction force (10^7 N) of the rigid walls vs displacement (cm)

In contrast to the typical load-displacement curve of the foam material at low impact velocity, the densification stage does not emerge after the plateau stage. This occurs because the failure pattern of the foam material at hypervelocity is erosion rather than crushing compaction. There is no densification during the impact.

For the quasi-static compression of the foam material, the reaction force at the fixed end is approximately equal to that at the moving end. In the low velocity impact case, the reaction force at the fixed end is similar to that at the moving end, but there is a time delay before the rise of the force in the fixed end because of the inertia effect. In the hypervelocity impact case, the inertia effect is dominant, and the rise of the force at the fixed end does not emerge until the moving end nearly reaches the fixed end.

3.4. Debris cloud spreading velocity

Assuming the spreading velocity of the particles on the edge of the debris cloud is v(r) where r is the radius from the center, the distribution of v(r) along r can be calculated from the SPH simulation output. The particles are divided into groups according to r, in which the increment of r is 0.01 cm. The average spreading velocity of the particles in one group is calculated as:

$$v_{average} = \frac{1}{n}\sum_{i=1}^{n}(\vec{v}_i \cdot \frac{\vec{r}_i}{|\vec{r}_i|}) \tag{3}$$

where "*i*" indicates the *i*th particle in the group, and $\vec{r}$ and $\vec{v}_i$ are the position and velocity vector of particle i. Only the X and Y components are considered in this calculation; the component along the impact direction is omitted.

These velocity distributions are measured at the time when the moving rigid wall reaches the unified displacement of 1.5 cm for all different impact velocities (Figure 9). The rise of the curves at about r = 0.635 cm is due to that location being the boundary of the foam cylinder. The location at r = 2.5 cm is about the edge of the debris cloud.

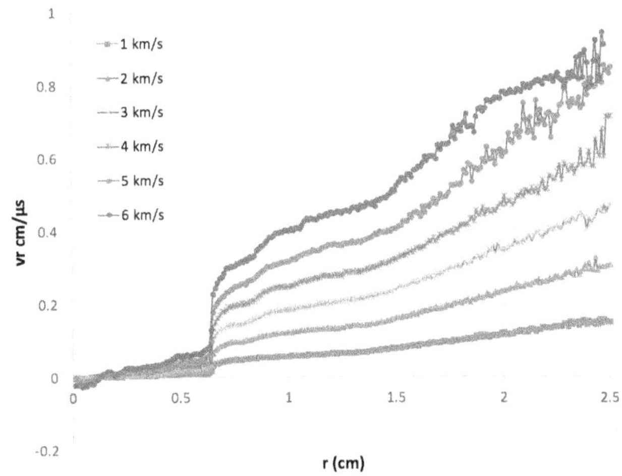

Figure 9 v(r) distribution along the radius at different impact velocities

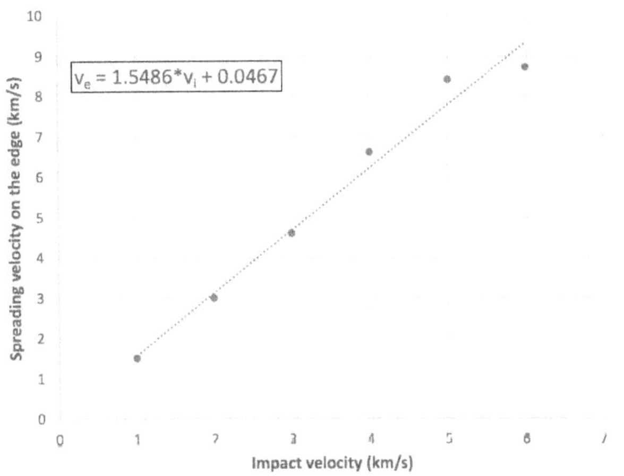

Figure 10 Spreading velocities on the edge of the debris cloud at different impact velocities

Spreading velocities on the edge of the debris cloud at different impact velocities are plotted in Figure 10. The linear fitting is applied to the data, and the equation for the dependence of the spreading velocity on impact velocity is also given in Figure 10.

4. Conclusions

The dynamic behavior of aluminum foam material subject to high- to hypervelocity impact is investigated in this paper. The cell-based numerical model of the foam material is built for the forward impact setup. The numerical results are analyzed from the aspect of overall failure pattern, base material failure and reaction force on the rigid wall.

The foam material is eroded for the high- to hypervelocity impact, while the upsetting and crushing do not occur as they occur for low velocity impact. A plateau stage can be seen in the reaction force curve before the moving end nearly reaches the fixed end. There is no densification stage because the plateau stage is caused by the erosion mechanism. Spreading velocity distributions along the radius at different impact velocities are provided, and a linear relationship is found between the spreading velocity on the edge of the debris cloud and the impact velocity.

References

[1] Betke U., Klaus M., Eggebrecht J. G., et al. Mofs meet macropores: dynamic direct crystallization of the microporous aluminum isophthalate cau-10 on reticulated open-cellular alumina foams[J]. Microporous and Mesoporous Materials. 2018, 265: 43-56.

[2] Ryan S., Hedman T., Christiansen E. L. Honeycomb vs. Foam: evaluating potential upgrades to iss module shielding[J]. Acta Astronautica. 2010, 67(7-8): 818-825.

[3] Ryan S., Christiansen E. L. Hypervelocity impact testing of advanced materials and structures for micrometeoroid and orbital debris shielding[J]. Acta Astronautica. 2013, 83: 216-231.

[4] Destefanis R., Schäfer F., Lambert M., et al. Enhanced space debris shields for manned spacecraft[J]. International Journal of Impact Engineering. 2003, 29(1 – 10): 215-226.

[5] Thoma K., Schäfer F., Hiermaier S., et al. An approach to achieve progress in spacecraft shielding[J]. Advances in Space Research. 2004, 34(5): 1063-1075.

[6] Destefanis R., Schafer F., Lambert M., et al. Selecting enhanced space debris shields for manned spacecraft[J]. International Journal of Impact Engineering. 2006, 33(1-12): 219-230.

[7] Ma Z., Jia B., Pang B. J., et al. Numerical simulation investigation in behavior of al-foams structure under hypervelocity impact[C]. Proceedings of the 4th European Conference on Space Debris. Darmstadt, Germany: 2005.

[8] Zhu F., Chou C. C., Yang K. H. Shock enhancement effect of lightweight composite structures and materials[J]. Composites Part B-Engineering. 2011, 42(5): 1202-1211.

[9] Sun Y., Li Q. M. Dynamic compressive behaviour of cellular materials: a review of phenomenon, mechanism and modelling[J]. International Journal of Impact Engineering. 2018, 112: 74-115.

[10] Zhang X., Wang R., Li Q. M. Modeling of hypervelocity impact of sandwiched open cell aluminum foam[J]. Procedia Engineering. 2017, 204: 262-269.

[11] Xi C. Q., Li Q. M. Meso-scale mechanism of compaction shock propagation in cellular materials[J]. International Journal of Impact Engineering. 2017, 109: 321-334.

[12] Selected hugoniots[R]. LA-4167-MS. Los Alamos Report, 1969.

[13] Kay G. Failure modeling of titanium 6al-4v and aluminum 2024-t3 with the johnson-cook material model[R]. FAA Report DOT/FAA/AR-03/57. 2003.

[14] Zhang X., Jia G., Huang H. Finite element reconstruction approach for on-orbit spacecraft breakup dynamics simulation and fragment analysis[J]. Advances in Space Research. 2013, 51(3): 423-433.

Proceedings of the 2019 Hypervelocity Impact Symposium
HVIS2019
April 14-19, 2019, Destin, FL, USA
HVIS2019-099

Explosion of hydrazine tanks due to space debris impacts

Jérôme Limido [a]*, Christian Puillet [b], Jean-Paul Vila [c]

a IMPETUS AFEA, 6 rue du Cers, 31330 Grenade sur Garonne, France
b CNES, 18 avenue Edouard Belin, 31401 Toulouse, France
c Institut de Mathématiques de Toulouse, Institut National des Sciences Appliquées de Toulouse, 135 Avenue de Rangueil, 31400 Toulouse, France,

Abstract

The impact of space debris on a spacecraft can result in a catastrophic event that not only destroys the structure but can also create more space debris. The design of any spacecraft requires understanding the potential damage that can be inflicted by such an event. We consider the case of a satellite hydrazine fuel tank and the consequences of a hypervelocity impact from space debris. The purpose of this study is to better understand the mechanisms of detonation of hydrazine vapor during a hypervelocity impact on a low-pressure reservoir. The IMPETUS Afea Solver® and Explo5 software were used to perform numerical simulations of operational impact configurations (v = 14 km / s). The multi-scale calculations are performed using the Next Generation IMPETUS ɣSPH solver. A numerical assessment of the impact performance at the scale of the reactive fluid reservoir representative of a real configuration (global and local scale) was carried out. The model includes the detonation in the fluid and the transfer of the momentum to the structure which includes capturing the perforation of the structure that results from the explosive loading. A first diagram (impact velocity, diameter / perforation-explosion) was constructed on a reference tank of diameter 50cm and thickness 1mm in titanium.

Keywords: Hydrazine tank, detonation, meshless

1. Introduction

The problem of the proliferation of space debris is becoming an important issue. Indeed, the Inter-Agency Space Debris Coordination Committee is an international governmental forum for global coordination of activities related to space debris issues. In June 2007, a document entitled "Space Debris Mitigation Guidelines" laid the groundwork for international space debris legislation.

It is therefore necessary to determine the potential damage that may occur on the most vulnerable structures of a spacecraft. One of such a structure found on satellites are pressurized tank by their nature are in a state of change as the liquid fuel is used over time. It is important to assess the risk over the life of the structure as part of the design criteria.

A CNES research study was conducted by IMPETUS and Jean-Paul Vila for better understanding of the mechanisms of detonation of hydrazine vapor during a hypervelocity impact on a low-pressure reservoir (liquid hydrazine prompt detonation is supposed unlikely [1]). IMPETUS Solver ® and Explo5 software were used to perform numerical simulations of operational impact configurations (v = 14 km / s). These multi-scale calculations are based on the use of the advanced numerical methods of the meshless IMPETUS AFea ɣSPH solver.

A numerical assessment of the impact performance at the scale of the reactive fluid reservoir representative of a real configuration (global and local scale) was carried out. We have studied the perforation-explosion transition of these reservoirs.

* Corresponding author. Tel.: +33561748682.
E-mail address: jerome@impetus.fr

The model considers the detonation in the fluid and the transfer of the momentum to the structure. Other potential catastrophic failure mode could exit but are not assessed as part of this study.

The study is organized as follows:

- Identification of the data needed to build the numerical models. We have analyzed reference studies from Bergh and Garcia [1]. This allowed us to identify the state equation of liquid hydrazine, the equation of state of the hydrazine vapor
- Local meshless model (dedicated to early stage impact till explosion initiation)
- Global meshless model (dedicated to explosive behavior of the full tank)

2. Hydrazine Behavior

This part is focused on the identification and validation of the equations of state necessary for the construction of the complete 3D model of reactive fluid. We therefore present first the reference works whose data will be used.

2.1. Bergh and Garcia Reference Study

The reference publication of Bergh and Garcia [1] proposes a modeling process for hydrazine reservoirs. We recall that these works were carried out in 4 phases:

- Hydrazine vapor study with thermochemical code CHEETAH
- Dynamic molecular hydrazine vapor and liquid study
- Local impact model of an axisymmetric HVI tank (Titanium, thickness 1 mm, diameter 500 mm)
- Global explosion model of a 2D axisymmetric HVI tank (Titanium, thickness 1 mm, diameter 500 mm)

The first two steps are used to identify the parameters of the state law of vapor and liquid hydrazine. The local model determines whether the critical state (explosion) of hydrazine is reached and its location. The global model implements an explosion simulation using the detonation point identified by the local model.

This study is based on several results of the literature that we present hereunder. The authors studied 3 debris diameters (0.5, 3.5 and 13mm) for 3 temperature ranges (T <90 ° C, 90 ° C <T <150 ° C, T> 150 ° C) for a velocity impact of 14 km / s.

The conclusions were as follows:

- Debris Diam 13mm Temperature> 90 ° C → explosive area
- Debris Diam 13mm Temperature <90 ° C → explosive area
- Debris Diam 3,5mm Temperature> 150 ° C → explosive area
- Debris Diam 3.5mm 90 ° c <Temperature <150 ° C → transition zone
- Debris Diam 3.5mm Temperature <90 ° C → "safe" zone
- Debris Diam 0.5mm → area "safe" Non-perforating

We will study the same configurations in this paper.

2.2. Liquid hydrazine

The first step required to construct a numerical model of liquid hydrazine is the identification of a state equation. We have identified in the literature the works of Garcia and Persson [3]. They present Hugoniot of liquid hydrazine. This identifies the parameters of the Mie- Gruneisen following equation:

$$P = \rho_0 C_0^2 \frac{\chi}{(1-S\chi)^2} \quad (1)$$

With $C_0 \sim 2632$ m / s and $S \sim 1.315$

We are now interested in the explosive behavior of liquid hydrazine. The study of Bergh and Garcia [1] on the mesoscopic scale (molecular dynamics) shows that liquid hydrazine can only explode under very severe conditions. These results agree with the experimental conclusions of Garcia and Chavez [4], in which, for example, a 100g cylinder impacted a 100mm diameter aluminum spherical tank filled with liquid hydrazine at a speed of 5000 m / s. Some reactions occurred in hydrazine but nothing to conclude a detonation. We shall see in the following that these reaction thresholds are much higher than those of the hydrazine vapor. We therefore consider in the following that the hydrazine film is not the critical mechanism of detonation.

2.3. Hydrazine vapor detonation

Few baseline studies are available on the behavior of hydrazine vapor. Pedley et al. [5] measured detonation velocity about 2400m / s (pressure 0.2 bars).

Bergh and Garcia [1] use a JWL state equation to model the explosive part. They based their analysis on the CHEETAH thermochemical code of the LLNL (USA). This code is not available to the public. We therefore chose to use the Explo5 Thermochemical Code to reproduce the approach of Bergh and Garcia. This approach is based on intermolecular potentials and statistical mechanics (Implemented in EXPLO5). The results obtained with Explo5 are very close to those determined experimentally by Pedley et al. [5] (see following table).

Table 1. Comparison Pedley Trials vs. Explo5 Prediction

	ρ, g/cm3	D experim., m/s	D, m/s	P, bar	Polytropic exponent, γ	Particle velocity, m/s	Sound velocity, m/s
Pedley							
T=393 K, p=0.6 bar	0.000596	2406	2513 (4.44%)	16.4			
T=393 K, p=0.328 bar	0.000326	2524	2512 (0,48%)	8.9			
T=393 K, p=0.2 bar	0.000199	2395	2511 (-4,84%)	5.5			
EXPLO5							
T=393 K, p=0.6 bar	0.000596	2406	2537 (5,4%)	16.97	1.298	1082	92.6
T=393 K, p=0.328 bar	0.000326	2524	2535 (0,44%)	9.25	1.302	1079	92.6
T=393 K, p=0.2 bar	0.000199	2395	2533 (5,76%)	5.64	1.310	1074	92.6

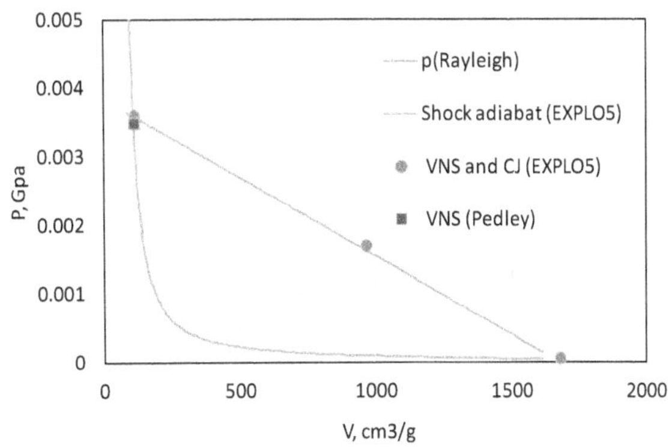

Fig. 1. EXPLO5 adiabatic shock calculation of unreacted hydrazine vapor, Rayleigh line, CJ and VNS vs. calculations Pedley (T = 393K, p = 0.6 bar)

After the validation step EXPLO5 was used to deduce JWL EOS parameters for Hydrazine Vapor.

3. Local Meshless model

We propose to build a local model based on γSPH method [2]. Indeed, the approach without a Lagrangian mesh allows total energy to be conserved by construction which is very important in the case of hypervelocity impact. We emphasize this point because it is very difficult to conserve energy with classical finite element ALE schemes. Contact management is indeed very complex and often non-conservative. The meshless methods make it possible to completely get rid of this problem. Moreover, the main limitations of conventional SPH methods for this type of application are the prohibitive calculation times and the lack of stability and precision. For ten years now, IMPETUS Afea and Vila Jean Paul have been developing innovative GPU-based algorithms that make it possible to overcome these two problems [10].

We present here a comparison with [1] in the impact case of an axisymmetric 2D model for an aluminum sphere, Diameter = 13 mm, Velocity = 14 km / s Pressure = 5 bar. The thickness of the tank is 1 mm, the liquid hydrazine film is 0.5 mm, and hydrazine vapor fills the rest of the field.

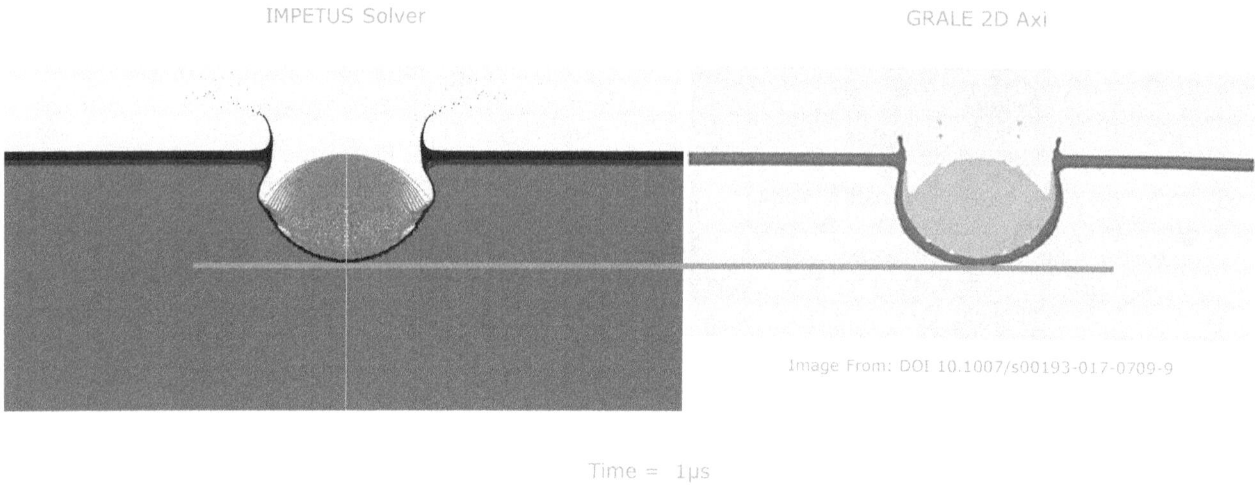

Fig. 2. Comparison of 2D Grale (ALE Solver) [1] / IMPETUS Solver®

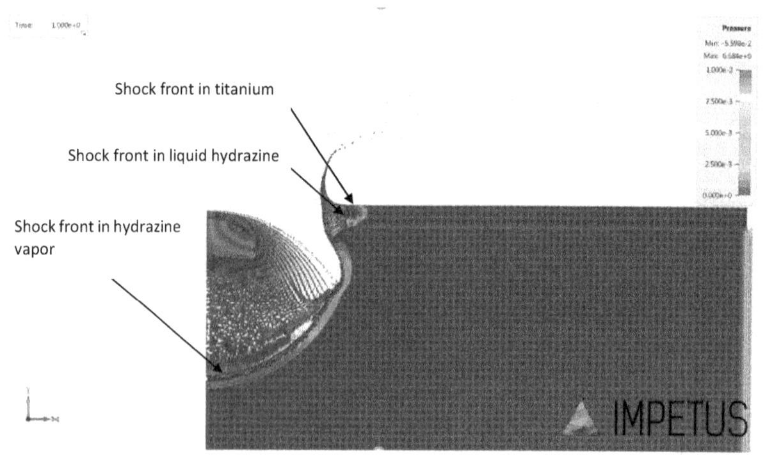

Fig. 3. Meshless local model, pressure visualization at 1µs

The results obtained are relatively similar both from the point of view of deformations and temporality. The expected results are observed (see Figure 3): fast elastic wave in the titanium skin, wave front glued in front of the impactor (the speed of sound in the hydrazine vapor is much lower than the speed of perforation).

Note that the computation times observed are very low due to the GPU parallelization of the code. Thus, the presented calculation takes about 1min using 1 Nvidia P6000 GPU Processor.

We used the same approach to model the impact case for 14 km / s of the 3.5 mm diameter aluminum sphere with a pressurized tank of 5 bar. This case does not lead to a trivial tank explosion without hydrazine. Indeed, the initial kinetic energy is in this case about Ek ~ 5 KJ (see Schäfer criteria [6]).

We have identified with this model a probable area of detonation reached at about 50 mm under the skin of the tank. It is the location of the detonation point that will be used for the global model (see next section).

4. Global Meshless model

The reference configuration is based on [1] is: titanium tank diameter 50 cm, aluminum spherical debris D = 3.5 mm V = 14 km / s, hydrazine vapor 5 bars.

As in the study [1], we consider the following hypotheses:

- Weak interaction hole + fragments with the waves of detonation
- Localization of the detonation point deduced from the local model

Note, however, that we do not expect to obtain results strictly equivalent to those of Bergh and Garcia [1] because we obtained substantially different equation of state parameters for the hydrazine vapor.

We propose a coupled 3D model FE / SPH:

- Tank is modeled by 3rd order solid elements + elastoplastic behavior model + Mie- Gruneisen equation of state
- Hydrazine vapor is represented by a ɣSPH approach (2,000,000 particles) + JWL state equation

We present a comparison of results obtained with Impetus Solver® with those of Ls-Dyna 2D Euler [1] in the following figure. Note that the position of the detonation point is not identical. This is quite normal because the equation of state of the hydrazine vapor is different in the 2 models. The critical point of detonation is reached more quickly in our study (position of the detonation point closer to the wall). The propagation velocity of the shock wave in hydrazine vapor is lower in the case of Ls-Dyna 2D than in the case of the Impetus Solver®. The wave goes further but is faster; the results for later time (> 100μs) are therefore relatively close.

Note also that the proposed approach allows for achieving directly a deformable and damageable model of the reservoir shell coupled with the detonation of hydrazine vapor. Lastly, the 3D approach developed makes it possible to consider extending the work to the non-isotropic (composite), non-spherical and off-axis symmetry reservoir study with associated computing times of approximately 1 hour using a P6000 Nvidia GPU Processor.

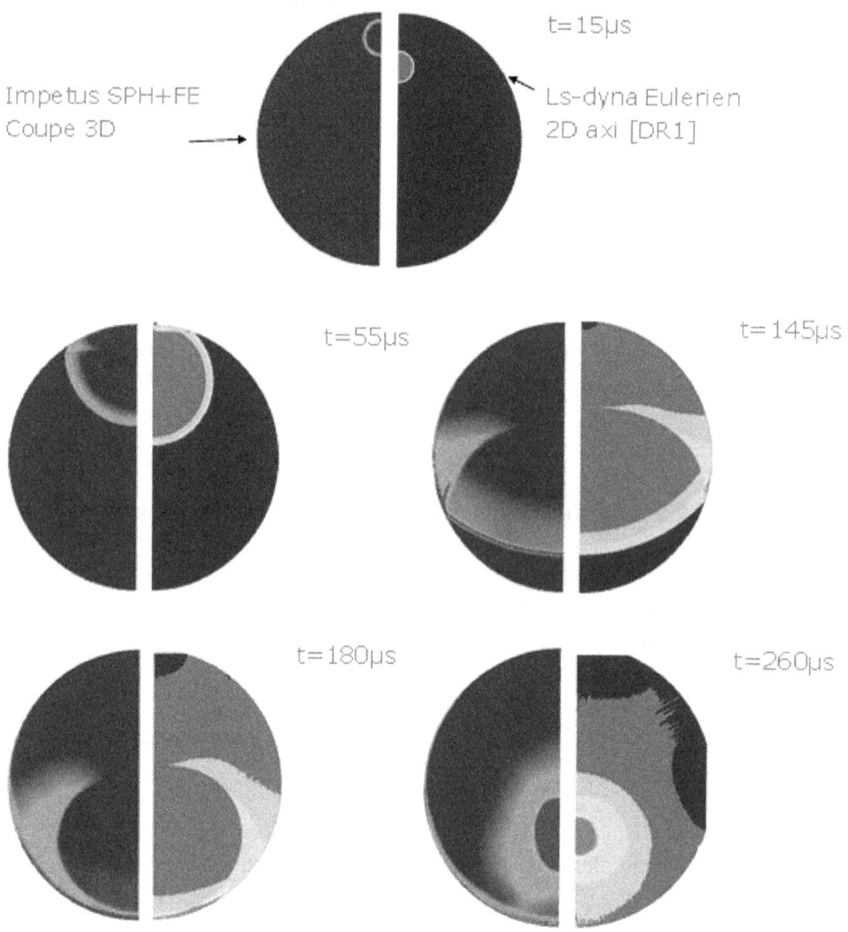

Fig. 4. Comparison Impetus 3 D / Grale 2D axi [1] cross section ¼ model

The following figure presents a 3D model ¼ visualizations for which the particles are made transparent for pressures below a given threshold (here 60bars). This allows for observing the pressure wave coming to lick the wall and converge towards the rear face of the tank. This mechanism involves a significant overpressure at the rear face (> 250bar).

Now let's look at the effect on the deformation of the reservoir. Indeed, the developed 3D model being coupled makes it possible to represent the stresses and deformations in the titanium envelope of the tank. The evolution of the plastic deformation of the reservoir is shown in Figure 6. This clearly shows the intense loading on the rear face during the convergence of pressure waves in hydrazine. It is obvious that the values reached are well above the admissible thresholds (plastic deformations greater than 5%). This is same conclusion as that of Bergh and Garcia [1], namely the rupture of the reservoir for the configuration D = 3.5 km / s, P = 5 bar and V = 14 km / s.

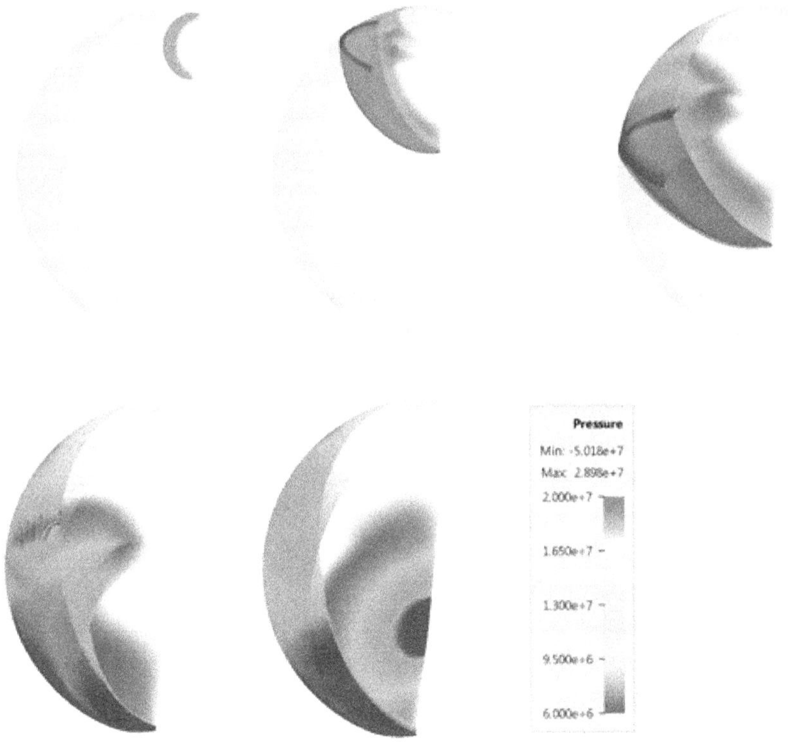

Fig. 5. 3D visualization of the pressure field

Fig. 6. Plastic deformation visualization of the titanium shell (red> 5% deformation)

It may also be interesting to put these results into perspective with the conventional criteria of spherical tank sizing and with the rupture pressure. Reference document JERG-0-001D [8] is used to evaluate the rupture pressure of a spherical tank via the following formula:

$P_{burst} = 2.0 * MEOP$ with $MEOP = (t* 4 * S) / (D + t* 0.4)$

D = internal diameter (0.5m), t = skin thickness (1mm), S = breaking strength of the material / safety factor (safety factor = 2.0 in our case), The rupture limit of Ti6Al4V is approximately 860 MPa

MEOP ~ 34 bars therefore the breaking pressure ~ 68 bar
Another method of calculation proposed by Pasquale M Sforza [9] defines the breaking pressure as follows:
Pburst = m / (0.004 x D^3) (in kPa)
m = mass of the tank (3.48 kg)
The rupture pressure in this case is ~ 69 bars

The pressures obtained with the digital reservoir model are much higher (~ 200 bar) which confirms the previous conclusions.

5. Conclusion

We have taken the approach of Bergh and Garcia [1] which breaks down the problem of modeling the explosion of reactive reservoirs in 2 steps: local model (localization point of detonation) and global (explosion).

A first diagram (impact velocity, diameter / perforation-explosion) was constructed on a reference tank of diameter 50cm thickness 1mm in titanium. The study included 9 configurations for impact speeds of 14 km / s, the results are summarized in the following table but not detailed for sake of brevity.

Impactor Diameter (mm)	Pressure (bars)	Damage mode	Tank state	Hydrazine vapour detonation
13	0,2	Back face debris impact	Rupture	Yes
13	1	Back face debris impact	Rupture	Yes
13	5	Back face debris impact	Rupture	Yes
3,5	0,2	Limited to impact zone	Safe	Yes
3,5	1	Overpressure hydrazine detonation	Transition	Yes
3,5	5	Overpressure hydrazine detonation	Rupture	Yes
0,5	0,2	Non perforating impact	Safe	No
0,5	1	Non perforating impact	Safe	No
0,5	5	Non perforating impact	Safe	No

Our approach allows for implementation of a 3D model with fluid coupling with a deformable structure in the framework of the γSPH method [2]. We obtained results qualitatively similar to those of Bergh and Garcia even if quantitative differences are observed due to the differences in the identification of the equation of state parameters. GPU implementation resulted in reduced computation time which opens the door for the analysis of off axis impact with non-spherical debris.

Acknowledgements

The authors would like to thank CNES for funding this research activity.

References

[1] M. Bergh, V. Garcia, A computational model for assessing high-velocity debris impact in space applications, Shock Waves, DOI 10.1007/s00193-017-0709-9, Jan. 2017

[2] Anthony Collé, Jérôme Limido, Jean-Paul Vila, An Accurate SPH Scheme for Dynamic Fragmentation modelling, EPJ Web Conf. 183 01030 (2018), DOI: 10.1051/epjconf/201818301030

[3] Garcia, Persson, The shock Hugoniot of liquid hydrazine in the pressure range of 3.1 to 21.4GPa, Int. Conf. « Shock waves in condensed matter », 1996

[4] Garcia, Chavez, Shock compression of liquid hydrazine, shock compression conference APS, 1995

[5] Pedley, M.D., Bishop, C.V., Benz, F.J., Bennett, C.A., McLenagan, R.D., Fenton, D.L., Knystautas, R., Lee, J.H., Peraldi, O., Dupre, G., Shepherd, J.E.: Hydrazine vapor detonations. In: Dynamics of Explosions, Progress in Astronautics and Aeronautics, vol. 114, pp. 45–63 (1988)

[6] Schäfer, F., Schneider, E., & Lambert, M., An Experimental Study to Investigate Hypervelocity Impacts on Pressure Vessels, Second European Conference on Space Debris, Organised by ESA, held 17-19 March, 1997, ESOC, Darmstadt, Germany (1997), ESA-SP 393., p. 435

[8] JERG-0-001D, Technical Standard for High Pressure Gas Equipment for Space Use, JAXA, March 25, 2016

[9] Pasquale M Sforza, Theory of Aerospace Propulsion, Butterworth-Heinemann, 2011 ISBN: 9780123848895 2009.

[10] A. Collé, J. Limido, J.-P. Vila, An accurate multi-regime SPH scheme for barotropic flows, Journal of Computational Physics, Volume 388, 2019, Pages 561-600,

Proceedings of the 2019 Hypervelocity Impact Symposium
HVIS2019
April 14-19, 2019, Destin, FL, USA
HVIS2019-118

Extended Abstract+

Hypervelocity impact of PrintCast A356/316L composites

Lauren L. Poole[a], Manny Gonzales[b], Matthew R. French[a],
William A. Yarberry III[a], Abdel R. Moustafa[a], Zachary C. Cordero[a,*]

[a] *Materials Science and NanoEngineering, Rice University, 6100 Main Street, Houston, TX 77005, USA*
[b] *Air Force Research Laboratory, Materials Manufacturing Directorate, 5135 Pearson Road, Wright-Patterson AFB, OH 45433, USA*

Shielding elements used to protect against micrometeoroids and orbital debris (MMOD) (e.g., Whipple shields, multi-shock shields, stuffed Whipple shields) typically incorporate thin bumper sheets that intercept and vaporize incident MMOD traveling at speeds in excess of several km/s. In some applications, however, space limitations prevent the use of large stand-offs, and components must instead be protected by a single monolithic shielding element. Electronics, for example, are often only protected by their housing. With such applications in mind, we describe a class of spatially efficient composite shielding elements fabricated using a hybrid additive manufacturing approach termed PrintCasting. The PrintCast process consists of two steps: First selective laser melting is used to fabricate a lattice preform in the shape of the final component. Next this preform is infiltrated with a liquid metal that has a melting point lower than that of the lattice. The resulting solidified part is a periodic interpenetrating composite in which each constituent forms a continuous network. Using a combination of hypervelocity impact experiments and shock transmission calculations, we demonstrate that these interpenetrating composite shielding elements mitigate spallation and other failure modes through multiple internal shock reflections at the buried heterophase interfaces.

A two-stage light-gas gun was used to perform normal incidence hypervelocity impact tests on monolithic 316L stainless steel, monolithic A356, layered A356/316L composites, and PrintCast interpenetrating A356/316L composites. The projectile was a 1.9 mm Al sphere with a striking velocity of 5.9±0.1 km/s. All the targets had an areal density of 3.5 g/cm^2. The layered and interpenetrating composites were 7.5 mm thick with 40 vol% 316L. The interpenetrating composite was prepared using PrintCasting [1] and possessed a body centered cubic (BCC) lattice symmetry. The structures of the recovered test specimens were characterized using optical and electron microscopy.

The PrintCast composite had the least rear surface deformation of all the impact specimens. The crater in the PrintCast composite was square, reflecting the cubic symmetry of the composite and the dissimilar mechanical properties of the constituent materials. The crater volume was only slightly larger than the volume of the crater the monolithic 316L target.

To assess the effects of mesostructure and material properties on the impact behavior of the different shielding elements, we computed the peak pressure of a spherical shock as it travels through monolithic A356, monolithic 316L, and layered composites of A356 and 316L with interface densities that matched our impact specimens. The stress profile in a twelve-layer composite, corresponding to the 7.5 mm thick interpenetrating composite, exhibited periodic pressure oscillations that accelerated shock attenuation [2, 3] and resulted in a rear surface pressure similar in magnitude to the spall strength of 316L and A356.

The PrintCasting process used to create these interpenetrating composite shielding elements can be easily extended to other material systems and lattice geometries. These capabilities, combined with the excellent shielding performance of the interpenetrating composites in the present work, make PrintCasting a promising approach for manufacturing lightweight and spatially efficient shielding elements.

+ Full paper can found at https://www.sciencedirect.com/journal/international-journal-of-impact-engineering/special-issue/10MTGP5W4VJ

* Corresponding author. Tel.: +1-713-348-2904.
E-mail address: zachary.cordero@rice.edu.

Acknowledgements

The authors gratefully acknowledge support from the Propulsion Materials Program, DOE Vehicle Technologies Office for additive manufacturing of the lattice preforms. We thank Robert McCandless of Jacobs Technology for their assistance with optical profilometry, Michael Orelley of Carl Zeiss Microscopy for imaging assistance, and Tom Seabury of ExxonMobil and Logan Ware of Rice University for their support with our metallographic investigations.

References

[1] Pawlowski, A., Cordero, Z., French, M., Muth, T., Carver, J., Dinwiddie, R., Elliott, A., Shyam, A., Splitter, D., 2017. Damage-tolerant metallic composites via melt infiltration of additively manufactured preforms, Materials & Design 127, p. 346.

[2] Han, W., Cerreta, N., Mara, N., Beyerlein, I., Carpenter, J., Zheng, J., Trujillo, C., Dickerson, P., Misra, A., 2013. Deformation and failure of shocked bulk Cu–Nb nanolaminates, Acta Materialia 63, p. 150.

[3] Zhuang, S., Ravichandran, G., Grady, D., 2003. An experimental investigation of shock wave propagation in periodically layered composites, Journal of the Mechanics and Physics of Solids 51, p. 245.

www.ingramcontent.com/pod-product-compliance
Lightning Source LLC
LaVergne TN
LVHW082011200926
848481LV00010B/2840